STATCRUNCH EXAMPLES AND EXERCISES INTEGRATED THROUGHOUT

StatCrunch® is online statistical software available for purchase that enables users to perform complex analyses, share data sets, and generate compelling reports of their data. Interactive graphics are embedded to help users understand statistical concepts; graphics are available for export to enrich reports with visual representations of data.

STATCRUNCH EXAMPLES

Examples with a SC icon indicate that you can log on to StatCrunch to learn how the example problem can be solved using the StatCrunch software. Note: accessing this content online requires a MyStatLab or StatCrunch account.

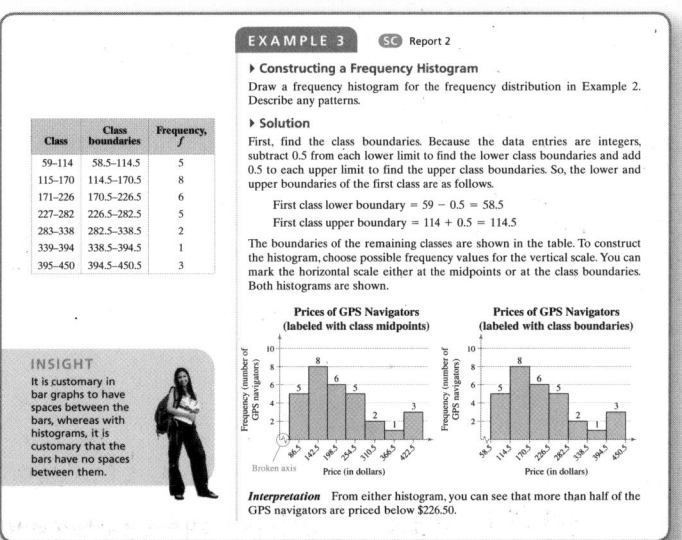

page 42

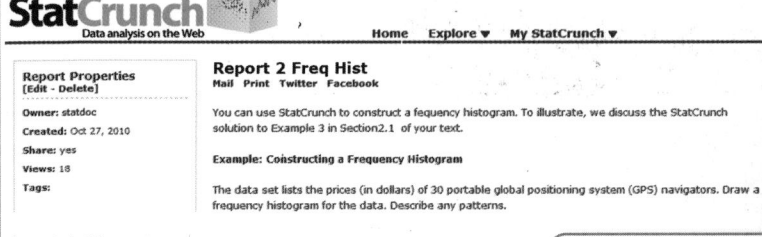

STATCRUNCH REPORTS

Data sets for 50 of the text's examples are provided in StatCrunch. Interactive, step-by-step instructions guide you through how to use StatCrunch to solve the example problem.

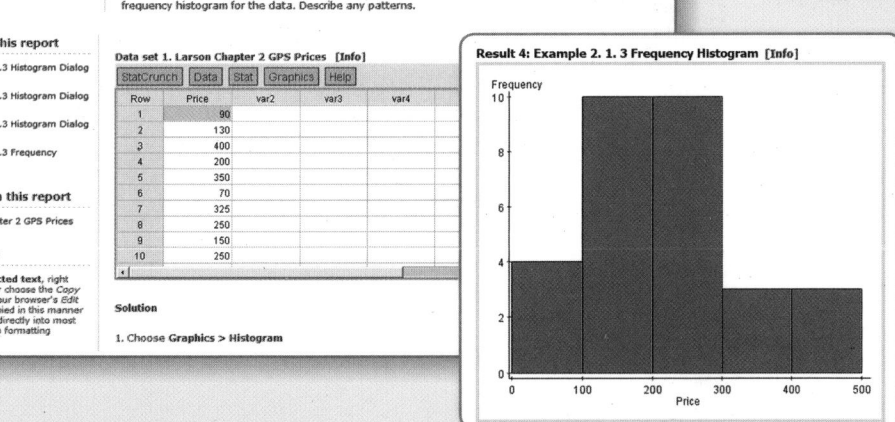

STATCRUNCH EXERCISES

Specific SC icon exercises direct you to solve the problem using StatCrunch, which allows you to practice using statistical software.

SC *In Exercises 27 and 28, use StatCrunch to construct 90%, 95%, and 99% confidence intervals for the population proportion. Interpret the results and compare the widths of the confidence intervals.*

27. Congress In a survey of 1025 U.S. adults, 802 disapprove of the job Congress is doing. *(Adapted from The Gallup Poll)*

28. UFOs In a survey of 2303 U.S. adults, 734 believe in UFOs. *(Adapted from Harris Interactive)*

page 335

ELEMENTARY STATISTICS

PICTURING THE WORLD

Fifth Edition

Ron Larson
The Pennsylvania State University
The Behrend College

Betsy Farber
Bucks County Community College

Prentice Hall

Boston   Columbus   Indianapolis   New York   San Francisco   Upper Saddle River
Amsterdam   Cape Town   Dubai   London   Madrid   Milan   Munich   Paris   Montréal   Toronto
Delhi   Mexico City   São Paulo   Sydney   Hong Kong   Seoul   Singapore   Taipei   Tokyo

Editor-in-Chief: *Deirdre Lynch*
Acquisitions Editor: *Marianne Stepanian*
Senior Content Editor: *Chere Bemelmans*
Editorial Assistant: *Sonia Ashraf*
Senior Managing Editor: *Karen Wernholm*
Associate Managing Editor: *Tamela Ambush*
Supplements Production Coordinator: *Katherine Roz*
Media Producer: *Audra Walsh*
MathXL Project Supervisor: *Bob Carroll*
TestGen QA Manager Assessment Content: *Marty Wright*
Senior Marketing Manager: *Alex Gay*
Marketing Assistant: *Kathleen DeChavez*
Senior Author Support/Technology Specialist: *Joe Vetere*
Rights and Permissions Advisor: *Michael Joyce*
Image Manager: *Rachel Youdelman*
Senior Manufacturing Buyer: *Carol Melville*
Senior Media Buyer: *Ginny Michaud*
Design Manager: *Andrea Nix*
Senior Designer: *Beth Paquin*
Text Design: *Lisa Kuhn, Curio Press LLC*
Composition: *Larson Texts, Inc.*
Illustrations: *Larson Texts, Inc.*

Cover Design: *Rokusek Design*

Cover Image: *Alice/Getty Images*

For permission to use copyrighted material, grateful acknowledgment is made to the copyright holders on the page following the Index, which is hereby made part of this copyright page.

Many of the designations used by manufacturers and sellers to distinguish their products are claimed as trademarks. Where those designations appear in this book, and Pearson Education was aware of a trademark claim, the designations have been printed in initial caps or all caps.

Library of Congress Cataloging-in-Publication Data

Larson, Ron, 1941-
  Elementary statistics : picturing the world / Ron Larson, Betsy Farber. -- 5th ed.
     p. cm.
  ISBN 978-0-321-69362-4
  1. Statistics--Textbooks. I. Farber, Elizabeth. II. Title.
  QA276.12.L373 2012
  519.5--dc22

                                          2010000454

6 7 8 9 10—DOW—14 13

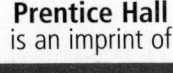
Prentice Hall
is an imprint of

www.pearsonhighered.com

ISBN 10: 0-321-69362-0
ISBN 13: 978-0-321-69362-4

ABOUT THE AUTHORS

Ron Larson
The Pennsylvania State University
The Behrend College

Ron Larson received his Ph.D. in mathematics from the University of Colorado in 1970. At that time he accepted a position with Penn State University, and he currently holds the rank of professor of mathematics at the university. Larson is the lead author of more than two dozen mathematics textbooks that range from sixth grade through calculus levels. Many of his texts, such as the eighth edition of his calculus text, are leaders in their markets. Larson is also one of the pioneers in the use of multimedia and the Internet to enhance the learning of mathematics. He has authored multimedia programs, extending from the elementary school through calculus levels. Larson is a member of several professional groups and is a frequent speaker at national and regional mathematics meetings.

Betsy Farber
Bucks County Community College

Betsy Farber received her Bachelor's degree in mathematics from Penn State University and her Master's degree in mathematics from the College of New Jersey. Since 1976, she has been teaching all levels of mathematics at Bucks County Community College in Newtown, Pennsylvania, where she currently holds the rank of professor. She is particularly interested in developing new ways to make statistics relevant and interesting to her students and has been teaching statistics in many different modes—with the TI-83 Plus, with MINITAB, and by distance learning as well as in the traditional classroom. A member of the American Mathematical Association of Two-Year Colleges (AMATYC), she is an author of *The Student Edition to MINITAB* and *A Guide to MINITAB*. She served as consulting editor for *Statistics, A First Course* and has written computer tutorials for the CD-ROM correlating to the texts in the Streeter Series in mathematics.

CONTENTS

PART ONE DESCRIPTIVE STATISTICS

1

INTRODUCTION TO STATISTICS

2

DESCRIPTIVE STATISTICS   36

PART TWO PROBABILITY AND PROBABILITY DISTRIBUTIONS

PART THREE STATISTICAL INFERENCE

HYPOTHESIS TESTING WITH ONE SAMPLE   354

HYPOTHESIS TESTING WITH TWO SAMPLES   426

PART FOUR MORE STATISTICAL INFERENCE

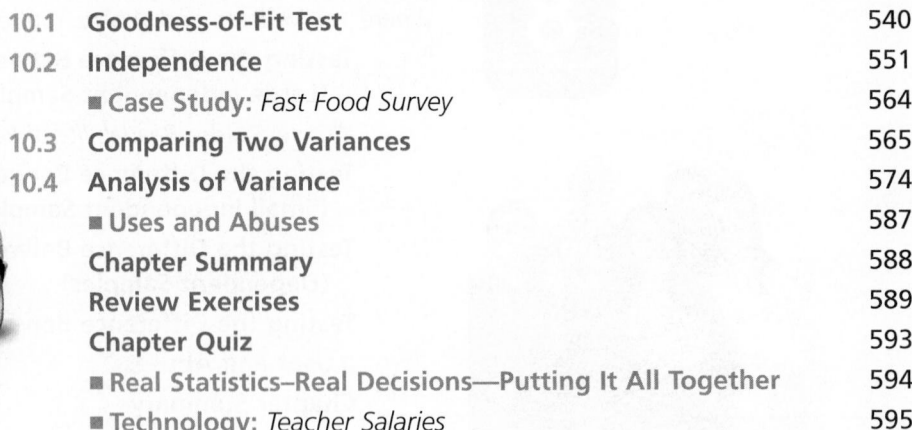

11 NONPARAMETRIC TESTS   596

APPENDICES

PREFACE

Welcome to *Elementary Statistics: Picturing the World,* Fifth Edition. You will find that this textbook is written with a balance of rigor and simplicity. It combines step-by-step instruction, real-life examples and exercises, carefully developed features, and technology that makes statistics accessible to all.

We are grateful for the overwhelming acceptance of the first four editions. It is gratifying to know that our vision of combining theory, pedagogy, and design to exemplify how statistics is used to picture and describe the world has helped students learn about statistics and make informed decisions.

WHAT'S NEW IN THIS EDITION

The goal of the Fifth Edition was a thorough update of the key features, examples, and exercises:

Examples This edition includes more than 210 examples, approximately 50% of which are new or revised.

Exercises Approximately 50% of the more than 2100 exercises are new or revised. We've also added 75 conceptual and critical thinking exercises throughout the text.

StatCrunch® Examples New to this edition are more than 50 StatCrunch Reports. These interactive reports, called out in the book with the **SC** icon, provide step-by-step instructions for how to use the online statistical software StatCrunch to solve the examples. *Note:* Accessing these reports requires a MyStatLab or StatCrunch account.

StatCrunch Exercises New to this edition are more than 80 exercises that instruct students to solve the exercise using StatCrunch. This allows students to practice the software skills learned in the StatCrunch Examples. *Note:* Solving the exercises using StatCrunch requires a MyStatLab or StatCrunch account.

Extensive Feature Updates Approximately 50% of the following key features have been replaced, making this edition fresh and relevant to today's students:

- Chapter Openers
- Case Studies
- Putting It All Together: Real Statistics—Real Decisions

Revised Content The following sections have been changed:

- **Section 2.2, More Graphs and Displays,** now defines misleading graphs.
- **Section 2.5, Measures of Position,** now defines the modified boxplot.

- **Section 9.1, Correlation,** now defines perfect positive linear correlation and perfect negative linear correlation.

FEATURES OF THE FIFTH EDITION

Guiding Student Learning

Where You've Been and **Where You're Going** Each chapter begins with a two-page visual description of a real-life problem. *Where You've Been* shows students how the chapter fits into the bigger picture of statistics by connecting it to topics learned in earlier chapters. *Where You're Going* gives students an overview of the chapter, exploring concepts in the context of real-world settings.

What You Should Learn Each section is organized by learning objectives, presented in everyday language in *What You Should Learn.* The same objectives are then used as subsection titles throughout the section.

Definitions and Formulas are clearly presented in easy-to-locate boxes. They are often followed by **Guidelines,** which explain *In Words* and *In Symbols* how to apply the formula or understand the definition.

Margin Features help reinforce understanding:

- **Study Tips** show how to read a table, use technology, or interpret a result or a graph. **Round-off Rules** guide the student during calculations.
- **Insights** help drive home an important interpretation or connect different concepts.
- **Picturing the World** Each section contains a real-life "mini case study" called *Picturing the World* illustrating important concepts in the section. Each feature concludes with a question and can be used for general class discussion or group work. The answers to these questions are included in the *Annotated Instructor's Edition.*

Examples and Exercises

Examples Every concept in the text is clearly illustrated with one or more step-by-step examples. Most examples have an interpretation step that shows the student how the solution may be interpreted within the real-life context of the example and promotes critical thinking and writing skills. Each example, which is numbered and titled for easy reference, is followed by a similar exercise called **Try It Yourself** so students can immediately practice the skill learned. The answers to these exercises are given in the back of the book, and the worked-out solutions are given in the *Student's Solutions Manual.* The Videos on DVD show clips of an instructor working out each *Try It Yourself* exercise.

StatCrunch Examples New to this edition are more than 50 StatCrunch Reports. These interactive reports, called out in the book with the SC icon, provide step-by-step instructions for how to use the online statistical software StatCrunch to solve the examples. Go to *www.statcrunch.com*, choose **Explore ▼ Groups,** and search for "Larson Elementary Statistics 5/e" to access the StatCrunch Reports. *Note:* Accessing these reports requires a MyStatLab or StatCrunch account.

Technology Examples Many sections contain a worked example that shows how technology can be used to calculate formulas, perform tests, or display data. Screen displays from MINITAB®, Excel®, and the TI-83/84 Plus graphing calculator are given. Additional screen displays are presented at the ends of selected chapters, and detailed instructions are given in separate technology manuals available with the book.

Exercises The Fifth Edition includes more than 2100 exercises, giving students practice in performing calculations, making decisions, providing explanations, and applying results to a real-life setting. Approximately 50% of these exercises are new or revised. The exercises at the end of each section are divided into three parts:

- **Building Basic Skills and Vocabulary** are short answer, true or false, and vocabulary exercises carefully written to nurture student understanding.

- **Using and Interpreting Concepts** are skill or word problems that move from basic skill development to more challenging and interpretive problems.

- **Extending Concepts** go beyond the material presented in the section. They tend to be more challenging and are not required as prerequisites for subsequent sections.

For the sections that contain StatCrunch examples, there are corresponding **StatCrunch exercises** that direct students to use StatCrunch to solve the exercises. *Note:* Using StatCrunch requires a MyStatLab or StatCrunch account.

Technology Answers Answers in the back of the book are found using tables. Answers found using technology are also included when there are discrepancies due to rounding.

Review and Assessment

Chapter Summary Each chapter concludes with a Chapter Summary that answers the question *What did you learn?* The objectives listed are correlated to Examples in the section as well as to the Review Exercises.

Chapter Review Exercises A set of Review Exercises follows each Chapter Summary. The order of the exercises follows the chapter organization. Answers to all odd-numbered exercises are given in the back of the book.

Chapter Quizzes Each chapter ends with a Chapter Quiz. The answers to all quiz questions are provided in the back of the book. For additional help, see the step-by-step video solutions on the companion DVD-ROM.

Cumulative Review A Cumulative Review at the end of Chapters 2, 5, 8, and 11 concludes each part of the text. Exercises in the Cumulative Review are in random order and may incorporate multiple ideas. Answers to all odd-numbered exercises are given in the back of the book.

Statistics in the Real World

Uses and Abuses: Statistics in the Real World Each chapter features a discussion on how statistical techniques should be used, while cautioning students about common abuses. The discussion includes ethics, where appropriate. Exercises help students apply their knowledge.

Applet Activities Selected sections contain activities that encourage interactive investigation of concepts in the lesson with exercises that ask students to draw conclusions. The accompanying applets are contained on the DVD that accompanies new copies of the text.

Chapter Case Study Each chapter has a full-page Case Study featuring actual data from a real-world context and questions that illustrate the important concepts of the chapter.

Putting It All Together: Real Statistics–Real Decisions This feature encourages students to think critically and make informed decisions about real-world data. Exercises guide students from interpretation to drawing of conclusions.

Chapter Technology Project Each chapter has a Technology project using MINITAB, Excel, and the TI-83/84 Plus that gives students insight into how technology is used to handle large data sets or real-life questions.

CONTINUED STRONG PEDAGOGY FROM THE FOURTH EDITION

Versatile Course Coverage The table of contents was developed to give instructors many options. For instance, the *Extending Concepts* exercises, applet activities, Real Statistics–Real Decisions, and Uses and Abuses provide sufficient content for the text to be used in a two-semester course. More commonly, we expect the text to be used in a three-credit semester course or a four-credit semester course that includes a lab component. In such cases, instructors will have to pare down the text's 46 sections.

Graphical Approach As with most introductory statistics texts, we begin the descriptive statistics chapter (Chapter 2) with a survey of different ways to display data graphically. A difference between this text and many others is that **we continue to incorporate the graphical display of data throughout the text.** For example, see the use of stem-and-leaf plots to display data on pages 385 and 386. This emphasis on graphical displays is beneficial to all students, especially those utilizing visual learning strategies.

Balanced Approach The text strikes a **balance among computation, decision making, and conceptual understanding.** We have provided many Examples, Exercises, and Try It Yourself exercises that go beyond mere computation.

Variety of Real-Life Applications We have chosen real-life applications that are representative of the majors of students taking introductory statistics courses. We want statistics to come alive and appear relevant to students so they understand the importance of and rationale for studying statistics. We wanted the applications to be **authentic**—but they also need to be **accessible**. See the Index of Applications on page XVI.

Data Sets and Source Lines The data sets in the book were chosen for interest, variety, and their ability to illustrate concepts. Most of the **240-plus data sets** contain real data with source lines. The remaining data sets contain simulated data that are representative of real-life situations. All data sets containing 20 or more entries are available in a variety of formats; they are available electronically on the DVD and Internet. In the exercise sets, the data sets that are available electronically are indicated by the icon ⊙.

Flexible Technology Although most formulas in the book are illustrated with "hand" calculations, we assume that most students have access to some form of technology tool, such as MINITAB, Excel, the TI-83 Plus, or the TI-84 Plus. Because the use of technology varies widely, we have made the text flexible. **It can be used in courses with no more technology than a scientific calculator—or it can be used in courses that require sophisticated technology tools.** Whatever your use of technology, we are sure you agree with us that the goal of the course is not computation. Rather, it is to help students gain an understanding of the basic concepts and uses of statistics.

Prerequisites Algebraic manipulations are kept to a minimum—often we display informal versions of formulas using words in place of or in addition to variables.

Choice of Tables Our experience has shown that students find a **cumulative density function** (CDF) table easier to use than a "0-to-z" table. Using the CDF table to find the area under a normal curve is a topic of Section 5.1 on pages 239–243. Because we realize that some teachers prefer to use the "0-to-z" table, we have provided an alternative presentation of this topic using the "0-to-z" table in Appendix A.

Page Layout Statistics is more accessible when it is carefully formatted on each page with a consistent open layout. This text is the first college-level statistics book to be written so that its features are not split from one page to the next. Although this process requires extra planning, the result is a presentation that is clean and clear.

MEETING THE STANDARDS

MAA, AMATYC, NCTM Standards This text answers the call for a **student-friendly text that emphasizes the uses of statistics.** Our job as introductory instructors is not to produce statisticians but to produce informed consumers of statistical reports. For this reason, we have included exercises that require students to interpret results, provide written explanations, find patterns, and make decisions.

GAISE Recommendations Funded by the American Statistical Association, the Guidelines for Assessment and Instruction in Statistics Education (GAISE) Project developed six recommendations for teaching introductory statistics in a college course. These recommendations are:

- Emphasize statistical literacy and develop statistical thinking.
- Use real data.
- Stress conceptual understanding rather than mere knowledge of procedures.
- Foster active learning in the classroom.
- Use technology for developing conceptual understanding and analyzing data.
- Use assessments to improve and evaluate student learning.

The examples, exercises, and features in this text embrace all of these recommendations.

SUPPLEMENTS

STUDENT RESOURCES

Student Solutions Manual Includes complete worked-out solutions to all of the *Try It Yourself* exercises, the odd-numbered exercises, and all of the *Chapter Quiz* exercises.
(ISBN-13: 978-0-321-69373-0; ISBN-10: 0-321-69373-6)

Videos on DVD-ROM A comprehensive set of videos tied to the textbook, containing short video clips of an instructor working every *Try It Yourself* exercise. New to this edition are section lecture videos.
(ISBN-13: 978-0-321-69374-7; ISBN-10: 0-321-69374-4)

A **Companion DVD-ROM** is bound in new copies of *Elementary Statistics: Picturing the World*. The DVD holds a number of supporting materials, including:

- **Chapter Quiz Prep:** video solutions to Chapter Quiz questions in the text, with English and Spanish captions
- **Data Sets:** selected data sets from the text, available in Excel, MINITAB (v.14), TI-83 / TI-84 and txt (tab delimited)
- **Applets:** 15 applets by Webster West
- **DDXL:** an Excel add-in

Graphing Calculator Manual Tutorial instruction and worked out examples for the TI-83/84 Plus graphing calculator. (ISBN-13: 978-0-321-69379-2; ISBN-10: 0-321-69379-5)

Excel Manual Tutorial instruction and worked-out examples for Excel. (ISBN-13: 978-0-321-69380-8; ISBN-10: 0-321-69380-9)

Minitab Manual Tutorial instruction and worked-out examples for Minitab. (ISBN-13: 978-0-321-69377-8; ISBN-10: 0-321-69377-9)

Study Cards for the following statistical software products are available: Minitab, Excel, SPSS, JMP, R, StatCrunch, and the TI-83/84 Plus graphing calculator.

INSTRUCTOR RESOURCES

Annotated Instructor's Edition Includes suggested activities, additional ways to present material, common pitfalls, alternative formats or approaches, and other helpful teaching tips. All answers to the section and review exercises are provided with short answers appearing in the margin next to the exercise. (ISBN-13: 978-0-321-69365-5; ISBN-10: 0-321-69365-5)

Instructor Solutions Manual Includes complete solutions to all of the exercises, *Try It Yourself* exercises, Case Studies, Technology pages, Uses and Abuses exercises, and Real Statistics–Real Decisions exercises. (ISBN-13: 978-0-321-69366-2; ISBN-10: 0-321-69366-3)

TestGen® (*www.pearsoned.com/testgen*) Enables instructors to build, edit, and print, and administer tests using a computerized bank of questions developed to cover all the objectives of the text. TestGen is algorithmically based, allowing instructors to create multiple but equivalent versions of the same question or test with the click of a button. Instructors can also modify test bank questions or add new questions. The software and testbank are available for download from Pearson Education's online catalog (*www.pearsonhighered.com/irc*).

Online Test Bank A test bank derived from TestGen® available for download at *www.pearsonhighered.com/irc*.

PowerPoint Lecture Slides Fully editable and printable slides that follow the textbook. Use during lecture or post to a website in an online course. Most slides include notes offering suggestions for how the material may effectively be presented in class. These slides are available within MyStatLab or at *www.pearsonhighered.com/irc*.

Active Learning Questions Prepared in PowerPoint®, these questions are intended for use with classroom response systems. Several multiple-choice questions are available for each chapter of the book, allowing instructors to quickly assess mastery of material in class. The Active Learning Questions are available to download from within MyStatLab or at *www.pearsonhighered.com/irc*.

TECHNOLOGY SUPPLEMENTS

MyStatLab™ Online Course (access code required)

MyStatLab is a series of text-specific, easily customizable online courses for Pearson Education's textbooks in statistics. For students, MyStatLab™ provides students with a personalized interactive learning environment that adapts to each student's learning style and gives them immediate feedback and help. Because MyStatLab is delivered over the Internet, students can learn at their own pace and work whenever they want. MyStatLab provides instructors with a rich and flexible set of text-specific resources, including course management tools, to support online, hybrid, or traditional courses. MyStatLab is available to qualified adopters and includes access to **StatCrunch**. For more information, visit *www.mystatlab.com* or contact your Pearson representative.

MathXL® for Statistics Online Course (access code required)

MathXL® for Statistics is a powerful online homework, tutorial, and assessment system that accompanies Pearson textbooks in statistics. With MathXL for Statistics, instructors can:

- Create, edit, and assign online homework and tests using algorithmically generated exercises correlated at the objective level to the textbook.
- Create and assign their own online exercises and import TestGen tests for added flexibility.
- Maintain records of all student work, tracked in MathXL's online gradebook.

With MathXL for Statistics, students can:

- Take chapter tests in MathXL and receive personalized study plans and/or personalized homework assignments based on their test results.
- Use the study plan and/or the homework to link directly to tutorial exercises for the objectives they need to study.
- Students can also access supplemental animations and video clips directly from selected exercises.

MathXL for Statistics is available to qualified adopters. For more information, visit our website at *www.mathxl.com*, or contact your Pearson representative.

StatCrunch®

StatCrunch® is an online statistical software website that allows users to perform complex analyses, share data sets, and generate compelling reports of their data. Developed by programmers and statisticians, StatCrunch currently has more than twelve thousand data sets available for students to analyze, covering almost any topic of interest. Interactive graphics are embedded to help users understand statistical concepts and are available for export to enrich reports with visual representations of data. Additional features include:

- A full range of numerical and graphical methods that allow users to analyze and gain insights from any data set.
- Flexible upload options that allow users to work with their .txt or Excel® files, both online and offline.
- Reporting options that help users create a wide variety of visually appealing representations of their data.

StatCrunch is available to qualified adopters. For more information, visit our website at *www.statcrunch.com*, or contact your Pearson representative.

ACKNOWLEDGMENTS

We owe a debt of gratitude to the many reviewers who helped us shape and refine *Elementary Statistics: Picturing the World,* Fifth Edition.

REVIEWERS OF THE CURRENT EDITION

Dawn Dabney, Northeast State Community College
David Gilbert, Santa Barbara City College
Donna Gorton, Butler Community College
Dr. Larry Green, Lake Tahoe Community College
Lloyd Jaisingh, Morehead State
Austin Lovenstein, Pulaski Technical College
Lyn A. Noble, Florida Community College at Jacksonville—
    South Campus
Nishant Patel, Northwest Florida State
Jack Plaggemeyer, Little Big Horn College
Abdullah Shuaibi, Truman College
Cathleen Zucco-Teveloff, Rowan University

REVIEWERS OF THE PREVIOUS EDITIONS

Rosalie Abraham, Florida Community College at
    Jacksonville
Ahmed Adala, Metropolitan Community College
Olcay Akman, College of Charleston
Polly Amstutz, University of Nebraska, Kearney
John J. Avioli, Christopher Newport University
David P. Benzel, Montgomery College
John Bernard, University of Texas—Pan American
G. Andy Chang, Youngstown State University
Keith J. Craswell, Western Washington University
Carol Curtis, Fresno City College
Cara DeLong, Fayetteville Technical Community College
Ginger Dewey, York Technical College
David DiMarco, Neumann College
Gary Egan, Monroe Community College
Charles Ehler, Anne Arundel Community College
Harold W. Ellingsen, Jr., SUNY—Potsdam
Michael Eurgubian, Santa Rosa Jr. College
Jill Fanter, Walters State Community College
Douglas Frank, Indiana University of Pennsylvania
Frieda Ganter, California State University
Sonja Hensler, St. Petersburg Jr. College
Sandeep Holay, Southeast Community College, Lincoln Campus

Nancy Johnson, Manatee Community College
Martin Jones, College of Charleston
David Kay, Moorpark College
Mohammad Kazemi, University of North Carolina—Charlotte
Jane Keller, Metropolitan Community College
Susan Kellicut, Seminole Community College
Hyune-Ju Kim, Syracuse University
Rita Kolb, Cantonsville Community College
Rowan Lindley, Westchester Community College
Jeffrey Linek, St. Petersburg Jr. College
Benny Lo, DeVry University, Fremont
Diane Long, College of DuPage
Rhonda Magel, North Dakota State University
Mike McGann, Ventura Community College
Vicki McMillian, Ocean County College
Lynn Meslinsky, Erie Community College
Lyn A. Noble, Florida Community College at Jacksonville—
    South Campus
Julie Norton, California State University—Hayward
Lynn Onken, San Juan College
Lindsay Packer, College of Charleston
Eric Preibisius, Cuyamaca Community College
Melonie Rasmussen, Pierce College
Neal Rogness, Grand Valley State University
Elisabeth Schuster, Benedictine University
Jean Sells, Sacred Heart University
John Seppala, Valdosta State University
Carole Shapero, Oakton Community College
Aileen Solomon, Trident Technical College
Sandra L. Spain, Thomas Nelson Community College
Michelle Strager-McCarney, Penn State—Erie, The Behrend
    College
Deborah Swiderski, Macomb Community College
William J. Thistleton, SUNY—Institute of Technology, Utica
Agnes Tuska, California State University—Fresno
Clark Vangilder, DeVry University
Ting-Xiu Wang, Oakton Community College
Dex Whittinghall, Rowan University

We also give special thanks to the people at Pearson Education who worked with us in the development of *Elementary Statistics: Picturing the World,* Fifth Edition: Marianne Stepanian, Dana Jones, Chere Bemelmans, Alex Gay, Kathleen DeChavez, Audra Walsh, Tamela Ambush, Joyce Kneuer, Courtney Marsh, Andrea Sheehan and Rich Williams. We also thank the staff of Larson Texts, Inc., who assisted with the development and production of the book. On a personal level, we are grateful to our spouses, Deanna Gilbert Larson and Richard Farber, for their love, patience, and support. Also, a special thanks goes to R. Scott O'Neil.

We have worked hard to make *Elementary Statistics: Picturing the World,* Fifth Edition, a clean, clear, and enjoyable text from which to teach and learn statistics. Despite our best efforts to ensure accuracy and ease of use, many users will undoubtedly have suggestions for improvement. We welcome your suggestions.

Ron Larson

Ron Larson, odx@psu.edu

Betsy Farber

Betsy Farber, farberb@bucks.edu

HOW TO STUDY STATISTICS

FOR EXTRA HELP:
MyStatLab

STUDY STRATEGIES

Congratulations! You are about to begin your study of statistics. As you progress through the course, you should discover how to use statistics in your everyday life and in your career. The prerequisites for this course are two years of algebra, an open mind, and a willingness to study. When you are studying statistics, the material you learn each day builds on material you learned previously. There are no shortcuts—you must keep up with your studies every day. Before you begin, read through the following hints that will help you succeed.

Making a Plan Make your own course plan right now! A good rule of thumb is to study at least two hours for every hour in class. After your first major exam, you will know if your efforts were sufficient. If you did not get the grade you wanted, then you should increase your study time, improve your study efficiency, or both.

Preparing for Class Before every class, review your notes from the previous class and read the portion of the text that is to be covered. Pay special attention to the definitions and rules that are highlighted. Read the examples and work through the Try It Yourself exercises that accompany each example. These steps take self-discipline, but they will pay off because you will benefit much more from your instructor's presentation.

Attending Class Attend every class. Arrive on time with your text, materials for taking notes, and your calculator. If you must miss a class, get the notes from another student, go to a tutor or your instructor for help, or view the appropriate Video on DVD. Try to learn the material that was covered in the class you missed before attending the next class.

Participating in Class When reading the text before class, reviewing your notes from a previous class, or working on your homework, write down any questions you have about the material. Ask your instructor these questions during class. Doing so will help you (and others in your class) understand the material better.

Taking Notes
During class, be sure to take notes on definitions, examples, concepts, and rules. Focus on the instructor's cues to identify important material. Then, as soon after class as possible, review your notes and add any explanations that will help to make your notes more understandable to you.

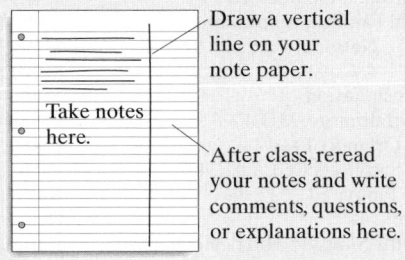

Draw a vertical line on your note paper.

Take notes here.

After class, reread your notes and write comments, questions, or explanations here.

Doing the Homework Learning statistics is like learning to play the piano or to play basketball. You cannot develop skills just by watching someone do it; you must do it yourself. The best time to do your homework is right after class, when the concepts are still fresh in your mind. Doing homework at this time increases your chances of retaining the information in long-term memory.

Finding a Study Partner When you get stuck on a problem, you may find that it helps to work with a partner. Even if you feel you are giving more help than you are getting, you will find that teaching others is an excellent way to learn.

Keeping Up with the Work Don't let yourself fall behind in this course. If you are having trouble, seek help immediately—from your instructor, a statistics tutor, your study partner, or additional study aids such as the Chapter Quiz Prep videos on DVD-ROM and the Try It Yourself video clips on the videos on DVD-ROM. Remember: If you have trouble with one section of your statistics text, there's a good chance that you will have trouble with later sections unless you take steps to improve your understanding.

Getting Stuck Every statistics student has had this experience: You work a problem and cannot solve it, or the answer you get does not agree with the one given in the text. When this happens, consider asking for help or taking a break to clear your thoughts. You might even want to sleep on it, or rework the problem, or reread the section in the text. Avoid getting frustrated or spending too much time on a single problem.

Preparing for Tests Cramming for a statistics test seldom works. If you keep up with the work and follow the suggestions given here, you should be almost ready for the test. To prepare for the chapter test, review the Chapter Summary and work the Review Exercises and the Cumulative Review Exercises. Then set aside some time to take the sample Chapter Quiz. Analyze the results of your Chapter Quiz to locate and correct test-taking errors.

Taking a Test Most instructors do not recommend studying right up to the minute the test begins. Doing so tends to make people anxious. The best cure for test-taking anxiety is to prepare well in advance. Once the test begins, read the directions carefully and work at a reasonable pace. (You might want to read the entire test first, then work the problems in the order in which you feel most comfortable.) Don't rush! People who hurry tend to make careless errors. If you finish early, take a few moments to clear your thoughts and then go over your work.

Learning from Mistakes After your test is returned to you, go over any errors you might have made. Doing so will help you avoid repeating some systematic or conceptual errors. Don't dismiss any error as just a "dumb mistake." Take advantage of any mistakes by hunting for ways to improve your test-taking skills.

INDEX OF APPLICATIONS

1 CHAPTER

INTRODUCTION TO STATISTICS

In 2008, the population of New Orleans, Louisiana grew faster than any other large city in the United States. Despite the increase, the population of 311,853 was still well below the pre-Hurricane Katrina population of 484,674.

You are already familiar with many of the practices of statistics, such as taking surveys, collecting data, and describing populations. What you may not know is that collecting accurate statistical data is often difficult and costly. Consider, for instance, the monumental task of counting and describing the entire population of the United States. If you were in charge of such a census, how would you do it? How would you ensure that your results are accurate? These and many more concerns are the responsibility of the United States Census Bureau, which conducts the census every decade.

In Chapter 1, you will be introduced to the basic concepts and goals of statistics. For instance, statistics were used to construct the following graphs, which show the fastest growing U.S. cities (population over 100,000) in 2008 by percent increase in population, U.S. cities with the largest numerical increases in population, and the regions where the cities are located.

For the 2010 Census, the Census Bureau sent short forms to every household. Short forms ask all members of every household such things as their gender, age, race, and ethnicity. Previously, a long form, which covered additional topics, was sent to about 17% of the population. But for the first time since 1940, the long form is being replaced by the American Community Survey, which will survey about 3 million households a year throughout the decade. These 3 million households will form a sample. In this course, you will learn how the data collected from a sample are used to infer characteristics about the entire population.

Fastest Growing U.S. Cities (Population over 100,000)

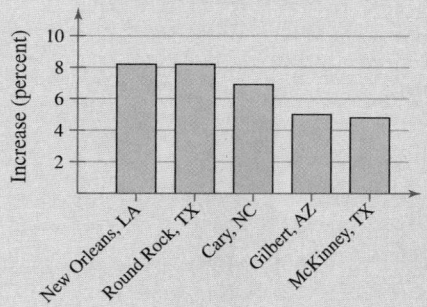

Location of the 25 Fastest Growing U.S. Cities

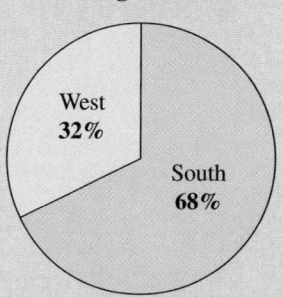

U.S. Cities with Largest Numerical Increases

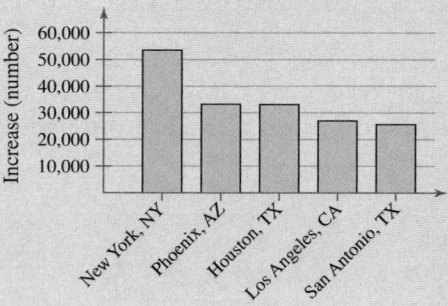

Location of the 25 U.S. Cities with Largest Numerical Increases

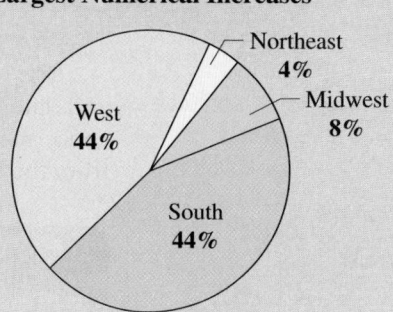

1.1   An Overview of Statistics

A Definition of Statistics ▸ Data Sets ▸ Branches of Statistics

WHAT YOU SHOULD LEARN

▸ The definition of statistics

▸ How to distinguish between a population and a sample and between a parameter and a statistic

▸ How to distinguish between descriptive statistics and inferential statistics

Note to Instructor*

To initiate a class discussion, bring a recent newspaper article that describes the results of a statistical study.

*Annotations such as this appear in this Annotated Instructor's Edition, but not in the student text.

▸ A DEFINITION OF STATISTICS

As you begin this course, you may wonder: *What is statistics? Why should I study statistics? How can studying statistics help me in my profession?* Almost every day you are exposed to statistics. For instance, consider the following.

- "The number of Americans with diabetes will nearly double in the next 25 years." *(Source: Diabetes Care)*

- "The NRF expects holiday sales to decline 1% versus a 3.4% drop in holiday sales the previous year." *(Source: National Retail Federation)*

- "EIA projects total U.S. natural gas consumption will decline by 2.6 percent in 2009 and increase by 0.5 percent in 2010." *(Source: Energy Information Administration)*

The three statements you just read are based on the collection of *data*.

DEFINITION

Data consist of information coming from observations, counts, measurements, or responses.

Sometimes data are presented graphically. If you have ever read *USA TODAY*, you have certainly seen one of that newspaper's most popular features, *USA TODAY Snapshots*. Graphics such as this present information in a way that is easy to understand.

Job seekers need a keen eye

How many typos in a résumé does it take for you to decide not to consider a job candidate for a position with your company?

One 40%
Two 36%
Three 14%
Don't know 3%
Four or more 7%

Source: Accountemps

The use of statistics dates back to census taking in ancient Babylonia, Egypt, and later in the Roman Empire, when data were collected about matters concerning the state, such as births and deaths. In fact, the word *statistics* is derived from the Latin word *status*, meaning "state." So, what is statistics?

DEFINITION

Statistics is the science of collecting, organizing, analyzing, and interpreting data in order to make decisions.

▶ DATA SETS

There are two types of data sets you will use when studying statistics. These data sets are called *populations* and *samples*.

DEFINITION

A **population** is the collection of *all* outcomes, responses, measurements, or counts that are of interest.

A **sample** is a subset, or part, of a population.

A sample should be representative of a population so that sample data can be used to form conclusions about that population. Sample data must be collected using an appropriate method, such as *random sampling*. (You will learn more about random sampling in Section 1.3.) If they are not collected using an appropriate method, the data are of no value.

EXAMPLE 1

▶ Identifying Data Sets

In a recent survey, 1500 adults in the United States were asked if they thought there was solid evidence of global warming. Eight hundred fifty-five of the adults said yes. Identify the population and the sample. Describe the sample data set. *(Adapted from Pew Research Center)*

▶ Solution

The population consists of the responses of all adults in the United States, and the sample consists of the responses of the 1500 adults in the United States in the survey. The sample is a subset of the responses of all adults in the United States. The sample data set consists of 855 yes's and 645 no's.

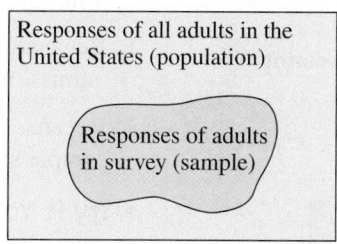

Responses of all adults in the United States (population)

Responses of adults in survey (sample)

▶ Try It Yourself 1

The U.S. Department of Energy conducts weekly surveys of approximately 900 gasoline stations to determine the average price per gallon of regular gasoline. On January 11, 2010, the average price was $2.75 per gallon. Identify the population and the sample. Describe the sample data set. *(Source: Energy Information Administration)*

a. Identify the *population* and the *sample*.
b. What does the sample data set consist of?   *Answer: Page A30*

Whether a data set is a population or a sample usually depends on the context of the real-life situation. For instance, in Example 1, the population was the set of responses of all adults in the United States. Depending on the purpose of the survey, the population could have been the set of responses of all adults who live in California or who have cellular phones or who read a particular magazine.

Two important terms that are used throughout this course are *parameter* and *statistic*.

DEFINITION

A **parameter** is a numerical description of a *population* characteristic.

A **statistic** is a numerical description of a *sample* characteristic.

It is important to note that a sample statistic can differ from sample to sample whereas a population parameter is constant for a population.

EXAMPLE 2

▶ **Distinguishing Between a Parameter and a Statistic**

Decide whether the numerical value describes a population parameter or a sample statistic. Explain your reasoning.

1. A recent survey of 200 college career centers reported that the average starting salary for petroleum engineering majors is $83,121. *(Source: National Association of Colleges and Employers)*

2. The 2182 students who accepted admission offers to Northwestern University in 2009 have an average SAT score of 1442. *(Source: Northwestern University)*

3. In a random check of a sample of retail stores, the Food and Drug Administration found that 34% of the stores were not storing fish at the proper temperature.

▶ **Solution**

1. Because the average of $83,121 is based on a subset of the population, it is a sample statistic.

2. Because the SAT score of 1442 is based on all the students who accepted admission offers in 2009, it is a population parameter.

3. Because the percent of 34% is based on a subset of the population, it is a sample statistic.

▶ **Try It Yourself 2**

In 2009, Major League Baseball teams spent a total of $2,655,395,194 on players' salaries. Does this numerical value describe a population parameter or a sample statistic? *(Source: USA Today)*

a. Decide whether the numerical value is from a *population* or a *sample*.
b. Specify whether the numerical value is a *parameter* or a *statistic*.

Answer: Page A30

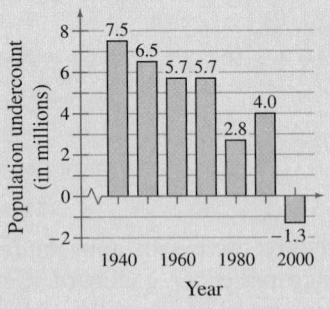
In this course, you will see how the use of statistics can help you make informed decisions that affect your life. Consider the census that the U.S. government takes every decade. When taking the census, the Census Bureau attempts to contact everyone living in the United States. Although it is impossible to count everyone, it is important that the census be as accurate as it can be, because public officials make many decisions based on the census information. Data collected in the 2010 census will determine how to assign congressional seats and how to distribute public funds.

▸ BRANCHES OF STATISTICS

The study of statistics has two major branches: *descriptive statistics* and *inferential statistics*.

DEFINITION

Descriptive statistics is the branch of statistics that involves the organization, summarization, and display of data.

Inferential statistics is the branch of statistics that involves using a sample to draw conclusions about a population. A basic tool in the study of inferential statistics is probability.

EXAMPLE 3

▸ **Descriptive and Inferential Statistics**

Decide which part of the study represents the descriptive branch of statistics. What conclusions might be drawn from the study using inferential statistics?

1. A large sample of men, aged 48, was studied for 18 years. For unmarried men, approximately 70% were alive at age 65. For married men, 90% were alive at age 65. *(Source: The Journal of Family Issues)*

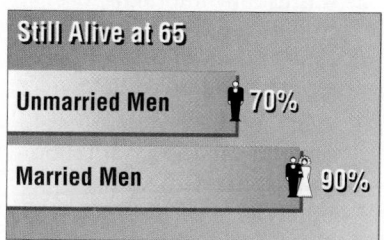

Still Alive at 65

Unmarried Men　70%

Married Men　90%

2. In a sample of Wall Street analysts, the percentage who incorrectly forecasted high-tech earnings in a recent year was 44%. *(Source: Bloomberg News)*

▸ **Solution**

1. Descriptive statistics involves statements such as "For unmarried men, approximately 70% were alive at age 65" and "For married men, 90% were alive at 65." A possible inference drawn from the study is that being married is associated with a longer life for men.

2. The part of this study that represents the descriptive branch of statistics involves the statement "the percentage [of Wall Street analysts] who incorrectly forecasted high-tech earnings in a recent year was 44%." A possible inference drawn from the study is that the stock market is difficult to forecast, even for professionals.

▸ **Try It Yourself 3**

A survey conducted among 1017 men and women by Opinion Research Corporation International found that 76% of women and 60% of men had a physical examination within the previous year. *(Source: Men's Health)*

a. Identify the *descriptive* aspect of the survey.
b. What *inferences* could be drawn from this survey?   *Answer: Page A30*

Note to Instructor

Ask students to identify the sample and the population in Example 3. Also ask students to suggest other inferences that might be drawn. This would be a good time to tell students that one can never be 100% sure about inferences.

Throughout this course you will see applications of both branches. A major theme in this course will be how to use sample statistics to make inferences about unknown population parameters.

1.1 EXERCISES

FOR EXTRA HELP:
MyStatLab

1. A sample is a subset of a population.

2. It is usually impractical (too expensive and/or time-consuming) to obtain all the population data.

3. A parameter is a numerical description of a population characteristic. A statistic is a numerical description of a sample characteristic.

4. The two main branches of statistics are descriptive statistics and inferential statistics.

5. False. A statistic is a numerical measure that describes a sample characteristic.

6. True   7. True

8. False. Inferential statistics involves using a sample to draw conclusions about a population.

9. False. A population is the collection of *all* outcomes, responses, measurements, or counts that are of interest.

10. See Selected Answers, page A109.

11. Population, because it is a collection of the heights of all the players on the school's basketball team.

12. See Selected Answers, page A109.

13. Sample, because the collection of the 500 spectators is a subset of the population.

14. See Selected Answers, page A109.

15. Sample, because the collection of the 20 patients is a subset of the population.

■ BUILDING BASIC SKILLS AND VOCABULARY

1. How is a sample related to a population?

2. Why is a sample used more often than a population?

3. What is the difference between a parameter and a statistic?

4. What are the two main branches of statistics?

True or False? *In Exercises 5–10, determine whether the statement is true or false. If it is false, rewrite it as a true statement.*

5. A statistic is a measure that describes a population characteristic.

6. A sample is a subset of a population.

7. It is impossible for the Census Bureau to obtain all the census data about the population of the United States.

8. Inferential statistics involves using a population to draw a conclusion about a corresponding sample.

9. A population is the collection of some outcomes, responses, measurements, or counts that are of interest.

10. A sample statistic will not change from sample to sample.

Classifying a Data Set *In Exercises 11–20, determine whether the data set is a population or a sample. Explain your reasoning.*

11. The height of each player on a school's basketball team

12. The amount of energy collected from every wind turbine on a wind farm

13. A survey of 500 spectators from a stadium with 42,000 spectators

14. The annual salary of each pharmacist at a pharmacy

15. The cholesterol levels of 20 patients in a hospital with 100 patients

16. The number of televisions in each U.S. household

17. The final score of each golfer in a tournament

18. The age of every third person entering a clothing store

19. The political party of every U.S. president

20. The soil contamination levels at 10 locations near a landfill

Graphical Analysis *In Exercises 21–24, use the Venn diagram to identify the population and the sample.*

21.
Parties of registered voters in Warren County

Parties of Warren County voters who respond to online survey

22.
Number of students who donate at a blood drive

Number of students who donate that have type O⁺ blood

16. Population, because it is a collection of the number of televisions in all U.S. households.

17. Population, because it is a collection of all the golfers' scores in the tournament.

18. Sample, because only the age of every third person entering the clothing store is recorded.

19. Population, because it is a collection of all the U.S. presidents' political parties.

20. Sample, because the collection of the 10 soil contamination levels is a subset of the population.

21. Population: Parties of registered voters in Warren County

    Sample: Parties of Warren County voters who respond to online survey

22. See Selected Answers, page A109.

23. See Odd Answers, page A50.

24. See Selected Answers, page A109.

25. Population: Collection of the responses of all adults in the United States

    Sample: Collection of the responses of the 1000 adults surveyed

26. See Selected Answers, page A109.

27. See Odd Answers, page A50.

28. See Selected Answers, page A109.

29. Population: Collection of the opinions of all registered voters

    Sample: Collection of the opinions of the 800 registered voters surveyed

30. See Selected Answers, page A109.

31. See Odd Answers, page A50.

32. See Selected Answers, page A109.

33. See Odd Answers, page A50.

34. Population: Collection of the quality of all light bulbs from the day's production

    Sample: Collection of the quality of the 20 light bulbs selected from the day's production

35. Statistic. The value $68,000 is a numerical description of a sample of annual salaries.

36. Statistic. 43% is a numerical description of a sample of high school students.

23.

Ages of adults in the United States who own cellular phones

Ages of adults in the U.S. who own Samsung cellular phones

24.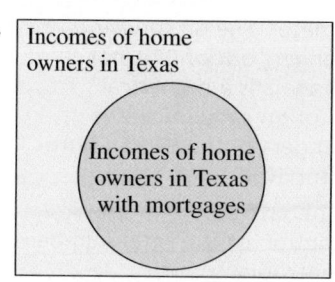

Incomes of home owners in Texas

Incomes of home owners in Texas with mortgages

■ USING AND INTERPRETING CONCEPTS

Identifying Populations and Samples *In Exercises 25–34, identify the population and the sample.*

25. A survey of 1000 U.S. adults found that 59% think buying a home is the best investment a family can make. *(Source: Rasmussen Reports)*

26. A study of 33,043 infants in Italy was conducted to find a link between a heart rhythm abnormality and sudden infant death syndrome. *(Source: New England Journal of Medicine)*

27. A survey of 1442 U.S. adults found that 36% received an influenza vaccine for the current flu season. *(Source: Zogby International)*

28. A survey of 1600 people found that 76% plan on using the Microsoft Windows 7™ operating system at their businesses. *(Source: Information Technology Intelligence Corporation and Sunbelt Software)*

29. A survey of 800 registered voters found that 50% think economic stimulus is the most important issue to consider when voting for Congress. *(Source: Diageo/Hotline Poll)*

30. A survey of 496 students at a college found that 10% planned on traveling out of the country during spring break.

31. A survey of 546 U.S. women found that more than 56% are the primary investors in their households. *(Adapted from Roper Starch Worldwide for Intuit)*

32. A survey of 791 vacationers from the United States found that they planned on spending at least $2000 for their next vacation.

33. A magazine mails questionnaires to each company in Fortune magazine's top 100 best companies to work for and receives responses from 85 of them.

34. At the end of the day, a quality control inspector selects 20 light bulbs from the day's production and tests them.

Distinguishing Between a Parameter and a Statistic *In Exercises 35–42, determine whether the numerical value is a parameter or a statistic. Explain your reasoning.*

35. The average annual salary for 35 of a company's 1200 accountants is $68,000.

36. In a survey of a sample of high school students, 43% said that their mothers had taught them the most about managing money. *(Source: Harris Poll for Girls Incorporated)*

37. Parameter. The 62 surviving passengers out of 97 total passengers is a numerical description of all of the passengers of the Hindenburg that survived.

38. Parameter. 52% is a numerical description of the total number of governors.

39. Statistic. 8% is a numerical description of a sample of computer users.

40. Parameter. 12% is a numerical description of all new magazines.

41. Statistic. 44% is a numerical description of a sample of all people.

42. Parameter. The score 21.0 is a numerical description of ACT scores for all graduates.

43. The statement "more than 56% are the primary investors in their households" is an example of descriptive statistics.

An inference drawn from the sample is that an association exists between U.S. women and being the primary investors in their households.

44. The statement "spending at least $2000 for their next vacation" is an example of descriptive statistics.

An inference drawn from the sample is that vacations in the United States cost more than $2000.

45. Answers will vary.

46. See Selected Answers, page A109.

47. (a) An inference drawn from the sample is that senior citizens who live in Florida have better memories than senior citizens who do not live in Florida.

(b) This inference may incorrectly imply that if you live in Florida you will have a better memory.

48. (a) An inference drawn from the sample is that the obesity rate among boys ages 2 to 19 is increasing.

(b) This inference may incorrectly imply that the trend will continue in future years.

49. Answers will vary.

37. Sixty-two of the 97 passengers aboard the Hindenburg airship survived its explosion.

38. In January 2010, 52% of the governors of the 50 states in the United States were Democrats.

39. In a survey of 300 computer users, 8% said their computers had malfunctions that needed to be repaired by service technicians.

40. In a recent year, the interest category for 12% of all new magazines was sports. *(Source: Oxbridge Communications)*

41. In a recent survey of 2000 people, 44% said China is the world's leading economic power. *(Source: Pew Research Center)*

42. In a recent year, the average math scores for all graduates on the ACT was 21.0. *(Source: ACT, Inc.)*

43. Which part of the survey described in Exercise 31 represents the descriptive branch of statistics? Make an inference based on the results of the survey.

44. Which part of the survey described in Exercise 32 represents the descriptive branch of statistics? Make an inference based on the results of the survey.

■ EXTENDING CONCEPTS

45. Identifying Data Sets in Articles Find a newspaper or magazine article that describes a survey.

(a) Identify the sample used in the survey.

(b) What is the sample's population?

(c) Make an inference based on the results of the survey.

46. Sleep Deprivation In a recent study, volunteers who had 8 hours of sleep were three times more likely to answer questions correctly on a math test than were sleep-deprived participants. *(Source: CBS News)*

(a) Identify the sample used in the study.

(b) What is the sample's population?

(c) Which part of the study represents the descriptive branch of statistics?

(d) Make an inference based on the results of the study.

47. Living in Florida A study shows that senior citizens who live in Florida have better memories than senior citizens who do not live in Florida.

(a) Make an inference based on the results of this study.

(b) What is wrong with this type of reasoning?

48. Increase in Obesity Rates A study shows that the obesity rate among boys ages 2 to 19 has increased over the past several years. *(Source: Washington Post)*

(a) Make an inference based on the results of this study.

(b) What is wrong with this type of reasoning?

49. Writing Write an essay about the importance of statistics for one of the following.

- A study on the effectiveness of a new drug
- An analysis of a manufacturing process
- Making conclusions about voter opinions using surveys

1.2 Data Classification

Types of Data ▸ Levels of Measurement

WHAT YOU SHOULD LEARN

▸ How to distinguish between qualitative data and quantitative data

▸ How to classify data with respect to the four levels of measurement: nominal, ordinal, interval, and ratio

▸ TYPES OF DATA

When doing a study, it is important to know the kind of data involved. The nature of the data you are working with will determine which statistical procedures can be used. In this section, you will learn how to classify data by type and by level of measurement. Data sets can consist of two types of data: *qualitative data* and *quantitative data*.

DEFINITION

Qualitative data consist of attributes, labels, or nonnumerical entries.

Quantitative data consist of numerical measurements or counts.

EXAMPLE 1

▸ **Classifying Data by Type**

The suggested retail prices of several Ford vehicles are shown in the table. Which data are qualitative data and which are quantitative data? Explain your reasoning. *(Source: Ford Motor Company)*

Model	Suggested retail price
Focus Sedan	$15,995
Fusion	$19,270
Mustang	$20,995
Edge	$26,920
Flex	$28,495
Escape Hybrid	$32,260
Expedition	$35,085
F-450	$44,145

▸ **Solution**

The information shown in the table can be separated into two data sets. One data set contains the names of vehicle models, and the other contains the suggested retail prices of vehicle models. The names are nonnumerical entries, so these are qualitative data. The suggested retail prices are numerical entries, so these are quantitative data.

▸ **Try It Yourself 1**

The populations of several U.S. cities are shown in the table. Which data are qualitative data and which are quantitative data? *(Source: U.S. Census Bureau)*

a. *Identify* the two data sets.
b. Decide whether each data set consists of *numerical* or *nonnumerical* entries.
c. Specify the *qualitative* data and the *quantitative* data. *Answer: Page A30*

City	Population
Baltimore, MD	636,919
Jacksonville, FL	807,815
Memphis, TN	669,651
Pasadena, CA	143,080
San Antonio, TX	1,351,305
Seattle, WA	598,541

▶ LEVELS OF MEASUREMENT

Another characteristic of data is its level of measurement. The level of measurement determines which statistical calculations are meaningful. The four levels of measurement, in order from lowest to highest, are *nominal*, *ordinal*, *interval*, and *ratio*.

DEFINITION

Data at the **nominal level of measurement** are qualitative only. Data at this level are categorized using names, labels, or qualities. No mathematical computations can be made at this level.

Data at the **ordinal level of measurement** are qualitative or quantitative. Data at this level can be arranged in order, or ranked, but differences between data entries are not meaningful.

When numbers are at the nominal level of measurement, they simply represent a label. Examples of numbers used as labels include Social Security numbers and numbers on sports jerseys. For instance, it would not make sense to add the numbers on the players' jerseys for the Chicago Bears.

EXAMPLE 2

▶ Classifying Data by Level

Two data sets are shown. Which data set consists of data at the nominal level? Which data set consists of data at the ordinal level? Explain your reasoning. *(Source: The Nielsen Company)*

Top Five TV Programs (from 5/4/09 to 5/10/09)
1. American Idol–Wednesday
2. American Idol–Tuesday
3. Dancing with the Stars
4. NCIS
5. The Mentalist

Network Affiliates in Pittsburgh, PA	
WTAE	(ABC)
WPXI	(NBC)
KDKA	(CBS)
WPGH	(FOX)

▶ Solution

The first data set lists the ranks of five TV programs. The data set consists of the ranks 1, 2, 3, 4, and 5. Because the ranks can be listed in order, these data are at the ordinal level. Note that the difference between a rank of 1 and 5 has no mathematical meaning. The second data set consists of the call letters of each network affiliate in Pittsburgh. The call letters are simply the names of network affiliates, so these data are at the nominal level.

▶ Try It Yourself 2

Consider the following data sets. For each data set, decide whether the data are at the nominal level or at the ordinal level.

1. The final standings for the Pacific Division of the National Basketball Association

2. A collection of phone numbers

a. *Identify* what each data set represents.

b. Specify the *level of measurement* and justify your answer.

Answer: Page A30

PICTURING THE WORLD

In 2009, Forbes Magazine chose the 75 best business schools in the United States. Forbes based their rankings on the return on investment achieved by the graduates from the class of 2004. Graduates of the top five M.B.A. programs typically earn more than $200,000 within five years. (Source: Forbes)

Forbes Top Five U.S. Business Schools
1. Stanford
2. Dartmouth
3. Harvard
4. Chicago
5. Pennsylvania

In this list, what is the level of measurement?

The data is at the ordinal level of measurement.

The two highest levels of measurement consist of quantitative data only.

> ### DEFINITION
>
> Data at the **interval level of measurement** can be ordered, and meaningful differences between data entries can be calculated. At the interval level, a zero entry simply represents a position on a scale; the entry is not an inherent zero.
>
> Data at the **ratio level of measurement** are similar to data at the interval level, with the added property that a zero entry is an inherent zero. A ratio of two data values can be formed so that one data value can be meaningfully expressed as a multiple of another.

An *inherent zero* is a zero that implies "none." For instance, the amount of money you have in a savings account could be zero dollars. In this case, the zero represents no money; it is an inherent zero. On the other hand, a temperature of 0°C does not represent a condition in which no heat is present. The 0°C temperature is simply a position on the Celsius scale; it is not an inherent zero.

To distinguish between data at the interval level and at the ratio level, determine whether the expression "twice as much" has any meaning in the context of the data. For instance, $2 is twice as much as $1, so these data are at the ratio level. On the other hand, 2°C is not twice as warm as 1°C, so these data are at the interval level.

New York Yankees' World Series Victories (Years)
1923, 1927, 1928, 1932, 1936, 1937, 1938, 1939, 1941, 1943, 1947, 1949, 1950, 1951, 1952, 1953, 1956, 1958, 1961, 1962, 1977, 1978, 1996, 1998, 1999, 2000, 2009

2009 American League Home Run Totals (by Team)	
Baltimore	160
Boston	212
Chicago	184
Cleveland	161
Detroit	183
Kansas City	144
Los Angeles	173
Minnesota	172
New York	244
Oakland	135
Seattle	160
Tampa Bay	199
Texas	224
Toronto	209

EXAMPLE 3

▶ Classifying Data by Level

Two data sets are shown at the left. Which data set consists of data at the interval level? Which data set consists of data at the ratio level? Explain your reasoning. *(Source: Major League Baseball)*

▶ Solution

Both of these data sets contain quantitative data. Consider the dates of the Yankees' World Series victories. It makes sense to find differences between specific dates. For instance, the time between the Yankees' first and last World Series victories is

$$2009 - 1923 = 86 \text{ years.}$$

But it does not make sense to say that one year is a multiple of another. So, these data are at the interval level. However, using the home run totals, you can find differences *and* write ratios. From the data, you can see that Texas hit 63 more home runs than Cleveland hit and that New York hit about 1.5 times as many home runs as Seattle hit. So, these data are at the ratio level.

▶ Try It Yourself 3

Decide whether the data are at the interval level or at the ratio level.

1. The body temperatures (in degrees Fahrenheit) of an athlete during an exercise session

2. The heart rates (in beats per minute) of an athlete during an exercise session

a. *Identify* what each data set represents.
b. Specify the *level of measurement* and justify your answer.

Answer: Page A30

The following tables summarize which operations are meaningful at each of the four levels of measurement. When identifying a data set's level of measurement, use the highest level that applies.

Level of measurement	Put data in categories	Arrange data in order	Subtract data values	Determine if one data value is a multiple of another
Nominal	Yes	No	No	No
Ordinal	Yes	Yes	No	No
Interval	Yes	Yes	Yes	No
Ratio	Yes	Yes	Yes	Yes

Summary of Four Levels of Measurement

	Example of a Data Set	Meaningful Calculations
Nominal Level (Qualitative data)	*Types of Shows Televised by a Network* Comedy Documentaries Drama Cooking Reality Shows Soap Operas Sports Talk Shows	*Put in a category.* For instance, a show televised by the network could be put into one of the eight categories shown.
Ordinal Level (Qualitative or quantitative data)	*Motion Picture Association of America Ratings Description* G General Audiences PG Parental Guidance Suggested PG-13 Parents Strongly Cautioned R Restricted NC-17 No One Under 17 Admitted	Put in a category and *put in order.* For instance, a PG rating has a stronger restriction than a G rating.
Interval Level (Quantitative data)	*Average Monthly Temperatures (in degrees Fahrenheit) for Denver, CO* Jan 29.2 Jul 73.4 Feb 33.2 Aug 71.7 Mar 39.6 Sep 62.4 Apr 47.6 Oct 51.0 May 57.2 Nov 37.5 Jun 67.6 Dec 30.3 *(Source: National Climatic Data Center)*	Put in a category, put in order, and *find differences between values.* For instance, $57.2 - 47.6 = 9.6°F$. So, May is 9.6° warmer than April.
Ratio Level (Quantitative data)	*Average Monthly Precipitation (in inches) for Orlando, FL* Jan 2.4 Jul 7.2 Feb 2.4 Aug 6.3 Mar 3.5 Sep 5.8 Apr 2.4 Oct 2.7 May 3.7 Nov 2.3 Jun 7.4 Dec 2.3 *(Source: National Climatic Data Center)*	Put in a category, put in order, find differences between values, and *find ratios of values.* For instance, $\frac{7.4}{3.7} = 2$. So, there is twice as much rain in June as in May.

1.2 EXERCISES

■ BUILDING BASIC SKILLS AND VOCABULARY

1. Name each level of measurement for which data can be qualitative.

2. Name each level of measurement for which data can be quantitative.

True or False? *In Exercises 3–6, determine whether the statement is true or false. If it is false, rewrite it as a true statement.*

3. Data at the ordinal level are quantitative only.

4. For data at the interval level, you cannot calculate meaningful differences between data entries.

5. More types of calculations can be performed with data at the nominal level than with data at the interval level.

6. Data at the ratio level cannot be put in order.

■ USING AND INTERPRETING CONCEPTS

Classifying Data by Type *In Exercises 7–18, determine whether the data are qualitative or quantitative. Explain your reasoning.*

7. telephone numbers in a directory

8. heights of hot air balloons

9. body temperatures of patients

10. eye colors of models

11. lengths of songs on MP3 player

12. carrying capacities of pickups

13. player numbers for a soccer team

14. student ID numbers

15. weights of infants at a hospital

16. species of trees in a forest

17. responses on an opinion poll

18. wait times at a grocery store

Classifying Data by Level *In Exercises 19–24, determine whether the data are qualitative or quantitative, and identify the data set's level of measurement. Explain your reasoning.*

19. Football The top five teams in the final college football poll released in January 2010 are listed. *(Source: Associated Press)*

  1. Alabama   2. Texas   3. Florida   4. Boise State   5. Ohio State

20. Politics The three political parties in the 111th Congress are listed below.

  Republican   Democrat   Independent

21. Top Salespeople The regions representing the top salespeople in a corporation for the past six years are given.

  Southeast   Northwest   Northeast
  Southeast   Southwest   Southwest

22. Fish Lengths The lengths (in inches) of a sample of striped bass caught in Maryland waters are listed. *(Adapted from National Marine Fisheries Service, Fisheries Statistics and Economics Division)*

  16   17.25   19   18.75   21   20.3   19.8   24   21.82

Answers (left column):

1. Nominal and ordinal

2. Ordinal, interval, and ratio

3. False. Data at the ordinal level can be qualitative or quantitative.

4. See Selected Answers, page A109.

5. False. More types of calculations can be performed with data at the interval level than with data at the nominal level.

6. See Selected Answers, page A109.

7. Qualitative, because telephone numbers are labels.

8. See Selected Answers, page A109.

9. Quantitative, because body temperatures are numerical measurements.

10. See Selected Answers, page A109.

11. Quantitative, because song lengths are numerical measurements.

12. See Selected Answers, page A109.

13. Qualitative, because player numbers are labels.

14. See Selected Answers, page A109.

15. Quantitative, because infant weights are numerical measurements.

16. See Selected Answers, page A109.

17. Qualitative, because the poll responses are attributes.

18. See Selected Answers, page A109.

19. See Odd Answers, page A50.

20. See Selected Answers, page A109.

21. See Odd Answers, page A50.

22. See Selected Answers, page A109.

23. Qualitative. Ordinal. Data can be arranged in order, but the differences between data entries are not meaningful.

24. Quantitative. Ratio. A ratio of two data values can be formed, so one data value can be expressed as a multiple of another.

25. Ordinal

26. Ratio

27. Nominal

28. Ratio

29. (a) Interval (b) Nominal
    (c) Ratio (d) Ordinal

30. (a) Interval (b) Nominal
    (c) Interval (d) Ratio

31. An inherent zero is a zero that implies "none." Answers will vary.

32. Answers will vary.

---

23. **Best Seller List** The top five hardcover nonfiction books on *The New York Times* Best Seller List on January 19, 2010 are shown. *(Source: The New York Times)*

    1. Committed 2. Have a Little Faith 3. The Checklist Manifesto
    4. Going Rogue 5. Stones Into Schools

24. **Ticket Prices** The average ticket prices for 10 Broadway shows in 2009 are listed. *(Adapted from The Broadway League)*

    $149 $128 $124 $91 $96 $106 $112 $95 $86 $74

Graphical Analysis *In Exercises 25–28, identify the level of measurement of the data listed on the horizontal axis in the graph.*

25. **Over the Next Few Years, How Likely Is It That the United States Will Enter a 1930s-Like Depression?**

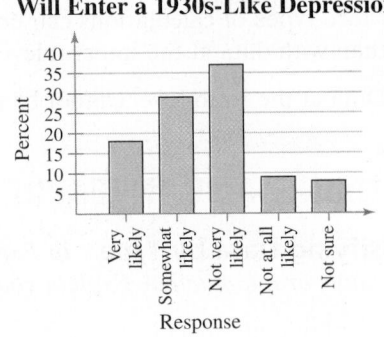

(Source: Rasmussen Reports)

26. **Average January Snowfall for 15 Cities**

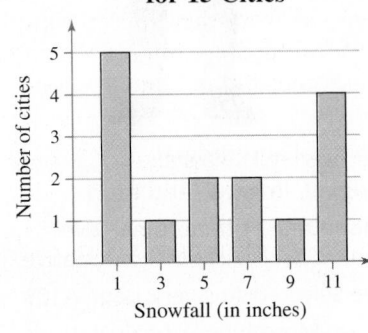

(Source: National Climatic Data Center)

27. **Gender Profile of the 111th Congress**

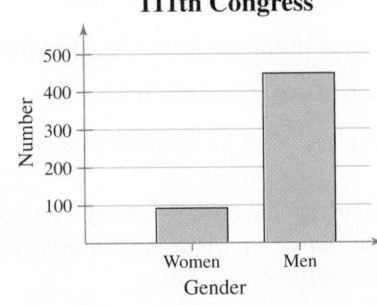

(Source: Congressional Research Service)

28. **Motor Vehicle Accidents by Year**

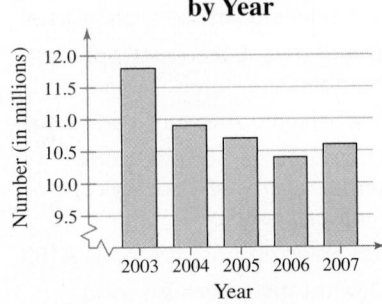

(Source: National Safety Council)

29. The following items appear on a physician's intake form. Identify the level of measurement of the data.

    a. Temperature **b.** Allergies
    c. Weight **d.** Pain level (scale of 0 to 10)

30. The following items appear on an employment application. Identify the level of measurement of the data.

    a. Highest grade level completed **b.** Gender
    c. Year of college graduation **d.** Number of years at last job

■ **EXTENDING CONCEPTS**

31. **Writing** What is an inherent zero? Describe three examples of data sets that have inherent zeros and three that do not.

32. **Writing** Describe two examples of data sets for each of the four levels of measurement. Justify your answer.

Rating Television Shows in the United States

The Nielsen Company has been rating television programs for more than 60 years. Nielsen uses several sampling procedures, but its main one is to track the viewing patterns of 20,000 households. These contain more than 45,000 people and are chosen to form a cross section of the overall population. The households represent various locations, ethnic groups, and income brackets. The data gathered from the Nielsen sample of 20,000 households are used to draw inferences about the population of all households in the United States.

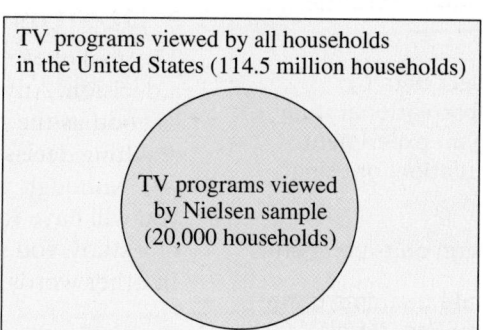

TV programs viewed by all households in the United States (114.5 million households)

TV programs viewed by Nielsen sample (20,000 households)

Top-Ranked Programs in Overall Viewing for the Week of 11/23/09–11/29/09

Rank	Rank Last Week	Program Name	Network	Day, Time	Rating	Share	Audience
1	2	Dancing with the Stars	ABC	Mon., 8:00 P.M.	12.9	19	20,411,000
2	1	NCIS	CBS	Tues., 8:00 P.M.	12.3	20	20,348,000
3	4	Dancing with the Stars Results	ABC	Tues., 9:00 P.M.	12.0	20	19,294,000
4	3	NBC Sunday Night Football	NBC	Sun., 8:15 P.M.	11.5	18	19,210,000
5	8	NCIS: Los Angeles	CBS	Tues., 9:00 P.M.	10.4	16	17,221,000
6	5	60 Minutes	CBS	Sun., 7:00 P.M.	9.0	14	14,377,000
7	15	The Big Bang Theory	CBS	Mon., 9:30 P.M.	8.4	13	14,129,000
8	16	Sunday Night NFL Pre-Kick	NBC	Sun., 8:00 P.M.	8.4	13	13,927,000
9	12	Two and a Half Men	CBS	Mon., 9:00 P.M.	8.3	12	13,877,000
10	11	Criminal Minds	CBS	Wed., 9:00 P.M.	8.2	14	13,605,000

Copyrighted information of The Nielsen Company, licensed for use herein.

■ EXERCISES

1. **Rating Points** Each rating point represents 1,145,000 households, or 1% of the households in the United States. Does a program with a rating of 8.4 have twice the number of households as a program with a rating of 4.2? Explain your reasoning.

2. **Sampling Percent** What percentage of the total number of U.S. households is used in the Nielsen sample?

3. **Nominal Level of Measurement** Which columns in the table contain data at the nominal level?

4. **Ordinal Level of Measurement** Which columns in the table contain data at the ordinal level? Describe two ways that the data can be ordered.

5. **Interval Level of Measurement** Which column in the table contains data at the interval level? How can these data be ordered?

6. **Ratio Level of Measurement** Which columns contain data at the ratio level?

7. **Rankings** The column listed as "Share" gives the percentage of televisions in use at a given time. The 11th ranked program for this week is CSI: Miami with a rating of 8.4 and share of 14. Using this information, how does Nielsen rank the programs? Why do you think they do it this way? Explain your reasoning.

8. **Inferences** What decisions (inferences) can be made on the basis of the Nielsen ratings?

1.3 Data Collection and Experimental Design

Design of a Statistical Study ▸ Data Collection ▸ Experimental Design
▸ Sampling Techniques

WHAT YOU SHOULD LEARN

▸ How to design a statistical study

▸ How to collect data by doing an observational study, performing an experiment, using a simulation, or using a survey

▸ How to design an experiment

▸ How to create a sample using random sampling, simple random sampling, stratified sampling, cluster sampling, and systematic sampling and how to identify a biased sample

▸ DESIGN OF A STATISTICAL STUDY

The goal of every statistical study is to collect data and then use the data to make a decision. Any decision you make using the results of a statistical study is only as good as the process used to obtain the data. If the process is flawed, then the resulting decision is questionable.

Although you may never have to develop a statistical study, it is likely that you will have to interpret the results of one. And before you interpret the results of a study, you should determine whether the results are valid, as well as reliable. In other words, you should be familiar with how to design a statistical study.

GUIDELINES

Designing a Statistical Study

1. Identify the variable(s) of interest (the focus) and the population of the study.
2. Develop a detailed plan for collecting data. If you use a sample, make sure the sample is representative of the population.
3. Collect the data.
4. Describe the data, using descriptive statistics techniques.
5. Interpret the data and make decisions about the population using inferential statistics.
6. Identify any possible errors.

▸ DATA COLLECTION

There are several ways you can collect data. Often, the focus of the study dictates the best way to collect data. The following is a brief summary of four methods of data collection.

- ***Do an observational study*** In an **observational study,** a researcher observes and measures characteristics of interest of part of a population but does not change existing conditions. For instance, an observational study was performed in which researchers observed and recorded the mouthing behavior on nonfood objects of children up to three years old. *(Source: Pediatrics Magazine)*

- ***Perform an experiment*** In performing an **experiment,** a **treatment** is applied to part of a population and responses are observed. Another part of the population may be used as a **control group,** in which no treatment is applied. In many cases, subjects (sometimes called **experimental units**) in the control group are given a **placebo,** which is a harmless, unmedicated treatment, that is made to look like the real treatment. The responses of the treatment group and control group can then be compared and studied. In most cases, it is a good idea to use the same number of subjects for each treatment. For instance, an experiment was performed in which diabetics took cinnamon extract daily while a control group took none. After 40 days, the diabetics who took the cinnamon reduced their risk of heart disease while the control group experienced no change. *(Source: Diabetes Care)*

INSIGHT

In an observational study, a researcher does not influence the responses. In an experiment, a researcher deliberately applies a treatment before observing the responses.

Note to Instructor

Students will be given the opportunity to use simulations when they study probability. Refer to page 187 to see an interesting example of a simulation.

- *Use a simulation*   A **simulation** is the use of a mathematical or physical model to reproduce the conditions of a situation or process. Collecting data often involves the use of computers. Simulations allow you to study situations that are impractical or even dangerous to create in real life, and often they save time and money. For instance, automobile manufacturers use simulations with dummies to study the effects of crashes on humans. Throughout this course, you will have the opportunity to use applets that simulate statistical processes on a computer.

- *Use a survey*   A **survey** is an investigation of one or more characteristics of a population. Most often, surveys are carried out on *people* by asking them questions. The most common types of surveys are done by interview, mail, or telephone. In designing a survey, it is important to word the questions so that they do not lead to biased results, which are not representative of a population. For instance, a survey is conducted on a sample of female physicians to determine whether the primary reason for their career choice is financial stability. In designing the survey, it would be acceptable to make a list of reasons and ask each individual in the sample to select her first choice.

PICTURING THE WORLD

The Gallup Organization conducts many polls (or surveys) regarding the president, Congress, and political and nonpolitical issues. A commonly cited Gallup poll is the public approval rating of the president. For instance, the approval ratings for President Barack Obama throughout 2009 are shown in the following graph. (The rating is from the poll conducted at the end of each month.)

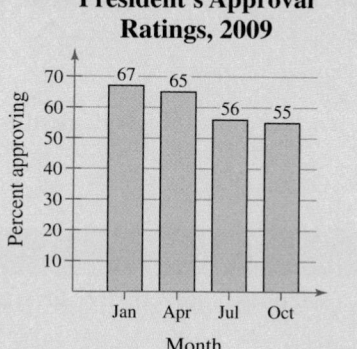

President's Approval Ratings, 2009

Discuss some ways that Gallup could select a biased sample to conduct a poll. How could Gallup select a sample that is unbiased?

Gallup may have surveyed only people who like President Obama. An unbiased sample could be obtained from random sampling by telephone.

EXAMPLE 1

▸ **Deciding on Methods of Data Collection**

Consider the following statistical studies. Which method of data collection would you use to collect data for each study? Explain your reasoning.

1. A study of the effect of changing flight patterns on the number of airplane accidents

2. A study of the effect of eating oatmeal on lowering blood pressure

3. A study of how fourth grade students solve a puzzle

4. A study of U.S. residents' approval rating of the U.S. president

▸ **Solution**

1. Because it is impractical to create this situation, use a simulation.

2. In this study, you want to measure the effect a treatment (eating oatmeal) has on patients. So, you would want to perform an experiment.

3. Because you want to observe and measure certain characteristics of part of a population, you could do an observational study.

4. You could use a survey that asks, "Do you approve of the way the president is handling his job?"

▸ **Try It Yourself 1**

Consider the following statistical studies. Which method of data collection would you use to collect data for each study?

1. A study of the effect of exercise on relieving depression

2. A study of the success of graduates of a large university in finding a job within one year of graduation

a. Identify the *focus* of the study.
b. Identify the *population* of the study.
c. Choose an appropriate *method of data collection*.   *Answer: Page A30*

▸ EXPERIMENTAL DESIGN

In order to produce meaningful unbiased results, experiments should be carefully designed and executed. It is important to know what steps should be taken to make the results of an experiment valid. Three key elements of a well-designed experiment are *control*, *randomization*, and *replication*.

Because experimental results can be ruined by a variety of factors, being able to control these influential factors is important. One such factor is a *confounding variable*.

> ### DEFINITION
>
> A **confounding variable** occurs when an experimenter cannot tell the difference between the effects of different factors on a variable.

For instance, to attract more customers, a coffee shop owner experiments by remodeling her shop using bright colors. At the same time, a shopping mall nearby has its grand opening. If business at the coffee shop increases, it cannot be determined whether it is because of the new colors or the new shopping mall. The effects of the colors and the shopping mall have been confounded.

Another factor that can affect experimental results is the *placebo effect*. The **placebo effect** occurs when a subject reacts favorably to a placebo when in fact the subject has been given no medicated treatment at all. To help control or minimize the placebo effect, a technique called *blinding* can be used.

> ### DEFINITION
>
> **Blinding** is a technique where the subjects do not know whether they are receiving a treatment or a placebo. In a **double-blind experiment,** neither the experimenter nor the subjects know if the subjects are receiving a treatment or a placebo. The experimenter is informed after all the data have been collected. This type of experimental design is preferred by researchers.

Another technique that can be used to obtain unbiased results is *randomization*.

> ### DEFINITION
>
> **Randomization** is a process of randomly assigning subjects to different treatment groups.

In a **completely randomized design,** subjects are assigned to different treatment groups through random selection. In some experiments, it may be necessary for the experimenter to use **blocks,** which are groups of subjects with similar characteristics. A commonly used experimental design is a **randomized block design.** To use a randomized block design, you should divide subjects with similar characteristics into blocks, and then, within each block, randomly assign subjects to treatment groups. For instance, an experimenter who is testing the effects of a new weight loss drink may first divide the subjects into age categories such as 30–39 years old, 40–49 years old, and over 50 years old, and then, within each age group, randomly assign subjects to either the treatment group or the control group as shown.

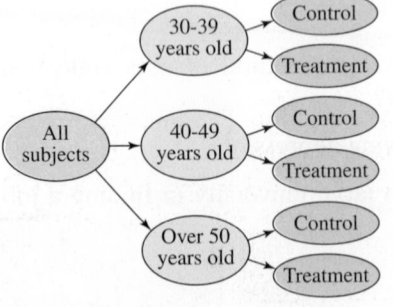

Randomized Block Design

Another type of experimental design is a **matched-pairs design,** where subjects are paired up according to a similarity. One subject in the pair is randomly selected to receive one treatment while the other subject receives a different treatment. For instance, two subjects may be paired up because of their age, geographical location, or a particular physical characteristic.

Sample size, which is the number of subjects, is another important part of experimental design. To improve the validity of experimental results, *replication* is required.

DEFINITION

Replication is the repetition of an experiment under the same or similar conditions.

For instance, suppose an experiment is designed to test a vaccine against a strain of influenza. In the experiment, 10,000 people are given the vaccine and another 10,000 people are given a placebo. Because of the sample size, the effectiveness of the vaccine would most likely be observed. But, if the subjects in the experiment are not selected so that the two groups are similar (according to age and gender), the results are of less value.

EXAMPLE 2

▶ **Analyzing an Experimental Design**

A company wants to test the effectiveness of a new gum developed to help people quit smoking. Identify a potential problem with the given experimental design and suggest a way to improve it.

1. The company identifies ten adults who are heavy smokers. Five of the subjects are given the new gum and the other five subjects are given a placebo. After two months, the subjects are evaluated and it is found that the five subjects using the new gum have quit smoking.

2. The company identifies one thousand adults who are heavy smokers. The subjects are divided into blocks according to gender. Females are given the new gum and males are given the placebo. After two months, a significant number of the female subjects have quit smoking.

▶ **Solution**

1. The sample size being used is not large enough to validate the results of the experiment. The experiment must be replicated to improve the validity.

2. The groups are not similar. The new gum may have a greater effect on women than on men, or vice versa. The subjects can be divided into blocks according to gender, but then, within each block, they must be randomly assigned to be in the treatment group or in the control group.

▶ **Try It Yourself 2**

Using the information in Example 2, suppose the company identifies 240 adults who are heavy smokers. The subjects are randomly assigned to be in a treatment group or in a control group. Each subject is also given a DVD featuring the dangers of smoking. After four months, most of the subjects in the treatment group have quit smoking.

a. Identify a *potential problem* with the experimental design.
b. How could the design be *improved*? *Answer: Page A30*

▶ SAMPLING TECHNIQUES

A **census** is a count or measure of an *entire* population. Taking a census provides complete information, but it is often costly and difficult to perform. A **sampling** is a count or measure of *part* of a population, and is more commonly used in statistical studies. To collect unbiased data, a researcher must ensure that the sample is representative of the population. Appropriate sampling techniques must be used to ensure that inferences about the population are valid. Remember that when a study is done with faulty data, the results are questionable. Even with the best methods of sampling, a **sampling error** may occur. A sampling error is the difference between the results of a sample and those of the population. When you learn about inferential statistics, you will learn techniques of controlling sampling errors.

A **random sample** is one in which every member of the population has an equal chance of being selected. A **simple random sample** is a sample in which every possible sample of the same size has the same chance of being selected. One way to collect a simple random sample is to assign a different number to each member of the population and then use a random number table like the one in Appendix B. Responses, counts, or measures for members of the population whose numbers correspond to those generated using the table would be in the sample. Calculators and computer software programs are also used to generate random numbers (see page 34).

INSIGHT

A **biased sample** is one that is not representative of the population from which it is drawn. For instance, a sample consisting of only 18- to 22-year-old college students would not be representative of the entire 18- to 22-year-old population in the country.

To explore this topic further, see Activity 1.3 on page 26.

Table 1—Random Numbers

92630	78240	19267	95457	53497	23894	37708	79862
79445	78735	71549	44843	26104	67318	00701	34986
59654	71966	27386	50004	05358	94031	29281	18544
31524	49587	76612	39789	13537	48086	59483	60680
06348	76938	90379	51392	55887	71015	09209	79157

Portion of Table 1 found in Appendix B

Consider a study of the number of people who live in West Ridge County. To use a simple random sample to count the number of people who live in West Ridge County households, you could assign a different number to each household, use a technology tool or table of random numbers to generate a sample of numbers, and then count the number of people living in each selected household.

STUDY TIP

Here are instructions for using the random integer generator on a TI-83/84 Plus for Example 3.

⟨MATH⟩

Choose the PRB menu.

5: randInt(

⟨1⟩ ⟨,⟩ ⟨7⟩ ⟨3⟩ ⟨1⟩ ⟨,⟩ ⟨8⟩ ⟨)⟩

⟨ENTER⟩

```
randInt(1,731,8)
⟨537 33 249 728…
```

Continuing to press ⟨ENTER⟩ will generate more random samples of 8 integers.

EXAMPLE 3   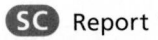 SC Report 1

▶ Using a Simple Random Sample

There are 731 students currently enrolled in a statistics course at your school. You wish to form a sample of eight students to answer some survey questions. Select the students who will belong to the simple random sample.

▶ Solution

Assign numbers 1 to 731 to the students in the course. In the table of random numbers, choose a starting place at random and read the digits in groups of three (because 731 is a three-digit number). For instance, if you started in the third row of the table at the beginning of the second column, you would group the numbers as follows:

$$719|66 \quad 2|738|6 \quad 50|004| \quad 053|58 \quad 9|403|1 \quad 29|281| \quad 185|44$$

Ignoring numbers greater than 731, the first eight numbers are 719, 662, 650, 4, 53, 589, 403, and 129. The students assigned these numbers will make up the sample. To find the sample using a TI-83/84 Plus, follow the instructions in the margin.

▶ **Try It Yourself 3**

A company employs 79 people. Choose a simple random sample of five to survey.

a. In the table in Appendix B, randomly choose a *starting place*.
b. *Read the digits* in groups of two.
c. Write the five random numbers.   *Answer: Page A30*

When you choose members of a sample, you should decide whether it is acceptable to have the same population member selected more than once. If it is acceptable, then the sampling process is said to be *with replacement*. If it is not acceptable, then the sampling process is said to be *without replacement*.

There are several other commonly used sampling techniques. Each has advantages and disadvantages.

• **Stratified Sample** When it is important for the sample to have members from each segment of the population, you should use a stratified sample. Depending on the focus of the study, members of the population are divided into two or more subsets, called *strata*, that share a similar characteristic such as age, gender, ethnicity, or even political preference. A sample is then randomly selected from each of the strata. Using a stratified sample ensures that each segment of the population is represented. For instance, to collect a stratified sample of the number of people who live in West Ridge County households, you could divide the households into socioeconomic levels, and then randomly select households from each level.

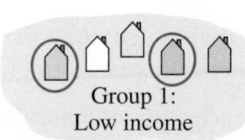

Group 1:
Low income

Group 2:
Middle income

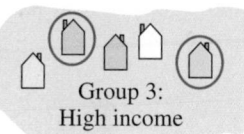

Group 3:
High income

Stratified Sampling

• **Cluster Sample** When the population falls into naturally occurring subgroups, each having similar characteristics, a cluster sample may be the most appropriate. To select a cluster sample, divide the population into groups, called *clusters*, and select all of the members in one or more (but not all) of the clusters. Examples of clusters could be different sections of the same course or different branches of a bank. For instance, to collect a cluster sample of the number of people who live in West Ridge County households, divide the households into groups according to zip codes, then select all the households in one or more, but not all, zip codes and count the number of people living in each household. In using a cluster sample, care must be taken to ensure that all clusters have similar characteristics. For instance, if one of the zip code clusters has a greater proportion of high-income people, the data might not be representative of the population.

Zip Code Zones in West Ridge County

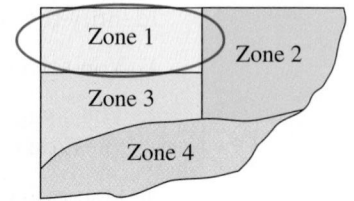

Cluster Sampling

- *Systematic Sample*   A systematic sample is a sample in which each member of the population is assigned a number. The members of the population are ordered in some way, a starting number is randomly selected, and then sample members are selected at regular intervals from the starting number. (For instance, every 3rd, 5th, or 100th member is selected.) For instance, to collect a systematic sample of the number of people who live in West Ridge County households, you could assign a different number to each household, randomly choose a starting number, select every 100th household, and count the number of people living in each. An advantage of systematic sampling is that it is easy to use. In the case of any regularly occurring pattern in the data, however, this type of sampling should be avoided.

Systematic Sampling

A type of sample that often leads to biased studies (so it is not recommended) is a **convenience sample.** A convenience sample consists only of available members of the population.

Note to Instructor

Ask students whether they have ever been part of a sampling process. Discuss the types of samples they have been part of. For students who don't think they have ever participated in a sampling, ask them if they have ever had a blood sample drawn or sampled a food.

EXAMPLE 4

▶ Identifying Sampling Techniques

You are doing a study to determine the opinions of students at your school regarding stem cell research. Identify the sampling technique you are using if you select the samples listed. Discuss potential sources of bias (if any). Explain.

1. You divide the student population with respect to majors and randomly select and question some students in each major.

2. You assign each student a number and generate random numbers. You then question each student whose number is randomly selected.

3. You select students who are in your biology class.

▶ Solution

1. Because students are divided into strata (majors) and a sample is selected from each major, this is a stratified sample.

2. Each sample of the same size has an equal chance of being selected and each student has an equal chance of being selected, so this is a simple random sample.

3. Because the sample is taken from students that are readily available, this is a convenience sample. The sample may be biased because biology students may be more familiar with stem cell research than other students and may have stronger opinions.

▶ Try It Yourself 4

You want to determine the opinions of students regarding stem cell research. Identify the sampling technique you are using if you select the samples listed.

1. You select a class at random and question each student in the class.

2. You assign each student a number and, after choosing a starting number, question every 25th student.

a. Determine *how* the sample is *selected* and identify the corresponding *sampling technique*.

b. Discuss potential sources of *bias* (if any). Explain.   *Answer: Page A30*

1.3 EXERCISES

FOR EXTRA HELP:
MyStatLab

1. In an experiment, a treatment is applied to part of a population and responses are observed. In an observational study, a researcher measures characteristics of interest of a part of a population but does not change existing conditions.

2. See Selected Answers, page A109.

3. See Odd Answers, page A51.

4. See Selected Answers, page A109.

5. True

6. See Selected Answers, page A109.

7. False. Using stratified sampling guarantees that members of each group within a population will be sampled.

8. See Selected Answers, page A109.

9. See Odd Answers, page A51.

10. True

11. Use a census because all the patients are accessible and the number of patients is not too large.

12. See Selected Answers, page A109.

13. Perform an experiment because you want to measure the effect of a treatment on the human digestive system.

14. See Selected Answers, page A109.

15. Use a simulation because the situation is impractical and dangerous to create in real life.

16. See Selected Answers, page A109.

17. See Odd Answers, page A51.

■ BUILDING BASIC SKILLS AND VOCABULARY

1. What is the difference between an observational study and an experiment?

2. What is the difference between a census and a sampling?

3. What is the difference between a random sample and a simple random sample?

4. What is replication in an experiment, and why is it important?

True or False?   *In Exercises 5–10, determine whether the statement is true or false. If it is false, rewrite it as a true statement.*

5. In a randomized block design, subjects with similar characteristics are divided into blocks, and then, within each block, randomly assigned to treatment groups.

6. A double-blind experiment is used to increase the placebo effect.

7. Using a systematic sample guarantees that members of each group within a population will be sampled.

8. A census is a count of part of a population.

9. The method for selecting a stratified sample is to order a population in some way and then select members of the population at regular intervals.

10. To select a cluster sample, divide a population into groups and then select all of the members in at least one (but not all) of the groups.

Deciding on the Method of Data Collection   *In Exercises 11–16, explain which method of data collection you would use to collect data for the study.*

11. A study of the health of 168 kidney transplant patients at a hospital

12. A study of motorcycle helmet usage in a city without a helmet law

13. A study of the effect on the human digestive system of potato chips made with a fat substitute

14. A study of the effect of a product's warning label to determine whether consumers will still buy the product

15. A study of how fast a virus would spread in a metropolitan area

16. A study of how often people wash their hands in public restrooms

■ USING AND INTERPRETING CONCEPTS

17. Allergy Drug   A pharmaceutical company wants to test the effectiveness of a new allergy drug. The company identifies 250 females 30–35 years old who suffer from severe allergies. The subjects are randomly assigned into two groups. One group is given the new allergy drug and the other is given a placebo that looks exactly like the new allergy drug. After six months, the subjects' symptoms are studied and compared.

(a) Identify the experimental units and treatments used in this experiment.

(b) Identify a potential problem with the experimental design being used and suggest a way to improve it.

(c) How could this experiment be designed to be double-blind?

18. See Selected Answers, page A109.

19. Simple random sampling is used because each telephone number has an equal chance of being dialed, and all samples of 1400 phone numbers have an equal chance of being selected. The sample may be biased because only homes with telephones will be sampled.

20. Stratified sampling is used because the persons are divided into strata (rural and urban), and a sample is selected from each stratum.

21. Convenience sampling is used because the students are chosen due to their convenience of location. Bias may enter into the sample because the students sampled may not be representative of the population of students.

22. Cluster sampling is used because the disaster area is divided into grids, and 30 grids are then entirely selected. A possible source of bias is that certain grids may have been much more severely damaged than others.

23. Simple random sampling is used because each customer has an equal chance of being contacted, and all samples of 580 customers have an equal chance of being selected.

24. Systematic sampling is used because every tenth person entering the shopping mall is sampled. It is possible for bias to enter into the sample if, for some reason, there is a regular pattern to the people entering the shopping mall.

25. Stratified sampling is used because a sample is taken from each one-acre subplot.

26. See Selected Answers, page A109.

27. Answers will vary.

28. Answers will vary.

29. Answers will vary. *Sample answer:* Treatment group: Jake, Maria, Lucy, Adam, Bridget, Vanessa, Rick, Dan, and Mary. Control group: Mike, Ron, Carlos, Steve, Susan, Kate, Pete, Judy, and Connie. A random number table is used.

30. Answers will vary.

18. **Sneakers** Nike developed a new type of sneaker designed to help delay the onset of arthritis in the knee. Eighty people with early signs of arthritis volunteered for a study. One-half of the volunteers wore the experimental sneakers and the other half wore regular Nike sneakers that looked exactly like the experimental sneakers. The individuals wore the sneakers every day. At the conclusion of the study, their symptoms were evaluated and MRI tests were performed on their knees. *(Source: Washington Post)*

   (a) Identify the experimental units and treatments used in this experiment.

   (b) Identify a potential problem with the experimental design being used and suggest a way to improve it.

   (c) The experiment is described as a placebo-controlled, double-blind study. Explain what this means.

   (d) Of the 80 volunteers, suppose 40 are men and 40 are women. How could blocking be used in designing this experiment?

Identifying Sampling Techniques   *In Exercises 19–26, identify the sampling technique used, and discuss potential sources of bias (if any). Explain.*

19. Using random digit dialing, researchers call 1400 people and ask what obstacles (such as childcare) keep them from exercising.

20. Chosen at random, 500 rural and 500 urban persons age 65 or older are asked about their health and their experience with prescription drugs.

21. Questioning students as they leave a university library, a researcher asks 358 students about their drinking habits.

22. After a hurricane, a disaster area is divided into 200 equal grids. Thirty of the grids are selected, and every occupied household in the grid is interviewed to help focus relief efforts on what residents require the most.

23. Chosen at random, 580 customers at a car dealership are contacted and asked their opinions of the service they received.

24. Every tenth person entering a mall is asked to name his or her favorite store.

25. Soybeans are planted on a 48-acre field. The field is divided into one-acre subplots. A sample is taken from each subplot to estimate the harvest.

26. From calls made with randomly generated telephone numbers, 1012 respondents are asked if they rent or own their residences.

27. **Random Number Table** Use the seventh row of Table 1 in Appendix B to generate 12 random numbers between 1 and 99.

28. **Random Number Table** Use the twelfth row of Table 1 in Appendix B to generate 10 random numbers between 1 and 920.

29. **Sleep Deprivation** A researcher wants to study the effects of sleep deprivation on motor skills. Eighteen people volunteer for the experiment: Jake, Maria, Mike, Lucy, Ron, Adam, Bridget, Carlos, Steve, Susan, Vanessa, Rick, Dan, Kate, Pete, Judy, Mary, and Connie. Use a random number generator to choose nine subjects for the treatment group. The other nine subjects will go into the control group. List the subjects in each group. Tell which method you would use to generate the random numbers.

30. **Random Number Generation** Volunteers for an experiment are numbered from 1 to 70. The volunteers are to be randomly assigned to two different treatment groups. Use a random number generator different from the one you used in Exercise 29 to choose 35 subjects for the treatment group. The other 35 subjects will go into the control group. List the subjects, according to number, in each group. Tell which method you used to generate the random numbers.

31. Census, because it is relatively easy to obtain the ages of the 115 residents.

32. See Selected Answers, page A109.

33. The question is biased because it already suggests that eating whole-grain foods improves your health. The question might be rewritten as "How does eating whole-grain foods affect your health?"

34. The question is biased because it already suggests that text messaging while driving increases the risk of a crash. The question might be rewritten as "Does text messaging while driving affect the risk of a crash?"

35. The survey question is unbiased.

36. The question is biased because it already suggests that the media have a negative effect on teen girls' dieting habits. The question might be rewritten as "Do you think the media have an effect on teen girls' dieting habits?"

37. The households sampled represent various locations, ethnic groups, and income brackets. Each of these variables is considered a stratum. Stratified sampling ensures that each segment of the population is represented.

38. (a)–(d) Answers will vary.

39. Observational studies may be referred to as natural experiments because they involve observing naturally occurring events that are not influenced by the study.

40. See Selected Answers, page A110.

41. See Odd Answers, page A51.

42. *Sample answer:* Confounding would occur in a study of the effectiveness of a billboard as an advertisement for a hotel if the billboard were put up just prior to a busy travel holiday.

43. If blinding is not used, then the placebo effect is more likely to occur.

44. See Selected Answers, page A110.

45. Both a randomized block design and a stratified sample split their members into groups based on similar characteristics.

Choosing Between a Census and a Sampling   *In Exercises 31 and 32, determine whether you would take a census or use a sampling. If you would use a sampling, decide what sampling technique you would use. Explain your reasoning.*

31. The average age of the 115 residents of a retirement community

32. The most popular type of movie among 100,000 online movie rental subscribers

Recognizing a Biased Question   *In Exercises 33–36, determine whether the survey question is biased. If the question is biased, suggest a better wording.*

33. Why does eating whole-grain foods improve your health?

34. Why does text messaging while driving increase the risk of a crash?

35. How much do you exercise during an average week?

36. Why do you think the media have a negative effect on teen girls' dieting habits?

37. Writing   A sample of television program ratings by The Nielsen Company is described on page 15. Discuss the strata used in the sample. Why is it important to have a stratified sample for these ratings?

SC 38. Use StatCrunch to generate the following random numbers.
   a. 8 numbers between 1 and 50
   b. 15 numbers between 1 and 150
   c. 16 numbers between 1 and 325
   d. 20 numbers between 1 and 1000

■ EXTENDING CONCEPTS

39. Observational studies are sometimes referred to as *natural experiments*. Explain, in your own words, what this means.

40. Open and Closed Questions   Two types of survey questions are open questions and closed questions. An open question allows for any kind of response; a closed question allows for only a fixed response. An open question, and a closed question with its possible choices, are given below. List an advantage and a disadvantage of each question.

Open Question   What can be done to get students to eat healthier foods?

Closed Question   How would you get students to eat healthier foods?
   1. Mandatory nutrition course
   2. Offer only healthy foods in the cafeteria and remove unhealthy foods
   3. Offer more healthy foods in the cafeteria and raise the prices on unhealthy foods

41. Who Picked These People?   Some polling agencies ask people to call a telephone number and give their response to a question. (a) List an advantage and a disadvantage of a survey conducted in this manner. (b) What sampling technique is used in such a survey?

42. Give an example of an experiment where confounding may occur.

43. Why is it important to use blinding in an experiment?

44. How are the placebo effect and the Hawthorne effect similar? How are they different?

45. How is a randomized block design in experiments similar to a stratified sample?

ACTIVITY 1.3   Random Numbers

APPLET

The *random numbers* applet is designed to allow you to generate random numbers from a range of values. You can specify integer values for the minimum value, maximum value, and the number of samples in the appropriate fields. You should not use decimal points when filling in the fields. When SAMPLE is clicked, the applet generates random values, which are displayed as a list in the text field.

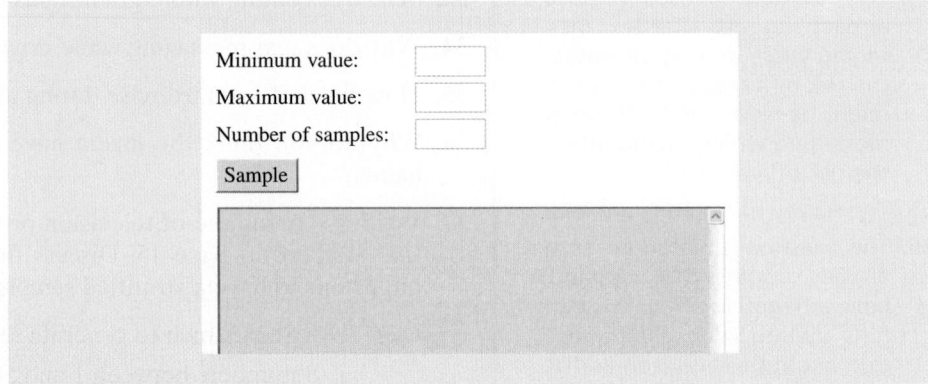

■ Explore

Step 1   Specify a minimum value.
Step 2   Specify a maximum value.
Step 3   Specify the number of samples.
Step 4   Click SAMPLE to generate a list of random values.

■ Draw Conclusions

APPLET

1. Specify the minimum, maximum, and number of samples to be 1, 20, and 8, respectively, as shown. Run the applet. Continue generating lists until you obtain one that shows that the random sample is taken with replacement. Write down this list. How do you know that the list is a random sample taken with replacement?

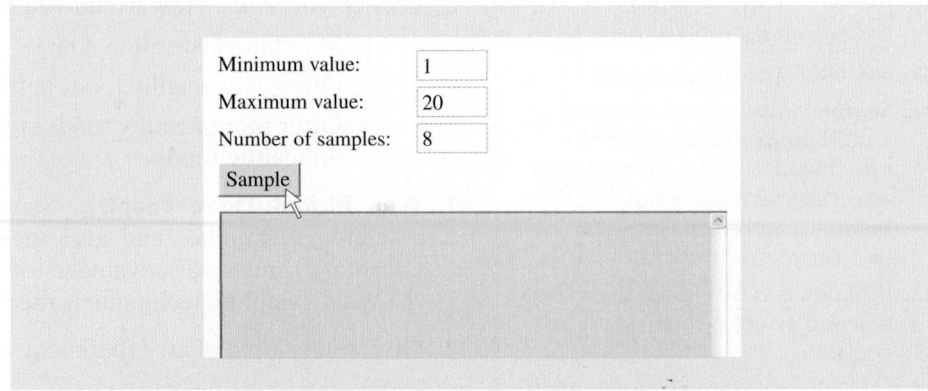

2. Use the applet to repeat Example 3 on page 20. What values did you use for the minimum, maximum, and number of samples? Which method do you prefer? Explain.

USES AND ABUSES

Uses

Experiments with Favorable Results An experiment that began in March 2003 studied 321 women with advanced breast cancer. All of the women had been previously treated with other drugs, but the cancer had stopped responding to the medications. The women were then given the opportunity to take a new drug combined with a particular chemotherapy drug.

The subjects were divided into two groups, one that took the new drug combined with a chemotherapy drug, and one that took only the chemotherapy drug. After three years, results showed that the new drug in combination with the chemotherapy drug delayed the progression of cancer in the subjects. The results were so significant that the study was stopped, and the new drug was offered to all women in the study. The Food and Drug Administration has since approved use of the new drug in conjunction with a chemotherapy drug.

Abuses

Experiments with Unfavorable Results From 1988 to 1991, one hundred eighty thousand teenagers in Norway were used as subjects to test a new vaccine against the deadly bacteria *meningococcus b*. A brochure describing the possible effects of the vaccine stated, "it is unlikely to expect serious complications," while information provided to the Norwegian Parliament stated, "serious side effects can not be excluded." The vaccine trial had some disastrous results: More than 500 side effects were reported, with some considered serious, and several of the subjects developed serious neurological diseases. The results showed that the vaccine was providing immunity in only 57% of the cases. This result was not sufficient for the vaccine to be added to Norway's vaccination program. Compensations have since been paid to the vaccine victims.

Ethics

Experiments help us further understand the world that surrounds us. But, in some cases, they can do more harm than good. In the Norwegian experiments, several ethical questions arise. Was the Norwegian experiment unethical if the best interests of the subjects were neglected? When should the experiment have been stopped? Should it have been conducted at all? If serious side effects are not reported and are withheld from subjects, there is no ethical question here, it is just wrong.

On the other hand, the breast cancer researchers would not want to deny the new drug to a group of patients with a life-threatening disease. But again, questions arise. How long must a researcher continue an experiment that shows better-than-expected results? How soon can a researcher conclude a drug is safe for the subjects involved?

■ EXERCISES

1. *Unfavorable Results* Find an example of a real-life experiment that had unfavorable results. What could have been done to avoid the outcome of the experiment?

2. *Stopping an Experiment* In your opinion, what are some problems that may arise if clinical trials of a new experimental drug or vaccine are stopped early and then the drug or vaccine is distributed to other subjects or patients?

1   CHAPTER SUMMARY

What did you **learn**?	EXAMPLE(S)	REVIEW EXERCISES
Section 1.1		
■ How to distinguish between a population and a sample	*1*	*1–4*
■ How to distinguish between a parameter and a statistic	*2*	*5–8*
■ How to distinguish between descriptive statistics and inferential statistics	*3*	*9, 10*
Section 1.2		
■ How to distinguish between qualitative data and quantitative data	*1*	*11–16*
■ How to classify data with respect to the four levels of measurement: nominal, ordinal, interval, and ratio	*2, 3*	*17–20*
Section 1.3		
■ How data are collected: by doing an observational study, performing an experiment, using a simulation, or using a survey	*1*	*21–24*
■ How to design an experiment	*2*	*25, 26*
■ How to create a sample using random sampling, simple random sampling, stratified sampling, cluster sampling, and systematic sampling	*3, 4*	*27–34*
■ How to identify a biased sample	*4*	*35–38*

1 REVIEW EXERCISES

1. Population: Collection of the opinions of all U.S. adults about credit cards

   Sample: Collection of the opinions of the 1000 U.S. adults surveyed about credit cards

2. See Selected Answers, page A110.

3. Population: Collection of the average annual percentage rates of all credit cards

   Sample: Collection of the average annual percentage rates of the 39 credit cards sampled

4. Population: Collection of the considerations of leaving practice of all physicians in the United States

   Sample: Collection of the considerations of leaving practice of the 1205 physicians surveyed

5. Parameter   6. Statistic

7. Parameter   8. Statistic

9. The statement "the average annual percentage rate (APR) [charged by credit cards] is 12.83%" is an example of descriptive statistics.

   An inference drawn from the sample is that all credit cards have an annual percentage rate of 12.83%.

10. See Selected Answers, page A110.

11. Quantitative, because monthly salaries are numerical measurements.

12. Qualitative, because Social Security numbers are labels for employees.

13. Quantitative, because ages are numerical measurements.

14. Qualitative, because zip codes are labels for customers.

15. Quantitative, because revenues are numerical measurements.

16. Qualitative, because marital statuses are attributes.

17. Interval. The data can be ordered and meaningful differences can be calculated, but it makes no sense to say that 100 degrees is twice as hot as 50 degrees.

18. Ordinal. The data are qualitative but could be arranged in order of threat level.

■ SECTION 1.1

In Exercises 1–4, identify the population and the sample.

1. A survey of 1000 U.S. adults found that 83% think credit cards tempt people to buy things they cannot afford. *(Source: Rasmussen Reports)*

2. Thirty-eight nurses working in the San Francisco area were surveyed concerning their opinions of managed health care.

3. A survey of 39 credit cards found that the average annual percentage rate (APR) is 12.83%. *(Source: Consumer Action)*

4. A survey of 1205 physicians found that about 60% had considered leaving the practice of medicine because they were discouraged over the state of U.S. health care. *(Source: The Physician Executive Journal of Medical Management)*

In Exercises 5–8, determine whether the numerical value describes a parameter or a statistic.

5. The 2009 team payroll of the Philadelphia Phillies was $113,004,046. *(Source: USA Today)*

6. In a survey of 752 adults in the United States, 42% think there should be a law that prohibits people from talking on cell phones in public places. *(Source: University of Michigan)*

7. In a recent study of math majors at a university, 10 students were minoring in physics.

8. Fifty percent of a sample of 1508 U.S. adults say they oppose drilling for oil and gas in the Arctic National Wildlife Refuge. *(Source: Pew Research Center)*

9. Which part of the study described in Exercise 3 represents the descriptive branch of statistics? Make an inference based on the results of the study.

10. Which part of the survey described in Exercise 4 represents the descriptive branch of statistics? Make an inference based on the results of the survey.

■ SECTION 1.2

In Exercises 11–16, determine which data are qualitative data and which are quantitative data. Explain your reasoning.

11. The monthly salaries of the employees at an accounting firm

12. The Social Security numbers of the employees at an accounting firm

13. The ages of a sample of 350 employees of a software company

14. The zip codes of a sample of 350 customers at a sporting goods store

15. The 2010 revenues of the companies on the Fortune 500 list

16. The marital statuses of all professional golfers

In Exercises 17–20, identify the data set's level of measurement. Explain your reasoning.

17. The daily high temperatures (in degrees Fahrenheit) for Mohave, Arizona for a week in June are listed. *(Source: Arizona Meteorological Network)*

    93 91 86 94 103 104 103

18. The levels of the Homeland Security Advisory System are listed.

    Severe High Elevated Guarded Low

19. Nominal. The data are qualitative and cannot be arranged in a meaningful order.
20. See Selected Answers, page A110.
21. Take a census because CEOs keep accurate records of charitable donations.
22. Perform a simulation because it is impractical to create this situation.
23. Perform an experiment because you want to measure the effect of training dogs from animal shelters on inmates.
24. See Selected Answers, page A110.
25. The subjects could be split into male and female and then be randomly assigned to each of the five treatment groups.
26. See Selected Answers, page A110.
27. Answers will vary.
28. See Selected Answers, page A110.
29. Simple random sampling is used because random telephone numbers were generated and called.
30. Convenience sampling is used because the student sampled a convenient group of friends.
31. Cluster sampling is used because each community is considered a cluster and every pregnant woman in a selected community is surveyed.
32. Systematic sampling is used because every third car is checked.
33. Stratified sampling is used because 25 students are randomly selected from each grade level.
34. See Selected Answers, page A110.
35. Telephone sampling samples only individuals who have telephones, who are available, and who are willing to respond.
36. The study may be biased toward the opinions of the student's friends.
37. The selected communities may not be representative of the entire area.
38. The street the law enforcement official is using may be outside a bar.

19. The four departments of a printing company are listed.

    Administration   Sales   Production   Billing

20. The total compensations (in millions of dollars) of the top ten female CEOs in the United States are listed. *(Source: Forbes)*

    9.4   5.3   11.8   11.1   9.4   4.1   6.6   5.7   4.6   4.5

■ SECTION 1.3

In Exercises 21–24, decide which method of data collection you would use to collect data for the study. Explain your reasoning.

21. A study of charitable donations of the CEOs in Syracuse, New York
22. A study of the effect of koalas on the ecosystem of Kangaroo Island, Australia
23. A study of how training dogs from animal shelters affects inmates at a prison
24. A study of college professors' opinions on teaching classes online

In Exercises 25 and 26, an experiment is being performed to test the effects of sleep deprivation on memory recall. Two hundred students volunteer for the experiment. The students will be placed in one of five different treatment groups, including the control group.

25. Explain how you could design an experiment so that it uses a randomized block design.
26. Explain how you could design an experiment so that it uses a completely randomized design.
27. **Random Number Table** Use the fifth row of Table 1 in Appendix B to generate 8 random numbers between 1 and 650.
28. **Census or Sampling?** You want to know the favorite spring break destination among 15,000 students at a university. Decide whether you would take a census or use a sampling. If you would use a sampling, decide what technique you would use. Explain your reasoning.

In Exercises 29–34, identify the sampling technique used in the study. Explain your reasoning.

29. Using random digit dialing, researchers ask 1003 U.S. adults their plans on working during retirement. *(Source: Princeton Survey Research Associates International)*
30. A student asks 18 friends to participate in a psychology experiment.
31. A pregnancy study in Cebu, Philippines randomly selects 33 communities from the Cebu metropolitan area, then interviews all available pregnant women in these communities. *(Adapted from Cebu Longitudinal Health and Nutrition Survey)*
32. Law enforcement officials stop and check the driver of every third vehicle for blood alcohol content.
33. Twenty-five students are randomly selected from each grade level at a high school and surveyed about their study habits.
34. A journalist interviews 154 people waiting at an airport baggage claim and asks them how safe they feel during air travel.

In Exercises 35–38, identify a bias or error that might occur in the indicated survey or study.

35. study in Exercise 29
36. experiment in Exercise 30
37. study in Exercise 31
38. sampling in Exercise 32

1 CHAPTER QUIZ

1. Population: Collection of the prostate conditions of all men

Sample: Collection of the prostate conditions of 20,000 men in study

2. (a) Statistic

(b) Parameter

(c) Statistic

3. (a) Qualitative

(b) Quantitative

4. (a) Ordinal, because badge numbers can be ordered and often indicate seniority of service, but no meaningful mathematical computation can be performed.

(b) Ratio, because one data value can be expressed as a multiple of another.

(c) Ordinal, because data can be arranged in order, but the differences between data entries make no sense.

(d) Interval, because meaningful differences between entries can be calculated but a zero entry is not an inherent zero.

5. (a) Perform an experiment because you want to measure the effect of a treatment on lead levels in adults.

(b) Use a survey because it would be impossible to question everyone in the population.

6. Randomized block design

7. (a) Convenience sampling, because all of the people sampled are in one convenient location.

(b) Systematic sampling, because every tenth machine part is sampled.

(c) Stratified sampling, because the population is first stratified and then a sample is collected from each stratum.

8. Convenience sampling

Take this quiz as you would take a quiz in class. After you are done, check your work against the answers given in the back of the book.

1. Identify the population and the sample in the following study.

A study of the dietary habits of 20,000 men was conducted to find a link between high intakes of dairy products and prostate cancer. *(Source: Harvard School of Public Health)*

2. Determine whether the numerical value is a parameter or a statistic.

(a) In a survey of 2253 Internet users, 19% use Twitter or another service to share social updates. *(Source: Pew Internet Project)*

(b) At a college, 90% of the Board of Trustees members approved the contract of the new president.

(c) A survey of 846 chief financial officers and senior comptrollers shows that 55% of U.S. companies are reducing bonuses. *(Source: Grant Thornton International)*

3. Determine whether the data are qualitative or quantitative.

(a) A list of debit card pin numbers

(b) The final scores on a video game

4. Identify each data set's level of measurement. Explain your reasoning.

(a) A list of badge numbers of police officers at a precinct

(b) The horsepowers of racing car engines

(c) The top 10 grossing films released in 2010

(d) The years of birth for the runners in the Boston marathon

5. Decide which method of data collection you would use to gather data for each study. Explain your reasoning.

(a) A study on the effect of low dietary intake of vitamin C and iron on lead levels in adults

(b) The ages of people living within 500 miles of your home

6. An experiment is being performed to test the effects of a new drug on high blood pressure. The experimenter identifies 320 people ages 35–50 years old with high blood pressure for participation in the experiment. The subjects are divided into equal groups according to age. Within each group, subjects are then randomly selected to be in either the treatment group or the control group. What type of experimental design is being used for this experiment?

7. Identify the sampling technique used in each study. Explain your reasoning.

(a) A journalist goes to a campground to ask people how they feel about air pollution.

(b) For quality assurance, every tenth machine part is selected from an assembly line and measured for accuracy.

(c) A study on attitudes about smoking is conducted at a college. The students are divided by class (freshman, sophomore, junior, and senior). Then a random sample is selected from each class and interviewed.

8. Which sampling technique used in Exercise 7 could lead to a biased study?

PUTTING IT ALL TOGETHER

Real Statistics — Real Decisions

You are a researcher for a professional research firm. Your firm has won a contract to do a study for an air travel industry publication. The editors of the publication would like to know their readers' thoughts on air travel factors such as ticket purchase, services, safety, comfort, economic growth, and security. They would also like to know the thoughts of adults who use air travel for business as well as for recreation.

The editors have given you their readership database and 20 questions they would like to ask (two sample questions from a previous study are given at the right). You know that it is too expensive to contact all of the readers, so you need to determine a way to contact a representative sample of the entire readership population.

■ EXERCISES

1. *How Would You Do It?*

(a) What sampling technique would you use to select the sample for the study? Why?

(b) Will the technique you choose in part (a) give you a sample that is representative of the population?

(c) Describe the method for collecting data.

(d) Identify possible flaws or biases in your study.

2. *Data Classification*

(a) What type of data do you expect to collect: qualitative, quantitative, or both? Why?

(b) At what levels of measurement do you think the data in the study will be? Why?

(c) Will the data collected for the study represent a population or a sample?

(d) Will the numerical descriptions of the data be parameters or statistics?

3. *How They Did It*

When the *Resource Systems Group* did a similar study, they used an Internet survey. They sent out 1000 invitations to participate in the survey and received 621 completed surveys.

(a) Describe some possible errors in collecting data by Internet surveys.

(b) Compare your method for collecting data in Exercise 1 to this method.

How did you acquire your ticket?

Response	Percent
Travel agent	35.1%
Directly from airline	20.9%
Online, using the airline's website	21.0%
Online, from a travel site other than the airline	18.5%
Other	4.5%

(Source: Resource Systems Group)

How many associates, friends, or family members traveled together in your party?

Response	Percent
1 (traveled alone)	48.7%
2 (traveled with one other person)	29.7%
3 (traveled with 2 others)	7.1%
4 (traveled with 3 others)	7.7%
5 (traveled with 4 others)	3.0%
6 or more (traveled with 5 or more others)	3.8%

(Source: Resource Systems Group)

HISTORY OF STATISTICS - TIMELINE

CONTRIBUTOR	TIME	CONTRIBUTION
John Graunt (1620–1674)	**17th century**	Studied records of deaths in London in the early 1600s. The first to make extensive statistical observations from massive amounts of data (Chapter 2), his work laid the foundation for modern statistics.
Blaise Pascal (1623–1662) **Pierre de Fermat** (1601–1665)		Pascal and Fermat corresponded about basic probability problems (Chapter 3)—especially those dealing with gaming and gambling.
Pierre Laplace (1749–1827)	**18th century**	Studied probability (Chapter 3) and is credited with putting probability on a sure mathematical footing.
Carl Friedrich Gauss (1777–1855)		Studied regression and the method of least squares (Chapter 9) through astronomy. In his honor, the normal distribution is sometimes called the Gaussian distribution.
Lambert Quetelet (1796–1874)	**19th century**	Used descriptive statistics (Chapter 2) to analyze crime and mortality data and studied census techniques. Described normal distributions (Chapter 5) in connection with human traits such as height.
Francis Galton (1822–1911)		Used regression and correlation (Chapter 9) to study genetic variation in humans. He is credited with the discovery of the Central Limit Theorem (Chapter 5).
Karl Pearson (1857–1936) 	**20th century**	Studied natural selection using correlation (Chapter 9). Formed first academic department of statistics and helped develop chi-square analysis (Chapter 6).
William Gosset (1876–1937)		Studied process of brewing and developed t-test to correct problems connected with small sample sizes (Chapter 6).
Charles Spearman (1863–1945)		British psychologist who was one of the first to develop intelligence testing using factor analysis (Chapter 10).
Ronald Fisher (1890–1962) 		Studied biology and natural selection and developed ANOVA (Chapter 10), stressed the importance of experimental design (Chapter 1), and was the first to identify the null and alternative hypotheses (Chapter 7).
Frank Wilcoxon (1892–1965)	**20th century (later)**	Biochemist who used statistics to study plant pathology. He introduced two-sample tests (Chapter 8), which led the way to the development of nonparametric statistics.
John Tukey (1915–2000) 		Worked at Princeton during World War II. Introduced exploratory data analysis techniques such as stem-and-leaf plots (Chapter 2). Also, worked at Bell Laboratories and is best known for his work in inferential statistics (Chapters 6–11).
David Kendall (1918–2007)		Worked at Princeton and Cambridge. Was a leading authority on applied probability and data analysis (Chapters 2 and 3).

TECHNOLOGY

Note to Instructor

As an instructor, you can decide the extent to which you want to use technology with this text. Supplements with detailed instructions are available for each of the technology tools shown.

USING TECHNOLOGY IN STATISTICS

With large data sets, you will find that calculators or computer software programs can help perform calculations and create graphics. Of the many calculators and statistical software programs that are available, we have chosen to incorporate the TI-83/84 Plus graphing calculator, and MINITAB and Excel software into this text.

The following example shows how to use these three technologies to generate a list of random numbers. This list of random numbers can be used to select sample members or perform simulations.

EXAMPLE

▸ **Generating a List of Random Numbers**

A quality control department inspects a random sample of 15 of the 167 cars that are assembled at an auto plant. How should the cars be chosen?

▸ **Solution**

One way to choose the sample is to first number the cars from 1 to 167. Then you can use technology to form a list of random numbers from 1 to 167. Each of the technology tools shown requires different steps to generate the list. Each, however, does require that you identify the minimum value as 1 and the maximum value as 167. Check your user's manual for specific instructions.

MINITAB	
↓	**C1**
1	167
2	11
3	74
4	160
5	18
6	70
7	80
8	56
9	37
10	6
11	82
12	126
13	98
14	104
15	137

EXCEL	
	A
1	41
2	16
3	91
4	58
5	151
6	36
7	96
8	154
9	2
10	113
11	157
12	103
13	64
14	135
15	90

TI-83/84 PLUS

```
randInt(1, 167, 15)
{17 42 152 59 5 116 125
64 122 55 58 60 82 152
105}
```

Note to Instructor

In Exercises 6 and 8, students may argue that these results are reasonable because there is a relatively small number of trials in each case. This is a reasonable answer because of the law of large numbers, which will be discussed in Chapter 3.

Recall that when you generate a list of random numbers, you should decide whether it is acceptable to have numbers that repeat. If it is acceptable, then the sampling process is said to be with replacement. If it is not acceptable, then the sampling process is said to be without replacement.

With each of the three technology tools shown on page 34, you have the capability of sorting the list so that the numbers appear in order. Sorting helps you see whether any of the numbers in the list repeat. If it is not acceptable to have repeats, you should specify that the tool generate more random numbers than you need.

■ EXERCISES

1. The SEC (Securities and Exchange Commission) is investigating a financial services company. The company being investigated has 86 brokers. The SEC decides to review the records for a random sample of 10 brokers. Describe how this investigation could be done. Then use technology to generate a list of 10 random numbers from 1 to 86 and order the list.

2. A quality control department is testing 25 smartphones from a shipment of 300 smartphones. Describe how this test could be done. Then use technology to generate a list of 25 random numbers from 1 to 300 and order the list.

3. Consider the population of ten digits: 0, 1, 2, 3, 4, 5, 6, 7, 8, and 9. Select three random samples of five digits from this list. Find the average of each sample. Compare your results with the average of the entire population. Comment on your results. (*Hint:* To find the average, sum the data entries and divide the sum by the number of entries.)

4. Consider the population of 41 whole numbers from 0 to 40. What is the average of these numbers? Select three random samples of seven numbers from this list. Find the average of each sample. Compare your results with the average of the entire population. Comment on your results. (*Hint:* To find the average, sum the data entries and divide the sum by the number of entries.)

5. Use random numbers to simulate rolling a six-sided die 60 times. How many times did you obtain each number from 1 to 6? Are the results what you expected?

6. You rolled a six-sided die 60 times and got the following tally.

20 ones	20 twos	15 threes
3 fours	2 fives	0 sixes

    Does this seem like a reasonable result? What inference might you draw from the result?

7. Use random numbers to simulate tossing a coin 100 times. Let 0 represent heads, and let 1 represent tails. How many times did you obtain each number? Are the results what you expected?

8. You tossed a coin 100 times and got 77 heads and 23 tails. Does this seem like a reasonable result? What inference might you draw from the result?

9. A political analyst would like to survey a sample of the registered voters in a county. The county has 47 election districts. How could the analyst use random numbers to obtain a cluster sample?

Extended solutions are given in the *Technology Supplement*.
Technical instruction is provided for MINITAB, Excel, and the TI-83/84 Plus.

2
CHAPTER

DESCRIPTIVE STATISTICS

Brothers Sam and Bud Walton opened the first Wal-Mart store in 1962. Today, the Walton family is one of the richest families in the world. Members of the Walton family held four spots in the top 50 richest people in the world in 2009.

In Chapter 1, you learned that there are many ways to collect data. Usually, researchers must work with sample data in order to analyze populations, but occasionally it is possible to collect all the data for a given population. For instance, the following represents the ages of the 50 richest people in the world in 2009.

89, 89, 87, 86, 86, 85, 83, 83, 82, 81, 80, 78, 78, 77, 76, 73, 73, 73, 72, 69, 69, 68, 67, 66, 66, 65, 65, 64, 63, 61, 61, 60, 59, 58, 57, 56, 54, 54, 53, 53, 51, 51, 49, 47, 46, 44, 43, 42, 36, 35

WHERE YOU'RE GOING ▶▶

In Chapter 2, you will learn ways to organize and describe data sets. The goal is to make the data easier to understand by describing trends, averages, and variations. For instance, in the raw data showing the ages of the 50 richest people in the world in 2009, it is not easy to see any patterns or special characteristics. Here are some ways you can organize and describe the data.

Make a frequency distribution table.

Class	Frequency, f
35–41	2
42–48	5
49–55	7
56–62	7
63–69	10
70–76	5
77–83	8
84–90	6

Draw a histogram.

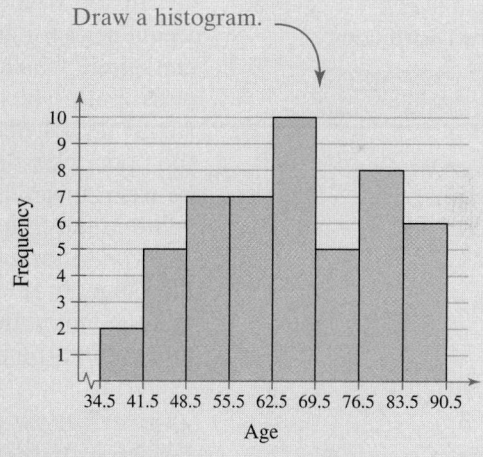

$$\text{Mean} = \frac{89 + 89 + 87 + 86 + 86 + \cdots + 43 + 42 + 36 + 35}{50}$$

$$= \frac{3263}{50}$$

$$= 65.26 \text{ years old}$$

Find an average.

$$\text{Range} = 89 - 35$$

$$= 54 \text{ years}$$

Find how the data vary.

2.1 Frequency Distributions and Their Graphs

Frequency Distributions ▸ Graphs of Frequency Distributions

WHAT YOU SHOULD LEARN

▸ How to construct a frequency distribution including limits, midpoints, relative frequencies, cumulative frequencies, and boundaries

▸ How to construct frequency histograms, frequency polygons, relative frequency histograms, and ogives

▸ FREQUENCY DISTRIBUTIONS

You will learn that there are many ways to organize and describe a data set. Important characteristics to look for when organizing and describing a data set are its **center,** its **variability** (or spread), and its **shape.** Measures of center and shapes of distributions are covered in Section 2.3.

When a data set has many entries, it can be difficult to see patterns. In this section, you will learn how to organize data sets by grouping the data into *intervals* called *classes* and forming a *frequency distribution.* You will also learn how to use frequency distributions to construct graphs.

DEFINITION

A **frequency distribution** is a table that shows **classes** or **intervals** of data entries with a count of the number of entries in each class. The **frequency** f of a class is the number of data entries in the class.

Example of a Frequency Distribution

Class	Frequency, f
1–5	5
6–10	8
11–15	6
16–20	8
21–25	5
26–30	4

In the frequency distribution shown at the left there are six classes. The frequencies for each of the six classes are 5, 8, 6, 8, 5, and 4. Each class has a **lower class limit,** which is the least number that can belong to the class, and an **upper class limit,** which is the greatest number that can belong to the class. In the frequency distribution shown, the lower class limits are 1, 6, 11, 16, 21, and 26, and the upper class limits are 5, 10, 15, 20, 25, and 30. The **class width** is the distance between lower (or upper) limits of consecutive classes. For instance, the class width in the frequency distribution shown is $6 - 1 = 5$.

The difference between the maximum and minimum data entries is called the **range.** In the frequency table shown, suppose the maximum data entry is 29, and the minimum data entry is 1. The range then is $29 - 1 = 28$. You will learn more about the range of a data set in Section 2.4.

STUDY TIP

In a frequency distribution, it is best if each class has the same width. Answers shown will use the minimum data value for the lower limit of the first class. Sometimes it may be more convenient to choose a lower limit that is slightly lower than the minimum value. The frequency distribution produced will vary slightly.

GUIDELINES

Constructing a Frequency Distribution from a Data Set

1. Decide on the number of classes to include in the frequency distribution. The number of classes should be between 5 and 20; otherwise, it may be difficult to detect any patterns.

2. Find the class width as follows. Determine the range of the data, divide the range by the number of classes, and *round up to the next convenient number.*

3. Find the class limits. You can use the minimum data entry as the lower limit of the first class. To find the remaining lower limits, add the class width to the lower limit of the preceding class. Then find the upper limit of the first class. Remember that classes cannot overlap. Find the remaining upper class limits.

4. Make a tally mark for each data entry in the row of the appropriate class.

5. Count the tally marks to find the total frequency f for each class.

Note to Instructor

Let students know that there can be many correct versions of a frequency distribution. To make it easy to check answers, however, students should follow the conventions for constructing frequency distributions shown in the text.

INSIGHT

If you obtain a whole number when calculating the class width of a frequency distribution, use the next whole number as the class width. Doing this ensures that you will have enough space in your frequency distribution for all the data values.

Lower limit	Upper limit
59	114
115	170
171	226
227	282
283	338
339	394
395	450

STUDY TIP

The uppercase Greek letter sigma (Σ) is used throughout statistics to indicate a summation of values.

Note to Instructor

Be sure that students interpret the class width correctly as the distance between lower (or upper) limits of consecutive classes. A common error is to use a class width of 55 for the class 59–114. Students should be shown that this class actually has a width of 56.

EXAMPLE 1

▸ **Constructing a Frequency Distribution from a Data Set**

The following sample data set lists the prices (in dollars) of 30 portable global positioning system (GPS) navigators. Construct a frequency distribution that has seven classes.

90	130	400	200	350	70	325	250	150	250
275	270	150	130	59	200	160	450	300	130
220	100	200	400	200	250	95	180	170	150

▸ **Solution**

1. The number of classes (7) is stated in the problem.

2. The minimum data entry is 59 and the maximum data entry is 450, so the range is $450 - 59 = 391$. Divide the range by the number of classes and round up to find the class width.

$$\text{Class width} = \frac{391}{7} \qquad \frac{\text{Range}}{\text{Number of classes}}$$
$$\approx 55.86 \qquad \text{Round up to 56.}$$

3. The minimum data entry is a convenient lower limit for the first class. To find the lower limits of the remaining six classes, add the class width of 56 to the lower limit of each previous class. The upper limit of the first class is 114, which is one less than the lower limit of the second class. The upper limits of the other classes are $114 + 56 = 170$, $170 + 56 = 226$, and so on. The lower and upper limits for all seven classes are shown.

4. Make a tally mark for each data entry in the appropriate class. For instance, the data entry 130 is in the 115–170 class, so make a tally mark in that class. Continue until you have made a tally mark for each of the 30 data entries.

5. The number of tally marks for a class is the frequency of that class.

The frequency distribution is shown in the following table. The first class, 59–114, has five tally marks. So, the frequency of this class is 5. Notice that the sum of the frequencies is 30, which is the number of entries in the sample data set. The sum is denoted by Σf, where Σ is the uppercase Greek letter **sigma**.

**Frequency Distribution for
Prices (in dollars) of GPS Navigators**

Prices →   ← Number of GPS navigators

Class	Tally	Frequency, f
59–114	$\|\|\|\|$	5
115–170	$\|\|\|\|$ $\|\|\|$	8
171–226	$\|\|\|\|$ $\|$	6
227–282	$\|\|\|\|$	5
283–338	$\|\|$	2
339–394	$\|$	1
395–450	$\|\|\|$	3
		$\Sigma f = 30$

Check that the sum of the frequencies equals the number in the sample.

▶ **Try It Yourself 1**

Construct a frequency distribution using the ages of the 50 richest people data set listed in the Chapter Opener on page 37. Use eight classes.

a. State the *number* of classes.
b. Find the minimum and maximum values and the *class width*.
c. Find the *class limits*.
d. *Tally* the data entries.
e. Write the *frequency f* of each class.

Answer: Page A30

After constructing a standard frequency distribution such as the one in Example 1, you can include several additional features that will help provide a better understanding of the data. These features (the *midpoint, relative frequency*, and *cumulative frequency* of each class) can be included as additional columns in your table.

DEFINITION

The **midpoint** of a class is the sum of the lower and upper limits of the class divided by two. The midpoint is sometimes called the *class mark*.

$$\text{Midpoint} = \frac{(\text{Lower class limit}) + (\text{Upper class limit})}{2}$$

The **relative frequency** of a class is the portion or percentage of the data that falls in that class. To find the relative frequency of a class, divide the frequency *f* by the sample size *n*.

$$\text{Relative frequency} = \frac{\text{Class frequency}}{\text{Sample size}} = \frac{f}{n}$$

The **cumulative frequency** of a class is the sum of the frequencies of that class and all previous classes. The cumulative frequency of the last class is equal to the sample size *n*.

After finding the first midpoint, you can find the remaining midpoints by adding the class width to the previous midpoint. For instance, if the first midpoint is 86.5 and the class width is 56, then the remaining midpoints are

$$86.5 + 56 = 142.5$$

$$142.5 + 56 = 198.5$$

$$198.5 + 56 = 254.5$$

$$254.5 + 56 = 310.5$$

and so on.

You can write the relative frequency as a fraction, decimal, or percent. The sum of the relative frequencies of all the classes should be equal to 1, or 100%. Due to rounding, the sum may be slightly less than or greater than 1. So, values such as 0.99 and 1.01 are sufficient.

EXAMPLE 2

▶ **Finding Midpoints, Relative Frequencies, and Cumulative Frequencies**

Using the frequency distribution constructed in Example 1, find the midpoint, relative frequency, and cumulative frequency of each class. Identify any patterns.

▶ **Solution**

The midpoints, relative frequencies, and cumulative frequencies of the first three classes are calculated as follows.

Class	f	Midpoint	Relative frequency	Cumulative frequency
59–114	5	$\dfrac{59 + 114}{2} = 86.5$	$\dfrac{5}{30} \approx 0.17$	5
115–170	8	$\dfrac{115 + 170}{2} = 142.5$	$\dfrac{8}{30} \approx 0.27$	$5 + 8 = 13$
171–226	6	$\dfrac{171 + 226}{2} = 198.5$	$\dfrac{6}{30} = 0.2$	$13 + 6 = 19$

The remaining midpoints, relative frequencies, and cumulative frequencies are shown in the following expanded frequency distribution.

Frequency Distribution for Prices (in dollars) of GPS Navigators

Prices

Number of GPS navigators

Portion of GPS navigators

Class	Frequency, f	Midpoint	Relative frequency	Cumulative frequency
59–114	5	86.5	0.17	5
115–170	8	142.5	0.27	13
171–226	6	198.5	0.2	19
227–282	5	254.5	0.17	24
283–338	2	310.5	0.07	26
339–394	1	366.5	0.03	27
395–450	3	422.5	0.1	30
	$\Sigma f = 30$		$\Sigma \dfrac{f}{n} \approx 1$	

Interpretation   There are several patterns in the data set. For instance, the most common price range for GPS navigators was \$115 to \$170.

▶ **Try It Yourself 2**

Using the frequency distribution constructed in Try It Yourself 1, find the midpoint, relative frequency, and cumulative frequency of each class. Identify any patterns.

a. Use the formulas to find each *midpoint*, *relative frequency*, and *cumulative frequency*.
b. *Organize* your results in a frequency distribution.
c. *Identify* patterns that emerge from the data.   *Answer: Page A31*

▶ GRAPHS OF FREQUENCY DISTRIBUTIONS

Sometimes it is easier to identify patterns of a data set by looking at a graph of the frequency distribution. One such graph is a *frequency histogram*.

DEFINITION

A **frequency histogram** is a bar graph that represents the frequency distribution of a data set. A histogram has the following properties.

1. The horizontal scale is quantitative and measures the data values.
2. The vertical scale measures the frequencies of the classes.
3. Consecutive bars must touch.

Because consecutive bars of a histogram must touch, bars must begin and end at class boundaries instead of class limits. **Class boundaries** are the numbers that separate classes *without* forming gaps between them. If data entries are integers, subtract 0.5 from each lower limit to find the lower class boundaries. To find the upper class boundaries, add 0.5 to each upper limit. The upper boundary of a class will equal the lower boundary of the next higher class.

EXAMPLE 3　　SC　Report 2

▶ Constructing a Frequency Histogram

Draw a frequency histogram for the frequency distribution in Example 2. Describe any patterns.

▶ Solution

First, find the class boundaries. Because the data entries are integers, subtract 0.5 from each lower limit to find the lower class boundaries and add 0.5 to each upper limit to find the upper class boundaries. So, the lower and upper boundaries of the first class are as follows.

> First class lower boundary = 59 − 0.5 = 58.5
> First class upper boundary = 114 + 0.5 = 114.5

The boundaries of the remaining classes are shown in the table. To construct the histogram, choose possible frequency values for the vertical scale. You can mark the horizontal scale either at the midpoints or at the class boundaries. Both histograms are shown.

Class	Class boundaries	Frequency, f
59–114	58.5–114.5	5
115–170	114.5–170.5	8
171–226	170.5–226.5	6
227–282	226.5–282.5	5
283–338	282.5–338.5	2
339–394	338.5–394.5	1
395–450	394.5–450.5	3

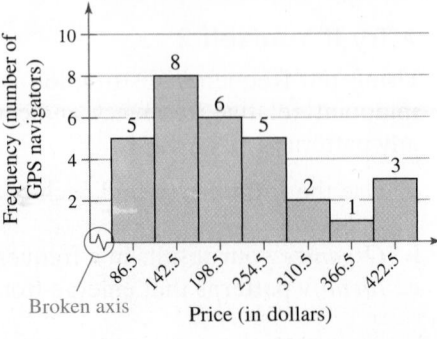

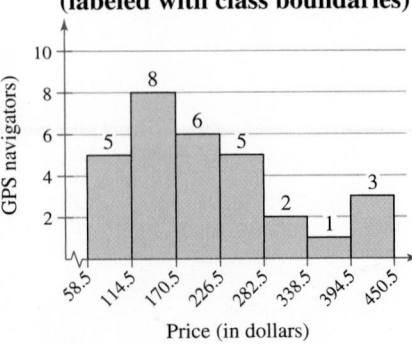

Interpretation From either histogram, you can see that more than half of the GPS navigators are priced below $226.50.

▶ **Try It Yourself 3**

Use the frequency distribution from Try It Yourself 2 to construct a frequency histogram that represents the ages of the 50 richest people. Describe any patterns.

a. Find the *class boundaries*.
b. Choose appropriate *horizontal and vertical scales*.
c. Use the frequency distribution to *find the height of each bar*.
d. *Describe* any patterns in the data.   *Answer: Page A31*

Another way to graph a frequency distribution is to use a frequency polygon. A **frequency polygon** is a line graph that emphasizes the continuous change in frequencies.

EXAMPLE 4

▶ **Constructing a Frequency Polygon**

Draw a frequency polygon for the frequency distribution in Example 2. Describe any patterns.

▶ **Solution**

To construct the frequency polygon, use the same horizontal and vertical scales that were used in the histogram labeled with class midpoints in Example 3. Then plot points that represent the midpoint and frequency of each class and connect the points in order from left to right. Because the graph should begin and end on the horizontal axis, extend the left side to one class width before the first class midpoint and extend the right side to one class width after the last class midpoint.

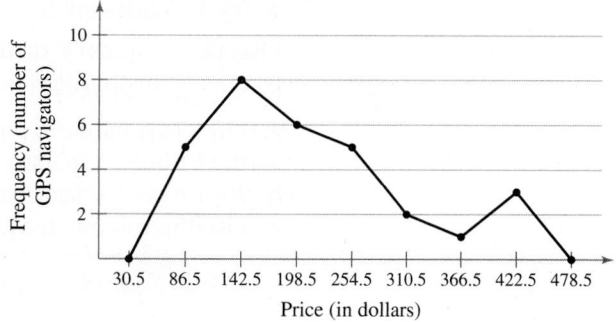

Prices of GPS Navigators

Interpretation   You can see that the frequency of GPS navigators increases up to $142.50 and then decreases.

▶ **Try It Yourself 4**

Use the frequency distribution from Try It Yourself 2 to construct a frequency polygon that represents the ages of the 50 richest people. Describe any patterns.

a. Choose appropriate *horizontal and vertical scales*.
b. *Plot points* that represent the midpoint and frequency of each class.
c. *Connect the points* and extend the sides as necessary.
d. *Describe* any patterns in the data.   *Answer: Page A31*

A **relative frequency histogram** has the same shape and the same horizontal scale as the corresponding frequency histogram. The difference is that the vertical scale measures the *relative* frequencies, not frequencies.

EXAMPLE 5   SC   Report 3

▶ Constructing a Relative Frequency Histogram

Draw a relative frequency histogram for the frequency distribution in Example 2.

▶ Solution

The relative frequency histogram is shown. Notice that the shape of the histogram is the same as the shape of the frequency histogram constructed in Example 3. The only difference is that the vertical scale measures the relative frequencies.

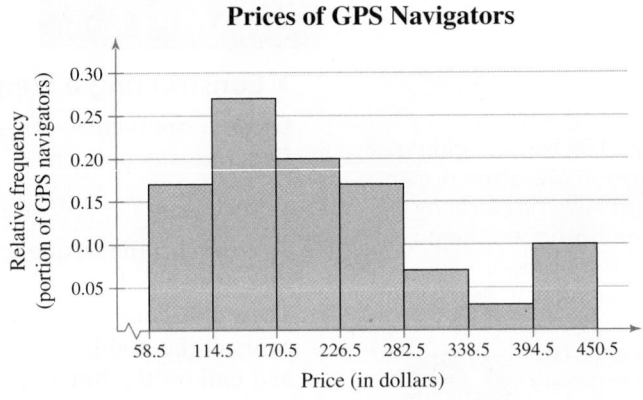

Interpretation   From this graph, you can quickly see that 0.27 or 27% of the GPS navigators are priced between \$114.50 and \$170.50, which is not as immediately obvious from the frequency histogram.

▶ Try It Yourself 5

Use the frequency distribution in Try It Yourself 2 to construct a relative frequency histogram that represents the ages of the 50 richest people.

a. *Use the same horizontal scale* that was used in the frequency histogram in the Chapter Opener.
b. *Revise the vertical scale* to reflect relative frequencies.
c. Use the relative frequencies to *find the height of each bar*.

Answer: Page A31

If you want to describe the number of data entries that are equal to or below a certain value, you can easily do so by constructing a *cumulative frequency graph*.

DEFINITION

A **cumulative frequency graph**, or **ogive** (pronounced ō′jīve), is a line graph that displays the cumulative frequency of each class at its upper class boundary. The upper boundaries are marked on the horizontal axis, and the cumulative frequencies are marked on the vertical axis.

> ## GUIDELINES
>
> **Constructing an Ogive (Cumulative Frequency Graph)**
>
> 1. Construct a frequency distribution that includes cumulative frequencies as one of the columns.
> 2. Specify the horizontal and vertical scales. The horizontal scale consists of upper class boundaries, and the vertical scale measures cumulative frequencies.
> 3. Plot points that represent the upper class boundaries and their corresponding cumulative frequencies.
> 4. Connect the points in order from left to right.
> 5. The graph should start at the lower boundary of the first class (cumulative frequency is zero) and should end at the upper boundary of the last class (cumulative frequency is equal to the sample size).

EXAMPLE 6

▶ Constructing an Ogive

Draw an ogive for the frequency distribution in Example 2. Estimate how many GPS navigators cost $300 or less. Also, use the graph to estimate when the greatest increase in price occurs.

▶ Solution

Using the cumulative frequencies, you can construct the ogive shown. The upper class boundaries, frequencies, and cumulative frequencies are shown in the table. Notice that the graph starts at 58.5, where the cumulative frequency is 0, and the graph ends at 450.5, where the cumulative frequency is 30.

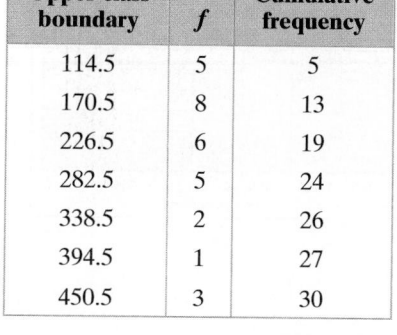

Upper class boundary	f	Cumulative frequency
114.5	5	5
170.5	8	13
226.5	6	19
282.5	5	24
338.5	2	26
394.5	1	27
450.5	3	30

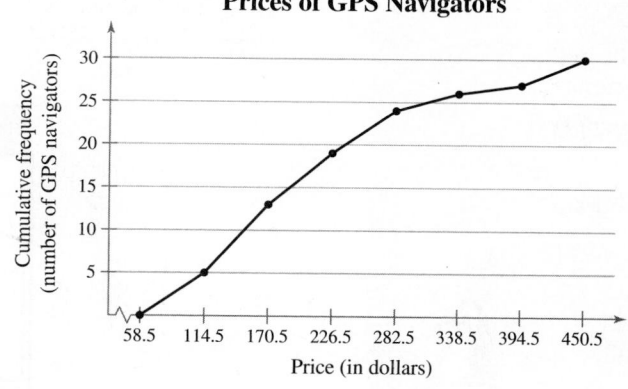

Prices of GPS Navigators

Interpretation   From the ogive, you can see that about 25 GPS navigators cost $300 or less. It is evident that the greatest increase occurs between $114.50 and $170.50, because the line segment is steepest between these two class boundaries.

Another type of ogive uses percent as the vertical axis instead of frequency (see Example 5 in Section 2.5).

▶ **Try It Yourself 6**

Use the frequency distribution from Try It Yourself 2 to construct an ogive that represents the ages of the 50 richest people. Estimate the number of people who are 80 years old or younger.

a. Specify the *horizontal and vertical scales*.
b. *Plot* the points given by the upper class boundaries and the cumulative frequencies.
c. *Construct* the graph.
d. *Estimate* the number of people who are 80 years old or younger.
e. *Interpret* the results in the context of the data.   *Answer: Page A31*

EXAMPLE 7

▶ **Using Technology to Construct Histograms**

Use a calculator or a computer to construct a histogram for the frequency distribution in Example 2.

▶ **Solution**

MINITAB, Excel, and the TI-83/84 Plus each have features for graphing histograms. Try using this technology to draw the histograms as shown.

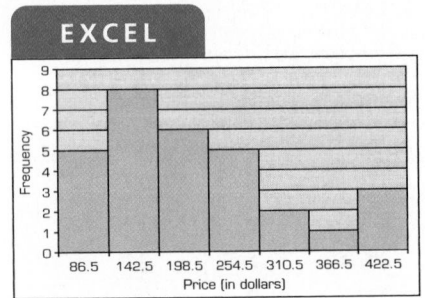

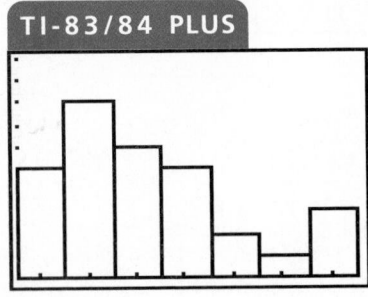

STUDY TIP

Detailed instructions for using MINITAB, Excel, and the TI-83/84 Plus are shown in the Technology Guide that accompanies this text. For instance, here are instructions for creating a histogram on a TI-83/84 Plus.

STAT   ENTER

Enter midpoints in L1.
Enter frequencies in L2.

2nd   STATPLOT

Turn on Plot 1.
Highlight Histogram.

Xlist: L1
Freq: L2

ZOOM   9

WINDOW

Xscl=56

GRAPH

▶ **Try It Yourself 7**

Use a calculator or a computer and the frequency distribution from Try It Yourself 2 to construct a frequency histogram that represents the ages of the 50 richest people.

a. *Enter* the data
b. *Construct* the histogram.   *Answer: Page A31*

2.1 EXERCISES

1. Organizing the data into a frequency distribution may make patterns within the data more evident. Sometimes it is easier to identify patterns of a data set by looking at a graph of the frequency distribution.

2. See Selected Answers, page A110.

3. See Odd Answers, page A52.

4. See Selected Answers, page A110.

5. The sum of the relative frequencies must be 1 or 100% because it is the sum of all portions or percentages of the data.

6. See Selected Answers, page A110.

7. False. Class width is the difference between lower or upper limits of consecutive classes.

8. True

9. False. An ogive is a graph that displays cumulative frequencies.

10. True

11. Class width = 8; Lower class limits: 9, 17, 25, 33, 41, 49, 57; Upper class limits: 16, 24, 32, 40, 48, 56, 64

12. See Selected Answers, page A110.

13. Class width = 15; Lower class limits: 17, 32, 47, 62, 77, 92, 107, 122; Upper class limits: 31, 46, 61, 76, 91, 106, 121, 136

14. See Selected Answers, page A110.

15. See Odd Answers, page A52.

16. See Selected Answers, page A110.

17. See Odd Answers, page A53.

18. See Selected Answers, page A110.

■ BUILDING BASIC SKILLS AND VOCABULARY

1. What are some benefits of representing data sets using frequency distributions? What are some benefits of using graphs of frequency distributions?

2. Why should the number of classes in a frequency distribution be between 5 and 20?

3. What is the difference between class limits and class boundaries?

4. What is the difference between relative frequency and cumulative frequency?

5. After constructing an expanded frequency distribution, what should the sum of the relative frequencies be? Explain.

6. What is the difference between a frequency polygon and an ogive?

True or False? *In Exercises 7–10, determine whether the statement is true or false. If it is false, rewrite it as a true statement.*

7. In a frequency distribution, the class width is the distance between the lower and upper limits of a class.

8. The midpoint of a class is the sum of its lower and upper limits divided by two.

9. An ogive is a graph that displays relative frequencies.

10. Class boundaries are used to ensure that consecutive bars of a histogram touch.

In Exercises 11–14, use the given minimum and maximum data entries and the number of classes to find the class width, the lower class limits, and the upper class limits.

11. min = 9, max = 64, 7 classes

12. min = 12, max = 88, 6 classes

13. min = 17, max = 135, 8 classes

14. min = 54, max = 247, 10 classes

Reading a Frequency Distribution *In Exercises 15 and 16, use the given frequency distribution to find the (a) class width, (b) class midpoints, and (c) class boundaries.*

15.

Cleveland, OH High Temperatures (°F)

Class	Frequency, f
20–30	19
31–41	43
42–52	68
53–63	69
64–74	74
75–85	68
86–96	24

16.

Travel Time to Work (in minutes)

Class	Frequency, f
0–9	188
10–19	372
20–29	264
30–39	205
40–49	83
50–59	76
60–69	32

17. Use the frequency distribution in Exercise 15 to construct an expanded frequency distribution, as shown in Example 2.

18. Use the frequency distribution in Exercise 16 to construct an expanded frequency distribution, as shown in Example 2.

19. (a) Number of classes = 7

(b) Least frequency ≈ 10

(c) Greatest frequency ≈ 300

(d) Class width = 10

20. (a) Number of classes = 7

(b) Least frequency ≈ 100

(c) Greatest frequency ≈ 900

(d) Class width = 5

21. (a) 50

(b) 22.5–23.5 pounds

22. (a) 50

(b) 64–66 inches

23. (a) 42

(b) 29.5 pounds

(c) 35

(d) 2

24. (a) 44

(b) 66 inches

(c) 20

(d) 6

Graphical Analysis   *In Exercises 19 and 20, use the frequency histogram to*

(a) *determine the number of classes.*

(b) *estimate the frequency of the class with the least frequency.*

(c) *estimate the frequency of the class with the greatest frequency.*

(d) *determine the class width.*

19.   **Employee Salaries**   **20.**   **Tree Heights**

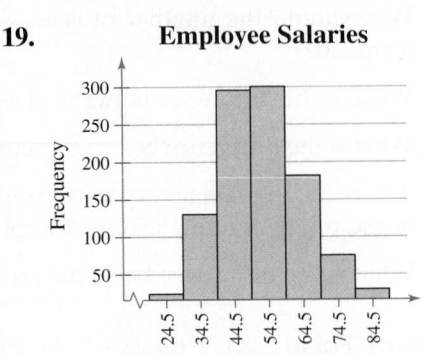

Salary (in thousands of dollars)

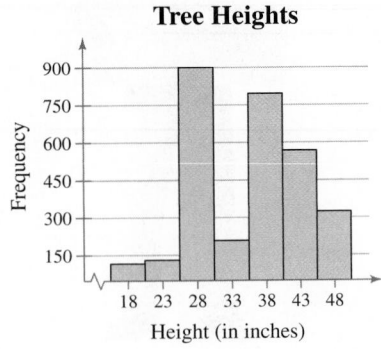

Height (in inches)

Graphical Analysis   *In Exercises 21 and 22, use the ogive to approximate*

(a) *the number in the sample.*

(b) *the location of the greatest increase in frequency.*

21.   **Male Beagles**   **22.**   **Adult Females, Ages 20–29**

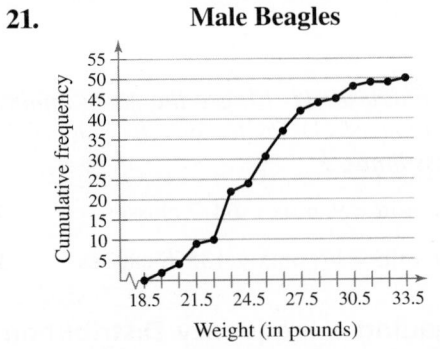

Weight (in pounds)

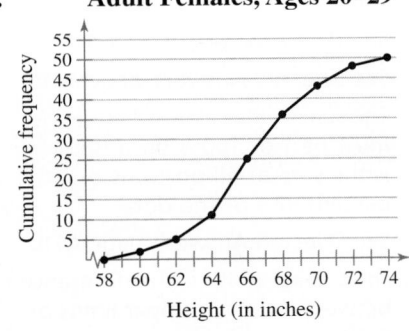

Height (in inches)

23. Use the ogive in Exercise 21 to approximate

(a) the cumulative frequency for a weight of 27.5 pounds.

(b) the weight for which the cumulative frequency is 45.

(c) the number of beagles that weigh between 22.5 pounds and 29.5 pounds.

(d) the number of beagles that weigh more than 30.5 pounds.

24. Use the ogive in Exercise 22 to approximate

(a) the cumulative frequency for a height of 72 inches.

(b) the height for which the cumulative frequency is 25.

(c) the number of adult females that are between 62 and 66 inches tall.

(d) the number of adult females that are taller than 70 inches.

25. (a) Class with greatest relative frequency: 8–9 inches

Class with least relative frequency: 17–18 inches

(b) Greatest relative frequency ≈ 0.195

Least relative frequency ≈ 0.005

(c) Approximately 0.01

26. (a) Class with greatest relative frequency: 19–20 minutes

Class with least relative frequency: 21–22 minutes

(b) Greatest relative frequency ≈ 40%

Least relative frequency ≈ 2%

(c) Approximately 33%

27. Class with greatest frequency: 29.5–32.5

Classes with least frequency: 11.5–14.5 and 38.5–41.5

28. Class with greatest frequency: 7.75–8.25

Class with least frequency: 6.25–6.75

29. See Odd Answers, page A53.

30. See Selected Answers, page A110.

Graphical Analysis   *In Exercises 25 and 26, use the relative frequency histogram to*

(a) *identify the class with the greatest, and the class with the least, relative frequency.*

(b) *approximate the greatest and least relative frequencies.*

(c) *approximate the relative frequency of the second class.*

25.   Atlantic Croaker Fish

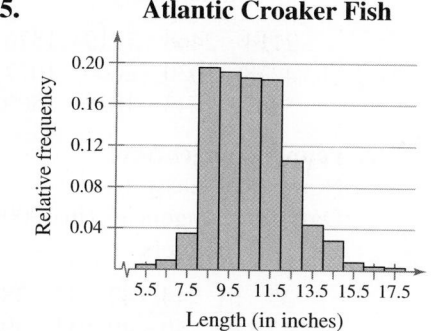

26.   Emergency Response Times

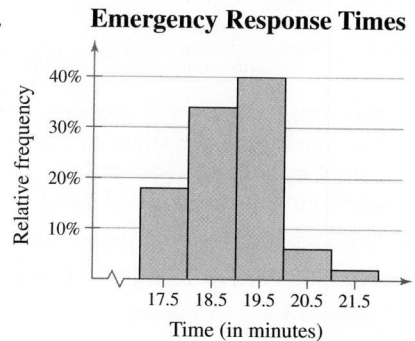

Graphical Analysis   *In Exercises 27 and 28, use the frequency polygon to identify the class with the greatest, and the class with the least, frequency.*

27.   Raw MCAT Scores for 60 Applicants

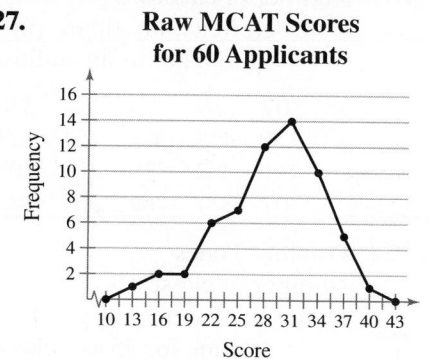

28.   Shoe Sizes for 50 Females

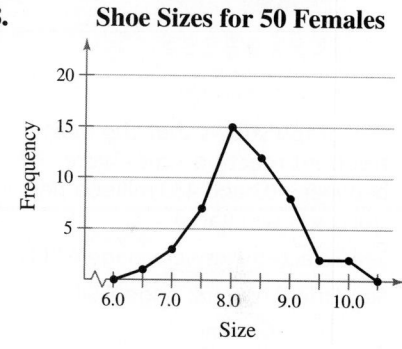

■ **USING AND INTERPRETING CONCEPTS**

Constructing a Frequency Distribution   *In Exercises 29 and 30, construct a frequency distribution for the data set using the indicated number of classes. In the table, include the midpoints, relative frequencies, and cumulative frequencies. Which class has the greatest frequency and which has the least frequency?*

29. Political Blog Reading Times

Number of classes: 5

Data set: Time (in minutes) spent reading a political blog in a day

7	39	13	9	25	8	22	0	2	18	2	30	7
35	12	15	8	6	5	29	0	11	39	16	15	

30. Book Spending

Number of classes: 6

Data set: Amount (in dollars) spent on books for a semester

91	472	279	249	530	376	188	341	266	199
142	273	189	130	489	266	248	101	375	486
190	398	188	269	43	30	127	354	84	

indicates that the data set for this exercise is available electronically.

31. See Odd Answers, page A53.

July Sales for
Representatives

The graph shows that most of
the sales representatives at the
company sold between $1000
and $2019. (Answers will vary.)

32. See Selected Answers, page A111.

33. See Odd Answers, page A53.

Reaction Times for Females

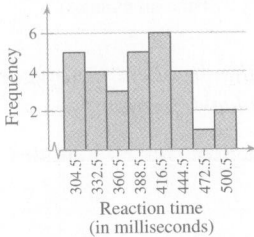

The graph shows that the most
frequent reaction times were
between 403 and 430 milliseconds.
(Answers will vary.)

34. See Selected Answers, page A111.

35. See Odd Answers, page A54.

Gasoline
Consumption

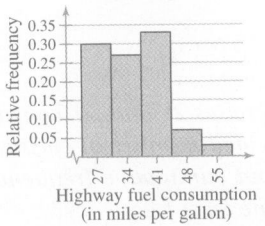

Class with greatest relative
frequency: 38–44
Class with least relative
frequency: 52–58

36. See Selected Answers, page A111.

Constructing a Frequency Distribution and a Frequency Histogram

In Exercises 31–34, construct a frequency distribution and a frequency histogram for the data set using the indicated number of classes. Describe any patterns.

31. Sales
Number of classes: 6
Data set: July sales (in dollars) for all sales representatives at
a company

2114	2468	7119	1876	4105	3183	1932	1355
4278	1030	2000	1077	5835	1512	1697	2478
3981	1643	1858	1500	4608	1000

32. Pepper Pungencies
Number of classes: 5
Data set: Pungencies (in 1000s of Scoville units) of 24 tabasco
peppers

35	51	44	42	37	38	36	39
44	43	40	40	32	39	41	38
42	39	40	46	37	35	41	39

33. Reaction Times
Number of classes: 8
Data set: Reaction times (in milliseconds) of a sample of 30 adult
females to an auditory stimulus

507	389	305	291	336	310	514	442
373	428	387	454	323	441	388	426
411	382	320	450	309	416	359	388
307	337	469	351	422	413

34. Fracture Times
Number of classes: 5
Data set: Amounts of pressure (in pounds per square inch) at fracture
time for 25 samples of brick mortar

2750	2862	2885	2490	2512	2456	2554
2872	2601	2877	2721	2692	2888	2755
2867	2718	2641	2834	2466	2596	2519
2532	2885	2853	2517

Constructing a Frequency Distribution and a Relative Frequency Histogram

In Exercises 35–38, construct a frequency distribution and a relative frequency histogram for the data set using five classes. Which class has the greatest relative frequency and which has the least relative frequency?

35. Gasoline Consumption
Data set: Highway fuel consumptions (in miles per gallon) for a sample
of cars

32	35	28	40	30	42	55	40	45	24
28	34	40	36	34	40	30	25	28	32
40	35	25	44	26	39	38	42	45	32

36. ATM Withdrawals
Data set: A sample of ATM withdrawals (in dollars)

35	10	30	25	75	10	30	20	20	10	40
50	40	30	60	70	25	40	10	60	20	80
40	25	20	10	20	25	30	50	80	20

37. See Odd Answers, page A54.

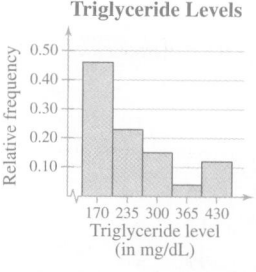

Triglyceride Levels

Class with greatest relative frequency: 138–202

Class with least relative frequency: 333–397

38. See Selected Answers, page A111.

39. See Odd Answers, page A54.

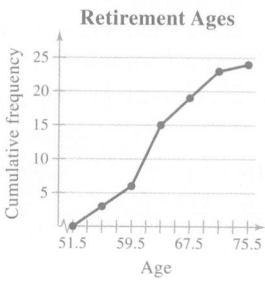

Retirement Ages

Location of the greatest increase in frequency: 60–63

40. See Selected Answers, page A112.

41. See Odd Answers, page A54.

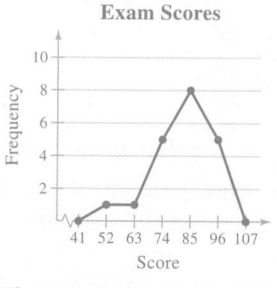

Exam Scores

The graph shows that the most frequent exam scores were between 80 and 90. (Answers will vary.)

42. See Selected Answers, page A112.

43. See Odd Answers, page A54.

37. Triglyceride Levels

Data set: Triglyceride levels (in milligrams per deciliter of blood) of a sample of patients

209	140	155	170	265	138	180	295	250
320	270	225	215	390	420	462	150	200
400	295	240	200	190	145	160	175	

38. Years of Service

Data set: Years of service of a sample of New York state troopers

12	7	9	8	9	8	12	10	9
10	6	8	13	12	10	11	7	14
12	9	8	10	9	11	13	8	

Constructing a Cumulative Frequency Distribution and an Ogive

In Exercises 39 and 40, construct a cumulative frequency distribution and an ogive for the data set using six classes. Then describe the location of the greatest increase in frequency.

39. Retirement Ages

Data set: Retirement ages for a sample of doctors

70	54	55	71	57	58	63	65
60	66	57	62	63	60	63	60
66	60	67	69	69	52	61	73

40. Saturated Fat Intakes

Data set: Daily saturated fat intakes (in grams) of a sample of people

38	32	34	39	40	54	32	17	29	33
57	40	25	36	33	24	42	16	31	33

Constructing a Frequency Distribution and a Frequency Polygon

In Exercises 41 and 42, construct a frequency distribution and a frequency polygon for the data set. Describe any patterns.

41. Exam Scores

Number of classes: 5

Data set: Exam scores for all students in a statistics class

83	92	94	82	73	98	78	85	72	90
89	92	96	89	75	85	63	47	75	82

42. Children of the Presidents

Number of classes: 6

Data set: Number of children of the U.S. presidents

(Source: presidentschildren.com)

0	5	6	0	3	4	0	4	10	15	0	6	2	3	0
4	5	4	8	7	3	5	3	2	6	3	3	1	2	
2	6	1	2	3	2	2	4	4	4	6	1	2	2	

In Exercises 43 and 44, use the data set to construct (a) an expanded frequency distribution, (b) a frequency histogram, (c) a frequency polygon, (d) a relative frequency histogram, and (e) an ogive.

43. Pulse Rates

Number of classes: 6

Data set: Pulse rates of students in a class

68	105	95	80	90	100	75	70	84	98	102	70
65	88	90	75	78	94	110	120	95	80	76	108

44. See Selected Answers, page A112.

45. See Odd Answers, page A55.

46. See Selected Answers, page A112.

In general, a greater number of classes better preserves the actual values of the data set but is not as helpful for observing general trends and making conclusions. In choosing the number of classes, an important consideration is the size of the data set. For instance, you would not want to use 20 classes if your data set contained 20 entries. In this particular example, as the number of classes increases, the histogram shows more fluctuation. The histograms with 10 and 20 classes have classes with zero frequencies. Not much is gained by using more than five classes. Therefore, it appears that five classes would be best.

47. (a) See Odd Answers, page A55.

(b) 16.7%, because the sum of the relative frequencies for the last three classes is 0.167.

(c) $9600, because the sum of the relative frequencies for the last two classes is 0.10.

48. (a) See Selected Answers, page A112.

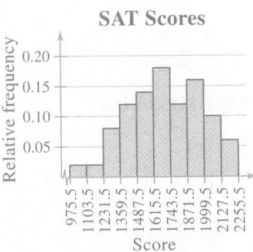

SAT Scores

(b) 62%; The sum of the relative frequencies of the class starting with 1616 and all classes with higher scores is 0.62.

(c) A score of 1360 or above, because the sum of the relative frequencies of the class starting with 1360 and all classes with higher scores is 0.88.

44. Hospitals

Number of classes: 8

Data set: Number of hospitals in each state *(Source: American Hospital Directory)*

15	100	56	74	360	53	34	8	213	116
15	38	21	143	97	59	76	110	83	51
23	116	55	91	75	19	108	14	25	14
73	40	30	213	154	97	36	181	12	63
29	121	378	36	91	7	61	71	40	15

SC **45.** Use StatCrunch to construct a frequency histogram and a relative frequency histogram for the following data set that shows the finishing times (in minutes) for 25 runners in a marathon. Use seven classes.

159	164	165	170	215	200	167	225	192	185	235	240	225
191	194	175	167	234	158	172	180	240	176	159	231	

46. Writing   What happens when the number of classes is increased for a frequency histogram? Use the data set listed and a technology tool to create frequency histograms with 5, 10, and 20 classes. Which graph displays the data best?

2	7	3	2	11	3	15	8	4	9	10	13	9
7	11	10	1	2	12	5	6	4	2	9	15	

■ EXTENDING CONCEPTS

47. What Would You Do?   You work at a bank and are asked to recommend the amount of cash to put in an ATM each day. You don't want to put in too much (security) or too little (customer irritation). Here are the daily withdrawals (in 100s of dollars) for 30 days.

72	84	61	76	104	76	86	92	80	88
98	76	97	82	84	67	70	81	82	89
74	73	86	81	85	78	82	80	91	83

(a) Construct a relative frequency histogram for the data using 8 classes.

(b) If you put $9000 in the ATM each day, what percent of the days in a month should you expect to run out of cash? Explain your reasoning.

(c) If you are willing to run out of cash for 10% of the days, how much cash should you put in the ATM each day? Explain your reasoning.

48. What Would You Do?   You work in the admissions department for a college and are asked to recommend the minimum SAT scores that the college will accept for a position as a full-time student. Here are the SAT scores for a sample of 50 applicants.

1760	1502	1375	1310	1601	1942	1380	2211	1622	1771
1150	1351	1682	1618	2051	1742	1463	1395	1860	1918
1882	1996	1525	1510	2120	1700	1818	1869	1440	1235
976	1513	1790	2250	2102	1905	1979	1588	1420	1730
2175	1930	1965	1658	2005	2125	1260	1560	1635	1620

(a) Construct a relative frequency histogram for the data using 10 classes.

(b) If you set the minimum score at 1616, what percent of the applicants will meet this requirement? Explain your reasoning.

(c) If you want to accept the top 88% of the applicants, what should the minimum score be? Explain your reasoning.

2.2 More Graphs and Displays

WHAT YOU SHOULD LEARN

▸ How to graph and interpret quantitative data sets using stem-and-leaf plots and dot plots

▸ How to graph and interpret qualitative data sets using pie charts and Pareto charts

▸ How to graph and interpret paired data sets using scatter plots and time series charts

Graphing Quantitative Data Sets ▸ Graphing Qualitative Data Sets ▸ Graphing Paired Data Sets

▸ GRAPHING QUANTITATIVE DATA SETS

In Section 2.1, you learned several traditional ways to display quantitative data graphically. In this section, you will learn a newer way to display quantitative data, called a **stem-and-leaf plot.** Stem-and-leaf plots are examples of **exploratory data analysis (EDA),** which was developed by John Tukey in 1977.

In a stem-and-leaf plot, each number is separated into a **stem** (for instance, the entry's leftmost digits) and a **leaf** (for instance, the rightmost digit). You should have as many leaves as there are entries in the original data set and the leaves should be single digits. A stem-and-leaf plot is similar to a histogram but has the advantage that the graph still contains the original data values. Another advantage of a stem-and-leaf plot is that it provides an easy way to sort data.

EXAMPLE 1 Report 4

▸ Constructing a Stem-and-Leaf Plot

The following are the numbers of text messages sent last week by the cellular phone users on one floor of a college dormitory. Display the data in a stem-and-leaf plot. What can you conclude?

155	159	144	129	105	145	126	116	130	114	122	112	112	142
126	118	118	108	122	121	109	140	126	119	113	117	118	109
109	119	139	139	122	78	133	126	123	145	121	134	124	119
132	133	124	129	112	126	148	147						

▸ **Solution** Because the data entries go from a low of 78 to a high of 159, you should use stem values from 7 to 15. To construct the plot, list these stems to the left of a vertical line. For each data entry, list a leaf to the right of its stem. For instance, the entry 155 has a stem of 15 and a leaf of 5. The resulting stem-and-leaf plot will be unordered. To obtain an ordered stem-and-leaf plot, rewrite the plot with the leaves in increasing order from left to right. Be sure to include a key.

STUDY TIP

It is important to include a key for a stem-and-leaf plot to identify the values of the data. This is done by showing a value represented by a stem and one leaf.

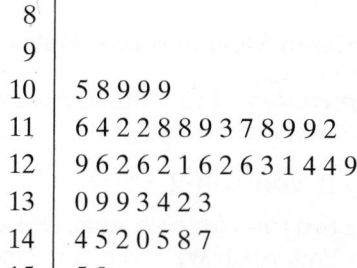

Number of Text Messages Sent

```
 7 | 8          Key: 15|5 = 155
 8 |
 9 |
10 | 5 8 9 9 9
11 | 6 4 2 2 8 8 9 3 7 8 9 9 2
12 | 9 6 2 6 2 1 6 2 6 3 1 4 4 9 6
13 | 0 9 9 3 4 2 3
14 | 4 5 2 0 5 8 7
15 | 5 9
```

Unordered Stem-and-Leaf Plot

Number of Text Messages Sent

```
 7 | 8          Key: 15|5 = 155
 8 |
 9 |
10 | 5 8 9 9 9
11 | 2 2 2 3 4 6 7 8 8 8 9 9 9
12 | 1 1 2 2 2 3 4 4 6 6 6 6 6 9 9
13 | 0 2 3 3 4 9 9
14 | 0 2 4 5 5 7 8
15 | 5 9
```

Ordered Stem-and-Leaf Plot

Interpretation From the display, you can conclude that more than 50% of the cellular phone users sent between 110 and 130 text messages.

▶ **Try It Yourself 1**

Use a stem-and-leaf plot to organize the ages of the 50 richest people data set listed in the Chapter Opener on page 37. What can you conclude?

a. List all possible *stems*.
b. List the *leaf* of each data entry to the right of its stem and include a *key*.
c. *Rewrite* the stem-and-leaf plot so that the leaves are ordered.
d. Use the plot to make a *conclusion*.   *Answer: Page A31*

Note to Instructor

If you are using MINITAB or Excel, ask students to use this technology to construct a stem-and-leaf plot.

EXAMPLE 2

▶ **Constructing Variations of Stem-and-Leaf Plots**

Organize the data given in Example 1 using a stem-and-leaf plot that has two rows for each stem. What can you conclude?

▶ **Solution**

Use the stem-and-leaf plot from Example 1, except now list each stem twice. Use the leaves 0, 1, 2, 3, and 4 in the first stem row and the leaves 5, 6, 7, 8, and 9 in the second stem row. The revised stem-and-leaf plot is shown. Notice that by using two rows per stem, you obtain a more detailed picture of the data.

INSIGHT

You can use stem-and-leaf plots to identify unusual data values called *outliers*. In Examples 1 and 2, the data value 78 is an outlier. You will learn more about outliers in Section 2.3.

Number of Text Messages Sent		Number of Text Messages Sent	
7	Key: 15│5 = 155	7	Key: 15│5 = 155
7	8	7	8
8		8	
8		8	
9		9	
9		9	
10		10	
10	5 8 9 9 9	10	5 8 9 9 9
11	4 2 2 3 2	11	2 2 2 3 4
11	6 8 8 9 7 8 9 9	11	6 7 8 8 8 9 9 9
12	2 2 1 2 3 1 4 4	12	1 1 2 2 2 3 4 4
12	9 6 6 6 6 9 6	12	6 6 6 6 6 9 9
13	0 3 4 2 3	13	0 2 3 3 4
13	9 9	13	9 9
14	4 2 0	14	0 2 4
14	5 5 8 7	14	5 5 7 8
15		15	
15	5 9	15	5 9
Unordered Stem-and-Leaf Plot		**Ordered Stem-and-Leaf Plot**	

Interpretation From the display, you can conclude that most of the cellular phone users sent between 105 and 135 text messages.

▶ **Try It Yourself 2**

Using two rows for each stem, revise the stem-and-leaf plot you constructed in Try It Yourself 1. What can you conclude?

a. List each stem *twice*.
b. List all leaves *using the appropriate stem row*.
c. Use the plot to make a *conclusion*.   *Answer: Page A32*

You can also use a dot plot to graph quantitative data. In a **dot plot,** each data entry is plotted, using a point, above a horizontal axis. Like a stem-and-leaf plot, a dot plot allows you to see how data are distributed, determine specific data entries, and identify unusual data values.

EXAMPLE 3   SC Report 5

▸ Constructing a Dot Plot

Use a dot plot to organize the text messaging data given in Example 1. What can you conclude from the graph?

155	159	144	129	105	145	126	116	130	114	122	112
112	142	126	118	118	108	122	121	109	140	126	119
113	117	118	109	109	119	139	139	122	78	133	126
123	145	121	134	124	119	132	133	124	129	112	126
148	147										

▸ Solution

So that each data entry is included in the dot plot, the horizontal axis should include numbers between 70 and 160. To represent a data entry, plot a point above the entry's position on the axis. If an entry is repeated, plot another point above the previous point.

Number of Text Messages Sent

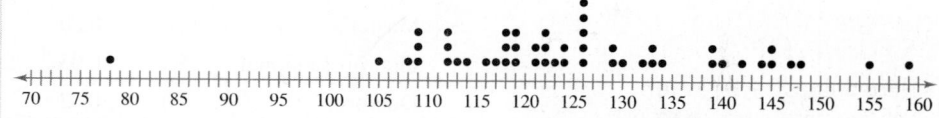

Interpretation From the dot plot, you can see that most values cluster between 105 and 148 and the value that occurs the most is 126. You can also see that 78 is an unusual data value.

▸ Try It Yourself 3

Use a dot plot to organize the ages of the 50 richest people data set listed in the Chapter Opener on page 37. What can you conclude from the graph?

a. Choose an appropriate scale for the *horizontal axis*.
b. Represent each data entry by *plotting a point*.
c. *Describe* any patterns in the data.   *Answer: Page A32*

Technology can be used to construct stem-and-leaf plots and dot plots. For instance, a MINITAB dot plot for the text messaging data is shown below.

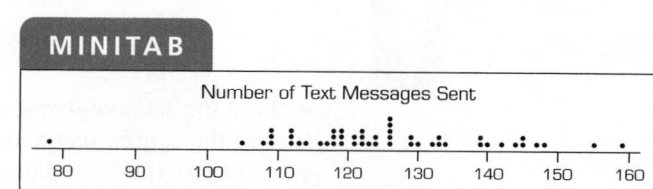

▶ GRAPHING QUALITATIVE DATA SETS

Pie charts provide a convenient way to present qualitative data graphically as percents of a whole. A **pie chart** is a circle that is divided into sectors that represent categories. The area of each sector is proportional to the frequency of each category. In most cases, you will be interpreting a pie chart or constructing one using technology. Example 4 shows how to construct a pie chart by hand.

Earned Degrees Conferred in 2007

Type of degree	Number (thousands)
Associate's	728
Bachelor's	1525
Master's	604
First professional	90
Doctoral	60

EXAMPLE 4   SC Report 6

▶ Constructing a Pie Chart

The numbers of earned degrees conferred (in thousands) in 2007 are shown in the table. Use a pie chart to organize the data. What can you conclude? *(Source: U.S. National Center for Education Statistics)*

▶ Solution

Begin by finding the relative frequency, or percent, of each category. Then construct the pie chart using the central angle that corresponds to each category. To find the central angle, multiply 360° by the category's relative frequency. For instance, the central angle for associate's degrees is $360°(0.24) \approx 86°$. To construct a pie chart in Excel, follow the instructions in the margin.

Type of degree	f	Relative frequency	Angle
Associate's	728	0.24	86°
Bachelor's	1525	0.51	184°
Master's	604	0.20	72°
First professional	90	0.03	11°
Doctoral	60	0.02	7°

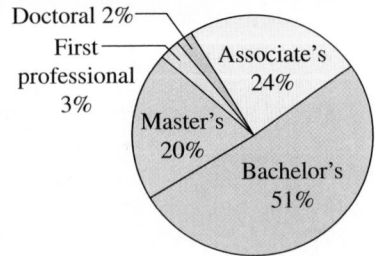

Earned Degrees Conferred in 2007

Interpretation   From the pie chart, you can see that over one half of the degrees conferred in 2007 were bachelor's degrees.

▶ Try It Yourself 4

The numbers of earned degrees conferred (in thousands) in 1990 are shown in the table. Use a pie chart to organize the data. Compare the 1990 data with the 2007 data. *(Source: U.S. National Center for Education Statistics)*

Earned Degrees Conferred in 1990

Type of degree	Number (thousands)
Associate's	455
Bachelor's	1052
Master's	325
First professional	71
Doctoral	38

a. Find the *relative frequency* and *central angle* of each category.
b. Use the *central angle* to find the portion that corresponds to each category.
c. *Compare* the 1990 data with the 2007 data.   *Answer: Page A32*

STUDY TIP

Here are instructions for constructing a pie chart using Excel. First, enter the degree types and their corresponding frequencies or relative frequencies in two separate columns. Then highlight the two columns, click on the Chart Wizard, and select *Pie* as your chart type. Click *Next* throughout the Chart Wizard while constructing your pie chart.

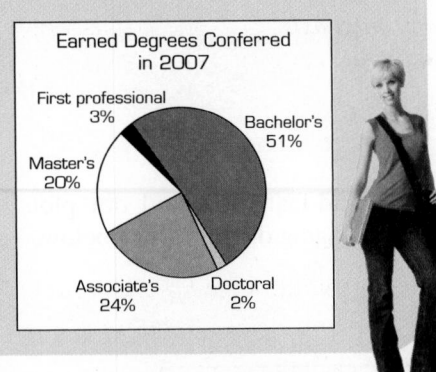

Another way to graph qualitative data is to use a Pareto chart. A **Pareto chart** is a vertical bar graph in which the height of each bar represents frequency or relative frequency. The bars are positioned in order of decreasing height, with the tallest bar positioned at the left. Such positioning helps highlight important data and is used frequently in business.

EXAMPLE 5   SC Report 7

▶ Constructing a Pareto Chart

In a recent year, the retail industry lost $36.5 billion in inventory shrinkage. Inventory shrinkage is the loss of inventory through breakage, pilferage, shoplifting, and so on. The main causes of inventory shrinkage are administrative error ($5.4 billion), employee theft ($15.9 billion), shoplifting ($12.7 billion), and vendor fraud ($1.4 billion). If you were a retailer, which causes of inventory shrinkage would you address first? (*Source: National Retail Federation and the University of Florida*)

▶ Solution

Using frequencies for the vertical axis, you can construct the Pareto chart as shown.

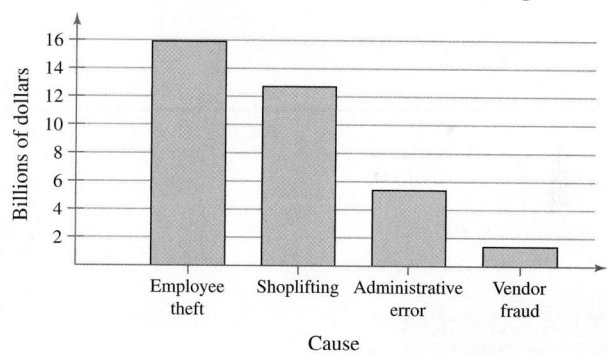

Main Causes of Inventory Shrinkage

Interpretation   From the graph, it is easy to see that the causes of inventory shrinkage that should be addressed first are employee theft and shoplifting.

▶ Try It Yourself 5

Every year, the Better Business Bureau (BBB) receives complaints from customers. In a recent year, the BBB received the following complaints.

7792 complaints about home furnishing stores
5733 complaints about computer sales and service stores
14,668 complaints about auto dealers
9728 complaints about auto repair shops
4649 complaints about dry cleaning companies

Use a Pareto chart to organize the data. What source is the greatest cause of complaints? (*Source: Council of Better Business Bureaus*)

a. Find the *frequency or relative frequency* for each data entry.
b. *Position the bars in decreasing order* according to frequency or relative frequency.
c. *Interpret* the results in the context of the data.   *Answer: Page A32*

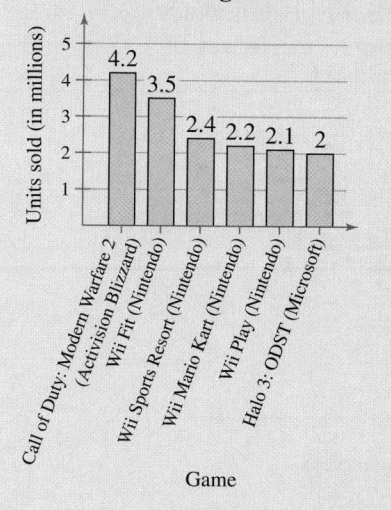

▶ GRAPHING PAIRED DATA SETS

When each entry in one data set corresponds to one entry in a second data set, the sets are called **paired data sets.** For instance, suppose a data set contains the costs of an item and a second data set contains sales amounts for the item at each cost. Because each cost corresponds to a sales amount, the data sets are paired. One way to graph paired data sets is to use a **scatter plot,** where the ordered pairs are graphed as points in a coordinate plane. A scatter plot is used to show the relationship between two quantitative variables.

EXAMPLE 6

▶ Interpreting a Scatter Plot

The British statistician Ronald Fisher (see page 33) introduced a famous data set called Fisher's Iris data set. This data set describes various physical characteristics, such as petal length and petal width (in millimeters), for three species of iris. In the scatter plot shown, the petal lengths form the first data set and the petal widths form the second data set. As the petal length increases, what tends to happen to the petal width? *(Source: Fisher, R. A., 1936)*

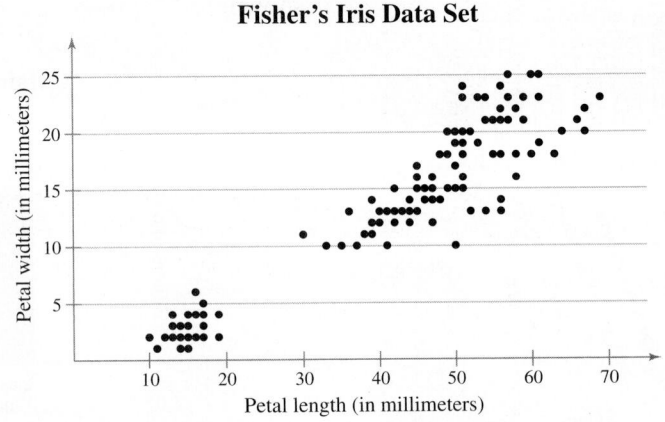

Fisher's Iris Data Set

▶ Solution

The horizontal axis represents the petal length, and the vertical axis represents the petal width. Each point in the scatter plot represents the petal length and petal width of one flower.

Interpretation   From the scatter plot, you can see that as the petal length increases, the petal width also tends to increase.

▶ Try It Yourself 6

The lengths of employment and the salaries of 10 employees are listed in the table at the left. Graph the data using a scatter plot. What can you conclude?

a. Label the *horizontal and vertical axes.*
b. *Plot* the paired data.
c. *Describe* any trends.   *Answer: Page A32*

Length of employment (in years)	Salary (in dollars)
5	32,000
4	32,500
8	40,000
4	27,350
2	25,000
10	43,000
7	41,650
6	39,225
9	45,100
3	28,000

You will learn more about scatter plots and how to analyze them in Chapter 9.

A data set that is composed of quantitative entries taken at regular intervals over a period of time is called a **time series.** For instance, the amount of precipitation measured each day for one month is a time series. You can use a **time series chart** to graph a time series.

See MINITAB and TI-83/84 Plus steps on pages 122 and 123.

Note to Instructor

Consider asking students to find a time series plot in a magazine or newspaper and bring it to class for discussion.

EXAMPLE 7   **SC** Report 8

▶ Constructing a Time Series Chart

The table lists the number of cellular telephone subscribers (in millions) and subscribers' average local monthly bills for service (in dollars) for the years 1998 through 2008. Construct a time series chart for the number of cellular subscribers. What can you conclude? *(Source: Cellular Telecommunications & Internet Association)*

Year	Subscribers (in millions)	Average bill (in dollars)
1998	69.2	39.43
1999	86.0	41.24
2000	109.5	45.27
2001	128.4	47.37
2002	140.8	48.40
2003	158.7	49.91
2004	182.1	50.64
2005	207.9	49.98
2006	233.0	50.56
2007	255.4	49.79
2008	270.3	50.07

▶ Solution

Let the horizontal axis represent the years and let the vertical axis represent the number of subscribers (in millions). Then plot the paired data and connect them with line segments.

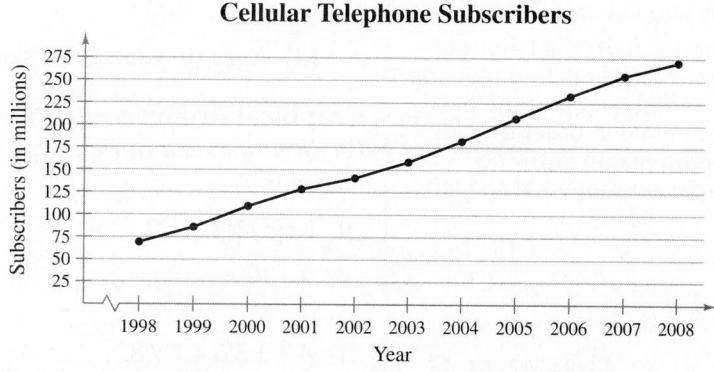

Cellular Telephone Subscribers

Interpretation The graph shows that the number of subscribers has been increasing since 1998.

▶ Try It Yourself 7

Use the table in Example 7 to construct a time series chart for subscribers' average local monthly cellular telephone bills for the years 1998 through 2008. What can you conclude?

a. Label the *horizontal and vertical axes.*
b. *Plot* the paired data and *connect* them with line segments.
c. *Describe* any patterns you see.   *Answer: Page A32*

2.2 EXERCISES

FOR EXTRA HELP:
MyStatLab

1. Quantitative: stem-and-leaf plot, dot plot, histogram, scatter plot, time series chart

   Qualitative: pie chart, Pareto chart

2. Unlike the histogram, the stem-and-leaf plot still contains the original data values. However, some data are difficult to organize in a stem-and-leaf plot.

3. Both the stem-and-leaf plot and the dot plot allow you to see how data are distributed, to determine specific data entries, and to identify unusual data values.

4. In a Pareto chart, the height of each bar represents frequency or relative frequency and the bars are positioned in order of decreasing height with the tallest bar positioned at the left.

5. b

6. d

7. a

8. c

9. 27, 32, 41, 43, 43, 44, 47, 47, 48, 50, 51, 51, 52, 53, 53, 53, 54, 54, 54, 54, 55, 56, 56, 58, 59, 68, 68, 68, 73, 78, 78, 85

   Max: 85; Min: 27

10. 12.9, 13.3, 13.6, 13.7, 13.7, 14.1, 14.1, 14.1, 14.1, 14.3, 14.4, 14.4, 14.6, 14.9, 14.9, 15.0, 15.0, 15.0, 15.1, 15.2, 15.4, 15.6, 15.7, 15.8, 15.8, 15.8, 15.9, 16.1, 16.6, 16.7

    Max: 16.7; Min: 12.9

11. 13, 13, 14, 14, 14, 15, 15, 15, 15, 15, 16, 17, 17, 18, 19

    Max: 19; Min: 13

12. See Selected Answers, page A112.

■ BUILDING BASIC SKILLS AND VOCABULARY

1. Name some ways to display quantitative data graphically. Name some ways to display qualitative data graphically.

2. What is an advantage of using a stem-and-leaf plot instead of a histogram? What is a disadvantage?

3. In terms of displaying data, how is a stem-and-leaf plot similar to a dot plot?

4. How is a Pareto chart different from a standard vertical bar graph?

Putting Graphs in Context *In Exercises 5–8, match the plot with the description of the sample.*

5.
```
0 | 8          Key: 0|8 = 0.8
1 | 5 6 8
2 | 1 3 4 5
3 | 0 9
4 | 0 0
```

6.
```
6 | 7 8          Key: 6|7 = 67
7 | 4 5 5 8 8 8
8 | 1 3 5 5 8 8 9
9 | 0 0 0 2 4
```

7.
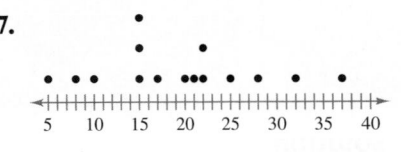

8.
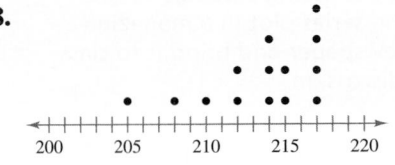

(a) Time (in minutes) it takes a sample of employees to drive to work

(b) Grade point averages of a sample of students with finance majors

(c) Top speeds (in miles per hour) of a sample of high-performance sports cars

(d) Ages (in years) of a sample of residents of a retirement home

Graphical Analysis *In Exercises 9–12, use the stem-and-leaf plot or dot plot to list the actual data entries. What is the maximum data entry? What is the minimum data entry?*

9. Key: 2|7 = 27
```
2 | 7
3 | 2
4 | 1 3 3 4 7 7 8
5 | 0 1 1 2 3 3 3 4 4 4 4 5 6 6 8 9
6 | 8 8 8
7 | 3 8 8
8 | 5
```

10. Key: 12|9 = 12.9
```
12 |
12 | 9
13 | 3
13 | 6 7 7
14 | 1 1 1 1 3 4 4
14 | 6 9 9
15 | 0 0 0 1 2 4
15 | 6 7 8 8 8 9
16 | 1
16 | 6 7
```

11.
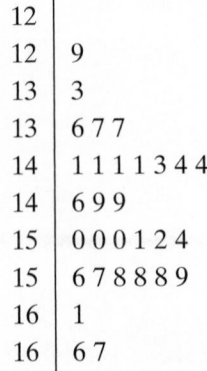

12.

13. Answers will vary. *Sample answer:* Users spend the most amount of time on MySpace and the least amount of time on Twitter.

14. Answers will vary. *Sample answer:* Motor vehicle thefts decreased between 2003 and 2008.

15. Answers will vary. *Sample answer:* Tailgaters irk drivers the most, and too-cautious drivers irk drivers the least.

16. Answers will vary. *Sample answer:* Twice as many people "sped up" than "cut off a car."

17. Key: 6|7 = 67

    6 | 7 8
    7 | 3 5 5 6 9
    8 | 0 0 2 3 5 5 7 7 8
    9 | 0 1 1 1 2 4 5 5

    It appears that most grades for the biology midterm were in the 80s or 90s. (Answers will vary.)

18. Key: 4|9 = 49

    4 | 8 9
    5 | 0 0 0 1 2
    5 | 5 5 5 6 7 7 8 9 9
    6 | 0 0 0 1 2 2 2 3 3 4 4 4
    6 | 7
    7 | 4

    It appears that most of the highest paid CEOs are between 55 and 65 years old. (Answers will vary.)

19. Key: 4|3 = 4.3

    4 | 3 9
    5 | 1 8 8 8 9
    6 | 4 8 9 9 9
    7 | 0 0 2 2 2 5
    8 | 0 1

    It appears that most ice had a thickness of 5.8 centimeters to 7.2 centimeters. (Answers will vary.)

20. Key: 17|5 = 17.5

    16 | 4 8
    17 | 1 1 3 4 5 5 6 7 9
    18 | 1 3 4 4 6 6 6 9
    19 | 0 0 2 3 3 5 6
    20 | 1 8

    It appears that most farmers charge 17 to 19 cents per pound of apples. (Answers will vary.)

■ USING AND INTERPRETING CONCEPTS

Graphical Analysis *In Exercises 13–16, give three conclusions that can be drawn from the graph.*

13.

Average Time Spent on Top 5 Social Networking Sites

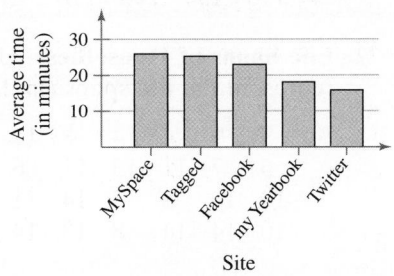

(Source: Experian Hitwise)

14.

Motor Vehicle Thefts in U.S.

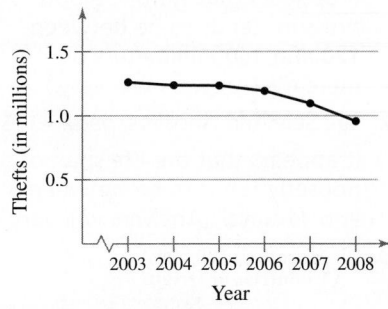

(Source: Federal Bureau of Investigation)

15.

How Other Drivers Irk Us

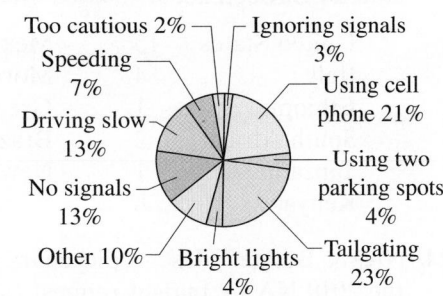

(Adapted from Reuters/Zogby)

16.

Driving and Cell Phone Use

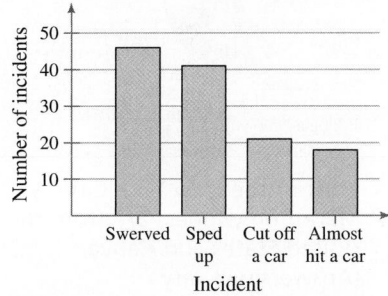

(Adapted from USA Today)

Graphing Data Sets *In Exercises 17–30, organize the data using the indicated type of graph. What can you conclude about the data?*

17. Exam Scores Use a stem-and-leaf plot to display the data. The data represent the scores of a biology class on a midterm exam.

    75 85 90 80 87 67 82 88 95 91 73 80
    83 92 94 68 75 91 79 95 87 76 91 85

18. Highest Paid CEOs Use a stem-and-leaf plot that has two rows for each stem to display the data. The data represent the ages of the top 30 highest paid CEOs. *(Source: Forbes)*

    64 74 55 55 62 63 50 67 51 59 50
    52 50 59 62 64 57 61 49 63 62 60
    55 56 48 58 64 60 60 57

19. Ice Thickness Use a stem-and-leaf plot to display the data. The data represent the thicknesses (in centimeters) of ice measured at 20 different locations on a frozen lake.

    5.8 6.4 6.9 7.2 5.1 4.9 4.3 5.8 7.0 6.8
    8.1 7.5 7.2 6.9 5.8 7.2 8.0 7.0 6.9 5.9

20. Apple Prices Use a stem-and-leaf plot to display the data. The data represent the prices (in cents per pound) paid to 28 farmers for apples.

    19.2 19.6 16.4 17.1 19.0 17.4 17.3
    20.1 19.0 17.5 17.6 18.6 18.4 17.7
    19.5 18.4 18.9 17.5 19.3 20.8 19.3
    18.6 18.6 18.3 17.1 18.1 16.8 17.9

21.   Systolic Blood Pressures

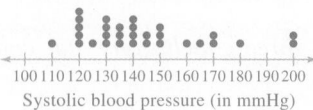

It appears that systolic blood pressure tends to be between 120 and 150 millimeters of mercury. (Answers will vary.)

22. See Selected Answers, page A113.

It appears that the life span of a housefly tends to be between 8 and 11 days. (Answers will vary.)

23.   Marathon Winners' Countries of Origin

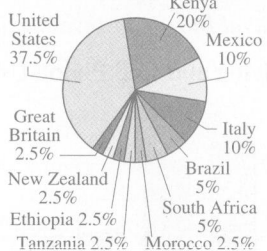

Most of the New York City Marathon winners are from the United States and Kenya. (Answers will vary.)

24. See Selected Answers, page A113.

25. See Odd Answers, page A56.

26. See Selected Answers, page A113.

27. See Odd Answers, page A56.

Hours	Hourly wage
33	12.16
37	9.98
34	10.79
40	11.71
35	11.80
33	11.51
40	13.65
33	12.05
28	10.54
45	10.33
37	11.57
28	10.17

TABLE FOR EXERCISE 27

21. Systolic Blood Pressures   Use a dot plot to display the data. The data represent the systolic blood pressures (in millimeters of mercury) of 30 patients at a doctor's office.

120	135	140	145	130	150	120	170	145	125
130	110	160	180	200	150	200	135	140	120
120	130	140	170	120	165	150	130	135	140

22. Life Spans of Houseflies   Use a dot plot to display the data. The data represent the life spans (in days) of 40 houseflies.

9	9	4	4	8	11	10	5	8	13	9
6	7	11	13	11	6	9	8	14	10	6
10	10	8	7	14	11	7	8	6	11	13
10	14	14	8	13	14	10				

23. New York City Marathon   Use a pie chart to display the data. The data represent the number of men's New York City Marathon winners from each country through 2009. *(Source: New York Road Runners)*

United States	15	Mexico	4
Italy	4	Morocco	1
Ethiopia	1	Great Britain	1
South Africa	2	Brazil	2
Tanzania	1	New Zealand	1
Kenya	8		

24. NASA Budget   Use a pie chart to display the data. The data represent the 2010 NASA budget request (in millions of dollars) divided among five categories. *(Source: NASA)*

Science, aeronautics, exploration	8947
Space operations	6176
Education	126
Cross-agency support	3401
Inspector general	36

25. Barrel of Oil   Use a Pareto chart to display the data. The data represent how a 42-gallon barrel of crude oil is distributed. *(Adapted from American Petroleum Institute)*

Gasoline	43%
Kerosene-type jet fuel	9%
Distillate fuel oil (home heating, diesel fuel, etc.)	24%
Coke	5%
Residual fuel oil (industry, marine transportation, etc.)	4%
Liquefied refinery gases	3%
Other	12%

26. UV Index   Use a Pareto chart to display the data. The data represent the ultraviolet indices for five cities at noon on a recent date. *(Source: National Oceanic and Atmospheric Administration)*

Atlanta, GA	Boise, ID	Concord, NH	Denver, CO	Miami, FL
9	7	8	7	10

27. Hourly Wages   Use a scatter plot to display the data shown in the table. The data represent the number of hours worked and the hourly wages (in dollars) for a sample of 12 production workers. Describe any trends shown.

Number of students per teacher	Average teacher's salary
17.1	28.7
17.5	47.5
18.9	31.8
17.1	28.1
20.0	40.3
18.6	33.8
14.4	49.8
16.5	37.5
13.3	42.5
18.4	31.9

TABLE FOR EXERCISE 28

28. See Selected Answers, page A113.

29.

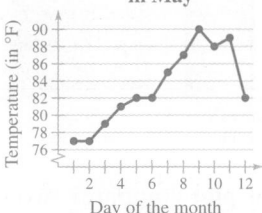

Daily High Temperatures in May

It appears that it was hottest from May 7th to May 11th. (Answers will vary.)

30. See Selected Answers, page A113.

31. Variable: Scores

Decimal point is 1 digit(s) to the right of the colon.

5 : 5

6 : 2

6 : 8

7 : 0 1

7 : 5 6

8 : 0 2 3

8 : 5 6 7 8 8 9

9 : 0 3 3

9 : 5 5 8 9

10 : 0

It appears that most scores on the final exam in economics were in the 80s and 90s. (Answers will vary.)

32. See Selected Answers, page A113.

It appears that most screen sizes are between 2.5 and 3 inches. (Answers will vary.)

33. See Odd Answers, page A56.

34. See Selected Answers, page A113.

28. Salaries Use a scatter plot to display the data shown in the table. The data represent the number of students per teacher and the average teacher salaries (in thousands of dollars) for a sample of 10 school districts. Describe any trends shown.

29. Daily High Temperatures Use a time series chart to display the data. The data represent the daily high temperatures for a city for a period of 12 days.

May 1	May 2	May 3	May 4	May 5	May 6
77°	77°	79°	81°	82°	82°

May 7	May 8	May 9	May 10	May 11	May 12
85°	87°	90°	ˉ88°	89°	82°

30. Manufacturing Use a time series chart to display the data. The data represent the percentages of the U.S. gross domestic product (GDP) that come from the manufacturing sector. *(Source: U.S. Bureau of Economic Analysis)*

1997	1998	1999	2000	2001	2002
16.6%	15.4%	14.8%	14.5%	13.2%	12.9%

2003	2004	2005	2006	2007	2008
12.5%	12.2%	11.9%	12.0%	11.7%	11.5%

SC *In Exercises 31–34, use StatCrunch to organize the data using the indicated type of graph. What can you conclude about the data?*

31. Use a stem-and-leaf plot to display the data. The data represent the scores of an economics class on a final exam.

82 93 95 75 68 90 98 71 85 88 100 93
70 80 89 62 55 95 83 86 88 76   99 87

32. Use a dot plot to display the data. The data represent the screen sizes (in inches) of 20 DVD camcorders.

3.0 2.7 3.2 2.7 1.8 2.7 2.7 3.0 2.7 3.0
2.5 3.2 2.7 2.7 3.0 2.7 2.0 2.7 3.0 2.5

33. Use (a) a pie chart and (b) a Pareto chart to display the data. The data represent the results of an online survey that asked adults which type of investment they would focus on in 2010. *(Adapted from CNN)*

U.S. stocks	11,521	Emerging markets	5267
Bonds	3292	Commodities	1975
Bank accounts	10,533		

34. The data represent the number of motor vehicles (in millions) registered in the U.S. and the number of crashes (in millions). *(Source: U.S. National Highway Safety Traffic Administration)*

Year	2000	2001	2002	2003	2004	2005	2006	2007
Registrations	221	230	230	231	237	241	244	247
Crashes	6.4	6.3	6.3	6.3	6.2	6.2	6.0	6.0

(a) Use a scatter plot to display the number of registrations.

(b) Use a scatter plot to display the number of crashes.

(c) Construct a time series chart for the number of registrations.

(d) Construct a time series chart for the number of crashes.

35. (a) The graph is misleading because the large gap from 0 to 90 makes it appear that the sales for the 3rd quarter are disproportionately larger than the other quarters. (Answers will vary.)

    (b) See Odd Answers, page A56.

36. See Selected Answers, page A114.

37. See Odd Answers, page A56.

38. (a) The graph is misleading because the "OPEC countries" bar is wider than the "non-OPEC countries" bar.

    (b) See Selected Answers, page A114.

Law Firm A		Law Firm B
5 0	9	0 3
8 5 2 2 2	10	5 7
9 9 7 0 0	11	0 0 5
1 1	12	0 3 3 5
	13	2 2 5 9
	14	1 3 3 3 9
	15	5 5 5 6
	16	4 9 9
9 9 5 1 0	17	1 2 5
5 5 5 2 1	18	9
9 9 8 7 5	19	0
3	20	

Key: 5|19|0 = $195,000 for Law Firm A and $190,000 for Law Firm B

FIGURE FOR EXERCISE 39

39. (a) At Law Firm A, the lowest salary was $90,000 and the highest salary was $203,000; at Law Firm B, the lowest salary was $90,000 and the highest salary was $190,000.

    (b) There are 30 lawyers at Law Firm A and 32 lawyers at Law Firm B.

    (c) At Law Firm A, the salaries tend to be clustered at the far ends of the distribution range and at Law Firm B, the salaries tend to fall in the middle of the distribution range.

40. See Selected Answers, page A114.

■ EXTENDING CONCEPTS

A Misleading Graph? *A misleading graph is a statistical graph that is not drawn appropriately. This type of graph can misrepresent data and lead to false conclusions. In Exercises 35–38, (a) explain why the graph is misleading, and (b) redraw the graph so that it is not misleading.*

35.
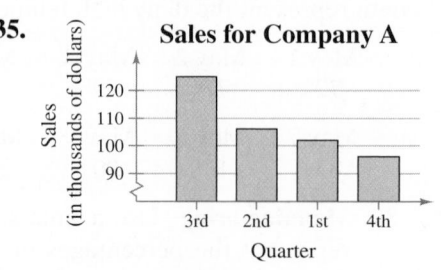
Sales for Company A

36.
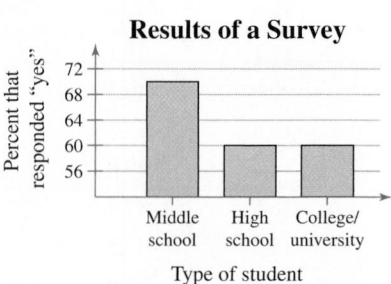
Results of a Survey

37.
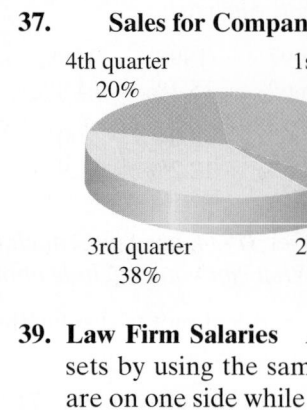
Sales for Company B

38.
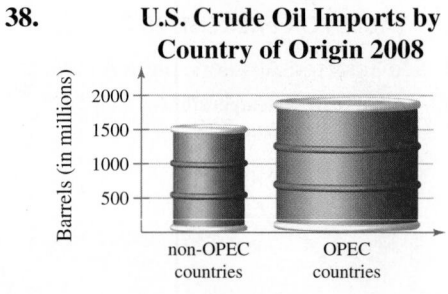
U.S. Crude Oil Imports by Country of Origin 2008

39. **Law Firm Salaries** A **back-to-back stem-and-leaf plot** compares two data sets by using the same stems for each data set. Leaves for the first data set are on one side while leaves for the second data set are on the other side. The back-to-back stem-and-leaf plot shows the salaries (in thousands of dollars) of all lawyers at two small law firms.

    (a) What are the lowest and highest salaries at Law Firm A? at Law Firm B?

    (b) How many lawyers are in each firm?

    (c) Compare the distribution of salaries at each law firm. What do you notice?

40. **Yoga Classes** The data sets show the ages of all participants in two yoga classes.

3:00 P.M. Class						8:00 P.M. Class					
40	60	73	77	51	68	19	18	20	29	39	43
68	35	68	53	64	75	71	56	44	44	18	19
76	69	59	55	38	57	19	18	18	20	25	29
68	84	75	62	73	75	25	22	31	24	24	23
85	77					19	19	18	28	20	31

    (a) Make a back-to-back stem-and-leaf plot to display the data.

    (b) What are the lowest and highest ages of participants in the 3:00 P.M. class? in the 8:00 P.M. class?

    (c) How many participants are in each class?

    (d) Compare the distribution of ages in each class. What conclusion(s) can you make based on your observations?

2.3   Measures of Central Tendency

WHAT YOU SHOULD LEARN

- How to find the mean, median, and mode of a population and of a sample

- How to find a weighted mean of a data set and the mean of a frequency distribution

- How to describe the shape of a distribution as symmetric, uniform, or skewed and how to compare the mean and median for each

Mean, Median, and Mode ▸ Weighted Mean and Mean of Grouped Data ▸ The Shapes of Distributions

▸ MEAN, MEDIAN, AND MODE

In Sections 2.1 and 2.2, you learned about the graphical representations of quantitative data. In Sections 2.3 and 2.4, you will learn how to supplement graphical representations with numerical statistics that describe the center and variability of a data set.

A **measure of central tendency** is a value that represents a typical, or central, entry of a data set. The three most commonly used measures of central tendency are the *mean*, the *median*, and the *mode*.

DEFINITION

The **mean** of a data set is the sum of the data entries divided by the number of entries. To find the mean of a data set, use one of the following formulas.

$$\text{Population Mean: } \mu = \frac{\sum x}{N} \qquad \text{Sample Mean: } \bar{x} = \frac{\sum x}{n}$$

The lowercase Greek letter μ (pronounced mu) represents the population mean and $\bar{x}$ (read as "x bar") represents the sample mean. Note that N represents the number of entries in a *population* and n represents the number of entries in a *sample*. Recall that the uppercase Greek letter sigma (Σ) indicates a summation of values.

EXAMPLE 1   SC   Report 9

▸ **Finding a Sample Mean**

The prices (in dollars) for a sample of round-trip flights from Chicago, Illinois to Cancun, Mexico are listed. What is the mean price of the flights?

872   432   397   427   388   782   397

▸ **Solution**

The sum of the flight prices is

$$\sum x = 872 + 432 + 397 + 427 + 388 + 782 + 397 = 3695.$$

To find the mean price, divide the sum of the prices by the number of prices in the sample.

$$\bar{x} = \frac{\sum x}{n} = \frac{3695}{7} \approx 527.9$$

So, the mean price of the flights is about $527.90.

▸ **Try It Yourself 1**

The heights (in inches) of the players on the 2009–2010 Cleveland Cavaliers basketball team are listed. What is the mean height?

a. *Find the sum* of the data entries.
b. *Divide the sum* by the number of data entries.
c. *Interpret* the results in the context of the data.

Answer: Page A32

STUDY TIP

Notice that the mean in Example 1 has one more decimal place than the original set of data values. This *round-off rule* will be used throughout the text. Another important *round-off rule* is that rounding should not be done until the final answer of a calculation.

Heights of Players							
74	78	81	87	81	80	77	80
85	78	80	83	75	81	73	

DEFINITION

The **median** of a data set is the value that lies in the middle of the data when the data set is ordered. The median measures the center of an ordered data set by dividing it into two equal parts. If the data set has an odd number of entries, the median is the middle data entry. If the data set has an even number of entries, the median is the mean of the two middle data entries.

EXAMPLE 2 · SC Report 10

▶ **Finding the Median**

Find the median of the flight prices given in Example 1.

▶ **Solution**

To find the median price, first order the data.

388   397   397   427   432   782   872

Because there are seven entries (an odd number), the median is the middle, or fourth, data entry. So, the median flight price is $427.

▶ **Try It Yourself 2**

The ages of a sample of fans at a rock concert are listed. Find the median age.

24   27   19   21   18   23   21   20   19   33   30   29   21
18   24   26   38   19   35   34   33   30   21   27   30

a. *Order* the data entries.
b. *Find the middle* data entry.
c. *Interpret* the results in the context of the data.   *Answer: Page A32*

STUDY TIP

In a data set, there are the same number of data values above the median as there are below the median. For instance, in Example 2, three of the prices are below $427 and three are above $427.

EXAMPLE 3

▶ **Finding the Median**

In Example 2, the flight priced at $432 is no longer available. What is the median price of the remaining flights?

▶ **Solution**

The remaining prices, in order, are 388, 397, 397, 427, 782, and 872.

Because there are six entries (an even number), the median is the mean of the two middle entries.

$$\text{Median} = \frac{397 + 427}{2} = 412$$

So, the median price of the remaining flights is $412.

▶ **Try It Yourself 3**

The prices (in dollars) of a sample of digital photo frames are listed. Find the median price of the digital photo frames.

25   100   130   60   140   200   220   80   250   97

a. *Order* the data entries.
b. *Find the mean* of the two middle data entries.
c. *Interpret* the results in the context of the data.   *Answer: Page A32*

DEFINITION

The **mode** of a data set is the data entry that occurs with the greatest frequency. A data set can have one mode, more than one mode, or no mode. If no entry is repeated, the data set has no mode. If two entries occur with the same greatest frequency, each entry is a mode and the data set is called **bimodal.**

EXAMPLE 4   SC Report 11

> ▶ **Finding the Mode**

Find the mode of the flight prices given in Example 1.

> ▶ **Solution**

Ordering the data helps to find the mode.

   388   397   397   427   432   782   872

From the ordered data, you can see that the entry 397 occurs twice, whereas the other data entries occur only once. So, the mode of the flight prices is $397.

> ▶ **Try It Yourself 4**

The prices (in dollars per square foot) for a sample of South Beach (Miami Beach, FL) condominiums are listed. Find the mode of the prices.

   324   462   540   450   638   564   670   618   624   825
   540   980   1650   1420   670   830   912   750   1260   450
   975   670   1100   980   750   723   705   385   475   720

a. Write the data in *order*.
b. Identify the entry, or entries, that occur with the *greatest frequency*.
c. *Interpret* the results in the context of the data.   *Answer: Page A32*

EXAMPLE 5

> ▶ **Finding the Mode**

At a political debate, a sample of audience members were asked to name the political party to which they belonged. Their responses are shown in the table. What is the mode of the responses?

> ▶ **Solution**

The response occurring with the greatest frequency is Republican. So, the mode is Republican.

Interpretation   In this sample, there were more Republicans than people of any other single affiliation.

> ▶ **Try It Yourself 5**

In a survey, 1000 U.S. adults were asked if they thought public cellular phone conversations were rude. Of those surveyed, 510 responded "Yes," 370 responded "No," and 120 responded "Not sure." What is the mode of the responses? *(Adapted from Fox TV/Rasmussen Reports)*

a. Identify the entry that occurs with the *greatest frequency*.
b. *Interpret* the results in the context of the data.   *Answer: Page A32*

INSIGHT

The mode is the only measure of central tendency that can be used to describe data at the nominal level of measurement. But when working with quantitative data, the mode is rarely used.

Political party	Frequency, f
Democrat	34
Republican	56
Other	21
Did not respond	9

Although the mean, the median, and the mode each describe a typical entry of a data set, there are advantages and disadvantages of using each. The mean is a reliable measure because it takes into account every entry of a data set. However, the mean can be greatly affected when the data set contains *outliers*.

Ages in a Class						
20	20	20	20	20	20	21
21	21	21	22	22	22	23
23	23	23	24	24	65	

Outlier ⟍↗

DEFINITION

An **outlier** is a data entry that is far removed from the other entries in the data set.

A data set can have one or more outliers, causing **gaps** in a distribution. Conclusions that are drawn from a data set that contains outliers may be flawed.

EXAMPLE 6

▶ **Comparing the Mean, the Median, and the Mode**

Find the mean, the median, and the mode of the sample ages of students in a class shown at the left. Which measure of central tendency best describes a typical entry of this data set? Are there any outliers?

▶ **Solution**

Mean: $\bar{x} = \dfrac{\sum x}{n} = \dfrac{475}{20} \approx 23.8$ years

Median: Median $= \dfrac{21 + 22}{2} = 21.5$ years

Mode: The entry occurring with the greatest frequency is 20 years.

Interpretation   The mean takes every entry into account but is influenced by the outlier of 65. The median also takes every entry into account, and it is not affected by the outlier. In this case the mode exists, but it doesn't appear to represent a typical entry. Sometimes a graphical comparison can help you decide which measure of central tendency best represents a data set. The histogram shows the distribution of the data and the locations of the mean, the median, and the mode. In this case, it appears that the median best describes the data set.

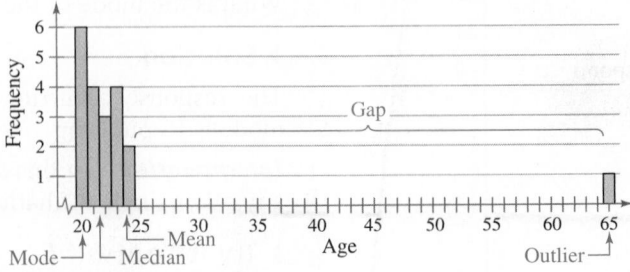

Ages of Students in a Class

▶ **Try It Yourself 6**

Remove the data entry 65 from the data set in Example 6. Then rework the example. How does the absence of this outlier change each of the measures?

a. Find the *mean*, the *median*, and the *mode*.

b. *Compare* these measures of central tendency with those found in Example 6.

Answer: Page A33

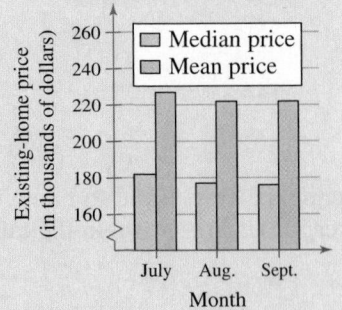

▶ WEIGHTED MEAN AND MEAN OF GROUPED DATA

Sometimes data sets contain entries that have a greater effect on the mean than do other entries. To find the mean of such a data set, you must find the *weighted mean*.

DEFINITION

A **weighted mean** is the mean of a data set whose entries have varying weights. A weighted mean is given by

$$\overline{x} = \frac{\sum(x \cdot w)}{\sum w}$$

where w is the weight of each entry x.

EXAMPLE 7

▶ Finding a Weighted Mean

You are taking a class in which your grade is determined from five sources: 50% from your test mean, 15% from your midterm, 20% from your final exam, 10% from your computer lab work, and 5% from your homework. Your scores are 86 (test mean), 96 (midterm), 82 (final exam), 98 (computer lab), and 100 (homework). What is the weighted mean of your scores? If the minimum average for an A is 90, did you get an A?

▶ Solution

Begin by organizing the scores and the weights in a table.

Source	Score, x	Weight, w	xw
Test mean	86	0.50	43.0
Midterm	96	0.15	14.4
Final exam	82	0.20	16.4
Computer lab	98	0.10	9.8
Homework	100	0.05	5.0
		$\sum w = 1$	$\sum(x \cdot w) = 88.6$

$$\overline{x} = \frac{\sum(x \cdot w)}{\sum w}$$

$$= \frac{88.6}{1}$$

$$= 88.6$$

Your weighted mean for the course is 88.6. So, you did not get an A.

▶ Try It Yourself 7

An error was made in grading your final exam. Instead of getting 82, you scored 98. What is your new weighted mean?

a. Multiply each score by its weight and *find the sum of these products*.
b. Find the *sum of the weights*.
c. Find the *weighted mean*.
d. *Interpret* the results in the context of the data.

Answer: Page A33

If data are presented in a frequency distribution, you can approximate the mean as follows.

DEFINITION

The **mean of a frequency distribution** for a sample is approximated by

$$\bar{x} = \frac{\sum (x \cdot f)}{n} \qquad \text{Note that } n = \sum f.$$

where x and f are the midpoints and frequencies of a class, respectively.

GUIDELINES

Finding the Mean of a Frequency Distribution

IN WORDS	IN SYMBOLS
1. Find the midpoint of each class.	$x = \dfrac{(\text{Lower limit}) + (\text{Upper limit})}{2}$
2. Find the sum of the products of the midpoints and the frequencies.	$\sum (x \cdot f)$
3. Find the sum of the frequencies.	$n = \sum f$ inconsistence
4. Find the mean of the frequency distribution.	$\bar{x} = \dfrac{\sum (x \cdot f)}{n}$

EXAMPLE 8

▶ **Finding the Mean of a Frequency Distribution**

Use the frequency distribution at the left to approximate the mean number of minutes that a sample of Internet subscribers spent online during their most recent session.

Class midpoint, x	Frequency, f	xf
12.5	6	75.0
24.5	10	245.0
36.5	13	474.5
48.5	8	388.0
60.5	5	302.5
72.5	6	435.0
84.5	2	169.0
	$n = 50$	$\sum = 2089.0$

▶ **Solution**

$$\bar{x} = \frac{\sum (x \cdot f)}{n}$$

$$= \frac{2089.0}{50}$$

$$\approx 41.8$$

So, the mean time spent online was approximately 41.8 minutes.

▶ **Try It Yourself 8**

Use a frequency distribution to approximate the mean age of the 50 richest people. (See Try It Yourself 2 on page 41.)

a. Find the *midpoint* of each class.
b. Find the *sum of the products* of each midpoint and corresponding frequency.
c. Find the *sum of the frequencies*.
d. Find the *mean of the frequency distribution*. *Answer: Page A33*

▶ THE SHAPES OF DISTRIBUTIONS

A graph reveals several characteristics of a frequency distribution. One such characteristic is the shape of the distribution.

DEFINITION

A frequency distribution is **symmetric** when a vertical line can be drawn through the middle of a graph of the distribution and the resulting halves are approximately mirror images.

A frequency distribution is **uniform** (or **rectangular**) when all entries, or classes, in the distribution have equal or approximately equal frequencies. A uniform distribution is also symmetric.

A frequency distribution is skewed if the "tail" of the graph elongates more to one side than to the other. A distribution is **skewed left (negatively skewed)** if its tail extends to the left. A distribution is **skewed right (positively skewed)** if its tail extends to the right.

To explore this topic further, see Activity 2.3 on page 79.

When a distribution is symmetric and unimodal, the mean, median, and mode are equal. If a distribution is skewed left, the mean is less than the median and the median is usually less than the mode. If a distribution is skewed right, the mean is greater than the median and the median is usually greater than the mode. Examples of these commonly occurring distributions are shown.

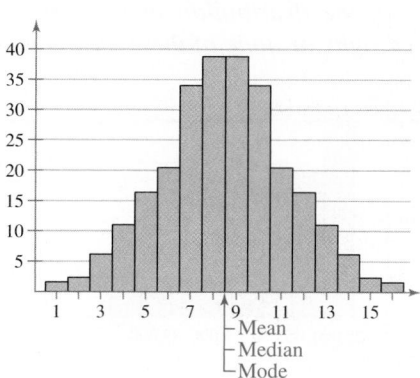

Symmetric Distribution

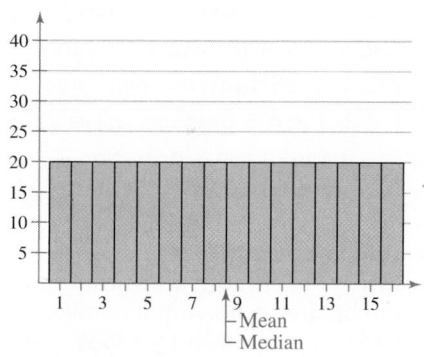

Uniform Distribution

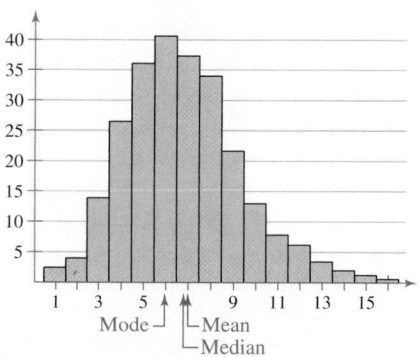

Skewed Left Distribution

Skewed Right Distribution

INSIGHT

Be aware that there are many different shapes of distributions. In some cases, the shape cannot be classified as symmetric, uniform, or skewed. A distribution can have several gaps caused by outliers, or **clusters** of data. Clusters occur when several types of data are included in the one data set.

The mean will always fall in the direction in which the distribution is skewed. For instance, when a distribution is skewed left, the mean is to the left of the median.

2.3 EXERCISES

1. True
2. False. Every quantitative data set has a median.
3. True
4. True
5. 1, 2, 2, 2, 3 (Answers will vary.)
6. 2, 4, 5, 5, 6, 8 (Answers will vary.)
7. 2, 5, 7, 9, 35 (Answers will vary.)
8. 1, 2, 3, 3, 3, 4, 5 (Answers will vary.)
9. The shape of the distribution is skewed right because the bars have a "tail" to the right.
10. The shape of the distribution is symmetric because a vertical line can be drawn down the middle, creating two halves that look approximately the same.
11. The shape of the distribution is uniform because the bars are approximately the same height.
12. The shape of the distribution is skewed left because the bars have a "tail" to the left.
13. (11), because the distribution of values ranges from 1 to 12 and has (approximately) equal frequencies.
14. (9), because the distribution has values in the thousands of dollars and is skewed right due to the few executives that make much higher salaries than the majority of the employees.
15. (12), because the distribution has a maximum value of 90 and is skewed left due to a few students scoring much lower than the majority of the students.
16. See Selected Answers, page A114.

■ BUILDING BASIC SKILLS AND VOCABULARY

True or False? *In Exercises 1–4, determine whether the statement is true or false. If it is false, rewrite it as a true statement.*

1. The mean is the measure of central tendency most likely to be affected by an outlier.

2. Some quantitative data sets do not have medians.

3. A data set can have the same mean, median, and mode.

4. When each data class has the same frequency, the distribution is symmetric.

Constructing Data Sets *In Exercises 5–8, construct the described data set. The values in the data set cannot all be the same.*

5. Median and mode are the same.   6. Mean and mode are the same.

7. Mean is *not* representative of a typical number in the data set.

8. Mean, median, and mode are the same.

Graphical Analysis *In Exercises 9–12, determine whether the approximate shape of the distribution in the histogram is symmetric, uniform, skewed left, skewed right, or none of these. Justify your answer.*

9.
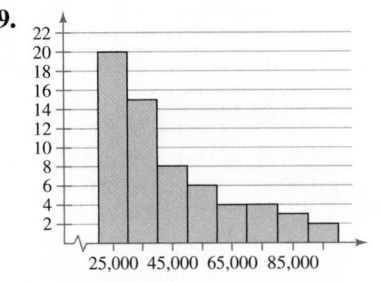

10.
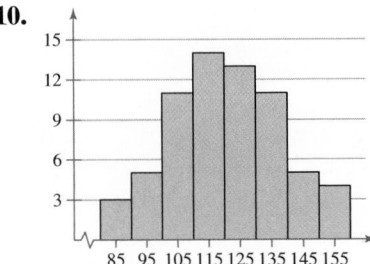

11.
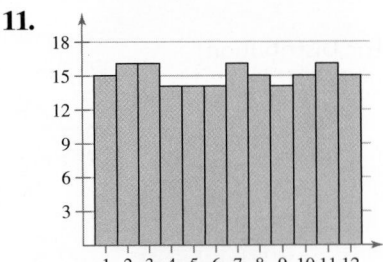

12.
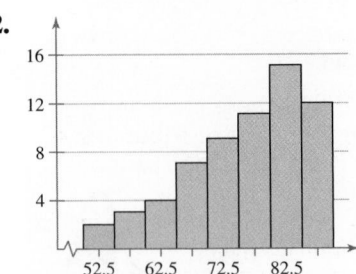

Matching *In Exercises 13–16, match the distribution with one of the graphs in Exercises 9–12. Justify your decision.*

13. The frequency distribution of 180 rolls of a dodecagon (a 12-sided die)

14. The frequency distribution of salaries at a company where a few executives make much higher salaries than the majority of employees

15. The frequency distribution of scores on a 90-point test where a few students scored much lower than the majority of students

16. The frequency distribution of weights for a sample of seventh grade boys

17. $\bar{x} \approx 4.9$; median = 5; mode = 4

18. $\bar{x} = 39.6$; median = 39; mode = 39

19. $\bar{x} \approx 11.0$; median = 11.0; mode = 11.7; The mode does not represent the center of the data because 11.7 is the largest number in the data set.

20. $\bar{x} = 200.4$; median = 186; mode = none; The mode cannot be found because no data point is repeated.

21. $\bar{x} \approx 21.46$; median = 21.95; mode = 20.4

22. $\bar{x} \approx 61.2$; median = 55; mode = 80, 125; The modes do not represent the center of the data set because they are large values compared to the rest of the data.

23. $\bar{x}$ = not possible; median = not possible; mode = "Eyeglasses"; The mean and median cannot be found because the data are at the nominal level of measurement.

24. See Selected Answers, page A114.

25. $\bar{x} \approx 170.63$; median = 169.3; mode = none; There is no mode because no data point is repeated.

26. See Selected Answers, page A114.

Type of lenses	Frequency, f
Contacts	40
Eyeglasses	570
Contacts and eyeglasses	180
None	210

TABLE FOR EXERCISE 23

How Do People Eat Their Potatoes?

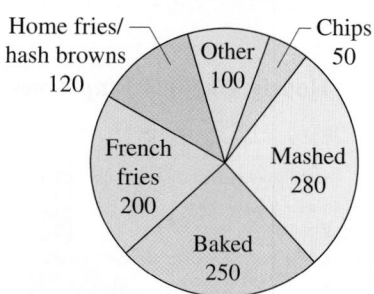

FIGURE FOR EXERCISE 26

■ USING AND INTERPRETING CONCEPTS

Finding and Discussing the Mean, Median, and Mode *In Exercises 17–34, find the mean, median, and mode of the data, if possible. If any of these measures cannot be found or a measure does not represent the center of the data, explain why.*

17. Concert Tickets The number of concert tickets purchased online for the last 13 purchases

4 2 5 8 6 6 4 3 2 4 7 8 5

18. Tuition The 2009–2010 tuition and fees (in thousands of dollars) for the top 10 liberal arts colleges *(Source: U.S. News and World Report)*

39 39 38 51 38 40 37 40 35 39

19. MCAT Scores The average medical college admission test (MCAT) scores for a sample of seven medical schools *(Source: Association of American Medical Colleges)*

11.0 11.7 10.3 11.7 11.7 10.7 9.7

20. Cholesterol The cholesterol levels of a sample of 10 female employees

154 240 171 188 235 203 184 173 181 275

21. NFL The average points per game scored by each NFL team during the 2009 regular season *(Source: National Football League)*

20.4 19.7 17.5 26.7 22.7 21.8 16.6 29.4
26.0 22.5 28.8 19.1 18.1 12.3 16.4 15.2
16.1 23.4 20.6 18.4 23.0 25.1 26.8 31.9
24.4 28.4 20.4 22.1 15.3 10.9 24.2 22.6

22. Power Failures The durations (in minutes) of power failures at a residence in the last 10 years

18 26 45 75 125 80 33 40 44 49
89 80 96 125 12 61 31 63 103 28

23. Eyeglasses and Contacts The responses of a sample of 1000 adults who were asked what type of corrective lenses they wore are shown in the table at the left. *(Adapted from American Optometric Association)*

24. Living on Your Own The responses of a sample of 1177 young adults who were asked what surprised them the most as they began to live on their own *(Adapted from Charles Schwab)*

Amount of first salary: 63   Trying to find a job: 125
Number of decisions: 163   Money needed: 326
Paying bills: 150   Trying to save: 275
How hard it is breaking away from parents: 75

25. Top Speeds The top speeds (in miles per hour) for a sample of seven sports cars

187.3 181.8 180.0 169.3 162.2 158.1 155.7

26. Potatoes The pie chart at the left shows the responses of a sample of 1000 adults who were asked their favorite way to eat potatoes. *(Adapted from Idaho Potato Commission)*

27. $\bar{x}$ = 168.7; median = 162.5; mode = 125; The mode does not represent the center of the data because 125 is the smallest number in the data set.

28. $\bar{x}$ = 16.6; median = 15; mode = none; The mode cannot be found because no data point is repeated.

29. $\bar{x} \approx$ 14.11; median = 14.25; mode = 2.5; The mode does not represent the center of the data because 2.5 is much smaller than most of the data in the set.

30. $\bar{x} \approx$ 240.3; median = 136; mode = 28; The mode does not represent the center of the data set because it is the second smallest number in the data set.

31. $\bar{x} \approx$ 29.82; median = 32; mode = 24, 35

32. $\bar{x} \approx$ 2.49; median = 2.35; mode = 4.0; The mode does not represent the center of the data set because it is the largest value in the data set.

33. $\bar{x} \approx$ 19.5; median = 20; mode = 15

34. $\bar{x} \approx$ 207.33; median = 210; mode = 210

35. The data are skewed right.

    A = mode, because it is the data entry that occurred most often.

    B = median, because the median is to the left of the mean in a skewed right distribution.

    C = mean, because the mean is to the right of the median in a skewed right distribution.

36. The data are skewed left.

    A = mean, because the mean is to the left of the median in a skewed left distribution.

    B = median, because the median is to the right of the mean in a skewed left distribution.

    C = mode, because it is the data entry that occurred most often.

27. **Typing Speeds**   The typing speeds (in words per minute) for several stenographers

    125 140 170 155 132 175 225 210 125 230

28. **Eating Disorders**   The number of weeks it took to reach a target weight for a sample of five patients with eating disorders treated by psychodynamic psychotherapy *(Source: The Journal of Consulting and Clinical Psychology)*

    15.0 31.5 10.0 25.5 1.0

29. **Eating Disorders**   The number of weeks it took to reach a target weight for a sample of 14 patients with eating disorders treated by psychodynamic psychotherapy and cognitive behavior techniques *(Source: The Journal of Consulting and Clinical Psychology)*

    2.5 20.0 11.0 10.5 17.5 16.5 13.0
    15.5 26.5   2.5 27.0 28.5   1.5   5.0

30. **Aircraft**   The number of aircraft that 15 airlines have in their fleets *(Source: Airline Transport Association)*

    136 110   38 625 350 755   52   32
    142   9 537   28 409 354   28

31. **Weights (in pounds) of Carry-On Luggage on a Plane**

0	6 7	Key: 3|2 = 32
1	2 5 8 9	
2	0 4 4 4 5 8 9	
3	2 2 3 5 5 5 6 8 9	
4	0 1 2 7 8	
5	1	

32. **Grade Point Averages of Students in a Class**

0	8	Key: 0|8 = 0.8
1	5 6 8	
2	1 3 4 5	
3	0 9	
4	0 0	

33. **Time (in minutes) It Takes Employees to Drive to Work**

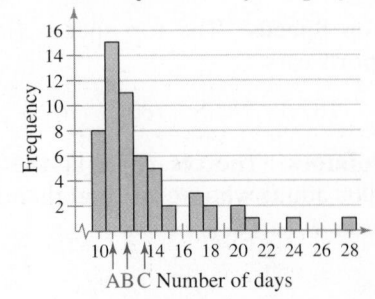

34. **Prices (in dollars per night) of Hotel Rooms in a City**

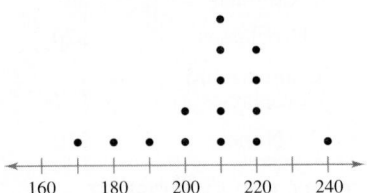

Graphical Analysis   *In Exercises 35 and 36, the letters A, B, and C are marked on the horizontal axis. Describe the shape of the data. Then determine which is the mean, which is the median, and which is the mode. Justify your answers.*

35. **Sick Days Used by Employees**

36. **Hourly Wages of Employees**

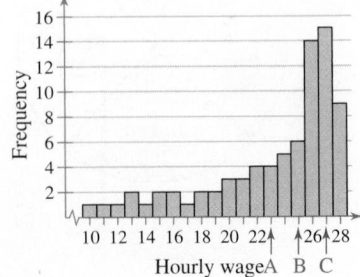

37. Mode, because the data are at the nominal level of measurement.

38. Mean, because the data are symmetric.

39. Mean, because there are no outliers.

40. Median, because there is an outlier.

41. 89

42. $75,840

43. $612.73

44. $982.19

45. 2.8

In Exercises 37–40, without performing any calculations, determine which measure of central tendency best represents the graphed data. Explain your reasoning.

37. **What Do You Think About "Green" Products?**

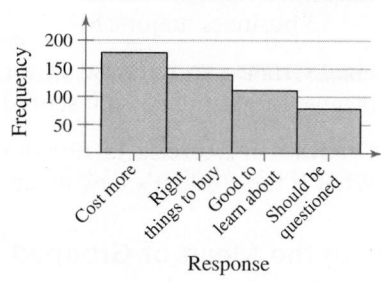

(Adapted from Green Home Furnishings Consumer Study)

38. **Heights of Players on Two Opposing Volleyball Teams**

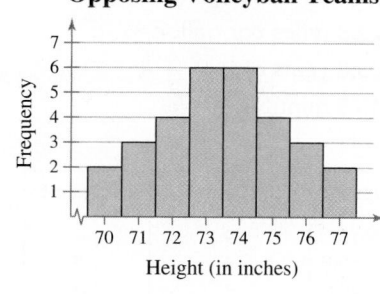

39. **Heart Rate of a Sample of Adults**

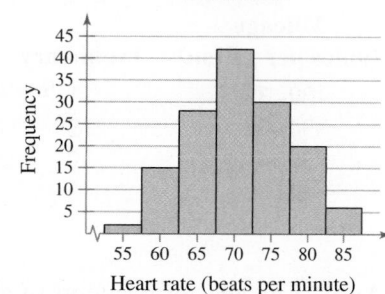

40. **Body Mass Index (BMI) of People in a Gym**

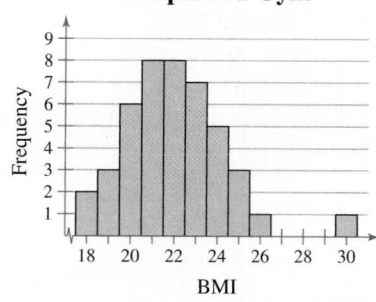

Finding the Weighted Mean   *In Exercises 41–46, find the weighted mean of the data.*

41. **Final Grade**   The scores and their percents of the final grade for a statistics student are given. What is the student's mean score?

	Score	Percent of final grade
Homework	85	5%
Quizzes	80	35%
Project	100	20%
Speech	90	15%
Final exam	93	25%

42. **Salaries**   The average starting salaries (by degree attained) for 25 employees at a company are given. What is the mean starting salary for these employees?

8 with MBAs: $92,500   17 with BAs in business: $68,000

43. **Account Balance**   For the month of April, a checking account has a balance of $523 for 24 days, $2415 for 2 days, and $250 for 4 days. What is the account's mean daily balance for April?

44. **Account Balance**   For the month of May, a checking account has a balance of $759 for 15 days, $1985 for 5 days, $1410 for 5 days, and $348 for 6 days. What is the account's mean daily balance for May?

45. **Grades**   A student receives the following grades, with an A worth 4 points, a B worth 3 points, a C worth 2 points, and a D worth 1 point. What is the student's mean grade point score?

B in 2 three-credit classes   D in 1 two-credit class
A in 1 four-credit class   C in 1 three-credit class

46. 84

47. 87

48. 3.0

49. 36.2 miles per gallon

50. 29.3 miles per gallon

51. 35.8 years old

52. 15.3 minutes

53.

Class	Frequency, f	Midpoint
127–161	9	144
162–196	8	179
197–231	3	214
232–266	3	249
267–301	1	284
	$\Sigma f = 24$	

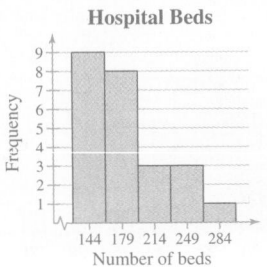

Hospital Beds

Positively skewed

54.

Class	Frequency, f	Midpoint
3–4	3	3.5
5–6	8	5.5
7–8	4	7.5
9–10	2	9.5
11–12	2	11.5
13–14	1	13.5
	$\Sigma f = 20$	

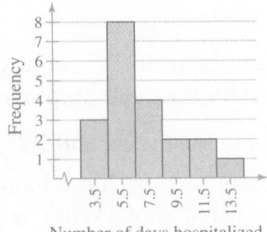

Hospitalization

Positively skewed

46. Scores The mean scores for students in a statistics course (by major) are given. What is the mean score for the class?

   9 engineering majors: 85
   5 math majors: 90
13 business majors: 81

47. Final Grade In Exercise 41, an error was made in grading your final exam. Instead of getting 93, you scored 85. What is your new weighted mean?

48. Grades In Exercise 45, one of the student's B grades gets changed to an A. What is the student's new mean grade point score?

Finding the Mean of Grouped Data

In Exercises 49–52, approximate the mean of the grouped data.

49. Fuel Economy The highway mileage (in miles per gallon) for 30 small cars

Mileage (miles per gallon)	Frequency
29–33	11
34–38	12
39–43	2
44–48	5

50. Fuel Economy The city mileage (in miles per gallon) for 24 family sedans

Mileage (miles per gallon)	Frequency
22–27	16
28–33	2
34–39	2
40–45	3
46–51	1

51. Ages The ages of residents of a town

Age	Frequency
0–9	55
10–19	70
20–29	35
30–39	56
40–49	74
50–59	42
60–69	38
70–79	17
80–89	10

52. Phone Calls The lengths of calls (in minutes) made by a salesperson in one week

Length of call	Number of calls
1–5	12
6–10	26
11–15	20
16–20	7
21–25	11
26–30	7
31–35	4
36–40	4
41–45	1

Identifying the Shape of a Distribution

In Exercises 53–56, construct a frequency distribution and a frequency histogram of the data using the indicated number of classes. Describe the shape of the histogram as symmetric, uniform, negatively skewed, positively skewed, or none of these.

53. Hospital Beds
Number of classes: 5
Data set: The number of beds in a sample of 24 hospitals

149   167   162   127   130   180   160   167
221   145   137   194   207   150   254   262
244   297   137   204   166   174   180   151

55.

Class	Frequency, f	Midpoint
62–64	3	63
65–67	7	66
68–70	9	69
71–73	8	72
74–76	3	75
	$\Sigma f = 30$	

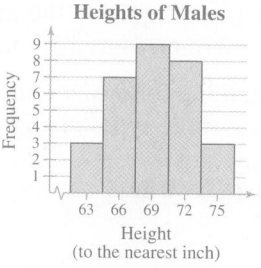

Heights of Males

Symmetric

56. See Selected Answers, page A114.

57. (a) $\bar{x} = 6.005$

median = 6.01

(b) $\bar{x} = 5.945$

median = 6.01

(c) Mean

U.S. Exports (in billions of dollars)	
Canada: 261.1	Japan: 65.1
Mexico: 151.2	South Korea: 34.7
Germany: 54.5	Singapore: 27.9
Taiwan: 24.9	France: 28.8
Netherlands: 39.7	Brazil: 32.3
China: 69.7	Belgium: 28.9
Australia: 22.2	Italy: 15.5
Malaysia: 12.9	Thailand: 9.1
Switzerland: 22.0	
Saudi Arabia: 12.5	
United Kingdom: 53.6	

TABLE FOR EXERCISE 58

58. (a) $\bar{x} \approx 50.87$, median = 28.9

(b) $\bar{x} \approx 39.19$, median = 28.85; the mean

(c) $\bar{x} \approx 49.22$, median = 28.85; the mean

59. See Odd Answers, page A57.

60. See Selected Answers, page A114.

54. Hospitalization
Number of classes: 6
Data set: The number of days 20 patients remained hospitalized

6	9	7	14	4	5	6	8	4	11
10	6	8	6	5	7	6	6	3	11

55. Heights of Males
Number of classes: 5
Data set: The heights (to the nearest inch) of 30 males

67	76	69	68	72	68	65	63	75	69
66	72	67	66	69	73	64	62	71	73
68	72	71	65	69	66	74	72	68	69

56. Six-Sided Die
Number of classes: 6
Data set: The results of rolling a six-sided die 30 times

1	4	6	1	5	3	2	5	4	6	1	2	4	3	5
6	3	2	1	1	5	6	2	4	4	3	1	6	2	4

57. Coffee Contents During a quality assurance check, the actual coffee contents (in ounces) of six jars of instant coffee were recorded as 6.03, 5.59, 6.40, 6.00, 5.99, and 6.02.

(a) Find the mean and the median of the coffee content.

(b) The third value was incorrectly measured and is actually 6.04. Find the mean and median of the coffee content again.

(c) Which measure of central tendency, the mean or the median, was affected more by the data entry error?

58. U.S. Exports The table at the left shows the U.S. exports (in billions of dollars) to 19 countries for a recent year. *(Source: U.S. Department of Commerce)*

(a) Find the mean and median.

(b) Find the mean and median without the U.S. exports to Canada. Which measure of central tendency, the mean or the median, was affected more by the elimination of the Canadian exports?

(c) The U.S. exports to India were $17.7 billion. Find the mean and median with the Indian exports added to the original data set. Which measure of central tendency was affected more by adding the Indian exports?

SC *In Exercises 59 and 60, use StatCrunch to find the sample size, mean, median, minimum data value, and maximum data value of the data.*

59. The data represent the amounts (in dollars) made by several families during a community yard sale.

95	120	125.50	105.25	82	102.75	130	151.50	145.25	79	97

60. The data represent the prices (in dollars) of the stocks in the Dow Jones Industrial Average during a recent session. *(Source: CNN Money)*

83.62	15.90	42.61	26.35	16.89	61.46	62.07	79.53	24.99	34.05
69.62	16.77	52.69	21.46	132.39	65.10	44.56	29.08	62.54	39.92
31.07	19.46	57.19	28.30	61.49	49.28	72.77	31.38	54.33	31.06

61. (a) $\bar{x} = 358$, median $= 375$

(b) $\bar{x} = 1074$, median $= 1125$

(c) The mean and median in part (b) are three times the mean and median in part (a).

(d) If you multiply the mean and median from part (b) by 12, you will get the mean and median of the data set in inches.

62. (a) Mean, because Car A has the highest mean of the three.

(b) Median, because Car B has the highest median of the three.

(c) Mode, because Car C has the highest mode of the three.

63. Car A, because the midrange is the largest.

64. (a) $\bar{x} \approx 49.2$, median $= 46.5$

(b) Key: $3|6 = 36$

```
1 | 1 3
2 | 2 8
3 | 6 6 6 7 7 7 8
4 | 1 3 4 6 7 —— mean
5 | 1 1 1 3
6 | 1 2 3 4      median
7 | 2 2 4 6
8 | 5
9 | 0
```

(c) Positively skewed

65. (a) 49.2

(b) $\bar{x} \approx 49.2$; median $= 46.5$; mode $= 36, 37, 51$; midrange $= 50.5$

(c) Using the trimmed mean eliminates potential outliers that could affect the mean of the entries.

■ EXTENDING CONCEPTS

61. Golf   The distances (in yards) for nine holes of a golf course are listed.

336   393   408   522   147   504   177   375   360

(a) Find the mean and median of the data.

(b) Convert the distances to feet. Then rework part (a).

(c) Compare the measures you found in part (b) with those found in part (a). What do you notice?

(d) Use your results from part (c) to explain how to find quickly the mean and median of the given data set if the distances are measured in inches.

62. Data Analysis   A consumer testing service obtained the following mileages (in miles per gallon) in five test runs performed with three types of compact cars.

	Run 1	Run 2	Run 3	Run 4	Run 5
Car A:	28	32	28	30	34
Car B:	31	29	31	29	31
Car C:	29	32	28	32	30

(a) The manufacturer of Car A wants to advertise that its car performed best in this test. Which measure of central tendency—mean, median, or mode—should be used for its claim? Explain your reasoning.

(b) The manufacturer of Car B wants to advertise that its car performed best in this test. Which measure of central tendency—mean, median, or mode—should be used for its claim? Explain your reasoning.

(c) The manufacturer of Car C wants to advertise that its car performed best in this test. Which measure of central tendency—mean, median, or mode—should be used for its claim? Explain your reasoning.

63. Midrange   Another measure of central tendency that is rarely used but is easy to calculate is the **midrange.** It can be found by the formula

$$\frac{(\text{Maximum data entry}) + (\text{Minimum data entry})}{2}.$$

Which of the manufacturers in Exercise 62 would prefer to use the midrange statistic in their ads? Explain your reasoning.

64. Data Analysis   Students in an experimental psychology class did research on depression as a sign of stress. A test was administered to a sample of 30 students. The scores are given.

44 51 11 90 76 36 64 37 43 72 53 62 36 74 51
72 37 28 38 61 47 63 36 41 22 37 51 46 85 13

(a) Find the mean and median of the data.

(b) Draw a stem-and-leaf plot for the data using one row per stem. Locate the mean and median on the display.

(c) Describe the shape of the distribution.

65. Trimmed Mean   To find the 10% **trimmed mean** of a data set, order the data, delete the lowest 10% of the entries and the highest 10% of the entries, and find the mean of the remaining entries.

(a) Find the 10% trimmed mean for the data in Exercise 64.

(b) Compare the four measures of central tendency, including the midrange.

(c) What is the benefit of using a trimmed mean versus using a mean found using all data entries? Explain your reasoning.

ACTIVITY 2.3   Mean Versus Median

APPLET

The *mean versus median* applet is designed to allow you to investigate interactively the mean and the median as measures of the center of a data set. Points can be added to the plot by clicking the mouse above the horizontal axis. The mean of the points is shown as a green arrow and the median is shown as a red arrow. If the two values are the same, then a single yellow arrow is displayed. Numeric values for the mean and median are shown above the plot. Points on the plot can be removed by clicking on the point and then dragging the point into the trash can. All of the points on the plot can be removed by simply clicking inside the trash can. The range of values for the horizontal axis can be specified by inputting lower and upper limits and then clicking UPDATE.

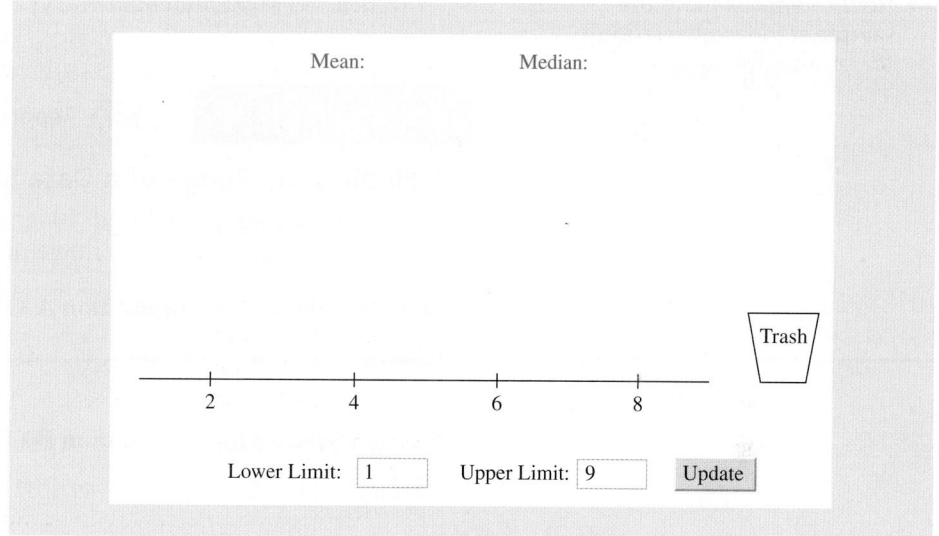

■ Explore

Step 1   Specify a lower limit.
Step 2   Specify an upper limit.
Step 3   Add 15 points to the plot.
Step 4   Remove all of the points from the plot.

■ Draw Conclusions

APPLET

1. Specify the lower limit to be 1 and the upper limit to be 50. Add at least 10 points that range from 20 to 40 so that the mean and the median are the same. What is the shape of the distribution? What happens at first to the mean and median when you add a few points that are less than 10? What happens over time as you continue to add points that are less than 10?

2. Specify the lower limit to be 0 and the upper limit to be 0.75. Place 10 points on the plot. Then change the upper limit to 25. Add 10 more points that are greater than 20 to the plot. Can the mean be any one of the points that were plotted? Can the median be any one of the points that were plotted? Explain.

2.4 Measures of Variation

Range ▸ Deviation, Variance, and Standard Deviation ▸ Interpreting Standard Deviation ▸ Standard Deviation for Grouped Data

▸ RANGE

In this section, you will learn different ways to measure the variation of a data set. The simplest measure is the *range* of the set.

DEFINITION

The **range** of a data set is the difference between the maximum and minimum data entries in the set. To find the range, the data must be quantitative.

Range = (Maximum data entry) − (Minimum data entry)

EXAMPLE 1   **SC** Report 12

▸ **Finding the Range of a Data Set**

Two corporations each hired 10 graduates. The starting salaries for each graduate are shown. Find the range of the starting salaries for Corporation A.

Starting Salaries for Corporation A (1000s of dollars)

Salary	41	38	39	45	47	41	44	41	37	42

Starting Salaries for Corporation B (1000s of dollars)

Salary	40	23	41	50	49	32	41	29	52	58

▸ **Solution**

Ordering the data helps to find the least and greatest salaries.

$$37 \quad 38 \quad 39 \quad 41 \quad 41 \quad 41 \quad 42 \quad 44 \quad 45 \quad 47$$

Minimum ⤸   Maximum

Range = (Maximum salary) − (Minimum salary)

= 47 − 37

= 10

So, the range of the starting salaries for Corporation A is 10, or $10,000.

▸ **Try It Yourself 1**

Find the range of the starting salaries for Corporation B.

a. Identify the *minimum* and *maximum* salaries.
b. Find the *range*.
c. *Compare* your answer with that for Example 1.

Answer: Page A33

▶ DEVIATION, VARIANCE, AND STANDARD DEVIATION

As a measure of variation, the range has the advantage of being easy to compute. Its disadvantage, however, is that it uses only two entries from the data set. Two measures of variation that use all the entries in a data set are the *variance* and the *standard deviation*. However, before you learn about these measures of variation, you need to know what is meant by the *deviation* of an entry in a data set.

DEFINITION

The **deviation** of an entry x in a population data set is the difference between the entry and the mean μ of the data set.

$$\text{Deviation of } x = x - \mu$$

Deviations of Starting Salaries for Corporation A

Salary (1000s of dollars) x	Deviation (1000s of dollars) $x - \mu$
41	−0.5
38	−3.5
39	−2.5
45	3.5
47	5.5
41	−0.5
44	2.5
41	−0.5
37	−4.5
42	0.5
$\Sigma x = 415$	$\Sigma(x - \mu) = 0$

EXAMPLE 2

▶ Finding the Deviations of a Data Set

Find the deviation of each starting salary for Corporation A given in Example 1.

▶ Solution

The mean starting salary is $\mu = 415/10 = 41.5$, or $41,500. To find out how much each salary deviates from the mean, subtract 41.5 from the salary. For instance, the deviation of 41, or $41,000 is

$$41 - 41.5 = -0.5, \text{ or } -\$500. \qquad \text{Deviation of } x = x - \mu$$

The table at the left lists the deviations of each of the 10 starting salaries.

▶ Try It Yourself 2

Find the deviation of each starting salary for Corporation B given in Example 1.

a. Find the *mean* of the data set.
b. *Subtract* the mean from each salary.   *Answer: Page A33*

In Example 2, notice that the sum of the deviations is zero. Because this is true for any data set, it doesn't make sense to find the average of the deviations. To overcome this problem, you can square each deviation. When you add the squares of the deviations, you compute a quantity called the **sum of squares,** denoted SS_x. In a population data set, the mean of the squares of the deviations is called the **population variance.**

DEFINITION

The **population variance** of a population data set of N entries is

$$\text{Population variance} = \sigma^2 = \frac{\Sigma(x - \mu)^2}{N}.$$

The symbol σ is the lowercase Greek letter sigma.

Note to Instructor

We have used the formulas here that are derived from the definition of the population variance and standard deviation because we feel they are easier to remember than the shortcut formula. If you prefer to use the shortcut formula, we have included it on page 96.

DEFINITION

The **population standard deviation** of a population data set of N entries is the square root of the population variance.

$$\text{Population standard deviation} = \sigma = \sqrt{\sigma^2} = \sqrt{\frac{\sum (x - \mu)^2}{N}}$$

GUIDELINES

Finding the Population Variance and Standard Deviation

IN WORDS	IN SYMBOLS
1. Find the mean of the population data set.	$\mu = \dfrac{\sum x}{N}$
2. Find the deviation of each entry.	$x - \mu$
3. Square each deviation.	$(x - \mu)^2$
4. Add to get the **sum of squares.**	$SS_x = \sum (x - \mu)^2$
5. Divide by N to get the **population variance.**	$\sigma^2 = \dfrac{\sum (x - \mu)^2}{N}$
6. Find the square root of the variance to get the **population standard deviation.**	$\sigma = \sqrt{\dfrac{\sum (x - \mu)^2}{N}}$

**Sum of Squares of Starting Salaries
for Corporation A**

Salary x	Deviation $x - \mu$	Squares $(x - \mu)^2$
41	−0.5	0.25
38	−3.5	12.25
39	−2.5	6.25
45	3.5	12.25
47	5.5	30.25
41	−0.5	0.25
44	2.5	6.25
41	−0.5	0.25
37	−4.5	20.25
42	0.5	0.25
	$\Sigma = 0$	$SS_x = 88.5$

EXAMPLE 3

▸ **Finding the Population Standard Deviation**

Find the population standard deviation of the starting salaries for Corporation A given in Example 1.

▸ **Solution**

The table at the left summarizes the steps used to find SS_x.

$$SS_x = 88.5, \qquad N = 10, \qquad \sigma^2 = \frac{88.5}{10} \approx 8.9, \qquad \sigma = \sqrt{\frac{88.5}{10}} \approx 3.0$$

So, the population variance is about 8.9, and the population standard deviation is about 3.0, or $3000.

▸ **Try It Yourself 3**

Find the population variance and standard deviation of the starting salaries for Corporation B given in Example 1.

a. Find the *mean* and each *deviation*, as you did in Try It Yourself 2.
b. *Square* each deviation and *add* to get the sum of squares.
c. *Divide* by N to get the population variance.
d. Find the *square root* of the population variance to get the population standard deviation.
e. *Interpret* the results by giving the population standard deviation in dollars.

Answer: Page A33

STUDY TIP

Notice that the variance and standard deviation in Example 3 have one more decimal place than the original set of data values has. This is the same *round-off rule* that was used to calculate the mean.

DEFINITION

The **sample variance** and **sample standard deviation** of a sample data set of n entries are listed below.

$$\text{Sample variance} = s^2 = \frac{\sum (x - \bar{x})^2}{n - 1}$$

$$\text{Sample standard deviation} = s = \sqrt{s^2} = \sqrt{\frac{\sum (x - \bar{x})^2}{n - 1}}$$

GUIDELINES

Finding the Sample Variance and Standard Deviation

IN WORDS	IN SYMBOLS
1. Find the mean of the sample data set.	$\bar{x} = \dfrac{\sum x}{n}$
2. Find the deviation of each entry.	$x - \bar{x}$
3. Square each deviation.	$(x - \bar{x})^2$
4. Add to get the **sum of squares**.	$SS_x = \sum (x - \bar{x})^2$
5. Divide by $n - 1$ to get the **sample variance**.	$s^2 = \dfrac{\sum (x - \bar{x})^2}{n - 1}$
6. Find the square root of the variance to get the **sample standard deviation**.	$s = \sqrt{\dfrac{\sum (x - \bar{x})^2}{n - 1}}$

Symbols in Variance and Standard Deviation Formulas

	Population	Sample
Variance	σ^2	s^2
Standard deviation	σ	s
Mean	μ	$\bar{x}$
Number of entries	N	n
Deviation	$x - \mu$	$x - \bar{x}$
Sum of squares	$\sum (x - \mu)^2$	$\sum (x - \bar{x})^2$

See MINITAB and TI-83/84 Plus steps on pages 122 and 123.

EXAMPLE 4   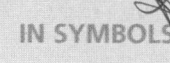 Report 13

▶ **Finding the Sample Standard Deviation**

The starting salaries given in Example 1 are for the Chicago branches of Corporations A and B. Each corporation has several other branches, and you plan to use the starting salaries of the Chicago branches to estimate the starting salaries for the larger populations. Find the *sample* standard deviation of the starting salaries for the Chicago branch of Corporation A.

▶ **Solution**

$$SS_x = 88.5, \qquad n = 10, \qquad s^2 = \frac{88.5}{9} \approx 9.8, \qquad s = \sqrt{\frac{88.5}{9}} \approx 3.1$$

So, the sample variance is about 9.8, and the sample standard deviation is about 3.1, or $3100.

▶ **Try It Yourself 4**

Find the sample standard deviation of the starting salaries for the Chicago branch of Corporation B.

a. Find the *sum of squares*, as you did in Try It Yourself 3.
b. *Divide* by $n - 1$ to get the sample variance.
c. Find the *square root* of the sample variance to get the sample standard deviation.
d. *Interpret* the results by giving the sample standard deviation in dollars.

Answer: Page A33

Office Rental Rates		
35.00	33.50	37.00
23.75	26.50	31.25
36.50	40.00	32.00
39.25	37.50	34.75
37.75	37.25	36.75
27.00	35.75	26.00
37.00	29.00	40.50
24.50	33.00	38.00

EXAMPLE 5

▶ Using Technology to Find the Standard Deviation

Sample office rental rates (in dollars per square foot per year) for Miami's central business district are shown in the table. Use a calculator or a computer to find the mean rental rate and the sample standard deviation. *(Adapted from Cushman & Wakefield Inc.)*

▶ Solution

MINITAB, Excel, and the TI-83/84 Plus each have features that automatically calculate the means and the standard deviations of data sets. Try using this technology to find the mean and the standard deviation of the office rental rates. From the displays, you can see that $\bar{x} \approx 33.73$ and $s \approx 5.09$.

MINITAB

Descriptive Statistics: Rental Rates

Variable	N	Mean	SE Mean	StDev	Minimum
Rental Rates	24	33.73	1.04	5.09	23.75

Variable	Q1	Median	Q3	Maximum
Rental Rates	29.56	35.38	37.44	40.50

EXCEL

	A	B
1	Mean	33.72917
2	Standard Error	1.038864
3	Median	35.375
4	Mode	37
5	Standard Deviation	5.089373
6	Sample Variance	25.90172
7	Kurtosis	-0.74282
8	Skewness	-0.70345
9	Range	16.75
10	Minimum	23.75
11	Maximum	40.5
12	Sum	809.5
13	Count	24

TI-83/84 PLUS

1-Var Stats
$\bar{x}=33.72916667$
$\Sigma x=809.5$
$\Sigma x^2=27899.5$
$Sx=5.089373342$
$\sigma x=4.982216639$
$n=24$

Sample Mean
Sample Standard Deviation

▶ Try It Yourself 5

Sample office rental rates (in dollars per square foot per year) for Seattle's central business district are listed. Use a calculator or a computer to find the mean rental rate and the sample standard deviation. *(Adapted from Cushman & Wakefield Inc.)*

40.00	43.00	46.00	40.50	35.75	39.75	32.75
36.75	35.75	38.75	38.75	36.75	38.75	39.00
29.00	35.00	42.75	32.75	40.75	35.25	

a. *Enter* the data.
b. *Calculate* the sample mean and the sample standard deviation.

Answer: Page A33

▶ INTERPRETING STANDARD DEVIATION

When interpreting the standard deviation, remember that it is a measure of the typical amount an entry deviates from the mean. The more the entries are spread out, the greater the standard deviation.

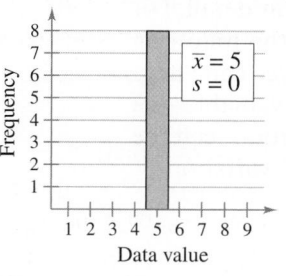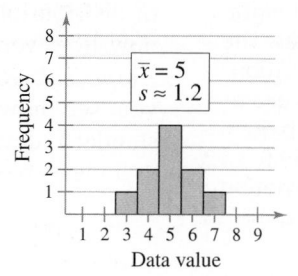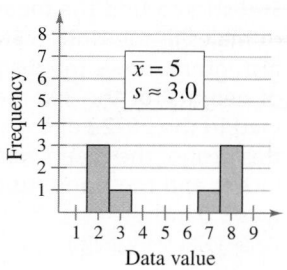

To explore this topic further, see Activity 2.4 on page 98.

EXAMPLE 6

▶ Estimating Standard Deviation

Without calculating, estimate the population standard deviation of each data set.

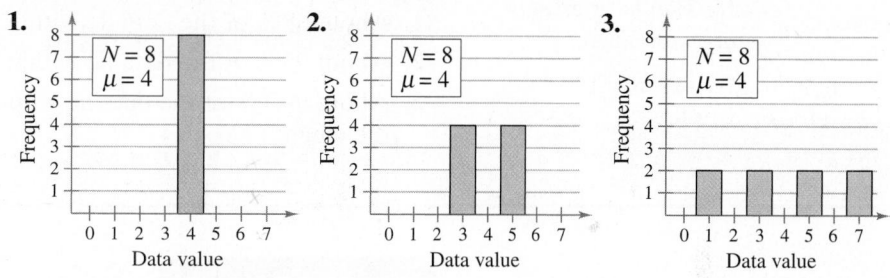

▶ Solution

1. Each of the eight entries is 4. So, each deviation is 0, which implies that

$$\sigma = 0.$$

2. Each of the eight entries has a deviation of ± 1. So, the population standard deviation should be 1. By calculating, you can see that

$$\sigma = 1.$$

3. Each of the eight entries has a deviation of ± 1 or ± 3. So, the population standard deviation should be about 2. By calculating, you can see that

$$\sigma \approx 2.24.$$

▶ Try It Yourself 6

Write a data set that has 10 entries, a mean of 10, and a population standard deviation that is approximately 3. (There are many correct answers.)

a. *Write* a data set that has five entries that are three units less than 10 and five entries that are three units more than 10.

b. *Calculate* the population standard deviation to check that σ is approximately 3. *Answer: Page A33*

PICTURING THE WORLD

A survey was conducted by the National Center for Health Statistics to find the mean height of males in the United States. The histogram shows the distribution of heights for the 808 men examined in the 20–29 age group. In this group, the mean was 69.9 inches and the standard deviation was 3.0 inches. (Adapted from National Center for Health Statistics)

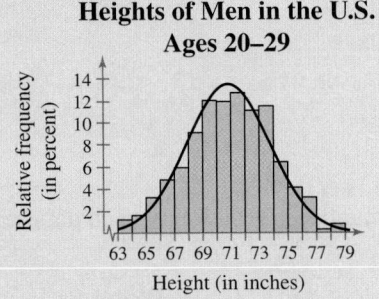

Heights of Men in the U.S. Ages 20–29

Roughly which two heights contain the middle 95% of the data?

The heights 63.9 inches and 75.9 inches are each two standard deviations from the mean, so they contain about 95% of the data.

Heights of Women in the U.S. Ages 20–29

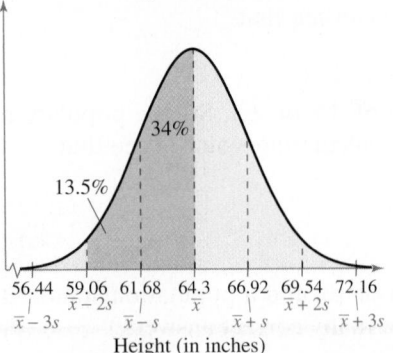

INSIGHT

Data values that lie more than two standard deviations from the mean are considered unusual. Data values that lie more than three standard deviations from the mean are very unusual.

Many real-life data sets have distributions that are approximately symmetric and bell-shaped. Later in the text, you will study this type of distribution in detail. For now, however, the following *Empirical Rule* can help you see how valuable the standard deviation can be as a measure of variation.

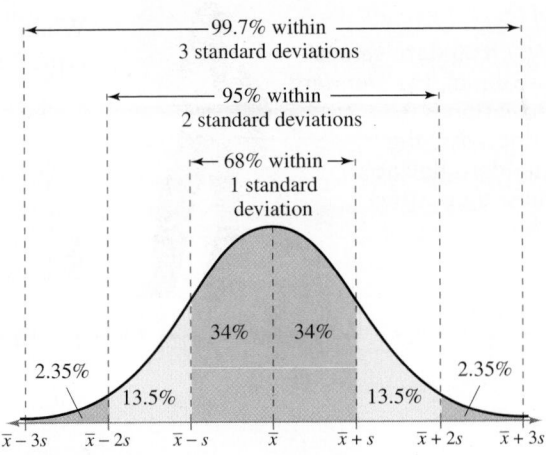

Bell-Shaped Distribution

EMPIRICAL RULE (OR 68–95–99.7 RULE)

For data with a (symmetric) bell-shaped distribution, the standard deviation has the following characteristics.

1. About 68% of the data lie within one standard deviation of the mean.
2. About 95% of the data lie within two standard deviations of the mean.
3. About 99.7% of the data lie within three standard deviations of the mean.

EXAMPLE 7

▶ **Using the Empirical Rule**

In a survey conducted by the National Center for Health Statistics, the sample mean height of women in the United States (ages 20–29) was 64.3 inches, with a sample standard deviation of 2.62 inches. Estimate the percent of women whose heights are between 59.06 inches and 64.3 inches. *(Adapted from National Center for Health Statistics)*

▶ **Solution**

The distribution of women's heights is shown. Because the distribution is bell-shaped, you can use the Empirical Rule. The mean height is 64.3, so when you subtract two standard deviations from the mean height, you get

$$\overline{x} - 2s = 64.3 - 2(2.62) = 59.06.$$

Because 59.06 is two standard deviations below the mean height, the percent of the heights between 59.06 and 64.3 inches is 13.5% + 34% = 47.5%.

Interpretation So, 47.5% of women are between 59.06 and 64.3 inches tall.

▶ **Try It Yourself 7**

Estimate the percent of women's heights that are between 64.3 and 66.92 inches tall.

a. How many *standard deviations* is 66.92 to the right of 64.3?
b. Use the *Empirical Rule* to estimate the percent of the data between $\overline{x}$ and $\overline{x} + s$.
c. *Interpret* the result in the context of the data.

Answer: Page A33

The Empirical Rule applies only to (symmetric) bell-shaped distributions. What if the distribution is not bell-shaped, or what if the shape of the distribution is not known? The following theorem gives an inequality statement that applies to *all* distributions. It is named after the Russian statistician Pafnuti Chebychev (1821–1894).

Note to Instructor

Explain that k represents the number of standard deviations from the mean. Ask students to calculate the percents for $k = 4$ and $k = 5$. Then ask them what happens as k increases. Point out that it is helpful to draw a number line and mark it in units of standard deviations.

CHEBYCHEV'S THEOREM

The portion of any data set lying within k standard deviations $(k > 1)$ of the mean is at least

$$1 - \frac{1}{k^2}.$$

- $k = 2$: In any data set, at least $1 - \frac{1}{2^2} = \frac{3}{4}$, or 75%, of the data lie within 2 standard deviations of the mean.

- $k = 3$: In any data set, at least $1 - \frac{1}{3^2} = \frac{8}{9}$, or 88.9%, of the data lie within 3 standard deviations of the mean.

EXAMPLE 8

▶ Using Chebychev's Theorem

The age distributions for Alaska and Florida are shown in the histograms. Decide which is which. Apply Chebychev's Theorem to the data for Florida using $k = 2$. What can you conclude?

INSIGHT

In Example 8, Chebychev's Theorem gives you an inequality statement that says that at least 75% of the population of Florida is under the age of 88.8. This is a true statement, but it is not nearly as strong a statement as could be made from reading the histogram.

In general, Chebychev's Theorem gives the minimum percent of data values that fall within the given number of standard deviations of the mean. Depending on the distribution, there is probably a higher percent of data falling in the given range.

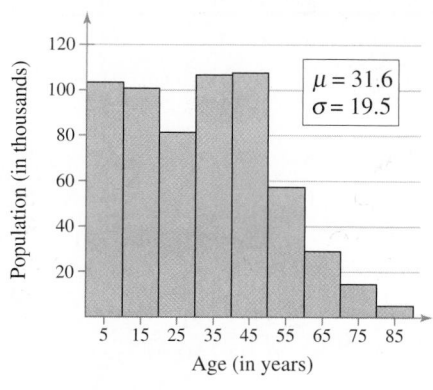

$\mu = 31.6$
$\sigma = 19.5$

Population (in thousands) vs. Age (in years)

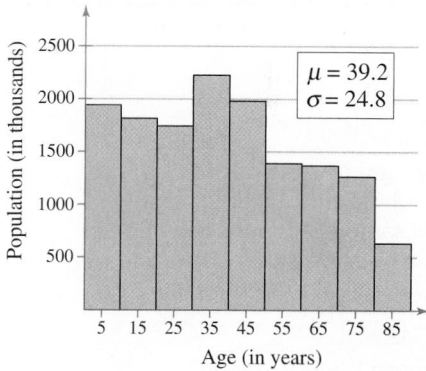

$\mu = 39.2$
$\sigma = 24.8$

Population (in thousands) vs. Age (in years)

▶ Solution

The histogram on the right shows Florida's age distribution. You can tell because the population is greater and older. Moving two standard deviations to the left of the mean puts you below 0, because $\mu - 2\sigma = 39.2 - 2(24.8) = -10.4$. Moving two standard deviations to the right of the mean puts you at $\mu + 2\sigma = 39.2 + 2(24.8) = 88.8$. By Chebychev's Theorem, you can say that at least 75% of the population of Florida is between 0 and 88.8 years old.

▶ Try It Yourself 8

Apply Chebychev's Theorem to the data for Alaska using $k = 2$. What can you conclude?

a. *Subtract* two standard deviations from the mean.
b. *Add* two standard deviations to the mean.
c. *Apply* Chebychev's Theorem for $k = 2$ and *interpret* the results.

Answer: Page A33

▶ STANDARD DEVIATION FOR GROUPED DATA

In Section 2.1, you learned that large data sets are usually best represented by frequency distributions. The formula for the sample standard deviation for a frequency distribution is

$$\text{Sample standard deviation} = s = \sqrt{\frac{\Sigma (x - \bar{x})^2 f}{n - 1}}$$

where $n = \Sigma f$ is the number of entries in the data set.

EXAMPLE 9

▶ Finding the Standard Deviation for Grouped Data

You collect a random sample of the number of children per household in a region. The results are shown at the left. Find the sample mean and the sample standard deviation of the data set.

▶ Solution

These data could be treated as 50 individual entries, and you could use the formulas for mean and standard deviation. Because there are so many repeated numbers, however, it is easier to use a frequency distribution.

Number of Children in 50 Households

1	3	1	1	1
1	2	2	1	0
1	1	0	0	0
1	5	0	3	6
3	0	3	1	1
1	1	6	0	1
3	6	6	1	2
2	3	0	1	1
4	1	1	2	2
0	3	0	2	4

x	f	xf	$x - \bar{x}$	$(x - \bar{x})^2$	$(x - \bar{x})^2 f$
0	× 10	0	−1.8	3.24	32.40
1	× 19	19	−0.8	0.64	12.16
2	× 7	14	0.2	0.04	0.28
3	7	21	1.2	1.44	10.08
4	2	8	2.2	4.84	9.68
5	1	5	3.2	10.24	10.24
6	4	24	4.2	17.64	70.56
	$\Sigma = 50$	$\Sigma = 91$			$\Sigma = 145.40$

$$\bar{x} = \frac{\Sigma xf}{n} = \frac{91}{50} \approx 1.8 \qquad \text{Sample mean}$$

Use the sum of squares to find the sample standard deviation.

$$s = \sqrt{\frac{\Sigma (x - \bar{x})^2 f}{n - 1}} = \sqrt{\frac{145.4}{49}} \approx 1.7 \qquad \text{Sample standard deviation}$$

So, the sample mean is about 1.8 children, and the sample standard deviation is about 1.7 children.

▶ Try It Yourself 9

Change three of the 6's in the data set to 4's. How does this change affect the sample mean and sample standard deviation?

a. Write the first three columns of a *frequency distribution*.
b. Find the *sample mean*.
c. Complete the *last three columns* of the frequency distribution.
d. Find the *sample standard deviation*.   *Answer: Page A33*

When a frequency distribution has classes, you can estimate the sample mean and the sample standard deviation by using the midpoint of each class.

EXAMPLE 10

▶ Using Midpoints of Classes

The circle graph at the right shows the results of a survey in which 1000 adults were asked how much they spend in preparation for personal travel each year. Make a frequency distribution for the data. Then use the table to estimate the sample mean and the sample standard deviation of the data set. *(Adapted from Travel Industry Association of America)*

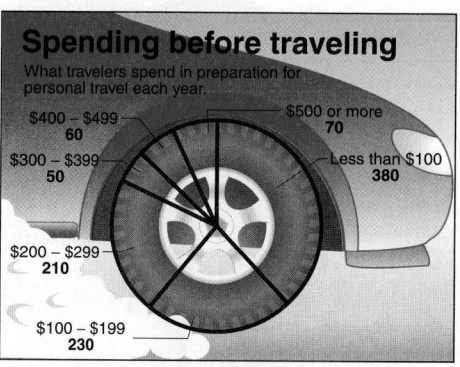

Spending before traveling
What travelers spend in preparation for personal travel each year.

- $400 – $499: 60
- $500 or more: 70
- $300 – $399: 50
- Less than $100: 380
- $200 – $299: 210
- $100 – $199: 230

▶ Solution

Begin by using a frequency distribution to organize the data.

Class	x	f	xf	$x - \bar{x}$	$(x - \bar{x})^2$	$(x - \bar{x})^2 f$
0–99	49.5	380	18,810	−142.5	20,306.25	7,716,375.0
100–199	149.5	230	34,385	−42.5	1806.25	415,437.5
200–299	249.5	210	52,395	57.5	3306.25	694,312.5
300–399	349.5	50	17,475	157.5	24,806.25	1,240,312.5
400–499	449.5	60	26,970	257.5	66,306.25	3,978,375.0
500+	599.5	70	41,965	407.5	166,056.25	11,623,937.5
		$\Sigma = 1000$	$\Sigma = 192,000$			$\Sigma = 25,668,750.0$

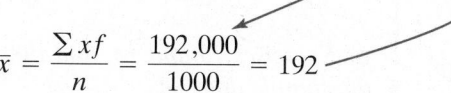

$$\bar{x} = \frac{\Sigma xf}{n} = \frac{192,000}{1000} = 192 \qquad \text{Sample mean}$$

Use the sum of squares to find the sample standard deviation.

$$s = \sqrt{\frac{\Sigma (x - \bar{x})^2 f}{n - 1}} = \sqrt{\frac{25,668,750}{999}} \approx 160.3 \qquad \text{Sample standard deviation}$$

So, the sample mean is $192 per year, and the sample standard deviation is about $160.30 per year.

▶ Try It Yourself 10

In the frequency distribution, 599.5 was chosen to represent the class of $500 or more. How would the sample mean and standard deviation change if you used 650 to represent this class?

a. Write the first four columns of a *frequency distribution*.
b. Find the *sample mean*.
c. Complete the *last three columns* of the frequency distribution.
d. Find the *sample standard deviation*. *Answer: Page A34*

STUDY TIP

When a class is open, as in the last class, you must assign a single value to represent the midpoint. For this example, we selected 599.5.

2.4 EXERCISES

FOR EXTRA HELP:
MyStatLab

1. The range is the difference between the maximum and minimum values of a data set. The advantage of the range is that it is easy to calculate. The disadvantage is that it uses only two entries from the data set.

2. A deviation $(x - \mu)$ is the difference between an entry x and the mean of the data μ. The sum of the deviations is always zero.

3. The units of variance are squared. Its units are meaningless. (Example: dollars^2)

4. The standard deviation is the positive square root of the variance. The standard deviation and variance can never be negative. Squared deviations can never be negative.

5. $\{9, 9, 9, 9, 9, 9, 9\}$

6. $\{3, 3, 3, 7, 7, 7\}$

7. When calculating the population standard deviation, you divide the sum of the squared deviations by N, then take the square root of that value. When calculating the sample standard deviation, you divide the sum of the squared deviations by $n - 1$, then take the square root of that value.

■ BUILDING BASIC SKILLS AND VOCABULARY

1. Explain how to find the range of a data set. What is an advantage of using the range as a measure of variation? What is a disadvantage?

2. Explain how to find the deviation of an entry in a data set. What is the sum of all the deviations in any data set?

3. Why is the standard deviation used more frequently than the variance? (*Hint:* Consider the units of the variance.)

4. Explain the relationship between variance and standard deviation. Can either of these measures be negative? Explain.

5. Construct a sample data set for which $n = 7$, $\bar{x} = 9$, and $s = 0$.

6. Construct a population data set for which $N = 6$, $\mu = 5$, and $\sigma = 2$.

7. Describe the difference between the calculation of population standard deviation and that of sample standard deviation.

8. Given a data set, how do you know whether to calculate σ or s?

9. Discuss the similarities and the differences between the Empirical Rule and Chebychev's Theorem.

10. What must you know about a data set before you can use the Empirical Rule?

In Exercises 11 and 12, find the range, mean, variance, and standard deviation of the population data set.

11. 9 5 9 10 11 12 7 7 8 12

12. 18 20 19 21 19 17 15
    17 25 22 19 20 16 18

In Exercises 13 and 14, find the range, mean, variance, and standard deviation of the sample data set.

13. 4 15 9 12 16 8 11 19 14

14. 28 25 21 15 7 14 9
    27 21 24 14 17 16

Graphical Reasoning   *In Exercises 15–18, find the range of the data set represented by the display or graph.*

15.
2	3 9
3	0 0 2 3 6 7
4	0 1 2 3 3 8
5	0 1 1 9
6	1 2 9 9
7	5 9
8	4 8
9	0 2 5 6

Key: 2|3 = 23

16. **Bride's Age at First Marriage**

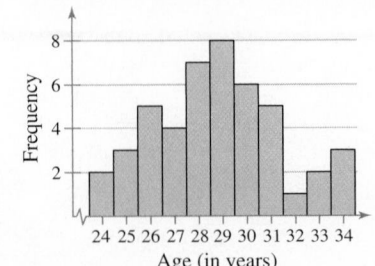

8. When given a data set, one would have to determine if it represented the population or if it was a sample taken from the population. If the data are a population, then σ is calculated. If the data are a sample, then s is calculated.

9. Similarity: Both estimate proportions of the data contained within k standard deviations of the mean.

Difference: The Empirical Rule assumes the distribution is bell-shaped; Chebychev's Theorem makes no such assumption.

10. You must know that the distribution is bell-shaped.

11. Range = 7, $\mu = 9$, $\sigma^2 = 4.8$, $\sigma \approx 2.2$

12. Range = 10, $\mu = 19$, $\sigma^2 \approx 6.1$, $\sigma \approx 2.5$

13. Range = 15, $\bar{x} = 12$, $s^2 \approx 21$, $s \approx 4.6$

14. Range = 21, $\bar{x} \approx 18.3$, $s^2 \approx 44.2$, $s \approx 6.7$

15. 73

16. 10

17. 24

18. 6.2

19. (a) Range = 17.8
(b) Range = 39.8

20. Changing the maximum value of the data set greatly affects the range.

21. The data set in (a) has a standard deviation of 24 and the data set in (b) has a standard deviation of 16, because the data in (a) have more variability.

22. The data set in (a) has a standard deviation of 2.4 and the data set in (b) has a standard deviation of 5, because the data in (b) have more variability.

23. Company B; An offer of $33,000 is two standard deviations from the mean of Company A's starting salaries, which makes it unlikely. The same offer is within one standard deviation of the mean of Company B's starting salaries, which makes the offer likely.

17.

```
          •   •
          •   •   •
      •   •   •• • ••
  • •  •• • •• • ••  ••••  •
 ◄++++++++++++++++++++++++++++►
   75    80   85   90   95
```

18.

```
0 | 5 5 9        Key: 0|5 = 0.5
1 | 1 3 4 6 9
2 | 2 5 7 9 9
3 | 0 1 5 5 5 5
4 | 7 7 9
5 |
6 | 3 4 7
```

19. Archaeology The depths (in inches) at which 10 artifacts are found are given below.

20.7 24.8 30.5 26.2 36.0 34.3 30.3 29.5 27.0 38.5

(a) Find the range of the data set.

(b) Change 38.5 to 60.5 and find the range of the new data set.

20. In Exercise 19, compare your answer to part (a) with your answer to part (b). How do outliers affect the range of a data set?

■ USING AND INTERPRETING CONCEPTS

21. Graphical Reasoning Both data sets have a mean of 165. One has a standard deviation of 16, and the other has a standard deviation of 24. By looking at the graphs, which is which? Explain your reasoning.

(a)
```
12 | 8 9        Key: 12|8 = 128
13 | 5 5 8
14 | 1 2
15 | 0 0 6 7
16 | 4 5 9
17 | 1 3 6 8
18 | 0 8 9
19 | 6
20 | 3 5 7
```

(b)
```
12 |              Key: 13|1 = 131
13 | 1
14 | 2 3 5
15 | 0 4 5 6 8
16 | 1 1 2 3 3 3
17 | 1 5 8 8
18 | 2 3 4 5
19 | 0 2
20 |
```

22. Graphical Reasoning Both data sets represented below have a mean of 50. One has a standard deviation of 2.4, and the other has a standard deviation of 5. By looking at the graphs, which is which? Explain your reasoning.

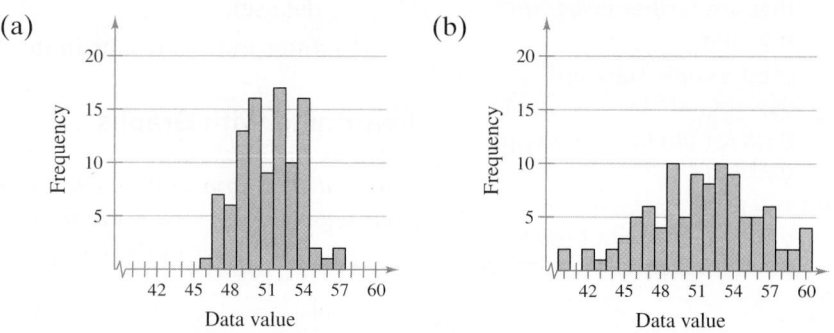

23. Salary Offers You are applying for jobs at two companies. Company A offers starting salaries with $\mu = \$31{,}000$ and $\sigma = \$1000$. Company B offers starting salaries with $\mu = \$31{,}000$ and $\sigma = \$5000$. From which company are you more likely to get an offer of $33,000 or more? Explain your reasoning.

24. Player B; A smaller standard deviation means that Player B's scores tend to fall within a smaller interval of values than Player A's scores.

25. (a) Dallas: $\bar{x} \approx 44.28$;
median = 44.7; range = 11.3;
$s^2 \approx 18.33$; $s \approx 4.28$

New York City: $\bar{x} \approx 50.91$;
median = 50.6; range = 17.8;
$s^2 \approx 50.36$; $s \approx 7.10$

(b) It appears from the data that the annual salaries in New York City are more variable than the annual salaries in Dallas. The annual salaries in Dallas have a lower mean and a lower median than the annual salaries in New York City.

26. See Selected Answers, page A114.

27. (a) Males: $\bar{x} \approx 1643$;
median = 1679.5;
range = 1087;
$s^2 \approx 116{,}477.4$; $s \approx 341.3$

Females: $\bar{x} \approx 1709.1$;
median = 1686.5;
range = 947; $s^2 \approx 91{,}625.0$;
$s \approx 302.7$

(b) It appears from the data that the SAT scores for males are more variable than the SAT scores for females. The SAT scores for males have a lower mean and median than the SAT scores for females.

28. See Selected Answers, page A114.

29. (a) Greatest sample standard deviation: (ii)

Data set (ii) has more entries that are farther away from the mean.

Least sample standard deviation: (iii)

Data set (iii) has more entries that are close to the mean.

(b) The three data sets have the same mean but have different standard deviations.

---

24. Golf Strokes An Internet site compares the strokes per round for two professional golfers. Which golfer is more consistent: Player A with $\mu = 71.5$ strokes and $\sigma = 2.3$ strokes, or Player B with $\mu = 70.1$ strokes and $\sigma = 1.2$ strokes? Explain your reasoning.

Comparing Two Data Sets   *In Exercises 25–28, you are asked to compare two data sets and interpret the results.*

25. Annual Salaries Sample annual salaries (in thousands of dollars) for accountants in Dallas and New York City are listed.

Dallas: 41.6   50.0   49.5   38.7   39.9   45.8   44.7   47.8   40.5
New York City: 45.6   41.5   57.6   55.1   59.3   59.0   50.6   47.2   42.3

(a) Find the mean, median, range, variance, and standard deviation of each data set.

(b) Interpret the results in the context of the real-life setting.

26. Annual Salaries Sample annual salaries (in thousands of dollars) for electrical engineers in Boston and Chicago are listed.

Boston: 70.4   84.2   58.5   64.5   71.6   79.9   88.3   80.1   69.9
Chicago: 69.4   71.5   65.4   59.9   70.9   68.5   62.9   70.1   60.9

(a) Find the mean, median, range, variance, and standard deviation of each data set.

(b) Interpret the results in the context of the real-life setting.

27. SAT Scores Sample SAT scores for eight males and eight females are listed.

Male SAT scores: 1520   1750   2120   1380   1982   1645   1033   1714
Female SAT scores: 1785   1507   1497   1952   2210   1871   1263   1588

(a) Find the mean, median, range, variance, and standard deviation of each data set.

(b) Interpret the results in the context of the real-life setting.

28. Batting Averages Sample batting averages for baseball players from two opposing teams are listed.

Team A: 0.295   0.310   0.325   0.272   0.256   0.297   0.320   0.384   0.235
Team B: 0.285   0.305   0.315   0.270   0.292   0.330   0.335   0.268   0.290

(a) Find the mean, median, range, variance, and standard deviation of each data set.

(b) Interpret the results in the context of the real-life setting.

Reasoning with Graphs   *In Exercises 29–32, you are asked to compare three data sets. (a) Without calculating, determine which data set has the greatest sample standard deviation and which has the least sample standard deviation. Explain your reasoning. (b) How are the data sets the same? How do they differ?*

29. (i)   (ii)   (iii)

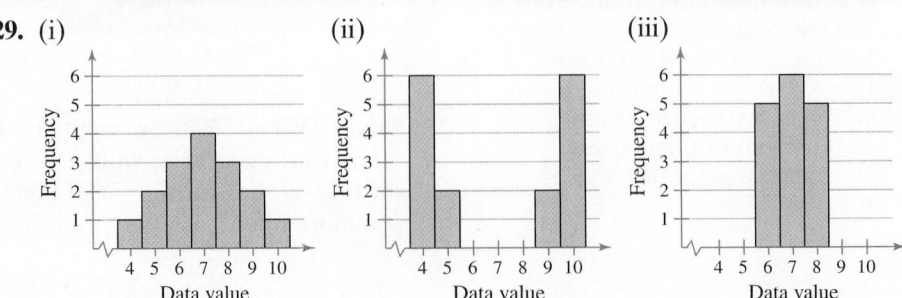

30. (a) Greatest sample standard deviation: (i)

Data set (i) has more entries that are farther away from the mean.

Least sample standard deviation: (iii)

Data set (iii) has more entries that are close to the mean.

(b) The three data sets have the same mean, median, and mode but have different standard deviations.

31. (a) Greatest sample standard deviation: (ii)

Data set (ii) has more entries that are farther away from the mean.

Least sample standard deviation: (iii)

Data set (iii) has more entries that are close to the mean.

(b) The three data sets have the same mean, median, and mode but have different standard deviations.

32. (a) Greatest sample standard deviation: (iii)

Data set (iii) has more entries that are farther away from the mean.

Least sample standard deviation: (i)

Data set (i) has more entries that are close to the mean.

(b) The three data sets have the same mean and median but have different modes and standard deviations.

33. 68%

34. Between $1500 and $3300

35. (a) 51   (b) 17

36. (a) 38   (b) 19

37. $2180, $1000, $2000, $950; $2180 is very unusual because it is more than 3 standard deviations from the mean.

30.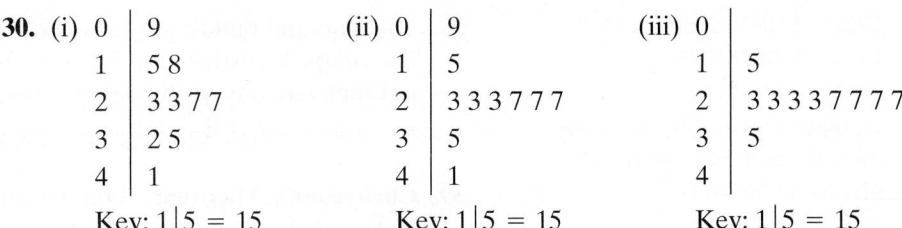

31. (i)   (ii)   (iii)

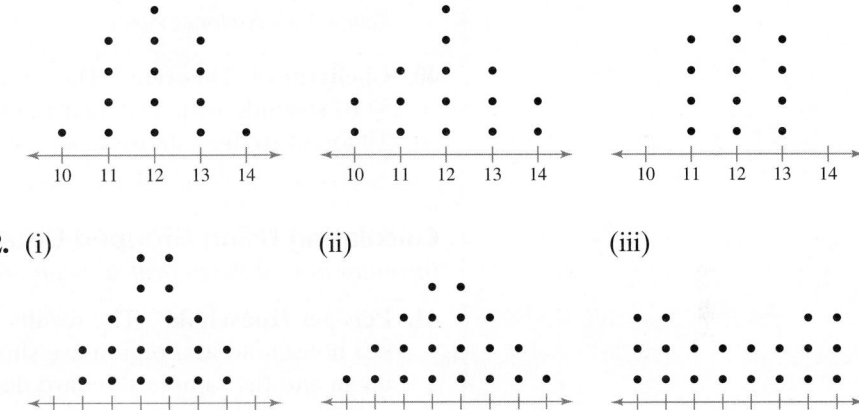

32. (i)   (ii)   (iii)

Using the Empirical Rule *In Exercises 33–38, you are asked to use the Empirical Rule.*

33. The mean value of land and buildings per acre from a sample of farms is $1500, with a standard deviation of $200. Estimate the percent of farms whose land and building values per acre are between $1300 and $1700. (Assume the data set has a bell-shaped distribution.)

34. The mean value of land and buildings per acre from a sample of farms is $2400, with a standard deviation of $450. Between what two values do about 95% of the data lie? (Assume the data set has a bell-shaped distribution.)

35. Using the sample statistics from Exercise 33, do the following. (Assume the number of farms in the sample is 75.)

(a) Estimate the number of farms whose land and building values per acre are between $1300 and $1700.

(b) If 25 additional farms were sampled, about how many of these farms would you expect to have land and building values between $1300 per acre and $1700 per acre?

36. Using the sample statistics from Exercise 34, do the following. (Assume the number of farms in the sample is 40.)

(a) Estimate the number of farms whose land and building values per acre are between $1500 and $3300.

(b) If 20 additional farms were sampled, about how many of these farms would you expect to have land and building values between $1500 per acre and $3300 per acre?

37. The land and building values per acre for eight more farms are listed. Using the sample statistics from Exercise 33, determine which of the data values are unusual. Are any of the data values very unusual? Explain.

$1150, $1775, $2180, $1000, $1475, $2000, $1850, $950

38. $3325, $1045, $3800, $1490; $3800 is very unusual

39. 24

40. At least 75% of the 400-meter dash times lie between 54.97 and 59.17 seconds.

41. $\bar{x} \approx 2.1, s \approx 1.3$

42. $\bar{x} \approx 1.7, s \approx 0.8$

43. See Odd Answers, page A58.

38. The land and building values per acre for eight more farms are listed. Using the sample statistics from Exercise 34, determine which of the data values are unusual. Are any of the data values very unusual? Explain.

$3325, $1045, $2450, $3200, $3800, $1490, $1675, $2950

39. Chebychev's Theorem Old Faithful is a famous geyser at Yellowstone National Park. From a sample with $n = 32$, the mean duration of Old Faithful's eruptions is 3.32 minutes and the standard deviation is 1.09 minutes. Using Chebychev's Theorem, determine at least how many of the eruptions lasted between 1.14 minutes and 5.5 minutes. *(Source: Yellowstone National Park)*

40. Chebychev's Theorem The mean time in a women's 400-meter dash is 57.07 seconds, with a standard deviation of 1.05 seconds. Apply Chebychev's Theorem to the data using $k = 2$. Interpret the results.

Calculating Using Grouped Data *In Exercises 41–48, use the grouped data formulas to find the indicated mean and standard deviation.*

41. Pets per Household The results of a random sample of the number of pets per household in a region are shown in the histogram. Estimate the sample mean and the sample standard deviation of the data set.

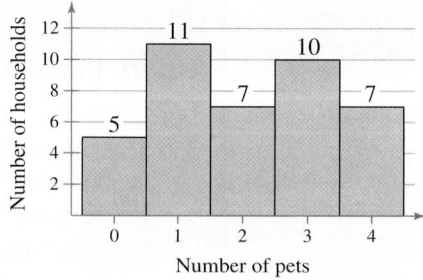

42. Cars per Household The results of a random sample of the number of cars per household in a region are shown in the histogram. Estimate the sample mean and the sample standard deviation of the data set.

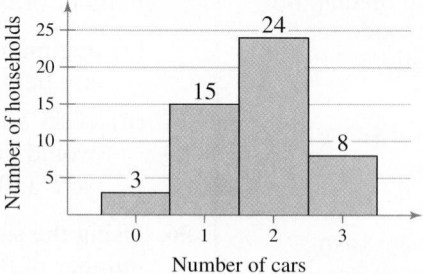

43. Football Wins The number of regular season wins for each National Football League team in 2009 are listed. Make a frequency distribution (using five classes) for the data set. Then approximate the population mean and the population standard deviation of the data set. *(Source: National Football League)*

```
10  9  7  6  10   9  9  5  14   9  8  7
13  8  5  4  11  11  8  4  12  11  7  2
13  9  8  3  10   8  5  1
```

44. See Selected Answers, page A114.
45. See Odd Answers, page A59.
46. See Selected Answers, page A115.
47. See Odd Answers, page A59.

44. Water Consumption The number of gallons of water consumed per day by a small village are listed. Make a frequency distribution (using five classes) for the data set. Then approximate the population mean and the population standard deviation of the data set.

| 167 | 180 | 192 | 173 | 145 | 151 | 174 | 175 | 178 | 160 |
| 195 | 224 | 244 | 146 | 162 | 146 | 177 | 163 | 149 | 188 |

45. Amounts of Caffeine The amounts of caffeine in a sample of five-ounce servings of brewed coffee are shown in the histogram. Make a frequency distribution for the data. Then use the table to estimate the sample mean and the sample standard deviation of the data set.

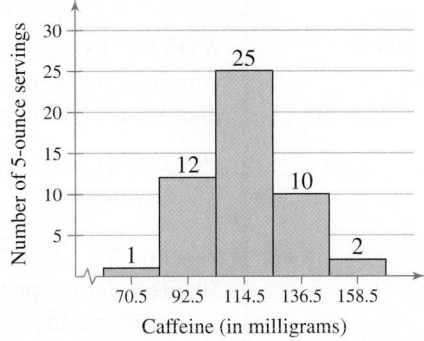

46. Supermarket Trips Thirty people were randomly selected and asked how many trips to the supermarket they had made in the past week. The responses are shown in the histogram. Make a frequency distribution for the data. Then use the table to estimate the sample mean and the sample standard deviation of the data set.

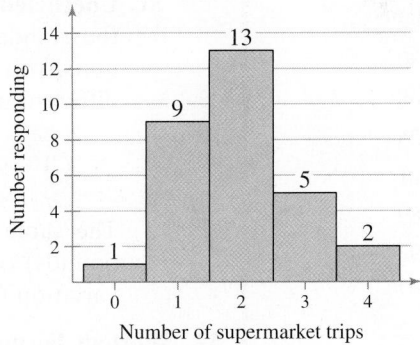

47. U.S. Population The estimated distribution (in millions) of the U.S. population by age for the year 2015 is shown in the pie chart. Make a frequency distribution for the data. Then use the table to estimate the sample mean and the sample standard deviation of the data set. Use 70 as the midpoint for "65 years and over." *(Source: Population Division, U.S. Census Bureau)*

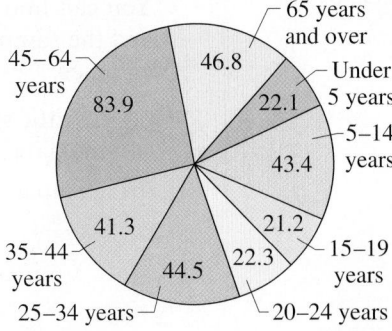

48. See Selected Answers, page A115.

49. See Odd Answers, page A59.

50. See Selected Answers, page A115.

51. $CV_{heights} = \dfrac{3.29}{72.75} \cdot 100\% \approx 4.5\%$

$CV_{weights} = \dfrac{17.69}{187.83} \cdot 100\% \approx 9.4\%$

It appears that weight is more variable than height.

52. (a) Male: 341.3

Female: 302.7

(b) The sample standard deviations are equal.

Heights	Weights
72	180
74	168
68	225
76	201
74	189
69	192
72	197
79	162
70	174
69	171
77	185
73	210

TABLE FOR EXERCISE 51

48. Brazil's Population
Brazil's estimated population for the year 2015 is shown in the histogram. Make a frequency distribution for the data. Then use the table to estimate the sample mean and the sample standard deviation of the data set. (*Adapted from U.S. Census Bureau, International Data Base*)

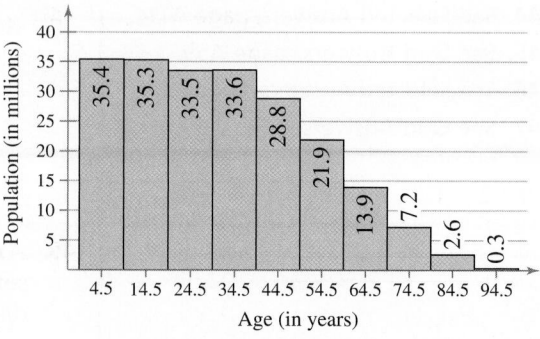

SC *In Exercises 49 and 50, use StatCrunch to find the sample size, mean, variance, standard deviation, median, range, minimum data value, and maximum data value of the data.*

49. The data represent the total amounts (in dollars) spent by several families at a restaurant.

49 56 75 64 55 49 62 89 30 34 60 52 60 72 75

50. The data represent the prices (in dollars) of several Hewlett-Packard office printers. (*Source: Hewlett-Packard*)

199.99 499.99 149.99 119.99 129.99 229.99
179.99 89.99 299.99 249.99 349.99 99.99

■ EXTENDING CONCEPTS

51. Coefficient of Variation The **coefficient of variation** *CV* describes the standard deviation as a percent of the mean. Because it has no units, you can use the coefficient of variation to compare data with different units.

$$CV = \frac{\text{Standard deviation}}{\text{Mean}} \times 100\%$$

The table at the left shows the heights (in inches) and weights (in pounds) of the members of a basketball team. Find the coefficient of variation for each data set. What can you conclude?

52. Shortcut Formula You used $SS_x = \sum (x - \bar{x})^2$ when calculating variance and standard deviation. An alternative formula that is sometimes more convenient for hand calculations is

$$SS_x = \sum x^2 - \frac{(\sum x)^2}{n}.$$

You can find the sample variance by dividing the sum of squares by $n - 1$ and the sample standard deviation by finding the square root of the sample variance.

(a) Use the shortcut formula to calculate the sample standard deviations for the data sets given in Exercise 27.

(b) Compare your results with those obtained in Exercise 27.

53. (a) $\bar{x} \approx 41.5$, $s \approx 5.3$

(b) $\bar{x} \approx 43.6$, $s \approx 5.6$

(c) $\bar{x} \approx 3.5$, $s \approx 0.4$

(d) When each entry is multiplied by a constant k, the new sample mean is $k \cdot \bar{x}$, and the new sample standard deviation is $k \cdot s$.

54. (a) $\bar{x} \approx 41.7$, $s \approx 6.0$

(b) $\bar{x} \approx 42.7$, $s \approx 6.0$

(c) $\bar{x} \approx 39.7$, $s \approx 6.0$

(d) Adding a constant k to, or subtracting it from, each entry makes the new sample mean $\bar{x} + k$ with the sample standard deviation being unaffected.

55. (a) Males: 249, Females: 245.4; The mean absolute deviation is less than the sample standard deviation.

(b) Team A: 0.0315, Team B: 0.0199; The mean absolute deviation is less than the sample standard deviation.

56. 10

Set $1 - \dfrac{1}{k^2} = 0.99$ and solve for k.

57. (a) $P \approx -2.61$

The data are skewed left.

(b) $P \approx 4.12$

The data are skewed right.

(c) $P = 0$

The data are symmetric.

(d) $P = 1$

The data are skewed right.

53. Scaling Data Sample annual salaries (in thousands of dollars) for employees at a company are listed.

42   36   48   51   39   39   42   36   48   33   39   42   45

(a) Find the sample mean and sample standard deviation.

(b) Each employee in the sample is given a 5% raise. Find the sample mean and sample standard deviation for the revised data set.

(c) To calculate the monthly salary, divide each original salary by 12. Find the sample mean and sample standard deviation for the revised data set.

(d) What can you conclude from the results of (a), (b), and (c)?

54. Shifting Data Sample annual salaries (in thousands of dollars) for employees at a company are listed.

40   35   49   53   38   39   40   37   49   34   38   43   47

(a) Find the sample mean and sample standard deviation.

(b) Each employee in the sample is given a $1000 raise. Find the sample mean and sample standard deviation for the revised data set.

(c) Each employee in the sample takes a pay cut of $2000 from their original salary. Find the sample mean and sample standard deviation for the revised data set.

(d) What can you conclude from the results of (a), (b), and (c)?

55. Mean Absolute Deviation Another useful measure of variation for a data set is the **mean absolute deviation** (*MAD*). It is calculated by the formula

$$\frac{\sum |x - \bar{x}|}{n}.$$

(a) Find the mean absolute deviations of the data sets in Exercise 27. Compare your results with the sample standard deviation.

(b) Find the mean absolute deviations of the data sets in Exercise 28. Compare your results with the sample standard deviation.

56. Chebychev's Theorem At least 99% of the data in any data set lie within how many standard deviations of the mean? Explain how you obtained your answer.

57. Pearson's Index of Skewness The English statistician Karl Pearson (1857–1936) introduced a formula for the skewness of a distribution.

$$P = \frac{3(\bar{x} - \text{median})}{s} \qquad \text{Pearson's index of skewness}$$

Most distributions have an index of skewness between -3 and 3. When $P > 0$, the data are skewed right. When $P < 0$, the data are skewed left. When $P = 0$, the data are symmetric. Calculate the coefficient of skewness for each distribution. Describe the shape of each.

(a) $\bar{x} = 17$, $s = 2.3$, median $= 19$

(b) $\bar{x} = 32$, $s = 5.1$, median $= 25$

(c) $\bar{x} = 9.2$, $s = 1.8$, median $= 9.2$

(d) $\bar{x} = 42$, $s = 6.0$, median $= 40$

ACTIVITY 2.4 Standard Deviation

APPLET

The *standard deviation* applet is designed to allow you to investigate interactively the standard deviation as a measure of spread for a data set. Points can be added to the plot by clicking the mouse above the horizontal axis. The mean of the points is shown as a green arrow. A numeric value for the standard deviation is shown above the plot. Points on the plot can be removed by clicking on the point and then dragging the point into the trash can. All of the points on the plot can be removed by simply clicking inside the trash can. The range of values for the horizontal axis can be specified by inputting lower and upper limits and then clicking UPDATE.

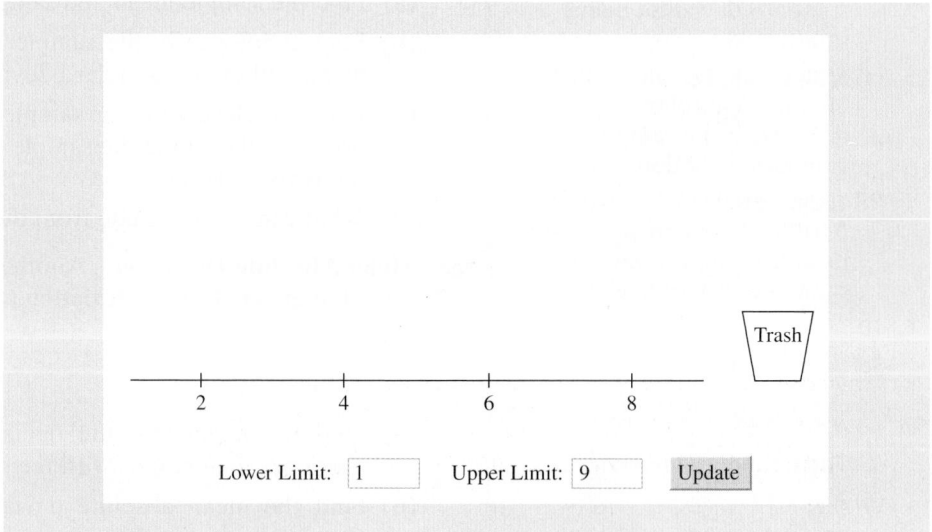

■ Explore

Step 1 Specify a lower limit.
Step 2 Specify an upper limit.
Step 3 Add 15 points to the plot.
Step 4 Remove all of the points from the plot.

■ Draw Conclusions

APPLET

1. Specify the lower limit to be 10 and the upper limit to be 20. Plot 10 points that have a mean of about 15 and a standard deviation of about 3. Write the estimates of the values of the points. Plot a point with a value of 15. What happens to the mean and standard deviation? Plot a point with a value of 20. What happens to the mean and standard deviation?

2. Specify the lower limit to be 30 and the upper limit to be 40. How can you plot eight points so that the points have the largest possible standard deviation? Use the applet to plot the set of points and then use the formula for standard deviation to confirm the value given in the applet. How can you plot eight points so that the points have the lowest possible standard deviation? Explain.

Earnings of Athletes

The earnings of professional athletes in different sports can vary. An athlete can be paid a base salary, earn signing bonuses upon signing a new contract, or even earn money by finishing in a certain position in a race or tournament. The data shown below are the earnings (for performance only, no endorsements) from Major League Baseball (MLB), Major League Soccer (MLS), the National Basketball Association (NBA), the National Football League (NFL), the National Hockey League (NHL), the National Association for Stock Car Auto Racing (NASCAR), and the Professional Golf Association Tour (PGA) for a recent year.

Organization	Number of players
MLB	858
MLS	410
NBA	463
NFL	1861
NHL	722
NASCAR	76
PGA	262

Number of Players Separated into Earnings Ranges

Organization	$0–$500,000	$500,001–$2,000,000	$2,000,001–$6,000,000	$6,000,001–$10,000,000	$10,000,001 +
MLB	353	182	164	85	74
MLS	403	5	1	1	0
NBA	35	157	137	77	57
NFL	554	746	438	85	38
NHL	42	406	237	37	0
NASCAR	23	16	31	6	0
PGA	110	115	36	1	0

■ EXERCISES

1. **Revenue** Which organization had the greatest total player earnings? Explain your reasoning.

2. **Mean Earnings** Estimate the mean earnings of a player in each organization. Use $19,000,000 as the midpoint for $10,000,001+.

3. **Revenue** Which organization had the greatest earnings per player? Explain your reasoning.

4. **Standard Deviation** Estimate the standard deviation for the earnings of a player in each organization. Use $19,000,000 as the midpoint for $10,000,001+.

5. **Standard Deviation** Which organization had the greatest standard deviation? Explain your reasoning.

6. **Bell-Shaped Distribution** Of the seven organizations, which is most bell-shaped? Explain your reasoning.

2.5   Measures of Position

Quartiles ▸ Percentiles and Other Fractiles ▸ The Standard Score

WHAT YOU SHOULD LEARN

▸ How to find the first, second, and third quartiles of a data set

▸ How to find the interquartile range of a data set

▸ How to represent a data set graphically using a box-and-whisker plot

▸ How to interpret other fractiles such as percentiles

▸ How to find and interpret the standard score (z-score)

▸ QUARTILES

In this section, you will learn how to use fractiles to specify the position of a data entry within a data set. **Fractiles** are numbers that partition, or divide, an ordered data set into equal parts. For instance, the median is a fractile because it divides an ordered data set into two equal parts.

DEFINITION

The three **quartiles,** Q_1, Q_2, and Q_3, approximately divide an ordered data set into four equal parts. About one quarter of the data fall on or below the **first quartile** Q_1. About one half of the data fall on or below the **second quartile** Q_2 (the second quartile is the same as the median of the data set). About three quarters of the data fall on or below the **third quartile** Q_3.

EXAMPLE 1   Report 14

▸ Finding the Quartiles of a Data Set

The number of nuclear power plants in the top 15 nuclear power-producing countries in the world are listed. Find the first, second, and third quartiles of the data set. What can you conclude? *(Source: International Atomic Energy Agency)*

7   18   11   6   59   17   18   54   104   20   31   8   10   15   19

▸ Solution

First, order the data set and find the median Q_2. Once you find Q_2, divide the data set into two halves. The first and third quartiles are the medians of the lower and upper halves of the data set.

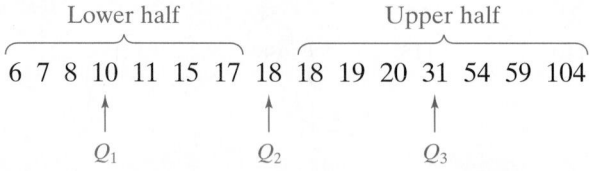

Interpretation   About one fourth of the countries have 10 or fewer nuclear power plants; about one half have 18 or fewer; and about three fourths have 31 or fewer.

▸ Try It Yourself 1

Find the first, second, and third quartiles for the ages of the 50 richest people using the data set listed in the Chapter Opener on page 37. What can you conclude?

a. *Order* the data set.
b. Find the *median* Q_2.
c. Find the *first and third quartiles*, Q_1 and Q_3.
d. *Interpret* the results in the context of the data.   *Answer: Page A34*

EXAMPLE 2

▶ Using Technology to Find Quartiles

The tuition costs (in thousands of dollars) for 25 liberal arts colleges are listed. Use a calculator or a computer to find the first, second, and third quartiles. What can you conclude?

23 25 30 23 20 22 21 15 25 24 30 25 30
20 23 29 20 19 22 23 29 23 28 22 28

▶ Solution

MINITAB, Excel, and the TI-83/84 Plus each have features that automatically calculate quartiles. Try using this technology to find the first, second, and third quartiles of the tuition data. From the displays, you can see that $Q_1 = 21.5$, $Q_2 = 23$, and $Q_3 = 28$.

STUDY TIP

There are several ways to find the quartiles of a data set. Regardless of how you find the quartiles, the results are rarely off by more than one data entry. For instance, in Example 2, the first quartile, as determined by Excel, is 22 instead of 21.5.

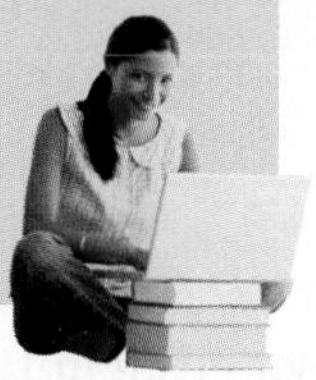

MINITAB

Descriptive Statistics: Tuition

Variable	N	Mean	SE Mean	StDev	Minimum
Tuition	25	23.960	0.788	3.942	15.000

Variable	Q1	Median	Q3	Maximum
Tuition	21.500	23.000	28.000	30.000

EXCEL

	A	B	C	D
1	23			
2	25		Quartile(A1:A25,1)	
3	30		22	
4	23			
5	20		Quartile(A1:A25,2)	
6	22		23	
7	21			
8	15		Quartile(A1:A25,3)	
9	25		28	
10	24			
11	30			
12	25			
13	30			
14	20			
15	23			
16	29			
17	20			
18	19			
19	22			
20	23			
21	29			
22	23			
23	28			
24	22			
25	28			

TI-83/84 PLUS

1-Var Stats
↑n=25
minX=15
Q_1=21.5
Med=23
Q_3=28
maxX=30

Note to Instructor

For MINITAB and the TI-83/84 Plus, quartiles are found with the following ranks.

Q_1: $\frac{1(n+1)}{4}$

Q_2: $\frac{2(n+1)}{4}$

Q_3: $\frac{3(n+1)}{4}$

Interpretation About one quarter of these colleges charge tuition of $21,500 or less; one half charge $23,000 or less; and about three quarters charge $28,000 or less.

▶ **Try It Yourself 2**

The tuition costs (in thousands of dollars) for 25 universities are listed. Use a calculator or a computer to find the first, second, and third quartiles. What can you conclude?

20 26 28 25 31 14 23 15 12 26 29 24 31
19 31 17 15 17 20 31 32 16 21 22 28

a. *Enter* the data.
b. Calculate the *first, second, and third quartiles.*
c. *Interpret* the results in the context of the data.   *Answer: Page A34*

After finding the quartiles of a data set, you can find the *interquartile range.*

DEFINITION

The **interquartile range (IQR)** of a data set is a measure of variation that gives the range of the middle 50% of the data. It is the difference between the third and first quartiles.

$$\text{Interquartile range (IQR)} = Q_3 - Q_1$$

EXAMPLE 3

▶ **Finding the Interquartile Range**

Find the interquartile range of the data set given in Example 1. What can you conclude from the result?

▶ **Solution**

From Example 1, you know that $Q_1 = 10$ and $Q_3 = 31$. So, the interquartile range is

$$\text{IQR} = Q_3 - Q_1 = 31 - 10 = 21.$$

Interpretation   The number of power plants in the middle portion of the data set vary by at most 21.

▶ **Try It Yourself 3**

Find the interquartile range for the ages of the 50 richest people listed in the Chapter Opener on page 37.

a. Find the *first and third quartiles,* Q_1 and Q_3.
b. *Subtract* Q_1 from Q_3.
c. *Interpret* the result in the context of the data.   *Answer: Page A34*

The IQR can also be used to identify outliers. First, multiply the IQR by 1.5. Then subtract that value from Q_1, and add that value to Q_3. Any data value that is smaller than $Q_1 - 1.5(\text{IQR})$ or larger than $Q_3 + 1.5(\text{IQR})$ is an outlier. For instance, the IQR in Example 1 is $31 - 10 = 21$ and $1.5(21) = 31.5$. So, adding 31.5 to Q_3 gives $Q_3 + 31.5 = 31 + 31.5 = 62.5$. Because $104 > 62.5$, 104 is an outlier.

Another important application of quartiles is to represent data sets using box-and-whisker plots. A **box-and-whisker plot** (or **boxplot**) is an exploratory data analysis tool that highlights the important features of a data set. To graph a box-and-whisker plot, you must know the following values.

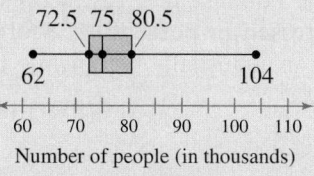
1. The minimum entry
2. The first quartile Q_1
3. The median Q_2
4. The third quartile Q_3
5. The maximum entry

These five numbers are called the **five-number summary** of the data set.

GUIDELINES

Drawing a Box-and-Whisker Plot

1. Find the five-number summary of the data set.
2. Construct a horizontal scale that spans the range of the data.
3. Plot the five numbers above the horizontal scale.
4. Draw a box above the horizontal scale from Q_1 to Q_3 and draw a vertical line in the box at Q_2.
5. Draw whiskers from the box to the minimum and maximum entries.

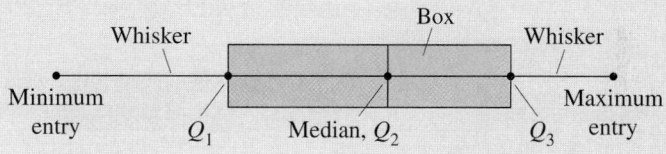

EXAMPLE 4   SC Report 15

▶ **Drawing a Box-and-Whisker Plot**

Draw a box-and-whisker plot that represents the data set given in Example 1. What can you conclude from the display?

See MINITAB and TI-83/84 Plus steps on pages 122 and 123.

▶ **Solution**

The five-number summary of the data set is displayed below. Using these five numbers, you can construct the box-and-whisker plot shown.

Min = 6, Q_1 = 10, Q_2 = 18, Q_3 = 31,   Max = 104,

Number of Power Plants

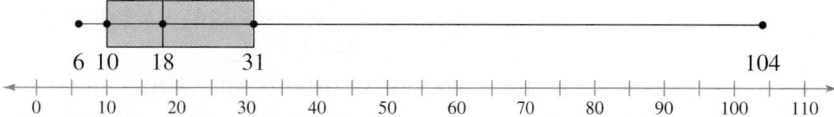

Interpretation   You can make several conclusions from the display. One is that about half the data values are between 10 and 31. By looking at the length of the right whisker, you can also conclude that the data value of 104 is a possible outlier.

▶ **Try It Yourself 4**

Draw a box-and-whisker plot that represents the ages of the 50 richest people listed in the Chapter Opener on page 37. What can you conclude?

a. Find the *five-number summary* of the data set.
b. Construct a *horizontal scale* and *plot* the five numbers above it.
c. Draw the *box,* the *vertical line*, and the *whiskers*.
d. Make some *conclusions*.

Answer: Page A34

▸ PERCENTILES AND OTHER FRACTILES

In addition to using quartiles to specify a measure of position, you can also use percentiles and deciles. These common fractiles are summarized as follows.

Fractiles	Summary	Symbols
Quartiles	Divide a data set into 4 equal parts.	Q_1, Q_2, Q_3
Deciles	Divide a data set into 10 equal parts.	$D_1, D_2, D_3, \ldots, D_9$
Percentiles	Divide a data set into 100 equal parts.	$P_1, P_2, P_3, \ldots, P_{99}$

Percentiles are often used in education and health-related fields to indicate how one individual compares with others in a group. They can also be used to identify unusually high or unusually low values. For instance, test scores and children's growth measurements are often expressed in percentiles. Scores or measurements in the 95th percentile and above are unusually high, while those in the 5th percentile and below are unusually low.

EXAMPLE 5

▸ Interpreting Percentiles

The ogive at the right represents the cumulative frequency distribution for SAT test scores of college-bound students in a recent year. What test score represents the 62nd percentile? How should you interpret this?
(Source: The College Board)

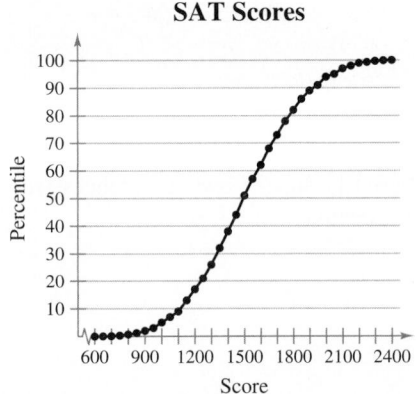

▸ Solution

From the ogive, you can see that the 62nd percentile corresponds to a test score of 1600.

Interpretation This means that approximately 62% of the students had an SAT score of 1600 or less.

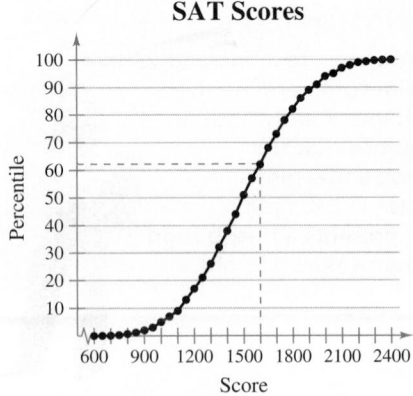

▸ Try It Yourself 5

The ages of the 50 richest people are represented in the cumulative frequency graph at the left. At what percentile is someone who is 66 years old? How should you interpret this?

a. *Use the graph* to find the percentile that corresponds to the given age.
b. *Interpret* the results in the context of the data.

Answer: Page A34

Ages of the 50 Richest People

▸ **THE STANDARD SCORE**

When you know the mean and standard deviation of a data set, you can measure a data value's position in the data set with a *standard score*, or *z-score*.

DEFINITION

The **standard score,** or **z-score,** represents the number of standard deviations a given value x falls from the mean μ. To find the z-score for a given value, use the following formula.

$$z = \frac{\text{Value} - \text{Mean}}{\text{Standard deviation}} = \frac{x - \mu}{\sigma}$$

A z-score can be negative, positive, or zero. If z is negative, the corresponding x-value is less than the mean. If z is positive, the corresponding x-value is greater than the mean. And if $z = 0$, the corresponding x-value is equal to the mean. A z-score can be used to identify an unusual value of a data set that is approximately bell-shaped.

EXAMPLE 6

▸ **Finding z-Scores**

The mean speed of vehicles along a stretch of highway is 56 miles per hour with a standard deviation of 4 miles per hour. You measure the speeds of three cars traveling along this stretch of highway as 62 miles per hour, 47 miles per hour, and 56 miles per hour. Find the z-score that corresponds to each speed. What can you conclude?

▸ **Solution**

The z-score that corresponds to each speed is calculated below.

$$x = 62 \text{ mph} \qquad\qquad x = 47 \text{ mph} \qquad\qquad x = 56 \text{ mph}$$

$$z = \frac{62 - 56}{4} = 1.5 \qquad z = \frac{47 - 56}{4} = -2.25 \qquad z = \frac{56 - 56}{4} = 0$$

Interpretation   From the z-scores, you can conclude that a speed of 62 miles per hour is 1.5 standard deviations above the mean; a speed of 47 miles per hour is 2.25 standard deviations below the mean; and a speed of 56 miles per hour is equal to the mean. If the distribution of the speeds is approximately bell-shaped, the car traveling 47 miles per hour is said to be traveling unusually slowly, because its speed corresponds to a z-score of -2.25.

▸ **Try It Yourself 6**

The monthly utility bills in a city have a mean of $70 and a standard deviation of $8. Find the z-scores that correspond to utility bills of $60, $71, and $92. What can you conclude?

a. *Identify* μ and σ. *Transform* each value to a z-score.
b. *Interpret* the results.   *Answer: Page A34*

When a distribution is approximately bell-shaped, you know from the Empirical Rule that about 95% of the data lie within 2 standard deviations of the mean. So, when this distribution's values are transformed to z-scores, about 95% of the z-scores should fall between -2 and 2. A z-score outside of this range will occur about 5% of the time and would be considered unusual. So, according to the Empirical Rule, a z-score less than -3 or greater than 3 would be very unusual, with such a score occurring about 0.3% of the time.

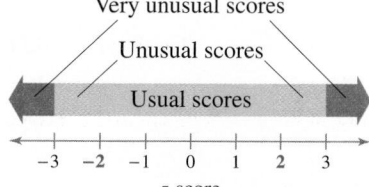

In Example 6, you used z-scores to compare data values within the same data set. You can also use z-scores to compare data values from different data sets.

EXAMPLE 7

▶ **Comparing z-Scores from Different Data Sets**

In 2009, Heath Ledger won the Oscar for Best Supporting Actor at age 29 for his role in the movie *The Dark Knight*. Penelope Cruz won the Oscar for Best Supporting Actress at age 34 for her role in *Vicky Cristina Barcelona*. The mean age of all Best Supporting Actor winners is 49.5, with a standard deviation of 13.8. The mean age of all Best Supporting Actress winners is 39.9, with a standard deviation of 14.0. Find the z-scores that correspond to the ages of Ledger and Cruz. Then compare your results.

▶ **Solution**

The z-scores that correspond to the ages of the two performers are calculated below.

$$\textit{Heath Ledger} \qquad z = \frac{x - \mu}{\sigma}$$

$$= \frac{29 - 49.5}{13.8}$$

$$\approx -1.49$$

$$\textit{Penelope Cruz} \qquad z = \frac{x - \mu}{\sigma}$$

$$= \frac{34 - 39.9}{14.0}$$

$$\approx -0.42$$

The age of Heath Ledger was 1.49 standard deviations below the mean, and the age of Penelope Cruz was 0.42 standard deviation below the mean.

Interpretation Compared with other Best Supporting Actor winners, Heath Ledger was relatively younger, whereas the age of Penelope Cruz was only slightly lower than the average age of other Best Supporting Actress winners. Both z-scores fall between -2 and 2, so neither score would be considered unusual.

▶ **Try It Yourself 7**

In 2009, Sean Penn won the Oscar for Best Actor at age 48 for his role in the movie *Milk*. Kate Winslet won the Oscar for Best Actress at age 33 for her role in *The Reader*. The mean age of all Best Actor winners is 43.7, with a standard deviation of 8.7. The mean age of all Best Actress winners is 35.9, with a standard deviation of 11.4. Find the z-scores that correspond to the ages of Penn and Winslet. Then compare your results.

a. *Identify* μ and σ for each data set.
b. *Transform* each value to a z-score.
c. *Compare* your results.   *Answer: Page A34*

2.5 EXERCISES

FOR EXTRA HELP:
MyStatLab

1. The soccer team scored fewer points per game than 75% of the teams in the league.

2. See Selected Answers, page A115.

3. The student scored higher than 78% of the students who took the actuarial exam.

4. See Selected Answers, page A115.

5. The interquartile range of a data set can be used to identify outliers because data values that are greater than $Q_3 + 1.5(IQR)$ or less than $Q_1 - 1.5(IQR)$ are considered outliers.

6. See Selected Answers, page A115.

7. False. The median of a data set is a fractile, but the mean may or may not be a fractile depending on the distribution of the data.

8. True

9. True

10. See Selected Answers, page A115.

11. False. The 50th percentile is equivalent to Q_2.

12. False. It is possible to have a z-score of 0.

13. False. A z-score of −2.5 is considered unusual.

14. True

15. See Odd Answers, page A59.

16. See Selected Answers, page A115.

17. See Odd Answers, page A59.

18. See Selected Answers, page A115.

19. See Odd Answers, page A59.

20. See Selected Answers, page A115.

■ BUILDING BASIC SKILLS AND VOCABULARY

1. The goals scored per game by a soccer team represent the first quartile for all teams in a league. What can you conclude about the team's goals scored per game?

2. A salesperson at a company sold $6,903,435 of hardware equipment last year, a figure that represented the eighth decile of sales performance at the company. What can you conclude about the salesperson's performance?

3. A student's score on an actuarial exam is in the 78th percentile. What can you conclude about the student's exam score?

4. A counselor tells a child's parents that their child's IQ is in the 93rd percentile for the child's age group. What can you conclude about the child's IQ?

5. Explain how the interquartile range of a data set can be used to identify outliers.

6. Describe the relationship between quartiles and percentiles.

True or False?   *In Exercises 7–14, determine whether the statement is true or false. If it is false, rewrite it as a true statement.*

7. The mean and median of a data set are both fractiles.

8. About one quarter of a data set falls below Q_1.

9. The second quartile is the median of an ordered data set.

10. The five numbers you need to graph a box-and-whisker plot are the minimum, the maximum, Q_1, Q_3, and the mean.

11. The 50th percentile is equivalent to Q_1.

12. It is impossible to have a z-score of 0.

13. A z-score of −2.5 is considered very unusual.

14. A z-score of 1.99 is considered usual.

■ USING AND INTERPRETING CONCEPTS

Graphical Analysis   *In Exercises 15–20, use the box-and-whisker plot to identify (a) the five-number summary, and (b) the interquartile range.*

15.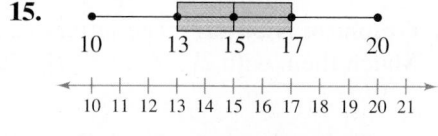

16.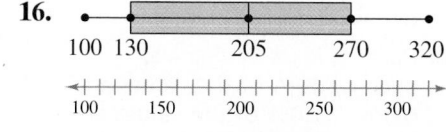

17.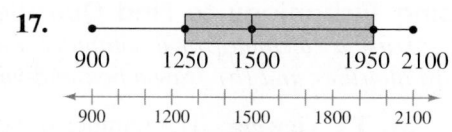

18.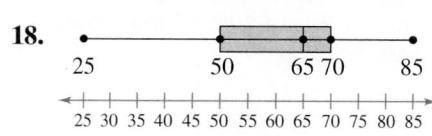

19.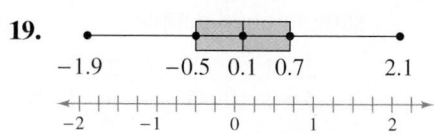

20.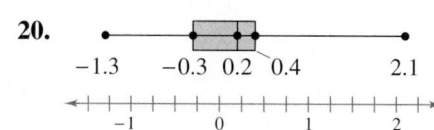

21. (a) Min = 24, Q_1 = 28, Q_2 = 35, Q_3 = 41, Max = 60

(b)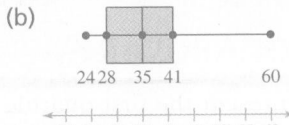

22. (a) Min = 150, Q_1 = 172, Q_2 = 177, Q_3 = 180, Max = 182

(b)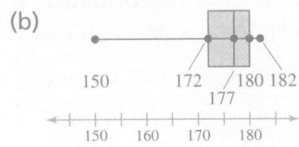

23. (a) Min = 1, Q_1 = 4.5, Q_2 = 6, Q_3 = 7.5, Max = 9

(b)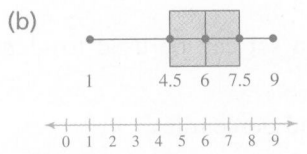

24. (a) Min = 1, Q_1 = 3, Q_2 = 5, Q_3 = 8, Max = 9

(b)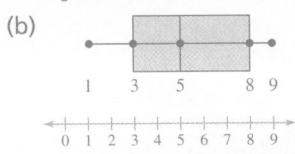

25. None. The data are not skewed or symmetric.

26. Skewed right. Most of the data lie to the left on the box plot.

27. Skewed left. Most of the data lie to the right on the box plot.

28. Symmetric. The data are evenly spaced to the left and to the right of the median.

29. Q_1 = B, Q_2 = A, Q_3 = C, because about one quarter of the data fall on or below 17, 18.5 is the median of the entire data set, and about three quarters of the data fall on or below 20.

30. P_{10} = T, P_{50} = R, P_{80} = S, because 10% of the values are below T, 50% of the values are below R, and 80% of the values are below S.

In Exercises 21–24, (a) find the five-number summary, and (b) draw a box-and-whisker plot of the data.

21. 39 36 30 27 26 24 28 35 39 60 50 41 35 32 51

22. 171 176 182 150 178 180 173 170 174 178 181 180

23. 4 7 7 5 2 9 7 6 8 5 8 4 1 5 2 8 7 6 6 9

24. 2 7 1 3 1 2 8 9 9 2 5 4 7 3 7 5 4 7
     2 3 5 9 5 6 3 9 3 4 9 8 8 2 3 9 5

Interpreting Graphs *In Exercises 25–28, use the box-and-whisker plot to determine if the shape of the distribution represented is symmetric, skewed left, skewed right, or none of these. Justify your answer.*

25.

26.

27.

28.

29. Graphical Analysis The letters A, B, and C are marked on the histogram. Match them with Q_1, Q_2 (the median), and Q_3. Justify your answer.

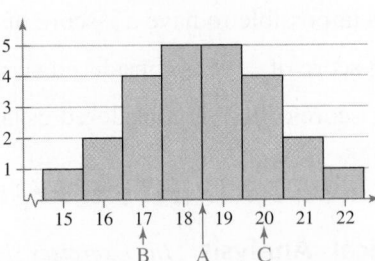

FIGURE FOR EXERCISE 29

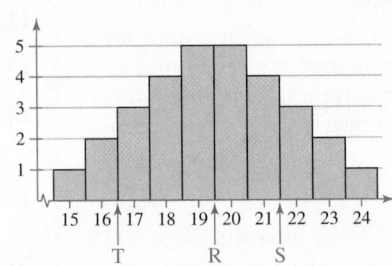

FIGURE FOR EXERCISE 30

30. Graphical Analysis The letters R, S, and T are marked on the histogram. Match them with P_{10}, P_{50}, and P_{80}. Justify your answer.

Using Technology to Find Quartiles and Draw Graphs *In Exercises 31–34, use a calculator or a computer to (a) find the data set's first, second, and third quartiles, and (b) draw a box-and-whisker plot that represents the data set.*

31. TV Viewing The number of hours of television watched per day by a sample of 28 people

2 4 1 5 7 2 5 4 4 2 3 6 4 3
5 2 0 3 5 9 4 5 2 1 3 6 7 2

31. (a) $Q_1 = 2$, $Q_2 = 4$, $Q_3 = 5$

(b)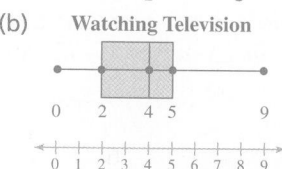

Watching Television

Number of hours

32. (a) $Q_1 = 2$, $Q_2 = 4.5$, $Q_3 = 6.5$

(b)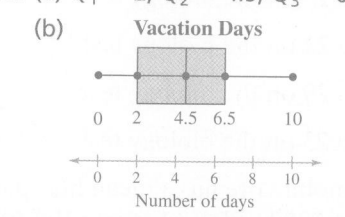

Vacation Days

Number of days

33. (a) $Q_1 = 3$, $Q_2 = 3.85$, $Q_3 = 5.2$

(b)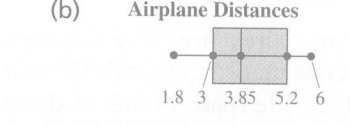

Airplane Distances

Distance (in miles)

34. (a) $Q_1 = 15.125$, $Q_2 = 15.8$, $Q_3 = 17.65$

(b)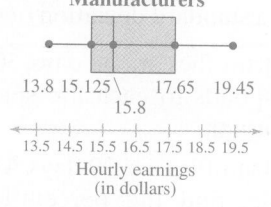

Railroad Equipment Manufacturers

Hourly earnings (in dollars)

35. (a) 5 (b) 50% (c) 25%

36. (a) $17.65 (b) 50% (c) 50%

37. $A \rightarrow z = -1.43$

$B \rightarrow z = 0$

$C \rightarrow z = 2.14$

A z-score of 2.14 would be unusual.

38. $B \rightarrow z = 0.77$

$C \rightarrow z = 1.54$

$A \rightarrow z = -1.54$

None of the z-scores is unusual.

32. **Vacation Days** The number of vacation days used by a sample of 20 employees in a recent year

3 9 2 1 7 5 3 2 2 6
4 0 10 0 3 5 7 8 6 5

33. **Airplane Distances** The distances (in miles) from an airport of a sample of 22 inbound and outbound airplanes

2.8 2.0 3.0 3.0 3.2 5.9 3.5 3.6
1.8 5.5 3.7 5.2 3.8 3.9 6.0 2.5
4.0 4.1 4.6 5.0 5.5 6.0

34. **Hourly Earnings** The hourly earnings (in dollars) of a sample of 25 railroad equipment manufacturers

15.60 18.75 14.60 15.80 14.35 13.90 17.50 17.55 13.80
14.20 19.05 15.35 15.20 19.45 15.95 16.50 16.30 15.25
15.05 19.10 15.20 16.22 17.75 18.40 15.25

35. **TV Viewing** Refer to the data set given in Exercise 31 and the box-and-whisker plot you drew that represents the data set.

(a) About 75% of the people watched no more than how many hours of television per day?

(b) What percent of the people watched more than 4 hours of television per day?

(c) If you randomly selected one person from the sample, what is the likelihood that the person watched less than 2 hours of television per day? Write your answer as a percent.

36. **Manufacturer Earnings** Refer to the data set given in Exercise 34 and the box-and-whisker plot you drew that represents the data set.

(a) About 75% of the manufacturers made less than what amount per hour?

(b) What percent of the manufacturers made more than $15.80 per hour?

(c) If you randomly selected one manufacturer from the sample, what is the likelihood that the manufacturer made less than $15.80 per hour? Write your answer as a percent.

Graphical Analysis *In Exercises 37 and 38, the midpoints A, B, and C are marked on the histogram. Match them with the indicated z-scores. Which z-scores, if any, would be considered unusual?*

37. $z = 0$

$z = 2.14$

$z = -1.43$

38. $z = 0.77$

$z = 1.54$

$z = -1.54$

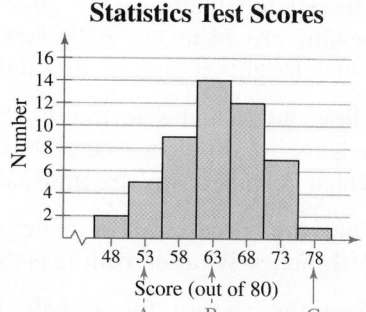

Statistics Test Scores

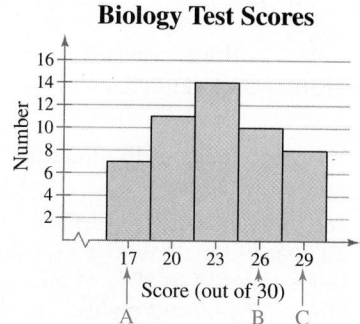

Biology Test Scores

39. (a) Statistics: $z = \dfrac{75 - 63}{7} \approx 1.71$

Biology: $z = \dfrac{25 - 23}{3.9} \approx 0.51$

(b) The student did better on the statistics test.

40. See Selected Answers, page A115.

41. (a) Statistics: $z = \dfrac{78 - 63}{7} \approx 2.14$

Biology: $z = \dfrac{29 - 23}{3.9} \approx 1.54$

(b) The student did better on the statistics test.

42. (a) Statistics: $z = \dfrac{63 - 63}{7} = 0$

Biology: $z = \dfrac{23 - 23}{3.9} = 0$

(b) The student performed equally on both tests.

43. (a) See Odd Answers, page A60.

(b) For 30,500, 2.5th percentile

For 37,250, 84th percentile

For 35,000, 50th percentile

44. (a) $z_1 = \dfrac{34 - 33}{4} = 0.25$,

$z_2 = \dfrac{30 - 33}{4} = -0.75$,

$z_3 = \dfrac{42 - 33}{4} = 2.25$

The life span of 42 days is unusual.

(b) For 29, 16th percentile; For 41, 97.5th percentile; For 25, 2.5th percentile

Adult Males Ages 20–29

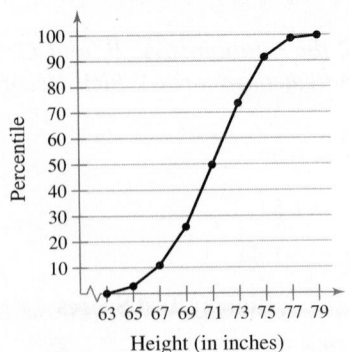

Height (in inches)

FIGURE FOR EXERCISES 45–50

45. 72 inches; 60% of the heights are below 72 inches.

46. 98th percentile; A 20- to 29-year-old male who is 77 inches tall is taller than 98% of all 20- to 29-year-old males.

Comparing Test Scores *For the statistics test scores in Exercise 37, the mean is 63 and the standard deviation is 7.0, and for the biology test scores in Exercise 38, the mean is 23 and the standard deviation is 3.9. In Exercises 39–42, you are given the test scores of a student who took both tests.*

   (a) Transform each test score to a z-score.

   (b) Determine on which test the student had a better score.

39. A student gets a 75 on the statistics test and a 25 on the biology test.

40. A student gets a 60 on the statistics test and a 22 on the biology test.

41. A student gets a 78 on the statistics test and a 29 on the biology test.

42. A student gets a 63 on the statistics test and a 23 on the biology test.

43. Life Spans of Tires A certain brand of automobile tire has a mean life span of 35,000 miles, with a standard deviation of 2250 miles. (Assume the life spans of the tires have a bell-shaped distribution.)

   (a) The life spans of three randomly selected tires are 34,000 miles, 37,000 miles, and 30,000 miles. Find the z-score that corresponds to each life span. According to the z-scores, would the life spans of any of these tires be considered unusual?

   (b) The life spans of three randomly selected tires are 30,500 miles, 37,250 miles, and 35,000 miles. Using the Empirical Rule, find the percentile that corresponds to each life span.

44. Life Spans of Fruit Flies The life spans of a species of fruit fly have a bell-shaped distribution, with a mean of 33 days and a standard deviation of 4 days.

   (a) The life spans of three randomly selected fruit flies are 34 days, 30 days, and 42 days. Find the z-score that corresponds to each life span and determine if any of these life spans are unusual.

   (b) The life spans of three randomly selected fruit flies are 29 days, 41 days, and 25 days. Using the Empirical Rule, find the percentile that corresponds to each life span.

Interpreting Percentiles *In Exercises 45–50, use the cumulative frequency distribution to answer the questions. The cumulative frequency distribution represents the heights of males in the United States in the 20–29 age group. The heights have a bell-shaped distribution (see Picturing the World, page 86) with a mean of 69.9 inches and a standard deviation of 3.0 inches. (Adapted from National Center for Health Statistics)*

45. What height represents the 60th percentile? How should you interpret this?

46. What percentile is a height of 77 inches? How should you interpret this?

47. Three adult males in the 20–29 age group are randomly selected. Their heights are 74 inches, 62 inches, and 80 inches. Use z-scores to determine which heights, if any, are unusual.

48. Three adult males in the 20–29 age group are randomly selected. Their heights are 70 inches, 66 inches, and 68 inches. Use z-scores to determine which heights, if any, are unusual.

49. Find the z-score for a male in the 20–29 age group whose height is 71.1 inches. What percentile is this?

50. Find the z-score for a male in the 20–29 age group whose height is 66.3 inches. What percentile is this?

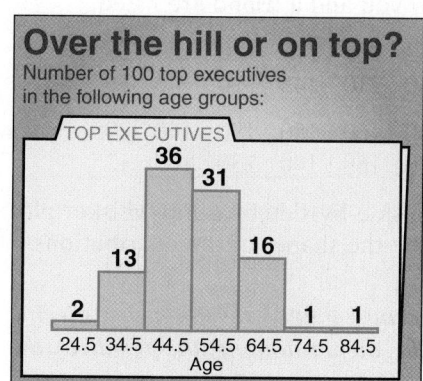

Over the hill or on top?

Number of 100 top executives in the following age groups:

TOP EXECUTIVES

36 31 13 16 2 1 1

24.5 34.5 44.5 54.5 64.5 74.5 84.5
Age

FIGURE FOR EXERCISE 51

47. $z_1 = \dfrac{74 - 69.9}{3.0} \approx 1.37$

$z_2 = \dfrac{62 - 69.9}{3.0} \approx -2.63$

$z_3 = \dfrac{80 - 69.9}{3.0} \approx 3.37$

The heights of 62 and 80 inches are unusual.

48. $z_1 = \dfrac{70 - 69.9}{3.0} \approx 0.03,$

$z_2 = \dfrac{66 - 69.9}{3.0} \approx -1.3,$

$z_3 = \dfrac{68 - 69.9}{3.0} \approx -0.63$

None of the heights is unusual.

49. $z = \dfrac{71.1 - 69.9}{3.0} = 0.4$

About the 50th percentile

50. $z = \dfrac{66.3 - 69.9}{3.0} = -1.2$

About the 10th percentile

51. (a) Min = 27, Q_1 = 42, Q_2 = 49, Q_3 = 56, Max = 82

(b) See Odd Answers, page A60.

(c) Half of the executives are between 42 and 56 years old.

(d) 49, because half of the executives are older and half are younger.

(e) The age groups 20–29, 70–79, and 80–89 would all be considered unusual because they lie more than two standard deviations from the mean.

52. 5

53. 33.75

54. 10.975

55. 19.8

56. See Selected Answers, page A115.

EXTENDING CONCEPTS

51. Ages of Executives   The ages of a sample of 100 executives are listed.

31	62	51	44	61	47	49	45	40	52	60	51	67
47	63	54	59	43	63	52	50	54	61	41	48	49
51	54	39	54	47	52	36	53	74	33	53	68	44
40	60	42	50	48	42	42	36	57	42	48	56	51
54	42	27	43	43	41	54	49	49	47	51	28	54
36	36	41	60	55	42	59	35	65	48	56	82	39
54	49	61	56	57	32	38	48	64	51	45	46	62
63	59	63	32	47	40	37	49	57				

(a) Find the five-number summary.

(b) Draw a box-and-whisker plot that represents the data set.

(c) Interpret the results in the context of the data.

(d) On the basis of this sample, at what age would you expect to be an executive? Explain your reasoning.

(e) Which age groups, if any, can be considered unusual? Explain your reasoning.

Midquartile   *Another measure of position is called the* **midquartile.** *You can find the midquartile of a data set by using the following formula.*

$$Midquartile = \frac{Q_1 + Q_3}{2}$$

In Exercises 52–55, find the midquartile of the given data set.

52. 5 7 1 2 3 10 8 7 5 3

53. 23 36 47 33 34 40 39 24 32 22 38 41

54. 12.3 9.7 8.0 15.4 16.1 11.8 12.7 13.4
12.2 8.1 7.9 10.3 11.2

55. 21.4 20.8 19.7 15.2 31.9 18.7 15.6 16.7
19.8 13.4 22.9 28.7 19.8 17.2 30.1

56. Song Lengths   **Side-by-side box-and-whisker plots** can be used to compare two or more different data sets. Each box-and-whisker plot is drawn on the same number line to compare the data sets more easily. The lengths (in seconds) of songs played at two different concerts are shown.

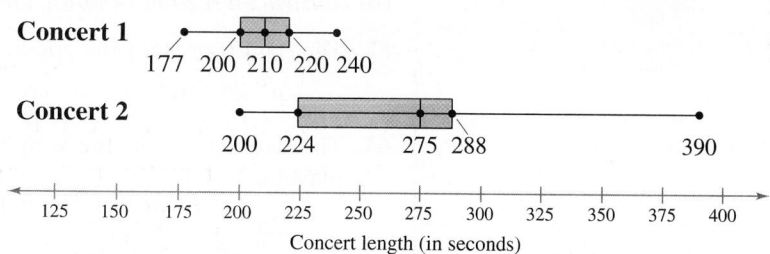

Concert 1
177 200 210 220 240

Concert 2
200 224 275 288 390

125 150 175 200 225 250 275 300 325 350 375 400
Concert length (in seconds)

(a) Describe the shape of each distribution. Which concert has less variation in song lengths?

(b) Which distribution is more likely to have outliers? Explain your reasoning.

(c) Which concert do you think has a standard deviation of 16.3? Explain your reasoning.

(d) Can you determine which concert lasted longer? Explain.

57. Credit Card Purchases

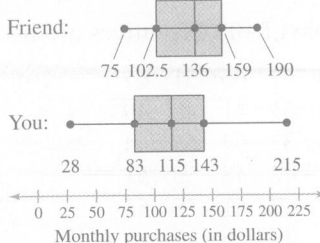

Friend:
75 102.5 136 159 190

You:
28 83 115 143 215

0 25 50 75 100 125 150 175 200 225
Monthly purchases (in dollars)

The shape of your bill is symmetric, and the shape of your friend's bill is uniform.

58. 4th percentile

59. 40th percentile

60. (a) 24, 2

(b)
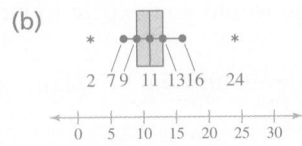

2 7 9 11 13 16   24

0   5   10   15   20   25   30

61. (a) 62, 95

(b)
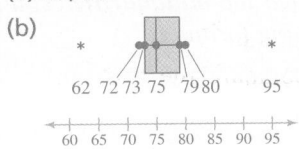

62 72 73 75 79 80   95

60   65   70   75   80   85   90   95

62. See Selected Answers, page A115.

63. (a) See Odd Answers, page A61.

(b) Weights of Professional Football Players

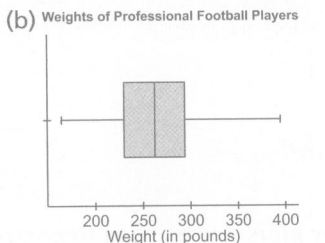

200   250   300   350   400
Weight (in pounds)

(c) Weights of Professional Football Players

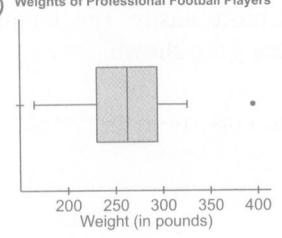

200   250   300   350   400
Weight (in pounds)

57. Credit Card Purchases The monthly credit card purchases (rounded to the nearest dollar) over the last two years for you and a friend are listed.

You: 60 95 102 110 130 130 162 200 215 120 124 28
           58 40 102 105 141 160 130 210 145   90   46 76

Friend: 100 125 132 90 85   75 140 160 180 190 160 105
           145 150 151 82 78 115 170 158 140 130 165 125

Use a calculator or a computer to draw a side-by-side box-and-whisker plot that represents the data sets. Then describe the shapes of the distributions.

Finding Percentiles *You can find the percentile that corresponds to a specific data value x by using the following formula, then rounding the result to the nearest whole number.*

$$\text{Percentile of } x = \frac{\text{number of data values less than } x}{\text{total number of data values}} \cdot 100$$

In Exercises 58 and 59, use the information from Example 7 and the fact that there have been 73 Oscars for Best Supporting Actor and 73 Oscars for Best Supporting Actress awarded.

58. Only three winners were younger than Heath Ledger when they won the Oscar for Best Supporting Actor. Find the percentile that corresponds to Heath Ledger's age.

59. Forty-three winners were older than Penelope Cruz when they won the Oscar for Best Supporting Actress. Find the percentile that corresponds to Penelope Cruz's age.

Modified Boxplot *A **modified boxplot** is a boxplot that uses symbols to identify outliers. The horizontal line of a modified boxplot extends as far as the minimum data value that is not an outlier and the maximum data value that is not an outlier. In Exercises 60 and 61, (a) identify any outliers (using the 1.5 × IQR rule), and (b) draw a modified boxplot that represents the data set. Use asterisks (*) to identify outliers.*

60. 16 9 11 12 8 10 12 13 11 10 24 9 2 15 7

61. 75 78 80 75 62 72 74 75 80 95 76 72

SC *In Exercises 62 and 63, use StatCrunch to (a) find the five-number summary, (b) construct a regular boxplot, and (c) construct a modified boxplot for the data.*

62. The data represent the speeds (in miles per hour) of several vehicles.

68 88 70 72 70 69 72 62 65 70 75 52 65

63. The data represent the weights (in pounds) of several professional football players.

225 250 305 285 275 265 290 310 290 250 210 225
308 325 260 165 195 245 235 298 395 255 268 190

USES AND ABUSES

Uses

Descriptive statistics help you see trends or patterns in a set of raw data. A good description of a data set consists of (1) a measure of the center of the data, (2) a measure of the variability (or spread) of the data, and (3) the shape (or distribution) of the data. When you read reports, news items, or advertisements prepared by other people, you are seldom given the raw data used for a study. Instead you see graphs, measures of central tendency, and measures of variability. To be a discerning reader, you need to understand the terms and techniques of descriptive statistics.

Procter & Gamble's Stock Price

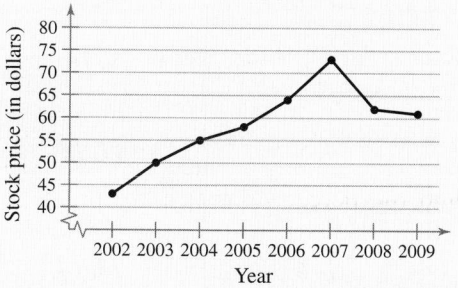

Abuses

Knowing how statistics are calculated can help you analyze questionable statistics. For instance, suppose you are interviewing for a sales position and the company reports that the average yearly commission earned by the five people in its sales force is $60,000. This is a misleading statement if it is based on four commissions of $25,000 and one of $200,000. The median would more accurately describe the yearly commission, but the company used the mean because it is a greater amount.

Statistical graphs can also be misleading. Compare the two time series charts at the left, which show the year-end stock prices for the Procter & Gamble Corporation. The data are the same for each chart. The first graph, however, has a cropped vertical axis, which makes it appear that the stock price increased greatly from 2002 to 2007, then decreased greatly from 2007 to 2009. In the second graph, the scale on the vertical axis begins at zero. This graph correctly shows that the stock price changed modestly during this time period. *(Source: Procter & Gamble Corporation)*

Procter & Gamble's Stock Price

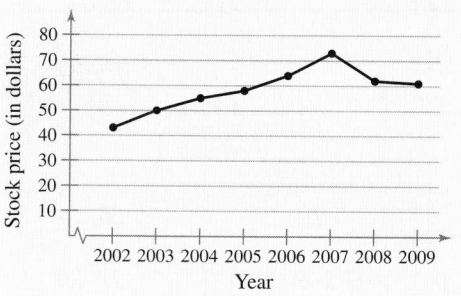

Ethics

Mark Twain helped popularize the saying, "There are three kinds of lies: lies, damned lies, and statistics." In short, even the most accurate statistics can be used to support studies or statements that are incorrect. Unscrupulous people can use misleading statistics to "prove" their point. Being informed about how statistics are calculated and questioning the data are ways to avoid being misled.

■ EXERCISES

1. Use the Internet or some other resource to find an example of a graph that might lead to incorrect conclusions.

2. You are publishing an article that discusses how eating oatmeal can help lower cholesterol. Because eating oatmeal might help people with high cholesterol, you include a graph that exaggerates the effects of eating oatmeal on lowering cholesterol. Do you think it is ethical to publish this graph? Explain.

2 CHAPTER SUMMARY

What did you **learn?**	EXAMPLE(S)	REVIEW EXERCISES
Section 2.1		
■ How to construct a frequency distribution including limits, midpoints, relative frequencies, cumulative frequencies, and boundaries	*1, 2*	*1*
■ How to construct frequency histograms, frequency polygons, relative frequency histograms, and ogives	*3–7*	*2–6*
Section 2.2		
■ How to graph quantitative data sets using stem-and-leaf plots and dot plots	*1–3*	*7, 8*
■ How to graph and interpret paired data sets using scatter plots and time series charts	*6, 7*	*9, 10*
■ How to graph qualitative data sets using pie charts and Pareto charts	*4, 5*	*11, 12*
Section 2.3		
■ How to find the mean, median, and mode of a population and a sample	*1–6*	*13, 14*
■ How to find a weighted mean of a data set and the mean of a frequency distribution	*7, 8*	*15–18*
■ How to describe the shape of a distribution as symmetric, uniform, or skewed and how to compare the mean and median for each		*19–24*
Section 2.4		
■ How to find the range of a data set	*1*	*25, 26*
■ How to find the variance and standard deviation of a population and a sample	*2–5*	*27–30*
■ How to use the Empirical Rule and Chebychev's Theorem to interpret standard deviation	*6–8*	*31–34*
■ How to approximate the sample standard deviation for grouped data	*9, 10*	*35, 36*
Section 2.5		
■ How to find the quartiles and interquartile range of a data set	*1–3*	*37, 38, 41*
■ How to draw a box-and-whisker plot	*4*	*39, 40, 42*
■ How to interpret other fractiles such as percentiles	*5*	*43, 44*
■ How to find and interpret the standard score (z-score)	*6, 7*	*45–48*

2 REVIEW EXERCISES

1. See Odd Answers, page A61.

2. See Selected Answers, page A116.

3.

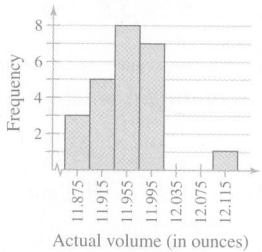

Liquid Volume 12-oz Cans

4. See Selected Answers, page A116.

5.

Class	Midpoint	Frequency, f
79–93	86	9
94–108	101	12
109–123	116	5
124–138	131	3
139–153	146	2
154–168	161	1
		$\sum f = 32$

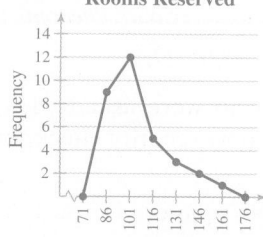

Rooms Reserved

6. See Selected Answers, page A116.

7.
```
1 | 0 0                Key: 1|0 = 10
2 | 0 0 2 5 5
3 | 0 3 4 5 5 8
4 | 1 2 4 4 7 8
5 | 2 3 3 7 9
6 | 1 1 5
7 | 1 5
8 | 9
```

8. See Selected Answers, page A116.

9.

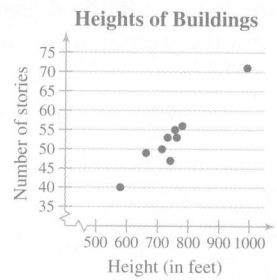

Heights of Buildings

The number of stories appears to increase with height.

■ SECTION 2.1

In Exercises 1 and 2, use the following data set. The data set represents the number of students per faculty member for 20 public colleges. (Source: Kiplinger)

13 15 15   8 16 20 28 19 18 15
21 23 30 17 10 16 15 16 20 15

1. Make a frequency distribution of the data set using five classes. Include the class limits, midpoints, boundaries, frequencies, relative frequencies, and cumulative frequencies.

2. Make a relative frequency histogram using the frequency distribution in Exercise 1. Then determine which class has the greatest relative frequency and which has the least relative frequency.

In Exercises 3 and 4, use the following data set. The data represent the actual liquid volumes (in ounces) in 24 twelve-ounce cans.

11.95 11.91 11.86 11.94 12.00 11.93 12.00 11.94
12.10 11.95 11.99 11.94 11.89 12.01 11.99 11.94
11.92 11.98 11.88 11.94 11.98 11.92 11.95 11.93

3. Make a frequency histogram of the data set using seven classes.

4. Make a relative frequency histogram of the data set using seven classes.

In Exercises 5 and 6, use the following data set. The data represent the number of rooms reserved during one night's business at a sample of hotels.

153 104 118 166   89 104 100   79
 93   96 116   94 140   84   81   96
108 111   87 126 101 111 122 108
126   93 108   87 103   95 129   93

5. Make a frequency distribution of the data set with six classes and draw a frequency polygon.

6. Make an ogive of the data set using six classes.

■ SECTION 2.2

In Exercises 7 and 8, use the following data set. The data represent the air quality indices for 30 U.S. cities. (Source: AIRNow)

25 35 20 75 10 10 61 89 44 22
34 33 38 30 47 53 44 57 71 20
42 52 48 41 35 59 53 61 65 25

7. Make a stem-and-leaf plot of the data set. Use one line per stem.

8. Make a dot plot of the data set.

9. The following are the heights (in feet) and the number of stories of nine notable buildings in Houston. Use the data to construct a scatter plot. What type of pattern is shown in the scatter plot? *(Source: Emporis Corporation)*

Height (in feet)	992	780	762	756	741	732	714	662	579
Number of stories	71	56	53	55	47	53	50	49	40

10.

U.S. Unemployment Rate

11.

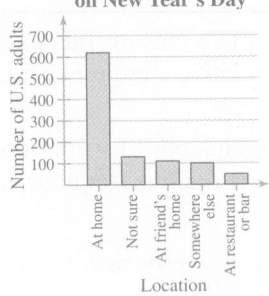

Location at Midnight on New Year's Day

12.

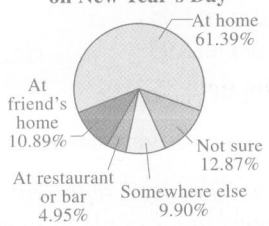

Location at Midnight on New Year's Day
At home 61.39%
At friend's home 10.89%
Not sure 12.87%
At restaurant or bar 4.95%
Somewhere else 9.90%

13. $\bar{x}$ = 29.15; median = 29.5; mode = 29.5

14. $\bar{x}$ = not possible; median = not possible; mode = "Approved"; The mean and median cannot be found because the data are at the nominal level of measurement.

15. 17.8

16. 2.1

17. 82.1

18. 88.8

19. Skewed

20. Skewed

10. The U.S. unemployment rate over a 12-year period is given. Use the data to construct a time series chart. *(Source: U.S. Bureau of Labor Statistics)*

Year	1998	1999	2000	2001	2002	2003
Unemployment rate	4.5	4.2	4.0	4.7	5.8	6.0

Year	2004	2005	2006	2007	2008	2009
Unemployment rate	5.5	5.1	4.6	4.6	5.8	9.3

In Exercises 11 and 12, use the following data set. The data set represents the results of a survey that asked U.S. adults where they would be at midnight when the new year arrived. *(Adapted from Rasmussen Reports)*

Response	At home	At friend's home	At restaurant or bar	Somewhere else	Not sure
Number	620	110	50	100	130

11. Make a Pareto chart of the data set.

12. Make a pie chart of the data set.

■ SECTION 2.3

In Exercises 13 and 14, find the mean, median, and mode of the data, if possible. If any of these measures cannot be found or a measure does not represent the center of the data, explain why.

13. Vertical Jumps The vertical jumps (in inches) of a sample of 10 college basketball players at the 2009 NBA Draft Combine *(Source: Sports Phenoms, Inc.)*

26.0　29.5　27.0　30.5　29.5　25.0　31.5　33.0　32.0　27.5

14. Airport Scanners The responses of 542 adults who were asked whether they approved the use of full-body scanners at airport security checkpoints *(Adapted from USA Today/Gallup Poll)*

Approved: 423　　　Did not approve: 108　　　No opinion: 11

15. Estimate the mean of the frequency distribution you made in Exercise 1.

16. The following frequency distribution shows the number of magazine subscriptions per household for a sample of 60 households. Find the mean number of subscriptions per household.

Number of magazines	0	1	2	3	4	5	6
Frequency	13	9	19	8	5	2	4

17. Six test scores are given. The first 5 test scores are 15% of the final grade, and the last test score is 25% of the final grade. Find the weighted mean of the test scores.

78　72　86　91　87　80

18. Four test scores are given. The first 3 test scores are 20% of the final grade, and the last test score is 40% of the final grade. Find the weighted mean of the test scores.

96　85　91　86

19. Describe the shape of the distribution in the histogram you made in Exercise 3. Is the distribution symmetric, uniform, or skewed?

20. Describe the shape of the distribution in the histogram you made in Exercise 4. Is the distribution symmetric, uniform, or skewed?

21. Skewed left

22. Skewed right

23. Median; When a distribution is skewed left, the mean is to the left of the median.

24. Mean; When a distribution is skewed right, the mean is to the right of the median.

25. $2.80

26. $3.84

27. $\mu \approx 6.9$, $\sigma \approx 4.6$

28. $\mu = 68.0$, $\sigma \approx 10.7$

29. $\bar{x} = 2453.4$, $s \approx 306.1$

30. $\bar{x} \approx 52{,}082.4$, $s \approx 3234.3$

31. Between $41.50 and $56.50

32. 68%

In Exercises 21 and 22, determine whether the approximate shape of the distribution in the histogram is symmetric, uniform, skewed left, skewed right, or none of these. Justify your answer.

21.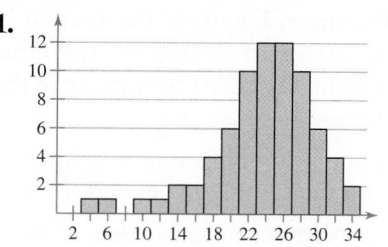

22.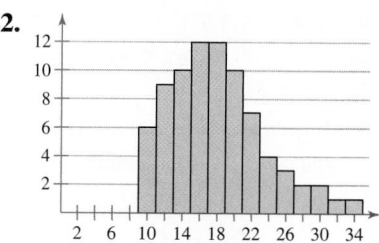

23. For the histogram in Exercise 21, which is greater, the mean or the median? Explain your reasoning.

24. For the histogram in Exercise 22, which is greater, the mean or the median? Explain your reasoning.

■ SECTION 2.4

25. The data set represents the mean prices of movie tickets (in U.S. dollars) for a sample of 12 U.S. cities. Find the range of the data set.

7.82   7.38   6.42   6.76   6.34   7.44   6.15   5.46   7.92   6.58   8.26   7.17

26. The data set represents the mean prices of movie tickets (in U.S. dollars) for a sample of 12 Japanese cities. Find the range of the data set.

19.73   16.48   19.10   18.56   17.68   17.19
16.63   15.99   16.66   19.59   15.89   16.49

27. The mileages (in thousands of miles) for a rental car company's fleet are listed. Find the population mean and the population standard deviation of the data.

4   2   9   12   15   3   6   8   1   4   14   12   3   3

28. The ages of the Supreme Court justices as of January 27, 2010 are listed. Find the population mean and the population standard deviation of the data. *(Source: Supreme Court of the United States)*

55   89   73   73   61   76   71   59   55

29. Dormitory room prices (in dollars) for one school year for a sample of four-year universities are listed. Find the sample mean and the sample standard deviation of the data.

2445   2940   2399   1960   2421   2940   2657   2153
2430   2278   1947   2383   2710   2761   2377

30. Sample salaries (in dollars) of high school teachers are listed. Find the sample mean and the sample standard deviation of the data.

49,632   54,619   58,298   48,250   51,842   50,875   53,219   49,924

31. The mean rate for satellite television for a sample of households was $49.00 per month, with a standard deviation of $2.50 per month. Between what two values do 99.7% of the data lie? (Assume the data set has a bell-shaped distribution.)

32. The mean rate for satellite television for a sample of households was $49.50 per month, with a standard deviation of $2.75 per month. Estimate the percent of satellite television rates between $46.75 and $52.25. (Assume the data set has a bell-shaped distribution.)

33. 30 customers

34. 15 flights

35. $\bar{x} \approx 2.5$, $s \approx 1.2$

36. $\bar{x} \approx 2.4$, $s \approx 1.7$

37. Min = 42, Q_1 = 47.5, Q_2 = 53, Q_3 = 54, Max = 60

38. 6.5 highway miles per gallon

39.

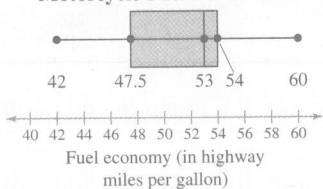

Motorcycle Fuel Economies

42   47.5   53 54   60

40 42 44 46 48 50 52 54 56 58 60
Fuel economy (in highway miles per gallon)

40. 17

41. 4.5

42.

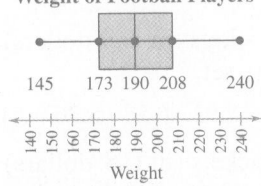

Weight of Football Players

145   173 190 208   240

140 150 160 170 180 190 200 210 220 230 240
Weight

The distribution of the weights is symmetric.

43. 35% scored higher than 75.

44. 86th percentile

45. Not unusual

46. Unusual

47. Unusual

48. Not unusual

---

33. The mean sale per customer for 40 customers at a gas station is $36.00, with a standard deviation of $8.00. Using Chebychev's Theorem, determine at least how many of the customers spent between $20.00 and $52.00.

34. The mean length of the first 20 space shuttle flights was about 7 days, and the standard deviation was about 2 days. Using Chebychev's Theorem, determine at least how many of the flights lasted between 3 days and 11 days. *(Source: NASA)*

35. From a random sample of households, the number of televisions are listed. Find the sample mean and the sample standard deviation of the data.

Number of televisions	0	1	2	3	4	5
Number of households	1	8	13	10	5	3

36. From a random sample of airplanes, the number of defects found in their fuselages are listed. Find the sample mean and the sample standard deviation of the data.

Number of defects	0	1	2	3	4	5	6
Number of airplanes	4	5	2	9	1	3	1

■ SECTION 2.5

In Exercises 37–40, use the following data set. The data represent the fuel economies (in highway miles per gallon) of several Harley-Davidson motorcycles. (Source: Total Motorcycle)

53 57 60 57 54 53 54 53 54 42 48
53 47 47 50 48 42 42 54 54 60

37. Find the five-number summary of the data set.

38. Find the interquartile range.

39. Make a box-and-whisker plot of the data.

40. About how many motorcycles fall on or below the third quartile?

41. Find the interquartile range of the data from Exercise 13.

42. The weights (in pounds) of the defensive players on a high school football team are given. Draw a box-and-whisker plot of the data and describe the shape of the distribution.

173 145 205 192 197 227 156 240 172 185
208 185 190 167 212 228 190 184 195

43. A student's test grade of 75 represents the 65th percentile of the grades. What percent of students scored higher than 75?

44. As of January 2010, there were 755 "oldies" radio stations in the United States. If one station finds that 104 stations have a larger daily audience than it has, what percentile does this station come closest to in the daily audience rankings? *(Source: Radio-locator.com)*

In Exercises 45–48, use the following information. The towing capacities (in pounds) of 25 four-wheel drive pickup trucks have a bell-shaped distribution, with a mean of 11,830 pounds and a standard deviation of 2370 pounds. Use z-scores to determine if the towing capacities of the following randomly selected four-wheel drive pickup trucks are unusual.

45. 16,500 pounds

46. 5500 pounds

47. 18,000 pounds

48. 11,300 pounds

2 CHAPTER QUIZ

1. See Odd Answers, page A62.

2. 125.2, 13.0

3. (a)   **U.S. Sporting Goods**

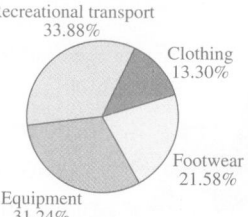

Recreational transport
33.88%

Clothing
13.30%

Footwear
21.58%

Equipment
31.24%

(b)   **U.S. Sporting Goods**

4. (a) $\bar{x} \approx 751.6$; median = 784.5; mode = none

The mean best describes a typical salary because there are no outliers.

(b) Range = 575; $s^2 \approx 48{,}135.1$; $s \approx 219.4$

5. Between $125,000 and $185,000

6. (a) $z = 3.0$, unusual

(b) $z \approx -6.67$, very unusual

(c) $z \approx 1.33$

(d) $z = -2.2$, unusual

7. (a) Min = 59, Q_1 = 74, Q_2 = 83.5, Q_3 = 88, Max = 103

(b) 14

(c)   **Wins for Each Team**

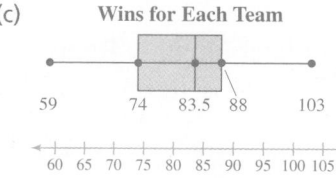

59   74 83.5 88   103

60 65 70 75 80 85 90 95 100 105
Number of wins

Take this quiz as you would take a quiz in class. After you are done, check your work against the answers given in the back of the book.

1. The data set represents the number of minutes a sample of 25 people exercise each week.

108	139	120	123	120	132	123	131	131
157	150	124	111	101	135	119	116	117
127	128	139	119	118	114	127		

(a) Make a frequency distribution of the data set using five classes. Include class limits, midpoints, boundaries, frequencies, relative frequencies, and cumulative frequencies.

(b) Display the data using a frequency histogram and a frequency polygon on the same axes.

(c) Display the data using a relative frequency histogram.

(d) Describe the distribution's shape as symmetric, uniform, or skewed.

(e) Display the data using a stem-and-leaf plot. Use one line per stem.

(f) Display the data using a box-and-whisker plot.

(g) Display the data using an ogive.

2. Use frequency distribution formulas to approximate the sample mean and the sample standard deviation of the data set in Exercise 1.

3. U.S. sporting goods sales (in billions of dollars) can be classified in four areas: clothing (10.6), footwear (17.2), equipment (24.9), and recreational transport (27.0). Display the data using (a) a pie chart and (b) a Pareto chart. (*Source: National Sporting Goods Association*)

4. Weekly salaries (in dollars) for a sample of registered nurses are listed.

774   446   1019   795   908   667   444   960

(a) Find the mean, median, and mode of the salaries. Which best describes a typical salary?

(b) Find the range, variance, and standard deviation of the data set. Interpret the results in the context of the real-life setting.

5. The mean price of new homes from a sample of houses is $155,000 with a standard deviation of $15,000. The data set has a bell-shaped distribution. Between what two prices do 95% of the houses fall?

6. Refer to the sample statistics from Exercise 5 and use z-scores to determine which, if any, of the following house prices is unusual.

(a) $200,000   (b) $55,000   (c) $175,000   (d) $122,000

7. The number of regular season wins for each Major League Baseball team in 2009 are listed. (*Source: Major League Baseball*)

| 103 | 95 | 84 | 75 | 64 | 87 | 86 | 79 | 65 | 65 | 97 | 87 | 85 | 75 | 93 |
| 87 | 86 | 70 | 59 | 91 | 83 | 80 | 78 | 74 | 62 | 95 | 92 | 88 | 75 | 70 |

(a) Find the five-number summary of the data set.

(b) Find the interquartile range.

(c) Display the data using a box-and-whisker plot.

PUTTING IT ALL TOGETHER

Real Statistics — Real Decisions

You are a member of your local apartment association. The association represents rental housing owners and managers who operate residential rental property throughout the greater metropolitan area. Recently, the association has received several complaints from tenants in a particular area of the city who feel that their monthly rental fees are much higher compared to other parts of the city.

You want to investigate the rental fees. You gather the data shown in the table at the right. Area A represents the area of the city where tenants are unhappy about their monthly rents. The data represent the monthly rents paid by a random sample of tenants in Area A and three other areas of similar size. Assume all the apartments represented are approximately the same size with the same amenities.

AMERICA'S LEADING ADVOCATE FOR QUALITY RENTAL HOUSING

 **EXERCISES**

1. *How Would You Do It?*

(a) How would you investigate the complaints from renters who are unhappy about their monthly rents?

(b) Which statistical measure do you think would best represent the data sets for the four areas of the city?

(c) Calculate the measure from part (b) for each of the four areas.

2. *Displaying the Data*

(a) What type of graph would you choose to display the data? Explain your reasoning.

(b) Construct the graph from part (a).

(c) Based on your data displays, does it appear that the monthly rents in Area A are higher than the rents in the other areas of the city? Explain.

3. *Measuring the Data*

(a) What other statistical measures in this chapter could you use to analyze the monthly rent data?

(b) Calculate the measures from part (a).

(c) Compare the measures from part (b) with the graph you constructed in Exercise 2. Do the measurements support your conclusion in Exercise 2? Explain.

4. *Discussing the Data*

(a) Do you think the complaints in Area A are legitimate? How do you think they should be addressed?

(b) What reasons might you give as to why the rents vary among different areas of the city?

The Monthly Rents (in dollars) Paid by 12 Randomly Selected Apartment Tenants in 4 Areas of Your City

Area A	Area B	Area C	Area D
1275	1124	1085	928
1110	954	827	1096
975	815	793	862
862	1078	1170	735
1040	843	919	798
997	745	943	812
1119	796	756	1232
908	816	765	1036
890	938	809	998
1055	1082	1020	914
860	750	710	1005
975	703	775	930

Highest Monthly Rents

AVERAGE PER CITY

New York, NY	$2922
San Francisco, CA	$1904
Boston, MA	$1658
San Jose, CA	$1612
Los Angeles, CA	$1452

(Source: Forbes)

TECHNOLOGY

MINITAB   EXCEL   TI-83/84 PLUS

Dairy Farmers of America is an association that provides help to dairy farmers. Part of this help is gathering and distributing statistics on milk production.

MONTHLY MILK PRODUCTION

The following data set was supplied by a dairy farmer. It lists the monthly milk productions (in pounds) for 50 Holstein dairy cows. *(Source: Matlink Dairy, Clymer, NY)*

2825	2072	2733	2069	2484
4285	2862	3353	1449	2029
1258	2982	2045	1677	1619
2597	3512	2444	1773	2284
1884	2359	2046	2364	2669
3109	2804	1658	2207	2159
2207	2882	1647	2051	2202
3223	2383	1732	2230	1147
2711	1874	1979	1319	2923
2281	1230	1665	1294	2936

www.dfamilk.com

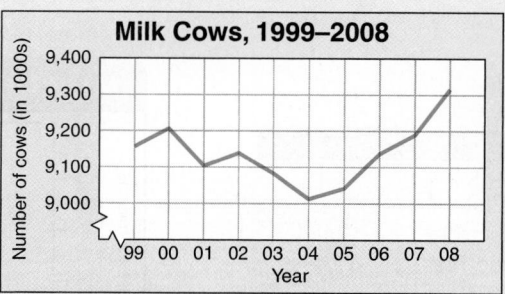

(Source: National Agricultural Statistics Service)

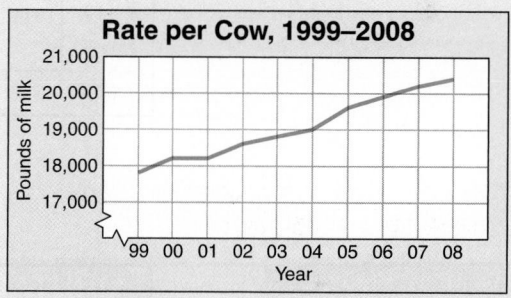

(Source: National Agricultural Statistics Service)

From 1999 to 2008, the number of dairy cows in the United States increased by only 1.7% while the yearly milk production per cow increased by almost 15%.

■ EXERCISES

In Exercises 1–4, use a computer or calculator. If possible, print your results.

1. Find the sample mean of the data.

2. Find the sample standard deviation of the data.

3. Make a frequency distribution for the data. Use a class width of 500.

4. Draw a histogram for the data. Does the distribution appear to be bell-shaped?

5. What percent of the distribution lies within one standard deviation of the mean? Within two standard deviations of the mean? How do these results agree with the Empirical Rule?

In Exercises 6–8, use the frequency distribution found in Exercise 3.

6. Use the frequency distribution to estimate the sample mean of the data. Compare your results with Exercise 1.

7. Use the frequency distribution to find the sample standard deviation for the data. Compare your results with Exercise 2.

8. **Writing** Use the results of Exercises 6 and 7 to write a general statement about the mean and standard deviation for grouped data. Do the formulas for grouped data give results that are as accurate as the individual entry formulas?

Extended solutions are given in the *Technology Supplement*.
Technical instruction is provided for MINITAB, Excel, and the TI-83/84 Plus.

USING TECHNOLOGY TO DETERMINE DESCRIPTIVE STATISTICS

Here are some MINITAB and TI-83/84 Plus printouts for three examples in this chapter.

(See Example 7, page 59.)

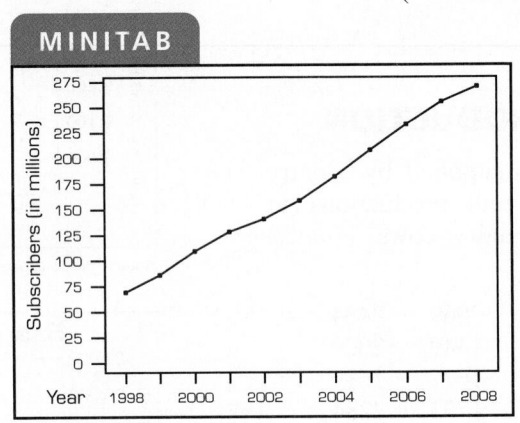

(See Example 4, page 83.)

MINITAB

Descriptive Statistics: Salaries

Variable	N	Mean	SE Mean	StDev	Minimum
Salaries	10	41.500	0.992	3.136	37.000

Variable	Q1	Median	Q3	Maximum
Salaries	38.750	41.000	44.250	47.000

(See Example 4, page 103.)

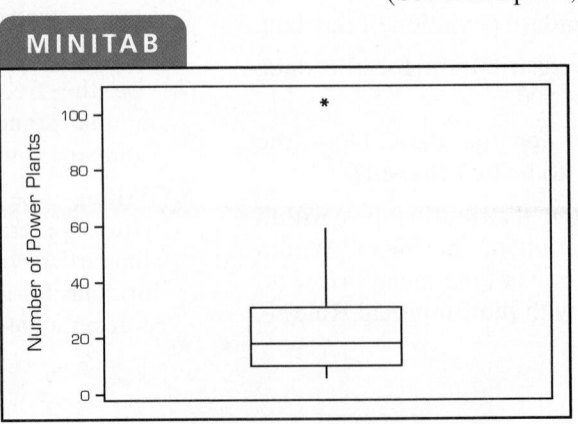

(See Example 7, page 59.)

(See Example 4, page 83.)

(See Example 4, page 103.)

TI-83/84 PLUS

STAT PLOTS
1: Plot1...Off
     L1 L2 □
2: Plot2...Off
    L1 L2 □
3: Plot3...Off
    L1 L2 □

TI-83/84 PLUS

EDIT **CALC** TESTS
1: 1-Var Stats
2: 2-Var Stats
3: Med-Med
4: LinReg(ax+b)
5: QuadReg
6: CubicReg
7↓ QuartReg

TI-83/84 PLUS

STAT PLOTS
1: Plot1...Off
    L1 L2 □
2: Plot2...Off
    L1 L2 □
3: Plot3...Off
    L1 L2 □
4↓ PlotsOff

TI-83/84 PLUS

Plot1 Plot2 Plot3
On Off
Type:
Xlist: L1
Ylist: L2
Mark: ■ + .

TI-83/84 PLUS

1-Var Stats L1

TI-83/84 PLUS

Plot1 Plot2 Plot3
On Off
Type:
Xlist: L1
Freq: 1

TI-83/84 PLUS

ZOOM MEMORY
4↑ ZDecimal
5: ZSquare
6: ZStandard
7: ZTrig
8: ZInteger
9: ZoomStat
0: ZoomFit

TI-83/84 PLUS

1-Var Stats
$\bar{x}$= 41.5
Σx= 415
Σx^2= 17311
Sx= 3.13581462
σx= 2.974894956
↓n= 10

TI-83/84 PLUS

ZOOM MEMORY
4↑ ZDecimal
5: ZSquare
6: ZStandard
7: ZTrig
8: ZInteger
9: ZoomStat
0: ZoomFit

TI-83/84 PLUS

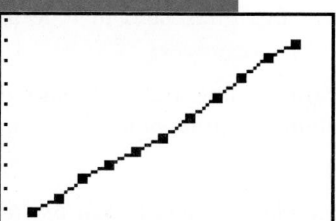

TI-83/84 PLUS

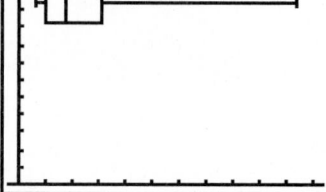

CUMULATIVE REVIEW
Chapters 1 and 2

In Exercises 1 and 2, identify the sampling technique used and discuss potential sources of bias (if any). Explain.

Section 1.3

1. For quality assurance, every fortieth toothbrush is taken from each of four assembly lines and tested to make sure the bristles stay in the toothbrush.

2. Using random digit dialing, researchers asked 1200 U.S. adults their thoughts on health care reform.

Section 2.2

3. In 2008, a worldwide study of all airlines found that baggage delays were caused by transfer baggage mishandling (49%), failure to load at originating airport (16%), arrival station mishandling (8%), space-weight restriction (6%), loading/offloading error (5%), tagging error (3%), and ticketing error/bag switch/security/other (13%). Use a Pareto chart to organize the data. *(Source: Société International de Télécommunications Aéronautiques)*

In Exercises 4 and 5, determine whether the numerical value is a parameter or a statistic. Explain your reasoning.

Section 1.1

4. In 2009, the average salary of a Major League Baseball player was $2,996,106. *(Source: Major League Baseball)*

5. In a recent survey of 1000 voters, 19% said that First Lady of the United States Michelle Obama will be very involved in policy decisions. *(Source: Rasmussen Reports)*

Sections 2.4 and 2.5

6. The mean annual salary for a sample of electrical engineers is $83,500, with a standard deviation of $1500. The data set has a bell-shaped distribution.

   (a) Use the Empirical Rule to estimate the number of electrical engineers whose annual salaries are between $80,500 and $86,500.

   (b) If 40 additional electrical engineers were sampled, about how many of these electrical engineers would you expect to have annual salaries between $80,500 and $86,500?

   (c) The salaries of three randomly selected electrical engineers are $90,500, $79,750, and $82,600. Find the z-score that corresponds to each salary. According to the z-scores, would the salaries of any of these engineers be considered unusual?

In Exercises 7 and 8, identify the population and the sample.

Section 1.1

7. A survey of career counselors at 195 colleges and universities found that 90% of the students working with their offices were interested in federal jobs or internships. *(Source: Partnership for Public Service Survey)*

8. A study of 232,606 people was conducted to find a link between taking antioxidant vitamins and living a longer life. *(Source: Journal of the American Medical Association)*

In Exercises 9 and 10, decide which method of data collection you would use to collect data for the study. Explain.

Section 1.3

9. A study of the years of service of the 100 members of the Senate

10. A study of the effects of removing recess from schools

Section 1.2 *In Exercises 11 and 12, determine whether the data are qualitative or quantitative and identify the data set's level of measurement.*

11. The number of games started by pitchers with at least one start for the New York Yankees in 2009 are listed. *(Source: Major League Baseball)*

9 34 1 33 32 31 7 9 6

12. The five top-earning states in 2008 by median income are listed. *(Source: U.S. Census Bureau)*

1. Maryland 2. New Jersey 3. Connecticut 4. Alaska 5. Hawaii

Section 2.5 **13.** The number of tornadoes by state in a recent year is listed. (a) Find the data set's five-number summary, (b) draw a box-and-whisker plot that represents the data set, and (c) describe the shape of the distribution. *(Source: National Climatic Data Center)*

81	1	8	69	30	34	0	0	56	54
2	6	21	14	46	136	17	23	2	0
1	5	71	105	39	10	40	1	0	7
4	0	23	53	4	27	1	11	0	14
19	23	105	4	0	24	4	0	63	6

Section 2.3 **14.** Five test scores are given. The first four test scores are 15% of the final grade, and the last test score is 40% of the final grade. Find the weighted mean of the test scores.

85 92 84 89 91

Sections 2.3 and 2.4 **15.** Tail lengths (in feet) for a sample of American alligators are listed.

6.5 3.4 4.2 7.1 5.4 6.8 7.5 3.9 4.6

(a) Find the mean, median, and mode of the tail lengths. Which best describes a typical American alligator tail length? Explain your reasoning.

(b) Find the range, variance, and standard deviation of the data set. Interpret the results in the context of the real-life setting.

Section 1.1 **16.** A study shows that the number of deaths due to heart disease for women has decreased every year for the past five years.

(a) Make an inference based on the results of the study.

(b) What is wrong with this type of reasoning?

Sections 2.1 and 2.3 *In Exercises 17–19, use the following data set. The data represent the points scored by each player on the Montreal Canadiens in a recent NHL season. (Source: National Hockey League)*

5	64	50	1	41	0	39	23	32	28
26	23	33	23	22	1	17	18	12	11
11	9	65	3	2	41	21	1	0	39

17. Make a frequency distribution using eight classes. Include the class limits, midpoints, boundaries, frequencies, relative frequencies, and cumulative frequencies.

18. Describe the shape of the distribution.

19. Make a relative frequency histogram using the frequency distribution in Exercise 17. Then determine which class has the greatest relative frequency and which has the least relative frequency.

3 PROBABILITY

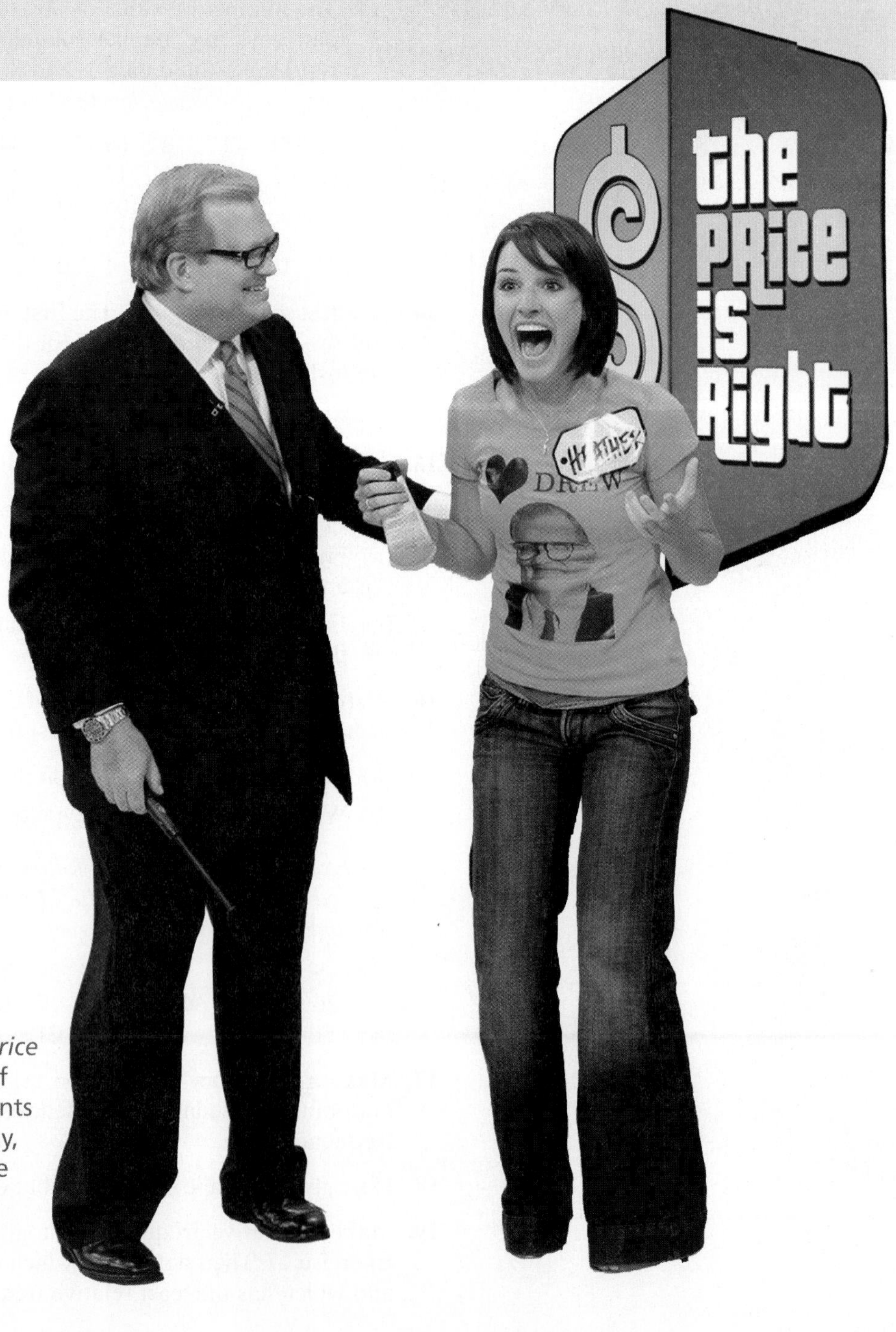

The television game show *The Price Is Right* presents a wide range of pricing games in which contestants compete for prizes using strategy, probability, and their knowledge of prices. One popular game is *Spelling Bee*.

In Chapters 1 and 2, you learned how to collect and describe data. Once the data are collected and described, you can use the results to write summaries, form conclusions, and make decisions. For instance, in *Spelling Bee*, contestants have a chance to win a car by choosing lettered cards that spell CAR or by choosing a single card that displays the entire word CAR. By collecting and analyzing data, you can determine the chances of winning the car.

To play *Spelling Bee*, contestants choose from 30 cards. Eleven cards display the letter C, eleven cards display A, six cards display R, and two cards display CAR. Depending on how well contestants play the game, they can choose two, three, four, or five cards.

Before the chosen cards are displayed, contestants are offered $1000 for each card. If contestants choose the money, the game is over. If contestants choose to try to win the car, the host displays one card. After a card is displayed, contestants are offered $1000 for each remaining card. If they do not accept the money, the host continues displaying cards. Play continues until contestants take the money, spell the word CAR, display the word CAR, or display all cards and do not spell CAR.

WHERE YOU'RE GOING ▶▶

In Chapter 3, you will learn how to determine the probability of an event. For instance, the following table shows the four ways that contestants on *Spelling Bee* can win a car and the corresponding probabilities.

You can see from the table that choosing more cards gives you a better chance of winning. These probabilities can be found using *combinations*, which will be discussed in Section 3.4.

Event	Probability
Winning by selecting two cards	$\frac{57}{435} \approx 0.131$
Winning by selecting three cards	$\frac{151}{406} \approx 0.372$
Winning by selecting four cards	$\frac{1067}{1827} \approx 0.584$
Winning by selecting five cards	$\frac{52,363}{71,253} \approx 0.735$

3.1 Basic Concepts of Probability and Counting

WHAT YOU SHOULD LEARN

▸ How to identify the sample space of a probability experiment and how to identify simple events

▸ How to use the Fundamental Counting Principle to find the number of ways two or more events can occur

▸ How to distinguish among classical probability, empirical probability, and subjective probability

▸ How to find the probability of the complement of an event

▸ How to use a tree diagram and the Fundamental Counting Principle to find more probabilities

▸ PROBABILITY EXPERIMENTS

When weather forecasters say that there is a 90% chance of rain or a physician says there is a 35% chance for a successful surgery, they are stating the likelihood, or *probability*, that a specific event will occur. Decisions such as "should you go golfing" or "should you proceed with surgery" are often based on these probabilities. In the previous chapter, you learned about the role of the descriptive branch of statistics. Because probability is the foundation of inferential statistics, it is necessary to learn about probability before proceeding to the second branch—inferential statistics.

DEFINITION

A **probability experiment** is an action, or trial, through which specific results (counts, measurements, or responses) are obtained. The result of a single trial in a probability experiment is an **outcome**. The set of all possible outcomes of a probability experiment is the **sample space**. An **event** is a subset of the sample space. It may consist of one or more outcomes.

STUDY TIP

Here is a simple example of the use of the terms *probability experiment*, *sample space*, *event*, and *outcome*.

Probability Experiment:
  Roll a six-sided die.

Sample Space:
  {1, 2, 3, 4, 5, 6}

Event:
  Roll an even number, {2, 4, 6}.

Outcome:
  Roll a 2, {2}.

Note to Instructor

Bring dice, a deck of cards, and some coins to class to illustrate examples. Differently colored gaming chips work well also.

EXAMPLE 1

▸ **Identifying the Sample Space of a Probability Experiment**

A probability experiment consists of tossing a coin and then rolling a six-sided die. Determine the number of outcomes and identify the sample space.

▸ **Solution**

There are two possible outcomes when tossing a coin: a head (H) or a tail (T). For each of these, there are six possible outcomes when rolling a die: 1, 2, 3, 4, 5, or 6. A **tree diagram** gives a visual display of the outcomes of a probability experiment by using branches that originate from a starting point. It can be used to find the number of possible outcomes in a sample space as well as individual outcomes.

Tree Diagram for Coin and Die Experiment

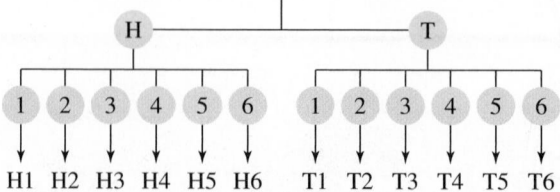

From the tree diagram, you can see that the sample space has 12 outcomes.

{H1, H2, H3, H4, H5, H6, T1, T2, T3, T4, T5, T6}

Note to Instructor

Ask students to use a tree diagram to develop the sample space that results from rolling two six-sided dice. Emphasize that the outcome {3, 2} is different from the outcome {2, 3} for a statistician, but not for a player. Point out that the event {3, 2} is a simple event. The event "roll a sum of 5" is not simple because it consists of the four outcomes {1, 4}, {2, 3}, {3, 2}, and {4, 1}.

▶ **Try It Yourself 1**

For each probability experiment, determine the number of outcomes and identify the sample space.

1. A probability experiment consists of recording a response to the survey statement at the left *and* the gender of the respondent.
2. A probability experiment consists of recording a response to the survey statement at the left *and* the geographic location (Northeast, South, Midwest, West) of the respondent.

a. Start a *tree diagram* by forming a branch for each possible response to the survey.
b. At the end of each survey response branch, draw a *new branch* for each possible outcome.
c. Find the *number of outcomes* in the sample space.
d. List the *sample space*.   *Answer: Page A34*

In the rest of this chapter, you will learn how to calculate the probability or likelihood of an event. Events are often represented by uppercase letters, such as A, B, and C. An event that consists of a single outcome is called a **simple event**. In Example 1, the event "tossing heads and rolling a 3" is a simple event and can be represented as $A = \{H3\}$. In contrast, the event "tossing heads and rolling an even number" is not simple because it consists of three possible outcomes $B = \{H2, H4, H6\}$.

EXAMPLE 2

▶ **Identifying Simple Events**

Determine the number of outcomes in each event. Then decide whether each event is simple or not. Explain your reasoning.

1. For quality control, you randomly select a machine part from a batch that has been manufactured that day. Event A is selecting a specific defective machine part.

2. You roll a six-sided die. Event B is rolling at least a 4.

▶ **Solution**

1. Event A has only one outcome: choosing the specific defective machine part. So, the event is a simple event.

2. Event B has three outcomes: rolling a 4, a 5, or a 6. Because the event has more than one outcome, it is not simple.

▶ **Try It Yourself 2**

You ask for a student's age at his or her last birthday. Determine the number of outcomes in each event. Then decide whether each event is simple or not. Explain your reasoning.

1. Event C: The student's age is between 18 and 23, inclusive.
2. Event D: The student's age is 20.

a. Determine the number of *outcomes* in the event.
b. State whether the event is *simple* or not. Explain your reasoning.
                                              Answer: Page A34

▶ **THE FUNDAMENTAL COUNTING PRINCIPLE**

In some cases, an event can occur in so many different ways that it is not practical to write out all the outcomes. When this occurs, you can rely on the Fundamental Counting Principle. The Fundamental Counting Principle can be used to find the number of ways two or more events can occur in sequence.

THE FUNDAMENTAL COUNTING PRINCIPLE

If one event can occur in m ways and a second event can occur in n ways, the number of ways the two events can occur in sequence is $m \cdot n$. This rule can be extended to any number of events occurring in sequence.

In words, the number of ways that events can occur in sequence is found by multiplying the number of ways one event can occur by the number of ways the other event(s) can occur.

EXAMPLE 3

▶ **Using the Fundamental Counting Principle**

You are purchasing a new car. The possible manufacturers, car sizes, and colors are listed.

Manufacturer:	Ford, GM, Honda
Car size:	compact, midsize
Color:	white (W), red (R), black (B), green (G)

How many different ways can you select one manufacturer, one car size, and one color? Use a tree diagram to check your result.

▶ **Solution**

There are three choices of manufacturers, two choices of car sizes, and four choices of colors. Using the Fundamental Counting Principle, you can conclude that the number of ways to select one manufacturer, one car size, and one color is

$3 \cdot 2 \cdot 4 = 24$ ways.

Using a tree diagram, you can see why there are 24 options.

Tree Diagram for Car Selections

▶ **Try It Yourself 3**

Your choices now include a Toyota and a tan car. How many different ways can you select one manufacturer, one car size, and one color? Use a tree diagram to check your result.

a. Find the *number of ways* each event can occur.
b. Use the *Fundamental Counting Principle*.
c. Use a *tree diagram* to check your result.

Answer: Page A35

EXAMPLE 4

▶ **Using the Fundamental Counting Principle**

The access code for a car's security system consists of four digits. Each digit can be any number from 0 through 9.

Access Code

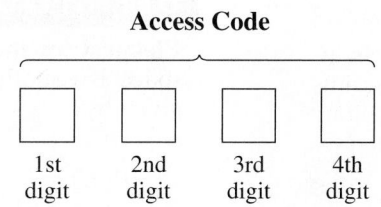

| 1st | 2nd | 3rd | 4th |
| digit | digit | digit | digit |

How many access codes are possible if

1. each digit can be used only once and not repeated?

2. each digit can be repeated?

3. each digit can be repeated but the first digit cannot be 0 or 1?

▶ **Solution**

1. Because each digit can be used only once, there are 10 choices for the first digit, 9 choices left for the second digit, 8 choices left for the third digit, and 7 choices left for the fourth digit. Using the Fundamental Counting Principle, you can conclude that there are

$$10 \cdot 9 \cdot 8 \cdot 7 = 5040$$

possible access codes.

2. Because each digit can be repeated, there are 10 choices for each of the four digits. So, there are

$$10 \cdot 10 \cdot 10 \cdot 10 = 10^4$$
$$= 10{,}000$$

possible access codes.

3. Because the first digit cannot be 0 or 1, there are 8 choices for the first digit. Then there are 10 choices for each of the other three digits. So, there are

$$8 \cdot 10 \cdot 10 \cdot 10 = 8000$$

possible access codes.

▶ **Try It Yourself 4**

How many license plates can you make if a license plate consists of

1. six (out of 26) alphabetical letters each of which can be repeated?
2. six (out of 26) alphabetical letters each of which cannot be repeated?
3. six (out of 26) alphabetical letters each of which can be repeated but the first letter cannot be A, B, C, or D?

a. *Identify* each event and the *number of ways* each event can occur.
b. Use the *Fundamental Counting Principle*.   *Answer: Page A35*

▶ **TYPES OF PROBABILITY**

The method you will use to calculate a probability depends on the type of probability. There are three types of probability: *classical probability*, *empirical probability*, and *subjective probability*. The probability that event E will occur is written as $P(E)$ and is read "the probability of event E."

DEFINITION

Classical (or **theoretical**) **probability** is used when each outcome in a sample space is equally likely to occur. The classical probability for an event E is given by

$$P(E) = \frac{\text{Number of outcomes in event } E}{\text{Total number of outcomes in sample space}}.$$

EXAMPLE 5

▶ **Finding Classical Probabilities**

You roll a six-sided die. Find the probability of each event.

1. Event A: rolling a 3

2. Event B: rolling a 7

3. Event C: rolling a number less than 5

▶ **Solution**

When a six-sided die is rolled, the sample space consists of six outcomes: $\{1, 2, 3, 4, 5, 6\}$.

1. There is one outcome in event $A = \{3\}$. So,

$$P(\text{rolling a 3}) = \frac{1}{6} \approx 0.167.$$

2. Because 7 is not in the sample space, there are no outcomes in event B. So,

$$P(\text{rolling a 7}) = \frac{0}{6} = 0.$$

3. There are four outcomes in event $C = \{1, 2, 3, 4\}$. So,

$$P(\text{rolling a number less than 5}) = \frac{4}{6} = \frac{2}{3} \approx 0.667.$$

▶ **Try It Yourself 5**

You select a card from a standard deck. Find the probability of each event.

1. Event D: Selecting a nine of clubs
2. Event E: Selecting a heart
3. Event F: Selecting a diamond, heart, club, or spade

a. Identify the *total number of outcomes* in the sample space.
b. Find the *number of outcomes* in the event.
c. Use the *classical probability formula*.

Answer: Page A35

Standard Deck of Playing Cards

Hearts	Diamonds	Spades	Clubs
A ♥	A ♦	A ♠	A ♣
K ♥	K ♦	K ♠	K ♣
Q ♥	Q ♦	Q ♠	Q ♣
J ♥	J ♦	J ♠	J ♣
10 ♥	10 ♦	10 ♠	10 ♣
9 ♥	9 ♦	9 ♠	9 ♣
8 ♥	8 ♦	8 ♠	8 ♣
7 ♥	7 ♦	7 ♠	7 ♣
6 ♥	6 ♦	6 ♠	6 ♣
5 ♥	5 ♦	5 ♠	5 ♣
4 ♥	4 ♦	4 ♠	4 ♣
3 ♥	3 ♦	3 ♠	3 ♣
2 ♥	2 ♦	2 ♠	2 ♣

When an experiment is repeated many times, regular patterns are formed. These patterns make it possible to find empirical probability. Empirical probability can be used even if each outcome of an event is not equally likely to occur.

DEFINITION

Empirical (or **statistical**) **probability** is based on observations obtained from probability experiments. The empirical probability of an event E is the relative frequency of event E.

$$P(E) = \frac{\text{Frequency of event } E}{\text{Total frequency}}$$

$$= \frac{f}{n}$$

EXAMPLE 6

▸ **Finding Empirical Probabilities**

A company is conducting a telephone survey of randomly selected individuals to get their overall impressions of the past decade (2000s). So far, 1504 people have been surveyed. The frequency distribution shows the results. What is the probability that the next person surveyed has a positive overall impression of the 2000s? *(Adapted from Princeton Survey Research Associates International)*

Response	Number of times, f
Positive	406
Negative	752
Neither	316
Don't know	30
	$\Sigma f = 1504$

▸ **Solution**

The event is a response of "positive." The frequency of this event is 406. Because the total of the frequencies is 1504, the empirical probability of the next person having a positive overall impression of the 2000s is

$$P(\text{positive}) = \frac{406}{1504}$$

$$\approx 0.270.$$

▸ **Try It Yourself 6**

An insurance company determines that in every 100 claims, 4 are fraudulent. What is the probability that the next claim the company processes will be fraudulent?

a. *Identify* the event. Find the *frequency* of the event.
b. Find the *total frequency* for the experiment.
c. Find the *empirical probability* of the event.   *Answer: Page A35*

PICTURING THE WORLD

It seems as if no matter how strange an event is, somebody wants to know the probability that it will occur. The following table lists the probabilities that some intriguing events will happen. (Adapted from Life: The Odds)

Event	Probability
Being audited by the IRS	0.6%
Writing a *New York Times* best seller	0.0045
Winning an Academy Award	0.000087
Having your identity stolen	0.5%
Spotting a UFO	0.0000003

Which of these events is most likely to occur? Least likely?

The event most likely to occur is being audited by the IRS. The event least likely to occur is spotting a UFO.

To explore this topic further, see Activity 3.1 on page 144.

As you increase the number of times a probability experiment is repeated, the empirical probability (relative frequency) of an event approaches the theoretical probability of the event. This is known as the **law of large numbers.**

LAW OF LARGE NUMBERS

As an experiment is repeated over and over, the empirical probability of an event approaches the theoretical (actual) probability of the event.

Probability of Tossing a Head

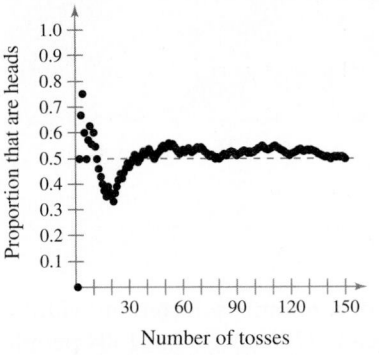

Number of tosses

As an example of this law, suppose you want to determine the probability of tossing a head with a fair coin. If you toss the coin 10 times and get only 3 heads, you obtain an empirical probability of $\frac{3}{10}$. Because you tossed the coin only a few times, your empirical probability is not representative of the theoretical probability, which is $\frac{1}{2}$. If, however, you toss the coin several thousand times, then the law of large numbers tells you that the empirical probability will be very close to the theoretical or actual probability.

The scatter plot at the left shows the results of simulating a coin toss 150 times. Notice that, as the number of tosses increases, the probability of tossing a head gets closer and closer to the theoretical probability of 0.5.

EXAMPLE 7

▶ **Using Frequency Distributions to Find Probabilities**

You survey a sample of 1000 employees at a company and record the age of each. The results are shown in the frequency distribution at the left. If you randomly select another employee, what is the probability that the employee will be between 25 and 34 years old?

Employee ages	Frequency, f
15 to 24	54
25 to 34	366
35 to 44	233
45 to 54	180
55 to 64	125
65 and over	42
	$\Sigma f = 1000$

▶ **Solution**

The event is selecting an employee who is between 25 and 34 years old. The frequency of this event is 366. Because the total of the frequencies is 1000, the empirical probability of selecting an employee between the ages of 25 and 34 years old is

$$P(\text{age 25 to 34}) = \frac{366}{1000}$$
$$= 0.366.$$

▶ **Try It Yourself 7**

Find the probability that an employee chosen at random will be between 15 and 24 years old.

a. Find the *frequency* of the event.
b. Find the *total of the frequencies*.
c. Find the *empirical probability* of the event.   *Answer: Page A35*

The third type of probability is **subjective probability.** Subjective probabilities result from intuition, educated guesses, and estimates. For instance, given a patient's health and extent of injuries, a doctor may feel that the patient has a 90% chance of a full recovery. Or a business analyst may predict that the chance of the employees of a certain company going on strike is 0.25.

EXAMPLE 8

▶ **Classifying Types of Probability**

Classify each statement as an example of classical probability, empirical probability, or subjective probability. Explain your reasoning.

1. The probability that you will get the flu this year is 0.1.

2. The probability that a voter chosen at random will be younger than 35 years old is 0.3.

3. The probability of winning a 1000-ticket raffle with one ticket is $\frac{1}{1000}$.

▶ **Solution**

1. This probability is most likely based on an educated guess. It is an example of subjective probability.

2. This statement is most likely based on a survey of a sample of voters, so it is an example of empirical probability.

3. Because you know the number of outcomes and each is equally likely, this is an example of classical probability.

▶ **Try It Yourself 8**

Based on previous counts, the probability of a salmon successfully passing through a dam on the Columbia River is 0.85. Is this statement an example of classical probability, empirical probability, or subjective probability? *(Source: Army Corps of Engineers)*

a. Identify the *event*.

b. Decide whether the probability is *determined* by knowing all possible outcomes, whether the probability is *estimated* from the results of an experiment, or whether the probability is an *educated guess*.

c. Make a *conclusion*.   *Answer: Page A35*

A probability cannot be negative or greater than 1. So, the probability of an event E is between 0 and 1, inclusive, as stated in the following rule.

RANGE OF PROBABILITIES RULE

The probability of an event E is between 0 and 1, inclusive. That is,

$$0 \leq P(E) \leq 1.$$

If the probability of an event is 1, the event is certain to occur. If the probability of an event is 0, the event is impossible. A probability of 0.5 indicates that an event has an even chance of occurring.

The following graph shows the possible range of probabilities and their meanings.

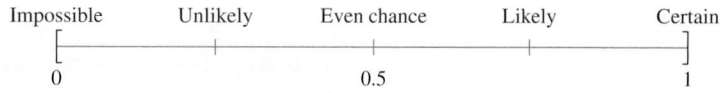

An event that occurs with a probability of 0.05 or less is typically considered unusual. Unusual events are highly unlikely to occur. Later in this course you will identify unusual events when studying inferential statistics.

▶ **COMPLEMENTARY EVENTS**

The sum of the probabilities of all outcomes in a sample space is 1 or 100%. An important result of this fact is that if you know the probability of an event E, you can find the probability of the *complement of event E*.

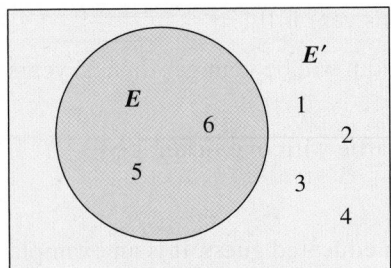

The area of the rectangle represents the total probability of the sample space (1 = 100%). The area of the circle represents the probability of event E, and the area outside the circle represents the probability of the complement of event E.

DEFINITION

The **complement of event E** is the set of all outcomes in a sample space that are not included in event E. The complement of event E is denoted by E' and is read as "E prime."

For instance, if you roll a die and let E be the event "the number is at least 5," then the complement of E is the event "the number is less than 5." In symbols, $E = \{5, 6\}$ and $E' = \{1, 2, 3, 4\}$.

Using the definition of the complement of an event and the fact that the sum of the probabilities of all outcomes is 1, you can determine the following formulas.

$$P(E) + P(E') = 1 \qquad P(E) = 1 - P(E') \qquad P(E') = 1 - P(E)$$

The Venn diagram at the left illustrates the relationship between the sample space, an event E, and its complement E'.

EXAMPLE 9

▶ **Finding the Probability of the Complement of an Event**

Use the frequency distribution in Example 7 to find the probability of randomly choosing an employee who is not between 25 and 34 years old.

▶ **Solution**

From Example 7, you know that

$$P(\text{age 25 to 34}) = \frac{366}{1000}$$
$$= 0.366.$$

So, the probability that an employee is not between 25 and 34 years old is

$$P(\text{age is not 25 to 34}) = 1 - \frac{366}{1000}$$
$$= \frac{634}{1000}$$
$$= 0.634.$$

▶ **Try It Yourself 9**

Use the frequency distribution in Example 7 to find the probability of randomly choosing an employee who is not between 45 and 54 years old.

a. Find the *probability* of randomly choosing an employee who is between 45 and 54 years old.
b. *Subtract* the resulting probability from 1.
c. *State the probability* as a fraction and as a decimal.   *Answer: Page A35*

▶ **PROBABILITY APPLICATIONS**

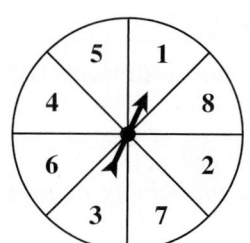

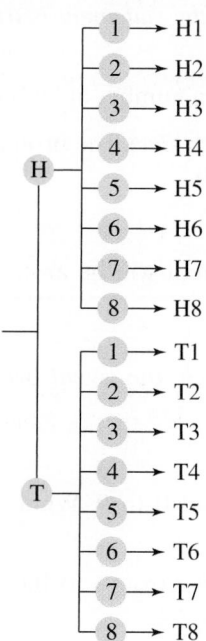

Tree Diagram for Coin and Spinner Experiment

H ─┬─ 1 ──▶ H1
   ├─ 2 ──▶ H2
   ├─ 3 ──▶ H3
   ├─ 4 ──▶ H4
   ├─ 5 ──▶ H5
   ├─ 6 ──▶ H6
   ├─ 7 ──▶ H7
   └─ 8 ──▶ H8

T ─┬─ 1 ──▶ T1
   ├─ 2 ──▶ T2
   ├─ 3 ──▶ T3
   ├─ 4 ──▶ T4
   ├─ 5 ──▶ T5
   ├─ 6 ──▶ T6
   ├─ 7 ──▶ T7
   └─ 8 ──▶ T8

EXAMPLE 10

▶ **Using a Tree Diagram**

A probability experiment consists of tossing a coin and spinning the spinner shown at the left. The spinner is equally likely to land on each number. Use a tree diagram to find the probability of each event.

1. Event A: tossing a tail and spinning an odd number

2. Event B: tossing a head or spinning a number greater than 3

▶ **Solution**   From the tree diagram at the left, you can see that there are 16 outcomes.

1. There are four outcomes in event $A = \{T1, T3, T5, T7\}$. So,

$$P(\text{tossing a tail and spinning an odd number}) = \frac{4}{16} = \frac{1}{4} = 0.25.$$

2. There are 13 outcomes in event $B = \{H1, H2, H3, H4, H5, H6, H7, H8, T4, T5, T6, T7, T8\}$. So,

$$P(\text{tossing a head or spinning a number greater than 3}) = \frac{13}{16} \approx 0.813.$$

▶ **Try It Yourself 10**

Find the probability of tossing a tail and spinning a number less than 6.

a. Find the *number of outcomes* in the event.
b. Find the *probability of the event*.   *Answer: Page A35*

EXAMPLE 11

▶ **Using the Fundamental Counting Principle**

Your college identification number consists of eight digits. Each digit can be 0 through 9 and each digit can be repeated. What is the probability of getting your college identification number when randomly generating eight digits?

▶ **Solution**   Because each digit can be repeated, there are 10 choices for each of the 8 digits. So, using the Fundamental Counting Principle, there are $10 \cdot 10 \cdot 10 \cdot 10 \cdot 10 \cdot 10 \cdot 10 \cdot 10 = 10^8 = 100,000,000$ possible   identification numbers. But only one of those numbers corresponds to your college identification number. So, the probability of randomly generating 8 digits and getting your college identification number is 1/100,000,000.

▶ **Try It Yourself 11**

Your college identification number consists of nine digits. The first two digits of each number will be the last two digits of the year you are scheduled to graduate. The other digits can be any number from 0 through 9, and each digit can be repeated. What is the probability of getting your college identification number when randomly generating the other seven digits?

a. Find the *total number* of possible identification numbers. Assume that you are scheduled to graduate in 2015.
b. Find the *probability* of randomly generating your identification number.
                    Answer: Page A35

3.1 EXERCISES

1. An outcome is the result of a single trial in a probability experiment, whereas an event is a set of one or more outcomes.

2. See Selected Answers, page A116.

3. The probability of an event cannot exceed 100%.

4. The Fundamental Counting Principle counts the number of ways that two or more events can occur in sequence.

5. The law of large numbers states that as an experiment is repeated over and over, the probabilities found in the experiment will approach the actual probabilities of the event. Example will vary.

6. $P(E) + P(E') = 1$
$P(E) = 1 - P(E')$
$P(E') = 1 - P(E)$

7. False. If you roll a six-sided die six times, the probability of rolling an even number at least once is approximately 0.984.

8. False. You flip a fair coin nine times and it lands tails up each time. The probability it will land heads up on the tenth flip is 0.5.

9. False. A probability of less than 0.05 indicates an unusual event.

10. True

11. b

12. d

13. c

14. a

■ BUILDING BASIC SKILLS AND VOCABULARY

1. What is the difference between an outcome and an event?

2. Determine which of the following numbers could not represent the probability of an event. Explain your reasoning.

   (a) 33.3%   (b) -1.5 (c) 0.0002   (d) 0   (e) $\frac{320}{1058}$ (f) $\frac{64}{25}$

3. Explain why the following statement is incorrect: *The probability of rain tomorrow is 150%.*

4. When you use the Fundamental Counting Principle, what are you counting?

5. Use your own words to describe the law of large numbers. Give an example.

6. List the three formulas that can be used to describe complementary events.

True or False?   *In Exercises 7–10, determine whether the statement is true or false. If it is false, rewrite it as a true statement.*

7. If you roll a six-sided die six times, you will roll an even number at least once.

8. You toss a fair coin nine times and it lands tails up each time. The probability it will land heads up on the tenth flip is greater than 0.5.

9. A probability of $\frac{1}{10}$ indicates an unusual event.

10. If an event is almost certain to happen, its complement will be an unusual event.

Matching Probabilities   *In Exercises 11–14, match the event with its probability.*

(a) 0.95   (b) 0.05   (c) 0.25   (d) 0

11. You toss a coin and randomly select a number from 0 to 9. What is the probability of getting tails and selecting a 3?

12. A random number generator is used to select a number from 1 to 100. What is the probability of selecting the number 153?

13. A game show contestant must randomly select a door. One door doubles her money while the other three doors leave her with no winnings. What is the probability she selects the door that doubles her money?

14. Five of the 100 digital video recorders (DVRs) in an inventory are known to be defective. What is the probability you randomly select an item that is not defective?

■ USING AND INTERPRETING CONCEPTS

Identifying a Sample Space   *In Exercises 15–20, identify the sample space of the probability experiment and determine the number of outcomes in the sample space. Draw a tree diagram if it is appropriate.*

15. Guessing the initial of a student's middle name

16. Guessing a student's letter grade (A, B, C, D, F) in a class

17. Drawing one card from a standard deck of cards

15. {A, B, C, D, E, F, G, H, I, J, K, L, M, N, O, P, Q, R, S, T, U, V, W, X, Y, Z}; 26

16. {A, B, C, D, F}; 5

17. See Odd Answers, page A64.

18. See Selected Answers, page A116.

19.

{(A, +), (A, −), (B, +), (B, −), (AB, +), (AB, −), (O, +), (O, −)}, where (A, +) represents positive Rh-factor with blood type A and (A, −) represents negative Rh-factor with blood type A; 8.

20. See Selected Answers, page A116.

21. 1; Simple event because it is an event that consists of a single outcome.

22. 499; Not a simple event because it is an event that consists of more than a single outcome.

23. 4; Not a simple event because it is an event that consists of more than a single outcome.

24. 1; Simple event because it is an event that consists of a single outcome.

25. 204

26. 200

27. 4500

28. 64

29. 0.083

30. 0.083

31. 0.667

32. 0.5

33. 0.417

34. 0.167

35. Empirical probability because company records were used to calculate the frequency of a washing machine breaking down.

36. Classical probability because each outcome in the sample space is equally likely to occur.

18. Tossing three coins

19. Determining a person's blood type (A, B, AB, O) and Rh-factor (positive, negative)

20. Rolling a pair of six-sided dice

Recognizing Simple Events *In Exercises 21–24, determine the number of outcomes in each event. Then decide whether the event is a simple event or not. Explain your reasoning.*

21. A computer is used to randomly select a number between 1 and 4000. Event *A* is selecting 253.

22. A computer is used to randomly select a number between 1 and 4000. Event *B* is selecting a number less than 500.

23. You randomly select one card from a standard deck. Event *A* is selecting an ace.

24. You randomly select one card from a standard deck. Event *B* is selecting a ten of diamonds.

25. Job Openings A software company is hiring for two positions: a software development engineer and a sales operations manager. How many ways can these positions be filled if there are 12 people applying for the engineering position and 17 people applying for the managerial position?

26. Menu A restaurant offers a $12 dinner special that has 5 choices for an appetizer, 10 choices for entrées, and 4 choices for dessert. How many different meals are available if you select an appetizer, an entrée, and a dessert?

27. Realty A realtor uses a lock box to store the keys for a house that is for sale. The access code for the lock box consists of four digits. The first digit cannot be zero and the last digit must be even. How many different codes are available?

28. True or False Quiz Assuming that no questions are left unanswered, in how many ways can a six-question true-false quiz be answered?

Classical Probabilities *In Exercises 29–34, a probability experiment consists of rolling a 12-sided die. Find the probability of each event.*

29. Event *A*: rolling a 2

30. Event *B*: rolling a 10

31. Event *C*: rolling a number greater than 4

32. Event *D*: rolling an even number

33. Event *E*: rolling a prime number

34. Event *F*: rolling a number divisible by 5

Classifying Types of Probability *In Exercises 35 and 36, classify the statement as an example of classical probability, empirical probability, or subjective probability. Explain your reasoning.*

35. According to company records, the probability that a washing machine will need repairs during a six-year period is 0.10.

36. The probability of choosing 6 numbers from 1 to 40 that match the 6 numbers drawn by a state lottery is 1/3,838,380 ≈ 0.00000026.

37. 0.159
38. 0.841
39. 0.000953
40. 0.999
41. 0.042; Yes
42. 0.125; No
43. 0.208; No
44. 0.042; Yes

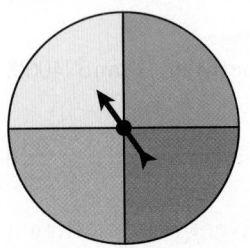

FIGURE FOR EXERCISES 41–44

45. (a) 1000   (b) 0.001   (c) 0.999
46. (a) 117,000
    (b) 0.00000855
    (c) 0.99999
47. {(SSS), (SSR), (SRS), (SRR), (RSS), (RSR), (RRS), (RRR)}
48. {(RRR)}
49. {(SSR), (SRS), (RSS)}
50. {(SSR), (SRS), (SRR), (RSS), (RSR), (RRS), (RRR)}
51. See Odd Answers, page A65.

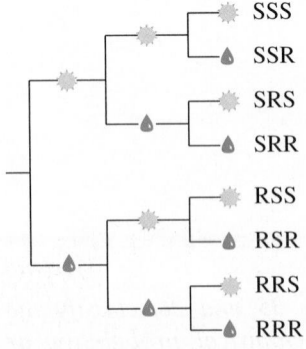

FIGURE FOR EXERCISES 47–50

Finding Probabilities *In Exercises 37–40, consider a company that selects employees for random drug tests. The company uses a computer to randomly select employee numbers that range from 1 to 6296.*

37. Find the probability of selecting a number less than 1000.

38. Find the probability of selecting a number greater than 1000.

39. Find the probability of selecting a number divisible by 1000.

40. Find the probability of selecting a number that is not divisible by 1000.

Probability Experiment *In Exercises 41–44, a probability experiment consists of rolling a six-sided die and spinning the spinner shown at the left. The spinner is equally likely to land on each color. Use a tree diagram to find the probability of each event. Then tell whether the event can be considered unusual.*

41. Event *A*: rolling a 5 and the spinner landing on blue

42. Event *B*: rolling an odd number and the spinner landing on green

43. Event *C*: rolling a number less than 6 and the spinner landing on yellow

44. Event *D*: not rolling a number less than 6 and the spinner landing on yellow

45. Security System The access code for a garage door consists of three digits. Each digit can be any number from 0 through 9, and each digit can be repeated.

(a) Find the number of possible access codes.

(b) What is the probability of randomly selecting the correct access code on the first try?

(c) What is the probability of not selecting the correct access code on the first try?

46. Security System An access code consists of a letter followed by four digits. Any letter can be used, the first digit cannot be 0, and the last digit must be even.

(a) Find the number of possible access codes.

(b) What is the probability of randomly selecting the correct access code on the first try?

(c) What is the probability of not selecting the correct access code on the first try?

Wet or Dry? *You are planning a three-day trip to Seattle, Washington in October. In Exercises 47–50, use the tree diagram shown at the left to answer each question.*

47. List the sample space.

48. List the outcome(s) of the event "It rains all three days."

49. List the outcome(s) of the event "It rains on exactly one day."

50. List the outcome(s) of the event "It rains on at least one day."

51. Sunny and Rainy Days You are planning a four-day trip to Seattle, Washington in October.

(a) Make a sunny day/rainy day tree diagram for your trip.

(b) List the sample space.

(c) List the outcome(s) of the event "It rains on exactly one day."

52.

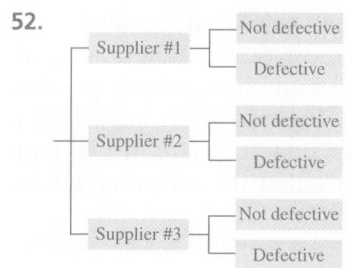

53. 0.399
54. 0.471
55. 0.040
56. 0.174
57. 0.936
58. 0.624
59. 0.033
60. 0.253
61. 0.275
62. 0.044

Ages of voters	Frequency (in millions)
18 to 20	5.8
21 to 24	9.3
25 to 34	22.7
35 to 44	25.4
45 to 64	54.9
65 and over	28.1

TABLE FOR EXERCISES 55–58

63. Yes; The event in Exercise 55 can be considered unusual because its probability is 0.05 or less.

64. Yes; The events in Exercises 59 and 62 can be considered unusual because their probabilities are 0.05 or less.

---

52. Machine Part Suppliers   Your company buys machine parts from three different suppliers. Make a tree diagram that shows the three suppliers and whether the parts they supply are defective.

Graphical Analysis   *In Exercises 53 and 54, use the diagram to answer the question.*

53. What is the probability that a registered voter in Virginia voted in the 2009 gubernatorial election? *(Source: Commonwealth of Virginia State Board of Elections)*

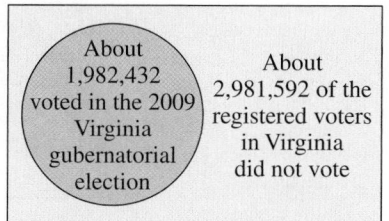

FIGURE FOR EXERCISE 53

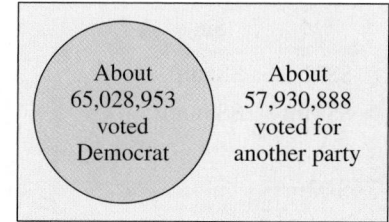

FIGURE FOR EXERCISE 54

54. What is the probability that a voter chosen at random did not vote for a Democratic representative in the 2008 election? *(Source: Federal Election Commission)*

Using a Frequency Distribution to Find Probabilities   *In Exercises 55–58, use the frequency distribution at the left, which shows the number of American voters (in millions) according to age, to find the probability that a voter chosen at random is in the given age range. (Source: U.S. Census Bureau)*

55. between 18 and 20 years old   **56.** between 35 and 44 years old

57. not between 21 and 24 years old   **58.** not between 45 and 64 years old

Using a Bar Graph to Find Probabilities   *In Exercises 59–62, use the following bar graph, which shows the highest level of education received by employees of a company.*

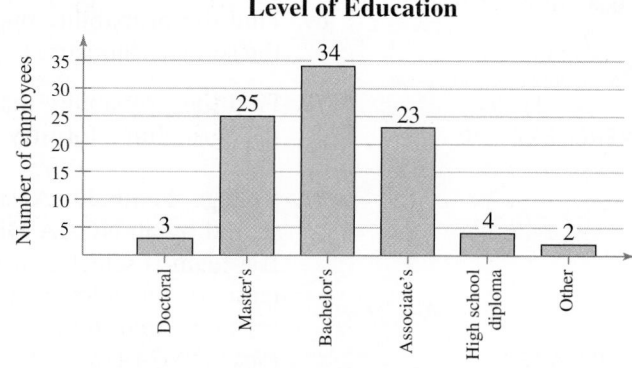

Find the probability that the highest level of education for an employee chosen at random is

59. a doctorate.   **60.** an associate's degree.

61. a master's degree.   **62.** a high school diploma.

63. Can any of the events in Exercises 55–58 be considered unusual? Explain.

64. Can any of the events in Exercises 59–62 be considered unusual? Explain.

65. (a) 0.5
    (b) 0.25
    (c) 0.25
66. 0.5

Parents
Ssmm and SsMm

	SM	Sm
Sm	SSMm	SSmm
Sm	SSMm	SSmm
sm	SsMm	Ssmm
sm	SsMm	Ssmm

	sM	sm
Sm	SsMm	Ssmm
Sm	SsMm	Ssmm
sm	ssMm	ssmm
sm	ssMm	ssmm

TABLE FOR EXERCISE 66

Workers (in thousands) by Industry for the U.S.

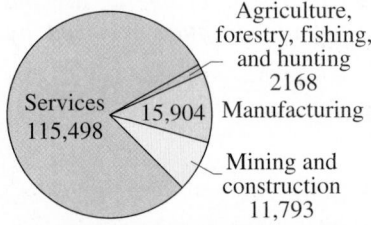

FIGURE FOR EXERCISES 67–70

67. 0.795
68. 0.109
69. 0.205
70. 0.985
71. (a) 0.225   (b) 0.133
    (c) 0.017; This event is unusual because its probability is 0.05 or less.

65. Genetics A *Punnett square* is a diagram that shows all possible gene combinations in a cross of parents whose genes are known. When two pink snapdragon flowers (RW) are crossed, there are four equally likely possible outcomes for the genetic makeup of the offspring: red (RR), pink (RW), pink (WR), and white (WW), as shown in the Punnett square. If two pink snapdragons are crossed, what is the probability that the offspring will be (a) pink, (b) red, and (c) white?

66. Genetics There are six basic types of coloring in registered collies: sable (SSmm), tricolor (ssmm), trifactored sable (Ssmm), blue merle (ssMm), sable merle (SSMm), and trifactored sable merle (SsMm). The Punnett square at the left shows the possible coloring of the offspring of a trifactored sable merle collie and a trifactored sable collie. What is the probability that the offspring will have the same coloring as one of its parents?

Using a Pie Chart to Find Probabilities *In Exercises 67–70, use the pie chart at the left, which shows the number of workers (in thousands) by industry for the United States.* (Source: U.S. Bureau of Labor Statistics)

67. Find the probability that a worker chosen at random was employed in the services industry.

68. Find the probability that a worker chosen at random was employed in the manufacturing industry.

69. Find the probability that a worker chosen at random was not employed in the services industry.

70. Find the probability that a worker chosen at random was not employed in the agriculture, forestry, fishing, and hunting industry.

71. College Football A stem-and-leaf plot for the number of touchdowns scored by all NCAA Division I Football Bowl Subdivision teams is shown. If a team is selected at random, find the probability the team scored (a) at least 51 touchdowns, (b) between 20 and 30 touchdowns, inclusive, and (c) more than 69 touchdowns. Are any of these events unusual? Explain. *(Source: NCAA)*

```
1 | 8 8 9                    Key: 1|8 = 18
2 | 1 1 3 4 4 5 5 6 6 7 7 8 8 9 9
3 | 0 1 1 2 3 3 3 3 3 4 4 4 4 4 5 5 5 5 5 7 7 7 7 8 8 9 9 9
4 | 0 0 0 0 1 2 2 2 2 3 3 4 4 4 4 4 4 4 4 5 5 5 5 5 5 6 6 6 6 6 7 7 7 8 8 8 8 8 9 9 9
5 | 0 0 0 0 1 1 2 2 2 2 4 5 5 5 6 6 7 9
6 | 0 0 1 2 2 3 4 5 6 8 9
7 | 6 7
```

72. (a) 0.25   (b) 0.5   (c) 0.5

73. The probability of randomly choosing a tea drinker who does not have a college degree

74. The probability of randomly choosing a smoker whose mother did not smoke

75. (a)

Sum	Probability
2	0.028
3	0.056
4	0.083
5	0.111
6	0.139
7	0.167
8	0.139
9	0.111
10	0.083
11	0.056
12	0.028

(b) Answers will vary.

(c) Answers will vary.

76. No. The odds of winning a prize are 1 : 6 (one winning cap and six losing caps). So, the statement should read, "one in seven game pieces wins a prize."

77. The first game; The probability of winning the second game is $\frac{1}{11} \approx 0.091$, which is less than $\frac{1}{10}$.

78. (a) 0.444
(b) 0.556

79. 13 : 39 = 1 : 3

80. 39 : 13 = 3 : 1

81. See Odd Answers, page A65.

---

72. Individual Stock Price   An individual stock is selected at random from the portfolio represented by the box-and-whisker plot shown. Find the probability that the stock price is (a) less than $21, (b) between $21 and $50, and (c) $30 or more.

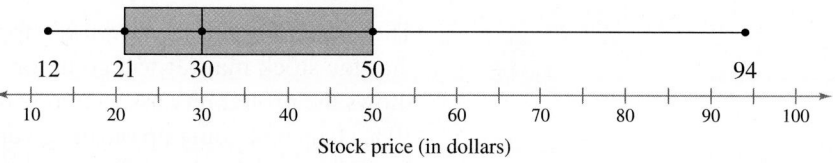

Stock price (in dollars)

Writing   *In Exercises 73 and 74, write a statement that represents the complement of the given probability.*

73. The probability of randomly choosing a tea drinker who has a college degree (Assume that you are choosing from the population of all tea drinkers.)

74. The probability of randomly choosing a smoker whose mother also smoked (Assume that you are choosing from the population of all smokers.)

■ EXTENDING CONCEPTS

75. Rolling a Pair of Dice   You roll a pair of six-sided dice and record the sum.

(a) List all of the possible sums and determine the probability of rolling each sum.

(b) Use a technology tool to simulate rolling a pair of dice and recording the sum 100 times. Make a tally of the 100 sums and use these results to list the probability of rolling each sum.

(c) Compare the probabilities in part (a) with the probabilities in part (b). Explain any similarities or differences.

Odds   *In Exercises 76–81, use the following information. The chances of winning are often written in terms of odds rather than probabilities. The **odds of winning** is the ratio of the number of successful outcomes to the number of unsuccessful outcomes. The **odds of losing** is the ratio of the number of unsuccessful outcomes to the number of successful outcomes. For example, if the number of successful outcomes is 2 and the number of unsuccessful outcomes is 3, the odds of winning are 2 : 3 (read "2 to 3") or $\frac{2}{3}$.*

76. A beverage company puts game pieces under the caps of its drinks and claims that one in six game pieces wins a prize. The official rules of the contest state that the odds of winning a prize are 1 : 6. Is the claim "one in six game pieces wins a prize" correct? Why or why not?

77. The probability of winning an instant prize game is $\frac{1}{10}$. The odds of winning a different instant prize game are 1 : 10. If you want the best chance of winning, which game should you play? Explain your reasoning.

78. The odds of an event occurring are 4 : 5. Find (a) the probability that the event will occur and (b) the probability that the event will not occur.

79. A card is picked at random from a standard deck of 52 playing cards. Find the odds that it is a spade.

80. A card is picked at random from a standard deck of 52 playing cards. Find the odds that it is not a spade.

81. The odds of winning an event A are $p : q$. Show that the probability of event A is given by $P(A) = \dfrac{p}{p + q}$.

ACTIVITY 3.1   Simulating the Stock Market

APPLET

The *simulating the stock market* applet allows you to investigate the probability that the stock market will go up on any given day. The plot at the top left corner shows the probability associated with each outcome. In this case, the market has a 50% chance of going up on any given day. When SIMULATE is clicked, outcomes for *n* days are simulated. The results of the simulations are shown in the frequency plot. If the animate option is checked, the display will show each outcome dropping into the frequency plot as the simulation runs. The individual outcomes are shown in the text field at the far right of the applet. The center plot shows in red the cumulative proportion of times that the market went up. The green line in the plot reflects the true probability of the market going up. As the experiment is conducted over and over, the cumulative proportion should converge to the true value.

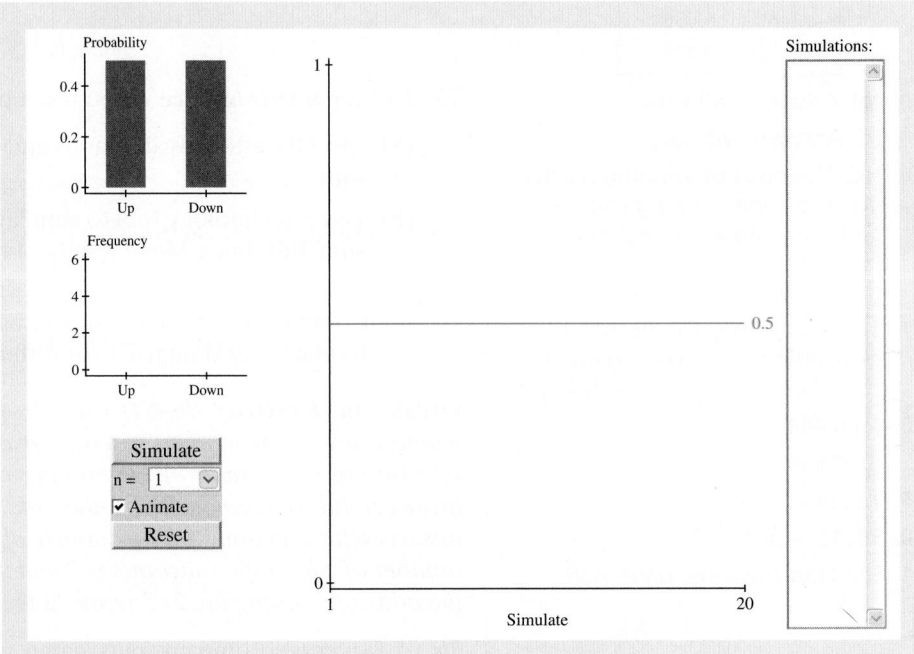

■ Explore

Step 1   Specify a value for *n*.
Step 2   Click SIMULATE four times.
Step 3   Click RESET.
Step 4   Specify another value for *n*.
Step 5   Click SIMULATE.

■ Draw Conclusions

APPLET

1. Run the simulation using *n* = 1 without clicking RESET. How many days did it take until there were three straight days on which the stock market went up? three straight days on which the stock market went down?

2. Run the applet to simulate the stock market activity over the last 35 business days. Find the empirical probability that the market goes up on day 36.

Conditional Probability and the Multiplication Rule

WHAT YOU SHOULD LEARN

▸ How to find the probability of an event given that another event has occurred

▸ How to distinguish between independent and dependent events

▸ How to use the Multiplication Rule to find the probability of two events occurring in sequence

▸ How to use the Multiplication Rule to find conditional probabilities

Conditional Probability ▸ Independent and Dependent Events ▸ The Multiplication Rule

▸ CONDITIONAL PROBABILITY

In this section, you will learn how to find the probability that two events occur in sequence. Before you can find this probability, however, you must know how to find *conditional probabilities*.

DEFINITION

A **conditional probability** is the probability of an event occurring, given that another event has already occurred. The conditional probability of event B occurring, given that event A has occurred, is denoted by $P(B|A)$ and is read as "probability of B, given A."

EXAMPLE 1

▸ **Finding Conditional Probabilities**

1. Two cards are selected in sequence from a standard deck. Find the probability that the second card is a queen, given that the first card is a king. (Assume that the king is not replaced.)

2. The table at the left shows the results of a study in which researchers examined a child's IQ and the presence of a specific gene in the child. Find the probability that a child has a high IQ, given that the child has the gene.

	Gene present	Gene not present	Total
High IQ	33	19	52
Normal IQ	39	11	50
Total	72	30	102

Note to Instructor

Point out to students that the table in Example 1 (part 2) is an example of a contingency table (or two-way table). They will learn more about contingency tables in Section 10.2.

Sample Space

	Gene present
High IQ	33
Normal IQ	39
Total	72

▸ **Solution**

1. Because the first card is a king and is not replaced, the remaining deck has 51 cards, 4 of which are queens. So,

$$P(B|A) = \frac{4}{51} \approx 0.078.$$

So, the probability that the second card is a queen, given that the first card is a king, is about 0.078.

2. There are 72 children who have the gene. So, the sample space consists of these 72 children, as shown at the left. Of these, 33 have a high IQ. So,

$$P(B|A) = \frac{33}{72} \approx 0.458.$$

So, the probability that a child has a high IQ, given that the child has the gene, is about 0.458.

▸ **Try It Yourself 1**

1. Find the probability that a child does not have the gene.
2. Find the probability that a child does not have the gene, given that the child has a normal IQ.

a. Find the *number of outcomes* in the event and in the sample space.
b. *Divide* the number of outcomes in the event by the number of outcomes in the sample space.

Answer: Page A35

PICTURING THE WORLD

Truman Collins, a probability and statistics enthusiast, wrote a program that finds the probability of landing on each square of a Monopoly board during a game. Collins explored various scenarios, including the effects of the Chance and Community Chest cards and the various ways of landing in or getting out of jail. Interestingly, Collins discovered that the length of each jail term affects the probabilities.

Monopoly square	Probability given short jail term	Probability given long jail term
Go	0.0310	0.0291
Chance	0.0087	0.0082
In Jail	0.0395	0.0946
Free Parking	0.0288	0.0283
Park Place	0.0219	0.0206
B&O RR	0.0307	0.0289
Water Works	0.0281	0.0265

Why do the probabilities depend on how long you stay in jail?

The probabilities depend on how long you stay in jail, because the more turns you spend in jail, the fewer chances you have to land on any other spot.

▶ INDEPENDENT AND DEPENDENT EVENTS

In some experiments, one event does not affect the probability of another. For instance, if you roll a die and toss a coin, the outcome of the roll of the die does not affect the probability of the coin landing on heads. These two events are *independent*. The question of the independence of two or more events is important to researchers in fields such as marketing, medicine, and psychology. You can use conditional probabilities to determine whether events are *independent*.

DEFINITION

Two events are **independent** if the occurrence of one of the events does not affect the probability of the occurrence of the other event. Two events A and B are independent if

$$P(B|A) = P(B) \quad \text{or if} \quad P(A|B) = P(A).$$

Events that are not independent are **dependent.**

To determine if A and B are independent, first calculate $P(B)$, the probability of event B. Then calculate $P(B|A)$, the probability of B, given A. If the values are equal, the events are independent. If $P(B) \neq P(B|A)$, then A and B are dependent events.

EXAMPLE 2

▶ **Classifying Events as Independent or Dependent**

Decide whether the events are independent or dependent.

1. Selecting a king from a standard deck (A), not replacing it, and then selecting a queen from the deck (B)

2. Tossing a coin and getting a head (A), and then rolling a six-sided die and obtaining a 6 (B)

3. Driving over 85 miles per hour (A), and then getting in a car accident (B)

▶ **Solution**

1. $P(B|A) = \frac{4}{51}$ and $P(B) = \frac{4}{52}$. The occurrence of A changes the probability of the occurrence of B, so the events are dependent.

2. $P(B|A) = \frac{1}{6}$ and $P(B) = \frac{1}{6}$. The occurrence of A does not change the probability of the occurrence of B, so the events are independent.

3. If you drive over 85 miles per hour, the chances of getting in a car accident are greatly increased, so these events are dependent.

▶ **Try It Yourself 2**

Decide whether the events are independent or dependent.

1. Smoking a pack of cigarettes per day (A) and developing emphysema, a chronic lung disease (B)

2. Exercising frequently (A) and having a 4.0 grade point average (B)

a. *Decide* whether the occurrence of the first event affects the probability of the second event.
b. State if the events are *independent* or *dependent*. *Answer: Page A35*

▸ **THE MULTIPLICATION RULE**

To find the probability of two events occurring in sequence, you can use the Multiplication Rule.

STUDY TIP

In words, to use the Multiplication Rule,

1. find the probability that the first event occurs,

2. find the probability that the second event occurs given that the first event has occurred, and

3. multiply these two probabilities.

THE MULTIPLICATION RULE FOR THE PROBABILITY OF *A* AND *B*

The probability that two events *A* and *B* will occur in sequence is

$$P(A \text{ and } B) = P(A) \cdot P(B|A).$$

If events *A* and *B* are independent, then the rule can be simplified to $P(A \text{ and } B) = P(A) \cdot P(B)$. This simplified rule can be extended to any number of independent events.

EXAMPLE 3

▸ **Using the Multiplication Rule to Find Probabilities**

1. Two cards are selected, without replacing the first card, from a standard deck. Find the probability of selecting a king and then selecting a queen.

2. A coin is tossed and a die is rolled. Find the probability of tossing a head and then rolling a 6.

▸ **Solution**

1. Because the first card is not replaced, the events are dependent.

$$P(K \text{ and } Q) = P(K) \cdot P(Q|K)$$

$$= \frac{4}{52} \cdot \frac{4}{51}$$

$$= \frac{16}{2652}$$

$$\approx 0.006$$

So, the probability of selecting a king and then a queen is about 0.006.

2. The events are independent.

$$P(H \text{ and } 6) = P(H) \cdot P(6)$$

$$= \frac{1}{2} \cdot \frac{1}{6}$$

$$= \frac{1}{12}$$

$$\approx 0.083$$

So, the probability of tossing a head and then rolling a 6 is about 0.083.

▸ **Try It Yourself 3**

1. The probability that a salmon swims successfully through a dam is 0.85. Find the probability that two salmon swim successfully through the dam.

2. Two cards are selected from a standard deck without replacement. Find the probability that they are both hearts.

a. Decide if the events are *independent* or *dependent*.

b. Use the *Multiplication Rule* to find the probability.   *Answer: Page A35*

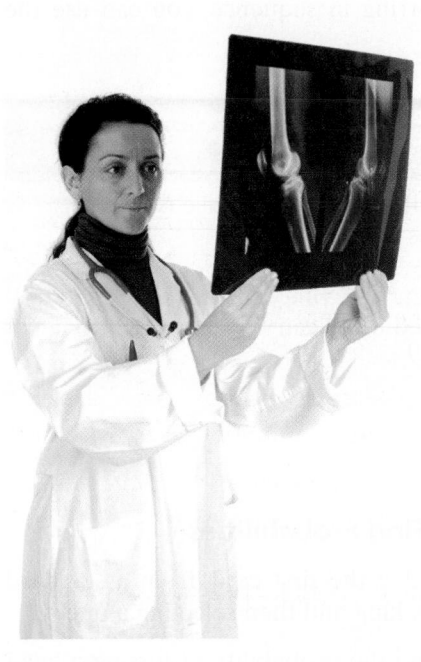

EXAMPLE 4

▸ **Using the Multiplication Rule to Find Probabilities**

The probability that a particular knee surgery is successful is 0.85.

1. Find the probability that three knee surgeries are successful.

2. Find the probability that none of the three knee surgeries are successful.

3. Find the probability that at least one of the three knee surgeries is successful.

▸ **Solution**

1. The probability that each knee surgery is successful is 0.85. The chance of success for one surgery is independent of the chances for the other surgeries.

$$P(\text{three surgeries are successful}) = (0.85)(0.85)(0.85)$$

$$\approx 0.614$$

So, the probability that all three surgeries are successful is about 0.614.

2. Because the probability of success for one surgery is 0.85, the probability of failure for one surgery is $1 - 0.85 = 0.15$.

$$P(\text{none of the three are successful}) = (0.15)(0.15)(0.15)$$

$$\approx 0.003$$

So, the probability that none of the surgeries are successful is about 0.003. Because 0.003 is less than 0.05, this can be considered an unusual event.

3. The phrase "at least one" means one or more. The complement to the event "at least one is successful" is the event "none are successful." Use the complement to find the probability.

$$P(\text{at least one is successful}) = 1 - P(\text{none are successful})$$

$$\approx 1 - 0.003$$

$$= 0.997.$$

So, the probability that at least one of the three surgeries is successful is about 0.997.

▸ **Try It Yourself 4**

The probability that a particular rotator cuff surgery is successful is 0.9. *(Source: The Orthopedic Center of St. Louis)*

1. Find the probability that three rotator cuff surgeries are successful.
2. Find the probability that none of the three rotator cuff surgeries are successful.
3. Find the probability that at least one of the three rotator cuff surgeries is successful.

a. Decide whether to find the probability of the event or its complement.
b. Use the *Multiplication Rule* to find the probability. If necessary, use the *complement*.
c. Determine if the event is *unusual*. Explain.   *Answer: Page A35*

In Example 4, you were asked to find a probability using the phrase "at least one." Notice that it was easier to find the probability of its complement, "none," and then subtract the probability of its complement from 1.

EXAMPLE 5

▶ **Using the Multiplication Rule to Find Probabilities**

More than 15,000 U.S. medical school seniors applied to residency programs in 2009. Of those, 93% were matched with residency positions. Eighty-two percent of the seniors matched with residency positions were matched with one of their top three choices. Medical students electronically rank the residency programs in their order of preference, and program directors across the United States do the same. The term "match" refers to the process whereby a student's preference list and a program director's preference list overlap, resulting in the placement of the student in a residency position. *(Source: National Resident Matching Program)*

1. Find the probability that a randomly selected senior was matched with a residency position *and* it was one of the senior's top three choices.

2. Find the probability that a randomly selected senior who was matched with a residency position did *not* get matched with one of the senior's top three choices.

3. Would it be unusual for a randomly selected senior to be matched with a residency position *and* that it was one of the senior's top three choices?

▶ **Solution**

Let A = {matched with residency position} and B = {matched with one of top three choices}. So, $P(A) = 0.93$ and $P(B|A) = 0.82$.

1. The events are dependent.

$$P(A \text{ and } B) = P(A) \cdot P(B|A) = (0.93) \cdot (0.82) \approx 0.763$$

So, the probability that a randomly selected senior was matched with one of the senior's top three choices is about 0.763.

2. To find this probability, use the complement.

$$P(B'|A) = 1 - P(B|A) = 1 - 0.82 = 0.18$$

So, the probability that a randomly selected senior was matched with a residency position that was not one of the senior's top three choices is 0.18.

3. It is not unusual because the probability of a senior being matched with a residency position that was one of the senior's top three choices is about 0.763, which is greater than 0.05.

▶ **Try It Yourself 5**

In a jury selection pool, 65% of the people are female. Of these 65%, one out of four works in a health field.

1. Find the probability that a randomly selected person from the jury pool is female and works in a health field.

2. Find the probability that a randomly selected person from the jury pool is female and does not work in a health field.

a. Determine *events A* and *B*.

b. Use the *Multiplication Rule* to write a formula to find the probability. If necessary, use the *complement*.

c. *Calculate* the probability.   *Answer: Page A35*

Medical School

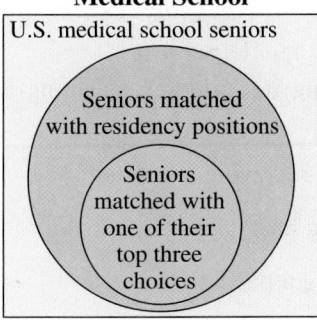

Jury Selection

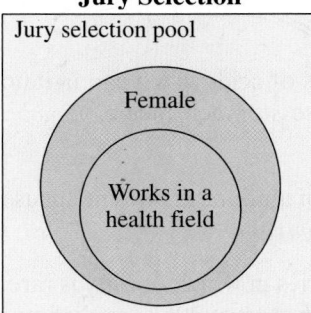

3.2 EXERCISES

1. Two events are independent if the occurrence of one of the events does not affect the probability of the occurrence of the other event, whereas two events are dependent if the occurrence of one of the events does affect the probability of the occurrence of the other event.

2. (a) Roll a die twice. The outcome of the second roll is independent of the outcome of the first roll.

   (b) Draw two cards (without replacement) from a standard 52-card deck. The outcome of the second draw is dependent on the outcome of the first draw.

3. The notation $P(B|A)$ means the probability of B, given A.

4. The complement of "at least one" is "none." So, the probability of getting at least one item is equal to $1 - P(\text{none of the items})$.

5. False. If two events are independent, then $P(A|B) = P(A)$.

6. False. If events A and B are independent, then $P(A \text{ and } B) = P(A) \cdot P(B)$.

7. Independent. The outcome of the first draw does not affect the outcome of the second draw.

■ BUILDING BASIC SKILLS AND VOCABULARY

1. What is the difference between independent and dependent events?

2. List examples of

   (a) two events that are independent.

   (b) two events that are dependent.

3. What does the notation $P(B|A)$ mean?

4. Explain how the complement can be used to find the probability of getting at least one item of a particular type.

True or False? *In Exercises 5 and 6, determine whether the statement is true or false. If it is false, rewrite it as a true statement.*

5. If two events are independent, $P(A|B) = P(B)$.

6. If events A and B are dependent, then $P(A \text{ and } B) = P(A) \cdot P(B)$.

Classifying Events *In Exercises 7–12, decide whether the events are independent or dependent. Explain your reasoning.*

7. Selecting a king from a standard deck, replacing it, and then selecting a queen from the deck

8. Returning a rented movie after the due date and receiving a late fee

9. A father having hazel eyes and a daughter having hazel eyes

10. Not putting money in a parking meter and getting a parking ticket

11. Rolling a six-sided die and then rolling the die a second time so that the sum of the two rolls is five

12. A ball numbered from 1 through 52 is selected from a bin, replaced, and then a second numbered ball is selected from the bin.

Classifying Events Based on Studies *In Exercises 13–16, identify the two events described in the study. Do the results indicate that the events are independent or dependent? Explain your reasoning.*

13. A study found that people who suffer from moderate to severe sleep apnea are at increased risk of having high blood pressure. *(Source: Journal of the American Medical Association)*

14. Stress causes the body to produce higher amounts of acid, which can irritate already existing ulcers. But, stress does not cause stomach ulcers. *(Source: Baylor College of Medicine)*

15. Studies found that exposure to everyday sources of aluminum does not cause Alzheimer's disease. *(Source: Alzheimer's Association)*

16. According to researchers, diabetes is rare in societies in which obesity is rare. In societies in which obesity has been common for at least 20 years, diabetes is also common. *(Source: American Diabetes Association)*

8. Dependent. The outcome of returning a movie after its due date affects the outcome of receiving a late fee.

9. Dependent. The outcome of a father having hazel eyes affects the outcome of a daughter having hazel eyes.

10. Dependent. The outcome of not putting money in a parking meter affects the outcome of getting a parking ticket.

11. Dependent. The sum of the rolls depends on which numbers came up on the first and second rolls.

12. Independent. The outcome of the first selection does not affect the outcome of the second selection.

13. Events: moderate to severe sleep apnea, high blood pressure; Dependent. People with moderate to severe sleep apnea are more likely to have high blood pressure.

14. Events: stress, ulcers; Independent. Stress only irritates already existing ulcers.

15. Events: exposure to aluminum, Alzheimer's disease; Independent. Exposure to everyday sources of aluminum does not cause Alzheimer's disease.

16. Events: diabetes, obesity; Dependent. In societies, the number of cases of diabetes increases as the number of cases of obesity increases.

17. (a) 0.6
    (b) 0.001
    (c) See Odd Answers, page A65.

18. (a) 0.222
    (b) 0.067
    (c) See Selected Answers, page A116.

19. (a) 0.308
    (b) 0.788
    (c) 0.757
    (d) 0.596
    (e) See Odd Answers, page A65.

■ USING AND INTERPRETING CONCEPTS

17. BRCA Gene In the general population, one woman in eight will develop breast cancer. Research has shown that approximately 1 woman in 600 carries a mutation of the BRCA gene. About 6 out of 10 women with this mutation develop breast cancer. *(Adapted from Susan G. Komen Breast Cancer Foundation)*

(a) Find the probability that a randomly selected woman will develop breast cancer, given that she has a mutation of the BRCA gene.

(b) Find the probability that a randomly selected woman will carry the mutation of the BRCA gene and will develop breast cancer.

(c) Are the events "carrying this mutation" and "developing breast cancer" independent or dependent? Explain.

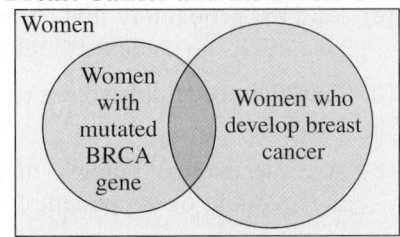

Breast Cancer and the BRCA Gene

FIGURE FOR EXERCISE 17

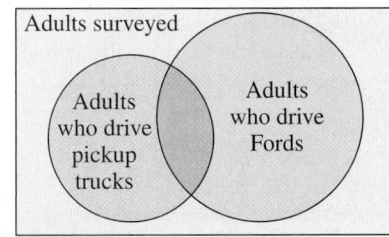

What Do You Drive?

FIGURE FOR EXERCISE 18

18. Pickup Trucks In a survey, 510 adults were asked if they drive a pickup truck and if they drive a Ford. The results showed that one in six adults surveyed drives a pickup truck, and three in ten adults surveyed drive a Ford. Of the adults surveyed that drive Fords, two in nine drive a pickup truck.

(a) Find the probability that a randomly selected adult drives a pickup truck, given that the adult drives a Ford.

(b) Find the probability that a randomly selected adult drives a Ford and drives a pickup truck.

(c) Are the events "driving a Ford" and "driving a pickup truck" independent or dependent? Explain.

19. Summer Vacation The table shows the results of a survey in which 146 families were asked if they own a computer and if they will be taking a summer vacation during the current year.

		Summer Vacation This Year		
		Yes	No	Total
Own a Computer	Yes	87	28	115
	No	14	17	31
	Total	101	45	146

(a) Find the probability that a randomly selected family is not taking a summer vacation this year.

(b) Find the probability that a randomly selected family owns a computer.

(c) Find the probability that a randomly selected family is taking a summer vacation this year, given that they own a computer.

(d) Find the probability that a randomly selected family is taking a summer vacation this year and owns a computer.

(e) Are the events "owning a computer" and "taking a summer vacation this year" independent or dependent events? Explain.

20. (a) 0.294

(b) 0.317

(c) 0.120

(d) 0.038

(e) See Selected Answers, page A116.

21. (a) 0.093

(b) 0.75

(c) No, the probability is not unusual because it is not less than or equal to 0.05.

22. (a) 0.688

(b) 0.2

(c) No, this is not unusual because the probability is not less than or equal to 0.05.

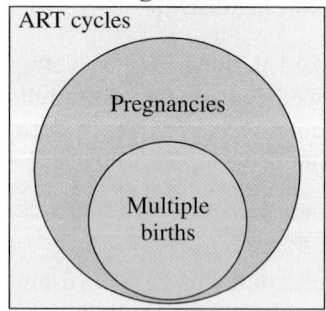

Pregnancies

FIGURE FOR EXERCISE 21

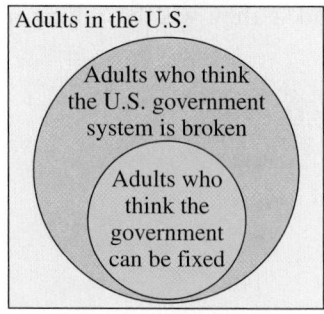

Government

FIGURE FOR EXERCISE 22

23. 0.745

24. 0.3

---

20. **Nursing Majors** The table shows the number of male and female students enrolled in nursing at the University of Oklahoma Health Sciences Center for a recent semester. *(Source: University of Oklahoma Health Sciences Center Office of Institutional Research)*

	Nursing majors	Non-nursing majors	Total
Males	151	1104	1255
Females	1016	1693	2709
Total	1167	2797	3964

(a) Find the probability that a randomly selected student is a nursing major.

(b) Find the probability that a randomly selected student is male.

(c) Find the probability that a randomly selected student is a nursing major, given that the student is male.

(d) Find the probability that a randomly selected student is a nursing major and male.

(e) Are the events "being a male student" and "being a nursing major" independent or dependent events? Explain.

21. **Assisted Reproductive Technology** A study found that 37% of the assisted reproductive technology (ART) cycles resulted in pregnancies. Twenty-five percent of the ART pregnancies resulted in multiple births. *(Source: National Center for Chronic Disease Prevention and Health Promotion)*

(a) Find the probability that a randomly selected ART cycle resulted in a pregnancy *and* produced a multiple birth.

(b) Find the probability that a randomly selected ART cycle that resulted in a pregnancy did *not* produce a multiple birth.

(c) Would it be unusual for a randomly selected ART cycle to result in a pregnancy and produce a multiple birth? Explain.

22. **Government** According to a survey, 86% of adults in the United States think the U.S. government system is broken. Of these 86%, about 8 out of 10 think the government can be fixed. *(Adapted from CNN/Opinion Research Corporation)*

(a) Find the probability that a randomly selected adult thinks the U.S. government system is broken *and* thinks the government can be fixed.

(b) Given that a randomly selected adult thinks the U.S. government system is broken, find the probability that he or she thinks the government *cannot* be fixed.

(c) Would it be unusual for a randomly selected adult to think the U.S. government system is broken and think the government can be fixed? Explain.

23. **Computers and Internet Access** A study found that 81% of households in the United States have computers. Of those 81%, 92% have Internet access. Find the probability that a U.S. household selected at random has a computer and has Internet access. *(Source: The Nielsen Company)*

24. **Surviving Surgery** A doctor gives a patient a 60% chance of surviving bypass surgery after a heart attack. If the patient survives the surgery, he has a 50% chance that the heart damage will heal. Find the probability that the patient survives surgery and the heart damage heals.

25. (a) 0.017
    (b) 0.757
    (c) 0.243
    (d) The event in part (a) is unusual because its probability is less than or equal to 0.05.
26. (a) 0.05
    (b) 0.55
    (c) 0.45
    (d) The event in part (a) is unusual because its probability is less than or equal to 0.05.
27. (a) 0.481
    (b) 0.465
    (c) 0.449
    (d) See Odd Answers, page A65.
28. (a) 0.556
    (b) 0.525
    (c) 0.167
    (d) See Selected Answers, page A116.

25. People Who Can Wiggle Their Ears   In a sample of 1000 people, 130 can wiggle their ears. Two unrelated people are selected at random without replacement.

(a) Find the probability that both people can wiggle their ears.

(b) Find the probability that neither person can wiggle his or her ears.

(c) Find the probability that at least one of the two people can wiggle his or her ears.

(d) Which of the events can be considered unusual? Explain.

26. Batteries   Sixteen batteries are tested to see if they last as long as the manufacturer claims. Four batteries fail the test. Two batteries are selected at random without replacement.

(a) Find the probability that both batteries fail the test.

(b) Find the probability that both batteries pass the test.

(c) Find the probability that at least one battery fails the test.

(d) Which of the events can be considered unusual? Explain.

27. Emergency Savings   The table shows the results of a survey in which 142 male and 145 female workers ages 25 to 64 were asked if they had at least one month's income set aside for emergencies.

	Male	Female	Total
Less than one month's income	66	83	149
One month's income or more	76	62	138
Total	142	145	287

(a) Find the probability that a randomly selected worker has one month's income or more set aside for emergencies.

(b) Given that a randomly selected worker is a male, find the probability that the worker has less than one month's income.

(c) Given that a randomly selected worker has one month's income or more, find the probability that the worker is a female.

(d) Are the events "having less than one month's income saved" and "being male" independent or dependent? Explain.

28. Health Care for Dogs   The table shows the results of a survey in which 90 dog owners were asked how much they had spent in the last year for their dog's health care, and whether their dogs were purebred or mixed breeds.

		Type of Dog		
		Purebred	Mixed breed	Total
Health Care	Less than $100	19	21	40
	$100 or more	35	15	50
	Total	54	36	90

(a) Find the probability that $100 or more was spent on a randomly selected dog's health care in the last year.

(b) Given that a randomly selected dog owner spent less than $100, find the probability that the dog was a mixed breed.

(c) Find the probability that a randomly selected dog owner spent $100 or more on health care and the dog was a mixed breed.

(d) Are the events "spending $100 or more on health care" and "having a mixed breed dog" independent or dependent? Explain.

29. (a) 0.00000590

    (b) 0.624

    (c) 0.376

30. (a) 0.030

    (b) 0.329

    (c) 0.671

31. (a) 0.25

    (b) 0.063

    (c) 0.000977

    (d) 0.237

    (e) 0.763

32. (a) 0.985

    (b) 0.015

    (c) 0.000000125

33. (a) 0.011

    (b) 0.458

34. (a) 0.00000751

    (b) 0.992

35. 0.444

---

29. Blood Types   The probability that a person in the United States has type B^+ blood is 9%. Five unrelated people in the United States are selected at random. *(Source: American Association of Blood Banks)*

(a) Find the probability that all five have type B^+ blood.

(b) Find the probability that none of the five have type B^+ blood.

(c) Find the probability that at least one of the five has type B^+ blood.

30. Blood Types   The probability that a person in the United States has type A^+ blood is 31%. Three unrelated people in the United States are selected at random. *(Source: American Association of Blood Banks)*

(a) Find the probability that all three have type A^+ blood.

(b) Find the probability that none of the three have type A^+ blood.

(c) Find the probability that at least one of the three has type A^+ blood.

31. Guessing   A multiple-choice quiz has five questions, each with four answer choices. Only one of the choices is correct. You have no idea what the answer is to any question and have to guess each answer.

(a) Find the probability of answering the first question correctly.

(b) Find the probability of answering the first two questions correctly.

(c) Find the probability of answering all five questions correctly.

(d) Find the probability of answering none of the questions correctly.

(e) Find the probability of answering at least one of the questions correctly.

32. Bookbinding Defects   A printing company's bookbinding machine has a probability of 0.005 of producing a defective book. This machine is used to bind three books.

(a) Find the probability that none of the books are defective.

(b) Find the probability that at least one of the books is defective.

(c) Find the probability that all of the books are defective.

33. Warehouses   A distribution center receives shipments of a product from three different factories in the following quantities: 50, 35, and 25. Three times a product is selected at random, each time without replacement. Find the probability that (a) all three products came from the third factory and (b) none of the three products came from the third factory.

34. Birthdays   Three people are selected at random. Find the probability that (a) all three share the same birthday and (b) none of the three share the same birthday. Assume 365 days in a year.

■ EXTENDING CONCEPTS

*According to **Bayes' Theorem**, the probability of event A, given that event B has occurred, is*

$$P(A|B) = \frac{P(A) \cdot P(B|A)}{P(A) \cdot P(B|A) + P(A') \cdot P(B|A')}.$$

In Exercises 35–38, use Bayes' Theorem to find $P(A|B)$.

35. $P(A) = \frac{2}{3}$, $P(A') = \frac{1}{3}$, $P(B|A) = \frac{1}{5}$, and $P(B|A') = \frac{1}{2}$

36. 0.4

37. 0.167

38. 0.797

39. (a) 0.074

(b) 0.999

40. (a) 0.462

(b) 0.538

(c) Yes

(d) Answers will vary.

41. 0.954

42. 0.933

36. $P(A) = \frac{3}{8}$, $P(A') = \frac{5}{8}$, $P(B|A) = \frac{2}{3}$, and $P(B|A') = \frac{3}{5}$

37. $P(A) = 0.25$, $P(A') = 0.75$, $P(B|A) = 0.3$, and $P(B|A') = 0.5$

38. $P(A) = 0.62$, $P(A') = 0.38$, $P(B|A) = 0.41$, and $P(B|A') = 0.17$

39. Reliability of Testing   A certain virus infects one in every 200 people. A test used to detect the virus in a person is positive 80% of the time if the person has the virus and 5% of the time if the person does not have the virus. (This 5% result is called a *false positive*.) Let A be the event "the person is infected" and B be the event "the person tests positive."

(a) Using Bayes' Theorem, if a person tests positive, determine the probability that the person is infected.

(b) Using Bayes' Theorem, if a person tests negative, determine the probability that the person is *not* infected.

40. Birthday Problem   You are in a class that has 24 students. You want to find the probability that at least two of the students share the same birthday.

(a) First, find the probability that each student has a different birthday.

$$P(\text{different birthdays}) = \overbrace{\frac{365}{365} \cdot \frac{364}{365} \cdot \frac{363}{365} \cdot \frac{362}{365} \cdots \frac{343}{365} \cdot \frac{342}{365}}^{24 \text{ factors}}$$

(b) The probability that at least two students have the same birthday is the complement of the probability in part (a). What is this probability?

(c) We used a technology tool to generate 24 random numbers between 1 and 365. Each number represents a birthday. Did we get at least two people with the same birthday?

228	348	181	317	81	183
52	346	177	118	315	273
252	168	281	266	285	13
118	360	8	193	57	107

(d) Use a technology tool to simulate the "Birthday Problem." Repeat the simulation 10 times. How many times did you get at least two people with the same birthday?

The Multiplication Rule and Conditional Probability   *By rewriting the formula for the Multiplication Rule, you can write a formula for finding conditional probabilities. The conditional probability of event B occurring, given that event A has occurred, is*

$$P(B|A) = \frac{P(A \text{ and } B)}{P(A)}.$$

In Exercises 41 and 42, use the following information.

- *The probability that an airplane flight departs on time is 0.89.*

- *The probability that a flight arrives on time is 0.87.*

- *The probability that a flight departs and arrives on time is 0.83.*

41. Find the probability that a flight departed on time given that it arrives on time.

42. Find the probability that a flight arrives on time given that it departed on time.

3.3   The Addition Rule

Mutually Exclusive Events ‣ The Addition Rule ‣ A Summary of Probability

‣ MUTUALLY EXCLUSIVE EVENTS

In Section 3.2, you learned how to find the probability of two events, *A* and *B*, occurring in sequence. Such probabilities are denoted by *P*(*A* and *B*). In this section, you will learn how to find the probability that at least one of two events will occur. Probabilities such as these are denoted by *P*(*A* or *B*) and depend on whether the events are *mutually exclusive*.

DEFINITION

Two events *A* and *B* are **mutually exclusive** if *A* and *B* cannot occur at the same time.

The Venn diagrams show the relationship between events that are mutually exclusive and events that are not mutually exclusive.

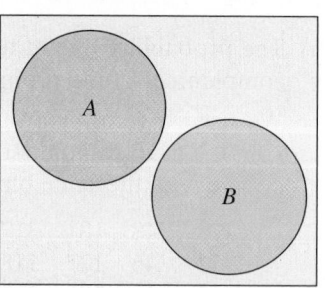

A and *B* are mutually exclusive.

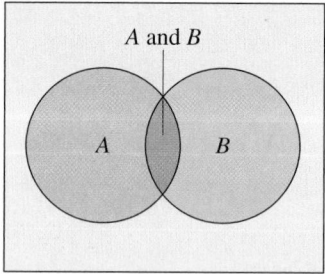

A and *B* are not mutually exclusive.

EXAMPLE 1

‣ **Mutually Exclusive Events**

Decide if the events are mutually exclusive. Explain your reasoning.

1. Event *A*: Roll a 3 on a die.
   Event *B*: Roll a 4 on a die.

2. Event *A*: Randomly select a male student.
   Event *B*: Randomly select a nursing major.

3. Event *A*: Randomly select a blood donor with type O blood.
   Event *B*: Randomly select a female blood donor.

‣ **Solution**

1. The first event has one outcome, a 3. The second event also has one outcome, a 4. These outcomes cannot occur at the same time, so the events are mutually exclusive.

2. Because the student can be a male nursing major, the events are not mutually exclusive.

3. Because the donor can be a female with type O blood, the events are not mutually exclusive.

▶ **Try It Yourself 1**

Decide if the events are mutually exclusive. Explain your reasoning.

1. Event *A*: Randomly select a jack from a standard deck of cards.
   Event *B*: Randomly select a face card from a standard deck of cards.
2. Event *A*: Randomly select a 20-year-old student.
   Event *B*: Randomly select a student with blue eyes.
3. Event *A*: Randomly select a vehicle that is a Ford.
   Event *B*: Randomly select a vehicle that is a Toyota.

a. Decide if one of the following statements is true.
   - Events *A* and *B* cannot occur at the same time.
   - Events *A* and *B* have no outcomes in common.
   - $P(A \text{ and } B) = 0$

b. Make a *conclusion*.   *Answer: Page A35*

▶ THE ADDITION RULE

To explore this topic further, see Activity 3.3 on page 166.

THE ADDITION RULE FOR THE PROBABILITY OF *A* OR *B*

The probability that events *A* or *B* will occur, $P(A \text{ or } B)$, is given by

$$P(A \text{ or } B) = P(A) + P(B) - P(A \text{ and } B).$$

If events *A* and *B* are mutually exclusive, then the rule can be simplified to $P(A \text{ or } B) = P(A) + P(B)$. This simplified rule can be extended to any number of mutually exclusive events.

In words, to find the probability that one event or the other will occur, add the individual probabilities of each event and subtract the probability that they both occur.

EXAMPLE 2

▶ **Using the Addition Rule to Find Probabilities**

1. You select a card from a standard deck. Find the probability that the card is a 4 or an ace.

2. You roll a die. Find the probability of rolling a number less than 3 or rolling an odd number.

▶ **Solution**

1. If the card is a 4, it cannot be an ace. So, the events are mutually exclusive, as shown in the Venn diagram. The probability of selecting a 4 or an ace is

$$P(4 \text{ or ace}) = P(4) + P(\text{ace}) = \frac{4}{52} + \frac{4}{52} = \frac{8}{52} = \frac{2}{13} \approx 0.154.$$

2. The events are not mutually exclusive because 1 is an outcome of both events, as shown in the Venn diagram. So, the probability of rolling a number less than 3 or an odd number is

$$P(\text{less than 3 or odd}) = P(\text{less than 3}) + P(\text{odd})$$
$$- P(\text{less than 3 and odd})$$
$$= \frac{2}{6} + \frac{3}{6} - \frac{1}{6} = \frac{4}{6} = \frac{2}{3} \approx 0.667.$$

Deck of 52 Cards

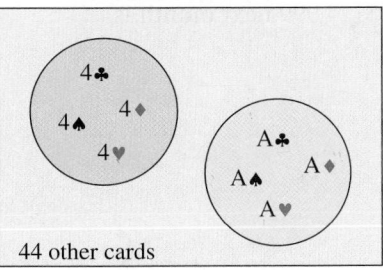

44 other cards

Roll a Die

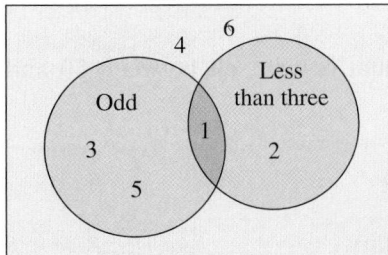

Note to Instructor

Point out to students that the notations $P(A \cap B)$ and $P(A \cup B)$ are sometimes used instead of $P(A$ and $B)$ and $P(A$ or $B)$, respectively. The symbol $\cap$ represents the intersection of two sets and the symbol $\cup$ represents the union of two sets. So, the Addition Rule can be written as

$$P(A \cup B) = P(A) + P(B) - P(A \cap B).$$

▶ **Try It Yourself 2**

1. A die is rolled. Find the probability of rolling a 6 or an odd number.
2. A card is selected from a standard deck. Find the probability that the card is a face card or a heart.

a. Decide whether the events are *mutually exclusive*.
b. Find $P(A)$, $P(B)$, and, if necessary, $P(A$ and $B)$.
c. Use the *Addition Rule* to find the probability.   *Answer: Page A35*

EXAMPLE 3

▶ **Finding Probabilities of Mutually Exclusive Events**

The frequency distribution shows volumes of sales (in dollars) and the number of months in which a sales representative reached each sales level during the past three years. If this sales pattern continues, what is the probability that the sales representative will sell between $75,000 and $124,999 next month?

Sales volume ($)	Months
0–24,999	3
25,000–49,999	5
50,000–74,999	6
75,000–99,999	7
100,000–124,999	9
125,000–149,999	2
150,000–174,999	3
175,000–199,999	1

▶ **Solution**

To solve this problem, define events A and B as follows.

$A = \{$monthly sales between $75,000 and $99,999$\}$

$B = \{$monthly sales between $100,000 and $124,999$\}$

Because events A and B are mutually exclusive, the probability that the sales representative will sell between $75,000 and $124,999 next month is

$$P(A \text{ or } B) = P(A) + P(B)$$

$$= \frac{7}{36} + \frac{9}{36}$$

$$= \frac{16}{36}$$

$$= \frac{4}{9} \approx 0.444.$$

▶ **Try It Yourself 3**

Find the probability that the sales representative will sell between $0 and $49,999.

a. Identify *events A* and *B*.
b. Decide if the events are *mutually exclusive*.
c. Find the *probability* of each event.
d. Use the *Addition Rule* to find the probability.   *Answer: Page A35*

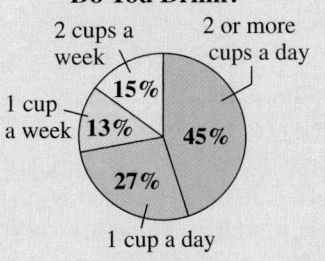
EXAMPLE 4

▶ Using the Addition Rule to Find Probabilities

A blood bank catalogs the types of blood, including positive or negative Rh-factor, given by donors during the last five days. The number of donors who gave each blood type is shown in the table. A donor is selected at random.

1. Find the probability that the donor has type O or type A blood.

2. Find the probability that the donor has type B blood or is Rh-negative.

		Blood Type				
		O	**A**	**B**	**AB**	**Total**
Rh-factor	**Positive**	156	139	37	12	344
	Negative	28	25	8	4	65
	Total	184	164	45	16	409

▶ Solution

1. Because a donor cannot have type O blood and type A blood, these events are mutually exclusive. So, using the Addition Rule, the probability that a randomly chosen donor has type O or type A blood is

$$P(\text{type O or type A}) = P(\text{type O}) + P(\text{type A})$$

$$= \frac{184}{409} + \frac{164}{409}$$

$$= \frac{348}{409}$$

$$\approx 0.851.$$

2. Because a donor can have type B blood and be Rh-negative, these events are not mutually exclusive. So, using the Addition Rule, the probability that a randomly chosen donor has type B blood or is Rh-negative is

$$P(\text{type B or Rh-neg}) = P(\text{type B}) + P(\text{Rh-neg}) - P(\text{type B and Rh-neg})$$

$$= \frac{45}{409} + \frac{65}{409} - \frac{8}{409}$$

$$= \frac{102}{409}$$

$$\approx 0.249.$$

▶ Try It Yourself 4

1. Find the probability that the donor has type B or type AB blood.
2. Find the probability that the donor has type O blood or is Rh-positive.

a. Identify *events A* and *B*.
b. Decide if the events are *mutually exclusive*.
c. Find the *probability* of each event.
d. Use the *Addition Rule* to find the probability.   *Answer: Page A35*

▸ A SUMMARY OF PROBABILITY

Type of Probability and Probability Rules	In Words	In Symbols
Classical Probability	The number of outcomes in the sample space is known and each outcome is equally likely to occur.	$P(E) = \dfrac{\text{Number of outcomes in event } E}{\text{Number of outcomes in sample space}}$
Empirical Probability	The frequency of outcomes in the sample space is estimated from experimentation.	$P(E) = \dfrac{\text{Frequency of event } E}{\text{Total frequency}} = \dfrac{f}{n}$
Range of Probabilities Rule	The probability of an event is between 0 and 1, inclusive.	$0 \le P(E) \le 1$
Complementary Events	The complement of event E is the set of all outcomes in a sample space that are not included in E, denoted by E'.	$P(E') = 1 - P(E)$
Multiplication Rule	The Multiplication Rule is used to find the probability of two events occurring in a sequence.	$P(A \text{ and } B) = P(A) \cdot P(B \mid A)$ $P(A \text{ and } B) = P(A) \cdot P(B)$ *Independent events*
Addition Rule	The Addition Rule is used to find the probability of at least one of two events occurring.	$P(A \text{ or } B) = P(A) + P(B) - P(A \text{ and } B)$ $P(A \text{ or } B) = P(A) + P(B)$ *Mutually exclusive events*

EXAMPLE 5

▸ Combining Rules to Find Probabilities

Use the graph at the right to find the probability that a randomly selected draft pick is not a running back or a wide receiver.

▸ Solution

Define events A and B.

    A: Draft pick is a running back.
    B: Draft pick is a wide receiver.

These events are mutually exclusive, so the probability that the draft pick is a running back or wide receiver is

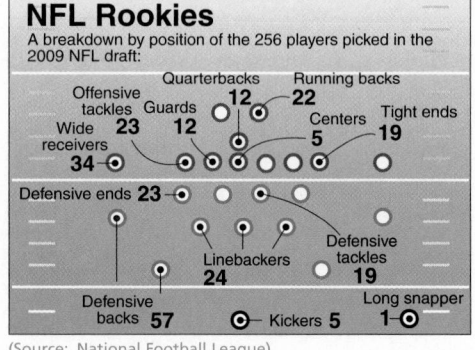

NFL Rookies
A breakdown by position of the 256 players picked in the 2009 NFL draft:

Quarterbacks **12**   Running backs **22**
Offensive tackles   Guards
Wide receivers **34**   **23**   **12**   Centers **5**   Tight ends **19**
Defensive ends **23**
Linebackers **24**   Defensive tackles **19**
Defensive backs **57**   Kickers **5**   Long snapper **1**

(Source: National Football League)

$$P(A \text{ or } B) = P(A) + P(B) = \tfrac{22}{256} + \tfrac{34}{256} = \tfrac{56}{256} = \tfrac{7}{32} \approx 0.219.$$

By taking the complement of $P(A \text{ or } B)$, you can determine that the probability of randomly selecting a draft pick who is not a running back or wide receiver is

$$1 - P(A \text{ or } B) = 1 - \tfrac{7}{32} = \tfrac{25}{32} \approx 0.781.$$

▸ Try It Yourself 5

Find the probability that a randomly selected draft pick is not a linebacker or a quarterback.

a. Find the *probability* that the draft pick is a linebacker or a quarterback.
b. Find the *complement* of the event.
                                                   Answer: Page A35

3.3 EXERCISES

FOR EXTRA HELP:
MyStatLab

1. $P(A \text{ and } B) = 0$ because A and B cannot occur at the same time.

2. Answers will vary. Sample answers are given.

   (a) Toss coin once: $A = \{\text{head}\}$ and $B = \{\text{tail}\}$

   (b) Draw one card: $A = \{\text{ace}\}$ and $B = \{\text{spade}\}$

3. True

4. False. Two events being independent does not imply they are mutually exclusive.

5. False. The probability that event A or event B will occur is $P(A \text{ or } B) = P(A) + P(B) - P(A \text{ and } B)$.

6. True

7. Not mutually exclusive. A student can be an athlete and on the Dean's list.

8. Mutually exclusive. A movie cannot have two ratings.

9. Not mutually exclusive. A public school teacher can be female and 25 years old.

10. Not mutually exclusive. A member of Congress can be a male senator.

11. Mutually exclusive. A student cannot have a birthday in both months.

12. Not mutually exclusive. A person can be between 18 and 24 years old and drive a convertible.

■ BUILDING BASIC SKILLS AND VOCABULARY

1. If two events are mutually exclusive, why is $P(A \text{ and } B) = 0$?

2. List examples of

   (a) two events that are mutually exclusive.

   (b) two events that are not mutually exclusive.

True or False? *In Exercises 3–6, determine whether the statement is true or false. If it is false, explain why.*

3. If two events are mutually exclusive, they have no outcomes in common.

4. If two events are independent, then they are also mutually exclusive.

5. The probability that event A or event B will occur is

$$P(A \text{ or } B) = P(A) + P(B) - P(A \text{ or } B).$$

6. If events A and B are mutually exclusive, then

$$P(A \text{ or } B) = P(A) + P(B).$$

Graphical Analysis *In Exercises 7 and 8, decide if the events shown in the Venn diagram are mutually exclusive. Explain your reasoning.*

7.

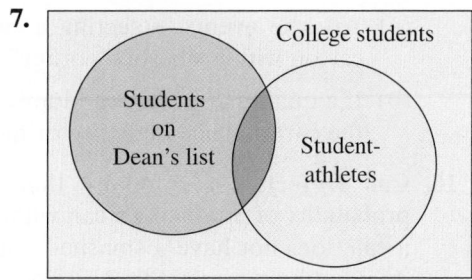

8.

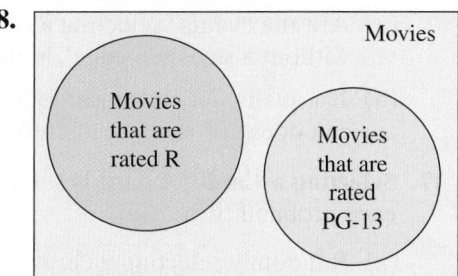

Recognizing Mutually Exclusive Events *In Exercises 9–12, decide if the events are mutually exclusive. Explain your reasoning.*

9. Event A: Randomly select a female public school teacher.
   Event B: Randomly select a public school teacher who is 25 years old.

10. Event A: Randomly select a member of the U.S. Congress.
    Event B: Randomly select a male U.S. Senator.

11. Event A: Randomly select a student with a birthday in April.
    Event B: Randomly select a student with a birthday in May.

12. Event A: Randomly select a person between 18 and 24 years old.
    Event B: Randomly select a person who drives a convertible.

13. (a) Not mutually exclusive. For five weeks the events overlapped.
    (b) 0.423
14. (a) Not mutually exclusive. An attendee can be a female and be a college professor.
    (b) 0.753
15. (a) Not mutually exclusive. A carton can have a puncture and a smashed corner.
    (b) 0.126
16. (a) Not mutually exclusive. A can may have no punctures and no smashed edges.
    (b) 0.997
17. (a) 0.308
    (b) 0.538
    (c) 0.308
18. (a) 0.5
    (b) 0.833
    (c) 0.667

■ USING AND INTERPRETING CONCEPTS

13. **Audit** During a 52-week period, a company paid overtime wages for 18 weeks and hired temporary help for 9 weeks. During 5 weeks, the company paid overtime *and* hired temporary help.

    (a) Are the events "selecting a week in which overtime wages were paid" and "selecting a week in which temporary help wages were paid" mutually exclusive? Explain.

    (b) If an auditor randomly examined the payroll records for only one week, what is the probability that the payroll for that week contained overtime wages or temporary help wages?

14. **Conference** A math conference has an attendance of 4950 people. Of these, 2110 are college professors and 2575 are female. Of the college professors, 960 are female.

    (a) Are the events "selecting a female" and "selecting a college professor" mutually exclusive? Explain.

    (b) The conference selects people at random to win prizes. Find the probability that a selected person is a female or a college professor.

15. **Carton Defects** A company that makes cartons finds that the probability of producing a carton with a puncture is 0.05, the probability that a carton has a smashed corner is 0.08, and the probability that a carton has a puncture and has a smashed corner is 0.004.

    (a) Are the events "selecting a carton with a puncture" and "selecting a carton with a smashed corner" mutually exclusive? Explain.

    (b) If a quality inspector randomly selects a carton, find the probability that the carton has a puncture or has a smashed corner.

16. **Can Defects** A company that makes soda pop cans finds that the probability of producing a can without a puncture is 0.96, the probability that a can does not have a smashed edge is 0.93, and the probability that a can does not have a puncture and does not have a smashed edge is 0.893.

    (a) Are the events "selecting a can without a puncture" and "selecting a can without a smashed edge" mutually exclusive? Explain.

    (b) If a quality inspector randomly selects a can, find the probability that the can does not have a puncture or does not have a smashed edge.

17. **Selecting a Card** A card is selected at random from a standard deck. Find each probability.

    (a) Randomly selecting a club or a 3

    (b) Randomly selecting a red suit or a king

    (c) Randomly selecting a 9 or a face card

18. **Rolling a Die** You roll a die. Find each probability.

    (a) Rolling a 5 or a number greater than 3

    (b) Rolling a number less than 4 or an even number

    (c) Rolling a 2 or an odd number

19. (a) 0.067
    (b) 0.84
    (c) 0.199
20. (a) 0.298
    (b) 0.445
    (c) 0.435
21. (a) 0.949
    (b) 0.388
22. (a) 0.19
    (b) 0.01
    (c) 0.48

19. U.S. Age Distribution The estimated percent distribution of the U.S. population for 2020 is shown in the pie chart. Find each probability. *(Source: U.S. Census Bureau)*

(a) Randomly selecting someone who is under 5 years old

(b) Randomly selecting someone who is not 65 years or over

(c) Randomly selecting someone who is between 20 and 34 years old

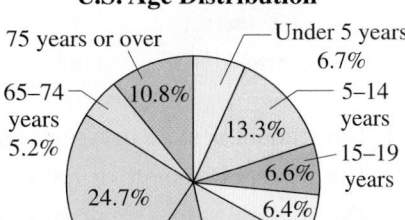

FIGURE FOR EXERCISE 19

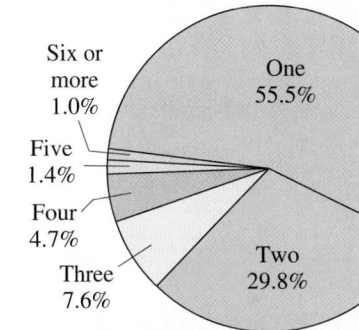

FIGURE FOR EXERCISE 20

20. Tacoma Narrows Bridge The percent distribution of the number of occupants in vehicles crossing the Tacoma Narrows Bridge in Washington is shown in the pie chart. Find each probability. *(Source: Washington State Department of Transportation)*

(a) Randomly selecting a car with two occupants

(b) Randomly selecting a car with two or more occupants

(c) Randomly selecting a car with between two and five occupants, inclusive

21. Education The number of responses to a survey are shown in the Pareto chart. The survey asked 1026 U.S. adults how they would grade the quality of public schools in the United States. Each person gave one response. Find each probability. *(Adapted from CBS News Poll)*

(a) Randomly selecting a person from the sample who did not give the public schools an A

(b) Randomly selecting a person from the sample who gave the public schools a D or an F

22. Olympics The number of responses to a survey are shown in the Pareto chart. The survey asked 1000 U.S. adults if they would watch a large portion of the 2010 Winter Olympics. Each person gave one response. Find each probability. *(Adapted from Rasmussen Reports)*

(a) Randomly selecting a person from the sample who is not at all likely to watch a large portion of the Winter Olympics

(b) Randomly selecting a person from the sample who is not sure whether they will watch a large portion of the Winter Olympics

(c) Randomly selecting a person from the sample who is neither somewhat likely nor very likely to watch a large portion of the Winter Olympics

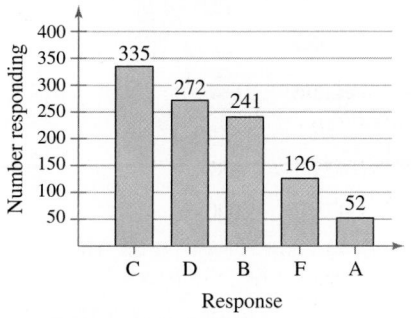

How Would You Grade the Quality of Public Schools in the U.S.?

FIGURE FOR EXERCISE 21

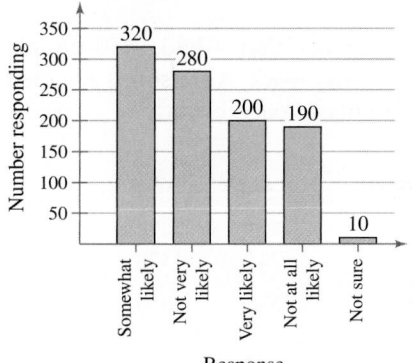

Will You Watch a Large Portion of the Winter Olympics?

FIGURE FOR EXERCISE 22

23. (a) 0.573
    (b) 0.962
    (c) 0.573
    (d) Not mutually exclusive. A male can be a nursing major.
24. (a) 0.538
    (b) 0.950
    (c) 0.575
    (d) 0.425
    (e) Not mutually exclusive. A right-handed person can be a female.
25. (a) 0.461
    (b) 0.762
    (c) 0.589
    (d) 0.922
    (e) Not mutually exclusive. A female can be frequently involved in charity work.

23. Nursing Majors The table shows the number of male and female students enrolled in nursing at the University of Oklahoma Health Sciences Center for a recent semester. A student is selected at random. Find the probability of each event. *(Adapted from University of Oklahoma Health Sciences Center Office of Institutional Research)*

	Nursing majors	Non-nursing majors	Total
Males	151	1104	1255
Females	1016	1693	2709
Total	1167	2797	3964

(a) The student is male or a nursing major.

(b) The student is female or not a nursing major.

(c) The student is not female or is a nursing major.

(d) Are the events "being male" and "being a nursing major" mutually exclusive? Explain.

24. Left-Handed People In a sample of 1000 people (525 men and 475 women), 113 are left-handed (63 men and 50 women). The results of the sample are shown in the table. A person is selected at random from the sample. Find the probability of each event.

		Gender		
		Male	Female	Total
Dominant Hand	Left	63	50	113
	Right	462	425	887
	Total	525	475	1000

(a) The person is left-handed or female.

(b) The person is right-handed or male.

(c) The person is not right-handed or is a male.

(d) The person is right-handed and is a female.

(e) Are the events "being right-handed" and "being female" mutually exclusive? Explain.

25. Charity The table shows the results of a survey that asked 2850 people whether they were involved in any type of charity work. A person is selected at random from the sample. Find the probability of each event.

	Frequently	Occasionally	Not at all	Total
Male	221	456	795	1472
Female	207	430	741	1378
Total	428	886	1536	2850

(a) The person is frequently or occasionally involved in charity work.

(b) The person is female or not involved in charity work at all.

(c) The person is male or frequently involved in charity work.

(d) The person is female or not frequently involved in charity work.

(e) Are the events "being female" and "being frequently involved in charity work" mutually exclusive? Explain.

26. (a) 0.475

(b) 0.595

(c) 0.662

(d) 0.752

(e) Mutually exclusive. A person who wears only contacts cannot wear both contacts and glasses.

27. Answers will vary.

28. 0.63

29. 0.55

30. If events *A*, *B*, and *C* are not mutually exclusive, $P(A \text{ and } B \text{ and } C)$ must be added because $P(A) + P(B) + P(C)$ counts the intersection of all three events three times and $-P(A \text{ and } B) - P(A \text{ and } C) - P(B \text{ and } C)$ subtracts the intersection of all three events three times. So, if $P(A \text{ and } B \text{ and } C)$ is not added at the end, it will not be counted.

26. Eye Survey   The table shows the results of a survey that asked 3203 people whether they wore contacts or glasses. A person is selected at random from the sample. Find the probability of each event.

	Only contacts	**Only glasses**	**Both**	**Neither**	**Total**
Male	64	841	177	456	1538
Female	189	427	368	681	1665
Total	253	1268	545	1137	3203

(a) The person wears only contacts or only glasses.

(b) The person is male or wears both contacts and glasses.

(c) The person is female or wears neither contacts nor glasses.

(d) The person is male or does not wear glasses.

(e) Are the events "wearing only contacts" and "wearing both contacts and glasses" mutually exclusive? Explain.

■ EXTENDING CONCEPTS

27. Writing   Is there a relationship between independence and mutual exclusivity? To decide, find examples of the following, if possible.

(a) Describe two events that are dependent and mutually exclusive.

(b) Describe two events that are independent and mutually exclusive.

(c) Describe two events that are dependent and not mutually exclusive.

(d) Describe two events that are independent and not mutually exclusive.

Use your results to write a conclusion about the relationship between independence and mutual exclusivity.

Addition Rule for Three Events   *The Addition Rule for the probability that event A or B or C will occur, $P(A \text{ or } B \text{ or } C)$, is given by*

$$P(A \text{ or } B \text{ or } C) = P(A) + P(B) + P(C) - P(A \text{ and } B) - P(A \text{ and } C) \\ - P(B \text{ and } C) + P(A \text{ and } B \text{ and } C).$$

In the Venn diagram shown, $P(A \text{ or } B \text{ or } C)$ is represented by the blue areas.

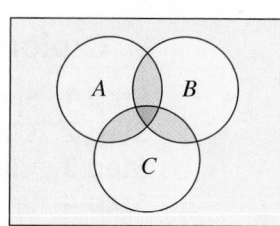

In Exercises 28 and 29, find $P(A \text{ or } B \text{ or } C)$ for the given probabilities.

28. $P(A) = 0.40$, $P(B) = 0.10$, $P(C) = 0.50$,
$P(A \text{ and } B) = 0.05$, $P(A \text{ and } C) = 0.25$, $P(B \text{ and } C) = 0.10$,
$P(A \text{ and } B \text{ and } C) = 0.03$

29. $P(A) = 0.38$, $P(B) = 0.26$, $P(C) = 0.14$,
$P(A \text{ and } B) = 0.12$, $P(A \text{ and } C) = 0.03$, $P(B \text{ and } C) = 0.09$,
$P(A \text{ and } B \text{ and } C) = 0.01$

30. Explain, in your own words, why in the Addition Rule for $P(A \text{ or } B \text{ or } C)$, $P(A \text{ and } B \text{ and } C)$ is added at the end of the formula.

ACTIVITY 3.3 Simulating the Probability of Rolling a 3 or 4

APPLET

The *simulating the probability of rolling a 3 or 4* applet allows you to investigate the probability of rolling a 3 or 4 on a fair die. The plot at the top left corner shows the probability associated with each outcome of a die roll. When ROLL is clicked, *n* simulations of the experiment of rolling a die are performed. The results of the simulations are shown in the frequency plot. If the animate option is checked, the display will show each outcome dropping into the frequency plot as the simulation runs. The individual outcomes are shown in the text field at the far right of the applet. The center plot shows in blue the cumulative proportion of times that an event of rolling a 3 or 4 occurs. The green line in the plot reflects the true probability of rolling a 3 or 4. As the experiment is conducted over and over, the cumulative proportion should converge to the true value.

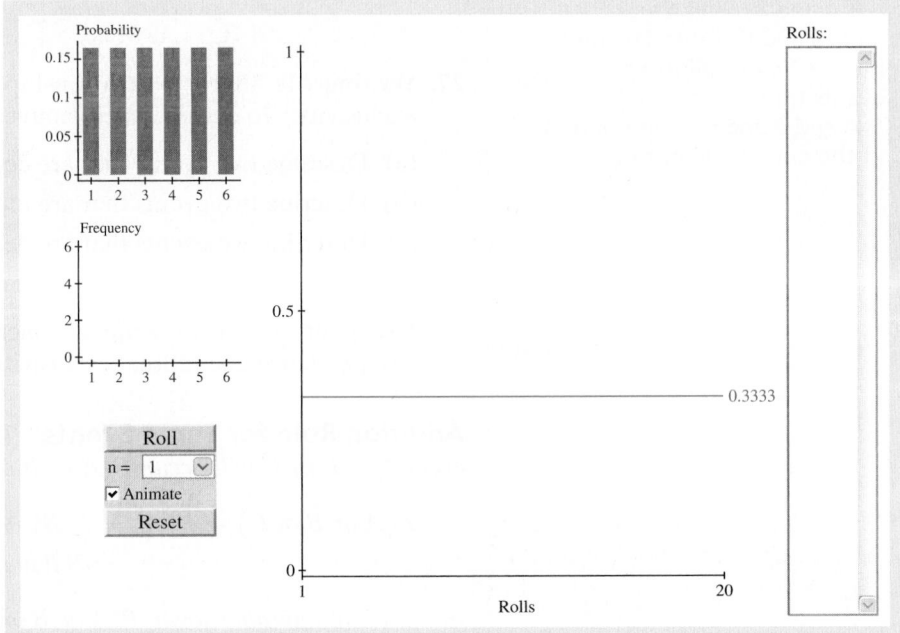

■ Explore

Step 1 Specify a value for *n*.

Step 2 Click ROLL four times.

Step 3 Click RESET.

Step 4 Specify another value for *n*.

Step 5 Click ROLL.

■ Draw Conclusions

APPLET

1. What is the theoretical probability of rolling a 3 or 4?

2. Run the simulation using each value of *n* one time. Clear the results after each trial. Compare the cumulative proportion of rolling a 3 or 4 for each trial with the theoretical probability of rolling a 3 or 4.

3. Suppose you want to modify the applet so you can find the probability of rolling a number less than 4. Describe the placement of the green line.

United States Congress

Congress is made up of the House of Representatives and the Senate. Members of the House of Representatives serve two-year terms and represent a district in a state. The number of representatives each state has is determined by population. States with larger populations have more representatives than states with smaller populations. The total number of representatives is set by law at 435 members. Members of the Senate serve six-year terms and represent a state. Each state has 2 senators, for a total of 100. The tables show the makeup of the 111th Congress by gender and political party. There are two vacant seats in the House of Representatives.

House of Representatives

		Political Party			
		Republican	Democrat	Independent	Total
Gender	**Male**	161	196	0	357
	Female	17	59	0	76
	Total	178	255	0	433

Senate

		Political Party			
		Republican	Democrat	Independent	Total
Gender	**Male**	37	44	2	83
	Female	4	13	0	17
	Total	41	57	2	100

■ EXERCISES

1. Find the probability that a randomly selected representative is female. Find the probability that a randomly selected senator is female.

2. Compare the probabilities from Exercise 1.

3. A representative is selected at random. Find the probability of each event.

   (a) The representative is male.

   (b) The representative is a Republican.

   (c) The representative is male given that the representative is a Republican.

   (d) The representative is female and a Democrat.

   (e) Are the events "being female" and "being a Democrat" independent or dependent events? Explain.

4. A senator is selected at random. Find the probability of each event.

   (a) The senator is male.

   (b) The senator is not a Democrat.

   (c) The senator is female or a Republican.

   (d) The senator is male or a Democrat.

   (e) Are the events "being female" and "being an Independent" mutually exclusive? Explain.

5. Using the same row and column headings as the tables above, create a combined table for Congress.

6. A member of Congress is selected at random. Use the table from Exercise 5 to find the probability of each event.

   (a) The member is Independent.

   (b) The member is female and a Republican.

   (c) The member is male or a Democrat.

3.4 Additional Topics in Probability and Counting

Permutations ▸ Combinations ▸ Applications of Counting Principles

WHAT YOU SHOULD LEARN

▸ How to find the number of ways a group of objects can be arranged in order

▸ How to find the number of ways to choose several objects from a group without regard to order

▸ How to use counting principles to find probabilities

▸ PERMUTATIONS

In Section 3.1, you learned that the Fundamental Counting Principle is used to find the number of ways two or more events can occur in sequence. In this section, you will study several other techniques for counting the number of ways an event can occur. An important application of the Fundamental Counting Principle is determining the number of ways that n objects can be arranged in order or in a *permutation*.

DEFINITION

A **permutation** is an ordered arrangement of objects. The number of different permutations of n distinct objects is $n!$.

The expression $n!$ is read as n **factorial** and is defined as follows.

$$n! = n \cdot (n - 1) \cdot (n - 2) \cdot (n - 3) \cdots 3 \cdot 2 \cdot 1$$

As a special case, $0! = 1$. Here are several other values of $n!$.

$$1! = 1, \ 2! = 2 \cdot 1 = 2, \ 3! = 3 \cdot 2 \cdot 1 = 6, \ 4! = 4 \cdot 3 \cdot 2 \cdot 1 = 24$$

STUDY TIP

Notice at the right that as n increases, $n!$ becomes very large. Take some time now to learn how to use the factorial key on your calculator.

EXAMPLE 1

▸ Finding the Number of Permutations of n Objects

The objective of a 9×9 Sudoku number puzzle is to fill the grid so that each row, each column, and each 3×3 grid contain the digits 1 to 9. How many different ways can the first row of a blank 9×9 Sudoku grid be filled?

▸ Solution

The number of permutations is $9! = 9 \cdot 8 \cdot 7 \cdot 6 \cdot 5 \cdot 4 \cdot 3 \cdot 2 \cdot 1 = 362,880$. So, there are 362,880 different ways the first row can be filled.

▸ Try It Yourself 1

The women's hockey teams for the 2010 Olympics are Canada, Sweden, Switzerland, Slovakia, United States, Finland, Russia, and China. How many different final standings are possible?

a. Determine the total number of women's hockey teams n that are in the 2010 Olympics.

b. Evaluate $n!$.

Answer: Page A35

Sudoku Number Puzzle

6	7	1				2	4	9
8			7		2			1
2				6				3
	5		6		3		2	
		8				7		
	1		8		4		6	
9				1				6
1			5		9			7
5	8	7				9	1	2

Suppose you want to choose some of the objects in a group and put them in order. Such an ordering is called a **permutation of n objects taken r at a time.**

PERMUTATIONS OF n OBJECTS TAKEN r AT A TIME

The number of permutations of n distinct objects taken r at a time is

$$_nP_r = \frac{n!}{(n - r)!}, \text{ where } r \leq n.$$

EXAMPLE 2

▶ Finding $_nP_r$

Find the number of ways of forming four-digit codes in which no digit is repeated.

▶ Solution

To form a four-digit code with no repeating digits, you need to select 4 digits from a group of 10, so $n = 10$ and $r = 4$.

$$_nP_r = {_{10}P_4} = \frac{10!}{(10 - 4)!}$$

$$= \frac{10!}{6!} = \frac{10 \cdot 9 \cdot 8 \cdot 7 \cdot 6 \cdot 5 \cdot 4 \cdot 3 \cdot 2 \cdot 1}{6 \cdot 5 \cdot 4 \cdot 3 \cdot 2 \cdot 1} = 5040$$

So, there are 5040 possible four-digit codes that do not have repeating digits.

▶ Try It Yourself 2

A psychologist shows a list of eight activities to her subject. How many ways can the subject pick a first, second, and third activity?

a. Find the quotient of $n!$ and $(n - r)!$. (List the factors and divide out.)
b. *Write* the result as a sentence.   *Answer: Page A35*

INSIGHT

Notice that the Fundamental Counting Principle can be used in Example 3 to obtain the same result. There are 43 choices for first place, 42 choices for second place, and 41 choices for third place. So, there are

$$43 \cdot 42 \cdot 41 = 74{,}046$$

ways the cars can finish first, second, and third.

EXAMPLE 3

▶ Finding $_nP_r$

Forty-three race cars started the 2010 Daytona 500. How many ways can the cars finish first, second, and third?

▶ Solution

You need to select three race cars from a group of 43, so $n = 43$ and $r = 3$. Because the order is important, the number of ways the cars can finish first, second, and third is

$$_nP_r = {_{43}P_3} = \frac{43!}{(43 - 3)!} = \frac{43!}{40!} = 43 \cdot 42 \cdot 41 = 74{,}046.$$

▶ Try It Yourself 3

The board of directors of a company has 12 members. One member is the president, another is the vice president, another is the secretary, and another is the treasurer. How many ways can these positions be assigned?

a. *Identify* the total number of objects n and the number of objects r being chosen in order.
b. *Evaluate* $_nP_r$.   *Answer: Page A35*

STUDY TIP

The letters *AAAABBC* can be rearranged in 7! orders, but many of these are not distinguishable. The number of distinguishable orders is

$$\frac{7!}{4! \cdot 2! \cdot 1!} = \frac{7 \cdot 6 \cdot 5}{2}$$

$$= 105.$$

You may want to order a group of n objects in which some of the objects are the same. For instance, consider a group of letters consisting of four As, two Bs, and one C. How many ways can you order such a group? Using the previous formula, you might conclude that there are $_7P_7 = 7!$ possible orders. However, because some of the objects are the same, not all of these permutations are *distinguishable*. How many distinguishable permutations are possible? The answer can be found using the formula for the number of distinguishable permutations.

DISTINGUISHABLE PERMUTATIONS

The number of **distinguishable permutations** of n objects, where n_1 are of one type, n_2 are of another type, and so on, is

$$\frac{n!}{n_1! \cdot n_2! \cdot n_3! \cdots n_k!}, \text{ where } n_1 + n_2 + n_3 + \cdots + n_k = n.$$

EXAMPLE 4

▶ **Finding the Number of Distinguishable Permutations**

A building contractor is planning to develop a subdivision. The subdivision is to consist of 6 one-story houses, 4 two-story houses, and 2 split-level houses. In how many distinguishable ways can the houses be arranged?

▶ **Solution**

There are to be 12 houses in the subdivision, 6 of which are of one type (one-story), 4 of another type (two-story), and 2 of a third type (split-level). So, there are

$$\frac{12!}{6! \cdot 4! \cdot 2!} = \frac{12 \cdot 11 \cdot 10 \cdot 9 \cdot 8 \cdot 7 \cdot 6!}{6! \cdot 4! \cdot 2!}$$

$$= 13{,}860 \text{ distinguishable ways.}$$

Interpretation There are 13,860 distinguishable ways to arrange the houses in the subdivision.

▶ **Try It Yourself 4**

The contractor wants to plant six oak trees, nine maple trees, and five poplar trees along the subdivision street. The trees are to be spaced evenly. In how many distinguishable ways can they be planted?

a. *Identify* the total number of objects n and the number of each type of object in the groups n_1, n_2, and n_3.

b. *Evaluate* $\dfrac{n!}{n_1! \cdot n_2! \cdots n_k!}$.

Answer: Page A36

▶ **COMBINATIONS**

You want to buy three DVDs from a selection of five DVDs labeled A, B, C, D, and E. There are 10 ways to make your selections.

ABC, ABD, ABE, ACD, ACE, ADE, BCD, BCE, BDE, CDE

In each selection, order does not matter (ABC is the same set as BAC). The number of ways to choose r objects from n objects without regard to order is called the number of **combinations of n objects taken r at a time.**

INSIGHT

You can think of a combination of n objects chosen r at a time as a permutation of n objects in which the r selected objects are alike and the remaining $n - r$ (not selected) objects are alike.

COMBINATIONS OF n OBJECTS TAKEN r AT A TIME

A combination is a selection of r objects from a group of n objects without regard to order and is denoted by $_nC_r$. The number of combinations of r objects selected from a group of n objects is

$$_nC_r = \frac{n!}{(n - r)!r!}.$$

STUDY TIP

Here are instructions for finding the number of combinations of n objects taken r at a time on a TI-83/84 Plus.

Enter the total number of objects n.

MATH

Choose the PRB menu.

3: nCr

Enter the number of objects r taken.

ENTER

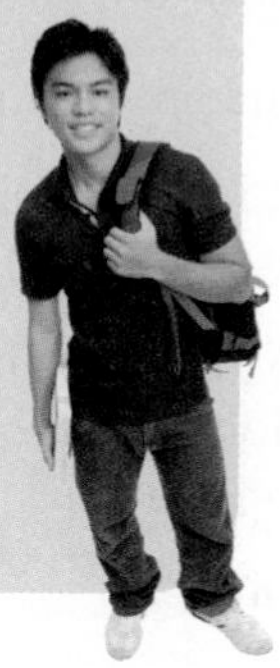

EXAMPLE 5

▸ Finding the Number of Combinations

A state's department of transportation plans to develop a new section of interstate highway and receives 16 bids for the project. The state plans to hire four of the bidding companies. How many different combinations of four companies can be selected from the 16 bidding companies?

▸ Solution

The state is selecting four companies from a group of 16, so $n = 16$ and $r = 4$. Because order is not important, there are

$$\begin{aligned} {}_nC_r = {}_{16}C_4 &= \frac{16!}{(16-4)!4!} \\ &= \frac{16!}{12!4!} \\ &= \frac{16 \cdot 15 \cdot 14 \cdot 13 \cdot 12!}{12! \cdot 4!} \\ &= 1820 \text{ different combinations.} \end{aligned}$$

Interpretation There are 1820 different combinations of four companies that can be selected from the 16 bidding companies.

▸ Try It Yourself 5

The manager of an accounting department wants to form a three-person advisory committee from the 20 employees in the department. In how many ways can the manager form this committee?

a. *Identify* the number of objects in the group n and the number of objects r to be selected.
b. *Evaluate* ${}_nC_r$.
c. *Write* the result as a sentence.

Answer: Page A36

The table summarizes the counting principles.

Principle	Description	Formula
Fundamental Counting Principle	If one event can occur in m ways and a second event can occur in n ways, the number of ways the two events can occur in sequence is $m \cdot n$.	$m \cdot n$
Permutations	The number of different ordered arrangements of n distinct objects	$n!$
	The number of permutations of n distinct objects taken r at a time, where $r \leq n$	${}_nP_r = \frac{n!}{(n-r)!}$
	The number of distinguishable permutations of n objects where n_1 are of one type, n_2 are of another type, and so on	$\frac{n!}{n_1! \cdot n_2! \cdots n_k!}$
Combinations	The number of combinations of r objects selected from a group of n objects without regard to order	${}_nC_r = \frac{n!}{(n-r)!r!}$

If you buy one ticket, the probability you win the Mega Millions lottery is

$$\frac{1}{175,711,536} \approx 5.69 \times 10^{-9}.$$

▶ **APPLICATIONS OF COUNTING PRINCIPLES**

EXAMPLE 6

▶ **Finding Probabilities**

A student advisory board consists of 17 members. Three members serve as the board's chair, secretary, and webmaster. Each member is equally likely to serve in any of the positions. What is the probability of selecting at random the three members who currently hold the three positions?

▶ **Solution**   There is one favorable outcome and there are

$$_{17}P_3 = \frac{17!}{(17-3)!} = \frac{17!}{14!} = \frac{17 \cdot 16 \cdot 15 \cdot 14!}{14!} = 17 \cdot 16 \cdot 15 = 4080$$

ways the three positions can be filled. So, the probability of correctly selecting the three members who hold each position is

$$P(\text{selecting the three members}) = \frac{1}{4080} \approx 0.0002.$$

▶ **Try It Yourself 6**

A student advisory board consists of 20 members. Two members serve as the board's chair and secretary. Each member is equally likely to serve in either of the positions. What is the probability of selecting at random the two members who currently hold the two positions?

a. Find the *number of ways* the two positions can be filled.
b. Find the *probability* of correctly selecting the two members.

Answer: Page A36

EXAMPLE 7

▶ **Finding Probabilities**

You have 11 letters consisting of one M, four I's, four S's, and two P's. If the letters are randomly arranged in order, what is the probability that the arrangement spells the word *Mississippi*?

▶ **Solution**   There is one favorable outcome and there are

$$\frac{11!}{1! \cdot 4! \cdot 4! \cdot 2!} = 34,650 \qquad \text{11 letters with 1, 4, 4, and 2 like letters}$$

distinguishable permutations of the given letters. So, the probability that the arrangement spells the word *Mississippi* is

$$P(\text{Mississippi}) = \frac{1}{34,650} \approx 0.00003.$$

▶ **Try It Yourself 7**

You have 6 letters consisting of one L, two E's, two T's, and one R. If the letters are randomly arranged in order, what is the probability that the arrangement spells the word *letter*?

a. Find the *number of favorable outcomes* and the *number of distinguishable permutations*.
b. Find the *probability* that the arrangement spells the word *letter*.

Answer: Page A36

EXAMPLE 8

▸ **Finding Probabilities**

Find the probability of picking five diamonds from a standard deck of playing cards.

▸ **Solution**

The possible number of ways of choosing 5 diamonds out of 13 is $_{13}C_5$. The number of possible five-card hands is $_{52}C_5$. So, the probability of being dealt 5 diamonds is

$$P(5 \text{ diamonds}) = \frac{_{13}C_5}{_{52}C_5} = \frac{1287}{2,598,960} \approx 0.0005.$$

▸ **Try It Yourself 8**

Find the probability of being dealt five diamonds from a standard deck of playing cards that also includes two jokers. In this case, the joker is considered to be a wild card that can be used to represent any card in the deck.

a. Find the *number of ways* of choosing 5 diamonds.
b. Find the *number of possible five-card hands*.
c. Find the *probability* of being dealt five diamonds.   *Answer: Page A36*

EXAMPLE 9

▸ **Finding Probabilities**

A food manufacturer is analyzing a sample of 400 corn kernels for the presence of a toxin. In this sample, three kernels have dangerously high levels of the toxin. If four kernels are randomly selected from the sample, what is the probability that exactly one kernel contains a dangerously high level of the toxin?

▸ **Solution**

The possible number of ways of choosing one toxic kernel out of three toxic kernels is $_3C_1$. The possible number of ways of choosing 3 nontoxic kernels from 397 nontoxic kernels is $_{397}C_3$. So, using the Fundamental Counting Principle, the number of ways of choosing one toxic kernel and three nontoxic kernels is

$$_3C_1 \cdot {_{397}C_3} = 3 \cdot 10,349,790$$
$$= 31,049,370.$$

The number of possible ways of choosing 4 kernels from 400 kernels is $_{400}C_4 = 1,050,739,900$. So, the probability of selecting exactly 1 toxic kernel is

$$P(1 \text{ toxic kernel}) = \frac{_3C_1 \cdot {_{397}C_3}}{_{400}C_4} = \frac{31,049,370}{1,050,739,900} \approx 0.030.$$

▸ **Try It Yourself 9**

A jury consists of five men and seven women. Three jury members are selected at random for an interview. Find the probability that all three are men.

a. *Find* the product of the number of ways to choose three men from five and the number of ways to choose zero women from seven.
b. Find the *number of ways* to choose 3 jury members from 12.
c. Find the *probability* that all three are men.   *Answer: Page A36*

3.4 EXERCISES

FOR EXTRA HELP:
MyStatLab

1. The number of ordered arrangements of *n* objects taken *r* at a time. An example of a permutation is the number of seating arrangements of you and three of your friends.

2. A selection of *r* of the *n* objects without regard to order. An example of a combination is the number of selections of different playoff teams from a volleyball tournament.

3. False. A permutation is an ordered arrangement of objects.

4. True

5. True

6. True

7. 15,120

8. 240

9. 56

10. 840

11. 203,490

12. 0.076

13. 0.030

14. 0.035

15. Permutation. The order of the eight cars in line matters.

16. Combination. The order of the committee members does not matter.

17. Combination. The order does not matter because the position of one captain is the same as the other.

18. Permutation. The order of the letters matters in the password.

■ BUILDING BASIC SKILLS AND VOCABULARY

1. When you calculate the number of permutations of *n* distinct objects taken *r* at a time, what are you counting? Give an example.

2. When you calculate the number of combinations of *r* objects taken from a group of *n* objects, what are you counting? Give an example.

True or False? *In Exercises 3–6, determine whether the statement is true or false. If it is false, rewrite it as a true statement.*

3. A combination is an ordered arrangement of objects.

4. The number of different ordered arrangements of *n* distinct objects is *n*!.

5. If you divide the number of permutations of 11 objects taken 3 at a time by 3!, you will get the number of combinations of 11 objects taken 3 at a time.

6. $_7C_5 = {_7C_2}$

In Exercises 7–14, perform the indicated calculation.

7. $_9P_5$

8. $_{16}P_2$

9. $_8C_3$

10. $_7P_4$

11. $_{21}C_8$

12. $\dfrac{_8C_4}{_{12}C_6}$

13. $\dfrac{_6P_2}{_{11}P_3}$

14. $\dfrac{_{10}C_7}{_{14}C_7}$

In Exercises 15–18, decide if the situation involves permutations, combinations, or neither. Explain your reasoning.

15. The number of ways eight cars can line up in a row for a car wash

16. The number of ways a four-member committee can be chosen from 10 people

17. The number of ways 2 captains can be chosen from 28 players on a lacrosse team

18. The number of four-letter passwords that can be created when no letter can be repeated

■ USING AND INTERPRETING CONCEPTS

19. **Video Games** You have seven different video games. How many different ways can you arrange the games side by side on a shelf?

20. **Skiing** Eight people compete in a downhill ski race. Assuming that there are no ties, in how many different orders can the skiers finish?

21. **Security Code** In how many ways can the letters A, B, C, D, E, and F be arranged for a six-letter security code?

22. **Starting Lineup** The starting lineup for a softball team consists of 10 players. How many different batting orders are possible using the starting lineup?

19. 5040

20. 40,320

21. 720

22. 3,628,800

23. 20,358,520

24. 24

25. 320,089,770

26. 4845

27. 50,400

28. 5,586,853,480

29. 6240

30. 3240

31. 86,296,950

32. 3,304,845,180

33. (a) 720

(b) sample

(c) 0.0014; Yes, the event can be considered unusual because its probability is less than or equal to 0.05.

34. (a) 60

(b) event

(c) 0.017; Yes, the event can be considered unusual because its probability is less than or equal to 0.05.

35. (a) 12

(b) tree

(c) 0.083; No, the event cannot be considered unusual because its probability is not less than or equal to 0.05.

36. (a) 360

(b) center

(c) 0.003; Yes, the event can be considered unusual because its probability is less than or equal to 0.05.

37. (a) 907,200

(b) population

(c) 0.000001; Yes, the event can be considered unusual because its probability is less than or equal to 0.05.

38. (a) 39,916,800

(b) distribution

(c) 0.00000003; Yes, the event can be considered unusual because its probability is less than or equal to 0.05.

23. Lottery Number Selection   A lottery has 52 numbers. In how many different ways can 6 of the numbers be selected? (Assume that order of selection is not important.)

24. Assembly Process   There are four processes involved in assembling a certain product. These processes can be performed in any order. Management wants to find which order is the least time-consuming. How many different orders will have to be tested?

25. Bracelets   You are putting 4 spacers, 10 gold charms, and 8 silver charms on a bracelet. In how many distinguishable ways can the spacers and charms be put on the bracelet?

26. Experimental Group   In order to conduct an experiment, 4 subjects are randomly selected from a group of 20 subjects. How many different groups of four subjects are possible?

27. Letters   In how many distinguishable ways can the letters in the word *statistics* be written?

28. Jury Selection   From a group of 40 people, a jury of 12 people is selected. In how many different ways can a jury of 12 people be selected?

29. Space Shuttle Menu   Space shuttle astronauts each consume an average of 3000 calories per day. One meal normally consists of a main dish, a vegetable dish, and two different desserts. The astronauts can choose from 10 main dishes, 8 vegetable dishes, and 13 desserts. How many different meals are possible? *(Source: NASA)*

30. Menu   A restaurant offers a dinner special that has 12 choices for entrées, 10 choices for side dishes, and 6 choices for dessert. For the special, you can choose one entrée, two side dishes, and one dessert. How many different meals are possible?

31. Water Samples   An environmental agency is analyzing water samples from 80 lakes for pollution. Five of the lakes have dangerously high levels of dioxin. If six lakes are randomly selected from the sample, how many ways could one polluted lake and five non-polluted lakes be chosen? Use a technology tool.

32. Soil Samples   An environmental agency is analyzing soil samples from 50 farms for lead contamination. Eight of the farms have dangerously high levels of lead. If 10 farms are randomly selected from the sample, how many ways could 2 contaminated farms and 8 noncontaminated farms be chosen? Use a technology tool.

Word Jumble   *In Exercises 33–38, do the following.*

(a) Find the number of distinguishable ways the letters can be arranged.

(b) There is one arrangement that spells an important term used throughout the course. Find the term.

(c) If the letters are randomly arranged in order, what is the probability that the arrangement spells the word from part (b)? Can this event be considered unusual? Explain.

33. palmes

34. nevte

35. etre

36. rnctee

37. unoppolati

38. sidtbitoiurn

39. 0.005
40. 0.012
41. (a) 0.016
    (b) 0.385
42. (a) 0.015
    (b) 0.070
43. (a) 70
    (b) 16
    (c) 0.086
44. (a) 6,760,000
    (b) 5,760,000
    (c) 0.5
45. (a) 67,600,000
    (b) 19,656,000
    (c) 0.000000015
46. (a) 1000
    (b) 800
    (c) 0.5
47. (a) 120
    (b) 12
    (c) 12
    (d) 0.4
48. (a) 56
    (b) 56
    (c) 112
    (d) 0.933

39. **Horse Race**   A horse race has 12 entries. Assuming that there are no ties, what is the probability that the three horses owned by one person finish first, second, and third?

40. **Pizza Toppings**   A pizza shop offers nine toppings. No topping is used more than once. What is the probability that the toppings on a three-topping pizza are pepperoni, onions, and mushrooms?

41. **Jukebox**   You look over the songs on a jukebox and determine that you like 15 of the 56 songs.

    (a) What is the probability that you like the next three songs that are played? (Assume a song cannot be repeated.)

    (b) What is the probability that you do not like the next three songs that are played? (Assume a song cannot be repeated.)

42. **Officers**   The offices of president, vice president, secretary, and treasurer for an environmental club will be filled from a pool of 14 candidates. Six of the candidates are members of the debate team.

    (a) What is the probability that all of the offices are filled by members of the debate team?

    (b) What is the probability that none of the offices are filled by members of the debate team?

43. **Employee Selection**   Four sales representatives for a company are to be chosen to participate in a training program. The company has eight sales representatives, two in each of four regions. In how many ways can the four sales representatives be chosen if (a) there are no restrictions and (b) the selection must include a sales representative from each region? (c) What is the probability that the four sales representatives chosen to participate in the training program will be from only two of the four regions if they are chosen at random?

44. **License Plates**   In a certain state, each automobile license plate number consists of two letters followed by a four-digit number. How many distinct license plate numbers can be formed if (a) there are no restrictions and (b) the letters O and I are not used? (c) What is the probability of selecting at random a license plate that ends in an even number?

45. **Password**   A password consists of two letters followed by a five-digit number. How many passwords are possible if (a) there are no restrictions and (b) none of the letters or digits can be repeated? (c) What is the probability of guessing the password in one trial if there are no restrictions?

46. **Area Code**   An area code consists of three digits. How many area codes are possible if (a) there are no restrictions and (b) the first digit cannot be a 1 or a 0? (c) What is the probability of selecting an area code at random that ends in an odd number if the first digit cannot be a 1 or a 0?

47. **Repairs**   In how many orders can three broken computers and two broken printers be repaired if (a) there are no restrictions, (b) the printers must be repaired first, and (c) the computers must be repaired first? (d) If the order of repairs has no restrictions and the order of repairs is done at random, what is the probability that a printer will be repaired first?

48. **Defective Units**   A shipment of 10 microwave ovens contains two defective units. In how many ways can a restaurant buy three of these units and receive (a) no defective units, (b) one defective unit, and (c) at least two nondefective units? (d) What is the probability of the restaurant buying at least two nondefective units?

Rate Your Financial Shape

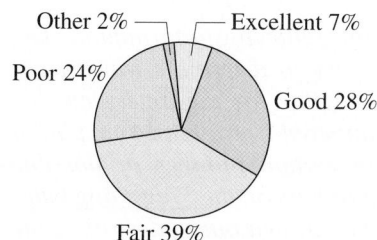

Other 2% — — Excellent 7%
Poor 24%
Good 28%
Fair 39%

FIGURE FOR EXERCISES 49–52

49. 0.000022
50. 0.000000562
51. 6.00×10^{-20}
52. 0.00000000401
53. (a) 658,008
    (b) 0.00000152
54. (a) 1.46×10^{22}
    (b) 8.53×10^{19}
    (c) 0.006
    (d) Yes, because the probability that no minorities are selected is very small.
55. (a) 0.0002
    (b) 0.0014
    (c) 0.0211
    (d) 0.0659
56. (a) 0.0077
    (b) 0.0003
    (c) 0.1916
    (d) 0.2647

Financial Shape *In Exercises 49–52, use the pie chart, which shows how U.S. adults rate their financial shape.* (*Source: Pew Research Center*)

49. Suppose 4 people are chosen at random from a group of 1200. What is the probability that all four would rate their financial shape as excellent? (Make the assumption that the 1200 people are represented by the pie chart.)

50. Suppose 10 people are chosen at random from a group of 1200. What is the probability that all 10 would rate their financial shape as poor? (Make the assumption that the 1200 people are represented by the pie chart.)

51. Suppose 80 people are chosen at random from a group of 500. What is the probability that none of the 80 people would rate their financial shape as fair? (Make the assumption that the 500 people are represented by the pie chart.)

52. Suppose 55 people are chosen at random from a group of 500. What is the probability that none of the 55 people would rate their financial shape as good? (Make the assumption that the 500 people are represented by the pie chart.)

53. **Probability**   In a state lottery, you must correctly select 5 numbers (in any order) out of 40 to win the top prize.

    (a) How many ways can 5 numbers be chosen from 40 numbers?

    (b) You purchase one lottery ticket. What is the probability that you will win the top prize?

54. **Probability**   A company that has 200 employees chooses a committee of 15 to represent employee retirement issues. When the committee is formed, none of the 56 minority employees are selected.

    (a) Use a technology tool to find the number of ways 15 employees can be chosen from 200.

    (b) Use a technology tool to find the number of ways 15 employees can be chosen from 144 nonminorities.

    (c) If the committee is chosen randomly (without bias), what is the probability that it contains no minorities?

    (d) Does your answer to part (c) indicate that the committee selection is biased? Explain your reasoning.

55. **Cards**   You are dealt a hand of five cards from a standard deck of playing cards. Find the probability of being dealt a hand consisting of

    (a) four-of-a-kind.

    (b) a full house, which consists of three of one kind and two of another kind.

    (c) three-of-a-kind. (The other two cards are different from each other.)

    (d) two clubs and one of each of the other three suits.

56. **Warehouse**   A warehouse employs 24 workers on first shift and 17 workers on second shift. Eight workers are chosen at random to be interviewed about the work environment. Find the probability of choosing

    (a) all first-shift workers.

    (b) all second-shift workers.

    (c) six first-shift workers.

    (d) four second-shift workers.

57. 1001; 1000
58. 24,024

59.

Team (worst team first)	1	2	3
Probability	0.250	0.199	0.156

Team (worst team first)	4	5	6
Probability	0.119	0.088	0.063

Team (worst team first)	7	8	9
Probability	0.043	0.028	0.017

Team (worst team first)	10	11	12
Probability	0.011	0.008	0.007

Team (worst team first)	13	14
Probability	0.006	0.005

Events in which any of Teams 7–14 win the first pick would be considered unusual because the probabilities are all less than or equal to 0.05.

60. 0.251
61. 0.314
62. 0.551

■ EXTENDING CONCEPTS

NBA Draft Lottery *In Exercises 57–62, use the following information. The National Basketball Association (NBA) uses a lottery to determine which team gets the first pick in its annual draft. The teams eligible for the lottery are the 14 non-playoff teams. Fourteen Ping-Pong balls numbered 1 through 14 are placed in a drum. Each of the 14 teams is assigned a certain number of possible four-number combinations that correspond to the numbers on the Ping-Pong balls, such as 3, 8, 10, and 12, as shown. Four balls are then drawn out to determine the first pick in the draft. The order in which the balls are drawn is not important. All of the four-number combinations are assigned to the 14 teams by computer except for one four-number combination. When this four-number combination is drawn, the balls are put back in the drum and another drawing takes place. For instance, if Team A has been assigned the four-number combination 3, 8, 10, 12 and the balls shown at the left are drawn, then Team A wins the first pick.*

After the first pick of the draft is determined, the process continues to choose the teams that will select second and third picks. A team may not win the lottery more than once. If the four-number combination belonging to a team that has already won is drawn, the balls are put back in the drum and another drawing takes place. The remaining order of the draft is determined by the number of losses of each team.

57. In how many ways can 4 of the numbers 1 to 14 be selected if order is not important? How many sets of 4 numbers are assigned to the 14 teams?

58. In how many ways can four of the numbers be selected if order is important?

In the Pareto chart, the number of combinations assigned to each of the 14 teams is shown. The team with the most losses (the worst team) gets the most chances to win the lottery. So, the worst team receives the greatest frequency of four-number combinations, 250. The team with the best record of the 14 non-playoff teams has the fewest chances, with 5 four-number combinations.

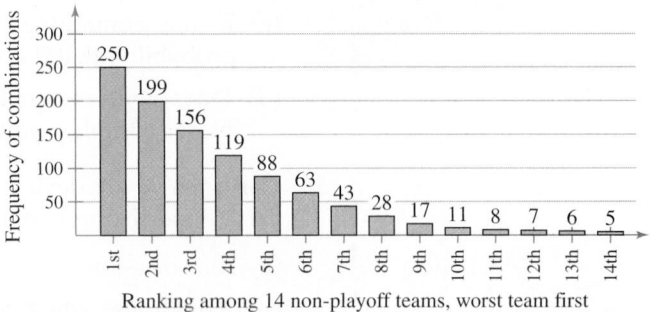

Frequency of Four-Number Combinations Assigned in the NBA Draft Lottery

59. For each team, find the probability that the team will win the first pick. Which of these events would be considered unusual? Explain.

60. What is the probability that the team with the worst record will win the second pick, given that the team with the best record, ranked 14th, wins the first pick?

61. What is the probability that the team with the worst record will win the third pick, given that the team with the best record, ranked 14th, wins the first pick and the team ranked 2nd wins the second pick?

62. What is the probability that neither the first- nor the second-worst team will get the first pick?

USES AND ABUSES

Uses

Probability affects decisions when the weather is forecast, when marketing strategies are determined, when medications are selected, and even when players are selected for professional sports teams. Although intuition is often used for determining probabilities, you will be better able to assess the likelihood that an event will occur by applying the rules of classical probability and empirical probability.

For instance, suppose you work for a real estate company and are asked to estimate the likelihood that a particular house will sell for a particular price within the next 90 days. You could use your intuition, but you could better assess the probability by looking at sales records for similar houses.

Abuses

One common abuse of probability is thinking that probabilities have "memories." For instance, if a coin is tossed eight times, the probability that it will land heads up all eight times is only about 0.004. However, if the coin has already been tossed seven times and has landed heads up each time, the probability that it will land heads up on the eighth time is 0.5. Each toss is independent of all other tosses. The coin does not "remember" that it has already landed heads up seven times.

Ethics

A human resources director for a company with 100 employees wants to show that her company is an equal opportunity employer of women and minorities. There are 40 women employees and 20 minority employees in the company. Nine of the women employees are minorities. Despite this fact, the director reports that 60% of the company is either a woman or a minority. If one employee is selected at random, the probability that the employee is a woman is 0.4 and the probability that the employee is a minority is 0.2. This does not mean, however, that the probability that a randomly selected employee is a woman or a minority is $0.4 + 0.2 = 0.6$, because nine employees belong to both groups. In this case, it would be ethically incorrect to omit this information from her report because these individuals would have been counted twice.

■ EXERCISES

1. *Assuming That Probability Has a "Memory"* A "Daily Number" lottery has a three-digit number from 000 to 999. You buy one ticket each day. Your number is 389.

   a. What is the probability of winning next Tuesday and Wednesday?

   b. You won on Tuesday. What's the probability of winning on Wednesday?

   c. You didn't win on Tuesday. What's the probability of winning on Wednesday?

2. *Adding Probabilities Incorrectly* A town has a population of 500 people. Suppose that the probability that a randomly chosen person owns a pickup truck is 0.25 and the probability that a randomly chosen person owns an SUV is 0.30. What can you say about the probability that a randomly chosen person owns a pickup or an SUV? Could this probability be 0.55? Could it be 0.60? Explain your reasoning.

3 CHAPTER SUMMARY

What did you learn?	EXAMPLE(S)	REVIEW EXERCISES
Section 3.1		
■ How to identify the sample space of a probability experiment and how to identify simple events	*1, 2*	*1–4*
■ How to use the Fundamental Counting Principle to find the number of ways two or more events can occur	*3, 4*	*5, 6*
■ How to distinguish among classical probability, empirical probability, and subjective probability	*5–8*	*7–12*
■ How to find the probability of the complement of an event and how to find other probabilities using the Fundamental Counting Principle	*9–11*	*13–16*
Section 3.2		
■ How to find conditional probabilities	*1*	*17, 18*
■ How to distinguish between independent and dependent events	*2*	*19–21*
■ How to use the Multiplication Rule to find the probability of two events occurring in sequence $P(A \text{ and } B) = P(A) \cdot P(B\|A)$ if events are dependent $P(A \text{ and } B) = P(A) \cdot P(B)$ if events are independent	*3–5*	*22–24*
Section 3.3		
■ How to determine if two events are mutually exclusive	*1*	*25–27*
■ How to use the Addition Rule to find the probability of two events $P(A \text{ or } B) = P(A) + P(B) - P(A \text{ and } B)$ $P(A \text{ or } B) = P(A) + P(B)$ if events are mutually exclusive	*2–5*	*28–40*
Section 3.4		
■ How to find the number of ways a group of objects can be arranged in order and the number of ways to choose several objects from a group without regard to order $_nP_r = \dfrac{n!}{(n-r)!}$ permutations of n objects taken r at a time $\dfrac{n!}{n_1! \cdot n_2! \cdot n_3! \cdots n_k!}$ distinguishable permutations $_nC_r = \dfrac{n!}{(n-r)!r!}$ combinations of n objects taken r at a time	*1–5*	*41–50*
■ How to use counting principles to find probabilities	*6–9*	*51–55*

3 ▶ REVIEW EXERCISES

1. Sample space:

{HHHH, HHHT, HHTH, HHTT,
HTHH, HTHT, HTTH, HTTT,
THHH, THHT, THTH, THTT,
TTHH, TTHT, TTTH, TTTT}; 4

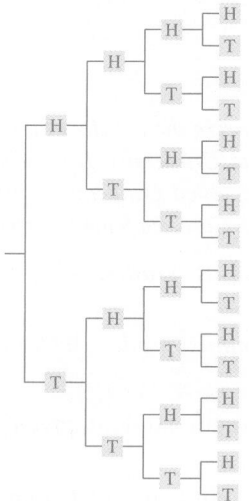

2. See Selected Answers, page A117.

3. Sample space:

{January, February, March,
April, May, June, July, August,
September, October, November,
December}; 3

4. See Selected Answers, page A117.

5. 84

6. 175,760,000

7. Empirical probability because it is based on observations obtained from probability experiments.

8. Classical probability because each outcome in the sample space is equally likely to occur.

9. Subjective probability because it is based on opinion.

10. Empirical probability because it is based on observations obtained from probability experiments.

11. Classical probability because all of the outcomes in the event and the sample space can be counted.

12. Empirical probability because it is based on observations obtained from probability experiments.

13. 0.215

14. 0.892

■ **SECTION 3.1**

In Exercises 1–4, identify the sample space of the probability experiment and determine the number of outcomes in the event. Draw a tree diagram if it is appropriate.

1. *Experiment:* Tossing four coins
Event: Getting three heads

2. *Experiment:* Rolling 2 six-sided dice
Event: Getting a sum of 4 or 5

3. *Experiment:* Choosing a month of the year
Event: Choosing a month that begins with the letter J

4. *Experiment:* Guessing the gender(s) of the three children in a family
Event: The family has two boys

In Exercises 5 and 6, use the Fundamental Counting Principle.

5. A student must choose from 7 classes to take at 8:00 A.M., 4 classes to take at 9:00 A.M., and 3 classes to take at 10:00 A.M. How many ways can the student arrange the schedule?

6. The state of Virginia's license plates have three letters followed by four digits. Assuming that any letter or digit can be used, how many different license plates are possible?

In Exercises 7–12, classify the statement as an example of classical probability, empirical probability, or subjective probability. Explain your reasoning.

7. On the basis of prior counts, a quality control officer says there is a 0.05 probability that a randomly chosen part is defective.

8. The probability of randomly selecting five cards of the same suit from a standard deck is about 0.0005.

9. The chance that Corporation A's stock price will fall today is 75%.

10. The probability that a person can roll his or her tongue is 70%.

11. The probability of rolling 2 six-sided dice and getting a sum greater than 9 is $\frac{1}{6}$.

12. The chance that a randomly selected person in the United States is between 15 and 29 years old is about 21%. *(Source: U.S. Census Bureau)*

In Exercises 13 and 14, the table shows the approximate distribution of the sizes of firms for a recent year. Use the table to determine the probability of the event.
(Adapted from U.S. Small Business Administration)

Number of employees	0 to 4	5 to 9	10 to 19	20 to 99	100 or more
Percent of firms	60.9%	17.6%	10.7%	9.0%	1.8%

13. What is the probability that a randomly selected firm will have at least 10 employees?

14. What is the probability that a randomly selected firm will have fewer than 20 employees?

15. 1.25×10^{-7}

16. 0.999999875

17. 0.92

18. 0.07

19. Independent. The outcomes of the first four coin tosses do not affect the outcome of the fifth coin toss.

20. Dependent. The outcome of taking a driver's education course affects the outcome of passing the driver's license exam.

21. Dependent. The outcome of getting high grades affects the outcome of being awarded an academic scholarship.

22. No; You do not know whether events A and B are independent or dependent.

23. 0.025; Yes, the event is unusual because its probability is less than or equal to 0.05.

24. 0.235; No, the event is not unusual because its probability is not less than or equal to 0.05.

25. Mutually exclusive. A jelly bean cannot be both completely red and completely yellow.

26. Not mutually exclusive. A person who loves cats can also own a dog.

Telephone Numbers   *The telephone numbers for a region of a state have an area code of 570. The next seven digits represent the local telephone numbers for that region. A local telephone number cannot begin with a 0 or 1. Your cousin lives within the given area code.*

15. What is the probability of randomly generating your cousin's telephone number?

16. What is the probability of not randomly generating your cousin's telephone number?

■ SECTION 3.2

For Exercises 17 and 18, the two statements below summarize the results of a study on the use of plus/minus grading at North Carolina State University. It shows the percents of graduate and undergraduate students who received grades with pluses and minuses (for example, C+, A−, etc.).   *(Source: North Carolina State University)*

- *Of all students who received one or more plus grades, 92% were undergraduates and 8% were graduates.*
- *Of all students who received one or more minus grades, 93% were undergraduates and 7% were graduates.*

17. Find the probability that a student is an undergraduate student, given that the student received a plus grade.

18. Find the probability that a student is a graduate student, given that the student received a minus grade.

In Exercises 19–21, decide whether the events are independent or dependent. Explain your reasoning.

19. Tossing a coin four times, getting four heads, and tossing it a fifth time and getting a head

20. Taking a driver's education course and passing the driver's license exam

21. Getting high grades and being awarded an academic scholarship

22. You are given that $P(A) = 0.35$ and $P(B) = 0.25$. Do you have enough information to find $P(A \text{ and } B)$? Explain.

In Exercises 23 and 24, find the probability of the sequence of events.

23. You are shopping, and your roommate has asked you to pick up toothpaste and dental rinse. However, your roommate did not tell you which brands to get. The store has eight brands of toothpaste and five brands of dental rinse. What is the probability that you will purchase the correct brands of both products? Is this an unusual event? Explain.

24. Your sock drawer has 18 folded pairs of socks, with 8 pairs of white, 6 pairs of black, and 4 pairs of blue. What is the probability, without looking in the drawer, that you will first select and remove a black pair, then select either a blue or a white pair? Is this an unusual event? Explain.

■ SECTION 3.3

In Exercises 25–27, decide if the events are mutually exclusive. Explain your reasoning.

25. Event A: Randomly select a red jelly bean from a jar.
Event B: Randomly select a yellow jelly bean from the same jar.

26. Event A: Randomly select a person who loves cats.
Event B: Randomly select a person who owns a dog.

27. Mutually exclusive. A person cannot be registered to vote in more than one state.

28. No; You do not know whether events A and B are mutually exclusive.

29. 0.60

30. 0.25

31. 0.538

32. 0.538

33. 0.583

34. 0.625

35. 0.291

36. 0.718

37. 0.188

38. 0.307

27. Event A: Randomly select a U.S. adult registered to vote in Illinois.
Event B: Randomly select a U.S. adult registered to vote in Florida.

28. You are given that $P(A) = 0.15$ and $P(B) = 0.40$. Do you have enough information to find $P(A$ or $B)$? Explain.

29. A random sample of 250 working adults found that 37% access the Internet at work, 44% access the Internet at home, and 21% access the Internet at both work and home. What is the probability that a person in this sample selected at random accesses the Internet at home or at work?

30. A sample of automobile dealerships found that 19% of automobiles sold are silver, 22% of automobiles sold are sport utility vehicles (SUVs), and 16% of automobiles sold are silver SUVs. What is the probability that a randomly chosen sold automobile from this sample is silver or an SUV?

In Exercises 31–34, determine the probability.

31. A card is randomly selected from a standard deck. Find the probability that the card is between 4 and 8, inclusive, or is a club.

32. A card is randomly selected from a standard deck. Find the probability that the card is red or a queen.

33. A 12-sided die, numbered 1 to 12, is rolled. Find the probability that the roll results in an odd number or a number less than 4.

34. An 8-sided die, numbered 1 to 8, is rolled. Find the probability that the roll results in an even number or a number greater than 6.

In Exercises 35 and 36, use the pie chart, which shows the percent distribution of the number of students in traditional U.S. elementary schools. (Source: U.S. National Center for Education Statistics)

35. Find the probability of randomly selecting a school with 600 or more students.

36. Find the probability of randomly selecting a school with between 300 and 999 students, inclusive.

In Exercises 37–40, use the Pareto chart, which shows the results of a survey in which 874 adults were asked which genre of movie they preferred. (Adapted from Rasmussen Reports)

Students in Elementary Schools

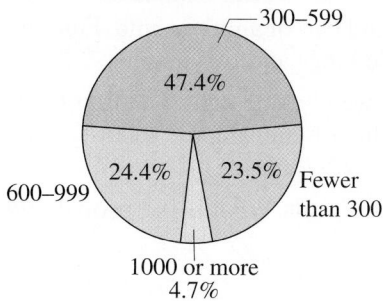

FIGURE FOR EXERCISES 35 AND 36

Which Genre of Movie Do You Prefer?

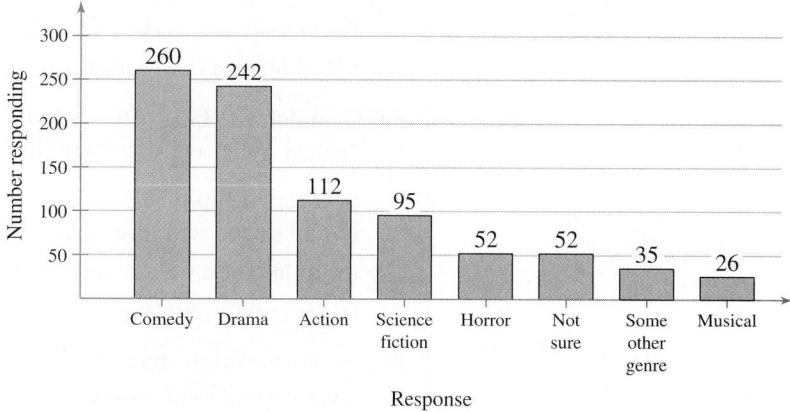

37. Find the probability of randomly selecting an adult from the sample who prefers an action movie or a horror movie.

38. Find the probability of randomly selecting an adult from the sample who prefers a drama or a musical.

39. 0.703
40. 0.763
41. 110
42. 20,160
43. 35
44. 0.083
45. 254,251,200
46. 5,414,950,296
47. 2730
48. 120
49. 2380
50. 78
51. 0.00000923; unusual
52. 0.0000064; unusual
53. (a) 0.955; not unusual
    (b) 0.000000761; unusual
    (c) 0.045; unusual
    (d) 0.999999239; not unusual
54. (a) 0.955; not unusual
    (b) 0.000000002; unusual
    (c) 0.045; unusual
    (d) 0.999999998; not unusual
55. (a) 0.071; not unusual
    (b) 0.005; unusual
    (c) 0.429; not unusual
    (d) 0.114; not unusual

39. Find the probability of randomly selecting an adult from the sample who does not prefer a comedy.

40. Find the probability of randomly selecting an adult from the sample who does not prefer a science fiction movie or an action movie.

■ SECTION 3.4

In Exercises 41–44, perform the indicated calculation.

41. $_{11}P_2$ **42.** $_8P_6$ **43.** $_7C_4$ **44.** $\dfrac{_5C_3}{_{10}C_3}$

45. Use a technology tool to find $_{50}P_5$.

46. Use a technology tool to find $_{38}C_{25}$.

In Exercises 47–50, use combinations and permutations.

47. Fifteen cyclists enter a race. In how many ways can they finish first, second, and third?

48. Five players on a basketball team must each choose a player on the opposing team to defend. In how many ways can they choose their defensive assignments?

49. A literary magazine editor must choose 4 short stories for this month's issue from 17 submissions. In how many ways can the editor choose this month's stories?

50. An employer must hire 2 people from a list of 13 applicants. In how many ways can the employer choose to hire the 2 people?

In Exercises 51–55, use counting principles to find the probability. Then tell whether the event can be considered unusual.

51. A full house consists of a three of one kind and two of another kind. Find the probability of a full house consisting of three kings and two queens.

52. A security code consists of three letters followed by one digit. The first letter cannot be an A, B, or C. What is the probability of guessing the security code in one trial?

53. A batch of 200 calculators contains 3 defective units. What is the probability that a sample of three calculators will have

(a) no defective calculators?

(b) all defective calculators?

(c) at least one defective calculator?

(d) at least one nondefective calculator?

54. A batch of 350 raffle tickets contains four winning tickets. You buy four tickets. What is the probability that you have

(a) no winning tickets?

(b) all of the winning tickets?

(c) at least one winning ticket?

(d) at least one nonwinning ticket?

55. A corporation has six male senior executives and four female senior executives. Four senior executives are chosen at random to attend a technology seminar. What is the probability of choosing

(a) four men?

(b) four women?

(c) two men and two women?

(d) one man and three women?

3 CHAPTER QUIZ

1. (a) 0.523
   (b) 0.508
   (c) 0.545
   (d) 0.772
   (e) 0.025
   (f) 0.673
   (g) 0.094
   (h) 0.574

2. The event in part (e) is unusual because its probability is less than or equal to 0.05.

3. Not mutually exclusive. A golfer can score the best round in a four-round tournament and still lose the tournament.

   Dependent. One event can affect the occurrence of the second event.

4. (a) 2,481,115
   (b) 1
   (c) 2,572,999

5. (a) 0.964
   (b) 0.000000389
   (c) 0.9999996

6. 450,000

7. 657,720

Take this quiz as you would take a quiz in class. After you are done, check your work against the answers given in the back of the book.

1. The table shows the number (in thousands) of earned degrees, by level and gender, conferred in the United States in a recent year. *(Source: U.S. National Center for Education Statistics)*

		Gender		
		Male	**Female**	**Total**
Level of Degree	**Associate's**	275	453	728
	Bachelor's	650	875	1525
	Master's	238	366	604
	Doctoral	30	30	60
	Total	1193	1724	2917

A person who earned a degree in the year is randomly selected. Find the probability of selecting someone who

(a) earned a bachelor's degree.

(b) earned a bachelor's degree given that the person is a female.

(c) earned a bachelor's degree given that the person is not a female.

(d) earned an associate's degree or a bachelor's degree.

(e) earned a doctorate given that the person is a male.

(f) earned a master's degree or is a female.

(g) earned an associate's degree and is a male.

(h) is a female given that the person earned a bachelor's degree.

2. Which event(s) in Exercise 1 can be considered unusual? Explain your reasoning.

3. Decide if the events are mutually exclusive. Then decide if the events are independent or dependent. Explain your reasoning.

   Event *A*: A golfer scoring the best round in a four-round tournament

   Event *B*: Losing the golf tournament

4. A shipment of 250 netbooks contains 3 defective units. Determine how many ways a vending company can buy three of these units and receive

   (a) no defective units.

   (b) all defective units.

   (c) at least one good unit.

5. In Exercise 4, find the probability of the vending company receiving

   (a) no defective units.

   (b) all defective units.

   (c) at least one good unit.

6. The access code for a warehouse's security system consists of six digits. The first digit cannot be 0 and the last digit must be even. How many different codes are available?

7. From a pool of 30 candidates, the offices of president, vice president, secretary, and treasurer will be filled. In how many different ways can the offices be filled?

PUTTING IT ALL TOGETHER

Real Statistics — Real Decisions

You work for the company that runs the Powerball® lottery. Powerball is a lottery game in which five white balls are chosen from a drum containing 59 balls and one red ball is chosen from a drum containing 39 balls. To win the jackpot, a player must match all five white balls and the red ball. Other winners and their prizes are also shown in the table.

Working in the public relations department, you handle many inquiries from the media and from lottery players. You receive the following e-mail.

> *You list the probability of matching only the red ball as 1/62. I know from my statistics class that the probability of winning is the ratio of the number of successful outcomes to the total number of outcomes. Could you please explain why the probability of matching only the red ball is 1/62?*

Your job is to answer this question, using the probability techniques you have learned in this chapter to justify your answer. In answering the question, assume only one ticket is purchased.

■ EXERCISES

1. *How Would You Do It?*

(a) How would you investigate the question about the probability of matching only the red ball?

(b) What statistical methods taught in this chapter would you use?

2. *Answering the Question*

Write an explanation that answers the question about the probability of matching only the red ball. Include in your explanation any probability formulas that justify your explanation.

3. *Another Question*

You receive another question asking how the overall probability of winning a prize in the Powerball lottery is determined. The overall probability of winning a prize in the Powerball lottery is 1/35. Write an explanation that answers the question and include any probability formulas that justify your explanation.

www.musl.com

Reprinted with permission from the MultiState Lottery Association.

Powerball Winners and Prizes

Match	Prize	Approximate probability
5 white, 1 red	Jackpot	1/195,249,054
5 white	$200,000	1/5,138,133
4 white, 1 red	$10,000	1/723,145
4 white	$100	1/19,030
3 white, 1 red	$100	1/13,644
3 white	$7	1/359
2 white, 1 red	$7	1/787
1 white, 1 red	$4	1/123
1 red	$3	1/62

(Source: Multi-State Lottery Association)

Where Is Powerball Played?

Powerball is played in 42 states, Washington, D.C., and the U.S. Virgin Islands

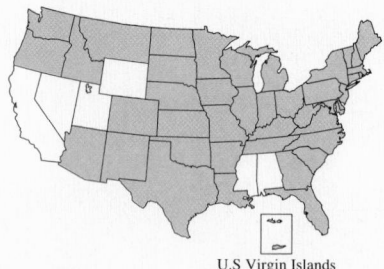

U.S Virgin Islands

(Source: Multi-State Lottery Association)

TECHNOLOGY

SIMULATION: COMPOSING MOZART VARIATIONS WITH DICE

Wolfgang Mozart (1756–1791) composed a wide variety of musical pieces. In his Musical Dice Game, he wrote a Wiener minuet with an almost endless number of variations. Each minuet has 16 bars. In the eighth and sixteenth bars, the player has a choice of two musical phrases. In each of the other 14 bars, the player has a choice of 11 phrases.

To create a minuet, Mozart suggested that the player toss 2 six-sided dice 16 times. For the eighth and sixteenth bars, choose Option 1 if the dice total is odd and Option 2 if it is even. For each of the other 14 bars, subtract 1 from the dice total. The following minuet is the result of the following sequence of numbers.

5	7	1	6	4	10	5	1
6	6	2	4	6	8	8	2

EXERCISES

1. How many phrases did Mozart write to create the Musical Dice Game minuet? Explain.

2. How many possible variations are there in Mozart's Musical Dice Game minuet? Explain.

3. Use technology to randomly select a number from 1 to 11.

   (a) What is the theoretical probability of each number from 1 to 11 occurring?

   (b) Use this procedure to select 100 integers from 1 to 11. Tally your results and compare them with the probabilities in part (a).

4. What is the probability of randomly selecting option 6, 7, or 8 for the first bar? For all 14 bars? Find each probability using (a) theoretical probability and (b) the results of Exercise 3(b).

5. Use technology to randomly select two numbers from 1, 2, 3, 4, 5, and 6. Find the sum and subtract 1 to obtain a total.

   (a) What is the theoretical probability of each total from 1 to 11?

   (b) Use this procedure to select 100 totals from 1 to 11. Tally your results and compare them with the probabilities in part (a).

6. Repeat Exercise 4 using the results of Exercise 5(b).

Extended solutions are given in the *Technology Supplement.*
Technical instruction is provided for MINITAB, Excel, and the TI-83/84 Plus.

4 DISCRETE PROBABILITY DISTRIBUTIONS

CHAPTER

The National Climatic Data Center (NCDC) is the world's largest active archive of weather data. NCDC archives weather data from the Coast Guard, Federal Aviation Administration, Military Services, the National Weather Service, and voluntary observers.

In Chapters 1 through 3, you learned how to collect and describe data and how to find the probability of an event. These skills are used in many different types of careers. For instance, data about climatic conditions are used to analyze and forecast the weather throughout the world. On a typical day, aircraft, National Weather Service cooperative observers, radar, remote sensing systems, satellites, ships, weather balloons, wind profilers, and a variety of other data-collection devices work together to provide meteorologists with data that are used to forecast the weather. Even with this much data, meteorologists cannot forecast the weather with certainty. Instead, they assign probabilities to certain weather conditions. For instance, a meteorologist might determine that there is a 40% chance of rain (based on the relative frequency of rain under similar weather conditions).

WHERE YOU'RE GOING ▶▶

In Chapter 4, you will learn how to create and use probability distributions. Knowing the shape, center, and variability of a probability distribution will enable you to make decisions in inferential statistics. You are a meteorologist working on a three-day forecast. Assuming that having rain on one day is independent of having rain on another day, you have determined that there is a 40% probability of rain

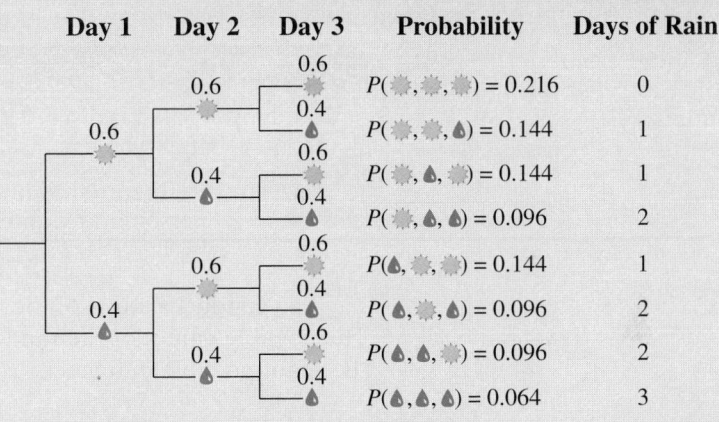

(and a 60% probability of no rain) on each of the three days. What is the probability that it will rain on 0, 1, 2, or 3 of the days? To answer this, you can create a probability distribution for the possible outcomes.

Using the *Addition Rule* with the probabilities in the tree diagram, you can determine the probabilities of having rain on various numbers of days. You can then use this information to graph a probability distribution.

Probability Distribution

Days of rain	Tally	Probability
0	1	0.216
1	3	0.432
2	3	0.288
3	1	0.064

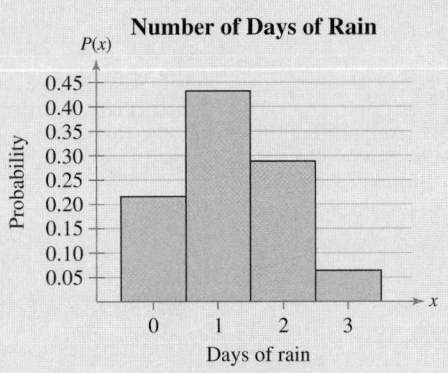

4.1 Probability Distributions

Random Variables ▸ Discrete Probability Distributions ▸ Mean, Variance, and Standard Deviation ▸ Expected Value

▸ RANDOM VARIABLES

The outcome of a probability experiment is often a count or a measure. When this occurs, the outcome is called a *random variable*.

> **DEFINITION**
>
> A **random variable** x represents a numerical value associated with each outcome of a probability experiment.

The word *random* indicates that x is determined by chance. There are two types of random variables: *discrete* and *continuous*.

> **DEFINITION**
>
> A random variable is **discrete** if it has a finite or countable number of possible outcomes that can be listed.
>
> A random variable is **continuous** if it has an uncountable number of possible outcomes, represented by an interval on the number line.

You conduct a study of the number of calls a telemarketer makes in one day. The possible values of the random variable x are 0, 1, 2, 3, 4, and so on. Because the set of possible outcomes

$$\{0, 1, 2, 3, \dots\}$$

can be listed, x is a discrete random variable. You can represent its values as points on a number line.

Number of Calls (Discrete)

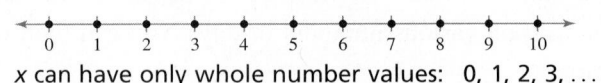

x can have only whole number values: 0, 1, 2, 3, ...

A different way to conduct the study would be to measure the time (in hours) a telemarketer spends making calls in one day. Because the time spent making calls can be any number from 0 to 24 (including fractions and decimals), x is a continuous random variable. You can represent its values with an interval on a number line.

Hours Spent on Calls (Continuous)

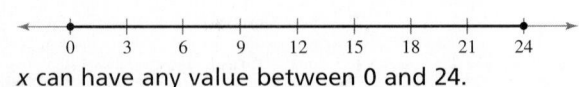

x can have any value between 0 and 24.

When a random variable is discrete, you can list the possible values it can assume. However, it is impossible to list all values for a continuous random variable.

Note to Instructor

In more advanced texts, the random variable is denoted X, while x represents particular values of the random variable. To simplify the discussion, we use x to represent both the random variable and its values.

EXAMPLE 1

▸ **Discrete Variables and Continuous Variables**

Decide whether the random variable x is discrete or continuous. Explain your reasoning.

1. Let x represent the number of Fortune 500 companies that lost money in the previous year.

2. Let x represent the volume of gasoline in a 21-gallon tank.

▸ **Solution**

1. The number of companies that lost money in the previous year can be counted.

   $$\{0, 1, 2, 3, \ldots, 500\}$$

   So, x is a *discrete* random variable.

2. The amount of gasoline in the tank can be any volume between 0 gallons and 21 gallons. So, x is a *continuous* random variable.

▸ **Try It Yourself 1**

Decide whether the random variable x is discrete or continuous. Explain your reasoning.

1. Let x represent the speed of a Space Shuttle.
2. Let x represent the number of calves born on a farm in one year.

a. Decide if x represents *counted* data or *measured* data.
b. Make a *conclusion* and *explain* your reasoning. *Answer: Page A36*

STUDY TIP

In most practical applications, discrete random variables represent counted data, while continuous random variables represent measured data.

INSIGHT

Values of variables such as age, height, and weight are usually rounded to the nearest year, inch, or pound. However, these values represent measured data, so they are continuous random variables.

It is important that you can distinguish between discrete and continuous random variables because different statistical techniques are used to analyze each. The remainder of this chapter focuses on discrete random variables and their probability distributions. You will study continuous distributions later.

▸ DISCRETE PROBABILITY DISTRIBUTIONS

Each value of a discrete random variable can be assigned a probability. By listing each value of the random variable with its corresponding probability, you are forming a *discrete probability distribution*.

DEFINITION

A **discrete probability distribution** lists each possible value the random variable can assume, together with its probability. A discrete probability distribution must satisfy the following conditions.

IN WORDS	IN SYMBOLS
1. The probability of each value of the discrete random variable is between 0 and 1, inclusive.	$0 \leq P(x) \leq 1$
2. The sum of all the probabilities is 1.	$\sum P(x) = 1$

Because probabilities represent relative frequencies, a discrete probability distribution can be graphed with a relative frequency histogram.

GUIDELINES

Constructing a Discrete Probability Distribution

Let x be a discrete random variable with possible outcomes $x_1, x_2, \ldots, x_n$.

1. Make a frequency distribution for the possible outcomes.
2. Find the sum of the frequencies.
3. Find the probability of each possible outcome by dividing its frequency by the sum of the frequencies.
4. Check that each probability is between 0 and 1, inclusive, and that the sum of all the probabilities is 1.

Frequency Distribution

Score, x	Frequency, f
1	24
2	33
3	42
4	30
5	21

EXAMPLE 2   SC   Report 16

▶ Constructing and Graphing a Discrete Probability Distribution

An industrial psychologist administered a personality inventory test for passive-aggressive traits to 150 employees. Each individual was given a score from 1 to 5, where 1 was extremely passive and 5 extremely aggressive. A score of 3 indicated neither trait. The results are shown at the left. Construct a probability distribution for the random variable x. Then graph the distribution using a histogram.

▶ Solution

Divide the frequency of each score by the total number of individuals in the study to find the probability for each value of the random variable.

$$P(1) = \frac{24}{150} = 0.16 \qquad P(2) = \frac{33}{150} = 0.22 \qquad P(3) = \frac{42}{150} = 0.28$$

$$P(4) = \frac{30}{150} = 0.20 \qquad P(5) = \frac{21}{150} = 0.14$$

The discrete probability distribution is shown in the following table.

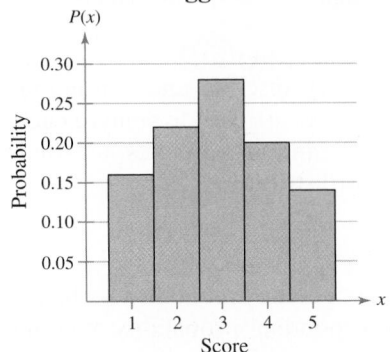

Passive-Aggressive Traits

x	1	2	3	4	5
$P(x)$	0.16	0.22	0.28	0.20	0.14

Note that $0 \le P(x) \le 1$ and $\sum P(x) = 1$.

The histogram is shown at the left. Because the width of each bar is one, the area of each bar is equal to the probability of a particular outcome. Also, the probability of an event corresponds to the sum of the areas of the outcomes included in the event. For instance, the probability of the event "having a score of 2 or 3" is equal to the sum of the areas of the second and third bars,

$$(1)(0.22) + (1)(0.28) = 0.22 + 0.28 = 0.50.$$

Interpretation   You can see that the distribution is approximately symmetric.

▶ Try It Yourself 2

A company tracks the number of sales new employees make each day during a 100-day probationary period. The results for one new employee are shown at the left. Construct and graph a probability distribution.

a. Find the *probability* of each outcome.
b. Organize the probabilities in a *probability distribution*.
c. Graph the probability distribution using a *histogram*.   *Answer: Page A36*

Frequency Distribution

Sales per day, x	Number of days, f
0	16
1	19
2	15
3	21
4	9
5	10
6	8
7	2

Probability Distribution

Days of rain, x	Probability, $P(x)$
0	0.216
1	0.432
2	0.288
3	0.064

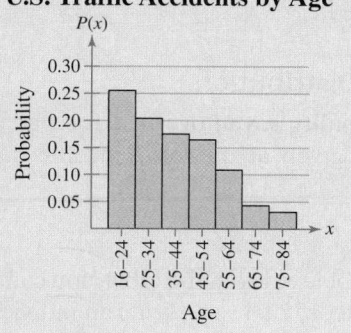

PICTURING THE WORLD

In a recent year in the United States, nearly 11 million traffic accidents were reported to the police. A histogram of traffic accidents for various age groups from 16 to 84 is shown. (Adapted from National Safety Council)

U.S. Traffic Accidents by Age

Estimate the probability that a randomly selected person involved in a traffic accident is in the 16 to 34 age group.

The probability that a randomly selected person involved in a traffic accident is in the 16 to 34 age group is about 0.45.

EXAMPLE 3

▸ Verifying Probability Distributions

Verify that the distribution at the left (see page 189) is a probability distribution.

▸ Solution

If the distribution is a probability distribution, then (1) each probability is between 0 and 1, inclusive, and (2) the sum of the probabilities equals 1.

1. Each probability is between 0 and 1.

2. $\sum P(x) = 0.216 + 0.432 + 0.288 + 0.064$
$= 1.$

Interpretation Because both conditions are met, the distribution is a probability distribution.

▸ Try It Yourself 3

Verify that the distribution you constructed in Try It Yourself 2 is a probability distribution.

a. Verify that the *probability* of each outcome is between 0 and 1, inclusive.
b. Verify that the *sum* of all the probabilities is 1.
c. Make a *conclusion*. *Answer: Page A36*

EXAMPLE 4

▸ Identifying Probability Distributions

Decide whether the distribution is a probability distribution. Explain your reasoning.

1.

x	5	6	7	8
$P(x)$	0.28	0.21	0.43	0.15

— more than 1

2.

x	1	2	3	4
$P(x)$	$\frac{1}{2}$	$\frac{1}{4}$	$\frac{5}{4}$	-1

neg not possible

▸ Solution

1. Each probability is between 0 and 1, but the sum of all the probabilities is 1.07, which is greater than 1. So, it is *not* a probability distribution.

2. The sum of all the probabilities is equal to 1, but $P(3)$ and $P(4)$ are not between 0 and 1. So, it is *not* a probability distribution. Probabilities can never be negative or greater than 1.

▸ Try It Yourself 4

Decide whether the distribution is a probability distribution. Explain your reasoning.

1.

x	5	6	7	8
$P(x)$	$\frac{1}{16}$	$\frac{5}{8}$	$\frac{1}{4}$	$\frac{1}{16}$

2.

x	1	2	3	4
$P(x)$	0.09	0.36	0.49	0.06

a. Verify that the *probability* of each outcome is between 0 and 1.
b. Verify that the *sum* of all the probabilities is 1.
c. Make a *conclusion*. *Answer: Page A36*

▶ **MEAN, VARIANCE, AND STANDARD DEVIATION**

You can measure the center of a probability distribution with its mean and measure the variability with its variance and standard deviation. The mean of a discrete random variable is defined as follows.

MEAN OF A DISCRETE RANDOM VARIABLE

The **mean** of a discrete random variable is given by

$$\mu = \Sigma x P(x).$$

Each value of x is multiplied by its corresponding probability and the products are added.

The mean of a random variable represents the "theoretical average" of a probability experiment and sometimes is not a possible outcome. If the experiment were performed many thousands of times, the mean of all the outcomes would be close to the mean of the random variable.

x	$P(x)$
1	0.16
2	0.22
3	0.28
4	0.20
5	0.14

Note to Instructor

Point out that 2.94 is not a possible score for one person. However, if the scores of the 150 employees were added up and divided by 150, the result would be 2.94.

STUDY TIP

Notice that the mean in Example 5 is rounded to one decimal place. This rounding was done because the mean of a probability distribution should be rounded to one more decimal place than was used for the random variable x. This *round-off rule* is also used for the variance and standard deviation of a probability distribution.

EXAMPLE 5

▶ **Finding the Mean of a Probability Distribution**

The probability distribution for the personality inventory test for passive-aggressive traits discussed in Example 2 is given at the left. Find the mean score. What can you conclude?

▶ **Solution**

Use a table to organize your work, as shown below. From the table, you can see that the mean score is approximately 2.9. A score of 3 represents an individual who exhibits neither passive nor aggressive traits. The mean is slightly under 3.

x	$P(x)$	$xP(x)$
1	0.16	$1(0.16) = 0.16$
2	0.22	$2(0.22) = 0.44$
3	0.28	$3(0.28) = 0.84$
4	0.20	$4(0.20) = 0.80$
5	0.14	$5(0.14) = 0.70$
	$\Sigma P(x) = 1$	$\Sigma xP(x) = 2.94$ ⟵ Mean

Interpretation You can conclude that the mean personality trait is neither extremely passive nor extremely aggressive, but is slightly closer to passive.

▶ **Try It Yourself 5**

Find the mean of the probability distribution you constructed in Try It Yourself 2. What can you conclude?

a. Find the *product* of each random outcome and its corresponding probability.
b. Find the *sum* of the products.
c. Make a *conclusion*.   *Answer: Page A36*

Although the mean of the random variable of a probability distribution describes a typical outcome, it gives no information about how the outcomes vary. To study the variation of the outcomes, you can use the variance and standard deviation of the random variable of a probability distribution.

STUDY TIP

A shortcut formula for the variance of a probability distribution is

$$\sigma^2 = [\Sigma x^2 P(x)] - \mu^2.$$

VARIANCE AND STANDARD DEVIATION OF A DISCRETE RANDOM VARIABLE

The **variance** of a discrete random variable is

$$\sigma^2 = \Sigma (x - \mu)^2 P(x).$$

The **standard deviation** is

$$\sigma = \sqrt{\sigma^2} = \sqrt{\Sigma (x - \mu)^2 P(x)}.$$

x	P(x)
1	0.16
2	0.22
3	0.28
4	0.20
5	0.14

EXAMPLE 6

▶ **Finding the Variance and Standard Deviation**

The probability distribution for the personality inventory test for passive-aggressive traits discussed in Example 2 is given at the left. Find the variance and standard deviation of the probability distribution.

▶ **Solution**

From Example 5, you know that before rounding, the mean of the distribution is $\mu = 2.94$. Use a table to organize your work, as shown below.

x	P(x)	x − μ	(x − μ)²	P(x)(x − μ)²
1	0.16	−1.94	3.764	0.602
2	0.22	−0.94	0.884	0.194
3	0.28	0.06	0.004	0.001
4	0.20	1.06	1.124	0.225
5	0.14	2.06	4.244	0.594
	$\Sigma P(x) = 1$			$\Sigma P(x)(x - \mu)^2 = 1.616$

 Variance

So, the variance is

$$\sigma^2 = 1.616 \approx 1.6$$

and the standard deviation is

$$\sigma = \sqrt{\sigma^2} = \sqrt{1.616} \approx 1.3.$$

Interpretation Most of the data values differ from the mean by no more than 1.3.

STUDY TIP

Detailed instructions for using MINITAB, Excel, and the TI-83/84 Plus are shown in the Technology Guide that accompanies this text. Here are instructions for finding the mean and standard deviation of a discrete random variable of a probability distribution on a TI-83/84 Plus.

STAT

Choose the EDIT menu.

1: Edit

Enter the possible values of the discrete random variable x in L1. Enter the probabilities P(x) in L2.

STAT

Choose the CALC menu.

1: 1–Var Stats

ENTER

2nd L1 , 2nd L2

ENTER

▶ **Try It Yourself 6**

Find the variance and standard deviation of the probability distribution constructed in Try It Yourself 2.

a. For each value of x, find the *square of the deviation* from the mean and multiply that value by the corresponding probability of x.
b. Find the *sum of the products* found in part (a) for the variance.
c. Take the *square root of the variance* to find the standard deviation.
d. *Interpret* the results.   *Answer: Page A36*

▶ **EXPECTED VALUE**

The mean of a random variable represents what you would expect to happen over thousands of trials. It is also called the *expected value*.

Note to Instructor
Tell students that expected value plays a role in decision theory.

DEFINITION

The **expected value** of a discrete random variable is equal to the mean of the random variable.

$$\text{Expected Value} = E(x) = \mu = \sum xP(x)$$

Although probabilities can never be negative, the expected value of a random variable can be negative.

INSIGHT

In most applications, an expected value of 0 has a practical interpretation. For instance, in games of chance, an expected value of 0 implies that a game is fair (an unlikely occurrence!). In a profit and loss analysis, an expected value of 0 represents the break-even point.

EXAMPLE 7

▶ **Finding an Expected Value**

At a raffle, 1500 tickets are sold at $2 each for four prizes of $500, $250, $150, and $75. You buy one ticket. What is the expected value of your gain?

▶ **Solution**

To find the gain for each prize, subtract the price of the ticket from the prize. For instance, your gain for the $500 prize is

$$\$500 - \$2 = \$498$$

and your gain for the $250 prize is

$$\$250 - \$2 = \$248.$$

Write a probability distribution for the possible gains (or outcomes).

Gain, x	$498	$248	$148	$73	−$2
Probability, $P(x)$	$\frac{1}{1500}$	$\frac{1}{1500}$	$\frac{1}{1500}$	$\frac{1}{1500}$	$\frac{1496}{1500}$

Then, using the probability distribution, you can find the expected value.

$$E(x) = \sum xP(x)$$
$$= \$498 \cdot \frac{1}{1500} + \$248 \cdot \frac{1}{1500} + \$148 \cdot \frac{1}{1500} + \$73 \cdot \frac{1}{1500} + (-\$2) \cdot \frac{1496}{1500}$$
$$= -\$1.35$$

Interpretation Because the expected value is negative, you can expect to lose an average of $1.35 for each ticket you buy.

▶ **Try It Yourself 7**

At a raffle, 2000 tickets are sold at $5 each for five prizes of $2000, $1000, $500, $250, and $100. You buy one ticket. What is the expected value of your gain?

a. Find the *gain* for each prize.
b. Write a *probability distribution* for the possible gains.
c. Find the *expected value*.
d. *Interpret* the results.   *Answer: Page A36*

4.1 EXERCISES

FOR EXTRA HELP:
MyStatLab

1. A random variable represents a numerical value associated with each outcome of a probability experiment.

   Examples: Answers will vary.

2. See Selected Answers, page A117.

3. No; Expected value may not be a possible value of x for one trial, but it represents the average value of x over a large number of trials.

4. See Selected Answers, page A117.

5. False. In most applications, discrete random variables represent counted data, while continuous random variables represent measured data.

6. True

7. True

8. See Selected Answers, page A117.

9. Discrete; Attendance is a random variable that is countable.

10. See Selected Answers, page A117.

11. Continuous; Distance traveled is a random variable that must be measured.

12. See Selected Answers, page A117.

13. See Odd Answers, page A67.

14. See Selected Answers, page A117.

15. See Odd Answers, page A67.

16. See Selected Answers, page A117.

17. See Odd Answers, page A67.

18. See Selected Answers, page A117.

19. See Odd Answers, page A67.

20. See Selected Answers, page A117.

■ BUILDING BASIC SKILLS AND VOCABULARY

1. What is a random variable? Give an example of a discrete random variable and a continuous random variable. Justify your answer.

2. What is a discrete probability distribution? What are the two conditions that determine a probability distribution?

3. Is the expected value of the probability distribution of a random variable always one of the possible values of x? Explain.

4. What is the significance of the mean of a probability distribution?

True or False? *In Exercises 5–8, determine whether the statement is true or false. If it is false, rewrite it as a true statement.*

5. In most applications, continuous random variables represent counted data, while discrete random variables represent measured data.

6. For a random variable x, the word *random* indicates that the value of x is determined by chance.

7. The mean of a random variable represents the "theoretical average" of a probability experiment and sometimes is not a possible outcome.

8. The expected value of a discrete random variable is equal to the standard deviation of the random variable.

Graphical Analysis *In Exercises 9–12, decide whether the graph represents a discrete random variable or a continuous random variable. Explain your reasoning.*

9. The attendance at concerts for a rock group

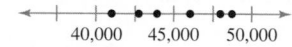

10. The length of time student-athletes practice each week

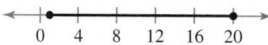

11. The distance a baseball travels after being hit

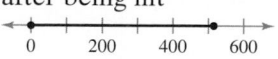

12. The annual traffic fatalities in the United States *(Source: U.S. National Highway Traffic Safety Administration)*

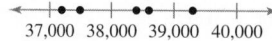

Distinguishing Between Discrete and Continuous Random Variables

In Exercises 13–20, decide whether the random variable x is discrete or continuous. Explain your reasoning.

13. Let x represent the number of books in a university library.

14. Let x represent the length of time it takes to get to work.

15. Let x represent the volume of blood drawn for a blood test.

16. Let x represent the number of tornadoes in the month of June in Oklahoma.

17. Let x represent the number of messages posted each month on a social networking website.

18. Let x represent the tension at which a randomly selected guitar's strings have been strung.

19. Let x represent the amount of snow (in inches) that fell in Nome, Alaska last winter.

20. Let x represent the total number of die rolls required for an individual to roll a five.

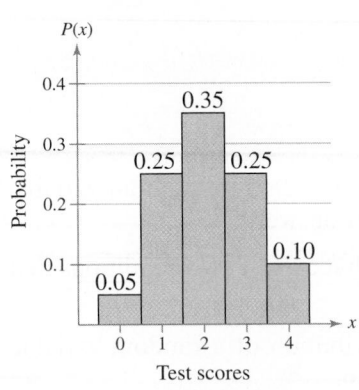

FIGURE FOR EXERCISE 21

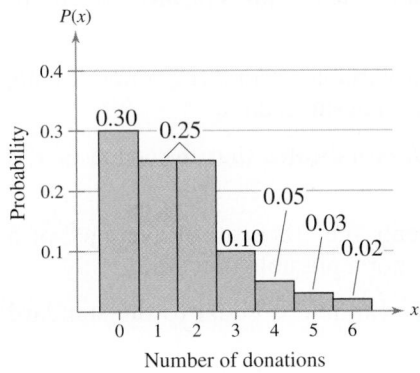

FIGURE FOR EXERCISE 22

21. (a) 0.35
    (b) 0.90
22. (a) 0.45
    (b) 0.80
23. 0.22
24. 0.15
25. Yes
26. No, $\sum P(x) = 0.97$.
27. (a) and (b) See Odd Answers, page A67.
    (c) 0.5, 0.8, 0.9
    (d) The mean is 0.5, so the average number of dogs per household is about 0 or 1 dog. The standard deviation is 0.9, so most of the households differ from the mean by no more than about 1 dog.

■ USING AND INTERPRETING CONCEPTS

21. Employee Testing A company gave psychological tests to prospective employees. The random variable x represents the possible test scores. Use the histogram to find the probability that a person selected at random from the survey's sample had a test score of (a) more than two and (b) less than four.

22. Blood Donations A survey asked a sample of people how many times they donate blood each year. The random variable x represents the number of donations in one year. Use the histogram to find the probability that a person selected at random from the survey's sample donated blood (a) more than once in a year and (b) less than three times in a year.

Determining a Missing Probability *In Exercises 23 and 24, determine the probability distribution's missing probability value.*

23.

x	0	1	2	3	4
$P(x)$	0.07	0.20	0.38	?	0.13

24.

x	0	1	2	3	4	5	6
$P(x)$	0.05	?	0.23	0.21	0.17	0.11	0.08

Identifying Probability Distributions *In Exercises 25 and 26, decide whether the distribution is a probability distribution. If it is not a probability distribution, identify the property (or properties) that are not satisfied.*

25. Tires A mechanic checked the tire pressures on each car that he worked on for one week. The random variable x represents the number of tires that were underinflated.

x	0	1	2	3	4
$P(x)$	0.30	0.25	0.25	0.15	0.05

26. Quality Control A quality inspector checked for imperfections in rolls of fabric for one week. The random variable x represents the number of imperfections found.

x	0	1	2	3	4	5
$P(x)$	$\frac{3}{4}$	$\frac{1}{10}$	$\frac{1}{20}$	$\frac{1}{25}$	$\frac{1}{50}$	$\frac{1}{100}$

Constructing Probability Distributions *In Exercises 27–32, (a) use the frequency distribution to construct a probability distribution, (b) graph the probability distribution using a histogram and describe its shape, (c) find the mean, variance, and standard deviation of the probability distribution, and (d) interpret the results in the context of the real-life situation.*

27. Dogs The number of dogs per household in a small town

Dogs	0	1	2	3	4	5
Households	1491	425	168	48	29	14

28. (a) and (b) See Selected Answers, page A117.

(c) 5.8, 1.4, 1.2

(d) The mean is 5.8, so the average number of games played per World Series was about 6. The standard deviation is 1.2, so most of the World Series differed from the mean by no more than about 1 game.

29. (a) and (b) See Odd Answers, page A68.

(c) 2.4, 0.6, 0.8

(d) The mean is 2.4, so the average household in the town has about 2 televisions. The standard deviation is 0.8, so most of the households differ from the mean by no more than about 1 television.

30. See Selected Answers, page A117.

31. (a) and (b) See Odd Answers, page A68.

(c) 3.4, 2.1, 1.5

(d) The mean is 3.4, so the average employee worked 3.4 hours of overtime. The standard deviation is 1.5, so the overtime worked by most of the employees differed from the mean by no more than 1.5 hours.

32. See Selected Answers, page A117.

33. An expected value of 0 means that the money gained is equal to the money spent, representing the break-even point.

34. A "fair bet" in a game of chance has an expected value of 0, which means that the chances of losing are equal to the chances of winning.

35. (a) 5.3

(b) 3.3

(c) 1.8

(d) 5.3

(e) The expected value is 5.3, so an average student is expected to answer about 5 questions correctly. The standard deviation is 1.8, so most of the students' quiz results differ from the expected value by no more than about 2 questions.

36. See Selected Answers, page A118.

28. Baseball   The number of games played in the World Series from 1903 to 2009 *(Source: Major League Baseball)*

Games played	4	5	6	7	8
Frequency	20	23	23	36	3

29. Televisions   The number of televisions per household in a small town

Televisions	0	1	2	3
Households	26	442	728	1404

30. Camping Chairs   The number of defects per batch of camping chairs inspected

Defects	0	1	2	3	4	5
Batches	95	113	87	64	13	8

31. Overtime Hours   The number of overtime hours worked in one week per employee

Overtime hours	0	1	2	3	4	5	6
Employees	6	12	29	57	42	30	16

32. Extracurricular Activities   The number of school-related extracurricular activities per student

Activities	0	1	2	3	4	5	6	7
Students	19	39	52	57	68	41	27	17

33. Writing   The expected value of an accountant's profit and loss analysis is 0. Explain what this means.

34. Writing   In a game of chance, what is the relationship between a "fair bet" and its expected value? Explain.

Finding Expected Value   *In Exercises 35–40, use the probability distribution or histogram to find the (a) mean, (b) variance, (c) standard deviation, and (d) expected value of the probability distribution, and (e) interpret the results.*

35. Quiz   Students in a class take a quiz with eight questions. The random variable x represents the number of questions answered correctly.

x	0	1	2	3	4	5	6	7	8
$P(x)$	0.02	0.02	0.06	0.06	0.08	0.22	0.30	0.16	0.08

36. 911 Calls   A 911 service center recorded the number of calls received per hour. The random variable x represents the number of calls per hour for one week.

x	0	1	2	3	4	5	6	7
$P(x)$	0.01	0.10	0.26	0.33	0.18	0.06	0.03	0.03

37. (a) 2.0

(b) 1.0

(c) 1.0

(d) 2.0

(e) The expected value is 2.0, so an average hurricane that hits the U.S. mainland is expected to be a category 2 hurricane. The standard deviation is 1.0, so most of the hurricanes differ from the expected value by no more than 1 category level.

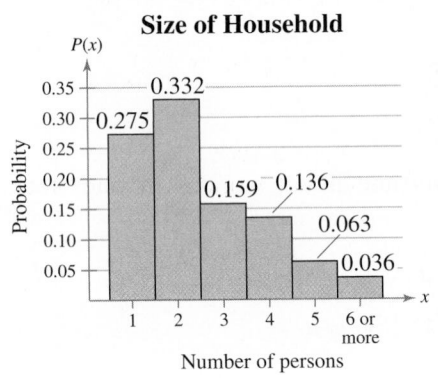

Size of Household

FIGURE FOR EXERCISE 39

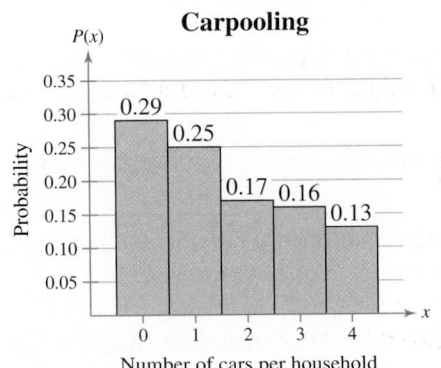

Carpooling

FIGURE FOR EXERCISE 40

38. (a) 1.7

(b) 1.0

(c) 1.0

(d) 1.7

(e) The expected value is 1.7, so an average car crossing the Tacoma Narrows Bridge is expected to have either 1 or 2 people in it. The standard deviation is 1.0, so most of the car occupancies differ from the expected value by no more than 1 occupant.

37. Hurricanes The histogram shows the distribution of hurricanes that have hit the U.S. mainland by category, with 1 the weakest level and 5 the strongest. *(Source: Weather Research Center)*

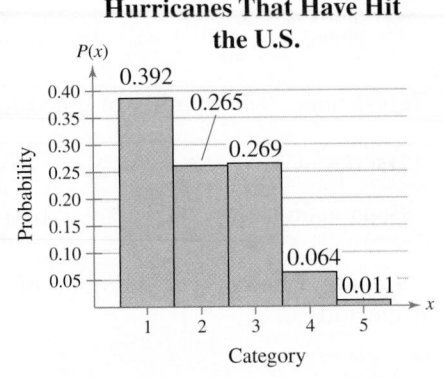

Hurricanes That Have Hit the U.S.

FIGURE FOR EXERCISE 37

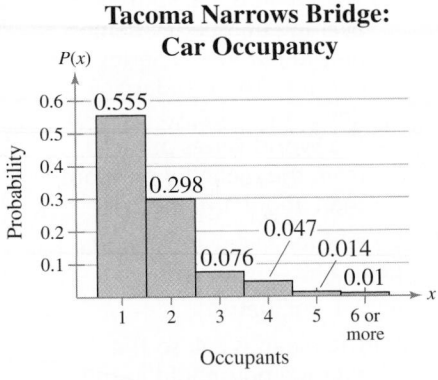

Tacoma Narrows Bridge: Car Occupancy

FIGURE FOR EXERCISE 38

38. Car Occupancy The histogram shows the distribution of occupants in cars crossing the Tacoma Narrows Bridge in Washington each week. *(Adapted from Washington State Department of Transportation)*

39. Household Size The histogram shows the distribution of household sizes in the United States for a recent year. *(Adapted from U.S. Census Bureau)*

40. Carpooling The histogram shows the distribution of carpooling by the number of cars per household. *(Adapted from Federal Highway Administration)*

41. Finding Probabilities Use the probability distribution you made for Exercise 27 to find the probability of randomly selecting a household that has (a) fewer than two dogs, (b) at least one dog, and (c) between one and three dogs, inclusive.

42. Finding Probabilities Use the probability distribution you made for Exercise 28 to find the probability of randomly selecting a World Series that consisted of (a) four games, (b) at least five games, and (c) between four and six games, inclusive.

43. Unusual Values A person lives in a household with three dogs and claims that having three dogs is not unusual. Use the information in Exercise 27 to determine if this person is correct. Explain your reasoning.

44. Unusual Values A person randomly chooses a World Series in which eight games were played and claims that this is an unusual event. Use the information in Exercise 28 to determine if this person is correct. Explain you reasoning.

Games of Chance *In Exercises 45 and 46, find the expected net gain to the player for one play of the game. If x is the net gain to a player in a game of chance, then E(x) is usually negative. This value gives the average amount per game the player can expect to lose.*

45. In American roulette, the wheel has the 38 numbers

$$00, 0, 1, 2, \ldots, 34, 35, \text{ and } 36$$

marked on equally spaced slots. If a player bets $1 on a number and wins, then the player keeps the dollar and receives an additional 35 dollars. Otherwise, the dollar is lost.

46. A charity organization is selling $5 raffle tickets as part of a fund-raising program. The first prize is a trip to Mexico valued at $3450, and the second prize is a weekend spa package valued at $750. The remaining 20 prizes are $25 gas cards. The number of tickets sold is 6000.

39. (a) 2.5

   (b) 1.9

   (c) 1.4

   (d) 2.5

   (e) The expected value is 2.5, so an average household is expected to have either 2 or 3 people. The standard deviation is 1.4, so most of the household sizes differ from the expected value by no more than 1 or 2 people.

40. (a) 1.6

   (b) 1.9

   (c) 1.4

   (d) 1.6

   (e) The expected value is 1.6, so an average household is expected to have either 1 or 2 cars. The standard deviation is 1.4, so most of the households differ from the expected value by no more than 1 or 2 cars.

41. (a) 0.881

   (b) 0.314

   (c) 0.294

42. (a) 0.19

   (b) 0.81

   (c) 0.628

43. A household with three dogs is unusual because the probability of this event is 0.022, which is less than 0.05.

44. A World Series in which eight games were played is unusual because the probability of this event is 0.029, which is less than 0.05.

45. −$0.05

46. −$4.22

47. See Odd Answers, page A68.

48. (a) See Selected Answers, page A118.

   (b) Approximately uniform

49. $38,800

50. $4054.96

51. 3020; 28

52. 441.29

SC *In Exercises 47 and 48, use StatCrunch to (a) construct and graph a probability distribution and (b) describe its shape.*

47. Computers   The number of computers per household in a small town

Computers	0	1	2	3
Households	300	280	95	20

48. Students   The enrollments (in thousands) for grades 1 through 8 in the United States for a recent year *(Source: U.S. National Center for Education Statistics)*

Grade	1	2	3	4	5	6	7	8
Enrollment	3750	3640	3627	3585	3601	3660	3715	3765

■ EXTENDING CONCEPTS

Linear Transformation of a Random Variable   *In Exercises 49 and 50, use the following information. For a random variable x, a new random variable y can be created by applying a **linear transformation** $y = a + bx$, where a and b are constants. If the random variable x has mean μ_x and standard deviation σ_x, then the mean, variance, and standard deviation of y are given by the following formulas.*

$$\mu_y = a + b\mu_x \qquad \sigma_y^2 = b^2\sigma_x^2 \qquad \sigma_y = |b|\sigma_x$$

49. The mean annual salary of employees at a company is $36,000. At the end of the year, each employee receives a $1000 bonus and a 5% raise (based on salary). What is the new mean annual salary (including the bonus and raise) of the employees?

50. The mean annual salary of employees at a company is $36,000 with a variance of 15,202,201. At the end of the year, each employee receives a $2000 bonus and a 4% raise (based on salary). What is the standard deviation of the new salaries?

Independent and Dependent Random Variables   *Two random variables x and y are **independent** if the value of x does not affect the value of y. If the variables are not independent, they are **dependent**. A new random variable can be formed by finding the sum or difference of random variables. If a random variable x has mean μ_x and a random variable y has mean μ_y, then the means of the sum and difference of the variables are given by the following equations.*

$$\mu_{x+y} = \mu_x + \mu_y \qquad\qquad \mu_{x-y} = \mu_x - \mu_y$$

If random variables are independent, then the variance and standard deviation of the sum or difference of the random variables can be found. So, if a random variable x has variance σ_x^2 and a random variable y has variance σ_y^2, then the variances of the sum and difference of the variables are given by the following equations. Note that the variance of the difference is the sum of the variances.

$$\sigma_{x+y}^2 = \sigma_x^2 + \sigma_y^2 \qquad\qquad \sigma_{x-y}^2 = \sigma_x^2 + \sigma_y^2$$

In Exercises 51 and 52, the distribution of SAT scores for college-bound male seniors has a mean of 1524 and a standard deviation of 317. The distribution of SAT scores for college-bound female seniors has a mean of 1496 and a standard deviation of 307. One male and one female are randomly selected. Assume their scores are independent. (Source: The College Board)

51. What is the average sum of their scores? What is the average difference of their scores?

52. What is the standard deviation of the difference in their scores?

4.2 Binomial Distributions

WHAT YOU SHOULD LEARN

▸ How to determine if a probability experiment is a binomial experiment

▸ How to find binomial probabilities using the binomial probability formula

▸ How to find binomial probabilities using technology, formulas, and a binomial probability table

▸ How to graph a binomial distribution

▸ How to find the mean, variance, and standard deviation of a binomial probability distribution

▸ BINOMIAL EXPERIMENTS

There are many probability experiments for which the results of each trial can be reduced to two outcomes: success and failure. For instance, when a basketball player attempts a free throw, he or she either makes the basket or does not. Probability experiments such as these are called *binomial experiments*.

DEFINITION

A **binomial experiment** is a probability experiment that satisfies the following conditions.

1. The experiment is repeated for a fixed number of trials, where each trial is independent of the other trials.
2. There are only two possible outcomes of interest for each trial. The outcomes can be classified as a success (S) or as a failure (F).
3. The probability of a success $P(S)$ is the same for each trial.
4. The random variable x counts the number of successful trials.

Note to Instructor

Emphasize that p represents the probability of success in a single trial. Students often have trouble with this concept.

NOTATION FOR BINOMIAL EXPERIMENTS

SYMBOL	DESCRIPTION
n	The number of times a trial is repeated
$p = P(S)$	The probability of success in a single trial
$q = P(F)$	The probability of failure in a single trial ($q = 1 - p$)
x	The random variable represents a count of the number of successes in n trials: $x = 0, 1, 2, 3, \ldots, n$.

Trial	Outcome	S or F?
1		F
2		S
3		F
4		F
5		S

There are two successful outcomes. So, $x = 2$.

Here is a simple example of a binomial experiment. From a standard deck of cards, you pick a card, note whether it is a club or not, and replace the card. You repeat the experiment five times, so $n = 5$. The outcomes of each trial can be classified in two categories: S = selecting a club and F = selecting another suit. The probabilities of success and failure are

$$p = P(S) = \frac{1}{4} \quad \text{and} \quad q = P(F) = \frac{3}{4}.$$

The random variable x represents the number of clubs selected in the five trials. So, the possible values of the random variable are

0, 1, 2, 3, 4, and 5.

For instance, if $x = 2$, then exactly two of the five cards are clubs and the other three are not clubs. An example of an experiment with $x = 2$ is shown at the left. Note that x is a discrete random variable because its possible values can be listed.

PICTURING THE WORLD

In a recent survey of U.S. adults who used the social networking website Twitter were asked if they had ever posted comments about their personal lives. The respondents' answers were either yes or no. (Adapted from Zogby International)

Survey question: Have you ever posted comments about your personal life on Twitter?

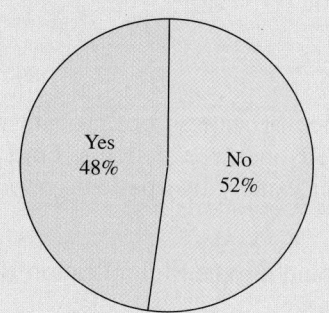

Why is this a binomial experiment? Identify the probability of success p. Identify the probability of failure q.

This is a binomial experiment because there are only two outcomes and the probability for success is the same for each trial, with $p = 0.48$ and $q = 0.52$.

EXAMPLE 1

▸ **Identifying and Understanding Binomial Experiments**

Decide whether the experiment is a binomial experiment. If it is, specify the values of n, p, and q, and list the possible values of the random variable x. If it is not, explain why.

1. A certain surgical procedure has an 85% chance of success. A doctor performs the procedure on eight patients. The random variable represents the number of successful surgeries.

2. A jar contains five red marbles, nine blue marbles, and six green marbles. You randomly select three marbles from the jar, *without replacement*. The random variable represents the number of red marbles.

▸ **Solution**

1. The experiment is a binomial experiment because it satisfies the four conditions of a binomial experiment. In the experiment, each surgery represents one trial. There are eight surgeries, and each surgery is independent of the others. There are only two possible outcomes for each surgery—either the surgery is a success or it is a failure. Also, the probability of success for each surgery is 0.85. Finally, the random variable x represents the number of successful surgeries.

$$n = 8$$
$$p = 0.85$$
$$q = 1 - 0.85$$
$$= 0.15$$
$$x = 0, 1, 2, 3, 4, 5, 6, 7, 8$$

2. The experiment is not a binomial experiment because it does not satisfy all four conditions of a binomial experiment. In the experiment, each marble selection represents one trial, and selecting a red marble is a success. When the first marble is selected, the probability of success is 5/20. However, because the marble is not replaced, the probability of success for subsequent trials is no longer 5/20. So, the trials are not independent, and the probability of a success is not the same for each trial.

▸ **Try It Yourself 1**

Decide whether the following is a binomial experiment. If it is, specify the values of n, p, and q, and list the possible values of the random variable x. If it is not, explain why.

> You take a multiple-choice quiz that consists of 10 questions. Each question has four possible answers, only one of which is correct. To complete the quiz, you randomly guess the answer to each question. The random variable represents the number of correct answers.

a. Identify a *trial* of the experiment and what is a success.
b. Decide if the experiment *satisfies the four conditions* of a binomial experiment.
c. Make a *conclusion* and *identify n, p, q*, and the possible values of x, if possible.

Answer: Page A36

▶ BINOMIAL PROBABILITY FORMULA

There are several ways to find the probability of x successes in n trials of a binomial experiment. One way is to use a tree diagram and the Multiplication Rule. Another way is to use the binomial probability formula.

BINOMIAL PROBABILITY FORMULA

In a binomial experiment, the probability of exactly x successes in n trials is

$$P(x) = {}_nC_x\, p^x q^{n-x} = \frac{n!}{(n-x)!\,x!}\, p^x q^{n-x}.$$

EXAMPLE 2   **SC**   Report 17

▶ Finding Binomial Probabilities

Microfracture knee surgery has a 75% chance of success on patients with degenerative knees. The surgery is performed on three patients. Find the probability of the surgery being successful on exactly two patients. *(Source: Illinois Sportsmedicine and Orthopedic Center)*

▶ **Solution**   **Method 1:** Draw a tree diagram and use the Multiplication Rule.

1st Surgery	2nd Surgery	3rd Surgery	Outcome	Number of Successes	Probability
		S	SSS	3	$\frac{3}{4}\cdot\frac{3}{4}\cdot\frac{3}{4}=\frac{27}{64}$
	S	F	SSF	2	$\frac{3}{4}\cdot\frac{3}{4}\cdot\frac{1}{4}=\frac{9}{64}$
S		S	SFS	2	$\frac{3}{4}\cdot\frac{1}{4}\cdot\frac{3}{4}=\frac{9}{64}$
	F	F	SFF	1	$\frac{3}{4}\cdot\frac{1}{4}\cdot\frac{1}{4}=\frac{3}{64}$
		S	FSS	2	$\frac{1}{4}\cdot\frac{3}{4}\cdot\frac{3}{4}=\frac{9}{64}$
	S	F	FSF	1	$\frac{1}{4}\cdot\frac{3}{4}\cdot\frac{1}{4}=\frac{3}{64}$
F		S	FFS	1	$\frac{1}{4}\cdot\frac{1}{4}\cdot\frac{3}{4}=\frac{3}{64}$
	F	F	FFF	0	$\frac{1}{4}\cdot\frac{1}{4}\cdot\frac{1}{4}=\frac{1}{64}$

There are three outcomes that have exactly two successes, and each has a probability of $\frac{9}{64}$. So, the probability of a successful surgery on exactly two patients is $3\left(\frac{9}{64}\right) \approx 0.422$.

Method 2: Use the binomial probability formula.

In this binomial experiment, the values of n, p, q, and x are $n = 3$, $p = \frac{3}{4}$, $q = \frac{1}{4}$, and $x = 2$. The probability of exactly two successful surgeries is

$$P(2\text{ successful surgeries}) = \frac{3!}{(3-2)!2!}\left(\frac{3}{4}\right)^2\left(\frac{1}{4}\right)^1$$

$$= 3\left(\frac{9}{16}\right)\left(\frac{1}{4}\right) = 3\left(\frac{9}{64}\right) = \frac{27}{64} \approx 0.422.$$

▶ **Try It Yourself 2**

A card is selected from a standard deck and replaced. This experiment is repeated a total of five times. Find the probability of selecting exactly three clubs.

a. *Identify* a trial, a success, and a failure.
b. *Identify* n, p, q, and x.
c. Use the *binomial probability formula*.   *Answer: Page A36*

By listing the possible values of x with the corresponding probabilities, you can construct a **binomial probability distribution.**

EXAMPLE 3

▶ **Constructing a Binomial Distribution**

Why Do We Like Texting?

- Convenient for basic information — 79%
- Works where talking won't do — 75%
- Quicker than calling — 56%
- Easier when facing arguments — 37%
- Dislike phone conversations — 27%
- Great for flirting — 25%

(Source: GfK Roper for Best Buy Mobile)

In a survey, U.S. adults were asked to give reasons why they liked texting on their cellular phones. The results are shown in the graph. Seven adults who participated in the survey are randomly selected and asked whether they like texting because it is quicker than calling. Create a binomial probability distribution for the number of adults who respond yes.

x	$P(x)$
0	0.0032
1	0.0284
2	0.1086
3	0.2304
4	0.2932
5	0.2239
6	0.0950
7	0.0173
	$\Sigma P(x) = 1$

▶ **Solution**

From the graph, you can see that 56% of adults like texting because it is quicker than calling. So, $p = 0.56$ and $q = 0.44$. Because $n = 7$, the possible values of x are 0, 1, 2, 3, 4, 5, 6, and 7.

$$P(0) = {_7}C_0(0.56)^0(0.44)^7 = 1(0.56)^0(0.44)^7 \approx 0.0032$$

$$P(1) = {_7}C_1(0.56)^1(0.44)^6 = 7(0.56)^1(0.44)^6 \approx 0.0284$$

$$P(2) = {_7}C_2(0.56)^2(0.44)^5 = 21(0.56)^2(0.44)^5 \approx 0.1086$$

$$P(3) = {_7}C_3(0.56)^3(0.44)^4 = 35(0.56)^3(0.44)^4 \approx 0.2304$$

$$P(4) = {_7}C_4(0.56)^4(0.44)^3 = 35(0.56)^4(0.44)^3 \approx 0.2932$$

$$P(5) = {_7}C_5(0.56)^5(0.44)^2 = 21(0.56)^5(0.44)^2 \approx 0.2239$$

$$P(6) = {_7}C_6(0.56)^6(0.44)^1 = 7(0.56)^6(0.44)^1 \approx 0.0950$$

$$P(7) = {_7}C_7(0.56)^7(0.44)^0 = 1(0.56)^7(0.44)^0 \approx 0.0173$$

Notice in the table at the left that all the probabilities are between 0 and 1 and that the sum of the probabilities is 1.

▶ **Try It Yourself 3**

Seven adults who participated in the survey are randomly selected and asked whether they like texting because it works where talking won't do. Create a binomial distribution for the number of adults who respond yes.

a. *Identify* a trial, a success, and a failure.
b. *Identify* n, p, q, and possible values for x.
c. Use the *binomial probability formula* for each value of x.
d. Use a *table* to show that the properties of a probability distribution are satisfied.   *Answer: Page A37*

STUDY TIP

When probabilities are rounded to a fixed number of decimal places, the sum of the probabilities may differ slightly from 1.

▸ **FINDING BINOMIAL PROBABILITIES**

In Examples 2 and 3, you used the binomial probability formula to find the probabilities. A more efficient way to find binomial probabilities is to use a calculator or a computer. For instance, you can find binomial probabilities using MINITAB, Excel, and the TI-83/84 Plus.

EXAMPLE 4

▸ **Finding a Binomial Probability Using Technology**

The results of a recent survey indicate that 67% of U.S. adults consider air conditioning a necessity. If you randomly select 100 adults, what is the probability that exactly 75 adults consider air conditioning a necessity? Use a technology tool to find the probability. *(Source: Opinion Research Corporation)*

▸ **Solution**

MINITAB, Excel, and the TI-83/84 Plus each have features that allow you to find binomial probabilities automatically. Try using these technologies. You should obtain results similar to the following.

MINITAB		TI-83/84 PLUS

Probability Distribution Function

Binomial with n = 100 and p = 0.67

x	P(X=x)
75	0.0201004

binompdf(100,.67,75)
  .0201004116

EXCEL			
	A	B	C	D
1	BINOMDIST(75,100,0.67,FALSE)			
2				0.020100412

Interpretation From these displays, you can see that the probability that exactly 75 adults consider air conditioning a necessity is about 0.02. Because 0.02 is less than 0.05, this can be considered an unusual event.

▸ **Try It Yourself 4**

The results of a recent survey indicate that 71% of people in the United States use more than one topping on their hot dogs. If you randomly select 250 people, what is the probability that exactly 178 of them will use more than one topping? Use a technology tool to find the probability. *(Source: ICR Survey Research Group for Hebrew International)*

a. *Identify n, p,* and *x.*
b. Calculate the *binomial probability.*
c. *Interpret* the results.
d. Determine if the event is *unusual.* Explain. *Answer: Page A37*

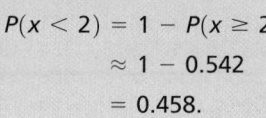

Using a TI-83/84 Plus, you can find the probability in part (1) automatically.

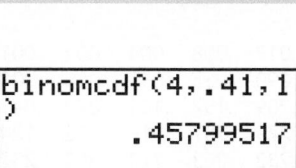

The cumulative distribution function (CDF) computes the probability of "x or fewer" successes. The CDF adds the areas for the given x-value and all those to its left.

Note to Instructor

Remind students to decide for each question whether it is easier to calculate a probability as given or to use the complement rule to get their answer.

EXAMPLE 5   SC Report 18

▸ Finding Binomial Probabilities Using Formulas

A survey indicates that 41% of women in the United States consider reading their favorite leisure-time activity. You randomly select four U.S. women and ask them if reading is their favorite leisure-time activity. Find the probability that (1) exactly two of them respond yes, (2) at least two of them respond yes, and (3) fewer than two of them respond yes. *(Source: Louis Harris & Associates)*

▸ Solution

1. Using $n = 4$, $p = 0.41$, $q = 0.59$, and $x = 2$, the probability that exactly two women will respond yes is

$$P(2) = {}_4C_2(0.41)^2(0.59)^2 = 6(0.41)^2(0.59)^2 \approx 0.351.$$

2. To find the probability that at least two women will respond yes, find the sum of $P(2)$, $P(3)$, and $P(4)$.

$$P(2) = {}_4C_2(0.41)^2(0.59)^2 = 6(0.41)^2(0.59)^2 \approx 0.351094$$
$$P(3) = {}_4C_3(0.41)^3(0.59)^1 = 4(0.41)^3(0.59)^1 \approx 0.162654$$
$$P(4) = {}_4C_4(0.41)^4(0.59)^0 = 1(0.41)^4(0.59)^0 \approx 0.028258$$

So, the probability that at least two will respond yes is

$$P(x \geq 2) = P(2) + P(3) + P(4)$$
$$\approx 0.351094 + 0.162654 + 0.028258$$
$$\approx 0.542.$$

3. To find the probability that fewer than two women will respond yes, find the sum of $P(0)$ and $P(1)$.

$$P(0) = {}_4C_0(0.41)^0(0.59)^4 = 1(0.41)^0(0.59)^4 \approx 0.121174$$
$$P(1) = {}_4C_1(0.41)^1(0.59)^3 = 4(0.41)^1(0.59)^3 \approx 0.336822$$

So, the probability that fewer than two will respond yes is

$$P(x < 2) = P(0) + P(1)$$
$$\approx 0.121174 + 0.336822$$
$$\approx 0.458.$$

▸ Try It Yourself 5

A survey indicates that 21% of men in the United States consider fishing their favorite leisure-time activity. You randomly select five U.S. men and ask them if fishing is their favorite leisure-time activity. Find the probability that (1) exactly two of them respond yes, (2) at least two of them respond yes, and (3) fewer than two of them respond yes. *(Source: Louis Harris & Associates)*

a. Determine the appropriate *value of x* for each situation.
b. Find the *binomial probability* for each value of x. Then find the *sum*, if necessary.
c. *Write* the result as a sentence.   *Answer: Page A37*

Finding binomial probabilities with the binomial probability formula can be a tedious process. To make this process easier, you can use a binomial probability table. Table 2 in Appendix B lists the binomial probabilities for selected values of n and p.

EXAMPLE 6

▸ Finding a Binomial Probability Using a Table

About ten percent of workers (16 years and over) in the United States commute to their jobs by carpooling. You randomly select eight workers. What is the probability that exactly four of them carpool to work? Use a table to find the probability. *(Source: American Community Survey)*

▸ Solution

A portion of Table 2 in Appendix B is shown here. Using the distribution for $n = 8$ and $p = 0.1$, you can find the probability that $x = 4$, as shown by the highlighted areas in the table.

													p	
n	x	.01	.05	.10	.15	.20	.25	.30	.35	.40	.45	.50	.55	.60
2	0	.980	.902	.810	.723	.640	.563	.490	.423	.360	.303	.250	.203	.160
	1	.020	.095	.180	.255	.320	.375	.420	.455	.480	.495	.500	.495	.480
	2	.000	.002	.010	.023	.040	.063	.090	.123	.160	.203	.250	.303	.360
3	0	.970	.857	.729	.614	.512	.422	.343	.275	.216	.166	.125	.091	.064
	1	.029	.135	.243	.325	.384	.422	.441	.444	.432	.408	.375	.334	.288
	2	.000	.007	.027	.057	.096	.141	.189	.239	.288	.334	.375	.408	.432
	3	.000	.000	.001	.003	.008	.016	.027	.043	.064	.091	.125	.166	.216
8	0	.923	.663	.430	.272	.168	.100	.058	.032	.017	.008	.004	.002	.001
	1	.075	.279	.383	.385	.336	.267	.198	.137	.090	.055	.031	.016	.008
	2	.003	.051	.149	.238	.294	.311	.296	.259	.209	.157	.109	.070	.041
	3	.000	.005	.033	.084	.147	.208	.254	.279	.279	.257	.219	.172	.124
	4	.000	.000	.005	.018	.046	.087	.136	.188	.232	.263	.273	.263	.232
	5	.000	.000	.000	.003	.009	.023	.047	.081	.124	.172	.219	.257	.279
	6	.000	.000	.000	.000	.001	.004	.010	.022	.041	.070	.109	.157	.209
	7	.000	.000	.000	.000	.000	.000	.001	.003	.008	.016	.031	.055	.090
	8	.000	.000	.000	.000	.000	.000	.000	.000	.001	.002	.004	.008	.017

To explore this topic further, see Activity 4.2 on page 216.

Interpretation So, the probability that exactly four of the eight workers carpool to work is 0.005. Because 0.005 is less than 0.05, this can be considered an unusual event.

▸ Try It Yourself 6

About fifty-five percent of all small businesses in the United States have a website. If you randomly select 10 small businesses, what is the probability that exactly four of them have a website? Use a table to find the probability. *(Adapted from Webvisible/Nielsen Online)*

a. *Identify* a trial, a success, and a failure.
b. Identify $n, p,$ and x.
c. Use Table 2 in Appendix B to find the *binomial probability*.
d. *Interpret* the results.
e. Determine if the event is *unusual*. Explain.

Answer: Page A37

▶ GRAPHING BINOMIAL DISTRIBUTIONS

In Section 4.1, you learned how to graph discrete probability distributions. Because a binomial distribution is a discrete probability distribution, you can use the same process.

EXAMPLE 7

▶ Graphing a Binomial Distribution

Sixty percent of households in the United States own a video game console. You randomly select six households and ask them if they own a video game console. Construct a probability distribution for the random variable x. Then graph the distribution. *(Source: Deloitte LLP)*

▶ Solution

To construct the binomial distribution, find the probability for each value of x. Using $n = 6$, $p = 0.6$, and $q = 0.4$, you can obtain the following.

x	0	1	2	3	4	5	6
$P(x)$	0.004	0.037	0.138	0.276	0.311	0.187	0.047

You can graph the probability distribution using a histogram as shown below.

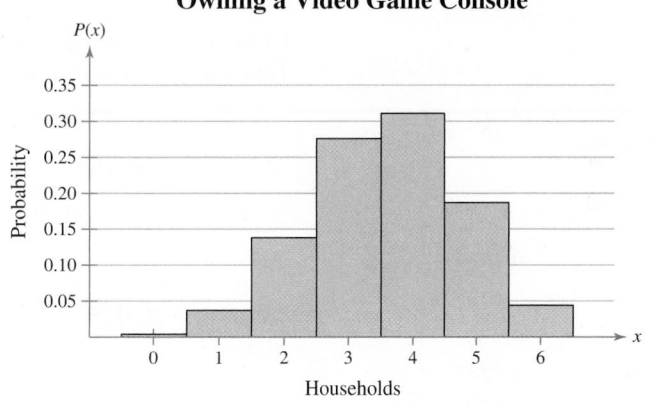

Owning a Video Game Console

Interpretation   From the histogram, you can see that it would be unusual if none, only one, or all six of the households owned a video game console because of the low probabilities.

▶ Try It Yourself 7

Eighty-one percent of households in the United States own a computer. You randomly select four households and ask if they own a computer. Construct a probability distribution for the random variable x. Then graph the distribution. *(Source: Nielsen)*

a. Find the *binomial probability* for each value of the random variable x.
b. *Organize* the values of x and their corresponding probabilities in a table.
c. Use a *histogram* to graph the binomial distribution. Then describe its shape.
d. Are any of the events *unusual*? Explain.   *Answer: Page A37*

Notice in Example 7 that the histogram is skewed left. The graph of a binomial distribution with $p > 0.5$ is skewed left, whereas the graph of a binomial distribution with $p < 0.5$ is skewed right. The graph of a binomial distribution with $p = 0.5$ is symmetric.

▸ MEAN, VARIANCE, AND STANDARD DEVIATION

Although you can use the formulas you learned in Section 4.1 for mean, variance, and standard deviation of a discrete probability distribution, the properties of a binomial distribution enable you to use much simpler formulas.

POPULATION PARAMETERS OF A BINOMIAL DISTRIBUTION

Mean: $\mu = np$

Variance: $\sigma^2 = npq$

Standard deviation: $\sigma = \sqrt{npq}$

Note to Instructor

It is useful to have students calculate the mean, variance, and standard deviation for a binomial experiment using the general formulas for discrete probability distributions given in Section 4.1. A quick example (and good review) would be to use the distribution for $n = 4$ and $p = 0.3$.

Point out that the formula $\mu = np$ gives the same result rounded to two decimal places.

x	$P(x)$	$xP(x)$
0	0.240	0
1	0.412	0.412
2	0.265	0.530
3	0.076	0.228
4	0.008	0.032
	$\Sigma = 1.001$	$\Sigma = 1.202$

EXAMPLE 8

▸ Finding and Interpreting Mean, Variance, and Standard Deviation

In Pittsburgh, Pennsylvania, about 56% of the days in a year are cloudy. Find the mean, variance, and standard deviation for the number of cloudy days during the month of June. Interpret the results and determine any unusual values. *(Source: National Climatic Data Center)*

▸ Solution

There are 30 days in June. Using $n = 30$, $p = 0.56$, and $q = 0.44$, you can find the mean, variance, and standard deviation as shown below.

$$\mu = np = 30 \cdot 0.56$$
$$= 16.8$$
$$\sigma^2 = npq = 30 \cdot 0.56 \cdot 0.44$$
$$\approx 7.4$$
$$\sigma = \sqrt{npq} = \sqrt{30 \cdot 0.56 \cdot 0.44}$$
$$\approx 2.7$$

Interpretation On average, there are 16.8 cloudy days during the month of June. The standard deviation is about 2.7 days. Values that are more than two standard deviations from the mean are considered unusual. Because $16.8 - 2(2.7) = 11.4$, a June with 11 cloudy days or less would be unusual. Similarly, because $16.8 + 2(2.7) = 22.2$, a June with 23 cloudy days or more would also be unusual.

▸ Try It Yourself 8

In San Francisco, California, 44% of the days in a year are clear. Find the mean, variance, and standard deviation for the number of clear days during the month of May. Interpret the results and determine any unusual events. *(Source: National Climatic Data Center)*

a. *Identify* a success and the values of n, p, and q.
b. Find the *product* of n and p to calculate the mean.
c. Find the *product* of n, p, and q for the variance.
d. Find the *square root* of the variance to find the standard deviation.
e. *Interpret* the results.
f. Determine any *unusual* events.

Answer: Page A37

4.2 EXERCISES

1. Each trial is independent of the other trials if the outcome of one trial does not affect the outcome of any of the other trials.

2. The random variable measures the number of successes in *n* trials.

3. (a) *p* = 0.50
   (b) *p* = 0.20
   (c) *p* = 0.80

4. (a) *p* = 0.75
   (b) *p* = 0.50
   (c) *p* = 0.25

5. (a) *n* = 12
   (b) *n* = 4
   (c) *n* = 8

   As *n* increases, the distribution becomes more symmetric.

6. (a) *n* = 10
   (b) *n* = 15
   (c) *n* = 5

   As *n* increases, the distribution becomes more symmetric.

■ BUILDING BASIC SKILLS AND VOCABULARY

1. In a binomial experiment, what does it mean to say that each trial is independent of the other trials?

2. In a binomial experiment with *n* trials, what does the random variable measure?

Graphical Analysis   *In Exercises 3 and 4, match each given probability with the correct graph. The histograms represent binomial distributions. Each distribution has the same number of trials n but different probabilities of success p.*

3. *p* = 0.20, *p* = 0.50, *p* = 0.80

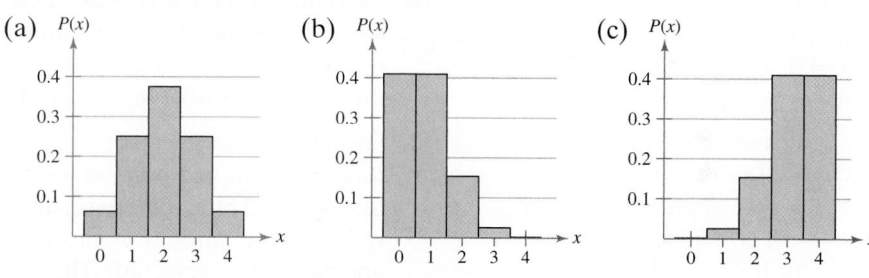

4. *p* = 0.25, *p* = 0.50, *p* = 0.75

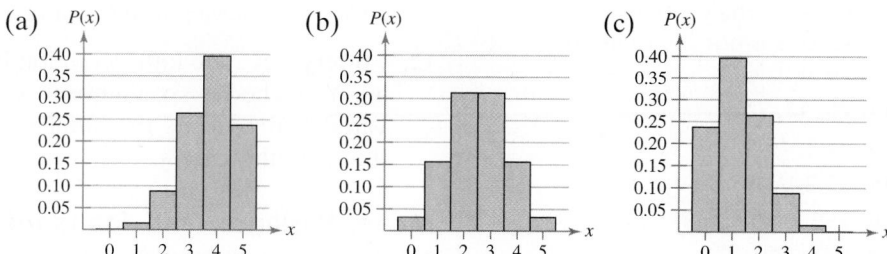

Graphical Analysis   *In Exercises 5 and 6, match each given value of n with the correct graph. Each histogram shown represents part of a binomial distribution. Each distribution has the same probability of success p but different numbers of trials n. What happens as the value of n increases and p remains the same?*

5. *n* = 4, *n* = 8, *n* = 12

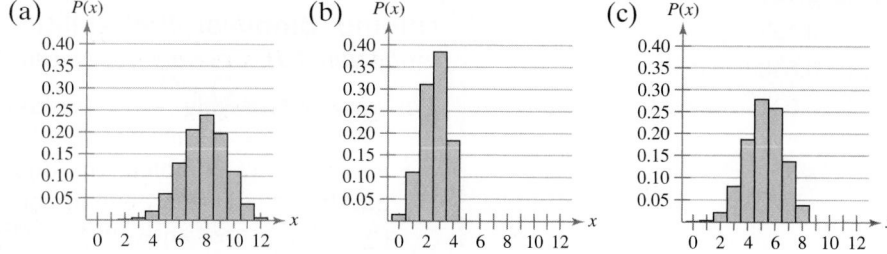

6. *n* = 5, *n* = 10, *n* = 15

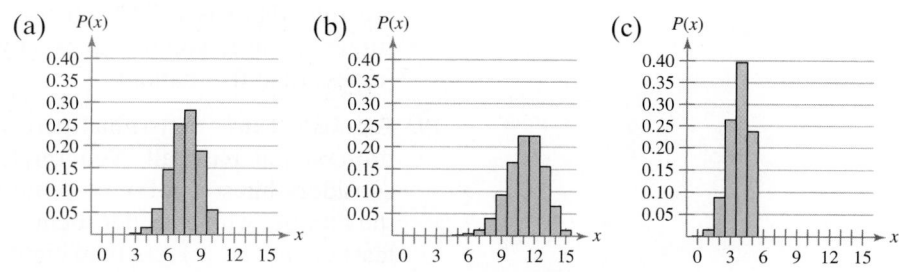

7. (a) $x = 0, 1, 2, 3, 4, 11, 12$
    (b) $x = 0$
    (c) $x = 0, 1, 2, 8$

8. (a) $x = 0, 1, 2, 3, 4$
    (b) $x = 0, 1, 2, 3, 4, 5, 6, 7, 8, 15$
    (c) $x = 0, 1$

9. Binomial experiment
    Success: baby recovers
    $n = 5, p = 0.80, q = 0.20,$
    $x = 0, 1, 2, 3, 4, 5$

10. Binomial experiment
    Success: person does not make a purchase
    $n = 18, p = 0.74, q = 0.26,$
    $x = 0, 1, 2, \ldots, 18$

11. Binomial experiment
    Success: selecting an officer who is postponing or reducing the amount of vacation
    $n = 20, p = 0.31, q = 0.69,$
    $x = 0, 1, 2, \ldots, 20$

12. Not a binomial experiment because the probability of a success is not the same for each trial.

13. 20, 12, 3.5

14. 54.6, 19.1, 4.4

15. 32.2, 23.9, 4.9

16. 259.1, 46.6, 6.8

17. (a) 0.088
    (b) 0.104
    (c) 0.896

18. (a) 0.318
    (b) 0.647
    (c) 0.353

19. (a) 0.111
    (b) 0.152
    (c) 0.848

20. (a) 0.021
    (b) 0.026
    (c) 0.974

7. Identify the unusual values of x in each histogram in Exercise 5.

8. Identify the unusual values of x in each histogram in Exercise 6.

Identifying and Understanding Binomial Experiments
In Exercises 9–12, decide whether the experiment is a binomial experiment. If it is, identify a success, specify the values of n, p, and q, and list the possible values of the random variable x. If it is not a binomial experiment, explain why.

9. **Cyanosis** Cyanosis is the condition of having bluish skin due to insufficient oxygen in the blood. About 80% of babies born with cyanosis recover fully. A hospital is caring for five babies born with cyanosis. The random variable represents the number of babies that recover fully. *(Source: The World Book Encyclopedia)*

10. **Clothing Store Purchases** From past records, a clothing store finds that 26% of the people who enter the store will make a purchase. During a one-hour period, 18 people enter the store. The random variable represents the number of people who do not make a purchase.

11. **Survey** A survey asks 1400 chief financial officers, "Has the economy forced you to postpone or reduce the amount of vacation you plan to take this year?" Thirty-one percent of those surveyed say they are postponing or reducing the amount of vacation. Twenty officers participating in the survey are randomly selected. The random variable represents the number of officers who are postponing or reducing the amount of vacation. *(Source: Robert Half Management Resources)*

12. **Lottery** A state lottery randomly chooses 6 balls numbered from 1 through 40. You choose six numbers and purchase a lottery ticket. The random variable represents the number of matches on your ticket to the numbers drawn in the lottery.

Mean, Variance, and Standard Deviation
In Exercises 13–16, find the mean, variance, and standard deviation of the binomial distribution with the given values of n and p.

13. $n = 50, p = 0.4$

14. $n = 84, p = 0.65$

15. $n = 124, p = 0.26$

16. $n = 316, p = 0.82$

■ USING AND INTERPRETING CONCEPTS

Finding Binomial Probabilities
In Exercises 17–26, find the indicated probabilities. If convenient, use technology to find the probabilities.

17. **Answer Guessing** You are taking a multiple-choice quiz that consists of five questions. Each question has four possible answers, only one of which is correct. To complete the quiz, you randomly guess the answer to each question. Find the probability of guessing (a) exactly three answers correctly, (b) at least three answers correctly, and (c) less than three answers correctly.

18. **Surgery Success** A surgical technique is performed on seven patients. You are told there is a 70% chance of success. Find the probability that the surgery is successful for (a) exactly five patients, (b) at least five patients, and (c) less than five patients.

19. **Baseball Fans** Fifty-nine percent of men consider themselves fans of professional baseball. You randomly select 10 men and ask each if he considers himself a fan of professional baseball. Find the probability that the number who consider themselves baseball fans is (a) exactly eight, (b) at least eight, and (c) less than eight. *(Source: Gallup Poll)*

21. (a) 0.257
    (b) 0.220
    (c) 0.780
22. (a) 1.627×10^{-9}
    (b) 0.9999999983
    (c) 1.662×10^{-9}
23. (a) 0.187
    (b) 0.605
    (c) 0.084
24. (a) 0.264
    (b) 8.555×10^{-5}
    (c) 0.999
25. (a) 0.255
    (b) 0.562
    (c) 0.783
26. (a) 0.194
    (b) 0.158
    (c) 0.351
27. (a) $n = 6$, $p = 0.63$

x	P(x)
0	0.003
1	0.026
2	0.112
3	0.253
4	0.323
5	0.220
6	0.063

(b)

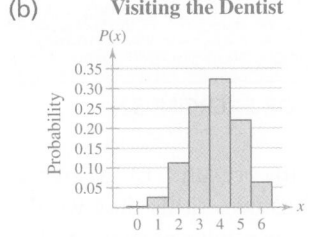

Visiting the Dentist

Skewed left

(c) 3.8, 1.4, 1.2

(d) On average, 3.8 out of 6 adults are visiting the dentist less because of the economy. The standard deviation is 1.2, so most samples of 6 adults would differ from the mean by no more than 1.2 people. The values $x = 0$ and $x = 1$ would be unusual because their probabilities are less than 0.05.

20. Favorite Cookie   Ten percent of adults say oatmeal raisin is their favorite cookie. You randomly select 12 adults and ask them to name their favorite cookie. Find the probability that the number who say oatmeal raisin is their favorite cookie is (a) exactly four, (b) at least four, and (c) less than four. *(Source: WEAREVER)*

21. Savings   Fifty-five percent of U.S. households say they would feel secure if they had $50,000 in savings. You randomly select 8 households and ask them if they would feel secure if they had $50,000 in savings. Find the probability that the number that say they would feel secure is (a) exactly five, (b) more than five, and (c) at most five. *(Source: HSBC Consumer Survey)*

22. Honeymoon Financing   Seventy percent of married couples paid for their honeymoon themselves. You randomly select 20 married couples and ask them if they paid for their honeymoon themselves. Find the probability that the number of couples who say they paid for their honeymoon themselves is (a) exactly one, (b) more than one, and (c) at most one. *(Source: Bride's Magazine)*

23. Financial Advice   Forty-three percent of adults say they get their financial advice from family members. You randomly select 14 adults and ask them if they get their financial advice from family members. Find the probability that the number who say they get their financial advice from family members is (a) exactly five, (b) at least six, and (c) at most three. *(Source: Sun Life Unretirement Index)*

24. Retirement   Fourteen percent of workers believe they will need less than $250,000 when they retire. You randomly select 10 workers and ask them how much money they think they will need for retirement. Find the probability that the number of workers who say they will need less than $250,000 when they retire is (a) exactly two, (b) more than six, and (c) at most five. *(Source: Retirement Corporation of America)*

25. Credit Cards   Twenty-eight percent of college students say they use credit cards because of the rewards program. You randomly select 10 college students and ask them to name the reason they use credit cards. Find the probability that the number of college students who say they use credit cards because of the rewards program is (a) exactly two, (b) more than two, and (c) between two and five, inclusive. *(Source: Experience.com)*

26. Movies on Phone   Twenty-five percent of adults say they would watch streaming movies on their phone at work. You randomly select 12 adults and ask them if they would watch streaming movies on their phone at work. Find the probability that the number who say they would watch streaming movies on their phone at work is (a) exactly four, (b) more than four, and (c) between four and eight, inclusive. *(Source: mSpot)*

Constructing Binomial Distributions   *In Exercises 27–30, (a) construct a binomial distribution, (b) graph the binomial distribution using a histogram and describe its shape, (c) find the mean, variance, and standard deviation of the binomial distribution, and (d) interpret the results in the context of the real-life situation. What values of the random variable x would you consider unusual? Explain your reasoning.*

27. Visiting the Dentist   Sixty-three percent of adults say they are visiting the dentist less because of the economy. You randomly select six adults and ask them if they are visiting the dentist less because of the economy. *(Source: American Optometric Association)*

28. (a) $n = 5$, $p = 0.25$

x	P(x)
0	0.237
1	0.396
2	0.264
3	0.088
4	0.015
5	0.001

(b) See Selected Answers, page A118.

(c) 1.3, 0.9, 1.0

(d) On average, 1.3 adults out of 5 have no trouble sleeping at night. The standard deviation is 1.0, so most samples of five adults would differ from the mean by no more than 1 person. The values $x = 4$ and $x = 5$ would be unusual because their probabilities are less than 0.05.

29. (a) $n = 4$, $p = 0.05$

x	P(x)
0	0.814506
1	0.171475
2	0.013537
3	0.000475
4	0.000006

(b) See Odd Answers, page A69.

(c) 0.2, 0.2, 0.4

(d) On average, 0.2 eligible adult out of every 4 gives blood. The standard deviation is 0.4, so most samples of four eligible adults would differ from the mean by at most 0.4 adult.

$x = 2$, 3, and 4 would be unusual because their probabilities are less than 0.05.

30. (a) and (b) See Selected Answers, page A118.

(c) 2.0, 1.2, 1.1

(d) On average, 2.0 adults out of 5 have O^+ blood. The standard deviation is 1.1, so most samples of 5 adults would differ from the mean by at most 1.1 adults. The value $x = 5$ would be unusual because the probability is less than 0.05.

28. **No Trouble Sleeping**   One in four adults claims to have no trouble sleeping at night. You randomly select five adults and ask them if they have any trouble sleeping at night. *(Source: Marist Institute for Public Opinion)*

29. **Blood Donors**   Five percent of people in the United States eligible to donate blood actually do. You randomly select four eligible blood donors and ask them if they donate blood. *(Source: MetLife Consumer Education Center)*

30. **Blood Types**   Thirty-nine percent of people in the United States have type O^+ blood. You randomly select five Americans and ask them if their blood type is O^+. *(Source: American Association of Blood Banks)*

31. **Annoying Flights**   The graph shows the results of a survey of travelers who were asked to name what they found most annoying on a flight. You randomly select six people who participated in the survey and ask them to name what they find most annoying on a flight. Let x represent the number who name crying kids as the most annoying. *(Source: USA Today)*

(a) Construct a binomial distribution.

(b) Find the probability that exactly two people name "crying kids."

(c) Find the probability that at least five people name "crying kids."

FIGURE FOR EXERCISE 31

FIGURE FOR EXERCISE 32

32. **Small-Business Owners**   The graph shows the results of a survey of small-business owners who were asked which business skills they would like to develop further. You randomly select five owners who participated in the survey and ask them which business skills they want to develop further. Let x represent the number who said financial management was the skill they wanted to develop further. *(Source: American Express)*

(a) Construct a binomial distribution.

(b) Find the probability that exactly two owners say "financial management."

(c) Find the probability that fewer than four owners say "financial management."

33. Find the mean and standard deviation of the binomial distribution in Exercise 31 and interpret the results in the context of the real-life situation. What values of x would you consider unusual? Explain your reasoning.

34. Find the mean and standard deviation of the binomial distribution in Exercise 32 and interpret the results in the context of the real-life situation. What values of x would you consider unusual? Explain your reasoning.

31. (a) $n = 6$, $p = 0.37$

x	$P(x)$
0	0.063
1	0.220
2	0.323
3	0.253
4	0.112
5	0.026
6	0.003

(b) 0.323

(c) 0.029

32. (a) $n = 5$, $p = 0.48$

x	$P(x)$
0	0.038
1	0.175
2	0.324
3	0.299
4	0.138
5	0.025

(b) 0.324

(c) 0.837

33. 2.2, 1.2

On average, 2.2 out of 6 travelers would name "crying kids" as the most annoying. The standard deviation is 1.2, so most samples of 6 travelers would differ from the mean by at most 1.2 travelers. The values $x = 5$ and $x = 6$ would be unusual because their probabilities are less than 0.05.

34. 2.4, 1.1

On average, if 5 owners are randomly selected, 2.4 owners will name financial management as the skill they want to develop further. The standard deviation is 1.1, so most samples of 5 owners would differ from the mean by at most 1.1 owners. The values $x = 0$ and $x = 5$ would be unusual because their probabilities are less than 0.05.

35. (a) 0.081

(b) 0.541

(c) 0.022; This event is unusual because its probability is less than 0.05.

36. (a) 0.270

(b) 0.431

(c) 0.024; This event is unusual because its probability is less than 0.05.

37. 0.033

38. 0.002

SC *In Exercises 35 and 36, use the StatCrunch binomial calculator to find the indicated probabilities. Then determine if the event is unusual. Explain your reasoning.*

35. Pet Owners   Sixty-six percent of pet owners say they consider their pet to be their best friend. You randomly select 10 pet owners and ask them if they consider their pet to be their best friend. Find the probability that the number who say their pet is their best friend is (a) exactly nine, (b) at least seven, and (c) at most three. *(Adapted from Kelton Research)*

36. Eco-Friendly Vehicles   Fifty-three percent of 18- to 30-year-olds say they would pay more for an eco-friendly vehicle. You randomly select eight 18- to 30-year-olds and ask each if they would pay more for an eco-friendly vehicle. Find the probability that the number who say they would pay more for an eco-friendly vehicle is (a) exactly four, (b) at least five, and (c) less than two. *(Source: Deloitte LLP and Michigan State University)*

■ EXTENDING CONCEPTS

Multinomial Experiments   *In Exercises 37 and 38, use the following information.*

A **multinomial experiment** is a probability experiment that satisfies the following conditions.

1. The experiment is repeated a fixed number of times n where each trial is independent of the other trials.

2. Each trial has k possible mutually exclusive outcomes: $E_1, E_2, E_3, \ldots, E_k$.

3. Each outcome has a fixed probability. So, $P(E_1) = p_1$, $P(E_2) = p_2$, $P(E_3) = p_3, \ldots, P(E_k) = p_k$. The sum of the probabilities for all outcomes is

$$p_1 + p_2 + p_3 + \cdots + p_k = 1.$$

4. x_1 is the number of times E_1 will occur, x_2 is the number of times E_2 will occur, x_3 is the number of times E_3 will occur, and so on.

5. The discrete random variable x counts the number of times $x_1, x_2, x_3, \ldots, x_k$ occurs in n independent trials where

$$x_1 + x_2 + x_3 + \cdots + x_k = n.$$

The probability that x will occur is

$$P(x) = \frac{n!}{x_1! x_2! x_3! \cdots x_k!} \, p_1^{x_1} p_2^{x_2} p_3^{x_3} \cdots p_k^{x_k}.$$

37. Genetics   According to a theory in genetics, if tall and colorful plants are crossed with short and colorless plants, four types of plants will result: tall and colorful, tall and colorless, short and colorful, and short and colorless, with corresponding probabilities of $\frac{9}{16}$, $\frac{3}{16}$, $\frac{3}{16}$, and $\frac{1}{16}$. If 10 plants are selected, find the probability that 5 will be tall and colorful, 2 will be tall and colorless, 2 will be short and colorful, and 1 will be short and colorless.

38. Genetics   Another proposed theory in genetics gives the corresponding probabilities for the four types of plants described in Exercise 37 as $\frac{5}{16}$, $\frac{4}{16}$, $\frac{1}{16}$, and $\frac{6}{16}$. If 10 plants are selected, find the probability that 5 will be tall and colorful, 2 will be tall and colorless, 2 will be short and colorful, and 1 will be short and colorless.

ACTIVITY 4.2 Binomial Distribution

APPLET

The *binomial distribution* applet allows you to simulate values from a binomial distribution. You can specify the parameters for the binomial distribution (*n* and *p*) and the number of values to be simulated (*N*). When you click SIMULATE, *N* values from the specified binomial distribution will be plotted at the right. The frequency of each outcome is shown in the plot.

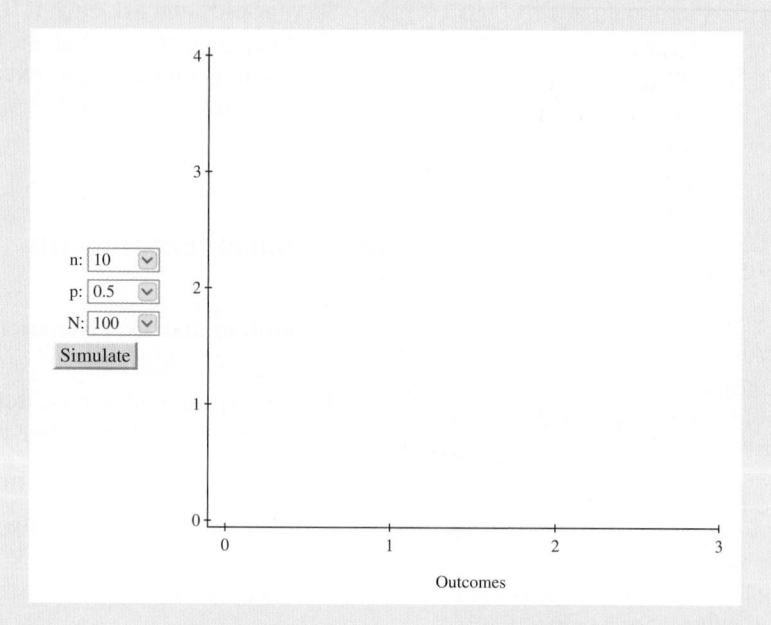

■ Explore

Step 1 Specify a value of *n*.
Step 2 Specify a value of *p*.
Step 3 Specify a value of *N*.
Step 4 Click SIMULATE.

■ Draw Conclusions

APPLET

1. During a presidential election year, 70% of a county's eligible voters actually vote. Simulate selecting *n* = 10 eligible voters *N* = 10 times (for 10 communities in the county). Use the results to estimate the probability that the number who voted in this election is (a) exactly 5, (b) at least 8, and (c) at most 7.

2. During a non-presidential election year, 20% of the eligible voters in the same county as in Exercise 1 actually vote. Simulate selecting *n* = 10 eligible voters *N* = 10 times (for 10 communities in the county). Use the results to estimate the probability that the number who voted in this election is (a) exactly 4, (b) at least 5, and (c) less than 4.

3. Suppose in Exercise 1 you select *n* = 10 eligible voters *N* = 100 times. Estimate the probability that the number who voted in this election is exactly 5. Compare this result with the result in Exercise 1 part (a). Which of these is closer to the probability found using the binomial probability formula?

Binomial Distribution of Airplane Accidents

The Air Transport Association of America (ATA) is a support organization for the principal U.S. airlines. Some of the ATA's activities include promoting the air transport industry and conducting industry-wide studies.

The ATA also keeps statistics about commercial airline flights, including those that involve accidents. From 1979 through 2008 for aircraft with 10 or more seats, there were 76 fatal commercial airplane accidents involving U.S. airlines. The distribution of these accidents is shown in the histogram at the right.

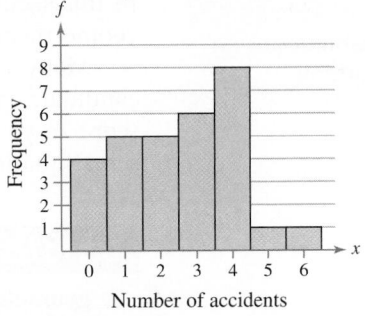

Fatal Commercial Airplane Accidents per Year (1979–2008)

Year	1979	1980	1981	1982	1983	1984	1985	1986	1987	1988	1989	1990	1991	1992	1993
Accidents	4	0	4	4	4	1	4	2	4	3	5	4	3	3	1

Year	1994	1995	1996	1997	1998	1999	2000	2001	2002	2003	2004	2005	2006	2007	2008
Accidents	4	1	3	3	1	2	2	6	0	2	1	3	2	0	0

■ EXERCISES

1. In 2006, there were about 11 million commercial flights in the United States. If one is selected at random, what is the probability that it involved a fatal accident?

2. Suppose that the probability of a fatal accident in a given year is 0.0000004. A binomial probability distribution for $n = 11,000,000$ and $p = 0.0000004$ with $x = 0$ to 12 is shown.

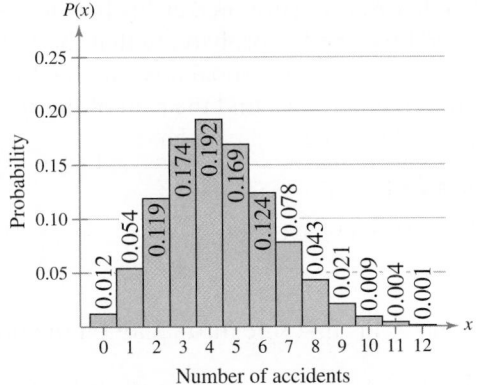

What is the probability that there will be (a) 4 fatal accidents in a year? (b) 10 fatal accidents? (c) between 1 and 5, inclusive?

3. Construct a binomial distribution for $n = 11,000,000$ and $p = 0.0000008$ with $x = 0$ to 12. Compare your results with the distribution in Exercise 2.

4. Is a binomial distribution a good model for determining the probabilities of various numbers of fatal accidents during a year? Explain your reasoning and include a discussion of the four criteria for a binomial experiment.

5. According to analysis by *USA TODAY*, air flight is so safe that a person "would have to fly every day for more than 64,000 years before dying in an accident." How can such a statement be justified?

4.3 More Discrete Probability Distributions

WHAT YOU SHOULD LEARN

▸ How to find probabilities using the geometric distribution

▸ How to find probabilities using the Poisson distribution

The Geometric Distribution ▸ The Poisson Distribution ▸ Summary of Discrete Probability Distributions

▸ THE GEOMETRIC DISTRIBUTION

In this section, you will study two more discrete probability distributions—the geometric distribution and the Poisson distribution.

Many actions in life are repeated until a success occurs. For instance, a CPA candidate might take the CPA exam several times before receiving a passing score, or you might have to send an e-mail several times before it is successfully sent. Situations such as these can be represented by a *geometric distribution*.

Note to Instructor

If you choose, this section can be omitted. None of the material included here will be needed in later sections.

DEFINITION

A **geometric distribution** is a discrete probability distribution of a random variable x that satisfies the following conditions.

1. A trial is repeated until a success occurs.
2. The repeated trials are independent of each other.
3. The probability of success p is constant for each trial.
4. The random variable x represents the number of the trial in which the first success occurs.

The probability that the first success will occur on trial number x is

$$P(x) = pq^{x-1}, \text{ where } q = 1 - p.$$

In other words, when the first success occurs on the third trial, the outcome is *FFS*, and the probability is $P(3) = q \cdot q \cdot p$, or $P(3) = p \cdot q^2$.

STUDY TIP

Here are instructions for finding a geometric probability on a TI-83/84 Plus.

2nd DISTR

D: geometpdf(

Enter the values of p and x separated by commas.

ENTER

```
geometpdf(.74,3)
            .050024
geometpdf(.74,4)
            .01300624
```

Using a TI-83/84 Plus, you can find the probabilities used in Example 1 automatically.

EXAMPLE 1

▸ **Finding Probabilities Using the Geometric Distribution**

Basketball player LeBron James makes a free throw shot about 74% of the time. Find the probability that the first free throw shot LeBron makes occurs on the third or fourth attempt. *(Source: ESPN)*

▸ **Solution**   To find the probability that LeBron makes his first free throw shot on the third or fourth attempt, first find the probability that the first shot he makes will occur on the third attempt and the probability that the first shot he makes will occur on the fourth attempt. Then, find the sum of the resulting probabilities. Using $p = 0.74$, $q = 0.26$, and $x = 3$, you have

$$P(3) = 0.74 \cdot (0.26)^2 = 0.050024.$$

Using $p = 0.74$, $q = 0.26$, and $x = 4$, you have

$$P(4) = 0.74 \cdot (0.26)^3 \approx 0.013006.$$

So, the probability that LeBron makes his first free throw shot on the third or fourth attempt is

$$P(\text{shot made on third or fourth attempt}) = P(3) + P(4)$$

$$\approx 0.050024 + 0.013006 \approx 0.063.$$

Note to Instructor
The sum of all the probabilities of a geometric distribution is a geometric series with an infinite number of terms. The sum of this geometric series is $s = a/(1 - r)$, where a is the first term of the series, and r is the common ratio of consecutive terms. In this case, $a = p$ and $r = q$, so $s = p/(1 - q) = p/p = 1$. So, each probability value is between 0 and 1, inclusive, and the sum of all the probabilities is 1.

▶ **Try It Yourself 1**

Find the probability that LeBron makes his first free throw shot before his third attempt.

a. Use the *geometric distribution* to find $P(1)$ and $P(2)$.
b. Find the *sum* of $P(1)$ and $P(2)$.
c. *Write* the result as a sentence.   *Answer: Page A37*

Even though theoretically a success may never occur, the geometric distribution is a discrete probability distribution because the values of x can be listed—1, 2, 3, ... Notice that as x becomes larger, $P(x)$ gets closer to zero. For instance,

$$P(15) = 0.74(0.26)^{14} \approx 0.0000000048.$$

▶ **THE POISSON DISTRIBUTION**

In a binomial experiment, you are interested in finding the probability of a specific number of successes in a given number of trials. Suppose instead that you want to know the probability that a specific number of occurrences takes place within a given unit of time or space. For instance, to determine the probability that an employee will take 15 sick days within a year, you can use the *Poisson distribution*.

STUDY TIP

Here are instructions for finding a Poisson probability on a TI-83/84 Plus.

2nd DISTR

B: poissonpdf(

Enter the values of μ and x separated by commas.

ENTER

```
Poissonpdf(3,4)
      .1680313557
```

Using a TI-83/84 Plus, you can find the probability in Example 2 automatically.

DEFINITION

The **Poisson distribution** is a discrete probability distribution of a random variable x that satisfies the following conditions.

1. The experiment consists of counting the number of times x an event occurs in a given interval. The interval can be an interval of time, area, or volume.

2. The probability of the event occurring is the same for each interval.

3. The number of occurrences in one interval is independent of the number of occurrences in other intervals.

The probability of exactly x occurrences in an interval is

$$P(x) = \frac{\mu^x e^{-\mu}}{x!}$$

where e is an irrational number approximately equal to 2.71828 and μ is the mean number of occurrences per interval unit.

EXAMPLE 2   (SC) Report 19

▶ **Using the Poisson Distribution**

The mean number of accidents per month at a certain intersection is three. What is the probability that in any given month four accidents will occur at this intersection?

▶ **Solution**

Using $x = 4$ and $\mu = 3$, the probability that 4 accidents will occur in any given month at the intersection is

$$P(4) \approx \frac{3^4 (2.71828)^{-3}}{4!} \approx 0.168.$$

▸ **Try It Yourself 2**

What is the probability that more than four accidents will occur in any given month at the intersection?

a. Use the *Poisson distribution* to find $P(0)$, $P(1)$, $P(2)$, $P(3)$, and $P(4)$.
b. Find the *sum* of $P(0)$, $P(1)$, $P(2)$, $P(3)$, and $P(4)$.
c. *Subtract* the sum from 1.
d. *Write* the result as a sentence.

Answer: Page A37

In Example 2, you used a formula to determine a Poisson probability. You can also use a table to find Poisson probabilities. Table 3 in Appendix B lists the Poisson probabilities for selected values of x and μ. You can use technology tools, such as MINITAB, Excel, and the TI-83/84 Plus, to find Poisson probabilities as well.

PICTURING THE WORLD

The first successful suspension bridge built in the United States, the Tacoma Narrows Bridge, spans the Tacoma Narrows in Washington State. The average occupancy of vehicles that travel across the bridge is 1.6. The following probability distribution represents the vehicle occupancy on the bridge during a five-day period. (Adapted from Washington State Department of Transportation)

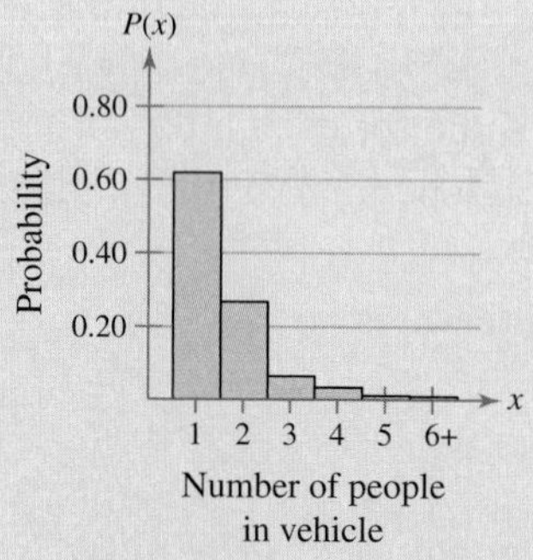

What is the probability that a randomly selected vehicle has two occupants or fewer?

The probability that a randomly selected vehicle has two occupants or fewer is about 0.85.

EXAMPLE 3

▸ Finding Poisson Probabilities Using a Table

A population count shows that the average number of rabbits per acre living in a field is 3.6. Use a table to find the probability that seven rabbits are found on any given acre of the field.

▸ Solution

A portion of Table 3 in Appendix B is shown here. Using the distribution for $\mu = 3.6$ and $x = 7$, you can find the Poisson probability as shown by the highlighted areas in the table.

	μ						
x	**3.1**	**3.2**	**3.3**	**3.4**	**3.5**	**3.6**	**3.7**
0	.0450	.0408	.0369	.0334	.0302	.0273	.0247
1	.1397	.1304	.1217	.1135	.1057	.0984	.0915
2	.2165	.2087	.2008	.1929	.1850	.1771	.1692
3	.2237	.2226	.2209	.2186	.2158	.2125	.2087
4	.1734	.1781	.1823	.1858	.1888	.1912	.1931
5	.1075	.1140	.1203	.1264	.1322	.1377	.1429
6	.0555	.0608	.0662	.0716	.0771	.0826	.0881
7	.0246	.0278	.0312	.0348	.0385	.0425	.0466
8	.0095	.0111	.0129	.0148	.0169	.0191	.0215
9	.0033	.0040	.0047	.0056	.0066	.0076	.0089
10	.0010	.0013	.0016	.0019	.0023	.0028	.0033

Interpretation So, the probability that seven rabbits are found on any given acre is 0.0425. Because 0.0425 is less than 0.05, this can be considered an unusual event.

▸ **Try It Yourself 3**

Two thousand brown trout are introduced into a small lake. The lake has a volume of 20,000 cubic meters. Use a table to find the probability that three brown trout are found in any given cubic meter of the lake.

a. Find the *average* number of brown trout per cubic meter.
b. *Identify* μ and x.
c. *Use* Table 3 in Appendix B to find the Poisson probability.
d. *Interpret* the results.
e. Determine if the event is *unusual*. Explain.

Answer: Page A37

▶ SUMMARY OF DISCRETE PROBABILITY DISTRIBUTIONS

The following table summarizes the discrete probability distributions discussed in this chapter.

Distribution	Summary	Formulas
Binomial Distribution	A binomial experiment satisfies the following conditions. **1.** The experiment is repeated for a fixed number n of independent trials. **2.** There are only two possible outcomes for each trial. Each outcome can be classified as a success or as a failure. **3.** The probability of a success must remain constant for each trial. **4.** The random variable x counts the number of successful trials. The parameters of a binomial distribution are n and p.	n = the number of times a trial repeats x = the number of successes in n trials p = probability of success in a single trial q = probability of failure in a single trial $q = 1 - p$ The probability of exactly x successes in n trials is $$P(x) = {}_nC_x\, p^x q^{n-x}$$ $$= \frac{n!}{(n-x)!x!}\, p^x q^{n-x}.$$
Geometric Distribution	A geometric distribution is a discrete probability distribution of a random variable x that satisfies the following conditions. **1.** A trial is repeated until a success occurs. **2.** The repeated trials are independent of each other. **3.** The probability of success p is constant for each trial. **4.** The random variable x represents the number of the trial in which the first success occurs. The parameter of a geometric distribution is p.	x = the number of the trial in which the first success occurs p = probability of success in a single trial q = probability of failure in a single trial $q = 1 - p$ The probability that the first success occurs on trial number x is $$P(x) = pq^{x-1}.$$
Poisson Distribution	The Poisson distribution is a discrete probability distribution of a random variable x that satisfies the following conditions. **1.** The experiment consists of counting the number of times x an event occurs over a specified interval of time, area, or volume. **2.** The probability of the event occurring is the same for each interval. **3.** The number of occurrences in one interval is independent of the number of occurrences in other intervals. The parameter of a Poisson distribution is μ.	x = the number of occurrences in the given interval μ = the mean number of occurrences in a given time or space unit The probability of exactly x occurrences in an interval is $$P(x) = \frac{\mu^x e^{-\mu}}{x!}.$$

4.3 EXERCISES

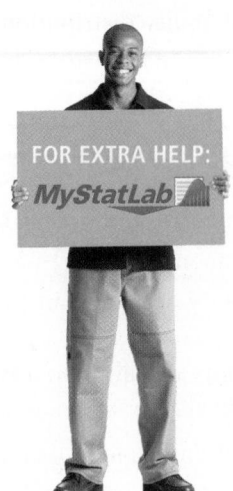

FOR EXTRA HELP:
MyStatLab

1. 0.080

2. 0.45

3. 0.062

4. 0.028

5. 0.175

6. 0.089

7. 0.251

8. 0.042

9. In a binomial distribution, the value of *x* represents the number of successes in *n* trials, and in a geometric distribution the value of *x* represents the first trial that results in a success.

10. In a binomial distribution, the value of *x* represents the number of successes in *n* trials, and in a Poisson distribution the value of *x* represents the number of occurrences in an interval.

11. Geometric. You are interested in counting the number of trials until the first success.

12. Poisson. You are interested in counting the number of occurrences that take place within a given unit of time.

13. Binomial. You are interested in counting the number of successes out of *n* trials.

14. Geometric. You are interested in counting the number of trials until the first success.

15. (a) 0.082
    (b) 0.469
    (c) 0.531

■ BUILDING BASIC SKILLS AND VOCABULARY

In Exercises 1–4, assume the geometric distribution applies. Use the given probability of success p to find the indicated probability.

1. Find $P(3)$ when $p = 0.65$.

2. Find $P(1)$ when $p = 0.45$.

3. Find $P(5)$ when $p = 0.09$.

4. Find $P(8)$ when $p = 0.28$.

In Exercises 5–8, assume the Poisson distribution applies. Use the given mean μ to find the indicated probability.

5. Find $P(4)$ when $\mu = 5$.

6. Find $P(3)$ when $\mu = 6$.

7. Find $P(2)$ when $\mu = 1.5$.

8. Find $P(5)$ when $\mu = 9.8$.

9. In your own words, describe the difference between the value of *x* in a binomial distribution and in a geometric distribution.

10. In your own words, describe the difference between the value of *x* in a binomial distribution and in a Poisson distribution.

Deciding on a Distribution *In Exercises 11–14, decide which probability distribution—binomial, geometric, or Poisson—applies to the question. You do not need to answer the question. Instead, justify your choice.*

11. Pilot's Test *Given:* The probability that a student passes the written test for a private pilot's license is 0.75. *Question:* What is the probability that a student will fail on the first attempt and pass on the second attempt?

12. Precipitation *Given:* In Tampa, Florida, the mean number of days in July with 0.01 inch or more precipitation is 16. *Question:* What is the probability that Tampa has 20 days with 0.01 inch or more precipitation next July? *(Source: National Climatic Data Center)*

13. Carry-On Luggage *Given:* Fifty-four percent of U.S. adults think Congress should place size limits on carry-on bags. In a survey of 110 randomly chosen adults, people are asked, "Do you think Congress should place size limits on carry-on bags?" *Question:* What is the probability that exactly 60 of the people answer yes? *(Source: TripAdvisor)*

14. Breaking Up *Given:* Twenty-nine percent of Americans ages 16 to 21 years old say that they would break up with their boyfriend/girlfriend for $10,000. You select at random twenty 16- to 21-year-olds. *Question:* What is the probability that the first person who says he or she would break up with their boyfriend/girlfriend for $10,000 is the fifth person selected? *(Source: Bank of America Student Banking & Seventeen)*

■ USING AND INTERPRETING CONCEPTS

Using a Distribution to Find Probabilities *In Exercises 15–22, find the indicated probabilities using the geometric distribution or the Poisson distribution. Then determine if the events are unusual. If convenient, use a Poisson probability table or technology to find the probabilities.*

15. Telephone Sales Assume the probability that you will make a sale on any given telephone call is 0.19. Find the probability that you (a) make your first sale on the fifth call, (b) make your first sale on the first, second, or third call, and (c) do not make a sale on the first three calls.

16. (a) 0.036; unusual
(b) 0.053
(c) 0.017; unusual

17. (a) 0.195
(b) 0.434 (*Tech:* 0.433)
(c) 0.566 (*Tech:* 0.567)

18. (a) 0.228
(b) 0.876
(c) 0.124

19. (a) 0.329
(b) 0.878
(c) 0.122

20. (a) 0.002; unusual
(b) 0.006; unusual
(c) 0.980

21. (a) 0.105
(b) 0.578
(c) 0.316

22. (a) 0.124
(b) 0.645
(c) 0.355

23. (a) 0.140
(b) 0.042; This event is unusual because its probability is less than 0.05.
(c) 0.064

24. (a) 0.043; This event is unusual because its probability is less than 0.05.
(b) 0.003; This event is unusual because its probability is less than 0.05.
(c) 0.256

16. Bankruptcies The mean number of bankruptcies filed per minute in the United States in a recent year was about two. Find the probability that (a) exactly five businesses will file bankruptcy in any given minute, (b) at least five businesses will file bankruptcy in any given minute, and (c) more than five businesses will file bankruptcy in any given minute. (*Source: Administrative Office of the U.S. Courts*)

17. Typographical Errors A newspaper finds that the mean number of typographical errors per page is four. Find the probability that (a) exactly three typographical errors are found on a page, (b) at most three typographical errors are found on a page, and (c) more than three typographical errors are found on a page.

18. Pass Completions Football player Peyton Manning completes a pass 64.8% of the time. Find the probability that (a) the first pass Peyton completes is the second pass, (b) the first pass Peyton completes is the first or second pass, and (c) Peyton does not complete his first two passes. (*Source: National Football League*)

19. Major Hurricanes A major hurricane is a hurricane with wind speeds of 111 miles per hour or greater. During the 20th century, the mean number of major hurricanes to strike the U.S. mainland per year was about 0.6. Find the probability that in a given year (a) exactly one major hurricane strikes the U.S. mainland, (b) at most one major hurricane strikes the U.S. mainland, and (c) more than one major hurricane strikes the U.S. mainland. (*Source: National Hurricane Center*)

20. Glass Manufacturer A glass manufacturer finds that 1 in every 500 glass items produced is warped. Find the probability that (a) the first warped glass item is the tenth item produced, (b) the first warped glass item is the first, second, or third item produced, and (c) none of the first 10 glass items produced are defective.

21. Winning a Prize A cereal maker places a game piece in each of its cereal boxes. The probability of winning a prize in the game is 1 in 4. Find the probability that you (a) win your first prize with your fourth purchase, (b) win your first prize with your first, second, or third purchase, and (c) do not win a prize with your first four purchases.

22. Precipitation The mean number of days with 0.01 inch or more precipitation per month in Baltimore, Maryland, is about 9.5. Find the probability that in a given month, (a) there are exactly 10 days with 0.01 inch or more precipitation, (b) there are at most 10 days with 0.01 inch or more precipitation, and (c) there are more than 10 days with 0.01 inch or more precipitation. (*Source: National Climatic Data Center*)

SC *In Exercises 23 and 24, use the StatCrunch Poisson calculator to find the indicated probabilities. Then determine if the events are unusual. Explain your reasoning.*

23. Oil Tankers The mean number of oil tankers at a port city is 8 per day. The port has facilities to handle up to 12 oil tankers in a day. Find the probability that on a given day, (a) eight oil tankers will arrive, (b) at most three oil tankers will arrive, and (c) too many oil tankers will arrive.

24. Kidney Transplants The mean number of kidney transplants performed per day in the United States in a recent year was about 45. Find the probability that on a given day, (a) exactly 50 kidney transplants will be performed, (b) at least 65 kidney transplants will be performed, and (c) no more than 40 kidney transplants will be performed. (*Source: U.S. Department of Health and Human Services*)

25. (a) 0.1254235482
    (b) 0.1254084986; The results are approximately the same.

26. (a) 0.629
    (b) 0.343
    (c) 0.029

27. (a) 1000, 999,000, 999.5
    On average you would have to play 1000 times in order to win the lottery. The standard deviation is 999.5 times.
    (b) 1000 times
    Lose money. On average you would win $500 once in every 1000 times you play the lottery. So, the net gain would be −$500.

28. (a) 200, 39,800, 199.5
    On average, 200 records will be examined before finding one that has been miscalculated. The standard deviation is 199.5 records.
    (b) 200

29. (a) 3.9, 2.0; The standard deviation is 2.0 strokes, so most of Phil's scores per hole differ from the mean by no more than 2.0 strokes.
    (b) 0.385

30. (a) 29.9, 5.5; The standard deviation is 5.5 inches, so most of the January snowfalls in Mount Shasta differ from the mean by no more than 5.5 inches.
    (b) 0.116

■ EXTENDING CONCEPTS

25. **Comparing Binomial and Poisson Distributions** An automobile manufacturer finds that 1 in every 2500 automobiles produced has a particular manufacturing defect. (a) Use a binomial distribution to find the probability of finding 4 cars with the defect in a random sample of 6000 cars. (b) The Poisson distribution can be used to approximate the binomial distribution for large values of n and small values of p. Repeat (a) using a Poisson distribution and compare the results.

26. **Hypergeometric Distribution** Binomial experiments require that any sampling be done with replacement because each trial must be independent of the others. The **hypergeometric distribution** also has two outcomes: success and failure. However, the sampling is done without replacement. Given a population of N items having k successes and $N − k$ failures, the probability of selecting a sample of size n that has x successes and $n − x$ failures is given by

$$P(x) = \frac{(_kC_x)(_{N-k}C_{n-x})}{_NC_n}.$$

In a shipment of 15 microchips, 2 are defective and 13 are not defective. A sample of three microchips is chosen at random. Find the probability that (a) all three microchips are not defective, (b) one microchip is defective and two are not defective, and (c) two microchips are defective and one is not defective.

Geometric Distribution: Mean and Variance *In Exercises 27 and 28, use the fact that the mean of a geometric distribution is $\mu = 1/p$ and the variance is $\sigma^2 = q/p^2$.*

27. **Daily Lottery** A daily number lottery chooses three balls numbered 0 to 9. The probability of winning the lottery is 1/1000. Let x be the number of times you play the lottery before winning the first time. (a) Find the mean, variance, and standard deviation. Interpret the results. (b) How many times would you expect to have to play the lottery before winning? Assume that it costs $1 to play and winners are paid $500. Would you expect to make or lose money playing this lottery? Explain.

28. **Paycheck Errors** A company assumes that 0.5% of the paychecks for a year were calculated incorrectly. The company has 200 employees and examines the payroll records from one month. (a) Find the mean, variance, and standard deviation. Interpret the results. (b) How many employee payroll records would you expect to examine before finding one with an error?

Poisson Distribution: Variance *In Exercises 29 and 30, use the fact that the variance of a Poisson distribution is $\sigma^2 = \mu$.*

29. **Golf** In a recent year, the mean number of strokes per hole for golfer Phil Mickelson was about 3.9. (a) Find the variance and standard deviation. Interpret the results. (b) How likely is Phil to play an 18-hole round and have more than 72 strokes? *(Source: PGATour.com)*

30. **Snowfall** The mean snowfall in January in Mount Shasta, California is 29.9 inches. (a) Find the variance and standard deviation. Interpret the results. (b) Find the probability that the snowfall in January in Mount Shasta, California will exceed 3 feet. *(Source: National Climatic Data Center)*

USES AND ABUSES

Uses

There are countless occurrences of binomial probability distributions in business, science, engineering, and many other fields.

For instance, suppose you work for a marketing agency and are in charge of creating a television ad for Brand A toothpaste. The toothpaste manufacturer claims that 40% of toothpaste buyers prefer its brand. To check whether the manufacturer's claim is reasonable, your agency conducts a survey. Of 100 toothpaste buyers selected at random, you find that only 35 (or 35%) prefer Brand A. Could the manufacturer's claim still be true? What if your random sample of 100 found only 25 people (or 25%) who express a preference for Brand A? Would you still be justified in running the advertisement?

Knowing the characteristics of binomial probability distributions will help you answer this type of question. By the time you have completed this course, you will be able make educated decisions about the reasonableness of the manufacturer's claim.

Ethics

Suppose the toothpaste manufacturer also claims that four out of five dentists recommend Brand A toothpaste. Your agency wants to mention this fact in the television ad, but when determining how the sample of dentists was formed, you find that the dentists were paid to recommend the toothpaste. Including this statement when running the advertisement would be unethical.

Abuses

Interpreting the "Most Likely" Outcome A common misuse of binomial probability distributions is to think that the "most likely" outcome is the outcome that will occur most of the time. For instance, suppose you randomly choose a committee of four from a large population that is 50% women and 50% men. The most likely composition of the committee will be two men and two women. Although this is the most likely outcome, the probability that it will occur is only 0.375. There is a 0.5 chance that the committee will contain one man and three women or three men and one woman. So, if either of these outcomes occurs, you should not assume that the selection was unusual or biased.

■ EXERCISES

In Exercises 1–4, suppose that the manufacturer's claim is true—40% of toothpaste buyers prefer Brand A toothpaste. Use the graph and technology to answer the questions. Explain your reasoning.

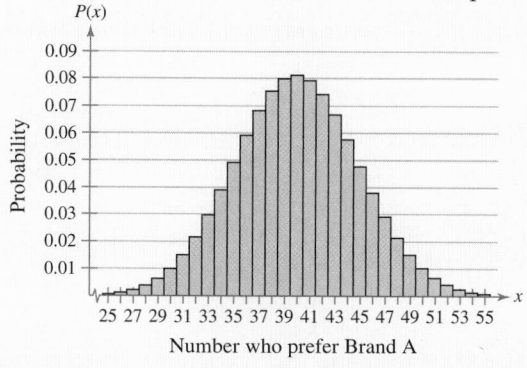

Number who prefer Brand A

1. ***Interpreting the "Most Likely" Outcome*** In a random sample of 100, what is the most likely outcome? How likely is it?

2. ***Interpreting the "Most Likely" Outcome*** In a random sample of 100, what is the probability that between 35 and 45 people, inclusive, prefer Brand A?

3. Suppose in a random sample of 100, you found 36 who prefer Brand A. Would the manufacturer's claim be believable?

4. Suppose in a random sample of 100, you found 25 who prefer Brand A. Would the manufacturer's claim be believable?

4 CHAPTER SUMMARY

What did you **learn?**	EXAMPLE(S)	REVIEW EXERCISES

Section 4.1

- How to distinguish between discrete random variables and continuous random variables — *1* — *1–6*

- How to determine if a distribution is a probability distribution — *3–4* — *7–10*

- How to construct a discrete probability distribution and its graph and find the mean, variance, and standard deviation of a discrete probability distribution — *2, 5, 6* — *11–14*

  $\mu = \Sigma x P(x)$ Mean of a discrete random variable

  $\sigma^2 = \Sigma (x - \mu)^2 P(x)$ Variance of a discrete random variable

  $\sigma = \sqrt{\sigma^2} = \sqrt{\Sigma (x - \mu)^2 P(x)}$ Standard deviation of a discrete random variable

- How to find the expected value of a discrete probability distribution — *7* — *15–16*

Section 4.2

- How to determine if a probability experiment is a binomial experiment — *1* — *17–20*

- How to find binomial probabilities using the binomial probability formula, a binomial probability table, and technology — *2, 4–6* — *21–24*

  $P(x) = {}_nC_x p^x q^{n-x} = \dfrac{n!}{(n-x)!x!} p^x q^{n-x}$ Binomial probability formula

- How to construct a binomial distribution and its graph and find the mean, variance, and standard deviation of a binomial probability distribution — *3, 7, 8* — *25–28*

  $\mu = np$ Mean of a binomial distribution

  $\sigma^2 = npq$ Variance of a binomial distribution

  $\sigma = \sqrt{npq}$ Standard deviation of a binomial distribution

Section 4.3

- How to find probabilities using the geometric distribution — *1* — *29, 30*

  $P(x) = pq^{x-1}$ Probability that the first success will occur on trial number x

- How to find probabilities using the Poisson distribution — *2, 3* — *31–33*

  $P(x) = \dfrac{\mu^x e^{-\mu}}{x!}$ Probability of exactly x occurrences in an interval

4 REVIEW EXERCISES

1. Continuous; The length of time spent sleeping is a random variable that cannot be counted.

2. Discrete; The number of fish caught is a random variable that is countable.

3. Discrete

4. Continuous

5. Continuous

6. Discrete

7. No, $\sum P(x) \neq 1$.

8. Yes

9. Yes

10. No, $P(5) > 1$ and $\sum P(x) \neq 1$.

■ SECTION 4.1

In Exercises 1 and 2, decide whether the graph represents a discrete random variable or a continuous random variable. Explain your reasoning.

1. The number of hours spent sleeping each day

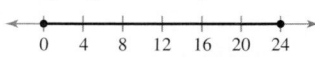

2. The number of fish caught during a fishing tournament

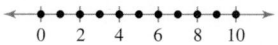

In Exercises 3–6, decide whether the random variable x is discrete or continuous.

3. Let *x* represent the number of pumps in use at a gas station.

4. Let *x* represent the weight of a truck at a weigh station.

5. Let *x* represent the amount of carbon dioxide emitted from a car's tailpipe each day.

6. Let *x* represent the number of people that activate a metal detector at an airport each hour.

In Exercises 7–10, decide whether the distribution is a probability distribution. If it is not, identify the property that is not satisfied.

7. The daily limit for catching bass at a lake is four. The random variable *x* represents the number of fish caught in a day.

x	0	1	2	3	4
P(x)	0.36	0.23	0.08	0.14	0.29

8. The random variable *x* represents the number of tickets a police officer writes out each shift.

x	0	1	2	3	4	5
P(x)	0.09	0.23	0.29	0.16	0.21	0.02

9. A greeting card shop keeps records of customers' buying habits. The random variable *x* represents the number of cards sold to an individual customer in a shopping visit.

x	1	2	3	4	5	6	7
P(x)	0.68	0.14	0.08	0.05	0.02	0.02	0.01

10. The random variable *x* represents the number of classes in which a student is enrolled in a given semester at a university.

x	1	2	3	4	5	6	7	8
P(x)	$\frac{1}{80}$	$\frac{2}{75}$	$\frac{1}{10}$	$\frac{12}{25}$	$\frac{27}{20}$	$\frac{1}{5}$	$\frac{2}{25}$	$\frac{1}{120}$

11. (a) and (b) See Odd Answers, page A70.

   (c) 6.4, 2.9, 1.7

   (d) The mean is 6.4, so the average number of pages per section is about 6 pages. The standard deviation is 1.7, so most of the sections differ from the mean by no more than about 2 pages.

12. (a) and (b) See Selected Answers, page A118.

   (c) 1.3, 1.0, 1.0

   (d) The mean is 1.3, so the average number of hits per game is about 1 hit. The standard deviation is 1.0, so most of the games differ from the mean by no more than 1 hit.

13. (a) and (b) See Odd Answers, page A70.

   (c) 2.8, 1.7, 1.3

   (d) The mean is 2.8, so the average number of cellular phones per household is about 3. The standard deviation is 1.3, so most of the households differ from the mean by no more than about 1 cell phone.

14. (a) and (b) See Selected Answers, page A118.

   (c) 31.7, 262.8, 16.2

   (d) The mean is 31.7, so the average length of an advertising block is 31.7 seconds. The standard deviation is 16.2, so most of the commercials differ from the mean by no more than 16.2 seconds.

15. 3.4

16. 2.5

In Exercises 11–14,

(a) use the frequency distribution table to construct a probability distribution,

(b) graph the probability distribution using a histogram and describe its shape,

(c) find the mean, variance, and standard deviation of the probability distribution, and

(d) interpret the results in the context of the real-life situation.

11. The number of pages in a section from 10 statistics texts

Pages	Sections
2	3
3	12
4	72
5	115
6	169
7	120
8	83
9	48
10	22
11	6

12. The number of hits per game played by a baseball player during a recent season

Hits	Games
0	29
1	62
2	33
3	12
4	3
5	1

13. The distribution of the number of cellular phones per household in a small town is given.

Cellphones	Families
0	5
1	35
2	68
3	73
4	42
5	19
6	8

14. A television station sells advertising in 15-, 30-, 60-, 90-, and 120-second blocks. The distribution of sales for one 24-hour day is given.

Length (in seconds)	Number
15	76
30	445
60	30
90	3
120	12

In Exercises 15 and 16, find the expected value of the random variable.

15. A person has shares of eight different stocks. The random variable x represents the number of stocks showing a loss on a selected day.

x	0	1	2	3	4	5	6	7	8
$P(x)$	0.02	0.11	0.18	0.32	0.15	0.09	0.05	0.05	0.03

16. A local pub has a chicken wing special on Tuesdays. The pub owners purchase wings in cases of 300. The random variable x represents the number of cases used during the special.

x	1	2	3	4
$P(x)$	$\frac{1}{9}$	$\frac{1}{3}$	$\frac{1}{2}$	$\frac{1}{18}$

Answers (left column)

17. No; In a binomial experiment, there are only two possible outcomes: success or failure.

18. Yes; The experiment has 20 independent trials, there are two possible outcomes (C and not C), the probability of success is the same for each trial, and the random variable x represents the number of outcomes of C. $p = 0.20$

19. Yes; $n = 12$, $p = 0.24$, $q = 0.76$, $x = 0, 1, \ldots, 12$

20. No; The experiment is not repeated for a fixed number of trials.

21. (a) 0.208
  (b) 0.322 (*Tech:* 0.321)
  (c) 0.114

22. (a) 0.232
  (b) 0.842
  (c) 0.609

23. (a) 0.196
  (b) 0.332
  (c) 0.137

24. (a) 0.316
  (b) 0.492
  (c) 0.177

25. (a)

x	$P(x)$
0	0.125
1	0.323
2	0.332
3	0.171
4	0.044
5	0.005

  (b) See Odd Answers, page A70.

  (c) 1.7, 1.1, 1.1; The mean is 1.7, so an average of 1.7 out of 5 women have spouses who never help with household chores. The standard deviation is 1.1, so most samples of 5 women differ from the mean by no more than 1.1 women.

  (d) The values $x = 4$ and $x = 5$ are unusual because their probabilities are less than 0.05.

■ SECTION 4.2

In Exercises 17 and 18, use the following information.

A probability experiment has n independent trials. Each trial has three possible outcomes: A, B, and C. For each trial, P(A) = 0.30, P(B) = 0.50, and P(C) = 0.20. There are 20 trials.

17. Can a binomial experiment be used to find the probability of 6 outcomes of A, 10 outcomes of B, and 4 outcomes of C? Explain your reasoning.

18. Can a binomial experiment be used to find the probability of 4 outcomes of C and 16 outcomes that are *not* C? Explain your reasoning. What is the probability of success for each trial?

In Exercises 19 and 20, decide whether the experiment is a binomial experiment. If it is not, identify the property that is not satisfied. If it is, list the values of n, p, and q and the values that x can assume.

19. Bags of plain M&M's contain 24% blue candies. One candy is selected from each of 12 bags. The random variable represents the number of blue candies selected. (*Source: Mars, Incorporated*)

20. A fair coin is tossed repeatedly until 15 heads are obtained. The random variable x counts the number of tosses.

In Exercises 21–24, find the indicated probabilities.

21. One in four adults is currently on a diet. You randomly select eight adults and ask them if they are currently on a diet. Find the probability that the number who say they are currently on a diet is (a) exactly three, (b) at least three, and (c) more than three. (*Source: Wirthlin Worldwide*)

22. One in four people in the United States owns individual stocks. You randomly select 12 people and ask them if they own individual stocks. Find the probability that the number who say they own individual stocks is (a) exactly two, (b) at least two, and (c) more than two. (*Source: Pew Research Center*)

23. Forty-three percent of businesses in the United States require a doctor's note when an employee takes sick time. You randomly select nine businesses and ask each if it requires a doctor's note when an employee takes sick time. Find the probability that the number who say they require a doctor's note is (a) exactly five, (b) at least five, and (c) more than five. (*Source: Harvard School of Public Health*)

24. In a typical day, 31% of people in the United States with Internet access go online to get news. You randomly select five people in the United States with Internet access and ask them if they go online to get news. Find the probability that the number who say they go online to get news is (a) exactly two, (b) at least two, and (c) more than two. (*Source: Pew Research Center*)

In Exercises 25–28,

(a) construct a binomial distribution,

(b) graph the binomial distribution using a histogram and describe its shape,

(c) find the mean, variance, and standard deviation of the binomial distribution and interpret the results in the context of the real-life situation, and

(d) determine the values of the random variable x that you would consider unusual.

25. Thirty-four percent of women in the United States say their spouses never help with household chores. You randomly select five U.S. women and ask if their spouses help with household chores. (*Source: Boston Consulting Group*)

26. (a) and (b) See Selected Answers, page A118.

   (c) 4.1, 1.3, 1.1; The mean is 4.1, so an average of 4.1 out of 6 families say that their children have an influence on their vacation destinations. The standard deviation is 1.1, so most samples of 6 families differ from the mean by no more than 1.1 families.

   (d) The values $x = 0$ and $x = 1$ are unusual because their probabilities are less than 0.05.

27. (a) and (b) See Odd Answers, page A70.

   (c) 1.6, 1.0, 1.0; The mean is 1.6, so an average of 1.6 out of 4 trucks have diesel engines. The standard deviation is 1.0, so most samples of 4 trucks differ from the mean by no more than 1 truck.

   (d) The value $x = 4$ is unusual because its probability is less than 0.05.

28. (a) and (b) See Selected Answers, page A119.

   (c) 3.2, 1.2, 1.1; The mean is 3.2, so an average of 3.2 out of 5 mothers choose fast food one to three times a week. The standard deviation is 1.1, so most samples of 5 mothers differ from the mean by no more than 1.1 mothers.

   (d) The value $x = 0$ is unusual because its probability is less than 0.05.

29. (a) 0.134

   (b) 0.186

   (c) 0.176

30. (a) 0.429

   (b) 0.245

   (c) 0.673

   (d) 0.813

31. (a) 0.765

   (b) 0.205

   (c) 0.997

   (d) 0.030; unusual

32. (a) 0.997

   (b) 0.875

   (c) 1 (*Tech:* 0.9999607)

33. The probability increases as the rate increases, and decreases as the rate decreases.

---

26. Sixty-eight percent of families say that their children have an influence on their vacation destinations. You randomly select six families and ask if their children have an influence on their vacation destinations. *(Source: YPB&R)*

27. In a recent year, forty percent of trucks sold by a company had diesel engines. You randomly select four trucks sold by the company and check if they have diesel engines.

28. Sixty-three percent of U.S. mothers with school-age children choose fast food as a dining option for their families one to three times a week. You randomly select five U.S. mothers with school-age children and ask if they choose fast food as a dining option for their families one to three times a week. *(Adapted from Market Day)*

■ SECTION 4.3

In Exercises 29–32, find the indicated probabilities using the geometric distribution or the Poisson distribution. Then determine if the events are unusual. If convenient, use a Poisson probability table or technology to find the probabilities.

29. Twenty-two percent of former smokers say they tried to quit four or more times before they were habit-free. You randomly select 10 former smokers. Find the probability that the first person who tried to quit four or more times is (a) the third person selected, (b) the fourth or fifth person selected, and (c) not one of the first seven people selected. *(Source: Porter Novelli Health Styles)*

30. In a recent season, hockey player Sidney Crosby scored 33 goals in 77 games he played. Assume that his goal production stayed at that level the following season. What is the probability that he would get his first goal

   (a) in the first game of the season?

   (b) in the second game of the season?

   (c) in the first or second game of the season?

   (d) within the first three games of the season? *(Source: ESPN)*

31. During a 69-year period, tornadoes killed 6755 people in the United States. Assume this rate holds true today and is constant throughout the year. Find the probability that tomorrow

   (a) no one in the U. S. is killed by a tornado,

   (b) one person in the U.S. is killed by a tornado,

   (c) at most two people in the U.S. are killed by a tornado, and

   (d) more than one person in the U.S. is killed by a tornado. *(Source: National Weather Service)*

32. It is estimated that sharks kill 10 people each year worldwide. Find the probability that at least 3 people are killed by sharks this year

   (a) assuming that this rate is true,

   (b) if the rate is actually 5 people a year, and

   (c) if the rate is actually 15 people a year. *(Source: International Shark Attack File)*

33. In Exercise 32, describe what happens to the probability of at least three people being killed by sharks this year as the rate increases and decreases.

4 CHAPTER QUIZ

1. (a) Discrete; The number of lightning strikes that occur in Wyoming during the month of June is a random variable that is countable.

   (b) Continuous; The fuel (in gallons) used by the Space Shuttle during takeoff is a random variable that has an infinite number of possible outcomes and cannot be counted.

Intensity	Number of hurricanes
1	114
2	74
3	76
4	18
5	3

TABLE FOR EXERCISE 2

2. See Odd Answers, page A71.

3. (a) and (b) See Odd Answers, page A71.

   (c) 5.1, 0.8, 0.9; The average number of successful surgeries is 5.1 out of 6. The standard deviation is 0.9, so most samples of 6 surgeries differ from the mean by no more than 0.9 surgery.

   (d) 0.041; Yes, this event is unusual because 0.041 < 0.05.

   (e) 0.047; Yes, this event is unusual because 0.041 < 0.05.

4. (a) 0.175

   (b) 0.440

   (c) 0.007

5. 0.038; Yes, this event is unusual because 0.038 < 0.05.

6. 0.335; No, this event is not unusual because 0.335 > 0.05.

Take this quiz as you would take a quiz in class. After you are done, check your work against the answers given in the back of the book.

1. Decide if the random variable x is discrete or continuous. Explain your reasoning.

   (a) Let x represent the number of lightning strikes that occur in Wyoming during the month of June.

   (b) Let x represent the amount of fuel (in gallons) used by the Space Shuttle during takeoff.

2. The table lists the number of U.S. mainland hurricane strikes (from 1851 to 2008) for various intensities according to the Saffir-Simpson Hurricane Scale. *(Source: National Oceanic and Atmospheric Administration)*

   (a) Construct a probability distribution of the data.

   (b) Graph the discrete probability distribution using a probability histogram. Then describe its shape.

   (c) Find the mean, variance, and standard deviation of the probability distribution and interpret the results.

   (d) Find the probability that a hurricane selected at random for further study has an intensity of at least four.

3. The success rate of corneal transplant surgery is 85%. The surgery is performed on six patients. *(Adapted from St. Luke's Cataract & Laser Institute)*

   (a) Construct a binomial distribution.

   (b) Graph the binomial distribution using a probability histogram. Then describe its shape.

   (c) Find the mean, variance, and standard deviation of the probability distribution and interpret the results.

   (d) Find the probability that the surgery is successful for exactly three patients. Is this an unusual event? Explain.

   (e) Find the probability that the surgery is successful for fewer than four patients. Is this an unusual event? Explain.

4. A newspaper finds that the mean number of typographical errors per page is five. Find the probability that

   (a) exactly five typographical errors will be found on a page,

   (b) fewer than five typographical errors will be found on a page, and

   (c) no typographical errors will be found on a page.

In Exercises 5 and 6, use the following information. Basketball player Dwight Howard makes a free throw shot about 60.2% of the time. (Source: ESPN)

5. Find the probability that the first free throw shot Dwight makes is the fourth shot. Is this an unusual event? Explain.

6. Find the probability that the first free throw shot Dwight makes is the second or third shot. Is this an unusual event? Explain.

PUTTING IT ALL TOGETHER

Real Statistics — Real Decisions

The Centers for Disease Control and Prevention (CDC) is required by law to publish a report on assisted reproductive technologies (ART). ART includes all fertility treatments in which both the egg and the sperm are used. These procedures generally involve removing eggs from a woman's ovaries, combining them with sperm in the laboratory, and returning them to the woman's body or giving them to another woman.

   You are helping to prepare the CDC report and select at random 10 ART cycles for a special review. None of the cycles resulted in a clinical pregnancy. Your manager feels it is impossible to select at random 10 ART cycles that did not result in a clinical pregnancy. Use the information provided at the right and your knowledge of statistics to determine if your manager is correct.

**Results of ART Cycles
Using Fresh Nondonor
Eggs or Embryos**

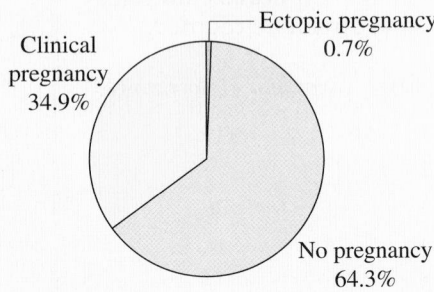

(Source: Centers for Disease Control and Prevention)

■ EXERCISES

1. *How Would You Do It?*

(a) How would you determine if your manager's view is correct, that it is impossible to select at random 10 ART cycles that did not result in a clinical pregnancy?

(b) What probability distribution do you think best describes the situation? Do you think the distribution of the number of clinical pregnancies is discrete or continuous? Why?

2. *Answering the Question*

Write an explanation that answers the question, "Is it possible to select at random 10 ART cycles that did not result in a clinical pregnancy?" Include in your explanation the appropriate probability distribution and your calculation of the probability of no clinical pregnancies in 10 ART cycles.

3. *Suspicious Samples?*

Which of the following samples would you consider suspicious if someone told you that the sample was selected at random? Would you believe that the samples were selected at random? Why or why not?

(a) Selecting at random 10 ART cycles among women of age 40, eight of which resulted in clinical pregnancies.

(b) Selecting at random 10 ART cycles among women of age 41, none of which resulted in clinical pregnancies.

**Pregnancy and Live Birth Rates for
ART Cycles Among Women
of Age 40 and Older**

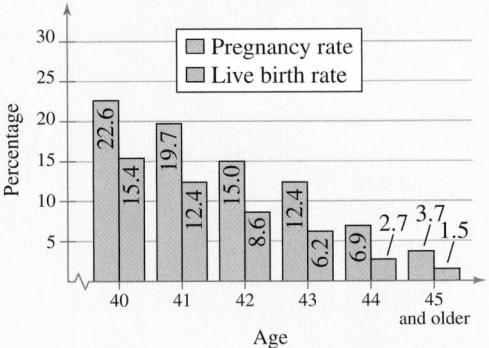

(Source: Centers for Disease Control and Prevention)

TECHNOLOGY

MINITAB   EXCEL   TI-83/84 PLUS

USING POISSON DISTRIBUTIONS AS QUEUING MODELS

Queuing means waiting in line to be served. There are many examples of queuing in everyday life: waiting at a traffic light, waiting in line at a grocery checkout counter, waiting for an elevator, holding for a telephone call, and so on.

Poisson distributions are used to model and predict the number of people (calls, computer programs, vehicles) arriving at the line. In the following exercises, you are asked to use Poisson distributions to analyze the queues at a grocery store checkout counter.

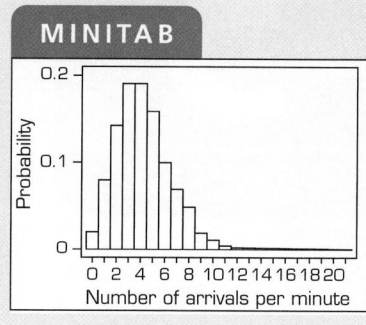

MINITAB

■ EXERCISES

In Exercises 1–7, consider a grocery store that can process a total of four customers at its checkout counters each minute.

1. Suppose that the mean number of customers who arrive at the checkout counters each minute is 4. Create a Poisson distribution with $\mu = 4$ for $x = 0$ to 20. Compare your results with the histogram shown at the upper right.

2. MINITAB was used to generate 20 random numbers with a Poisson distribution for $\mu = 4$. Let the random number represent the number of arrivals at the checkout counter each minute for 20 minutes.

   3 3 3 3 5 5 6 7 3 6
   3 5 6 3 4 6 2 2 4 1

   During each of the first four minutes, only three customers arrived. These customers could all be processed, so there were no customers waiting after four minutes.

   (a) How many customers were waiting after 5 minutes? 6 minutes? 7 minutes? 8 minutes?

   (b) Create a table that shows the number of customers waiting at the end of 1 through 20 minutes.

3. Generate a list of 20 random numbers with a Poisson distribution for $\mu = 4$. Create a table that shows the number of customers waiting at the end of 1 through 20 minutes.

4. Suppose that the mean increases to 5 arrivals per minute. You can still process only four per minute. How many would you expect to be waiting in line after 20 minutes?

5. Simulate the setting in Exercise 4. Do this by generating a list of 20 random numbers with a Poisson distribution for $\mu = 5$. Then create a table that shows the number of customers waiting at the end of 20 minutes.

6. Suppose that the mean number of arrivals per minute is 5. What is the probability that 10 customers will arrive during the first minute?

7. Suppose that the mean number of arrivals per minute is 4.

   (a) What is the probability that three, four, or five customers will arrive during the third minute?

   (b) What is the probability that more than four customers will arrive during the first minute?

   (c) What is the probability that more than four customers will arrive during each of the first four minutes?

Extended solutions are given in the *Technology Supplement.*
Technical instruction is provided for MINITAB, Excel, and the TI-83/84 Plus.

5 NORMAL PROBABILITY DISTRIBUTIONS

The bottom shell of an Eastern Box Turtle has hinges so the turtle can retract its head, tail, and legs into the shell. The shell can also regenerate if it has been damaged.

In Chapters 1 through 4, you learned how to collect and describe data, find the probability of an event, and analyze discrete probability distributions. You also learned that if a sample is used to make inferences about a population, then it is critical that the sample not be biased. Suppose, for instance, that you wanted to determine the rate of clinical mastitis (infections caused by bacteria that can alter milk production) in dairy herds. How would you organize the study? When the Animal Health Service performed this study, it used random sampling and then classified the results according to breed, housing, hygiene, health, milking management, and milking machine. One conclusion from the study was that herds with Red and White cows as the predominant breed had a higher rate of clinical mastitis than herds with Holstein-Friesian cows as the main breed.

In Chapter 5, you will learn how to recognize normal (bell-shaped) distributions and how to use their properties in real-life applications. Suppose that you worked for the North Carolina Zoo and were collecting data about various physical traits of Eastern Box Turtles at the zoo. Which of the following would you expect to have bell-shaped, symmetric distributions: carapace (top shell) length, plastral (bottom shell) length, carapace width, plastral width, weight, total length? For instance, the four graphs below show the carapace length and plastral length of male and female Eastern Box Turtles in the North Carolina Zoo. Notice that the male Eastern Box Turtle carapace length distribution is bell-shaped, but the other three distributions are skewed left.

Female Eastern Box Turtle Carapace Length

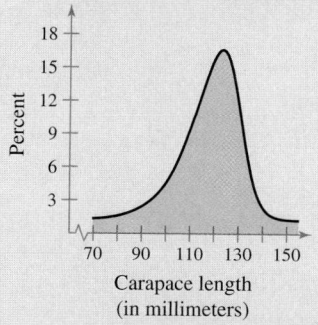

Male Eastern Box Turtle Carapace Length

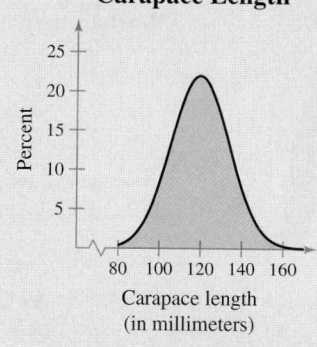

Female Eastern Box Turtle Plastral Length

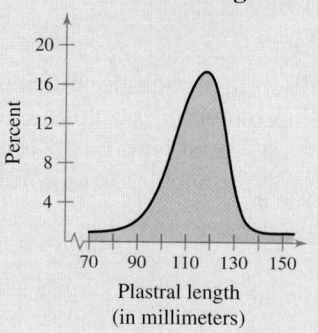

Male Eastern Box Turtle Plastral Length

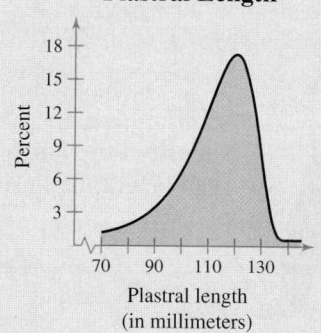

5.1 Introduction to Normal Distributions and the Standard Normal Distribution

▸ How to interpret graphs of normal probability distributions

▸ How to find areas under the standard normal curve

Note to Instructor

Draw several different continuous probability curves. Then point out that the normal (or Gaussian) curve is graphed using the formula shown at the bottom of the page. Have students discuss measures in nature that are normally distributed. Mention that often grades in a statistics class are not normally distributed.

INSIGHT

To learn how to determine if a random sample is taken from a normal distribution, see Appendix C.

INSIGHT

A probability density function has two requirements.

1. The total area under the curve is equal to 1.

2. The function can never be negative.

Properties of a Normal Distribution ▸ The Standard Normal Distribution

▸ PROPERTIES OF A NORMAL DISTRIBUTION

In Section 4.1, you distinguished between discrete and continuous random variables, and learned that a continuous random variable has an infinite number of possible values that can be represented by an interval on the number line. Its probability distribution is called a **continuous probability distribution.** In this chapter, you will study the most important continuous probability distribution in statistics—the *normal distribution*. Normal distributions can be used to model many sets of measurements in nature, industry, and business. For instance, the systolic blood pressures of humans, the lifetimes of plasma televisions, and even housing costs are all normally distributed random variables.

DEFINITION

A **normal distribution** is a continuous probability distribution for a random variable x. The graph of a normal distribution is called the **normal curve.** A normal distribution has the following properties.

1. The mean, median, and mode are equal.
2. The normal curve is bell-shaped and is symmetric about the mean.
3. The total area under the normal curve is equal to 1.
4. The normal curve approaches, but never touches, the x-axis as it extends farther and farther away from the mean.
5. Between $\mu - \sigma$ and $\mu + \sigma$ (in the center of the curve), the graph curves downward. The graph curves upward to the left of $\mu - \sigma$ and to the right of $\mu + \sigma$. The points at which the curve changes from curving upward to curving downward are called *inflection points*.

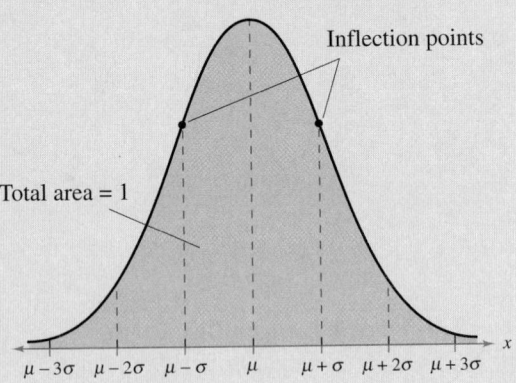

You have learned that a discrete probability distribution can be graphed with a histogram. For a continuous probability distribution, you can use a **probability density function (pdf).** A normal curve with mean μ and standard deviation σ can be graphed using the normal probability density function.

$$y = \frac{1}{\sigma\sqrt{2\pi}} e^{-(x-\mu)^2/2\sigma^2}.$$

A normal curve depends completely on the two parameters μ and σ because $e \approx 2.718$ and $\pi \approx 3.14$ are constants.

A normal distribution can have any mean and any positive standard deviation. These two parameters, μ and σ, completely determine the shape of the normal curve. The mean gives the location of the line of symmetry, and the standard deviation describes how much the data are spread out.

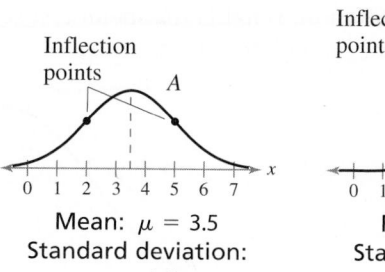

Mean: $\mu = 3.5$
Standard deviation: $\sigma = 1.5$

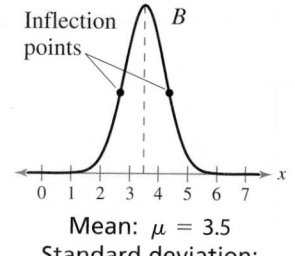

Mean: $\mu = 3.5$
Standard deviation: $\sigma = 0.7$

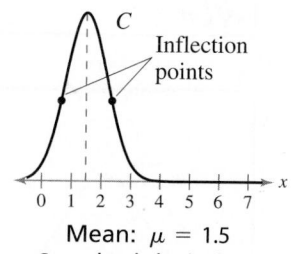

Mean: $\mu = 1.5$
Standard deviation: $\sigma = 0.7$

Notice that curve *A* and curve *B* above have the same mean, and curve *B* and curve *C* have the same standard deviation. The total area under each curve is 1.

EXAMPLE 1

▶ **Understanding Mean and Standard Deviation**

1. Which normal curve has a greater mean?

2. Which normal curve has a greater standard deviation?

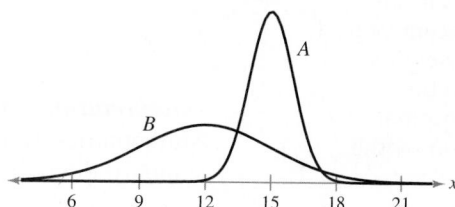

▶ **Solution**

1. The line of symmetry of curve *A* occurs at $x = 15$. The line of symmetry of curve *B* occurs at $x = 12$. So, curve *A* has a greater mean.

2. Curve *B* is more spread out than curve *A*. So, curve *B* has a greater standard deviation.

▶ **Try It Yourself 1**

Consider the normal curves shown at the right. Which normal curve has the greatest mean? Which normal curve has the greatest standard deviation?

a. Find the location of the *line of symmetry* of each curve. Make a conclusion about which mean is greatest.

b. Determine which normal curve is *more spread out*. Make a conclusion about which standard deviation is greatest.   *Answer: Page A37*

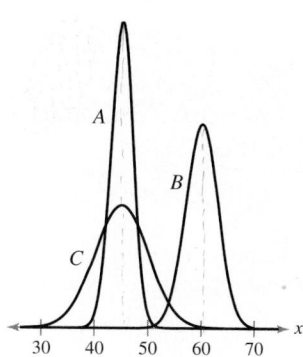

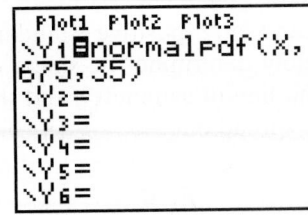

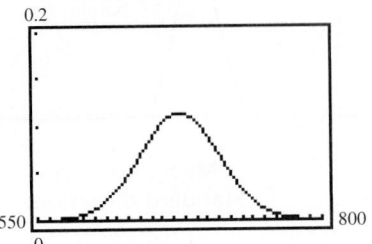

Once you determine the mean and standard deviation, you can use a TI-83/84 Plus to graph the normal curve in Example 2.

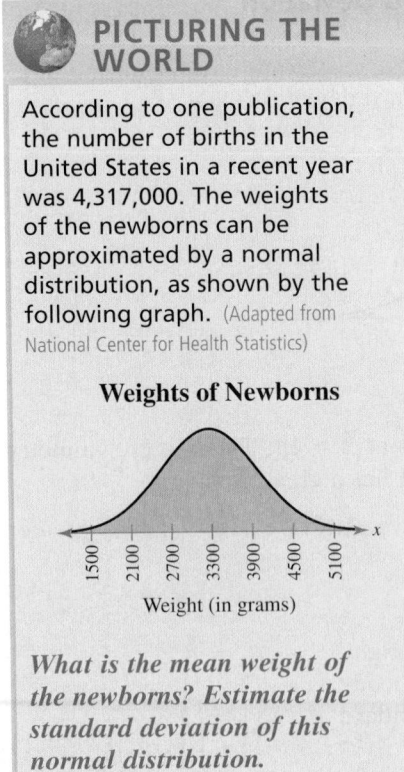

PICTURING THE WORLD

According to one publication, the number of births in the United States in a recent year was 4,317,000. The weights of the newborns can be approximated by a normal distribution, as shown by the following graph. (Adapted from National Center for Health Statistics)

Weights of Newborns

Weight (in grams)

What is the mean weight of the newborns? Estimate the standard deviation of this normal distribution.

The mean is about 3300 grams and the standard deviation is about 600 grams.

EXAMPLE 2

▶ Interpreting Graphs of Normal Distributions

The scaled test scores for the New York State Grade 8 Mathematics Test are normally distributed. The normal curve shown below represents this distribution. What is the mean test score? Estimate the standard deviation of this normal distribution. *(Adapted from New York State Education Department)*

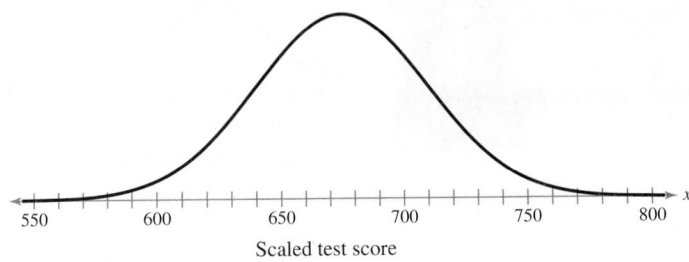

Scaled test score

▶ Solution

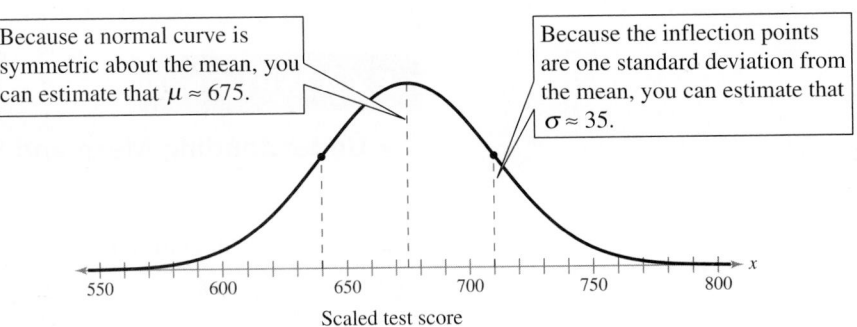

Because a normal curve is symmetric about the mean, you can estimate that $\mu \approx 675$.

Because the inflection points are one standard deviation from the mean, you can estimate that $\sigma \approx 35$.

Scaled test score

Interpretation The scaled test scores for the New York State Grade 8 Mathematics Test are normally distributed with a mean of about 675 and a standard deviation of about 35.

▶ Try It Yourself 2

The scaled test scores for the New York State Grade 8 English Language Arts Test are normally distributed. The normal curve shown below represents this distribution. What is the mean test score? Estimate the standard deviation of this normal distribution. *(Adapted from New York State Education Department)*

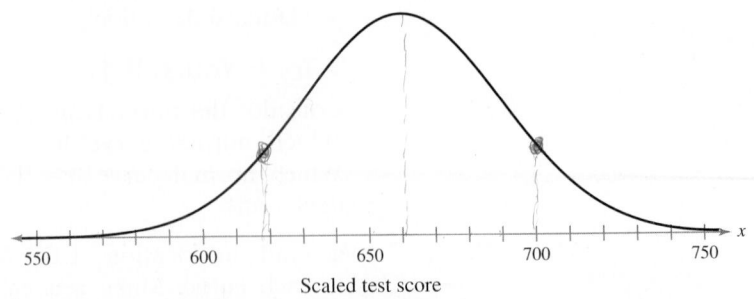

Scaled test score

a. Find the *line of symmetry* and identify the *mean*.
b. Estimate the *inflection points* and identify the *standard deviation*.

Answer: Page A37

▸ THE STANDARD NORMAL DISTRIBUTION

There are infinitely many normal distributions, each with its own mean and standard deviation. The normal distribution with a mean of 0 and a standard deviation of 1 is called the *standard normal distribution*. The horizontal scale of the graph of the standard normal distribution corresponds to z-scores. In Section 2.5, you learned that a z-score is a measure of position that indicates the number of standard deviations a value lies from the mean. Recall that you can transform an x-value to a z-score using the formula

$$z = \frac{\text{Value} - \text{Mean}}{\text{Standard deviation}}$$

$$= \frac{x - \mu}{\sigma}. \qquad \text{Round to the nearest hundredth.}$$

DEFINITION

The **standard normal distribution** is a normal distribution with a mean of 0 and a standard deviation of 1.

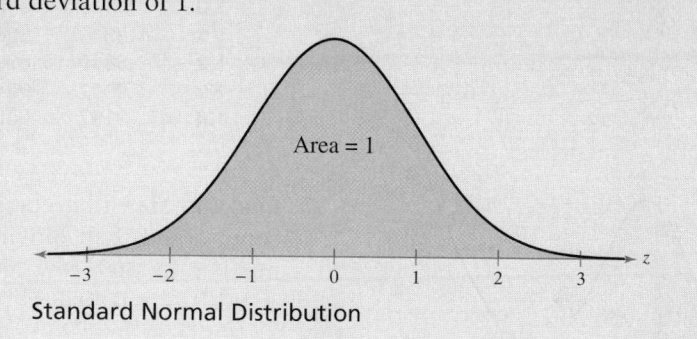

Standard Normal Distribution

If each data value of a normally distributed random variable x is transformed into a z-score, the result will be the standard normal distribution. When this transformation takes place, the area that falls in the interval under the nonstandard normal curve is the *same* as that under the standard normal curve within the corresponding z-boundaries.

In Section 2.4, you learned to use the Empirical Rule to approximate areas under a normal curve when the values of the random variable x corresponded to $-3, -2, -1, 0, 1, 2$, or 3 standard deviations from the mean. Now, you will learn to calculate areas corresponding to other x-values. After you use the formula given above to transform an x-value to a z-score, you can use the Standard Normal Table in Appendix B. The table lists the cumulative area under the standard normal curve to the left of z for z-scores from -3.49 to 3.49. As you examine the table, notice the following.

PROPERTIES OF THE STANDARD NORMAL DISTRIBUTION

1. The cumulative area is close to 0 for z-scores close to $z = -3.49$.
2. The cumulative area increases as the z-scores increase.
3. The cumulative area for $z = 0$ is 0.5000.
4. The cumulative area is close to 1 for z-scores close to $z = 3.49$.

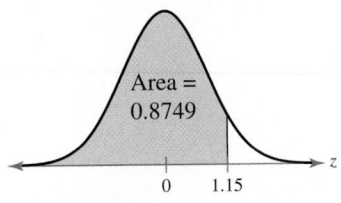

Area = 0.8749

0   1.15

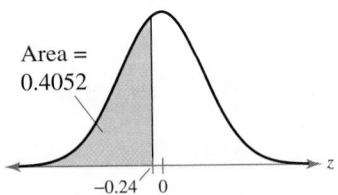

Area = 0.4052

-0.24 0

EXAMPLE 3

▸ Using the Standard Normal Table

1. Find the cumulative area that corresponds to a z-score of 1.15.

2. Find the cumulative area that corresponds to a z-score of -0.24.

▸ Solution

1. Find the area that corresponds to $z = 1.15$ by finding 1.1 in the left column and then moving across the row to the column under 0.05. The number in that row and column is 0.8749. So, the area to the left of $z = 1.15$ is 0.8749.

z	.00	.01	.02	.03	.04	.05	.06
0.0	.5000	.5040	.5080	.5120	.5160	.5199	.5239
0.1	.5398	.5438	.5478	.5517	.5557	.5596	.5636
0.2	.5793	.5832	.5871	.5910	.5948	.5987	.6026
0.9	.8159	.8186	.8212	.8238	.8264	.8289	.8315
1.0	.8413	.8438	.8461	.8485	.8508	.8531	.8554
1.1	.8643	.8665	.8686	.8708	.8729	.8749	.8770
1.2	.8849	.8869	.8888	.8907	.8925	.8944	.8962
1.3	.9032	.9049	.9066	.9082	.9099	.9115	.9131
1.4	.9192	.9207	.9222	.9236	.9251	.9265	.9279

2. Find the area that corresponds to $z = -0.24$ by finding -0.2 in the left column and then moving across the row to the column under 0.04. The number in that row and column is 0.4052. So, the area to the left of $z = -0.24$ is 0.4052.

z	.09	.08	.07	.06	.05	.04	.03
-3.4	.0002	.0003	.0003	.0003	.0003	.0003	.0003
-3.3	.0003	.0004	.0004	.0004	.0004	.0004	.0004
-3.2	.0005	.0005	.0005	.0006	.0006	.0006	.0006
-0.5	.2776	.2810	.2843	.2877	.2912	.2946	.2981
-0.4	.3121	.3156	.3192	.3228	.3264	.3300	.3336
-0.3	.3483	.3520	.3557	.3594	.3632	.3669	.3707
-0.2	.3859	.3897	.3936	.3974	.4013	.4052	.4090
-0.1	.4247	.4286	.4325	.4364	.4404	.4443	.4483
-0.0	.4641	.4681	.4721	.4761	.4801	.4840	.4880

You can also use a computer or calculator to find the cumulative area that corresponds to a z-score, as shown in the margin.

▸ Try It Yourself 3

1. Find the cumulative area that corresponds to a z-score of -2.19.

2. Find the cumulative area that corresponds to a z-score of 2.17.

Locate the given z-score and *find the area* that corresponds to it in the Standard Normal Table.   *Answer: Page A37*

When the z-score is not in the table, use the entry closest to it. If the given z-score is exactly midway between two z-scores, then use the area midway between the corresponding areas.

STUDY TIP

Here are instructions for finding the area that corresponds to $z = -0.24$ on a TI-83/84 Plus.

To specify the lower bound in this case, use $-10,000$.

> 2nd DISTR

> 2: normalcdf(

> $-10000, -.24$)

> ENTER

```
normalcdf(-10000
,-.24)
        .405165175
```

You can use the following guidelines to find various types of areas under the standard normal curve.

GUIDELINES

Finding Areas Under the Standard Normal Curve

1. Sketch the standard normal curve and shade the appropriate area under the curve.
2. Find the area by following the directions for each case shown.

    a. To find the area to the *left* of z, find the area that corresponds to z in the Standard Normal Table.

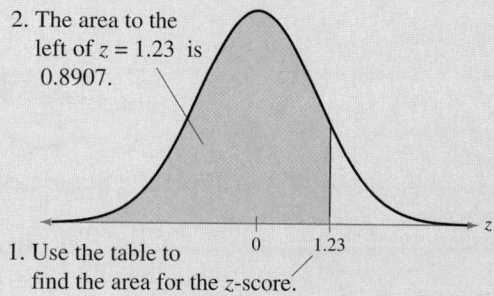

2. The area to the left of $z = 1.23$ is 0.8907.

1. Use the table to find the area for the z-score.

    b. To find the area to the *right* of z, use the Standard Normal Table to find the area that corresponds to z. Then subtract the area from 1.

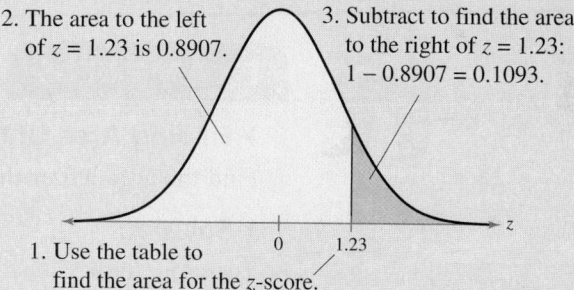

2. The area to the left of $z = 1.23$ is 0.8907.

3. Subtract to find the area to the right of $z = 1.23$:
$1 - 0.8907 = 0.1093$.

1. Use the table to find the area for the z-score.

    c. To find the area *between* two z-scores, find the area corresponding to each z-score in the Standard Normal Table. Then subtract the smaller area from the larger area.

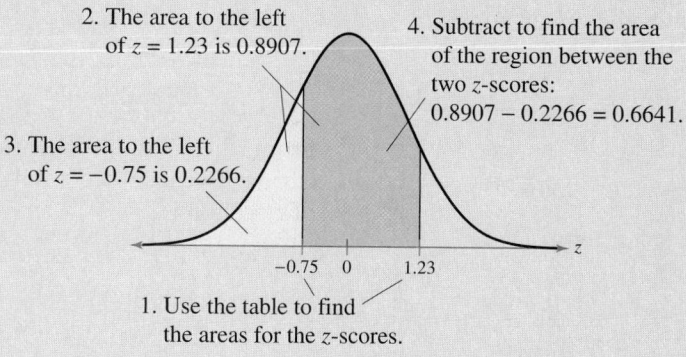

2. The area to the left of $z = 1.23$ is 0.8907.

4. Subtract to find the area of the region between the two z-scores:
$0.8907 - 0.2266 = 0.6641$.

3. The area to the left of $z = -0.75$ is 0.2266.

1. Use the table to find the areas for the z-scores.

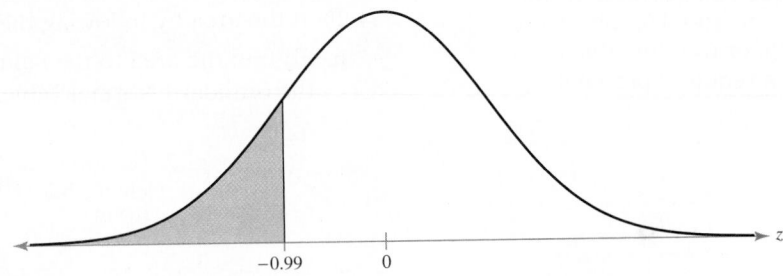

Using a TI-83/84 Plus, you can find the area automatically.

INSIGHT

Because the normal distribution is a continuous probability distribution, the area under the standard normal curve to the left of a z-score gives the probability that z is less than that z-score. For instance, in Example 4, the area to the left of $z = -0.99$ is 0.1611. So, $P(z < -0.99) = 0.1611$, which is read as "the probability that z is less than -0.99 is 0.1611."

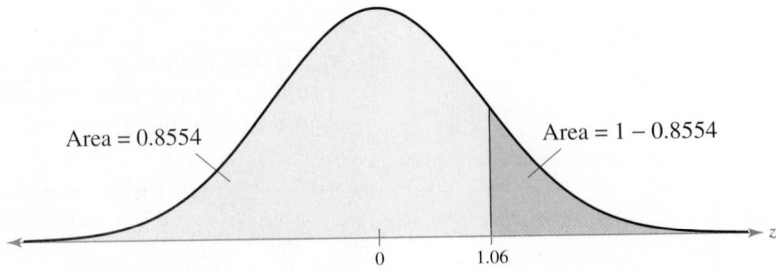

Use 10,000 for the upper bound.

EXAMPLE 4

▶ **Finding Area Under the Standard Normal Curve**

Find the area under the standard normal curve to the left of $z = -0.99$.

▶ **Solution**

The area under the standard normal curve to the left of $z = -0.99$ is shown.

From the Standard Normal Table, this area is equal to 0.1611.

▶ **Try It Yourself 4**

Find the area under the standard normal curve to the left of $z = 2.13$.

a. *Draw* the standard normal curve and shade the area under the curve and to the left of $z = 2.13$.
b. Use the Standard Normal Table to *find the area* that corresponds to $z = 2.13$.
Answer: Page A38

EXAMPLE 5

▶ **Finding Area Under the Standard Normal Curve**

Find the area under the standard normal curve to the right of $z = 1.06$.

▶ **Solution**

The area under the standard normal curve to the right of $z = 1.06$ is shown.

Area = 0.8554   Area = 1 − 0.8554

From the Standard Normal Table, the area to the left of $z = 1.06$ is 0.8554. Because the total area under the curve is 1, the area to the right of $z = 1.06$ is

$$\text{Area} = 1 - 0.8554$$
$$= 0.1446.$$

▶ **Try It Yourself 5**

Find the area under the standard normal curve to the right of $z = -2.16$.

a. *Draw* the standard normal curve and shade the area below the curve and to the right of $z = -2.16$.
b. Use the Standard Normal Table to *find the area* to the left of $z = -2.16$.
c. *Subtract* the area from 1.
Answer: Page A38

EXAMPLE 6

▶ **Finding Area Under the Standard Normal Curve**

Find the area under the standard normal curve between $z = -1.5$ and $z = 1.25$.

▶ **Solution**

The area under the standard normal curve between $z = -1.5$ and $z = 1.25$ is shown.

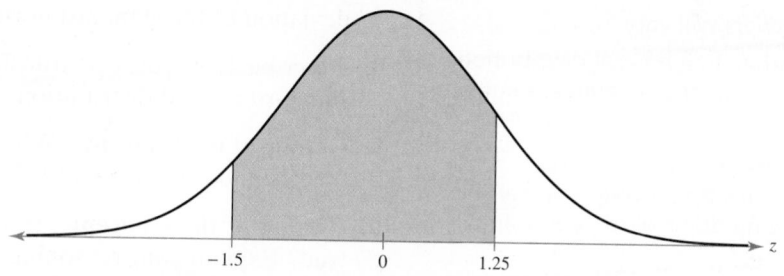

From the Standard Normal Table, the area to the left of $z = 1.25$ is 0.8944 and the area to the left of $z = -1.5$ is 0.0668. So, the area between $z = -1.5$ and $z = 1.25$ is

$$\text{Area} = 0.8944 - 0.0668$$
$$= 0.8276.$$

Interpretation   So, 82.76% of the area under the curve falls between $z = -1.5$ and $z = 1.25$.

▶ **Try It Yourself 6**

Find the area under the standard normal curve between $z = -2.165$ and $z = -1.35$.

a. Use the Standard Normal Table to *find the area* to the left of $z = -1.35$.
b. Use the Standard Normal Table to *find the area* to the left of $z = -2.165$.
c. *Subtract* the smaller area from the larger area.
d. *Interpret* the results.
Answer: Page A38

Recall that in Section 2.4 you learned, using the Empirical Rule, that values lying more than two standard deviations from the mean are considered unusual. Values lying more than three standard deviations from the mean are considered *very* unusual. So, if a z-score is greater than 2 or less than -2, it is unusual. If a z-score is greater than 3 or less than -3, it is *very* unusual.

```
normalcdf(-1.5,1
.25)
        .8275429323
```

When using technology, your answers may differ slightly from those found using the Standard Normal Table.

5.1 EXERCISES

FOR EXTRA HELP:
MyStatLab

1. Answers will vary.

2. Neither. In a normal distribution, the mean and median are equal.

3. 1

4. Points at which the curve changes from curving upward to curving downward; $\mu - \sigma$ and $\mu + \sigma$

5. Answers will vary.

   Similarities: The two curves will have the same line of symmetry.

   Differences: The curve with the larger standard deviation will be more spread out than the curve with the smaller standard deviation.

6. Answers will vary.

   Similarities: The two curves will have the same shape because they have equal standard deviations.

   Differences: The two curves will have different lines of symmetry.

7. $\mu = 0$, $\sigma = 1$

8. Transform each data value x into a z-score by subtracting the mean from x and dividing the result by the standard deviation. In symbols,
   $$z = \frac{x - \mu}{\sigma}.$$

■ BUILDING BASIC SKILLS AND VOCABULARY

1. Find three real-life examples of a continuous variable. Which do you think may be normally distributed? Why?

2. In a normal distribution, which is greater, the mean or the median? Explain.

3. What is the total area under the normal curve?

4. What do the inflection points on a normal distribution represent? Where do they occur?

5. Draw two normal curves that have the same mean but different standard deviations. Describe the similarities and differences.

6. Draw two normal curves that have different means but the same standard deviation. Describe the similarities and differences.

7. What is the mean of the standard normal distribution? What is the standard deviation of the standard normal distribution?

8. Describe how you can transform a nonstandard normal distribution to a standard normal distribution.

9. **Getting at the Concept** Why is it correct to say "a" normal distribution and "the" standard normal distribution?

10. **Getting at the Concept** If a z-score is 0, which of the following must be true? Explain your reasoning.

    (a) The mean is 0.

    (b) The corresponding x-value is 0.

    (c) The corresponding x-value is equal to the mean.

Graphical Analysis *In Exercises 11–16, determine whether the graph could represent a variable with a normal distribution. Explain your reasoning.*

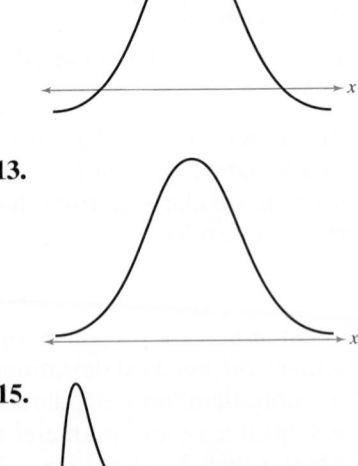

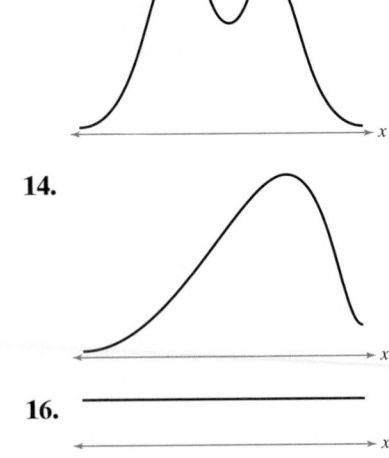

9. "The" standard normal distribution is used to describe one specific normal distribution ($\mu = 0$, $\sigma = 1$). "A" normal distribution is used to describe a normal distribution with any mean and standard deviation.

10. (c) is true because a z-score equal to 0 indicates that the corresponding x-value is equal to the mean.

11. No, the graph crosses the x-axis.

12. No, the graph is not symmetric.

13. Yes, the graph fulfills the properties of the normal distribution.

14. No, the graph is skewed left.

15. No, the graph is skewed right.

16. No, the graph is not bell-shaped.

17. It is normal because it is bell-shaped and nearly symmetric.

18. No, the graph is skewed right.

19. 0.0968

20. 0.9554

21. 0.0228

22. 0.9893

23. 0.4878

24. 0.6247

25. 0.5319

26. 0.9992

27. 0.005

28. 0.9139

29. 0.7422

30. 0.0006

31. 0.6387

32. 0.0532

33. 0.4979

34. 0.437

35. 0.95

36. 0.9802

37. 0.2006 (*Tech:* 0.2005)

38. 0.05

Graphical Analysis *In Exercises 17 and 18, determine whether the histogram represents data with a normal distribution. Explain your reasoning.*

17.

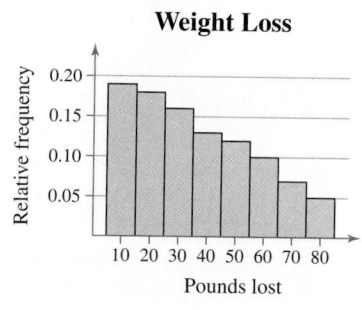

Waiting Time in a Dentist's Office

18.

Weight Loss

■ USING AND INTERPRETING CONCEPTS

Graphical Analysis *In Exercises 19–24, find the area of the indicated region under the standard normal curve. If convenient, use technology to find the area.*

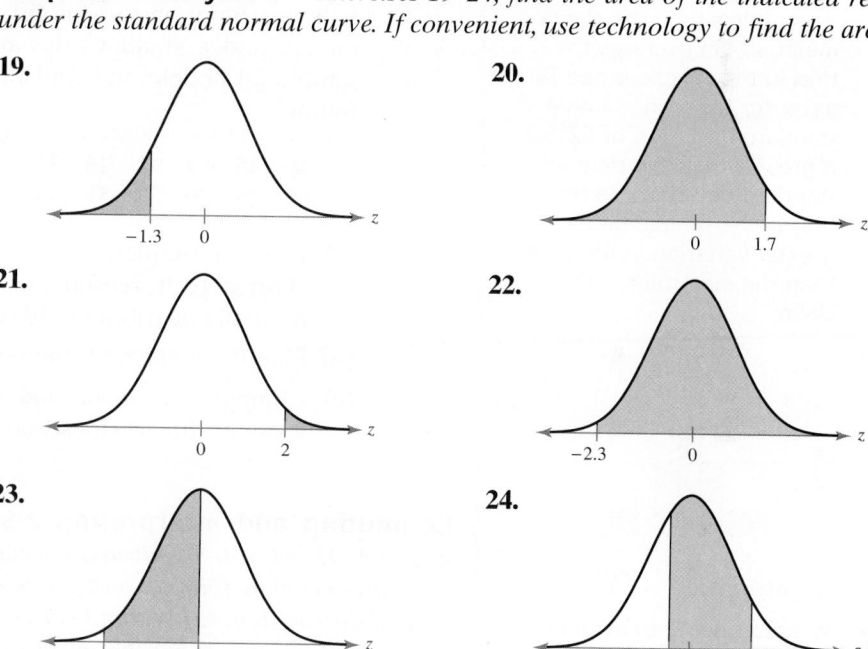

19.

20.

21.

22.

23.

24.

Finding Area *In Exercises 25–38, find the indicated area under the standard normal curve. If convenient, use technology to find the area.*

25. To the left of $z = 0.08$

26. To the right of $z = -3.16$

27. To the left of $z = -2.575$

28. To the left of $z = 1.365$

29. To the right of $z = -0.65$

30. To the right of $z = 3.25$

31. To the right of $z = -0.355$

32. To the right of $z = 1.615$

33. Between $z = 0$ and $z = 2.86$

34. Between $z = -1.53$ and $z = 0$

35. Between $z = -1.96$ and $z = 1.96$

36. Between $z = -2.33$ and $z = 2.33$

37. To the left of $z = -1.28$ and to the right of $z = 1.28$

38. To the left of $z = -1.96$ and to the right of $z = 1.96$

39. (a)

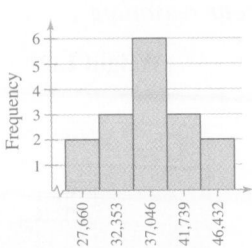

Life Spans of Tires

It is reasonable to assume that the life spans are normally distributed because the histogram is symmetric and bell-shaped.

(b) 37,234.7, 6259.2

(c) The sample mean of 37,234.7 hours is less than the claimed mean, so, on average, the tires in the sample lasted for a shorter time. The sample standard deviation of 6259.2 is greater than the claimed standard deviation, so the tires in the sample had a greater variation in life span than the manufacturer's claim.

40. (a)

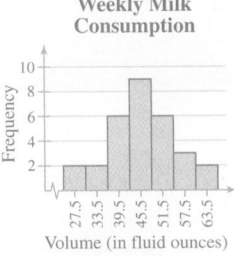

Weekly Milk Consumption

It is reasonable to assume that the weekly milk consumptions are normally distributed because the histogram is nearly symmetric and bell-shaped.

(b) 45.9, 9.5

(c) The mean of your sample is 2.8 fluid ounces less than that of the previous study, so the average milk consumption from the sample is less than in the previous study. The standard deviation is about 0.9 fluid ounce greater than that of the previous study, so the milk consumptions are slightly more spread out than in the previous study.

39. Manufacturer Claims You work for a consumer watchdog publication and are testing the advertising claims of a tire manufacturer. The manufacturer claims that the life spans of the tires are normally distributed, with a mean of 40,000 miles and a standard deviation of 4000 miles. You test 16 tires and get the following life spans.

48,778	41,046	29,083	36,394	32,302	42,787	41,972	37,229
25,314	31,920	38,030	38,445	30,750	38,886	36,770	46,049

(a) Draw a frequency histogram to display these data. Use five classes. Is it reasonable to assume that the life spans are normally distributed? Why?

(b) Find the mean and standard deviation of your sample.

(c) Compare the mean and standard deviation of your sample with those in the manufacturer's claim. Discuss the differences.

40. Milk Consumption You are performing a study about weekly per capita milk consumption. A previous study found weekly per capita milk consumption to be normally distributed, with a mean of 48.7 fluid ounces and a standard deviation of 8.6 fluid ounces. You randomly sample 30 people and find their weekly milk consumptions to be as follows.

40	45	54	41	43	31	47	30	33	37	48	57	52	45	38
65	25	39	53	51	58	52	40	46	44	48	61	47	49	57

(a) Draw a frequency histogram to display these data. Use seven classes. Is it reasonable to assume that the consumptions are normally distributed? Why?

(b) Find the mean and standard deviation of your sample.

(c) Compare the mean and standard deviation of your sample with those of the previous study. Discuss the differences.

Computing and Interpreting z-Scores of Normal Distributions *In Exercises 41–44, you are given a normal distribution, the distribution's mean and standard deviation, four values from that distribution, and a graph of the standard normal distribution. (a) Without converting to z-scores, match the values with the letters A, B, C, and D on the given graph of the standard normal distribution. (b) Find the z-score that corresponds to each value and check your answers to part (a). (c) Determine whether any of the values are unusual.*

41. Blood Pressure The systolic blood pressures of a sample of adults are normally distributed, with a mean pressure of 115 millimeters of mercury and a standard deviation of 3.6 millimeters of mercury. The systolic blood pressures of four adults selected at random are 121 millimeters of mercury, 113 millimeters of mercury, 105 millimeters of mercury, and 127 millimeters of mercury.

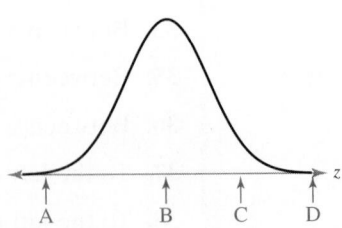

41. (a) A = 105; B = 113; C = 121;
D = 127

(b) −2.78; −0.56; 1.67; 3.33

(c) x = 105 is unusual because
its corresponding z-score
(−2.78) lies more than
2 standard deviations from
the mean, and x = 127 is
very unusual because its
corresponding z-score (3.33)
lies more than 3 standard
deviations from the mean.

42. (a) A = 11.92; B = 11.99;
C = 12.01; D = 12.12

(b) −1.6; −0.2; 0.2; 2.4

(c) x = 12.12 is unusual because
its corresponding z-score
(2.4) lies more than
2 standard deviations
from the mean.

43. (a) A = 1241; B = 1392;
C = 1924; D = 2202

(b) −0.86; −0.375; 1.33; 2.22

(c) x = 2202 is unusual because
its corresponding z-score
(2.22) lies more than
2 standard deviations from
the mean.

44. (a) A = 9; B = 15; C = 22;
D = 35

(b) −2.42; −1.22; 0.18; 2.78

(c) x = 9 is unusual because its
corresponding z-score
(−2.42) lies more than 2
standard deviations from the
mean. x = 35 is unusual
because its corresponding
z-score (2.78) lies more than
2 standard deviations from
the mean.

45. 0.9750

46. 0.1908

47. 0.9775

48. 0.1003

49. 0.84

50. 0.4382

42. Cereal Boxes The weights of the contents of cereal boxes are normally distributed, with a mean weight of 12 ounces and a standard deviation of 0.05 ounce. The weights of the contents of four cereal boxes selected at random are 12.01 ounces, 11.92 ounces, 12.12 ounces, and 11.99 ounces.

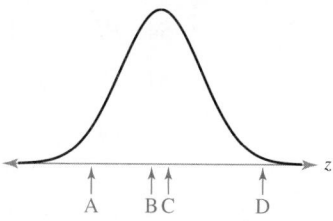

43. SAT Scores The SAT is an exam used by colleges and universities to evaluate undergraduate applicants. The test scores are normally distributed. In a recent year, the mean test score was 1509 and the standard deviation was 312. The test scores of four students selected at random are 1924, 1241, 2202, and 1392. *(Source: The College Board)*

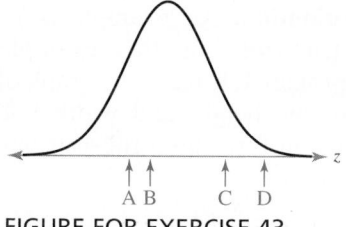 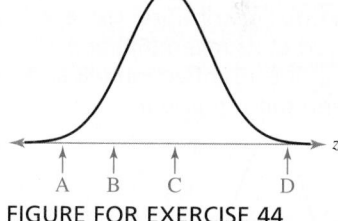

FIGURE FOR EXERCISE 43   FIGURE FOR EXERCISE 44

44. ACT Scores The ACT is an exam used by colleges and universities to evaluate undergraduate applicants. The test scores are normally distributed. In a recent year, the mean test score was 21.1 and the standard deviation was 5.0. The test scores of four students selected at random are 15, 22, 9, and 35. *(Source: ACT, Inc.)*

Graphical Analysis *In Exercises 45–50, find the probability of z occurring in the indicated region. If convenient, use technology to find the probability.*

45. 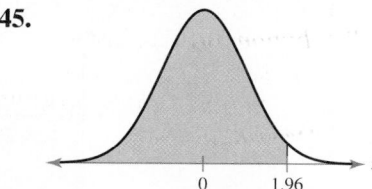   **46.**

47. 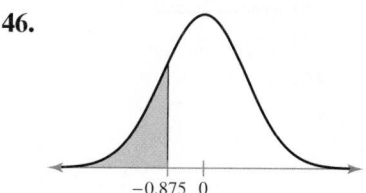   **48.**

49. 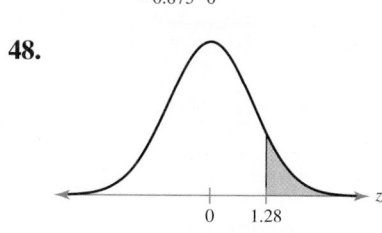   **50.**

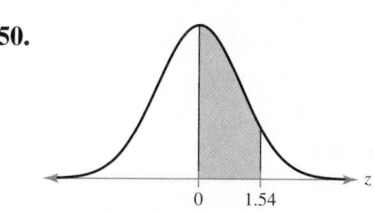

51. 0.9265

52. 0.4286

53. 0.0148

54. 0.9678

55. 0.3133

56. 0.2002

57. 0.901 (Tech: 0.9011)

58. 0.8764

59. 0.0098 (Tech: 0.0099)

60. 0.1236

61.

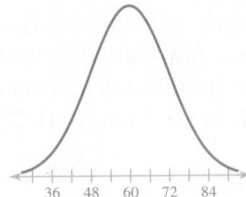

The normal distribution curve is centered at its mean (60) and has 2 points of inflection (48 and 72) representing $\mu \pm \sigma$.

62.

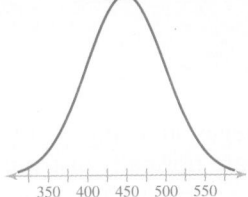

The normal distribution curve is centered at its mean (450) and has 2 points of inflection (400 and 500) representing $\mu \pm \sigma$.

63. (a) Area under curve

= area of square

= (1)(1) = 1

(b) 0.25

(c) 0.4

64. (a)

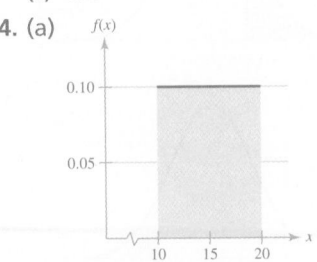

Area under curve

= area of rectangle

= (20 − 10) · (0.10)

= 1

(b) 0.3

(c) 0.5

Finding Probabilities   *In Exercises 51–60, find the indicated probability using the standard normal distribution. If convenient, use technology to find the probability.*

51. $P(z < 1.45)$ **52.** $P(z < -0.18)$ **53.** $P(z > 2.175)$

54. $P(z > -1.85)$ **55.** $P(-0.89 < z < 0)$ **56.** $P(0 < z < 0.525)$

57. $P(-1.65 < z < 1.65)$ **58.** $P(-1.54 < z < 1.54)$

59. $P(z < -2.58 \text{ or } z > 2.58)$ **60.** $P(z < -1.54 \text{ or } z > 1.54)$

■ EXTENDING CONCEPTS

61. Writing   Draw a normal curve with a mean of 60 and a standard deviation of 12. Describe how you constructed the curve and discuss its features.

62. Writing   Draw a normal curve with a mean of 450 and a standard deviation of 50. Describe how you constructed the curve and discuss its features.

63. Uniform Distribution   Another continuous distribution is the **uniform distribution**. An example is $f(x) = 1$ for $0 \le x \le 1$. The mean of the distribution for this example is 0.5 and the standard deviation is approximately 0.29. The graph of the distribution for this example is a square with the height and width both equal to 1 unit. In general, the density function for a uniform distribution on the interval from $x = a$ to $x = b$ is given by

$$f(x) = \frac{1}{b - a}.$$

The mean is

$$\frac{a + b}{2}$$

and the standard deviation is

$$\sqrt{\frac{(b - a)^2}{12}}.$$

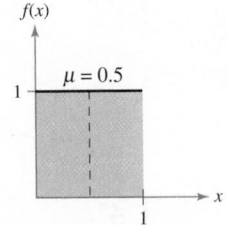

(a) Verify that the area under the curve is 1.

(b) Find the probability that x falls between 0.25 and 0.5.

(c) Find the probability that x falls between 0.3 and 0.7.

64. Uniform Distribution   Consider the uniform density function $f(x) = 0.1$ for $10 \le x \le 20$. The mean of this distribution is 15 and the standard deviation is about 2.89.

(a) Draw a graph of the distribution and show that the area under the curve is 1.

(b) Find the probability that x falls between 12 and 15.

(c) Find the probability that x falls between 13 and 18.

5.2 Normal Distributions: Finding Probabilities

WHAT YOU SHOULD LEARN

▸ How to find probabilities for normally distributed variables using a table and using technology

Probability and Normal Distributions

▸ PROBABILITY AND NORMAL DISTRIBUTIONS

If a random variable x is normally distributed, you can find the probability that x will fall in a given interval by calculating the area under the normal curve for the given interval. To find the area under any normal curve, you can first convert the upper and lower bounds of the interval to z-scores. Then use the standard normal distribution to find the area. For instance, consider a normal curve with $\mu = 500$ and $\sigma = 100$, as shown at the upper left. The value of x one standard deviation above the mean is $\mu + \sigma = 500 + 100 = 600$. Now consider the standard normal curve shown at the lower left. The value of z one standard deviation above the mean is $\mu + \sigma = 0 + 1 = 1$. Because a z-score of 1 corresponds to an x-value of 600, and areas are not changed with a transformation to a standard normal curve, the shaded areas in the graphs are equal.

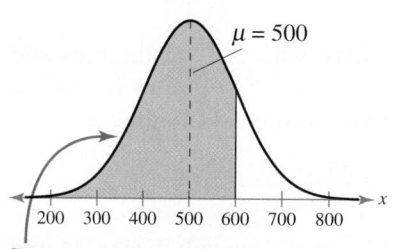

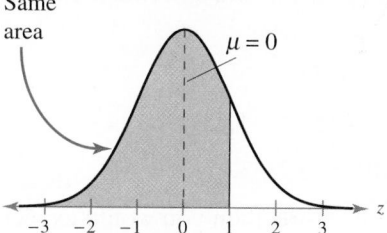

Same area

| **EXAMPLE 1** | **SC** Report 20 |

▸ Finding Probabilities for Normal Distributions

A survey indicates that people use their cellular phones an average of 1.5 years before buying a new one. The standard deviation is 0.25 year. A cellular phone user is selected at random. Find the probability that the user will use their current phone for less than 1 year before buying a new one. Assume that the variable x is normally distributed. *(Adapted from Fonebak)*

▸ Solution

The graph shows a normal curve with $\mu = 1.5$ and $\sigma = 0.25$ and a shaded area for x less than 1. The z-score that corresponds to 1 year is

$$z = \frac{x - \mu}{\sigma} = \frac{1 - 1.5}{0.25} = -2.$$

The Standard Normal Table shows that $P(z < -2) = 0.0228$. The probability that the user will use their cellular phone for less than 1 year before buying a new one is 0.0228.

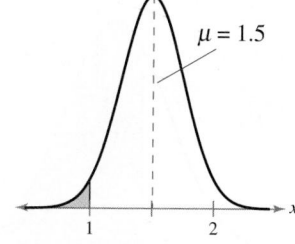

Age of cellular phone (in years)

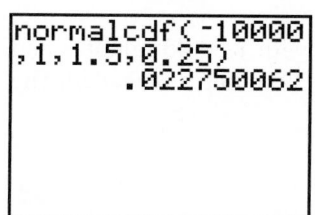

In Example 1, you can use a TI-83/84 Plus to find the probability automatically.

Interpretation   So, 2.28% of cellular phone users will use their cellular phone for less than 1 year before buying a new one. Because 2.28% is less than 5%, this is an unusual event.

▸ Try It Yourself 1

The average speed of vehicles traveling on a stretch of highway is 67 miles per hour with a standard deviation of 3.5 miles per hour. A vehicle is selected at random. What is the probability that it is violating the 70 mile per hour speed limit? Assume the speeds are normally distributed.

a. *Sketch* a graph.
b. *Find the z-score* that corresponds to 70 miles per hour.
c. *Find the area* to the right of that z-score.
d. *Interpret* the results.

Answer: Page A38

STUDY TIP

Another way to write the answer to Example 1 is $P(x < 1) = 0.0228$.

EXAMPLE 2

▶ **Finding Probabilities for Normal Distributions**

A survey indicates that for each trip to the supermarket, a shopper spends an average of 45 minutes with a standard deviation of 12 minutes in the store. The lengths of time spent in the store are normally distributed and are represented by the variable x. A shopper enters the store. (a) Find the probability that the shopper will be in the store for each interval of time listed below. (b) Interpret your answer if 200 shoppers enter the store. How many shoppers would you expect to be in the store for each interval of time listed below?

1. Between 24 and 54 minutes   **2.** More than 39 minutes

▶ **Solution**

1. (a) The graph at the left shows a normal curve with $\mu = 45$ minutes and $\sigma = 12$ minutes. The area for x between 24 and 54 minutes is shaded. The z-scores that correspond to 24 minutes and to 54 minutes are

$$z_1 = \frac{24 - 45}{12} = -1.75 \quad \text{and} \quad z_2 = \frac{54 - 45}{12} = 0.75.$$

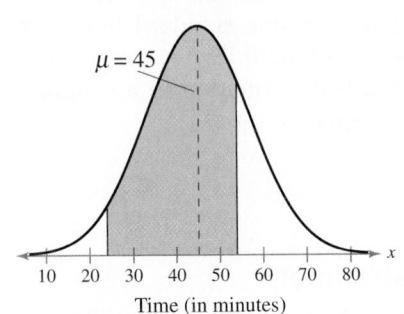

So, the probability that a shopper will be in the store between 24 and 54 minutes is

$$P(24 < x < 54) = P(-1.75 < z < 0.75)$$
$$= P(z < 0.75) - P(z < -1.75)$$
$$= 0.7734 - 0.0401 = 0.7333.$$

(b) **Interpretation** If 200 shoppers enter the store, then you would expect $200(0.7333) = 146.66$, or about 147, shoppers to be in the store between 24 and 54 minutes.

2. (a) The graph at the left shows a normal curve with $\mu = 45$ minutes and $\sigma = 12$ minutes. The area for x greater than 39 minutes is shaded. The z-score that corresponds to 39 minutes is

$$z = \frac{39 - 45}{12} = -0.5.$$

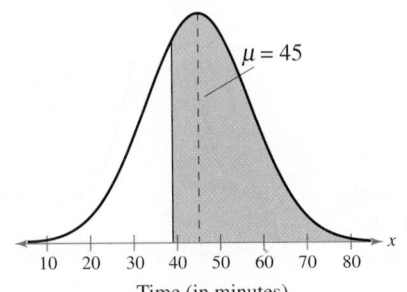

So, the probability that a shopper will be in the store more than 39 minutes is

$$P(x > 39) = P(z > -0.5) = 1 - P(z < -0.5) = 1 - 0.3085 = 0.6915.$$

(b) **Interpretation** If 200 shoppers enter the store, then you would expect $200(0.6915) = 138.3$, or about 138, shoppers to be in the store more than 39 minutes.

▶ **Try It Yourself 2**

What is the probability that the shopper in Example 2 will be in the supermarket between 33 and 60 minutes?

a. *Sketch* a graph.
b. *Find the z-scores* that correspond to 33 minutes and 60 minutes.
c. *Find the cumulative area* for each z-score and *subtract* the smaller area from the larger area.
d. *Interpret* your answer if 150 shoppers enter the store. How many shoppers would you expect to be in the store between 33 and 60 minutes?

Answer: Page A38

Another way to find normal probabilities is to use a calculator or a computer. You can find normal probabilities using MINITAB, Excel, and the TI-83/84 Plus.

EXAMPLE 3

▶ **Using Technology to Find Normal Probabilities**

Triglycerides are a type of fat in the bloodstream. The mean triglyceride level in the United States is 134 milligrams per deciliter. Assume the triglyceride levels of the population of the United States are normally distributed, with a standard deviation of 35 milligrams per deciliter. You randomly select a person from the United States. What is the probability that the person's triglyceride level is less than 80? Use a technology tool to find the probability. *(Adapted from University of Maryland Medical Center)*

▶ **Solution**

MINITAB, Excel, and the TI-83/84 Plus each have features that allow you to find normal probabilities without first converting to standard z-scores. For each, you must specify the mean and standard deviation of the population, as well as the x-value(s) that determine the interval.

MINITAB

Cumulative Distribution Function

Normal with mean = 134 and standard deviation = 35

x	P[X <= x]
80	0.0614327

EXCEL

	A	B	C
1	NORMDIST(80,134,35,TRUE)		
2			0.06143272

TI-83/84 PLUS

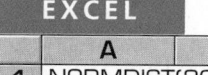

normalcdf(-10000,80,134,35)
        .0614327356

From the displays, you can see that the probability that the person's triglyceride level is less than 80 is about 0.0614, or 6.14%.

▶ **Try It Yourself 3**

A person from the United States is selected at random. What is the probability that the person's triglyceride level is between 100 and 150? Use a technology tool.

a. *Read the user's guide* for the technology tool you are using.
b. *Enter the appropriate data* to obtain the probability.
c. *Write* the result as a sentence. *Answer: Page A38*

Example 3 shows only one of several ways to find normal probabilities using MINITAB, Excel, and the TI-83/84 Plus.

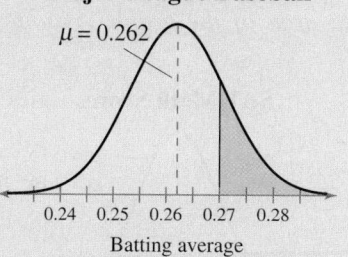

5.2 EXERCISES

FOR EXTRA HELP:
MyStatLab

1. 0.4207

2. 0.9032

3. 0.3446

4. 0.8289

5. 0.1787 (*Tech:* 0.1788)

6. 0.3557 (*Tech:* 0.3558)

7. 0.3442 (*Tech:* 0.3451)

8. 0.0832 (*Tech:* 0.0837)

9. 0.2747 (*Tech:* 0.2737)

10. 0.2597 (*Tech:* 0.2571)

11. 0.3387 (*Tech:* 0.3385)

12. 0.4735

■ BUILDING BASIC SKILLS AND VOCABULARY

Computing Probabilities *In Exercises 1–6, assume the random variable x is normally distributed with mean $\mu = 174$ and standard deviation $\sigma = 20$. Find the indicated probability.*

1. $P(x < 170)$

2. $P(x < 200)$

3. $P(x > 182)$

4. $P(x > 155)$

5. $P(160 < x < 170)$

6. $P(172 < x < 192)$

Graphical Analysis *In Exercises 7–12, assume a member is selected at random from the population represented by the graph. Find the probability that the member selected at random is from the shaded area of the graph. Assume the variable x is normally distributed.*

7.   **SAT Writing Scores**

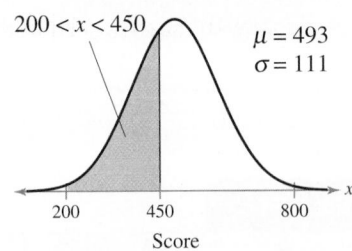

$200 < x < 450$
$\mu = 493$
$\sigma = 111$

Score

(*Source: The College Board*)

8.   **SAT Math Scores**

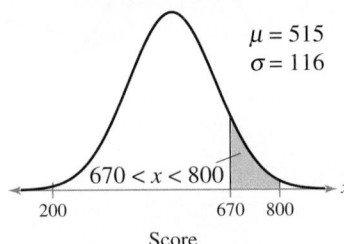

$670 < x < 800$
$\mu = 515$
$\sigma = 116$

Score

(*Source: The College Board*)

9. **U.S. Men Ages 35–44:**
Total Cholesterol

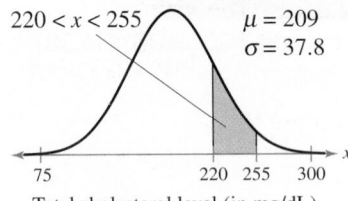

$220 < x < 255$
$\mu = 209$
$\sigma = 37.8$

Total cholesterol level (in mg/dL)

(*Adapted from National Center for Health Statistics*)

10. **U.S. Women Ages 35–44:**
Total Cholesterol

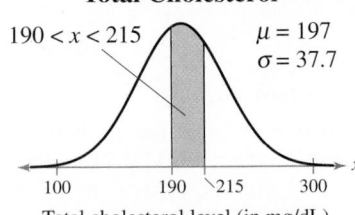

$190 < x < 215$
$\mu = 197$
$\sigma = 37.7$

Total cholesterol level (in mg/dL)

(*Adapted from National Center for Health Statistics*)

11.   **Ford Fusion:**
Braking Distance

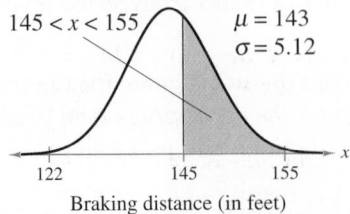

$145 < x < 155$
$\mu = 143$
$\sigma = 5.12$

Braking distance (in feet)

(*Adapted from Consumer Reports*)

12.   **Hyundai Elantra:**
Braking Distance

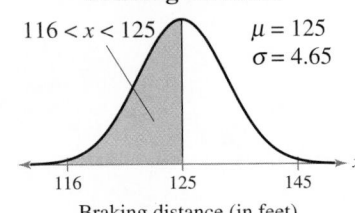

$116 < x < 125$
$\mu = 125$
$\sigma = 4.65$

Braking distance (in feet)

(*Adapted from Consumer Reports*)

13. (a) 0.0968
    (b) 0.6612
    (c) 0.2420
    (d) No, none of the events are unusual because their probabilities are greater than 0.05.

14. (a) 0.0013
    (b) 0.7488 (*Tech:* 0.7483)
    (c) 0.0087 (*Tech:* 0.0085)
    (d) Yes, the events in parts (a) and (c) are unusual because their probabilities are less than 0.05.

15. (a) 0.1867 (*Tech:* 0.1870)
    (b) 0.4171 (*Tech:* 0.4176)
    (c) 0.0166 (*Tech:* 0.0167)
    (d) Yes, the event in part (c) is unusual because its probability is less than 0.05.

16. (a) 0.2514 (*Tech:* 0.2525)
    (b) 0.0675 (*Tech:* 0.0662)
    (c) 0.0475 (*Tech:* 0.0478)
    (d) Yes, the event in part (c) is unusual because its probability is less than 0.05.

17. (a) 0.0228
    (b) 0.927
    (c) 0.0013

■ USING AND INTERPRETING CONCEPTS

Finding Probabilities *In Exercises 13–20, find the indicated probabilities. If convenient, use technology to find the probabilities.*

13. **Heights of Men** A survey was conducted to measure the heights of U.S. men. In the survey, respondents were grouped by age. In the 20–29 age group, the heights were normally distributed, with a mean of 69.9 inches and a standard deviation of 3.0 inches. A study participant is randomly selected. *(Adapted from U.S. National Center for Health Statistics)*

    (a) Find the probability that his height is less than 66 inches.
    (b) Find the probability that his height is between 66 and 72 inches.
    (c) Find the probability that his height is more than 72 inches.
    (d) Can any of these events be considered unusual? Explain your reasoning.

14. **Heights of Women** A survey was conducted to measure the heights of U.S. women. In the survey, respondents were grouped by age. In the 20–29 age group, the heights were normally distributed, with a mean of 64.3 inches and a standard deviation of 2.6 inches. A study participant is randomly selected. *(Adapted from U.S. National Center for Health Statistics)*

    (a) Find the probability that her height is less than 56.5 inches.
    (b) Find the probability that her height is between 61 and 67 inches.
    (c) Find the probability that her height is more than 70.5 inches.
    (d) Can any of these events be considered unusual? Explain your reasoning.

15. **ACT English Scores** In a recent year, the ACT scores for the English portion of the test were normally distributed, with a mean of 20.6 and a standard deviation of 6.3. A high school student who took the English portion of the ACT is randomly selected. *(Source: ACT, Inc.)*

    (a) Find the probability that the student's ACT score is less than 15.
    (b) Find the probability that the student's ACT score is between 18 and 25.
    (c) Find the probability that the student's ACT score is more than 34.
    (d) Can any of these events be considered unusual? Explain your reasoning.

16. **Beagles** The weights of adult male beagles are normally distributed, with a mean of 25 pounds and a standard deviation of 3 pounds. A beagle is randomly selected.

    (a) Find the probability that the beagle's weight is less than 23 pounds.
    (b) Find the probability that the weight is between 24.5 and 25 pounds.
    (c) Find the probability that the beagle's weight is more than 30 pounds.
    (d) Can any of these events be considered unusual? Explain your reasoning.

17. **Computer Usage** A survey was conducted to measure the number of hours per week adults in the United States spend on their computers. In the survey, the numbers of hours were normally distributed, with a mean of 7 hours and a standard deviation of 1 hour. A survey participant is randomly selected.

    (a) Find the probability that the number of hours spent on the computer by the participant is less than 5 hours per week.
    (b) Find the probability that the number of hours spent on the computer by the participant is between 5.5 and 9.5 hours per week.
    (c) Find the probability that the number of hours spent on the computer by the participant is more than 10 hours per week.

18. (a) 0.0062
    (b) 0.7492 (*Tech:* 0.7499)
    (c) 0.0004
19. (a) 0.0073
    (b) 0.7215 (*Tech:* 0.7218)
    (c) 0.0228
20. (a) 0.2743
    (b) 0.4452
    (c) 0.0228
21. (a) 83.15% (*Tech:* 83.25%)
    (b) 305 scores (*Tech:* 304 scores)
22. (a) 44.83% (*Tech:* 44.86%)
    (b) 349 scores (*Tech:* 348 scores)
23. (a) 66.28% (*Tech:* 66.4%)
    (b) 22 men

18. **Utility Bills**   The monthly utility bills in a city are normally distributed, with a mean of $100 and a standard deviation of $12. A utility bill is randomly selected.

    (a) Find the probability that the utility bill is less than $70.
    (b) Find the probability that the utility bill is between $90 and $120.
    (c) Find the probability that the utility bill is more than $140.

19. **Computer Lab Schedule**   The times per week a student uses a lab computer are normally distributed, with a mean of 6.2 hours and a standard deviation of 0.9 hour. A student is randomly selected.

    (a) Find the probability that the student uses a lab computer less than 4 hours per week.
    (b) Find the probability that the student uses a lab computer between 5 and 7 hours per week.
    (c) Find the probability that the student uses a lab computer more than 8 hours per week.

20. **Health Club Schedule**   The times per workout an athlete uses a stairclimber are normally distributed, with a mean of 20 minutes and a standard deviation of 5 minutes. An athlete is randomly selected.

    (a) Find the probability that the athlete uses a stairclimber for less than 17 minutes.
    (b) Find the probability that the athlete uses a stairclimber between 20 and 28 minutes.
    (c) Find the probability that the athlete uses a stairclimber for more than 30 minutes.

Using Normal Distributions   *In Exercises 21–28, answer the questions about the specified normal distribution.*

21. **SAT Writing Scores**   Use the normal distribution of SAT writing scores in Exercise 7 for which the mean is 493 and the standard deviation is 111.

    (a) What percent of the SAT writing scores are less than 600?
    (b) If 1000 SAT writing scores are randomly selected, about how many would you expect to be greater than 550?

22. **SAT Math Scores**   Use the normal distribution of SAT math scores in Exercise 8 for which the mean is 515 and the standard deviation is 116.

    (a) What percent of the SAT math scores are less than 500?
    (b) If 1500 SAT math scores are randomly selected, about how many would you expect to be greater than 600?

23. **Cholesterol**   Use the normal distribution of men's total cholesterol levels in Exercise 9 for which the mean is 209 milligrams per deciliter and the standard deviation is 37.8 milligrams per deciliter.

    (a) What percent of the men have a total cholesterol level less than 225 milligrams per deciliter of blood?
    (b) If 250 U.S. men in the 35–44 age group are randomly selected, about how many would you expect to have a total cholesterol level greater than 260 milligrams per deciliter of blood?

24. (a) 70.19% (*Tech:* 70.21%)

    (b) 125 women

25. (a) 99.87%

    (b) 1 adult

26. (a) 1.88% (*Tech:* 1.86%)

    (b) 61 bills

27. 1.5% (*Tech:* 1.51%); It is unusual for a battery to have a life span that is more than 2065 hours because the probability is less than 0.05.

28. 5.94%, It is not unusual for a person to consume less than 3.1 pounds of peanuts in a year because the probability is greater than 0.05.

29. (a) 0.3085

    (b) 0.1499

    (c) 0.0668; No, because 0.0668 > 0.05, this event is not unusual.

30. (a) 0.1587

    (b) 0.1587

    (c) 0.0013; Yes, because 0.0013 < 0.05, this event is unusual.

24. Cholesterol   Use the normal distribution of women's total cholesterol levels in Exercise 10 for which the mean is 197 milligrams per deciliter and the standard deviation is 37.7 milligrams per deciliter.

(a) What percent of the women have a total cholesterol level less than 217 milligrams per deciliter of blood?

(b) If 200 U.S. women in the 35–44 age group are randomly selected, about how many would you expect to have a total cholesterol level greater than 185 milligrams per deciliter of blood?

25. Computer Usage   Use the normal distribution of computer usage in Exercise 17 for which the mean is 7 hours and the standard deviation is 1 hour.

(a) What percent of the adults spend more than 4 hours per week on their computer?

(b) If 35 adults in the United States are randomly selected, about how many would you expect to say they spend less than 5 hours per week on their computer?

26. Utility Bills   Use the normal distribution of utility bills in Exercise 18 for which the mean is $100 and the standard deviation is $12.

(a) What percent of the utility bills are more than $125?

(b) If 300 utility bills are randomly selected, about how many would you expect to be less than $90?

27. Battery Life Spans   The life spans of batteries are normally distributed, with a mean of 2000 hours and a standard deviation of 30 hours. What percent of batteries have a life span that is more than 2065 hours? Would it be unusual for a battery to have a life span that is more than 2065 hours? Explain your reasoning.

28. Peanuts   Assume the mean annual consumptions of peanuts are normally distributed, with a mean of 5.9 pounds per person and a standard deviation of 1.8 pounds per person. What percent of people annually consume less than 3.1 pounds of peanuts per person? Would it be unusual for a person to consume less than 3.1 pounds of peanuts in a year? Explain your reasoning.

SC *In Exercises 29 and 30, use the StatCrunch normal calculator to find the indicated probabilities.*

29. Soft Drink Machine   The amounts a soft drink machine is designed to dispense for each drink are normally distributed, with a mean of 12 fluid ounces and a standard deviation of 0.2 fluid ounce. A drink is randomly selected.

(a) Find the probability that the drink is less than 11.9 fluid ounces.

(b) Find the probability that the drink is between 11.8 and 11.9 fluid ounces.

(c) Find the probability that the drink is more than 12.3 fluid ounces. Can this be considered an unusual event? Explain your reasoning.

30. Machine Parts   The thicknesses of washers produced by a machine are normally distributed, with a mean of 0.425 inch and a standard deviation of 0.005 inch. A washer is randomly selected.

(a) Find the probability that the washer is less than 0.42 inch thick.

(b) Find the probability that the washer is between 0.40 and 0.42 inch thick.

(c) Find the probability that the washer is more than 0.44 inch thick. Can this be considered an unusual event? Explain your reasoning.

31. Out of control, because there is a point more than three standard deviations beyond the mean.

32. Out of control, because two out of three consecutive points lie more than two standard deviations from the mean.

33. Out of control, because there are nine consecutive points below the mean, and two out of three consecutive points lie more than two standard deviations from the mean.

34. In control, because none of the three warning signals detected a change.

■ EXTENDING CONCEPTS

Control Charts *Statistical process control (SPC) is the use of statistics to monitor and improve the quality of a process, such as manufacturing an engine part. In SPC, information about a process is gathered and used to determine if a process is meeting all of the specified requirements. One tool used in SPC is a* **control chart.** *When individual measurements of a variable x are normally distributed, a control chart can be used to detect processes that are possibly out of statistical control. Three warning signals that a control chart uses to detect a process that may be out of control are as follows.*

(1) A point lies beyond three standard deviations of the mean.

(2) There are nine consecutive points that fall on one side of the mean.

(3) At least two of three consecutive points lie more than two standard deviations from the mean.

In Exercises 31–34, a control chart is shown. Each chart has horizontal lines drawn at the mean μ, at $\mu \pm 2\sigma$, and at $\mu \pm 3\sigma$. Determine if the process shown is in control or out of control. Explain.

31. A gear has been designed to have a diameter of 3 inches. The standard deviation of the process is 0.2 inch.

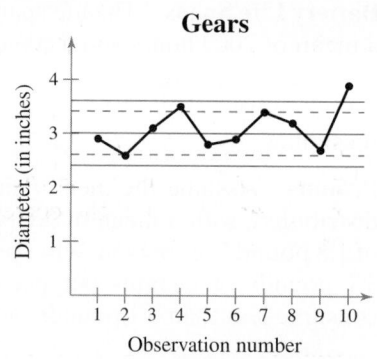

32. A nail has been designed to have a length of 4 inches. The standard deviation of the process is 0.12 inch.

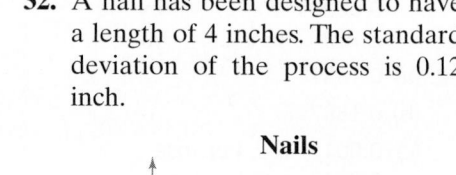
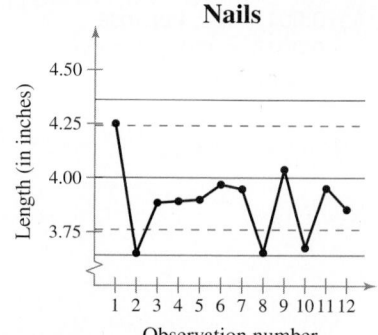

33. A liquid-dispensing machine has been designed to fill bottles with 1 liter of liquid. The standard deviation of the process is 0.1 liter.

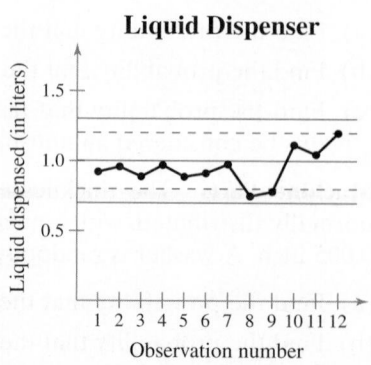

34. An engine part has been designed to have a diameter of 55 millimeters. The standard deviation of the process is 0.001 millimeter.

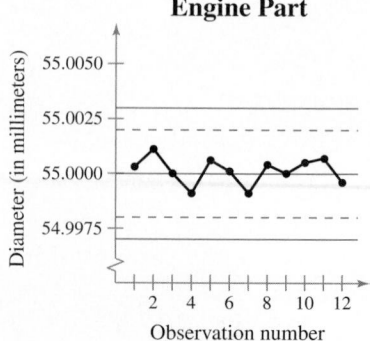

5.3   Normal Distributions: Finding Values

WHAT YOU SHOULD LEARN

▸ How to find a z-score given the area under the normal curve

▸ How to transform a z-score to an x-value

▸ How to find a specific data value of a normal distribution given the probability

Note to Instructor

Have students note that as the z-scores increase, the cumulative areas increase. The CDF is one-to-one and as such has an inverse function. Discuss how this inverse function (INVCDF) can be used when a cumulative area (percentile) is known and the z-score must be found.

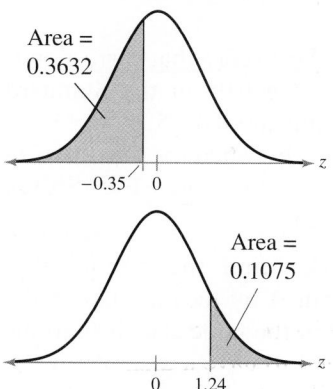

Area = 0.3632

Area = 0.1075

STUDY TIP

Here are instructions for finding the z-score that corresponds to a given area on a TI-83/84 Plus.

 2nd DISTR

3: invNorm(

Enter the cumulative area.

ENTER

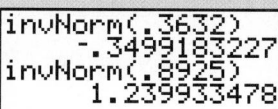
```
invNorm(.3632)
        -.3499183227
invNorm(.8925)
         1.239933478
```

Finding z-Scores ▸ Transforming a z-Score to an x-Value ▸ Finding a Specific Data Value for a Given Probability

▸ FINDING z-SCORES

In Section 5.2, you were given a normally distributed random variable x and you found the probability that x would fall in a given interval by calculating the area under the normal curve for the given interval.

But what if you are given a probability and want to find a value? For instance, a university might want to know the lowest test score a student can have on an entrance exam and still be in the top 10%, or a medical researcher might want to know the cutoff values for selecting the middle 90% of patients by age. In this section, you will learn how to find a value given an area under a normal curve (or a probability), as shown in the following example.

EXAMPLE 1

▸ Finding a z-Score Given an Area

1. Find the z-score that corresponds to a cumulative area of 0.3632.

2. Find the z-score that has 10.75% of the distribution's area to its right.

▸ Solution

1. Find the z-score that corresponds to an area of 0.3632 by locating 0.3632 in the Standard Normal Table. The values at the beginning of the corresponding row and at the top of the corresponding column give the z-score. For this area, the row value is -0.3 and the column value is 0.05. So, the z-score is -0.35.

z	.09	.08	.07	.06	.05	.04	.03
−3.4	.0002	.0003	.0003	.0003	.0003	.0003	.0003
−0.5	.2776	.2810	.2843	.2877	.2912	.2946	.2981
−0.4	.3121	.3156	.3192	.3228	.3264	.3300	.3336
−0.3	.3483	.3520	.3557	.3594	.3632	.3669	.3707
−0.2	.3859	.3897	.3936	.3974	.4013	.4052	.4090

2. Because the area to the right is 0.1075, the cumulative area is $1 - 0.1075 = 0.8925$. Find the z-score that corresponds to an area of 0.8925 by locating 0.8925 in the Standard Normal Table. For this area, the row value is 1.2 and the column value is 0.04. So, the z-score is 1.24.

z	.00	.01	.02	.03	.04	.05	.06
0.0	.5000	.5040	.5080	.5120	.5160	.5199	.5239
1.0	.8413	.8438	.8461	.8485	.8508	.8531	.8554
1.1	.8643	.8665	.8686	.8708	.8729	.8749	.8770
1.2	.8849	.8869	.8888	.8907	.8925	.8944	.8962
1.3	.9032	.9049	.9066	.9082	.9099	.9115	.9131

You can also use a computer or calculator to find the z-scores that correspond to the given cumulative areas, as shown in the margin.

STUDY TIP

In most cases, the given area will not be found in the table, so use the entry closest to it. If the given area is halfway between two area entries, use the z-score halfway between the corresponding z-scores.

▶ **Try It Yourself 1**

1. Find the z-score that has 96.16% of the distribution's area to the right.
2. Find the z-score for which 95% of the distribution's area lies between $-z$ and z.

   a. *Determine* the cumulative area.
   b. *Locate* the area in the Standard Normal Table.
   c. *Find the z-score* that corresponds to the area.   *Answer: Page A38*

In Section 2.5, you learned that percentiles divide a data set into 100 equal parts. To find a z-score that corresponds to a percentile, you can use the Standard Normal Table. Recall that if a value x represents the 83rd percentile P_{83}, then 83% of the data values are below x and 17% of the data values are above x.

EXAMPLE 2

▶ **Finding a z-Score Given a Percentile**

Find the z-score that corresponds to each percentile.

1. P_5

2. P_{50}

3. P_{90}

▶ **Solution**

1. To find the z-score that corresponds to P_5, find the z-score that corresponds to an area of 0.05 (see upper figure) by locating 0.05 in the Standard Normal Table. The areas closest to 0.05 in the table are 0.0495 ($z = -1.65$) and 0.0505 ($z = -1.64$). Because 0.05 is halfway between the two areas in the table, use the z-score that is halfway between -1.64 and -1.65. So, the z-score that corresponds to an area of 0.05 is -1.645.

2. To find the z-score that corresponds to P_{50}, find the z-score that corresponds to an area of 0.5 (see middle figure) by locating 0.5 in the Standard Normal Table. The area closest to 0.5 in the table is 0.5000, so the z-score that corresponds to an area of 0.5 is 0.

3. To find the z-score that corresponds to P_{90}, find the z-score that corresponds to an area of 0.9 (see lower figure) by locating 0.9 in the Standard Normal Table. The area closest to 0.9 in the table is 0.8997, so the z-score that corresponds to an area of 0.9 is about 1.28.

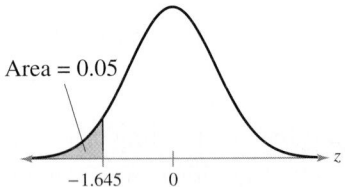

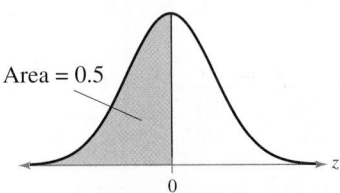

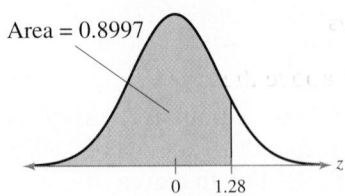

▶ **Try It Yourself 2**

Find the z-score that corresponds to each percentile.

1. P_{10} 2. P_{20} 3. P_{99}

   a. *Write* the percentile as an area. If necessary, draw a graph of the area to visualize the problem.
   b. *Locate* the area in the Standard Normal Table. If the area is not in the table, use the closest area. (See Study Tip above.)
   c. *Identify* the z-score that corresponds to the area.   *Answer: Page A38*

▸ TRANSFORMING A *z*-SCORE TO AN *x*-VALUE

Recall that to transform an *x*-value to a *z*-score, you can use the formula

$$z = \frac{x - \mu}{\sigma}.$$

This formula gives *z* in terms of *x*. If you solve this formula for *x*, you get a new formula that gives *x* in terms of *z*.

$$z = \frac{x - \mu}{\sigma} \qquad \text{Formula for } z \text{ in terms of } x$$

$$z\sigma = x - \mu \qquad \text{Multiply each side by } \sigma.$$

$$\mu + z\sigma = x \qquad \text{Add } \mu \text{ to each side.}$$

$$x = \mu + z\sigma \qquad \text{Interchange sides.}$$

TRANSFORMING A *z*-SCORE TO AN *x*-VALUE

To transform a standard *z*-score to a data value *x* in a given population, use the formula

$$x = \mu + z\sigma.$$

EXAMPLE 3

▸ Finding an *x*-Value Corresponding to a *z*-Score

A veterinarian records the weights of cats treated at a clinic. The weights are normally distributed, with a mean of 9 pounds and a standard deviation of 2 pounds. Find the weights *x* corresponding to *z*-scores of 1.96, −0.44, and 0. Interpret your results.

▸ Solution

The *x*-value that corresponds to each standard *z*-score is calculated using the formula $x = \mu + z\sigma$.

$z = 1.96$: $x = 9 + 1.96(2) = 12.92$ pounds

$z = -0.44$: $x = 9 + (-0.44)(2) = 8.12$ pounds

$z = 0$: $x = 9 + 1.96(0) = 9$ pounds

Interpretation   You can see that 12.92 pounds is above the mean, 8.12 pounds is below the mean, and 9 pounds is equal to the mean.

▸ Try It Yourself 3

A veterinarian records the weights of dogs treated at a clinic. The weights are normally distributed, with a mean of 52 pounds and a standard deviation of 15 pounds. Find the weights *x* corresponding to *z*-scores of −2.33, 3.10, and 0.58. Interpret your results.

a. *Identify* μ and σ of the normal distribution.
b. *Transform* each *z*-score to an *x*-value.
c. *Interpret* the results.   *Answer: Page A38*

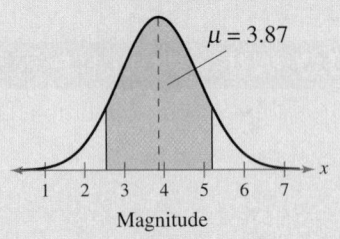
STUDY TIP

Here are instructions for finding a specific x-value for a given probability on a TI-83/84 Plus.

2nd DISTR

3: invNorm(

Enter the values for the area under the normal distribution, the specified mean, and the specified standard deviation separated by commas.

ENTER

```
invNorm(.9,50,10
)
       62.81551567
```

▶ FINDING A SPECIFIC DATA VALUE FOR A GIVEN PROBABILITY

You can also use the normal distribution to find a specific data value (*x*-value) for a given probability, as shown in Examples 4 and 5.

EXAMPLE 4   **SC** Report 21

▶ Finding a Specific Data Value

Scores for the California Peace Officer Standards and Training test are normally distributed, with a mean of 50 and a standard deviation of 10. An agency will only hire applicants with scores in the top 10%. What is the lowest score you can earn and still be eligible to be hired by the agency? *(Source: State of California)*

▶ Solution

Exam scores in the top 10% correspond to the shaded region shown.

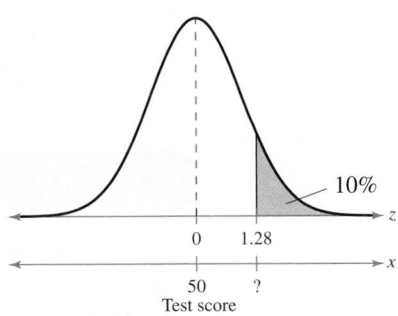

A test score in the top 10% is any score above the 90th percentile. To find the score that represents the 90th percentile, you must first find the *z*-score that corresponds to a cumulative area of 0.9. From the Standard Normal Table, you can find that the area closest to 0.9 is 0.8997. So, the *z*-score that corresponds to an area of 0.9 is $z = 1.28$. Using the equation $x = \mu + z\sigma$, you have

$$x = \mu + z\sigma$$
$$= 50 + 1.28(10)$$
$$\approx 62.8.$$

Interpretation   The lowest score you can earn and still be eligible to be hired by the agency is about 63.

▶ Try It Yourself 4

The braking distances of a sample of Nissan Altimas are normally distributed, with a mean of 129 feet and a standard deviation of 5.18 feet. What is the longest braking distance one of these Nissan Altimas could have and still be in the bottom 1%? *(Adapted from Consumer Reports)*

a. *Sketch* a graph.
b. *Find the z-score* that corresponds to the given area.
c. *Find x* using the equation $x = \mu + z\sigma$.
d. *Interpret* the result.

Answer: Page A38

Note to Instructor

Mention that to use the cumulative table to find a z-score, you must first express the area given as a cumulative area. It helps to explain cumulative areas in terms of percentiles. For instance, the score in the top 20% represents the 80th percentile. So, you must find the z-score that corresponds to a cumulative area of 0.8. Point out that if students are using a technology tool to find an x-value that corresponds to an area, it is not necessary first to find a z-score.

EXAMPLE 5   SC Report 22

▸ **Finding a Specific Data Value**

In a randomly selected sample of women ages 20–34, the mean total cholesterol level is 188 milligrams per deciliter with a standard deviation of 41.3 milligrams per deciliter. Assume the total cholesterol levels are normally distributed. Find the highest total cholesterol level a woman in this 20–34 age group can have and still be in the bottom 1%. *(Adapted from National Center for Health Statistics)*

▸ **Solution**

Total cholesterol levels in the lowest 1% correspond to the shaded region shown.

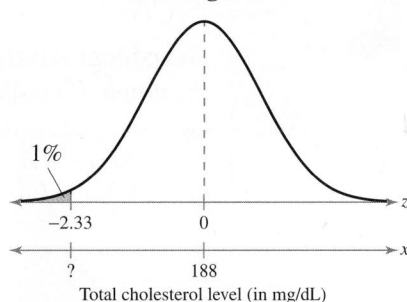

Total Cholesterol Levels in Women Ages 20–34

Total cholesterol level (in mg/dL)

A total cholesterol level in the lowest 1% is any level below the 1st percentile. To find the level that represents the 1st percentile, you must first find the z-score that corresponds to a cumulative area of 0.01. From the Standard Normal Table, you can find that the area closest to 0.01 is 0.0099. So, the z-score that corresponds to an area of 0.01 is $z = -2.33$. Using the equation $x = \mu + z\sigma$, you have

$$x = \mu + z\sigma$$

$$= 188 + (-2.33)(41.3)$$

$$\approx 91.77.$$

Interpretation   The value that separates the lowest 1% of total cholesterol levels for women in the 20–34 age group from the highest 99% is about 92 milligrams per deciliter.

▸ **Try It Yourself 5**

The lengths of time employees have worked at a corporation are normally distributed, with a mean of 11.2 years and a standard deviation of 2.1 years. In a company cutback, the lowest 10% in seniority are laid off. What is the maximum length of time an employee could have worked and still be laid off?

a. *Sketch* a graph.
b. *Find the z-score* that corresponds to the given area.
c. *Find x* using the equation $x = \mu + z\sigma$.
d. *Interpret* the result.   *Answer: Page A38*

```
invNorm(.01,188,
41.3)
          91.92183267
```

Using a TI-83/84 Plus, you can find the highest total cholesterol level automatically.

5.3 EXERCISES

1. −0.81
2. −0.16
3. 2.39
4. 0.84
5. −1.645
6. 1.04
7. 1.555
8. −2.605
9. −1.04
10. −0.52
11. 1.175
12. 0.44
13. −0.67
14. −0.25
15. 0.67
16. 0.84
17. −0.38
18. 0.25
19. −0.58
20. 1.99
21. −1.645, 1.645
22. −1.96, 1.96
23. −1.18
24. 0.79
25. 1.18
26. −0.79
27. −1.28, 1.28
28. −2.575, 2.575

■ BUILDING BASIC SKILLS AND VOCABULARY

In Exercises 1–16, use the Standard Normal Table to find the z-score that corresponds to the given cumulative area or percentile. If the area is not in the table, use the entry closest to the area. If the area is halfway between two entries, use the z-score halfway between the corresponding z-scores. If convenient, use technology to find the z-score.

1. 0.2090	**2.** 0.4364	**3.** 0.9916	**4.** 0.7995
5. 0.05	**6.** 0.85	**7.** 0.94	**8.** 0.0046
9. P_{15}	**10.** P_{30}	**11.** P_{88}	**12.** P_{67}
13. P_{25}	**14.** P_{40}	**15.** P_{75}	**16.** P_{80}

Graphical Analysis   *In Exercises 17–22, find the indicated z-score(s) shown in the graph. If convenient, use technology to find the z-score(s).*

17.
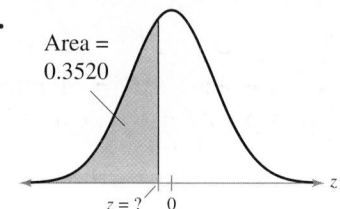

18.
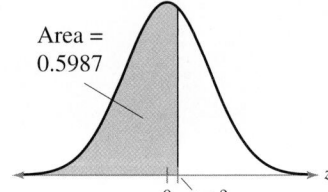

19.
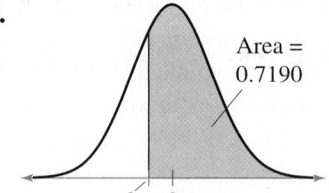

20.
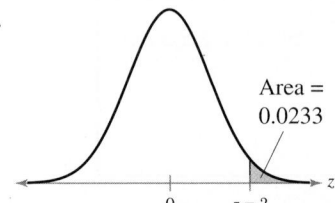

21.
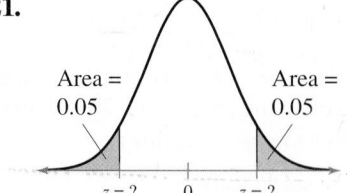

22.
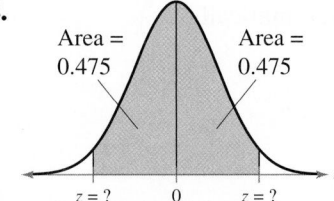

In Exercises 23–30, find the indicated z-score.

23. Find the z-score that has 11.9% of the distribution's area to its left.

24. Find the z-score that has 78.5% of the distribution's area to its left.

25. Find the z-score that has 11.9% of the distribution's area to its right.

26. Find the z-score that has 78.5% of the distribution's area to its right.

27. Find the z-score for which 80% of the distribution's area lies between −z and z.

28. Find the z-score for which 99% of the distribution's area lies between −z and z.

29. −0.06, 0.06

30. −0.15, 0.15

31. (a) 68.58 inches
   (b) 62.56 inches
      (*Tech:* 62.55 inches)

32. (a) 73.74 inches
   (b) 67.89 inches
      (*Tech:* 67.88 inches)

33. (a) 161.72 days
      (*Tech:* 161.73 days)
   (b) 221.22 days
      (*Tech:* 221.33 days)

34. (a) 1852.5 days
      (*Tech:* 1852.84 days)
   (b) 1531.63 days
      (*Tech:* 1530.67 days)

35. (a) 7.75 hours (*Tech:* 7.74 hours)
   (b) 5.43 hours and 6.77 hours

29. Find the z-score for which 5% of the distribution's area lies between $-z$ and z.

30. Find the z-score for which 12% of the distribution's area lies between $-z$ and z.

■ USING AND INTERPRETING CONCEPTS

Using Normal Distributions   *In Exercises 31–36, answer the questions about the specified normal distribution.*

31. Heights of Women   In a survey of women in the United States (ages 20–29), the mean height was 64.3 inches with a standard deviation of 2.6 inches. *(Adapted from National Center for Health Statistics)*

(a) What height represents the 95th percentile?

(b) What height represents the first quartile?

32. Heights of Men   In a survey of men in the United States (ages 20–29), the mean height was 69.9 inches with a standard deviation of 3.0 inches. *(Adapted from National Center for Health Statistics)*

(a) What height represents the 90th percentile?

(b) What height represents the first quartile?

33. Heart Transplant Waiting Times   The time spent (in days) waiting for a heart transplant for people ages 35–49 in a recent year can be approximated by a normal distribution, as shown in the graph. *(Adapted from Organ Procurement and Transplantation Network)*

(a) What waiting time represents the 5th percentile?

(b) What waiting time represents the third quartile?

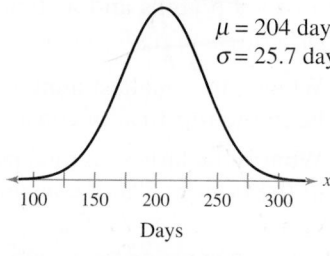

Time Spent Waiting for a Heart

$\mu = 204$ days
$\sigma = 25.7$ days

FIGURE FOR EXERCISE 33

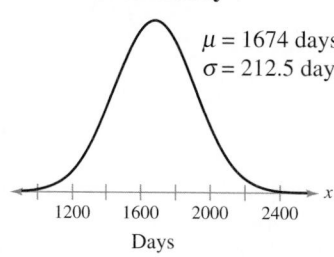

Time Spent Waiting for a Kidney

$\mu = 1674$ days
$\sigma = 212.5$ days

FIGURE FOR EXERCISE 34

34. Kidney Transplant Waiting Times   The time spent (in days) waiting for a kidney transplant for people ages 35–49 in a recent year can be approximated by a normal distribution, as shown in the graph. *(Adapted from Organ Procurement and Transplantation Network)*

(a) What waiting time represents the 80th percentile?

(b) What waiting time represents the first quartile?

35. Sleeping Times of Medical Residents   The average time spent sleeping (in hours) for a group of medical residents at a hospital can be approximated by a normal distribution, as shown in the graph. *(Source: National Institute of Occupational Safety and Health, Japan)*

(a) What is the shortest time spent sleeping that would still place a resident in the top 5% of sleeping times?

(b) Between what two values does the middle 50% of the sleep times lie?

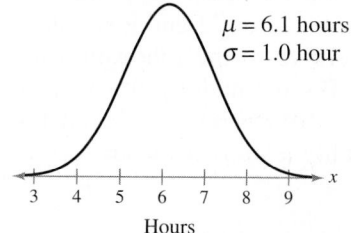

Sleeping Times of Medical Residents

$\mu = 6.1$ hours
$\sigma = 1.0$ hour

FIGURE FOR EXERCISE 35

Annual U.S. per Capita Ice Cream Consumption

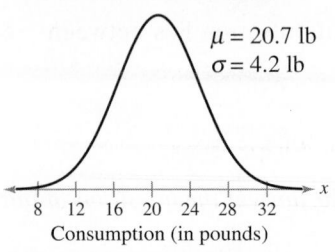

$\mu = 20.7$ lb
$\sigma = 4.2$ lb

Consumption (in pounds)

FIGURE FOR EXERCISE 36

36. (a) 15.32 pounds
    (b) 15.32 pounds and 26.08 pounds
37. 32.61 ounces
38. 7.93 ounces
39. (a) 18.88 pounds
        (*Tech:* 18.90 pounds)
    (b) 12.04 pounds
        (*Tech:* 12.05 pounds)
40. (a) 13.1 pounds
    (b) 5.79 pounds
41. Tires that wear out by 26,800 miles (*Tech:* 26,796 miles) will be replaced free of charge.
42. A = 83.52 (*Tech:* 83.53);
    B = 76.68 (*Tech:* 76.72);
    C = 67.32 (*Tech:* 67.28);
    D = 60.48 (*Tech:* 60.47)

36. Ice Cream   The annual per capita consumption of ice cream (in pounds) in the United States can be approximated by a normal distribution, as shown in the graph. *(Adapted from U.S. Department of Agriculture)*

(a) What is the largest annual per capita consumption of ice cream that can be in the bottom 10% of consumptions?

(b) Between what two values does the middle 80% of the consumptions lie?

37. Bags of Baby Carrots   The weights of bags of baby carrots are normally distributed, with a mean of 32 ounces and a standard deviation of 0.36 ounce. Bags in the upper 4.5% are too heavy and must be repackaged. What is the most a bag of baby carrots can weigh and not need to be repackaged?

38. Vending Machine   A vending machine dispenses coffee into an eight-ounce cup. The amounts of coffee dispensed into the cup are normally distributed, with a standard deviation of 0.03 ounce. You can allow the cup to overfill 1% of the time. What amount should you set as the mean amount of coffee to be dispensed?

SC   *In Exercises 39 and 40, use the StatCrunch normal calculator to find the indicated values.*

39. Apples   The annual per capita consumption of fresh apples (in pounds) in the United States can be approximated by a normal distribution, with a mean of 16.2 pounds and a standard deviation of 4 pounds. *(Adapted from U.S. Department of Agriculture)*

(a) What is the smallest annual per capita consumption of apples that can be in the top 25% of consumptions?

(b) What is the largest annual per capita consumption of apples that can be in the bottom 15% of consumptions?

40. Oranges   The annual per capita consumption of fresh oranges (in pounds) in the United States can be approximated by a normal distribution, with a mean of 9.9 pounds and a standard deviation of 2.5 pounds. *(Adapted from U.S. Department of Agriculture)*

(a) What is the smallest annual per capita consumption of oranges that can be in the top 10% of consumptions?

(b) What is the largest annual per capita consumption of oranges that can be in the bottom 5% of consumptions?

■ EXTENDING CONCEPTS

41. Writing a Guarantee   You sell a brand of automobile tire that has a life expectancy that is normally distributed, with a mean life of 30,000 miles and a standard deviation of 2500 miles. You want to give a guarantee for free replacement of tires that don't wear well. How should you word your guarantee if you are willing to replace approximately 10% of the tires?

42. Statistics Grades   In a large section of a statistics class, the points for the final exam are normally distributed, with a mean of 72 and a standard deviation of 9. Grades are to be assigned according to the following rule: the top 10% receive A's, the next 20% receive B's, the middle 40% receive C's, the next 20% receive D's, and the bottom 10% receive F's. Find the lowest score on the final exam that would qualify a student for an A, a B, a C, and a D.

Final Exam Grades

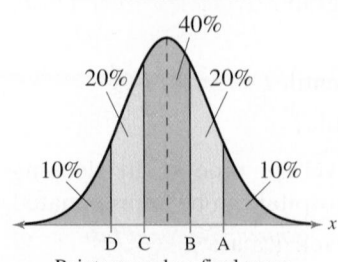

40%

20%   20%

10%   10%

D   C   B   A

Points scored on final exam

FIGURE FOR EXERCISE 42

Birth Weights in America

The National Center for Health Statistics (NCHS) keeps records of many health-related aspects of people, including the birth weights of all babies born in the United States.

The birth weight of a baby is related to its gestation period (the time between conception and birth). For a given gestation period, the birth weights can be approximated by a normal distribution. The means and standard deviations of the birth weights for various gestation periods are shown in the table below.

One of the many goals of the NCHS is to reduce the percentage of babies born with low birth weights. As you can see from the graph below, the problem of low birth weights increased from 1992 to 2006.

Gestation period	Mean birth weight	Standard deviation
Under 28 weeks	1.90 lb	1.22 lb
28 to 31 weeks	4.12 lb	1.87 lb
32 to 33 weeks	5.14 lb	1.57 lb
34 to 36 weeks	6.19 lb	1.29 lb
37 to 39 weeks	7.29 lb	1.08 lb
40 weeks	7.66 lb	1.04 lb
41 weeks	7.75 lb	1.07 lb
42 weeks and over	7.57 lb	1.11 lb

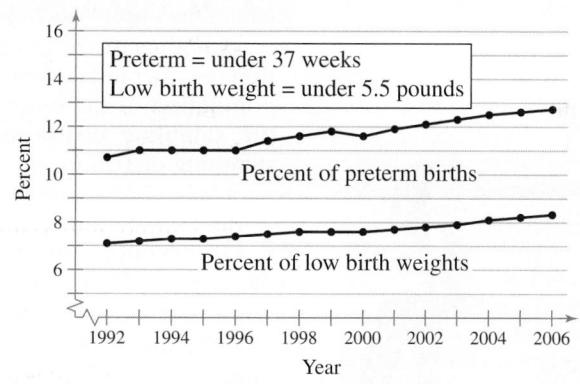

Preterm = under 37 weeks
Low birth weight = under 5.5 pounds

Percent of preterm births

Percent of low birth weights

▰ EXERCISES

1. The distributions of birth weights for three gestation periods are shown. Match the curves with the gestation periods. Explain your reasoning.

(a)

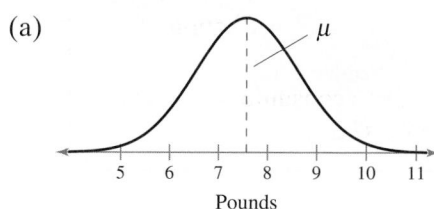

(b)

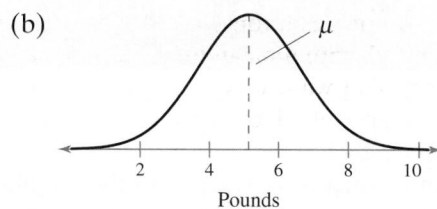

(c)

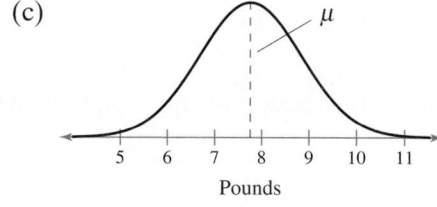

2. What percent of the babies born within each gestation period have a low birth weight (under 5.5 pounds)? Explain your reasoning.

(a) Under 28 weeks   (b) 32 to 33 weeks
(c) 40 weeks   (d) 42 weeks and over

3. Describe the weights of the top 10% of the babies born within each gestation period. Explain your reasoning.

(a) Under 28 weeks   (b) 34 to 36 weeks
(c) 41 weeks   (d) 42 weeks and over

4. For each gestation period, what is the probability that a baby will weigh between 6 and 9 pounds at birth?

(a) Under 28 weeks   (b) 28 to 31 weeks
(c) 34 to 36 weeks   (d) 37 to 39 weeks

5. A birth weight of less than 3.25 pounds is classified by the NCHS as a "very low birth weight." What is the probability that a baby has a very low birth weight for each gestation period?

(a) Under 28 weeks   (b) 28 to 31 weeks
(c) 32 to 33 weeks   (d) 37 to 39 weeks

5.4 Sampling Distributions and the Central Limit Theorem

Sampling Distributions ▸ The Central Limit Theorem ▸ Probability and the Central Limit Theorem

WHAT YOU SHOULD LEARN

▸ How to find sampling distributions and verify their properties

▸ How to interpret the Central Limit Theorem

▸ How to apply the Central Limit Theorem to find the probability of a sample mean

▸ SAMPLING DISTRIBUTIONS

In previous sections, you studied the relationship between the mean of a population and values of a random variable. In this section, you will study the relationship between a population mean and the means of samples taken from the population.

DEFINITION

A **sampling distribution** is the probability distribution of a sample statistic that is formed when samples of size n are repeatedly taken from a population. If the sample statistic is the sample mean, then the distribution is the **sampling distribution of sample means.** Every sample statistic has a sampling distribution.

For instance, consider the following Venn diagram. The rectangle represents a large population, and each circle represents a sample of size n. Because the sample entries can differ, the sample means can also differ. The mean of Sample 1 is $\bar{x}_1$; the mean of Sample 2 is $\bar{x}_2$; and so on. The sampling distribution of the sample means for samples of size n for this population consists of $\bar{x}_1$, $\bar{x}_2$, $\bar{x}_3$, and so on. If the samples are drawn with replacement, an infinite number of samples can be drawn from the population.

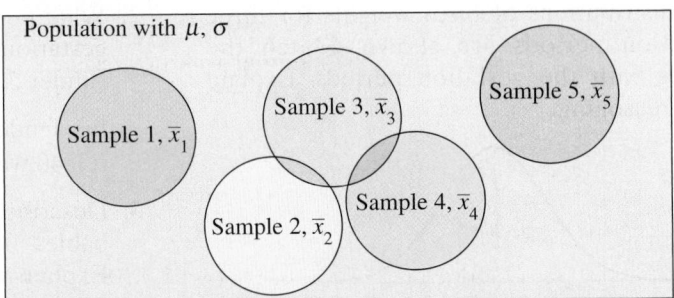

INSIGHT

Sample means can vary from one another and can also vary from the population mean. This type of variation is to be expected and is called *sampling error*.

Note to Instructor

A good exercise that can be used in conjunction with the Venn diagram is to have each student randomly select a place in the random number table and write down the next five digits horizontally. Students can verify that the population of digits {0, 1, 2, ..., 9} is uniform and has a mean of 4.5 and standard deviation of 2.87. Have each student calculate the mean of his or her sample and write that result on the board. Students can easily see that the sample means vary but are not dispersed as much as the population (range 0 to 9) is. Construct a histogram of the sample means; find the mean of these means and the standard deviation of the means. (With a TI-83/84 Plus, this takes little time even if only one student does the calculations.) Because the population standard deviation is known for this simulation, the results will be approximately normal.

PROPERTIES OF SAMPLING DISTRIBUTIONS OF SAMPLE MEANS

1. The mean of the sample means $\mu_{\bar{x}}$ is equal to the population mean μ.

$$\mu_{\bar{x}} = \mu$$

2. The standard deviation of the sample means $\sigma_{\bar{x}}$ is equal to the population standard deviation σ divided by the square root of the sample size n.

$$\sigma_{\bar{x}} = \frac{\sigma}{\sqrt{n}}$$

The standard deviation of the sampling distribution of the sample means is called the **standard error of the mean.**

**Probability Histogram
of Population of x**

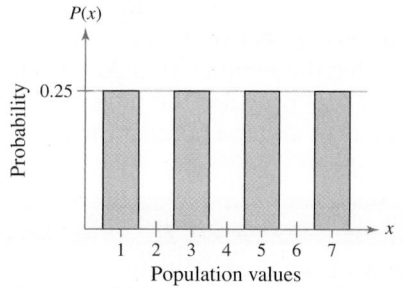

**Probability Distribution
of Sample Means**

$\bar{x}$	f	Probability
1	1	$1/16 = 0.0625$
2	2	$2/16 = 0.1250$
3	3	$3/16 = 0.1875$
4	4	$4/16 = 0.2500$
5	3	$3/16 = 0.1875$
6	2	$2/16 = 0.1250$
7	1	$1/16 = 0.0625$

**Probability Histogram of
Sampling Distribution of $\bar{x}$**

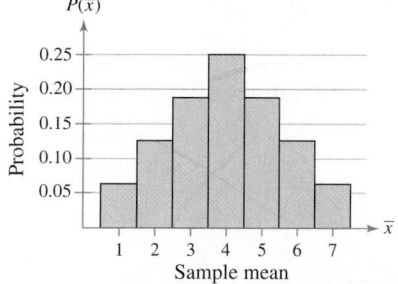

 To explore this topic further,
see Activity 5.4 on page 280.

EXAMPLE 1

▶ A Sampling Distribution of Sample Means

You write the population values $\{1, 3, 5, 7\}$ on slips of paper and put them in a box. Then you randomly choose two slips of paper, with replacement. List all possible samples of size $n = 2$ and calculate the mean of each. These means form the sampling distribution of the sample means. Find the mean, variance, and standard deviation of the sample means. Compare your results with the mean $\mu = 4$, variance $\sigma^2 = 5$, and standard deviation $\sigma = \sqrt{5} \approx 2.236$ of the population.

▶ Solution

List all 16 samples of size 2 from the population and the mean of each sample.

Sample	Sample mean, $\bar{x}$	Sample	Sample mean, $\bar{x}$
1, 1	1	5, 1	3
1, 3	2	5, 3	4
1, 5	3	5, 5	5
1, 7	4	5, 7	6
3, 1	2	7, 1	4
3, 3	3	7, 3	5
3, 5	4	7, 5	6
3, 7	5	7, 7	7

After constructing a probability distribution of the sample means, you can graph the sampling distribution using a probability histogram as shown at the left. Notice that the shape of the histogram is bell-shaped and symmetric, similar to a normal curve. The mean, variance, and standard deviation of the 16 sample means are

$$\mu_{\bar{x}} = 4$$

$$(\sigma_{\bar{x}})^2 = \frac{5}{2} = 2.5 \quad \text{and} \quad \sigma_{\bar{x}} = \sqrt{\frac{5}{2}} = \sqrt{2.5} \approx 1.581.$$

These results satisfy the properties of sampling distributions because

$$\mu_{\bar{x}} = \mu = 4 \quad \text{and} \quad \sigma_{\bar{x}} = \frac{\sigma}{\sqrt{n}} = \frac{\sqrt{5}}{\sqrt{2}} \approx 1.581.$$

▶ Try It Yourself 1

List all possible samples of $n = 3$, with replacement, from the population $\{1, 3, 5, 7\}$. Calculate the mean, variance, and standard deviation of the sample means. Compare these values with the corresponding population parameters.

a. Form all possible *samples of size 3* and find the *mean* of each.
b. Make a *probability distribution* of the sample means and find the *mean*, *variance*, and *standard deviation*.
c. *Compare* the mean, variance, and standard deviation of the sample means with those of the population.   *Answer: Page A38*

▶ THE CENTRAL LIMIT THEOREM

The Central Limit Theorem forms the foundation for the inferential branch of statistics. This theorem describes the relationship between the sampling distribution of sample means and the population that the samples are taken from. The Central Limit Theorem is an important tool that provides the information you'll need to use sample statistics to make inferences about a population mean.

Note to Instructor

The sample mean $\bar{x}$ varies from sample to sample and is a random variable. As a random variable, it has a probability distribution, called the sampling distribution of the mean. Mention that other sample statistics, such as s^2, s, and $\hat{p}$, have different sampling distributions that will be studied in the next chapter.

THE CENTRAL LIMIT THEOREM

1. If samples of size n, where $n \geq 30$, are drawn from any population with a mean μ and a standard deviation σ, then the sampling distribution of sample means approximates a normal distribution. The greater the sample size, the better the approximation.

2. If the population itself is normally distributed, then the sampling distribution of sample means is normally distributed for *any* sample size n.

In either case, the sampling distribution of sample means has a mean equal to the population mean.

$$\mu_{\bar{x}} = \mu \qquad \text{Mean}$$

The sampling distribution of sample means has a variance equal to $1/n$ times the variance of the population and a standard deviation equal to the population standard deviation divided by the square root of n.

$$\sigma_{\bar{x}}^2 = \frac{\sigma^2}{n} \qquad \text{Variance}$$

$$\sigma_{\bar{x}} = \frac{\sigma}{\sqrt{n}} \qquad \text{Standard deviation}$$

Recall that the standard deviation of the sampling distribution of the sample means, $\sigma_{\bar{x}}$, is also called the standard error of the mean.

INSIGHT

The distribution of sample means has the same mean as the population. But its standard deviation is less than the standard deviation of the population. This tells you that the distribution of sample means has the same center as the population, but it is not as spread out.

Moreover, the distribution of sample means becomes less and less spread out (tighter concentration about the mean) as the sample size n increases.

1. Any Population Distribution

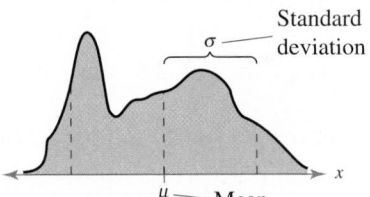

Distribution of Sample Means, $n \geq 30$

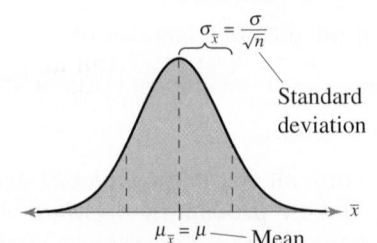

2. Normal Population Distribution

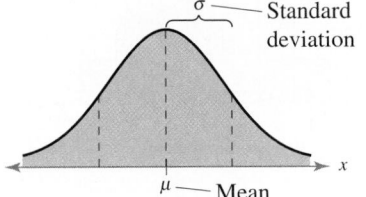

Distribution of Sample Means (any n)

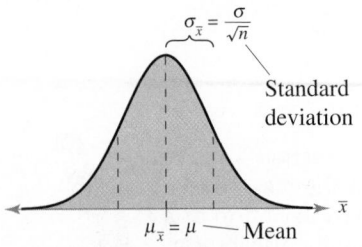

EXAMPLE 2

▶ Interpreting the Central Limit Theorem

Cellular phone bills for residents of a city have a mean of $63 and a standard deviation of $11, as shown in the following graph. Random samples of 100 cellular phone bills are drawn from this population and the mean of each sample is determined. Find the mean and standard error of the mean of the sampling distribution. Then sketch a graph of the sampling distribution of sample means. *(Adapted from JD Power and Associates)*

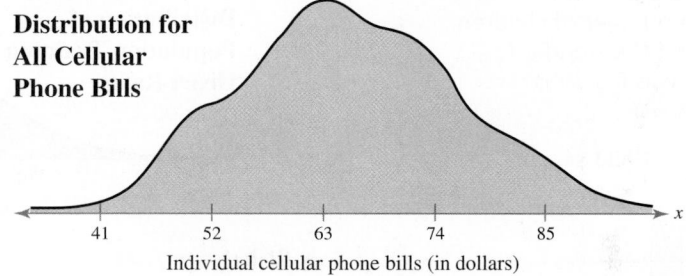

Distribution for All Cellular Phone Bills

Individual cellular phone bills (in dollars)

▶ Solution

The mean of the sampling distribution is equal to the population mean, and the standard error of the mean is equal to the population standard deviation divided by $\sqrt{n}$. So,

$$\mu_{\bar{x}} = \mu = 63 \quad \text{and} \quad \sigma_{\bar{x}} = \frac{\sigma}{\sqrt{n}} = \frac{11}{\sqrt{100}} = 1.1.$$

Interpretation   From the Central Limit Theorem, because the sample size is greater than 30, the sampling distribution can be approximated by a normal distribution with $\mu = \$63$ and $\sigma = \$1.10$, as shown in the graph below.

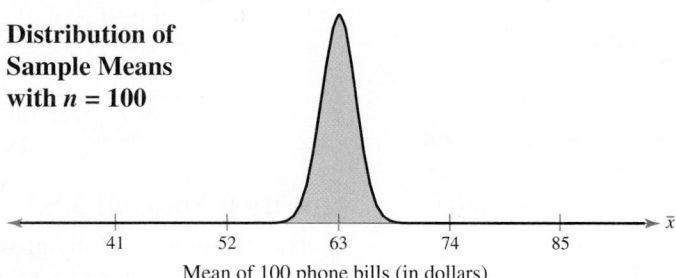

Distribution of Sample Means with $n = 100$

Mean of 100 phone bills (in dollars)

▶ Try It Yourself 2

Suppose random samples of size 64 are drawn from the population in Example 2. Find the mean and standard error of the mean of the sampling distribution. Sketch a graph of the sampling distribution and compare it with the sampling distribution in Example 2.

a. *Find* $\mu_{\bar{x}}$ and $\sigma_{\bar{x}}$.
b. *Identify* the sample size. If $n \geq 30$, *sketch* a normal curve with mean $\mu_{\bar{x}}$ and standard deviation $\sigma_{\bar{x}}$.
c. *Compare* the results with those in Example 2.   *Answer: Page A39*

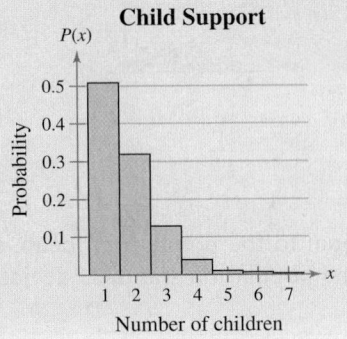
EXAMPLE 3

▶ **Interpreting the Central Limit Theorem**

Suppose the training heart rates of all 20-year-old athletes are normally distributed, with a mean of 135 beats per minute and standard deviation of 18 beats per minute, as shown in the following graph. Random samples of size 4 are drawn from this population, and the mean of each sample is determined. Find the mean and standard error of the mean of the sampling distribution. Then sketch a graph of the sampling distribution of sample means.

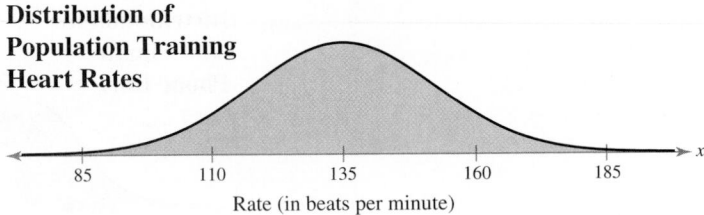

Distribution of Population Training Heart Rates

Rate (in beats per minute)

▶ **Solution**

The mean of the sampling distribution is equal to the population mean, and the standard error of the mean is equal to the population standard deviation divided by $\sqrt{n}$. So,

$$\mu_{\bar{x}} = \mu = 135 \text{ beats per minute} \quad \text{and} \quad \sigma_{\bar{x}} = \frac{\sigma}{\sqrt{n}} = \frac{18}{\sqrt{4}} = 9 \text{ beats per minute.}$$

Interpretation From the Central Limit Theorem, because the population is normally distributed, the sampling distribution of the sample means is also normally distributed, as shown in the graph below.

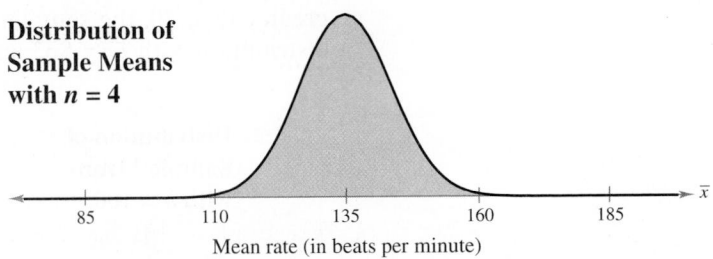

Distribution of Sample Means with *n* = 4

Mean rate (in beats per minute)

▶ **Try It Yourself 3**

The diameters of fully grown white oak trees are normally distributed, with a mean of 3.5 feet and a standard deviation of 0.2 foot, as shown in the graph below. Random samples of size 16 are drawn from this population, and the mean of each sample is determined. Find the mean and standard error of the mean of the sampling distribution. Then sketch a graph of the sampling distribution.

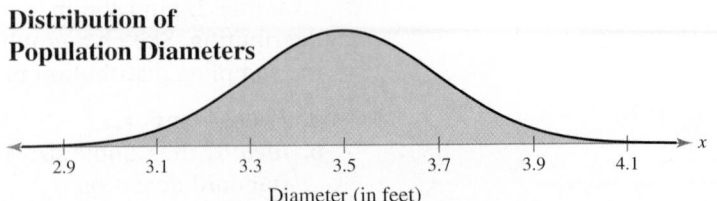

Distribution of Population Diameters

Diameter (in feet)

a. *Find* $\mu_{\bar{x}}$ *and* $\sigma_{\bar{x}}$.
b. *Sketch* a normal curve with mean $\mu_{\bar{x}}$ and standard deviation $\sigma_{\bar{x}}$.

Answer: Page A39

▶ PROBABILITY AND THE CENTRAL LIMIT THEOREM

In Section 5.2, you learned how to find the probability that a random variable x will fall in a given interval of population values. In a similar manner, you can find the probability that a sample mean $\bar{x}$ will fall in a given interval of the $\bar{x}$ sampling distribution. To transform $\bar{x}$ to a z-score, you can use the formula

$$z = \frac{\text{Value} - \text{Mean}}{\text{Standard error}} = \frac{\bar{x} - \mu_{\bar{x}}}{\sigma_{\bar{x}}} = \frac{\bar{x} - \mu}{\sigma/\sqrt{n}}.$$

EXAMPLE 4

▶ Finding Probabilities for Sampling Distributions

The graph at the right shows the lengths of time people spend driving each day. You randomly select 50 drivers ages 15 to 19. What is the probability that the mean time they spend driving each day is between 24.7 and 25.5 minutes? Assume that $\sigma = 1.5$ minutes.

▶ Solution

The sample size is greater than 30, so you can use the Central Limit Theorem to conclude that the distribution of sample means is approximately normal, with a mean and a standard deviation of

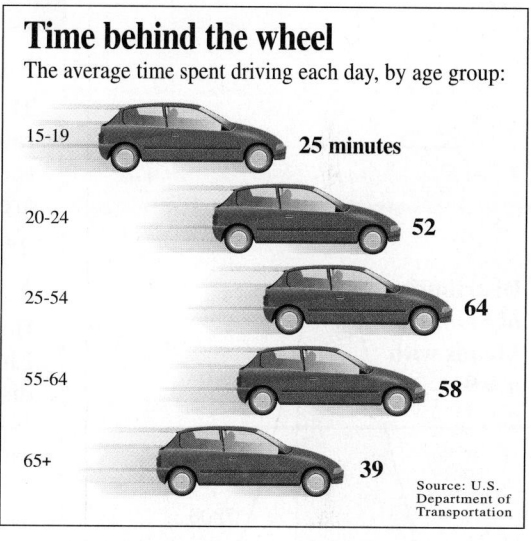

Time behind the wheel

The average time spent driving each day, by age group:

15-19   25 minutes
20-24   52
25-54   64
55-64   58
65+   39

Source: U.S. Department of Transportation

$$\mu_{\bar{x}} = \mu = 25 \text{ minutes} \qquad \text{and} \qquad \sigma_{\bar{x}} = \frac{\sigma}{\sqrt{n}} = \frac{1.5}{\sqrt{50}} \approx 0.21213 \text{ minute}.$$

The graph of this distribution is shown at the left with a shaded area between 24.7 and 25.5 minutes. The z-scores that correspond to sample means of 24.7 and 25.5 minutes are

$$z_1 = \frac{24.7 - 25}{1.5/\sqrt{50}} \approx \frac{-0.3}{0.21213} \approx -1.41 \qquad \text{and}$$

$$z_2 = \frac{25.5 - 25}{1.5/\sqrt{50}} \approx \frac{0.5}{0.21213} \approx 2.36.$$

So, the probability that the mean time the 50 people spend driving each day is between 24.7 and 25.5 minutes is

$$P(24.7 < \bar{x} < 25.5) = P(-1.41 < z < 2.36)$$
$$= P(z < 2.36) - P(z < -1.41)$$
$$= 0.9909 - 0.0793 = 0.9116.$$

Interpretation   Of the samples of 50 drivers ages 15 to 19, 91.16% will have a mean driving time that is between 24.7 and 25.5 minutes, as shown in the graph at the left. This implies that, assuming the value of $\mu = 25$ is correct, only 8.84% of such sample means will lie outside the given interval.

Distribution of Sample Means with $n = 50$

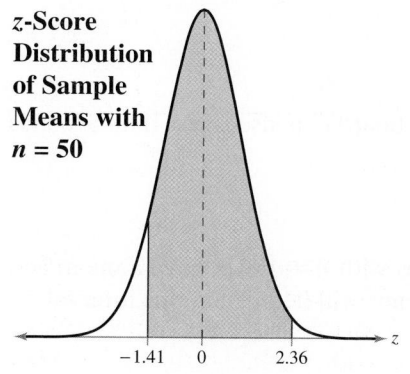

$\mu = 25$

24.2   24.6   25.0   25.4   25.8 $\bar{x}$
       24.7   25.5
Mean time (in minutes)

z-Score Distribution of Sample Means with $n = 50$

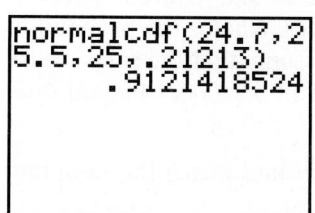

−1.41   0   2.36 z

```
normalcdf(24.7,2
5.5,25,.21213)
         .9121418524
```

In Example 4, you can use a TI-83/84 Plus to find the probability automatically once the standard error of the mean is calculated.

STUDY TIP

Before you find probabilities for intervals of the sample mean $\bar{x}$, use the Central Limit Theorem to determine the mean and the standard deviation of the sampling distribution of the sample means. That is, calculate $\mu_{\bar{x}}$ and $\sigma_{\bar{x}}$.

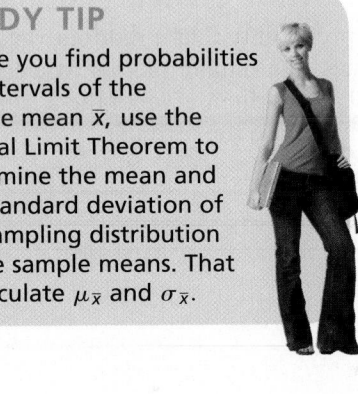

▶ **Try It Yourself 4**

You randomly select 100 drivers ages 15 to 19 from Example 4. What is the probability that the mean time they spend driving each day is between 24.7 and 25.5 minutes? Use $\mu = 25$ and $\sigma = 1.5$ minutes.

a. Use the Central Limit Theorem to *find* $\mu_{\bar{x}}$ and $\sigma_{\bar{x}}$ and *sketch* the sampling distribution of the sample means.
b. *Find the z-scores* that correspond to $\bar{x} = 24.7$ minutes and $\bar{x} = 25.5$ minutes.
c. *Find the cumulative area* that corresponds to each z-score and calculate the probability.
d. *Interpret* the results.   *Answer: Page A39*

EXAMPLE 5

▶ **Finding Probabilities for Sampling Distributions**

The mean room and board expense per year at four-year colleges is $7540. You randomly select 9 four-year colleges. What is the probability that the mean room and board is less than $7800? Assume that the room and board expenses are normally distributed with a standard deviation of $1245. *(Adapted from National Center for Education Statistics)*

▶ **Solution**

Because the population is normally distributed, you can use the Central Limit Theorem to conclude that the distribution of sample means is normally distributed, with a mean of $7540 and a standard deviation of $415.

$$\mu_{\bar{x}} = \mu = 7540 \quad \text{and} \quad \sigma_{\bar{x}} = \frac{\sigma}{\sqrt{n}} = \frac{1245}{\sqrt{9}} = 415$$

The graph of this distribution is shown at the left. The area to the left of $7800 is shaded. The z-score that corresponds to $7800 is

$$z = \frac{7800 - 7540}{1245/\sqrt{9}} = \frac{260}{415} \approx 0.63.$$

So, the probability that the mean room and board expense is less than $7800 is

$$P(\bar{x} < 7800) = P(z < 0.63)$$
$$= 0.7357.$$

Interpretation So, 73.57% of such samples with $n = 9$ will have a mean less than $7800 and 26.43% of these sample means will lie outside this interval.

▶ **Try It Yourself 5**

The average sales price of a single-family house in the United States is $290,600. You randomly select 12 single-family houses. What is the probability that the mean sales price is more than $265,000? Assume that the sales prices are normally distributed with a standard deviation of $36,000. *(Adapted from The U.S. Commerce Department)*

a. Use the Central Limit Theorem to *find* $\mu_{\bar{x}}$ and $\sigma_{\bar{x}}$ and *sketch* the sampling distribution of the sample means.
b. *Find the z-score* that corresponds to $\bar{x} = \$265,000$.
c. *Find the cumulative area* that corresponds to the z-score and calculate the probability.
d. *Interpret* the results.   *Answer: Page A39*

Distribution of Sample Means with $n = 9$

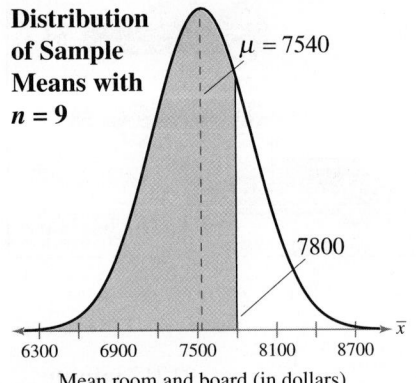

Mean room and board (in dollars)

```
normalcdf(-10000
,7800,7540,415)
          .7345085183
```

In Example 5, you can use a TI-83/84 Plus to find the probability automatically.

The Central Limit Theorem can also be used to investigate unusual events. An unusual event is one that occurs with a probability of less than 5%.

EXAMPLE 6

▶ **Finding Probabilities for *x* and $\bar{x}$**

An education finance corporation claims that the average credit card debts carried by undergraduates are normally distributed, with a mean of $3173 and a standard deviation of $1120. *(Adapted from Sallie Mae)*

1. What is the probability that a randomly selected undergraduate, who is a credit card holder, has a credit card balance less than $2700?

2. You randomly select 25 undergraduates who are credit card holders. What is the probability that their mean credit card balance is less than $2700?

3. Compare the probabilities from (1) and (2) and interpret your answer in terms of the corporation's claim.

▶ **Solution**

1. In this case, you are asked to find the probability associated with a certain value of the random variable *x*. The *z*-score that corresponds to *x* = $2700 is

$$z = \frac{x - \mu}{\sigma} = \frac{2700 - 3173}{1120} \approx -0.42.$$

So, the probability that the card holder has a balance less than $2700 is

$$P(x < 2700) = P(z < -0.42) = 0.3372.$$

2. Here, you are asked to find the probability associated with a sample mean $\bar{x}$. The *z*-score that corresponds to $\bar{x}$ = $2700 is

$$z = \frac{\bar{x} - \mu_{\bar{x}}}{\sigma_{\bar{x}}} = \frac{\bar{x} - \mu}{\sigma/\sqrt{n}} = \frac{2700 - 3173}{1120/\sqrt{25}} = \frac{-473}{224} \approx -2.11.$$

So, the probability that the mean credit card balance of the 25 card holders is less than $2700 is

$$P(\bar{x} < 2700) = P(z < -2.11) = 0.0174.$$

3. *Interpretation* Although there is about a 34% chance that an undergraduate will have a balance less than $2700, there is only about a 2% chance that the mean of a sample of 25 will have a balance less than $2700. Because there is only a 2% chance that the mean of a sample of 25 will have a balance less than $2700, this is an unusual event. So, it is possible that the corporation's claim that the mean is $3173 is incorrect.

▶ **Try It Yourself 6**

A consumer price analyst claims that prices for liquid crystal display (LCD) computer monitors are normally distributed, with a mean of $190 and a standard deviation of $48. (1) What is the probability that a randomly selected LCD computer monitor costs less than $200? (2) You randomly select 10 LCD computer monitors. What is the probability that their mean cost is less than $200? (3) Compare these two probabilities.

a. *Find the z-scores* that correspond to *x* and $\bar{x}$.
b. Use the Standard Normal Table to *find the probability* associated with each *z*-score.
c. *Compare* the probabilities and *interpret* your answer.   *Answer: Page A39*

STUDY TIP

To find probabilities for individual members of a population with a normally distributed random variable *x*, use the formula

$$z = \frac{x - \mu}{\sigma}.$$

To find probabilities for the mean $\bar{x}$ of a sample size *n*, use the formula

$$z = \frac{\bar{x} - \mu_{\bar{x}}}{\sigma_{\bar{x}}}.$$

Note to Instructor

You may want to tell students that the second formula can also be used to calculate *z*-scores for individual values. Consider a sample of *n* = 1 for an individual value.

5.4 EXERCISES

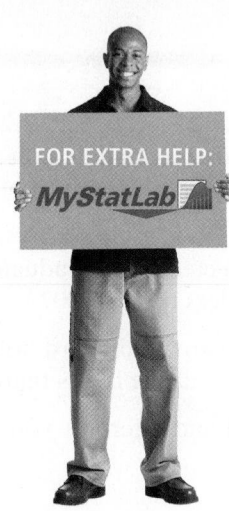

BUILDING BASIC SKILLS AND VOCABULARY

In Exercises 1–4, a population has a mean $\mu = 150$ and a standard deviation $\sigma = 25$. Find the mean and standard deviation of a sampling distribution of sample means with the given sample size n.

1. $n = 50$
2. $n = 100$
3. $n = 250$
4. $n = 1000$

True or False? In Exercises 5–8, determine whether the statement is true or false. If it is false, rewrite it as a true statement.

5. As the size of a sample increases, the mean of the distribution of sample means increases.

6. As the size of a sample increases, the standard deviation of the distribution of sample means increases.

7. A sampling distribution is normal only if the population is normal.

8. If the size of a sample is at least 30, you can use z-scores to determine the probability that a sample mean falls in a given interval of the sampling distribution.

Graphical Analysis In Exercises 9 and 10, the graph of a population distribution is shown with its mean and standard deviation. Assume that a sample size of 100 is drawn from each population. Decide which of the graphs labeled (a)–(c) would most closely resemble the sampling distribution of the sample means for each graph. Explain your reasoning.

9. The waiting time (in seconds) at a traffic signal during a red light

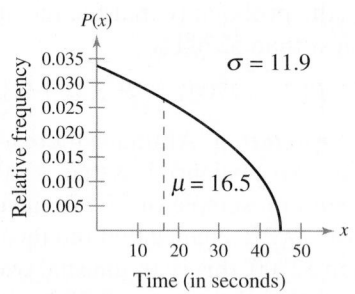

(a)

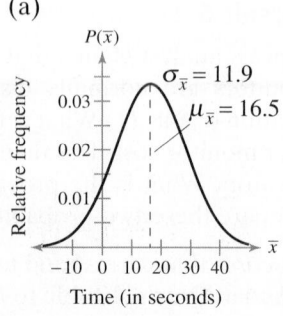

(b)

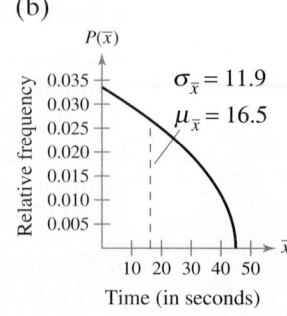

(c)

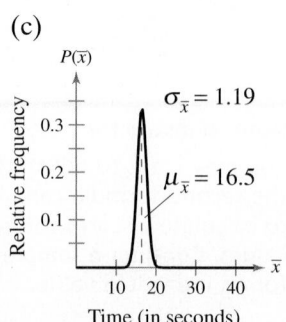

Answers

1. 150, 3.536
2. 150, 2.5
3. 150, 1.581
4. 150, 0.791
5. False. As the size of a sample increases, the mean of the distribution of sample means does not change.
6. False. As the size of the sample increases, the standard deviation of the distribution of sample means decreases.
7. False. A sampling distribution is normal if either $n \geq 30$ or the population is normal.
8. True
9. (c), because $\mu_{\bar{x}} = 16.5$, $\sigma_{\bar{x}} = 1.19$, and the graph approximates a normal curve.

10. (b), because $\mu_{\bar{x}} = 5.8$, $\sigma_{\bar{x}} = 0.23$, and the graph approximates a normal curve.

11. See Odd Answers, page A73.

$\mu = 7.5$, $\sigma \approx 5.36$

$\mu_{\bar{x}} = 7.5$, $\sigma_{\bar{x}} \approx 3.09$

The means are equal but the standard deviation of the sampling distribution is smaller.

12.

Sample	Mean
130, 130	130
130, 200	165
130, 230	180
130, 270	200
200, 130	165
200, 200	200
200, 230	215
200, 270	235
230, 130	180
230, 200	215
230, 230	230
230, 270	250
270, 130	200
270, 200	235
270, 230	250
270, 270	270

$\mu = 207.5$, $\sigma \approx 51.17$

$\mu_{\bar{x}} = 207.5$, $\sigma_{\bar{x}} \approx 36.18$

The means are equal but the standard deviation of the sampling distribution is smaller.

13. 0.9726; not unusual

14. 0.0082; unusual

15. 0.0351 (*Tech:* 0.0349); unusual

16. 0.5; not unusual

10. The annual snowfall (in feet) for a central New York state county

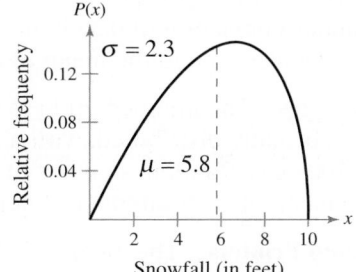

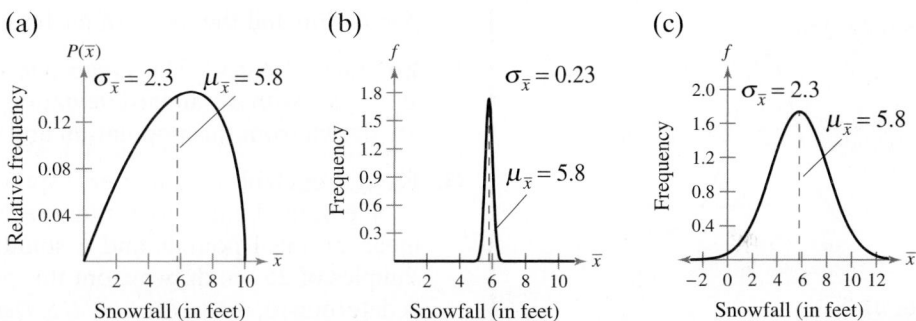

(a)   (b)   (c)

Verifying Properties of Sampling Distributions *In Exercises 11 and 12, find the mean and standard deviation of the population. List all samples (with replacement) of the given size from that population. Find the mean and standard deviation of the sampling distribution and compare them with the mean and standard deviation of the population.*

11. The number of DVDs rented by each of four families in the past month is 8, 4, 16, and 2. Use a sample size of 3.

12. Four friends paid the following amounts for their MP3 players: $200, $130, $270, and $230. Use a sample size of 2.

Finding Probabilities *In Exercises 13–16, the population mean and standard deviation are given. Find the required probability and determine whether the given sample mean would be considered unusual. If convenient, use technology to find the probability.*

13. For a sample of $n = 64$, find the probability of a sample mean being less than 24.3 if $\mu = 24$ and $\sigma = 1.25$.

14. For a sample of $n = 100$, find the probability of a sample mean being greater than 24.3 if $\mu = 24$ and $\sigma = 1.25$.

15. For a sample of $n = 45$, find the probability of a sample mean being greater than 551 if $\mu = 550$ and $\sigma = 3.7$.

16. For a sample of $n = 36$, find the probability of a sample mean being less than 12,750 or greater than 12,753 if $\mu = 12,750$ and $\sigma = 1.7$.

■ USING AND INTERPRETING CONCEPTS

Using the Central Limit Theorem *In Exercises 17–22, use the Central Limit Theorem to find the mean and standard error of the mean of the indicated sampling distribution. Then sketch a graph of the sampling distribution.*

17. 7.6, 0.101

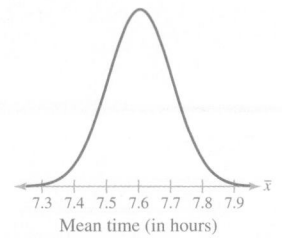

7.3 7.4 7.5 7.6 7.7 7.8 7.9
Mean time (in hours)

18. 800, 25.820

See Selected Answers, page A119.

19. 235, 13.864

207.3   235   262.7
Mean price (in dollars)

20. 47.2, 0.6

See Selected Answers, page A119.

21. 188.4, 10.9

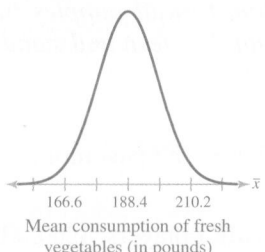

166.6   188.4   210.2
Mean consumption of fresh
vegetables (in pounds)

22. 24.2, 1.479

See Selected Answers, page A119.

23. $n = 24$: 7.6, 0.07;
$n = 36$: 7.6, 0.06

See Odd Answers, page A73.

As the sample size increases, the standard error decreases, while the mean of the sample means remains constant.

24. $n = 30$: 800, 18.257;
$n = 45$: 800, 14.907

See Selected Answers, page A119.

As the sample size increases, the standard error decreases, while the mean of the sample means remains the same.

25. 0.0003; Only 0.03% of samples of 35 specialists will have a mean salary less than $60,000. This is an extremely unusual event.

26. 0.2514 (*Tech:* 0.2503); About 25% of samples of 48 flight attendants will have a mean salary less than $56,100.

17. Employed Persons   The amounts of time employees at a large corporation work each day are normally distributed, with a mean of 7.6 hours and a standard deviation of 0.35 hour. Random samples of size 12 are drawn from the population and the mean of each sample is determined.

18. Fly Eggs   The numbers of eggs female house flies lay during their lifetimes are normally distributed, with a mean of 800 eggs and a standard deviation of 100 eggs. Random samples of size 15 are drawn from this population and the mean of each sample is determined.

19. Photo Printers   The mean price of photo printers on a website is $235 with a standard deviation of $62. Random samples of size 20 are drawn from this population and the mean of each sample is determined.

20. Employees' Ages   The mean age of employees at a large corporation is 47.2 years with a standard deviation of 3.6 years. Random samples of size 36 are drawn from this population and the mean of each sample is determined.

21. Fresh Vegetables   The per capita consumption of fresh vegetables by people in the United States in a recent year was normally distributed, with a mean of 188.4 pounds and a standard deviation of 54.5 pounds. Random samples of 25 are drawn from this population and the mean of each sample is determined. *(Adapted from U.S. Department of Agriculture)*

22. Coffee   The per capita consumption of coffee by people in the United States in a recent year was normally distributed, with a mean of 24.2 gallons and a standard deviation of 8.1 gallons. Random samples of 30 are drawn from this population and the mean of each sample is determined. *(Adapted from U.S. Department of Agriculture)*

23. Repeat Exercise 17 for samples of size 24 and 36. What happens to the mean and the standard deviation of the distribution of sample means as the size of the sample increases?

24. Repeat Exercise 18 for samples of size 30 and 45. What happens to the mean and the standard deviation of the distribution of sample means as the size of the sample increases?

Finding Probabilities   *In Exercises 25–30, find the probabilities and interpret the results. If convenient, use technology to find the probabilities.*

25. Salaries   The population mean annual salary for environmental compliance specialists is about $63,500. A random sample of 35 specialists is drawn from this population. What is the probability that the mean salary of the sample is less than $60,000? Assume $\sigma = \$6100$. *(Adapted from Salary.com)*

26. Salaries   The population mean annual salary for flight attendants is $56,275. A random sample of 48 flight attendants is selected from this population. What is the probability that the mean annual salary of the sample is less than $56,100? Assume $\sigma = \$1800$. *(Adapted from Salary.com)*

27. Gas Prices: New England   During a certain week the mean price of gasoline in the New England region was $2.714 per gallon. A random sample of 32 gas stations is drawn from this population. What is the probability that the mean price for the sample was between $2.695 and $2.725 that week? Assume $\sigma = \$0.045$. *(Adapted from U.S. Energy Information Administration)*

28. Gas Prices: California   During a certain week the mean price of gasoline in California was $2.999 per gallon. A random sample of 38 gas stations is drawn from this population. What is the probability that the mean price for the sample was between $3.010 and $3.025 that week? Assume $\sigma = \$0.049$.
(Adapted from U.S. Energy Information Administration)

27. 0.9078 (*Tech:* 0.9083); About 91% of samples of 32 gas stations that week will have a mean price between $2.695 and $2.725.

28. 0.0833 (*Tech:* 0.0827); About 8% of samples of 38 gas stations that week will have a mean price between $3.010 and $3.025.

29. ≈ 0 (*Tech:* 0.0000002); There is almost no chance that a random sample of 60 women will have a mean height greater than 66 inches. This event is almost impossible.

30. 0.3974 (*Tech:* 0.3981); About 40% of samples of 60 men will have a mean height greater than 70 inches.

31. It is more likely to select a sample of 20 women with a mean height less than 70 inches because the sample of 20 has a higher probability.

32. It is more likely to select 1 man with a height less than 65 inches because the probability is greater.

33. Yes, it is very unlikely that you would have randomly sampled 40 cans with a mean equal to 127.9 ounces because it is more than 2 standard deviations from the mean of the sample means.

34. Yes, it is very unlikely that you would have randomly sampled 40 containers with a mean equal to 64.05 ounces because it is more than 2 standard deviations away from the sample mean.

35. (a) 0.0008

    (b) Claim is inaccurate.

    (c) No, assuming the manufacturer's claim is true, because 96.25 is within 1 standard deviation of the mean for an individual board.

36. (a) 0.0179

    (b) Claim is inaccurate.

    (c) No, assuming the manufacturer's claim is true, because 10.21 is within 1 standard deviation of the mean for an individual carton.

29. **Heights of Women** The mean height of women in the United States (ages 20–29) is 64.3 inches. A random sample of 60 women in this age group is selected. What is the probability that the mean height for the sample is greater than 66 inches? Assume $\sigma = 2.6$ inches. (*Source: National Center for Health Statistics*)

30. **Heights of Men** The mean height of men in the United States (ages 20–29) is 69.9 inches. A random sample of 60 men in this age group is selected. What is the probability that the mean height for the sample is greater than 70 inches? Assume $\sigma = 3.0$ inches. (*Source: National Center for Health Statistics*)

31. **Which Is More Likely?** Assume that the heights given in Exercise 29 are normally distributed. Are you more likely to randomly select 1 woman with a height less than 70 inches or are you more likely to select a sample of 20 women with a mean height less than 70 inches? Explain.

32. **Which Is More Likely?** Assume that the heights given in Exercise 30 are normally distributed. Are you more likely to randomly select 1 man with a height less than 65 inches or are you more likely to select a sample of 15 men with a mean height less than 65 inches? Explain.

33. **Make a Decision** A machine used to fill gallon-sized paint cans is regulated so that the amount of paint dispensed has a mean of 128 ounces and a standard deviation of 0.20 ounce. You randomly select 40 cans and carefully measure the contents. The sample mean of the cans is 127.9 ounces. Does the machine need to be reset? Explain your reasoning.

34. **Make a Decision** A machine used to fill half-gallon-sized milk containers is regulated so that the amount of milk dispensed has a mean of 64 ounces and a standard deviation of 0.11 ounce. You randomly select 40 containers and carefully measure the contents. The sample mean of the containers is 64.05 ounces. Does the machine need to be reset? Explain your reasoning.

35. **Lumber Cutter** Your lumber company has bought a machine that automatically cuts lumber. The seller of the machine claims that the machine cuts lumber to a mean length of 8 feet (96 inches) with a standard deviation of 0.5 inch. Assume the lengths are normally distributed. You randomly select 40 boards and find that the mean length is 96.25 inches.

    (a) Assuming the seller's claim is correct, what is the probability that the mean of the sample is 96.25 inches or more?

    (b) Using your answer from part (a), what do you think of the seller's claim?

    (c) Would it be unusual to have an individual board with a length of 96.25 inches? Why or why not?

36. **Ice Cream Carton Weights** A manufacturer claims that the mean weight of its ice cream cartons is 10 ounces with a standard deviation of 0.5 ounce. Assume the weights are normally distributed. You test 25 cartons and find their mean weight is 10.21 ounces.

    (a) Assuming the manufacturer's claim is correct, what is the probability that the mean of the sample is 10.21 ounces or more?

    (b) Using your answer from part (a), what do you think of the manufacturer's claim?

    (c) Would it be unusual to have an individual carton with a weight of 10.21 ounces? Why or why not?

37. (a) 0.0002

(b) Claim is inaccurate.

(c) No, assuming the manufacturer's claim is true, because 49,721 is within 1 standard deviation of the mean for an individual tire.

38. (a) 0.0068 (*Tech:* 0.0067)

(b) Claim is inaccurate.

(c) No, assuming the manufacturer's claim is true, because 37,650 is within 1 standard deviation of the mean for an individual brake pad.

39. No, because the *z*-score (0.88) is not unusual.

40. It is very unlikely the machine is calibrated to produce a bolt with a mean diameter of 4 inches.

37. Life of Tires A manufacturer claims that the life span of its tires is 50,000 miles. You work for a consumer protection agency and you are testing this manufacturer's tires. Assume the life spans of the tires are normally distributed. You select 100 tires at random and test them. The mean life span is 49,721 miles. Assume $\sigma = 800$ miles.

(a) Assuming the manufacturer's claim is correct, what is the probability that the mean of the sample is 49,721 miles or less?

(b) Using your answer from part (a), what do you think of the manufacturer's claim?

(c) Would it be unusual to have an individual tire with a life span of 49,721 miles? Why or why not?

38. Brake Pads A brake pad manufacturer claims its brake pads will last for 38,000 miles. You work for a consumer protection agency and you are testing this manufacturer's brake pads. Assume the life spans of the brake pads are normally distributed. You randomly select 50 brake pads. In your tests, the mean life of the brake pads is 37,650 miles. Assume $\sigma = 1000$ miles.

(a) Assuming the manufacturer's claim is correct, what is the probability that the mean of the sample is 37,650 miles or less?

(b) Using your answer from part (a), what do you think of the manufacturer's claim?

(c) Would it be unusual to have an individual brake pad last for 37,650 miles? Why or why not?

▉ EXTENDING CONCEPTS

39. SAT Scores The mean critical reading SAT score is 501, with a standard deviation of 112. A particular high school claims that its students have unusually high critical reading SAT scores. A random sample of 50 students from this school was selected, and the mean critical reading SAT score was 515. Is the high school justified in its claim? Explain. *(Source: The College Board)*

40. Machine Calibrations A machine in a manufacturing plant is calibrated to produce a bolt that has a mean diameter of 4 inches and a standard deviation of 0.5 inch. An engineer takes a random sample of 100 bolts from this machine and finds the mean diameter is 4.2 inches. What are some possible consequences of these findings?

Finite Correction Factor *The formula for the standard error of the mean*

$$\sigma_{\bar{x}} = \frac{\sigma}{\sqrt{n}}$$

*given in the Central Limit Theorem is based on an assumption that the population has infinitely many members. This is the case whenever sampling is done with replacement (each member is put back after it is selected), because the sampling process could be continued indefinitely. The formula is also valid if the sample size is small in comparison with the population. However, when sampling is done without replacement and the sample size n is more than 5% of the finite population of size N ($n/N > 0.05$), there is a finite number of possible samples. A **finite correction factor**,*

$$\sqrt{\frac{N-n}{N-1}}$$

should be used to adjust the standard error. The sampling distribution of the sample means will be normal with a mean equal to the population mean, and the standard error of the mean will be

41. Yes, the finite correction factor should be used; 0.0003

42. Yes, the finite correction factor should be used; ≈ 1 (*Tech:* 0.9998)

43. See Odd Answers, page A73.

44. See Selected Answers, page A119.

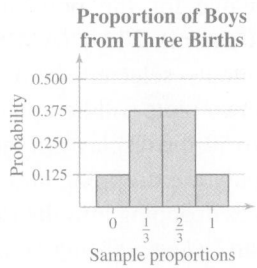

Proportion of Boys from Three Births

The spread of the histogram for each proportion is equal to the number of occurrences of 0, 1, 2, and 3 boys, respectively, from the binomial distribution.

45. See Odd Answers, page A73.

The sample means are equal to the proportions.

46. See Selected Answers, page A119.

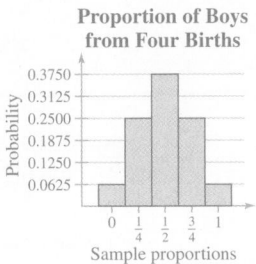

Proportion of Boys from Four Births

47. 0.0446 (*Tech:* 0.0441); About 4.5% (*Tech:* 4.4%) of samples of 105 female heart transplant patients will have a mean 3-year survival rate of less than 70%. Because the probability is less than 0.05, this is an unusual event.

$$\sigma_{\bar{x}} = \frac{\sigma}{\sqrt{n}} \sqrt{\frac{N-n}{N-1}}.$$

In Exercises 41 and 42, determine if the finite correction factor should be used. If so, use it in your calculations when you find the probability.

41. Gas Prices In a sample of 900 gas stations, the mean price of regular gasoline at the pump was $2.702 per gallon and the standard deviation was $0.009 per gallon. A random sample of size 55 is drawn from this population. What is the probability that the mean price per gallon is less than $2.698? *(Adapted from U.S. Department of Energy)*

42. Old Faithful In a sample of 500 eruptions of the Old Faithful geyser at Yellowstone National Park, the mean duration of the eruptions was 3.32 minutes and the standard deviation was 1.09 minutes. A random sample of size 30 is drawn from this population. What is the probability that the mean duration of eruptions is between 2.5 minutes and 4 minutes? *(Adapted from Yellowstone National Park)*

Sampling Distribution of Sample Proportions   *The sample mean is not the only statistic with a sampling distribution. Every sample statistic, such as the sample median, the sample standard deviation, and the sample proportion, has a sampling distribution. For a random sample of size n, the **sample proportion** is the number of individuals in the sample with a specified characteristic divided by the sample size. The **sampling distribution of sample proportions** is the distribution formed when sample proportions of size n are repeatedly taken from a population where the probability of an individual with a specified characteristic is p.*

In Exercises 43–46, suppose three births are randomly selected. There are two equally possible outcomes for each birth, a boy (b) or a girl (g). The number of boys can equal 0, 1, 2, or 3. These correspond to sample proportions of 0, 1/3, 2/3, and 1.

43. List the eight possible samples that can result from randomly selecting three births. For instance, let bbb represent a sample of three boys. Make a table that shows each sample, the number of boys in each sample, and the proportion of boys in each sample.

44. Use the table from Exercise 43 to construct the sampling distribution of the sample proportion of boys from three births. Graph the sampling distribution using a probability histogram. What do you notice about the spread of the histogram as compared to the binomial probability distribution for the number of boys in each sample?

45. Let $x = 1$ represent a boy and $x = 0$ represent a girl. Using these values, find the sample mean for each sample. What do you notice?

46. Construct a sampling distribution of the sample proportion of boys from four births.

47. Heart Transplants About 77% of all female heart transplant patients will survive for at least 3 years. One hundred five female heart transplant patients are randomly selected. What is the probability that the sample proportion surviving for at least 3 years will be less than 70%? Interpret your results. Assume the sampling distribution of sample proportions is a normal distribution. The mean of the sample proportion is equal to the population proportion p, and the standard deviation is equal to $\sqrt{\dfrac{pq}{n}}$. *(Source: American Heart Association)*

ACTIVITY 5.4   Sampling Distributions

APPLET

The *sampling distributions* applet allows you to investigate sampling distributions by repeatedly taking samples from a population. The top plot displays the distribution of a population. Several options are available for the population distribution (Uniform, Bell-shaped, Skewed, Binary, and Custom). When SAMPLE is clicked, N random samples of size n will be repeatedly selected from the population. The sample statistics specified in the bottom two plots will be updated for each sample. If N is set to 1 and n is less than or equal to 50, the display will show, in an animated fashion, the points selected from the population dropping into the second plot and the corresponding summary statistic values dropping into the third and fourth plots. Click RESET to stop an animation and clear existing results. Summary statistics for each plot are shown in the panel at the left of the plot.

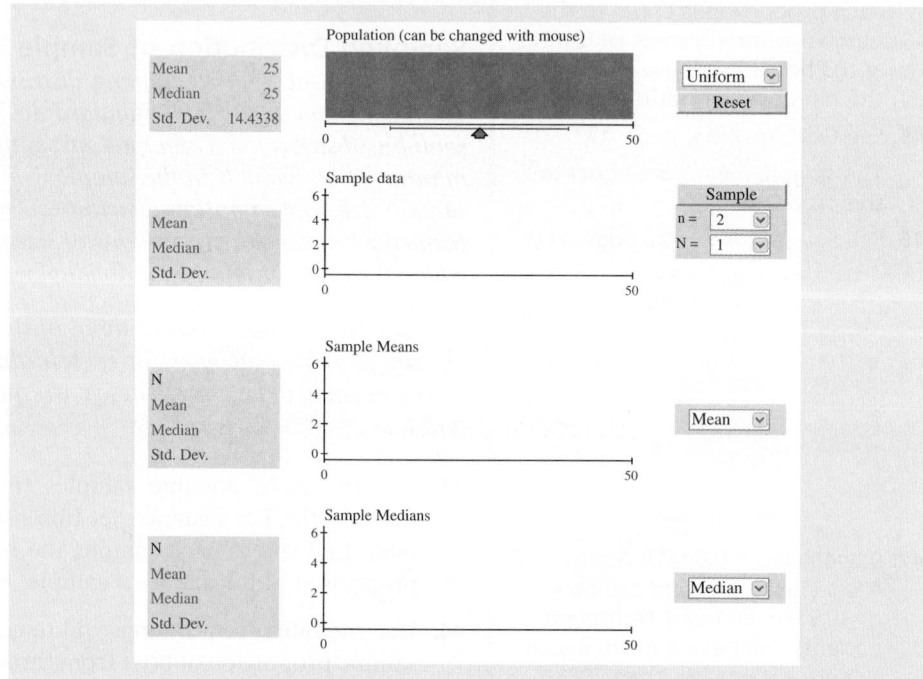

■ Explore

Step 1   Specify a distribution.
Step 2   Specify values of n and N.
Step 3   Specify what to display in the bottom two graphs.
Step 4   Click SAMPLE to generate the sampling distributions.

■ Draw Conclusions

APPLET

1. Run the simulation using $n = 30$ and $N = 10$ for a uniform, a bell-shaped, and a skewed distribution. What is the mean of the sampling distribution of the sample means for each distribution? For each distribution, is this what you would expect?

2. Run the simulation using $n = 50$ and $N = 10$ for a bell-shaped distribution. What is the standard deviation of the sampling distribution of the sample means? According to the formula, what should the standard deviation of the sampling distribution of the sample means be? Is this what you would expect?

5.5 Normal Approximations to Binomial Distributions

WHAT YOU SHOULD LEARN

▸ How to decide when a normal distribution can approximate a binomial distribution

▸ How to find the continuity correction

▸ How to use a normal distribution to approximate binomial probabilities

Approximating a Binomial Distribution ▸ Continuity Correction ▸ Approximating Binomial Probabilities

▸ APPROXIMATING A BINOMIAL DISTRIBUTION

In Section 4.2, you learned how to find binomial probabilities. For instance, if a surgical procedure has an 85% chance of success and a doctor performs the procedure on 10 patients, it is easy to find the probability of exactly two successful surgeries.

But what if the doctor performs the surgical procedure on 150 patients and you want to find the probability of *fewer than 100* successful surgeries? To do this using the techniques described in Section 4.2, you would have to use the binomial formula 100 times and find the sum of the resulting probabilities. This approach is not practical, of course. A better approach is to use a normal distribution to approximate the binomial distribution.

NORMAL APPROXIMATION TO A BINOMIAL DISTRIBUTION

If $np \geq 5$ and $nq \geq 5$, then the binomial random variable x is approximately normally distributed, with mean

$$\mu = np$$

and standard deviation

$$\sigma = \sqrt{npq}$$

where n is the number of independent trials, p is the probability of success in a single trial, and q is the probability of failure in a single trial.

To see why this result is valid, look at the following binomial distributions for $p = 0.25$, $q = 1 - 0.25 = 0.75$, and $n = 4$, $n = 10$, $n = 25$, and $n = 50$. Notice that as n increases, the histogram approaches a normal curve.

STUDY TIP

Properties of a binomial experiment

• n independent trials

• Two possible outcomes: success or failure

• Probability of success is p; probability of failure is $q = 1 - p$

• p is constant for each trial

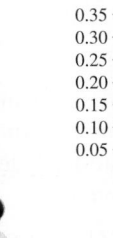

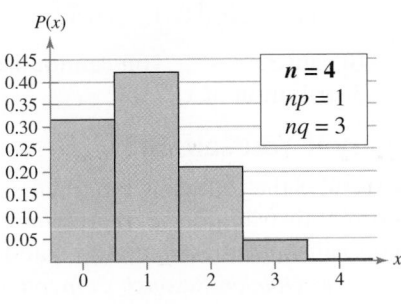

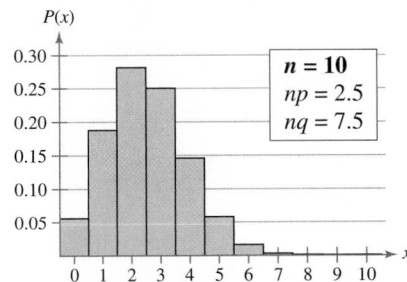

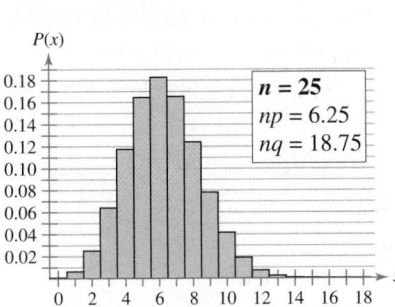

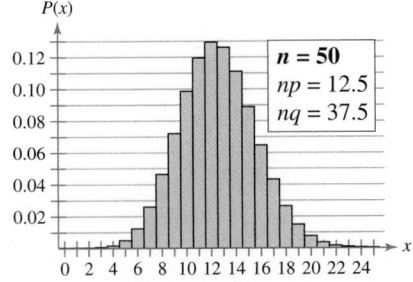

EXAMPLE 1

▶ Approximating a Binomial Distribution

Two binomial experiments are listed. Decide whether you can use the normal distribution to approximate x, the number of people who reply yes. If you can, find the mean and standard deviation. If you cannot, explain why. *(Source: Opinion Research Corporation)*

1. Sixty-two percent of adults in the United States have an HDTV in their home. You randomly select 45 adults in the United States and ask them if they have an HDTV in their home.

2. Twelve percent of adults in the United States who do not have an HDTV in their home are planning to purchase one in the next two years. You randomly select 30 adults in the United States who do not have an HDTV and ask them if they are planning to purchase one in the next two years.

▶ Solution

1. In this binomial experiment, $n = 45$, $p = 0.62$, and $q = 0.38$. So,

$$np = 45(0.62) = 27.9$$

and

$$nq = 45(0.38) = 17.1.$$

Because np and nq are greater than 5, you can use a normal distribution with

$$\mu = np = 27.9$$

and

$$\sigma = \sqrt{npq} = \sqrt{45 \cdot 0.62 \cdot 0.38} \approx 3.26$$

to approximate the distribution of x.

2. In this binomial experiment, $n = 30$, $p = 0.12$, and $q = 0.88$. So,

$$np = 30(0.12) = 3.6$$

and

$$nq = 30(0.88) = 26.4.$$

Because $np < 5$, you cannot use a normal distribution to approximate the distribution of x.

▶ Try It Yourself 1

Consider the following binomial experiment. Decide whether you can use the normal distribution to approximate x, the number of people who reply yes. If you can, find the mean and standard deviation. If you cannot, explain why. *(Source: Opinion Research Corporation)*

Five percent of adults in the United States are planning to purchase a 3D TV in the next two years. You randomly select 125 adults in the United States and ask them if they are planning to purchase a 3D TV in the next two years.

a. *Identify* n, p, and q.
b. *Find* the products np and nq.
c. *Decide* whether you can use a normal distribution to approximate x.
d. *Find* the mean μ and standard deviation σ, if appropriate.

Answer: Page A39

Note to Instructor

For technology users who are not limited to $n = 20$ in the table, many more binomial problems can be calculated without using a normal distribution approximation. However, students should be shown that even technology has limitations. The TI-83/84 Plus cannot calculate the cumulative binomial probability for $n = 10,000$, $p = 0.4$, and $x = 9000$, but that probability can be calculated using a normal approximation. Likewise, depending on the version of MINITAB or Excel you are using, there are memory limitations for the binomial distribution.

▶ **CONTINUITY CORRECTION**

A binomial distribution is discrete and can be represented by a probability histogram. To calculate *exact* binomial probabilities, you can use the binomial formula for each value of x and add the results. Geometrically, this corresponds to adding the areas of bars in the probability histogram. Remember that each bar has a width of one unit and x is the midpoint of the interval.

When you use a *continuous* normal distribution to approximate a binomial probability, you need to move 0.5 unit to the left and right of the midpoint to include all possible x-values in the interval. When you do this, you are making a **continuity correction.**

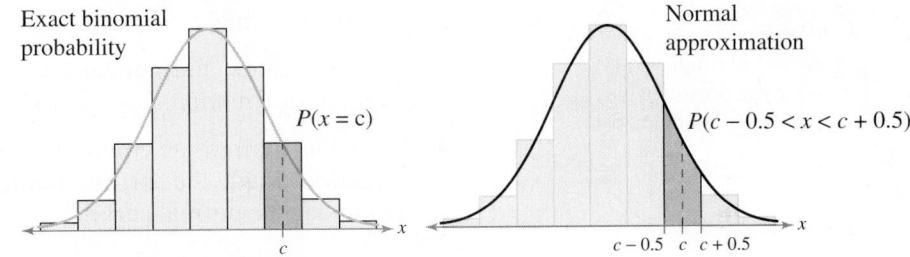

STUDY TIP

To use a continuity correction, simply subtract 0.5 from the lowest value and/or add 0.5 to the highest.

EXAMPLE 2

▶ **Using a Continuity Correction**

Use a continuity correction to convert each of the following binomial intervals to a normal distribution interval.

1. The probability of getting between 270 and 310 successes, inclusive

2. The probability of getting at least 158 successes

3. The probability of getting fewer than 63 successes

▶ **Solution**

1. The discrete midpoint values are $270, 271, \ldots, 310$. The corresponding interval for the continuous normal distribution is

$$269.5 < x < 310.5.$$

2. The discrete midpoint values are $158, 159, 160, \ldots$. The corresponding interval for the continuous normal distribution is

$$x > 157.5.$$

3. The discrete midpoint values are $\ldots, 60, 61, 62$. The corresponding interval for the continuous normal distribution is

$$x < 62.5.$$

▶ **Try It Yourself 2**

Use a continuity correction to convert each of the following binomial intervals to a normal distribution interval.

1. The probability of getting between 57 and 83 successes, inclusive
2. The probability of getting at most 54 successes

a. List the *midpoint values* for the binomial probability.
b. Use a *continuity correction* to write the normal distribution interval.

Answer: Page A39

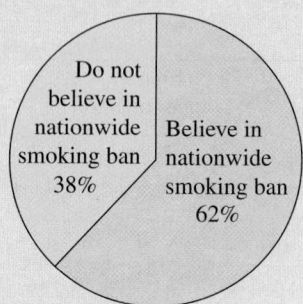

Do not believe in nationwide smoking ban 38% | Believe in nationwide smoking ban 62%

Assume that this survey is a true indication of the proportion of the population who say there should be a nationwide ban on smoking in all public places. If you sampled 50 adults at random, what is the probability that between 25 and 30, inclusive, would say there should be a nationwide ban on smoking in all public places?

If 50 adults were sampled at random, the probability that between 25 and 30, inclusive, would say there should be a nationwide ban on smoking in all public places is 0.4117 (*Tech:* 0.4130).

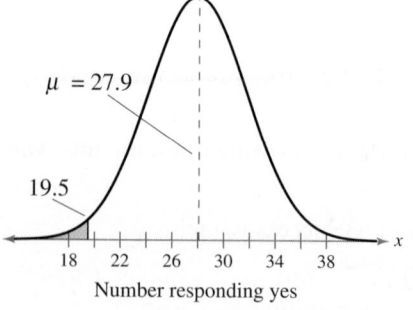

$\mu = 27.9$

19.5

18 22 26 30 34 38 → x

Number responding yes

▶ APPROXIMATING BINOMIAL PROBABILITIES

GUIDELINES

Using a Normal Distribution to Approximate Binomial Probabilities

IN WORDS | IN SYMBOLS

1. Verify that a binomial distribution applies. — Specify n, p, and q.

2. Determine if you can use a normal distribution to approximate x, the binomial variable. — Is $np \geq 5$? Is $nq \geq 5$?

3. Find the mean μ and standard deviation σ for the distribution. — $\mu = np$, $\sigma = \sqrt{npq}$

4. Apply the appropriate continuity correction. Shade the corresponding area under the normal curve. — Add or subtract 0.5 from endpoints.

5. Find the corresponding z-score(s). — $z = \dfrac{x - \mu}{\sigma}$

6. Find the probability. — Use the Standard Normal Table.

EXAMPLE 3

▶ **Approximating a Binomial Probability**

Sixty-two percent of adults in the United States have an HDTV in their home. You randomly select 45 adults in the United States and ask them if they have an HDTV in their home. What is the probability that fewer than 20 of them respond yes? (*Source: Opinion Research Corporation*)

▶ **Solution**

From Example 1, you know that you can use a normal distribution with $\mu = 27.9$ and $\sigma \approx 3.26$ to approximate the binomial distribution. Remember to apply the continuity correction for the value of x. In the binomial distribution, the possible midpoint values for "fewer than 20" are

... 17, 18, 19.

To use a normal distribution, add 0.5 to the right-hand boundary 19 to get $x = 19.5$. The graph at the left shows a normal curve with $\mu = 27.9$ and $\sigma \approx 3.26$ and a shaded area to the left of 19.5. The z-score that corresponds to $x = 19.5$ is

$$z = \frac{19.5 - 27.9}{3.26}$$

$$\approx -2.58.$$

Using the Standard Normal Table,

$$P(z < -2.58) = 0.0049.$$

Interpretation The probability that fewer than 20 people respond yes is approximately 0.0049, or about 0.49%.

▶ **Try It Yourself 3**

Five percent of adults in the United States are planning to purchase a 3D TV in the next two years. You randomly select 125 adults in the United States and ask them if they are planning to purchase a 3D TV in the next two years. What is the probability that more than 9 respond yes? (See Try It Yourself 1.) *(Source: Opinion Research Corporation)*

a. *Determine* whether you can use a normal distribution to approximate the binomial variable (see part (c) of Try It Yourself 1).
b. Find the *mean μ* and the *standard deviation σ* for the distribution (see part (d) of Try It Yourself 1).
c. Apply a *continuity correction* to rewrite $P(x > 9)$ and sketch a graph.
d. *Find* the corresponding *z*-score.
e. Use the Standard Normal Table to *find the area* to the left of *z* and *calculate the probability*.   *Answer: Page A39*

EXAMPLE 4

▶ **Approximating a Binomial Probability**

Fifty-eight percent of adults say that they never wear a helmet when riding a bicycle. You randomly select 200 adults in the United States and ask them if they wear a helmet when riding a bicycle. What is the probability that at least 120 adults will say they never wear a helmet when riding a bicycle? *(Source: Consumer Reports National Research Center)*

▶ **Solution**   Because $np = 200 \cdot 0.58 = 116$ and $nq = 200 \cdot 0.42 = 84$, the binomial variable x is approximately normally distributed, with

$$\mu = np = 116 \quad \text{and} \quad \sigma = \sqrt{npq} = \sqrt{200 \cdot 0.58 \cdot 0.42} \approx 6.98.$$

Using the continuity correction, you can rewrite the discrete probability $P(x \geq 120)$ as the continuous probability $P(x \geq 119.5)$. The graph shows a normal curve with $\mu = 116$, $\sigma = 6.98$, and a shaded area to the right of 119.5. The *z*-score that corresponds to 119.5 is

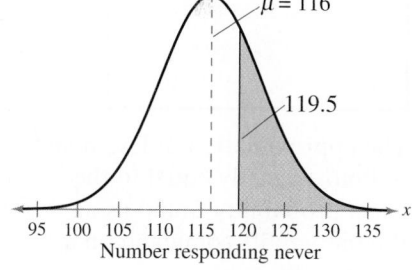

$$z = \frac{119.5 - 116}{6.98} \approx 0.50.$$

So, the probability that at least 120 will say yes is approximately

$$P(x \geq 119.5) = P(z \geq 0.50)$$
$$= 1 - P(z \leq 0.50) = 1 - 0.6915 = 0.3085.$$

▶ **Try It Yourself 4**

In Example 4, what is the probability that at most 100 adults will say they never wear a helmet when riding a bicycle?

a. *Determine* whether you can use a normal distribution to approximate the binomial variable (see Example 4).
b. Find the *mean μ* and the *standard deviation σ* for the distribution (see Example 4).
c. Apply a *continuity correction* to rewrite $P(x \leq 100)$ and sketch a graph.
d. *Find* the corresponding *z*-score.
e. Use the Standard Normal Table to *find the area* to the left of *z* and *calculate the probability*.   *Answer: Page A39*

STUDY TIP

In a discrete distribution, there is a difference between $P(x \geq c)$ and $P(x > c)$. This is true because the probability that *x* is exactly *c* is not 0. In a continuous distribution, however, there is no difference between $P(x \geq c)$ and $P(x > c)$ because the probability that *x* is exactly *c* is 0.

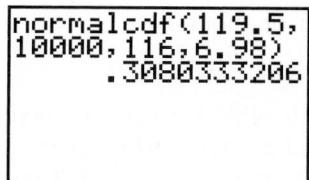

```
normalcdf(119.5,
10000,116,6.98)
         .3080333206
```

In Example 4, you can use a TI-83/84 Plus to find the probability once the mean, standard deviation, and continuity correction are calculated. Use 10,000 for the upper bound.

EXAMPLE 5

▶ **Approximating a Binomial Probability**

A survey reports that 62% of Internet users use Windows® Internet Explorer® as their browser. You randomly select 150 Internet users and ask them whether they use Internet Explorer® as their browser. What is the probability that exactly 96 will say yes? *(Source: Net Applications)*

▶ **Solution**

Because $np = 150 \cdot 0.62 = 93$ and $nq = 150 \cdot 0.38 = 57$, the binomial variable x is approximately normally distributed, with

$$\mu = np = 93 \quad \text{and} \quad \sigma = \sqrt{npq} = \sqrt{150 \cdot 0.62 \cdot 0.38} \approx 5.94.$$

Using the continuity correction, you can rewrite the discrete probability $P(x = 96)$ as the continuous probability $P(95.5 < x < 96.5)$. The graph shows a normal curve with $\mu = 93$, $\sigma = 5.94$, and a shaded area between 95.5 and 96.5.

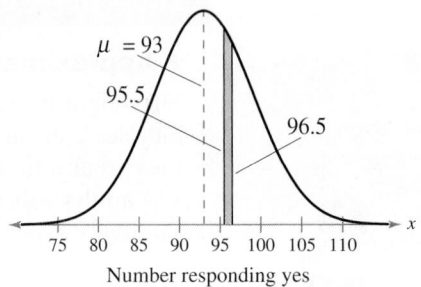

Number responding yes

The z-scores that correspond to 95.5 and 96.5 are

$$z_1 = \frac{95.5 - 93}{5.94} \approx 0.42 \quad \text{and} \quad z_2 = \frac{96.5 - 93}{5.94} \approx 0.59.$$

So, the probability that exactly 96 Internet users will say they use Internet Explorer® is

$$
\begin{aligned}
P(95.5 < x < 96.5) &= P(0.42 < z < 0.59) \\
&= P(z < 0.59) - P(z < 0.42) \\
&= 0.7224 - 0.6628 \\
&= 0.0596.
\end{aligned}
$$

Interpretation The probability that exactly 96 of the Internet users will say they use Internet Explorer® is approximately 0.0596, or about 6%.

▶ **Try It Yourself 5**

A survey reports that 24% of Internet users use Mozilla® Firefox® as their browser. You randomly select 150 Internet users and ask them whether they use Firefox® as their browser. What is the probability that exactly 27 will say yes? *(Source: Net Applications)*

a. *Determine* whether you can use a normal distribution to approximate the binomial variable.
b. Find the *mean* μ and the *standard deviation* σ for the distribution.
c. Apply a *continuity correction* to rewrite $P(x = 27)$ and sketch a graph.
d. *Find* the corresponding z-scores.
e. Use the Standard Normal Table to *find the area* to the left of each z-score and *calculate the probability*.

Answer: Page A39

```
binompdf(150,.62
,96)
    .0595828329
```

The approximation in Example 5 is almost exactly equal to the exact probability found using the binompdf(command on a TI-83/84 Plus.

5.5 EXERCISES

FOR EXTRA HELP:

MyStatLab

1. Properties of a binomial experiment:

(1) The experiment is repeated for a fixed number of independent trials.

(2) There are two possible outcomes: success or failure.

(3) The probability of success is the same for each trial.

(4) The random variable x counts the number of successful trials.

2. $np \geq 5$ and $nq \geq 5$

3. Cannot use normal distribution.

4. Cannot use normal distribution.

5. Cannot use normal distribution.

6. Can use normal distribution.

7. Cannot use normal distribution because $nq < 5$.

8. Can use normal distribution; $\mu = 12.6$, $\sigma \approx 2.16$

9. Can use normal distribution; $\mu = 27.5$, $\sigma \approx 3.52$

10. Can use normal distribution; $\mu = 5.7$, $\sigma \approx 2.15$

11. Cannot use normal distribution because $nq < 5$.

12. Can use normal distribution; $\mu = 9.15$, $\sigma \approx 1.89$

13. a

14. d

15. c

16. b

■ BUILDING BASIC SKILLS AND VOCABULARY

1. What are the properties of a binomial experiment?

2. What are the conditions for using a normal distribution to approximate a binomial distribution?

In Exercises 3–6, the sample size n, probability of success p, and probability of failure q are given for a binomial experiment. Decide whether you can use a normal distribution to approximate the random variable x.

3. $n = 24$, $p = 0.85$, $q = 0.15$

4. $n = 15$, $p = 0.70$, $q = 0.30$

5. $n = 18$, $p = 0.90$, $q = 0.10$

6. $n = 20$, $p = 0.65$, $q = 0.35$

Approximating a Binomial Distribution   *In Exercises 7–12, a binomial experiment is given. Decide whether you can use a normal distribution to approximate the binomial distribution. If you can, find the mean and standard deviation. If you cannot, explain why.*

7. **House Contract**   A survey of U.S. adults found that 85% read every word or at least enough to understand a contract for buying or selling a home before signing. You randomly select 10 adults and ask them if they read every word or at least enough to understand a contract for buying or selling a home before signing. *(Source: FindLaw.com)*

8. **Organ Donors**   A survey of U.S. adults found that 63% would want their organs transplanted into a patient who needs them if they were killed in an accident. You randomly select 20 adults and ask them if they would want their organs transplanted into a patient who needs them if they were killed in an accident. *(Source: USA Today)*

9. **Multivitamins**   A survey of U.S. adults found that 55% have used a multivitamin in the past 12 months. You randomly select 50 adults and ask them if they have used a multivitamin in the past 12 months. *(Source: Harris Interactive)*

10. **Happiness at Work**   A survey of U.S. adults found that 19% are happy with their current employer. You randomly select 30 adults and ask them if they are happy with their current employer. *(Source: Opinion Research Corporation)*

11. **Going Green**   A survey of U.S. adults found that 76% would pay more for an environmentally friendly product. You randomly select 20 adults and ask them if they would pay more for an environmentally friendly product. *(Source: Opinion Research Corporation)*

12. **Online Habits**   A survey of U.S. adults found that 61% look online for health information. You randomly select 15 adults and ask them if they look online for health information. *(Source: Pew Research Center)*

In Exercises 13–16, use a continuity correction and match the binomial probability statement with the corresponding normal distribution statement.

Binomial Probability	Normal Probability
13. $P(x > 109)$	(a) $P(x > 109.5)$
14. $P(x \geq 109)$	(b) $P(x < 108.5)$
15. $P(x \leq 109)$	(c) $P(x \leq 109.5)$
16. $P(x < 109)$	(d) $P(x \geq 108.5)$

17. The probability of getting fewer than 25 successes; $P(x < 24.5)$

18. The probability of getting at least 110 successes; $P(x > 109.5)$

19. The probability of getting exactly 33 successes; $P(32.5 < x < 33.5)$

20. The probability of getting more than 65 successes; $P(x > 65.5)$

21. The probability of getting at most 150 successes; $P(x < 150.5)$

22. The probability of getting between 55 and 60 successes; $P(55.5 < x < 59.5)$

23. Can use normal distribution.

(a) 0.0782 *(Tech:* 0.0785)

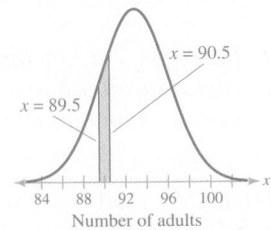

(b) 0.9147 *(Tech:* 0.9151)

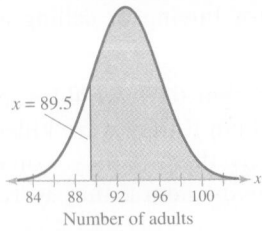

(c) 0.0853 *(Tech:* 0.0849)

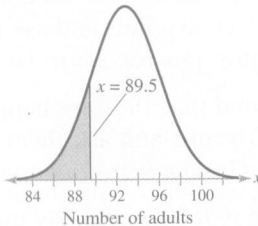

(d) No, none of the probabilities are less than 0.05.

24. See Selected Answers, page A119.

25. Can use normal distribution.

(a) ≈ 1

See Odd Answers, page A74.

(b) 0.9798 *(Tech:* 0.9801)

See Odd Answers, page A74.

(c) 0.6097 *(Tech:* 0.6109)

See Odd Answers, page A74.

(d) No, none of the probabilities are less than 0.05.

26. See Selected Answers, page A119.

In Exercises 17–22, a binomial probability is given. Write the probability in words. Then, use a continuity correction to convert the binomial probability to a normal distribution probability.

17. $P(x < 25)$ **18.** $P(x \geq 110)$ **19.** $P(x = 33)$

20. $P(x > 65)$ **21.** $P(x \leq 150)$ **22.** $P(55 < x < 60)$

■ USING AND INTERPRETING CONCEPTS

Approximating Binomial Probabilities *In Exercises 23–30, decide whether you can use a normal distribution to approximate the binomial distribution. If you can, use the normal distribution to approximate the indicated probabilities and sketch their graphs. If you cannot, explain why and use a binomial distribution to find the indicated probabilities.*

23. **Internet Use** A survey of U.S. adults ages 18–29 found that 93% use the Internet. You randomly select 100 adults ages 18–29 and ask them if they use the Internet. *(Source: Pew Research Center)*

(a) Find the probability that exactly 90 people say they use the Internet.

(b) Find the probability that at least 90 people say they use the Internet.

(c) Find the probability that fewer than 90 people say they use the Internet.

(d) Are any of the probabilities in parts (a)–(c) unusual? Explain.

24. **Internet Use** A survey of U.S. adults ages 50–64 found that 70% use the Internet. You randomly select 80 adults ages 50–64 and ask them if they use the Internet. *(Source: Pew Research Center)*

(a) Find the probability that at least 70 people say they use the Internet.

(b) Find the probability that exactly 50 people say they use the Internet.

(c) Find the probability that more than 60 people say they use the Internet.

(d) Are any of the probabilities in parts (a)–(c) unusual? Explain.

25. **Favorite Sport** A survey of U.S. adults found that 35% say their favorite sport is professional football. You randomly select 150 adults and ask them if their favorite sport is professional football. *(Source: Harris Interactive)*

(a) Find the probability that at most 75 people say their favorite sport is professional football.

(b) Find the probability that more than 40 people say their favorite sport is professional football.

(c) Find the probability that between 50 and 60 people, inclusive, say their favorite sport is professional football.

(d) Are any of the probabilities in parts (a)–(c) unusual? Explain.

26. **College Graduates** About 34% of workers in the United States are college graduates. You randomly select 50 workers and ask them if they are a college graduate. *(Source: U.S. Bureau of Labor Statistics)*

(a) Find the probability that exactly 12 workers are college graduates.

(b) Find the probability that more than 23 workers are college graduates.

(c) Find the probability that at most 18 workers are college graduates.

(d) A committee is looking for 30 working college graduates to volunteer at a career fair. The committee randomly selects 125 workers. What is the probability that there will not be enough college graduates?

27. Can use normal distribution.

(a) 0.0692 (*Tech:* 0.0691)

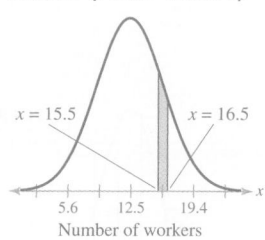

(b) 0.8770 (*Tech:* 0.8771)

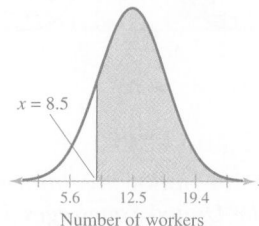

(c) 0.8078 (*Tech:* 0.8080)

See Odd Answers, page A74.

(d) 0.8212 (*Tech:* 0.8221)

See Odd Answers, page A74.

28. Cannot use normal distribution because $nq < 5$.

(a) 0.0036

(b) 0.641

(c) 0.943

(d) Yes, the event in part (a) is unusual because its probability is less than 0.05.

29. (a) Can use normal distribution. 0.0069

See Odd Answers, page A74.

(b) Can use normal distribution. 0.3557 (*Tech:* 0.3545)

See Odd Answers, page A75.

(c) Can use normal distribution. 0.0558 (*Tech:* 0.0595)

See Odd Answers, page A75.

(d) Cannot use normal distribution because $np < 5$ and $nq < 5$; 0.002

30. Cannot use normal distribution because $np < 5$.

(a) 0.999871

(b) 0.255

(c) 0.003

(d) 0.230

27. **Public Transportation**   Five percent of workers in the United States use public transportation to get to work. You randomly select 250 workers and ask them if they use public transportation to get to work. (*Source: U.S. Census Bureau*)

(a) Find the probability that exactly 16 workers will say yes.

(b) Find the probability that at least 9 workers will say yes.

(c) Find the probability that fewer than 16 workers will say yes.

(d) A transit authority offers discount rates to companies that have at least 30 employees who use public transportation to get to work. There are 500 employees in a company. What is the probability that the company will not get the discount?

28. **Concert Tickets**   A survey of U.S. adults who attend at least one music concert a year found that 67% say concert tickets are too expensive. You randomly select 12 adults who attend at least one music concert a year and ask them if concert tickets are too expensive. (*Source: Rasmussen Reports*)

(a) Find the probability that fewer than 4 people say that concert tickets are too expensive.

(b) Find the probability that between 7 and 9 people, inclusive, say that concert tickets are too expensive.

(c) Find the probability that at most 10 people say that concert tickets are too expensive.

(d) Are any of the probabilities in parts (a)–(c) unusual? Explain.

29. **News**   A survey of U.S. adults ages 18–24 found that 34% get no news on an average day. You randomly select 200 adults ages 18–24 and ask them if they get news on an average day. (*Source: Pew Research Center*)

(a) Find the probability that at least 85 people say they get no news on an average day.

(b) Find the probability that fewer than 66 people say they get no news on an average day.

(c) Find the probability that exactly 68 people say they get no news on an average day.

(d) A college English teacher wants students to discuss current events. The teacher randomly selects six students from the class. What is the probability that none of the students can talk about current events because they get no news on an average day.

30. **Long Work Weeks**   A survey of U.S. workers found that 2.9% work more than 70 hours per week. You randomly select 10 workers in the United States and ask them if they work more than 70 hours per week.

(a) Find the probability that at most 3 people say they work more than 70 hours per week.

(b) Find the probability that at least 1 person says he or she works more than 70 hours per week.

(c) Find the probability that more than 2 people say they work more than 70 hours per week.

(d) A large company is concerned about overworked employees who work more than 70 hours per week. The company randomly selects 50 employees. What is the probability there will be no employee working more than 70 hours?

31. Binomial: 0.549; Normal: 0.5463 (*Tech:* 0.5466); The results are about the same.

32. Binomial: 0.191; Normal: 0.1875 (*Tech:* 0.1886); The results are about the same.

33. Highly unlikely. Answers will vary.

34. Probable. Answers will vary.

35. 0.1020

36. 0.1736 (*Tech:* 0.1727)

Graphical Analysis *In Exercises 31 and 32, write the binomial probability and the normal probability for the shaded region of the graph. Find the value of each probability and compare the results.*

31.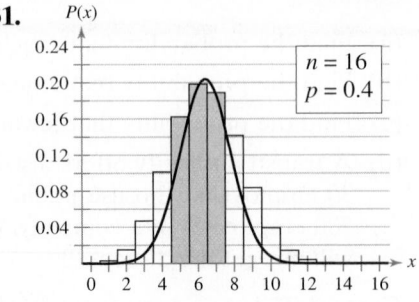

32.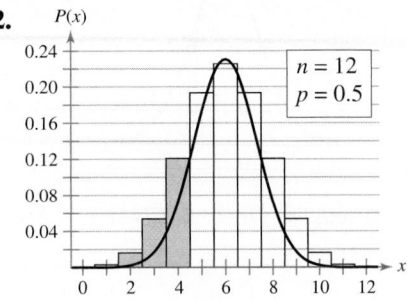

■ EXTENDING CONCEPTS

Getting Physical *In Exercises 33 and 34, use the following information. The graph shows the results of a survey of adults in the United States ages 33 to 51 who were asked if they participated in a sport. Seventy percent of adults said they regularly participated in at least one sport, and they gave their favorite sport.*

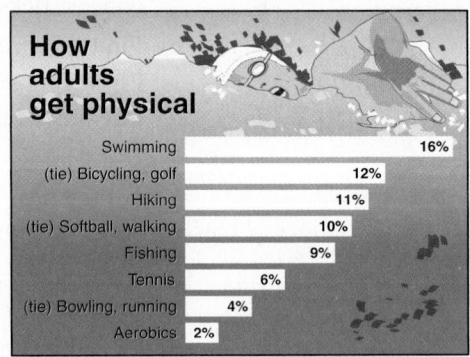

33. You randomly select 250 people in the United States ages 33 to 51 and ask them if they regularly participate in at least one sport. You find that 60% say no. How likely is this result? Do you think this sample is a good one? Explain your reasoning.

34. You randomly select 300 people in the United States ages 33 to 51 and ask them if they regularly participate in at least one sport. Of the 200 who say yes, 9% say they participate in hiking. How likely is this result? Do you think this sample is a good one? Explain your reasoning.

Testing a Drug *In Exercises 35 and 36, use the following information. A drug manufacturer claims that a drug cures a rare skin disease 75% of the time. The claim is checked by testing the drug on 100 patients. If at least 70 patients are cured, this claim will be accepted.*

35. Find the probability that the claim will be rejected assuming that the manufacturer's claim is true.

36. Find the probability that the claim will be accepted assuming that the actual probability that the drug cures the skin disease is 65%.

USES AND ABUSES

Uses

Normal Distributions   Normal distributions can be used to describe many real-life situations and are widely used in the fields of science, business, and psychology. They are the most important probability distributions in statistics and can be used to approximate other distributions, such as discrete binomial distributions.

The most incredible application of the normal distribution lies in the Central Limit Theorem. This theorem states that no matter what type of distribution a population may have, as long as the sample size is at least 30, the distribution of sample means will be approximately normal. If the population is itself normal, then the distribution of sample means will be normal no matter how small the sample is.

The normal distribution is essential to sampling theory. Sampling theory forms the basis of statistical inference, which you will begin to study in the next chapter.

Abuses

Unusual Events   Suppose a population is normally distributed, with a mean of 100 and standard deviation of 15. It would not be unusual for an individual value taken from this population to be 115 or more. In fact, this will happen almost 16% of the time. It *would* be, however, highly unusual to take random samples of 100 values from that population and obtain a sample with a mean of 115 or more. Because the population is normally distributed, the mean of the sample distribution will be 100, and the standard deviation will be 1.5. A sample mean of 115 lies 10 standard deviations above the mean. This would be an extremely unusual event. When an event this unusual occurs, it is a good idea to question the original claimed value of the mean.

Although normal distributions are common in many populations, people try to make *non-normal* statistics fit a normal distribution. The statistics used for normal distributions are often inappropriate when the distribution is obviously non-normal.

◼ EXERCISES

1. ***Is It Unusual?***   A population is normally distributed, with a mean of 100 and a standard deviation of 15. Determine if either of the following events is unusual. Explain your reasoning.

   a. The mean of a sample of 3 is 115 or more.

   b. The mean of a sample of 20 is 105 or more.

2. ***Find the Error***   The mean age of students at a high school is 16.5, with a standard deviation of 0.7. You use the Standard Normal Table to help you determine that the probability of selecting one student at random and finding his or her age to be more than 17.5 years is about 8%. What is the error in this problem?

3. Give an example of a distribution that might be non-normal.

5 CHAPTER SUMMARY

What did you **learn?**	EXAMPLE(S)	REVIEW EXERCISES
Section 5.1		
■ How to interpret graphs of normal probability distributions	*1, 2*	*1–6*
■ How to find areas under the standard normal curve	*3–6*	*7–28*
Section 5.2		
■ How to find probabilities for normally distributed variables	*1–3*	*29–38*
Section 5.3		
■ How to find a *z*-score given the area under the normal curve	*1, 2*	*39–46*
■ How to transform a *z*-score to an *x*-value $$x = \mu + z\sigma$$	*3*	*47, 48*
■ How to find a specific data value of a normal distribution given the probability	*4, 5*	*49–52*
Section 5.4		
■ How to find sampling distributions and verify their properties	*1*	*53, 54*
■ How to interpret the Central Limit Theorem $$\mu_{\bar{x}} = \mu \quad \text{Mean}$$ $$\sigma_{\bar{x}} = \frac{\sigma}{\sqrt{n}} \quad \text{Standard deviation}$$	*2, 3*	*55, 56*
■ How to apply the Central Limit Theorem to find the probability of a sample mean	*4–6*	*57–62*
Section 5.5		
■ How to decide when a normal distribution can approximate a binomial distribution $$\mu = np \quad \text{Mean}$$ $$\sigma = \sqrt{npq} \quad \text{Standard deviation}$$	*1*	*63, 64*
■ How to find the continuity correction	*2*	*65–68*
■ How to use a normal distribution to approximate binomial probabilities	*3–5*	*69, 70*

5 REVIEW EXERCISES

■ SECTION 5.1

In Exercises 1 and 2, use the graph to estimate μ and σ.

1.
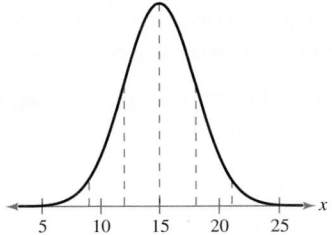

2.
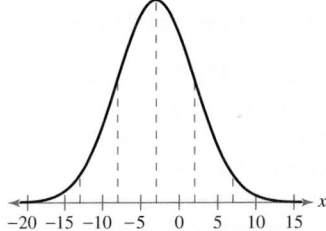

In Exercises 3 and 4, use the normal curves shown.

3. Which normal curve has the greatest mean? Explain your reasoning.

4. Which normal curve has the greatest standard deviation? Explain your reasoning.

In Exercises 5 and 6, use the following information and standard scores to investigate observations about a normal population. A batch of 2500 resistors is normally distributed, with a mean resistance of 1.5 ohms and a standard deviation of 0.08 ohm. Four resistors are randomly selected and tested. Their resistances are measured at 1.32, 1.54, 1.66, and 1.78 ohms.

5. How many standard deviations from the mean are these observations?

6. Are there any unusual observations?

In Exercises 7 and 8, find the area of the indicated region under the standard normal curve. If convenient, use technology to find the area.

7.
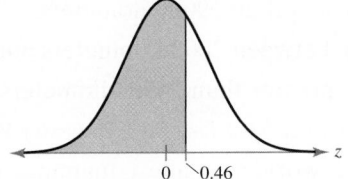

8.
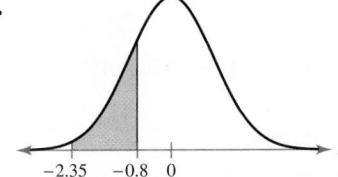

In Exercises 9–20, find the indicated area under the standard normal curve. If convenient, use technology to find the area.

9. To the left of $z = 0.33$

10. To the left of $z = -1.95$

11. To the right of $z = -0.57$

12. To the right of $z = 3.22$

13. To the left of $z = -2.825$

14. To the right of $z = 0.015$

15. Between $z = -1.64$ and the mean

16. Between $z = -1.55$ and $z = 1.04$

17. Between $z = 0.05$ and $z = 1.71$

18. Between $z = -2.68$ and $z = 2.68$

19. To the left of $z = -1.5$ and to the right of $z = 1.5$

20. To the left of $z = 0.64$ and to the right of $z = 3.415$

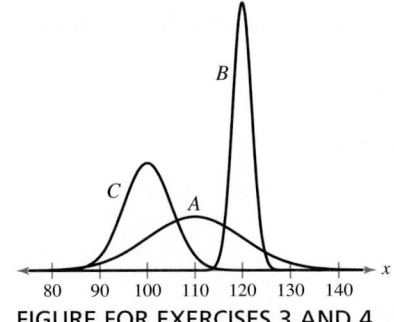

FIGURE FOR EXERCISES 3 AND 4

1. $\mu = 15$, $\sigma = 3$

2. $\mu = -3$, $\sigma = 5$

3. Curve *B* has the greatest mean because its line of symmetry occurs the farthest to the right.

4. Curve *A* has the greatest standard deviation because it is the most spread out.

5. −2.25; 0.5; 2; 3.5

6. 1.32 and 1.78 are unusual.

7. 0.6772

8. 0.2025

9. 0.6293

10. 0.0256

11. 0.7157

12. 0.0006

13. 0.00235 (*Tech:* 0.00236)

14. 0.4940

15. 0.4495

16. 0.7902 (*Tech:* 0.7903)

17. 0.4365 (*Tech:* 0.4364)

18. 0.9926

19. 0.1336

20. 0.7392

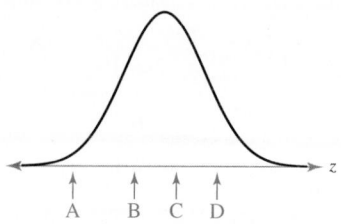

FIGURE FOR EXERCISES 21 AND 22

21. A = 8; B = 17; C = 23; D = 29

22. A: −2.16; B: −0.71; C: 0.26;
D: 1.23; The test score of 8
is unusual because its
corresponding z-score is more
than 2 standard deviations from
the mean.

23. 0.8997

24. 0.7704

25. 0.9236 (Tech: 0.9237)

26. 0.3364

27. 0.0124

28. 0.5465

29. 0.8944

30. 0.0088

31. 0.2266

32. 0.6179

33. 0.2684 (Tech: 0.2685)

34. 0.4400 (Tech: 0.4401)

35. (a) 0.3156
(b) 0.3099
(c) 0.3446

36. (a) 0.9544 (Tech: 0.9545)
(b) 0.3420
(c) 0.0026

37. No, none of the events are
unusual because their
probabilities are greater
than 0.05.

38. Yes, the event in part (c) is
unusual because its probability
is less than 0.05.

39. −0.07

40. −1.28

41. 1.13

42. −2.05

43. 1.04

44. −0.10

In Exercises 21 and 22, use the following information. In a recent year, the ACT scores for the reading portion of the test were normally distributed, with a mean of 21.4 and a standard deviation of 6.2. The test scores of four students selected at random are 17, 29, 8, and 23. (Source: ACT, Inc.)

21. Without converting to z-scores, match the values with the letters A, B, C, and D on the given graph.

22. Find the z-score that corresponds to each value and check your answers in Exercise 21. Are any of the values unusual? Explain.

In Exercises 23–28, find the indicated probabilities. If convenient, use technology to find the probability.

23. $P(z < 1.28)$

24. $P(z > -0.74)$

25. $P(-2.15 < z < 1.55)$

26. $P(0.42 < z < 3.15)$

27. $P(z < -2.50 \text{ or } z > 2.50)$

28. $P(z < 0 \text{ or } z > 1.68)$

■ SECTION 5.2

In Exercises 29–34, assume the random variable x is normally distributed, with mean $\mu = 74$ and standard deviation $\sigma = 8$. Find the indicated probability.

29. $P(x < 84)$

30. $P(x < 55)$

31. $P(x > 80)$

32. $P(x > 71.6)$

33. $P(60 < x < 70)$

34. $P(72 < x < 82)$

In Exercises 35 and 36, find the indicated probabilities.

35. A study found that the mean migration distance of the green turtle was 2200 kilometers and the standard deviation was 625 kilometers. Assuming that the distances are normally distributed, find the probability that a randomly selected green turtle migrates a distance of

(a) less than 1900 kilometers.

(b) between 2000 kilometers and 2500 kilometers.

(c) greater than 2450 kilometers.

(Adapted from Dorling Kindersley Visual Encyclopedia)

36. The world's smallest mammal is the Kitti's hog-nosed bat, with a mean weight of 1.5 grams and a standard deviation of 0.25 gram. Assuming that the weights are normally distributed, find the probability of randomly selecting a bat that weighs

(a) between 1.0 gram and 2.0 grams.

(b) between 1.6 grams and 2.2 grams.

(c) more than 2.2 grams.

(Adapted from Dorling Kindersley Visual Encyclopedia)

37. Can any of the events in Exercise 35 be considered unusual? Explain your reasoning.

38. Can any of the events in Exercise 36 be considered unusual? Explain your reasoning.

■ SECTION 5.3

In Exercises 39–44, use the Standard Normal Table to find the z-score that corresponds to the given cumulative area or percentile. If the area is not in the table, use the entry closest to the area. If convenient, use technology to find the z-score.

45. 0.51

46. 1.88

47. 42.5 meters

48. 50.6 meters

Braking Distance of a Cadillac Escalade

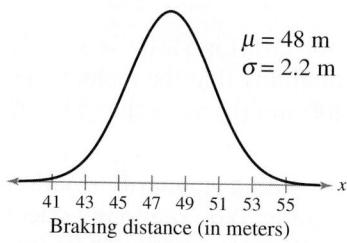

$\mu = 48$ m
$\sigma = 2.2$ m

41 43 45 47 49 51 53 55
Braking distance (in meters)

FIGURE FOR EXERCISES 47–52

49. 51.6 meters

50. 49.5 meters

51. 50.8 meters

52. 44.4 meters

53. $\mu = 145$, $\sigma = 45$

$\mu_{\bar{x}} = 145$, $\sigma_{\bar{x}} \approx 25.98$

The means are the same, but $\sigma_{\bar{x}}$ is less than σ.

54. $\mu = 1.5$, $\sigma = 1.118$; $\mu_{\bar{x}} = 1.5$, $\sigma_{\bar{x}} \approx 0.791$

The means are the same, but $\sigma_{\bar{x}}$ is less than σ.

55. 76, 3.465

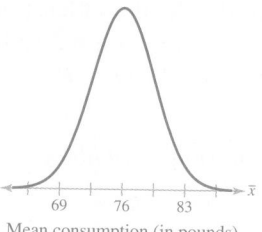

69   76   83
Mean consumption (in pounds)

56. 108.3, 5.550

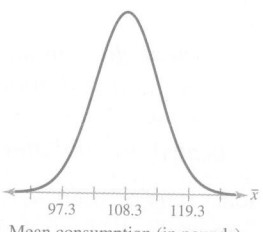

97.3   108.3   119.3
Mean consumption (in pounds)

57. (a) 0.0485 (*Tech:* 0.0482)

(b) 0.8180

(c) 0.0823 (*Tech:* 0.0829)

(a) and (c) are smaller, (b) is larger. This is to be expected because the standard error of the sample means is smaller.

39. 0.4721   **40.** 0.1   **41.** 0.8708

42. P_2 **43.** P_{85} **44.** P_{46}

45. Find the z-score that has 30.5% of the distribution's area to its right.

46. Find the z-score for which 94% of the distribution's area lies between $-z$ and z.

In Exercises 47–52, use the following information. On a dry surface, the braking distance (in meters) of a Cadillac Escalade can be approximated by a normal distribution, as shown in the graph at the left. (Adapted from Consumer Reports)

47. Find the braking distance of a Cadillac Escalade that corresponds to $z = -2.5$.

48. Find the braking distance of a Cadillac Escalade that corresponds to $z = 1.2$.

49. What braking distance of a Cadillac Escalade represents the 95th percentile?

50. What braking distance of a Cadillac Escalade represents the third quartile?

51. What is the shortest braking distance of a Cadillac Escalade that can be in the top 10% of braking distances?

52. What is the longest braking distance of a Cadillac Escalade that can be in the bottom 5% of braking distances?

■ SECTION 5.4

In Exercises 53 and 54, use the given population to find the mean and standard deviation of the population and the mean and standard deviation of the sampling distribution. Compare the values.

53. A corporation has four executives. The number of minutes of overtime per week reported by each is 90, 120, 160, and 210. Draw three executives' names from this population, with replacement.

54. There are four residents sharing a house. The number of times each washes their car each month is 1, 2, 0, and 3. Draw two names from this population, with replacement.

In Exercises 55 and 56, use the Central Limit Theorem to find the mean and standard error of the mean of the indicated sampling distribution. Then sketch a graph of the sampling distribution.

55. The per capita consumption of citrus fruits by people in the United States in a recent year was normally distributed, with a mean of 76.0 pounds and a standard deviation of 20.5 pounds. Random samples of 35 people are drawn from this population and the mean of each sample is determined. *(Adapted from U.S. Department of Agriculture)*

56. The per capita consumption of red meat by people in the United States in a recent year was normally distributed, with a mean of 108.3 pounds and a standard deviation of 35.1 pounds. Random samples of 40 people are drawn from this population and the mean of each sample is determined. *(Adapted from U.S. Department of Agriculture)*

In Exercises 57–62, find the probabilities for the sampling distributions. Interpret the results.

57. Refer to Exercise 35. A sample of 12 green turtles is randomly selected. Find the probability that the sample mean of the distance migrated is (a) less than 1900 kilometers, (b) between 2000 kilometers and 2500 kilometers, and (c) greater than 2450 kilometers. Compare your answers with those in Exercise 35.

58. (a) $\approx$ 1 (*Tech:* 0.9999999)

(b) 0.1446 (*Tech:* 0.1450)

(c) $\approx$ 0

(a) is larger and (b) and (c) are smaller.

59. (a) 0.1867 (*Tech:* 0.1855)

(b) $\approx$ 0

60. (a) 0.9918

(b) $\approx$ 1 (*Tech:* 0.9998)

61. 0.0019 (*Tech:* 0.0018)

62. 0.0006

63. Cannot use normal distribution because $nq < 5$.

64. Can use normal distribution; $\mu = 22.5$, $\sigma \approx 2.37$

65. $P(x > 24.5)$

66. $P(x < 36.5)$

67. $P(44.5 < x < 45.5)$

68. $P(49.5 < x < 50.5)$

69. Can use normal distribution.

$\approx$ 0 (*Tech:* 0.0002)

70. Cannot use normal distribution because $np < 5$.

0.019

58. Refer to Exercise 36. A sample of seven Kitti's hog-nosed bats is randomly selected. Find the probability that the sample mean is (a) between 1.0 gram and 2.0 grams, (b) between 1.6 grams and 2.2 grams, and (c) more than 2.2 grams. Compare your answers with those in Exercise 36.

59. The mean annual salary for chauffeurs is $29,200. A sample of 45 chauffeurs is randomly selected. What is the probability that the mean annual salary is (a) less than $29,000 and (b) more than $31,000? Assume $\sigma = \$1500$. (*Source: Salary.com*)

60. The mean value of land and buildings per acre for farms is $1300. A sample of 36 farms is randomly selected. What is the probability that the mean value of land and buildings per acre is (a) less than $1400 and (b) more than $1150? Assume $\sigma = \$250$.

61. The mean price of houses in a city is $1.5 million with a standard deviation of $500,000. The house prices are normally distributed. You randomly select 15 houses in this city. What is the probability that the mean price will be less than $1.125 million?

62. Mean rent in a city is $500 per month with a standard deviation of $30. The rents are normally distributed. You randomly select 15 apartments in this city. What is the probability that the mean rent will be more than $525?

■ SECTION 5.5

In Exercises 63 and 64, a binomial experiment is given. Decide whether you can use a normal distribution to approximate the binomial distribution. If you can, find the mean and standard deviation. If you cannot, explain why.

63. In a recent year, the American Cancer Society said that the five-year survival rate for new cases of stage 1 kidney cancer is 96%. You randomly select 12 men who were new stage 1 kidney cancer cases this year and calculate the five-year survival rate of each. (*Source: American Cancer Society, Inc.*)

64. A survey indicates that 75% of U.S. adults who go to the theater at least once a month think movie tickets are too expensive. You randomly select 30 adults and ask them if they think movie tickets are too expensive. (*Source: Rasmussen Reports*)

In Exercises 65–68, write the binomial probability as a normal probability using the continuity correction.

65. $P(x \geq 25)$

66. $P(x \leq 36)$

67. $P(x = 45)$

68. $P(x = 50)$

In Exercises 69 and 70, decide whether you can use a normal distribution to approximate the binomial distribution. If you can, use the normal distribution to approximate the indicated probabilities and sketch their graphs. If you cannot, explain why and use a binomial distribution to find the indicated probabilities.

69. Seventy percent of children ages 12 to 17 keep at least part of their savings in a savings account. You randomly select 45 children and ask them if they keep at least part of their savings in a savings account. Find the probability that at most 20 children will say yes. (*Source: International Communications Research for Merrill Lynch*)

70. Thirty-one percent of people in the United States have type A⁺ blood. You randomly select 15 people in the United States and ask them if their blood type is A⁺. Find the probability that more than 8 adults say they have A⁺ blood. (*Source: American Association of Blood Banks*)

5 CHAPTER QUIZ

Answers

1. (a) 0.9945
   (b) 0.9990
   (c) 0.6212
   (d) 0.83685 (*Tech:* 0.83692)

2. (a) 0.9198 (*Tech:* 0.9199)
   (b) 0.1940 (*Tech:* 0.1938)
   (c) 0.0456 (*Tech:* 0.0455)

3. 0.0475 (*Tech:* 0.0478); Yes, the event is unusual because its probability is less than 0.05.

4. 0.2586 (*Tech:* 0.2611); No, the event is not unusual because its probability is greater than 0.05.

5. 21.19%

6. 503 students (*Tech:* 505 students)

7. 125

8. 80

9. 0.0049; About 0.5% of samples of 60 students will have a mean IQ score greater than 105. This is a very unusual event.

10. More likely to select one student with an IQ score greater than 105 because the standard error of the mean is less than the standard deviation.

11. Can use normal distribution.
    $\mu = 28.35, \sigma \approx 2.32$

12. 0.0004; This event is extremely unusual because its probability is much less than 0.05.

Take this quiz as you would take a quiz in class. After you are done, check your work against the answers given in the back of the book.

1. Find each standard normal probability.
   (a) $P(z > -2.54)$
   (b) $P(z < 3.09)$
   (c) $P(-0.88 < z < 0.88)$
   (d) $P(z < -1.445 \text{ or } z > -0.715)$

2. Find each normal probability for the given parameters.
   (a) $\mu = 5.5, \sigma = 0.08, P(5.36 < x < 5.64)$
   (b) $\mu = -8.2, \sigma = 7.84, P(-5.00 < x < 0)$
   (c) $\mu = 18.5, \sigma = 9.25, P(x < 0 \text{ or } x > 37)$

In Exercises 3–10, use the following information. Students taking a standardized IQ test had a mean score of 100 with a standard deviation of 15. Assume that the scores are normally distributed. (*Adapted from Audiblox*)

3. Find the probability that a student had a score higher than 125. Is this an unusual event? Explain.

4. Find the probability that a student had a score between 95 and 105. Is this an unusual event? Explain.

5. What percent of the students had an IQ score that is greater than 112?

6. If 2000 students are randomly selected, how many would be expected to have an IQ score that is less than 90?

7. What is the lowest score that would still place a student in the top 5% of the scores?

8. What is the highest score that would still place a student in the bottom 10% of the scores?

9. A random sample of 60 students is drawn from this population. What is the probability that the mean IQ score is greater than 105? Interpret your result.

10. Are you more likely to randomly select one student with an IQ score greater than 105 or are you more likely to randomly select a sample of 15 students with a mean IQ score greater than 105? Explain.

In Exercises 11 and 12, use the following information. In a survey of adults under age 65, 81% say they are concerned about the amount and security of personal online data that can be accessed by cybercriminals and hackers. You randomly select 35 adults and ask them if they are concerned about the amount and security of personal online data that can be accessed by cybercriminals and hackers. (*Source: Financial Times/Harris Poll*)

11. Decide whether you can use a normal distribution to approximate the binomial distribution. If you can, find the mean and standard deviation. If you cannot, explain why.

12. Find the probability that at most 20 adults say they are concerned about the amount and security of personal online data that can be accessed by cybercriminals and hackers. Interpret the result.

PUTTING IT ALL TOGETHER

Real Statistics — Real Decisions

You are the human resources director for a corporation and want to implement a health improvement program for employees to decrease employee medical absences. You perform a six-month study with a random sample of employees. Your goal is to decrease absences by 50%. (Assume all data are normally distributed.)

■ EXERCISES

1. *Preliminary Thoughts*

   You got the idea for this health improvement program from a national survey in which 75% of people who responded said they would participate in such a program if offered by their employer. You randomly select 60 employees and ask them whether they would participate in such a program.

   (a) Find the probability that exactly 35 will say yes.

   (b) Find the probability that at least 40 will say yes.

   (c) Find the probability that fewer than 20 will say yes.

   (d) Based on the results in parts (a)–(c), explain why you chose to perform the study.

2. *Before the Program*

   Before the study, the mean number of absences during a six-month period of the participants was 6, with a standard deviation of 1.5. An employee is randomly selected.

   (a) Find the probability that the employee's number of absences is less than 5.

   (b) Find the probability that the employee's number of absences is between 5 and 7.

   (c) Find the probability that the employee's number of absences is more than 7.

3. *After the Program*

   The graph at the right represents the results of the study.

   (a) What is the mean number of absences for employees? Explain how you know.

   (b) Based on the results, was the goal of decreasing absences by 50% reached?

   (c) Describe how you would present your results to the board of directors of the corporation.

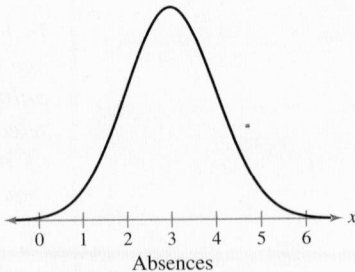

FIGURE FOR EXERCISE 3

TECHNOLOGY

| MINITAB | EXCEL | TI-83/84 PLUS |

U.S. Census Bureau

www.census.gov

AGE DISTRIBUTION IN THE UNITED STATES

One of the jobs of the U.S. Census Bureau is to keep track of the age distribution in the country. The age distribution in 2009 is shown below.

Age Distribution in the U.S.

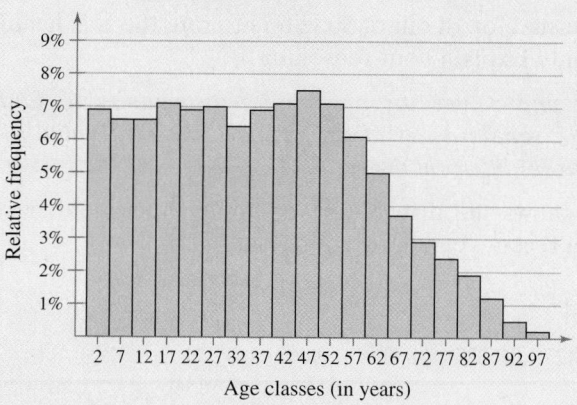

Age classes (in years)

Class	Class midpoint	Relative frequency
0–4	2	6.9%
5–9	7	6.6%
10–14	12	6.6%
15–19	17	7.1%
20–24	22	6.9%
25–29	27	7.0%
30–34	32	6.4%
35–39	37	6.9%
40–44	42	7.1%
45–49	47	7.5%
50–54	52	7.1%
55–59	57	6.1%
60–64	62	5.0%
65–69	67	3.7%
70–74	72	2.9%
75–79	77	2.4%
80–84	82	1.9%
85–89	87	1.2%
90–94	92	0.5%
95–99	97	0.2%

■ EXERCISES

We used a technology tool to select random samples with n = 40 from the age distribution of the United States. The means of the 36 samples were as follows.

28.14, 31.56, 36.86, 32.37, 36.12, 39.53,
36.19, 39.02, 35.62, 36.30, 34.38, 32.98,
36.41, 30.24, 34.19, 44.72, 38.84, 42.87,
38.90, 34.71, 34.13, 38.25, 38.04, 34.07,
39.74, 40.91, 42.63, 35.29, 35.91, 34.36,
36.51, 36.47, 32.88, 37.33, 31.27, 35.80

1. Enter the age distribution of the United States into a technology tool. Use the tool to find the mean age in the United States.

2. Enter the set of sample means into a technology tool. Find the mean of the set of sample means. How does it compare with the mean age in the United States? Does this agree with the result predicted by the Central Limit Theorem?

3. Are the ages of people in the United States normally distributed? Explain your reasoning.

4. Sketch a relative frequency histogram for the 36 sample means. Use nine classes. Is the histogram approximately bell-shaped and symmetric? Does this agree with the result predicted by the Central Limit Theorem?

5. Use a technology tool to find the standard deviation of the ages of people in the United States.

6. Use a technology tool to find the standard deviation of the set of 36 sample means. How does it compare with the standard deviation of the ages? Does this agree with the result predicted by the Central Limit Theorem?

Extended solutions are given in the *Technology Supplement*.
Technical instruction is provided for MINITAB, Excel, and the TI-83/84 Plus.

CUMULATIVE REVIEW
Chapters 3 – 5

Section 5.5

1. A survey of voters in the United States found that 15% rate the U.S. health care system as excellent. You randomly select 50 voters and ask them how they rate the U.S. health care system. *(Source: Rasmussen Reports)*

   (a) Verify that the normal distribution can be used to approximate the binomial distribution.

   (b) Find the probability that at most 14 voters rate the U.S. health care system as excellent.

   (c) Is it unusual for 14 out of 50 voters to rate the U.S. health care system as excellent? Explain your reasoning.

In Exercises 2 and 3, use the probability distribution to find the (a) mean, (b) variance, (c) standard deviation, and (d) expected value of the probability distribution, and (e) interpret the results.

Section 4.1

2. The table shows the distribution of family household sizes in the United States for a recent year. *(Source: U.S. Census Bureau)*

x	2	3	4	5	6	7
$P(x)$	0.427	0.227	0.200	0.093	0.034	0.018

3. The table shows the distribution of fouls per game for a player in a recent NBA season. *(Source: NBA.com)*

x	0	1	2	3	4	5	6
$P(x)$	0.012	0.049	0.159	0.256	0.244	0.195	0.085

Section 4.1

4. Use the probability distribution in Exercise 3 to find the probability of randomly selecting a game in which the player had (a) fewer than four fouls, (b) at least three fouls, and (c) between two and four fouls, inclusive.

Section 3.4

5. From a pool of 16 candidates, 9 men and 7 women, the offices of president, vice president, secretary, and treasurer will be filled. (a) In how many different ways can the offices be filled? (b) What is the probability that all four of the offices are filled by women?

Section 5.1

In Exercises 6–11, use the Standard Normal Table to find the indicated area under the standard normal curve.

6. To the left of $z = 0.72$

7. To the left of $z = -3.08$

8. To the right of $z = -0.84$

9. Between $z = 0$ and $z = 2.95$

10. Between $z = -1.22$ and $z = -0.26$

11. To the left of $z = 0.12$ or to the right of $z = 1.72$

Section 4.2

12. Forty-five percent of adults say they are interested in regularly measuring their carbon footprint. You randomly select 11 adults and ask them if they are interested in regularly measuring their carbon footprint. Find the probability that the number of adults who say they are interested is (a) exactly eight, (b) at least five, and (c) less than two. Are any of these events unusual? Explain your reasoning. *(Source: Sacred Heart University Polling)*

Section 4.3 **13.** An auto parts seller finds that 1 in every 200 parts sold is defective. Use the geometric distribution to find the probability that (a) the first defective part is the tenth part sold, (b) the first defective part is the first, second, or third part sold, and (c) none of the first 10 parts sold are defective.

Sections 3.1, 3.2, and 3.3 **14.** The table shows the results of a survey in which 2,944,100 public and 401,900 private school teachers were asked about their full-time teaching experience. *(Adapted from U.S. National Center for Education Statistics)*

	Public	Private	Total
Less than 3 years	177,300	27,600	204,900
3 to 9 years	995,800	154,500	1,150,300
10 to 20 years	906,300	111,600	1,017,900
20 years or more	864,700	108,200	972,900
Total	2,944,100	401,900	3,346,000

(a) Find the probability that a randomly selected private school teacher has 10 to 20 years of full-time teaching experience.

(b) Given that a randomly selected teacher has 3 to 9 years of full-time experience, find the probability that the teacher is at a public school.

(c) Are the events "being a public school teacher" and "having 20 years or more of full-time teaching experience" independent? Explain.

(d) Find the probability that a randomly selected teacher is either at a public school or has less than 3 years of full-time teaching experience.

(e) Find the probability that a randomly selected teacher has 3 to 9 years of full-time teaching experience or is at a private school.

Section 5.4 **15.** The initial pressures for bicycle tires when first filled are normally distributed, with a mean of 70 pounds per square inch (psi) and a standard deviation of 1.2 psi.

(a) Random samples of size 40 are drawn from this population and the mean of each sample is determined. Use the Central Limit Theorem to find the mean and standard error of the mean of the sampling distribution. Then sketch a graph of the sampling distribution of sample means.

(b) A random sample of 15 tires is drawn from this population. What is the probability that the mean tire pressure of the sample $\bar{x}$ is less than 69 psi?

Sections 5.2 and 5.3 **16.** The life spans of car batteries are normally distributed, with a mean of 44 months and a standard deviation of 5 months.

(a) A car battery is selected at random. Find the probability that the life span of the battery is less than 36 months.

(b) A car battery is selected at random. Find the probability that the life span of the battery is between 42 and 60 months.

(c) What is the shortest life expectancy a car battery can have and still be in the top 5% of life expectancies?

Section 3.4 **17.** A florist has 12 different flowers from which floral arrangements can be made. (a) If a centerpiece is to be made using four different flowers, how many different centerpieces can be made? (b) What is the probability that the four flowers in the centerpiece are roses, gerbers, hydrangeas, and callas?

Section 4.2 **18.** About fifty percent of adults say they feel vulnerable to identity theft. You randomly select 16 adults and ask them if they feel vulnerable to identity theft. Find the probability that the number who say they feel vulnerable is (a) exactly 12, (b) no more than 6, and (c) more than 7. Are any of these events unusual? Explain your reasoning. *(Adapted from KRC Research for Fellowes)*

6

CHAPTER

CONFIDENCE INTERVALS

David Wechsler was one of the most influential psychologists of the 20th century. He is known for developing intelligence tests, such as the Wechsler Adult Intelligence Scale and the Wechsler Intelligence Scale for Children.

In Chapters 1 through 5, you studied descriptive statistics (how to collect and describe data) and probability (how to find probabilities and analyze discrete and continuous probability distributions). For instance, psychologists use descriptive statistics to analyze the data collected during experiments and trials.

One of the most commonly administered psychological tests is the Wechsler Adult Intelligence Scale. It is an intelligence quotient (IQ) test that is standardized to have a normal distribution with a mean of 100 and a standard deviation of 15.

In this chapter, you will begin your study of inferential statistics—the second major branch of statistics. For instance, a chess club wants to estimate the mean IQ of its members. The mean of a random sample of members is 115. Because this estimate consists of a single number represented by a point on a number line, it is called a point estimate. The problem with using a point estimate is that it is rarely equal to the exact parameter (mean, standard deviation, or proportion) of the population.

In this chapter, you will learn how to make a more meaningful estimate by specifying an interval of values on a number line, together with a statement of how confident you are that your interval contains the population parameter. Suppose the club wants to be 90% confident of its estimate for the mean IQ of its members. Here is an overview of how to construct an interval estimate.

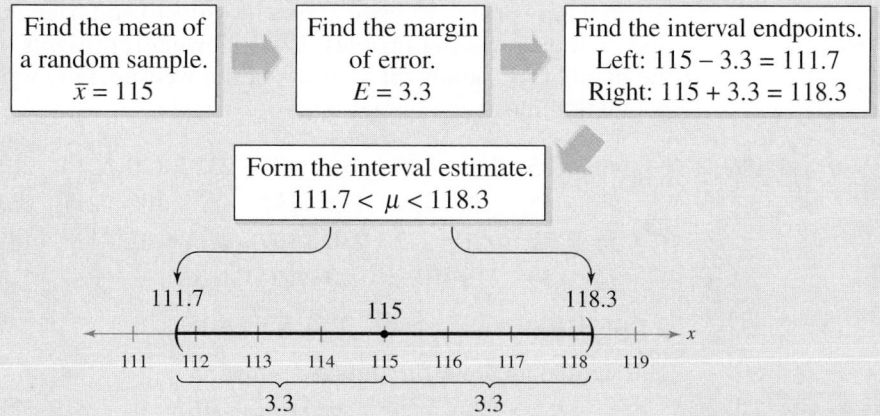

So, the club can be 90% confident that the mean IQ of its members is between 111.7 and 118.3.

6.1 Confidence Intervals for the Mean (Large Samples)

Estimating Population Parameters ▸ Confidence Intervals for the Population Mean ▸ Sample Size

WHAT YOU SHOULD LEARN

▸ How to find a point estimate and a margin of error

▸ How to construct and interpret confidence intervals for the population mean

▸ How to determine the minimum sample size required when estimating μ

▸ ESTIMATING POPULATION PARAMETERS

In this chapter, you will learn an important technique of statistical inference—to use sample statistics to estimate the value of an unknown population parameter. In this section, you will learn how to use sample statistics to make an estimate of the population parameter μ when the sample size is at least 30 or when the population is normally distributed and the standard deviation σ is known. To make such an inference, begin by finding a *point estimate*.

DEFINITION

A **point estimate** is a single value estimate for a population parameter. The most unbiased point estimate of the population mean μ is the sample mean $\overline{x}$.

The validity of an estimation method is increased if a sample statistic is unbiased and has low variability. A statistic is unbiased if it does not overestimate or underestimate the population parameter. In Chapter 5, you learned that the mean of all possible sample means of the same size equals the population mean. As a result, $\overline{x}$ is an unbiased estimator of μ. When the standard error $\sigma/\sqrt{n}$ of a sample mean is decreased by increasing n, it becomes less variable.

EXAMPLE 1

▸ Finding a Point Estimate

A social networking website allows its users to add friends, send messages, and update their personal profiles. The following represents a random sample of the number of friends for 40 users of the website. Find a point estimate of the population mean μ. *(Adapted from Facebook)*

140	105	130	97	80	165	232	110	214	201	122
98	65	88	154	133	121	82	130	211	153	114
58	77	51	247	236	109	126	132	125	149	122
74	59	218	192	90	117	105				

▸ Solution

The sample mean of the data is

$$\overline{x} = \frac{\Sigma x}{n} = \frac{5232}{40} = 130.8.$$

So, the point estimate for the mean number of friends for all users of the website is 130.8 friends.

▸ Try It Yourself 1

Another random sample of the number of friends for 30 users of the website is shown at the left. Use this sample to find another point estimate for μ.

a. *Find* the sample mean.

b. *Estimate* the mean number of friends of the population.   *Answer: Page A39*

Sample Data

Number of Friends					
162	114	131	87	108	63
249	135	172	196	127	100
146	214	80	55	71	130
95	156	201	227	137	125
145	179	74	215	137	124

In Example 1, the probability that the population mean is ex... virtually zero. So, instead of estimating μ to be exactly 130.8 us... estimate, you can estimate that μ lies in an interval. This is called... *interval estimate.*

DEFINITION

An **interval estimate** is an interval, or range of values, used to estimate a population parameter.

Although you can assume that the point estimate in Example 1 is not equal to the actual population mean, it is probably close to it. To form an interval estimate, use the point estimate as the center of the interval, then add and subtract a margin of error. For instance, if the margin of error is 15.7, then an interval estimate would be given by 130.8 ± 15.7 or $115.1 < \mu < 146.5$. The point estimate and interval estimate are as follows.

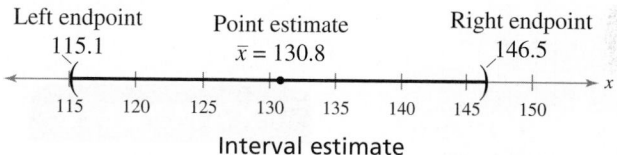

Interval estimate

Before finding a margin of error for an interval estimate, you should first determine how confident you need to be that your interval estimate contains the population mean μ.

DEFINITION

The **level of confidence** c is the probability that the interval estimate contains the population parameter.

You know from the Central Limit Theorem that when $n \geq 30$, the sampling distribution of sample means is a normal distribution. The level of confidence c is the area under the standard normal curve between the *critical values*, $-z_c$ and z_c. **Critical values** are values that separate sample statistics that are probable from sample statistics that are improbable, or unusual. You can see from the graph that c is the percent of the area under the normal curve between $-z_c$ and z_c. The area remaining is $1 - c$, so the area in each tail is $\frac{1}{2}(1 - c)$. For instance, if $c = 90\%$, then 5% of the area lies to the left of $-z_c = -1.645$ and 5% lies to the right of $z_c = 1.645$.

STUDY TIP

In this course, you will usually use 90%, 95%, and 99% levels of confidence. The following z-scores correspond to these levels of confidence.

Level of Confidence	z_c
90%	1.645
95%	1.96
99%	2.575

Note to Instructor

Work through the details of finding z_c with different values of c. Point out to students that there is a natural variation from one sample to another, but c is the percent of these sample means that will be between $-z_c$ and z_c.

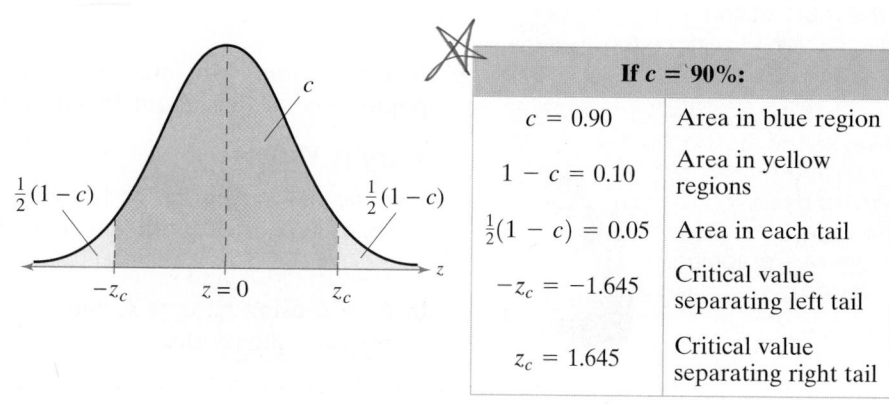

	If $c = 90\%$:
$c = 0.90$	Area in blue region
$1 - c = 0.10$	Area in yellow regions
$\frac{1}{2}(1 - c) = 0.05$	Area in each tail
$-z_c = -1.645$	Critical value separating left tail
$z_c = 1.645$	Critical value separating right tail

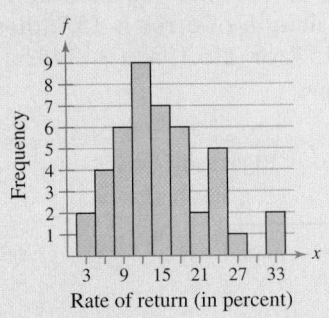

The difference between the point estimate and the actual parameter value is called the **sampling error.** When μ is estimated, the sampling error is the difference $\bar{x} - \mu$. In most cases, of course, μ is unknown, and $\bar{x}$ varies from sample to sample. However, you can calculate a maximum value for the error if you know the level of confidence and the sampling distribution.

DEFINITION

Given a level of confidence c, the **margin of error E** (sometimes also called the maximum error of estimate or error tolerance) is the greatest possible distance between the point estimate and the value of the parameter it is estimating.

$$E = z_c \sigma_{\bar{x}} = z_c \frac{\sigma}{\sqrt{n}}$$

In order to use this technique, it is assumed that the population standard deviation is known. This is rarely the case, but when $n \geq 30$, the sample standard deviation s can be used in place of σ.

EXAMPLE 2

▶ **Finding the Margin of Error**

Use the data given in Example 1 and a 95% confidence level to find the margin of error for the mean number of friends for all users of the website. Assume that the sample standard deviation is about 53.0.

▶ **Solution**

The z-score that corresponds to a 95% confidence level is 1.96. This implies that 95% of the area under the standard normal curve falls within 1.96 standard deviations of the mean. (You can approximate the distribution of the sample means with a normal curve by the Central Limit Theorem because $n = 40 \geq 30$.) You don't know the population standard deviation σ. But because $n \geq 30$, you can use s in place of σ.

Using the values $z_c = 1.96$, $\sigma \approx s \approx 53.0$, and $n = 40$,

$$E = z_c \frac{\sigma}{\sqrt{n}}$$

$$\approx 1.96 \cdot \frac{53.0}{\sqrt{40}}$$

$$\approx 16.4.$$

Interpretation   You are 95% confident that the margin of error for the population mean is about 16.4 friends.

▶ **Try It Yourself 2**

Use the data given in Try It Yourself 1 and a 95% confidence level to find the margin of error for the mean number of friends for all users of the website.

a. *Identify* z_c, n, and s.
b. *Find* E using z_c, $\sigma \approx s$, and n.
c. *Interpret* the results.

Answer: Page A39

▶ CONFIDENCE INTERVALS FOR THE POPULATION MEAN

Using a point estimate and a margin of error, you can construct an interval estimate of a population parameter such as μ. This interval estimate is called a *confidence interval*.

DEFINITION

A *c*-confidence interval for the population mean μ is

$$\bar{x} - E < \mu < \bar{x} + E.$$

The probability that the confidence interval contains μ is c.

GUIDELINES

Finding a Confidence Interval for a Population Mean ($n \geq 30$ or σ known with a normally distributed population)

IN WORDS	IN SYMBOLS
1. Find the sample statistics n and $\bar{x}$.	$\bar{x} = \dfrac{\sum x}{n}$
2. Specify σ, if known. Otherwise, if $n \geq 30$, find the sample standard deviation s and use it as an estimate for σ.	$s = \sqrt{\dfrac{\sum (x - \bar{x})^2}{n - 1}}$
3. Find the critical value z_c that corresponds to the given level of confidence.	Use the Standard Normal Table or technology.
4. Find the margin of error E.	$E = z_c \dfrac{\sigma}{\sqrt{n}}$
5. Find the left and right endpoints and form the confidence interval.	Left endpoint: $\bar{x} - E$ Right endpoint: $\bar{x} + E$ Interval: $\bar{x} - E < \mu < \bar{x} + E$

EXAMPLE 3   **SC** Report 23

| | See MINITAB steps on page 352. |

▶ **Constructing a Confidence Interval**

Use the data given in Example 1 to construct a 95% confidence interval for the mean number of friends for all users of the website.

▶ **Solution**   In Examples 1 and 2, you found that $\bar{x} = 130.8$ and $E \approx 16.4$. The confidence interval is as follows.

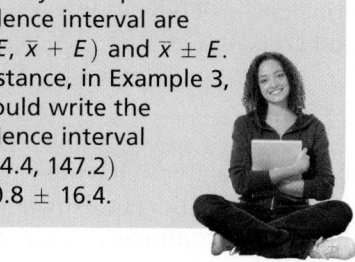

      Left Endpoint   Right Endpoint

$\bar{x} - E \approx 130.8 - 16.4 = 114.4$ $\bar{x} + E \approx 130.8 + 16.4 = 147.2$

$$114.4 < \mu < 147.2$$

Interpretation   With 95% confidence, you can say that the population mean number of friends is between 114.4 and 147.2.

▶ **Try It Yourself 3**

Use the data given in Try It Yourself 1 to construct a 95% confidence interval for the mean number of friends for all users of the website. Compare your result with the interval found in Example 3.

a. *Find* $\bar{x}$ *and E.*
b. Find the *left* and *right endpoints* of the confidence interval.
c. *Interpret* the results and compare them with Example 3. *Answer: Page A39*

EXAMPLE 4

▶ **Constructing a Confidence Interval Using Technology**

Use a technology tool to construct a 99% confidence interval for the mean number of friends for all users of the website using the sample in Example 1.

▶ **Solution**

To use a technology tool to solve the problem, enter the data and recall that the sample standard deviation is $s \approx 53.0$. Then, use the confidence interval command to calculate the confidence interval (*1-Sample Z* for MINITAB). The display should look like the one shown below. To construct a confidence interval using a TI-83/84 Plus, follow the instructions in the margin.

MINITAB

One-Sample Z: Friends

The assumed standard deviation = 53

Variable	N	Mean	StDev	SE Mean	99% CI
Friends	40	130.80	52.63	8.38	(109.21, 152.39)

So, a 99% confidence interval for μ is $(109.2, 152.4)$.

Interpretation With 99% confidence, you can say that the population mean number of friends is between 109.2 and 152.4.

▶ **Try It Yourself 4**

Use the sample data in Example 1 and a technology tool to construct 75%, 85%, and 99% confidence intervals for the mean number of friends for all users of the website. How does the width of the confidence interval change as the level of confidence increases?

a. *Enter* the data.
b. *Use* the appropriate command to construct each confidence interval.
c. *Compare* the widths of the confidence intervals for $c = 0.75, 0.85,$ and 0.99.
Answer: Page A39

In Example 4 and Try It Yourself 4, the same sample data were used to construct confidence intervals with different levels of confidence. Notice that as the level of confidence increases, the width of the confidence interval also increases. In other words, when the same sample data are used, *the greater the level of confidence, the wider the interval.*

If the population is normally distributed and the population standard deviation σ is known, you may use the normal sampling distribution for any sample size, as shown in Example 5.

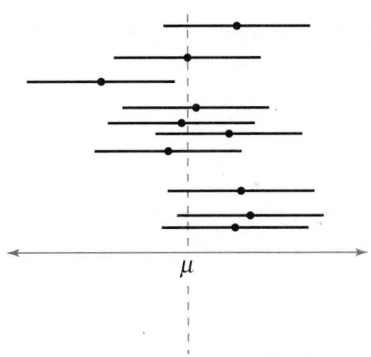

The horizontal segments represent 90% confidence intervals for different samples of the same size. In the long run, 9 of every 10 such intervals will contain μ.

EXAMPLE 5

See TI-83/84 Plus steps on page 353.

▶ Constructing a Confidence Interval, σ Known

A college admissions director wishes to estimate the mean age of all students currently enrolled. In a random sample of 20 students, the mean age is found to be 22.9 years. From past studies, the standard deviation is known to be 1.5 years, and the population is normally distributed. Construct a 90% confidence interval of the population mean age.

▶ Solution

Using $n = 20$, $\overline{x} = 22.9$, $\sigma = 1.5$, and $z_c = 1.645$, the margin of error at the 90% confidence level is

$$E = z_c \frac{\sigma}{\sqrt{n}}$$

$$= 1.645 \cdot \frac{1.5}{\sqrt{20}} \approx 0.6.$$

The 90% confidence interval can be written as $\overline{x} \pm E \approx 22.9 \pm 0.6$ or as follows.

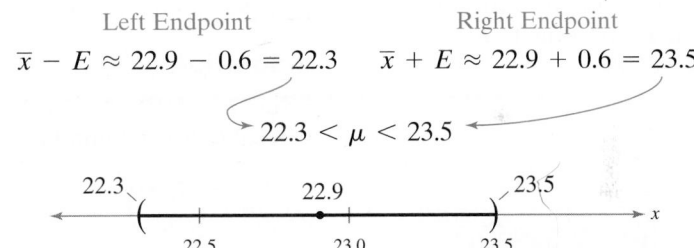

Left Endpoint   Right Endpoint
$\overline{x} - E \approx 22.9 - 0.6 = 22.3$ $\overline{x} + E \approx 22.9 + 0.6 = 23.5$

$22.3 < \mu < 23.5$

Interpretation   With 90% confidence, you can say that the mean age of all the students is between 22.3 and 23.5 years.

▶ Try It Yourself 5

Construct a 90% confidence interval of the population mean age for the college students in Example 5 with the sample size increased to 30 students. Compare your answer with Example 5.

a. *Identify* n, $\overline{x}$, σ, and z_c, and *find* E.
b. Find the *left* and *right endpoints* of the confidence interval.
c. *Interpret* the results and compare them with Example 5.   *Answer: Page A39*

After constructing a confidence interval, it is important that you interpret the results correctly. Consider the 90% confidence interval constructed in Example 5. Because μ is a fixed value predetermined by the population, it is either in the interval or not. It is *not* correct to say "There is a 90% probability that the actual mean will be in the interval (22.3, 23.5)." This statement is wrong because it suggests that the value of μ can vary, which is not true. The correct way to interpret your confidence interval is "If a large number of samples is collected and a confidence interval is created for each sample, approximately 90% of these intervals will contain μ."

Note to Instructor

Students often have trouble identifying the value of E. Remind them that E represents the now given margin of error (or error tolerance). Also point out that the precision is not related to the size of the population—only to the size of the sample.

▸ SAMPLE SIZE

For the same sample statistics, as the level of confidence increases, the confidence interval widens. As the confidence interval widens, the precision of the estimate decreases. One way to improve the precision of an estimate without decreasing the level of confidence is to increase the sample size. But how large a sample size is needed to guarantee a certain level of confidence for a given margin of error?

FIND A MINIMUM SAMPLE SIZE TO ESTIMATE μ

Given a c-confidence level and a margin of error E, the minimum sample size n needed to estimate the population mean μ is

$$n = \left(\frac{z_c \sigma}{E} \right)^2.$$

If σ is unknown, you can estimate it using s, provided you have a preliminary sample with at least 30 members.

EXAMPLE 6

▸ Determining a Minimum Sample Size

You want to estimate the mean number of friends for all users of the website. How many users must be included in the sample if you want to be 95% confident that the sample mean is within seven friends of the population mean?

▸ Solution

Using $c = 0.95$, $z_c = 1.96$, $\sigma \approx s \approx 53.0$ (from Example 2), and $E = 7$, you can solve for the minimum sample size n.

$$n = \left(\frac{z_c \sigma}{E} \right)^2$$

$$\approx \left(\frac{1.96 \cdot 53.0}{7} \right)^2$$

$$\approx 220.23$$

When necessary, round up to obtain a whole number. So, you should include at least 221 users in your sample.

Interpretation You already have 40, so you need 181 more. Note that 221 is the *minimum* number of users to include in the sample. You could include more, if desired.

▸ Try It Yourself 6

How many users must be included in the sample if you want to be 95% confident that the sample mean is within 10 users of the population mean? Compare your answer with Example 6.

a. *Identify* z_c, E, and s.
b. *Use* z_c, E, and $\sigma \approx s$ to find the *minimum sample size n*.
c. *Interpret* the results and compare them with Example 6.

Answer: Page A40

6.1 EXERCISES

FOR EXTRA HELP:
MyStatLab

1. You are more likely to be correct using an interval estimate because it is unlikely that a point estimate will exactly equal the population mean.

2. The percent of the population who fall under the "yes" category probably lies between 40% and 50%.

3. d; As the level of confidence increases, z_c increases, causing wider intervals.

4. See Selected Answers, page A120.

5. 1.28

6. 1.44

7. 1.15

8. 2.17

9. −0.47

10. 0.74

11. 1.76

12. −1.56

13. 1.861

14. 0.675

15. 0.192

16. 1.030

17. c

18. d

19. b

20. a

21. (12.0, 12.6)

22. (31.22, 31.56)

23. (9.7, 11.31)

24. (20.0, 21.2)

■ BUILDING BASIC SKILLS AND VOCABULARY

1. When estimating a population mean, are you more likely to be correct if you use a point estimate or an interval estimate? Explain your reasoning.

2. A news reporter reports the results of a survey and states that 45% of those surveyed responded "yes" with a margin of error of "plus or minus 5%." Explain what this means.

3. Given the same sample statistics, which level of confidence would produce the widest confidence interval? Explain your reasoning.

    (a) 90%   (b) 95%   (c) 98%   (d) 99%

4. You construct a 95% confidence interval for a population mean using a random sample. The confidence interval is $24.9 < \mu < 31.5$. Is the probability that μ is in this interval 0.95? Explain.

In Exercises 5–8, find the critical value z_c necessary to construct a confidence interval at the given level of confidence.

5. $c = 0.80$ **6.** $c = 0.85$ **7.** $c = 0.75$ **8.** $c = 0.97$

Graphical Analysis *In Exercises 9–12, use the values on the number line to find the sampling error.*

9. $\bar{x} = 3.8$ $\mu = 4.27$

    3.4 3.6 3.8 4.0 4.2 4.4 4.6

10. $\mu = 8.76$ $\bar{x} = 9.5$

    8.6 8.8 9.0 9.2 9.4 9.6 9.8

11. $\mu = 24.67$ $\bar{x} = 26.43$

    24   25   26   27

12. $\bar{x} = 46.56$ $\mu = 48.12$

    46   47   48   49

In Exercises 13–16, find the margin of error for the given values of c, s, and n.

13. $c = 0.95, s = 5.2, n = 30$ **14.** $c = 0.90, s = 2.9, n = 50$

15. $c = 0.80, s = 1.3, n = 75$ **16.** $c = 0.975, s = 4.6, n = 100$

Matching *In Exercises 17–20, match the level of confidence c with its representation on the number line, given $\bar{x} = 57.2$, $s = 7.1$, and $n = 50$.*

17. $c = 0.88$ **18.** $c = 0.90$ **19.** $c = 0.95$ **20.** $c = 0.98$

(a)   54.9   57.2   59.5

    54 55 56 57 58 59 60

(b)   55.2   57.2   59.2

    54 55 56 57 58 59 60

(c)   55.6   57.2   58.8

    54 55 56 57 58 59 60

(d)   55.5   57.2   58.9

    54 55 56 57 58 59 60

In Exercises 21–24, construct the indicated confidence interval for the population mean μ. If convenient, use technology to construct the confidence interval.

21. $c = 0.90, \bar{x} = 12.3, s = 1.5, n = 50$

22. $c = 0.95, \bar{x} = 31.39, s = 0.8, n = 82$

23. $c = 0.99, \bar{x} = 10.5, s = 2.14, n = 45$

24. $c = 0.80, \bar{x} = 20.6, s = 4.7, n = 100$

25. 1.4, 13.4

26. 4.27, 25.88

27. 0.17, 1.88

28. 0.016, 3.16

29. 126

30. 25

31. 7

32. 139

33. 1.95, 28.15

34. 18.45, 62.52

35. (428.68, 476.92); (424.06, 481.54)

With 90% confidence, you can say that the population mean price is between $428.68 and $476.92; with 95% confidence, you can say that the population mean price is between $424.06 and $481.54. The 95% CI is wider.

36. (2.26, 2.42); (2.25, 2.43)

With 90% confidence, you can say that the population mean price is between $2.26 and $2.42; with 95% confidence, you can say that the population mean price is between $2.25 and $2.43. The 95% CI is wider.

37. (87.0, 111.6); (84.7, 113.9)

With 90% confidence, you can say that the population mean is between 87.0 and 111.6 calories; with 95% confidence, you can say that the population mean is between 84.7 and 113.9 calories. The 95% CI is wider.

38. (21, 25); (21, 25)

With 90% confidence and with 95% confidence, you can say that the population mean concentration is between 21 and 25 cubic centimeters per cubic meter. When rounded to the nearest whole number, both confidence intervals have the same width.

39. (2532.20, 2767.80)

With 95% confidence, you can say that the population mean cost is between $2532.20 and $2767.80.

40. (144.85, 155.15)

With 99% confidence, you can say that the population mean repair cost is between $144.85 and $155.15.

In Exercises 25–28, use the given confidence interval to find the margin of error and the sample mean.

25. $(12.0, 14.8)$ **26.** $(21.61, 30.15)$

27. $(1.71, 2.05)$ **28.** $(3.144, 3.176)$

In Exercises 29–32, determine the minimum sample size n needed to estimate μ for the given values of c, s, and E.

29. $c = 0.90$, $s = 6.8$, $E = 1$ **30.** $c = 0.95$, $s = 2.5$, $E = 1$

31. $c = 0.80$, $s = 4.1$, $E = 2$ **32.** $c = 0.98$, $s = 10.1$, $E = 2$

■ USING AND INTERPRETING CONCEPTS

Finding the Margin of Error *In Exercises 33 and 34, use the given confidence interval to find the estimated margin of error. Then find the sample mean.*

33. Commute Times A government agency reports a confidence interval of $(26.2, 30.1)$ when estimating the mean commute time (in minutes) for the population of workers in a city.

34. Book Prices A store manager reports a confidence interval of $(44.07, 80.97)$ when estimating the mean price (in dollars) for the population of textbooks.

Constructing Confidence Intervals *In Exercises 35–38, you are given the sample mean and the sample standard deviation. Use this information to construct the 90% and 95% confidence intervals for the population mean. Interpret the results and compare the widths of the confidence intervals. If convenient, use technology to construct the confidence intervals.*

35. Home Theater Systems A random sample of 34 home theater systems has a mean price of $452.80 and a standard deviation of $85.50.

36. Gasoline Prices From a random sample of 48 days in a recent year, U.S. gasoline prices had a mean of $2.34 and a standard deviation of $0.32. *(Source: U.S. Energy Information Administration)*

37. Juice Drinks A random sample of 31 eight-ounce servings of different juice drinks has a mean of 99.3 calories and a standard deviation of 41.5 calories. *(Adapted from The Beverage Institute for Health and Wellness)*

38. Sodium Chloride Concentration In 36 randomly selected seawater samples, the mean sodium chloride concentration was 23 cubic centimeters per cubic meter and the standard deviation was 6.7 cubic centimeters per cubic meter. *(Adapted from Dorling Kindersley Visual Encyclopedia)*

39. Replacement Costs: Transmissions You work for a consumer advocate agency and want to estimate the population mean cost of replacing a car's transmission. As part of your study, you randomly select 50 replacement costs and find the mean to be $2650.00. The sample standard deviation is $425.00. Construct a 95% confidence interval for the population mean replacement cost. Interpret the results. *(Adapted from CostHelper)*

40. Repair Costs: Refrigerators In a random sample of 60 refrigerators, the mean repair cost was $150.00 and the standard deviation was $15.50. Construct a 99% confidence interval for the population mean repair cost. Interpret the results. *(Adapted from Consumer Reports)*

41. (2556.87, 2743.13)
[*Tech:* (2556.90, 2743.10)]

The $n = 50$ CI is wider because a smaller sample is taken, giving less information about the population.

42. (143.69, 156.31)

The $n = 40$ CI is wider because a smaller sample is taken, giving less information about the population.

43. (3.09, 3.15)

With 95% confidence, you can say that the population mean time is between 3.09 and 3.15 minutes.

44. (140.77, 167.57)
[*Tech:* (140.76, 167.58)]

With 99% confidence, you can say that the population mean nightly cost is between $140.77 (*Tech:* $140.76) and $167.57 (*Tech:* $167.58).

45. (3.10, 3.14)

The $s = 0.09$ CI is wider because of the increased variability within the sample.

46. (139.41, 168.93)

The $s = 42.50$ CI is wider because of the increased variability within the sample.

47. (a) An increase in the level of confidence will widen the confidence interval.

(b) An increase in the sample size will narrow the confidence interval.

(c) An increase in the standard deviation will widen the confidence interval.

48. Answers will vary.

49. (14.6, 15.6); (14.4, 15.8)

With 90% confidence, you can say that the population mean length of time is between 14.6 and 15.6 minutes; with 99% confidence, you can say that the population mean length of time is between 14.4 and 15.8 minutes. The 99% CI is wider.

50. (27.6, 30.4); (26.8, 31.2)

With 90% confidence, you can say that the population mean length of time is between 27.6 and 30.4 minutes; with 99% confidence, you can say that the population mean length of time is between 26.8 and 31.2 minutes. The 99% CI is wider.

51. 89

---

41. Repeat Exercise 39, changing the sample size to $n = 80$. Which confidence interval is wider? Explain.

42. Repeat Exercise 40, changing the sample size to $n = 40$. Which confidence interval is wider? Explain.

43. Swimming Times   A random sample of forty-eight 200-meter swims has a mean time of 3.12 minutes and a standard deviation of 0.09 minute. Construct a 95% confidence interval for the population mean time. Interpret the results.

44. Hotels   A random sample of 55 standard hotel rooms in the Philadelphia, PA area has a mean nightly cost of $154.17 and a standard deviation of $38.60. Construct a 99% confidence interval for the population mean cost. Interpret the results.

45. Repeat Exercise 43, using a standard deviation of $s = 0.06$ minute. Which confidence interval is wider? Explain.

46. Repeat Exercise 44, using a standard deviation of $s = \$42.50$. Which confidence interval is wider? Explain.

47. If all other quantities remain the same, how does the indicated change affect the width of a confidence interval?

(a) Increase in the level of confidence

(b) Increase in the sample size

(c) Increase in the standard deviation

48. Describe how you would construct a 90% confidence interval to estimate the population mean age for students at your school.

Constructing Confidence Intervals   *In Exercises 49 and 50, use the given information to construct the 90% and 99% confidence intervals for the population mean. Interpret the results and compare the widths of the confidence intervals. If convenient, use technology to construct the confidence intervals.*

49. DVRs   A research council wants to estimate the mean length of time (in minutes) the average U.S. adult spends watching TVs using digital video recorders (DVRs) each day. To determine this estimate, the research council takes a random sample of 20 U.S. adults and obtains the following results.

15, 18, 17, 20, 24, 12, 9, 15, 14, 25, 8, 6, 10, 14, 16, 20, 27, 10, 9, 13

From past studies, the research council assumes that σ is 1.3 minutes and that the population of times is normally distributed. *(Adapted from the Council for Research Excellence)*

50. Text Messaging   A telecommunications company wants to estimate the mean length of time (in minutes) that 18- to 24-year-olds spend text messaging each day. In a random sample of twenty-seven 18- to 24-year-olds, the mean length of time spent text messaging was 29 minutes. From past studies, the company assumes that σ is 4.5 minutes and that the population of times is normally distributed. *(Adapted from the Council for Research Excellence)*

51. Minimum Sample Size   Determine the minimum required sample size if you want to be 95% confident that the sample mean is within one unit of the population mean given $\sigma = 4.8$. Assume the population is normally distributed.

52. 4

53. (a) 121 servings

 (b) 208 servings

 (c) The 99% CI requires a larger sample because more information is needed from the population to be 99% confident.

54. (a) 4 students

 (b) 10 students

 (c) The 99% CI requires a larger sample because more information is needed from the population to be 99% confident.

55. (a) 32 cans

 (b) 87 cans

 $E = 0.15$ requires a larger sample size. As the error size decreases, a larger sample must be taken to obtain enough information from the population to ensure the desired accuracy.

Error tolerance = 0.25 oz

FIGURE FOR EXERCISE 55

Error tolerance = 1 mL

FIGURE FOR EXERCISE 56

56. See Selected Answers, page A120.

57. (a) 16 sheets

 (b) 62 sheets

 $E = 0.0625$ requires a larger sample size. As the error size decreases, a larger sample must be taken to obtain enough information from the population to ensure the desired accuracy.

52. **Minimum Sample Size** Determine the minimum required sample size if you want to be 99% confident that the sample mean is within two units of the population mean given $\sigma = 1.4$. Assume the population is normally distributed.

53. **Cholesterol Contents of Cheese** A cheese processing company wants to estimate the mean cholesterol content of all one-ounce servings of cheese. The estimate must be within 0.5 milligram of the population mean.

 (a) Determine the minimum required sample size to construct a 95% confidence interval for the population mean. Assume the population standard deviation is 2.8 milligrams.

 (b) Repeat part (a) using a 99% confidence interval.

 (c) Which level of confidence requires a larger sample size? Explain.

54. **Ages of College Students** An admissions director wants to estimate the mean age of all students enrolled at a college. The estimate must be within 1 year of the population mean. Assume the population of ages is normally distributed.

 (a) Determine the minimum required sample size to construct a 90% confidence interval for the population mean. Assume the population standard deviation is 1.2 years.

 (b) Repeat part (a) using a 99% confidence interval.

 (c) Which level of confidence requires a larger sample size? Explain.

55. **Paint Can Volumes** A paint manufacturer uses a machine to fill gallon cans with paint (see figure).

 (a) The manufacturer wants to estimate the mean volume of paint the machine is putting in the cans within 0.25 ounce. Determine the minimum sample size required to construct a 90% confidence interval for the population mean. Assume the population standard deviation is 0.85 ounce.

 (b) Repeat part (a) using an error tolerance of 0.15 ounce. Which error tolerance requires a larger sample size? Explain.

56. **Water Dispensing Machine** A beverage company uses a machine to fill one-liter bottles with water (see figure). Assume that the population of volumes is normally distributed.

 (a) The company wants to estimate the mean volume of water the machine is putting in the bottles within 1 milliliter. Determine the minimum sample size required to construct a 95% confidence interval for the population mean. Assume the population standard deviation is 3 milliliters.

 (b) Repeat part (a) using an error tolerance of 2 milliliters. Which error tolerance requires a larger sample size? Explain.

57. **Plastic Sheet Cutting** A machine cuts plastic into sheets that are 50 feet (600 inches) long. Assume that the population of lengths is normally distributed.

 (a) The company wants to estimate the mean length of the sheets within 0.125 inch. Determine the minimum sample size required to construct a 95% confidence interval for the population mean. Assume the population standard deviation is 0.25 inch.

 (b) Repeat part (a) using an error tolerance of 0.0625 inch. Which error tolerance requires a larger sample size? Explain.

58. (a) 34 units

(b) 135 units

$E = 0.02125$ requires a larger sample size. As the error decreases, a larger sample must be taken to obtain enough information from the population to ensure the desired accuracy.

59. (a) 42 soccer balls

(b) 60 soccer balls

$\sigma = 0.3$ requires a larger sample size. Due to the increased variability in the population, a larger sample size is needed to ensure the desired accuracy.

60. See Selected Answers, page A120.

61. (a) An increase in the level of confidence will increase the minimum sample size required.

(b) An increase (larger E) in the error tolerance will decrease the minimum sample size required.

(c) An increase in the population standard deviation will increase the minimum sample size required.

62. *Sample answer:* A 99% CI may not be practical to use in all situations. It may produce a CI so wide that it has no practical application.

63. (212.8, 221.4)

With 95% confidence, you can say that the population mean airfare price is between $212.8 and $221.4.

```
18 | 3 3
19 | 7
20 | 9 9
21 | 2 2 2 3 3 3 3 3 3 6 6
22 | 2 2 2 3 6 6 8 8 8 8 8 9
23 | 8 8
```

Key: 18|3 = 183

FIGURE FOR EXERCISE 63

64. (19.153, 20.768)

With 95% confidence, you can say that the population mean closing stock price is between $19.153 and $20.768.

58. Paint Sprayer A company uses an automated sprayer to apply paint to metal furniture. The company sets the sprayer to apply the paint one mil (1/1000 of an inch) thick.

(a) The company wants to estimate the mean thickness of paint the sprayer is applying within 0.0425 mil. Determine the minimum sample size required to construct a 90% confidence interval for the population mean. Assume the population standard deviation is 0.15 mil.

(b) Repeat part (a) using an error tolerance of 0.02125 mil. Which error tolerance requires a larger sample size? Explain.

59. Soccer Balls A soccer ball manufacturer wants to estimate the mean circumference of soccer balls within 0.1 inch.

(a) Determine the minimum sample size required to construct a 99% confidence interval for the population mean. Assume the population standard deviation is 0.25 inch.

(b) Repeat part (a) using a standard deviation of 0.3 inch. Which standard deviation requires a larger sample size? Explain.

60. Mini-Soccer Balls A soccer ball manufacturer wants to estimate the mean circumference of mini-soccer balls within 0.15 inch. Assume that the population of circumferences is normally distributed.

(a) Determine the minimum sample size required to construct a 99% confidence interval for the population mean. Assume the population standard deviation is 0.20 inch.

(b) Repeat part (a) using a standard deviation of 0.10 inch. Which standard deviation requires a larger sample size? Explain.

61. If all other quantities remain the same, how does the indicated change affect the minimum sample size requirement?

(a) Increase in the level of confidence

(b) Increase in the error tolerance

(c) Increase in the standard deviation

62. When estimating the population mean, why not construct a 99% confidence interval every time?

Using Technology *In Exercises 63 and 64, you are given a data sample. Use a technology tool to construct a 95% confidence interval for the population mean. Interpret your answer.*

63. Airfare The stem-and-leaf plot shows the results of a random sample of airfare prices (in dollars) for a one-way ticket from Boston, MA to Chicago, IL *(Adapted from Expedia, Inc.)*

64. Stock Prices A random sample of the closing stock prices for the Oracle Corporation for a recent year *(Source: Yahoo! Inc.)*

18.41	16.91	16.83	17.72	15.54	15.56	18.01	19.11	19.79
18.32	18.65	20.71	20.66	21.04	21.74	22.13	21.96	22.16
22.86	20.86	20.74	22.05	21.42	22.34	22.83	24.34	17.97
14.47	19.06	18.42	20.85	21.43	21.97	21.81		

65. **80% confidence interval results:**

μ : population mean

Standard deviation = 344.9

See Odd Answers, page A77.

90% confidence interval results:

μ : population mean

Standard deviation = 344.9

See Odd Answers, page A77.

95% confidence interval results:

μ : population mean

Standard deviation = 344.9

See Odd Answers, page A77.

With 80% confidence, you can say that the population mean sodium content is between 962.0 and 1123.4 milligrams; with 90% confidence, you can say it is between 939.1 and 1146.3 milligrams; with 95% confidence, you can say it is between 919.3 and 1166.1 milligrams.

66. See Selected Answers, page A120.

67. (a) 0.707

(b) 0.949

(c) 0.962

(d) 0.975

(e) The finite population correction factor approaches 1 as the sample size decreases and the population size remains the same.

68. (a) 0.711

(b) 0.937

(c) 0.964

(d) 0.975

(e) The finite population correction factor approaches 1 as the population size increases and the sample size remains the same.

69. See Odd Answers, page A77.

SC *In Exercises 65 and 66, use StatCrunch to construct the 80%, 90%, and 95% confidence intervals for the population mean. Interpret the results.*

65. **Sodium** A random sample of 30 sandwiches from a fast food restaurant has a mean of 1042.7 milligrams of sodium and a standard deviation of 344.9 milligrams of sodium. *(Source: McDonald's Corporation)*

66. **Carbohydrates** The following represents a random sample of the amounts of carbohydrates (in grams) for 30 sandwiches from a fast food restaurant. *(Source: McDonald's Corporation)*

31	33	34	33	37	40	40	45	37	38	63	61	59	38	40
44	51	59	52	60	54	62	39	33	26	34	27	35	28	26

■ EXTENDING CONCEPTS

Finite Population Correction Factor *In Exercises 67 and 68, use the following information.*

In this section, you studied the construction of a confidence interval to estimate a population mean when the population is large or infinite. When a population is finite, the formula that determines the standard error of the mean $\sigma_{\bar{x}}$ needs to be adjusted. If N is the size of the population and n is the size of the sample (where $n \geq 0.05N$), the standard error of the mean is

$$\sigma_{\bar{x}} = \frac{\sigma}{\sqrt{n}} \sqrt{\frac{N-n}{N-1}}.$$

The expression $\sqrt{(N-n)/(N-1)}$ is called the *finite population correction factor*. The margin of error is

$$E = z_c \frac{\sigma}{\sqrt{n}} \sqrt{\frac{N-n}{N-1}}.$$

67. Determine the finite population correction factor for each of the following.

(a) $N = 1000$ and $n = 500$ (b) $N = 1000$ and $n = 100$

(c) $N = 1000$ and $n = 75$ (d) $N = 1000$ and $n = 50$

(e) What happens to the finite population correction factor as the sample size n decreases but the population size N remains the same?

68. Determine the finite population correction factor for each of the following.

(a) $N = 100$ and $n = 50$ (b) $N = 400$ and $n = 50$

(c) $N = 700$ and $n = 50$ (d) $N = 1000$ and $n = 50$

(e) What happens to the finite population correction factor as the population size N increases but the sample size n remains the same?

69. **Sample Size** The equation for determining the sample size

$$n = \left(\frac{z_c \sigma}{E}\right)^2$$

can be obtained by solving the equation for the margin of error

$$E = \frac{z_c \sigma}{\sqrt{n}}$$

for n. Show that this is true and justify each step.

Marathon Training

A marathon is a foot race with a distance of 26.22 miles. It was one of the original events of the modern Olympics, where it was a men's-only event. The women's marathon did not become an Olympic event until 1984. The Olympic record for the men's marathon was set during the 2008 Olympics by Samuel Kamau Wanjiru of Kenya, with a time of 2 hours, 6 minutes, 32 seconds. The Olympic record for the women's marathon was set during the 2000 Olympics by Naoko Takahashi of Japan, with a time of 2 hours, 23 minutes, 14 seconds.

Training for a marathon typically lasts at least 6 months. The training is gradual, with increases in distance about every 2 weeks. About 1 to 3 weeks before the race, the distance run is decreased slightly. The stem-and-leaf plots below show the marathon training times (in minutes) for a sample of 30 male runners and 30 female runners.

**Training Times (in minutes)
of Male Runners**

```
15 | 5 8 9 9 9              Key: 15|5 = 155
16 | 0 0 0 0 1 2 3 4 4 5 8 9
17 | 0 1 1 3 5 6 6 7 7 9
18 | 0 1 5
```

**Training Times (in minutes)
of Female Runners**

```
17 | 8 9 9                  Key: 17|8 = 178
18 | 0 0 0 0 1 2 3 4 6 6 7 9
19 | 0 0 0 1 3 4 5 5 6 6
20 | 0 0 1 2 3
```

■ EXERCISES

1. Use the sample to find a point estimate for the mean training time of the

   (a) male runners.

   (b) female runners.

2. Find the standard deviation of the training times for the

   (a) male runners.

   (b) female runners.

3. Use the sample to construct a 95% confidence interval for the population mean training time of the

   (a) male runners.

   (b) female runners.

4. Interpret the results of Exercise 3.

5. Use the sample to construct a 95% confidence interval for the population mean training time of all runners. How do your results differ from those in Exercise 3? Explain.

6. A trainer wants to estimate the population mean running times for both male and female runners within 2 minutes. Determine the minimum sample size required to construct a 99% confidence interval for the population mean training time of

   (a) male runners. Assume the population standard deviation is 8.9 minutes.

   (b) female runners. Assume the population standard deviation is 8.4 minutes.

6.2 Confidence Intervals for the Mean (Small Samples)

The *t*-Distribution ▸ Confidence Intervals and *t*-Distributions

WHAT YOU SHOULD LEARN

- ▸ How to interpret the *t*-distribution and use a *t*-distribution table

- ▸ How to construct confidence intervals when $n < 30$, the population is normally distributed, and σ is unknown

▸ THE *t*-DISTRIBUTION

In many real-life situations, the population standard deviation is unknown. Moreover, because of various constraints such as time and cost, it is often not practical to collect samples of size 30 or more. So, how can you construct a confidence interval for a population mean given such circumstances? If the random variable is normally distributed (or approximately normally distributed), you can use a *t-distribution*.

DEFINITION

If the distribution of a random variable x is approximately normal, then

$$t = \frac{\bar{x} - \mu}{\frac{s}{\sqrt{n}}}$$

follows a ***t*-distribution.**

Critical values of t are denoted by t_c. Several properties of the *t*-distribution are as follows.

1. The *t*-distribution is bell-shaped and symmetric about the mean.
2. The *t*-distribution is a family of curves, each determined by a parameter called the *degrees of freedom*. The **degrees of freedom** are the number of free choices left after a sample statistic such as $\bar{x}$ is calculated. When you use a *t*-distribution to estimate a population mean, the degrees of freedom are equal to one less than the sample size.

   $$\text{d.f.} = n - 1 \qquad \text{Degrees of freedom}$$

3. The total area under a *t*-curve is 1 or 100%.
4. The mean, median, and mode of the *t*-distribution are equal to 0.
5. As the degrees of freedom increase, the *t*-distribution approaches the normal distribution. After 30 d.f., the *t*-distribution is very close to the standard normal *z*-distribution.

INSIGHT

The following example illustrates the concept of degrees of freedom. Suppose the number of chairs in a classroom equals the number of students: 25 chairs and 25 students. Each of the first 24 students to enter the classroom has a choice as to which chair he or she will sit in. There is no freedom of choice, however, for the 25th student who enters the room.

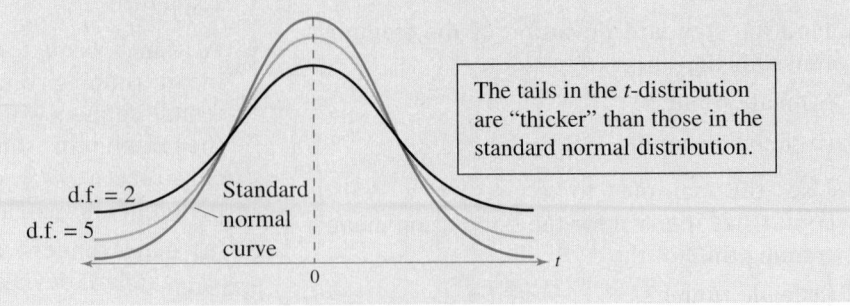

The tails in the *t*-distribution are "thicker" than those in the standard normal distribution.

d.f. = 2
d.f. = 5
Standard normal curve

Table 5 in Appendix B lists critical values of t for selected confidence intervals and degrees of freedom.

EXAMPLE 1

▶ **Finding Critical Values of *t***

Find the critical value t_c for a 95% confidence level when the sample size is 15.

▶ **Solution**

Because $n = 15$, the degrees of freedom are

$$\text{d.f.} = n - 1$$
$$= 15 - 1$$
$$= 14.$$

A portion of Table 5 is shown. Using d.f. $= 14$ and $c = 0.95$, you can find the critical value t_c, as shown by the highlighted areas in the table.

	Level of confidence, c	0.50	0.80	0.90	0.95	0.98
	One tail, α	0.25	0.10	0.05	0.025	0.01
d.f.	**Two tails, α**	0.50	0.20	0.10	0.05	0.02
1		1.000	3.078	6.314	12.706	31.821
2		.816	1.886	2.920	4.303	6.965
3		.765	1.638	2.353	3.182	4.541
12		.695	1.356	1.782	2.179	2.681
13		.694	1.350	1.771	2.160	2.650
14		.692	1.345	1.761	2.145	2.624
15		.691	1.341	1.753	2.131	2.602
16		.690	1.337	1.746	2.120	2.583
28		.683	1.313	1.701	2.048	2.467
29		.683	1.311	1.699	2.045	2.462
∞		.674	1.282	1.645	1.960	2.326

From the table, you can see that $t_c = 2.145$. The graph shows the *t*-distribution for 14 degrees of freedom, $c = 0.95$, and $t_c = 2.145$.

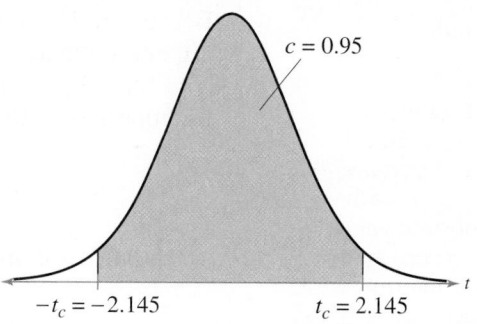

$c = 0.95$

$-t_c = -2.145$ $t_c = 2.145$

Interpretation   So, 95% of the area under the *t*-distribution curve with 14 degrees of freedom lies between $t = \pm 2.145$.

▶ **Try It Yourself 1**

Find the critical value t_c for a 90% confidence level when the sample size is 22.

a. Identify the *degrees of freedom*.
b. Identify the *level of confidence c*.
c. *Use* Table 5 in Appendix B to find t_c.   *Answer: Page A40*

▶ **CONFIDENCE INTERVALS AND *t*-DISTRIBUTIONS**

Constructing a confidence interval using the *t*-distribution is similar to constructing a confidence interval using the normal distribution—both use a point estimate $\bar{x}$ and a margin of error E.

GUIDELINES

Constructing a Confidence Interval for the Mean: *t*-Distribution

IN WORDS	IN SYMBOLS
1. Find the sample statistics n, $\bar{x}$, and s.	$\bar{x} = \dfrac{\sum x}{n}, \quad s = \sqrt{\dfrac{\sum (x - \bar{x})^2}{n - 1}}$
2. Identify the degrees of freedom, the level of confidence c, and the critical value t_c.	d.f. $= n - 1$
3. Find the margin of error E.	$E = t_c \dfrac{s}{\sqrt{n}}$
4. Find the left and right endpoints and form the confidence interval.	Left endpoint: $\bar{x} - E$ Right endpoint: $\bar{x} + E$ Interval: $\bar{x} - E < \mu < \bar{x} + E$

EXAMPLE 2    Report 24

See MINITAB steps on page 352.

▶ **Constructing a Confidence Interval**

You randomly select 16 coffee shops and measure the temperature of the coffee sold at each. The sample mean temperature is 162.0°F with a sample standard deviation of 10.0°F. Construct a 95% confidence interval for the population mean temperature. Assume the temperatures are approximately normally distributed.

▶ **Solution**

Because the sample size is less than 30, σ is unknown, and the temperatures are approximately normally distributed, you can use the *t*-distribution. Using $n = 16$, $\bar{x} = 162.0$, $s = 10.0$, $c = 0.95$, and d.f. $= 15$, you can use Table 5 to find that $t_c = 2.131$. The margin of error at the 95% confidence level is

$$E = t_c \frac{s}{\sqrt{n}} = 2.131 \cdot \frac{10.0}{\sqrt{16}} \approx 5.3.$$

The confidence interval is as follows.

Left Endpoint   Right Endpoint
$\bar{x} - E \approx 162 - 5.3 = 156.7$ $\bar{x} + E \approx 162 + 5.3 = 167.3$

$156.7 < \mu < 167.3$

Interpretation   With 95% confidence, you can say that the population mean temperature of coffee sold is between 156.7°F and 167.3°F.

▶ **Try It Yourself 2**

Construct 90% and 99% confidence intervals for the population mean temperature.

a. *Find* t_c *and* E *for each level of confidence.*
b. *Use* $\bar{x}$ *and* E *to find the* **left** *and* **right endpoints** *of the confidence interval.*
c. *Interpret the results.*   *Answer: Page A40*

EXAMPLE 3   **SC** Report 25

See TI-83/84 Plus
steps on page 353.

▶ **Constructing a Confidence Interval**

To explore this topic further, see Activity 6.2 on page 326.

You randomly select 20 cars of the same model that were sold at a car dealership and determine the number of days each car sat on the dealership's lot before it was sold. The sample mean is 9.75 days, with a sample standard deviation of 2.39 days. Construct a 99% confidence interval for the population mean number of days the car model sits on the dealership's lot. Assume the days on the lot are normally distributed.

▶ **Solution**

Because the sample size is less than 30, σ is unknown, and the days on the lot are normally distributed, you can use the *t*-distribution. Using $n = 20$, $\bar{x} = 9.75$, $s = 2.39$, $c = 0.99$, and d.f. $= 19$, you can use Table 5 to find that $t_c = 2.861$. The margin of error at the 99% confidence level is

$$E = t_c \frac{s}{\sqrt{n}}$$

$$= 2.861 \cdot \frac{2.39}{\sqrt{20}}$$

$$\approx 1.53.$$

The confidence interval is as follows.

HISTORICAL REFERENCE

William S. Gosset (1876–1937)

Developed the *t*-distribution while employed by the Guinness Brewing Company in Dublin, Ireland. Gosset published his findings using the pseudonym Student. The *t*-distribution is sometimes referred to as Student's *t*-distribution. (See page 33 for others who were important in the history of statistics.)

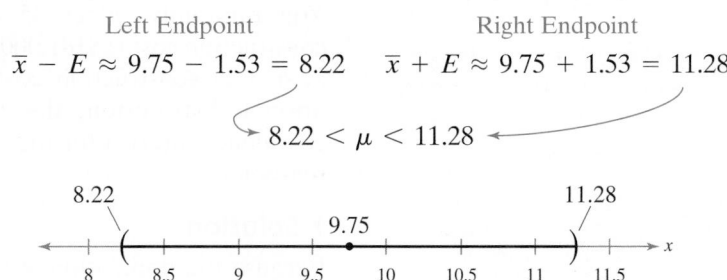

Left Endpoint   Right Endpoint
$\bar{x} - E \approx 9.75 - 1.53 = 8.22$ $\bar{x} + E \approx 9.75 + 1.53 = 11.28$

$8.22 < \mu < 11.28$

Interpretation   With 99% confidence, you can say that the population mean number of days the car model sits on the dealership's lot is between 8.22 and 11.28.

▶ **Try It Yourself 3**

Construct 90% and 95% confidence intervals for the population mean number of days the car model sits on the dealership's lot. Compare the widths of the confidence intervals.

a. *Find* t_c *and* E *for each level of confidence.*
b. *Use* $\bar{x}$ *and* E *to find the* **left** *and* **right endpoints** *of the confidence interval.*
c. *Interpret the results and compare the widths of the confidence intervals.*
                                                        Answer: Page A40

The following flowchart describes when to use the normal distribution and when to use a *t*-distribution to construct a confidence interval for the population mean.

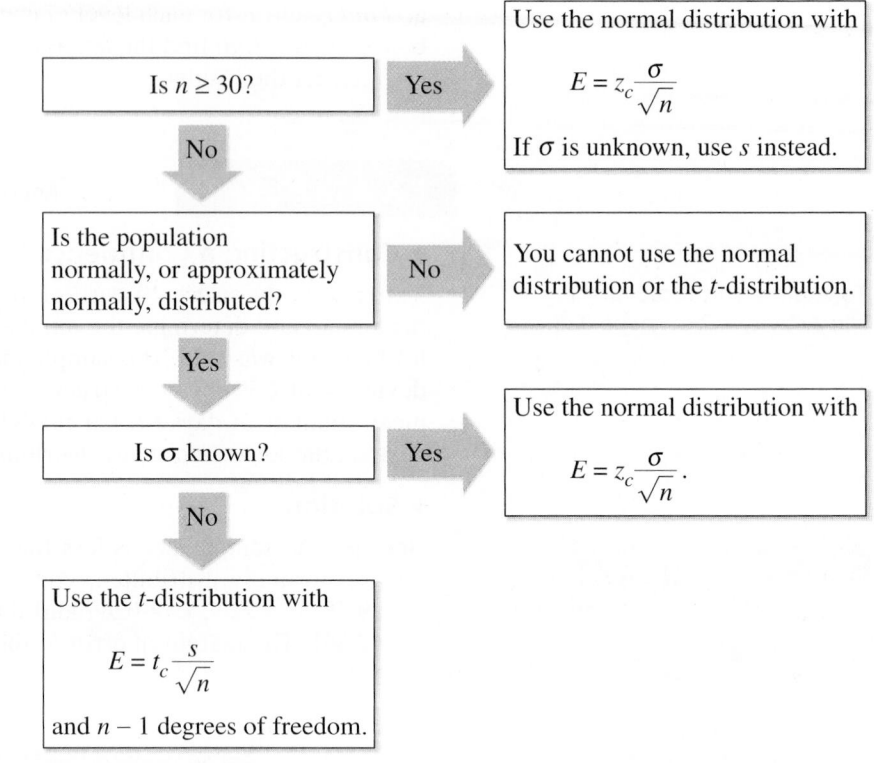

▶ **Choosing the Normal Distribution or the *t*-Distribution**

You randomly select 25 newly constructed houses. The sample mean construction cost is $181,000 and the population standard deviation is $28,000. Assuming construction costs are normally distributed, should you use the normal distribution, the *t*-distribution, or neither to construct a 95% confidence interval for the population mean construction cost? Explain your reasoning.

▶ **Solution**

Because the population is normally distributed and the population standard deviation is known, you should use the normal distribution.

▶ **Try It Yourself 4**

You randomly select 18 adult male athletes and measure the resting heart rate of each. The sample mean heart rate is 64 beats per minute, with a sample standard deviation of 2.5 beats per minute. Assuming the heart rates are normally distributed, should you use the normal distribution, the *t*-distribution, or neither to construct a 90% confidence interval for the population mean heart rate? Explain your reasoning.

Use the flowchart above to determine which distribution you should use to construct the 90% confidence interval for the population mean heart rate.

Answer: Page A40

6.2 EXERCISES

FOR EXTRA HELP:
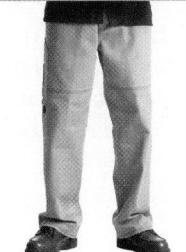

1. 1.833
2. 2.201
3. 2.947
4. 2.539
5. 2.7
6. 4.9
7. 1.2
8. 4.5
9. (a) (10.9, 14.1)
   (b) The *t*-CI is wider.
10. (a) (12.7, 14.1)
    (b) The *t*-CI is wider.
11. (a) (4.1, 4.5)
    (b) When rounded to the nearest tenth, the normal CI and the *t*-CI have the same width.
12. (a) (20.0, 29.4)
    (b) The *t*-CI is wider.
13. 3.7, 18.4
14. 1.18, 7.35
15. 9.5, 74.1
16. 6.8, 23
17. 6.0; (29.5, 41.5); With 95% confidence, you can say that the population mean commute time is between 29.5 and 41.5 minutes.
18. See Selected Answers, page A120.
19. 6.4; (29.1, 41.9); With 95% confidence, you can say that the population mean commute time is between 29.1 and 41.9 minutes. This confidence interval is slightly wider than the one found in Exercise 17.
20. See Selected Answers, page A120.

■ BUILDING BASIC SKILLS AND VOCABULARY

In Exercises 1–4, find the critical value t_c for the given confidence level c and sample size n.

1. $c = 0.90$, $n = 10$

2. $c = 0.95$, $n = 12$

3. $c = 0.99$, $n = 16$

4. $c = 0.98$, $n = 20$

In Exercises 5–8, find the margin of error for the given values of c, s, and n.

5. $c = 0.95$, $s = 5$, $n = 16$

6. $c = 0.99$, $s = 3$, $n = 6$

7. $c = 0.90$, $s = 2.4$, $n = 12$

8. $c = 0.98$, $s = 4.7$, $n = 9$

In Exercises 9–12, (a) construct the indicated confidence interval for the population mean μ using a t-distribution. (b) If you had incorrectly used a normal distribution, which interval would be wider?

9. $c = 0.90$, $\bar{x} = 12.5$, $s = 2.0$, $n = 6$

10. $c = 0.95$, $\bar{x} = 13.4$, $s = 0.85$, $n = 8$

11. $c = 0.98$, $\bar{x} = 4.3$, $s = 0.34$, $n = 14$

12. $c = 0.99$, $\bar{x} = 24.7$, $s = 4.6$, $n = 10$

In Exercises 13–16, use the given confidence interval to find the margin of error and the sample mean.

13. (14.7, 22.1)

14. (6.17, 8.53)

15. (64.6, 83.6)

16. (16.2, 29.8)

■ USING AND INTERPRETING CONCEPTS

Constructing Confidence Intervals *In Exercises 17 and 18, you are given the sample mean and the sample standard deviation. Assume the random variable is normally distributed and use a t-distribution to find the margin of error and construct a 95% confidence interval for the population mean. Interpret the results. If convenient, use technology to construct the confidence interval.*

17. Commute Time to Work In a random sample of eight people, the mean commute time to work was 35.5 minutes and the sample standard deviation was 7.2 minutes.

18. Driving Distance to Work In a random sample of five people, the mean driving distance to work was 22.2 miles and the sample standard deviation was 5.8 miles.

19. You research commute times to work and find that the population standard deviation was 9.3 minutes. Repeat Exercise 17, using a normal distribution with the appropriate calculations for a standard deviation that is known. Compare the results.

20. You research driving distances to work and find that the population standard deviation was 5.2 miles. Repeat Exercise 18, using a normal distribution with the appropriate calculations for a standard deviation that is known. Compare the results.

21. (a) (3.80, 5.20)

    (b) (4.41, 4.59); The *t*-CI in part (a) is wider.

22. (a) (1.35, 1.65)

    (b) (1.48, 1.52); The *t*-CI in part (a) is wider.

23. (a) 90,182.9

    (b) 3724.9

    (c) (87,438.6, 92,927.2)

24. (a) 68,555.6

    (b) 3243.5

    (c) (65,944.6, 71,166.6)
       [*Tech:* (65,944, 71,167)]

25. (a) 1767.7

    (b) 252.2

    (c) (1541.5, 1993.8)

26. (a) 2.35

    (b) 1.03

    (c) (1.56, 3.14) [*Tech:* (1.56, 3.15)]

27. Use a normal distribution because $n \geq 30$.

    (26.0, 29.4); With 95% confidence, you can say that the population mean BMI is between 26.0 and 29.4.

28. Use a *t*-distribution because $n < 30$, the interest rates are normally distributed, and σ is unknown.

    (4.79, 5.19); With 95% confidence, you can say that the population mean interest rate is between 4.79% and 5.19%.

29. Use a *t*-distribution because $n < 30$, the miles per gallon are normally distributed, and σ is unknown.

    (20.5, 23.3) [*Tech:* (20.5, 23.4)]; With 95% confidence, you can say that the population mean is between 20.5 and 23.3 miles per gallon.

30. Use a normal distribution because although $n < 30$, the yards per carry are normally distributed and σ is known.

    (3.71, 4.89); With 95% confidence, you can say that the population mean is between 3.71 and 4.89 yards per carry.

31. Cannot use normal or *t*-distribution because $n < 30$ and the times are not normally distributed.

Constructing Confidence Intervals *In Exercises 21 and 22, you are given the sample mean and the sample standard deviation. Assume the random variable is normally distributed and use a normal distribution or a t-distribution to construct a 90% confidence interval for the population mean. If convenient, use technology to construct the confidence interval.*

21. **Waste Generated**   (a) In a random sample of 10 adults from the United States, the mean waste generated per person per day was 4.50 pounds and the standard deviation was 1.21 pounds. (b) Repeat part (a), assuming the same statistics came from a sample size of 500. Compare the results. *(Adapted from U.S. Environmental Protection Agency)*

22. **Waste Recycled**   (a) In a random sample of 12 adults from the United States, the mean waste recycled per person per day was 1.50 pounds and the standard deviation was 0.28 pound. (b) Repeat part (a), assuming the same statistics came from a sample size of 600. Compare the results. *(Adapted from U.S. Environmental Protection Agency)*

Constructing Confidence Intervals *In Exercises 23–26, a data set is given. For each data set, (a) find the sample mean, (b) find the sample standard deviation, and (c) construct a 99% confidence interval for the population mean. Assume the population of each data set is normally distributed. If convenient, use a technology tool.*

23. **Earnings**   The annual earnings of 16 randomly selected computer software engineers *(Adapted from U.S. Bureau of Labor Statistics)*

    | 92,184 | 86,919 | 90,176 | 91,740 | 95,535 | 90,108 | 94,815 | 88,114 |
    | 85,406 | 90,197 | 89,944 | 93,950 | 84,116 | 96,054 | 85,119 | 88,549 |

24. **Earnings**   The annual earnings of 14 randomly selected physical therapists *(Adapted from U.S. Bureau of Labor Statistics)*

    | 63,118 | 65,740 | 72,899 | 68,500 | 66,726 | 65,554 | 69,247 |
    | 64,963 | 68,627 | 70,448 | 71,842 | 66,873 | 74,103 | 71,138 |

25. **SAT Scores**   The SAT scores of 12 randomly selected high school seniors

    | 1704 | 1940 | 1518 | 2005 | 1432 | 1872 |
    | 1998 | 1658 | 1825 | 1670 | 2210 | 1380 |

26. **GPA**   The grade point averages (GPA) of 15 randomly selected college students

    | 2.3 | 3.3 | 2.6 | 1.8 | 0.2 | 3.1 | 4.0 | 0.7 |
    | 2.3 | 2.0 | 3.1 | 3.4 | 1.3 | 2.6 | 2.6 |

Choosing a Distribution *In Exercises 27–32, use a normal distribution or a t-distribution to construct a 95% confidence interval for the population mean. Justify your decision. If neither distribution can be used, explain why. Interpret the results. If convenient, use technology to construct the confidence interval.*

27. **Body Mass Index**   In a random sample of 50 people, the mean body mass index (BMI) was 27.7 and the standard deviation was 6.12. Assume the body mass indexes are normally distributed. *(Adapted from Centers for Disease Control)*

28. **Mortgages**   In a random sample of 15 mortgage institutions, the mean interest rate was 4.99% and the standard deviation was 0.36%. Assume the interest rates are normally distributed. *(Adapted from Federal Reserve)*

32. Use a *t*-distribution because
n < 30, the lengths of stay are
normally distributed, and *σ* is
unknown.

(5.3, 7.3); With 95% confidence,
you can say that the population
mean length of stay is between
5.3 and 7.3 days.

33. **90% confidence interval results:**

μ : mean of Variable

See Odd Answers, page A78.

95% confidence interval results:

μ : mean of Variable

See Odd Answers, page A78.

99% confidence interval results:

μ : mean of Variable

See Odd Answers, page A78.

With 90% confidence, you can
say that the population mean
time spent on homework is
between 11.5 and 12.9 hours;
with 95% confidence, you can
say it is between 11.3 and 13.1
hours; and with 99% confidence,
you can say it is between 11.0
and 13.4 hours. As the level of
confidence increases, the
intervals get wider.

34. **90% confidence interval results:**

μ : population mean

See Selected Answers, page A120.

95% confidence interval results:

μ : population mean

See Selected Answers, page A120.

99% confidence interval results:

μ : population mean

See Selected Answers, page A120.

With 90% confidence, you can
say that the population mean
weekly time spent weight lifting
is between 6.2 and 8.2 hours;
with 95% confidence, you can
say it is between 5.9 and 8.5
hours; and with 99% confidence,
you can say it is between 5.4
and 9.0 hours. As the level of
confidence increases, the
intervals get wider.

35. No; They are not making good
tennis balls because the desired
bounce height of 55.5 inches is
not between 55.9 and 56.1 inches.

36. Yes; They are making good light
bulbs because the desired bulb
life of 1000 hours is between
997 and 1033 hours.

29. **Sports Cars: Miles per Gallon** You take a random survey of 25 sports
cars and record the miles per gallon for each. The data are listed below.
Assume the miles per gallon are normally distributed.

15 27 24 24 20 21 24 14 21 25 21 13 21
25 22 21 25 24 22 24 24 22 21 24 24

30. **Yards Per Carry** In a recent season, the standard deviation of the
yards per carry for all running backs was 1.34. The yards per carry of 20
randomly selected running backs are listed below. Assume the yards per
carry are normally distributed. *(Source: National Football League)*

5.6 4.4 3.8 4.5 3.3 5.0 3.6 3.7 4.8 3.5
5.6 3.0 6.8 4.7 2.2 3.3 5.7 3.0 5.0 4.5

31. **Hospital Waiting Times** In a random sample of 19 patients at a hospital's
minor emergency department, the mean waiting time before seeing a
medical professional was 23 minutes and the standard deviation was
11 minutes. Assume the waiting times are not normally distributed.

32. **Hospital Length of Stay** In a random sample of 13 people, the mean length
of stay at a hospital was 6.3 days and the standard deviation was 1.7 days.
Assume the lengths of stay are normally distributed. *(Adapted from American
Hospital Association)*

SC *In Exercises 33 and 34, use StatCrunch to construct the 90%, 95%, and 99%
confidence intervals for the population mean. Interpret the results and compare the
widths of the confidence intervals. Assume the random variable is normally distributed.*

33. **Homework** The weekly time spent (in hours) on homework for 18 randomly
selected high school students

12.0 11.3 13.5 11.7 12.0 13.0 15.5 10.8 12.5
12.3 14.0 9.5 8.8 10.0 12.8 15.0 11.8 13.0

34. **Weight Lifting** In a random sample of 11 college football players, the mean
weekly time spent weight lifting was 7.2 hours and the standard deviation
was 1.9 hours.

■ EXTENDING CONCEPTS

35. **Tennis Ball Manufacturing** A company manufactures tennis balls. When
its tennis balls are dropped onto a concrete surface from a height of
100 inches, the company wants the mean height the balls bounce upward to
be 55.5 inches. This average is maintained by periodically testing random
samples of 25 tennis balls. If the *t*-value falls between $-t_{0.99}$ and $t_{0.99}$, the
company will be satisfied that it is manufacturing acceptable tennis balls. A
sample of 25 balls is randomly selected and tested. The mean bounce height
of the sample is 56.0 inches and the standard deviation is 0.25 inch. Assume
the bounce heights are approximately normally distributed. Is the company
making acceptable tennis balls? Explain your reasoning.

36. **Light Bulb Manufacturing** A company manufactures light bulbs. The
company wants the bulbs to have a mean life span of 1000 hours. This
average is maintained by periodically testing random samples of 16 light
bulbs. If the *t*-value falls between $-t_{0.99}$ and $t_{0.99}$, the company will be
satisfied that it is manufacturing acceptable light bulbs. A sample of 16 light
bulbs is randomly selected and tested. The mean life span of the sample is
1015 hours and the standard deviation is 25 hours. Assume the life spans are
approximately normally distributed. Is the company making acceptable light
bulbs? Explain your reasoning.

ACTIVITY 6.2

Confidence Intervals for a Mean (the impact of not knowing the standard deviation)

APPLET

The *confidence intervals for a mean (the impact of not knowing the standard deviation)* applet allows you to visually investigate confidence intervals for a population mean. You can specify the sample size *n*, the shape of the distribution (Normal or Right-skewed), the population mean (Mean), and the true population standard deviation (Std. Dev.). When you click SIMULATE, 100 separate samples of size *n* will be selected from a population with these population parameters. For each of the 100 samples, a 95% Z confidence interval (known standard deviation) and a 95% T confidence interval (unknown standard deviation) are displayed in the plot at the right. The 95% Z confidence interval is displayed in green and the 95% T confidence interval is displayed in blue. If an interval does not contain the population mean, it is displayed in red. Additional simulations can be carried out by clicking SIMULATE multiple times. The cumulative number of times that each type of interval contains the population mean is also shown. Press CLEAR to clear existing results and start a new simulation.

■ Explore

Step 1 Specify a value for *n*.
Step 2 Specify a distribution.
Step 3 Specify a value for the mean.
Step 4 Specify a value for the standard deviation.
Step 5 Click SIMULATE to generate the confidence intervals.

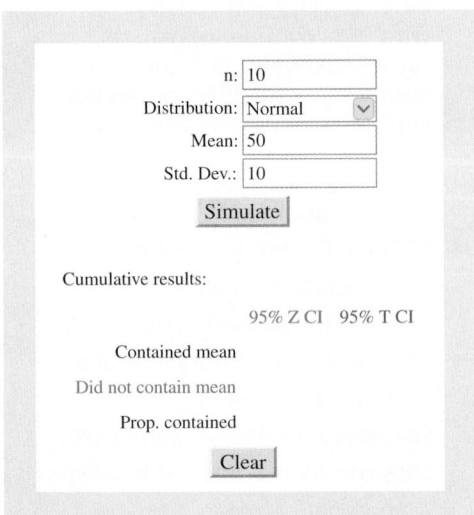

■ Draw Conclusions

APPLET

1. Set *n* = 30, Mean = 25, Std. Dev. = 5, and the distribution to Normal. Run the simulation so that at least 1000 confidence intervals are generated. Compare the proportion of the 95% Z confidence intervals and 95% T confidence intervals that contain the population mean. Is this what you would expect? Explain.

2. In a random sample of 24 high school students, the mean number of hours of sleep per night during the school week was 7.26 hours and the standard deviation was 1.19 hours. Assume the sleep times are normally distributed. Run the simulation for *n* = 10 so that at least 500 confidence intervals are generated. What proportion of the 95% Z confidence intervals and 95% T confidence intervals contain the population mean? Should you use a Z confidence interval or a T confidence interval for the mean number of hours of sleep? Explain.

6.3 Confidence Intervals for Population Proportions

Point Estimate for a Population Proportion ▸ Confidence Intervals for a Population Proportion ▸ Finding a Minimum Sample Size

▸ POINT ESTIMATE FOR A POPULATION PROPORTION

Recall from Section 4.2 that the probability of success in a single trial of a binomial experiment is p. This probability is a **population proportion.** In this section, you will learn how to estimate a population proportion p using a confidence interval. As with confidence intervals for μ, you will start with a point estimate.

DEFINITION

The **point estimate for p,** the population proportion of successes, is given by the proportion of successes in a sample and is denoted by

$$\hat{p} = \frac{x}{n} \qquad \text{Sample proportion}$$

where x is the number of successes in the sample and n is the sample size. The point estimate for the population proportion of failures is $\hat{q} = 1 - \hat{p}$. The symbols $\hat{p}$ and $\hat{q}$ are read as "p hat" and "q hat."

EXAMPLE 1

▸ Finding a Point Estimate for p

In a survey of 1000 U.S. adults, 662 said that it is acceptable to check personal e-mail while at work. Find a point estimate for the population proportion of U.S. adults who say it is acceptable to check personal e-mail while at work. *(Adapted from Liberty Mutual)*

▸ Solution

Using $n = 1000$ and $x = 662$,

$$\hat{p} = \frac{x}{n}$$
$$= \frac{662}{1000}$$
$$= 0.662$$
$$= 66.2\%$$

So, the point estimate for the population proportion of U.S. adults who say it is acceptable to check personal e-mail while at work is 66.2%.

▸ Try It Yourself 1

In a survey of 1006 U.S. adults, 181 said that Abraham Lincoln was the greatest president. Find a point estimate for the population proportion of U.S. adults who say Abraham Lincoln was the greatest president. *(Adapted from The Gallup Poll)*

a. *Identify* x and n.
b. *Use* x and n to find $\hat{p}$.   *Answer: Page A40*

INSIGHT

In the first two sections, estimates were made for quantitative data. In this section, sample proportions are used to make estimates for qualitative data.

In a recent year, there were about 9600 bird-aircraft collisions reported. A poll surveyed 2138 people about bird-aircraft collisions. Of those surveyed, 667 said that they are worried about bird-aircraft collisions. *(Adapted from TripAdvisor)*

Find a 90% confidence interval for the population proportion of people that are worried about bird-aircraft collisions.

(0.296, 0.328)

▶ **CONFIDENCE INTERVALS FOR A POPULATION PROPORTION**

Constructing a confidence interval for a population proportion p is similar to constructing a confidence interval for a population mean. You start with a point estimate and calculate a margin of error.

DEFINITION

A *c*-**confidence interval for a population proportion** p is

$$\hat{p} - E < p < \hat{p} + E$$

where

$$E = z_c \sqrt{\frac{\hat{p}\hat{q}}{n}}.$$

The probability that the confidence interval contains p is c.

In Section 5.5, you learned that a binomial distribution can be approximated by a normal distribution if $np \geq 5$ and $nq \geq 5$. When $n\hat{p} \geq 5$ and $n\hat{q} \geq 5$, the sampling distribution of $\hat{p}$ is approximately normal with a mean of

$$\mu_{\hat{p}} = p$$

and a standard error of

$$\sigma_{\hat{p}} = \sqrt{\frac{pq}{n}}.$$

GUIDELINES

Constructing a Confidence Interval for a Population Proportion

IN WORDS	IN SYMBOLS
1. Identify the sample statistics n and x.	
2. Find the point estimate $\hat{p}$.	$\hat{p} = \dfrac{x}{n}$
3. Verify that the sampling distribution of $\hat{p}$ can be approximated by a normal distribution.	$n\hat{p} \geq 5,\ n\hat{q} \geq 5$
4. Find the critical value z_c that corresponds to the given level of confidence c.	Use the Standard Normal Table or technology.
5. Find the margin of error E.	$E = z_c \sqrt{\dfrac{\hat{p}\hat{q}}{n}}$
6. Find the left and right endpoints and form the confidence interval.	Left endpoint: $\hat{p} - E$ Right endpoint: $\hat{p} + E$ Interval: $\hat{p} - E < p < \hat{p} + E$

MINITAB and TI-83/84 Plus steps are shown on pages 352 and 353.

EXAMPLE 2  Report 26

▶ Constructing a Confidence Interval for *p*

Use the data given in Example 1 to construct a 95% confidence interval for the population proportion of U.S. adults who say that it is acceptable to check personal e-mail while at work.

▶ Solution

From Example 1, $\hat{p} = 0.662$. So,

$$\hat{q} = 1 - 0.662 = 0.338.$$

Using $n = 1000$, you can verify that the sampling distribution of $\hat{p}$ can be approximated by a normal distribution.

$$n\hat{p} = 1000 \cdot 0.662 = 662 > 5$$

$$n\hat{q} = 1000 \cdot 0.338 = 338 > 5$$

Using $z_c = 1.96$, the margin of error is

$$E = z_c \sqrt{\frac{\hat{p}\hat{q}}{n}} = 1.96\sqrt{\frac{(0.662)(0.338)}{1000}} \approx 0.029.$$

The 95% confidence interval is as follows.

Left Endpoint   Right Endpoint

$\hat{p} - E \approx 0.662 - 0.029 = 0.633$ $\hat{p} + E \approx 0.662 + 0.029 = 0.691$

$$0.633 < p < 0.691$$

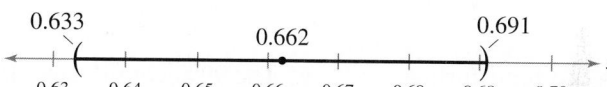

Interpretation With 95% confidence, you can say that the population proportion of U.S. adults who say that it is acceptable to check personal e-mail while at work is between 63.3% and 69.1%.

▶ Try It Yourself 2

Use the data given in Try It Yourself 1 to construct a 90% confidence interval for the population proportion of U.S. adults who say that Abraham Lincoln was the greatest president.

a. *Find* $\hat{p}$ *and* $\hat{q}$.
b. *Verify* that the sampling distribution of $\hat{p}$ can be approximated by a normal distribution.
c. *Find* z_c *and* E.
d. Use $\hat{p}$ and E to find the *left* and *right endpoints* of the confidence interval.
e. *Interpret* the results.   *Answer: Page A40*

The confidence level of 95% used in Example 2 is typical of opinion polls. The result, however, is usually not stated as a confidence interval. Instead, the result of Example 2 would be stated as "66.2% with a margin of error of ±2.9%."

Note to Instructor

Point out that the value of *E* is calculated by multiplying the *z*-score (the number of standard deviations from the mean) by the standard error of $\hat{p}$. The *z*-scores are found the same way they were found in Section 6.1.

STUDY TIP

Notice in Example 2 that the confidence interval for the population proportion *p* is rounded to three decimal places. This *round-off rule* will be used throughout the text.

EXAMPLE 3 SC Report 27

▶ **Constructing a Confidence Interval for p**

The graph shown at the right is from a survey of 498 U.S. adults. Construct a 99% confidence interval for the population proportion of U.S. adults who think that teenagers are the more dangerous drivers. (*Source: The Gallup Poll*)

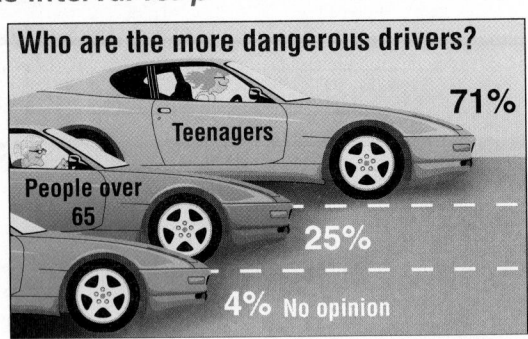

Who are the more dangerous drivers?

Teenagers **71%**

People over 65 **25%**

4% No opinion

INSIGHT

In Example 3, note that $n\hat{p} \geq 5$ and $n\hat{q} \geq 5$. So, the sampling distribution of $\hat{p}$ is approximately normal.

To explore this topic further, see Activity 6.3 on page 336.

▶ **Solution**

From the graph, $\hat{p} = 0.71$. So,

$$\hat{q} = 1 - 0.71$$
$$= 0.29.$$

Using these values and the values $n = 498$ and $z_c = 2.575$, the margin of error is

$$E = z_c \sqrt{\frac{\hat{p}\hat{q}}{n}}$$

$$\approx 2.575\sqrt{\frac{(0.71)(0.29)}{498}}$$ Use Table 4 in Appendix B to estimate that z_c is halfway between 2.57 and 2.58.

$$\approx 0.052.$$

The 99% confidence interval is as follows.

Left Endpoint

Right Endpoint

$$\hat{p} - E \approx 0.71 - 0.052 = 0.658 \qquad \hat{p} + E \approx 0.71 + 0.052 = 0.762$$

$$0.658 < p < 0.762$$

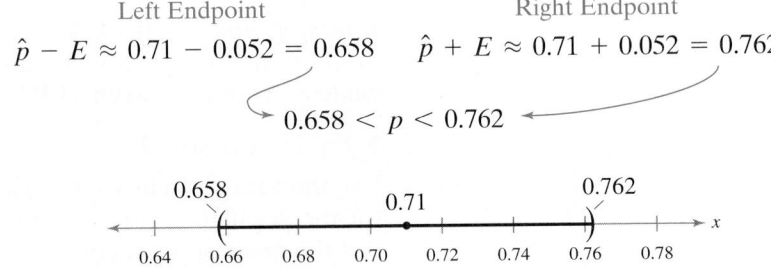

Interpretation With 99% confidence, you can say that the population proportion of U.S. adults who think that teenagers are the more dangerous drivers is between 65.8% and 76.2%.

▶ **Try It Yourself 3**

Use the data given in Example 3 to construct a 99% confidence interval for the population proportion of adults who think that people over 65 are the more dangerous drivers.

a. *Find* $\hat{p}$ *and* $\hat{q}$.
b. *Verify* that the sampling distribution of $\hat{p}$ can be approximated by a normal distribution.
c. *Find* z_c *and* E.
d. *Use* $\hat{p}$ *and* E to find the *left* and *right endpoints* of the confidence interval.
e. *Interpret* the results.

Answer: Page A40

▶ FINDING A MINIMUM SAMPLE SIZE

One way to increase the precision of a confidence interval without decreasing the level of confidence is to increase the sample size.

FINDING A MINIMUM SAMPLE SIZE TO ESTIMATE p

Given a c-confidence level and a margin of error E, the minimum sample size n needed to estimate p is

$$n = \hat{p}\hat{q}\left(\frac{z_c}{E}\right)^2.$$

This formula assumes that you have preliminary estimates of $\hat{p}$ and $\hat{q}$. If not, use $\hat{p} = 0.5$ and $\hat{q} = 0.5$.

EXAMPLE 4

▶ Determining a Minimum Sample Size

You are running a political campaign and wish to estimate, with 95% confidence, the population proportion of registered voters who will vote for your candidate. Your estimate must be accurate within 3% of the population proportion. Find the minimum sample size needed if (1) no preliminary estimate is available and (2) a preliminary estimate gives $\hat{p} = 0.31$. Compare your results.

▶ Solution

1. Because you do not have a preliminary estimate of $\hat{p}$, use $\hat{p} = 0.5$ and $\hat{q} = 0.5$. Using $z_c = 1.96$ and $E = 0.03$, you can solve for n.

$$n = \hat{p}\hat{q}\left(\frac{z_c}{E}\right)^2 = (0.5)(0.5)\left(\frac{1.96}{0.03}\right)^2 \approx 1067.11$$

Because n is a decimal, round up to the nearest whole number, 1068.

2. You have a preliminary estimate of $\hat{p} = 0.31$. So, $\hat{q} = 0.69$. Using $z_c = 1.96$ and $E = 0.03$, you can solve for n.

$$n = \hat{p}\hat{q}\left(\frac{z_c}{E}\right)^2 = (0.31)(0.69)\left(\frac{1.96}{0.03}\right)^2 \approx 913.02$$

Because n is a decimal, round up to the nearest whole number, 914.

Interpretation With no preliminary estimate, the minimum sample size should be at least 1068 registered voters. With a preliminary estimate of $\hat{p} = 0.31$, the sample size should be at least 914 registered voters. So, you will need a larger sample size if no preliminary estimate is available.

▶ Try It Yourself 4

You wish to estimate, with 90% confidence, the population proportion of females who refuse to eat leftovers. Your estimate must be accurate within 2% of the population proportion. Find the minimum sample size needed if (1) no preliminary estimate is available and (2) a previous survey found that 11% of females refuse to eat leftovers. *(Source: Consumer Reports National Research Center)*

a. *Identify* $\hat{p}$, $\hat{q}$, z_c, and E. If $\hat{p}$ is unknown, use 0.5.
b. *Use* $\hat{p}$, $\hat{q}$, z_c, and E to find the *minimum sample size n*.
c. *Determine* how many females should be included in the sample.

Answer: Page A40

6.3 EXERCISES

1. False. To estimate the value of *p*, the population proportion of successes, use the point estimate $\hat{p} = x/n$.

2. True

3. 0.750, 0.250

4. 0.830, 0.170

5. 0.423, 0.577

6. 0.110, 0.890

7. $E = 0.014, \hat{p} = 0.919$

8. $E = 0.115, \hat{p} = 0.360$

9. $E = 0.042, \hat{p} = 0.554$

10. $E = 0.088, \hat{p} = 0.175$

11. (0.557, 0.619)
[*Tech:* (0.556, 0.619)];
(0.551, 0.625)
[*Tech:* (0.550, 0.625)];
With 90% confidence, you can say that the population proportion of U.S. males ages 18–64 who say they have gone to the dentist in the past year is between 55.7% (*Tech:* 55.6%) and 61.9%; with 95% confidence, you can say it is between 55.1% (*Tech:* 55.0%) and 62.5%. The 95% confidence interval is slightly wider.

■ BUILDING BASIC SKILLS AND VOCABULARY

True or False? *In Exercises 1 and 2, determine whether the statement is true or false. If it is false, rewrite it as a true statement.*

1. To estimate the value of *p*, the population proportion of successes, use the point estimate *x*.

2. The point estimate for the proportion of failures is $1 - \hat{p}$.

Finding $\hat{p}$ and $\hat{q}$ *In Exercises 3–6, let p be the population proportion for the given condition. Find point estimates of p and q.*

3. **Recycling** In a survey of 1002 U.S. adults, 752 say they recycle. (*Adapted from ABC News Poll*)

4. **Charity** In a survey of 2939 U.S. adults, 2439 say they have contributed to a charity in the past 12 months. (*Adapted from Harris Interactive*)

5. **Computers** In a survey of 11,605 parents, 4912 think that the government should subsidize the costs of computers for lower-income families. (*Adapted from DisneyFamily.com*)

6. **Vacation** In a survey of 1003 U.S. adults, 110 say they would go on vacation to Europe if cost did not matter. (*Adapted from The Gallup Poll*)

In Exercises 7–10, use the given confidence interval to find the margin of error and the sample proportion.

7. (0.905, 0.933)

8. (0.245, 0.475)

9. (0.512, 0.596)

10. (0.087, 0.263)

■ USING AND INTERPRETING CONCEPTS

Constructing Confidence Intervals *In Exercises 11 and 12, construct 90% and 95% confidence intervals for the population proportion. Interpret the results and compare the widths of the confidence intervals. If convenient, use technology to construct the confidence intervals.*

11. **Dental Visits** In a survey of 674 U.S. males ages 18–64, 396 say they have gone to the dentist in the past year. (*Adapted from National Center for Health Statistics*)

12. **Dental Visits** In a survey of 420 U.S. females ages 18–64, 279 say they have gone to the dentist in the past year. (*Adapted from National Center for Health Statistics*)

Constructing Confidence Intervals *In Exercises 13 and 14, construct a 99% confidence interval for the population proportion. Interpret the results.*

13. **Going Green** In a survey of 3110 U.S. adults, 1435 say they have started paying bills online in the last year. (*Adapted from Harris Interactive*)

14. **Seen a Ghost** In a survey of 4013 U.S. adults, 722 say they have seen a ghost. (*Adapted from Pew Research Center*)

12. (0.626, 0.702); (0.619, 0.709);
With 90% confidence, you can say that the population proportion of U.S. females ages 18–64 who say they have gone to the dentist in the past year is between 62.6% and 70.2%; with 95% confidence, you can say it is between 61.9% and 70.9%. The 95% confidence interval is slightly wider.

13. (0.438, 0.484);
With 99% confidence, you can say that the population proportion of U.S. adults who say they have started paying bills online in the last year is between 43.8% and 48.4%.

14. (0.164, 0.196);
With 99% confidence, you can say that the population proportion of U.S. adults who say they have seen a ghost is between 16.4% and 19.6%.

15. (0.622, 0.644)

16. (0.185, 0.229)
[*Tech:* (0.184, 0.229)]

17. (a) 601 adults
(b) 413 adults
(c) Having an estimate of the population proportion reduces the minimum sample size needed.

18. (a) 4145 adults
(b) 4130 adults
(c) Having an estimate of the population proportion reduces the minimum sample size needed.

19. (a) 752 adults
(b) 483 adults
(c) Having an estimate of the population proportion reduces the minimum sample size needed.

20. (a) 385 adults
(b) 303 adults
(c) Having an estimate of the population proportion reduces the minimum sample size needed.

15. Nail Polish   In a survey of 7000 women, 4431 say they change their nail polish once a week. Construct a 95% confidence interval for the population proportion of women who change their nail polish once a week. (*Adapted from Essie Cosmetics*)

16. World Series   In a survey of 891 U.S. adults who follow baseball in a recent year, 184 said that the Boston Red Sox would win the World Series. Construct a 90% confidence interval for the population proportion of U.S. adults who follow baseball who in a recent year said that the Boston Red Sox would win the World Series. (*Adapted from Harris Interactive*)

17. Alternative Energy   You wish to estimate, with 95% confidence, the population proportion of U.S. adults who want more funding for alternative energy. Your estimate must be accurate within 4% of the population proportion.

(a) No preliminary estimate is available. Find the minimum sample size needed.

(b) Find the minimum sample size needed, using a prior study that found that 78% of U.S. adults want more funding for alternative energy. (*Source: Pew Research Center*)

(c) Compare the results from parts (a) and (b).

18. Reading Fiction   You wish to estimate, with 99% confidence, the population proportion of U.S. adults who read fiction books. Your estimate must be accurate within 2% of the population proportion.

(a) No preliminary estimate is available. Find the minimum sample size needed.

(b) Find the minimum sample size needed, using a prior study that found that 47% of U.S. adults read fiction books. (*Source: National Endowment for the Arts*)

(c) Compare the results from parts (a) and (b).

19. Emergency Room Visits   You wish to estimate, with 90% confidence, the population proportion of U.S. adults who made one or more emergency room visits in the past year. Your estimate must be accurate within 3% of the population proportion.

(a) No preliminary estimate is available. Find the minimum sample size needed.

(b) Find the minimum sample size needed, using a prior study that found that 20.1% of U.S. adults made one or more emergency room visits in the past year. (*Source: National Center for Health Statistics*)

(c) Compare the results from parts (a) and (b).

20. Ice Cream   You wish to estimate, with 95% confidence, the population proportion of U.S. adults who say chocolate is their favorite ice cream flavor. Your estimate must be accurate within 5% of the population proportion.

(a) No preliminary estimate is available. Find the minimum sample size needed.

(b) Find the minimum sample size needed, using a prior study that found that 27% of U.S. adults say that chocolate is their favorite ice cream flavor. (*Source: Harris Interactive*)

(c) Compare the results from parts (a) and (b).

21. (a) (0.234, 0.306)
    (b) (0.450, 0.530)
    (c) (0.275, 0.345)

22. (a) No, the confidence intervals do not overlap.
    (b) No, the confidence intervals do not overlap.
    (c) Yes, the confidence intervals overlap.

23. (a) (0.274, 0.366)
    (b) (0.511, 0.609)

24. (a) (0.313, 0.407)
    (b) (0.217, 0.303)

25. No, it is unlikely that the two proportions are equal because the confidence intervals estimating the proportions do not overlap. The 99% confidence intervals are (0.260, 0.380) and (0.496, 0.624). Although these intervals are wider, they still do not overlap.

26. No, it is unlikely that the two proportions are equal because the confidence intervals estimating the proportions do not overlap. The 99% confidence intervals are (0.298, 0.422) and (0.204, 0.316). Using these intervals, it is possible that the two proportions are equal because the confidence intervals overlap.

Constructing Confidence Intervals

In Exercises 21 and 22, use the following information. The graph shows the results of a survey in which 1017 adults from the United States, 1060 adults from Italy, and 1126 adults from Great Britain were asked if they believe climate change poses a large threat to the world. (Source: Harris Interactive)

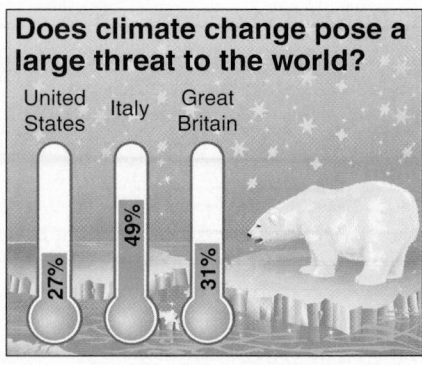

Does climate change pose a large threat to the world?

21. **Global Warming** Construct a 99% confidence interval for

    (a) the population proportion of adults from the United States who say that climate change poses a large threat to the world.

    (b) the population proportion of adults from Italy who say that climate change poses a large threat to the world.

    (c) the population proportion of adults from Great Britain who say that climate change poses a large threat to the world.

22. **Global Warming** Determine whether it is possible that the following proportions are equal and explain your reasoning.

    (a) The proportion of adults from Exercise 21(a) and the proportion of adults from Exercise 21(b).

    (b) The proportion of adults from Exercise 21(b) and the proportion of adults from Exercise 21(c).

    (c) The proportion of adults from Exercise 21(a) and the proportion of adults from Exercise 21(c).

Constructing Confidence Intervals

In Exercises 23 and 24, use the following information. The table shows the results of a survey in which separate samples of 400 adults each from the East, South, Midwest, and West were asked if traffic congestion is a serious problem in their community. (Adapted from Harris Interactive)

Bad Traffic Congestion?
Adults who say that traffic congestion is a serious problem

East	36%
South	32%
Midwest	26%
West	56%

23. **South and West** Construct a 95% confidence interval for the population proportion of adults

    (a) from the South who say traffic congestion is a serious problem.

    (b) from the West who say traffic congestion is a serious problem.

24. **East and Midwest** Construct a 95% confidence interval for the population proportion of adults

    (a) from the East who say traffic congestion is a serious problem.

    (b) from the Midwest who say traffic congestion is a serious problem.

25. **Writing** Is it possible that the proportions in Exercise 23 are equal? What if you used a 99% confidence interval? Explain your reasoning.

26. **Writing** Is it possible that the proportions in Exercise 24 are equal? What if you used a 99% confidence interval? Explain your reasoning.

27. See Odd Answers, page A78.

28. See Selected Answers, page A120.

29. (0.304, 0.324) is approximately a 97.6% CI.

30. (0.160, 0.220) is approximately a 99.2% CI.

31. If $n\hat{p} < 5$ or $n\hat{q} < 5$, the sampling distribution of $\hat{p}$ may not be normally distributed, so z_c cannot be used to calculate the confidence interval.

32. See Selected Answers, page A121.

33. See Odd Answers, page A79.

SC *In Exercises 27 and 28, use StatCrunch to construct 90%, 95%, and 99% confidence intervals for the population proportion. Interpret the results and compare the widths of the confidence intervals.*

27. **Congress**   In a survey of 1025 U.S. adults, 802 disapprove of the job Congress is doing. *(Adapted from The Gallup Poll)*

28. **UFOs**   In a survey of 2303 U.S. adults, 734 believe in UFOs. *(Adapted from Harris Interactive)*

■ EXTENDING CONCEPTS

Newspaper Surveys   *In Exercises 29 and 30, translate the newspaper excerpt into a confidence interval for p. Approximate the level of confidence.*

29. In a survey of 8451 U.S. adults, 31.4% said they were taking vitamin E as a supplement. The survey's margin of error is plus or minus 1%. *(Source: Decision Analyst, Inc.)*

30. In a survey of 1000 U.S. adults, 19% are concerned that their taxes will be audited by the Internal Revenue Service. The survey's margin of error is plus or minus 3%. *(Source: Rasmussen Reports)*

31. **Why Check It?**   Why is it necessary to check that $n\hat{p} \geq 5$ and $n\hat{q} \geq 5$?

32. **Sample Size**   The equation for determining the sample size

$$n = \hat{p}\hat{q}\left(\frac{z_c}{E}\right)^2$$

can be obtained by solving the equation for the margin of error

$$E = z_c \sqrt{\frac{\hat{p}\hat{q}}{n}}$$

for n. Show that this is true and justify each step.

33. **Maximum Value of $\hat{p}\hat{q}$**   Complete the tables for different values of $\hat{p}$ and $\hat{q} = 1 - \hat{p}$. From the tables, which value of $\hat{p}$ appears to give the maximum value of the product $\hat{p}\hat{q}$?

$\hat{p}$	$\hat{q} = 1 - \hat{p}$	$\hat{p}\hat{q}$	$\hat{p}$	$\hat{q} = 1 - \hat{p}$	$\hat{p}\hat{q}$
0.0	1.0	0.00	0.45		
0.1	0.9	0.09	0.46		
0.2	0.8		0.47		
0.3			0.48		
0.4			0.49		
0.5			0.50		
0.6			0.51		
0.7			0.52		
0.8			0.53		
0.9			0.54		
1.0			0.55		

ACTIVITY 6.3 Confidence Intervals for a Proportion

APPLET

The *confidence intervals for a proportion* applet allows you to visually investigate confidence intervals for a population proportion. You can specify the sample size n and the population proportion p. When you click SIMULATE, 100 separate samples of size n will be selected from a population with a proportion of successes equal to p. For each of the 100 samples, a 95% confidence interval (in green) and a 99% confidence interval (in blue) are displayed in the plot at the right. Each of these intervals is computed using the standard normal approximation. If an interval does not contain the population proportion, it is displayed in red. Note that the 99% confidence interval is always wider than the 95% confidence interval. Additional simulations can be carried out by clicking SIMULATE multiple times. The cumulative number of times that each type of interval contains the population proportion is also shown. Press CLEAR to clear existing results and start a new simulation.

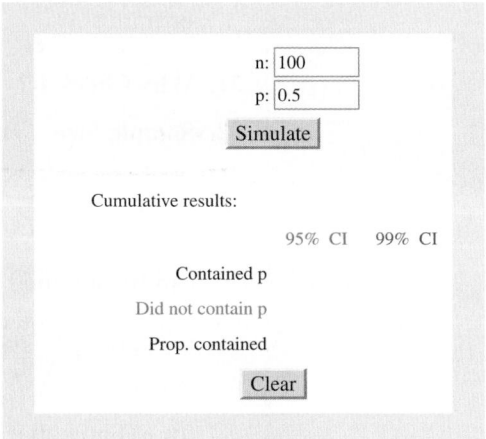

■ Explore

Step 1 Specify a value for n.
Step 2 Specify a value for p.
Step 3 Click SIMULATE to generate the confidence intervals.

■ Draw Conclusions

APPLET

1. Run the simulation for $p = 0.6$ and $n = 10, 20, 40,$ and 100. Clear the results after each trial. What proportion of the confidence intervals for each confidence level contains the population proportion? What happens to the proportion of confidence intervals that contains the population proportion for each confidence level as the sample size increases?

2. Run the simulation for $p = 0.4$ and $n = 100$ so that at least 1000 confidence intervals are generated. Compare the proportion of confidence intervals that contains the population proportion for each confidence level. Is this what you would expect? Explain.

6.4 Confidence Intervals for Variance and Standard Deviation

The Chi-Square Distribution ▸ Confidence Intervals for σ^2 and σ

▶ THE CHI-SQUARE DISTRIBUTION

In manufacturing, it is necessary to control the amount that a process varies. For instance, an automobile part manufacturer must produce thousands of parts to be used in the manufacturing process. It is important that the parts vary little or not at all. How can you measure, and consequently control, the amount of variation in the parts? You can start with a point estimate.

DEFINITION

The **point estimate for σ^2** is s^2 and the **point estimate for σ** is s. The most unbiased estimate for σ^2 is s^2.

You can use a *chi-square distribution* to construct a confidence interval for the variance and standard deviation.

STUDY TIP

The Greek letter χ is pronounced "$k\bar{i}$," which rhymes with the more familiar Greek letter π.

DEFINITION

If a random variable x has a normal distribution, then the distribution of

$$\chi^2 = \frac{(n-1)s^2}{\sigma^2}$$

forms a **chi-square distribution** for samples of any size $n > 1$. Four properties of the chi-square distribution are as follows.

1. All chi-square values χ^2 are greater than or equal to 0.
2. The chi-square distribution is a family of curves, each determined by the degrees of freedom. To form a confidence interval for σ^2, use the χ^2-distribution with degrees of freedom equal to one less than the sample size.

   $$\text{d.f.} = n - 1 \qquad \text{Degrees of freedom}$$

3. The area under each curve of the chi-square distribution equals 1.
4. Chi-square distributions are positively skewed.

Note to Instructor

If you are short of time, this section can be omitted. Or, if you prefer, it can be covered with Chapter 10 when additional chi-square applications are presented.

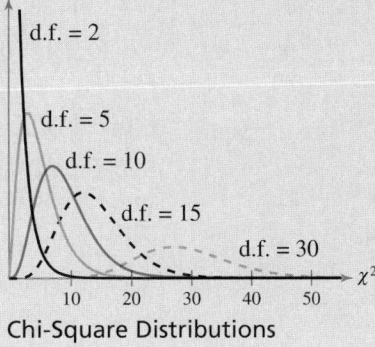

Chi-Square Distributions

STUDY TIP

For chi-square critical values with a c-confidence level, the following values are what you look up in Table 6 in Appendix B.

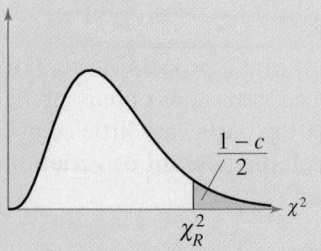

Area to the right of χ_R^2

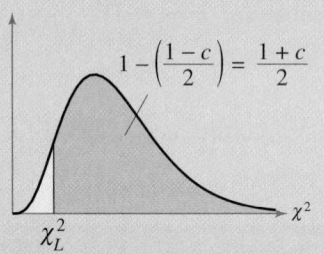

Area to the right of χ_L^2

The result is that you can conclude that the area between the left and right critical values is c.

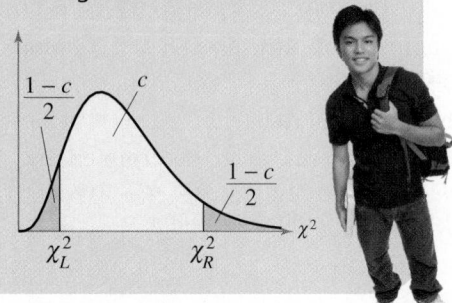

There are two critical values for each level of confidence. The value χ_R^2 represents the right-tail critical value and χ_L^2 represents the left-tail critical value. Table 6 in Appendix B lists critical values of χ^2 for various degrees of freedom and areas. Each area in the table represents the region under the chi-square curve to the *right* of the critical value.

EXAMPLE 1

▶ Finding Critical Values for χ^2

Find the critical values χ_R^2 and χ_L^2 for a 95% confidence interval when the sample size is 18.

▶ Solution

Because the sample size is 18, there are

$$\text{d.f.} = n - 1 = 18 - 1 = 17 \text{ degrees of freedom.}$$

The areas to the right of χ_R^2 and χ_L^2 are

$$\text{Area to right of } \chi_R^2 = \frac{1 - c}{2} = \frac{1 - 0.95}{2} = 0.025$$

and

$$\text{Area to right of } \chi_L^2 = \frac{1 + c}{2} = \frac{1 + 0.95}{2} = 0.975.$$

Part of Table 6 is shown. Using d.f. = 17 and the areas 0.975 and 0.025, you can find the critical values, as shown by the highlighted areas in the table.

Degrees of freedom	α							
	0.995	0.99	0.975	0.95	0.90	0.10	0.05	0.025
1	—	—	0.001	0.004	0.016	2.706	3.841	5.024
2	0.010	0.020	0.051	0.103	0.211	4.605	5.991	7.378
3	0.072	0.115	0.216	0.352	0.584	6.251	7.815	9.348
15	4.601	5.229	6.262	7.261	8.547	22.307	24.996	27.488
16	5.142	5.812	6.908	7.962	9.312	23.542	26.296	28.845
17	5.697	6.408	7.564	8.672	10.085	24.769	27.587	30.191
18	6.265	7.015	8.231	9.390	10.865	25.989	28.869	31.526
19	6.844	7.633	8.907	10.117	11.651	27.204	30.144	32.852
20	7.434	8.260	9.591	10.851	12.443	28.412	31.410	34.170

χ_L^2 $\quad$ χ_R^2

From the table, you can see that $\chi_R^2 = 30.191$ and $\chi_L^2 = 7.564$.

Interpretation So, 95% of the area under the curve lies between 7.564 and 30.191.

▶ Try It Yourself 1

Find the critical values χ_R^2 and χ_L^2 for a 90% confidence interval when the sample size is 30.

a. Identify the *degrees of freedom* and the *level of confidence.*
b. *Find the areas* to the right of χ_R^2 and χ_L^2.
c. *Use* Table 6 in Appendix B to find χ_R^2 and χ_L^2.
d. *Interpret* the results.   *Answer: Page A40*

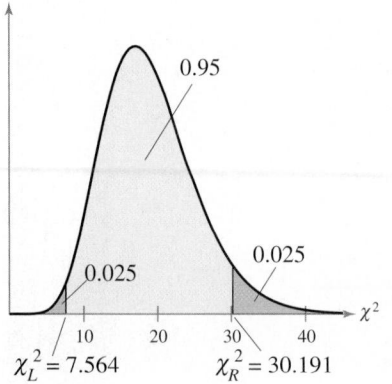

▶ CONFIDENCE INTERVALS FOR σ^2 AND σ

You can use the critical values χ_R^2 and χ_L^2 to construct confidence intervals for a population variance and standard deviation. The best point estimate for the variance is s^2 and the best point estimate for the standard deviation is s.

DEFINITION

The c-confidence intervals for the population variance and standard deviation are as follows.

Confidence Interval for σ^2:

$$\frac{(n-1)s^2}{\chi_R^2} < \sigma^2 < \frac{(n-1)s^2}{\chi_L^2}$$

Confidence Interval for σ:

$$\sqrt{\frac{(n-1)s^2}{\chi_R^2}} < \sigma < \sqrt{\frac{(n-1)s^2}{\chi_L^2}}$$

The probability that the confidence intervals contain σ^2 or σ is c.

GUIDELINES

Constructing a Confidence Interval for a Variance and Standard Deviation

IN WORDS	IN SYMBOLS
1. Verify that the population has a normal distribution.	
2. Identify the sample statistic n and the degrees of freedom.	d.f. $= n - 1$
3. Find the point estimate s^2.	$s^2 = \dfrac{\Sigma(x - \overline{x})^2}{n - 1}$
4. Find the critical values χ_R^2 and χ_L^2 that correspond to the given level of confidence c.	Use Table 6 in Appendix B.

	Left Endpoint	Right Endpoint
5. Find the left and right endpoints and form the confidence interval for the population variance.	$\dfrac{(n-1)s^2}{\chi_R^2} < \sigma^2 <$	$\dfrac{(n-1)s^2}{\chi_L^2}$

6. Find the confidence interval for the population standard deviation by taking the square root of each endpoint.	$\sqrt{\dfrac{(n-1)s^2}{\chi_R^2}} < \sigma <$	$\sqrt{\dfrac{(n-1)s^2}{\chi_L^2}}$

PICTURING THE WORLD

The Florida panther is one of the most endangered mammals on Earth. In the southeastern United States, the only breeding population (about 100) can be found on the southern tip of Florida. Most of the panthers live in (1) the Big Cypress National Preserve, (2) Everglades National Park, and (3) the Florida Panther National Wildlife Refuge, as shown on the map. In a recent study of 19 female panthers, it was found that the mean litter size was 2.4 kittens, with a standard deviation of 0.9.
(Source: U.S. Fish & Wildlife Service)

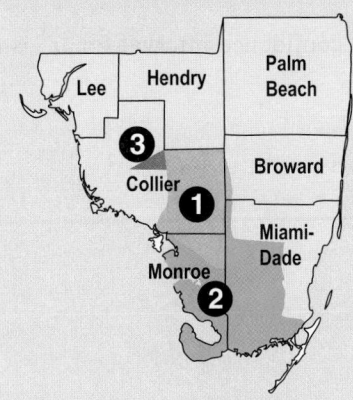

Construct a 90% confidence interval for the standard deviation of the litter size for female Florida panthers. Assume the litter sizes are normally distributed.

(0.71, 1.25)

▶ **Constructing a Confidence Interval**

You randomly select and weigh 30 samples of an allergy medicine. The sample standard deviation is 1.20 milligrams. Assuming the weights are normally distributed, construct 99% confidence intervals for the population variance and standard deviation.

▶ **Solution**

The areas to the right of χ_R^2 and χ_L^2 are

$$\text{Area to right of } \chi_R^2 = \frac{1-c}{2} = \frac{1-0.99}{2} = 0.005$$

and

$$\text{Area to right of } \chi_L^2 = \frac{1+c}{2} = \frac{1+0.99}{2} = 0.995.$$

Using the values $n = 30$, d.f. $= 29$, and $c = 0.99$, the critical values χ_R^2 and χ_L^2 are

$$\chi_R^2 = 52.336 \qquad \text{and} \qquad \chi_L^2 = 13.121.$$

Using these critical values and $s = 1.20$, the confidence interval for σ^2 is as follows.

Left Endpoint

$$\frac{(n-1)s^2}{\chi_R^2} = \frac{(30-1)(1.20)^2}{52.336}$$
$$\approx 0.80$$

Right Endpoint

$$\frac{(n-1)s^2}{\chi_L^2} = \frac{(30-1)(1.20)^2}{13.121}$$
$$\approx 3.18$$

$$0.80 < \sigma^2 < 3.18$$

The confidence interval for σ is

$$\sqrt{\frac{(30-1)(1.20)^2}{52.336}} < \sigma < \sqrt{\frac{(30-1)(1.20)^2}{13.121}}$$

$$0.89 < \sigma < 1.78.$$

Interpretation   With 99% confidence, you can say that the population variance is between 0.80 and 3.18, and the population standard deviation is between 0.89 and 1.78 milligrams.

▶ **Try It Yourself 2**

Find the 90% and 95% confidence intervals for the population variance and standard deviation of the medicine weights.

a. Find the *critical values* χ_R^2 and χ_L^2 for each confidence interval.
b. Use n, s, χ_R^2, and χ_L^2 to find the *left* and *right endpoints* for each confidence interval for the population variance.
c. Find the *square roots* of the endpoints of each confidence interval.
d. *Specify* the 90% and 95% confidence intervals for the population variance and standard deviation. *Answer: Page A40*

Note to Instructor

Point out that the left endpoint requires using χ_R^2 and the right endpoint requires using χ_L^2. This is true because $\chi_R^2 > \chi_L^2$ and dividing the same numerator by a larger value will produce a smaller quotient.

STUDY TIP

When a confidence interval for a population variance or standard deviation is computed, the general *round-off rule* is to round off to the same number of decimal places given for the sample variance or standard deviation.

6.4 EXERCISES

FOR EXTRA HELP:
MyStatLab

1. Yes

2. It approaches the shape of a normal curve.

3. 14.067, 2.167

4. 31.319, 4.075

5. 32.852, 8.907

6. 44.314, 11.524

7. 52.336, 13.121

8. 63.167, 37.689

9. (a) (0.0000413, 0.000157)

   (b) (0.00643, 0.0125)

   With 90% confidence, you can say that the population variance is between 0.0000413 and 0.000157, and the population standard deviation is between 0.00643 and 0.0125 milligram.

10. (a) (0.000610, 0.00220)

    (b) (0.0247, 0.0469)

    With 90% confidence, you can say that the population variance is between 0.000610 and 0.00220, and the population standard deviation is between 0.0247 and 0.0469 fluid ounce.

11. (a) (0.0305, 0.191)

    (b) (0.175, 0.438)

    With 99% confidence, you can say that the population variance is between 0.0305 and 0.191, and the population standard deviation is between 0.175 and 0.438 hour.

12. See Selected Answers, page A121.

■ BUILDING BASIC SKILLS AND VOCABULARY

1. Does a population have to be normally distributed in order to use the chi-square distribution?

2. What happens to the shape of the chi-square distribution as the degrees of freedom increase?

In Exercises 3–8, find the critical values χ_R^2 and χ_L^2 for the given confidence level c and sample size n.

3. $c = 0.90$, $n = 8$

4. $c = 0.99$, $n = 15$

5. $c = 0.95$, $n = 20$

6. $c = 0.98$, $n = 26$

7. $c = 0.99$, $n = 30$

8. $c = 0.80$, $n = 51$

■ USING AND INTERPRETING CONCEPTS

Constructing Confidence Intervals *In Exercises 9–24, assume each sample is taken from a normally distributed population and construct the indicated confidence intervals for (a) the population variance σ^2 and (b) the population standard deviation σ. Interpret the results.*

9. Vitamins To analyze the variation in weights of vitamin supplement tablets, you randomly select and weigh 14 tablets. The results (in milligrams) are shown. Use a 90% level of confidence.

500.000	499.995	500.010	499.997	500.015
499.988	500.000	499.996	500.020	500.002
499.998	499.996	500.003	500.000	

10. Cough Syrup You randomly select and measure the volumes of the contents of 15 bottles of cough syrup. The results (in fluid ounces) are shown. Use a 90% level of confidence.

4.211	4.246	4.269	4.241	4.260
4.293	4.189	4.248	4.220	4.239
4.253	4.209	4.300	4.256	4.290

11. Car Batteries The reserve capacities (in hours) of 18 randomly selected automotive batteries are shown. Use a 99% level of confidence. *(Adapted from Consumer Reports)*

1.70	1.60	1.94	1.58	1.74	1.60
1.86	1.72	1.38	1.46	1.64	1.49
1.55	1.70	1.75	0.88	1.77	2.07

12. Bolts You randomly select and measure the lengths of 17 bolts. The results (in inches) are shown. Use a 95% level of confidence.

1.286	1.138	1.240	1.132	1.381	1.137
1.300	1.167	1.240	1.401	1.241	1.171
1.217	1.360	1.302	1.331	1.383	

13. (a) (6.63, 55.46)

(b) (2.58, 7.45)

With 99% confidence, you can say that the population variance is between 6.63 and 55.46, and the population standard deviation is between 2.58 and 7.45 dollars per year.

14. (a) (7432, 24,421)

(b) (86, 156)

With 80% confidence, you can say that the population variance is between 7432 and 24,421, and the population standard deviation is between $86 and $156.

15. (a) (380.0, 3942.6)

(b) (19.5, 62.8)

With 98% confidence, you can say that the population variance is between 380.0 and 3942.6, and the population standard deviation is between $19.5 and $62.8.

16. (a) (0.33, 2.79)

(b) (0.58, 1.67)

With 99% confidence, you can say that the population variance is between 0.33 and 2.79, and the population standard deviation is between 0.58 and 1.67 pounds.

17. (a) (22.5, 98.7)

(b) (4.7, 9.9)

With 95% confidence, you can say that the population variance is between 22.5 and 98.7, and the population standard deviation is between 4.7 and 9.9 beats per minute.

18. (a) (10,901.06, 40,789.17)

(b) (104, 202)

With 98% confidence, you can say that the population variance is between 10,901.06 and 40,789.17, and the population standard deviation is between $104 and $202.

19. (a) (128, 492)

(b) (11, 22)

With 95% confidence, you can say that the population variance is between 128 and 492, and the population standard deviation is between 11 and 22 grains per gallon.

13. LCD TVs A magazine includes a report on the energy costs per year for 32-inch liquid crystal display (LCD) televisions. The article states that 14 randomly selected 32-inch LCD televisions have a sample standard deviation of $3.90. Use a 99% level of confidence. *(Adapted from Consumer Reports)*

14. Digital Cameras A magazine includes a report on the prices of subcompact digital cameras. The article states that 11 randomly selected subcompact digital cameras have a sample standard deviation of $109. Use an 80% level of confidence. *(Adapted from Consumer Reports)*

15. Spring Break As part of your spring break planning, you randomly select 10 hotels in Cancun, Mexico, and record the room rate for each hotel. The results are shown in the stem-and-leaf plot. Use a 98% level of confidence. *(Source: Expedia, Inc.)*

```
 6 | 9       Key: 7|4 = 74
 7 | 4
 8 |
 9 | 0 9 9
10 |
11 | 2
12 |
13 | 6 9
14 | 9
15 | 0
```

16. Cordless Drills The weights (in pounds) of a random sample of 14 cordless drills are shown in the stem-and-leaf plot. Use a 99% level of confidence. *(Adapted from Consumer Reports)*

```
3 | 4 6 9       Key: 3|4 = 3.4
4 | 6 8 9
5 | 1 3 4 5 7 9
6 | 0 1
```

17. Pulse Rates The pulse rates of a random sample of 16 adults are shown in the dot plot. Use a 95% level of confidence.

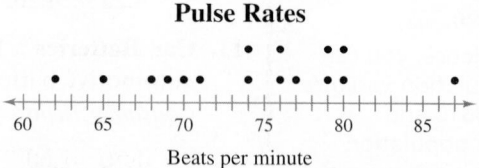

Pulse Rates

Beats per minute

18. Blu-Ray™ Players The prices of a random sample of 27 Blu-ray™ players are shown in the dot plot. Use a 98% level of confidence. *(Adapted from Consumer Reports)*

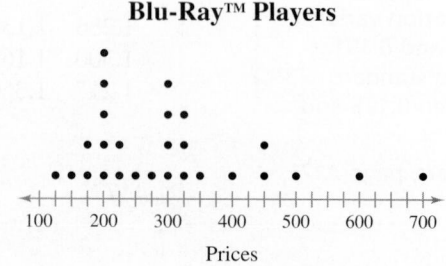

Blu-Ray™ Players

Prices

20. (a) (8,831,450, 21,224,305)

(b) (2972, 4607)

With 90% confidence, you can say that the population variance is between 8,831,450 and 21,224,305, and the population standard deviation is between $2972 and $4607.

21. (a) (9,104,741, 25,615,326)

(b) (3017, 5061)

With 80% confidence, you can say that the population variance is between 9,104,741 and 25,615,326, and the population standard deviation is between $3017 and $5061.

22. (a) (39.13, 136.11)

(b) (6.26, 11.67)

With 98% confidence, you can say that the population variance is between 39.13 and 136.11, and the population standard deviation is between 6.26 and 11.67 inches.

23. (a) (7.0, 30.6)

(b) (2.6, 5.5)

With 98% confidence, you can say that the population variance is between 7.0 and 30.6, and the population standard deviation is between 2.6 and 5.5 minutes.

24. (a) (9,586,982, 28,564,792)

(b) (3096, 5345)

With 90% confidence, you can say that the population variance is between 9,586,982 and 28,564,792, and the population standard deviation is between $3096 and $5345.

25. See Odd Answers, page A79.

26. See Selected Answers, page A121.

27. See Odd Answers, page A79.

28. See Selected Answers, page A121.

29. Yes, because all of the values in the confidence interval are less than 0.015.

30. No, because 0.025 is contained in the confidence interval.

31. Answers will vary. *Sample answer:* Unlike a confidence interval for a population mean or proportion, a confidence interval for a population variance does not have a margin of error. The left and right endpoints must be calculated separately.

19. Water Quality As part of a water quality survey, you test the water hardness in several randomly selected streams. The results are shown in the figure. Use a 95% level of confidence.

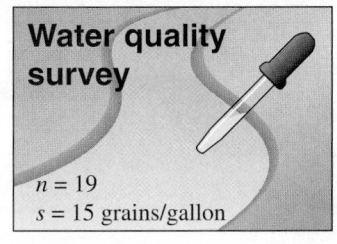

Water quality survey

$n = 19$

$s = 15$ grains/gallon

20. Website Costs As part of a survey, you ask a random sample of business owners how much they would be willing to pay for a website for their company. The results are shown in the figure. Use a 90% level of confidence.

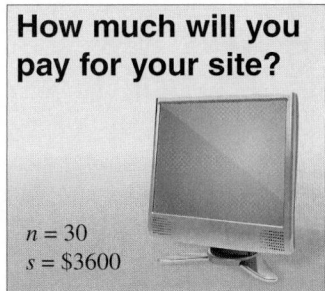

How much will you pay for your site?

$n = 30$

$s = \$3600$

21. Annual Earnings The annual earnings of 14 randomly selected computer software engineers have a sample standard deviation of $3725. Use an 80% level of confidence. *(Adapted from U.S. Bureau of Labor Statistics)*

22. Annual Precipitation The average annual precipitations (in inches) of a random sample of 30 years in San Francisco, California have a sample standard deviation of 8.18 inches. Use a 98% level of confidence. *(Source: Golden Gate Weather Services)*

23. Waiting Times The waiting times (in minutes) of a random sample of 22 people at a bank have a sample standard deviation of 3.6 minutes. Use a 98% level of confidence.

24. Motorcycles The prices of a random sample of 20 new motorcycles have a sample standard deviation of $3900. Use a 90% level of confidence.

SC *In Exercises 25–28, use StatCrunch to help you construct the indicated confidence intervals for the population variance σ^2 and the population standard deviation σ. Assume each sample is taken from a normally distributed population.*

25. $c = 0.95, s^2 = 11.56, n = 30$ **26.** $c = 0.99, s^2 = 0.64, n = 7$

27. $c = 0.90, s = 35, n = 18$ **28.** $c = 0.97, s = 278.1, n = 45$

■ **EXTENDING CONCEPTS**

29. Vitamin Tablet Weights You are analyzing the sample of vitamin supplement tablets in Exercise 9. The population standard deviation of the tablets' weights should be less than 0.015 milligram. Does the confidence interval you constructed for σ suggest that the variation in the tablets' weights is at an acceptable level? Explain your reasoning.

30. Cough Syrup Bottle Contents You are analyzing the sample of cough syrup bottles in Exercise 10. The population standard deviation of the volumes of the bottles' contents should be less than 0.025 fluid ounce. Does the confidence interval you constructed for σ suggest that the variation in the volumes of the bottles' contents is at an acceptable level? Explain your reasoning.

31. In your own words, explain how finding a confidence interval for a population variance is different from finding a confidence interval for a population mean or proportion.

USES AND ABUSES

Statistics in the Real World

Uses

By now, you know that complete information about population parameters is often not available. The techniques of this chapter can be used to make interval estimates of these parameters so that you can make informed decisions.

From what you learned in this chapter, you know that point estimates (sample statistics) of population parameters are usually close but rarely equal to the actual values of the parameters they are estimating. Remembering this can help you make good decisions in your career and in everyday life. For instance, suppose the results of a survey tell you that 52% of the population plans to vote in favor of the rezoning of a portion of a town from residential to commercial use. You know that this is only a point estimate of the actual proportion that will vote in favor of rezoning. If the interval estimate is $0.49 < p < 0.55$, then you know this means it is possible that the item will not receive a majority vote.

Abuses

Unrepresentative Samples There are many ways that surveys can result in incorrect predictions. When you read the results of a survey, remember to question the sample size, the sampling technique, and the questions asked. For instance, suppose you want to know the proportion of people who will vote in favor of rezoning. From the diagram below, you can see that even if your sample is large enough, it may not consist of actual voters.

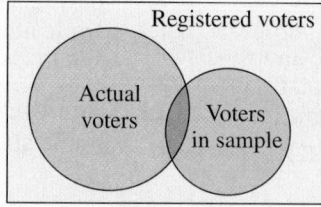

Using a small sample might be the only way to make an estimate, but be aware that a change in one data value may completely change the results. Generally, the larger the sample size, the more accurate the results will be.

Biased Survey Questions In surveys, it is also important to analyze the wording of the questions. For instance, the question about rezoning might be presented as: "Knowing that rezoning will result in more businesses contributing to school taxes, would you support the rezoning?"

■ EXERCISES

1. ***Unrepresentative Samples*** Find an example of a survey that is reported in a newspaper, magazine, or on a website. Describe different ways that the sample could have been unrepresentative of the population.

2. ***Biased Survey Questions*** Find an example of a survey that is reported in a newspaper, magazine, or on a website. Describe different ways that the survey questions could have been biased.

6 CHAPTER SUMMARY

What did you **learn?**	EXAMPLE(S)	REVIEW EXERCISES
Section 6.1		
■ How to find a point estimate and a margin of error $$E = z_c \frac{\sigma}{\sqrt{n}} \qquad \text{Margin of error}$$	1, 2	1, 2
■ How to construct and interpret confidence intervals for the population mean $$\bar{x} - E < \mu < \bar{x} + E$$	3–5	3–6
■ How to determine the minimum sample size required when estimating μ	6	7–10
Section 6.2		
■ How to interpret the t-distribution and use a t-distribution table $$t = \frac{(\bar{x} - \mu)}{(s/\sqrt{n})}$$	1	11–16
■ How to construct confidence intervals when $n < 30$, the population is normally distributed, and σ is unknown $$\bar{x} - E < \mu < \bar{x} + E, \quad E = t_c \frac{s}{\sqrt{n}}$$	2–4	17–26
Section 6.3		
■ How to find a point estimate for a population proportion $$\hat{p} = \frac{x}{n}$$	1	27–34
■ How to construct a confidence interval for a population proportion $$\hat{p} - E < p < \hat{p} + E, \quad E = z_c \sqrt{\frac{\hat{p}\hat{q}}{n}}$$	2, 3	35–42
■ How to determine the minimum sample size required when estimating a population proportion	4	43, 44
Section 6.4		
■ How to interpret the chi-square distribution and use a chi-square distribution table $$\chi^2 = \frac{(n-1)s^2}{\sigma^2}$$	1	45–48
■ How to use the chi-square distribution to construct a confidence interval for the variance and standard deviation $$\frac{(n-1)s^2}{\chi_R^2} < \sigma^2 < \frac{(n-1)s^2}{\chi_L^2}, \quad \sqrt{\frac{(n-1)s^2}{\chi_R^2}} < \sigma < \sqrt{\frac{(n-1)s^2}{\chi_L^2}}$$	2	49–52

6 REVIEW EXERCISES

1. (a) 103.5
   (b) 9.0
2. (a) 9.5
   (b) 2.1
3. (15.6, 16.0)
4. (7.647, 7.703)
5. 1.675, 22.425
6. 0.067, 7.495
7. 47 people
8. 1992 people
9. 49 people
10. 1095 people
11. 1.383
12. 2.069
13. 2.624
14. 2.756
15. $n = 20$
16. $n = 20$
17. 11.2
18. 0.5
19. 0.7
20. 10.6
21. (60.9, 83.3)
22. (3, 4)

■ SECTION 6.1

In Exercises 1 and 2, find (a) the point estimate of the population mean μ and (b) the margin of error for a 90% confidence interval.

1. Waking times of 40 people who start work at 8:00 A.M. (in minutes past 5:00 A.M.)

135	145	95	140	135	95	110	50
90	165	110	125	80	125	130	110
25	75	65	100	60	125	115	135
95	90	140	40	75	50	130	85
100	160	135	45	135	115	75	130

2. Lengths of commutes to work of 32 people (in miles)

12	9	7	2	8	7	3	27
21	10	13	3	7	2	30	7
6	13	6	14	4	1	10	3
13	6	2	9	2	12	16	18

In Exercises 3 and 4, construct the indicated confidence interval for the population mean μ. If convenient, use technology to construct the confidence interval.

3. $c = 0.99, \overline{x} = 15.8, s = 0.85, n = 80$

4. $c = 0.95, \overline{x} = 7.675, s = 0.105, n = 55$

In Exercises 5 and 6, use the given confidence interval to find the margin of error and the sample mean.

5. (20.75, 24.10)   **6.** (7.428, 7.562)

In Exercises 7–10, determine the minimum sample size n needed to estimate μ.

7. Use the results of Exercise 1. Determine the minimum survey size that is necessary to be 95% confident that the sample mean waking time is within 10 minutes of the actual mean waking time.

8. Use the results of Exercise 1. Now suppose you want 99% confidence with a margin of error of 2 minutes. How many people would you need to survey?

9. Use the results of Exercise 2. Determine the minimum survey size that is necessary to be 95% confident that the sample mean length of commutes to work is within 2 miles of the actual mean length of commutes to work.

10. Use the results of Exercise 2. Now suppose you want 98% confidence with a margin of error of 0.5 mile. How many people would you need to survey?

■ SECTION 6.2

In Exercises 11–14, find the critical value t_c for the given confidence level c and sample size n.

11. $c = 0.80, n = 10$ **12.** $c = 0.95, n = 24$

13. $c = 0.98, n = 15$ **14.** $c = 0.99, n = 30$

15. Consider a 90% confidence interval for μ. Assume σ is not known. For which sample size, $n = 20$ or $n = 30$, is the critical value t_c larger?

23. (6.1, 7.5)
24. (14.6, 35.8)
25. (2050, 2386)
26. (1944, 2492)
27. 0.81, 0.19
28. 0.85, 0.15
29. 0.540, 0.460
30. 0.113, 0.887
31. 0.140, 0.860
32. 0.251, 0.749
33. 0.490, 0.510
34. 0.520, 0.480
35. (0.790, 0.830)

   With 95% confidence, you can say that the population proportion of U.S. adults who say they will participate in the 2010 Census is between 79.0% and 83.0%.

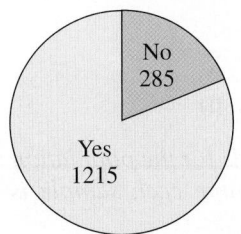

FIGURE FOR EXERCISE 27

36. (0.809, 0.891)

   With 99% confidence, you can say that the population proportion of U.S. adults who say they would trust doctors to tell the truth is between 80.9% and 89.1%.

37. (0.514, 0.566)
   [*Tech:* (0.514, 0.565)]

   With 90% confidence, you can say that the population proportion of U.S. adults who say they have worked the night shift at some point in their lives is between 51.4% and 56.6% (*Tech:* 56.5%).

38. (0.086, 0.139)
   [*Tech:* (0.087, 0.138)]

   With 98% confidence, you can say that the population proportion of U.S. adults who say they are making the minimum payment(s) on their credit card(s) is between 8.6% (*Tech:* 8.7%) and 13.9% (*Tech:* 13.8%).

16. Consider a 90% confidence interval for μ. Assume σ is not known. For which sample size, $n = 20$ or $n = 30$, is the confidence interval wider?

In Exercises 17–20, find the margin of error for μ.

17. $c = 0.90$, $s = 25.6$, $n = 16$, $\bar{x} = 72.1$
18. $c = 0.95$, $s = 1.1$, $n = 25$, $\bar{x} = 3.5$
19. $c = 0.98$, $s = 0.9$, $n = 12$, $\bar{x} = 6.8$
20. $c = 0.99$, $s = 16.5$, $n = 20$, $\bar{x} = 25.2$

In Exercises 21–24, construct the confidence interval for μ using the statistics from the given exercise. If convenient, use technology to construct the confidence interval.

21. Exercise 17   22. Exercise 18
23. Exercise 19   24. Exercise 20

25. In a random sample of 28 sports cars, the average annual fuel cost was $2218 and the standard deviation was $523. Construct a 90% confidence interval for μ. Assume the annual fuel costs are normally distributed. (*Adapted from U.S. Department of Energy*)

26. Repeat Exercise 25 using a 99% confidence interval.

■ SECTION 6.3

In Exercises 27–34, let p be the proportion of the population who respond yes. Use the given information to find $\hat{p}$ and $\hat{q}$.

27. A survey asks 1500 U.S. adults if they will participate in the 2010 Census. The results are shown in the pie chart. (*Adapted from Pew Research Center*)

28. In a survey of 500 U.S. adults, 425 say they would trust doctors to tell the truth. (*Adapted from Harris Interactive*)

29. In a survey of 1023 U.S. adults, 552 say they have worked the night shift at some point in their lives. (*Adapted from CNN/Opinion Research*)

30. In a survey of 800 U.S. adults, 90 are making the minimum payment(s) on their credit card(s). (*Adapted from Cambridge Consumer Credit Index*)

31. In a survey of 1008 U.S. adults, 141 say the cost of health care is the most important financial problem facing their family today. (*Adapted from Gallup, Inc.*)

32. In a survey of 938 U.S. adults, 235 say the phrase "you know" is the most annoying conversational phrase. (*Adapted from Marist Poll*)

33. In a survey of 706 parents with kids 4 to 8 years old, 346 say that they know their state booster seat law. (*Adapted from Knowledge Networks, Inc.*)

34. In a survey of 2365 U.S. adults, 1230 say they worry most about missing deductions when filing their taxes. (*Adapted from USA TODAY*)

In Exercises 35–42, construct the indicated confidence interval for the population proportion p. If convenient, use technology to construct the confidence interval. Interpret the results.

35. Use the sample in Exercise 27 with $c = 0.95$.
36. Use the sample in Exercise 28 with $c = 0.99$.
37. Use the sample in Exercise 29 with $c = 0.90$.
38. Use the sample in Exercise 30 with $c = 0.98$.

39. (0.112, 0.168)

With 99% confidence, you can say that the population proportion of U.S. adults who say that the cost of healthcare is the most important financial problem facing their family today is between 11.2% and 16.8%.

40. (0.228, 0.274) [Tech: (0.227, 0.274)]

With 90% confidence, you can say that the population proportion of U.S. adults who say the phrase "you know" is the most annoying conversational phrase is between 22.8% (Tech: 22.7%) and 27.4%.

41. (0.466, 0.514)

With 80% confidence, you can say that the population proportion of parents with kids 4 to 8 years old who say they know their state booster seat law is between 46.6% and 51.4%.

42. (0.496, 0.544)

With 98% confidence, you can say that the population proportion of U.S. adults who say they worry most about missing deductions when filing their taxes is between 49.6% and 54.4%.

43. (a) 385 adults

(b) 359 adults

(c) Having an estimate of the population proportion reduces the minimum sample size needed.

44. 2473 adults; The sample size is larger.

45. 23.337, 4.404

46. 42.980, 10.856

47. 14.067, 2.167

48. 23.589, 1.735

49. (27.2, 113.5); (5.2, 10.7)

50. (22.9, 152.5); (4.8, 12.3)

With 99% confidence, you can say that the population variance is between 22.9 and 152.5 and the population standard deviation is between 4.8 and 12.3 ounces. The confidence intervals in Exercise 50 are wider than those in Exercise 49.

51. (0.80, 3.07); (0.89, 1.75)

52. See Selected Answers, page A121.

39. Use the sample in Exercise 31 with $c = 0.99$.

40. Use the sample in Exercise 32 with $c = 0.90$.

41. Use the sample in Exercise 33 with $c = 0.80$.

42. Use the sample in Exercise 34 with $c = 0.98$.

43. You wish to estimate, with 95% confidence, the population proportion of U.S. adults who think they should be saving more money. Your estimate must be accurate within 5% of the population proportion.

(a) No preliminary estimate is available. Find the minimum sample size needed.

(b) Find the minimum sample size needed, using a prior study that found that 63% of U.S. adults think that they should be saving more money. *(Source: Pew Research Center)*

(c) Compare the results from parts (a) and (b).

44. Repeat Exercise 43 part (b), using a 99% confidence level and a margin of error of 2.5%. How does this sample size compare with your answer from Exercise 43 part (b)?

■ SECTION 6.4

In Exercises 45–48, find the critical values χ_R^2 and χ_L^2 for the given confidence level c and sample size n.

45. $c = 0.95, n = 13$

46. $c = 0.98, n = 25$

47. $c = 0.90, n = 8$

48. $c = 0.99, n = 10$

In Exercises 49–52, construct the indicated confidence intervals for the population variance σ^2 and the population standard deviation σ. Assume each sample is taken from a normally distributed population.

49. A random sample of the weights (in ounces) of 17 superzoom digital cameras is shown in the stem-and-leaf plot. Use a 95% level of confidence. *(Adapted from Consumer Reports)*

```
0 | 7 8 8 9          Key: 1|3 = 13
1 | 0 1 3 4 5 5 5 7 7 9
2 | 1 4
3 | 5
```

50. Repeat Exercise 49 using a 99% level of confidence. Interpret the results and compare with Exercise 49.

51. A random sample of the acceleration times (in seconds) from 0 to 60 miles per hour for 26 sedans is shown in the dot plot. Use a 98% level of confidence. *(Adapted from Consumer Reports)*

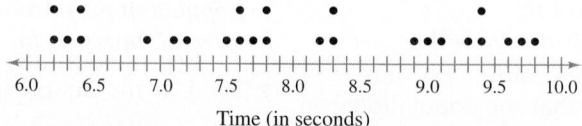

Acceleration Times From 0–60 Miles Per Hour for Sedans

Time (in seconds)

52. Repeat Exercise 51 using a 90% level of confidence. Interpret the results and compare with Exercise 51.

6 CHAPTER QUIZ

1. (a) 6.85

(b) 0.65; You are 95% confident that the margin of error for the population mean is about 0.65 minute.

(c) (6.20, 7.50)

With 95% confidence, you can say that the population mean amount of time is between 6.20 and 7.50 minutes.

2. 39 college students

3. (a) 33.11; 2.38

(b) (31.73, 34.49)

With 90% confidence, you can say that the population mean time played in the season is between 31.73 and 34.49 minutes.

(c) (30.38, 35.84)

With 90% confidence, you can say that the population mean time played in the season is between 30.38 and 35.84 minutes. This confidence interval is wider than the one found in part (b).

4. (6510, 7138)

5. (a) 0.780

(b) (0.762, 0.798)
[*Tech:* (0.762, 0.799)]

(c) 712 adults

6. (a) (2.10, 5.99)

(b) (1.45, 2.45)

Take this quiz as you would take a quiz in class. After you are done, check your work against the answers given in the back of the book.

1. The following data set represents the amounts of time (in minutes) spent watching online videos each day for a random sample of 30 college students. *(Adapted from the Council for Research Excellence)*

5.0	6.25	8.0	5.5	4.75	4.5	7.2	6.6	5.8	5.5
4.2	5.4	6.75	9.8	8.2	6.4	7.8	6.5	5.5	6.0
3.8	6.75	9.25	10.0	9.6	7.2	6.4	6.8	9.8	10.2

(a) Find the point estimate of the population mean.

(b) Find the margin of error for a 95% level of confidence. Interpret the result.

(c) Construct a 95% confidence interval for the population mean. Interpret the results.

2. You want to estimate the mean time college students spend watching online videos each day. The estimate must be within 1 minute of the population mean. Determine the required sample size to construct a 99% confidence interval for the population mean. Assume the population standard deviation is 2.4 minutes.

3. The following data set represents the average number of minutes played for a random sample of professional basketball players in a recent season. *(Source: ESPN)*

35.9 33.8 34.7 31.5 33.2 29.1 30.7 31.2 36.1 34.9

(a) Find the sample mean and the sample standard deviation.

(b) Construct a 90% confidence interval for the population mean and interpret the results. Assume the population of the data set is normally distributed.

(c) Repeat part (b), assuming $\sigma = 5.25$ minutes per game. Interpret and compare the results.

4. In a random sample of seven aerospace engineers, the mean monthly income was $6824 and the standard deviation was $340. Assume the monthly incomes are normally distributed and construct a 95% confidence interval for the population mean monthly income for aerospace engineers. *(Adapted from U.S. Bureau of Labor Statistics)*

5. In a survey of 1383 U.S. adults, 1079 favor increasing federal funding for research on wind, solar, and hydrogen technology. *(Adapted from Pew Research Center)*

(a) Find a point estimate for the population proportion p of those in favor of increasing federal funding for research on wind, solar, and hydrogen technology.

(b) Construct a 90% confidence interval for the population proportion.

(c) Find the minimum sample size needed to estimate the population proportion at the 99% confidence level in order to ensure that the estimate is accurate within 4% of the population proportion.

6. Refer to the data set in Exercise 1. Assume the population of times spent watching online videos each day is normally distributed.

(a) Construct a 95% confidence interval for the population variance.

(b) Construct a 95% confidence interval for the population standard deviation.

PUTTING IT ALL TOGETHER

Real Statistics — Real Decisions

In 1974, the Safe Drinking Water Act was passed "to protect public health by regulating the nation's public drinking water supply." In accordance with the act, the Environmental Protection Agency (EPA) has regulations that limit the levels of contaminants in drinking water supplied by water utilities. These utilities are required to supply water quality reports to their customers annually. These reports discuss the source of the water, its treatment, and the results of water quality monitoring that is performed daily. The results of this monitoring indicate whether or not drinking water is healthy enough for consumption.

A water department tests for contaminants at water treatment plants and at customers' taps. These regulated parameters include microorganisms, organic chemicals, and inorganic chemicals. For instance, cyanide is an inorganic chemical that is regulated. Its presence in drinking water is the result of discharges from steel, plastics, and fertilizer factories. The maximum contaminant level for cyanide is set at 0.2 part per million.

You work for a city's water department and are interpreting the results shown in the graph at the right. The graph shows the point estimates for the population mean concentration and the 95% confidence intervals for μ for cyanide over a three-year period. The data are based on random water samples taken by the city's three water treatment plants.

Cyanide

(Graph: Mean concentration level (in parts per million) vs. Year, showing point estimates and 95% confidence intervals for Year 1, Year 2, and Year 3.)

■ EXERCISES

1. **Interpreting the Results**

   Use the graph to decide if there has been a change in the mean concentration level of cyanide for the given years. Explain your reasoning.

   (a) From Year 1 to Year 2   (b) From Year 2 to Year 3

   (c) From Year 1 to Year 3

2. **What Can You Conclude?**

   Using the results of Exercise 1, what can you conclude about the concentrations of cyanide in the drinking water?

3. **What Do You Think?**

   The confidence interval for Year 2 is much larger than the other years. What do you think may have caused this larger confidence level?

4. **How Do You Think They Did It?**

   How do you think the water department constructed the 95% confidence intervals for the population mean concentration of cyanide in the water? Do the following to answer the question. (You do not need to make any calculations.)

   (a) What sampling distribution do you think they used? Why?

   (b) Do you think they used the population standard deviation in calculating the margin of error? Why or why not? If not, what could they have used?

THE GALLUP ORGANIZATION

WWW.GALLUP.COM

MOST ADMIRED POLLS

Since 1946, the Gallup Organization has conducted a "most admired" poll. The methodology for the 2009 poll is described at the right.

> ### Survey Question
> *What man* that you have heard or read about, living today in any part of the world, do you admire most? And who is your second choice?*

Reprinted with permission from GALLUP.

*Survey respondents are asked an identical question about most admired woman.

"Results are based on telephone interviews with 1,025 national adults, aged 18 and older, conducted Dec. 11–13, 2009. For results based on the total sample of national adults, one can say with 95% confidence that the maximum margin of sampling error is ±4 percentage points. Interviews are conducted with respondents on land-line telephones (for respondents with a land-line telephone) and cellular phones (for respondents who are cell-phone only). In addition to sampling error, question wording and practical difficulties in conducting surveys can introduce error or bias into the findings of public opinion polls."

■ EXERCISES

1. In 2009, the most named man was Barack Obama at 30%. Use a technology tool to find a 95% confidence interval for the population proportion that would have chosen Barack Obama.

2. In 2009, the most named woman was Hillary Clinton at 16%. Use a technology tool to find a 95% confidence interval for the population proportion that would have chosen Hillary Clinton.

3. Do the confidence intervals you obtained in Exercises 1 and 2 agree with the statement issued by the Gallup Organization that the margin of error is ±4%? Explain.

4. The second most named woman was Sarah Palin, who was named by 15% of the people in the sample. Use a technology tool to find a 95% confidence interval for the population proportion that would have chosen Sarah Palin.

5. Use a technology tool to simulate a most admired poll. Assume that the actual population proportion who most admire Sarah Palin is 18%. Run the simulation several times using $n = 1025$.

   (a) What was the least value you obtained for $\hat{p}$?

   (b) What was the greatest value you obtained for $\hat{p}$?

   ### MINITAB

Number of rows of data to generate:	200
Store in column(s):	C1
Number of trials:	1025
Event probability:	0.18

6. Is it probable that the population proportion who most admire Sarah Palin is 18% or greater? Explain your reasoning.

Extended solutions are given in the *Technology Supplement*.
Technical instruction is provided for MINITAB, Excel, and the TI-83/84 Plus.

USING TECHNOLOGY TO CONSTRUCT CONFIDENCE INTERVALS

Here are some MINITAB and TI-83/84 Plus printouts for some examples in this chapter. Answers may be slightly different because of rounding.

(See Example 3, page 307.)

140	105	130	97	80	165	232	110	214	201	122	98	65	88
154	133	121	82	130	211	153	114	58	77	51	247	236	109
126	132	125	149	122	74	59	218	192	90	117	105		

Display Descriptive Statistics...
Store Descriptive Statistics...
Graphical Summary...

1-Sample Z...
1-Sample t...
2-Sample t...
Paired t...

1 Proportion...
2 Proportions...

MINITAB

One-Sample Z: Friends

The assumed standard deviation = 53

Variable	N	Mean	StDev	SE Mean	95% CI
Friends	40	130.80	52.63	8.38	[114.38, 147.22]

(See Example 2, page 320.)

Display Descriptive Statistics...
Store Descriptive Statistics...
Graphical Summary...

1-Sample Z...
1-Sample t...
2-Sample t...
Paired t...

1 Proportion...
2 Proportions...

MINITAB

One-Sample T

N	Mean	StDev	SE Mean	95% CI
16	162.00	10.00	2.50	[156.67, 167.33]

(See Example 2, page 329.)

Display Descriptive Statistics...
Store Descriptive Statistics...
Graphical Summary...

1-Sample Z...
1-Sample t...
2-Sample t...
Paired t...

1 Proportion...
2 Proportions...

MINITAB

Test and CI for One Proportion

Sample	X	N	Sample p	95% CI
1	662	1000	0.662000	[0.631738, 0.691305]

(See Example 5, page 309.)

TI-83/84 PLUS

EDIT CALC TESTS
1: Z–Test...
2: T–Test...
3: 2–SampZTest...
4: 2–SampTTest...
5: 1–PropZTest...
6: 2–PropZTest...
7↓ ZInterval...

⬇

TI-83/84 PLUS

ZInterval
 Inpt: Data **Stats**
 s: 1.5
 x̄: 22.9
 n: 20
 C–Level: .9
 Calculate

⬇

TI-83/84 PLUS

ZInterval
 (22.348, 23.452)
 x̄= 22.9
 n= 20

(See Example 3, page 321.)

TI-83/84 PLUS

EDIT CALC **TESTS**
2↑ T–Test...
3: 2–SampZTest...
4: 2–SampTTest...
5: 1–PropZTest...
6: 2–PropZTest...
7: ZInterval...
8↓ TInterval...

⬇

TI-83/84 PLUS

TInterval
 Inpt: Data **Stats**
 x̄: 9.75
 Sx: 2.39
 n: 20
 C–Level: .99
 Calculate

⬇

TI-83/84 PLUS

TInterval
 (8.2211, 11.279)
 x̄= 9.75
 Sx= 2.39
 n= 20

(See Example 2, page 329.)

TI-83/84 PLUS

EDIT CALC **TESTS**
5↑ 1–PropZTest...
6: 2–PropZTest...
7: ZInterval...
8: TInterval...
9: 2–SampZInt...
0: 2–SampTInt...
A↓ 1–PropZInt...

⬇

TI-83/84 PLUS

1–PropZInt
 x: 662
 n: 1000
 C–Level: .95
 Calculate

⬇

TI-83/84 PLUS

1–PropZInt
 (.63268, .69132)
 p̂= 0.662
 n= 1000

7 HYPOTHESIS TESTING WITH ONE SAMPLE

CHAPTER

Computer software is protected by federal copyright laws. Each year, software companies lose billions of dollars because of pirated software. Federal criminal penalties for software piracy can include fines of up to $250,000 and jail terms of up to five years.

In Chapter 6, you began your study of inferential statistics. There, you learned how to form a confidence interval estimate about a population parameter, such as the proportion of people in the United States who agree with a certain statement. For instance, in a nationwide poll conducted by *Harris Interactive* on behalf of the Business Software Alliance (BSA), U.S. students ages 8 to 18 years were asked several questions about their attitudes toward copyright law and Internet behavior. Here are some of the results.

Survey Question	Number Surveyed	Number Who Said Yes
Have you ever downloaded music from the Internet without paying for it?	1196	361
Have you ever downloaded movies from the Internet without paying for them?	1196	95
Have you ever downloaded software from the Internet without paying for it?	1196	133

In this chapter, you will continue your study of inferential statistics. But now, instead of making an estimate about a population parameter, you will learn how to test a claim about a parameter.

For instance, suppose that you work for *Harris Interactive* and are asked to test a claim that the proportion of U.S. students ages 8 to 18 who download music without paying for it is $p = 0.25$. To test the claim, you take a random sample of $n = 1196$ students and find that 361 of them download music without paying for it. Your sample statistic is $\hat{p} \approx 0.302$.

Is your sample statistic different enough from the claim ($p = 0.25$) to decide that the claim is false? The answer lies in the sampling distribution of sample proportions taken from a population in which $p = 0.25$. The graph below shows that your sample statistic is more than 4 standard errors from the claimed value. If the claim is true, the probability of the sample statistic being 4 standard errors or more from the claimed value is extremely small. Something is wrong! If your sample was truly random, then you can conclude that the actual proportion of the student population is not 0.25. In other words, you tested the original claim (hypothesis), and you decided to reject it.

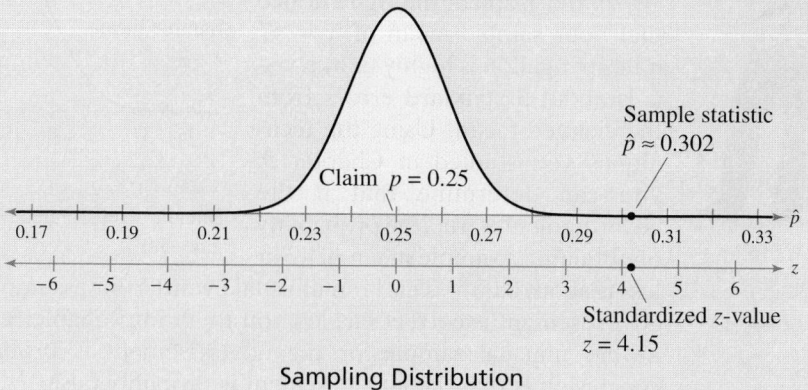

Sampling Distribution

7.1 Introduction to Hypothesis Testing

Hypothesis Tests ▸ Stating a Hypothesis ▸ Types of Errors and Level of Significance ▸ Statistical Tests and *P*-Values ▸ Making a Decision and Interpreting the Decision ▸ Strategies for Hypothesis Testing

▸ HYPOTHESIS TESTS

Throughout the remainder of this course, you will study an important technique in inferential statistics called hypothesis testing. A **hypothesis test** is a process that uses sample statistics to test a claim about the value of a population parameter. Researchers in fields such as medicine, psychology, and business rely on hypothesis testing to make informed decisions about new medicines, treatments, and marketing strategies.

For instance, suppose an automobile manufacturer advertises that its new hybrid car has a mean gas mileage of 50 miles per gallon. If you suspect that the mean mileage is not 50 miles per gallon, how could you show that the advertisement is false?

Obviously, you cannot test *all* the vehicles, but you can still make a reasonable decision about the mean gas mileage by taking a random sample from the population of vehicles and measuring the mileage of each. If the sample mean differs enough from the advertisement's mean, you can decide that the advertisement is wrong.

For instance, to test that the mean gas mileage of all hybrid vehicles of this type is $\mu = 50$ miles per gallon, you could take a random sample of $n = 30$ vehicles and measure the mileage of each. Suppose you obtain a sample mean of $\bar{x} = 47$ miles per gallon with a sample standard deviation of $s = 5.5$ miles per gallon. Does this indicate that the manufacturer's advertisement is false?

To decide, you do something unusual—*you assume the advertisement is correct!* That is, you assume that $\mu = 50$. Then, you examine the sampling distribution of sample means (with $n = 30$) taken from a population in which $\mu = 50$ and $\sigma = 5.5$. From the Central Limit Theorem, you know this sampling distribution is normal with a mean of 50 and standard error of

$$\frac{5.5}{\sqrt{30}} \approx 1.$$

Sampling Distribution of $\bar{x}$

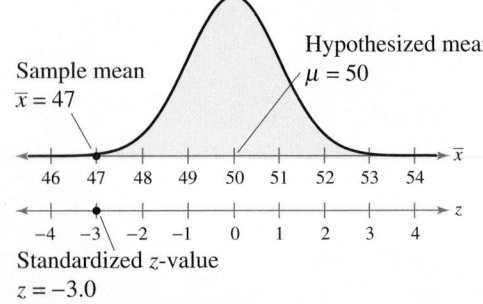

In the graph at the right, notice that your sample mean of $\bar{x} = 47$ miles per gallon is highly unlikely— it is about 3 standard errors from the claimed mean! Using the techniques you studied in Chapter 5, you can determine that if the advertisement is true, the probability of obtaining a sample mean of 47 or less is about 0.0013. This is an unusual event! Your assumption that the company's advertisement is correct has led you to an improbable result. So, either you had a very unusual sample, or the advertisement is probably false. The logical conclusion is that the advertisement is probably false.

▶ STATING A HYPOTHESIS

A statement about a population parameter is called a **statistical hypothesis.** To test a population parameter, you should carefully state a pair of hypotheses—one that represents the claim and the other, its complement. When one of these hypotheses is false, the other must be true. Either hypothesis—the *null hypothesis* or the *alternative hypothesis*—may represent the original claim.

INSIGHT

The term *null hypothesis* was introduced by Ronald Fisher (see page 33). If the statement in the null hypothesis is not true, then the alternative hypothesis must be true.

DEFINITION

1. A **null hypothesis** H_0 is a statistical hypothesis that contains a statement of equality, such as $\leq$, $=$, or $\geq$.
2. The **alternative hypothesis** H_a is the complement of the null hypothesis. It is a statement that must be true if H_0 is false and it contains a statement of strict inequality, such as $>$, $\neq$, or $<$.

H_0 is read as "H sub-zero" or "H naught" and H_a is read as "H sub-a."

To write the null and alternative hypotheses, translate the claim made about the population parameter from a verbal statement to a mathematical statement. Then, write its complement. For instance, if the claim value is k and the population parameter is μ, then some possible pairs of null and alternative hypotheses are

$$\begin{cases} H_0: \mu \leq k \\ H_a: \mu > k \end{cases} \qquad \begin{cases} H_0: \mu \geq k \\ H_a: \mu < k \end{cases} \qquad \begin{cases} H_0: \mu = k \\ H_a: \mu \neq k \end{cases}.$$

Regardless of which of the three pairs of hypotheses you use, you always assume $\mu = k$ and examine the sampling distribution on the basis of this assumption. Within this sampling distribution, you will determine whether or not a sample statistic is unusual.

The following table shows the relationship between possible verbal statements about the parameter μ and the corresponding null and alternative hypotheses. Similar statements can be made to test other population parameters, such as p, σ, or σ^2.

PICTURING THE WORLD

A sample of 25 randomly selected patients with early-stage high blood pressure underwent a special chiropractic adjustment to help lower their blood pressure. After eight weeks, the mean drop in the patients' systolic blood pressure was 14 millimeters of mercury. So, it is claimed that the mean drop in systolic blood pressure of all patients who undergo this special chiropractic adjustment is 14 millimeters of mercury. (Adapted from Journal of Human Hypertension)

Determine a null hypothesis and alternative hypothesis for this claim.

$H_0: \mu = 14$; $H_a: \mu \neq 14$

Verbal Statement H_0 The mean is . . .	Mathematical Statements	Verbal Statement H_a The mean is . . .
. . . greater than or equal to k. . . . at least k. . . . not less than k.	$\begin{cases} H_0: \mu \geq k \\ H_a: \mu < k \end{cases}$	. . . less than k. . . . below k. . . . fewer than k.
. . . less than or equal to k. . . . at most k. . . . not more than k.	$\begin{cases} H_0: \mu \leq k \\ H_a: \mu > k \end{cases}$	. . . greater than k. . . . above k. . . . more than k.
. . . equal to k. . . . k. . . . exactly k.	$\begin{cases} H_0: \mu = k \\ H_a: \mu \neq k \end{cases}$	. . . not equal to k. . . . different from k. . . . not k.

EXAMPLE 1

▶ Stating the Null and Alternative Hypotheses

Write the claim as a mathematical sentence. State the null and alternative hypotheses, and identify which represents the claim.

1. A school publicizes that the proportion of its students who are involved in at least one extracurricular activity is 61%.

2. A car dealership announces that the mean time for an oil change is less than 15 minutes.

3. A company advertises that the mean life of its furnaces is more than 18 years.

▶ Solution

1. The claim "the proportion ... is 61%" can be written as $p = 0.61$. Its complement is $p \neq 0.61$. Because $p = 0.61$ contains the statement of equality, it becomes the null hypothesis. In this case, the null hypothesis represents the claim.

$$H_0: p = 0.61 \text{ (Claim)}$$

$$H_a: p \neq 0.61$$

2. The claim "the mean ... is less than 15 minutes" can be written as $\mu < 15$. Its complement is $\mu \geq 15$. Because $\mu \geq 15$ contains the statement of equality, it becomes the null hypothesis. In this case, the alternative hypothesis represents the claim.

$$H_0: \mu \geq 15 \text{ minutes}$$

$$H_a: \mu < 15 \text{ minutes (Claim)}$$

3. The claim "the mean ... is more than 18 years" can be written as $\mu > 18$. Its complement is $\mu \leq 18$. Because $\mu \leq 18$ contains the statement of equality, it becomes the null hypothesis. In this case, the alternative hypothesis represents the claim.

$$H_0: \mu \leq 18 \text{ years}$$

$$H_a: \mu > 18 \text{ years (Claim)}$$

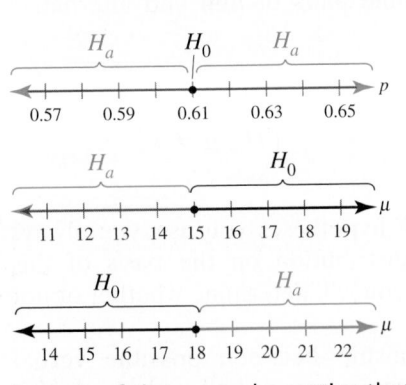

In each of these graphs, notice that each point on the number line is in H_0 or H_a, but no point is in both.

▶ Try It Yourself 1

Write the claim as a mathematical sentence. State the null and alternative hypotheses, and identify which represents the claim.

1. A consumer analyst reports that the mean life of a certain type of automobile battery is not 74 months.
2. An electronics manufacturer publishes that the variance of the life of its home theater systems is less than or equal to 2.7.
3. A realtor publicizes that the proportion of homeowners who feel their house is too small for their family is more than 24%.

a. Identify the *verbal claim* and write it as a *mathematical statement*.
b. Write the *complement* of the claim.
c. Identify the *null* and *alternative hypotheses* and determine which one represents the claim.

Answer: Page A40

▶ TYPES OF ERRORS AND LEVEL OF SIGNIFICANCE

No matter which hypothesis represents the claim, you always begin a hypothesis test by assuming that the equality condition in the null hypothesis is true. So, when you perform a hypothesis test, you make one of two decisions:

1. reject the null hypothesis or

2. fail to reject the null hypothesis.

Because your decision is based on a sample rather than the entire population, there is always the possibility you will make the wrong decision.

For instance, suppose you claim that a certain coin is not fair. To test your claim, you flip the coin 100 times and get 49 heads and 51 tails. You would probably agree that you do not have enough evidence to support your claim. Even so, it is possible that the coin is actually not fair and you had an unusual sample.

But what if you flip the coin 100 times and get 21 heads and 79 tails? It would be a rare occurrence to get only 21 heads out of 100 tosses with a fair coin. So, you probably have enough evidence to support your claim that the coin is not fair. However, you can't be 100% sure. It is possible that the coin is fair and you had an unusual sample.

If p represents the proportion of heads, the claim that "the coin is not fair" can be written as the mathematical statement $p \neq 0.5$. Its complement, "the coin is fair," is written as $p = 0.5$. So, your null hypothesis and alternative hypothesis are

$$H_0: p = 0.5$$

and

$$H_a: p \neq 0.5. \quad \text{(Claim)}$$

Remember, the only way to be absolutely certain of whether H_0 is true or false is to test the entire population. Because your decision—to reject H_0 or to fail to reject H_0—is based on a sample, you must accept the fact that your decision might be incorrect. You might reject a null hypothesis when it is actually true. Or, you might fail to reject a null hypothesis when it is actually false.

DEFINITION

A **type I error** occurs if the null hypothesis is rejected when it is true.

A **type II error** occurs if the null hypothesis is not rejected when it is false.

The following table shows the four possible outcomes of a hypothesis test.

Decision	Truth of H_0	
	H_0 is true.	H_0 is false.
Do not reject H_0.	Correct decision	Type II error
Reject H_0.	Type I error	Correct decision

Hypothesis testing is sometimes compared to the legal system used in the United States. Under this system, the following steps are used.

1. A carefully worded accusation is written.

2. The defendant is assumed innocent (H_0) until proven guilty. The burden of proof lies with the prosecution. If the evidence is not strong enough, there is no conviction. A "not guilty" verdict does not prove that a defendant is innocent.

3. The evidence needs to be conclusive beyond a reasonable doubt. The system assumes that more harm is done by convicting the innocent (type I error) than by not convicting the guilty (type II error).

	Truth About Defendant	
Verdict	**Innocent**	**Guilty**
Not guilty	Justice	Type II error
Guilty	Type I error	Justice

EXAMPLE 2

▶ Identifying Type I and Type II Errors

The USDA limit for salmonella contamination for chicken is 20%. A meat inspector reports that the chicken produced by a company exceeds the USDA limit. You perform a hypothesis test to determine whether the meat inspector's claim is true. When will a type I or type II error occur? Which is more serious? *(Source: U.S. Department of Agriculture)*

▶ Solution

Let p represent the proportion of the chicken that is contaminated. The meat inspector's claim is "more than 20% is contaminated." You can write the null and alternative hypotheses as follows.

H_0: $p \le 0.2$ The proportion is less than or equal to 20%.

H_a: $p > 0.2$ (Claim)   The proportion is greater than 20%.

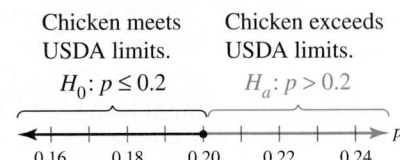

A type I error will occur if the actual proportion of contaminated chicken is less than or equal to 0.2, but you reject H_0. A type II error will occur if the actual proportion of contaminated chicken is greater than 0.2, but you do not reject H_0. With a type I error, you might create a health scare and hurt the sales of chicken producers who were actually meeting the USDA limits. With a type II error, you could be allowing chicken that exceeded the USDA contamination limit to be sold to consumers. A type II error is more serious because it could result in sickness or even death.

▶ Try It Yourself 2

A company specializing in parachute assembly states that its main parachute failure rate is not more than 1%. You perform a hypothesis test to determine whether the company's claim is false. When will a type I or type II error occur? Which is more serious?

a. State the *null* and *alternative hypotheses*.
b. Write the possible *type I* and *type II* errors.
c. *Determine* which error is more serious.   *Answer: Page A40*

You will reject the null hypothesis when the sample statistic from the sampling distribution is unusual. You have already identified unusual events to be those that occur with a probability of 0.05 or less. When statistical tests are used, an unusual event is sometimes required to have a probability of 0.10 or less, 0.05 or less, or 0.01 or less. Because there is variation from sample to sample, there is always a possibility that you will reject a null hypothesis when it is actually true. In other words, although the null hypothesis is true, your sample statistic is determined to be an unusual event in the sampling distribution. You can decrease the probability of this happening by lowering the *level of significance*.

INSIGHT

When you decrease α (the maximum allowable probability of making a type I error), you are likely to be increasing β. The value $1 - \beta$ is called the **power of the test**. It represents the probability of rejecting the null hypothesis when it is false. The value of the power is difficult (and sometimes impossible) to find in most cases.

Note to Instructor

You can use an example of "false positive" and "false negative" results for a medical test (say, diabetes) to discuss type I and type II errors. You might also want to point out that computation of β is beyond the scope of an introductory statistics text.

> ## DEFINITION
>
> In a hypothesis test, the **level of significance** is your maximum allowable probability of making a type I error. It is denoted by α, the lowercase Greek letter alpha.
>
> The probability of a type II error is denoted by β, the lowercase Greek letter beta.

By setting the level of significance at a small value, you are saying that you want the probability of rejecting a true null hypothesis to be small. Three commonly used levels of significance are $\alpha = 0.10$, $\alpha = 0.05$, and $\alpha = 0.01$.

▶ STATISTICAL TESTS AND *P*-VALUES

After stating the null and alternative hypotheses and specifying the level of significance, the next step in a hypothesis test is to obtain a random sample from the population and calculate sample statistics such as the mean and the standard deviation. The statistic that is compared with the parameter in the null hypothesis is called the **test statistic.** The type of test used and the sampling distribution are based on the test statistic.

In this chapter, you will learn about several one-sample statistical tests. The following table shows the relationships between population parameters and their corresponding test statistics and standardized test statistics.

Population parameter	Test statistic	Standardized test statistic
μ	$\overline{x}$	z (Section 7.2, $n \geq 30$), t (Section 7.3, $n < 30$)
p	$\hat{p}$	z (Section 7.4)
σ^2	s^2	χ^2 (Section 7.5)

One way to decide whether to reject the null hypothesis is to determine whether the probability of obtaining the standardized test statistic (or one that is more extreme) is less than the level of significance.

> ## DEFINITION
>
> If the null hypothesis is true, a **P-value** (or **probability value**) of a hypothesis test is the probability of obtaining a sample statistic with a value as extreme or more extreme than the one determined from the sample data.

The *P*-value of a hypothesis test depends on the nature of the test. There are three types of hypothesis tests—left-tailed, right-tailed, and two-tailed. The type of test depends on the location of the region of the sampling distribution that favors a rejection of H_0. This region is indicated by the alternative hypothesis.

DEFINITION

1. If the alternative hypothesis H_a contains the less-than inequality symbol ($<$), the hypothesis test is a **left-tailed test.**

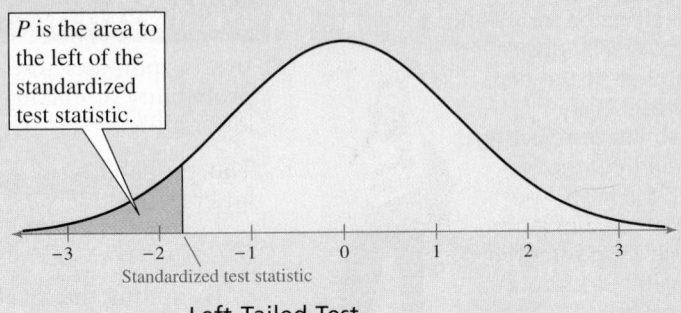

$H_0: \mu \geq k$
$H_a: \mu < k$

P is the area to the left of the standardized test statistic.

Standardized test statistic

Left-Tailed Test

2. If the alternative hypothesis H_a contains the greater-than inequality symbol ($>$), the hypothesis test is a **right-tailed test.**

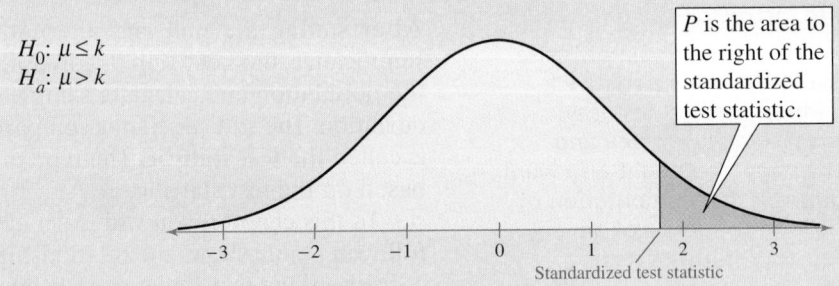

$H_0: \mu \leq k$
$H_a: \mu > k$

P is the area to the right of the standardized test statistic.

Standardized test statistic

Right-Tailed Test

3. If the alternative hypothesis H_a contains the not-equal-to symbol ($\neq$), the hypothesis test is a **two-tailed test.** In a two-tailed test, each tail has an area of $\frac{1}{2}P$.

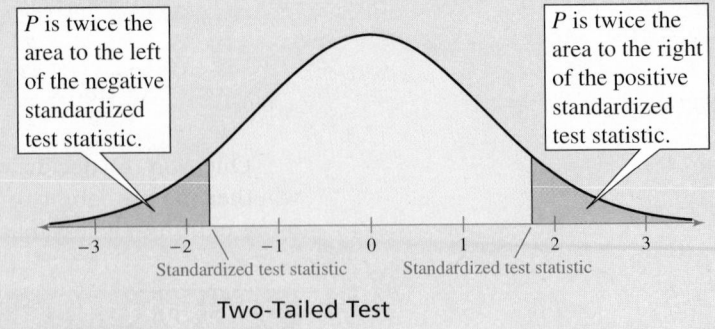

$H_0: \mu = k$
$H_a: \mu \neq k$

P is twice the area to the left of the negative standardized test statistic.

P is twice the area to the right of the positive standardized test statistic.

Standardized test statistic   Standardized test statistic

Two-Tailed Test

STUDY TIP

The third type of test is called a two-tailed test because evidence that would support the alternative hypothesis could lie in either tail of the sampling distribution.

The smaller the *P*-value of the test, the more evidence there is to reject the null hypothesis. A very small *P*-value indicates an unusual event. Remember, however, that even a very low *P*-value does not constitute proof that the null hypothesis is false, only that it is probably false.

EXAMPLE 3

▶ **Identifying the Nature of a Hypothesis Test**

For each claim, state H_0 and H_a in words and in symbols. Then determine whether the hypothesis test is a left-tailed test, right-tailed test, or two-tailed test. Sketch a normal sampling distribution and shade the area for the P-value.

1. A school publicizes that the proportion of its students who are involved in at least one extracurricular activity is 61%.

2. A car dealership announces that the mean time for an oil change is less than 15 minutes.

3. A company advertises that the mean life of its furnaces is more than 18 years.

▶ **Solution**

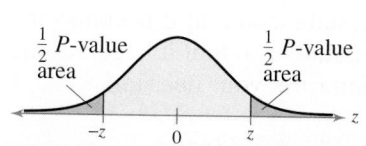

In Symbols	*In Words*
1. H_0: $p = 0.61$	The proportion of students who are involved in at least one extracurricular activity is 61%.
H_a: $p \neq 0.61$	The proportion of students who are involved in at least one extracurricular activity is not 61%.

Because H_a contains the $\neq$ symbol, the test is a two-tailed hypothesis test. The graph of the normal sampling distribution at the left shows the shaded area for the P-value.

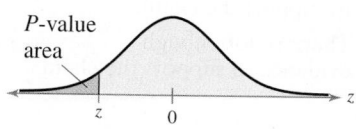

2. H_0: $\mu \geq 15$ min	The mean time for an oil change is greater than or equal to 15 minutes.
H_a: $\mu < 15$ min	The mean time for an oil change is less than 15 minutes.

Because H_a contains the $<$ symbol, the test is a left-tailed hypothesis test. The graph of the normal sampling distribution at the left shows the shaded area for the P-value.

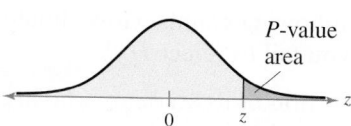

3. H_0: $\mu \leq 18$ yr	The mean life of the furnaces is less than or equal to 18 years.
H_a: $\mu > 18$ yr	The mean life of the furnaces is more than 18 years.

Because H_a contains the $>$ symbol, the test is a right-tailed hypothesis test. The graph of the normal sampling distribution at the left shows the shaded area for the P-value.

▶ **Try It Yourself 3**

For each claim, state H_0 and H_a in words and in symbols. Then determine whether the hypothesis test is a left-tailed test, right-tailed test, or two-tailed test. Sketch a normal sampling distribution and shade the area for the P-value.

1. A consumer analyst reports that the mean life of a certain type of automobile battery is not 74 months.
2. An electronics manufacturer publishes that the variance of the life of its home theater systems is less than or equal to 2.7.
3. A realtor publicizes that the proportion of homeowners who feel their house is too small for their family is more than 24%.

a. *Write* H_0 and H_a in words and in symbols.
b. *Determine* whether the test is *left-tailed*, *right-tailed*, or *two-tailed*.
c. *Sketch* the sampling distribution and *shade* the area for the P-value.

Answer: Page A40

▸ MAKING A DECISION AND INTERPRETING THE DECISION

To conclude a hypothesis test, you make a decision and interpret that decision. There are only two possible outcomes to a hypothesis test: (1) reject the null hypothesis and (2) fail to reject the null hypothesis.

INSIGHT

In this chapter, you will learn that there are two basic types of **decision rules** for deciding whether to reject H_0 or fail to reject H_0. The decision rule described on this page is based on *P*-values. The second basic type of decision rule is based on rejection regions. When the standardized test statistic falls in the rejection region, the observed probability (*P*-value) of a type I error is less than α. You will learn more about rejection regions in the next section.

DECISION RULE BASED ON *P*-VALUE

To use a *P*-value to make a conclusion in a hypothesis test, compare the *P*-value with α.

1. If $P \leq \alpha$, then reject H_0.
2. If $P > \alpha$, then fail to reject H_0.

Failing to reject the null hypothesis does not mean that you have accepted the null hypothesis as true. It simply means that there is not enough evidence to reject the null hypothesis. If you want to support a claim, state it so that it becomes the alternative hypothesis. If you want to reject a claim, state it so that it becomes the null hypothesis. The following table will help you interpret your decision.

	Claim	
Decision	**Claim is H_0.**	**Claim is H_a.**
Reject H_0.	There is enough evidence to reject the claim.	There is enough evidence to support the claim.
Fail to reject H_0.	There is not enough evidence to reject the claim.	There is not enough evidence to support the claim.

EXAMPLE 4

▸ Interpreting a Decision

You perform a hypothesis test for each of the following claims. How should you interpret your decision if you reject H_0? If you fail to reject H_0?

1. H_0 (Claim): A school publicizes that the proportion of its students who are involved in at least one extracurricular activity is 61%.
2. H_a (Claim): A car dealership announces that the mean time for an oil change is less than 15 minutes.

▸ Solution

1. The claim is represented by H_0. If you reject H_0, then you should conclude "there is enough evidence to reject the school's claim that the proportion of students who are involved in at least one extracurricular activity is 61%." If you fail to reject H_0, then you should conclude "there is not enough evidence to reject the school's claim that the proportion of students who are involved in at least one extracurricular activity is 61%."
2. The claim is represented by H_a, so the null hypothesis is "the mean time for an oil change is greater than or equal to 15 minutes." If you reject H_0, then you should conclude "there is enough evidence to support the dealership's claim that the mean time for an oil change is less than 15 minutes." If you fail to reject H_0, then you should conclude "there is not enough evidence to support the dealership's claim that the mean time for an oil change is less than 15 minutes."

▶ **Try It Yourself 4**

You perform a hypothesis test for the following claim. How should you interpret your decision if you reject H_0? If you fail to reject H_0?

H_a (Claim): A realtor publicizes that the proportion of homeowners who feel their house is too small for their family is more than 24%. *P > 24%*

a. Interpret your decision if you *reject* the null hypothesis.
b. Interpret your decision if you *fail to reject* the null hypothesis.

Answer: Page A41

The general steps for a hypothesis test using *P*-values are summarized below.

STEPS FOR HYPOTHESIS TESTING

1. State the claim mathematically and verbally. Identify the null and alternative hypotheses.

$$H_0: \quad ? \qquad H_a: \quad ?$$

2. Specify the level of significance.

$$\alpha = \quad ?$$

3. Determine the standardized sampling distribution and sketch its graph.

> This sampling distribution is based on the assumption that H_0 is true.

4. Calculate the test statistic and its corresponding standardized test statistic. Add it to your sketch.

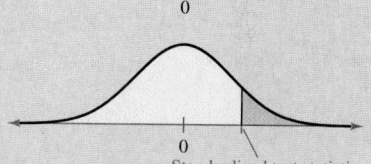

Standardized test statistic

5. Find the *P*-value.
6. Use the following decision rule.

| Is the *P*-value less than or equal to the level of significance? | **No** → | Fail to reject H_0. |

Yes ↓

Reject H_0.

7. Write a statement to interpret the decision in the context of the original claim.

In the steps above, the graphs show a right-tailed test. However, the same basic steps also apply to left-tailed and two-tailed tests.

▶ STRATEGIES FOR HYPOTHESIS TESTING

In a courtroom, the strategy used by an attorney depends on whether the attorney is representing the defense or the prosecution. In a similar way, the strategy that you will use in hypothesis testing should depend on whether you are trying to support or reject a claim. Remember that you cannot use a hypothesis test to support your claim if your claim is the null hypothesis. So, as a researcher, if you want a conclusion that supports your claim, word your claim so it is the alternative hypothesis. If you want to reject a claim, word it so it is the null hypothesis.

EXAMPLE 5

▶ Writing the Hypotheses

A medical research team is investigating the benefits of a new surgical treatment. One of the claims is that the mean recovery time for patients after the new treatment is less than 96 hours. How would you write the null and alternative hypotheses if (1) you are on the research team and want to support the claim? (2) you are on an opposing team and want to reject the claim?

▶ Solution

1. To answer the question, first think about the context of the claim. Because you want to support this claim, make the alternative hypothesis state that the mean recovery time for patients is less than 96 hours. So, $H_a: \mu < 96$ hours. Its complement, $\mu \geq 96$ hours, would be the null hypothesis.

$$H_0: \mu \geq 96$$
$$H_a: \mu < 96 \quad \text{(Claim)}$$

2. First think about the context of the claim. As an opposing researcher, you do not want the recovery time to be less than 96 hours. Because you want to reject this claim, make it the null hypothesis. So, $H_0: \mu \leq 96$ hours. Its complement, $\mu > 96$ hours, would be the alternative hypothesis.

$$H_0: \mu \leq 96 \quad \text{(Claim)}$$
$$H_a: \mu > 96$$

▶ Try It Yourself 5

1. You represent a chemical company that is being sued for paint damage to automobiles. You want to support the claim that the mean repair cost per automobile is less than $650. How would you write the null and alternative hypotheses?
2. You are on a research team that is investigating the mean temperature of adult humans. The commonly accepted claim is that the mean temperature is about 98.6°F. You want to show that this claim is false. How would you write the null and alternative hypotheses?

   a. *Determine* whether you want to support or reject the claim.
   b. Write the *null* and *alternative hypotheses*.   *Answer: Page A41*

7.1 EXERCISES

■ BUILDING BASIC SKILLS AND VOCABULARY

1. What are the two types of hypotheses used in a hypothesis test? How are they related?

2. Describe the two types of error possible in a hypothesis test decision.

3. What are the two decisions that you can make from performing a hypothesis test?

4. Does failing to reject the null hypothesis mean that the null hypothesis is true? Explain.

True or False? *In Exercises 5–10, determine whether the statement is true or false. If it is false, rewrite it as a true statement.*

5. In a hypothesis test, you assume the alternative hypothesis is true.

6. A statistical hypothesis is a statement about a sample.

7. If you decide to reject the null hypothesis, you can support the alternative hypothesis.

8. The level of significance is the maximum probability you allow for rejecting a null hypothesis when it is actually true.

9. A large P-value in a test will favor rejection of the null hypothesis.

10. If you want to support a claim, write it as your null hypothesis.

Stating Hypotheses *In Exercises 11–16, use the given statement to represent a claim. Write its complement and state which is H_0 and which is H_a.*

11. $\mu \leq 645$

12. $\mu < 128$

13. $\sigma \neq 5$

14. $\sigma^2 \geq 1.2$

15. $p < 0.45$

16. $p = 0.21$

Graphical Analysis *In Exercises 17–20, match the alternative hypothesis with its graph. Then state the null hypothesis and sketch its graph.*

17. $H_a: \mu > 3$

18. $H_a: \mu < 3$

19. $H_a: \mu \neq 3$

20. $H_a: \mu > 2$

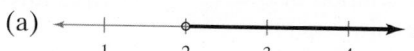

(a)

(b)

(c)

(d)

Identifying Tests *In Exercises 21–24, determine whether the hypothesis test with the given null and alternative hypotheses is left-tailed, right-tailed, or two-tailed.*

21. $H_0: \mu \leq 8.0$
$H_a: \mu > 8.0$

22. $H_0: \sigma \geq 5.2$
$H_a: \sigma < 5.2$

23. $H_0: \sigma^2 = 142$
$H_a: \sigma^2 \neq 142$

24. $H_0: p = 0.25$
$H_a: p \neq 0.25$

---

25. $\mu > 750$

  $H_0: \mu \le 750$; $H_a: \mu > 750$ (claim)

26. $\sigma < 3$

  $H_0: \sigma \ge 3$; $H_a: \sigma < 3$ (claim)

27. $\sigma \le 320$

  $H_0: \sigma \le 320$ (claim); $H_a: \sigma > 320$

28. $\mu \ge 85$

  $H_0: \mu \ge 85$ (claim); $H_a: \mu < 85$

29. $\mu < 45$

  $H_0: \mu \ge 45$; $H_a: \mu < 45$ (claim)

30. $p = 0.74$

  $H_0: p = 0.74$ (claim); $H_a: p \ne 0.74$

31. A type I error will occur if the actual proportion of new customers who return to buy their next piece of furniture is at least 0.60, but you reject $H_0: p \ge 0.60$.

  A type II error will occur if the actual proportion of new customers who return to buy their next piece of furniture is less than 0.60, but you fail to reject $H_0: p \ge 0.60$.

32. A type I error will occur if the actual mean flow rate of the hose is 16 gallons per minute, but you reject $H_0: \mu = 16$.

  A type II error will occur if the actual mean flow rate of the hose is not 16 gallons per minute, but you fail to reject $H_0: \mu = 16$.

33. A type I error will occur if the actual standard deviation of the length of time to play a game is less than or equal to 12 minutes, but you reject $H_0: \sigma \le 12$.

  A type II error will occur if the actual standard deviation of the length of time to play a game is greater than 12 minutes, but you fail to reject $H_0: \sigma \le 12$.

34. See Selected Answers, page A121.

35. A type I error will occur if the actual proportion of applicants who become police officers is at most 0.20, but you reject $H_0: p \le 0.20$.

  A type II error will occur if the actual proportion of applicants who become police officers is greater than 0.20, but you fail to reject $H_0: p \le 0.20$.

36. See Selected Answers, page A121.

37. See Odd Answers, page A81.

38. See Selected Answers, page A121.

■ USING AND INTERPRETING CONCEPTS

Stating the Hypotheses *In Exercises 25–30, write the claim as a mathematical sentence. State the null and alternative hypotheses, and identify which represents the claim.*

25. **Light Bulbs** A light bulb manufacturer claims that the mean life of a certain type of light bulb is more than 750 hours.

26. **Shipping Errors** As stated by a company's shipping department, the number of shipping errors per million shipments has a standard deviation that is less than 3.

27. **Base Price of an ATV** The standard deviation of the base price of a certain type of all-terrain vehicle is no more than $320.

28. **Oak Trees** A state park claims that the mean height of the oak trees in the park is at least 85 feet.

29. **Drying Time** A company claims that its brands of paint have a mean drying time of less than 45 minutes.

30. **MP3 Players** According to a recent survey, 74% of college students own an MP3 player. *(Source: Harris Interactive)*

Identifying Errors *In Exercises 31–36, write sentences describing type I and type II errors for a hypothesis test of the indicated claim.*

31. **Repeat Buyers** A furniture store claims that at least 60% of its new customers will return to buy their next piece of furniture.

32. **Flow Rate** A garden hose manufacturer advertises that the mean flow rate of a certain type of hose is 16 gallons per minute.

33. **Chess** A local chess club claims that the length of time to play a game has a standard deviation of more than 12 minutes.

34. **Video Game Systems** A researcher claims that the proportion of adults in the United States who own a video game system is not 26%.

35. **Police** A police station publicizes that at most 20% of applicants become police officers.

36. **Computers** A computer repairer advertises that the mean cost of removing a virus infection is less than $100.

Identifying Tests *In Exercises 37–42, state H_0 and H_a in words and in symbols. Then determine whether the hypothesis test is left-tailed, right-tailed, or two-tailed. Explain your reasoning.*

37. **Security Alarms** At least 14% of all homeowners have a home security alarm.

38. **Clocks** A manufacturer of grandfather clocks claims that the mean time its clocks lose is no more than 0.02 second per day.

39. **Golf** The standard deviation of the 18-hole scores for a golfer is less than 2.1 strokes.

40. **Lung Cancer** A government report claims that the proportion of lung cancer cases that are due to smoking is 87%. *(Source: LungCancer.org)*

39. H_0: The standard deviation of the 18-hole scores for a golfer is greater than or equal to 2.1 strokes.

H_a: The standard deviation of the 18-hole scores for a golfer is less than 2.1 strokes.

H_0: $\sigma \geq 2.1$; H_a: $\sigma < 2.1$

Left-tailed because the alternative hypothesis contains $<$.

40. See Selected Answers, page A121.

41. H_0: The mean length of the baseball team's games is greater than or equal to 2.5 hours.

H_a: The mean length of the baseball team's games is less than 2.5 hours.

H_0: $\mu \geq 2.5$; H_a: $\mu < 2.5$

Left-tailed because the alternative hypothesis contains $<$.

42. See Selected Answers, page A121.

43. (a) There is enough evidence to support the scientist's claim that the mean incubation period for swan eggs is less than 40 days.

(b) There is not enough evidence to support the scientist's claim that the mean incubation period for swan eggs is less than 40 days.

44. See Selected Answers, page A121.

45. See Odd Answers, page A81.

46. See Selected Answers, page A122.

47. (a) There is enough evidence to support the researcher's claim that the proportion of people who have had no health care visits in the past year is less than 17%.

(b) There is not enough evidence to support the researcher's claim that the proportion of people who have had no health care visits in the past year is less than 17%.

48. See Selected Answers, page A122.

49. H_0: $\mu \geq 60$; H_a: $\mu < 60$

50. H_0: $\mu = 21$; H_a: $\mu \neq 21$

51. (a) H_0: $\mu \geq 15$; H_a: $\mu < 15$
(b) H_0: $\mu \leq 15$; H_a: $\mu > 15$

52. (a) H_0: $\mu \leq 28$; H_a: $\mu > 28$
(b) H_0: $\mu \geq 28$; H_a: $\mu < 28$

41. Baseball A baseball team claims that the mean length of its games is less than 2.5 hours.

42. Tuition A state claims that the mean tuition of its universities is no more than $25,000 per year.

Interpreting a Decision *In Exercises 43–48, consider each claim. If a hypothesis test is performed, how should you interpret a decision that*

(a) rejects the null hypothesis?

(b) fails to reject the null hypothesis?

43. Swans A scientist claims that the mean incubation period for swan eggs is less than 40 days.

44. Lawn Mowers The standard deviation of the life of a certain type of lawn mower is at most 2.8 years.

45. Hourly Wages The U.S. Department of Labor claims that the proportion of full-time workers earning over $450 per week is greater than 75%. *(Adapted from U.S. Bureau of Labor Statistics)*

46. Gas Mileage An automotive manufacturer claims the standard deviation for the gas mileage of its models is 3.9 miles per gallon.

47. Health Care Visits A researcher claims that the proportion of people who have had no health care visits in the past year is less than 17%. *(Adapted from National Center for Health Statistics)*

48. Calories A sports drink maker claims the mean calorie content of its beverages is 72 calories per serving.

49. Writing Hypotheses: Medicine Your medical research team is investigating the mean cost of a 30-day supply of a certain heart medication. A pharmaceutical company thinks that the mean cost is less than $60. You want to support this claim. How would you write the null and alternative hypotheses?

50. Writing Hypotheses: Taxicab Company A taxicab company claims that the mean travel time between two destinations is about 21 minutes. You work for the bus company and want to reject this claim. How would you write the null and alternative hypotheses?

51. Writing Hypotheses: Refrigerator Manufacturer A refrigerator manufacturer claims that the mean life of its competitor's refrigerators is less than 15 years. You are asked to perform a hypothesis test to test this claim. How would you write the null and alternative hypotheses if

(a) you represent the manufacturer and want to support the claim?

(b) you represent the competitor and want to reject the claim?

52. Writing Hypotheses: Internet Provider An Internet provider is trying to gain advertising deals and claims that the mean time a customer spends online per day is greater than 28 minutes. You are asked to test this claim. How would you write the null and alternative hypotheses if

(a) you represent the Internet provider and want to support the claim?

(b) you represent a competing advertiser and want to reject the claim?

53. If you decrease α, you are decreasing the probability that you will reject H_0. Therefore, you are increasing the probability of failing to reject H_0. This could increase β, the probability of failing to reject H_0 when H_0 is false.

54. If $\alpha = 0$, the null hypothesis cannot be rejected and the hypothesis test is useless.

55. Yes; If the P-value is less than $\alpha = 0.05$, it is also less than $\alpha = 0.10$.

56. Not necessarily; A P-value less than $\alpha = 0.10$ may or may not also be less than $\alpha = 0.05$.

57. (a) Fail to reject H_0 because the confidence interval includes values greater than 70.

   (b) Reject H_0 because the confidence interval is located entirely to the left of 70.

   (c) Fail to reject H_0 because the confidence interval includes values greater than 70.

58. (a) Fail to reject H_0 because the confidence interval includes values less than 54.

   (b) Fail to reject H_0 because the confidence interval includes values less than 54.

   (c) Reject H_0 because the confidence interval is located entirely to the right of 54.

59. (a) Reject H_0 because the confidence interval is located entirely to the right of 0.20.

   (b) Fail to reject H_0 because the confidence interval includes values less than 0.20.

   (c) Fail to reject H_0 because the confidence interval includes values less than 0.20.

60. (a) Fail to reject H_0 because the confidence interval includes values greater than 0.73.

   (b) Reject H_0 because the confidence interval is located entirely to the left of 0.73.

   (c) Fail to reject H_0 because the confidence interval includes values greater than 0.73.

■ EXTENDING CONCEPTS

53. Getting at the Concept   Why can decreasing the probability of a type I error cause an increase in the probability of a type II error?

54. Getting at the Concept   Explain why a level of significance of $\alpha = 0$ is not used.

55. Writing   A null hypothesis is rejected with a level of significance of 0.05. Is it also rejected at a level of significance of 0.10? Explain.

56. Writing   A null hypothesis is rejected with a level of significance of 0.10. Is it also rejected at a level of significance of 0.05? Explain.

Graphical Analysis   *In Exercises 57–60, you are given a null hypothesis and three confidence intervals that represent three samplings. Decide whether each confidence interval indicates that you should reject H_0. Explain your reasoning.*

57. H_0: $\mu \geq 70$

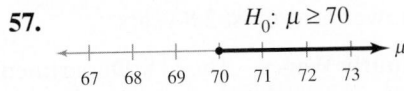

(a) $67 < \mu < 71$

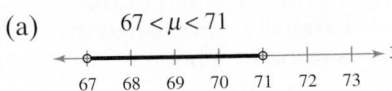

(b) $67 < \mu < 69$

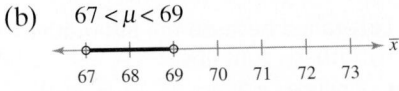

(c) $69.5 < \mu < 72.5$

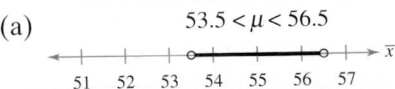

58. H_0: $\mu \leq 54$

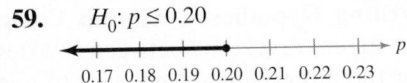

(a) $53.5 < \mu < 56.5$

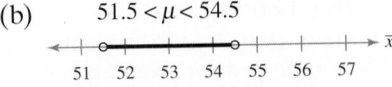

(b) $51.5 < \mu < 54.5$

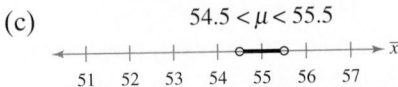

(c) $54.5 < \mu < 55.5$

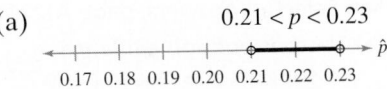

59. H_0: $p \leq 0.20$

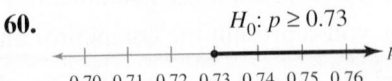

(a) $0.21 < p < 0.23$

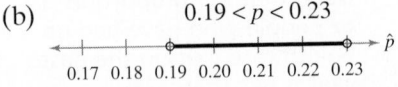

(b) $0.19 < p < 0.23$

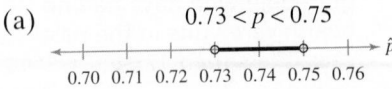

(c) $0.175 < p < 0.205$

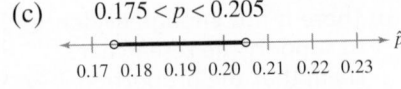

60. H_0: $p \geq 0.73$

(a) $0.73 < p < 0.75$

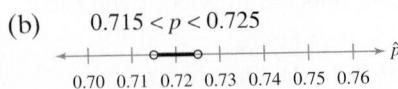

(b) $0.715 < p < 0.725$

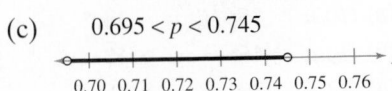

(c) $0.695 < p < 0.745$

7.2 Hypothesis Testing for the Mean (Large Samples)

WHAT YOU SHOULD LEARN

▸ How to find *P*-values and use them to test a mean μ

▸ How to use *P*-values for a *z*-test

▸ How to find critical values and rejection regions in a normal distribution

▸ How to use rejection regions for a *z*-test

▸ USING *P*-VALUES TO MAKE DECISIONS

In Chapter 5, you learned that when the sample size is at least 30, the sampling distribution for $\overline{x}$ (the sample mean) is normal. In Section 7.1, you learned that a way to reach a conclusion in a hypothesis test is to use a *P*-value for the sample statistic, such as $\overline{x}$. Recall that when you assume the null hypothesis is true, a *P*-value (or probability value) of a hypothesis test is the probability of obtaining a sample statistic with a value as extreme or more extreme than the one determined from the sample data. The decision rule for a hypothesis test based on a *P*-value is as follows.

DECISION RULE BASED ON *P*-VALUE

To use a *P*-value to make a conclusion in a hypothesis test, compare the *P*-value with α.

1. If $P \leq \alpha$, then reject H_0.
2. If $P > \alpha$, then fail to reject H_0.

Note to Instructor

If a *P*-value is less than 0.01, the null hypothesis will be rejected at the common levels of $\alpha = 0.01$, $\alpha = 0.05$, and $\alpha = 0.10$. If the *P*-value is greater than 0.10, then you would fail to reject H_0 for these common levels. Make sure students know that the same conclusion will be reached regardless of whether they use the critical value method or the *P*-value method.

EXAMPLE 1

▸ **Interpreting a *P*-Value**

The *P*-value for a hypothesis test is $P = 0.0237$. What is your decision if the level of significance is (1) $\alpha = 0.05$ and (2) $\alpha = 0.01$?

▸ **Solution**

1. Because $0.0237 < 0.05$, you should reject the null hypothesis.

2. Because $0.0237 > 0.01$, you should fail to reject the null hypothesis.

▸ **Try It Yourself 1**

The *P*-value for a hypothesis test is $P = 0.0347$. What is your decision if the level of significance is (1) $\alpha = 0.01$ and (2) $\alpha = 0.05$?

a. *Compare* the *P*-value with the level of significance.
b. *Make* a decision.

Answer: Page A41

INSIGHT

The lower the *P*-value, the more evidence there is in favor of rejecting H_0. The *P*-value gives you the lowest level of significance for which the sample statistic allows you to reject the null hypothesis. In Example 1, you would reject H_0 at any level of significance greater than or equal to 0.0237.

FINDING THE *P*-VALUE FOR A HYPOTHESIS TEST

After determining the hypothesis test's standardized test statistic and the test statistic's corresponding area, do one of the following to find the *P*-value.

a. For a left-tailed test, $P = $ (Area in left tail).
b. For a right-tailed test, $P = $ (Area in right tail).
c. For a two-tailed test, $P = 2$(Area in tail of test statistic).

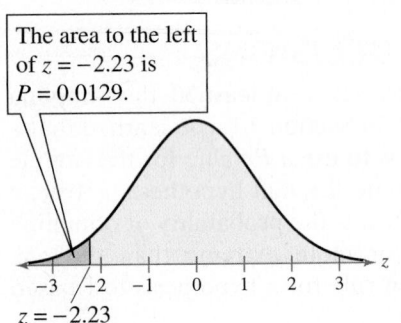

The area to the left of $z = -2.23$ is $P = 0.0129$.

$z = -2.23$

Left-Tailed Test

EXAMPLE 2

▶ Finding a *P*-Value for a Left-Tailed Test

Find the *P*-value for a left-tailed hypothesis test with a test statistic of $z = -2.23$. Decide whether to reject H_0 if the level of significance is $\alpha = 0.01$.

▶ Solution

The graph shows a standard normal curve with a shaded area to the left of $z = -2.23$. For a left-tailed test,

$$P = (\text{Area in left tail}).$$

From Table 4 in Appendix B, the area corresponding to $z = -2.23$ is 0.0129, which is the area in the left tail. So, the *P*-value for a left-tailed hypothesis test with a test statistic of $z = -2.23$ is $P = 0.0129$.

Interpretation Because the *P*-value of 0.0129 is greater than 0.01, you should fail to reject H_0.

▶ Try It Yourself 2

Find the *P*-value for a left-tailed hypothesis test with a test statistic of $z = -1.71$. Decide whether to reject H_0 if the level of significance is $\alpha = 0.05$.

a. *Use* Table 4 in Appendix B to find the area that corresponds to $z = -1.71$.
b. *Calculate* the *P*-value for a left-tailed test, the area in the left tail.
c. *Compare* the *P*-value with α and *decide* whether to reject H_0.

Answer: Page A41

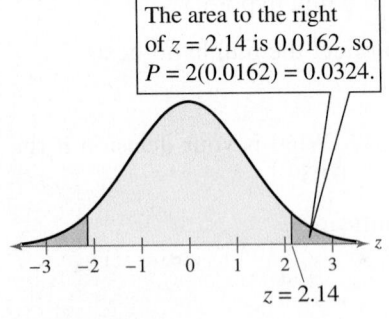

The area to the right of $z = 2.14$ is 0.0162, so $P = 2(0.0162) = 0.0324$.

$z = 2.14$

Two-Tailed Test

EXAMPLE 3

▶ Finding a *P*-Value for a Two-Tailed Test

Find the *P*-value for a two-tailed hypothesis test with a test statistic of $z = 2.14$. Decide whether to reject H_0 if the level of significance is $\alpha = 0.05$.

▶ Solution

The graph shows a standard normal curve with shaded areas to the left of $z = -2.14$ and to the right of $z = 2.14$. For a two-tailed test,

$$P = 2(\text{Area in tail of test statistic}).$$

From Table 4, the area corresponding to $z = 2.14$ is 0.9838. The area in the right tail is $1 - 0.9838 = 0.0162$. So, the *P*-value for a two-tailed hypothesis test with a test statistic of $z = 2.14$ is

$$P = 2(0.0162) = 0.0324.$$

Interpretation Because the *P*-value of 0.0324 is less than 0.05, you should reject H_0.

▶ Try It Yourself 3

Find the *P*-value for a two-tailed hypothesis test with a test statistic of $z = 1.64$. Decide whether to reject H_0 if the level of significance is $\alpha = 0.10$.

a. *Use* Table 4 to find the area that corresponds to $z = 1.64$.
b. *Calculate* the *P*-value for a two-tailed test, twice the area in the tail of the test statistic.
c. *Compare* the *P*-value with α and *decide* whether to reject H_0.

Answer: Page A41

▶ USING *P*-VALUES FOR A *z*-TEST

The *z*-test for the mean is used in populations for which the sampling distribution of sample means is normal. To use the *z*-test, you need to find the standardized value for your test statistic $\bar{x}$.

$$z = \frac{(\text{Sample mean}) - (\text{Hypothesized mean})}{\text{Standard error}}$$

STUDY TIP

With all hypothesis tests, it is helpful to sketch the sampling distribution. Your sketch should include the standardized test statistic.

z-TEST FOR A MEAN μ

The **z-test for a mean** is a statistical test for a population mean. The *z*-test can be used when the population is normal and σ is known, or for any population when the sample size n is at least 30. The **test statistic** is the sample mean $\bar{x}$ and the **standardized test statistic** is

$$z = \frac{\bar{x} - \mu}{\sigma/\sqrt{n}}.$$

Recall that $\dfrac{\sigma}{\sqrt{n}}$ = standard error = $\sigma_{\bar{x}}$.

When $n \geq 30$, you can use the sample standard deviation s in place of σ.

INSIGHT

When the sample size is at least 30, you know the following about the sampling distribution of sample means.

(1) The shape is normal.

(2) The mean is the hypothesized mean.

(3) The standard error is $s/\sqrt{n}$, where s is used in place of σ.

Note to Instructor

We use the same format for all hypothesis testing throughout the text. Using the same format makes it easier for students to understand the logic of the test. Emphasize that the sampling distribution and, consequently, the logic of the test are based on the assumption that the equality condition of the null hypothesis is true.

GUIDELINES

Using *P*-Values for a z-Test for Mean μ

IN WORDS	IN SYMBOLS
1. State the claim mathematically and verbally. Identify the null and alternative hypotheses.	State H_0 and H_a.
2. Specify the level of significance.	Identify α.
3. Determine the standardized test statistic.	$z = \dfrac{\bar{x} - \mu}{\sigma/\sqrt{n}}$ or, if $n \geq 30$, use $\sigma \approx s$.
4. Find the area that corresponds to z.	Use Table 4 in Appendix B.

5. Find the *P*-value.

   a. For a left-tailed test, $P = (\text{Area in left tail})$.

   b. For a right-tailed test, $P = (\text{Area in right tail})$.

   c. For a two-tailed test, $P = 2(\text{Area in tail of test statistic})$.

6. Make a decision to reject or fail to reject the null hypothesis.	Reject H_0 if *P*-value is less than or equal to α. Otherwise, fail to reject H_0.

7. Interpret the decision in the context of the original claim.

EXAMPLE 4

▶ Hypothesis Testing Using *P*-Values

In auto racing, a pit stop is where a racing vehicle stops for new tires, fuel, repairs, and other mechanical adjustments. The efficiency of a pit crew that makes these adjustments can affect the outcome of a race. A pit crew claims that its mean pit stop time (for 4 new tires and fuel) is less than 13 seconds. A random selection of 32 pit stop times has a sample mean of 12.9 seconds and a standard deviation of 0.19 second. Is there enough evidence to support the claim at $\alpha = 0.01$? Use a *P*-value.

▶ Solution

The claim is "the mean pit stop time is less than 13 seconds." So, the null and alternative hypotheses are

$$H_0:\ \mu \geq 13 \text{ seconds} \qquad \text{and} \qquad H_a:\ \mu < 13 \text{ seconds. (Claim)}$$

The level of significance is $\alpha = 0.01$. The standardized test statistic is

$$z = \frac{\bar{x} - \mu}{\sigma/\sqrt{n}} \qquad \text{Because } n \geq 30, \text{ use the } z\text{-test.}$$

$$\approx \frac{12.9 - 13}{0.19/\sqrt{32}} \qquad \text{Because } n \geq 30, \text{ use } \sigma \approx s = 0.19. \text{ Assume } \mu = 13.$$

$$\approx -2.98.$$

In Table 4 in Appendix B, the area corresponding to $z = -2.98$ is 0.0014. Because this test is a left-tailed test, the *P*-value is equal to the area to the left of $z = -2.98$. So, $P = 0.0014$. Because the *P*-value is less than $\alpha = 0.01$, you should decide to reject the null hypothesis.

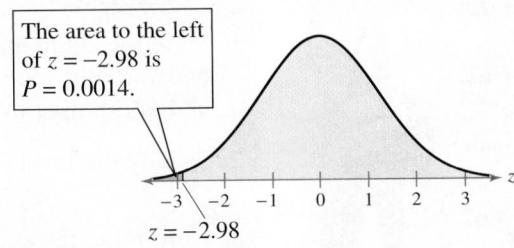

The area to the left of $z = -2.98$ is $P = 0.0014$.

$z = -2.98$

Left-Tailed Test

Interpretation   There is enough evidence at the 1% level of significance to support the claim that the mean pit stop time is less than 13 seconds.

▶ Try It Yourself 4

Homeowners claim that the mean speed of automobiles traveling on their street is greater than the speed limit of 35 miles per hour. A random sample of 100 automobiles has a mean speed of 36 miles per hour and a standard deviation of 4 miles per hour. Is there enough evidence to support the claim at $\alpha = 0.05$? Use a *P*-value.

a. Identify the *claim*. Then state the *null* and *alternative hypotheses*.
b. Identify the *level of significance*.
c. Find the *standardized test statistic z*.
d. Find the *P-value*.
e. *Decide* whether to reject the null hypothesis.
f. *Interpret* the decision in the context of the original claim.

Answer: Page A41

EXAMPLE 5   **SC** Report 29

See MINITAB
steps on page 424.

▶ **Hypothesis Testing Using *P*-Values**

The National Institute of Diabetes and Digestive and Kidney Diseases reports that the average cost of bariatric (weight loss) surgery is about $22,500. You think this information is incorrect. You randomly select 30 bariatric surgery patients and find that the average cost for their surgeries is $21,545 with a standard deviation of $3015. Is there enough evidence to support your claim at $\alpha = 0.05$? Use a *P*-value. *(Adapted from National Institute of Diabetes and Digestive and Kidney Diseases)*

▶ **Solution**

The claim is "the mean is different from $22,500." So, the null and alternative hypotheses are

$$H_0: \mu = \$22,500$$

and

$$H_a: \mu \neq \$22,500. \text{ (Claim)}$$

The level of significance is $\alpha = 0.05$. The standardized test statistic is

$$z = \frac{\bar{x} - \mu}{\sigma/\sqrt{n}} \qquad \text{Because } n \geq 30, \text{ use the } z\text{-test.}$$

$$\approx \frac{21,545 - 22,500}{3015/\sqrt{30}} \qquad \begin{array}{l} \text{Because } n \geq 30, \text{ use } \sigma \approx s = 3015. \\ \text{Assume } \mu = 22,500. \end{array}$$

$$\approx -1.73.$$

The area to the left of $z = -1.73$ is 0.0418, so $P = 2(0.0418) = 0.0836$.

Two-Tailed Test

In Table 4, the area corresponding to $z = -1.73$ is 0.0418. Because the test is a two-tailed test, the *P*-value is equal to twice the area to the left of $z = -1.73$. So,

$$P = 2(0.0418)$$

$$= 0.0836.$$

Because the *P*-value is greater than α, you should fail to reject the null hypothesis.

Interpretation   There is not enough evidence at the 5% level of significance to support the claim that the mean cost of bariatric surgery is different from $22,500.

▶ **Try It Yourself 5**

One of your distributors reports an average of 150 sales per day. You suspect that this average is not accurate, so you randomly select 35 days and determine the number of sales each day. The sample mean is 143 daily sales with a standard deviation of 15 sales. At $\alpha = 0.01$, is there enough evidence to doubt the distributor's reported average? Use a *P*-value.

a. Identify the *claim*. Then state the *null* and *alternative hypotheses*.
b. Identify the *level of significance*.
c. Find the *standardized test statistic z*.
d. Find the *P-value*.
e. *Decide* whether to reject the null hypothesis.
f. *Interpret* the decision in the context of the original claim.

Answer: Page A41

EXAMPLE 6

▶ **Using a Technology Tool to Find a *P*-Value**

What decision should you make for the following TI-83/84 Plus displays, using a level of significance of $\alpha = 0.05$?

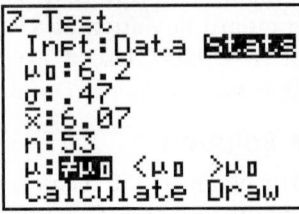

TI-83/84 Plus

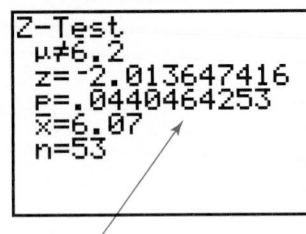

TI-83/84 Plus

▶ **Solution**

The *P*-value for this test is given as 0.0440464253. Because the *P*-value is less than 0.05, you should reject the null hypothesis.

▶ **Try It Yourself 6**

For the TI-83/84 Plus hypothesis test shown in Example 6, make a decision at the $\alpha = 0.01$ level of significance.

a. *Compare* the *P*-value with the level of significance.
b. *Make* your decision.   *Answer: Page A41*

▶ REJECTION REGIONS AND CRITICAL VALUES

Another method to decide whether to reject the null hypothesis is to determine whether the standardized test statistic falls within a range of values called the rejection region of the sampling distribution.

DEFINITION

A **rejection region** (or **critical region**) of the sampling distribution is the range of values for which the null hypothesis is not probable. If a test statistic falls in this region, the null hypothesis is rejected. A **critical value** z_0 separates the rejection region from the nonrejection region.

GUIDELINES

Finding Critical Values in a Normal Distribution

1. Specify the level of significance α.
2. Decide whether the test is left-tailed, right-tailed, or two-tailed.
3. Find the critical value(s) z_0. If the hypothesis test is
   a. *left-tailed*, find the *z*-score that corresponds to an area of α.
   b. *right-tailed*, find the *z*-score that corresponds to an area of $1 - \alpha$.
   c. *two-tailed*, find the *z*-scores that correspond to $\frac{1}{2}\alpha$ and $1 - \frac{1}{2}\alpha$.
4. Sketch the standard normal distribution. Draw a vertical line at each critical value and shade the rejection region(s).

If you cannot find the exact area in Table 4, use the area that is closest. When the area is exactly midway between two areas in the table, use the z-score midway between the corresponding z-scores.

EXAMPLE 7

▶ Finding a Critical Value for a Left-Tailed Test

Find the critical value and rejection region for a left-tailed test with $\alpha = 0.01$.

▶ Solution

The graph shows a standard normal curve with a shaded area of 0.01 in the left tail. In Table 4, the z-score that is closest to an area of 0.01 is -2.33. So, the critical value is $z_0 = -2.33$. The rejection region is to the left of this critical value.

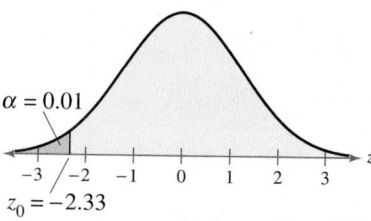

1% Level of Significance

▶ Try It Yourself 7

Find the critical value and rejection region for a left-tailed test with $\alpha = 0.10$.

a. *Draw* a graph of the standard normal curve with an area of α in the left tail.
b. *Use* Table 4 to find the area that is closest to α.
c. *Find* the z-score that corresponds to this area.
d. *Identify* the rejection region.

Answer: Page A41

EXAMPLE 8

▶ Finding a Critical Value for a Two-Tailed Test

Find the critical values and rejection regions for a two-tailed test with $\alpha = 0.05$.

▶ Solution

The graph shows a standard normal curve with shaded areas of $\frac{1}{2}\alpha = 0.025$ in each tail. The area to the left of $-z_0$ is $\frac{1}{2}\alpha = 0.025$, and the area to the left of z_0 is $1 - \frac{1}{2}\alpha = 0.975$. In Table 4, the z-scores that correspond to the areas 0.025 and 0.975 are -1.96 and 1.96, respectively. So, the critical values are $-z_0 = -1.96$ and $z_0 = 1.96$. The rejection regions are to the left of -1.96 and to the right of 1.96.

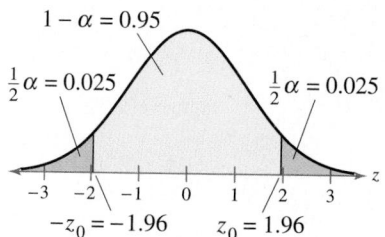

5% Level of Significance

▶ Try It Yourself 8

Find the critical values and rejection regions for a two-tailed test with $\alpha = 0.08$.

a. *Draw* a graph of the standard normal curve with an area of $\frac{1}{2}\alpha$ in each tail.
b. *Use* Table 4 to find the areas that are closest to $\frac{1}{2}\alpha$ and $1 - \frac{1}{2}\alpha$.
c. *Find* the z-scores that correspond to these areas.
d. *Identify* the rejection regions.

Answer: Page A41

STUDY TIP

Notice in Example 8 that the critical values are opposites. This is always true for two-tailed z-tests.

The table lists the critical values for commonly used levels of significance.

Alpha	Tail	z
0.10	Left	-1.28
	Right	1.28
	Two	± 1.645
0.05	Left	-1.645
	Right	1.645
	Two	± 1.96
0.01	Left	-2.33
	Right	2.33
	Two	± 2.575

▸ **USING REJECTION REGIONS FOR A z-TEST**

To conclude a hypothesis test using rejection region(s), you make a decision and interpret the decision as follows.

DECISION RULE BASED ON REJECTION REGION

To use a rejection region to conduct a hypothesis test, calculate the standardized test statistic z. If the standardized test statistic

1. is in the rejection region, then reject H_0.
2. is *not* in the rejection region, then fail to reject H_0.

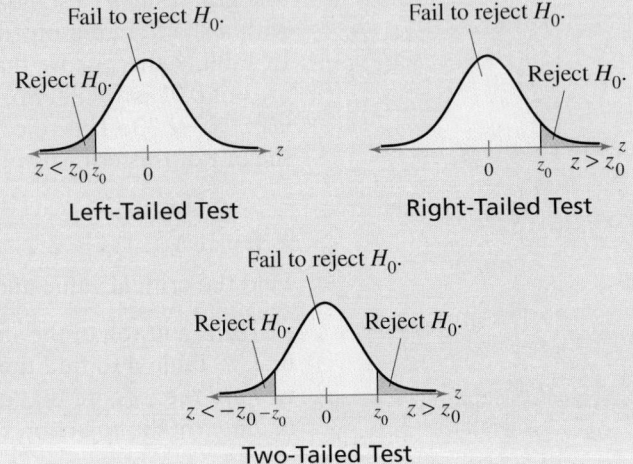

Left-Tailed Test

Right-Tailed Test

Two-Tailed Test

Failing to reject the null hypothesis does not mean that you have accepted the null hypothesis as true. It simply means that there is not enough evidence to reject the null hypothesis.

GUIDELINES

Using Rejection Regions for a z-Test for a Mean μ

IN WORDS	IN SYMBOLS
1. State the claim mathematically and verbally. Identify the null and alternative hypotheses.	State H_0 and H_a.
2. Specify the level of significance.	Identify α.
3. Determine the critical value(s).	Use Table 4 in Appendix B.
4. Determine the rejection region(s).	
5. Find the standardized test statistic and sketch the sampling distribution.	$z = \dfrac{\bar{x} - \mu}{\sigma/\sqrt{n}}$, or, if $n \geq 30$, use $\sigma \approx s$.
6. Make a decision to reject or fail to reject the null hypothesis.	If z is in the rejection region, reject H_0. Otherwise, fail to reject H_0.
7. Interpret the decision in the context of the original claim.	

See TI-83/84 Plus
steps on page 425.

EXAMPLE 9

▶ Testing μ with a Large Sample

Employees at a construction and mining company claim that the mean salary of the company's mechanical engineers is less than that of one of its competitors, which is $68,000. A random sample of 30 of the company's mechanical engineers has a mean salary of $66,900 with a standard deviation of $5500. At $\alpha = 0.05$, test the employees' claim.

▶ Solution

The claim is "the mean salary is less than $68,000." So, the null and alternative hypotheses can be written as

$$H_0: \mu \geq \$68,000 \qquad \text{and} \qquad H_a: \mu < \$68,000. \text{ (Claim)}$$

Because the test is a left-tailed test and the level of significance is $\alpha = 0.05$, the critical value is $z_0 = -1.645$ and the rejection region is $z < -1.645$. The standardized test statistic is

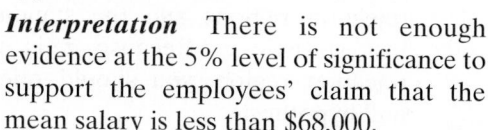

$$z = \frac{\bar{x} - \mu}{\sigma/\sqrt{n}} \qquad \text{Because } n \geq 30, \text{ use the } z\text{-test.}$$

$$\approx \frac{66,900 - 68,000}{5500/\sqrt{30}} \qquad \begin{array}{l} \text{Because } n \geq 30, \text{ use } \sigma \approx s = 5500. \\ \text{Assume } \mu = 68,000. \end{array}$$

$$\approx -1.10.$$

The graph shows the location of the rejection region and the standardized test statistic z. Because z is not in the rejection region, you fail to reject the null hypothesis.

Interpretation There is not enough evidence at the 5% level of significance to support the employees' claim that the mean salary is less than $68,000.

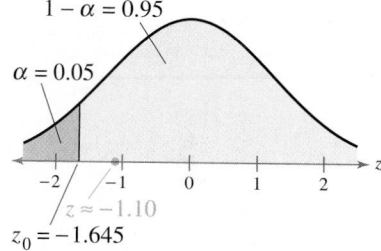

5% Level of Significance

Be sure you understand the decision made in this example. Even though your sample has a mean of $66,900, you cannot (at a 5% level of significance) support the claim that the mean of all the mechanical engineers' salaries is less than $68,000. The difference between your test statistic and the hypothesized mean is probably due to sampling error.

▶ Try It Yourself 9

The CEO of the company claims that the mean work day of the company's mechanical engineers is less than 8.5 hours. A random sample of 35 of the company's mechanical engineers has a mean work day of 8.2 hours with a standard deviation of 0.5 hour. At $\alpha = 0.01$, test the CEO's claim.

a. Identify the *claim* and state H_0 and H_a.
b. Identify the *level of significance* α.
c. Find the *critical value* z_0 and identify the *rejection region*.
d. Find the *standardized test statistic z. Sketch* a graph.
e. *Decide* whether to reject the null hypothesis.
f. *Interpret* the decision in the context of the original claim.

Answer: Page A41

EXAMPLE 10

▶ **Testing μ with a Large Sample**

The U.S. Department of Agriculture claims that the mean cost of raising a child from birth to age 2 by husband-wife families in the United States is $13,120. A random sample of 500 children (age 2) has a mean cost of $12,925 with a standard deviation of $1745. At $\alpha = 0.10$, is there enough evidence to reject the claim? *(Adapted from U.S. Department of Agriculture Center for Nutrition Policy and Promotion)*

▶ **Solution**

The claim is "the mean cost is $13,120." So, the null and alternative hypotheses are

$$H_0: \mu = \$13,120 \quad \text{(Claim)}$$

and

$$H_a: \mu \neq \$13,120.$$

Because the test is a two-tailed test and the level of significance is $\alpha = 0.10$, the critical values are $-z_0 = -1.645$ and $z_0 = 1.645$. The rejection regions are $z < -1.645$ and $z > 1.645$. The standardized test statistic is

$$z = \frac{\bar{x} - \mu}{\sigma/\sqrt{n}} \qquad \text{Because } n \geq 30, \text{ use the } z\text{-test.}$$

$$\approx \frac{12,925 - 13,120}{1745/\sqrt{500}} \qquad \begin{array}{l} \text{Because } n \geq 30, \text{ use } \sigma \approx s = 1745. \\ \text{Assume } \mu = 13,120. \end{array}$$

$$\approx -2.50.$$

The graph shows the location of the rejection regions and the standardized test statistic z. Because z is in the rejection region, you should reject the null hypothesis.

Interpretation   There is enough evidence at the 10% level of significance to reject the claim that the mean cost of raising a child from birth to age 2 by husband-wife families in the United States is $13,120.

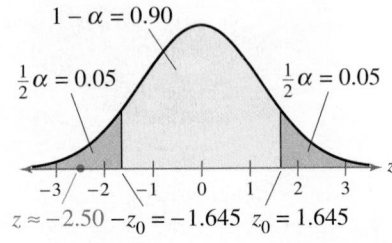

5% Level of Significance

▶ **Try It Yourself 10**

Using the information and results of Example 10, determine whether there is enough evidence to reject the claim that the mean cost of raising a child from birth to age 2 by husband-wife families in the United States is $13,120. Use $\alpha = 0.01$.

a. Identify the *level of significance α*.
b. Find the *critical values* $-z_0$ and z_0 and identify the *rejection regions*.
c. *Sketch* a graph. *Decide* whether to reject the null hypothesis.
d. *Interpret* the decision in the context of the original claim.

Answer: Page A41

Z-Test
μ≠13120
z=-2.498757912
p=.0124629702
x̄=12925
n=500

Using a TI-83/84 Plus, you can find the standardized test statistic automatically.

7.2 EXERCISES

FOR EXTRA HELP:
MyStatLab

1. In the z-test using rejection region(s), the test statistic is compared with critical values. The z-test using a *P*-value compares the *P*-value with the level of significance α.

2. No; Both involve comparing the test statistic's probability with the level of significance. The *P*-value method converts the standardized test statistic to a probability (*P*-value) and compares this with the level of significance, whereas the critical value method converts the level of significance to a z-score and compares this with the standardized test statistic.

3. $P = 0.0934$; Reject H_0.

4. $P = 0.0606$; Fail to reject H_0.

5. $P = 0.0069$; Reject H_0.

6. $P = 0.1093$; Fail to reject H_0.

7. $P = 0.0930$; Fail to reject H_0.

8. $P = 0.0214$; Fail to reject H_0.

9. b

10. d

11. c

12. a

13. (a) Fail to reject H_0.
(b) Reject H_0.

14. (a) Fail to reject H_0.
(b) Fail to reject H_0.

■ BUILDING BASIC SKILLS AND VOCABULARY

1. Explain the difference between the z-test for μ using rejection region(s) and the z-test for μ using a *P*-value.

2. In hypothesis testing, does choosing between the critical value method or the *P*-value method affect your conclusion? Explain.

In Exercises 3–8, find the P-value for the indicated hypothesis test with the given standardized test statistic z. Decide whether to reject H_0 for the given level of significance α.

3. Left-tailed test, $z = -1.32$, $\alpha = 0.10$

4. Left-tailed test, $z = -1.55$, $\alpha = 0.05$

5. Right-tailed test, $z = 2.46$, $\alpha = 0.01$

6. Right-tailed test, $z = 1.23$, $\alpha = 0.10$

7. Two-tailed test, $z = -1.68$, $\alpha = 0.05$

8. Two-tailed test, $z = 2.30$, $\alpha = 0.01$

Graphical Analysis   *In Exercises 9–12, match each P-value with the graph that displays its area. The graphs are labeled (a)–(d).*

9. $P = 0.0089$

10. $P = 0.3050$

11. $P = 0.0688$

12. $P = 0.0287$

(a)
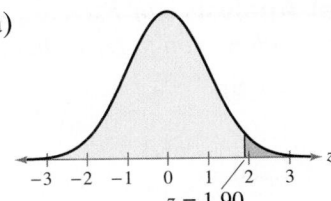
$z = 1.90$

(b)
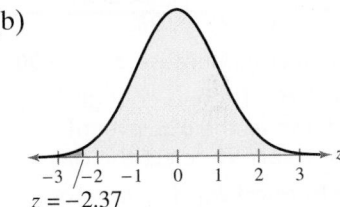
$z = -2.37$

(c)
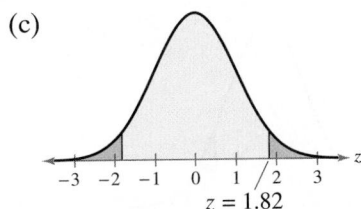
$z = 1.82$

(d)
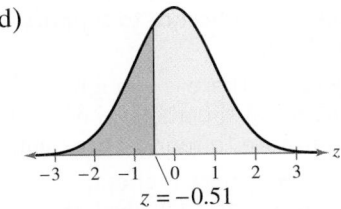
$z = -0.51$

13. Given $H_0: \mu = 100$, $H_a: \mu \neq 100$, and $P = 0.0461$.

    (a) Do you reject or fail to reject H_0 at the 0.01 level of significance?
    (b) Do you reject or fail to reject H_0 at the 0.05 level of significance?

14. Given $H_0: \mu \geq 8.5$, $H_a: \mu < 8.5$, and $P = 0.0691$.

    (a) Do you reject or fail to reject H_0 at the 0.01 level of significance?
    (b) Do you reject or fail to reject H_0 at the 0.05 level of significance?

15. Fail to reject H_0.

16. Reject H_0.

17. 1.645

See Odd Answers, page A81.

18. 1.41

See Selected Answers, page A122.

19. −1.88

See Odd Answers, page A81.

20. −1.34

See Selected Answers, page A122.

21. −2.33, 2.33

See Odd Answers, page A81.

22. −1.645, 1.645

See Selected Answers, page A122.

23. (a) Fail to reject H_0 because $z < 1.285$.

(b) Fail to reject H_0 because $z < 1.285$.

(c) Fail to reject H_0 because $z < 1.285$.

(d) Reject H_0 because $z > 1.285$.

24. (a) Reject H_0 because $z > 1.96$.

(b) Fail to reject H_0 because $-1.96 < z < 1.96$.

(c) Fail to reject H_0 because $-1.96 < z < 1.96$.

(d) Reject H_0 because $z < -1.96$.

25. Reject H_0. There is enough evidence at the 5% level of significance to reject the claim.

26. Fail to reject H_0. There is not enough evidence at the 10% level of significance to support the claim.

27. Reject H_0. There is enough evidence at the 2% level of significance to support the claim.

28. Reject H_0. There is enough evidence at the 1% level of significance to reject the claim.

29. (a) $H_0: \mu \le 30$; $H_a: \mu > 30$ (claim)

(b) 2.83; 0.9977

(c) 0.0023

(d) Reject H_0.

(e) There is enough evidence at the 1% level of significance to support the student's claim that the mean raw score for the school's applicants is more than 30.

In Exercises 15 and 16, use the TI-83/84 Plus displays to make a decision to reject or fail to reject the null hypothesis at the given level of significance.

15. $\alpha = 0.05$

16. $\alpha = 0.01$

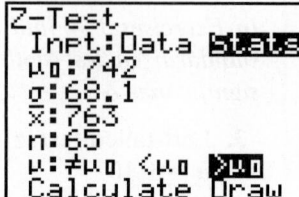

Finding Critical Values *In Exercises 17–22, find the critical value(s) for the indicated type of test and level of significance α. Include a graph with your answer.*

17. Right-tailed test, $\alpha = 0.05$

18. Right-tailed test, $\alpha = 0.08$

19. Left-tailed test, $\alpha = 0.03$

20. Left-tailed test, $\alpha = 0.09$

21. Two-tailed test, $\alpha = 0.02$

22. Two-tailed test, $\alpha = 0.10$

Graphical Analysis *In Exercises 23 and 24, state whether each standardized test statistic z allows you to reject the null hypothesis. Explain your reasoning.*

23. (a) $z = -1.301$

(b) $z = 1.203$

(c) $z = 1.280$

(d) $z = 1.286$

24. (a) $z = 1.98$

(b) $z = -1.89$

(c) $z = 1.65$

(d) $z = -1.99$

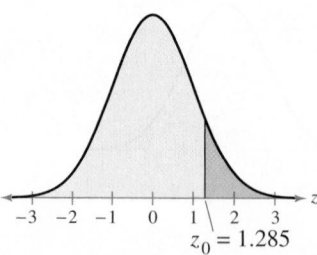

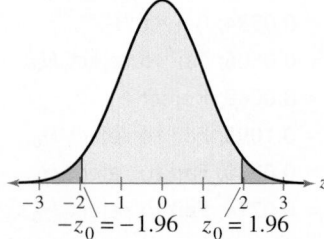

In Exercises 25–28, test the claim about the population mean μ at the given level of significance α using the given sample statistics.

25. Claim: $\mu = 40$; $\alpha = 0.05$.
Sample statistics: $\overline{x} = 39.2$, $s = 3.23$, $n = 75$

26. Claim: $\mu > 1745$; $\alpha = 0.10$.
Sample statistics: $\overline{x} = 1752$, $s = 38$, $n = 44$

27. Claim: $\mu \ne 8550$; $\alpha = 0.02$.
Sample statistics: $\overline{x} = 8420$, $s = 314$, $n = 38$

28. Claim: $\mu \le 22{,}500$; $\alpha = 0.01$.
Sample statistics: $\overline{x} = 23{,}250$, $s = 1200$, $n = 45$

30. (a) H_0: $\mu \geq 135$ (claim);
H_a: $\mu < 135$

(b) -3.43; 0.0003

(c) 0.0003

(d) Reject H_0.

(e) There is enough evidence at the 10% level of significance to reject the manufacturer's claim that the average activating temperature is at least 135°F.

31. (a) H_0: $\mu = 28.5$ (claim);
H_a: $\mu \neq 28.5$

(b) -1.71; 0.0436

(c) 0.0872 (*Tech:* 0.0878)

(d) Fail to reject H_0.

(e) There is not enough evidence at the 8% level of significance to reject the U.S. Department of Agriculture's claim that the mean consumption of bottled water by a person in the United States is 28.5 gallons per year.

32. (a) H_0: $\mu = 24.2$ (claim);
H_a: $\mu \neq 24.2$

(b) -2.40; 0.0082

(c) 0.0164 (*Tech:* 0.0166)

(d) Reject H_0.

(e) There is enough evidence at the 5% level of significance to reject the U.S. Department of Agriculture's claim that the mean consumption of coffee by a person in the United States is 24.2 gallons per year.

33. (a) H_0: $\mu = 15$ (claim); H_a: $\mu \neq 15$

(b) -0.22; 0.4129 (*Tech:* 0.4135)

(c) 0.8258 (*Tech:* 0.8270)

(d) Fail to reject H_0.

(e) There is not enough evidence at the 5% level of significance to reject the claim that the mean time it takes smokers to quit smoking permanently is 15 years.

34. (a) H_0: $\mu \leq 66{,}200$;
H_a: $\mu > 66{,}200$ (claim)

(b) 0.97; 0.8340

(c) 0.1660 (*Tech:* 0.1651)

(d) Fail to reject H_0.

(e) See Selected Answers, page A122.

USING AND INTERPRETING CONCEPTS

Testing Claims Using P-Values   *In Exercises 29–34,*

(a) write the claim mathematically and identify H_0 and H_a.

(b) find the standardized test statistic z and its corresponding area. If convenient, use technology.

(c) find the P-value. If convenient, use technology.

(d) decide whether to reject or fail to reject the null hypothesis.

(e) interpret the decision in the context of the original claim.

29. MCAT Scores A random sample of 50 medical school applicants at a university has a mean raw score of 31 with a standard deviation of 2.5 on the multiple choice portions of the Medical College Admission Test (MCAT). A student says that the mean raw score for the school's applicants is more than 30. At $\alpha = 0.01$, is there enough evidence to support the student's claim? *(Adapted from Association of American Medical Colleges)*

30. Sprinkler Systems A manufacturer of sprinkler systems designed for fire protection claims that the average activating temperature is at least 135°F. To test this claim, you randomly select a sample of 32 systems and find the mean activation temperature to be 133°F with a standard deviation of 3.3°F. At $\alpha = 0.10$, do you have enough evidence to reject the manufacturer's claim?

31. Bottled Water Consumption The U.S. Department of Agriculture claims that the mean consumption of bottled water by a person in the United States is 28.5 gallons per year. A random sample of 100 people in the United States has a mean bottled water consumption of 27.8 gallons per year with a standard deviation of 4.1 gallons. At $\alpha = 0.08$, can you reject the claim? *(Adapted from U.S. Department of Agriculture)*

32. Coffee Consumption The U.S. Department of Agriculture claims that the mean consumption of coffee by a person in the United States is 24.2 gallons per year. A random sample of 120 people in the United States shows that the mean coffee consumption is 23.5 gallons per year with a standard deviation of 3.2 gallons. At $\alpha = 0.05$, can you reject the claim? *(Adapted from U.S. Department of Agriculture)*

33. Quitting Smoking The lengths of time (in years) it took a random sample of 32 former smokers to quit smoking permanently are listed. At $\alpha = 0.05$, is there enough evidence to reject the claim that the mean time it takes smokers to quit smoking permanently is 15 years? *(Adapted from The Gallup Organization)*

15.7	13.2	22.6	13.0	10.7	18.1	14.7	7.0	17.3	7.5	21.8
12.3	19.8	13.8	16.0	15.5	13.1	20.7	15.5	9.8	11.9	16.9
7.0	19.3	13.2	14.6	20.9	15.4	13.3	11.6	10.9	21.6

34. Salaries An analyst claims that the mean annual salary for advertising account executives in Denver, Colorado is more than the national mean, $66,200. The annual salaries (in dollars) for a random sample of 35 advertising account executives in Denver are listed. At $\alpha = 0.09$, is there enough evidence to support the analyst's claim? *(Adapted from Salary.com)*

69,450	65,910	68,780	66,724	64,125	67,561	62,419
70,375	65,835	62,653	65,090	67,997	65,176	64,936
66,716	69,832	63,111	64,550	63,512	65,800	66,150
68,587	68,276	65,902	63,415	64,519	70,275	70,102
67,230	65,488	66,225	69,879	69,200	65,179	69,755

35. (a) H_0: $\mu = 40$ (claim); H_a: $\mu \neq 40$

(b) $-z_0 = -2.575$, $z_0 = 2.575$;

Rejection regions:
$z < -2.575$, $z > 2.575$

(c) -0.584

(d) Fail to reject H_0.

(e) There is not enough evidence at the 1% level of significance to reject the company's claim that the mean caffeine content per 12-ounce bottle of cola is 40 milligrams.

36. See Selected Answers, page A122.

37. (a) H_0: $\mu \geq 750$ (claim);
H_a: $\mu < 750$

(b) $z_0 = -2.05$;
Rejection region: $z < -2.05$

(c) -0.5

(d) Fail to reject H_0.

(e) There is not enough evidence at the 2% level of significance to reject the light bulb manufacturer's claim that the mean life of the bulb is at least 750 hours.

38. (a) H_0: $\mu \leq 920$ (claim);
H_a: $\mu > 920$

(b) $z_0 = 1.28$;
Rejection region: $z > 1.28$

(c) 1.84   (d) Reject H_0.

(e) There is enough evidence at the 10% level of significance to reject the restaurant's claim that the mean sodium content in one of their breakfast sandwiches is no more than 920 milligrams.

**Weight Loss (in pounds)
after One Month**

5	7 7 Key: 5|7 = 5.7
6	6 7
7	0 1 9
8	2 2 7 9
9	0 3 5 6 8
10	2 5 6 6
11	1 2 5 7 8
12	0 7 8
13	8
14	
15	0

FIGURE FOR EXERCISE 41

Testing Claims Using Critical Values   *In Exercises 35–42, (a) write the claim mathematically and identify H_0 and H_a, (b) find the critical values and identify the rejection regions, (c) find the standardized test statistic, (d) decide whether to reject or fail to reject the null hypothesis, and (e) interpret the decision in the context of the original claim.*

35. Caffeine Content in Colas   A company that makes cola drinks states that the mean caffeine content per 12-ounce bottle of cola is 40 milligrams. You want to test this claim. During your tests, you find that a random sample of thirty 12-ounce bottles of cola has a mean caffeine content of 39.2 milligrams with a standard deviation of 7.5 milligrams. At $\alpha = 0.01$, can you reject the company's claim? *(Adapted from American Beverage Association)*

36. Electricity Consumption   The U.S. Energy Information Association claims that the mean monthly residential electricity consumption in your town is 874 kilowatt-hours (kWh). You want to test this claim. You find that a random sample of 64 residential customers has a mean monthly electricity consumption of 905 kWh and a standard deviation of 125 kWh. At $\alpha = 0.05$, do you have enough evidence to reject the association's claim? *(Adapted from U.S. Energy Information Association)*

37. Light Bulbs   A light bulb manufacturer guarantees that the mean life of a certain type of light bulb is at least 750 hours. A random sample of 36 light bulbs has a mean life of 745 hours with a standard deviation of 60 hours. At $\alpha = 0.02$, do you have enough evidence to reject the manufacturer's claim?

38. Fast Food   A fast food restaurant estimates that the mean sodium content in one of its breakfast sandwiches is no more than 920 milligrams. A random sample of 44 breakfast sandwiches has a mean sodium content of 925 with a standard deviation of 18 milligrams. At $\alpha = 0.10$, do you have enough evidence to reject the restaurant's claim?

39. Nitrogen Dioxide Levels   A scientist estimates that the mean nitrogen dioxide level in Calgary is greater than 32 parts per billion. You want to test this estimate. To do so, you determine the nitrogen dioxide levels for 34 randomly selected days. The results (in parts per billion) are listed below. At $\alpha = 0.06$, can you support the scientist's estimate? *(Adapted from Clean Air Strategic Alliance)*

24	36	44	35	44	34	29	40	39	43	41	32
33	29	29	43	25	39	25	42	29	22	22	25
14	15	14	29	25	27	22	24	18	17		

40. Fluorescent Lamps   A fluorescent lamp manufacturer guarantees that the mean life of a certain type of lamp is at least 10,000 hours. You want to test this guarantee. To do so, you record the lives of a random sample of 32 fluorescent lamps. The results (in hours) are shown below. At $\alpha = 0.09$, do you have enough evidence to reject the manufacturer's claim?

8,800	9,155	13,001	10,250	10,002	11,413	8,234	10,402
10,016	8,015	6,110	11,005	11,555	9,254	6,991	12,006
10,420	8,302	8,151	10,980	10,186	10,003	8,814	11,445
6,277	8,632	7,265	10,584	9,397	11,987	7,556	10,380

41. Weight Loss   A weight loss program claims that program participants have a mean weight loss of at least 10 pounds after 1 month. You work for a medical association and are asked to test this claim. A random sample of 30 program participants and their weight losses (in pounds) after 1 month is listed in the stem-and-leaf plot at the left. At $\alpha = 0.03$, do you have enough evidence to reject the program's claim?

Evacuation Time (in seconds)

```
 0 | 7 9                Key: 0|7 = 7
 1 | 1 9 9
 2 | 2 6 7 9 9
 3 | 1 1 6 7 7 9 9
 4 | 1 1 3 3 3 4 6 6 7
 5 | 2 3 4 5 7 8 8 8 9 9
 6 | 1 3 3 4 6 6 7
 7 | 4 6 9
 8 | 4 6
 9 | 4
10 | 2
```

FIGURE FOR EXERCISE 42

39. (a) H_0: $\mu \le 32$; H_a: $\mu > 32$ (claim)

(b) $z_0 = 1.555$; Rejection region: $z > 1.555$

(c) -1.478

(d) Fail to reject H_0.

(e) There is not enough evidence at the 6% level of significance to support the scientist's claim that the mean nitrogen dioxide level in Calgary is greater than 32 parts per billion.

40. See Selected Answers, page A122.

41. See Odd Answers, page A82.

42. See Selected Answers, page A122.

43. See Odd Answers, page A82.

44. See Selected Answers, page A122.

45. See Odd Answers, page A82.

46. See Selected Answers, page A122.

47. Fail to reject H_0 because the standardized test statistic $z = -1.86$ is not in the rejection region ($z < -2.33$).

48. Fail to reject H_0 because the standardized test statistic $z = 1.548$ is not in the rejection region ($z > 1.645$).

49. b, d; If $\alpha = 0.05$, the rejection region is $z < -1.645$; because $z = -1.86$ is in the rejection region, you can reject H_0. If $n = 50$, the standardized test statistic is $z = -2.40$; because $z = -2.40$ is in the rejection region ($z < -2.33$), you can reject H_0.

50. See Selected Answers, page A122.

42. Fire Drill An engineering company claims that the mean time it takes an employee to evacuate a building during a fire drill is less than 60 seconds. You want to test this claim. A random sample of 50 employees and their evacuation times (in seconds) is listed in the stem-and-leaf plot at the left. At $\alpha = 0.01$, can you support the company's claim?

SC *In Exercises 43–46, use StatCrunch to help you test the claim about the population mean μ at the given level of significance α using the given sample statistics. For each claim, assume the population is normally distributed.*

43. Claim: $\mu = 58$; $\alpha = 0.10$. Sample statistics: $\bar{x} = 57.6$, $s = 2.35$, $n = 80$

44. Claim: $\mu > 495$; $\alpha = 0.05$. Sample statistics: $\bar{x} = 498.4$, $s = 17.8$, $n = 65$

45. Claim: $\mu \le 1210$; $\alpha = 0.08$. Sample statistics: $\bar{x} = 1234.21$, $s = 205.87$, $n = 250$

46. Claim: $\mu \ne 28{,}750$; $\alpha = 0.01$. Sample statistics: $\bar{x} = 29{,}130$, $s = 3200$, $n = 600$

■ EXTENDING CONCEPTS

47. Water Usage You believe the mean annual water usage of U.S. households is less than 127,400 gallons. You find that a random sample of 30 households has a mean water usage of 125,270 gallons with a standard deviation of 6275 gallons. You conduct a statistical experiment where H_0: $\mu \ge 127{,}400$ and H_a: $\mu < 127{,}400$. At $\alpha = 0.01$, explain why you cannot reject H_0. *(Adapted from American Water Works Association)*

48. Vehicle Miles of Travel You believe the annual mean vehicle miles of travel (VMT) per U.S. household is greater than 22,000 miles. You do some research and find that a random sample of 36 U.S. households has a mean annual VMT of 22,200 miles with a standard deviation of 775 miles. You conduct a statistical experiment where H_0: $\mu \le 22{,}000$ and H_a: $\mu > 22{,}000$. At $\alpha = 0.05$, explain why you cannot reject H_0. *(Adapted from U.S. Federal Highway Administration)*

49. Using Different Values of α and n In Exercise 47, you believe that H_0 is not valid. Which of the following allows you to reject H_0? Explain your reasoning.

(a) Use the same values but increase α from 0.01 to 0.02.

(b) Use the same values but increase α from 0.01 to 0.05.

(c) Use the same values but increase n from 30 to 40.

(d) Use the same values but increase n from 30 to 50.

50. Using Different Values of α and n In Exercise 48, you believe that H_0 is not valid. Which of the following allows you to reject H_0? Explain your reasoning.

(a) Use the same values but increase α from 0.05 to 0.06.

(b) Use the same values but increase α from 0.05 to 0.07.

(c) Use the same values but increase n from 36 to 40.

(d) Use the same values but increase n from 36 to 80.

Human Body Temperature: What's Normal?

In an article in the *Journal of Statistics Education* (vol. 4, no. 2), Allen Shoemaker describes a study that was reported in the Journal of the American Medical Association (JAMA).* It is generally accepted that the mean body temperature of an adult human is 98.6°F. In his article, Shoemaker uses the data from the JAMA article to test this hypothesis. Here is a summary of his test.

Claim: The body temperature of adults is 98.6°F.

$$H_0: \mu = 98.6°F \text{ (Claim)} \qquad H_a: \mu \neq 98.6°F$$

Sample Size: $n = 130$

Population: Adult human temperatures (Fahrenheit)

Distribution: Approximately normal

Test Statistics: $\bar{x} = 98.25$, $s = 0.73$

* Data for the JAMA article were collected from healthy men and women, ages 18 to 40, at the University of Maryland Center for Vaccine Development, Baltimore.

Men's Temperatures (in degrees Fahrenheit)

96	3
96	7 9
97	0 1 1 1 2 3 4 4 4 4
97	5 5 6 6 6 7 8 8 8 8 9 9
98	0 0 0 0 0 0 1 1 2 2 2 2 3 3 4 4 4 4
98	5 5 6 6 6 6 6 6 7 7 8 8 8 9
99	0 0 0 1 2 3 4
99	5
100	
100	

Key: 96|3 = 96.3

Women's Temperatures (in degrees Fahrenheit)

96	4
96	7 8
97	2 2 4
97	6 7 7 8 8 8 9 9 9
98	0 0 0 0 0 1 2 2 2 2 2 2 3 3 3 4 4 4 4 4
98	5 6 6 6 6 7 7 7 7 7 7 8 8 8 8 8 8 8 9
99	0 0 1 1 2 2 3 4
99	9
100	0
100	8

Key: 96|4 = 96.4

EXERCISES

1. Complete the hypothesis test for all adults (men and women) by performing the following steps. Use a level of significance of $\alpha = 0.05$.

   (a) Sketch the sampling distribution.

   (b) Determine the critical values and add them to your sketch.

   (c) Determine the rejection regions and shade them in your sketch.

   (d) Find the standardized test statistic. Add it to your sketch.

   (e) Make a decision to reject or fail to reject the null hypothesis.

   (f) Interpret the decision in the context of the original claim.

2. If you lower the level of significance to $\alpha = 0.01$, does your decision change? Explain your reasoning.

3. Test the hypothesis that the mean temperature of men is 98.6°F. What can you conclude at a level of significance of $\alpha = 0.01$?

4. Test the hypothesis that the mean temperature of women is 98.6°F. What can you conclude at a level of significance of $\alpha = 0.01$?

5. Use the sample of 130 temperatures to form a 99% confidence interval for the mean body temperature of adult humans.

6. The conventional "normal" body temperature was established by Carl Wunderlich over 100 years ago. What were possible sources of error in Wunderlich's sampling procedure?

7.3   Hypothesis Testing for the Mean (Small Samples)

WHAT YOU SHOULD LEARN

▸ How to find critical values in a *t*-distribution

▸ How to use the *t*-test to test a mean μ

▸ How to use technology to find *P*-values and use them with a *t*-test to test a mean μ

▸ CRITICAL VALUES IN A *t*-DISTRIBUTION

In Section 7.2, you learned how to perform a hypothesis test for a population mean when the sample size was at least 30. In real life, it is often not practical to collect samples of size 30 or more. However, if the population has a normal, or nearly normal, distribution, you can still test the population mean μ. To do so, you can use the *t*-sampling distribution with $n - 1$ degrees of freedom.

GUIDELINES

Finding Critical Values in a *t*-Distribution

1. Identify the level of significance α.

2. Identify the degrees of freedom d.f. $= n - 1$.

3. Find the critical value(s) using Table 5 in Appendix B in the row with $n - 1$ degrees of freedom. If the hypothesis test is

    a. *left-tailed*, use the "One Tail, α" column with a negative sign.

    b. *right-tailed*, use the "One Tail, α" column with a positive sign.

    c. *two-tailed*, use the "Two Tails, α" column with a negative and a positive sign.

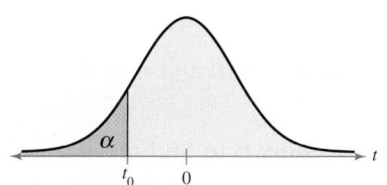

Left-Tailed Test

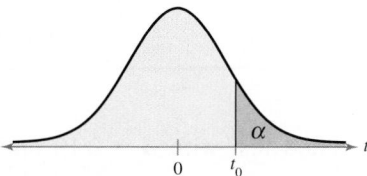

Right-Tailed Test

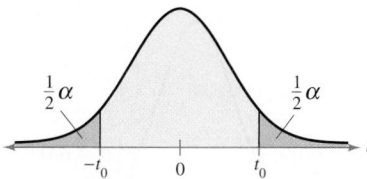

Two-Tailed Test

EXAMPLE 1

▸ **Finding Critical Values for *t***

Find the critical value t_0 for a left-tailed test with $\alpha = 0.05$ and $n = 21$.

▸ **Solution**

The degrees of freedom are

    d.f. $= n - 1$

            $= 21 - 1$

            $= 20.$

To find the critical value, use Table 5 in Appendix B with d.f. $= 20$ and $\alpha = 0.05$ in the "One Tail, α" column. Because the test is a left-tailed test, the critical value is negative. So,

    $t_0 = -1.725.$

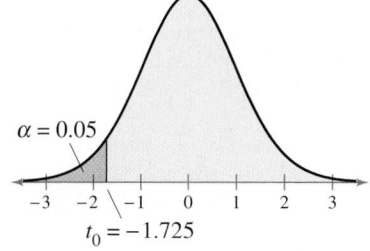

5% Level of Significance

▸ **Try It Yourself 1**

Find the critical value t_0 for a left-tailed test with $\alpha = 0.01$ and $n = 14$.

a. Identify the *degrees of freedom*.

b. *Use* the "One Tail, α" column in Table 5 in Appendix B to find t_0.

Answer: Page A41

Note to Instructor

A thoughtful student might ask what should be done if the sample size is small, the standard deviation is not known, and you cannot assume that the population is normally distributed. Chapter 11 will cover this case (see nonparametric tests). You can cover these tests immediately after this section if desirable.

EXAMPLE 2

▸ **Finding Critical Values for *t***

Find the critical value t_0 for a right-tailed test with $\alpha = 0.01$ and $n = 17$.

▸ **Solution**

The degrees of freedom are

$$\begin{aligned}
\text{d.f.} &= n - 1 \\
&= 17 - 1 \\
&= 16.
\end{aligned}$$

To find the critical value, use Table 5 with d.f. = 16 and $\alpha = 0.01$ in the "One Tail, α" column. Because the test is right-tailed, the critical value is positive. So,

$$t_0 = 2.583.$$

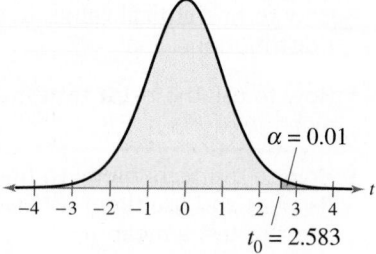

1% Level of Significance

▸ **Try It Yourself 2**

Find the critical value t_0 for a right-tailed test with $\alpha = 0.10$ and $n = 9$.

a. *Identify* the *degrees of freedom.*
b. *Use* the "One Tail, α" column in Table 5 in Appendix B to find t_0.

Answer: Page A41

EXAMPLE 3

▸ **Finding Critical Values for *t***

Find the critical values $-t_0$ and t_0 for a two-tailed test with $\alpha = 0.10$ and $n = 26$.

▸ **Solution**

The degrees of freedom are

$$\begin{aligned}
\text{d.f.} &= n - 1 \\
&= 26 - 1 \\
&= 25.
\end{aligned}$$

To find the critical values, use Table 5 with d.f. = 25 and $\alpha = 0.10$ in the "Two Tails, α" column. Because the test is two-tailed, one critical value is negative and one is positive. So,

$$-t_0 = -1.708 \quad \text{and} \quad t_0 = 1.708.$$

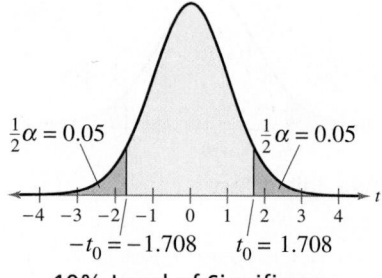

10% Level of Significance

▸ **Try It Yourself 3**

Find the critical values $-t_0$ and t_0 for a two-tailed test with $\alpha = 0.05$ and $n = 16$.

a. *Identify* the *degrees of freedom.*
b. *Use* the "Two Tails, α" column in Table 5 in Appendix B to find t_0.

Answer: Page A41

▶ THE t-TEST FOR A MEAN μ ($n < 30$, σ UNKNOWN)

To test a claim about a mean μ using a small sample ($n < 30$) from a normal, or nearly normal, distribution when σ is unknown, you can use a t-sampling distribution.

$$t = \frac{(\text{Sample mean}) - (\text{Hypothesized mean})}{\text{Standard error}}$$

t-TEST FOR A MEAN μ

The **t-test for a mean** is a statistical test for a population mean. The t-test can be used when the population is normal or nearly normal, σ is unknown, and $n < 30$. The **test statistic** is the sample mean $\bar{x}$ and the **standardized test statistic** is

$$t = \frac{\bar{x} - \mu}{s/\sqrt{n}}.$$

The degrees of freedom are

$$\text{d.f.} = n - 1.$$

PICTURING THE WORLD

On the basis of a t-test, a decision was made whether to send truckloads of waste contaminated with cadmium to a sanitary landfill or a hazardous waste landfill. The trucks were sampled to determine if the mean level of cadmium exceeded the allowable amount of 1 milligram per liter for a sanitary landfill. Assume the null hypothesis was $\mu \leq 1$.
(Adapted from Pacific Northwest National Laboratory)

	H_0 True	H_0 False
Fail to reject H_0.		
Reject H_0.		

Describe the possible type I and type II errors of this situation.

A type I error will occur if the actual amount of cadmium is less than or equal to 1 milligram per liter, but you reject H_0: $\mu \leq 1$. So, the trucks will be filled with sanitary waste, but will be sent to the hazardous waste landfill.

A type II error will occur if the actual amount of cadmium is greater than 1 milligram per liter, but you fail to reject H_0: $\mu \leq 1$. So, the trucks will be filled with hazardous waste, but will be sent to the sanitary landfill.

GUIDELINES

Using the t-Test for a Mean μ (Small Sample)

IN WORDS	IN SYMBOLS
1. State the claim mathematically and verbally. Identify the null and alternative hypotheses.	State H_0 and H_a.
2. Specify the level of significance.	Identify α.
3. Identify the degrees of freedom.	d.f. $= n - 1$
4. Determine the critical value(s).	Use Table 5 in Appendix B.
5. Determine the rejection region(s).	
6. Find the standardized test statistic and sketch the sampling distribution.	$t = \dfrac{\bar{x} - \mu}{s/\sqrt{n}}$
7. Make a decision to reject or fail to reject the null hypothesis.	If t is in the rejection region, reject H_0. Otherwise, fail to reject H_0.
8. Interpret the decision in the context of the original claim.	

Remember that when you make a decision, the possibility of a type I or a type II error exists.

If you prefer using P-values, turn to page 392 to learn how to use P-values for a t-test for a mean μ (small sample).

EXAMPLE 4

See MINITAB
steps on page 424.

▶ **Testing μ with a Small Sample**

A used car dealer says that the mean price of a 2008 Honda CR-V is at least $20,500. You suspect this claim is incorrect and find that a random sample of 14 similar vehicles has a mean price of $19,850 and a standard deviation of $1084. Is there enough evidence to reject the dealer's claim at $\alpha = 0.05$? Assume the population is normally distributed. *(Adapted from Kelley Blue Book)*

▶ **Solution**

The claim is "the mean price is at least $20,500." So, the null and alternative hypotheses are

$$H_0: \mu \geq \$20,500 \text{ (Claim)}$$

and

$$H_a: \mu < \$20,500.$$

The test is a left-tailed test, the level of significance is $\alpha = 0.05$, and the degrees of freedom are d.f. $= 14 - 1 = 13$. So, the critical value is $t_0 = -1.771$. The rejection region is $t < -1.771$. The standardized test statistic is

$$t = \frac{\bar{x} - \mu}{s/\sqrt{n}}$$ Because $n < 30$, use the t-test.

$$= \frac{19,850 - 20,500}{1084/\sqrt{14}}$$ Assume $\mu = 20,500$.

$$\approx -2.244.$$

To explore this topic further, see Activity 7.3 on page 397.

The graph shows the location of the rejection region and the standardized test statistic t. Because t is in the rejection region, you should reject the null hypothesis.

Interpretation There is enough evidence at the 5% level of significance to reject the claim that the mean price of a 2008 Honda CR-V is at least $20,500.

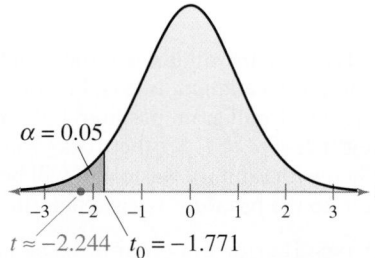

$t \approx -2.244 \quad t_0 = -1.771$

5% Level of Significance

▶ **Try It Yourself 4**

An insurance agent says that the mean cost of insuring a 2008 Honda CR-V is less than $1200. A random sample of 7 similar insurance quotes has a mean cost of $1125 and a standard deviation of $55. Is there enough evidence to support the agent's claim at $\alpha = 0.10$? Assume the population is normally distributed.

a. Identify the *claim* and state H_0 and H_a.
b. Identify the *level of significance* α and the *degrees of freedom*.
c. Find the *critical value t_0* and identify the *rejection region*.
d. Find the *standardized test statistic t. Sketch* a graph.
e. *Decide* whether to reject the null hypothesis.
f. *Interpret* the decision in the context of the original claim.

Answer: Page A41

EXAMPLE 5

See TI-83/84 Plus steps on page 425.

▶ Testing μ with a Small Sample

An industrial company claims that the mean pH level of the water in a nearby river is 6.8. You randomly select 19 water samples and measure the pH of each. The sample mean and standard deviation are 6.7 and 0.24, respectively. Is there enough evidence to reject the company's claim at $\alpha = 0.05$? Assume the population is normally distributed.

▶ Solution

The claim is "the mean pH level is 6.8." So, the null and alternative hypotheses are

$$H_0: \mu = 6.8 \ (\text{Claim})$$

and

$$H_a: \mu \neq 6.8.$$

The test is a two-tailed test, the level of significance is $\alpha = 0.05$, and the degrees of freedom are d.f. $= 19 - 1 = 18$. So, the critical values are $-t_0 = -2.101$ and $t_0 = 2.101$. The rejection regions are $t < -2.101$ and $t > 2.101$. The standardized test statistic is

$$t = \frac{\bar{x} - \mu}{s/\sqrt{n}} \qquad \text{Because } n < 30, \text{ use the } t\text{-test.}$$

$$= \frac{6.7 - 6.8}{0.24/\sqrt{19}} \qquad \text{Assume } \mu = 6.8.$$

$$\approx -1.816.$$

The graph shows the location of the rejection region and the standardized test statistic t. Because t is not in the rejection region, you fail to reject the null hypothesis.

Interpretation There is not enough evidence at the 5% level of significance to reject the claim that the mean pH is 6.8.

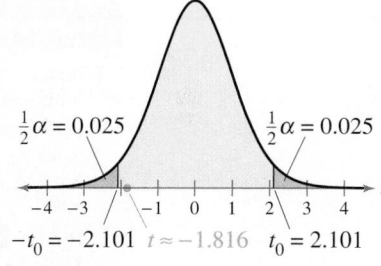

$\frac{1}{2}\alpha = 0.025$ $\frac{1}{2}\alpha = 0.025$

$-t_0 = -2.101$ $t \approx -1.816$ $t_0 = 2.101$

5% Level of Significance

▶ Try It Yourself 5

The company also claims that the mean conductivity of the river is 1890 milligrams per liter. The conductivity of a water sample is a measure of the total dissolved solids in the sample. You randomly select 19 water samples and measure the conductivity of each. The sample mean and standard deviation are 2500 milligrams per liter and 700 milligrams per liter, respectively. Is there enough evidence to reject the company's claim at $\alpha = 0.01$? Assume the population is normally distributed.

a. Identify the *claim* and state H_0 and H_a.
b. Identify the *level of significance* α and the *degrees of freedom*.
c. Find the *critical values* $-t_0$ and t_0 and identify the *rejection region*.
d. Find the *standardized test statistic t. Sketch* a graph.
e. *Decide* whether to reject the null hypothesis.
f. *Interpret* the decision in the context of the original claim.

Answer: Page A42

▶ USING *P*-VALUES WITH *t*-TESTS

Suppose you wanted to find a *P*-value given $t = 1.98$, 15 degrees of freedom, and a right-tailed test. Using Table 5 in Appendix B, you can determine that *P* falls between $\alpha = 0.025$ and $\alpha = 0.05$, but you cannot determine an exact value for *P*. In such cases, you can use technology to perform a hypothesis test and find exact *P*-values.

EXAMPLE 6    Report 30

▶ Using *P*-Values with a *t*-Test

A Department of Motor Vehicles office claims that the mean wait time is less than 14 minutes. A random sample of 10 people has a mean wait time of 13 minutes with a standard deviation of 3.5 minutes. At $\alpha = 0.10$, test the office's claim. Assume the population is normally distributed.

▶ Solution

The claim is "the mean wait time is less than 14 minutes." So, the null and alternative hypotheses are

$$H_0: \mu \geq 14 \text{ minutes}$$

and

$$H_a: \mu < 14 \text{ minutes. (Claim)}$$

The TI-83/84 Plus display at the far left shows how to set up the hypothesis test. The two displays on the right show the possible results, depending on whether you select "Calculate" or "Draw."

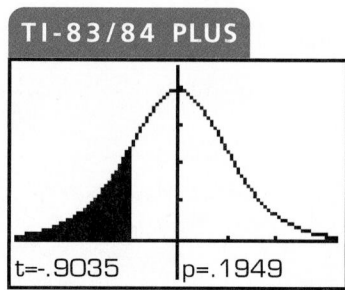

From the displays, you can see that $P \approx 0.1949$. Because the *P*-value is greater than $\alpha = 0.10$, you fail to reject the null hypothesis.

Interpretation   There is not enough evidence at the 10% level of significance to support the office's claim that the mean wait time is less than 14 minutes.

▶ Try It Yourself 6

Another Department of Motor Vehicles office claims that the mean wait time is at most 18 minutes. A random sample of 12 people has a mean wait time of 15 minutes with a standard deviation of 2.2 minutes. At $\alpha = 0.05$, test the office's claim. Assume the population is normally distributed.

a. Identify the *claim* and state H_0 and H_a.
b. *Use* a TI-83/84 Plus to find the *P*-value.
c. *Compare* the *P*-value with the level of significance α and *make* a decision.
d. *Interpret* the decision in the context of the original claim.

Answer: Page A42

7.3 EXERCISES

1. Identify the level of significance α and the degrees of freedom, d.f. $= n - 1$. Find the critical value(s) using the t-distribution table in the row with $n - 1$ d.f. If the hypothesis test is

  (1) left-tailed, use the "One Tail, α" column with a negative sign.

  (2) right-tailed, use the "One Tail, α" column with a positive sign.

  (3) two-tailed, use the "Two Tails, α" column with a negative and a positive sign.

2. See Selected Answers, page A122.

3. 1.717   **4.** 2.764   **5.** −1.328

6. −2.473   **7.** −2.056, 2.056

8. −1.721, 1.721

9. See Odd Answers, page A82.

10. See Selected Answers, page A122.

11. (a) Fail to reject H_0 because $-2.602 < t < 2.602$.

  (b) Fail to reject H_0 because $-2.602 < t < 2.602$.

  (c) Reject H_0 because $t > 2.602$.

  (d) Reject H_0 because $t < -2.602$.

12. See Selected Answers, page A122.

13. See Odd Answers, page A83.

14. See Selected Answers, page A122.

15. See Odd Answers, page A83.

16. See Selected Answers, page A123.

BUILDING BASIC SKILLS AND VOCABULARY

1. Explain how to find critical values for a t-sampling distribution.

2. Explain how to use a t-test to test a hypothesized mean μ given a small sample ($n < 30$). What assumption about the population is necessary?

In Exercises 3–8, find the critical value(s) for the indicated t-test, level of significance α, and sample size n.

3. Right-tailed test, $\alpha = 0.05$, $n = 23$

4. Right-tailed test, $\alpha = 0.01$, $n = 11$

5. Left-tailed test, $\alpha = 0.10$, $n = 20$

6. Left-tailed test, $\alpha = 0.01$, $n = 28$

7. Two-tailed test, $\alpha = 0.05$, $n = 27$

8. Two-tailed test, $\alpha = 0.10$, $n = 22$

Graphical Analysis *In Exercises 9–12, state whether the standardized test statistic t indicates that you should reject the null hypothesis. Explain.*

9. (a) $t = 2.091$
  (b) $t = 0$
  (c) $t = -1.08$
  (d) $t = -2.096$

10. (a) $t = 1.308$
  (b) $t = -1.389$
  (c) $t = 1.650$
  (d) $t = -0.998$

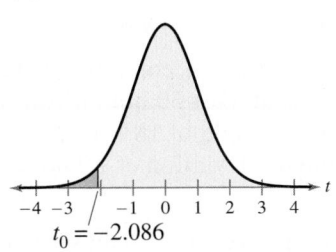

11. (a) $t = -2.502$
  (b) $t = 2.203$
  (c) $t = 2.680$
  (d) $t = -2.703$

12. (a) $t = 1.705$
  (b) $t = -1.755$
  (c) $t = -1.585$
  (d) $t = 1.745$

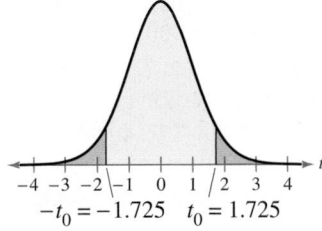

In Exercises 13–16, use a t-test to test the claim about the population mean μ at the given level of significance α using the given sample statistics. For each claim, assume the population is normally distributed.

13. Claim: $\mu = 15$; $\alpha = 0.01$. Sample statistics: $\bar{x} = 13.9$, $s = 3.23$, $n = 6$

14. Claim: $\mu > 25$; $\alpha = 0.05$. Sample statistics: $\bar{x} = 26.2$, $s = 2.32$, $n = 17$

15. Claim: $\mu \geq 8000$; $\alpha = 0.01$. Sample statistics: $\bar{x} = 7700$, $s = 450$, $n = 25$

16. Claim: $\mu \neq 52{,}200$; $\alpha = 0.10$. Sample statistics: $\bar{x} = 53{,}220$, $s = 2700$, $n = 18$

17. (a) $H_0: \mu = 18{,}000$ (claim);
$H_a: \mu \neq 18{,}000$

(b) $-t_0 = -2.145$, $t_0 = 2.145$;
Rejection regions:
$t < -2.145$, $t > 2.145$

(c) 1.21   (d) Fail to reject H_0.

(e) There is not enough
evidence at the 5% level
of significance to reject the
dealer's claim that the mean
price of a 2008 Subaru
Forester is $18,000.

18. See Selected Answers, page A123.

19. (a) $H_0: \mu \leq 60$; $H_a: \mu > 60$ (claim)

(b) $t_0 = 1.943$; Rejection region:
$t > 1.943$

(c) 2.12   (d) Reject H_0.

(e) There is enough evidence at
the 5% level of significance
to support the board's claim
that the mean number of
hours worked per week by
surgical faculty who teach
at an academic institution
is more than 60 hours.

20. See Selected Answers, page A123.

21. (a) $H_0: \mu \leq 1$; $H_a: \mu > 1$ (claim)

(b) $t_0 = 1.356$; Rejection region:
$t > 1.356$

(c) 6.44   (d) Reject H_0.

(e) There is enough evidence at
the 10% level of significance
to support the environmentalist's claim that the mean
amount of waste recycled
by adults in the United States
is more than 1 pound per
person per day.

22. See Selected Answers, page A123.

23. (a) $H_0: \mu = \$26{,}000$ (claim);
$H_a: \mu \neq \$26{,}000$

(b) $-t_0 = -2.262$, $t_0 = 2.262$;
Rejection regions:
$t < -2.262$, $t > 2.262$

(c) -0.15 (d) Fail to reject H_0.

(e) There is not enough evidence
at the 5% level of significance to reject the employment information service's
claim that the mean salary
for full-time male workers
over age 25 without a high
school diploma is $26,000.

■ USING AND INTERPRETING CONCEPTS

Testing Claims *In Exercises 17–24, (a) write the claim mathematically and identify H_0 and H_a, (b) find the critical value(s) and identify the rejection region(s), (c) find the standardized test statistic t, (d) decide whether to reject or fail to reject the null hypothesis, and (e) interpret the decision in the context of the original claim. If convenient, use technology. For each claim, assume the population is normally distributed.*

17. **Used Car Cost** A used car dealer says that the mean price of a 2008 Subaru Forester is $18,000. You suspect this claim is incorrect and find that a random sample of 15 similar vehicles has a mean price of $18,550 and a standard deviation of $1767. Is there enough evidence to reject the claim at $\alpha = 0.05$? *(Adapted from Kelley Blue Book)*

18. **IRS Wait Times** The Internal Revenue Service claims that the mean wait time for callers during a recent tax filing season was at most 7 minutes. A random sample of 11 callers has a mean wait time of 8.7 minutes and a standard deviation of 2.7 minutes. Is there enough evidence to reject the claim at $\alpha = 0.10$? *(Adapted from Internal Revenue Service)*

19. **Work Hours** A medical board claims that the mean number of hours worked per week by surgical faculty who teach at an academic institution is more than 60 hours. The hours worked include teaching hours as well as regular working hours. A random sample of 7 surgical faculty has a mean hours worked per week of 70 hours and a standard deviation of 12.5 hours. At $\alpha = 0.05$, do you have enough evidence to support the board's claim? *(Adapted from Journal of the American College of Surgeons)*

20. **Battery Life** A company claims that the mean battery life of their MP3 player is at least 30 hours. You suspect this claim is incorrect and find that a random sample of 18 MP3 players has a mean battery life of 28.5 hours and a standard deviation of 1.7 hours. Is there enough evidence to reject the claim at $\alpha = 0.01$?

21. **Waste Recycled** An environmentalist estimates that the mean amount of waste recycled by adults in the United States is more than 1 pound per person per day. You want to test this claim. You find that the mean waste recycled per person per day for a random sample of 13 adults in the United States is 1.50 pounds and the standard deviation is 0.28 pound. At $\alpha = 0.10$, can you support the claim? *(Adapted from U.S. Environmental Protection Agency)*

22. **Waste Generated** As part of your work for an environmental awareness group, you want to test a claim that the mean amount of waste generated by adults in the United States is more than 4 pounds per day. In a random sample of 22 adults in the United States, you find that the mean waste generated per person per day is 4.50 pounds with a standard deviation of 1.21 pounds. At $\alpha = 0.01$, can you support the claim? *(Adapted from U.S. Environmental Protection Agency)*

23. **Annual Pay** An employment information service claims the mean annual salary for full-time male workers over age 25 and without a high school diploma is $26,000. The annual salaries for a random sample of 10 full-time male workers without a high school diploma are listed. At $\alpha = 0.05$, test the claim that the mean salary is $26,000. *(Adapted from U.S. Bureau of Labor Statistics)*

26,185	23,814	22,374	25,189	26,318
20,767	30,782	29,541	24,597	28,955

24. (a) H_0: $\mu \le \$18{,}500$;
H_a: $\mu > \$18{,}500$ (claim)

(b) $t_0 = 1.363$; Rejection region: $t > 1.363$

(c) 1.43   (d) Reject H_0.

(e) There is enough evidence at the 10% level of significance to support the employment information service's claim that the mean annual salary for full-time female workers over age 25 without a high school diploma is more than $18,500.

25. (a) H_0: $\mu \le 45$; H_a: $\mu > 45$ (claim)

(b) 0.0052   (c) Reject H_0.

(d) There is enough evidence at the 10% level of significance to support the county's claim that the mean speed of the vehicles is greater than 45 miles per hour.

26. (a) H_0: $\mu \le 3500$;
H_a: $\mu > 3500$ (claim)

(b) 0.9200   (c) Fail to reject H_0.

(d) There is not enough evidence at the 5% level of significance to support the shop's claim that people travel more than 3500 miles between oil changes.

27. (a) H_0: $\mu = \$105$ (claim);
H_a: $\mu \ne \$105$

(b) 0.0165   (c) Fail to reject H_0.

(d) There is not enough evidence at the 1% level of significance to reject the travel association's claim that the mean daily meal cost for two adults traveling together on vacation in San Francisco is $105.

28. See Selected Answers, page A123.

29. (a) H_0: $\mu \ge 32$; H_a: $\mu < 32$ (claim)

(b) 0.0344   (c) Reject H_0.

(d) There is enough evidence at the 5% level of significance to support the brochure's claim that the mean class size for full-time faculty is fewer than 32 students.

30. See Selected Answers, page A123.

24. Annual Pay   An employment information service claims the mean annual salary for full-time female workers over age 25 and without a high school diploma is more than $18,500. The annual salaries for a random sample of 12 full-time female workers without a high school diploma are listed. At $\alpha = 0.10$, is there enough evidence to support the claim that the mean salary is more than $18,500? *(Adapted from U.S. Bureau of Labor Statistics)*

| 18,665 | 16,312 | 18,794 | 19,403 | 20,864 | 19,177 |
| 17,328 | 21,445 | 20,354 | 19,143 | 18,316 | 19,237 |

Testing Claims Using P-Values   *In Exercises 25–30, (a) write the claim mathematically and identify H_0 and H_a, (b) use technology to find the P-value, (c) decide whether to reject or fail to reject the null hypothesis, and (d) interpret the decision in the context of the original claim. Assume the population is normally distributed.*

25. Speed Limit   A county is considering raising the speed limit on a road because they claim that the mean speed of vehicles is greater than 45 miles per hour. A random sample of 25 vehicles has a mean speed of 48 miles per hour and a standard deviation of 5.4 miles per hour. At $\alpha = 0.10$, do you have enough evidence to support the county's claim?

26. Oil Changes   A repair shop believes that people travel more than 3500 miles between oil changes. A random sample of 8 cars getting an oil change has a mean distance of 3375 miles since having an oil change with a standard deviation of 225 miles. At $\alpha = 0.05$, do you have enough evidence to support the shop's claim?

27. Meal Cost   A travel association claims that the mean daily meal cost for two adults traveling together on vacation in San Francisco is $105. A random sample of 20 such groups of adults has a mean daily meal cost of $110 and a standard deviation of $8.50. Is there enough evidence to reject the claim at $\alpha = 0.01$? *(Adapted from American Automobile Association)*

28. Lodging Cost   A travel association claims that the mean daily lodging cost for two adults traveling together on vacation in San Francisco is at least $240. A random sample of 24 such groups of adults has a mean daily lodging cost of $233 and a standard deviation of $12.50. Is there enough evidence to reject the claim at $\alpha = 0.10$? *(Adapted from American Automobile Association)*

29. Class Size   You receive a brochure from a large university. The brochure indicates that the mean class size for full-time faculty is fewer than 32 students. You want to test this claim. You randomly select 18 classes taught by full-time faculty and determine the class size of each. The results are listed below. At $\alpha = 0.05$, can you support the university's claim?

35 28 29 33 32 40 26 25 29
28 30 36 33 29 27 30 28 25

30. Faculty Classroom Hours   The dean of a university estimates that the number of classroom hours per week for full-time faculty is 1 member of the student council, you want to test this claim. A rand of the number of classroom hours for eight full-time faculty for listed below. At $\alpha = 0.01$, can you reject the dean's claim?

11.8 8.6 12.6 7.9 6.4 10.4 13.6 9.1

31. See Odd Answers, page A83.

32. See Selected Answers, page A123.

33. See Odd Answers, page A83.

34. See Selected Answers, page A123.

35. Because the P-value = 0.0748 > 0.05, fail to reject H_0.

36. b, c, d; If $\alpha = 0.10$, the P-value ($P = 0.0748$) is less than $\alpha = 0.10$, so you can reject H_0. If $n = 8$, the P-value becomes $P = 0.0451$; because $P = 0.0451$ is less than $\alpha = 0.05$, you can reject H_0. If $n = 24$, the P-value becomes $P = 0.0012$; because $P = 0.0012$ is less than $\alpha = 0.05$, you can reject H_0.

37. Use the t-distribution because the population is normal, $n < 30$, and σ is unknown.

    Fail to reject H_0. There is not enough evidence at the 5% level of significance to reject the car company's claim that the mean gas mileage for the luxury sedan is at least 23 miles per gallon.

38. Use the z-distribution because σ is unknown and $n \geq 30$.

    Fail to reject H_0. There is not enough evidence at the 1% level of significance to support the education publication's claim that the average in-state tuition for one year of law school at a private institution is more than $35,000.

39. More likely; For degrees of freedom less than 30, the tails of a t-distribution curve are thicker than those of a standard normal distribution curve. So, if you incorrectly use a standard normal sampling distribution instead of a t-sampling distribution, the area under the curve at the tails will be smaller than what it would be for the t-test, meaning the critical value(s) will lie closer to the mean. This makes it more likely for the test statistic to be in the rejection region(s). This result is the same regardless of whether the test is left-tailed, right-tailed, or two-tailed; in each case, the tail thickness affects the location of the critical value(s).

SC *In Exercises 31–34, use StatCrunch and a t-test to help you test the claim about the population mean μ at the given level of significance α using the given sample statistics. For each claim, assume the population is normally distributed.*

31. Claim: $\mu \leq 75$; $\alpha = 0.05$. Sample statistics: $\bar{x} = 73.6$, $s = 3.2$, $n = 26$

32. Claim: $\mu \neq 27$; $\alpha = 0.01$. Sample statistics: $\bar{x} = 31.5$, $s = 4.7$, $n = 12$

33. Claim: $\mu < 188$; $\alpha = 0.05$. Sample statistics: $\bar{x} = 186$, $s = 12$, $n = 9$

34. Claim: $\mu \geq 2118$; $\alpha = 0.10$. Sample statistics: $\bar{x} = 1787$, $s = 384$, $n = 17$

■ EXTENDING CONCEPTS

35. **Credit Card Balances** To test the claim that the mean credit card debt for individuals is greater than $5000, you do some research and find that a random sample of 6 cardholders has a mean credit card balance of $5434 with a standard deviation of $625. You conduct a statistical experiment where $H_0: \mu \leq \$5000$ and $H_a: \mu > \$5000$. At $\alpha = 0.05$, explain why you cannot reject H_0. Assume the population is normally distributed. *(Adapted from TransUnion)*

36. **Using Different Values of α and n** In Exercise 35, you believe that H_0 is not valid. Which of the following allows you to reject H_0? Explain your reasoning.

    (a) Use the same values but decrease α from 0.05 to 0.01.

    (b) Use the same values but increase α from 0.05 to 0.10.

    (c) Use the same values but increase n from 6 to 8.

    (d) Use the same values but increase n from 6 to 24.

Deciding on a Distribution *In Exercises 37 and 38, decide whether you should use a normal sampling distribution or a t-sampling distribution to perform the hypothesis test. Justify your decision. Then use the distribution to test the claim. Write a short paragraph about the results of the test and what you can conclude about the claim.*

37. **Gas Mileage** A car company says that the mean gas mileage for its luxury sedan is at least 23 miles per gallon (mpg). You believe the claim is incorrect and find that a random sample of 5 cars has a mean gas mileage of 22 mpg and a standard deviation of 4 mpg. At $\alpha = 0.05$, test the company's claim. Assume the population is normally distributed.

38. **Private Law School** An education publication claims that the average in-state tuition for one year of law school at a private institution is more than $35,000. A random sample of 50 private law schools has a mean in-state tuition of $34,967 and a standard deviation of $5933 for one year. At $\alpha = 0.01$, test the publication's claim. Assume the population is normally distributed. *(Adapted from U.S. News and World Report)*

39. **Writing** You are testing a claim and incorrectly use the normal sampling distribution instead of the t-sampling distribution. Does this make it more or less likely to reject the null hypothesis? Is this result the same no matter whether the test is left-tailed, right-tailed, or two-tailed? Explain your reasoning.

ACTIVITY 7.3   Hypothesis Tests for a Mean

APPLET

The *hypothesis tests for a mean* applet allows you to visually investigate hypothesis tests for a mean. You can specify the sample size n, the shape of the distribution (Normal or Right skewed), the true population mean (Mean), the true population standard deviation (Std. Dev.), the null value for the mean (Null mean), and the alternative for the test (Alternative). When you click SIMULATE, 100 separate samples of size n will be selected from a population with these population parameters. For each of the 100 samples, a hypothesis test based on the T statistic is performed, and the results from each test are displayed in the plots at the right. The test statistic for each test is shown in the top plot and the P-value is shown in the bottom plot. The green and blue lines represent the cutoffs for rejecting the null hypothesis with the 0.05 and 0.01 level tests, respectively. Additional simulations can be carried out by clicking SIMULATE multiple times. The cumulative number of times that each test rejects the null hypothesis is also shown. Press CLEAR to clear existing results and start a new simulation.

■ Explore

Step 1 Specify a value for n.
Step 2 Specify a distribution.
Step 3 Specify a value for the mean.
Step 4 Specify a value for the standard deviation.
Step 5 Specify a value for the null mean.
Step 6 Specify an alternative hypothesis.
Step 7 Click SIMULATE to generate the hypothesis tests.

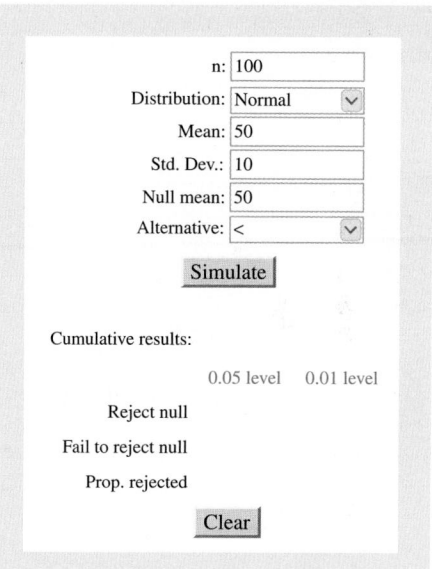

■ Draw Conclusions

APPLET

1. Set $n = 15$, Mean $= 40$, Std. Dev. $= 5$, Null mean $= 40$, alternative hypothesis to "not equal," and the distribution to "Normal." Run the simulation so that at least 1000 hypothesis tests are run. Compare the proportion of null hypothesis rejections for the 0.05 level and the 0.01 level. Is this what you would expect? Explain.

2. Suppose a null hypothesis is rejected at the 0.01 level. Will it be rejected at the 0.05 level? Explain. Suppose a null hypothesis is rejected at the 0.05 level. Will it be rejected at the 0.01 level? Explain.

3. Set $n = 25$, Mean $= 25$, Std. Dev. $= 3$, Null mean $= 27$, alternative hypothesis to "$<$," and the distribution to "Normal." What is the null hypothesis? Run the simulation so that at least 1000 hypothesis tests are run. Compare the proportion of null hypothesis rejections for the 0.05 level and the 0.01 level. Is this what you would expect? Explain.

7.4 Hypothesis Testing for Proportions

Hypothesis Test for Proportions

▶ HYPOTHESIS TEST FOR PROPORTIONS

In Sections 7.2 and 7.3, you learned how to perform a hypothesis test for a population mean. In this section, you will learn how to test a population proportion p.

Hypothesis tests for proportions can be used when politicians want to know the proportion of their constituents who favor a certain bill or when quality assurance engineers test the proportion of parts that are defective.

If $np \geq 5$ and $nq \geq 5$ for a binomial distribution, then the sampling distribution for $\hat{p}$ is approximately normal with a mean of

$$\mu_{\hat{p}} = p$$

and a standard error of

$$\sigma_{\hat{p}} = \sqrt{pq/n}.$$

z-TEST FOR A PROPORTION p

The **z-test for a proportion** is a statistical test for a population proportion p. The z-test can be used when a binomial distribution is given such that $np \geq 5$ and $nq \geq 5$. The **test statistic** is the sample proportion $\hat{p}$ and the **standardized test statistic** is

$$z = \frac{\hat{p} - \mu_{\hat{p}}}{\sigma_{\hat{p}}} = \frac{\hat{p} - p}{\sqrt{pq/n}}.$$

GUIDELINES

Using a z-Test for a Proportion p

Verify that $np \geq 5$ and $nq \geq 5$.

IN WORDS	IN SYMBOLS
1. State the claim mathematically and verbally. Identify the null and alternative hypotheses.	State H_0 and H_a.
2. Specify the level of significance.	Identify α.
3. Determine the critical value(s).	Use Table 4 in Appendix B.
4. Determine the rejection region(s).	
5. Find the standardized test statistic and sketch the sampling distribution.	$z = \dfrac{\hat{p} - p}{\sqrt{pq/n}}$
6. Make a decision to reject or fail to reject the null hypothesis.	If z is in the rejection region, reject H_0. Otherwise, fail to reject H_0.
7. Interpret the decision in the context of the original claim.	

INSIGHT

A hypothesis test for a proportion p can also be performed using P-values. Use the guidelines on page 373 for using P-values for a z-test for a mean μ, but in Step 3 find the standardized test statistic by using the formula

$$z = \frac{\hat{p} - p}{\sqrt{pq/n}}.$$

The other steps in the test are the same.

To explore this topic further, see Activity 7.4 on page 403.

EXAMPLE 1

See TI-83/84 Plus steps on page 425.

▶ **Hypothesis Test for a Proportion**

A research center claims that less than 50% of U.S. adults have accessed the Internet over a wireless network with a laptop computer. In a random sample of 100 adults, 39% say they have accessed the Internet over a wireless network with a laptop computer. At $\alpha = 0.01$, is there enough evidence to support the researcher's claim? *(Adapted from Pew Research Center)*

▶ **Solution**   The products $np = 100(0.50) = 50$ and $nq = 100(0.50) = 50$ are both greater than 5. So, you can use a z-test. The claim is "less than 50% have accessed the Internet over a wireless network with a laptop computer." So, the null and alternative hypotheses are

$$H_0: p \geq 0.5 \quad \text{and} \quad H_a: p < 0.5. \ (\text{Claim})$$

Because the test is a left-tailed test and the level of significance is $\alpha = 0.01$, the critical value is $z_0 = -2.33$ and the rejection region is $z < -2.33$. The standardized test statistic is

STUDY TIP

Remember that if the sample proportion is not given, you can find it using

$$\hat{p} = \frac{x}{n}$$

where x is the number of successes in the sample and n is the sample size.

$$z = \frac{\hat{p} - p}{\sqrt{pq/n}} \qquad \text{Because } np \geq 5 \text{ and } nq \geq 5, \text{ you can use the } z\text{-test.}$$

$$= \frac{0.39 - 0.5}{\sqrt{(0.5)(0.5)/100}} \qquad \text{Assume } p = 0.5.$$

$$= -2.2.$$

The graph shows the location of the rejection region and the standardized test statistic z. Because z is not in the rejection region, you should fail to reject the null hypothesis.

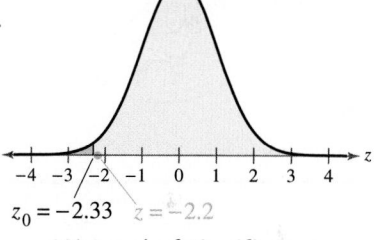

$z_0 = -2.33$ $z = -2.2$

1% Level of Significance

Interpretation   There is not enough evidence at the 1% level of significance to support the claim that less than 50% of U.S. adults have accessed the Internet over a wireless network with a laptop computer.

STUDY TIP

Remember that when you fail to reject H_0, a type II error is possible. For instance, in Example 1 the null hypothesis, $p \geq 0.5$, may be false.

▶ **Try It Yourself 1**

A research center claims that more than 25% of U.S. adults have used a cellular phone to access the Internet. In a random sample of 125 adults, 32% say they have used a cellular phone to access the Internet. At $\alpha = 0.05$, is there enough evidence to support the researcher's claim? *(Adapted from Pew Research Center)*

a. *Verify* that $np \geq 5$ and $nq \geq 5$.
b. *Identify* the *claim* and state H_0 and H_a.
c. *Identify* the *level of significance* α.
d. *Find* the *critical value* z_0 and identify the *rejection region*.
e. *Find* the *standardized test statistic z*. *Sketch* a graph.
f. *Decide* whether to reject the null hypothesis.
g. *Interpret* the decision in the context of the original claim.

Answer: Page A42

To use a P-value to perform the hypothesis test in Example 1, use Table 4 to find the area corresponding to $z = -2.2$. The area is 0.0139. Because this is a left-tailed test, the P-value is equal to the area to the left of $z = -2.2$. So, $P = 0.0139$. Because the P-value is greater than $\alpha = 0.01$, you should fail to reject the null hypothesis. Note that this is the same result obtained in Example 1.

See MINITAB steps on page 424.

PICTURING THE WORLD

A recent survey claimed that at least 70% of U.S. adults believe that cloning animals is morally wrong. To test this claim, you conduct a random telephone survey of 300 U.S. adults. In the survey, you find that 189 adults believe that cloning animals is morally wrong. (Adapted from The Gallup Poll)

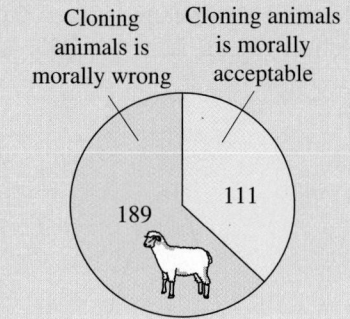

At $\alpha = 0.05$, is there enough evidence to reject the claim?

Yes, there is enough evidence at the 5% level of significance to reject the claim that at least 70% of U.S. adults believe that cloning animals is morally wrong.

EXAMPLE 2

▸ **Hypothesis Test for a Proportion**

A research center claims that 25% of college graduates think a college degree is not worth the cost. You decide to test this claim and ask a random sample of 200 college graduates whether they think a college degree is not worth the cost. Of those surveyed, 21% reply yes. At $\alpha = 0.10$, is there enough evidence to reject the claim? *(Adapted from Zogby International)*

▸ **Solution**

The products $np = 200(0.25) = 50$ and $nq = 200(0.75) = 150$ are both greater than 5. So, you can use a z-test. The claim is "25% of college graduates think a college degree is not worth the cost." So, the null and alternative hypotheses are

$$H_0: p = 0.25 \ \text{(Claim)} \quad \text{and} \quad H_a: p \neq 0.25.$$

Because the test is a two-tailed test and the level of significance is $\alpha = 0.10$, the critical values are $-z_0 = -1.645$ and $z_0 = 1.645$. The rejection regions are $z < -1.645$ and $z > 1.645$. The standardized test statistic is

$$z = \frac{\hat{p} - p}{\sqrt{pq/n}} \qquad \text{Because } np \geq 5 \text{ and } nq \geq 5, \text{ you can use the } z\text{-test.}$$

$$= \frac{0.21 - 0.25}{\sqrt{(0.25)(0.75)/200}} \qquad \text{Assume } p = 0.25.$$

$$= -1.31.$$

The graph shows the location of the rejection regions and the standardized test statistic z. Because z is not in the rejection region, you should fail to reject the null hypothesis.

Interpretation There is not enough evidence at the 10% level of significance to reject the claim that 25% of college graduates think a college degree is not worth the cost.

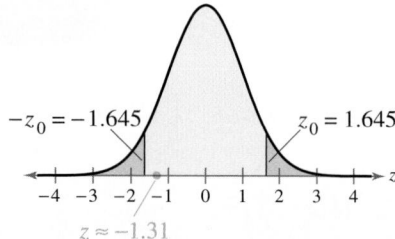

10% Level of Significance

▸ **Try It Yourself 2**

A research center claims that 30% of U.S. adults have not purchased a certain brand because they found the advertisements distasteful. You decide to test this claim and ask a random sample of 250 U.S. adults whether they have not purchased a certain brand because they found the advertisements distasteful. Of those surveyed, 36% reply yes. At $\alpha = 0.10$, is there enough evidence to reject the claim? *(Adapted from Harris Interactive)*

a. *Verify* that $np \geq 5$ and $nq \geq 5$.
b. *Identify* the *claim* and state H_0 and H_a.
c. *Identify* the *level of significance* α.
d. *Find* the *critical values* $-z_0$ and z_0 and identify the *rejection regions*.
e. *Find* the *standardized test statistic z*. *Sketch* a graph.
f. *Decide* whether to reject the null hypothesis.
g. *Interpret* the decision in the context of the original claim.

Answer: Page A42

7.4 EXERCISES

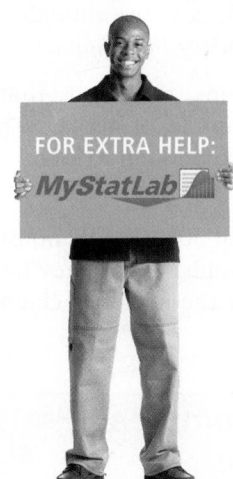

1. If $np \geq 5$ and $nq \geq 5$, the normal distribution can be used.

2. See Selected Answers, page A123.

3. Cannot use normal distribution.

4. Can use normal distribution.

Reject H_0. There is enough evidence at the 8% level of significance to reject the claim.

5. Can use normal distribution.

Fail to reject H_0. There is not enough evidence at the 5% level of significance to support the claim.

6. See Selected Answers, page A123.

7. Can use normal distribution.

Fail to reject H_0. There is not enough evidence at the 5% level of significance to reject the claim.

8. Cannot use normal distribution.

9. (a) H_0: $p \geq 0.25$;
    H_a: $p < 0.25$ (claim)

(b) $z_0 = -1.645$;
    Rejection region: $z < -1.645$

(c) -2.12 (d) Reject H_0.

(e) There is enough evidence at the 5% level of significance to support the researcher's claim that less than 25% of U.S. adults are smokers.

10. See Selected Answers, page A123.

11. See Odd Answers, page A84.

12. See Selected Answers, page A123.

13. See Odd Answers, page A84.

■ BUILDING BASIC SKILLS AND VOCABULARY

1. Explain how to decide when a normal distribution can be used to approximate a binomial distribution.

2. Explain how to test a population proportion p.

In Exercises 3–8, decide whether the normal sampling distribution can be used. If it can be used, test the claim about the population proportion p at the given level of significance α using the given sample statistics.

3. Claim: $p < 0.12$; $\alpha = 0.01$. Sample statistics: $\hat{p} = 0.10$, $n = 40$

4. Claim: $p \geq 0.48$; $\alpha = 0.08$. Sample statistics: $\hat{p} = 0.40$, $n = 90$

5. Claim: $p \neq 0.15$; $\alpha = 0.05$. Sample statistics: $\hat{p} = 0.12$, $n = 500$

6. Claim: $p > 0.70$; $\alpha = 0.04$. Sample statistics: $\hat{p} = 0.64$, $n = 225$

7. Claim: $p \leq 0.45$; $\alpha = 0.05$. Sample statistics: $\hat{p} = 0.52$, $n = 100$

8. Claim: $p = 0.95$; $\alpha = 0.10$. Sample statistics: $\hat{p} = 0.875$, $n = 50$

■ USING AND INTERPRETING CONCEPTS

Testing Claims *In Exercises 9–16, (a) write the claim mathematically and identify H_0 and H_a, (b) find the critical value(s) and identify the rejection region(s), (c) find the standardized test statistic z, (d) decide whether to reject or fail to reject the null hypothesis, and (e) interpret the decision in the context of the original claim. If convenient, use technology to find the standardized test statistic.*

9. Smokers   A medical researcher says that less than 25% of U.S. adults are smokers. In a random sample of 200 U.S. adults, 18.5% say that they are smokers. At $\alpha = 0.05$, is there enough evidence to reject the researcher's claim? *(Adapted from National Center for Health Statistics)*

10. Census   A research center claims that at least 40% of U.S. adults think the Census count is accurate. In a random sample of 600 U.S. adults, 35% say that the Census count is accurate. At $\alpha = 0.02$, is there enough evidence to reject the center's claim? *(Adapted from Rasmussen Reports)*

11. Cellular Phones and Driving   A research center claims that at most 50% of people believe that drivers should be allowed to use cellular phones with hands-free devices while driving. In a random sample of 150 U.S. adults, 58% say that drivers should be allowed to use cellular phones with hands-free devices while driving. At $\alpha = 0.01$, is there enough evidence to reject the center's claim? *(Adapted from Rasmussen Reports)*

12. Asthma   A medical researcher claims that 5% of children under 18 years of age have asthma. In a random sample of 250 children under 18 years of age, 9.6% say they have asthma. At $\alpha = 0.08$, is there enough evidence to reject the researcher's claim? *(Adapted from National Center for Health Statistics)*

13. Female Height   A research center claims that more than 75% of females ages 20–29 are taller than 62 inches. In a random sample of 150 females ages 20–29, 82% are taller than 62 inches. At $\alpha = 0.10$, is there enough evidence to support the center's claim? *(Adapted from National Center for Health Statistics)*

14. (a) H_0: $p = 0.16$ (claim);
H_a: $p \neq 0.16$

(b) $-z_0 = -1.96$, $z_0 = 1.96$;
Rejection regions:
$z < -1.96$, $z > 1.96$

(c) 1.89

(d) Fail to reject H_0.

(e) There is not enough evidence at the 5% level of significance to reject the research center's claim that 16% of U.S. adults say that curling is the Winter Olympic sport they would like to try the most.

15. (a) H_0: $p \geq 0.35$;
H_a: $p < 0.35$ (claim)

(b) $z_0 = -1.28$;
Rejection region: $z < -1.28$

(c) 1.68   (d) Fail to reject H_0.

(e) There is not enough evidence at the 10% level of significance to support the humane society's claim that less than 35% of U.S. households own a dog.

16. See Selected Answers, page A123.

17. Fail to reject H_0. There is not enough evidence at the 5% level of significance to reject the claim that at least 52% of adults are more likely to buy a product when there are free samples.

18. Answers will vary. *Sample answer:* The company should continue the use of giveaways because there is not enough evidence to say that less than 52% of adults are more likely to buy a product when there are free samples.

19. (a) H_0: $p \geq 0.35$;
H_a: $p < 0.35$ (claim)

(b) $z_0 = -1.28$;
Rejection region: $z < -1.28$

(c) 1.68   (d) Fail to reject H_0.

(e) There is not enough evidence at the 10% level of significance to support the humane society's claim that less than 35% of U.S. households own a dog.

The results are the same.

20. Answers will vary.

---

14. Curling A research center claims that 16% of U.S. adults say that curling is the Winter Olympic sport they would like to try the most. In a random sample of 300 U.S. adults, 20% say that curling is the Winter Olympic sport they would like to try the most. At $\alpha = 0.05$, is there enough evidence to reject the researcher's claim? *(Adapted from Zogby International)*

15. Dog Ownership A humane society claims that less than 35% of U.S. households own a dog. In a random sample of 400 U.S. households, 156 say they own a dog. At $\alpha = 0.10$, is there enough evidence to support the society's claim? *(Adapted from The Humane Society of the United States)*

16. Cat Ownership A humane society claims that 30% of U.S. households own a cat. In a random sample of 200 U.S. households, 72 say they own a cat. At $\alpha = 0.05$, is there enough evidence to reject the society's claim? *(Adapted from The Humane Society of the United States)*

Free Samples *In Exercises 17 and 18, use the graph, which shows what adults think about the effectiveness of free samples.*

17. Do Free Samples Work? You interview a random sample of 50 adults. The results of the survey show that 48% of the adults said they were more likely to buy a product when there are free samples. At $\alpha = 0.05$, can you reject the claim that at least 52% of adults are more likely to buy a product when there are free samples?

Free Samples Work
How effective adults say free samples are:

More likely to buy a product **52%**
Shouldn't do it **3%**
Nice, but not necessary **25%**
More likely to remember a product **20%**

Take One (free)

18. Should Free Samples Be Used? Use your conclusion from Exercise 17 to write a paragraph on the use of free samples. Do you think a company should use free samples to get people to buy a product? Explain.

■ EXTENDING CONCEPTS

Alternative Formula *In Exercises 19 and 20, use the following information. When you know the number of successes x, the sample size n, and the population proportion p, it can be easier to use the formula*

$$z = \frac{x - np}{\sqrt{npq}}$$

to find the standardized test statistic when using a z-test for a population proportion p.

19. Rework Exercise 15 using the alternative formula and compare the results.

20. The alternative formula is derived from the formula

$$z = \frac{\hat{p} - p}{\sqrt{pq/n}} = \frac{(x/n) - p}{\sqrt{pq/n}}.$$

Use this formula to derive the alternative formula. Justify each step.

ACTIVITY 7.4   Hypothesis Tests for a Proportion

APPLET

The *hypothesis tests for a proportion* applet allows you to visually investigate hypothesis tests for a population proportion. You can specify the sample size *n*, the population proportion (True *p*), the null value for the proportion (Null *p*), and the alternative for the test (Alternative). When you click SIMULATE, 100 separate samples of size *n* will be selected from a population with a proportion of successes equal to True *p*. For each of the 100 samples, a hypothesis test based on the Z statistic is performed, and the results from each test are displayed in plots at the right. The standardized test statistic for each test is shown in the top plot and the *P*-value is shown in the bottom plot. The green and blue lines represent the cutoffs for rejecting the null hypothesis with the 0.05 and 0.01 level tests, respectively. Additional simulations can be carried out by clicking SIMULATE multiple times. The cumulative number of times that each test rejects the null hypothesis is also shown. Press CLEAR to clear existing results and start a new simulation.

n:	100
True p:	0.5
Null p:	0.5
Alternative:	<

Simulate

Cumulative results:

	0.05 level	0.01 level
Reject null		
Fail to reject null		
Prop. rejected		

Clear

■ Explore

Step 1   Specify a value for *n*.
Step 2   Specify a value for True *p*.
Step 3   Specify a value for Null *p*.
Step 4   Specify an alternative hypothesis.
Step 5   Click SIMULATE to generate the hypothesis tests.

■ Draw Conclusions

APPLET

1. Set *n* = 25, True *p* = 0.35, Null *p* = 0.35, and the alternative hypothesis to "not equal." Run the simulation so that at least 1000 hypothesis tests are run. Compare the proportion of null hypothesis rejections for the 0.05 level and the 0.01 level. Is this what you would expect? Explain.

2. Set *n* = 50, True *p* = 0.6, Null *p* = 0.4, and the alternative hypothesis to "<." What is the null hypothesis? Run the simulation so that at least 1000 hypothesis tests are run. Compare the proportion of null hypothesis rejections for the 0.05 level and the 0.01 level. Perform a hypothesis test for each level. Use the results of the hypothesis tests to explain the results of the simulation.

7.5 Hypothesis Testing for Variance and Standard Deviation

Critical Values for a χ^2-Test ▶ The Chi-Square Test

▶ CRITICAL VALUES FOR A χ^2-TEST

In real life, it is often important to produce consistent predictable results. For instance, consider a company that manufactures golf balls. The manufacturer must produce millions of golf balls, each having the same size and the same weight. There is a very low tolerance for variation. If the population is normal, you can test the variance and standard deviation of the process using the chi-square distribution with $n - 1$ degrees of freedom.

> **GUIDELINES**
>
> **Finding Critical Values for the χ^2-Test**
> 1. Specify the level of significance α.
> 2. Determine the degrees of freedom d.f. $= n - 1$.
> 3. The critical values for the χ^2-distribution are found in Table 6 in Appendix B. To find the critical value(s) for a
>   a. *right-tailed test*, use the value that corresponds to d.f. and α.
>   b. *left-tailed test*, use the value that corresponds to d.f. and $1 - \alpha$.
>   c. *two-tailed test*, use the values that correspond to d.f. and $\frac{1}{2}\alpha$, and d.f. and $1 - \frac{1}{2}\alpha$.

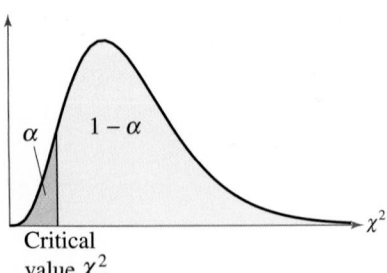

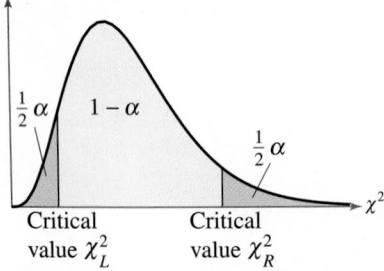

> **EXAMPLE 1**
>
> ▶ **Finding Critical Values for χ^2**
>
> Find the critical χ^2-value for a right-tailed test when $n = 26$ and $\alpha = 0.10$.
>
> ▶ **Solution**
>
> The degrees of freedom are
>
> $$\text{d.f.} = n - 1 = 26 - 1 = 25.$$
>
> The graph at the right shows a χ^2-distribution with 25 degrees of freedom and a shaded area of $\alpha = 0.10$ in the right tail. In Table 6 in Appendix B with d.f. $= 25$ and $\alpha = 0.10$, the critical value is
>
> $$\chi_0^2 = 34.382.$$

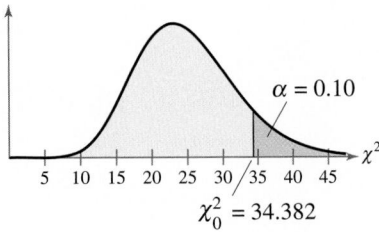

> ▶ **Try It Yourself 1**
>
> Find the critical χ^2-value for a right-tailed test when $n = 18$ and $\alpha = 0.01$.
>
> a. *Identify* the degrees of freedom and the level of significance.
> b. *Use* Table 6 in Appendix B to find the critical χ^2-value. *Answer: Page A42*

Note to Instructor

This section can be omitted or covered later (with Chapter 10) without loss of continuity.

EXAMPLE 2

▶ **Finding Critical Values for χ^2**

Find the critical χ^2-value for a left-tailed test when $n = 11$ and $\alpha = 0.01$.

▶ **Solution**

The degrees of freedom are

> d.f. $= n - 1 = 11 - 1 = 10$.

The graph shows a χ^2-distribution with 10 degrees of freedom and a shaded area of $\alpha = 0.01$ in the left tail. The area to the right of the critical value is

> $1 - \alpha = 1 - 0.01 = 0.99$.

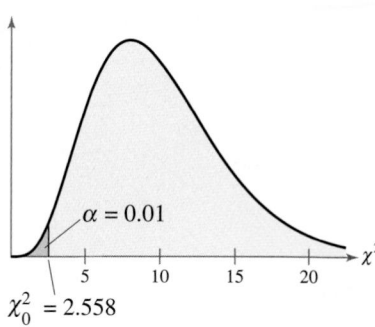

$\chi_0^2 = 2.558$

In Table 6 with d.f. $= 10$ and the area $1 - \alpha = 0.99$, the critical value is $\chi_0^2 = 2.558$.

▶ **Try It Yourself 2**

Find the critical χ^2-value for a left-tailed test when $n = 30$ and $\alpha = 0.05$.

a. *Identify* the degrees of freedom and the level of significance.
b. *Use* Table 6 in Appendix B to find the critical χ^2-value.   *Answer: Page A42*

Answer: Page A42

STUDY TIP

Note that because chi-square distributions are not symmetric (like normal or *t*-distributions), in a two-tailed test the two critical values are not opposites. Each critical value must be calculated separately.

EXAMPLE 3

▶ **Finding Critical Values for χ^2**

Find the critical χ^2-values for a two-tailed test when $n = 9$ and $\alpha = 0.05$.

▶ **Solution**

The degrees of freedom are

> d.f. $= n - 1 = 9 - 1 = 8$.

The graph shows a χ^2-distribution with 8 degrees of freedom and a shaded area of $\frac{1}{2}\alpha = 0.025$ in each tail. The areas to the right of the critical values are

> $\frac{1}{2}\alpha = 0.025$

and

> $1 - \frac{1}{2}\alpha = 0.975$.

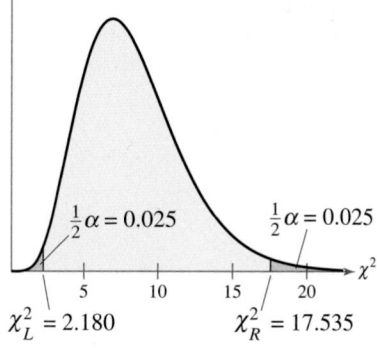

$\chi_L^2 = 2.180$ $\chi_R^2 = 17.535$

In Table 6 with d.f. $= 8$ and the areas 0.025 and 0.975, the critical values are $\chi_L^2 = 2.180$ and $\chi_R^2 = 17.535$.

▶ **Try It Yourself 3**

Find the critical χ^2-values for a two-tailed test when $n = 51$ and $\alpha = 0.01$.

a. *Identify* the degrees of freedom and the level of significance.
b. *Find* the first critical value χ_R^2 using Table 6 in Appendix B and the area $\frac{1}{2}\alpha$.
c. *Find* the second critical value χ_L^2 using Table 6 in Appendix B and the area $1 - \frac{1}{2}\alpha$.   *Answer: Page A4*

Answer: Page A4

THE CHI-SQUARE TEST

To test a variance σ^2 or a standard deviation σ of a population that is normally distributed, you can use the χ^2-test. The χ^2-test for a variance or standard deviation is not as robust as the tests for the population mean μ or the population proportion p. So, it is essential in performing a χ^2-test for a variance or standard deviation that the population be normally distributed. The results can be misleading if the population is not normal.

χ^2-TEST FOR A VARIANCE σ^2 OR STANDARD DEVIATION σ

The **χ^2-test for a variance or standard deviation** is a statistical test for a population variance or standard deviation. The χ^2-test can be used when the population is normal. The **test statistic** is s^2 and the **standardized test statistic**

$$\chi^2 = \frac{(n-1)s^2}{\sigma^2}$$

follows a chi-square distribution with degrees of freedom

d.f. $= n - 1$.

GUIDELINES

Note to Instructor

Review the properties of chi-square distributions. Tell students that this family of distributions will be used in later chapters, but the degrees of freedom for those tests are not necessarily $n - 1$.

Using the χ^2-Test for a Variance or Standard Deviation

IN WORDS	IN SYMBOLS
1. State the claim mathematically and verbally. Identify the null and alternative hypotheses.	State H_0 and H_a.
2. Specify the level of significance.	Identify α.
3. Determine the degrees of freedom.	d.f. $= n - 1$
4. Determine the critical value(s).	Use Table 6 in Appendix B.
5. Determine the rejection region(s).	
6. Find the standardized test statistic and sketch the sampling distribution.	$\chi^2 = \frac{(n-1)s^2}{\sigma^2}$
7. Make a decision to reject or fail to reject the null hypothesis.	If χ^2 is in the rejection region, reject H_0. Otherwise, fail to reject H_0.
8. Interpret the decision in the context of the original claim.	

A community center claims that the chlorine level in its pool has a standard deviation of 0.46 parts per million (ppm). A sampling of the pool's chlorine levels at 25 random times during a month yields a standard deviation of **0.61 ppm.** (Adapted from American Pool Supply)

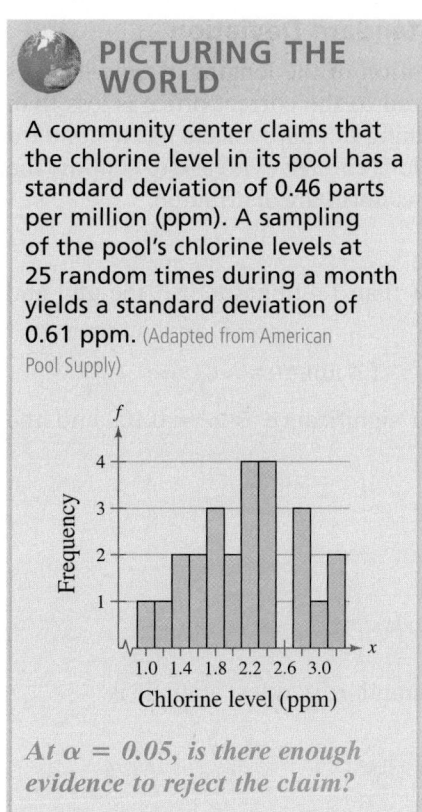

Chlorine level (ppm)

At α = 0.05, is there enough evidence to reject the claim?

Yes, there is enough evidence at the 5% level of significance to reject the claim that the chlorine level in the pool has a standard deviation of 0.46 parts per million.

EXAMPLE 4   SC Report 31

▶ Using a Hypothesis Test for the Population Variance

A dairy processing company claims that the variance of the amount of fat in the whole milk processed by the company is no more than 0.25. You suspect this is wrong and find that a random sample of 41 milk containers has a variance of 0.27. At $\alpha = 0.05$, is there enough evidence to reject the company's claim? Assume the population is normally distributed.

▶ Solution

The claim is "the variance is no more than 0.25." So, the null and alternative hypotheses are

$$H_0: \sigma^2 \leq 0.25 \text{ (Claim)} \quad \text{and} \quad H_a: \sigma^2 > 0.25.$$

The test is a right-tailed test, the level of significance is $\alpha = 0.05$, and the degrees of freedom are d.f. $= 41 - 1 = 40$. So, the critical value is

$$\chi_0^2 = 55.758.$$

The rejection region is $\chi^2 > 55.758$. The standardized test statistic is

$$
\begin{aligned}
\chi^2 &= \frac{(n-1)s^2}{\sigma^2} &&\text{Use the chi-square test.} \\
&= \frac{(41-1)(0.27)}{0.25} &&\text{Assume } \sigma^2 = 0.25. \\
&= 43.2.
\end{aligned}
$$

The graph shows the location of the rejection region and the standardized test statistic χ^2. Because χ^2 is not in the rejection region, you should fail to reject the null hypothesis.

Interpretation There is not enough evidence at the 5% level of significance to reject the company's claim that the variance of the amount of fat in the whole milk is no more than 0.25.

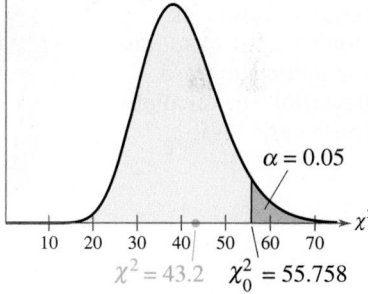

▶ Try It Yourself 4

A bottling company claims that the variance of the amount of sports drink in a 12-ounce bottle is no more than 0.40. A random sample of 31 bottles has a variance of 0.75. At $\alpha = 0.01$, is there enough evidence to reject the company's claim? Assume the population is normally distributed.

a. Identify the *claim* and state H_0 and H_a.
b. Identify the *level of significance* α and the *degrees of freedom*.
c. Find the *critical value* and identify the *rejection region*.
d. Find the *standardized test statistic* χ^2.
e. *Decide* whether to reject the null hypothesis. Use a graph if necessary.
f. *Interpret* the decision in the context of the original claim.

Answer: Page A42

EXAMPLE 5 SC Report 32

▶ Using a Hypothesis Test for the Standard Deviation

A company claims that the standard deviation of the lengths of time it takes an incoming telephone call to be transferred to the correct office is less than 1.4 minutes. A random sample of 25 incoming telephone calls has a standard deviation of 1.1 minutes. At $\alpha = 0.10$, is there enough evidence to support the company's claim? Assume the population is normally distributed.

▶ Solution

The claim is "the standard deviation is less than 1.4 minutes." So, the null and alternative hypotheses are

$$H_0: \sigma \geq 1.4 \text{ minutes} \quad \text{and} \quad H_a: \sigma < 1.4 \text{ minutes. (Claim)}$$

The test is a left-tailed test, the level of significance is $\alpha = 0.10$, and the degrees of freedom are

$$\text{d.f.} = 25 - 1$$
$$= 24.$$

So, the critical value is

$$\chi_0^2 = 15.659.$$

The rejection region is $\chi^2 < 15.659$. The standardized test statistic is

$$\chi^2 = \frac{(n-1)s^2}{\sigma^2} \qquad \text{Use the chi-square test.}$$
$$= \frac{(25-1)(1.1)^2}{1.4^2} \qquad \text{Assume } \sigma = 1.4.$$
$$\approx 14.816.$$

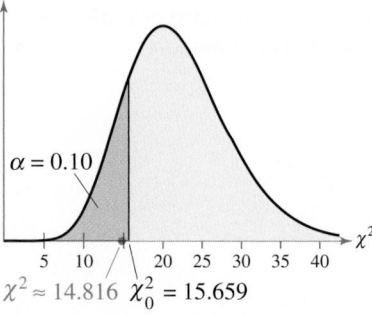

The graph shows the location of the rejection region and the standardized test statistic χ^2. Because χ^2 is in the rejection region, you should reject the null hypothesis.

Interpretation There is enough evidence at the 10% level of significance to support the claim that the standard deviation of the lengths of time it takes an incoming telephone call to be transferred to the correct office is less than 1.4 minutes.

▶ Try It Yourself 5

A police chief claims that the standard deviation of the lengths of response times is less than 3.7 minutes. A random sample of 9 response times has a standard deviation of 3.0 minutes. At $\alpha = 0.05$, is there enough evidence to support the police chief's claim? Assume the population is normally distributed.

a. Identify the *claim* and state H_0 and H_a.
b. Identify the *level of significance* α and the *degrees of freedom*.
c. Find the *critical value* and identify the *rejection region*.
d. Find the *standardized test statistic* χ^2.
e. *Decide* whether to reject the null hypothesis. Use a graph if necessary.
f. *Interpret* the decision in the context of the original claim.

Answer: Page A42

EXAMPLE 6

▶ **Using a Hypothesis Test for the Population Variance**

A sporting goods manufacturer claims that the variance of the strengths of a certain fishing line is 15.9. A random sample of 15 fishing line spools has a variance of 21.8. At $\alpha = 0.05$, is there enough evidence to reject the manufacturer's claim? Assume the population is normally distributed.

▶ **Solution**

The claim is "the variance is 15.9." So, the null and alternative hypotheses are

$$H_0\text{: } \sigma^2 = 15.9 \text{ (Claim)}$$

and

$$H_a\text{: } \sigma^2 \neq 15.9.$$

The test is a two-tailed test, the level of significance is $\alpha = 0.05$, and the degrees of freedom are

$$\text{d.f.} = 15 - 1$$
$$= 14.$$

So, the critical values are $\chi_L^2 = 5.629$ and $\chi_R^2 = 26.119$.

The rejection regions are $\chi^2 < 5.629$ and $\chi^2 > 26.119$. The standardized test statistic is

$$\chi^2 = \frac{(n-1)s^2}{\sigma^2} \qquad \text{Use the chi-square test.}$$

$$= \frac{(15-1)(21.8)}{15.9} \qquad \text{Assume } \sigma^2 = 15.9.$$

$$\approx 19.195.$$

The graph shows the location of the rejection regions and the standardized test statistic χ^2. Because χ^2 is not in the rejection regions, you should fail to reject the null hypothesis.

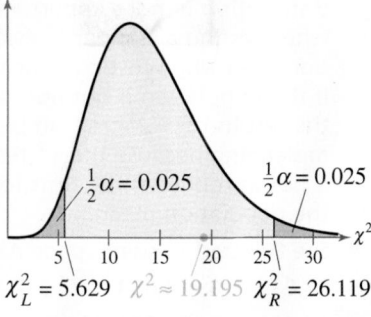

$$\chi_L^2 = 5.629 \quad \chi^2 \approx 19.195 \quad \chi_R^2 = 26.119$$

Interpretation There is not enough evidence at the 5% level of significance to reject the claim that the variance of the strengths of the fishing line is 15.9.

▶ **Try It Yourself 6**

A company that offers dieting products and weight loss services claims that the variance of the weight losses of their users is 25.5. A random sample of 13 users has a variance of 10.8. At $\alpha = 0.10$, is there enough evidence to reject the company's claim? Assume the population is normally distributed.

a. Identify the *claim* and state H_0 and H_a.
b. Identify the *level of significance* α and the *degrees of freedom*.
c. Find the *critical values* and identify the *rejection regions*.
d. Find the *standardized test statistic* χ^2.
e. *Decide* whether to reject the null hypothesis. Use a graph if necessary.
f. *Interpret* the decision in the context of the original claim.

Answer: Page A47

7.5 EXERCISES

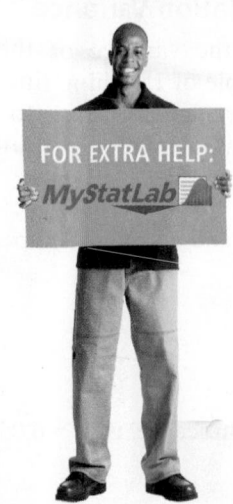

1. Specify the level of significance α. Determine the degrees of freedom. Determine the critical values using the χ^2-distribution. For a right-tailed test, use the value that corresponds to d.f. and α; for a left-tailed test, use the value that corresponds to d.f. and $1 - \alpha$; for a two-tailed test, use the values that correspond to d.f. and $\frac{1}{2}\alpha$, and d.f. and $1 - \frac{1}{2}\alpha$.

2. See Selected Answers, page A123.

3. The requirement of a normal distribution is more important when testing a standard deviation than when testing a mean. If the population is not normal, the results of a χ^2-test can be misleading because the χ^2-test is not as robust as the tests for the population mean.

4. See Selected Answers, page A124.

5. 38.885   6. 14.684   7. 0.872

8. 13.091   9. 60.391, 101.879

10. 35.534, 91.952

11. (a) Fail to reject H_0.
   (b) Fail to reject H_0.
   (c) Fail to reject H_0.
   (d) Reject H_0.

12. (a) Fail to reject H_0.
   (b) Fail to reject H_0.
   (c) Reject H_0.
   (d) Reject H_0.

13. (a) Fail to reject H_0.
   (b) Reject H_0.
   (c) Reject H_0.
   (d) Fail to reject H_0.

■ BUILDING BASIC SKILLS AND VOCABULARY

1. Explain how to find critical values in a χ^2-sampling distribution.

2. Can a critical value for the χ^2-test be negative? Explain.

3. When testing a claim about a population mean or a population standard deviation, a requirement is that the sample is from a population that is normally distributed. How is this requirement different between the two tests?

4. Explain how to test a population variance or a population standard deviation.

In Exercises 5–10, find the critical value(s) for the indicated test for a population variance, sample size n, and level of significance α.

5. Right-tailed test,
   $n = 27, \alpha = 0.05$

6. Right-tailed test,
   $n = 10, \alpha = 0.10$

7. Left-tailed test,
   $n = 7, \alpha = 0.01$

8. Left-tailed test,
   $n = 24, \alpha = 0.05$

9. Two-tailed test,
   $n = 81, \alpha = 0.10$

10. Two-tailed test,
    $n = 61, \alpha = 0.01$

Graphical Analysis *In Exercises 11–14, state whether the standardized test statistic χ^2 allows you to reject the null hypothesis.*

11. (a) $\chi^2 = 2.091$
    (b) $\chi^2 = 0$
    (c) $\chi^2 = 1.086$
    (d) $\chi^2 = 6.3471$

12. (a) $\chi^2 = 0.771$
    (b) $\chi^2 = 9.486$
    (c) $\chi^2 = 0.701$
    (d) $\chi^2 = 9.508$

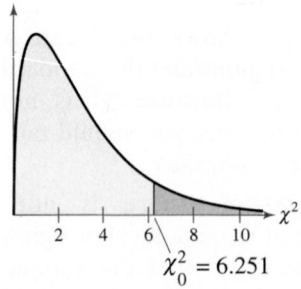

$\chi_0^2 = 6.251$ $\chi_L^2 = 0.711$ $\chi_R^2 = 9.488$

13. (a) $\chi^2 = 22.302$
    (b) $\chi^2 = 23.309$
    (c) $\chi^2 = 8.457$
    (d) $\chi^2 = 8.577$

14. (a) $\chi^2 = 10.065$
    (b) $\chi^2 = 10.075$
    (c) $\chi^2 = 10.585$
    (d) $\chi^2 = 10.745$

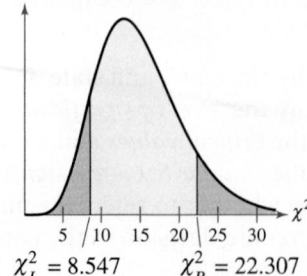

$\chi_L^2 = 8.547$ $\chi_R^2 = 22.307$

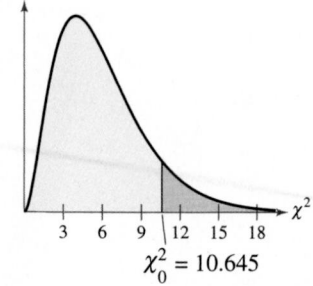

$\chi_0^2 = 10.645$

14. (a) Fail to reject H_0.
    (b) Fail to reject H_0.
    (c) Fail to reject H_0.
    (d) Reject H_0.

15. Fail to reject H_0. There is not enough evidence at the 5% level of significance to reject the claim.

16. Fail to reject H_0. There is not enough evidence at the 5% level of significance to reject the claim.

17. Reject H_0. There is enough evidence at the 10% level of significance to reject the claim.

18. See Selected Answers, page A124.

19. (a) H_0: $\sigma^2 = 1.25$ (claim);
     H_a: $\sigma^2 \neq 1.25$

    (b) $\chi_L^2 = 10.283$, $\chi_R^2 = 35.479$;

    Rejection regions:
     $\chi^2 < 10.283$, $\chi^2 > 35.479$

    (c) 22.68   (d) Fail to reject H_0.

    (e) There is not enough evidence at the 5% level of significance to reject the manufacturer's claim that the variance of the number of grams of carbohydrates in servings of its tortilla chips is 1.25.

20. See Selected Answers, page A124.

21. (a) H_0: $\sigma \geq 36$; H_a: $\sigma < 36$ (claim)

    (b) $\chi_0^2 = 13.240$;

    Rejection region: $\chi^2 < 13.240$

    (c) 18.076   (d) Fail to reject H_0.

    (e) There is not enough evidence at the 10% level of significance to support the test administrator's claim that the standard deviation for eighth graders on the examination is less than 36 points.

22. See Selected Answers, page A124.

23. (a) H_0: $\sigma \leq 25$ (claim); H_a: $\sigma > 25$

    (b) $\chi_0^2 = 36.741$;

    Rejection region: $\chi^2 > 36.741$

    (c) 41.515   (d) Reject H_0.

    (e) There is enough evidence at the 10% level of significance to reject the weather service's claim that the standard deviation of the number of fatalities per year from tornadoes is no more than 25.

24. See Selected Answers, page A124.

In Exercises 15–18, use a χ^2-test to test the claim about the population variance σ^2 or standard deviation σ at the given level of significance α using the given sample statistics. For each claim, assume the population is normally distributed.

15. Claim: $\sigma^2 = 0.52$; $\alpha = 0.05$. Sample statistics: $s^2 = 0.508$, $n = 18$

16. Claim: $\sigma^2 \geq 8.5$; $\alpha = 0.05$. Sample statistics: $s^2 = 7.45$, $n = 23$

17. Claim: $\sigma = 24.9$; $\alpha = 0.10$. Sample statistics: $s = 29.1$, $n = 51$

18. Claim: $\sigma < 40$; $\alpha = 0.01$. Sample statistics: $s = 40.8$, $n = 12$

■ USING AND INTERPRETING CONCEPTS

Testing Claims   *In Exercises 19–28, (a) write the claim mathematically and identify H_0 and H_a, (b) find the critical value(s) and identify the rejection region(s), (c) find the standardized test statistic χ^2, (d) decide whether to reject or fail to reject the null hypothesis, and (e) interpret the decision in the context of the original claim. For each claim, assume the population is normally distributed.*

19. Carbohydrates   A snack food manufacturer estimates that the variance of the number of grams of carbohydrates in servings of its tortilla chips is 1.25. A dietician is asked to test this claim and finds that a random sample of 22 servings has a variance of 1.35. At $\alpha = 0.05$, is there enough evidence to reject the manufacturer's claim?

20. Hybrid Vehicle Gas Mileage   An auto manufacturer believes that the variance of the gas mileages of its hybrid vehicles is 1.0. You work for an energy conservation agency and want to test this claim. You find that a random sample of the gas mileages of 25 of the manufacturer's hybrid vehicles has a variance of 1.65. At $\alpha = 0.05$, do you have enough evidence to reject the manufacturer's claim? *(Adapted from Green Hybrid)*

21. Science Assessment Tests   On a science assessment test, the scores of a random sample of 22 eighth grade students have a standard deviation of 33.4 points. This result prompts a test administrator to claim that the standard deviation for eighth graders on the examination is less than 36 points. At $\alpha = 0.10$, is there enough evidence to support the administrator's claim? *(Adapted from National Center for Educational Statistics)*

22. U.S. History Assessment Tests   A state school administrator says that the standard deviation of test scores for eighth grade students who took a U.S. history assessment test is less than 30 points. You work for the administrator and are asked to test this claim. You randomly select 18 tests and find that the tests have a standard deviation of 33.6 points. At $\alpha = 0.01$, is there enough evidence to support the administrator's claim? *(Adapted from National Center for Educational Statistics)*

23. Tornadoes   A weather service claims that the standard deviation of the number of fatalities per year from tornadoes is no more than 25. A random sample of the number of deaths for 28 years has a standard deviation of 31 fatalities. At $\alpha = 0.10$, is there enough evidence to reject the weather service's claim? *(Source: NOAA Weather Partners)*

24. Lengths of Stay   A doctor says the standard deviation of the lengths of stay for patients involved in a crash in which the vehicle struck a tree is 6.14 days. A random sample of 20 lengths of stay for patients involved in this type of crash has a standard deviation of 6.5 days. At $\alpha = 0.05$, can you reject the doctor's claim? *(Adapted from National Highway Traffic Safety Administration)*

25. (a) H_0: $\sigma \geq \$3500$;
H_a: $\sigma < \$3500$ (claim)

(b) $\chi_0^2 = 18.114$;
Rejection region: $\chi^2 < 18.114$

(c) 37.051   (d) Fail to reject H_0.

(e) There is not enough evidence at the 10% level of significance to support the insurance agent's claim that the standard deviation of the total charges for patients involved in a crash in which the vehicle struck a construction barricade is less than \$3500.

26. See Selected Answers, page A124.

27. (a) H_0: $\sigma \leq 6100$;
H_a: $\sigma > 6100$ (claim)

(b) $\chi_0^2 = 27.587$;
Rejection region: $\chi^2 > 27.587$

(c) 27.897   (d) Reject H_0.

(e) There is enough evidence at the 5% level of significance to support the claim that the standard deviation of the annual salaries of environmental engineers is greater than \$6100.

28. See Selected Answers, page A124.

29. See Odd Answers, page A85.

30. See Selected Answers, page A124.

31. See Odd Answers, page A85.

32. See Selected Answers, page A124.

33. P-value = 0.9059
Fail to reject H_0.

34. P-value = 0.1189
Fail to reject H_0.

TI-83/84 PLUS

χ^2 cdf [0, 43.2, 40]
.6637768667

---

35. P-value = 0.0462
Reject H_0.

36. P-value = 0.0983
Reject H_0.

---

25. **Total Charges**   An insurance agent says the standard deviation of the total hospital charges for patients involved in a crash in which the vehicle struck a construction barricade is less than \$3500. A random sample of 28 total hospital charges for patients involved in this type of crash has a standard deviation of \$4100. At $\alpha = 0.10$, can you support the agent's claim? *(Adapted from National Highway Traffic Safety Administration)*

26. **Hotel Room Rates**   A travel agency estimates that the standard deviation of the room rates of hotels in a certain city is no more than \$30. You work for a consumer advocacy group and are asked to test this claim. You find that a random sample of 21 hotels has a standard deviation of \$35.25. At $\alpha = 0.01$, do you have enough evidence to reject the agency's claim?

27. **Salaries**   The annual salaries (in dollars) of 18 randomly chosen environmental engineers are listed. At $\alpha = 0.05$, can you conclude that the standard deviation of the annual salaries is greater than \$6100? *(Adapted from Salary.com)*

63,125	59,749	52,369	55,979	61,550	54,644	50,420
47,291	51,357	56,901	53,499	49,998	69,712	64,575
45,850	46,297	63,770	71,589			

28. **Salaries**   A staffing organization states that the standard deviation of the annual salaries of commodity buyers is at least \$10,600. The annual salaries (in dollars) of 20 randomly chosen commodity buyers are listed. At $\alpha = 0.10$, can you reject the organization's claim? *(Adapted from Salary.com)*

79,319	68,825	65,129	75,899	85,070	76,270	68,750
70,982	69,237	63,470	79,025	55,880	80,985	75,264
66,918	65,459	70,598	86,579	71,225	57,311	

SC   *In Exercises 29–32, use StatCrunch to help you test the claim about the population variance σ^2 or standard deviation σ at the given level of significance α using the given sample statistics. For each claim, assume the population is normally distributed.*

29. Claim: $\sigma^2 \geq 9$; $\alpha = 0.01$. Sample statistics: $s^2 = 2.03$, $n = 10$

30. Claim: $\sigma^2 = 14.85$; $\alpha = 0.05$. Sample statistics: $s^2 = 28.75$, $n = 17$

31. Claim: $\sigma > 4.5$; $\alpha = 0.05$. Sample statistics: $s = 5.8$, $n = 15$

32. Claim: $\sigma \neq 418$; $\alpha = 0.10$. Sample statistics: $s = 305$, $n = 24$

■ EXTENDING CONCEPTS

P-Values   *You can calculate the P-value for a χ^2-test using technology. After calculating the χ^2-test value, you can use the cumulative density function (CDF) to calculate the area under the curve. From Example 4 on page 407, $\chi^2 = 43.2$. Using a TI-83/84 Plus (choose 7 from the DISTR menu), enter 0 for the lower bound, 43.2 for the upper bound, and 40 for the degrees of freedom, as shown at the left.*

The P-value is approximately $1 - 0.6638 = 0.3362$. Because $P > \alpha = 0.05$, the conclusion is to fail to reject H_0.

In Exercises 33–36, use the P-value method to perform the hypothesis test for the indicated exercise.

33. Exercise 25   34. Exercise 26

35. Exercise 27   36. Exercise 28

Statistics in the Real World

USES AND ABUSES

Uses

Hypothesis Testing   Hypothesis testing is important in many different fields because it gives a scientific procedure for assessing the validity of a claim about a population. Some of the concepts in hypothesis testing are intuitive, but some are not. For instance, the *American Journal of Clinical Nutrition* suggests that eating dark chocolate can help prevent heart disease. A random sample of healthy volunteers were assigned to eat 3.5 ounces of dark chocolate each day for 15 days. After 15 days, the mean systolic blood pressure of the volunteers was 6.4 millimeters of mercury lower. A hypothesis test could show if this drop in systolic blood pressure is significant or simply due to sampling error.

Careful inferences must be made concerning the results. In another part of the study, it was found that white chocolate did not result in similar benefits. So, the inference of health benefits cannot be extended to all types of chocolate. You also would not infer that you should eat large quantities of chocolate because the benefits must be weighed against known risks, such as weight gain, acne, and acid reflux.

Abuses

Not Using a Random Sample   The entire theory of hypothesis testing is based on the fact that the sample is randomly selected. If the sample is not random, then you cannot use it to infer anything about a population parameter.

Attempting to Prove the Null Hypothesis   If the *P*-value for a hypothesis test is greater than the level of significance, you have not proven the null hypothesis is true—only that there is not enough evidence to reject it. For instance, with a *P*-value higher than the level of significance, a researcher could not prove that there is no benefit to eating dark chocolate—only that there is not enough evidence to support the claim that there is a benefit.

Making Type I or Type II Errors   Remember that a type I error is rejecting a null hypothesis that is true and a type II error is failing to reject a null hypothesis that is false. You can decrease the probability of a type I error by lowering the level of significance. Generally, if you decrease the probability of making a type I error, you increase the probability of making a type II error. You can decrease the chance of making both types of errors by increasing the sample size.

■ EXERCISES

In Exercises 1–4, assume that you work in a transportation department. You are asked to write a report about the claim that 73% of U.S. adults who fly at least once a year favor full-body scanners at airports. (Adapted from Rasmussen Reports)

1. ***Not Using a Random Sample***   How could you choose a random sample to test this hypothesis?

2. ***Attempting to Prove the Null Hypothesis***   What is the null hypothesis in this situation? Describe how your report could be incorrect by trying to prove the null hypothesis.

3. ***Making a Type I Error***   Describe how your report could make a type I error.

4. ***Making a Type II Error***   Describe how your report could make a type II error.

Do You Favor the Use of Full-Body Scanners at Airports in the U.S.?

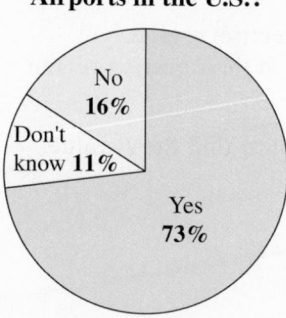

7 A SUMMARY OF HYPOTHESIS TESTING

With hypothesis testing, perhaps more than any other area of statistics, it can be difficult to see the forest for all the trees. To help you see the forest—the overall picture—a summary of what you studied in this chapter is provided.

Writing the Hypotheses

- You are given a claim about a population parameter μ, p, σ^2, or σ.
- Rewrite the claim and its complement using $\underbrace{\leq, \geq, =}_{H_0}$ and $\underbrace{>, <, \neq}_{H_a}$.
- Identify the claim. Is it H_0 or H_a?

Specifying a Level of Significance

- Specify α, the maximum acceptable probability of rejecting a valid H_0 (a type I error).

Specifying the Sample Size

- Specify your sample size n.

Choosing the Test   ■ Any population   ■ Normally distributed population

- **Mean:** H_0 describes a hypothesized population mean μ.
  - Use a **z-test** for *any* population if $n \geq 30$.
  - Use a **z-test** if the population is normal and σ is known for any n.
  - Use a **t-test** if the population is normal and $n < 30$, but σ is unknown.
- **Proportion:** H_0 describes a hypothesized population proportion p.
  - Use a **z-test** for any population if $np \geq 5$ and $nq \geq 5$.
- **Variance or Standard Deviation:** H_0 describes a hypothesized population variance σ^2 or standard deviation σ.
  - Use a **χ^2-test** if the population is normal.

Sketching the Sampling Distribution

- Use H_a to decide if the test is left-tailed, right-tailed, or two-tailed.

Finding the Standardized Test Statistic

- Take a random sample of size n from the population.
- Compute the test statistic $\bar{x}$, $\hat{p}$, or s^2.
- Find the standardized test statistic z, t, or χ^2.

Making a Decision

Option 1. Decision based on rejection region

- Use α to find the critical value(s) z_0, t_0, or χ_0^2 and rejection region(s).
- **Decision Rule:**

  Reject H_0 if the standardized test statistic is in the rejection region.
  Fail to reject H_0 if the standardized test statistic is not in the rejection region.

Option 2. Decision based on *P*-value

- Use the standardized test statistic or a technology tool to find the *P*-value.
- **Decision Rule:**

  Reject H_0 if $P \leq \alpha$.
  Fail to reject H_0 if $P > \alpha$.

INSIGHT

Large sample sizes will usually increase the cost and effort of testing a hypothesis, but they also tend to make your decision more reliable.

z-Test for a Hypothesized Mean μ *(Section 7.2)*

Test statistic: $\overline{x}$

Standardized test statistic: z

Critical value: z_0 (Use Table 4.)

If $n \geq 30$, s can be used in place of σ. Sampling distribution of sample means is a normal distribution.

$$z = \frac{\overline{x} - \mu}{\sigma / \sqrt{n}}$$

Sample mean ⟶ (numerator) ⟵ Hypothesized mean

Population standard deviation ⟶ (denominator) ⟵ Sample size

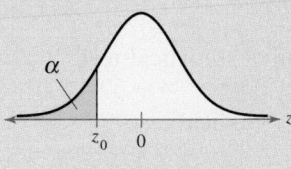
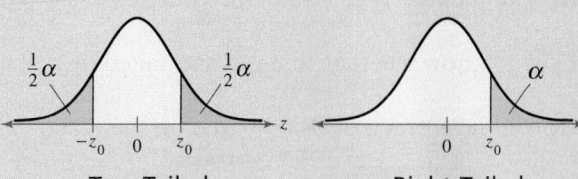

Left-Tailed   Two-Tailed   Right-Tailed

z-Test for a Hypothesized Proportion p *(Section 7.4)*

Test statistic: $\hat{p}$

Standardized test statistic: z

Critical value: z_0 (Use Table 4.)

Sampling distribution of sample proportions is a normal distribution.

$$z = \frac{\hat{p} - p}{\sqrt{pq/n}}$$

Sample proportion ⟶ (numerator) ⟵ Hypothesized proportion

$q = 1 - p$ ⟶ (denominator) ⟵ Sample size

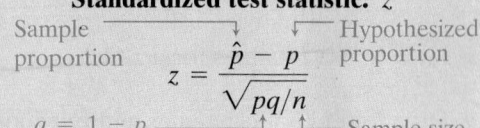

t-Test for a Hypothesized Mean μ *(Section 7.3)*

Test statistic: $\overline{x}$

Standardized test statistic: t

Critical value: t_0 (Use Table 5.)

Sampling distribution of sample means is approximated by a t-distribution with d.f. $= n - 1$.

$$t = \frac{\overline{x} - \mu}{s / \sqrt{n}}$$

Sample mean ⟶ (numerator) ⟵ Hypothesized mean

Sample standard deviation ⟶ (denominator) ⟵ Sample size

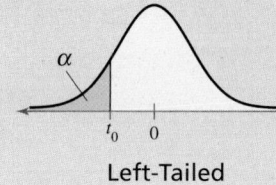
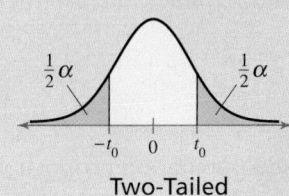
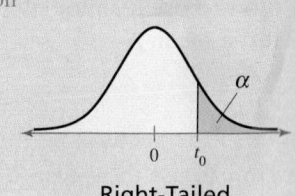

Left-Tailed   Two-Tailed   Right-Tailed

χ^2-Test for a Hypothesized Variance σ^2 or Standard Deviation σ *(Section 7.5)*

Test statistic: s^2

Standardized test statistic: χ^2

Critical value: χ_0^2 (Use Table 6.)

Sampling distribution is approximated by a chi-square distribution with d.f. $= n - 1$.

$$\chi^2 = \frac{(n - 1)s^2}{\sigma^2}$$

Sample size ⟶ (numerator) ⟵ Sample variance

⟵ Hypothesized variance

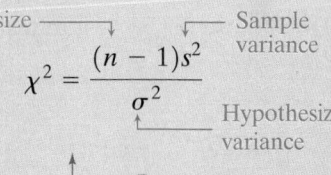
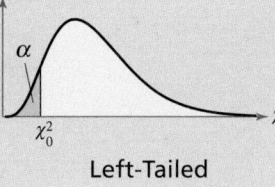
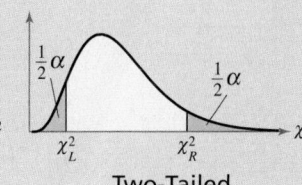
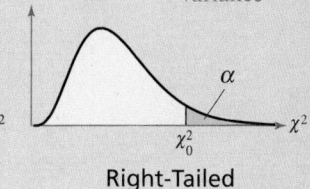

Left-Tailed   Two-Tailed   Right-Tailed

7 CHAPTER SUMMARY

What did you learn?	EXAMPLE(S)	REVIEW EXERCISES
Section 7.1		
■ How to state a null hypothesis and an alternative hypothesis	*1*	*1–6*
■ How to identify type I and type II errors	*2*	*7–10*
■ How to know whether to use a one-tailed or a two-tailed statistical test	*3*	*7–10*
■ How to interpret a decision based on the results of a statistical test	*4*	*7–10*
Section 7.2		
■ How to find P-values and use them to test a mean μ	*1–3*	*11, 12*
■ How to use P-values for a z-test	*4–6*	*13, 14, 23–28*
■ How to find critical values and rejection regions in a normal distribution	*7, 8*	*15–18*
■ How to use rejection regions for a z-test	*9, 10*	*19–28*
Section 7.3		
■ How to find critical values in a t-distribution	*1–3*	*29–32*
■ How to use the t-test to test a mean μ	*4, 5*	*33–40*
■ How to use technology to find P-values and use them with a t-test to test a mean μ	*6*	*41, 42*
Section 7.4		
■ How to use the z-test to test a population proportion p	*1, 2*	*43–52*
Section 7.5		
■ How to find critical values for a χ^2-test	*1–3*	*53–56*
■ How to use the χ^2-test to test a variance or a standard deviation	*4–6*	*57–63*

7 REVIEW EXERCISES

1. H_0: $\mu \leq 375$ (claim); H_a: $\mu > 375$

2. H_0: $\mu = 82$ (claim); H_a: $\mu \neq 82$

3. H_0: $p \geq 0.205$;
H_a: $p < 0.205$ (claim)

4. H_0: $\mu = 150{,}020$;
H_a: $\mu \neq 150{,}020$ (claim)

5. H_0: $\sigma \leq 1.9$; H_a: $\sigma > 1.9$ (claim)

6. H_0: $p \geq 0.64$ (claim); H_a: $p < 0.64$

7. (a) H_0: $p = 0.71$ (claim);
H_a: $p \neq 0.71$

(b) A type I error will occur if the actual proportion of Americans who support plans to order deep cuts in executive compensation at companies that have received federal bailout funds is 71%, but you reject H_0: $p = 0.71$.

A type II error will occur if the actual proportion is not 71%, but you fail to reject H_0: $p = 0.71$.

(c) Two-tailed because the alternative hypothesis contains $\neq$.

(d) There is enough evidence to reject the news outlet's claim that the proportion of Americans who support plans to order deep cuts in executive compensation at companies that have received federal bailout funds is 71%.

(e) There is not enough evidence to reject the news outlet's claim that the proportion of Americans who support plans to order deep cuts in executive compensation at companies that have received federal bailout funds is 71%.

8. See Selected Answers, page A124.

9. See Odd Answers, page A85.

10. See Selected Answers, page A124.

11. 0.1736; Fail to reject H_0.

12. 0.0102; Reject H_0.

13. See Odd Answers, page A85.

14. See Selected Answers, page A125.

15. See Odd Answers, page A85.

16. See Selected Answers, page A125.

17. See Odd Answers, page A85.

18. See Selected Answers, page A125.

■ SECTION 7.1

In Exercises 1–6, use the given statement to represent a claim. Write its complement and state which is H_0 and which is H_a.

1. $\mu \leq 375$

2. $\mu = 82$

3. $p < 0.205$

4. $\mu \neq 150{,}020$

5. $\sigma > 1.9$

6. $p \geq 0.64$

In Exercises 7–10, do the following.

(a) *State the null and alternative hypotheses, and identify which represents the claim.*

(b) *Determine when a type I or type II error occurs for a hypothesis test of the claim.*

(c) *Determine whether the hypothesis test is left-tailed, right-tailed, or two-tailed. Explain your reasoning.*

(d) *Explain how you should interpret a decision that rejects the null hypothesis.*

(e) *Explain how you should interpret a decision that fails to reject the null hypothesis.*

7. A news outlet reports that the proportion of Americans who support plans to order deep cuts in executive compensation at companies that have received federal bailout funds is 71%. *(Source: ABC News)*

8. An agricultural cooperative guarantees that the mean shelf life of a certain type of dried fruit is at least 400 days.

9. A soup maker says that the standard deviation of the sodium content in one serving of a certain soup is no more than 50 milligrams. *(Adapted from Consumer Reports)*

10. An energy bar maker claims that the mean number of grams of carbohydrates in one bar is less than 25.

■ SECTION 7.2

In Exercises 11 and 12, find the P-value for the indicated hypothesis test with the given standardized test statistic z. Decide whether to reject H_0 for the given level of significance α.

11. Left-tailed test, $z = -0.94$, $\alpha = 0.05$

12. Two-tailed test, $z = 2.57$, $\alpha = 0.10$

In Exercises 13 and 14, use a P-value to test the claim about the population mean μ using the given sample statistics. State your decision for $\alpha = 0.10$, $\alpha = 0.05$, and $\alpha = 0.01$ levels of significance. If convenient, use technology.

13. Claim: $\mu \leq 0.05$; Sample statistics: $\bar{x} = 0.057$, $s = 0.018$, $n = 32$

14. Claim: $\mu \neq 230$; Sample statistics: $\bar{x} = 216.5$, $s = 17.3$, $n = 48$

In Exercises 15–18, find the critical value(s) for the indicated z-test and level of significance α. Include a graph with your answer.

15. Left-tailed test, $\alpha = 0.02$

16. Two-tailed test, $\alpha = 0.005$

17. Right-tailed test, $\alpha = 0.025$

18. Two-tailed test, $\alpha = 0.08$

19. Fail to reject H_0 because $-1.645 < z < 1.645$.

20. Reject H_0 because $z > 1.645$.

21. Fail to reject H_0 because $-1.645 < z < 1.645$.

22. Reject H_0 because $z < -1.645$.

23. Reject H_0. There is enough evidence at the 5% level of significance to reject the claim.

24. Reject H_0. There is enough evidence at the 3% level of significance to support the claim.

25. Fail to reject H_0. There is not enough evidence at the 1% level of significance to support the claim.

26. Fail to reject H_0. There is not enough evidence at the 10% level of significance to reject the claim.

27. Fail to reject H_0. There is not enough evidence at the 1% level of significance to reject the U.S. Department of Agriculture's claim that the mean cost of raising a child from birth to age 2 by husband-wife families in rural areas is $10,380.

28. See Selected Answers, page A125.

29. $-2.093, 2.093$ **30.** 2.998

31. -2.977 **32.** $-2.718, 2.718$

33. Fail to reject H_0. There is not enough evidence at the 5% level of significance to support the claim.

34. Reject H_0. There is enough evidence at the 0.5% level of significance to support the claim.

35. Fail to reject H_0. There is not enough evidence at the 10% level of significance to reject the claim.

36. Reject H_0. There is enough evidence at the 2% level of significance to reject the claim.

37. Reject H_0. There is enough evidence at the 1% level of significance to reject the claim.

38. Fail to reject H_0. There is not enough evidence at the 2.5% level of significance to support the claim.

In Exercises 19–22, state whether each standardized test statistic z allows you to reject the null hypothesis. Explain your reasoning.

19. $z = 1.631$

20. $z = 1.723$

21. $z = -1.464$

22. $z = -1.655$

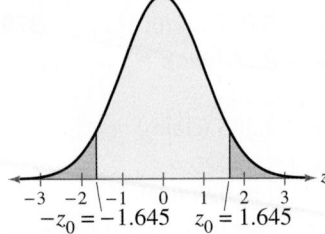

In Exercises 23–26, use a z-test to test the claim about the population mean μ at the given level of significance α using the given sample statistics. If convenient, use technology.

23. Claim: $\mu \le 45$; $\alpha = 0.05$. Sample statistics: $\bar{x} = 47.2$, $s = 6.7$, $n = 42$

24. Claim: $\mu \ne 8.45$; $\alpha = 0.03$. Sample statistics: $\bar{x} = 7.88$, $s = 1.75$, $n = 60$

25. Claim: $\mu < 5.500$; $\alpha = 0.01$. Sample statistics: $\bar{x} = 5.497$, $s = 0.011$, $n = 36$

26. Claim: $\mu = 7450$; $\alpha = 0.10$. Sample statistics: $\bar{x} = 7495$, $s = 243$, $n = 57$

In Exercises 27 and 28, test the claim about the population mean μ using rejection region(s) or a P-value. Interpret your decision in the context of the original claim. If convenient, use technology.

27. The U.S. Department of Agriculture claims that the mean cost of raising a child from birth to age 2 by husband-wife families in rural areas is $10,380. A random sample of 800 children (age 2) has a mean cost of $10,240 with a standard deviation of $1561. At $\alpha = 0.01$, is there enough evidence to reject the claim? *(Adapted from U.S. Department of Agriculture Center for Nutrition Policy and Promotion)*

28. A tourist agency in Hawaii claims the mean daily cost of meals and lodging for a family of 4 traveling in Hawaii is at most $650. You work for a consumer protection advocate and want to test this claim. In a random sample of 45 families of 4 traveling in Hawaii, the mean daily cost of meals and lodging is $657 with a standard deviation of $40. At $\alpha = 0.05$, do you have enough evidence to reject the tourist agency's claim? *(Adapted from American Automobile Association)*

■ SECTION 7.3

In Exercises 29–32, find the critical value(s) for the indicated t-test, level of significance α, and sample size n.

29. Two-tailed test, $\alpha = 0.05$, $n = 20$ **30.** Right-tailed test, $\alpha = 0.01$, $n = 8$

31. Left-tailed test, $\alpha = 0.005$, $n = 15$ **32.** Two-tailed test, $\alpha = 0.02$, $n = 12$

In Exercises 33–38, use a t-test to test the claim about the population mean μ at the given level of significance α using the given sample statistics. For each claim, assume the population is normally distributed. If convenient, use technology.

33. Claim: $\mu \ne 95$; $\alpha = 0.05$. Sample statistics: $\bar{x} = 94.1$, $s = 1.53$, $n = 12$

34. Claim: $\mu > 12,700$; $\alpha = 0.005$. Sample statistics: $\bar{x} = 12,855$, $s = 248$, $n = 21$

35. Claim: $\mu \ge 0$; $\alpha = 0.10$. Sample statistics: $\bar{x} = -0.45$, $s = 1.38$, $n = 16$

36. Claim: $\mu = 4.20$; $\alpha = 0.02$. Sample statistics: $\bar{x} = 4.61$, $s = 0.33$, $n = 9$

37. Claim: $\mu \le 48$; $\alpha = 0.01$. Sample statistics: $\bar{x} = 52$, $s = 2.5$, $n = 7$

38. Claim: $\mu < 850$; $\alpha = 0.025$. Sample statistics: $\bar{x} = 875$, $s = 25$, $n = 14$

39. Fail to reject H_0. There is not enough evidence at the 10% level of significance to reject the advertisement's claim that the mean monthly cost of joining a health club is $25.

40. Reject H_0. There is enough evidence at the 2.5% level of significance to reject the fitness magazine's claim that the mean cost of a yoga session is no more than $14.

41. There is not enough evidence at the 1% level of significance to reject the education publication's claim that the mean expenditure per student in public elementary and secondary schools is at least $10,200.

42. See Selected Answers, page A125.

43. Can use normal distribution.

Fail to reject H_0. There is not enough evidence at the 5% level of significance to reject the claim.

44. Can use normal distribution.

Reject H_0. There is enough evidence at the 1% level of significance to support the claim.

45. Can use normal distribution.

Fail to reject H_0. There is not enough evidence at the 8% level of significance to support the claim.

46. Can use normal distribution.

Reject H_0. There is enough evidence at the 3% level of significance to reject the claim.

47. Cannot use normal distribution.

48. Can use normal distribution.

Fail to reject H_0. There is not enough evidence at the 1% level of significance to support the claim.

49. Can use normal distribution.

Fail to reject H_0. There is not enough evidence at the 2% level of significance to support the claim.

50. Can use normal distribution.

Fail to reject H_0. There is not enough evidence at the 10% level of significance to reject the claim.

In Exercises 39 and 40, use a t-test to test the claim. Interpret your decision in the context of the original claim. For each claim, assume the population is normally distributed. If convenient, use technology.

39. A fitness magazine advertises that the mean monthly cost of joining a health club is $25. You work for a consumer advocacy group and are asked to test this claim. You find that a random sample of 18 clubs has a mean monthly cost of $26.25 and a standard deviation of $3.23. At $\alpha = 0.10$, do you have enough evidence to reject the advertisement's claim?

40. A fitness magazine claims that the mean cost of a yoga session is no more than $14. You work for a consumer advocacy group and are asked to test this claim. You find that a random sample of 29 yoga sessions has a mean cost of $15.59 and a standard deviation of $2.60. At $\alpha = 0.025$, do you have enough evidence to reject the magazine's claim?

In Exercises 41 and 42, use a t-statistic and its P-value to test the claim about the population mean μ using the given data. Interpret your decision in the context of the original claim. For each claim, assume the population is normally distributed. If convenient, use technology.

41. An education publication claims that the mean expenditure per student in public elementary and secondary schools is at least $10,200. You want to test this claim. You randomly select 16 school districts and find the average expenditure per student. The results are listed below. At $\alpha = 0.01$, can you reject the publication's claim? *(Adapted from National Center for Education Statistics)*

9,242	10,857	10,377	8,935	9,545	9,974
9,847	10,641	9,364	10,157	9,784	9,962
10,065	9,851	9,763	9,969		

42. A restaurant association says the typical household in the United States spends a mean amount of $2698 per year on food away from home. You are a consumer reporter for a national publication and want to test this claim. A random sample of 28 U.S. households has a mean amount spent on food away from home of $2764 and a standard deviation of $322. At $\alpha = 0.05$, do you have enough evidence to reject the association's claim? *(Adapted from U.S. Bureau of Labor Statistics)*

■ SECTION 7.4

In Exercises 43–50, decide whether the normal sampling distribution can be used to approximate the binomial distribution. If it can, use the z-test to test the claim about the population proportion p at the given level of significance α using the given sample statistics. If convenient, use technology.

43. Claim: $p = 0.15$; $\alpha = 0.05$. Sample statistics: $\hat{p} = 0.09$, $n = 40$

44. Claim: $p < 0.70$; $\alpha = 0.01$. Sample statistics: $\hat{p} = 0.50$, $n = 68$

45. Claim: $p < 0.09$; $\alpha = 0.08$. Sample statistics: $\hat{p} = 0.07$, $n = 75$

46. Claim: $p = 0.65$; $\alpha = 0.03$. Sample statistics: $\hat{p} = 0.76$, $n = 116$

47. Claim: $p \geq 0.04$; $\alpha = 0.10$. Sample statistics: $\hat{p} = 0.03$, $n = 30$

48. Claim: $p \neq 0.34$; $\alpha = 0.01$. Sample statistics: $\hat{p} = 0.29$, $n = 60$

49. Claim: $p \neq 0.24$; $\alpha = 0.02$. Sample statistics: $\hat{p} = 0.32$, $n = 50$

50. Claim: $p \leq 0.80$; $\alpha = 0.10$. Sample statistics: $\hat{p} = 0.85$, $n = 43$

51. Reject H_0. There is enough evidence at the 2% level of significance to support the polling agency's claim that over 16% of U.S. adults are without health care coverage.

52. Fail to reject H_0. There is not enough evidence at the 5% level of significance to reject the medical researcher's claim that the rate of false positives for a Western blot assay is 2%.

53. 30.144

54. 3.565; 29.819

55. 63.167

56. 1.145

57. Reject H_0. There is enough evidence at the 10% level of significance to support the claim.

58. Fail to reject H_0. There is not enough evidence at the 2.5% level of significance to reject the claim.

59. Fail to reject H_0. There is not enough evidence at the 5% level of significance to reject the claim.

60. Fail to reject H_0. There is not enough evidence at the 1% level of significance to support the claim.

61. Reject H_0. There is enough evidence at the 0.5% level of significance to reject the bolt manufacturer's claim that the variance is at most 0.01.

62. Fail to reject H_0. There is not enough evidence at the 1% level of significance to reject the restaurant's claim that the standard deviation of the lengths of serving times is 3 minutes.

63. You can reject H_0 at the 5% level of significance because $\chi^2 = 43.94 > 41.923$.

In Exercises 51 and 52, test the claim about the population proportion p. Interpret your decision in the context of the original claim. If convenient, use technology.

51. A polling agency reports that over 16% of U.S. adults are without health care coverage. In a random survey of 1420 U.S. adults, 256 said they did not have health care coverage. At $\alpha = 0.02$, is there enough evidence to support the agency's claim? *(Source: The Gallup Poll)*

52. The Western blot assay is a blood test for the presence of HIV. It has been found that this test sometimes gives false positive results for HIV. A medical researcher claims that the rate of false positives is 2%. A recent study of 300 randomly selected U.S. blood donors who do not have HIV found that 3 received a false positive test result. At $\alpha = 0.05$, is there enough evidence to reject the researcher's claim? *(Adapted from Centers for Disease Control and Prevention)*

■ SECTION 7.5

In Exercises 53–56, find the critical value(s) for the indicated χ^2-test for a population variance, sample size n, and level of significance α.

53. Right-tailed test, $n = 20$, $\alpha = 0.05$

54. Two-tailed test, $n = 14$, $\alpha = 0.01$

55. Right-tailed test, $n = 51$, $\alpha = 0.10$

56. Left-tailed test, $n = 6$, $\alpha = 0.05$

In Exercises 57–60, use a χ^2-test to test the claim about the population variance σ^2 or standard deviation σ at the given level of significance α and using the given sample statistics. For each claim, assume the population is normally distributed.

57. Claim: $\sigma^2 > 2$; $\alpha = 0.10$. Sample statistics: $s^2 = 2.95$, $n = 18$

58. Claim: $\sigma^2 \leq 60$; $\alpha = 0.025$. Sample statistics: $s^2 = 72.7$, $n = 15$

59. Claim: $\sigma = 1.25$; $\alpha = 0.05$. Sample statistics: $s = 1.03$, $n = 6$

60. Claim: $\sigma \neq 0.035$; $\alpha = 0.01$. Sample statistics: $s = 0.026$, $n = 16$

In Exercises 61 and 62, test the claim about the population variance or standard deviation. Interpret your decision in the context of the original claim. For each claim, assume the population is normally distributed.

61. A bolt manufacturer makes a type of bolt to be used in airtight containers. The manufacturer needs to be sure that all of its bolts are very similar in width, so it sets an upper tolerance limit for the variance of bolt width at 0.01. A random sample of the widths of 28 bolts has a variance of 0.064. At $\alpha = 0.005$, is there enough evidence to reject the manufacturer's claim?

62. A restaurant claims that the standard deviation of the lengths of serving times is 3 minutes. A random sample of 27 serving times has a standard deviation of 3.9 minutes. At $\alpha = 0.01$, is there enough evidence to reject the restaurant's claim?

63. In Exercise 62, is there enough evidence to reject the restaurant's claim at the $\alpha = 0.05$ level? Explain.

7 CHAPTER QUIZ

1. (a) H_0: $\mu \geq 170$ (claim);
   H_a: $\mu < 170$

   (b) One-tailed because the alternative hypothesis contains $<$; z-test because $n \geq 30$

   (c) $z_0 = -1.88$;
   Rejection region: $z < -1.88$

   (d) -2.59

   (e) Reject H_0.

   (f) There is enough evidence at the 3% level of significance to reject the service's claim that the mean consumption of vegetables and melons by people in the United States is at least 170 pounds per person.

2. See Odd Answers, page A86.

3. (a) H_0: $p \leq 0.10$ (claim);
   H_a: $p > 0.10$

   (b) One-tailed because the alternative hypothesis contains $>$; z-test because $np > 5$ and $nq > 5$

   (c) $z_0 = 1.75$;
   Rejection region: $z > 1.75$

   (d) 0.75

   (e) Fail to reject H_0.

   (f) There is not enough evidence at the 4% level of significance to reject the microwave oven maker's claim that no more than 10% of its microwaves need repair during the first 5 years of use.

4. See Odd Answers, page A86.

5. See Odd Answers, page A86.

6. See Odd Answers, page A86.

Take this quiz as you would take a quiz in class. After you are done, check your work against the answers given in the back of the book. If convenient, use technology.

For this quiz, do the following.

(a) *Write the claim mathematically. Identify H_0 and H_a.*

(b) *Determine whether the hypothesis test is one-tailed or two-tailed and whether to use a z-test, a t-test, or a χ^2-test. Explain your reasoning.*

(c) *If necessary, find the critical value(s) and identify the rejection region(s).*

(d) *Find the appropriate test statistic. If necessary, find the P-value.*

(e) *Decide whether to reject or fail to reject the null hypothesis.*

(f) *Interpret the decision in the context of the original claim.*

1. A research service estimates that the mean annual consumption of vegetables and melons by people in the United States is at least 170 pounds per person. A random sample of 360 people in the United States has a mean consumption of vegetables and melons of 168.5 pounds per year and a standard deviation of 11 pounds. At $\alpha = 0.03$, is there enough evidence to reject the service's claim that the mean consumption of vegetables and melons by people in the United States is at least 170 pounds per person? *(Adapted from U.S. Department of Agriculture)*

2. A hat company states that the mean hat size for a male is at least 7.25. A random sample of 12 hat sizes has a mean of 7.15 and a standard deviation of 0.27. At $\alpha = 0.05$, can you reject the company's claim that the mean hat size for a male is at least 7.25? Assume the population is normally distributed.

3. A maker of microwave ovens advertises that no more than 10% of its microwaves need repair during the first 5 years of use. In a random sample of 57 microwaves that are 5 years old, 13% needed repairs. At $\alpha = 0.04$, can you reject the maker's claim that no more than 10% of its microwaves need repair during the first five years of use? *(Adapted from Consumer Reports)*

4. A state school administrator says that the standard deviation of SAT critical reading test scores is 112. A random sample of 19 SAT critical reading test scores has a standard deviation of 143. At $\alpha = 0.10$, test the administrator's claim. What can you conclude? Assume the population is normally distributed. *(Adapted from The College Board)*

5. A government agency reports that the mean amount of earnings for full-time workers ages 25 to 34 with a master's degree is $62,569. In a random sample of 15 full-time workers ages 25 to 34 with a master's degree, the mean amount of earnings is $59,231 and the standard deviation is $5945. Is there enough evidence to reject the agency's claim? Use a P-value and $\alpha = 0.05$. Assume the population is normally distributed. *(Adapted from U.S. Census Bureau)*

6. A tourist agency in Kansas claims the mean daily cost of meals and lodging for a family of 4 traveling in the state is $201. You work for a consumer protection advocate and want to test this claim. In a random sample of 35 families of 4 traveling in Kansas, the mean daily cost of meals and lodging is $216 and the standard deviation is $30. Do you have enough evidence to reject the agency's claim? Use a P-value and $\alpha = 0.05$. *(Adapted from American Automobile Association)*

PUTTING IT ALL TOGETHER

Real Statistics — Real Decisions

In the 1970s and 1980s, PepsiCo, maker of Pepsi®, began airing television commercials in which it claimed more cola drinkers preferred Pepsi® over Coca-Cola® in a blind taste test. The Coca-Cola Company, maker of Coca-Cola®, was the market leader in soda sales. After the television ads began airing, Pepsi® sales increased and began rivaling Coca-Cola® sales.

Assume the claim is that more than 50% of cola drinkers preferred Pepsi® over Coca-Cola®. You work for an independent market research firm and are asked to test this claim.

■ EXERCISES

1. How Would You Do It?

(a) When PepsiCo performed this challenge, PepsiCo representatives went to shopping malls to obtain their sample. Do you think this type of sampling is representative of the population? Explain.

(b) What sampling technique would you use to select the sample for your study?

(c) Identify possible flaws or biases in your study.

2. Testing a Proportion

In your study, 280 out of 560 cola drinkers prefer Pepsi® over Coca-Cola®. Using these results, test the claim that more than 50% of cola drinkers prefer Pepsi® over Coca-Cola®. Use $\alpha = 0.05$. Interpret your decision in the context of the original claim. Does the decision support PepsiCo's claim?

3. Labeling Influence

The Baylor College of Medicine decided to replicate this taste test by monitoring brain activity while conducting the test on participants. They also wanted to see if brand labeling would affect the results. When participants were shown which cola they were sampling, Coca-Cola® was preferred by 75% of the participants. What conclusions can you draw from this study?

4. Your Conclusions

(a) Why do you think PepsiCo used a blind taste test?

(b) Do you think brand image or taste has more influence on consumer preferences for cola?

(c) What other factors may influence consumer preferences besides taste and branding?

TECHNOLOGY

MINITAB   EXCEL   TI-83/84 PLUS

THE CASE OF THE VANISHING WOMEN

53% ➡ 29% ➡ 9% ➡ 0%

From 1966 to 1968, Dr. Benjamin Spock and others were tried for conspiracy to violate the Selective Service Act by encouraging resistance to the Vietnam War. By a series of three selections, no women ended up being on the jury. In 1969, Hans Zeisel wrote an article in *The University of Chicago Law Review* using statistics and hypothesis testing to argue that the jury selection was biased against Dr. Spock. Dr. Spock was a well-known pediatrician and author of books about raising children. Millions of mothers had read his books and followed his advice. Zeisel argued that, by keeping women off the jury, the court prejudiced the verdict.

The jury selection process for Dr. Spock's trial is shown at the right.

Stage 1. The clerk of the Federal District Court selected 350 people "at random" from the Boston City Directory. The directory contained several hundred names, 53% of whom were women. However, only 102 of the 350 people selected were women.

Stage 2. The trial judge, Judge Ford, selected 100 people "at random" from the 350 people. This group was called a venire and it contained only nine women.

Stage 3. The court clerk assigned numbers to the members of the venire and, one by one, they were interrogated by the attorneys for the prosecution and defense until 12 members of the jury were chosen. At this stage, only one potential female juror was questioned, and she was eliminated by the prosecutor under his quota of peremptory challenges (for which he did not have to give a reason).

■ EXERCISES

1. The MINITAB display below shows a hypothesis test for a claim that the proportion of women in the city directory is $p = 0.53$. In the test, $n = 350$ and $\hat{p} \approx 0.2914$. Should you reject the claim? What is the level of significance? Explain.

2. In Exercise 1, you rejected the claim that $p = 0.53$. But this claim was true. What type of error is this?

3. If you reject a true claim with a level of significance that is virtually zero, what can you infer about the randomness of your sampling process?

4. Describe a hypothesis test for Judge Ford's "random" selection of the venire. Use a claim of

$$p = \frac{102}{350} \approx 0.2914.$$

(a) Write the null and alternative hypotheses.

(b) Use a technology tool to perform the test.

(c) Make a decision.

(d) Interpret the decision in the context of the original claim. Could Judge Ford's selection of 100 venire members have been random?

MINITAB

Test and CI for One Proportion

Test of p = 0.53 vs p not = 0.53

Sample	X	N	Sample p	99 % CI	Z-Value	P-Value
1	102	350	0.291429	(0.228862, 0.353995)	-8.94	0.000

Using the normal approximation.

Extended solutions are given in the *Technology Supplement*.
Technical instruction is provided for MINITAB, Excel, and the TI-83/84 Plus.

7 USING TECHNOLOGY TO PERFORM HYPOTHESIS TESTS

Here are some MINITAB and TI-83/84 Plus printouts for some of the examples in this chapter.

(See Example 5, page 375.)

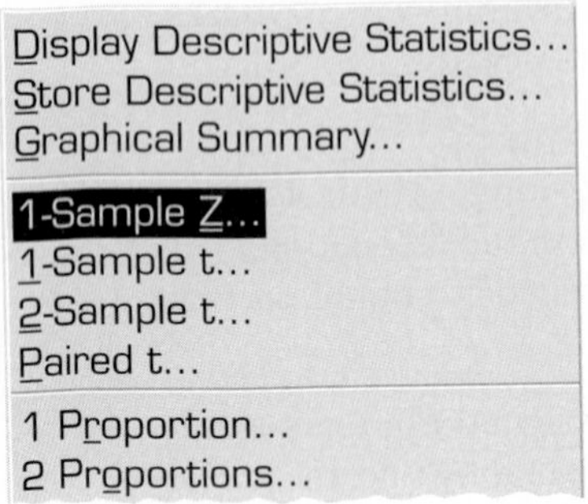

Display Descriptive Statistics...
Store Descriptive Statistics...
Graphical Summary...

1-Sample Z...
1-Sample t...
2-Sample t...
Paired t...

1 Proportion...
2 Proportions...

MINITAB

One-Sample Z

Test of mu = 22500 vs not = 22500
The assumed standard deviation = 3015

N	Mean	SE Mean	95% CI	Z	P
30	21545	550	(20466, 22624)	−1.73	0.083

(See Example 4, page 390.)

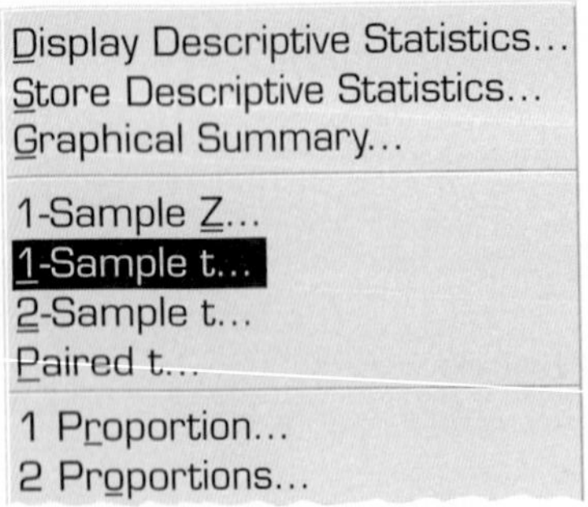

Display Descriptive Statistics...
Store Descriptive Statistics...
Graphical Summary...

1-Sample Z...
1-Sample t...
2-Sample t...
Paired t...

1 Proportion...
2 Proportions...

MINITAB

One-Sample T

Test of mu = 20500 vs < 20500

N	Mean	StDev	SE Mean	95% Upper Bound	T	P
14	19850	1084	290	20363	−2.24	0.021

(See Example 2, page 400.)

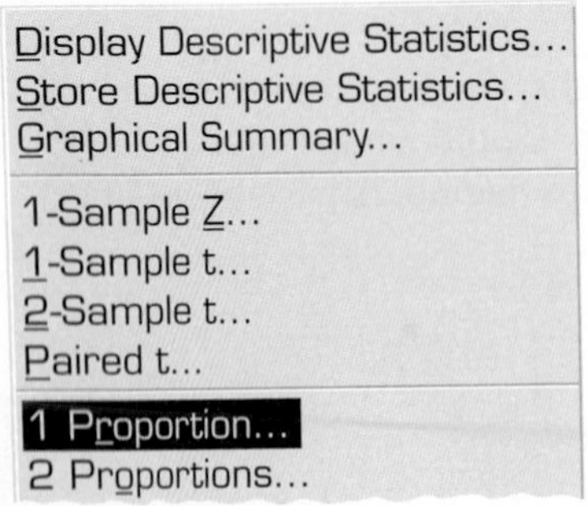

Display Descriptive Statistics...
Store Descriptive Statistics...
Graphical Summary...

1-Sample Z...
1-Sample t...
2-Sample t...
Paired t...

1 Proportion...
2 Proportions...

MINITAB

Test and CI for One Proportion

Test of p = 0.25 vs p not = 0.25

Sample	X	N	Sample p	90% CI	Z-Value	P-Value
1	42	200	0.210000	(0.162627, 0.257373)	−1.31	0.191

Using the normal approximation.

(See Example 9, page 379.)

(See Example 5, page 391.)

(See Example 1, page 399.)

TI-83/84 PLUS

EDIT CALC **TESTS**

1: Z–Test...
2: T–Test...
3: 2–SampZTest...
4: 2–SampTTest...
5: 1–PropZTest...
6: 2–PropZTest...
7↓ZInterval...

TI-83/84 PLUS

EDIT CALC **TESTS**

1: Z–Test...
2: T–Test...
3: 2–SampZTest...
4: 2–SampTTest...
5: 1–PropZTest...
6: 2–PropZTest...
7↓ZInterval...

TI-83/84 PLUS

EDIT CALC **TESTS**

1: Z–Test...
2: T–Test...
3: 2–SampZTest...
4: 2–SampTTest...
5: 1–PropZTest...
6: 2–PropZTest...
7↓ZInterval...

TI-83/84 PLUS

Z-Test

Inpt: Data **Stats**
μ_0: 68000
σ: 5500
$\bar{x}$: 66900
n: 30
μ: $\neq \mu_0$ **$<\mu_0$** $>\mu_0$
Calculate Draw

TI-83/84 PLUS

T-Test

Inpt: Data **Stats**
μ_0: 6.8
$\bar{x}$: 6.7
Sx: .24
n: 19
μ: **$\neq \mu_0$** $<\mu_0$ $>\mu_0$
Calculate Draw

TI-83/84 PLUS

1-PropZTest

p_0: .5
x: 39
n: 100
prop$\neq p_0$ **$<p_0$** $>p_0$
Calculate Draw

TI-83/84 PLUS

Z-Test

$\mu < 68000$
z= −1.095445115
p= .1366608782
$\bar{x}$= 66900
n= 30

TI-83/84 PLUS

T-Test

$\mu \neq 6.8$
t= −1.816207893
p= .0860316039
$\bar{x}$= 6.7
Sx= .24
n= 19

TI-83/84 PLUS

1-PropZTest

prop< .5
z= −2.2
p= .0139033989
$\hat{p}$= .39
n= 100

TI-83/84 PLUS

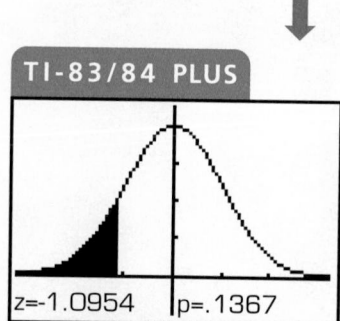

z=-1.0954   p=.1367

TI-83/84 PLUS

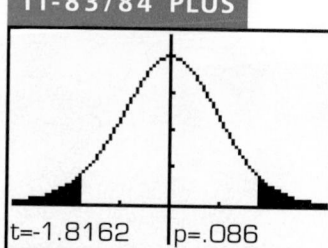

t=-1.8162   p=.086

TI-83/84 PLUS

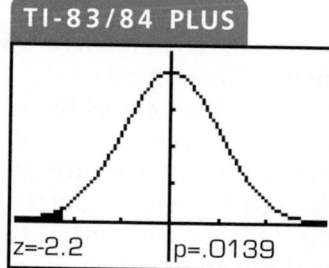

z=-2.2   p=.0139

8 HYPOTHESIS TESTING WITH TWO SAMPLES

CHAPTER

The *National Youth Tobacco Survey (NYTS)*, a report published by the Centers for Disease Control and Prevention, provides information on the most widely used tobacco products among U.S. students. One of the national health objectives is to reduce current cigarette use among high school students.

In Chapter 6, you were introduced to inferential statistics and you learned how to form confidence intervals to estimate a parameter. Then, in Chapter 7, you learned how to test a claim about a population parameter, basing your decision on sample statistics and their distributions.

The *National Youth Tobacco Survey (NYTS)* is a study conducted by the Centers for Disease Control and Prevention to provide information about student use of tobacco products. As part of this study in a recent year, a random sample of 7000 U.S. male high school students was surveyed. The following proportions were found.

Male High School Students ($n = 7000$)

Characteristic	Frequency	Proportion
Smoke cigarettes (at least one in the last 30 days)	1484	0.212
Smoke cigars (at least one in the last 30 days)	1162	0.166
Use smokeless tobacco (at least once in the last 30 days)	770	0.110

WHERE YOU'RE GOING ▶▶

In this chapter, you will continue your study of inferential statistics and hypothesis testing. Now, however, instead of testing a hypothesis about a single population, you will learn how to test a hypothesis that compares two populations.

For instance, in the *NYTS* study a random sample of 7489 U.S. female high school students was also surveyed. Here are the study's findings for this second group.

Female High School Students ($n = 7489$)

Characteristic	Frequency	Proportion
Smoke cigarettes (at least one in the last 30 days)	1378	0.184
Smoke cigars (at least one in the last 30 days)	539	0.072
Use smokeless tobacco (at least once in the last 30 days)	112	0.015

From these two samples, can you conclude that there is a difference in the proportion of high school students who smoke cigarettes, smoke cigars, or use smokeless tobacco among males and among females? Or, might the differences in the proportions be due to chance?

In this chapter, you will learn that you can answer these questions by testing the hypothesis that the two proportions are equal. For the proportions of students who use smokeless tobacco, for instance, you can conclude that the proportion of male high school students is different from the proportion of female high school students.

8.1 Testing the Difference Between Means (Large Independent Samples)

Independent and Dependent Samples ▸ An Overview of Two-Sample Hypothesis Testing ▸ Two-Sample z-Test for the Difference Between Means

▸ INDEPENDENT AND DEPENDENT SAMPLES

In Chapter 7, you studied methods for testing a claim about the value of a population parameter. In this chapter, you will learn how to test a claim comparing parameters from two populations. When you compare the means of two different populations, the method you use to sample as well as the sample sizes will determine the type of test you will use.

DEFINITION

Two samples are **independent** if the sample selected from one population is not related to the sample selected from the second population. Two samples are **dependent** if each member of one sample corresponds to a member of the other sample. Dependent samples are also called **paired samples** or **matched samples.**

Independent Samples

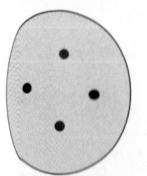 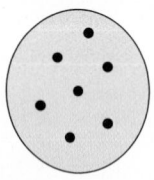

Sample 1   Sample 2

Dependent Samples

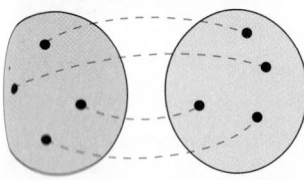

Sample 1   Sample 2

EXAMPLE 1

▸ Independent and Dependent Samples

Classify each pair of samples as independent or dependent and justify your answer.

1. Sample 1: Weights of 65 college students before their freshman year begins
   Sample 2: Weights of the same 65 college students after their freshman year

2. Sample 1: Scores for 38 adult males on a psychological screening test for attention-deficit hyperactivity disorder
   Sample 2: Scores for 50 adult females on a psychological screening test for attention-deficit hyperactivity disorder

▸ Solution

1. These samples are dependent. Because the weights of the same students are taken, the samples are related. The samples can be paired with respect to each student.

2. These samples are independent. It is not possible to form a pairing between the members of samples, the sample sizes are different, and the data represent scores for different individuals.

▸ Try It Yourself 1

Classify each pair of samples as independent or dependent.

1. Sample 1: Systolic blood pressures of 30 adult females
   Sample 2: Systolic blood pressures of 30 adult males
2. Sample 1: Midterm exam scores of 14 chemistry students
   Sample 2: Final exam scores of the same 14 chemistry students

a. Determine whether the samples are *independent* or *dependent*.
b. *Explain* your reasoning.

Answer: Page A43

▶ AN OVERVIEW OF TWO-SAMPLE HYPOTHESIS TESTING

In this section, you will learn how to test a claim comparing the means of two different populations using independent samples.

For instance, suppose you are developing a marketing plan for an Internet service provider and want to determine whether there is a difference in the amounts of time male and female college students spend online each day. The only way you can conclude with certainty that there is a difference is to take a census of all college students, calculate the mean daily times male students and female students spend online, and find the difference. Of course, it is not practical to take such a census. However, you can still determine with some degree of certainty whether such a difference exists.

You can begin by assuming that there is no difference in the mean times of the two populations. That is, $\mu_1 - \mu_2 = 0$. Then, by taking a random sample from each population, you can perform a two-sample hypothesis test using the test statistic $\overline{x}_1 - \overline{x}_2$. Suppose you obtain the following results.

INSIGHT

The members in the two samples are not matched or paired, so the samples are independent.

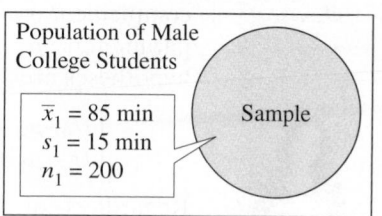

Population of Male College Students

$\overline{x}_1 = 85$ min
$s_1 = 15$ min
$n_1 = 200$

Sample

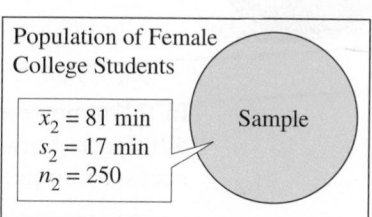

Population of Female College Students

$\overline{x}_2 = 81$ min
$s_2 = 17$ min
$n_2 = 250$

Sample

The graph below shows the sampling distribution of $\overline{x}_1 - \overline{x}_2$ for many similar samples taken from two populations for which $\mu_1 - \mu_2 = 0$. From the graph, you can see that it is quite unlikely to obtain sample means that differ by 4 minutes if the actual difference is 0. The difference of the sample means would be more than 2.5 standard errors from the hypothesized difference of 0! So, you can conclude that there is a significant difference in the amounts of time male college students and female college students spend online each day.

Sampling Distribution

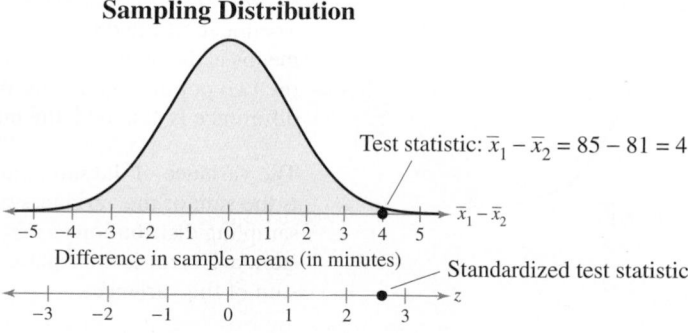

Test statistic: $\overline{x}_1 - \overline{x}_2 = 85 - 81 = 4$

$\overline{x}_1 - \overline{x}_2$

Difference in sample means (in minutes)

Standardized test statistic

z

It is important to remember that when you perform a two-sample hypothesis test using independent samples, you are testing a claim concerning the difference between the parameters in two populations, not the values of the parameters themselves.

DEFINITION

For a two-sample hypothesis test with independent samples,

1. the **null hypothesis H_0** is a statistical hypothesis that usually states there is no difference between the parameters of two populations. The null hypothesis always contains the symbol $\leq$, $=$, or $\geq$.
2. the **alternative hypothesis H_a** is a statistical hypothesis that is true when H_0 is false. The alternative hypothesis contains the symbol $>$, $\neq$, or $<$.

STUDY TIP

You can also write the null and alternative hypotheses as follows.

$$\begin{cases} H_0: \mu_1 - \mu_2 = 0 \\ H_a: \mu_1 - \mu_2 \neq 0 \end{cases}$$

$$\begin{cases} H_0: \mu_1 - \mu_2 \leq 0 \\ H_a: \mu_1 - \mu_2 > 0 \end{cases}$$

$$\begin{cases} H_0: \mu_1 - \mu_2 \geq 0 \\ H_a: \mu_1 - \mu_2 < 0 \end{cases}$$

To write the null and alternative hypotheses for a two-sample hypothesis test with independent samples, translate the claim made about the population parameters from a verbal statement to a mathematical statement. Then, write its complementary statement. For instance, if the claim is about two population parameters μ_1 and μ_2, then some possible pairs of null and alternative hypotheses are

$$\begin{cases} H_0: \mu_1 = \mu_2 \\ H_a: \mu_1 \neq \mu_2 \end{cases}, \quad \begin{cases} H_0: \mu_1 \leq \mu_2 \\ H_a: \mu_1 > \mu_2 \end{cases}, \quad \text{and} \quad \begin{cases} H_0: \mu_1 \geq \mu_2 \\ H_a: \mu_1 < \mu_2 \end{cases}.$$

Regardless of which hypotheses you use, you always assume there is no difference between the population means, or $\mu_1 = \mu_2$.

▶ TWO-SAMPLE z-TEST FOR THE DIFFERENCE BETWEEN MEANS

In the remainder of this section, you will learn how to perform a z-test for the difference between two population means μ_1 and μ_2. Three conditions are necessary to perform such a test.

1. The samples must be randomly selected.

2. The samples must be independent.

3. Each sample size must be at least 30 or, if not, each population must have a normal distribution with a known standard deviation.

If these requirements are met, then the **sampling distribution for $\bar{x}_1 - \bar{x}_2$, the difference of the sample means,** is a normal distribution with mean and standard error as follows.

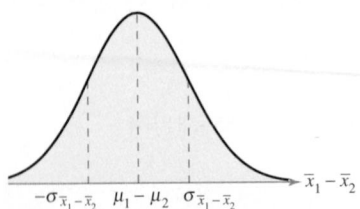

Sampling Distribution for $\bar{x}_1 - \bar{x}_2$

In Words	In Symbols
The mean of the difference of the sample means is the assumed difference between the two population means. When no difference is assumed, the mean is 0.	Mean $= \mu_{\bar{x}_1 - \bar{x}_2}$ $= \mu_{\bar{x}_1} - \mu_{\bar{x}_2}$ $= \mu_1 - \mu_2$
The variance of the sampling distribution is the sum of the variances of the individual sampling distributions for $\bar{x}_1$ and $\bar{x}_2$. The standard error is the square root of the sum of the variances.	Standard error $= \sigma_{\bar{x}_1 - \bar{x}_2}$ $= \sqrt{\sigma_{\bar{x}_1}^2 + \sigma_{\bar{x}_2}^2}$ $= \sqrt{\dfrac{\sigma_1^2}{n_1} + \dfrac{\sigma_2^2}{n_2}}$

Note to Instructor

Point out the similarities between
two-sample hypothesis tests and
the one-sample tests presented in
Chapter 7. Also, point out that
P-values can be used in this chapter
in much the same way as they were
used in Chapter 7.

Because the sampling distribution for $\bar{x}_1 - \bar{x}_2$ is a normal distribution, you can use the z-test to test the difference between two population means μ_1 and μ_2. Notice that the standardized test statistic takes the form of

$$z = \frac{(\text{Observed difference}) - (\text{Hypothesized difference})}{\text{Standard error}}.$$

TWO-SAMPLE z-TEST FOR THE DIFFERENCE BETWEEN MEANS

A **two-sample z-test** can be used to test the difference between two population means μ_1 and μ_2 when a large sample (at least 30) is randomly selected from each population and the samples are *independent*. The **test statistic** is $\bar{x}_1 - \bar{x}_2$, and the **standardized test statistic** is

$$z = \frac{(\bar{x}_1 - \bar{x}_2) - (\mu_1 - \mu_2)}{\sigma_{\bar{x}_1 - \bar{x}_2}} \quad \text{where} \quad \sigma_{\bar{x}_1 - \bar{x}_2} = \sqrt{\frac{\sigma_1^2}{n_1} + \frac{\sigma_2^2}{n_2}}.$$

When the samples are large, you can use s_1 and s_2 in place of σ_1 and σ_2. If the samples are not large, you can still use a two-sample z-test, provided the populations are normally distributed and the population standard deviations are known.

PICTURING THE WORLD

There are about 112,900 public elementary and secondary school teachers in Georgia and about 119,300 in Ohio. In a survey, 200 public elementary and secondary school teachers in each state were asked to report their salary. The results were as follows.
(Adapted from National Education Association)

Georgia
$\bar{x}_1 = \$49,900$
$s_1 = \$6935$

Ohio
$\bar{x}_2 = \$51,900$
$s_2 = \$6584$

Is there enough evidence to conclude that there is a difference in the mean salaries of public elementary and secondary school teachers in Georgia and Ohio using $\alpha = 0.01$?

There is enough evidence at the 1% level of significance to conclude that there is a difference in the mean salaries of public elementary and secondary school teachers in Georgia and Ohio.

If the null hypothesis states $\mu_1 = \mu_2$, $\mu_1 \leq \mu_2$, or $\mu_1 \geq \mu_2$, then $\mu_1 = \mu_2$ is assumed and the expression $\mu_1 - \mu_2$ is equal to 0 in the preceding test.

GUIDELINES

Using a Two-Sample z-Test for the Difference Between Means (Large Independent Samples)

IN WORDS	IN SYMBOLS
1. State the claim mathematically and verbally. Identify the null and alternative hypotheses.	State H_0 and H_a.
2. Specify the level of significance.	Identify α.
3. Determine the critical value(s).	Use Table 4 in Appendix B.
4. Determine the rejection region(s).	
5. Find the standardized test statistic and sketch the sampling distribution.	$z = \dfrac{(\bar{x}_1 - \bar{x}_2) - (\mu_1 - \mu_2)}{\sigma_{\bar{x}_1 - \bar{x}_2}}$
6. Make a decision to reject or fail to reject the null hypothesis.	If z is in the rejection region, reject H_0. Otherwise, fail to reject H_0.
7. Interpret the decision in the context of the original claim.	

A hypothesis test for the difference between means can also be performed using P-values. Use the guidelines above, skipping Steps 3 and 4. After finding the standardized test statistic, use the Standard Normal Table to calculate the P-value. Then make a decision to reject or fail to reject the null hypothesis. If P is less than or equal to α, reject H_0. Otherwise, fail to reject H_0.

Sample Statistics for Credit Card Debt

New York	Texas
$\bar{x}_1 = \$4446.25$	$\bar{x}_2 = \$4567.24$
$s_1 = \$1045.70$	$s_2 = \$1361.95$
$n_1 = 250$	$n_2 = 250$

EXAMPLE 2

See TI-83/84 Plus steps on page 479.

▶ A Two-Sample z-Test for the Difference Between Means

A credit card watchdog group claims that there is a difference in the mean credit card debts of households in New York and Texas. The results of a random survey of 250 households from each state are shown at the left. The two samples are independent. Do the results support the group's claim? Use $\alpha = 0.05$. *(Adapted from PlasticRewards.com)*

▶ Solution

The claim is "there is a difference in the mean credit card debts of households in New York and Texas." So, the null and alternative hypotheses are

$$H_0: \mu_1 = \mu_2 \quad \text{and} \quad H_a: \mu_1 \neq \mu_2. \text{ (Claim)}$$

Because the test is a two-tailed test and the level of significance is $\alpha = 0.05$, the critical values are $-z_0 = -1.96$ and $z_0 = 1.96$. The rejection regions are $z < -1.96$ and $z > 1.96$. Because both samples are large, s_1 and s_2 can be used in place of σ_1 and σ_2 to calculate the standard error.

$$\sigma_{\bar{x}_1 - \bar{x}_2} \approx \sqrt{\frac{s_1^2}{n_1} + \frac{s_2^2}{n_2}}$$

$$= \sqrt{\frac{1045.70^2}{250} + \frac{1361.95^2}{250}} \approx 108.5983$$

The standardized test statistic is

$$z = \frac{(\bar{x}_1 - \bar{x}_2) - (\mu_1 - \mu_2)}{\sigma_{\bar{x}_1 - \bar{x}_2}} \quad \text{Use the z-test.}$$

$$\approx \frac{(4446.25 - 4567.24) - 0}{108.5983} \quad \text{Assume } \mu_1 = \mu_2, \text{ so } \mu_1 - \mu_2 = 0.$$

$$\approx -1.11.$$

The graph at the left shows the location of the rejection regions and the standardized test statistic z. Because z is not in the rejection region, you should fail to reject the null hypothesis.

Interpretation   There is not enough evidence at the 5% level of significance to support the group's claim that there is a difference in the mean credit card debts of households in New York and Texas.

STUDY TIP

In Example 2, you can also use a *P*-value to perform the hypothesis test. For instance, the test is a two-tailed test, so the *P*-value is equal to twice the area to the left of $z = -1.11$, or

$$2(0.1335) = 0.267.$$

Because $0.267 > 0.05$, you should fail to reject H_0.

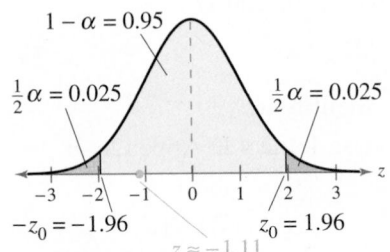

$1 - \alpha = 0.95$

$\frac{1}{2}\alpha = 0.025$ $\frac{1}{2}\alpha = 0.025$

$-z_0 = -1.96$ $z_0 = 1.96$

$z \approx -1.11$

▶ Try It Yourself 2

A survey indicates that the mean annual wages for forensic science technicians working for local and state governments are $53,300 and $51,910, respectively. The survey includes a randomly selected sample of size 100 from each government branch. The sample standard deviations are $6200 (local) and $5575 (state). The two samples are independent. At $\alpha = 0.10$, is there enough evidence to conclude that there is a difference in the mean annual wages? *(Adapted from U.S. Bureau of Labor Statistics)*

a. Identify the *claim* and state H_0 and H_a.
b. Identify the *level of significance* α.
c. Find the *critical values* and identify the *rejection regions*.
d. Use the *z*-test to find the *standardized test statistic z*. Sketch a graph.
e. *Decide* whether to reject the null hypothesis.
f. *Interpret* the decision in the context of the original claim.

Answer: Page A43

Sample Statistics for Daily Cost of Meals and Lodging for Two Adults

Texas	Virgina
$\bar{x}_1 = \$216$	$\bar{x}_2 = \$222$
$s_1 = \$18$	$s_2 = \$24$
$n_1 = 50$	$n_2 = 35$

EXAMPLE 3   Report 33

▶ Using Technology to Perform a Two-Sample *z*-Test

A travel agency claims that the average daily cost of meals and lodging for vacationing in Texas is less than the same average cost for vacationing in Virginia. The table at the left shows the results of a random survey of vacationers in each state. The two samples are independent. At $\alpha = 0.01$, is there enough evidence to support the claim? $[H_0: \mu_1 \geq \mu_2$ and $H_a: \mu_1 < \mu_2$ (claim)]   *(Adapted from American Automobile Association)*

▶ Solution

The top two displays show how to set up the hypothesis test using a TI-83/84 Plus. The remaining displays show the possible results, depending on whether you select *Calculate* or *Draw*.

STUDY TIP

Note that the TI-83/84 Plus displays $P \approx 0.1051$. Because $P > \alpha$, you should fail to reject the null hypothesis.

TI-83/84 PLUS

```
2-SampZTest
Inpt: Data  Stats
 σ1:18
 σ2:24
 x̄1:216
 n1:50
 x̄2:222
↓n2:35
```

TI-83/84 PLUS

```
2-SampZTest
↑σ2:24
 x̄1:216
 n1:50
 x̄2:222
 n2:35
 μ1:≠μ2  <μ2  >μ2
 Calculate Draw
```

Note to Instructor

Students might be confused about designating Texas as population 1 and Virginia as population 2. Explain that the final result of the test would be identical if the subscripts were reversed. Show that reversing the subscripts would produce a right-tailed test for which the test statistic is 1.25, the critical value is $z = 2.33$, and the rejection region is $z > 2.33$.

TI-83/84 PLUS

```
2-SampZTest
 μ1<μ2
 z= -1.252799556
 p= .1051393971
 x̄1= 216
 x̄2= 222
↓n1:50
```

TI-83/84 PLUS

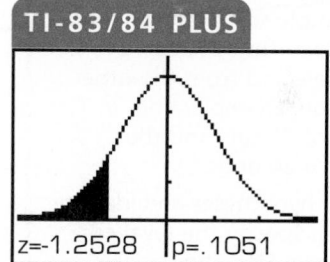

```
z=-1.2528  p=.1051
```

Because the test is a left-tailed test and $\alpha = 0.01$, the rejection region is $z < -2.33$. The standardized test statistic $z \approx -1.25$ is not in the rejection region, so you should fail to reject the null hypothesis.

Interpretation   There is not enough evidence at the 1% level of significance to support the travel agency's claim.

▶ Try It Yourself 3

A travel agency claims that the average daily cost of meals and lodging for vacationing in Alaska is greater than the same average cost for vacationing in Colorado. The table at the left shows the results of a random survey of vacationers in each state. The two samples are independent. At $\alpha = 0.05$, is there enough evidence to support the claim? *(Adapted from American Automobile Association)*

Sample Statistics for Daily Cost of Meals and Lodging for Two Adults

Alaska	Colorado
$\bar{x}_1 = \$274$	$\bar{x}_2 = \$271$
$s_1 = \$22$	$s_2 = \$18$
$n_1 = 150$	$n_2 = 200$

a. Use a TI-83/84 Plus to find the *test statistic* or the *P-value*.
b. *Decide* whether to reject the null hypothesis.
c. *Interpret* the decision in the context of the original claim.

Answer: Page A43

8.1 EXERCISES

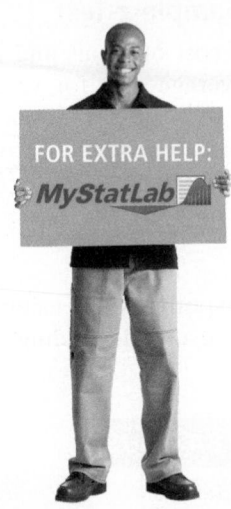

FOR EXTRA HELP:
MyStatLab

1. Two samples are dependent if each member of one sample corresponds to a member of the other sample. Example: The weights of 22 people before starting an exercise program and the weights of the same 22 people 6 weeks after starting the exercise program.

Two samples are independent if the sample selected from one population is not related to the sample selected from the other population. Example: The weights of 25 cats and the weights of 20 dogs.

2. State the hypotheses and identify the claim. Specify the level of significance. Find the critical value(s) and identify the rejection region(s). Find the standardized test statistic. Make a decision and interpret it in the context of the claim.

3. Use *P*-values.

4. (1) The samples must be randomly selected. (2) The samples must be independent. (3) $n_1 \geq 30$ and $n_2 \geq 30$, or each population must have a normal distribution with a known standard deviation.

5. Independent because different students were sampled.

6. Dependent because the same students were sampled.

7. Dependent because the same football players were sampled.

■ BUILDING BASIC SKILLS AND VOCABULARY

1. What is the difference between two samples that are dependent and two samples that are independent? Give an example of each.

2. Explain how to perform a two-sample *z*-test for the difference between the means of two populations using large independent samples.

3. Describe another way you can perform a hypothesis test for the difference between the means of two populations using large independent samples.

4. What conditions are necessary in order to use the *z*-test to test the difference between two population means?

In Exercises 5–12, classify the two given samples as independent or dependent. Explain your reasoning.

5. Sample 1: The SAT scores of 35 high school students who did not take an SAT preparation course
   Sample 2: The SAT scores of 40 high school students who did take an SAT preparation course

6. Sample 1: The SAT scores of 44 high school students
   Sample 2: The SAT scores of the same 44 high school students after taking an SAT preparation course

7. Sample 1: The maximum bench press weights for 53 football players
   Sample 2: The maximum bench press weights for the same football players after completing a weight lifting program

8. Sample 1: The IQ scores of 60 females
   Sample 2: The IQ scores of 60 males

9. Sample 1: The average speed of 23 powerboats using an old hull design
   Sample 2: The average speed of 14 powerboats using a new hull design

10. Sample 1: The commute times of 10 workers who use their own vehicles
   Sample 2: The commute times of the same 10 workers when they use public transportation

11. The table shows the braking distances (in feet) for each of four different sets of tires with the car's antilock braking system (ABS) on and with ABS off. The tests were done on ice with cars traveling at 15 miles per hour. *(Source: Consumer Reports)*

Tire set	1	2	3	4
Braking distance with ABS	42	55	43	61
Braking distance without ABS	58	67	59	75

12. The table shows the heart rates (in beats per minute) of five people before and after exercising.

Person	1	2	3	4	5
Heart rate before exercising	65	72	85	78	93
Heart rate after exercising	127	135	140	136	150

8. Independent because different individuals were sampled.

9. Independent because different boats were sampled.

10. Dependent because the same workers were sampled.

11. Dependent because the same tire sets were sampled.

12. Dependent because the same people were sampled.

13. (a) 2

(b) 2.95

(c) In the rejection region.

(d) Reject H_0. There is enough evidence at the 1% level of significance to reject the claim.

14. (a) 5

(b) 1.15

(c) Not in the rejection region.

(d) Fail to reject H_0. There is not enough evidence at the 10% level of significance to support the claim.

15. (a) 3

(b) 0.18

(c) Not in the rejection region.

(d) Fail to reject H_0. There is not enough evidence at the 5% level of significance to support the claim.

16. (a) 109

(b) 5.29

(c) In the rejection region.

(d) Reject H_0. There is enough evidence at the 3% level of significance to reject the claim.

17. Fail to reject H_0. There is not enough evidence at the 1% level of significance to support the claim.

18. Reject H_0. There is enough evidence at the 5% level of significance to support the claim.

19. Reject H_0.

20. Fail to reject H_0.

In Exercises 13–16, (a) find the test statistic, (b) find the standardized test statistic, (c) decide whether the standardized test statistic is in the rejection region, and (d) decide whether you should reject or fail to reject the null hypothesis. The samples are random and independent.

13. Claim: $\mu_1 = \mu_2$; $\alpha = 0.01$. Sample statistics: $\bar{x}_1 = 16$, $s_1 = 3.4$, $n_1 = 30$ and $\bar{x}_2 = 14$, $s_2 = 1.5$, $n_2 = 30$

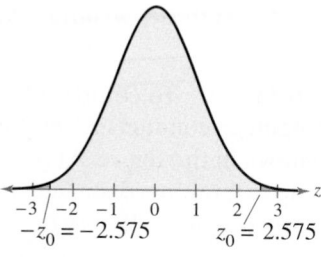

FIGURE FOR EXERCISE 13

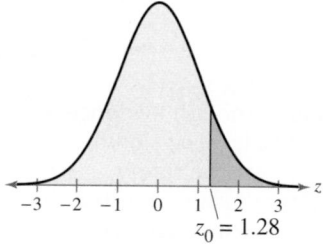

FIGURE FOR EXERCISE 14

14. Claim: $\mu_1 > \mu_2$; $\alpha = 0.10$. Sample statistics: $\bar{x}_1 = 500$, $s_1 = 40$, $n_1 = 100$ and $\bar{x}_2 = 495$, $s_2 = 15$, $n_2 = 75$

15. Claim: $\mu_1 < \mu_2$; $\alpha = 0.05$. Sample statistics: $\bar{x}_1 = 2435$, $s_1 = 75$, $n_1 = 35$ and $\bar{x}_2 = 2432$, $s_2 = 105$, $n_2 = 90$

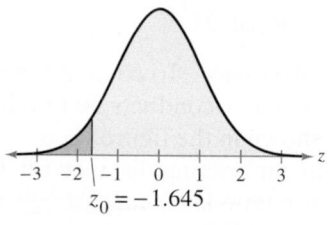

FIGURE FOR EXERCISE 15

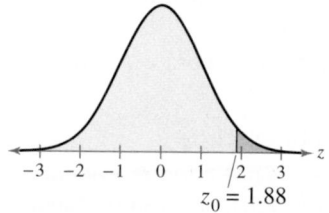

FIGURE FOR EXERCISE 16

16. Claim: $\mu_1 \leq \mu_2$; $\alpha = 0.03$. Sample statistics: $\bar{x}_1 = 5004$, $s_1 = 136$, $n_1 = 144$ and $\bar{x}_2 = 4895$, $s_2 = 215$, $n_2 = 156$

In Exercises 17 and 18, use the given sample statistics to test the claim about the difference between two population means μ_1 and μ_2 at the given level of significance α.

17. Claim: $\mu_1 > \mu_2$; $\alpha = 0.01$. Sample statistics: $\bar{x}_1 = 5.2$, $s_1 = 0.2$, $n_1 = 45$ and $\bar{x}_2 = 5.5$, $s_2 = 0.3$, $n_2 = 37$

18. Claim: $\mu_1 \neq \mu_2$; $\alpha = 0.05$. Sample statistics: $\bar{x}_1 = 52$, $s_1 = 2.5$, $n_1 = 70$ and $\bar{x}_2 = 45$, $s_2 = 5.5$, $n_2 = 60$

In Exercises 19 and 20, use the TI-83/84 Plus display to make a decision at the given level of significance. Make your decision using the standardized test statistic and using the P-value. Assume the sample sizes are equal.

19. $\alpha = 0.05$

```
2-SampZTest
μ1≠μ2
z=2.956485408
P=.0031118068
x̄1=2500
x̄2=2425
↓n1=120
```

20. $\alpha = 0.01$

```
2-SampZTest
μ1>μ2
z=1.941656065
P=.0260893059
x̄1=44
x̄2=42
↓n1=50
```

21. (a) The claim is "the mean braking distances are different for the two types of tires."

$H_0: \mu_1 = \mu_2;$
$H_a: \mu_1 \neq \mu_2$ (claim)

(b) $-z_0 = -1.645$, $z_0 = 1.645$;

Rejection regions:
$z < -1.645$, $z > 1.645$

(c) -2.786

(d) Reject H_0.

(e) There is enough evidence at the 10% level of significance to support the safety engineer's claim that the mean braking distances are different for the two types of tires.

22. See Selected Answers, page A125.

23. (a) The claim is "Region A's average wind speed is greater than Region B's."

$H_0: \mu_1 \leq \mu_2;$
$H_a: \mu_1 > \mu_2$ (claim)

(b) $z_0 = 1.645$;
Rejection region: $z > 1.645$

(c) 1.53

(d) Fail to reject H_0.

(e) There is not enough evidence at the 5% level of significance to conclude that Region A's average wind speed is greater than Region B's.

24. See Selected Answers, page A125.

25. (a) The claim is "male and female high school students have equal ACT scores."

$H_0: \mu_1 = \mu_2$ (claim);
$H_a: \mu_1 \neq \mu_2$

(b) $-z_0 = -2.575$, $z_0 = 2.575$;

Rejection regions:
$z < -2.575$, $z > 2.575$

(c) 0.202

(d) Fail to reject H_0.

(e) There is not enough evidence at the 1% level of significance to reject the claim that male and female high school students have equal ACT scores.

26. See Selected Answers, page A125.

27. (a) The claim is "the average home sales price in Dallas, Texas is the same as in Austin, Texas."

$H_0: \mu_1 = \mu_2$ (claim);
$H_a: \mu_1 \neq \mu_2$

▪ USING AND INTERPRETING CONCEPTS

Testing the Difference Between Two Means *In Exercises 21–34, (a) identify the claim and state H_0 and H_a, (b) find the critical value(s) and identify the rejection region(s), (c) find the standardized test statistic z, (d) decide whether to reject or fail to reject the null hypothesis, and (e) interpret the decision in the context of the original claim. If convenient, use technology to solve the problem. In each exercise, assume the samples are randomly selected and independent.*

21. **Braking Distances** To compare the braking distances for two types of tires, a safety engineer conducts 35 braking tests for each type. The results of the tests are shown in the figure. At $\alpha = 0.10$, can the engineer support the claim that the mean braking distances are different for the two types of tires? *(Adapted from Consumer Reports)*

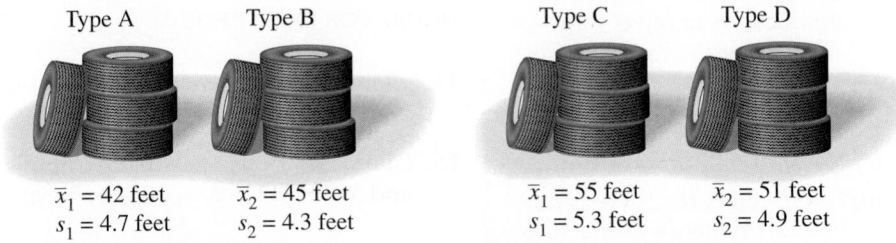

Type A   Type B   Type C   Type D

$\bar{x}_1 = 42$ feet $\bar{x}_2 = 45$ feet $\bar{x}_1 = 55$ feet $\bar{x}_2 = 51$ feet
$s_1 = 4.7$ feet $s_2 = 4.3$ feet $s_1 = 5.3$ feet $s_2 = 4.9$ feet

FIGURE FOR EXERCISE 21   FIGURE FOR EXERCISE 22

22. **Braking Distances** To compare the braking distances for two types of tires, a safety engineer conducts 50 braking tests for each type. The results of the tests are shown in the figure. At $\alpha = 0.10$, can the engineer support the claim that the mean braking distance for Type C is greater than the mean braking distance for Type D? *(Adapted from Consumer Reports)*

23. **Wind Energy** An energy company wants to choose between two regions in a state to install energy-producing wind turbines. The company will choose Region A if its average wind speed is greater than that of Region B. To test the regions, the average wind speed is calculated for 60 days in each region. The results of the company's research are shown in the figure. At $\alpha = 0.05$, should the company choose Region A?

Region A   Region B   Region C   Region D

$\bar{x}_1 = 13.2$ mph $\bar{x}_2 = 12.5$ mph $\bar{x}_1 = 14.0$ mph $\bar{x}_2 = 15.1$ mph
$s_1 = 2.3$ mph $s_2 = 2.7$ mph $s_1 = 2.9$ mph $s_2 = 3.3$ mph

FIGURE FOR EXERCISE 23   FIGURE FOR EXERCISE 24

24. **Wind Energy** An energy company wants to choose between two regions in a state to install energy-producing wind turbines. A researcher claims that the wind speeds in Region C and Region D are equal. The company tests the regions by calculating the average wind speed for 75 days in Region C and 80 days in Region D. The results of the company's research are shown in the figure. At $\alpha = 0.03$, can the company reject the researcher's claim?

(b) $-z_0 = -1.645$, $z_0 = 1.645$;
Rejection regions:
$z < -1.645$, $z > 1.645$

(c) -1.30

(d) Fail to reject H_0.

(e) There is not enough evidence at the 10% level of significance to reject the real estate agency's claim that the average home sales price in Dallas, Texas is the same as in Austin, Texas.

28. See Selected Answers, page A125.

29. (a) The claim is "the average home sales price in Dallas, Texas is the same as in Austin, Texas."

$H_0: \mu_1 = \mu_2$ (claim);
$H_a: \mu_1 \neq \mu_2$

(b) $-z_0 = -1.645$, $z_0 = 1.645$;
Rejection regions:
$z < -1.645$, $z > 1.645$

(c) 1.86

(d) Reject H_0.

(e) There is enough evidence at the 10% level of significance to reject the real estate agency's claim that the average home sales price in Dallas, Texas is the same as in Austin, Texas.

The new samples do lead to a different conclusion.

30. See Selected Answers, page A125.

31. (a) The claim is "children ages 6–17 spent more time watching television in 1981 than children ages 6–17 do today."

$H_0: \mu_1 \leq \mu_2$;
$H_a: \mu_1 > \mu_2$ (claim)

(b) $z_0 = 1.96$;
Rejection region: $z > 1.96$

(c) 3.01

(d) Reject H_0.

(e) There is enough evidence at the 2.5% level of significance to support the sociologist's claim that children ages 6–17 spent more time watching television in 1981 than children ages 6–17 do today.

25. **ACT Scores**   The mean ACT score for 43 male high school students is 21.1 and the standard deviation is 5.0. The mean ACT score for 56 female high school students is 20.9 and the standard deviation is 4.7. At $\alpha = 0.01$, can you reject the claim that male and female high school students have equal ACT scores? *(Adapted from ACT Inc.)*

26. **ACT Scores**   A guidance counselor claims that high school students in a college preparation program have higher ACT scores than those in a general program. The mean ACT score for 49 high school students who are in a college preparation program is 22.2 and the standard deviation is 4.8. The mean ACT score for 44 high school students who are in a general program is 20.0 and the standard deviation is 5.4. At $\alpha = 0.10$, can you support the guidance counselor's claim? *(Adapted from ACT Inc.)*

27. **Home Prices**   A real estate agency says that the average home sales price in Dallas, Texas is the same as in Austin, Texas. The average home sales price for 35 homes in Dallas is $240,993 and the standard deviation is $25,875. The average home sales price for 35 homes in Austin is $249,237 and the standard deviation is $27,110. At $\alpha = 0.10$, is there enough evidence to reject the agency's claim? *(Adapted from RealtyTrac Inc.)*

28. **Money Spent Eating Out**   A restaurant association says that households in the United States headed by people under the age of 25 spend less on food away from home than do households headed by people ages 65–74. The mean amount spent by 30 households headed by people under the age of 25 is $1876 and the standard deviation is $113. The mean amount spent by 30 households headed by people ages 65–74 is $1878 and the standard deviation is $85. At $\alpha = 0.05$, can you support the restaurant association's claim? *(Adapted from Bureau of Labor Statistics)*

29. **Home Prices**   Refer to Exercise 27. Two more samples are taken, one from Dallas and one from Austin. For 50 homes in Dallas, $\bar{x}_1 = \$247,245$ and $s_1 = \$22,740$. For 50 homes in Austin, $\bar{x}_2 = \$239,150$ and $s_2 = \$20,690$. Use $\alpha = 0.10$. Do the new samples lead to a different conclusion?

30. **Money Spent Eating Out**   Refer to Exercise 28. Two more samples are taken, one from each age group. For 40 households headed by people under the age of 25, $\bar{x}_1 = \$2015$ and $s_1 = \$124$. For 40 households headed by people ages 65–74, $\bar{x}_2 = \$2099$ and $s_2 = \$111$. Use $\alpha = 0.05$. Do the new samples lead to a different conclusion?

31. **Watching More TV?**   A sociologist claims that children ages 6–17 spent more time watching television in 1981 than children ages 6–17 do today. A study was conducted in 1981 to find the time that children ages 6–17 spent watching television on weekdays. The results (in hours per weekday) are shown below.

2.0	2.5	2.1	2.3	2.1	1.6	2.6	2.1	2.1	2.4
2.1	2.1	1.5	1.7	2.1	2.3	2.5	3.3	2.2	2.9
1.5	1.9	2.4	2.2	1.2	3.0	1.0	2.1	1.9	2.2

Recently, a similar study was conducted. The results are shown below.

2.9	1.8	0.9	1.6	2.0	1.7	2.5	1.1	1.6	2.0
1.4	1.7	1.7	1.9	1.6	1.7	1.2	2.0	2.6	1.6
1.5	2.5	1.6	2.1	1.7	1.8	1.1	1.4	1.2	2.3

At $\alpha = 0.025$, can you support the sociologist's claim? *(Adapted from University of Michigan's Institute for Social Research)*

32. See Selected Answers, page A125.

33. (a) The claim is "there is no difference in the mean washer diameter manufactured by two different methods."

$H_0: \mu_1 = \mu_2$ (claim);
$H_a: \mu_1 \neq \mu_2$

(b) $-z_0 = -2.575$, $z_0 = 2.575$;

Rejection regions:
$z < -2.575$, $z > 2.575$

(c) 64.978

(d) Reject H_0.

(e) There is enough evidence at the 1% level of significance to reject the production engineer's claim that there is no difference in the mean washer diameter manufactured by two different methods.

34. See Selected Answers, page A126.

35. They are equivalent through algebraic manipulation of the equation.

$\mu_1 = \mu_2 \Rightarrow \mu_1 - \mu_2 = 0$

36. See Selected Answers, page A126.

37. **Hypothesis test results:**

μ_1 : mean of population 1
(Std. Dev. = 5.4)

μ_2 : mean of population 2
(Std. Dev. = 7.5)

$\mu_1 - \mu_2$: mean difference

$H_0 : \mu_1 - \mu_2 = 0$

$H_A : \mu_1 - \mu_2 \neq 0$

See Odd Answers, page A88.

$P = 0.0031 < 0.01$, so reject H_0.

There is enough evidence at the 1% level of significance to support the claim.

38. See Selected Answers, page A126.

39. **Hypothesis test results:**

μ_1 : mean of population 1
(Std. Dev. = 0.92)

μ_2 : mean of population 2
(Std. Dev. = 0.73)

$\mu_1 - \mu_2$: mean difference

$H_0 : \mu_1 - \mu_2 = 0$

$H_A : \mu_1 - \mu_2 < 0$

See Odd Answers, page A88.

$P = 0.0492 < 0.05$, so reject H_0.

There is enough evidence at the 5% level of significance to reject the claim.

32. Spending More Time Studying? A sociologist thinks that middle school boys spent less time studying in 1981 than middle school boys do today. A study was conducted in 1981 to find the time that middle school boys spent studying on weekdays. The results (in minutes per weekday) are shown below.

31.9	35.4	28.0	39.1	30.5	31.9	33.0	29.6	35.7	30.2
38.8	35.9	37.1	36.2	32.6	36.9	24.2	28.5	28.7	41.1
33.8	32.1	28.7	35.4	36.6	34.3	35.5	34.2	33.8	25.3
27.7	21.9	30.0	36.8	26.9					

Recently, a similar study was conducted. The results are shown below.

44.7	54.6	41.1	46.7	43.0	46.6	42.9	48.7	50.0	47.9
47.2	58.0	51.0	41.1	49.6	51.3	39.0	45.6	49.8	54.4
47.1	45.5	52.8	49.4	47.2	54.8	40.2	45.4	48.6	50.0
51.5	55.0	44.7	42.2	52.0					

At $\alpha = 0.03$, can you support the sociologist's claim? *(Adapted from University of Michigan's Institute for Social Research)*

33. **Washer Diameters** A production engineer claims that there is no difference in the mean washer diameter manufactured by two different methods. The first method produces washers with the following diameters (in inches).

0.861	0.864	0.882	0.887	0.858	0.879	0.887	0.876	0.870
0.894	0.884	0.882	0.869	0.859	0.887	0.875	0.863	0.887
0.882	0.862	0.906	0.880	0.877	0.864	0.873	0.860	0.866
0.869	0.877	0.863	0.875	0.883	0.872	0.879	0.861	

The second method produces washers with these diameters (in inches).

0.705	0.703	0.715	0.711	0.690	0.720	0.702	0.686	0.704
0.712	0.718	0.695	0.708	0.695	0.699	0.715	0.691	0.696
0.680	0.703	0.697	0.694	0.714	0.694	0.672	0.688	0.700
0.715	0.709	0.698	0.696	0.700	0.706	0.695	0.715	

At $\alpha = 0.01$, can you reject the production engineer's claim?

34. **Nut Diameters** A production engineer claims that there is no difference in the mean nut diameter manufactured by two different methods. The first method produces nuts with the following diameters (in centimeters).

3.330	3.337	3.329	3.354	3.325	3.343	3.333	3.347	3.332
3.358	3.353	3.335	3.341	3.331	3.327	3.326	3.337	3.336
3.323	3.347	3.329	3.345	3.329	3.338	3.353	3.339	3.338
3.338	3.350	3.320	3.364	3.340	3.348	3.339	3.336	3.321
3.316	3.352	3.320	3.336					

The second method produces nuts with these diameters (in centimeters).

3.513	3.490	3.498	3.504	3.483	3.512	3.494	3.514	3.495
3.489	3.493	3.499	3.497	3.495	3.496	3.485	3.506	3.517
3.484	3.498	3.522	3.505	3.501	3.491	3.500	3.499	3.475
3.486	3.501	3.496	3.504	3.513	3.511	3.501	3.487	3.508
3.515	3.505	3.496	3.505					

At $\alpha = 0.04$, can you reject the production engineer's claim?

35. **Getting at the Concept** Explain why the null hypothesis $H_0: \mu_1 = \mu_2$ is equivalent to the null hypothesis $H_0: \mu_1 - \mu_2 = 0$.

40. Hypothesis test results:

μ_1: mean of population 1
(Std. Dev. = 193.8)

μ_2: mean of population 2
(Std. Dev. = 129.25)

$\mu_1 - \mu_2$: mean difference

$H_0: \mu_1 - \mu_2 = 0$

$H_A: \mu_1 - \mu_2 > 0$

See Selected Answers, page A126.

$P = 0.0278 < 0.03$, so reject H_0.

There is enough evidence at the 3% level of significance to reject the claim.

41. $H_0: \mu_1 - \mu_2 = -9$ (claim);
$H_a: \mu_1 - \mu_2 \neq -9$

Fail to reject H_0.

There is not enough evidence at the 1% level of significance to reject the claim that children spend 9 hours a week more in day care or preschool today than in 1981.

42. $H_0: \mu_1 - \mu_2 = -2$ (claim);
$H_a: \mu_1 - \mu_2 \neq -2$

Fail to reject H_0. There is not enough evidence at the 5% level of significance to reject the claim that the mean time per week children ages 6–8 watch television is 2 hours less than that of children ages 9–11.

43. $H_0: \mu_1 - \mu_2 \leq 10{,}000$;
$H_a: \mu_1 - \mu_2 > 10{,}000$ (claim)

Reject H_0. There is enough evidence at the 5% level of significance to support the claim that the difference in mean annual salaries of microbiologists in Maryland and California is more than $10,000.

Microbiologists in
Maryland

$\bar{x}_1 = \$94{,}980$
$s_1 = \$8795$
$n_1 = 42$

Microbiologists in
California

$\bar{x}_2 = \$80{,}830$
$s_2 = \$9250$
$n_2 = 38$

FIGURE FOR EXERCISE 43

36. Getting at the Concept   Explain why the null hypothesis $H_0: \mu_1 \geq \mu_2$ is equivalent to the null hypothesis $H_0: \mu_1 - \mu_2 \geq 0$.

SC   *In Exercises 37–40, use StatCrunch to help you test the claim about the difference between two population means μ_1 and μ_2 at the given level of significance α using the given sample statistics. Assume the samples are randomly selected and independent.*

37. Claim: $\mu_1 \neq \mu_2$; $\alpha = 0.01$. Sample statistics: $\bar{x}_1 = 64$, $s_1 = 5.4$, $n_1 = 50$ and $\bar{x}_2 = 60$, $s_2 = 7.5$, $n_2 = 45$

38. Claim: $\mu_1 > \mu_2$; $\alpha = 0.10$. Sample statistics: $\bar{x}_1 = 158.9$, $s_1 = 20.8$, $n_1 = 80$ and $\bar{x}_2 = 155.3$, $s_2 = 24.6$, $n_2 = 80$

39. Claim: $\mu_1 \geq \mu_2$; $\alpha = 0.05$. Sample statistics: $\bar{x}_1 = 4.75$, $s_1 = 0.92$, $n_1 = 35$ and $\bar{x}_2 = 5.07$, $s_2 = 0.73$, $n_2 = 40$

40. Claim: $\mu_1 \leq \mu_2$; $\alpha = 0.03$. Sample statistics: $\bar{x}_1 = 1740.28$, $s_1 = 193.80$, $n_1 = 100$ and $\bar{x}_2 = 1695.70$, $s_2 = 129.25$, $n_2 = 100$

■ EXTENDING CONCEPTS

Testing a Difference Other Than Zero   *Sometimes a researcher is interested in testing a difference in means other than zero. For instance, you may want to determine if children today spend an average of 9 hours a week more in day care (or preschool) than children did 20 years ago. In Exercises 41–44, you will test the difference between two means using a null hypothesis of $H_0: \mu_1 - \mu_2 = k$, $H_0: \mu_1 - \mu_2 \geq k$, or $H_0: \mu_1 - \mu_2 \leq k$. The standardized test statistic is still*

$$z = \frac{(\bar{x}_1 - \bar{x}_2) - (\mu_1 - \mu_2)}{\sigma_{\bar{x}_1 - \bar{x}_2}} \quad where \quad \sigma_{\bar{x}_1 - \bar{x}_2} = \sqrt{\frac{\sigma_1^2}{n_1} + \frac{\sigma_2^2}{n_2}}.$$

41. Time in Day Care or Preschool   In 1981, a study of 70 randomly selected children under 3 years old found that the mean length of time spent in day care or preschool per week was 11.5 hours with a standard deviation of 3.8 hours. A recent study of 65 randomly selected children under 3 years old found that the mean length of time spent in day care or preschool per week was 20 hours and the standard deviation was 6.7 hours. At $\alpha = 0.01$, test the claim that children spend 9 hours a week more in day care or preschool today than they did in 1981. *(Adapted from University of Michigan's Institute for Social Research)*

42. Time Watching TV   A recent study of 48 randomly selected children ages 6–8 found that the mean length of time spent watching television each week was 12.95 hours and the standard deviation was 4.31 hours. The mean time 56 randomly selected children ages 9–11 spent watching television each week was 15.02 hours and the standard deviation was 4.99 hours. At $\alpha = 0.05$, test the claim that the mean time per week children ages 6–8 watch television is 2 hours less than that of children ages 9–11. *(Adapted from University of Michigan's Institute for Social Research)*

43. Microbiologist Salaries   Is the difference between the mean annual salaries of microbiologists in Maryland and California more than $10,000? To decide, you select a random sample of microbiologists from each state. The results of each survey are shown in the figure. At $\alpha = 0.05$, what should you conclude? *(Adapted from U.S. Bureau of Labor Statistics)*

Physician Assistants in New Jersey

$\overline{x}_1 = \$88,540$
$s_1 = \$8225$
$n_1 = 30$

Registered Nurses in New Jersey

$\overline{x}_2 = \$72,870$
$s_2 = \$7640$
$n_2 = 32$

TABLES FOR EXERCISE 44

44. See Selected Answers, page A126.

45. $-3.6 < \mu_1 - \mu_2 < -0.2$

46. $1.5 < \mu_1 - \mu_2 < 2.1$

47. $H_0: \mu_1 \geq \mu_2$; $H_a: \mu_1 < \mu_2$ (claim)

Reject H_0. There is enough evidence at the 5% level of significance to support the claim. You should recommend the DASH diet and exercise program over the traditional diet and exercise program because the mean systolic blood pressure was significantly lower in the DASH program.

Irinotecan	Fluorouracil
$\overline{x}_1 = 10.3$	$\overline{x}_2 = 8.5$
$s_1 = 1.2$	$s_2 = 1.5$
$n_1 = 140$	$n_2 = 127$

TABLE FOR EXERCISE 46

48. See Selected Answers, page A126.

49. The 95% CI for $\mu_1 - \mu_2$ in Exercise 45 contained only values less than 0 and, as found in Exercise 47, there was enough evidence at the 5% level of significance to support the claim.

If the CI for $\mu_1 - \mu_2$ contains only negative numbers, you reject H_0 because the null hypothesis states that $\mu_1 - \mu_2$ is greater than or equal to 0.

50. See Selected Answers, page A126.

44. Registered Nurse and Physician Assistant Salaries At $\alpha = 0.10$, test the claim that the difference between the mean salary for physician assistants and the mean salary for registered nurses in New Jersey is greater than $15,000. The results of a survey of randomly selected physician assistants and registered nurses in New Jersey are shown at the left. (*Adapted from U.S. Bureau of Labor Statistics*)

Constructing Confidence Intervals for $\mu_1 - \mu_2$ *You can construct a confidence interval for the difference between two population means $\mu_1 - \mu_2$ by using the following if $n_1 \geq 30$ and $n_2 \geq 30$, or if both populations are normally distributed and both population standard deviations are known. Also, the samples must be randomly selected and independent.*

$$(\overline{x}_1 - \overline{x}_2) - z_c\sqrt{\frac{\sigma_1^2}{n_1} + \frac{\sigma_2^2}{n_2}} < \mu_1 - \mu_2 < (\overline{x}_1 - \overline{x}_2) + z_c\sqrt{\frac{\sigma_1^2}{n_1} + \frac{\sigma_2^2}{n_2}}$$

In Exercises 45 and 46, construct the indicated confidence interval for $\mu_1 - \mu_2$.

45. DASH Diet and Systolic Blood Pressure A study was conducted to see if a specific diet and exercise program called the DASH (Dietary Approaches to Stop Hypertension) program, which emphasizes the consumption of fruits, vegetables, and low-fat dairy products, can reduce systolic blood pressure more than a traditional diet and exercise program does. After 6 months, 269 people using the DASH diet had a mean systolic blood pressure of 123.1 mm Hg and a standard deviation of 9.9 mm Hg. After the same time period, 268 people using a traditional diet and exercise program had a mean systolic blood pressure of 125 mm Hg and a standard deviation of 10.1 mm Hg. Before the study, each group had the same mean systolic blood pressure. Construct a 95% confidence interval for $\mu_1 - \mu_2$, where μ_1 is the mean systolic blood pressure for the group using the DASH diet and exercise program and μ_2 is the mean systolic blood pressure for the group using the traditional diet and exercise program. (*Source: The Journal of the American Medical Association*)

46. Comparing Cancer Drugs In a study, two groups of patients with colorectal cancer are treated with different drugs. Group A is treated with the drug Irinotecan and Group B is treated with the drug Fluorouracil. The results of the study on the number of months in which the groups reported no cancer-related pain are shown at the left. Construct a 95% confidence interval for $\mu_1 - \mu_2$. (*Adapted from The Lancet*)

47. Make a Decision Refer to the study in Exercise 45. At $\alpha = 0.05$, test the claim that the mean systolic blood pressure for the group using the DASH diet and exercise program is less than the mean systolic blood pressure for the group using the traditional diet and exercise program. Would you recommend the DASH diet and exercise program over the traditional diet and exercise program? Explain.

48. Make a Decision Refer to the study in Exercise 46. At $\alpha = 0.05$, test the claim that the mean number of months of cancer-related pain relief obtained with Irinotecan is greater than the mean number of months of cancer-related pain relief obtained with Fluorouracil. Would you recommend Irinotecan over Fluorouracil to relieve cancer-related pain? Explain.

49. Compare the confidence interval you constructed in Exercise 45 with the hypothesis test result in Exercise 47. Explain why you would reject the null hypothesis if the confidence interval contains only negative numbers.

50. Compare the confidence interval you constructed in Exercise 46 with the hypothesis test result in Exercise 48. Explain why you would reject the null hypothesis if the confidence interval contains only positive numbers.

Readability of Patient Education Materials

Many patient education materials (PEMs) published by health organizations are written at above average readability levels, which may make it difficult for people to comprehend the information. Readability measures the grade level of education necessary to understand written material. According to the National Assessment of Adult Literacy, 14% of U.S. adults have below basic health literacy. Many studies are performed to determine the readability levels of different PEMs.

The table below shows the results of three different studies. Study 1 evaluated PEMs published by the American Cancer Society. Study 2 evaluated PEMs published by the National Clearinghouse for Alcohol and Drug Information. Study 3 evaluated PEMs from the American Academy of Family Physicians.

	Study 1 $n_1 = 51$	Study 2 $n_2 = 52$	Study 3 $n_3 = 171$
Readability level (by grade level)	$\bar{x}_1 = 11.9$ $s_1 = 2.2$	$\bar{x}_2 = 11.84$ $s_2 = 0.94$	$\bar{x}_3 = 9.43$ $s_3 = 1.31$

■ EXERCISES

In Exercises 1–3, perform a two-sample z-test to determine whether the mean readability levels of the two indicated studies are different. For each exercise, write your conclusion as a sentence. Use $\alpha = 0.05$.

1. Test the readability levels of PEMs in Study 1 against those in Study 2.

2. Test the readability levels of PEMs in Study 1 against those in Study 3.

3. Test the readability levels of PEMs in Study 2 against those in Study 3.

4. In which comparisons in Exercises 1–3 did you find a difference in readability levels? Write a summary of your findings.

5. Construct a 95% confidence interval for $\mu_1 - \mu_2$, where μ_1 is the mean readability level in Study 1 and μ_2 is the mean readability level in Study 2. Interpret the results. (See Extending Concepts in Section 8.1 Exercises.)

6. In a fourth study conducted by the Johns Hopkins Oncology Center, the mean readability level of 137 PEMs was 11.1, with a standard deviation of 1.67.

   (a) Test the mean readability level of this study against the level of Study 1. Use $\alpha = 0.01$.

   (b) Test the mean readability level of this study against the level of Study 2. Use $\alpha = 0.01$.

8.2 Testing the Difference Between Means (Small Independent Samples)

WHAT YOU SHOULD LEARN

▸ How to perform a *t*-test for the difference between two population means μ_1 and μ_2 using small independent samples

The Two-Sample *t*-Test for the Difference Between Means

▸ THE TWO-SAMPLE *t*-TEST FOR THE DIFFERENCE BETWEEN MEANS

As you have learned, in real life, it is often not practical to collect samples of size 30 or more from each of two populations. However, if both populations have a normal distribution, you can still test the difference between their means. In this section, you will learn how to use a *t*-test to test the difference between two population means μ_1 and μ_2 using independent samples from each population. The following conditions are necessary to use a *t*-test for small independent samples.

1. The samples must be randomly selected.

2. The samples must be independent.

3. Each population must have a normal distribution.

When these conditions are met, the sampling distribution for the difference between the sample means $\overline{x}_1 - \overline{x}_2$ is approximated by a *t*-distribution with mean $\mu_1 - \mu_2$. So, you can use a two-sample *t*-test to test the difference between the two population means μ_1 and μ_2. The standard error and the degrees of freedom of the sampling distribution depend on whether the population variances σ_1^2 and σ_2^2 are equal.

STUDY TIP

You will need to know whether the variances of two populations are equal. In this chapter, each example and exercise will state whether the variances are equal. You will learn to test for differences in variance of two populations in Chapter 10.

Note to Instructor

If you choose, you can cover tests for equal variances (Section 10.3) before doing these *t*-tests. In case you cover Section 10.3 later or do not have time to cover it at all, students will be informed whether to assume equal variances in each example, Try It Yourself, and exercise in this section. In each instance, we have run an *F*-test at the level of significance stated in the problem and have reported the outcome of the test.

TWO-SAMPLE *t*-TEST FOR THE DIFFERENCE BETWEEN MEANS

A **two-sample *t*-test** is used to test the difference between two population means μ_1 and μ_2 when a sample is randomly selected from each population. Performing this test requires each population to be normally distributed, and the samples should be independent. The **test statistic** is $\overline{x}_1 - \overline{x}_2$, and the **standardized test statistic** is

$$t = \frac{(\overline{x}_1 - \overline{x}_2) - (\mu_1 - \mu_2)}{s_{\overline{x}_1 - \overline{x}_2}}.$$

Variances are equal: If the population variances are equal, then information from the two samples is combined to calculate a **pooled estimate of the standard deviation** $\hat{\sigma}$.

$$\hat{\sigma} = \sqrt{\frac{(n_1 - 1)s_1^2 + (n_2 - 1)s_2^2}{n_1 + n_2 - 2}}$$

The standard error for the sampling distribution of $\overline{x}_1 - \overline{x}_2$ is

$$s_{\overline{x}_1 - \overline{x}_2} = \hat{\sigma} \cdot \sqrt{\frac{1}{n_1} + \frac{1}{n_2}} \qquad \text{Variances equal}$$

and d.f. $= n_1 + n_2 - 2$.

Variances are not equal: If the population variances are not equal, then the standard error is

$$s_{\overline{x}_1 - \overline{x}_2} = \sqrt{\frac{s_1^2}{n_1} + \frac{s_2^2}{n_2}} \qquad \text{Variances not equal}$$

and d.f. $=$ smaller of $n_1 - 1$ and $n_2 - 1$.

PICTURING THE WORLD

A study published by the American Psychological Association in the journal *Neuropsychology* reported that children with musical training showed better verbal memory than children with no musical training. The study also showed that the longer the musical training, the better the verbal memory. Suppose you tried to duplicate the results as follows. A verbal memory test with a possible 100 points was administered to 90 children. Half had musical training, while the other half had no training and acted as the control group. The 45 children with training had an average score of 83.12 with a standard deviation of 5.7. The 45 students in the control group had an average score of 79.9 with a standard deviation of 6.2.

At $\alpha = 0.05$, is there enough evidence to support the claim that children with musical training have better verbal memory test scores than those without training? Assume the populations are normally distributed and the population variances are equal.

There is enough evidence at the 5% level of significance to support the claim that children with musical training have better verbal memory test scores than those without training.

The requirements for the *z*-test described in Section 8.1 and the *t*-test described in this section are shown in the flowchart below.

Two-Sample Tests for Independent Samples

Are both sample sizes at least 30? — Yes → Use the *z*-test.

No ↓

Are both populations normal? — No → You cannot use the *z*-test or the *t*-test.

Yes ↓

Are both population standard deviations known? — No → Are the population variances equal? — Yes → Use the *t*-test with
$$s_{\bar{x}_1 - \bar{x}_2} = \hat{\sigma} \cdot \sqrt{\frac{1}{n_1} + \frac{1}{n_2}}$$
and d.f. $= n_1 + n_2 - 2$.

Yes ↓

Use the *z*-test.

(Are the population variances equal?) No ↓

Use the *t*-test with
$$s_{\bar{x}_1 - \bar{x}_2} = \sqrt{\frac{s_1^2}{n_1} + \frac{s_2^2}{n_2}}$$
and d.f. = smaller of $n_1 - 1$ and $n_2 - 1$.

GUIDELINES

Using a Two-Sample *t*-Test for the Difference Between Means (Small Independent Samples)

IN WORDS	IN SYMBOLS
1. State the claim mathematically and verbally. Identify the null and alternative hypotheses.	State H_0 and H_a.
2. Specify the level of significance.	Identify α.
3. Determine the degrees of freedom.	d.f. $= n_1 + n_2 - 2$ or d.f. $=$ smaller of $n_1 - 1$ and $n_2 - 1$
4. Determine the critical value(s).	Use Table 5 in Appendix B.
5. Determine the rejection region(s).	
6. Find the standardized test statistic and sketch the sampling distribution.	$t = \dfrac{(\bar{x}_1 - \bar{x}_2) - (\mu_1 - \mu_2)}{s_{\bar{x}_1 - \bar{x}_2}}$
7. Make a decision to reject or fail to reject the null hypothesis.	If t is in the rejection region, reject H_0. Otherwise, fail to reject H_0.
8. Interpret the decision in the context of the original claim.	

Sample Statistics for State Mathematics Test Scores

Teacher 1	Teacher 2
$\bar{x}_1 = 473$	$\bar{x}_2 = 459$
$s_1 = 39.7$	$s_2 = 24.5$
$n_1 = 8$	$n_2 = 18$

EXAMPLE 1 SC Report 34

See MINITAB steps on page 478.

▶ A Two-Sample *t*-Test for the Difference Between Means

The results of a state mathematics test for random samples of students taught by two different teachers at the same school are shown at the left. Can you conclude that there is a difference in the mean mathematics test scores for the students of the two teachers? Use $\alpha = 0.10$. Assume the populations are normally distributed and the population variances are not equal.

▶ Solution

The claim is "there is a difference in the mean mathematics test scores for the students of the two teachers." So, the null and alternative hypotheses are

$$H_0: \mu_1 = \mu_2 \quad \text{and} \quad H_a: \mu_1 \neq \mu_2. \text{ (Claim)}$$

Because the variances are not equal and the smaller sample size is 8, use d.f. $= 8 - 1 = 7$. Because the test is a two-tailed test with d.f. $= 7$ and $\alpha = 0.10$, the critical values are $-t_0 = -1.895$ and $t_0 = 1.895$. The rejection regions are $t < -1.895$ and $t > 1.895$. The standard error is

$$s_{\bar{x}_1 - \bar{x}_2} = \sqrt{\frac{s_1^2}{n_1} + \frac{s_2^2}{n_2}}$$

$$= \sqrt{\frac{39.7^2}{8} + \frac{24.5^2}{18}} \approx 15.1776.$$

The standardized test statistic is

$$t = \frac{(\bar{x}_1 - \bar{x}_2) - (\mu_1 - \mu_2)}{s_{\bar{x}_1 - \bar{x}_2}} \qquad \text{Use the } t\text{-test.}$$

$$\approx \frac{(473 - 459) - 0}{15.1776} \qquad \text{Assume } \mu_1 = \mu_2, \text{ so } \mu_1 - \mu_2 = 0.$$

$$\approx 0.922.$$

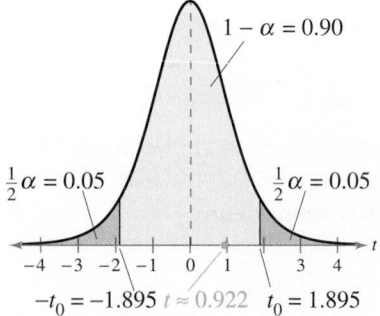

$1 - \alpha = 0.90$

$\frac{1}{2}\alpha = 0.05$ $\frac{1}{2}\alpha = 0.05$

$-t_0 = -1.895$ $t \approx 0.922$ $t_0 = 1.895$

The graph at the left shows the location of the rejection regions and the standardized test statistic *t*. Because *t* is not in the rejection region, you should fail to reject the null hypothesis.

Interpretation There is not enough evidence at the 10% level of significance to support the claim that the mean mathematics test scores for the students of the two teachers are different.

▶ Try It Yourself 1

The annual earnings of 15 people with a high school diploma and 12 people with a bachelor's degree or higher are shown at the left. Can you conclude that there is a difference in the mean annual earnings based on level of education? Use $\alpha = 0.01$. Assume the populations are normally distributed and the population variances are not equal.

Sample Statistics for Annual Earnings

High school diploma	Bachelor's degree or higher
$\bar{x}_1 = \$27,136$	$\bar{x}_2 = \$34,329$
$s_1 = \$2318$	$s_2 = \$4962$
$n_1 = 15$	$n_2 = 12$

a. Identify the *claim* and state H_0 and H_a.
b. Identify the *level of significance* α and the *degrees of freedom*.
c. Find the *critical values* and identify the *rejection regions*.
d. Find the *standardized test statistic t*. Sketch a graph.
e. *Decide* whether to reject the null hypothesis.
f. *Interpret* the decision in the context of the original claim.

Answer: Page A43

**Sample Statistics for
Calling Range**

Manufacturer	Competitor
$\bar{x}_1 = 1275$ ft	$\bar{x}_2 = 1250$ ft
$s_1 = 45$ ft	$s_2 = 30$ ft
$n_1 = 14$	$n_2 = 16$

EXAMPLE 2   Report 35

See TI-83/84 Plus
steps on page 479.

▶ A Two-Sample *t*-Test for the Difference Between Means

A manufacturer claims that the mean calling range (in feet) of its 2.4-GHz cordless telephones is greater than that of its leading competitor. You perform a study using 14 randomly selected phones from the manufacturer and 16 randomly selected similar phones from its competitor. The results are shown at the left. At $\alpha = 0.05$, can you support the manufacturer's claim? Assume the populations are normally distributed and the population variances are equal.

▶ Solution

The claim is "the mean calling range of the manufacturer's cordless phones is greater than that of its leading competitor." So, the null and alternative hypotheses are

$$H_0: \mu_1 \le \mu_2 \quad \text{and} \quad H_a: \mu_1 > \mu_2. \text{ (Claim)}$$

Because the variances are equal, d.f. $= n_1 + n_2 - 2 = 14 + 16 - 2 = 28$. Because the test is a right-tailed test with d.f. $= 28$ and $\alpha = 0.05$, the critical value is $t_0 = 1.701$. The rejection region is $t > 1.701$. The standard error is

$$s_{\bar{x}_1 - \bar{x}_2} = \sqrt{\frac{(n_1 - 1)s_1^2 + (n_2 - 1)s_2^2}{n_1 + n_2 - 2}} \cdot \sqrt{\frac{1}{n_1} + \frac{1}{n_2}}$$

$$= \sqrt{\frac{(13)(45^2) + (15)(30^2)}{14 + 16 - 2}} \cdot \sqrt{\frac{1}{14} + \frac{1}{16}} \approx 13.8018.$$

The standardized test statistic is

$$t = \frac{(\bar{x}_1 - \bar{x}_2) - (\mu_1 - \mu_2)}{s_{\bar{x}_1 - \bar{x}_2}} \qquad \text{Use the } t\text{-test.}$$

$$\approx \frac{(1275 - 1250) - 0}{13.8018} \qquad \text{Assume } \mu_1 = \mu_2, \text{ so } \mu_1 - \mu_2 = 0.$$

$$\approx 1.811.$$

The graph at the left shows the location of the rejection region and the standardized test statistic *t*. Because *t* is in the rejection region, you should decide to reject the null hypothesis.

Interpretation   There is enough evidence at the 5% level of significance to support the manufacturer's claim that its phones have a greater calling range than its competitor's.

▶ Try It Yourself 2

A manufacturer claims that the watt usage of its 17-inch flat panel monitors is less than that of its leading competitor. You perform a study and obtain the results shown at the left. At $\alpha = 0.10$, is there enough evidence to support the manufacturer's claim? Assume the populations are normally distributed and the population variances are equal.

a. Identify the *claim* and state H_0 and H_a.
b. Identify the *level of significance* α and the *degrees of freedom*.
c. Find the *critical value* and identify the *rejection region*.
d. Find the *standardized test statistic t. Sketch* a graph.
e. *Decide* whether to reject the null hypothesis.
f. *Interpret* the decision in the context of the original claim.

Answer: Page A43

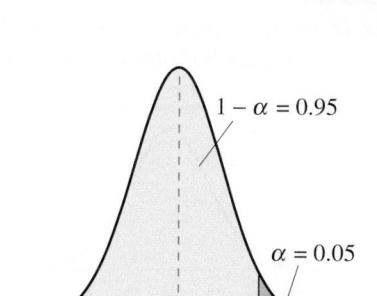

$1 - \alpha = 0.95$

$\alpha = 0.05$

$t_0 = 1.701 \quad t \approx 1.811$

**Sample Statistics for
Watt Usage**

Manufacturer	Competitor
$\bar{x}_1 = 32$	$\bar{x}_2 = 35$
$s_1 = 2.1$	$s_2 = 1.8$
$n_1 = 12$	$n_2 = 15$

8.2 EXERCISES

BUILDING BASIC SKILLS AND VOCABULARY

1. What conditions are necessary in order to use a t-test to test the difference between two population means?

2. Explain how to perform a two-sample t-test for the difference between the means of two populations.

In Exercises 3–8, use Table 5 in Appendix B to find the critical value(s) for the indicated alternative hypothesis, level of significance α, and sample sizes n_1 and n_2. Assume that the samples are independent, normal, and random, and that the population variances are (a) equal and (b) not equal.

3. $H_a\colon \mu_1 \neq \mu_2,\ \alpha = 0.10,\ n_1 = 11,\ n_2 = 14$

4. $H_a\colon \mu_1 > \mu_2,\ \alpha = 0.01,\ n_1 = 12,\ n_2 = 15$

5. $H_a\colon \mu_1 < \mu_2,\ \alpha = 0.05,\ n_1 = 7,\ n_2 = 11$

6. $H_a\colon \mu_1 \neq \mu_2,\ \alpha = 0.01,\ n_1 = 19,\ n_2 = 22$

7. $H_a\colon \mu_1 > \mu_2,\ \alpha = 0.05,\ n_1 = 13,\ n_2 = 8$

8. $H_a\colon \mu_1 < \mu_2,\ \alpha = 0.10,\ n_1 = 9,\ n_2 = 4$

In Exercises 9–12, (a) find the test statistic, (b) find the standardized test statistic, (c) decide whether the standardized test statistic is in the rejection region, and (d) decide whether you should reject or fail to reject the null hypothesis. Assume the populations are normally distributed.

9. Claim: $\mu_1 = \mu_2$; $\alpha = 0.01$.
Sample statistics: $\bar{x}_1 = 33.7,\ s_1 = 3.5$,
$n_1 = 12$ and $\bar{x}_2 = 35.5,\ s_2 = 2.2,\ n_2 = 17$.
Assume $\sigma_1^2 = \sigma_2^2$.

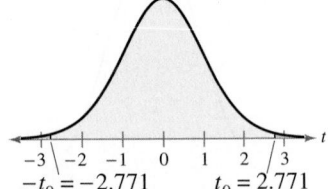

10. Claim: $\mu_1 < \mu_2$; $\alpha = 0.10$.
Sample statistics: $\bar{x}_1 = 0.345,\ s_1 = 0.305$,
$n_1 = 11$ and $\bar{x}_2 = 0.515,\ s_2 = 0.215,\ n_2 = 9$.
Assume $\sigma_1^2 = \sigma_2^2$.

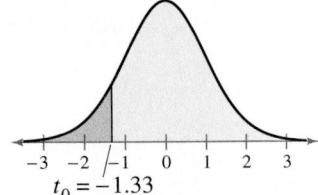

11. Claim: $\mu_1 \leq \mu_2$; $\alpha = 0.05$.
Sample statistics: $\bar{x}_1 = 2410,\ s_1 = 175$,
$n_1 = 13$ and $\bar{x}_2 = 2305,\ s_2 = 52,\ n_2 = 10$.
Assume $\sigma_1^2 \neq \sigma_2^2$.

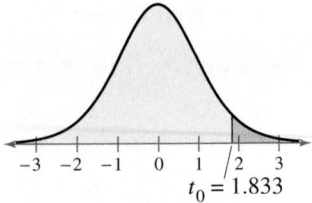

12. Claim: $\mu_1 > \mu_2$; $\alpha = 0.01$.
Sample statistics: $\bar{x}_1 = 52\ s_1 = 4.8$,
$n_1 = 16$ and $\bar{x}_2 = 50,\ s_2 = 1.2,\ n_2 = 14$.
Assume $\sigma_1^2 \neq \sigma_2^2$.

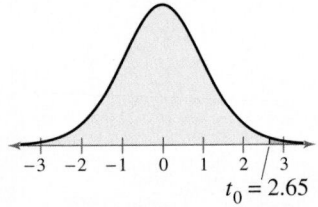

Selected Answers (margin)

1. (1) The samples must be randomly selected.

   (2) The samples must be independent.

   (3) Each population must have a normal distribution.

2. See Selected Answers, page A126.

3. (a) $-t_0 = -1.714,\ t_0 = 1.714$
   (b) $-t_0 = -1.812,\ t_0 = 1.812$

4. (a) $t_0 = 2.485$
   (b) $t_0 = 2.718$

5. (a) $t_0 = -1.746$
   (b) $t_0 = -1.943$

6. (a) $-t_0 = -2.576,\ t_0 = 2.576$
   (b) $-t_0 = -2.878,\ t_0 = 2.878$

7. (a) $t_0 = 1.729$
   (b) $t_0 = 1.895$

8. (a) $t_0 = -1.363$
   (b) $t_0 = -1.638$

9. (a) -1.8
   (b) -1.70
   (c) Not in the rejection region.
   (d) Fail to reject H_0.

10. (a) -0.17
    (b) -1.41
    (c) In the rejection region.
    (d) Reject H_0.

11. (a) 105
    (b) 2.05
    (c) In the rejection region.
    (d) Reject H_0.

12. (a) 2

(b) 1.61

(c) Not in the rejection region.

(d) Fail to reject H_0.

13. (a) The claim is "the mean annual costs of routine veterinarian visits for dogs and cats are the same."

$H_0: \mu_1 = \mu_2$ (claim);
$H_a: \mu_1 \neq \mu_2$

(b) $-t_0 = -1.943$, $t_0 = 1.943$;
Rejection regions:
$t < -1.943$, $t > 1.943$

(c) 1.90

(d) Fail to reject H_0.

(e) There is not enough evidence at the 10% level of significance to reject the pet association's claim that the mean annual costs of routine veterinarian visits for dogs and cats are the same.

14. (a) The claim is "athletes have a greater mean maximal oxygen consumption than non-athletes."

$H_0: \mu_1 \leq \mu_2$;
$H_a: \mu_1 > \mu_2$ (claim)

(b) $t_0 = 1.645$;
Rejection region: $t > 1.645$

(c) 7.20

(d) Reject H_0.

(e) There is enough evidence at the 5% level of significance to support the claim that athletes have a greater mean maximal oxygen consumption than non-athletes.

15. (a) The claim is "the mean bumper repair cost is less for mini cars than for midsize cars."

$H_0: \mu_1 \geq \mu_2$;
$H_a: \mu_1 < \mu_2$ (claim)

(b) $t_0 = -1.325$;
Rejection region: $t < -1.325$

(c) -0.93

(d) Fail to reject H_0.

(e) There is not enough evidence at the 10% level of significance to support the claim that the mean bumper repair cost is less for mini cars than for midsize cars.

16. See Selected Answers, page A126.

■ USING AND INTERPRETING CONCEPTS

Testing the Difference Between Two Means   *In Exercises 13–22, (a) identify the claim and state H_0 and H_a, (b) find the critical value(s) and identify the rejection region(s), (c) find the standardized test statistic t, (d) decide whether to reject or fail to reject the null hypothesis, and (e) interpret the decision in the context of the original claim. If convenient, use technology to solve the problem. In each exercise, assume the populations are normally distributed, and the samples are independent and random.*

13. Dogs and Cats   A pet association claims that the mean annual costs of routine veterinarian visits for dogs and cats are the same. The results for samples of the two types of pets are shown below. At $\alpha = 0.10$, can you reject the pet association's claim? Assume the population variances are not equal. *(Adapted from American Pet Products Association)*

Sample Statistics for Annual Routine Vet Visits

Dogs	Cats
$\bar{x}_1 = \$225$	$\bar{x}_2 = \$203$
$s_1 = \$28$	$s_2 = \$15$
$n_1 = 7$	$n_2 = 10$

14. Maximal Oxygen Consumption   The maximal oxygen consumption is a way to measure the physical fitness of an individual. It is the amount of oxygen in milliliters a person uses per kilogram of body weight per minute. A medical research center claims that athletes have a greater mean maximal oxygen consumption than non-athletes. The results for samples of the two groups are shown below. At $\alpha = 0.05$, can you support the research center's claim? Assume the population variances are equal.

Sample Statistics for Maximal Oxygen Consumptions

Athletes	Non-athletes
$\bar{x}_1 = 56$ ml/kg/min	$\bar{x}_2 = 47$ ml/kg/min
$s_1 = 4.9$ ml/kg/min	$s_2 = 3.1$ ml/kg/min
$n_1 = 23$	$n_2 = 21$

15. Bumper Repair Cost   In low speed crash tests, the mean bumper repair cost of 6 randomly selected mini cars is \$1621 with a standard deviation of \$493. In similar tests of 16 randomly selected midsize cars, the mean bumper repair cost is \$1895, with a standard deviation of \$648. At $\alpha = 0.10$, can you conclude that the mean bumper repair cost is less for mini cars than for midsize cars? Assume the population variances are equal. *(Adapted from Insurance Institute for Highway Safety)*

16. Footwell Intrusion   An insurance actuary claims that the mean footwell intrusions for small pickups and small SUVs are equal. Crash tests at 40 miles per hour were performed on 7 randomly selected small pickups and 13 randomly selected small SUVs. The amount that the footwell intruded on the driver's side was measured. The mean footwell intrusion for the small pickups was 11.18 centimeters with a standard deviation of 4.53. The mean footwell intrusion for the small SUVs was 9.52 centimeters with a standard deviation of 3.84. At $\alpha = 0.01$, can you reject the insurance actuary's claim? Assume the population variances are equal. *(Adapted from Insurance Institute for Highway Safety)*

17. (a) The claim is "the mean household income is greater in Allegheny County than it is in Erie County."

H_0: $\mu_1 \leq \mu_2$;
H_a: $\mu_1 > \mu_2$ (claim)

(b) $t_0 = 1.761$;
Rejection region: $t > 1.761$

(c) 1.99

(d) Reject H_0.

(e) There is enough evidence at the 5% level of significance to support the personnel director's claim that the mean household income is greater in Allegheny County than it is in Erie County.

18. (a) The claim is "the mean household income is greater in Hillsborough County than it is in Polk County."

H_0: $\mu_1 \leq \mu_2$;
H_a: $\mu_1 > \mu_2$ (claim)

(b) $t_0 = 2.583$;
Rejection region: $t > 2.583$

(c) 2.38

(d) Fail to reject H_0.

(e) There is not enough evidence at the 1% level of significance to support the personnel director's claim that the mean household income is greater in Hillsborough County than it is in Polk County.

19. (a) The claim is "the new treatment makes a difference in the tensile strength of steel bars."

H_0: $\mu_1 = \mu_2$
H_a: $\mu_1 \neq \mu_2$ (claim)

(b) $-t_0 = -2.831$, $t_0 = 2.831$;

Rejection regions:
$t < -2.831$, $t > 2.831$

(c) -2.76

(d) Fail to reject H_0.

(e) There is not enough evidence at the 1% level of significance to support the claim that the new treatment makes a difference in the tensile strength of steel bars.

20. See Selected Answers, page A126.

17. Annual Income A personnel director from Pennsylvania claims that the mean household income is greater in Allegheny County than it is in Erie County. In Allegheny County, a sample of 19 residents has a mean household income of $48,800 and a standard deviation of $8800. In Erie County, a sample of 15 residents has a mean household income of $44,000 and a standard deviation of $5100. At $\alpha = 0.05$, can you support the personnel director's claim? Assume the population variances are not equal. (*Adapted from U.S. Census Bureau*)

18. Annual Income A personnel director from Florida claims that the mean household income is greater in Hillsborough County than it is in Polk County. In Hillsborough County, a sample of 17 residents has a mean household income of $49,800 and a standard deviation of $4200. In Polk County, a sample of 18 residents has a mean household income of $44,400 and a standard deviation of $8600. At $\alpha = 0.01$, can you support the personnel director's claim? Assume the population variances are not equal. (*Adapted from U.S. Census Bureau*)

19. Tensile Strength The tensile strength of a metal is a measure of its ability to resist tearing when it is pulled lengthwise. A new experimental type of treatment produced steel bars with the following tensile strengths (in newtons per square millimeter).

Experimental Method:

| 391 | 383 | 333 | 378 | 368 |
| 401 | 339 | 376 | 366 | 348 |

The old method produced steel bars with the following tensile strengths (in newtons per square millimeter).

Old Method:

| 362 | 382 | 368 | 398 | 381 | 391 | 400 |
| 410 | 396 | 411 | 385 | 385 | 395 |

At $\alpha = 0.01$, does the new treatment make a difference in the tensile strength of steel bars? Assume the population variances are equal.

20. Tensile Strength An engineer wants to compare the tensile strengths of steel bars that are produced using a conventional method and an experimental method. (The tensile strength of a metal is a measure of its ability to resist tearing when pulled lengthwise.) To do so, the engineer randomly selects steel bars that are manufactured using each method and records the following tensile strengths (in newtons per square millimeter).

Experimental Method:

| 395 | 389 | 421 | 394 | 407 | 411 | 389 | 402 | 422 |
| 416 | 402 | 408 | 400 | 386 | 411 | 405 | 389 |

Conventional Method:

| 362 | 352 | 380 | 382 | 413 | 384 | 400 |
| 378 | 419 | 379 | 384 | 388 | 372 | 383 |

At $\alpha = 0.10$, can the engineer claim that the experimental method produces steel with greater mean tensile strength? Should the engineer recommend using the experimental method? Assume the population variances are not equal.

21. (a) The claim is "the new method of teaching reading produces higher reading test scores than the old method."

$H_0: \mu_1 \geq \mu_2$;
$H_a: \mu_1 < \mu_2$ (claim)

(b) $t_0 = -1.282$;
Rejection region: $t < -1.282$

(c) -4.295

(d) Reject H_0.

(e) There is enough evidence at the 10% level of significance to support the claim that the new method of teaching reading produces higher reading test scores than the old method and to recommend changing to the new method.

22. See Selected Answers, page A126.

23. Hypothesis test results:

μ_1 : mean of population 1

μ_2 : mean of population 2

$\mu_1 - \mu_2$: mean difference

$H_0 : \mu_1 - \mu_2 = 0$

$H_A : \mu_1 - \mu_2 > 0$

(with pooled variances)

See Odd Answers, page A89.

$P = 0.6789 > 0.10$, so fail to reject H_0.

There is not enough evidence at the 10% level of significance to support the claim.

24. See Selected Answers, page A126.

25. Hypothesis test results:

μ_1 : mean of population 1

μ_2 : mean of population 2

$\mu_1 - \mu_2$: mean difference

$H_0 : \mu_1 - \mu_2 = 0$

$H_A : \mu_1 - \mu_2 < 0$

(without pooled variances)

See Odd Answers, page A89.

$P = 0.0714 > 0.05$, so fail to reject H_0.

There is not enough evidence at the 5% level of significance to reject the claim.

26. See Selected Answers, page A127.

21. Teaching Methods   A new method of teaching reading is being tested on third grade students. A group of third grade students is taught using the new curriculum. A control group of third grade students is taught using the old curriculum. The reading test scores for the two groups are shown in the back-to-back stem-and-leaf plot.

Old Curriculum		New Curriculum
9	3	Key:
9 9	4	3 9|4 = 49 (old curriculum)
9 8 8 4 3 3 2 1	5	2 4 4|3 = 43 (new curriculum)
7 6 4 2 2 1 0 0	6	0 1 1 4 7 7 7 7 7 8 9 9
	7	0 1 1 2 3 3 4 9
	8	2 4

At $\alpha = 0.10$, is there enough evidence to conclude that the new method of teaching reading produces higher reading test scores than the old method does? Would you recommend changing to the new method? Assume the population variances are equal.

22. Teaching Methods   Two teaching methods and their effects on science test scores are being reviewed. A group of students is taught in traditional lab sessions. A second group of students is taught using interactive simulation software. The science test scores for the two groups are shown in the back-to-back stem-and-leaf plot.

Traditional Lab		Interactive Simulation Software
4	6	Key:
9 9 8 8 7 6 6 3 2 1 0	7	0 4 5 5 7 7 8 0|9 = 90 (traditional)
9 8 5 1 1 0 0	8	0 0 3 4 7 8 8 9 9 9|1 = 91 (interactive)
2 0	9	1 3 9

At $\alpha = 0.05$, can you support the claim that the mean science test score is lower for students taught using the traditional lab method than it is for students taught using the interactive simulation software? Assume the population variances are equal.

SC   *In Exercises 23–26, use StatCrunch to help you test the claim about the difference between two population means μ_1 and μ_2 at the given level of significance α using the given sample statistics. Assume that the populations are normally distributed, and the samples are independent and random.*

23. Claim: $\mu_1 \leq \mu_2$; $\alpha = 0.10$. Sample statistics: $\bar{x}_1 = 186$, $s_1 = 38$, $n_1 = 15$ and $\bar{x}_2 = 194$, $s_2 = 44$, $n_2 = 9$. Assume $\sigma_1^2 = \sigma_2^2$.

24. Claim: $\mu_1 \neq \mu_2$; $\alpha = 0.01$. Sample statistics: $\bar{x}_1 = 34$, $s_1 = 5.8$, $n_1 = 5$ and $\bar{x}_2 = 45$, $s_2 = 4.6$, $n_2 = 8$. Assume $\sigma_1^2 = \sigma_2^2$.

25. Claim: $\mu_1 < \mu_2$; $\alpha = 0.05$. Sample statistics: $\bar{x}_1 = 840$, $s_1 = 95$, $n_1 = 14$ and $\bar{x}_2 = 883$, $s_2 = 58$, $n_2 = 23$. Assume $\sigma_1^2 \neq \sigma_2^2$.

26. Claim: $\mu_1 = \mu_2$; $\alpha = 0.10$. Sample statistics: $\bar{x}_1 = 98.5$, $s_1 = 10.2$, $n_1 = 15$ and $\bar{x}_2 = 76.1$, $s_2 = 18.8$, $n_2 = 6$. Assume $\sigma_1^2 \neq \sigma_2^2$.

27. $45 < \mu_1 - \mu_2 < 307$

28. $-5 < \mu_1 - \mu_2 < 9$

29. $11 < \mu_1 - \mu_2 < 35$

30. $34 < \mu_1 - \mu_2 < 44$

**Sample Statistics for
Kidney Transplants**

35–49	50–64
$\bar{x}_1 = 1805$ days	$\bar{x}_2 = 1629$ days
$s_1 = 166$ days	$s_2 = 204$ days
$n_1 = 21$	$n_2 = 11$

TABLE FOR EXERCISE 27

**Sample Statistics for
Driving Distances**

Golfer 1	Golfer 2
$\bar{x}_1 = 267$ yd	$\bar{x}_2 = 244$ yd
$s_1 = 6$ yd	$s_2 = 12$ yd
$n_1 = 9$	$n_2 = 5$

TABLE FOR EXERCISE 29

**Sample Statistics for
African Elephant Lifespans**

Wild	Zoo
$\bar{x}_1 = 56.0$ yr	$\bar{x}_2 = 16.9$ yr
$s_1 = 8.6$ yr	$s_2 = 3.8$ yr
$n_1 = 20$	$n_2 = 12$

TABLE FOR EXERCISE 30

■ EXTENDING CONCEPTS

Constructing Confidence Intervals for $\mu_1 - \mu_2$ *If the sampling distribution for $\bar{x}_1 - \bar{x}_2$ is approximated by a t-distribution and the populations have equal variances, you can construct a confidence interval for $\mu_1 - \mu_2$ by using the following.*

$$(\bar{x}_1 - \bar{x}_2) - t_c \hat{\sigma} \cdot \sqrt{\frac{1}{n_1} + \frac{1}{n_2}} < \mu_1 - \mu_2 < (\bar{x}_1 - \bar{x}_2) + t_c \hat{\sigma} \cdot \sqrt{\frac{1}{n_1} + \frac{1}{n_2}}$$

where $\hat{\sigma} = \sqrt{\dfrac{(n_1 - 1)s_1^2 + (n_2 - 1)s_2^2}{n_1 + n_2 - 2}}$ and d.f. $= n_1 + n_2 - 2$

In Exercises 27 and 28, construct a confidence interval for $\mu_1 - \mu_2$. Assume the populations are approximately normal with equal variances.

27. **Kidney Transplant Waiting Times** To compare the mean times spent waiting for a kidney transplant for two age groups, you randomly select several people in each age group who have had a kidney transplant. The results are shown at the left. Construct a 95% confidence interval for the difference in mean times spent waiting for a kidney transplant for the two age groups. *(Adapted from Organ Procurement and Transplantation Network)*

28. **Heart Transplant Waiting Times** To compare the mean times spent waiting for a heart transplant for two age groups, you randomly select several people in each age group who have had a heart transplant. The results are shown below. Construct a 99% confidence interval for the difference in mean times spent waiting for a heart transplant for the two age groups. *(Adapted from Organ Procurement and Transplantation Network)*

**Sample Statistics for
Heart Transplants**

18–34	35–49
$\bar{x}_1 = 171$ days	$\bar{x}_2 = 169$ days
$s_1 = 8.5$ days	$s_2 = 11.5$ days
$n_1 = 26$	$n_2 = 24$

Constructing Confidence Intervals for $\mu_1 - \mu_2$ *If the sampling distribution for $\bar{x}_1 - \bar{x}_2$ is approximated by a t-distribution and the population variances are not equal, you can construct a confidence interval for $\mu_1 - \mu_2$ by using the following.*

$$(\bar{x}_1 - \bar{x}_2) - t_c \sqrt{\frac{s_1^2}{n_1} + \frac{s_2^2}{n_2}} < u_1 - u_2 < (\bar{x}_1 - \bar{x}_2) + t_c \sqrt{\frac{s_1^2}{n_1} + \frac{s_2^2}{n_2}}$$

and d.f. is the smaller of $n_1 - 1$ and $n_2 - 1$

In Exercises 29 and 30, construct the indicated confidence interval for $\mu_1 - \mu_2$. Assume the populations are approximately normal with unequal variances.

29. **Golf** To compare the mean driving distances for two golfers, you randomly select several drives from each golfer. The results are shown at the left. Construct a 90% confidence interval for the difference in mean driving distances for the two golfers.

30. **Elephants** To compare the mean lifespans of African elephants in the wild and in a zoo, you randomly select several lifespans from both locations. The results are shown at the left. Construct a 95% confidence interval for the difference in mean lifespans of elephants in the wild and in a zoo. *(Adapted from Science Magazine)*

8.3 Testing the Difference Between Means (Dependent Samples)

The *t*-Test for the Difference Between Means

▸ THE *t*-TEST FOR THE DIFFERENCE BETWEEN MEANS

In Sections 8.1 and 8.2, you performed two-sample hypothesis tests with independent samples using the test statistic $\bar{x}_1 - \bar{x}_2$ (the difference between the means of the two samples). To perform a two-sample hypothesis test with dependent samples, you will use a different technique. You will first find the difference *d* for each data pair:

$$d = x_1 - x_2.$$ Difference between entries for a data pair

The test statistic is the mean $\bar{d}$ of these differences

$$\bar{d} = \frac{\Sigma d}{n}.$$ Mean of the differences between paired data entries in the dependent samples

The following conditions are required to conduct the test.

1. The samples must be randomly selected.

2. The samples must be dependent (paired).

3. Both populations must be normally distributed.

If these requirements are met, then the **sampling distribution for $\bar{d}$, the mean of the differences of the paired data entries in the dependent samples,** is approximated by a *t*-distribution with $n - 1$ degrees of freedom, where *n* is the number of data pairs.

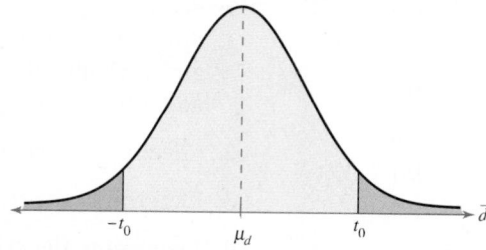

The following symbols are used for the *t*-test for μ_d. Although formulas are given for the mean and standard deviation of differences, you should use a technology tool to calculate these statistics.

Symbol	Description
n	The number of pairs of data
d	The difference between entries for a data pair, $d = x_1 - x_2$
μ_d	The hypothesized mean of the differences of paired data in the population
$\bar{d}$	The mean of the differences between the paired data entries in the dependent samples $$\bar{d} = \frac{\Sigma d}{n}$$
s_d	The standard deviation of the differences between the paired data entries in the dependent samples $$s_d = \sqrt{\frac{\Sigma (d - \bar{d})^2}{n - 1}}$$

STUDY TIP

You can also calculate the standard deviation of the differences between paired data entries using the shortcut formula

$$s_d = \sqrt{\frac{\Sigma d^2 - \left[\frac{(\Sigma d)^2}{n}\right]}{n - 1}}.$$

When you use a t-distribution to approximate the sampling distribution for $\bar{d}$, the mean of the differences between paired data entries, you can use a t-test to test a claim about the mean of the differences for a population of paired data.

t-TEST FOR THE DIFFERENCE BETWEEN MEANS

A t-test can be used to test the difference of two population means when a sample is randomly selected from each population. The requirements for performing the test are that each population must be normal and each member of the first sample must be paired with a member of the second sample. The **test statistic** is

$$\bar{d} = \frac{\Sigma d}{n}$$

and the **standardized test statistic** is

$$t = \frac{\bar{d} - \mu_d}{s_d/\sqrt{n}}.$$

The degrees of freedom are

$$\text{d.f.} = n - 1.$$

GUIDELINES

Using the t-Test for the Difference Between Means (Dependent Samples)

IN WORDS	IN SYMBOLS
1. State the claim mathematically and verbally. Identify the null and alternative hypotheses.	State H_0 and H_a.
2. Specify the level of significance.	Identify α.
3. Determine the degrees of freedom.	d.f. $= n - 1$
4. Determine the critical value(s).	Use Table 5 in Appendix B. If $n > 29$, use the last row (∞) in the t-distribution table.
5. Determine the rejection region(s).	
6. Calculate $\bar{d}$ and s_d.	$\bar{d} = \dfrac{\Sigma d}{n}$ $s_d = \sqrt{\dfrac{\Sigma (d - \bar{d})^2}{n - 1}}$
7. Find the standardized test statistic and sketch the sampling distribution.	$t = \dfrac{\bar{d} - \mu_d}{s_d/\sqrt{n}}$
8. Make a decision to reject or fail to reject the null hypothesis.	If t is in the rejection region, reject H_0. Otherwise, fail to reject H_0.
9. Interpret the decision in the context of the original claim.	

EXAMPLE 1

See MINITAB steps on page 478.

▶ The *t*-Test for the Difference Between Means

A shoe manufacturer claims that athletes can increase their vertical jump heights using the manufacturer's new Strength Shoes®. The vertical jump heights of eight randomly selected athletes are measured. After the athletes have used the Strength Shoes® for 8 months, their vertical jump heights are measured again. The vertical jump heights (in inches) for each athlete are shown in the table. At $\alpha = 0.10$, is there enough evidence to support the manufacturer's claim? Assume the vertical jump heights are normally distributed. *(Adapted from Coaches Sports Publishing)*

Athlete	1	2	3	4	5	6	7	8
Vertical jump height (before using shoes)	24	22	25	28	35	32	30	27
Vertical jump height (after using shoes)	26	25	25	29	33	34	35	30

▶ Solution

The claim is that "athletes can increase their vertical jump heights." In other words, the manufacturer claims that an athlete's vertical jump height before using the Strength Shoes® will be less than the athlete's vertical jump height after using the Strength Shoes®. Each difference is given by

$$d = (\text{jump height before shoes}) - (\text{jump height after shoes}).$$

The null and alternative hypotheses are

$$H_0: \mu_d \geq 0 \qquad \text{and} \qquad H_a: \mu_d < 0. \text{ (Claim)}$$

Because the test is a left-tailed test, $\alpha = 0.10$, and d.f. $= 8 - 1 = 7$, the critical value is $t_0 = -1.415$. The rejection region is $t < -1.415$. Using the table at the left, you can calculate $\overline{d}$ and s_d as follows. Notice that the shortcut formula is used to calculate the standard deviation.

Before	After	d	d^2
24	26	−2	4
22	25	−3	9
25	25	0	0
28	29	−1	1
35	33	2	4
32	34	−2	4
30	35	−5	25
27	30	−3	9
		$\Sigma = -14$	$\Sigma = 56$

$$\overline{d} = \frac{\Sigma d}{n} = \frac{-14}{8} = -1.75$$

$$s_d = \sqrt{\frac{\Sigma d^2 - \left[\frac{(\Sigma d)^2}{n}\right]}{n - 1}} = \sqrt{\frac{56 - \frac{(-14)^2}{8}}{8 - 1}} \approx 2.1213$$

The standardized test statistic is

$$t = \frac{\overline{d} - \mu_d}{s_d/\sqrt{n}} \qquad \text{Use the } t\text{-test.}$$

$$\approx \frac{-1.75 - 0}{2.1213/\sqrt{8}} \qquad \text{Assume } \mu_d = 0.$$

$$\approx -2.333.$$

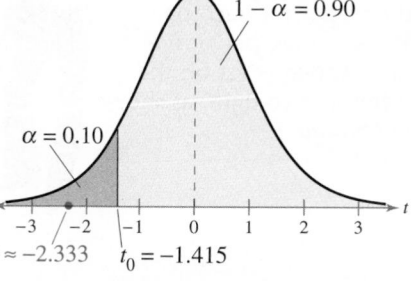

The graph at the right shows the location of the rejection region and the standardized test statistic t. Because t is in the rejection region, you should decide to reject the null hypothesis.

Interpretation There is enough evidence at the 10% level of significance to support the shoe manufacturer's claim that athletes can increase their vertical jump heights using the new Strength Shoes®.

STUDY TIP

You can also use technology and a *P*-value to perform a hypothesis test for the difference between means. For instance, in Example 1, you can enter the data in MINITAB (as shown on page 478) and find $P = 0.026$. Because $P < \alpha$, you should decide to reject the null hypothesis.

Before	After
4.85	4.78
4.90	4.90
5.08	5.05
4.72	4.65
4.62	4.64
4.54	4.50
5.25	5.24
5.18	5.27
4.81	4.75
4.57	4.43
4.63	4.61
4.77	4.82

▶ **Try It Yourself 1**

A shoe manufacturer claims that athletes can decrease their times in the 40-yard dash using the manufacturer's new Strength Shoes®. The 40-yard dash times of 12 randomly selected athletes are measured. After the athletes have used the Strength Shoes® for 8 months, their 40-yard dash times are measured again. The times (in seconds) are listed at the left. At $\alpha = 0.05$, is there enough evidence to support the manufacturer's claim? Assume the times are normally distributed. *(Adapted from Coaches Sports Publishing)*

a. Identify the *claim* and state H_0 and H_a.
b. Identify the *level of significance* α and the *degrees of freedom*.
c. Find the *critical value* t_0 and identify the *rejection region*.
d. *Calculate* $\overline{d}$ and s_d.
e. Use the *t*-test to find the *standardized test statistic t*. *Sketch* a graph.
f. *Decide* whether to reject the null hypothesis.
g. *Interpret* the decision in the context of the original claim.

Answer: Page A43

Note that in Example 1 it is possible that the vertical jump height improved because of other reasons. Many advertisements misuse statistical results by implying a cause-and-effect relationship that has not been substantiated by testing.

EXAMPLE 2   Report 36

▶ **The *t*-Test for the Difference Between Means**

A state legislator wants to determine whether her performance rating (0–100) has changed from last year to this year. The following table shows the legislator's performance ratings from the same 16 randomly selected voters for last year and this year. At $\alpha = 0.01$, is there enough evidence to conclude that the legislator's performance rating has changed? Assume the performance ratings are normally distributed.

Voter	1	2	3	4	5	6	7	8
Rating (last year)	60	54	78	84	91	25	50	65
Rating (this year)	56	48	70	60	85	40	40	55

Voter	9	10	11	12	13	14	15	16
Rating (last year)	68	81	75	45	62	79	58	63
Rating (this year)	80	75	78	50	50	85	53	60

▶ **Solution**

If there is a change in the legislator's rating, there will be a difference between "this year's" ratings and "last year's" ratings. Because the legislator wants to see if there is a difference, the null and alternative hypotheses are

$$H_0: \mu_d = 0 \quad \text{and} \quad H_a: \mu_d \neq 0. \text{ (Claim)}$$

Because the test is a two-tailed test, $\alpha = 0.01$, and d.f. $= 16 - 1 = 15$, the critical values are $-t_0 = -2.947$ and $t_0 = 2.947$. The rejection regions are $t < -2.947$ and $t > 2.947$.

Before	After	d	d^2
60	56	4	16
54	48	6	36
78	70	8	64
84	60	24	576
91	85	6	36
25	40	−15	225
50	40	10	100
65	55	10	100
68	80	−12	144
81	75	6	36
75	78	−3	9
45	50	−5	25
62	50	12	144
79	85	−6	36
58	53	5	25
63	60	3	9
		$\Sigma = 53$	$\Sigma = 1581$

Using the table at the left, you can calculate $\bar{d}$ and s_d as shown below.

$$\bar{d} = \frac{\Sigma d}{n} = \frac{53}{16} = 3.3125$$

$$s_d = \sqrt{\frac{\Sigma d^2 - \left[\frac{(\Sigma d)^2}{n}\right]}{n - 1}}$$

$$= \sqrt{\frac{1581 - \frac{53^2}{16}}{16 - 1}} \approx 9.6797$$

The standardized test statistic is

$$t = \frac{\bar{d} - \mu_d}{s_d/\sqrt{n}} \qquad \text{Use the } t\text{-test.}$$

$$\approx \frac{3.3125 - 0}{9.6797/\sqrt{16}} \qquad \text{Assume } \mu_d = 0.$$

$$\approx 1.369.$$

The graph at the right shows the location of the rejection region and the standardized test statistic t. Because t is not in the rejection region, you should fail to reject the null hypothesis.

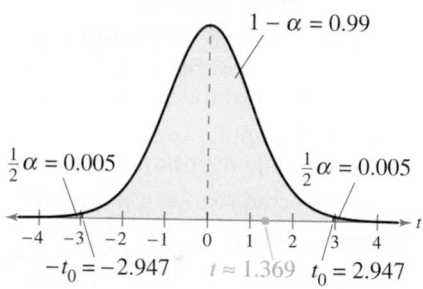

Interpretation There is not enough evidence at the 1% level of significance to conclude that the legislator's performance rating has changed.

▶ **Try It Yourself 2**

A medical researcher wants to determine whether a drug changes the body's temperature. Seven test subjects are randomly selected, and the body temperature (in degrees Fahrenheit) of each is measured. The subjects are then given the drug and, after 20 minutes, the body temperature of each is measured again. The results are listed below. At $\alpha = 0.05$, is there enough evidence to conclude that the drug changes the body's temperature? Assume the body temperatures are normally distributed.

Subject	1	2	3	4	5	6	7
Initial temperature	101.8	98.5	98.1	99.4	98.9	100.2	97.9
Second temperature	99.2	98.4	98.2	99	98.6	99.7	97.8

a. Identify the *claim* and state H_0 and H_a.
b. Identify the *level of significance* α and the *degrees of freedom*.
c. Find the *critical values* and identify the *rejection regions*.
d. Calculate $\bar{d}$ and s_d.
e. Use the t-test to find the *standardized test statistic t*. *Sketch* a graph.
f. *Decide* whether to reject the null hypothesis.
g. *Interpret* the decision in the context of the original claim.

Answer: Page A43

8.3 EXERCISES

BUILDING BASIC SKILLS AND VOCABULARY

1. What conditions are necessary in order to use the dependent samples t-test for the mean of the difference of two populations?

2. Explain what the symbols $\bar{d}$ and s_d represent.

In Exercises 3–8, test the claim about the mean of the difference of two populations. Use a t-test for dependent, random samples at the given level of significance with the given statistics. Is the test right-tailed, left-tailed, or two-tailed? Assume the populations are normally distributed.

3. Claim: $\mu_d < 0; \alpha = 0.05$. Statistics: $\bar{d} = 1.5, s_d = 3.2, n = 14$

4. Claim: $\mu_d = 0; \alpha = 0.01$. Statistics: $\bar{d} = 3.2, s_d = 8.45, n = 8$

5. Claim: $\mu_d \leq 0; \alpha = 0.10$. Statistics: $\bar{d} = 6.5, s_d = 9.54, n = 16$

6. Claim: $\mu_d > 0; \alpha = 0.05$. Statistics: $\bar{d} = 0.55, s_d = 0.99, n = 28$

7. Claim: $\mu_d \geq 0; \alpha = 0.01$. Statistics: $\bar{d} = -2.3, s_d = 1.2, n = 15$

8. Claim: $\mu_d \neq 0; \alpha = 0.10$. Statistics: $\bar{d} = -1, s_d = 2.75, n = 20$

USING AND INTERPRETING CONCEPTS

Testing the Difference Between Two Means *In Exercises 9–18, (a) identify the claim and state H_0 and H_a, (b) find the critical value(s) and identify the rejection region(s), (c) calculate $\bar{d}$ and s_d, (d) find the standardized test statistic t, (e) decide whether to reject or fail to reject the null hypothesis, and (f) interpret the decision in the context of the original claim. If convenient, use technology to solve the problem. For each randomly selected sample, assume the population is normally distributed.*

9. **Grammatical Errors**   A teacher claims that a grammar seminar will help students reduce the number of grammatical errors they make when writing a 1000-word essay. The table shows the number of grammatical errors made by seven students before participating in the seminar and after participating in the seminar. At $\alpha = 0.01$, is there enough evidence to conclude that the seminar reduced the number of errors?

Student	1	2	3	4	5	6	7
Errors (before)	15	10	12	8	5	4	9
Errors (after)	11	9	6	5	1	0	9

10. **SAT Scores**   An SAT preparation course claims to improve the test scores of students. The table shows the critical reading scores for 10 students the first two times they took the SAT. Before taking the SAT for the second time, the students took a course to try to improve their critical reading SAT scores. Test the claim at $\alpha = 0.01$.

Student	1	2	3	4	5	6	7	8	9	10
Score (first)	308	456	352	433	306	471	422	370	320	418
Score (second)	400	524	409	491	348	583	451	408	391	450

Answers (margin)

1. (1) Each sample must be randomly selected.

   (2) Each member of the first sample must be paired with a member of the second sample.

   (3) Both populations must be normally distributed.

2. See Selected Answers, page A127.

3. Left-tailed test; Fail to reject H_0.

4. Two-tailed test; Fail to reject H_0.

5. Right-tailed test; Reject H_0.

6. Right-tailed test; Reject H_0.

7. Left-tailed test; Reject H_0.

8. Two-tailed test; Fail to reject H_0.

9. (a) The claim is "a grammar seminar will help students reduce the number of grammatical errors."

   $H_0: \mu_d \leq 0$; $H_a: \mu_d > 0$ (claim)

   (b) $t_0 = 3.143$;
   Rejection region: $t > 3.143$

   (c) $\bar{d} \approx 3.143$; $s_d \approx 2.035$

   (d) 4.085

   (e) Reject H_0.

   (f) There is enough evidence at the 1% level of significance to support the teacher's claim that a grammar seminar will help students reduce the number of grammatical errors.

10. See Selected Answers, page A127.

11. (a) The claim is "a particular exercise program will help participants lose weight after one month."

$H_0: \mu_d \leq 0$; $H_a: \mu_d > 0$ (claim)

(b) $t_0 = 1.363$;
Rejection region: $t > 1.363$

(c) $\bar{d} = 3.75$; $s_d \approx 7.841$

(d) 1.657

(e) Reject H_0.

(f) There is enough evidence at the 10% level of significance to support the nutritionist's claim that the exercise program helps participants lose weight after one month.

12. (a) The claim is "a baseball clinic will help players raise their batting averages."

$H_0: \mu_d \geq 0$; $H_a: \mu_d < 0$ (claim)

(b) $t_0 = -1.771$;
Rejection region: $t < -1.771$

(c) $\bar{d} \approx -0.002$; $s_d \approx 0.015$

(d) -0.431

(e) Fail to reject H_0.

(f) There is not enough evidence at the 5% level of significance to support the claim that the baseball clinic will help players raise their batting averages.

13. (a) The claim is "soft tissue therapy and spinal manipulation help to reduce the length of time patients suffer from headaches."

$H_0: \mu_d \leq 0$; $H_a: \mu_d > 0$ (claim)

(b) $t_0 = 2.764$;
Rejection region: $t > 2.764$

(c) $\bar{d} \approx 1.255$; $s_d \approx 0.441$

(d) 9.429

(e) Reject H_0.

(f) There is enough evidence at the 1% level of significance to support the physical therapist's claim that soft tissue therapy and spinal manipulation help reduce the length of time patients suffer from headaches.

11. Losing Weight   A nutritionist claims that a particular exercise program will help participants lose weight after one month. The table shows the weights of 12 adults before participating in the exercise program and one month after participating in the exercise program. At $\alpha = 0.10$, can you conclude that the exercise program helps participants lose weight?

Participant	1	2	3	4	5	6
Weight before exercise program	157	185	120	212	230	165
Weight after exercise program	150	181	121	206	215	169

Participant	7	8	9	10	11	12
Weight before exercise program	207	251	196	140	137	172
Weight after exercise program	210	232	188	138	145	172

12. Batting Averages   A coach suggests that a baseball clinic will help players raise their batting averages. The table shows the batting averages of 14 players before participating in the clinic and two months after participating in the clinic. At $\alpha = 0.05$, is there enough evidence to conclude that the clinic helped the players raise their batting averages?

Player	1	2	3	4	5	6	7
Batting average before clinic	0.290	0.275	0.278	0.310	0.302	0.325	0.256
Batting average after clinic	0.295	0.320	0.280	0.300	0.298	0.330	0.260

Player	8	9	10	11	12	13	14
Batting average before clinic	0.350	0.380	0.316	0.270	0.300	0.330	0.340
Batting average after clinic	0.345	0.380	0.315	0.280	0.282	0.336	0.325

13. Headaches   A physical therapist suggests that soft tissue therapy and spinal manipulation help to reduce the lengths of time patients suffer from headaches. The table shows the number of hours per day 11 patients suffered from headaches before and after 7 weeks of receiving treatment. At $\alpha = 0.01$, is there enough evidence to support the therapist's claim? *(Adapted from The Journal of the American Medical Association)*

Patient	1	2	3	4	5	6	7	8	9	10	11
Hours (before)	2.8	2.4	2.8	2.6	2.7	2.9	3.2	2.9	4.1	1.6	2.5
Hours (after)	1.6	1.3	1.6	1.4	1.5	1.6	1.7	1.6	1.8	1.2	1.4

14. (a) The claim is "one 600-mg dose of Vitamin C will increase muscular endurance."

$H_0: \mu_d \geq 0$; $H_a: \mu_d < 0$ (claim)

(b) $t_0 = -1.761$;

Rejection region: $t < -1.761$

(c) $\bar{d} \approx 48.467$; $s_d \approx 239.005$

(d) 0.785

(e) Fail to reject H_0.

(f) There is not enough evidence at the 5% level of significance to support the physical therapist's claim that one 600-mg dose of Vitamin C will increase muscular endurance.

15. (a) The claim is "the new drug reduces systolic blood pressure."

$H_0: \mu_d \leq 0$; $H_a: \mu_d > 0$ (claim)

(b) $t_0 = 1.895$;

Rejection region: $t > 1.895$

(c) $\bar{d} = 14.75$; $s_d \approx 6.861$

(d) 6.081

(e) Reject H_0.

(f) There is enough evidence at the 5% level of significance to support the pharmaceutical company's claim that its new drug reduces systolic blood pressure.

16. (a) The claim is "garlic can reduce the thickness of plaque buildup in arteries."

$H_0: \mu_d \leq 0$; $H_a: \mu_d > 0$ (claim)

(b) $t_0 = 1.860$;

Rejection region: $t > 1.860$

(c) $\bar{d} = 0.020$; $s_d \approx 0.031$

(d) 1.947

(e) Reject H_0.

(f) There is enough evidence at the 5% level of significance to support the claim that garlic can reduce the thickness of plaque buildup in arteries.

17. (a) The claim is "the product ratings have changed from last year to this year."

$H_0: \mu_d = 0$; $H_a: \mu_d \neq 0$ (claim)

(b) $-t_0 = -2.365$, $t_0 = 2.365$

Rejection regions: $t < -2.365$, $t > 2.365$

14. Grip Strength   A physical therapist suggests that one 600-mg dose of Vitamin C will increase muscular endurance. The table shows the number of repetitions 15 males made on a hand dynamometer (measures grip strength) until the grip strengths in three consecutive trials were 50% of their maximum grip strength. At $\alpha = 0.05$, test the claim that Vitamin C will increase muscular endurance. *(Adapted from Journal of Sports Medicine and Physical Fitness)*

Participant	1	2	3	4	5	6	7	8
Repetitions using placebo	417	279	678	636	170	699	372	582
Repetitions using Vitamin C	145	185	387	593	248	245	349	902

Participant	9	10	11	12	13	14	15
Repetitions using placebo	363	258	288	526	180	172	278
Repetitions using Vitamin C	159	122	264	1052	218	117	185

15. Blood Pressure   A pharmaceutical company guarantees that its new drug reduces systolic blood pressure. The table shows the systolic blood pressures (in millimeters of mercury) of eight patients before taking the new drug and two hours after taking the drug. At $\alpha = 0.05$, can you conclude that the new drug reduces systolic blood pressure?

Patient	1	2	3	4
Systolic blood pressure (before)	201	171	186	162
Systolic blood pressure (after)	192	165	167	155

Patient	5	6	7	8
Systolic blood pressure (before)	165	167	175	148
Systolic blood pressure (after)	148	144	152	134

16. Plaque Thickness   A researcher believes that garlic can reduce the thickness of plaque buildup in arteries. The table shows the thicknesses (in millimeters) of plaque in the carotid arteries of nine patients with mild atherosclerosis before taking garlic on a daily basis and after four years of taking garlic on a daily basis. At $\alpha = 0.05$, can you conclude that garlic reduces the thickness of plaque buildup?

Patient	1	2	3	4	5	6	7	8	9
Thickness (before)	0.78	0.65	0.73	0.85	0.68	0.80	0.64	0.72	0.82
Thickness (after)	0.75	0.67	0.70	0.87	0.63	0.76	0.60	0.74	0.77

(c) $\bar{d} = -1$; $s_d \approx 1.309$

(d) -2.160

(e) Fail to reject H_0.

(f) There is not enough evidence at the 5% level of significance to support the claim that the product ratings have changed from last year to this year.

18. (a) The claim is "the scoring averages have changed."

   $H_0: \mu_d = 0$; $H_a: \mu_d \neq 0$ (claim)

   (b) $-t_0 = -1.833$, $t_0 = 1.833$;
   Rejection regions:
   $t < -1.833$, $t > 1.833$

   (c) $\bar{d} = -1.280$; $s_d \approx 3.051$

   (d) -1.327

   (e) Fail to reject H_0.

   (f) There is not enough evidence at the 10% level of significance to support the claim that the basketball players' scoring averages have changed.

19. Hypothesis test results:

   $\mu_1 - \mu_2$: mean of the paired difference between Cholesterol (before) and Cholesterol (after)

   $H_0 : \mu_1 - \mu_2 = 0$

   $H_A : \mu_1 - \mu_2 > 0$

   See Odd Answers, page A90.

   $P = 0.0702 > 0.05$, so fail to reject H_0.

   There is not enough evidence at the 5% level of significance to support the claim that the new cereal lowers total blood cholesterol levels.

20. Hypothesis test results:

   $\mu_1 - \mu_2$: mean of the paired difference between Time (beginning) and Time (end)

   $H_0 : \mu_1 - \mu_2 = 0$

   $H_A : \mu_1 - \mu_2 \neq 0$

   See Selected Answers, page A127.

   $P = 0.0029 < 0.01$, so reject H_0.

   There is enough evidence at the 1% level of significance to support the claim that the contestants' times have changed.

17. Product Ratings A company wants to determine whether its consumer product ratings (0–10) have changed from last year to this year. The table shows the company's product ratings from the same eight consumers for last year and this year. At $\alpha = 0.05$, is there enough evidence to conclude that the product ratings have changed?

Consumer	1	2	3	4	5	6	7	8
Rating (last year)	5	7	2	3	9	10	8	7
Rating (this year)	5	9	4	6	9	9	9	8

18. Points Per Game The scoring averages (in points per game) of 10 professional basketball players for their rookie and sophomore seasons are shown in the table below. At $\alpha = 0.10$, is there enough evidence to conclude that the scoring averages have changed? *(Source: National Basketball Association)*

Player	1	2	3	4	5	6	7	8	9	10
Points per game (rookie)	18.5	13.9	16.1	15.3	16.8	13.0	11.9	11.8	11.1	11.1
Points per game (sophomore)	17.5	14.7	16.9	16.3	20.4	18.9	14.6	6.3	14.2	12.5

SC *In Exercises 19 and 20, use StatCrunch to help you test the claim about the difference between two population means. For each randomly selected sample, assume the population is normally distributed.*

19. Cholesterol Levels A food manufacturer claims that eating its new cereal as part of a daily diet lowers total blood cholesterol levels. The table shows the total blood cholesterol levels (in milligrams per deciliter of blood) of seven patients before eating the cereal and after one year of eating the cereal as part of their diets. At $\alpha = 0.05$, can you conclude that the new cereal lowers total blood cholesterol levels?

Patient	1	2	3	4	5	6	7
Total blood cholesterol level (before)	210	225	240	250	255	270	235
Total blood cholesterol level (after)	200	220	245	248	252	268	232

20. Obstacle Course On a television show, eight contestants try to lose the highest percentage of weight in order to win a cash prize. As part of the show, the contestants are timed as they run an obstacle course. The table shows the times (in seconds) of the contestants at the beginning of the season and at the end of the season. At $\alpha = 0.01$, is there enough evidence to conclude that the contestants' times have changed?

Contestant	1	2	3	4	5	6	7	8
Time (beginning)	130.2	104.8	100.1	136.4	125.9	122.6	150.4	158.2
Time (end)	121.5	100.7	90.2	135.0	112.1	120.5	139.8	142.9

21. Yes; $P \approx 0.0003 < 0.05$, so you reject H_0.

22. Yes; $P \approx 0.2173 > 0.10$, so you fail to reject H_0.

23. $-1.76 < \mu_d < -1.29$

24. $-0.83 < \mu_d < -0.05$

■ EXTENDING CONCEPTS

21. In Exercise 15, use a P-value to perform the hypothesis test. Compare your result with the result obtained using rejection regions. Are they the same?

22. In Exercise 18, use a P-value to perform the hypothesis test. Compare your result with the result obtained using rejection regions. Are they the same?

Constructing Confidence Intervals for μ_d *To construct a confidence interval for μ_d, use the following inequality.*

$$\overline{d} - t_c \frac{s_d}{\sqrt{n}} < \mu_d < \overline{d} + t_c \frac{s_d}{\sqrt{n}}$$

In Exercises 23 and 24, construct the indicated confidence interval for μ_d. Assume the populations are normally distributed.

23. Drug Testing A sleep disorder specialist wants to test the effectiveness of a new drug that is reported to increase the number of hours of sleep patients get during the night. To do so, the specialist randomly selects 16 patients and records the number of hours of sleep each gets with and without the new drug. The results of the two-night study are listed below. Construct a 90% confidence interval for μ_d.

Patient	1	2	3	4	5	6	7	8	9
Hours of sleep without the drug	1.8	2.0	3.4	3.5	3.7	3.8	3.9	3.9	4.0
Hours of sleep using the drug	3.0	3.6	4.0	4.4	4.5	5.2	5.5	5.7	6.2

Patient	10	11	12	13	14	15	16
Hours of sleep without the drug	4.9	5.1	5.2	5.0	4.5	4.2	4.7
Hours of sleep using the drug	6.3	6.6	7.8	7.2	6.5	5.6	5.9

24. Herbal Medicine Testing An herbal medicine is tested on 14 randomly selected patients with sleeping disorders. The table shows the number of hours of sleep patients got during one night without using the herbal medicine and the number of hours of sleep the patients got on another night after the herbal medicine had been administered. Construct a 95% confidence interval for μ_d.

Patient	1	2	3	4	5	6	7
Hours of sleep without medicine	1.0	1.4	3.4	3.7	5.1	5.1	5.2
Hours of sleep using the herbal medicine	2.9	3.3	3.5	4.4	5.0	5.0	5.2

Patient	8	9	10	11	12	13	14
Hours of sleep without medicine	5.3	5.5	5.8	4.2	4.8	2.9	4.5
Hours of sleep using the herbal medicine	5.3	6.0	6.5	4.4	4.7	3.1	4.7

8.4 Testing the Difference Between Proportions

WHAT YOU SHOULD LEARN

▶ How to perform a z-test for the difference between two population proportions p_1 and p_2

Two-Sample z-Test for the Difference Between Proportions

▶ TWO-SAMPLE z-TEST FOR THE DIFFERENCE BETWEEN PROPORTIONS

In this section, you will learn how to use a z-test to test the difference between two population proportions p_1 and p_2 using a sample proportion from each population. If a claim is about two population parameters p_1 and p_2, then some possible pairs of null and alternative hypotheses are

$$\begin{cases} H_0: p_1 = p_2 \\ H_a: p_1 \neq p_2 \end{cases}, \quad \begin{cases} H_0: p_1 \leq p_2 \\ H_a: p_1 > p_2 \end{cases}, \quad \text{and} \quad \begin{cases} H_0: p_1 \geq p_2 \\ H_a: p_1 < p_2 \end{cases}.$$

Regardless of which hypotheses you use, you always assume there is no difference between the population proportions, or $p_1 = p_2$.

For instance, suppose you want to determine whether the proportion of female college students who earn a bachelor's degree in four years is different from the proportion of male college students who earn a bachelor's degree in four years. The following conditions are necessary to use a z-test to test such a difference.

1. The samples must be randomly selected.

2. The samples must be independent.

3. The samples must be large enough to use a normal sampling distribution. That is, $n_1 p_1 \geq 5$, $n_1 q_1 \geq 5$, $n_2 p_2 \geq 5$, and $n_2 q_2 \geq 5$.

If these conditions are met, then the **sampling distribution for $\hat{p}_1 - \hat{p}_2$, the difference between the sample proportions,** is a normal distribution with mean

$$\mu_{\hat{p}_1 - \hat{p}_2} = p_1 - p_2$$

and standard error

$$\sigma_{\hat{p}_1 - \hat{p}_2} = \sqrt{\frac{p_1 q_1}{n_1} + \frac{p_2 q_2}{n_2}}.$$

Notice that you need to know the population proportions to calculate the standard error. Because a hypothesis test for $p_1 - p_2$ is based on the assumption that $p_1 = p_2$, you can calculate a weighted estimate of p_1 and p_2 using

$$\overline{p} = \frac{x_1 + x_2}{n_1 + n_2}, \text{ where } x_1 = n_1 \hat{p}_1 \text{ and } x_2 = n_2 \hat{p}_2.$$

With the weighted estimate $\overline{p}$, the standard error of the sampling distribution for $\hat{p}_1 - \hat{p}_2$ is

$$\sigma_{\hat{p}_1 - \hat{p}_2} = \sqrt{\overline{p}\,\overline{q}\left(\frac{1}{n_1} + \frac{1}{n_2}\right)}, \text{ where } \overline{q} = 1 - \overline{p}.$$

Also observe that you need to know the population proportions to verify that the samples are large enough to be approximated by the normal distribution. But when determining whether the z-test can be used for the difference between proportions for a binomial experiment, you should use $\overline{p}$ in place of p_1 and p_2 and use $\overline{q}$ in place of q_1 and q_2.

STUDY TIP

The following symbols are used in the z-test for $p_1 - p_2$. See Sections 4.2 and 5.5 to review the binomial distribution.

Symbol	Description
p_1, p_2	Population proportions
x_1, x_2	Number of successes in each sample
n_1, n_2	Size of each sample
$\hat{p}_1, \hat{p}_2$	Sample proportions of successes
$\overline{p}$	Weighted estimate for p_1 and p_2

PICTURING THE WORLD

A medical research team conducted a study to test whether a drug lowers the chance of getting diabetes. In the study, 2623 people took the drug and 2646 people took a placebo. The results are shown below. (Source: The New England Journal of Medicine)

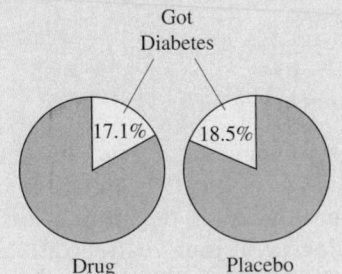

Got Diabetes

17.1%   18.5%

Drug   Placebo

At $\alpha = 0.05$, can you support the claim that the drug lowers the chance of getting diabetes?

There is not enough evidence at the 5% level of significance to support the claim that the drug lowers the chance of getting diabetes.

If the sampling distribution for $\hat{p}_1 - \hat{p}_2$ is normal, you can use a two-sample z-test to test the difference between two population proportions p_1 and p_2.

TWO-SAMPLE z-TEST FOR THE DIFFERENCE BETWEEN PROPORTIONS

A two-sample z-test is used to test the difference between two population proportions p_1 and p_2 when a sample is randomly selected from each population. The **test statistic** is $\hat{p}_1 - \hat{p}_2$, and the **standardized test statistic** is

$$z = \frac{(\hat{p}_1 - \hat{p}_2) - (p_1 - p_2)}{\sqrt{\overline{p}\,\overline{q}\left(\dfrac{1}{n_1} + \dfrac{1}{n_2}\right)}}$$

where

$$\overline{p} = \frac{x_1 + x_2}{n_1 + n_2} \quad \text{and} \quad \overline{q} = 1 - \overline{p}.$$

Note: $n_1\overline{p}$, $n_1\overline{q}$, $n_2\overline{p}$, and $n_2\overline{q}$ must be at least 5.

If the null hypothesis states $p_1 = p_2$, $p_1 \le p_2$, or $p_1 \ge p_2$, then $p_1 = p_2$ is assumed and the expression $p_1 - p_2$ is equal to 0 in the preceding test.

GUIDELINES

Using a Two-Sample z-Test for the Difference Between Proportions

IN WORDS	IN SYMBOLS
1. State the claim mathematically and verbally. Identify the null and alternative hypotheses.	State H_0 and H_a.
2. Specify the level of significance.	Identify α.
3. Determine the critical value(s).	Use Table 4 in Appendix B.
4. Determine the rejection region(s).	
5. Find the weighted estimate of p_1 and p_2. Verify that $n_1\overline{p}$, $n_1\overline{q}$, $n_2\overline{p}$, and $n_2\overline{q}$ are at least 5.	$\overline{p} = \dfrac{x_1 + x_2}{n_1 + n_2}$
6. Find the standardized test statistic and sketch the sampling distribution.	$z = \dfrac{(\hat{p}_1 - \hat{p}_2) - (p_1 - p_2)}{\sqrt{\overline{p}\,\overline{q}\left(\dfrac{1}{n_1} + \dfrac{1}{n_2}\right)}}$
7. Make a decision to reject or fail to reject the null hypothesis.	If z is in the rejection region, reject H_0. Otherwise, fail to reject H_0.
8. Interpret the decision in the context of the original claim.	

A hypothesis test for the difference between proportions can also be performed using *P*-values. Use the guidelines listed above, skipping Steps 3 and 4. After finding the standardized test statistic, use the Standard Normal Table to calculate the *P*-value. Then make a decision to reject or fail to reject the null hypothesis. If *P* is less than or equal to α, reject H_0. Otherwise, fail to reject H_0.

See TI-83/84 Plus steps on page 479.

EXAMPLE 1

▶ **A Two-Sample z-Test for the Difference Between Proportions**

A study of 150 randomly selected occupants in passenger cars and 200 randomly selected occupants in pickup trucks shows that 86% of occupants in passenger cars and 74% of occupants in pickup trucks wear seat belts. At $\alpha = 0.10$, can you reject the claim that the proportion of occupants who wear seat belts is the same for passenger cars and pickup trucks? (*Adapted from National Highway Traffic Safety Administration*)

▶ **Solution**

The claim is "the proportion of occupants who wear seat belts is the same for passenger cars and pickup trucks." So, the null and alternative hypotheses are

$$H_0: p_1 = p_2 \text{ (Claim)} \quad \text{and} \quad H_a: p_1 \neq p_2.$$

Because the test is two-tailed and the level of significance is $\alpha = 0.10$, the critical values are $-z_0 = -1.645$ and $z_0 = 1.645$. The rejection regions are $z < -1.645$ and $z > 1.645$. The weighted estimate of p_1 and p_2 is

$$\overline{p} = \frac{x_1 + x_2}{n_1 + n_2} = \frac{129 + 148}{150 + 200} = \frac{277}{350} \approx 0.7914$$

and

$$\overline{q} = 1 - \overline{p} \approx 1 - 0.7914 = 0.2086.$$

Because $n_1\overline{p} \approx 150(0.7914)$, $n_1\overline{q} \approx 150(0.2086)$, $n_2\overline{p} \approx 200(0.7914)$, and $n_2\overline{q} \approx 200(0.2086)$ are at least 5, you can use a two-sample z-test. The standardized test statistic is

$$z = \frac{(\hat{p}_1 - \hat{p}_2) - (p_1 - p_2)}{\sqrt{\overline{p}\,\overline{q}\left(\dfrac{1}{n_1} + \dfrac{1}{n_2}\right)}} \approx \frac{(0.86 - 0.74) - 0}{\sqrt{(0.7914)(0.2086)\left(\dfrac{1}{150} + \dfrac{1}{200}\right)}} \approx 2.73.$$

The graph at the left shows the location of the rejection regions and the standardized test statistic. Because z is in the rejection region, you should decide to reject the null hypothesis.

Interpretation There is enough evidence at the 10% level of significance to reject the claim that the proportion of occupants who wear seat belts is the same for passenger cars and pickup trucks.

▶ **Try It Yourself 1**

Consider the results of the *NYTS* study discussed in the Chapter Opener. At $\alpha = 0.05$, can you support the claim that there is a difference between the proportion of male high school students who smoke cigarettes and the proportion of female high school students who smoke cigarettes?

a. Identify the *claim* and state H_0 and H_a.
b. Identify the *level of significance* α.
c. Find the *critical values* and identify the *rejection regions*.
d. *Find* $\overline{p}$ and $\overline{q}$.
e. *Verify* that $n_1\overline{p}$, $n_1\overline{q}$, $n_2\overline{p}$, and $n_2\overline{q}$ are at least 5.
f. Find the *standardized test statistic z*. *Sketch* a graph.
g. *Decide* whether to reject the null hypothesis.
h. *Interpret* the decision in the context of the original claim.

Answer: Page A44

Sample Statistics for Vehicles

Passenger cars	Pickup trucks
$n_1 = 150$	$n_2 = 200$
$\hat{p}_1 = 0.86$	$\hat{p}_2 = 0.74$
$x_1 = 129$	$x_2 = 148$

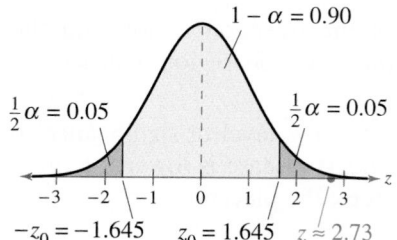

$1 - \alpha = 0.90$

$\frac{1}{2}\alpha = 0.05$ $\frac{1}{2}\alpha = 0.05$

$-z_0 = -1.645$ $z_0 = 1.645$ $z \approx 2.73$

Sample Statistics for Cholesterol-Reducing Medication

Received medication	Received placebo
$n_1 = 4700$	$n_2 = 4300$
$x_1 = 301$	$x_2 = 357$
$\hat{p}_1 = 0.064$	$\hat{p}_2 = 0.083$

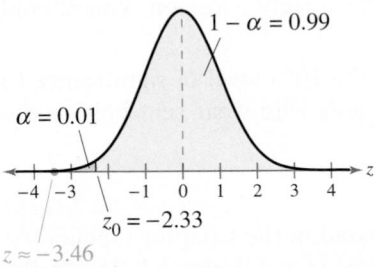

EXAMPLE 2

▸ **A Two-Sample z-Test for the Difference Between Proportions**

A medical research team conducted a study to test the effect of a cholesterol-reducing medication. At the end of the study, the researchers found that of the 4700 randomly selected subjects who took the medication, 301 died of heart disease. Of the 4300 randomly selected subjects who took a placebo, 357 died of heart disease. At $\alpha = 0.01$, can you support the claim that the death rate due to heart disease is lower for those who took the medication than for those who took the placebo? *(Adapted from The New England Journal of Medicine)*

▸ **Solution**

The claim is "the death rate due to heart disease is lower for those who took the medication than for those who took the placebo." So, the null and alternative hypotheses are

$$H_0: p_1 \geq p_2 \quad \text{and} \quad H_a: p_1 < p_2. \text{ (Claim)}$$

Because the test is left-tailed and the level of significance is $\alpha = 0.01$, the critical value is $z_0 = -2.33$. The rejection region is $z < -2.33$. The weighted estimate of p_1 and p_2 is

$$\overline{p} = \frac{x_1 + x_2}{n_1 + n_2} = \frac{301 + 357}{4700 + 4300} = \frac{658}{9000} \approx 0.0731$$

and

$$\overline{q} = 1 - \overline{p} \approx 1 - 0.0731 = 0.9269.$$

Because $n_1\overline{p} = 4700(0.0731)$, $n_1\overline{q} = 4700(0.9269)$, $n_2\overline{p} = 4300(0.0731)$, and $n_2\overline{q} = 4300(0.9269)$ are at least 5, you can use a two-sample z-test.

$$z = \frac{(\hat{p}_1 - \hat{p}_2) - (p_1 - p_2)}{\sqrt{\overline{p}\,\overline{q}\left(\frac{1}{n_1} + \frac{1}{n_2}\right)}} \approx \frac{(0.064 - 0.083) - 0}{\sqrt{(0.0731)(0.9269)\left(\frac{1}{4700} + \frac{1}{4300}\right)}} \approx -3.46$$

The graph at the left shows the location of the rejection region and the standardized test statistic. Because z is in the rejection region, you should decide to reject the null hypothesis.

Interpretation There is enough evidence at the 1% level of significance to support the claim that the death rate due to heart disease is lower for those who took the medication than for those who took the placebo.

▸ **Try It Yourself 2**

Consider the results of the *NYTS* study discussed in the Chapter Opener. At $\alpha = 0.05$, can you support the claim that the proportion of male high school students who smoke cigars is greater than the proportion of female high school students who smoke cigars?

a. Identify the *claim* and state H_0 and H_a.
b. Identify the *level of significance* α.
c. Find the *critical value* and identify the *rejection region*.
d. *Find* $\overline{p}$ and $\overline{q}$.
e. *Verify* that $n_1\overline{p}$, $n_1\overline{q}$, $n_2\overline{p}$, and $n_2\overline{q}$ are at least 5.
f. Find the *standardized test statistic z. Sketch* a graph.
g. *Decide* whether to reject the null hypothesis.
h. *Interpret* the decision in the context of the original claim.

Answer: Page A44

8.4 EXERCISES

1. (1) The samples must be randomly selected.

   (2) The samples must be independent.

   (3) $n_1\bar{p} \geq 5$, $n_1\bar{q} \geq 5$, $n_2\bar{p} \geq 5$, and $n_2\bar{q} \geq 5$

2. State the hypotheses and identify the claim. Specify the level of significance. Find the critical value(s) and rejection region(s). Find $\bar{p}$ and $\bar{q}$. Find the standardized test statistic. Make a decision and interpret it in the context of the claim.

   The test can also be performed by calculating the P-value and comparing it to α.

3. Can use normal sampling distribution; Fail to reject H_0.

4. Can use normal sampling distribution; Reject H_0.

5. Can use normal sampling distribution; Reject H_0.

6. Cannot use normal sampling distribution.

7. Can use normal sampling distribution; Fail to reject H_0.

8. Can use normal sampling distribution; Fail to reject H_0.

9. See Odd Answers, page A90.

■ BUILDING BASIC SKILLS AND VOCABULARY

1. What conditions are necessary in order to use the z-test to test the difference between two population proportions?

2. Explain how to perform a two-sample z-test for the difference between two population proportions.

In Exercises 3–8, decide whether the normal sampling distribution can be used. If it can be used, test the claim about the difference between two population proportions p_1 and p_2 at the given level of significance α using the given sample statistics. Assume the sample statistics are from independent, random samples.

3. Claim: $p_1 \neq p_2$; $\alpha = 0.01$.
   Sample statistics: $x_1 = 35$, $n_1 = 70$ and $x_2 = 36$, $n_2 = 60$

4. Claim: $p_1 < p_2$; $\alpha = 0.05$.
   Sample statistics: $x_1 = 471$, $n_1 = 785$ and $x_2 = 372$, $n_2 = 465$

5. Claim: $p_1 = p_2$; $\alpha = 0.10$.
   Sample statistics: $x_1 = 42$, $n_1 = 150$ and $x_2 = 76$, $n_2 = 200$

6. Claim: $p_1 > p_2$; $\alpha = 0.01$.
   Sample statistics: $x_1 = 6$, $n_1 = 20$ and $x_2 = 4$, $n_2 = 30$

7. Claim: $p_1 \leq p_2$; $\alpha = 0.10$.
   Sample statistics: $x_1 = 344$, $n_1 = 860$ and $x_2 = 304$, $n_2 = 800$

8. Claim: $p_1 = p_2$; $\alpha = 0.05$.
   Sample statistics: $x_1 = 29$, $n_1 = 45$ and $x_2 = 25$, $n_2 = 30$

■ USING AND INTERPRETING CONCEPTS

Testing the Difference Between Two Proportions   *In Exercises 9–18, (a) identify the claim and state H_0 and H_a, (b) find the critical value(s) and identify the rejection region(s), (c) find the standardized test statistic z, (d) decide whether to reject or fail to reject the null hypothesis, and (e) interpret the decision in the context of the original claim. If convenient, use technology to solve the problem. In each exercise, assume the samples are randomly selected and independent.*

9. **Plantar Heel Pain**   A medical research team conducted a study to test the effect of magnetic insoles for treating plantar heel pain. In the study, 54 subjects wore magnetic insoles and 41 subjects wore nonmagnetic insoles. All subjects wore their insoles for 4 weeks. The results are shown below. At $\alpha = 0.01$, can you support the claim that there is a difference in the proportion of subjects who feel all or mostly better after 4 weeks between subjects who used magnetic insoles and subjects who used nonmagnetic insoles? *(Adapted from The Journal of the American Medical Association)*

Do You Feel All or Mostly Better?

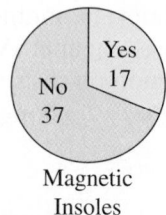

Magnetic Insoles

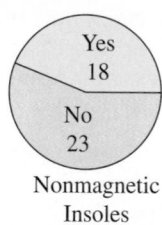

Nonmagnetic Insoles

10. See Selected Answers, page A127.

11. (a) The claim is "the proportion of males who enrolled in college is less than the proportion of females who enrolled in college."

$H_0: p_1 \geq p_2$;
$H_a: p_1 < p_2$ (claim)

(b) $z_0 = -1.645$;
Rejection region: $z < -1.645$

(c) -4.22

(d) Reject H_0.

(e) There is enough evidence at the 5% level of significance to support the claim that the proportion of males who enrolled in college is less than the proportion of females who enrolled in college.

12. See Selected Answers, page A127.

13. (a) The claim is "the proportion of subjects who are pain-free is the same for the two groups."

$H_0: p_1 = p_2$ (claim);
$H_a: p_1 \neq p_2$

(b) $-z_0 = -1.96$, $z_0 = 1.96$;
Rejection regions:
$z < -1.96$, $z > 1.96$

(c) 5.62 (*Tech:* 5.58)

(d) Reject H_0.

(e) There is enough evidence at the 5% level of significance to reject the claim that the proportion of subjects who are pain-free is the same for the two groups.

14. See Selected Answers, page A127.

15. (a) The claim is "the proportion of motorcyclists who wear a helmet is now greater."

$H_0: p_1 \leq p_2$;
$H_a: p_1 > p_2$ (claim)

(b) $z_0 = 1.645$;
Rejection region: $z > 1.645$

(c) 1.37

(d) Fail to reject H_0.

(e) There is not enough evidence at the 5% level of significance to support the claim that the proportion of motorcyclists who wear a helmet is now greater.

10. Cancer Drug A gastrointestinal stromal tumor is a rare form of cancer which develops in muscle tissue and blood vessels within the stomach or small intestine. A medical research team conducted a study to test the effect of a drug on this type of cancer. In the study, 300 subjects took the drug and 300 subjects took a placebo. All subjects had surgery to remove the tumor and then took the drug or placebo for one year. At $\alpha = 0.10$, can you support the claim that the proportion of subjects who are cancer-free after one year is greater for subjects who took the drug than for subjects who took a placebo? *(Adapted from American College of Surgeons Oncology Group)*

Are You Cancer-Free One Year After Surgery?

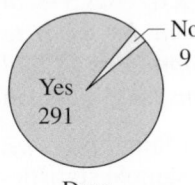

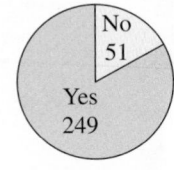

Drug   Placebo

11. Attending College In a survey of 875,000 males who completed high school during the past 12 months, 65.8% were enrolled in college. In a survey of 901,000 females who completed high school during the past 12 months, 66.1% were enrolled in college. At $\alpha = 0.05$, can you support the claim that the proportion of males who enrolled in college is less than the proportion of females who enrolled in college? *(Source: National Center for Education Statistics)*

12. Consumer Spending In a survey of 433 females, 72% have reduced the amount they spend on eating out. In a survey of 577 males, 65% have reduced the amount they spend on eating out. At $\alpha = 0.01$, can you reject the claim that there is no difference in the proportion of females who have reduced the amount they spend on eating out and the proportion of males who have reduced the amount they spend on eating out? *(Adapted from Morpace)*

13. Migraines A medical research team conducted a study to test the effect of a migraine drug. Of the 400 subjects who took the drug, 25% were pain-free after two hours. Of the 407 subjects who took a placebo, 10% were pain-free after two hours. At $\alpha = 0.05$, can you reject the claim that the proportion of subjects who are pain-free is the same for the two groups? *(Adapted from International Migraine Pain Assessment Clinical Trial)*

14. Migraines A medical research team conducted a study to test the effect of a migraine drug. Of the 400 subjects who took the drug, 65% were free of nausea after two hours. Of the 407 subjects who took a placebo, 53% were free of nausea after two hours. At $\alpha = 0.10$, can you support the claim that the proportion of subjects who are free of nausea is greater for subjects who took the drug than for subjects who took a placebo? *(Adapted from International Migraine Pain Assessment Clinical Trial)*

15. Motorcycle Helmet Use In a survey of 600 motorcyclists, 404 wear a helmet. In another survey of 500 motorcyclists taken one year before, 317 wore a helmet. At $\alpha = 0.05$, can you support the claim that the proportion of motorcyclists who wear a helmet is now greater? *(Adapted from National Highway Traffic Safety Administration)*

16. See Selected Answers, page A127.

17. (a) The claim is "the proportion of Internet users is the same for the two age groups."

$H_0: p_1 = p_2$ (claim);
$H_a: p_1 \neq p_2$

(b) $-z_0 = -2.575$, $z_0 = 2.575$;
Rejection regions:
$z < -2.575$, $z > 2.575$

(c) 5.31

(d) Reject H_0.

(e) There is enough evidence at the 1% level of significance to reject the claim that the proportion of Internet users is the same for the two age groups.

18. See Selected Answers, page A128.

19. There is enough evidence at the 5% level of significance to reject the claim that the proportion of customers who wait 20 minutes or less is the same at the Fairfax North and Fairfax South offices.

20. There is enough evidence at the 1% level of significance to support the claim that the proportion of customers who wait 20 minutes or less is greater at the Staunton office than it is at the Fairfax South office.

21. There is enough evidence at the 10% level of significance to support the claim that the proportion of customers who wait 20 minutes or less at the Roanoke office is less than the proportion of customers who wait 20 minutes or less at the Staunton office.

22. See Selected Answers, page A128.

23. No; When $\alpha = 0.01$, the rejection region becomes $z < -2.33$. Because $-2.02 > -2.33$, you fail to reject H_0. There is not enough evidence at the 1% level of significance to support the claim that the proportion of customers who wait 20 minutes or less at the Roanoke office is less than the proportion of customers who wait 20 minutes or less at the Staunton office.

24. See Selected Answers, page A128.

16. Motorcycle Helmet Use In a survey of 300 motorcyclists from the Northeast, 183 wear a helmet. In a survey of 300 motorcyclists from the Midwest, 201 wear a helmet. At $\alpha = 0.10$, can you support the claim that the proportion of motorcyclists who wear a helmet in the Northeast is less than the proportion of motorcyclists who wear a helmet in the Midwest? *(Adapted from National Highway Traffic Safety Administration)*

17. Internet Users In a survey of 450 adults 18 to 29 years of age, 419 said they use the Internet. In a survey of 400 adults 30 to 49 years of age, 324 said they use the Internet. At $\alpha = 0.01$, can you reject the claim that the proportion of Internet users is the same for the two age groups? *(Adapted from Pew Research Center)*

18. Internet Users In a survey of 485 adults who live in an urban area, 359 said they use the Internet. In a survey of 315 adults who live in a rural area, 221 said they use the Internet. At $\alpha = 0.10$, can you support the claim that the proportion of adults who use the Internet is greater for adults who live in an urban area than for adults who live in a rural area? *(Adapted from Pew Research Center)*

DMV Wait Times *In Exercises 19–22, refer to the figure, which shows the percentages of customers waiting 20 minutes or less at four district offices of the Department of Motor Vehicles (DMV) in Virginia. Assume the survey included 400 people from each district. (Adapted from Virginia Department of Motor Vehicles)*

19. Fairfax North and Fairfax South At $\alpha = 0.05$, can you reject the claim that the proportion of customers who wait 20 minutes or less is the same at the Fairfax North office and the Fairfax South office?

DMV Wait Times
Percentage of customers waiting 20 minutes or less

Fairfax North	60%
Fairfax South	51%
Roanoke	56%
Staunton	63%

20. Staunton and Fairfax South At $\alpha = 0.01$, can you support the claim that the proportion of customers who wait 20 minutes or less is greater at the Staunton office than at the Fairfax South office?

21. Roanoke and Staunton At $\alpha = 0.10$, can you support the claim that the proportion of customers who wait 20 minutes or less at the Roanoke office is less than the proportion of customers who wait 20 minutes or less at the Staunton office?

22. Roanoke and Fairfax North At $\alpha = 0.05$, can you support the claim that there is a difference between the proportion of customers who wait 20 minutes or less at the Roanoke office and the proportion of customers who wait 20 minutes or less at the Fairfax North office?

23. Writing Suppose you are testing Exercise 21 at $\alpha = 0.01$. Do you still make the same decision? Explain your reasoning.

24. Writing Suppose you are testing Exercise 22 at $\alpha = 0.10$. Do you still make the same decision? Explain your reasoning.

25. Hypothesis test results:

p_1 : proportion of successes for population 1

p_2 : proportion of successes for population 2

$p_1 - p_2$: difference in proportions

$H_0 : p_1 - p_2 = 0$

$H_A : p_1 - p_2 > 0$

See Odd Answers, page A91.

$P = 0.2518 > 0.05$, so fail to reject H_0.

There is not enough evidence at the 5% level of significance to support the claim that the proportion of men ages 18 to 24 living in their parents' homes was greater in 2000 than in 2009.

26. See Selected Answers, page A128.

27. Hypothesis test results:

p_1 : proportion of successes for population 1

p_2 : proportion of successes for population 2

$p_1 - p_2$: difference in proportions

$H_0 : p_1 - p_2 = 0$

$H_A : p_1 - p_2 \neq 0$

See Odd Answers, page A91.

$P < 0.0001 < 0.01$, so reject H_0.

There is enough evidence at the 1% level of significance to reject the claim that the proportion of 18- to 24-year-olds living in their parents' homes in 2000 was the same for men and women.

28. See Selected Answers, page A128.

29. $-0.028 < p_1 - p_2 < -0.012$

30. $0.021 < p_1 - p_2 < 0.039$

SC *In Exercises 25–28, refer to the figure and use StatCrunch to test the claim. Assume the survey included 13,300 men and 13,200 women in 2000 and 14,500 men and 14,200 women in 2009, and assume the samples are random and independent.* (Adapted from U.S. Census Bureau)

25. Men: Then and Now At $\alpha = 0.05$, can you support the claim that the proportion of men ages 18 to 24 living in their parents' homes was greater in 2000 than in 2009?

26. Women: Then and Now At $\alpha = 0.05$, can you support the claim that the proportion of women ages 18 to 24 living in their parents' homes was greater in 2000 than in 2009?

27. Then: Men and Women At $\alpha = 0.01$, can you reject the claim that the proportion of 18- to 24-year-olds living in their parents' homes in 2000 was the same for men and women?

Moving out of the nest

Percentage of 18- to 24-year-olds living in parents' homes in the USA:
☐ Men ■ Women

42.5%   44.8%

56.4%   56.0%

2000   2009

28. Now: Men and Women At $\alpha = 0.10$, can you reject the claim that the proportion of 18- to 24-year-olds living in their parents' homes in 2009 was the same for men and women?

■ EXTENDING CONCEPTS

Constructing Confidence Intervals for $p_1 - p_2$ *You can construct a confidence interval for the difference between two population proportions $p_1 - p_2$ by using the following inequality.*

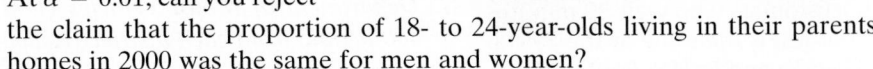
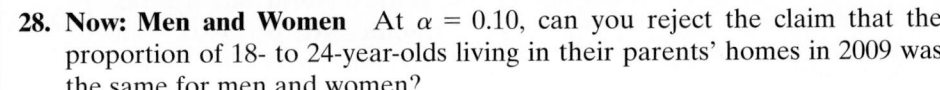

$$(\hat{p}_1 - \hat{p}_2) - z_c\sqrt{\frac{\hat{p}_1\hat{q}_1}{n_1} + \frac{\hat{p}_2\hat{q}_2}{n_2}} < p_1 - p_2 < (\hat{p}_1 - \hat{p}_2) + z_c\sqrt{\frac{\hat{p}_1\hat{q}_1}{n_1} + \frac{\hat{p}_2\hat{q}_2}{n_2}}$$

In Exercises 29 and 30, construct the indicated confidence interval for $p_1 - p_2$. Assume the samples are random and independent.

29. Students Planning to Study Education In a survey of 10,000 students taking the SAT, 7% were planning to study education in college. In another survey of 8000 students taken 10 years before, 9% were planning to study education in college. Construct a 95% confidence interval for $p_1 - p_2$, where p_1 is the proportion from the recent survey and p_2 is the proportion from the survey taken 10 years ago. (Adapted from The College Board)

30. Students Planning to Study Health-Related Fields In a survey of 10,000 students taking the SAT, 19% were planning to study health-related fields in college. In another survey of 8000 students taken 10 years before, 16% were planning to study health-related fields in college. Construct a 90% confidence interval for $p_1 - p_2$, where p_1 is the proportion from the recent survey and p_2 is the proportion from the survey taken 10 years ago. (Adapted from The College Board)

USES AND ABUSES

Uses

Hypothesis Testing with Two Samples Hypothesis testing enables you to decide whether differences in samples indicate actual differences in populations or are merely due to sampling error. For instance, a study conducted on 2 groups of 4-year-olds compared the behavior of the children who attended preschool with the behavior of those who stayed home with a parent. Aggressive behavior such as stealing toys, pushing other children, and starting fights was measured in both groups. The study showed that children who attended preschool were three times more likely to be aggressive than those who stayed home. These statistics were used to persuade parents to keep their children at home until they start school at age 5.

Abuses

Study Funding The study did not mention that it is normal for 4-year-olds to display aggressive behavior. Parents who keep their children at home but take them to play groups also observe their children being aggressive. Psychologists have suggested that this is the way children learn to interact with each other. The children who stayed home were less aggressive, but their behavior was considered abnormal. A follow-up study performed by a different group demonstrated that the children who stayed home before attending school ended up being more aggressive at a later age than those who had attended preschool.

The first study was funded by a mother support group who used the statistics to promote their own predetermined agenda. When dealing with statistics, always know who is paying for a study. (*Source: British Broadcasting Corporation*)

Using Nonrepresentative Samples In comparisons of data collected from two different samples, care should be taken to ensure that there are no confounding variables. For instance, suppose you are examining a claim that a new arthritis medication lessens joint pain.

If the group that is given the medication is over 60 years old and the group given the placebo is under 40, variables other than the medication might affect the outcome of the study. When you look for other abuses in a study, consider how the claim in the study was determined. What were the sample sizes? Were the samples random? Were they independent? Was the sampling conducted by an unbiased researcher?

People with arthritis

Group given new medication

Group given placebo

■ EXERCISES

1. ***Using Nonrepresentative Samples*** Assume that you work for the Food and Drug Administration. A pharmaceutical company has applied for approval to market a new arthritis medication. The research involved a test group that was given the medication and another test group that was given a placebo. Describe some ways that the test groups might not have been representative of the entire population of people with arthritis.

2. Medical research often involves blind and double-blind testing. Explain what these two terms mean.

8 CHAPTER SUMMARY

What did you **learn?**	EXAMPLE(S)	REVIEW EXERCISES

Section 8.1

- How to decide whether two samples are independent or dependent

  1   *1, 2*

- How to perform a two-sample z-test for the difference between two means μ_1 and μ_2 using large independent samples

  2, 3   *3–10*

  $$z = \frac{(\bar{x}_1 - \bar{x}_2) - (\mu_1 - \mu_2)}{\sigma_{\bar{x}_1 - \bar{x}_2}}$$

Section 8.2

- How to perform a t-test for the difference between two population means μ_1 and μ_2 using small independent samples

  1, 2   *11–18*

  $$t = \frac{(\bar{x}_1 - \bar{x}_2) - (\mu_1 - \mu_2)}{s_{\bar{x}_1 - \bar{x}_2}}$$

Section 8.3

- How to perform a t-test to test the mean of the differences for a population of paired data

  1, 2   *19–24*

  $$t = \frac{\bar{d} - \mu_d}{s_d / \sqrt{n}}$$

Two-Sample Hypothesis Testing for Population Means

Are the samples independent? → No → Are both populations normal? → No → Cannot use hypothesis tests discussed in this chapter.

Are both populations normal? → Yes → Use t-test for dependent samples (Section 8.3).

Are the samples independent? → Yes

Are both samples large? → No → Are both populations normal? → No → Cannot use hypothesis tests discussed in this chapter.

Are both samples large? → Yes → Use z-test for large independent samples (Section 8.1).

Are both populations normal? → Yes → Are both standard deviations known? → No → Use t-test for small independent samples (Section 8.2).

Are both standard deviations known? → Yes → Use z-test (Section 8.1).

Section 8.4

- How to perform a z-test for the difference between two population proportions p_1 and p_2

  1, 2   *25–32*

  $$z = \frac{(\hat{p}_1 - \hat{p}_2) - (p_1 - p_2)}{\sqrt{\bar{p}\,\bar{q}\left(\dfrac{1}{n_1} + \dfrac{1}{n_2}\right)}}$$

8 REVIEW EXERCISES

1. Dependent because the same cities were sampled.

2. Dependent because the same runners were sampled.

3. Fail to reject H_0. There is not enough evidence at the 5% level of significance to reject the claim.

4. Reject H_0. There is enough evidence at the 1% level of significance to reject the claim.

5. Reject H_0. There is enough evidence at the 10% level of significance to support the claim.

6. Fail to reject H_0. There is not enough evidence at the 5% level of significance to support the claim.

7. (a) The claim is "the Wendy's fish sandwich has less sodium than the Long John Silver's fish sandwich."

H_0: $\mu_1 \geq \mu_2$;
H_a: $\mu_1 < \mu_2$ (claim)

(b) $-z_0 = -1.645$;
Rejection region: $z < -1.645$

(c) -9.20

(d) Reject H_0.

(e) There is enough evidence at the 5% level of significance to support the claim that the Wendy's fish sandwich has less sodium than the Long John Silver's fish sandwich.

8. (a) The claim is "the mean annual salary of civilian federal employees is the same in California and Illinois."

H_0: $\mu_1 = \mu_2$ (claim);
H_a: $\mu_1 \neq \mu_2$

(b) $-z_0 = -1.645$, $z_0 = 1.645$;

Rejection regions:
$z < -1.645$, $z > 1.645$

(c) -1.81

(d) Reject H_0.

(e) See Selected Answers, page A128.

9. Yes; The new rejection region is $z < -2.33$, which contains $z = -9.20$, so you still reject H_0.

10. No; The new rejection regions are $z < -1.96$ and $z > 1.96$, which do not contain $z = -1.81$, so you fail to reject H_0.

■ SECTION 8.1

In Exercises 1 and 2, classify the two given samples as independent or dependent. Explain your reasoning.

1. Sample 1: Air pollution measurements for 15 cities

Sample 2: Air pollution measurements for those 15 cities five years after a law was passed restricting carbon emissions

2. Sample 1: Pulse rates of runners before a marathon

Sample 2: Pulse rates of the same runners after a marathon

In Exercises 3–6, use the given sample statistics to test the claim about the difference between two population means μ_1 and μ_2 at the given level of significance α. The samples are random and independent.

3. Claim: $\mu_1 \geq \mu_2$; $\alpha = 0.05$. Sample statistics: $\bar{x}_1 = 1.28$, $s_1 = 0.30$, $n_1 = 96$ and $\bar{x}_2 = 1.34$, $s_2 = 0.23$, $n_2 = 85$

4. Claim: $\mu_1 = \mu_2$; $\alpha = 0.01$. Sample statistics: $\bar{x}_1 = 5595$, $s_1 = 52$, $n_1 = 156$ and $\bar{x}_2 = 5575$, $s_2 = 68$, $n_2 = 216$

5. Claim: $\mu_1 < \mu_2$; $\alpha = 0.10$. Sample statistics: $\bar{x}_1 = 0.28$, $s_1 = 0.11$, $n_1 = 41$ and $\bar{x}_2 = 0.33$, $s_2 = 0.10$, $n_2 = 34$

6. Claim: $\mu_1 \neq \mu_2$; $\alpha = 0.05$. Sample statistics: $\bar{x}_1 = 87$, $s_1 = 14$, $n_1 = 410$ and $\bar{x}_2 = 85$, $s_2 = 15$, $n_2 = 340$

In Exercises 7 and 8, (a) identify the claim and state H_0 and H_a, (b) find the critical value(s) and identify the rejection region(s), (c) find the standardized test statistic z, (d) decide whether to reject or fail to reject the null hypothesis, and (e) interpret the decision in the context of the original claim. If convenient, use technology to solve the problem. In each exercise, assume the samples are randomly selected and independent.

7. In a fast food study, a researcher finds that the mean sodium content of 42 Wendy's fish sandwiches is 1010 milligrams with a standard deviation of 75 milligrams. The mean sodium content of 39 Long John Silver's fish sandwiches is 1180 milligrams with a standard deviation of 90 milligrams. At $\alpha = 0.05$, is there enough evidence for the researcher to conclude that the Wendy's fish sandwich has less sodium than the Long John Silver's fish sandwich? *(Adapted from Wendy's International Inc. and Long John Silver's Inc.)*

8. A government agency states that the mean annual salary of civilian federal employees is the same in California and Illinois. The mean annual salary for 180 civilian federal employees in California is \$66,210 and the standard deviation is \$6385. The mean annual salary for 180 civilian federal employees in Illinois is \$67,390 and the standard deviation is \$5998. At $\alpha = 0.10$, is there enough evidence to reject the agency's claim? *(Adapted from U.S. Office of Personnel Management)*

9. Suppose you are testing Exercise 7 at $\alpha = 0.01$. Do you still make the same decision? Explain your reasoning.

10. Suppose you are testing Exercise 8 at $\alpha = 0.05$. Do you still make the same decision? Explain your reasoning.

11. Reject H_0. There is enough evidence at the 5% level of significance to reject the claim.

12. Fail to reject H_0. There is not enough evidence at the 10% level of significance to reject the claim.

13. Fail to reject H_0. There is not enough evidence at the 5% level of significance to reject the claim.

14. Reject H_0. There is enough evidence at the 1% level of significance to reject the claim.

15. Reject H_0. There is enough evidence at the 1% level of significance to support the claim.

16. Fail to reject H_0. There is not enough evidence at the 10% level of significance to reject the claim.

17. (a) The claim is "third graders taught with the directed reading activities scored higher than those taught without the activities."

H_0: $\mu_1 \le \mu_2$;
H_a: $\mu_1 > \mu_2$ (claim)

(b) $t_0 = 1.645$;
Rejection region: $t > 1.645$

(c) 2.267

(d) Reject H_0.

(e) There is enough evidence at the 5% level of significance to support the claim that third graders taught with the directed reading activities scored higher than those taught without the activities.

■ SECTION 8.2

In Exercises 11–16, use the given sample statistics to test the claim about the difference between two population means μ_1 and μ_2 at the given level of significance α. Assume that the samples are random and independent and that the populations are approximately normally distributed.

11. Claim: $\mu_1 = \mu_2$; $\alpha = 0.05$. Sample statistics: $\bar{x}_1 = 228$, $s_1 = 27$, $n_1 = 20$ and $\bar{x}_2 = 207$, $s_2 = 25$, $n_2 = 13$. Assume equal variances.

12. Claim: $\mu_1 = \mu_2$; $\alpha = 0.10$. Sample statistics: $\bar{x}_1 = 0.015$, $s_1 = 0.011$, $n_1 = 8$ and $\bar{x}_2 = 0.019$, $s_2 = 0.004$, $n_2 = 6$. Assume variances are not equal.

13. Claim: $\mu_1 \le \mu_2$; $\alpha = 0.05$. Sample statistics: $\bar{x}_1 = 183.5$, $s_1 = 1.3$, $n_1 = 25$ and $\bar{x}_2 = 184.7$, $s_2 = 3.9$, $n_2 = 25$. Assume variances are not equal.

14. Claim: $\mu_1 \ge \mu_2$; $\alpha = 0.01$. Sample statistics: $\bar{x}_1 = 44.5$, $s_1 = 5.85$, $n_1 = 17$ and $\bar{x}_2 = 49.1$, $s_2 = 5.25$, $n_2 = 18$. Assume equal variances.

15. Claim: $\mu_1 \ne \mu_2$; $\alpha = 0.01$. Sample statistics: $\bar{x}_1 = 61$, $s_1 = 3.3$, $n_1 = 5$ and $\bar{x}_2 = 55$, $s_2 = 1.2$, $n_2 = 7$. Assume equal variances.

16. Claim: $\mu_1 \ge \mu_2$; $\alpha = 0.10$. Sample statistics: $\bar{x}_1 = 520$, $s_1 = 25$, $n_1 = 7$ and $\bar{x}_2 = 500$, $s_2 = 55$, $n_2 = 6$. Assume variances are not equal.

In Exercises 17 and 18, (a) identify the claim and state H_0 and H_a, (b) find the critical value(s) and identify the rejection region(s), (c) find the standardized test statistic t, (d) decide whether to reject or fail to reject the null hypothesis, and (e) interpret the decision in the context of the original claim. If convenient, use technology to solve the problem. In each exercise, assume the populations are normally distributed.

17. A study of methods for teaching reading in the third grade was conducted. A classroom of 21 students participated in directed reading activities for eight weeks. Another classroom, with 23 students, followed the same curriculum without the activities. Students in both classrooms then took the same reading test. The scores of the two groups are shown in the back-to-back stem-and-leaf plot.

Classroom With Activities		Classroom Without Activities
	1	0 7 9
4	2	0 6 8
3	3	3 7 7
9 9 6 4 3 3 3	4	1 2 2 2 3 6 8
9 8 7 7 6 4 3 2	5	3 4 5 5
7 2 1	6	0 2
1	7	
	8	5

Key: $4|2 = 24$ (classroom with activities)
  $2|0 = 20$ (classroom without activities)

At $\alpha = 0.05$, is there enough evidence to conclude that third graders taught with the directed reading activities scored higher than those taught without the activities? Assume the population variances are equal. *(Source: StatLib/Schmitt, Maribeth C., The Effects of an Elaborated Directed Reading Activity on the Metacomprehension Skills of Third Graders)*

18. (a) The claim is "there is no difference between the mean household incomes of two neighborhoods."

$H_0: \mu_1 = \mu_2$ (claim);
$H_a: \mu_1 \neq \mu_2$

(b) $-t_0 = -2.845$; $t_0 = 2.845$;
Rejection regions:
$t < -2.845$, $t > 2.845$

(c) 1.939

(d) Fail to reject H_0.

(e) There is not enough evidence at the 1% level of significance to reject the agent's claim that there is no difference between the mean household incomes of the two neighborhoods.

19. Two-tailed test; Reject H_0.

20. Left-tailed test; Fail to reject H_0.

21. Right-tailed test; Reject H_0.

22. Two-tailed test; Reject H_0.

23. (a) The claim is "the men's systolic blood pressure decreased."

$H_0: \mu_d \leq 0$; $H_a: \mu_d > 0$ (claim)

(b) $t_0 = 1.383$;
Rejection region: $t > 1.383$

(c) $\bar{d} = 5$; $s_d \approx 8.743$

(d) 1.808

(e) Reject H_0.

(f) There is enough evidence at the 10% level of significance to support the claim that the men's systolic blood pressure decreased.

24. (a) The claim is "the weight loss supplement will help users lose weight after two weeks."

$H_0: \mu_d \leq 0$; $H_a: \mu_d > 0$ (claim)

(b) $t_0 = 1.860$;
Rejection region: $t > 1.860$

(c) $\bar{d} \approx 1.444$; $s_d \approx 2.744$

(d) 1.579

(e) Fail to reject H_0.

(f) There is not enough evidence at the 5% level of significance to support the claim that the weight loss supplement will help users lose weight after two weeks.

18. A real estate agent claims that there is no difference between the mean household incomes of two neighborhoods. The mean income of 12 randomly selected households from the first neighborhood was $32,750 with a standard deviation of $1900. In the second neighborhood, 10 randomly selected households had a mean income of $31,200 with a standard deviation of $1825. At $\alpha = 0.01$, can you reject the real estate agent's claim? Assume the population variances are equal.

■ SECTION 8.3

In Exercises 19–22, using a test for dependent, random samples, test the claim about the mean of the difference of the two populations at the given level of significance α using the given statistics. Is the test right-tailed, left-tailed, or two-tailed? Assume the populations are normally distributed.

19. Claim: $\mu_d = 0$; $\alpha = 0.01$. Statistics: $\bar{d} = 8.5$, $s_d = 10.7$, $n = 16$

20. Claim: $\mu_d < 0$; $\alpha = 0.10$. Statistics: $\bar{d} = 3.2$, $s_d = 5.68$, $n = 25$

21. Claim: $\mu_d \leq 0$; $\alpha = 0.10$. Statistics: $\bar{d} = 10.3$, $s_d = 18.19$, $n = 33$

22. Claim: $\mu_d \neq 0$; $\alpha = 0.05$. Statistics: $\bar{d} = 17.5$, $s_d = 4.05$, $n = 37$

In Exercises 23 and 24, (a) identify the claim and state H_0 and H_a, (b) find the critical value(s) and identify the rejection region(s), (c) calculate $\bar{d}$ and s_d, (d) find the standardized test statistic t, (e) decide whether to reject or fail to reject the null hypothesis, and (f) interpret the decision in the context of the original claim. If convenient, use technology to solve the problem. For each sample, assume the population is normally distributed.

23. A medical researcher wants to test the effects of calcium supplements on men's systolic blood pressure. In part of the study, 10 randomly selected men are given a calcium supplement for 12 weeks. The researcher measures the men's systolic blood pressure (in millimeters of mercury) before and after the 12-week study and records the results shown below. At $\alpha = 0.10$, can the researcher claim that the men's systolic blood pressure decreased? *(Source: The Journal of the American Medical Association)*

Patient	1	2	3	4	5	6	7	8	9	10
Before	107	110	123	129	112	111	107	112	136	102
After	100	114	105	112	115	116	106	102	125	104

24. A physical fitness instructor claims that a particular weight loss supplement will help users lose weight after two weeks. The table shows the weights (in pounds) of 9 mildly overweight adults before using the supplement and two weeks after using the supplement. At $\alpha = 0.05$, can you conclude that the supplement helps users lose weight?

User	1	2	3	4	5	6	7	8	9
Weight before using supplement	228	210	245	272	203	198	256	217	240
Weight after using supplement	225	208	242	270	205	196	250	220	240

25. Can use normal sampling distribution; Fail to reject H_0.

26. Can use normal sampling distribution; Reject H_0.

27. Can use normal sampling distribution; Reject H_0.

28. Can use normal sampling distribution; Fail to reject H_0.

29. (a) The claim is "the proportions of U.S. adults who considered the amount of federal income tax they had to pay to be too high were the same for the two years."

H_0: $p_1 = p_2$ (claim);
H_a: $p_1 \neq p_2$

(b) $-z_0 = -2.575$, $z_0 = 2.575$;
Rejection regions:
$z < -2.575$, $z > 2.575$

(c) 2.65

(d) Reject H_0.

(e) There is enough evidence at the 1% level of significance to reject the claim that the proportions of U.S. adults who considered the amount of federal income tax they had to pay to be too high were the same for the two years.

30. See Selected Answers, page A128.

31. Yes; When $\alpha = 0.05$, the rejection regions become $z < -1.96$ and $z > 1.96$. Because 2.65 > 1.96, you still reject H_0. There is enough evidence at the 5% level of significance to reject the claim that the proportions of U.S. adults who considered the amount of federal income tax they had to pay to be too high were the same for the two years.

32. No; When $\alpha = 0.01$, the rejection region becomes $z > 2.33$. Because 1.80 < 2.33, you fail to reject H_0. There is not enough evidence at the 1% level of significance to support the claim that the proportion of U.S. adults who believe it is likely that life exists on other planets is less now than in 2007.

■ SECTION 8.4

In Exercises 25–28, decide whether the normal sampling distribution can be used. If it can be used, test the claim about the difference between two population proportions p_1 and p_2 at the given level of significance α using the given sample statistics. Assume the sample statistics are from independent, random samples.

25. Claim: $p_1 = p_2$; $\alpha = 0.05$. Sample statistics: $x_1 = 425$, $n_1 = 840$ and $x_2 = 410$, $n_2 = 760$

26. Claim: $p_1 \leq p_2$; $\alpha = 0.01$. Sample statistics: $x_1 = 36$, $n_1 = 100$ and $x_2 = 46$, $n_2 = 200$

27. Claim: $p_1 > p_2$; $\alpha = 0.10$. Sample statistics: $x_1 = 261$, $n_1 = 556$ and $x_2 = 207$, $n_2 = 483$

28. Claim: $p_1 < p_2$; $\alpha = 0.05$. Sample statistics: $x_1 = 86$, $n_1 = 900$ and $x_2 = 107$, $n_2 = 1200$

In Exercises 29 and 30, (a) identify the claim and state H_0 and H_a, (b) find the critical value(s) and identify the rejection region(s), (c) find the standardized test statistic z, (d) decide whether to reject or fail to reject the null hypothesis, and (e) interpret the decision in the context of the original claim. If convenient, use technology to solve the problem. In each exercise, assume the samples are randomly selected and that the samples are independent.

29. In a survey of 900 U.S. adults in 2008, 468 considered the amount of federal income tax they had to pay to be too high. In a recent year, in a survey of 1027 U.S. adults, 472 considered the amount too high. At $\alpha = 0.01$, can you reject the claim that the proportions of U.S. adults who considered the amount of federal income tax they had to pay to be too high were the same for the two years? *(Adapted from The Gallup Poll)*

Study Done in 2008 (900 U.S. Adults)

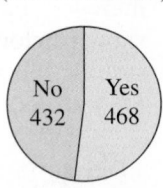

Study Done Recently (1027 U.S. Adults)

30. In a survey of 1000 U.S. adults in 2007, 57% said it is likely that life exists on other planets. In a recent year, in a survey of 1000 U.S. adults, 53% said it is likely that life exists on other planets. At $\alpha = 0.05$, can you support the claim that the proportion of U.S. adults who believe it is likely that life exists on other planets is less now than in 2007? *(Source: Rasmussen Reports)*

Survey Done in 2007 (1000 U.S. Adults)

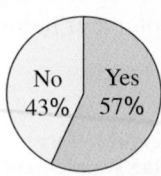

Survey Done Recently (1000 U.S. Adults)

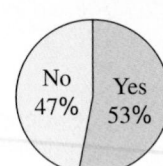

31. Suppose you are testing Exercise 29 at $\alpha = 0.05$. Do you still make the same decision? Explain your reasoning.

32. Suppose you are testing Exercise 30 at $\alpha = 0.01$. Do you still make the same decision? Explain your reasoning.

8 CHAPTER QUIZ

1. (a) H_0: $\mu_1 \leq \mu_2$;
   H_a: $\mu_1 > \mu_2$ (claim)

   (b) One-tailed because H_a contains >; z-test because n_1 and n_2 are each greater than 30.

   (c) $z_0 = 1.645$;
   Rejection region: $z > 1.645$

   (d) 0.585

   (e) Fail to reject H_0.

   (f) See Odd Answers, page A92.

2. (a) H_0: $\mu_1 = \mu_2$ (claim);
   H_a: $\mu_1 \neq \mu_2$

   (b) Two-tailed because H_a contains ≠; t-test because n_1 and n_2 are less than 30, the samples are independent, and the populations are normally distributed.

   (c) $-t_0 = -2.779$, $t_0 = 2.779$;
   Rejection regions:
   $t < -2.779$, $t > 2.779$

   (d) 0.341

   (e) Fail to reject H_0.

   (f) See Odd Answers, page A92.

3. (a) H_0: $p_1 = p_2$ (claim);
   H_a: $p_1 \neq p_2$

   (b) Two-tailed because H_a contains ≠; z-test because you are testing proportions and $n_1\bar{p}$, $n_1\bar{q}$, $n_2\bar{p}$, and $n_2\bar{q} \geq 5$.

   (c) $-z_0 = 1.645$, $z_0 = 1.645$;
   Rejection regions:
   $z < -1.645$, $z > 1.645$

   (d) 1.32

   (e) Fail to reject H_0.

   (f) See Odd Answers, page A92.

4. (a) H_0: $\mu_d \geq 0$; H_a: $\mu_d < 0$ (claim)

   (b) One-tailed because H_a contains <; t-test because both populations are normally distributed and the samples are dependent.

   (c) $t_0 = -2.718$;
   Rejection region: $t < -2.718$

   (d) −5.07

   (e) Reject H_0.

   (f) See Odd Answers, page A92.

Take this quiz as you would take a quiz in class. After you are done, check your work against the answers given in the back of the book.

For this quiz, do the following.

(a) Write the claim mathematically and identify H_0 and H_a.

(b) Determine whether the hypothesis test is a one-tailed test or a two-tailed test and whether to use a z-test or a t-test. Explain your reasoning.

(c) Find the critical value(s) and identify the rejection region(s).

(d) Use the appropriate test to find the appropriate standardized test statistic. If convenient, use technology.

(e) Decide whether to reject or fail to reject the null hypothesis.

(f) Interpret the decision in the context of the original claim.

1. The mean score on a science assessment for 49 randomly selected male high school students was 149 with a standard deviation of 35. The mean score on the same test for 50 randomly selected female high school students was 145 with a standard deviation of 33. At $\alpha = 0.05$, can you support the claim that the mean score on the science assessment for the male high school students was higher than for the female high school students? *(Adapted from National Center for Education Statistics)*

2. A science teacher claims that the mean scores on a science assessment test for fourth grade boys and girls are equal. The mean score for 13 randomly selected boys is 153 with a standard deviation of 32, and the mean score for 15 randomly selected girls is 149 with a standard deviation of 30. At $\alpha = 0.01$, can you reject the teacher's claim? Assume the populations are normally distributed and the variances are equal. *(Adapted from National Center for Education Statistics)*

3. In a random sample of 800 U.S. adults, 336 are worried that they or someone in their family will become a victim of terrorism. In another random sample of 1100 U.S. adults taken a month earlier, 429 were worried that they or someone in their family would become a victim of terrorism. At $\alpha = 0.10$, can you reject the claim that the proportion of U.S. adults who are worried that they or someone in their family will become a victim of terrorism has not changed? *(Adapted from The Gallup Poll)*

4. The table shows the credit scores for 12 randomly selected adults who are considered high-risk borrowers before and two years after they attend a personal finance seminar. At $\alpha = 0.01$, is there enough evidence to conclude that the seminar helps adults increase their credit scores?

Adult	1	2	3	4	5	6
Credit score (before seminar)	608	620	610	650	640	680
Credit score (after seminar)	646	692	715	669	725	786

Adult	7	8	9	10	11	12
Credit score (before seminar)	655	602	644	656	632	664
Credit score (after seminar)	700	650	660	650	680	702

PUTTING IT ALL TOGETHER

Real Statistics — Real Decisions

The National Hospital Discharge Survey (NHDS) is a national probability survey that has been conducted annually since 1965 by the Centers for Disease Control and Prevention's National Center for Health Statistics. From 1988 to 2007, the NHDS collected data from a sample of about 270,000 inpatient records provided by a national sample of about 500 hospitals. Beginning in 2007, this sample size was reduced to 239 hospitals. Only non-Federal short-stay hospitals, such as general hospitals and children's general hospitals, are included in the survey. The results of this survey provide information on the characteristics of inpatients discharged from these hospitals and are used to examine important topics of interest in public health.

You work for the National Center for Health Statistics. You want to test the claim that the mean length of stay for inpatients today is different than what it was a decade ago by analyzing data from a random selection of inpatient records. The results for several inpatients are shown in the histograms from a decade ago and the current year.

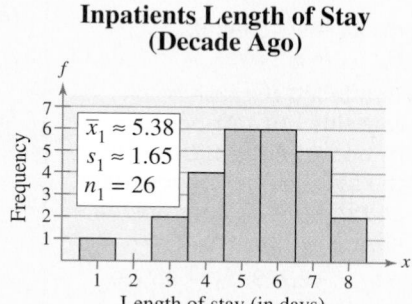

Inpatients Length of Stay (Decade Ago)

$\bar{x}_1 \approx 5.38$
$s_1 \approx 1.65$
$n_1 = 26$

■ EXERCISES

1. *How Could You Do It?*

Explain how you could use the given sampling technique to select the sample for the study.

(a) stratified sample   (b) cluster sample

(c) systematic sample   (d) simple random sample

2. *Choosing a Sampling Technique*

(a) Which sampling technique in Exercise 1 would you choose to implement for the study? Why?

(b) Identify possible flaws or biases in your study.

3. *Choosing a Test*

To test the claim that there is a difference in the mean length of hospital stays, should you use a *z*-test or a *t*-test? Are the samples independent or dependent? Do you need to know what each population's distribution is? Do you need to know anything about the population variances?

4. *Testing a Mean*

Test the claim that there is a difference in the mean length of hospital stays for inpatients. Assume the populations are normal and the population variances are not equal. Use $\alpha = 0.10$. Interpret the test's decision. Does the decision support the claim?

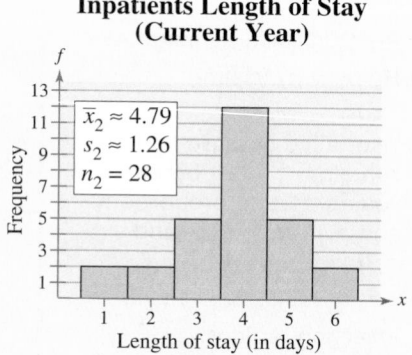

Inpatients Length of Stay (Current Year)

$\bar{x}_2 \approx 4.79$
$s_2 \approx 1.26$
$n_2 = 28$

TECHNOLOGY

TAILS OVER HEADS

In the article "Tails over Heads" in the Washington Post (Oct. 13, 1996), journalist William Casey describes one of his hobbies—keeping track of every coin he finds on the street! From January 1, 1985 until the article was written, Casey found 11,902 coins.

As each coin is found, Casey records the time, date, location, value, mint location, and whether the coin is lying heads up or tails up. In the article, Casey notes that 6130 coins were found tails up and 5772 were found heads up. Of the 11,902 coins found, 43 were minted in San Francisco, 7133 were minted in Philadelphia, and 4726 were minted in Denver.

A simulation of Casey's experiment can be done in MINITAB as shown below. A frequency histogram of one simulation's results is shown at the right.

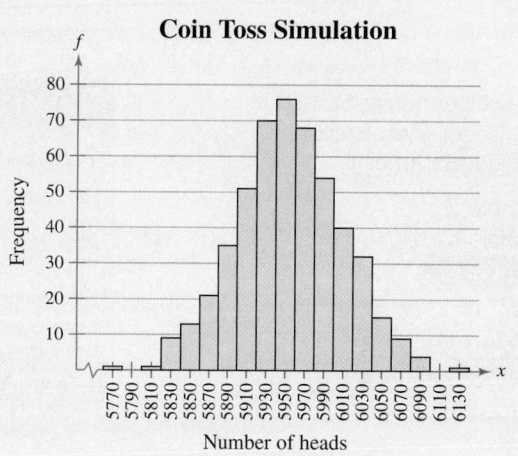

Coin Toss Simulation

Sample From Columns...
Chi-Square...
Normal...
Multivariate Normal...
F...
t...
Uniform...

Bernoulli...
Binomial...
Geometric...

MINITAB

Number of rows of data to generate: 500

Store in column(s): C1

Number of trials: 11902

Event probability: .5

◼ EXERCISES

1. Use a technology tool to perform a one-sample z-test to test the hypothesis that the probability that a "found coin" will be lying heads up is 0.5. Use $\alpha = 0.01$. Use Casey's data as your sample and write your conclusion as a sentence.

2. Do Casey's data differ significantly from chance? If so, what might be the reason?

3. In the simulation shown above, what percent of the trials had heads less than or equal to the number of heads in Casey's data? Use a technology tool to repeat the simulation. Are your results comparable?

In Exercises 4 and 5, use a technology tool to perform a two-sample z-test to decide whether there is a difference in the mint dates and in the values of coins found on a street from 1985 through 1996. Write your conclusion as a sentence. Use $\alpha = 0.05$.

4. Mint dates of coins (years)

    Philadelphia: $\bar{x}_1 = 1984.8$ $s_1 = 8.6$

    Denver: $\bar{x}_2 = 1983.4$ $s_2 = 8.4$

5. Value of coins (dollars)

    Philadelphia: $\bar{x}_1 = \$0.034$ $s_1 = \$0.054$

    Denver: $\bar{x}_2 = \$0.033$ $s_2 = \$0.052$

Extended solutions are given in the *Technology Supplement.* Technical instruction is provided for MINITAB, Excel, and the TI-83/84 Plus.

8 USING TECHNOLOGY TO PERFORM TWO-SAMPLE HYPOTHESIS TESTS

Here are some MINITAB and TI-83/84 Plus printouts for several examples in this chapter.

(See Example 1, page 444.)

Display Descriptive Statistics...
Store Descriptive Statistics...
Graphical Summary...

1-Sample Z...
1-Sample t...
2-Sample t...
Paired t...

1 Proportion...
2 Proportions...

MINITAB

Two-Sample T-Test and CI

Sample	N	Mean	StDev	SE Mean
1	8	473.0	39.7	14
2	18	459.0	24.5	5.8

Difference = mu (1) − mu (2)
Estimate for difference: 14.0
90% CI for difference: (−13.8, 41.8)
T-Test of difference = 0 (vs not =): T-Value = 0.92 P-Value = 0.380 DF = 9

(See Example 1, page 453.)

Vertical Jump Heights, Before and After Using Shoes

Athlete	1	2	3	4	5	6	7	8
Vertical jump height (before using shoes)	24	22	25	28	35	32	30	27
Vertical jump height (after using shoes)	26	25	25	29	33	34	35	30

Display Descriptive Statistics...
Store Descriptive Statistics...
Graphical Summary...

1-Sample Z...
1-Sample t...
2-Sample t...
Paired t...

1 Proportion...
2 Proportions...

MINITAB

Paired T-Test and CI: C1, C2

Paired T for C1 − C2

	N	Mean	StDev	SE Mean
C1	8	27.88	4.32	1.53
C2	8	29.63	4.07	1.44
Difference	8	−1.750	2.121	0.750

90% upper bound for mean difference: −0.689
T-Test of mean difference = 0 (vs < 0): T-Value = −2.33 P-Value = 0.026

(See Example 2, page 432.)

TI-83/84 PLUS

EDIT CALC **TESTS**
1: Z-Test...
2: T-Test...
3: 2-SampZTest...
4: 2-SampTTest...
5: 1-PropZTest...
6: 2-PropZTest...
7↓ZInterval...

↓

TI-83/84 PLUS

2-SampZTest
 Inpt: Data **Stats**
 σ1: 1045.7
 σ2: 1361.95
 x̄1: 4446.25
 n1: 250
 x̄2: 4567.24
↓n2: 250

↓

TI-83/84 PLUS

2-SampZTest
↑σ2: 1361.95
 x̄1: 4446.25
 n1: 250
 x̄2: 4567.24
 n2: 250
 μ1: **≠μ2** <μ2 >μ2
 Calculate Draw

↓

TI-83/84 PLUS

2-SampZTest
 μ1 ≠ μ2
 z= -1.114106102
 p= .2652337599
 x̄1= 4446.25
 x̄2= 4567.24
↓n1= 250

(See Example 2, page 445.)

TI-83/84 PLUS

EDIT CALC **TESTS**
1: Z-Test...
2: T-Test...
3: 2-SampZTest...
4: 2-SampTTest...
5: 1-PropZTest...
6: 2-PropZTest...
7↓ZInterval...

↓

TI-83/84 PLUS

2-SampTTest
 Inpt: Data **Stats**
 x̄1: 1275
 Sx1: 45
 n1: 14
 x̄2: 1250
 Sx2: 30
↓n2: 16

↓

TI-83/84 PLUS

2-SampTTest
↑n1: 14
 x̄2: 1250
 Sx2: 30
 n2: 16
 μ1: ≠μ2 <μ2 **>μ2**
 Pooled: No **Yes**
 Calculate Draw

↓

TI-83/84 PLUS

2-SampTTest
 μ1 >μ2
 t= 1.811358919
 p= .0404131295
 df= 28
 x̄1= 1275
↓x̄2= 1250

(See Example 1, page 463.)

TI-83/84 PLUS

EDIT CALC **TESTS**
1: Z-Test...
2: T-Test...
3: 2-SampZTest...
4: 2-SampTTest...
5: 1-PropZTest...
6: 2-PropZTest...
7↓ZInterval...

↓

TI-83/84 PLUS

2-PropZTest
 x1: 129
 n1: 150
 x2: 148
 n2: 200
 p1: **≠p2** <p2 >p2
 Calculate Draw

↓

TI-83/84 PLUS

2-PropZTest
 p1 ≠ p2
 z= 2.734478928
 p= .0062480166
 p̂1= .86
 p̂2= .74
↓p̂= .7914285714

CUMULATIVE REVIEW
Chapters 6 – 8

Sections 6.3 and 7.4

1. In a survey of 1000 people who attend community college, 13% are age 40 or older. *(Adapted from American Association of Community Colleges)*

   (a) Construct a 95% confidence interval for the proportion of people who attend community college that are age 40 or older.

   (b) A researcher claims that more than 10% of people who attend community college are age 40 or older. At $\alpha = 0.05$, can you support the researcher's claim? Interpret the decision in the context of the original claim.

Section 8.3

2. **Gas Mileage**   The table shows the gas mileages (in miles per gallon) of eight cars with and without using a fuel additive. At $\alpha = 0.10$, is there enough evidence to conclude that the additive improved gas mileage?

Car	1	2	3	4	5	6	7	8
Gas mileage without additive	23.1	25.4	21.9	24.3	19.9	21.2	25.9	24.8
Gas mileage with fuel additive	23.6	27.7	23.6	26.8	22.1	22.4	26.3	26.6

Sections 6.1 and 6.2

In Exercises 3–6, construct the indicated confidence interval for the population mean μ. Which distribution did you use to create the confidence interval?

3. $c = 0.95, \bar{x} = 26.97, s = 3.4, n = 42$

4. $c = 0.90, \bar{x} = 3.46, s = 1.63, n = 16$

5. $c = 0.99, \bar{x} = 12.1, s = 2.64, n = 26$

6. $c = 0.95, \bar{x} = 8.21, s = 0.62, n = 8$

Section 8.1

7. A pediatrician claims that the mean birth weight of a single-birth baby is greater than the mean birth weight of a baby that has a twin. The mean birth weight of a random sample of 85 single-birth babies is 3086 grams with a standard deviation of 563 grams. The mean birth weight of a random sample of 68 babies that have a twin is 2263 grams with a standard deviation of 624 grams. At $\alpha = 0.10$, can you support the pediatrician's claim? Interpret the decision in the context of the original claim.

Section 7.1

In Exercises 8–11, use the given statement to represent a claim. Write its complement and state which is H_0 and which is H_a.

8. $\mu < 33$

9. $p \geq 0.19$

10. $\sigma = 0.63$

11. $\mu \neq 2.28$

Sections 6.4 and 7.5

12. The mean number of chronic medications taken by a random sample of 26 elderly adults in a community has a sample standard deviation of 3.1 medications. Assume the population is normally distributed. *(Adapted from The Journal of the American Medical Association)*

(a) Construct a 99% confidence interval for the population variance.

(b) Construct a 99% confidence interval for the population standard deviation.

(c) A pharmacist believes that the standard deviation of the mean number of chronic medications taken by elderly adults in the community is less than 2.5 medications. At $\alpha = 0.01$, can you support the pharmacist's claim? Interpret the decision in the context of the original claim.

Section 8.2

13. An education organization claims that the mean SAT scores for male athletes and male non-athletes at a college are different. A random sample of 26 male athletes at the college has a mean SAT score of 1783 and a standard deviation of 218. A random sample of 18 male non-athletes at the college has a mean SAT score of 2064 and a standard deviation of 186. At $\alpha = 0.05$, can you support the organization's claim? Interpret the decision in the context of the original claim. Assume the populations are normally distributed and the population variances are equal.

Sections 6.2 and 7.3

14. The annual earnings for 26 randomly selected translators are shown below. Assume the population is normally distributed. *(Adapted from U.S. Bureau of Labor Statistics)*

39,023	36,340	40,517	43,351	43,136	44,504	33,873	39,204
42,853	36,864	37,952	35,207	34,777	37,163	37,724	34,033
38,288	38,738	40,217	38,844	38,949	38,831	43,533	39,613
39,336	38,438						

(a) Construct a 95% confidence interval for the population mean annual earnings for translators.

(b) A researcher claims that the mean annual earnings for translators is $40,000. At $\alpha = 0.05$, can you reject the researcher's claim? Interpret the decision in the context of the original claim.

Section 8.4

15. A medical research team studied the number of head and neck injuries sustained by hockey players. Of the 319 players who wore a full-face shield, 195 sustained an injury. Of the 323 players who wore a half-face shield, 204 sustained an injury. At $\alpha = 0.10$, can you reject the claim that the proportions of players sustaining head and neck injuries are the same for the two groups? Interpret the decision in the context of the original claim. *(Source: The Journal of the American Medical Association)*

Sections 6.1 and 7.2

16. A random sample of 40 ostrich eggs has a mean incubation period of 42 days and a standard deviation of 1.6 days.

(a) Construct a 95% confidence interval for the population mean incubation period.

(b) A zoologist claims that the mean incubation period for ostriches is at least 45 days. At $\alpha = 0.05$, can you reject the zoologist's claim? Interpret the decision in the context of the original claim.

9 CORRELATION AND REGRESSION

CHAPTER

In 2009, the New York Yankees had the highest team salary in Major League Baseball at $201.4 million and the Florida Marlins had the lowest team salary at $36.8 million. In the same year, the Los Angeles Dodgers had the highest average attendance at 46,440 and the Oakland Athletics had the lowest average attendance at 17,392.

◀◀ WHERE YOU'VE BEEN

In Chapters 1–8, you studied descriptive statistics, probability, and inferential statistics. One of the techniques you learned in descriptive statistics was graphing paired data with a scatter plot (Section 2.2). For instance, the salaries and average attendances at home games for the teams in Major League Baseball in 2009 are shown in graphical form at the right and in tabular form below.

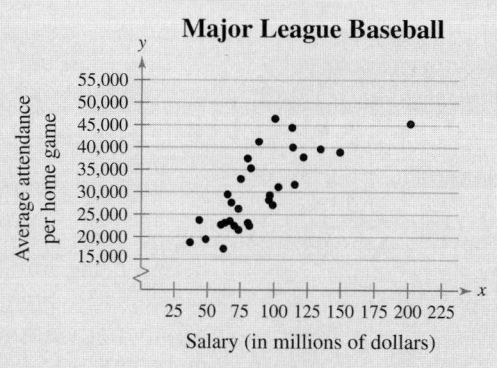

Major League Baseball

Salary (in millions of dollars)	73.5	96.7	67.1	121.7	134.8	96.1	73.6	81.6	75.2	115.1
Average Attendance Per Home Game	26,281	29,304	23,545	37,811	39,610	28,199	21,579	22,492	32,902	31,693

Salary (in millions of dollars)	36.8	103.0	70.5	113.7	100.4	80.2	65.3	149.4	201.4	62.3
Average Attendance Per Home Game	18,770	31,124	22,473	40,004	46,440	37,499	29,466	38,941	45,364	17,392

Salary (in millions of dollars)	113.0	48.7	43.7	82.6	98.9	88.5	63.3	68.2	80.5	60.3
Average Attendance Per Home Game	44,453	19,479	23,735	35,322	27,116	41,274	23,147	27,641	23,162	22,715

WHERE YOU'RE GOING ▶▶

In this chapter, you will study how to describe and test the significance of relationships between two variables when data are presented as ordered pairs. For instance, in the scatter plot above, it appears that higher team salaries tend to correspond to higher average attendances and lower team salaries tend to correspond to lower average attendances. This relationship is described by saying that the team salaries are positively correlated to the average attendances. Graphically, the relationship can be described by drawing a line, called a regression line, that fits the points as closely as possible, as shown below. The second scatter plot below shows the salaries and wins for the teams in Major League Baseball in 2009. From the scatter plot, it appears that there is a weak positive correlation between the team salaries and wins.

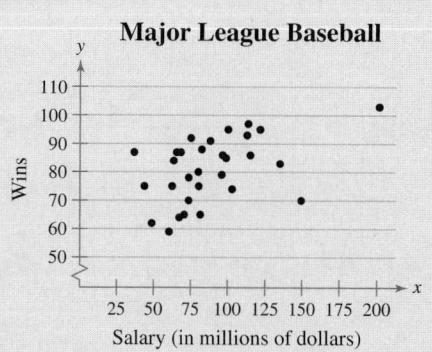

9.1 Correlation

An Overview of Correlation ▸ Correlation Coefficient ▸ Using a Table to Test a Population Correlation Coefficient ρ ▸ Hypothesis Testing for a Population Correlation Coefficient ρ ▸ Correlation and Causation

▸ AN OVERVIEW OF CORRELATION

Suppose a safety inspector wants to determine whether a relationship exists between the number of hours of training for an employee and the number of accidents involving that employee. Or suppose a psychologist wants to know whether a relationship exists between the number of hours a person sleeps each night and that person's reaction time. How would he or she determine if any relationship exists?

In this section, you will study how to describe what type of relationship, or correlation, exists between two quantitative variables and how to determine whether the correlation is significant.

DEFINITION

A **correlation** is a relationship between two variables. The data can be represented by the ordered pairs (x, y), where x is the **independent** (or **explanatory**) **variable** and y is the **dependent** (or **response**) **variable.**

In Section 2.2, you learned that the graph of ordered pairs (x, y) is called a *scatter plot*. In a scatter plot, the ordered pairs (x, y) are graphed as points in a coordinate plane. The independent (explanatory) variable x is measured by the horizontal axis, and the dependent (response) variable y is measured by the vertical axis. A scatter plot can be used to determine whether a linear (straight line) correlation exists between two variables. The following scatter plots show several types of correlation.

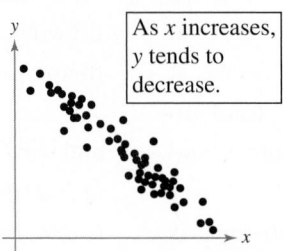

Negative Linear Correlation

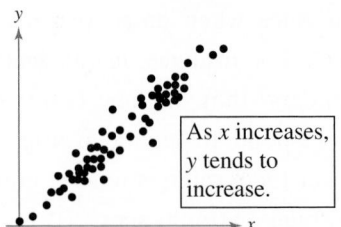

Positive Linear Correlation

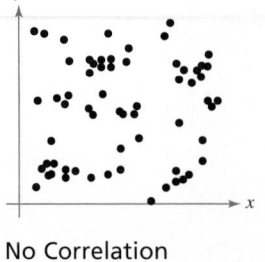

No Correlation

Nonlinear Correlation

GDP (trillions of $), x	CO_2 emissions (millions of metric tons), y
1.6	428.2
3.6	828.8
4.9	1214.2
1.1	444.6
0.9	264.0
2.9	415.3
2.7	571.8
2.3	454.9
1.6	358.7
1.5	573.5

EXAMPLE 1

▸ Constructing a Scatter Plot

An economist wants to determine whether there is a linear relationship between a country's gross domestic product (GDP) and carbon dioxide (CO_2) emissions. The data are shown in the table at the left. Display the data in a scatter plot and determine whether there appears to be a positive or negative linear correlation or no linear correlation. *(Source: World Bank and U.S. Energy Information Administration)*

▸ Solution

The scatter plot is shown at the right. From the scatter plot, it appears that there is a positive linear correlation between the variables.

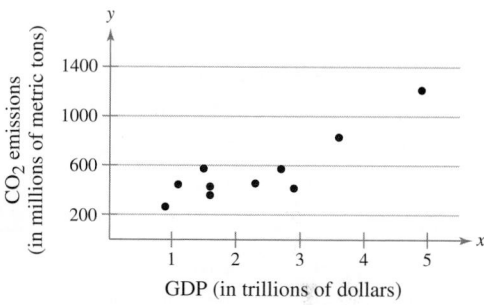

Interpretation Reading from left to right, as the gross domestic products increase, the carbon dioxide emissions tend to increase.

Number of years out of school, x	Annual contribution (1000s of $), y
1	12.5
10	8.7
5	14.6
15	5.2
3	9.9
24	3.1
30	2.7

▸ Try It Yourself 1

A director of alumni affairs at a small college wants to determine whether there is a linear relationship between the number of years alumni classes have been out of school and their annual contributions (in thousands of dollars). The data are shown in the table at the left. Display the data in a scatter plot and determine the type of correlation.

a. *Draw* and *label* the x- and y-axes.
b. *Plot* each ordered pair.
c. Does there appear to be a linear correlation? If so, *interpret* the correlation in the context of the data.
Answer: Page A44

Note to Instructor

Have students try to picture a line going through the points. If the line clearly has a positive slope, the correlation is positive. If the slope is clearly negative, the correlation is negative. If they have trouble picturing a line through the points, there is probably no linear correlation.

EXAMPLE 2

▸ Constructing a Scatter Plot

A student conducts a study to determine whether there is a linear relationship between the number of hours a student exercises each week and the student's grade point average (GPA). The data are shown in the following table. Display the data in a scatter plot and describe the type of correlation.

Hours of exercise, x	12	3	0	6	10	2	20	14	15	5
GPA, y	3.6	4.0	3.9	2.5	2.4	2.2	3.7	3.0	1.8	3.1

▸ Solution

The scatter plot is shown at the right. From the scatter plot, it appears that there is no linear correlation between the variables.

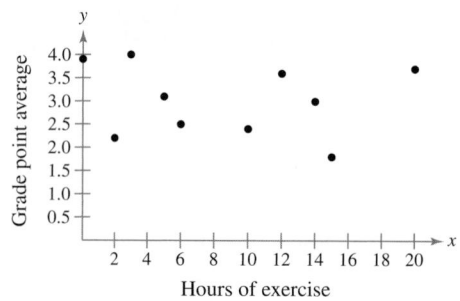

Interpretation The number of hours a student exercises each week does not appear to be related to the student's grade point average.

Duration, x	Time, y	Duration, x	Time, y
1.80	56	3.78	79
1.82	58	3.83	85
1.90	62	3.88	80
1.93	56	4.10	89
1.98	57	4.27	90
2.05	57	4.30	89
2.13	60	4.43	89
2.30	57	4.47	86
2.37	61	4.53	89
2.82	73	4.55	86
3.13	76	4.60	92
3.27	77	4.63	91
3.65	77		

▶ **Try It Yourself 2**

A researcher conducts a study to determine whether there is a linear relationship between a person's height (in inches) and pulse rate (in beats per minute). The data are shown in the following table. Display the data in a scatter plot and describe the type of correlation.

Height, x	68	72	65	70	62	75	78	64	68
Pulse rate, y	90	85	88	100	105	98	70	65	72

a. *Draw* and *label* the *x*- and *y*-axes.
b. *Plot* each ordered pair.
c. Does there appear to be a linear correlation? If so, *interpret* the correlation in the context of the data. *Answer: Page A44*

EXAMPLE 3   **SC** Report 37

▶ **Constructing a Scatter Plot Using Technology**

Old Faithful, located in Yellowstone National Park, is the world's most famous geyser. The durations (in minutes) of several of Old Faithful's eruptions and the times (in minutes) until the next eruption are shown in the table at the left. Using a TI-83/84 Plus, display the data in a scatter plot. Describe the type of correlation.

▶ **Solution**

Begin by entering the *x*-values into List 1 and the *y*-values into List 2. Use *Stat Plot* to construct the scatter plot. The plot should look similar to the one shown below. From the scatter plot, it appears that the variables have a positive linear correlation.

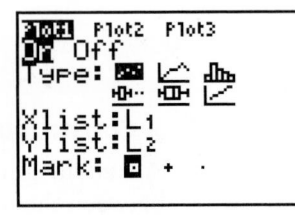 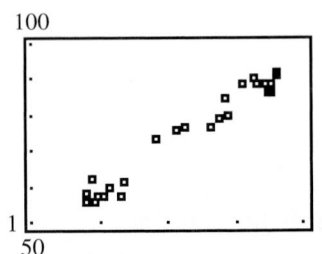

Interpretation You can conclude that the longer the duration of the eruption, the longer the time before the next eruption begins.

▶ **Try It Yourself 3**

Consider the data from the Chapter Opener on page 483 on the salaries and average attendances at home games for the teams in Major League Baseball. Use a technology tool to display the data in a scatter plot. Describe the type of correlation.

a. *Enter* the data into List 1 and List 2.
b. *Construct* the scatter plot.
c. Does there appear to be a linear correlation? If so, *interpret* the correlation in the context of the data. *Answer: Page A44*

▶ CORRELATION COEFFICIENT

Interpreting correlation using a scatter plot can be subjective. A more precise way to measure the type and strength of a linear correlation between two variables is to calculate the *correlation coefficient*. Although a formula for the sample correlation coefficient is given, it is more convenient to use a technology tool to calculate this value.

DEFINITION

The **correlation coefficient** is a measure of the strength and the direction of a linear relationship between two variables. The symbol *r* represents the sample correlation coefficient. A formula for *r* is

$$r = \frac{n\sum xy - (\sum x)(\sum y)}{\sqrt{n\sum x^2 - (\sum x)^2}\sqrt{n\sum y^2 - (\sum y)^2}}$$

where *n* is the number of pairs of data.

The population correlation coefficient is represented by ρ (the lowercase Greek letter rho, pronounced "row").

The range of the correlation coefficient is −1 to 1, inclusive. If *x* and *y* have a strong positive linear correlation, *r* is close to 1. If *x* and *y* have a strong negative linear correlation, *r* is close to −1. If *x* and *y* have perfect positive linear correlation or perfect negative linear correlation, *r* is equal to 1 or −1, respectively. If there is no linear correlation or a weak linear correlation, *r* is close to 0. It is important to remember that if *r* is close to 0, it does not mean that there is no relation between *x* and *y*, just that there is no linear relation. Several examples are shown below.

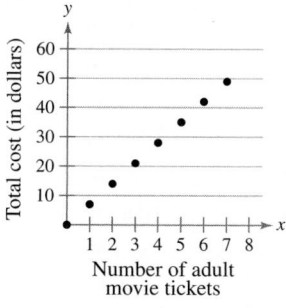

Perfect positive correlation
$r = 1$

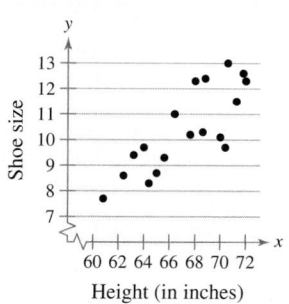

Strong positive correlation
$r = 0.81$

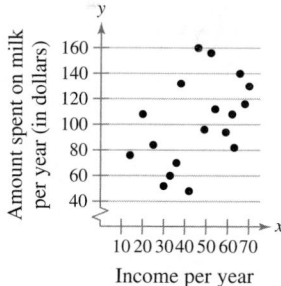

Weak positive correlation
$r = 0.45$

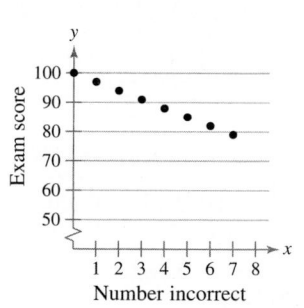

Perfect negative correlation
$r = -1$

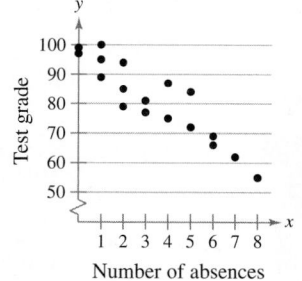

Strong negative correlation
$r = -0.92$

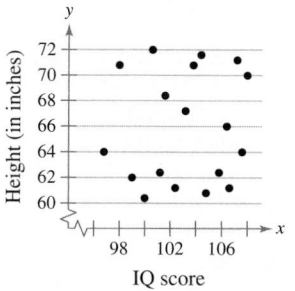

No correlation
$r = 0.04$

GUIDELINES

Calculating a Correlation Coefficient

IN WORDS	IN SYMBOLS
1. Find the sum of the x-values.	$\sum x$
2. Find the sum of the y-values.	$\sum y$
3. Multiply each x-value by its corresponding y-value and find the sum.	$\sum xy$
4. Square each x-value and find the sum.	$\sum x^2$
5. Square each y-value and find the sum.	$\sum y^2$
6. Use these five sums to calculate the correlation coefficient.	$r = \dfrac{n\sum xy - (\sum x)(\sum y)}{\sqrt{n\sum x^2 - (\sum x)^2}\sqrt{n\sum y^2 - (\sum y)^2}}$

EXAMPLE 4

▶ Finding the Correlation Coefficient

Calculate the correlation coefficient for the gross domestic products and carbon dioxide emissions data given in Example 1. What can you conclude?

▶ **Solution**   Use a table to help calculate the correlation coefficient.

GDP (trillions of \$), x	CO$_2$ emissions (millions of metric tons), y	xy	x^2	y^2
1.6	428.2	685.12	2.56	183,355.24
3.6	828.8	2983.68	12.96	686,909.44
4.9	1214.2	5949.58	24.01	1,474,281.64
1.1	444.6	489.06	1.21	197,669.16
0.9	264.0	237.6	0.81	69,696
2.9	415.3	1204.37	8.41	172,474.09
2.7	571.8	1543.86	7.29	326,955.24
2.3	454.9	1046.27	5.29	206,934.01
1.6	358.7	573.92	2.56	128,665.69
1.5	573.5	860.25	2.25	328,902.25
$\sum x = 23.1$	$\sum y = 5554$	$\sum xy = 15{,}573.71$	$\sum x^2 = 67.35$	$\sum y^2 = 3{,}775{,}842.76$

With these sums and $n = 10$, the correlation coefficient is

$$r = \frac{n\sum xy - (\sum x)(\sum y)}{\sqrt{n\sum x^2 - (\sum x)^2}\sqrt{n\sum y^2 - (\sum y)^2}}$$

$$= \frac{10(15{,}573.71) - (23.1)(5554)}{\sqrt{10(67.35) - 23.1^2}\sqrt{10(3{,}775{,}842.76) - 5554^2}}$$

$$= \frac{27{,}439.7}{\sqrt{139.89}\sqrt{6{,}911{,}511.6}} \approx 0.882.$$

The result $r \approx 0.882$ suggests a strong positive linear correlation.

Interpretation   As the gross domestic product increases, the carbon dioxide emissions also increase.

Number of years out of school, x	Annual contribution (1000s of \$), y
1	12.5
10	8.7
5	14.6
15	5.2
3	9.9
24	3.1
30	2.7

▶ **Try It Yourself 4**

Calculate the correlation coefficient for the number of years out of school and annual contribution data given in Try It Yourself 1. What can you conclude?

a. *Identify n and use a table* to help calculate $\sum x$, $\sum y$, $\sum xy$, $\sum x^2$, and $\sum y^2$.
b. *Use the resulting sums and n* to calculate r.
c. What can you conclude?   *Answer: Page A44*

EXAMPLE 5   Report 38

▶ **Using Technology to Find a Correlation Coefficient**

Use a technology tool to calculate the correlation coefficient for the Old Faithful data given in Example 3. What can you conclude?

▶ **Solution**

MINITAB, Excel, and the TI-83/84 Plus each have features that allow you to calculate a correlation coefficient for paired data sets. Try using this technology to find r. You should obtain results similar to the following.

MINITAB

Correlations: C1, C2

Pearson correlation of C1 and C2 = 0.979

EXCEL

	A	B	C
26	CORREL(A1:A25,B1:B25)		
27			0.978659

Before using the TI-83/84 Plus to calculate r, you must enter the Diagnostic On command. To do so, enter the following keystrokes:

[2nd] [0] cursor to *DiagnosticOn* [ENTER] [ENTER].

The following screens show how to find r using a TI-83/84 Plus with the data stored in List 1 and List 2. To begin, use the STAT keystroke.

To explore this topic further, see Activity 9.1 on page 500.

TI-83/84 PLUS

EDIT **CALC** TESTS
1: 1-Var Stats
2: 2-Var Stats
3: Med-Med
4: LinReg(ax+b)
5: QuadReg
6: CubicReg
7↓QuartReg

TI-83/84 PLUS

LinReg(ax+b) L1,
L2

TI-83/84 PLUS

LinReg
  y=ax+b
  a=12.48094391
  b=33.68290034
  r^2=.9577738551
  (r=.9786592129)

Correlation coefficient

The result $r \approx 0.979$ suggests a strong positive linear correlation.

▶ **Try It Yourself 5**

Calculate the correlation coefficient for the data from the Chapter Opener on page 483 on the salaries and average attendances at home games for the teams in Major League Baseball. What can you conclude?

a. *Enter* the data.
b. Use the appropriate feature to *calculate r*.
c. What can you conclude?   *Answer: Page A44*

▸ **USING A TABLE TO TEST A POPULATION CORRELATION COEFFICIENT ρ**

Once you have calculated r, the sample correlation coefficient, you will want to determine whether there is enough evidence to decide that the population correlation coefficient ρ is significant. In other words, based on a few pairs of data, can you make an inference about the population of all such data pairs? Remember that you are using sample data to make a decision about population data, so it is always possible that your inference may be wrong. In correlation studies, the small percentage of times when you decide that the correlation is significant when it is really not is called the *level of significance*. It is typically set at $\alpha = 0.01$ or 0.05. When $\alpha = 0.05$, you will probably decide that the population correlation coefficient is significant when it is really not 5% of the time. (Of course, 95% of the time, you will correctly determine that a correlation coefficient is significant.) When $\alpha = 0.01$, you will make this type of error only 1% of the time. When using a lower level of significance, however, you may fail to identify some significant correlations.

In order for a correlation coefficient to be significant, its absolute value must be close to 1. To determine whether the population correlation coefficient ρ is significant, use the critical values given in Table 11 in Appendix B. A portion of the table is shown below. If $|r|$ is greater than the critical value, there is enough evidence to decide that the correlation is significant. Otherwise, there is *not* enough evidence to say that the correlation is significant. For instance, to determine whether ρ is significant for five pairs of data ($n = 5$) at a level of significance of $\alpha = 0.01$, you need to compare $|r|$ with a critical value of 0.959, as shown in the table.

Number n of pairs of data in sample

Critical values for $\alpha = 0.05$ and $\alpha = 0.01$

n	$\alpha = 0.05$	$\alpha = 0.01$
4	0.950	0.990
5	0.878	(0.959)
6	0.811	0.917

If $|r| > 0.959$, the correlation is significant. Otherwise, there is *not* enough evidence to conclude that the correlation is significant. The guidelines for this process are as follows.

GUIDELINES

Using Table 11 for the Correlation Coefficient ρ

IN WORDS	IN SYMBOLS
1. Determine the number of pairs of data in the sample.	Determine n.
2. Specify the level of significance.	Identify α.
3. Find the critical value.	Use Table 11 in Appendix B.
4. Decide if the correlation is significant.	If $|r| >$ critical value, the correlation is significant. Otherwise, there is *not* enough evidence to conclude that the correlation is significant.
5. Interpret the decision in the context of the original claim.	

EXAMPLE 6

▶ Using Table 11 for a Correlation Coefficient

In Example 5, you used 25 pairs of data to find $r \approx 0.979$. Is the correlation coefficient significant? Use $\alpha = 0.05$.

▶ Solution

The number of pairs of data is 25, so $n = 25$. The level of significance is $\alpha = 0.05$. Using Table 11, find the critical value in the $\alpha = 0.05$ column that corresponds to the row with $n = 25$. The number in that column and row is 0.396.

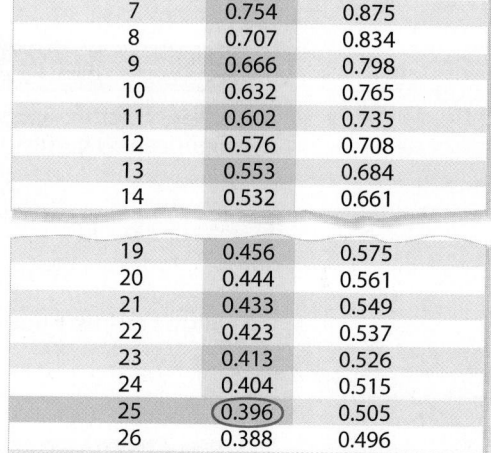

n	$\alpha = 0.05$	$\alpha = 0.01$
4	0.950	0.990
5	0.878	0.959
6	0.811	0.917
7	0.754	0.875
8	0.707	0.834
9	0.666	0.798
10	0.632	0.765
11	0.602	0.735
12	0.576	0.708
13	0.553	0.684
14	0.532	0.661
19	0.456	0.575
20	0.444	0.561
21	0.433	0.549
22	0.423	0.537
23	0.413	0.526
24	0.404	0.515
25	0.396	0.505
26	0.388	0.496
27	0.381	0.487
28	0.374	0.479
29	0.367	0.471

Because $|r| \approx 0.979 > 0.396$, you can decide that the population correlation is significant.

Interpretation   There is enough evidence at the 5% level of significance to conclude that there is a significant linear correlation between the duration of Old Faithful's eruptions and the time between eruptions.

▶ Try It Yourself 6

In Try It Yourself 4, you calculated the correlation coefficient of the number of years out of school and annual contribution data to be $r \approx -0.908$. Is the correlation coefficient significant? Use $\alpha = 0.01$.

a. Determine the *number of pairs of data* in the sample.
b. Identify the *level of significance*.
c. Find the *critical value*. Use Table 11 in Appendix B.
d. *Compare* $|r|$ with the critical value and *decide* if the correlation is significant.
e. *Interpret* the decision in the context of the original claim.

Answer: Page A44

Note to Instructor

The material on hypothesis testing requires previous coverage of Chapter 7. This material can be omitted if you cover correlation and regression before covering Chapters 7 and 8.

▶ **HYPOTHESIS TESTING FOR A POPULATION CORRELATION COEFFICIENT ρ**

You can also use a hypothesis test to determine whether the sample correlation coefficient r provides enough evidence to conclude that the population correlation coefficient ρ is significant. A hypothesis test for ρ can be one-tailed or two-tailed. The null and alternative hypotheses for these tests are as follows.

$$\begin{cases} H_0\text{: } \rho \geq 0 \text{ (no significant negative correlation)} \\ H_a\text{: } \rho < 0 \text{ (significant negative correlation)} \end{cases}$$ **Left-tailed test**

$$\begin{cases} H_0\text{: } \rho \leq 0 \text{ (no significant positive correlation)} \\ H_a\text{: } \rho > 0 \text{ (significant positive correlation)} \end{cases}$$ **Right-tailed test**

$$\begin{cases} H_0\text{: } \rho = 0 \text{ (no significant correlation)} \\ H_a\text{: } \rho \neq 0 \text{ (significant correlation)} \end{cases}$$ **Two-tailed test**

In this text, you will consider only two-tailed hypothesis tests for ρ.

Note to Instructor

Remind students that this is the same family of distributions that was used for hypothesis testing of means with small and/or dependent samples. There are $n - 2$ degrees of freedom because one degree of freedom is lost for each variable.

THE t-TEST FOR THE CORRELATION COEFFICIENT

A t-test can be used to test whether the correlation between two variables is significant. The **test statistic** is r and the **standardized test statistic**

$$t = \frac{r}{\sigma_r} = \frac{r}{\sqrt{\dfrac{1 - r^2}{n - 2}}}$$

follows a t-distribution with $n - 2$ degrees of freedom.

GUIDELINES

Using the t-Test for the Correlation Coefficient ρ

IN WORDS	IN SYMBOLS
1. Identify the null and alternative hypotheses.	State H_0 and H_a.
2. Specify the level of significance.	Identify α.
3. Identify the degrees of freedom.	d.f. $= n - 2$
4. Determine the critical value(s) and the rejection region(s).	Use Table 5 in Appendix B.
5. Find the standardized test statistic.	$t = \dfrac{r}{\sqrt{\dfrac{1 - r^2}{n - 2}}}$
6. Make a decision to reject or fail to reject the null hypothesis.	If t is in the rejection region, reject H_0. Otherwise, fail to reject H_0.
7. Interpret the decision in the context of the original claim.	

EXAMPLE 7   **SC** Report 39

▶ The *t*-Test for a Correlation Coefficient

In Example 4, you used 10 pairs of data to find $r \approx 0.882$. Test the significance of this correlation coefficient. Use $\alpha = 0.05$.

▶ Solution

The null and alternative hypotheses are

$$H_0: \rho = 0 \text{ (no correlation)} \quad \text{and} \quad H_a: \rho \neq 0 \text{ (significant correlation)}.$$

Because there are 10 pairs of data in the sample, there are $10 - 2 = 8$ degrees of freedom. Because the test is a two-tailed test, $\alpha = 0.05$, and d.f. $= 8$, the critical values are $-t_0 = -2.306$ and $t_0 = 2.306$. The rejection regions are $t < -2.306$ and $t > 2.306$. Using the *t*-test, the standardized test statistic is

$$t = \frac{r}{\sqrt{\dfrac{1 - r^2}{n - 2}}}$$

$$\approx \frac{0.882}{\sqrt{\dfrac{1 - (0.882)^2}{10 - 2}}} \approx 5.294.$$

The following graph shows the location of the rejection regions and the standardized test statistic.

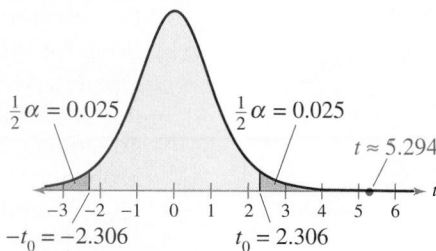

$\frac{1}{2}\alpha = 0.025$ $\frac{1}{2}\alpha = 0.025$

$t \approx 5.294$

$-t_0 = -2.306$ $t_0 = 2.306$

Because t is in the rejection region, you should decide to reject the null hypothesis.

Interpretation   There is enough evidence at the 5% level of significance to conclude that there is a significant linear correlation between gross domestic products and carbon dioxide emissions.

▶ Try It Yourself 7

In Try It Yourself 5, you calculated the correlation coefficient of the salaries and average attendances at home games for the teams in Major League Baseball to be $r \approx 0.74972$. Test the significance of this correlation coefficient. Use $\alpha = 0.01$.

a. State the *null* and *alternative hypotheses*.
b. Identify the *level of significance*.
c. Identify the *degrees of freedom*.
d. Determine the *critical values* and the *rejection regions*.
e. Find the *standardized test statistic*.
f. *Make a decision* to reject or fail to reject the null hypothesis.
g. *Interpret* the decision in the context of the original claim.

Answer: Page A45

INSIGHT

In Example 7, you can use Table 11 in Appendix B to test the population correlation coefficient ρ. Given $n = 10$ and $\alpha = 0.05$, the critical value from Table 11 is 0.632. Because

$$|r| \approx 0.882 > 0.632,$$

the correlation is significant. Note that this is the same result you obtained using a *t*-test for the population correlation coefficient ρ.

STUDY TIP

Be sure you see in Example 7 that rejecting the null hypothesis means that there is enough evidence that the correlation is significant.

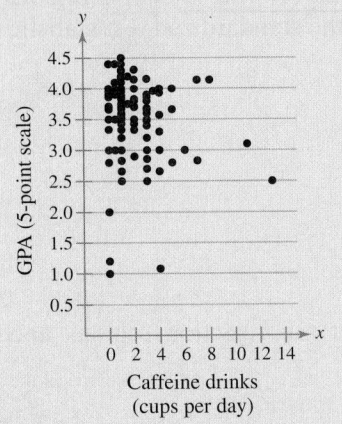
▶ CORRELATION AND CAUSATION

The fact that two variables are strongly correlated does not in itself imply a cause-and-effect relationship between the variables. More in-depth study is usually needed to determine whether there is a causal relationship between the variables.

If there is a significant correlation between two variables, a researcher should consider the following possibilities.

1. **Is there a direct cause-and-effect relationship between the variables?**

   That is, does *x* cause *y*? For instance, consider the relationship between gross domestic products and carbon dioxide emissions that has been discussed throughout this section. It is reasonable to conclude that an increase in a country's gross domestic product will result in higher carbon dioxide emissions.

2. **Is there a reverse cause-and-effect relationship between the variables?**

   That is, does *y* cause *x*? For instance, consider the Old Faithful data that have been discussed throughout this section. These variables have a positive linear correlation, and it is possible to conclude that the duration of an eruption affects the time before the next eruption. However, it is also possible that the time between eruptions affects the duration of the next eruption.

3. **Is it possible that the relationship between the variables can be caused by a third variable or perhaps a combination of several other variables?**

   For instance, consider the salaries and average attendances per home game for the teams in Major League Baseball listed in the Chapter Opener. Although these variables have a positive linear correlation, it is doubtful that just because a team's salary decreases, the average attendance per home game will also decrease. The relationship is probably due to several other variables, such as the economy, the players on the team, and whether or not the team is winning games.

4. **Is it possible that the relationship between two variables may be a coincidence?**

   For instance, although it may be possible to find a significant correlation between the number of animal species living in certain regions and the number of people who own more than two cars in those regions, it is highly unlikely that the variables are directly related. The relationship is probably due to coincidence.

Determining which of the cases above is valid for a data set can be difficult. For instance, consider the following example. Suppose a person breaks out in a rash each time he eats shrimp at a certain restaurant. The natural conclusion is that the person is allergic to shrimp. However, upon further study by an allergist, it is found that the person is not allergic to shrimp, but to a type of seasoning the chef is putting into the shrimp.

9.1 EXERCISES

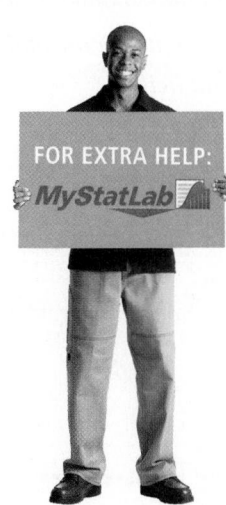

FOR EXTRA HELP:
MyStatLab

1. Increase

2. Decrease

3. The range of values for the correlation coefficient is −1 to 1, inclusive.

4. See Selected Answers, page A128.

5. Answers will vary. *Sample answer:*

   Perfect positive linear correlation: price per gallon of gasoline and total cost of gasoline

   Perfect negative linear correlation: distance from door and height of wheelchair ramp

6. A table can be used to compare *r* with a critical value, or a hypothesis test can be performed using a *t*-test.

7. *r* is the sample correlation coefficient, while *ρ* is the population correlation coefficient.

8. Answers will vary. *Sample answer:* The fact that two variables have a linear relationship does not necessarily imply that one variable is the cause of the other.

9. Negative linear correlation

10. No linear correlation

11. Perfect negative linear correlation

12. No linear correlation

13. Positive linear correlation

14. Perfect positive linear correlation

▪ BUILDING BASIC SKILLS AND VOCABULARY

1. Two variables have a positive linear correlation. Does the dependent variable increase or decrease as the independent variable increases?

2. Two variables have a negative linear correlation. Does the dependent variable increase or decrease as the independent variable increases?

3. Describe the range of values for the correlation coefficient.

4. What does the sample correlation coefficient r measure? Which value indicates a stronger correlation: $r = 0.918$ or $r = -0.932$? Explain your reasoning.

5. Give examples of two variables that have perfect positive linear correlation and two variables that have perfect negative linear correlation.

6. Explain how to decide whether a sample correlation coefficient indicates that the population correlation coefficient is significant.

7. Discuss the difference between r and ρ.

8. In your own words, what does it mean to say "correlation does not imply causation"?

Graphical Analysis *In Exercises 9–14, the scatter plots of paired data sets are shown. Determine whether there is a perfect positive linear correlation, a strong positive linear correlation, a perfect negative linear correlation, a strong negative linear correlation, or no linear correlation between the variables.*

9.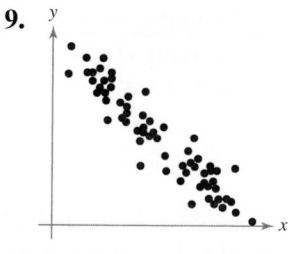

10.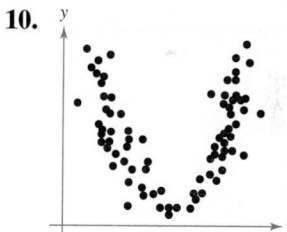

11.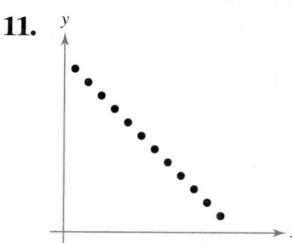

12.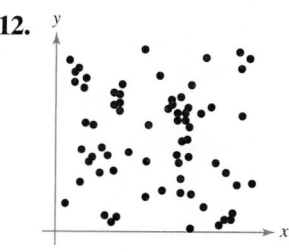

13.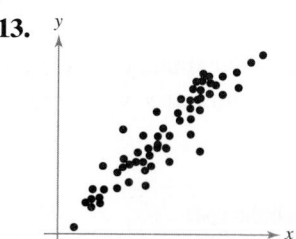

14.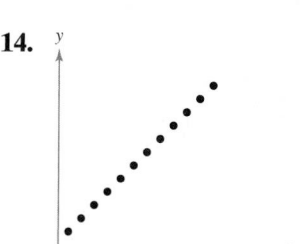

15. c; You would expect a positive linear correlation between age and income.

16. d; You would not expect age and height to be correlated.

17. b; You would expect a negative linear correlation between age and balance on student loans.

18. a; You would expect the relationship between age and body temperature to be fairly constant.

19. Explanatory variable: Amount of water consumed

    Response variable: Weight loss

20. Explanatory variable: Hours of safety classes

    Response variable: Number of accidents

21. (a)

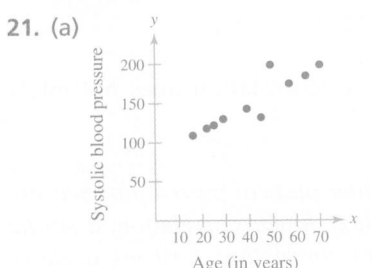

    (b) 0.908

    (c) Strong positive linear correlation

22. (a)

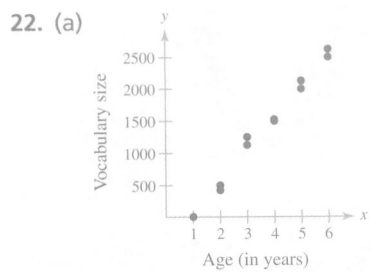

    (b) 0.996

    (c) Strong positive linear correlation

23. (a)

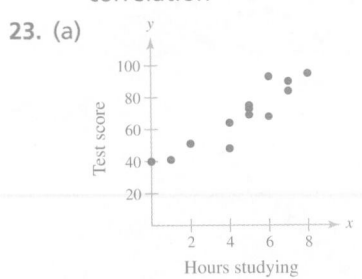

    (b) 0.923

    (c) Strong positive linear correlation

Graphical Analysis *In Exercises 15–18, the scatter plots show the results of a survey of 20 randomly selected males ages 24–35. Using age as the explanatory variable, match each graph with the appropriate description. Explain your reasoning.*

(a) *Age and body temperature*

(b) *Age and balance on student loans*

(c) *Age and income*

(d) *Age and height*

15.

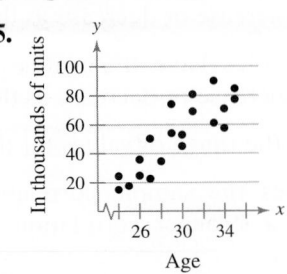

16.

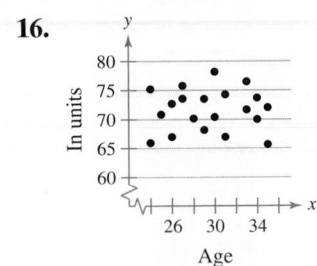

17.

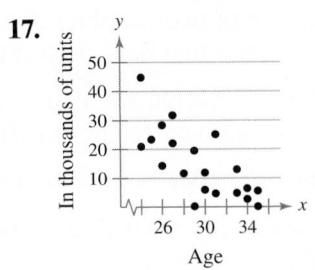

18.

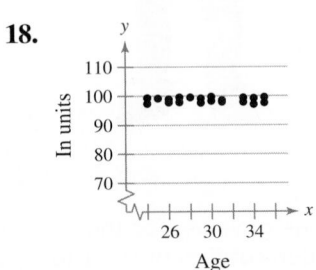

In Exercises 19 and 20, identify the explanatory variable and the response variable.

19. A nutritionist wants to determine if the amounts of water consumed each day by persons of the same weight and on the same diet can be used to predict individual weight loss.

20. An insurance company hires an actuary to determine whether the number of hours of safety driving classes can be used to predict the number of driving accidents for each driver.

■ USING AND INTERPRETING CONCEPTS

Constructing a Scatter Plot and Determining Correlation *In Exercises 21–28, (a) display the data in a scatter plot, (b) calculate the sample correlation coefficient r, and (c) make a conclusion about the type of correlation.*

21. **Age and Blood Pressure** The ages (in years) of 10 men and their systolic blood pressures

Age, x	16	25	39	45	49	64	70	29	57	22
Systolic blood pressure, y	109	122	143	132	199	185	199	130	175	118

22. **Age and Vocabulary** The ages (in years) of 11 children and the number of words in their vocabulary

Age, x	1	2	3	4	5	6
Vocabulary size, y	3	440	1200	1500	2100	2600

Age, x	3	5	2	4	6
Vocabulary size, y	1100	2000	500	1525	2500

24. (a)

(b) −0.831

(c) Strong negative linear correlation

25. (a)

(b) 0.604

(c) Weak positive linear correlation

26. (a)

(b) −0.972

(c) Strong negative linear correlation

27. (a)

(b) 0.828

(c) Strong positive linear correlation

28. (a) See Selected Answers, page A128.

(b) 0.981

(c) Strong positive linear correlation

23. Hours Studying and Test Scores The number of hours 13 students spent studying for a test and their scores on that test

Hours spent studying, x	0	1	2	4	4	5	5	5	6	6	7	7	8
Test score, y	40	41	51	48	64	69	73	75	68	93	84	90	95

24. Hours Online and Test Scores The number of hours 12 students spent online during the weekend and the scores of each student who took a test the following Monday

Hours spent online, x	0	1	2	3	3	5	5	5	6	7	7	10
Test score, y	96	85	82	74	95	68	76	84	58	65	75	50

25. Movie Budgets and Grosses The budget (in millions of dollars) and worldwide gross (in millions of dollars) for eight of the most expensive movies ever made *(Adapted from The Numbers)*

Budget, x	300	258	250	210	232	230	225	207
Gross, y	961	891	937	836	391	576	419	551

26. Speed of Sound The altitude (in thousands of feet) and speed of sound (in feet per second)

Altitude, x	0	5	10	15	20	25
Speed of sound, y	1116.3	1096.9	1077.3	1057.2	1036.8	1015.8

Altitude, x	30	35	40	45	50
Speed of sound, y	994.5	969.0	967.7	967.7	967.7

27. Earnings and Dividends The earnings per share and dividends per share for 12 medical supplies companies in a recent year *(Source: The Value Line Investment Survey)*

Earnings per share, x	6.00	1.44	4.44	3.38	3.63	4.46
Dividends per share, y	2.45	0.15	0.62	0.91	0.68	1.14

Earnings per share, x	3.80	1.43	1.88	4.57	4.28	2.92
Dividends per share, y	0.52	0.06	0.19	1.80	0.48	0.63

28. Crimes and Arrests The number of crimes reported (in millions) and the number of arrests reported (in millions) by the U.S. Department of Justice for 14 years *(Adapted from the National Crime Victimization Survey and Uniform Crime Reports)*

Crimes, x	1.60	1.55	1.44	1.40	1.32	1.23	1.22
Arrests, y	0.78	0.80	0.73	0.72	0.68	0.64	0.63

Crimes, x	1.23	1.22	1.18	1.16	1.19	1.21	1.20
Arrests, y	0.63	0.62	0.60	0.59	0.60	0.61	0.58

29. The correlation coefficient becomes $r \approx 0.621$. The new data entry is an outlier, so the linear correlation is weaker.

30. The correlation coefficient becomes $r \approx -0.343$. The new data entry is an outlier, so the linear correlation is weaker.

31. There is not enough evidence at the 1% level of significance to conclude that there is a significant linear correlation between vehicle weight and the variability in braking distance.

32. There is enough evidence at the 5% level of significance to conclude that there is a significant linear correlation between vehicle weight and the variability in braking distance.

33. There is enough evidence at the 1% level of significance to conclude that there is a significant linear correlation between the number of hours spent studying for a test and the score received on the test.

34. There is enough evidence at the 5% level of significance to conclude that there is a significant linear correlation between the number of hours spent online and the score received on the test.

35. There is enough evidence at the 1% level of significance to conclude that there is a significant linear correlation between earnings per share and dividends per share.

36. There is enough evidence at the 5% level of significance to conclude that there is a significant linear correlation between crimes and arrests.

29. A student spends 1 hour studying and gets a test score of 99. Add this data entry to the current data set in Exercise 23. Describe how adding this data entry changes the correlation coefficient r. Why do you think it changed?

30. A student spends 12 hours online during the weekend and gets a test score of 98. Add this data entry to the current data set in Exercise 24. Describe how adding this data entry changes the correlation coefficient r. Why do you think it changed?

Testing Claims   *In Exercises 31–36, use Table 11 in Appendix B as shown in Example 6, or perform a hypothesis test using Table 5 in Appendix B as shown in Example 7, to make a conclusion about the indicated correlation coefficient. If convenient, use technology to solve the problem.*

31. **Braking Distances: Dry Surface**   The weights (in pounds) of eight vehicles and the variability of their braking distances (in feet) when stopping on a dry surface are shown in the table. Can you conclude that there is a significant linear correlation between vehicle weight and variability in braking distance on a dry surface? Use $\alpha = 0.01$. *(Adapted from National Highway Traffic Safety Administration)*

Weight, x	5940	5340	6500	5100	5850	4800	5600	5890
Variability in braking distance, y	1.78	1.93	1.91	1.59	1.66	1.50	1.61	1.70

32. **Braking Distances: Wet Surface**   The weights (in pounds) of eight vehicles and the variability of their braking distances (in feet) when stopping on a wet surface are shown in the table. At $\alpha = 0.05$, can you conclude that there is a significant linear correlation between vehicle weight and variability in braking distance on a wet surface? *(Adapted from National Highway Traffic Safety Administration)*

Weight, x	5890	5340	6500	4800	5940	5600	5100	5850
Variability in braking distance, y	2.92	2.40	4.09	1.72	2.88	2.53	2.32	2.78

33. **Hours Studying and Test Scores**   The table in Exercise 23 shows the number of hours 13 students spent studying for a test and their scores on that test. At $\alpha = 0.01$, is there enough evidence to conclude that there is a significant linear correlation between the data? (Use the value of r found in Exercise 23.)

34. **Hours Online and Test Scores**   The table in Exercise 24 shows the number of hours spent online and the test scores for 12 randomly selected students. At $\alpha = 0.05$, is there enough evidence to conclude that there is a significant linear correlation between the data? (Use the value of r found in Exercise 24.)

35. **Earnings and Dividends**   The table in Exercise 27 shows the earnings per share and dividends per share for 12 medical supplies companies in a recent year. At $\alpha = 0.01$, can you conclude that there is a significant linear correlation between earnings per share and dividends per share? (Use the value of r found in Exercise 27.)

36. **Crimes and Arrests**   The table in Exercise 28 shows the number of crimes reported (in millions) and the number of arrests reported (in millions) by the U.S. Department of Justice for 14 years. At $\alpha = 0.05$, can you conclude that there is a significant linear correlation between the number of crimes and the number of arrests? (Use the value of r found in Exercise 28.)

37. (a) See Odd Answers, page A93.

(b) 0.848

(c) Reject H_0. There is enough evidence at the 1% level of significance to conclude that there is a significant linear correlation between the magnitudes of earthquakes and their depths below the surface at the epicenter.

38. (a) See Selected Answers, page A128.

(b) 0.603

(c) Fail to reject H_0. There is not enough evidence at the 5% level of significance to conclude that there is a significant linear correlation between family income level and percent of income donated to charities.

39. The correlation coefficient becomes $r \approx 0.085$. The new rejection regions are $t < -3.499$ and $t > 3.499$, and the new standardized test statistic is $t \approx 0.227$. So, you now fail to reject H_0.

40. The correlation coefficient becomes $r \approx -0.042$. The new rejection regions are $t < -2.447$ and $t > 2.447$, and the new standardized test statistic is $t \approx -0.102$. So, you still fail to reject H_0.

41. 0.883; 0.883; The correlation coefficient remains unchanged when the x-values and y-values are switched.

42. -0.779; -0.779; The correlation coefficient remains unchanged when the x-values and y-values are switched.

43. Answers will vary.

SC   *In Exercises 37 and 38, use StatCrunch to (a) display the data in a scatter plot, (b) calculate the correlation coefficient r, and (c) test the significance of the correlation coefficient.*

37. Earthquakes   A researcher wants to determine if there is a linear relationship between the magnitudes of earthquakes and their depths below the surface at the epicenter. The magnitudes and depths (in kilometers) of eight recent earthquakes are shown in the table. Use $\alpha = 0.01$. *(Source: U.S. Geological Survey)*

Magnitude, x	7.7	6.7	6.9	6.8	4.0	3.8	7.1	5.9
Depth, y	35	18	17	26	5	10	25	10

38. Income Level and Charitable Donations   A sociologist wants to determine if there is a linear relationship between family income level and percent of income donated to charities. The income levels (in thousands of dollars) and percents of income donated to charities for seven families are shown in the table. Use $\alpha = 0.05$.

Income level, x	50	65	48	42	59	72	60
Donating percent, y	4	8	5	5	10	7	6

39. An earthquake is recorded with a magnitude of 6.3 and a depth of 620 kilometers. Add this data entry to the current data set in Exercise 37. Describe how adding this data entry changes the correlation coefficient r and your decision to reject or fail to reject the null hypothesis.

40. A family has an income level of \$75,000 and donates 1% of their income to charities. Add this data entry to the current data set in Exercise 38. Describe how adding this data entry changes the correlation coefficient r and your decision to reject or fail to reject the null hypothesis.

■ **EXTENDING CONCEPTS**

Interchanging x and y   *In Exercises 41 and 42, calculate the correlation coefficient r, letting Row 1 represent the x-values and Row 2 the y-values. Then calculate the correlation coefficient r, letting Row 2 represent the x-values and Row 1 the y-values. What effect does switching the explanatory and response variables have on the correlation coefficient?*

41.

Row 1	16	25	39	45	49	64	70
Row 2	109	122	143	132	199	185	199

42.

Row 1	0	1	2	3	3	5	5	5	6	7
Row 2	96	85	82	74	95	68	76	84	58	65

43. Writing   Use your school's library, the Internet, or some other reference source to find a real-life data set with the indicated cause-and-effect relationship. Write a paragraph describing each variable and explain why you think the variables have the indicated cause-and-effect relationship.

(a) *Direct Cause-and-Effect:* Changes in one variable cause changes in the other variable.

(b) *Other Factors:* The relationship between the variables is caused by a third variable.

(c) *Coincidence:* The relationship between the variables is a coincidence.

ACTIVITY 9.1   Correlation by Eye

The *correlation by eye* applet allows you to guess the sample correlation coefficient *r* for a data set. When the applet loads, a data set consisting of 20 points is displayed. Points can be added to the plot by clicking the mouse.

Points on the plot can be removed by clicking on the point and then dragging the point into the trash can. All of the points on the plot can be removed by simply clicking inside the trash can. You can enter your guess for *r* in the "Guess" field, and then click SHOW R! to see if you are within 0.1 of the true value. When you click NEW DATA, a new data set is generated.

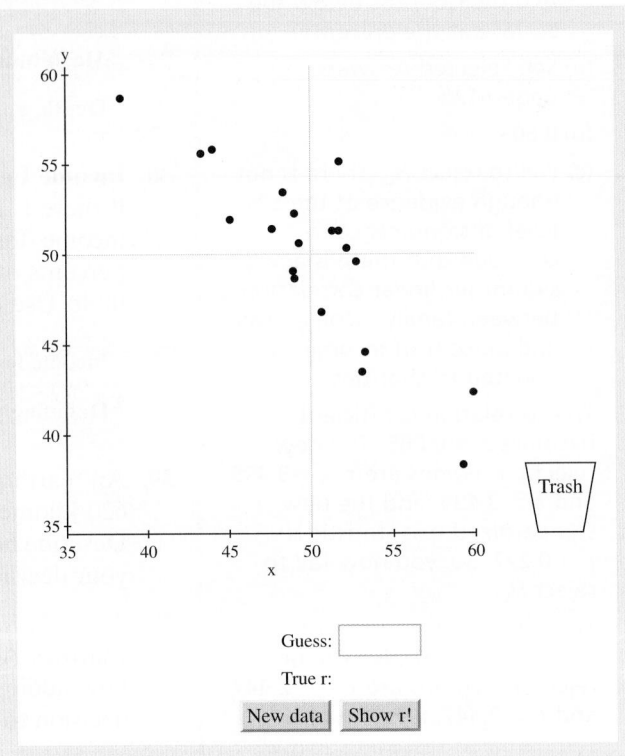

■ Explore

Step 1   Add five points to the plot.
Step 2   Enter a guess for *r*.
Step 3   Click SHOW R!.
Step 4   Click NEW DATA.
Step 5   Remove five points from the plot.
Step 6   Enter a guess for *r*.
Step 7   Click SHOW R!.

■ Draw Conclusions

1. Generate a new data set. Using your knowledge of correlation, try to guess the value of *r* for the data set. Repeat this 10 times. How many times were you correct? Describe how you chose each *r* value.

2. Describe how to create a data set with a value of *r* that is approximately 1.

3. Describe how to create a data set with a value of *r* that is approximately 0.

4. Try to create a data set with a value of *r* that is approximately −0.9. Then try to create a data set with a value of *r* that is approximately 0.9. What did you do differently to create the two data sets?

9.2   Linear Regression

Regression Lines ▸ Applications of Regression Lines

▸ REGRESSION LINES

After verifying that the linear correlation between two variables is significant, the next step is to determine the equation of the line that best models the data. This line is called a *regression line*, and its equation can be used to predict the value of y for a given value of x. Although many lines can be drawn through a set of points, a regression line is determined by specific criteria.

Consider the scatter plot and the line shown below. For each data point, d_i represents the difference between the observed y-value and the predicted y-value for a given x-value on the line. These differences are called **residuals** and can be positive, negative, or zero. When the point is above the line, d_i is positive. When the point is below the line, d_i is negative. If the observed y-value equals the predicted y-value, $d_i = 0$. Of all possible lines that can be drawn through a set of points, the regression line is the line for which the sum of the squares of all the residuals

$$\sum d_i^2$$

is a minimum.

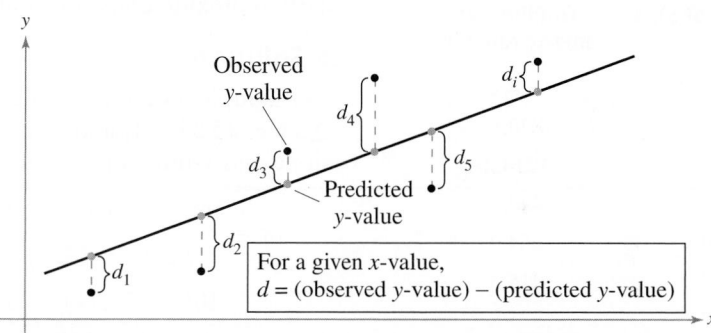

For a given x-value,
$d = $ (observed y-value) − (predicted y-value)

DEFINITION

A **regression line,** also called a **line of best fit,** is the line for which the sum of the squares of the residuals is a minimum.

When determining the equation of a regression line, it is helpful to construct a scatter plot of the data to check for outliers, which can greatly influence a regression line. You should also check for gaps and clusters in the data.

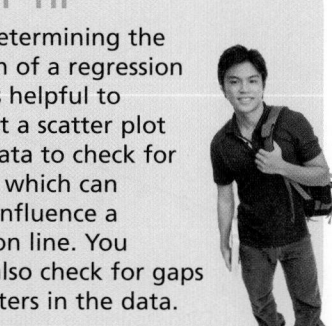

In algebra, you learned that you can write an equation of a line by finding its slope m and y-intercept b. The equation has the form

$$y = mx + b.$$

Recall that the slope of a line is the ratio of its rise over its run and the y-intercept is the y-value of the point at which the line crosses the y-axis. It is the y-value when $x = 0$.

In algebra, you used two points to determine the equation of a line. In statistics, you will use every point in the data set to determine the equation of the regression line.

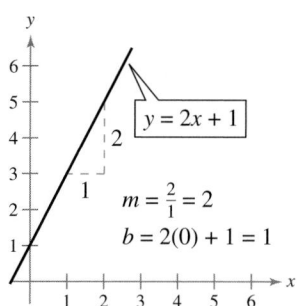

$y = 2x + 1$

$m = \frac{2}{1} = 2$

$b = 2(0) + 1 = 1$

The equation of a regression line allows you to use the independent (explanatory) variable x to make predictions for the dependent (response) variable y.

THE EQUATION OF A REGRESSION LINE

The equation of a regression line for an independent variable x and a dependent variable y is

$$\hat{y} = mx + b$$

where $\hat{y}$ is the predicted y-value for a given x-value. The slope m and y-intercept b are given by

$$m = \frac{n\sum xy - (\sum x)(\sum y)}{n\sum x^2 - (\sum x)^2} \quad \text{and} \quad b = \bar{y} - m\bar{x} = \frac{\sum y}{n} - m\frac{\sum x}{n}$$

where $\bar{y}$ is the mean of the y-values in the data set and $\bar{x}$ is the mean of the x-values. The regression line always passes through the point $(\bar{x}, \bar{y})$.

EXAMPLE 1

▶ **Finding the Equation of a Regression Line**

Find the equation of the regression line for the gross domestic products and carbon dioxide emissions data used in Section 9.1.

▶ **Solution**

In Example 4 of Section 9.1, you found that $n = 10$, $\sum x = 23.1$, $\sum y = 5554$, $\sum xy = 15{,}573.71$, and $\sum x^2 = 67.35$. You can use these values to calculate the slope and y-intercept of the regression line as shown.

$$m = \frac{n\sum xy - (\sum x)(\sum y)}{n\sum x^2 - (\sum x)^2}$$

$$= \frac{10(15{,}573.71) - (23.1)(5554)}{10(67.35) - 23.1^2}$$

$$= \frac{27{,}439.7}{139.89} \approx 196.151977$$

$$b = \bar{y} - m\bar{x} \approx \frac{5554}{10} - (196.151977)\frac{23.1}{10}$$

$$= 555.4 - (196.151977)(2.31)$$

$$\approx 102.2889$$

GDP (trillions of $), x	CO_2 emissions (millions of metric tons), y
1.6	428.2
3.6	828.8
4.9	1214.2
1.1	444.6
0.9	264.0
2.9	415.3
2.7	571.8
2.3	454.9
1.6	358.7
1.5	573.5

So, the equation of the regression line is

$$\hat{y} = 196.152x + 102.289.$$

To sketch the regression line, use any two x-values within the range of data and calculate their corresponding y-values from the regression line. Then draw a line through the two points. The regression line and scatter plot of the data are shown at the right. If you plot the point $(\bar{x}, \bar{y}) = (2.31, 555.4)$, you will notice that the line passes through this point.

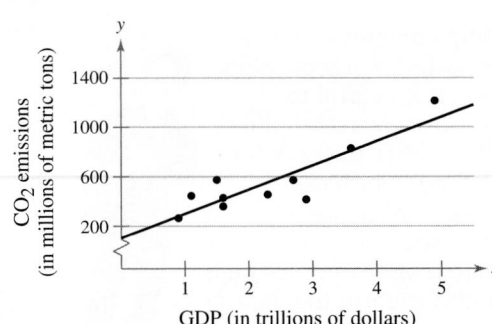

▶ **Try It Yourself 1**

Find the equation of the regression line for the number of years out of school and annual contribution data used in Section 9.1.

a. *Identify* n, $\sum x$, $\sum y$, $\sum xy$, and $\sum x^2$ from Try It Yourself 4 in Section 9.1.
b. Calculate the *slope m* and the *y-intercept b*.
c. Write the *equation* of the regression line.   *Answer: Page A45*

Duration, x	Time, y	Duration, x	Time, y
1.80	56	3.78	79
1.82	58	3.83	85
1.90	62	3.88	80
1.93	56	4.10	89
1.98	57	4.27	90
2.05	57	4.30	89
2.13	60	4.43	89
2.30	57	4.47	86
2.37	61	4.53	89
2.82	73	4.55	86
3.13	76	4.60	92
3.27	77	4.63	91
3.65	77		

Note to Instructor

Point out the different notation used by the TI-83/84 Plus, MINITAB, and Excel.

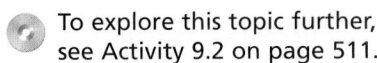

To explore this topic further, see Activity 9.2 on page 511.

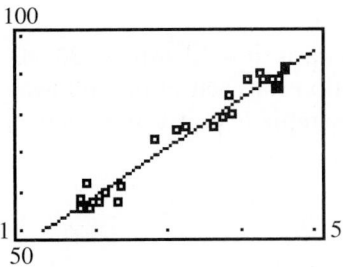

EXAMPLE 2   SC Report 40

▶ **Using Technology to Find a Regression Equation**

Use a technology tool to find the equation of the regression line for the Old Faithful data used in Section 9.1.

▶ **Solution**

MINITAB, Excel, and the TI-83/84 Plus each have features that automatically calculate a regression equation. Try using this technology to find the regression equation. You should obtain results similar to the following.

MINITAB

Regression Analysis: C2 versus C1

The regression equation is
C2 = 33.7 + 12.5 C1

Predictor	Coef	SE Coef	T	P
Constant	33.683	1.894	17.79	0.000
C1	12.4809	0.5464	22.84	0.000

S = 2.88153   R-Sq = 95.8%   R-Sq(adj) = 95.6%

EXCEL

	A	B	C	D
1	Slope:			
2	INDEX(LINEST(known_y's,known_x's),1)			
3				12.48094
4				
5	Y-intercept:			
6	INDEX(LINEST(known_y's,known_x's),2)			
7				33.6829

TI-83/84 PLUS

LinReg
y=ax+b
a=12.48094391
b=33.68290034
r^2=.9577738551
r=.9786592129

From the displays, you can see that the regression equation is

$$\hat{y} = 12.481x + 33.683.$$

The TI-83/84 Plus display at the left shows the regression line and a scatter plot of the data in the same viewing window. To do this, use *Stat Plot* to construct the scatter plot and enter the regression equation as y_1.

▶ **Try It Yourself 2**

Use a technology tool to find the equation of the regression line for the salaries and average attendances at home games for the teams in Major League Baseball given in the Chapter Opener on page 483.

a. *Enter* the data.
b. Perform the necessary steps to calculate the *slope* and *y-intercept*.
c. Specify the *regression equation*.   *Answer: Page A45*

PICTURING THE WORLD

The following scatter plot shows the relationship between the number of farms (in thousands) in a state and the total value of the farms (in billions of dollars).

(Source: U.S. Department of Agriculture and National Agriculture Statistics Service)

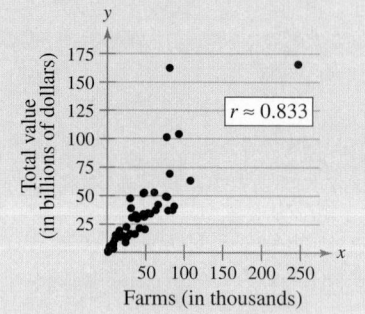

Describe the correlation between these two variables in words. Use the scatter plot to predict the total value of farms in a state that has 150,000 farms. The regression line for this scatter plot is $\hat{y} = 0.714x + 3.367$. Use this equation to make a prediction. (Assume x and y have a significant linear correlation.) How does your algebraic prediction compare with your graphical one?

$110 billion; $110.467 billion; The predictions are very close to each other.

▶ APPLICATIONS OF REGRESSION LINES

After finding the equation of a regression line, you can use the equation to predict *y*-values over the range of the data *if the correlation between x and y is significant*. For instance, an environmentalist could forecast carbon dioxide emissions on the basis of gross domestic products. To predict *y*-values, substitute the given *x*-value into the regression equation, then calculate $\hat{y}$, the predicted *y*-value.

EXAMPLE 3

▶ Predicting *y*-Values Using Regression Equations

The regression equation for the gross domestic products (in trillions of dollars) and carbon dioxide emissions (in millions of metric tons) data is

$$\hat{y} = 196.152x + 102.289.$$

Use this equation to predict the *expected* carbon dioxide emissions for the following gross domestic products. (Recall from Section 9.1, Example 7, that *x* and *y* have a significant linear correlation.)

1. 1.2 trillion dollars **2.** 2.0 trillion dollars **3.** 2.5 trillion dollars

▶ Solution

To predict the expected carbon dioxide emissions, substitute each gross domestic product for *x* in the regression equation. Then calculate $\hat{y}$.

1. $\hat{y} = 196.152x + 102.289$
$= 196.152(1.2) + 102.289$
≈ 337.671

Interpretation When the gross domestic product is \$1.2 trillion, the CO_2 emissions are about 337.671 million metric tons.

2. $\hat{y} = 196.152x + 102.289$
$= 196.152(2.0) + 102.289$
$= 494.593$

Interpretation When the gross domestic product is \$2.0 trillion, the CO_2 emissions are 494.593 million metric tons.

3. $\hat{y} = 196.152x + 102.289$
$= 196.152(2.5) + 102.289$
$= 592.669$

Interpretation When the gross domestic product is \$2.5 trillion, the CO_2 emissions are 592.669 million metric tons.

Prediction values are meaningful only for *x*-values in (or close to) the range of the data. The *x*-values in the original data set range from 0.9 to 4.9. So, it would not be appropriate to use the regression line $\hat{y} = 196.152x + 102.289$ to predict carbon dioxide emissions for gross domestic products such as \$0.2 or \$14.5 trillion dollars.

▶ Try It Yourself 3

The regression equation for the Old Faithful data is $\hat{y} = 12.481x + 33.683$. Use this to predict the time until the next eruption for each of the following eruption durations. (Recall from Section 9.1, Example 6, that *x* and *y* have a significant linear correlation.)

1. 2 minutes
2. 3.32 minutes

a. *Substitute* each value of *x* into the regression equation.
b. *Calculate* $\hat{y}$.
c. *Specify* the predicted time until the next eruption for each eruption duration.

Answer: Page A45

9.2 EXERCISES

FOR EXTRA HELP:
MyStatLab

1. A residual is the difference between the observed y-value of a data point and the predicted y-value on the regression line for the x-coordinate of the data point. A residual is positive when the data point is above the line, negative when the point is below the line, and zero when the observed y-value equals the predicted y-value.

2. Positive

3. Substitute a value of x into the equation of a regression line and solve for y.

4. Prediction values are meaningful only for x-values in (or close to) the range of the original data.

5. The correlation between variables must be significant.

6. Answers will vary. *Sample answer:* Because the regression line models the trend of the given data, and it is not known if the trend continues beyond the range of those data.

7. b

8. a

9. e

10. c

11. f

12. d

13. c

14. b

15. a

16. d

■ BUILDING BASIC SKILLS AND VOCABULARY

1. What is a residual? Explain when a residual is positive, negative, and zero.

2. Two variables have a positive linear correlation. Is the slope of the regression line for the variables positive or negative?

3. Explain how to predict y-values using the equation of a regression line.

4. Given a set of data and a corresponding regression line, describe all values of x that provide meaningful predictions for y.

5. In order to predict y-values using the equation of a regression line, what must be true about the correlation coefficient of the variables?

6. Why is it not appropriate to use a regression line to predict y-values for x-values that are not in (or close to) the range of x-values found in the data?

In Exercises 7–12, match the description in the left column with its symbol(s) in the right column.

7. The y-value of a data point corresponding to x_i **a.** $\hat{y}_i$

8. The y-value for a point on the regression line corresponding to x_i **b.** y_i

         c. b

9. Slope

10. y-intercept   **d.** $(\bar{x}, \bar{y})$

11. The mean of the y-values   **e.** m

         f. $\bar{y}$

12. The point a regression line always passes through

Graphical Analysis   *In Exercises 13–16, match the regression equation with the appropriate graph. (Note that the x- and y-axes are broken.)*

13. $\hat{y} = -1.04x + 50.3$

14. $\hat{y} = 1.662x + 83.34$

15. $\hat{y} = 0.00114x + 2.53$

16. $\hat{y} = -0.667x + 52.6$

a.

b.

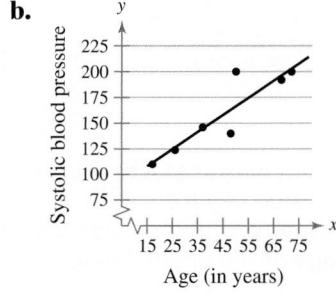

c.

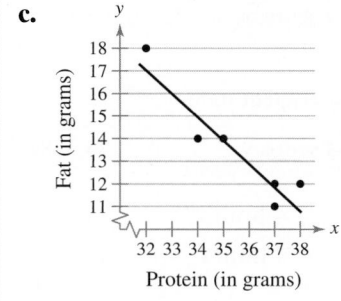

d.

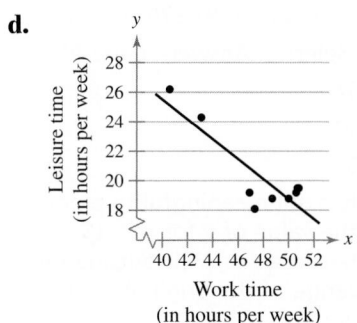

17. $\hat{y} = 0.065x + 0.465$

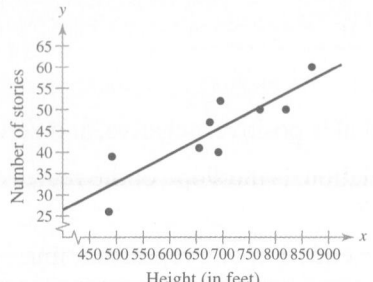

(a) 52 stories

(b) 49 stories

(c) It is not meaningful to predict the value of y for $x = 400$ because $x = 400$ is outside the range of the original data.

(d) 41 stories

18. $\hat{y} = 0.123x - 50.030$

See Selected Answers, page A129.

(a) 128.32

(b) It is not meaningful to predict the value of y for $x = 2720$ because $x = 2720$ is outside the range of the original data.

(c) 217.495

(d) 182.44

19. $\hat{y} = 7.350x + 34.617$

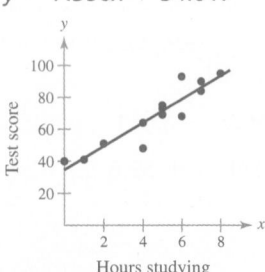

(a) 57

(b) 82

(c) It is not meaningful to predict the value of y for $x = 13$ because $x = 13$ is outside the range of the original data.

(d) 68

20. $\hat{y} = -4.067x + 93.970$

See Selected Answers, page A129.

(a) 78

(b) 61

(c) 57

(d) It is not meaningful to predict the value of y for $x = 15$ because $x = 15$ is outside the range of the original data.

■ USING AND INTERPRETING CONCEPTS

Finding the Equation of a Regression Line *In Exercises 17–24, find the equation of the regression line for the given data. Then construct a scatter plot of the data and draw the regression line. (Each pair of variables has a significant correlation.) Then use the regression equation to predict the value of y for each of the given x-values, if meaningful. If the x-value is not meaningful to predict the value of y, explain why not. If convenient, use technology to solve the problem.*

17. Atlanta Building Heights The heights (in feet) and the number of stories of nine notable buildings in Atlanta *(Source: Emporis Corporation)*

Height, x	869	820	771	696	692	676	656	492	486
Stories, y	60	50	50	52	40	47	41	39	26

(a) $x = 800$ feet   (b) $x = 750$ feet

(c) $x = 400$ feet   (d) $x = 625$ feet

18. Square Footages and Home Sale Prices The square footages and sale prices (in thousands of dollars) of seven homes *(Source: Howard Hanna)*

Square footage, x	1924	1592	2413	2332	1552	1312	1278
Sale price, y	174.9	136.9	275.0	219.9	120.0	99.9	145.0

(a) $x = 1450$ square feet   (b) $x = 2720$ square feet

(c) $x = 2175$ square feet   (d) $x = 1890$ square feet

19. Hours Studying and Test Scores The number of hours 13 students spent studying for a test and their scores on that test

Hours spent studying, x	0	1	2	4	4	5	5
Test score, y	40	41	51	48	64	69	73

Hours spent studying, x	5	6	6	7	7	8
Test score, y	75	68	93	84	90	95

(a) $x = 3$ hours   (b) $x = 6.5$ hours

(c) $x = 13$ hours   (d) $x = 4.5$ hours

20. Hours Online The number of hours 12 students spent online during the weekend and the scores of each student who took a test the following Monday

Hours spent online, x	0	1	2	3	3	5
Test score, y	96	85	82	74	95	68

Hours spent online, x	5	5	6	7	7	10
Test score, y	76	84	58	65	75	50

(a) $x = 4$ hours   (b) $x = 8$ hours

(c) $x = 9$ hours   (d) $x = 15$ hours

21. $\hat{y} = 2.472x + 80.813$

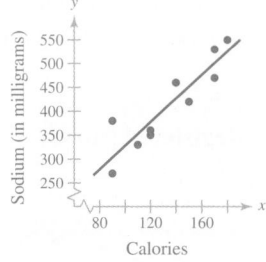

(a) 501.053 milligrams

(b) 328.013 milligrams

(c) 426.893 milligrams

(d) It is not meaningful to predict the value of y for x = 210 because x = 210 is outside the range of the original data.

22. $\hat{y} = 0.167x - 18.140$

See Selected Answers, page A129.

(a) 6.91 grams

(b) It is not meaningful to predict the value of y for x = 90 because x = 90 is outside the range of the original data.

(c) 11.085 grams

(d) 16.596 grams

23. $\hat{y} = 1.870x + 51.360$

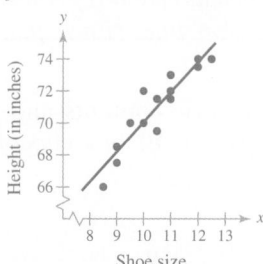

(a) 72.865 inches

(b) 66.32 inches

(c) It is not meaningful to predict the value of y for x = 15.5 because x = 15.5 is outside the range of the original data.

(d) 70.06 inches

24. $\hat{y} = -0.688x + 14.440$

See Selected Answers, page A129.

(a) 14.234 hours

(b) It is not meaningful to predict the value of y for x = 3.9 because x = 3.9 is outside the range of the original data.

(c) 14.027 hours

(d) 13.89 hours

21. Hot Dogs: Caloric and Sodium Content   The caloric contents and the sodium contents (in milligrams) of 10 beef hot dogs *(Source: Consumer Reports)*

Calories, x	150	170	120	120	90
Sodium, y	420	470	350	360	270

Calories, x	180	170	140	90	110
Sodium, y	550	530	460	380	330

(a) x = 170 calories   (b) x = 100 calories

(c) x = 140 calories   (d) x = 210 calories

22. High-Fiber Cereals: Caloric and Sugar Content   The caloric contents and the sugar contents (in grams) of 11 high-fiber breakfast cereals *(Source: Consumer Reports)*

Calories, x	140	200	160	170	170	190
Sugar, y	6	9	6	9	10	17

Calories, x	190	210	190	170	160
Sugar, y	13	18	19	10	10

(a) x = 150 calories   (b) x = 90 calories

(c) x = 175 calories   (d) x = 208 calories

23. Shoe Size and Height   The shoe sizes and heights (in inches) of 14 men

Shoe size, x	8.5	9.0	9.0	9.5	10.0	10.0	10.5
Height, y	66.0	68.5	67.5	70.0	70.0	72.0	71.5

Shoe size, x	10.5	11.0	11.0	11.0	12.0	12.0	12.5
Height, y	69.5	71.5	72.0	73.0	73.5	74.0	74.0

(a) x = size 11.5   (b) x = size 8.0

(c) x = size 15.5   (d) x = size 10.0

24. Age and Hours Slept   The ages (in years) of 10 infants and the number of hours each slept in a day

Age, x	0.1	0.2	0.4	0.7	0.6	0.9
Hours slept, y	14.9	14.5	13.9	14.1	13.9	13.7

Age, x	0.1	0.2	0.4	0.9
Hours slept, y	14.3	13.9	14.0	14.1

(a) x = 0.3 year   (b) x = 3.9 years

(c) x = 0.6 year   (d) x = 0.8 year

Years of experience, x	Annual salary (in thousands), y
0.5	40.2
2	42.9
4	45.1
5	46.7
7	50.2
9	53.6
10	54.0
12.5	58.4
13	61.8
16	63.9
18	67.5
20	64.3
22	60.1
25	59.9

TABLE FOR EXERCISES 25–29

25. Strong positive linear correlation; As the years of experience of the registered nurses increase, their salaries tend to increase.

26. $\hat{y} = 0.998x + 43.214$

See Selected Answers, page A129.

27. No, it is not meaningful to predict a salary for a registered nurse with 28 years of experience because $x = 28$ is outside the range of the original data.

28. $|r| \approx 0.880 > 0.661$ (the critical value), so the population has a significant correlation.

29. Answers will vary. *Sample answer:* Although it is likely that there is a cause-and-effect relationship between a registered nurse's years of experience and salary, you cannot use significant correlation to claim cause and effect. The relationship between the variables may also be influenced by other factors, such as work performance, level of education, or the number of years with an employer.

30. (a) $\hat{y} = 0.024x + 0.181$

(b) 0.773

(c) See Selected Answers, page A129.

31. (a) $\hat{y} = -0.159x + 5.827$

(b) −0.852

(c) See Odd Answers, page A95.

Registered Nurse Salaries *In Exercises 25–29, use the following information. You work for a salary analyst and gather the data shown in the table. The table shows the years of experience of 14 registered nurses and their annual salaries.* (Adapted from Payscale, Inc.)

25. Correlation 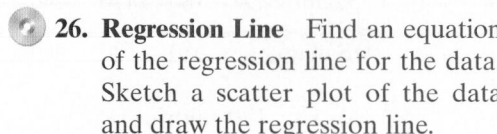 Using the scatter plot of the registered nurse salary data shown, what type of correlation, if any, do you think the data have? Explain.

26. Regression Line Find an equation of the regression line for the data. Sketch a scatter plot of the data and draw the regression line.

27. Using the Regression Line The analyst used the regression line you found in Exercise 26 to predict the annual salary for a registered nurse with 28 years of experience. Is this a valid prediction? Explain your reasoning.

28. Significant Correlation? The analyst claims that the population has a significant correlation for $\alpha = 0.01$. Verify this claim.

29. Cause and Effect Write a paragraph describing the cause-and-effect relationship between the years of experience and the annual salaries of registered nurses.

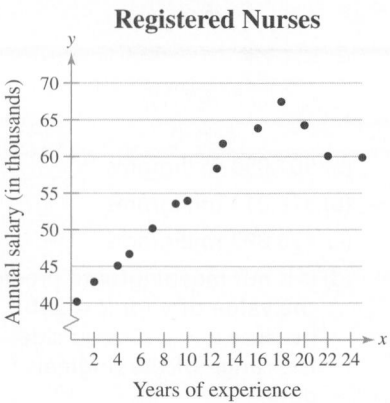

SC *In Exercises 30 and 31, use StatCrunch to (a) find the equation of the regression line for the data, (b) find the correlation coefficient r, and (c) construct the scatter plot of the data that also shows the regression line. (Each pair of variables has a significant correlation.)*

30. Hot Chocolates: Caloric and Fat Contents The caloric contents and the fat contents (in grams) of 6- to 8-ounce servings for 10 hot chocolate products *(Source: Consumer Reports)*

Calories, x	262	140	150	159	120	140	185	150	80	80
Fat, y	6.3	2.0	3.5	3.5	2.5	3.5	6.8	3	3	2.5

31. Wins and Earned Run Averages The number of wins and the earned run averages (mean number of earned runs allowed per nine innings pitched) for eight professional baseball pitchers in the 2009 regular season *(Source: ESPN)*

Wins, x	19	17	16	15	15	14	12	9
Earned run average, y	2.63	2.79	3.75	3.23	3.47	3.96	4.05	4.12

■ EXTENDING CONCEPTS

Interchanging x and y *In Exercises 32 and 33, do the following.*

(a) Find the equation of the regression line for the given data, letting Row 1 represent the x-values and Row 2 the y-values. Sketch a scatter plot of the data and draw the regression line.

32. (a) $\hat{y} = 1.724x + 79.733$

    See Selected Answers, page A129.

    (b) $\hat{y} = 0.453x - 26.448$

    See Selected Answers, page A129.

    (c) The slope of the line keeps the same sign, but the values of m and b change.

33. (a) $\hat{y} = -4.297x + 94.200$

    See Odd Answers, page A95.

    (b) $\hat{y} = -0.141x + 14.763$

    See Odd Answers, page A95.

    (c) The slope of the line keeps the same sign, but the values of m and b change.

34. (a) $\hat{y} = 1.711x + 3.912$

    (b) See Selected Answers, page A129.

    (c) See Selected Answers, page A129.

    (d) The residual plot shows no pattern in the residuals because the residuals fluctuate about 0. This suggests that the regression line is a good representation of the data.

35. (a) $\hat{y} = 0.139x + 21.024$

    (b) See Odd Answers, page A95.

    (c) See Odd Answers, page A95.

    (d) The residual plot shows a pattern because the residuals do not fluctuate about 0. This implies that the regression line is not a good representation of the relationship between the two variables.

36. See Selected Answers, page A129.

37. (a) See Odd Answers, page A95.

    (b) The point (44, 8) may be an outlier.

    (c) The point (44, 8) is not an influential point because the slopes and y-intercepts of the regression lines with the point included and without the point included are not significantly different.

(b) Find the equation of the regression line for the given data, letting Row 2 represent the x-values and Row 1 the y-values. Sketch a scatter plot of the data and draw the regression line.

(c) What effect does switching the explanatory and response variables have on the regression line?

32.

Row 1	16	25	39	45	49	64	70
Row 2	109	122	143	132	199	185	199

33.

Row 1	0	1	2	3	3	5	5	5	6	7
Row 2	96	85	82	74	95	68	76	84	58	65

Residual Plots *A **residual plot** allows you to assess correlation data and check for possible problems with a regression model. To construct a residual plot, make a scatter plot of (x, y − ŷ), where y − ŷ is the residual of each y-value. If the resulting plot shows any type of pattern, the regression line is not a good representation of the relationship between the two variables. If it does not show a pattern—that is, if the residuals fluctuate about 0—then the regression line is a good representation. Be aware that if a point on the residual plot appears to be outside the pattern of the other points, then it may be an outlier.*

In Exercises 34 and 35, (a) find the equation of the regression line, (b) construct a scatter plot of the data and draw the regression line, (c) construct a residual plot, and (d) determine if there are any patterns in the residual plot and explain what they suggest about the relationship between the variables.

34.

x	8	4	15	7	6	3	12	10	5
y	18	11	29	18	14	8	25	20	12

35.

x	38	34	40	46	43	48	60	55	52
y	24	22	27	32	30	31	27	26	28

Influential Points *An **influential point** is a point in a data set that can greatly affect the graph of a regression line. An outlier may or may not be an influential point. To determine if a point is influential, find two regression lines: one including all the points in the data set, and the other excluding the possible influential point. If the slope or y-intercept of the regression line shows significant changes, the point can be considered influential. An influential point can be removed from a data set only if there is proper justification.*

In Exercises 36 and 37, (a) construct a scatter plot of the data, (b) identify any possible outliers, and (c) determine if the point is influential. Explain your reasoning.

36.

x	1	3	6	8	12	14
y	4	7	10	9	15	3

37.

x	5	6	9	10	14	17	19	44
y	32	33	28	26	25	23	23	8

38. See Selected Answers, page A129.

39. See Odd Answers, page A95.

40. See Selected Answers, page A130.

41. See Odd Answers, page A95.

42. The exponential equation is a much better model for the data. The graph of the exponential equation fits the data better than the regression line.

Number of hours, x	Number of bacteria, y
1	165
2	280
3	468
4	780
5	1310
6	1920
7	4900

TABLE FOR EXERCISES 39–42

43. See Odd Answers, page A95.

44. See Selected Answers, page A130.

45. See Odd Answers, page A95.

x	y
1	695
2	410
3	256
4	110
5	80
6	75
7	68
8	74

TABLE FOR EXERCISES 43–46

46. The power equation is a much better model for the data. The graph of the power equation fits the data better than the regression line.

47. See Odd Answers, page A95.

48. See Selected Answers, page A130.

49. The logarithmic equation is a better model for the data. The graph of the logarithmic equation fits the data better than the regression line.

50. See Selected Answers, page A130.

---

38. **Chapter Opener** Consider the data from the Chapter Opener on page 483 on the salaries and average attendances at home games for the teams in Major League Baseball. Is the data point $(201.4, 45{,}364)$ an outlier? If so, is it influential? Explain.

Transformations to Achieve Linearity *When a linear model is not appropriate for representing data, other models can be used. In some cases, the values of x or y must be transformed to find an appropriate model. In a* **logarithmic transformation,** *the logarithms of the variables are used instead of the original variables when creating a scatter plot and calculating the regression line.*

In Exercises 39–42, use the data shown in the table, which shows the number of bacteria present after a certain number of hours.

39. Find the equation of the regression line for the data. Then construct a scatter plot of (x, y) and sketch the regression line with it.

40. Replace each y-value in the table with its logarithm, $\log y$. Find the equation of the regression line for the transformed data. Then construct a scatter plot of $(x, \log y)$ and sketch the regression line with it. What do you notice?

41. An **exponential equation** is a nonlinear regression equation of the form $y = ab^x$. Use a technology tool to find and graph the exponential equation for the original data. Include a scatter plot in your graph. Note that you can also find this model by solving the equation $\log y = mx + b$ from Exercise 40 for y.

42. Compare your results in Exercise 41 with the equation of the regression line and its graph in Exercise 39. Which equation is a better model for the data? Explain.

In Exercises 43–46, use the data shown in the table.

43. Find the equation of the regression line for the data. Then construct a scatter plot of (x, y) and sketch the regression line with it.

44. Replace each x-value and y-value in the table with its logarithm. Find the equation of the regression line for the transformed data. Then construct a scatter plot of $(\log x, \log y)$ and sketch the regression line with it. What do you notice?

45. A **power equation** is a nonlinear regression equation of the form $y = ax^b$. Use a technology tool to find and graph the power equation for the original data. Include a scatter plot in your graph. Note that you can also find this model by solving the equation $\log y = m(\log x) + b$ from Exercise 44 for y.

46. Compare your results in Exercise 45 with the equation of the regression line and its graph in Exercise 43. Which equation is a better model for the data? Explain.

Logarithmic Equation *In Exercises 47–50, use the following information and a technology tool. The* **logarithmic equation** *is a nonlinear regression equation of the form $y = a + b \ln x$.*

47. Find and graph the logarithmic equation for the data given in Exercise 23.

48. Find and graph the logarithmic equation for the data given in Exercise 24.

49. Compare your results in Exercise 47 with the equation of the regression line and its graph. Which equation is a better model for the data? Explain.

50. Compare your results in Exercise 48 with the equation of the regression line and its graph. Which equation is a better model for the data? Explain.

ACTIVITY 9.2   Regression by Eye

The *regression by eye* applet allows you to interactively estimate the regression line for a data set. When the applet loads, a data set consisting of 20 points is displayed. Points on the plot can be added to the plot by clicking the mouse. Points on the plot can be removed by clicking on the point and then dragging the point into the trash can. All of the points on the plot can be removed by simply clicking inside the trash can. You can move the green line on the plot by clicking and dragging the endpoints. You should try to move the line in order to minimize the sum of the squares of the residuals, also known as the sum of square error (SSE). Note that the regression line minimizes SSE. The SSE for the green line and for the regression line are given below the plot. The equations of each line are given above the plot. Click SHOW REGRESSION LINE! to see the regression line in the plot. Click NEW DATA to generate a new data set.

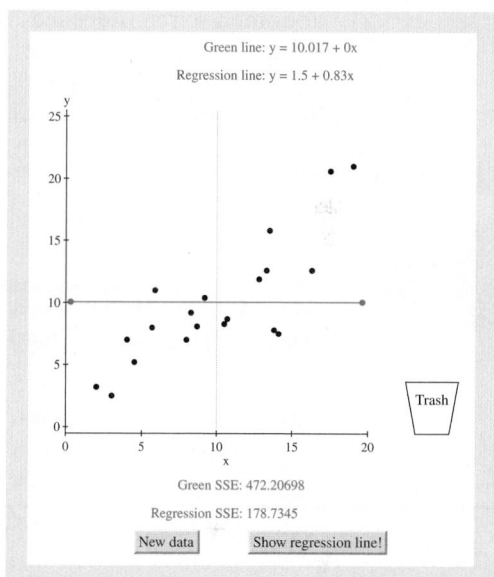

■ Explore

Step 1 Move the endpoints of the green line to try to approximate the regression line.

Step 2 Click SHOW REGRESSION LINE!.

■ Draw Conclusions

1. Click NEW DATA to generate a new data set. Try to move the green line to where the regression line should be. Then click SHOW REGRESSION LINE!. Repeat this five times. Describe how you moved each green line.

2. On a blank plot, place 10 points so that they have a strong positive correlation. Record the equation of the regression line. Then, add a point in the upper left corner of the plot and record the equation of the regression line. How does the regression line change?

3. Remove the point from the upper-left corner of the plot. Add 10 more points so that there is still a strong positive correlation. Record the equation of the regression line. Add a point in the upper-left corner of the plot and record the equation of the regression line. How does the regression line change?

4. Use the results of Exercises 2 and 3 to describe what happens to the slope of the regression line when an outlier is added as the sample size increases.

Correlation of Body Measurements

In a study published in *Medicine and Science in Sports and Exercise* (volume 17, no. 2, page 189) the measurements of 252 men (ages 22–81) are given. Of the 14 measurements taken of each man, some have significant correlations and others don't. For instance, the scatter plot at the right shows that the hip and abdomen circumferences of the men have a strong linear correlation ($r = 0.85$). The partial table shown here lists only the first nine rows of the data.

Hip and Abdomen Circumferences

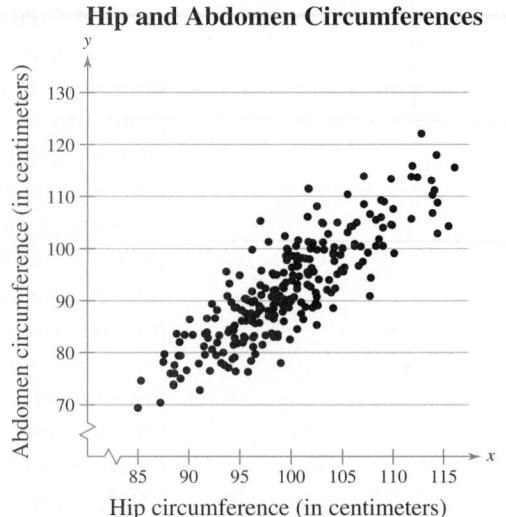

Age (yr)	Weight (lb)	Height (in.)	Neck (cm)	Chest (cm)	Abdom. (cm)	Hip (cm)	Thigh (cm)	Knee (cm)	Ankle (cm)	Bicep (cm)	Forearm (cm)	Wrist (cm)	Body fat %
22	173.25	72.25	38.5	93.6	83.0	98.7	58.7	37.3	23.4	30.5	28.9	18.2	6.1
22	154.00	66.25	34.0	95.8	87.9	99.2	59.6	38.9	24.0	28.8	25.2	16.6	25.3
23	154.25	67.75	36.2	93.1	85.2	94.5	59.0	37.3	21.9	32.0	27.4	17.1	12.3
23	198.25	73.50	42.1	99.6	88.6	104.1	63.1	41.7	25.0	35.6	30.0	19.2	11.7
23	159.75	72.25	35.5	92.1	77.1	93.9	56.1	36.1	22.7	30.5	27.2	18.2	9.4
23	188.15	77.50	38.0	96.6	85.3	102.5	59.1	37.6	23.2	31.8	29.7	18.3	10.3
24	184.25	71.25	34.4	97.3	100.0	101.9	63.2	42.2	24.0	32.2	27.7	17.7	28.7
24	210.25	74.75	39.0	104.5	94.4	107.8	66.0	42.0	25.6	35.7	30.6	18.8	20.9
24	156.00	70.75	35.7	92.7	81.9	95.3	56.4	36.5	22.0	33.5	28.3	17.3	14.2

Source: "Generalized Body Composition Prediction Equation for Men Using Simple Measurement Techniques" by K.W. Penrose et al. (1985). MEDICINE AND SCIENCE IN SPORTS AND EXERCISE, vol. 17, no.2, p. 189.

■ EXERCISES

1. Using your intuition, classify the following (x, y) pairs as having a weak correlation ($0 < r < 0.5$), a moderate correlation ($0.5 < r < 0.8$), or a strong correlation ($0.8 < r < 1.0$).

   (a) (weight, neck)   (b) (weight, height)
   (c) (age, body fat)   (d) (chest, hip)
   (e) (age, wrist)   (f) (ankle, wrist)
   (g) (forearm, height) (h) (bicep, forearm)
   (i) (weight, body fat) (j) (knee, thigh)
   (k) (hip, abdomen) (l) (abdomen, hip)

2. Now, use a technology tool to find the correlation coefficient for each pair in Exercise 1. Compare your results with those obtained by intuition.

3. Use a technology tool to find the regression line for each pair in Exercise 1 that has a strong correlation.

4. Use the results of Exercise 3 to predict the following.

   (a) The neck circumference of a man whose weight is 180 pounds

   (b) The abdomen circumference of a man whose hip circumference is 100 centimeters

5. Are there pairs of measurements that have stronger correlation coefficients than 0.85? Use a technology tool and intuition to reach a conclusion.

9.3   Measures of Regression and Prediction Intervals

WHAT YOU SHOULD LEARN

▸ How to interpret the three types of variation about a regression line

▸ How to find and interpret the coefficient of determination

▸ How to find and interpret the standard error of estimate for a regression line

▸ How to construct and interpret a prediction interval for y

▸ VARIATION ABOUT A REGRESSION LINE

In this section, you will study two measures used in correlation and regression studies—the coefficient of determination and the standard error of estimate. You will also learn how to construct a prediction interval for y using a regression line and a given value of x. Before studying these concepts, you need to understand the three types of variation about a regression line.

To find the total variation, the explained variation, and the unexplained variation about a regression line, you must first calculate the **total deviation,** the **explained deviation,** and the **unexplained deviation** for each ordered pair (x_i, y_i) in a data set. These deviations are shown in the graph.

$$\text{Total deviation} = y_i - \overline{y}$$
$$\text{Explained deviation} = \hat{y}_i - \overline{y}$$
$$\text{Unexplained deviation} = y_i - \hat{y}_i$$

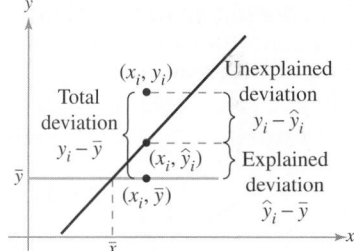

Note to Instructor

Remind students that without information about the regression line, the best predictor for y given a value of x is $\overline{y}$, the mean of the y-values.

After calculating the deviations for each data point (x_i, y_i), you can find the *total variation*, the *explained variation*, and the *unexplained variation*.

STUDY TIP

Consider the gross domestic products and carbon dioxide emissions data used throughout this chapter with a regression line of

$$\hat{y} = 196.152x + 102.289.$$

Using the data point (2.7, 571.8), you can find the total, explained, and unexplained deviations as follows.

Total deviation:

$y_i - \overline{y} = 571.8 - 555.4$

$\qquad = 16.4$

Explained deviation:

$\hat{y}_i - \overline{y} = 631.8994 - 555.4$

$\qquad = 76.4994$

Unexplained deviation:

$y_i - \hat{y}_i = 571.8 - 631.8994$

$\qquad = -60.0994$

DEFINITION

The **total variation** about a regression line is the sum of the squares of the differences between the y-value of each ordered pair and the mean of y.

$$\text{Total variation} = \sum (y_i - \overline{y})^2$$

The **explained variation** is the sum of the squares of the differences between each predicted y-value and the mean of y.

$$\text{Explained variation} = \sum (\hat{y}_i - \overline{y})^2$$

The **unexplained variation** is the sum of the squares of the differences between the y-value of each ordered pair and each corresponding predicted y-value.

$$\text{Unexplained variation} = \sum (y_i - \hat{y}_i)^2$$

The sum of the explained and unexplained variations is equal to the total variation.

$$\text{Total variation} = \text{Explained variation} + \text{Unexplained variation}$$

As its name implies, the *explained variation* can be explained by the relationship between x and y. The *unexplained variation* cannot be explained by the relationship between x and y and is due to chance or other variables.

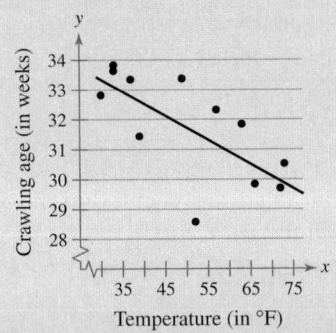
▶ THE COEFFICIENT OF DETERMINATION

You already know how to calculate the correlation coefficient r. The square of this coefficient is called the *coefficient of determination*. It can be shown that the coefficient of determination is equal to the ratio of the explained variation to the total variation.

DEFINITION

The **coefficient of determination** r^2 is the ratio of the explained variation to the total variation. That is,

$$r^2 = \frac{\text{Explained variation}}{\text{Total variation}}.$$

It is important that you interpret the coefficient of determination correctly. For instance, if the correlation coefficient is $r = 0.90$, then the coefficient of determination is

$$r^2 = 0.90^2$$
$$= 0.81.$$

This means that 81% of the variation in y can be explained by the relationship between x and y. The remaining 19% of the variation is unexplained and is due to other factors or to sampling error.

EXAMPLE 1    Report 41

▶ Finding the Coefficient of Determination

The correlation coefficient for the gross domestic products and carbon dioxide emissions data as calculated in Example 4 in Section 9.1 is $r \approx 0.882$. Find the coefficient of determination. What does this tell you about the explained variation of the data about the regression line? About the unexplained variation?

▶ Solution

The coefficient of determination is

$$r^2 \approx (0.882)^2$$
$$\approx 0.778.$$

Interpretation   About 77.8% of the variation in the carbon dioxide emissions can be explained by the variation in the gross domestic products. About 22.2% of the variation is unexplained and is due to chance or other variables.

▶ Try It Yourself 1

The correlation coefficient for the Old Faithful data as calculated in Example 5 in Section 9.1 is $r \approx 0.979$. Find the coefficient of determination. What does this tell you about the explained variation of the data about the regression line? About the unexplained variation?

a. Identify the *correlation coefficient r*.
b. Calculate the *coefficient of determination r^2*.
c. What percent of the variation in the times is *explained*? What percent is *unexplained*?
Answer: Page A45

▶ THE STANDARD ERROR OF ESTIMATE

When a $\hat{y}$-value is predicted from an x-value, the prediction is a point estimate. You can construct an interval estimate for $\hat{y}$, but first you need to calculate the *standard error of estimate*.

DEFINITION

The **standard error of estimate** s_e is the standard deviation of the observed y_i-values about the predicted $\hat{y}$-value for a given x_i-value. It is given by

$$s_e = \sqrt{\frac{\Sigma(y_i - \hat{y}_i)^2}{n - 2}}$$

where n is the number of ordered pairs in the data set.

From this formula, you can see that the standard error of estimate is the square root of the unexplained variation divided by $n - 2$. So, the closer the observed y-values are to the predicted y-values, the smaller the standard error of estimate will be.

GUIDELINES

Finding the Standard Error of Estimate s_e

IN WORDS	IN SYMBOLS
1. Make a table that includes the column headings shown at the right.	$x_i, y_i, \hat{y}_i, (y_i - \hat{y}_i),$ $(y_i - \hat{y}_i)^2$
2. Use the regression equation to calculate the predicted y-values.	$\hat{y}_i = mx_i + b$
3. Calculate the sum of the squares of the differences between each observed y-value and the corresponding predicted y-value.	$\Sigma(y_i - \hat{y}_i)^2$
4. Find the standard error of estimate.	$s_e = \sqrt{\dfrac{\Sigma(y_i - \hat{y}_i)^2}{n - 2}}$

Shoe size, x	Men's heights (in inches), y
10.0	70.0
10.5	71.0
9.5	70.0
11.0	72.0
12.0	74.0
8.5	66.0
9.0	68.5
13.0	76.0
10.5	71.5
10.5	70.5
10.0	72.0
9.5	70.0
10.0	71.0
10.5	69.5
11.0	71.5
12.0	73.5

You can also find the standard error of estimate using the following formula.

$$s_e = \sqrt{\frac{\Sigma y^2 - b\Sigma y - m\Sigma xy}{n - 2}}$$

This formula is easy to use if you have already calculated the slope m, the y-intercept b, and several of the sums. For instance, the regression line for the data set given at the left is $\hat{y} = 1.84247x + 51.77413$, and the values of the sums are $\Sigma y^2 = 80{,}877.5$, $\Sigma y = 1137$, and $\Sigma xy = 11{,}940.25$. When the alternative formula is used, the standard error of estimate is

$$s_e = \sqrt{\frac{\Sigma y^2 - b\Sigma y - m\Sigma xy}{n - 2}}$$

$$= \sqrt{\frac{80{,}877.5 - 51.77413(1137) - 1.84247(11{,}940.25)}{16 - 2}}$$

$$\approx 0.877.$$

CHAPTER 9

CORRELATION AND REGRESSION

516 CHAPTER 9 CORRELATION AND REGRESSION

EXAMPLE 2 (SC) Report 42

▶ Finding the Standard Error of Estimate

The regression equation for the gross domestic products and carbon dioxide emissions data as calculated in Example 1 in Section 9.2 is

$$\hat{y} = 196.152x + 102.289.$$

Find the standard error of estimate.

▶ Solution

Use a table to calculate the sum of the squared differences of each observed y-value and the corresponding predicted y-value.

x_i	y_i	$\hat{y}_i$	$y_i - \hat{y}_i$	$(y_i - \hat{y}_i)^2$
1.6	428.2	416.1322	12.0678	145.63179684
3.6	828.8	808.4362	20.3638	414.68435044
4.9	1214.2	1063.4338	150.7662	22,730.44706244
1.1	444.6	318.0562	126.5438	16,013.33331844
0.9	264.0	278.8258	−14.8258	219.80434564
2.9	415.3	671.1298	−255.8298	65,448.88656804
2.7	571.8	631.8994	−60.0994	3611.93788036
2.3	454.9	553.4386	−98.5386	9709.85568996
1.6	358.7	416.1322	−57.4322	3298.45759684
1.5	573.5	396.517	176.983	31,322.982289
				$\Sigma = 152,916.020898$

Unexplained variation

When $n = 10$ and $\Sigma (y_i - \hat{y}_i)^2 = 152,916.020898$ are used, the standard error of estimate is

$$s_e = \sqrt{\frac{\Sigma (y_i - \hat{y}_i)^2}{n - 2}}$$

$$= \sqrt{\frac{152,916.020898}{10 - 2}}$$

$$\approx 138.255.$$

Interpretation The standard error of estimate of the carbon dioxide emissions for a specific gross domestic product is about 138.255 million metric tons.

▶ Try It Yourself 2

A researcher collects the data shown at the left and concludes that there is a significant relationship between the amount of radio advertising time (in minutes per week) and the weekly sales of a product (in hundreds of dollars). Find the standard error of estimate. Use the regression equation

$$\hat{y} = 1.405x + 7.311.$$

a. *Use a table* to calculate the sum of the squared differences of each observed y-value and the corresponding predicted y-value.
b. Identify the *number n of ordered pairs* in the data set.
c. *Calculate* s_e.
d. *Interpret* the results.

Answer: Page A45

Radio ad time	Weekly sales
15	26
20	32
20	38
30	56
40	54
45	78
50	80
60	88

▶ PREDICTION INTERVALS

Two variables have a **bivariate normal distribution** if for any fixed values of x the corresponding values of y are normally distributed, and for any fixed values of y the corresponding values of x are normally distributed.

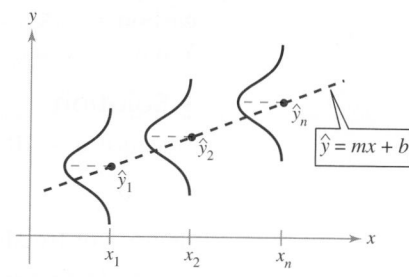

Bivariate Normal Distribution

Because regression equations are determined using sample data and because x and y are assumed to have a bivariate normal distribution, you can construct a *prediction interval* for the true value of y. To construct the prediction interval, use a *t*-distribution with $n - 2$ degrees of freedom.

DEFINITION

Given a linear regression equation $\hat{y} = mx + b$ and x_0, a specific value of x, a ***c*-prediction interval** for y is

$$\hat{y} - E < y < \hat{y} + E$$

where

$$E = t_c s_e \sqrt{1 + \frac{1}{n} + \frac{n(x_0 - \overline{x})^2}{n\sum x^2 - (\sum x)^2}}.$$

The point estimate is $\hat{y}$ and the margin of error is E. The probability that the prediction interval contains y is c.

GUIDELINES

Construct a Prediction Interval for *y* for a Specific Value of *x*

IN WORDS	IN SYMBOLS
1. Identify the number of ordered pairs in the data set n and the degrees of freedom.	d.f. $= n - 2$
2. Use the regression equation and the given x-value to find the point estimate $\hat{y}$.	$\hat{y}_i = mx_i + b$
3. Find the critical value t_c that corresponds to the given level of confidence c.	Use Table 5 in Appendix B.
4. Find the standard error of estimate s_e.	$s_e = \sqrt{\dfrac{\sum (y_i - \hat{y}_i)^2}{n - 2}}$
5. Find the margin of error E.	$E = t_c s_e \sqrt{1 + \dfrac{1}{n} + \dfrac{n(x_0 - \overline{x})^2}{n\sum x^2 - (\sum x)^2}}$
6. Find the left and right endpoints and form the prediction interval.	Left endpoint: $\hat{y} - E$ Right endpoint: $\hat{y} + E$ Interval: $\hat{y} - E < y < \hat{y} + E$

STUDY TIP

The formulas for s_e and E use the quantities $\sum(y_i - \hat{y}_i)^2$, $(\sum x)^2$, and $\sum x^2$. Use a table to calculate these quantities.

EXAMPLE 3   SC Report 43

▶ Constructing a Prediction Interval

Using the results of Example 2, construct a 95% prediction interval for the carbon dioxide emissions when the gross domestic product is $3.5 trillion. What can you conclude?

▶ Solution

Because $n = 10$, there are

$$10 - 2 = 8$$

degrees of freedom. Using the regression equation

$$\hat{y} = 196.152x + 102.289$$

and

$$x = 3.5,$$

the point estimate is

$$\hat{y} = 196.152x + 102.289$$

$$= 196.152(3.5) + 102.289$$

$$= 788.821.$$

From Table 5, the critical value is $t_c = 2.306$, and from Example 2, $s_e \approx 138.255$. Using these values, the margin of error is

$$E = t_c s_e \sqrt{1 + \frac{1}{n} + \frac{n(x_0 - \bar{x})^2}{n(\sum x^2) - (\sum x)^2}}$$

$$\approx (2.306)(138.255)\sqrt{1 + \frac{1}{10} + \frac{10(3.5 - 2.31)^2}{10(67.35) - (23.1)^2}}$$

$$\approx 349.424.$$

Using $\hat{y} = 788.821$ and $E = 349.424$, the prediction interval is

Left Endpoint	Right Endpoint
$788.821 - 349.424 = 439.397$	$788.821 + 349.424 = 1138.245$

$$439.397 < y < 1138.245.$$

Interpretation You can be 95% confident that when the gross domestic product is $3.5 trillion, the carbon dioxide emissions will be between 439.397 and 1138.245 million metric tons.

▶ Try It Yourself 3

Construct a 95% prediction interval for the carbon dioxide emissions when the gross domestic product is $4 trillion. What can you conclude?

a. *Specify n, d.f., t_c, s_e.*
b. *Calculate $\hat{y}$ when $x = 4$.*
c. Calculate the *margin of error E.*
d. Construct the *prediction interval.*
e. *Interpret* the results.

Answer: Page A45

INSIGHT

The greater the difference between x and $\bar{x}$, the wider the prediction interval is. For instance, in Example 3 the 95% prediction intervals for $0.9 < x < 4.9$ are shown below. Notice how the bands curve away from the regression line as x gets close to 0.9 and 4.9.

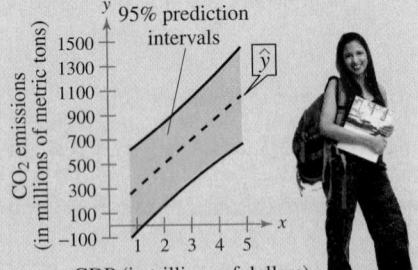

9.3 EXERCISES

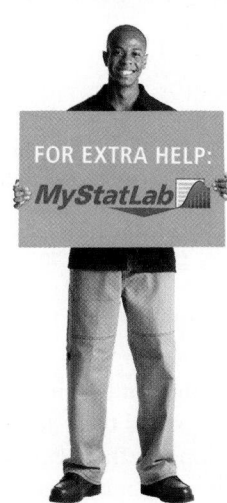

1. The total variation is the sum of the squares of the differences between the y-values of each ordered pair and the mean of the y-values of the ordered pairs, or $\Sigma(y_i - \bar{y})^2$.

2. See Selected Answers, page A130.

3. The unexplained variation is the sum of the squares of the differences between the observed y-values and the predicted y-values, or $\Sigma(y_i - \hat{y}_i)^2$.

4. See Selected Answers, page A130.

5. See Odd Answers, page A96.

6. See Selected Answers, page A130.

7. 0.216; About 21.6% of the variation is explained. About 78.4% of the variation is unexplained.

8. 0.108; About 10.8% of the variation is explained. About 89.2% of the variation is unexplained.

9. 0.916; About 91.6% of the variation is explained. About 8.4% of the variation is unexplained.

10. 0.776; About 77.6% of the variation is explained. About 22.4% of the variation is unexplained.

11. (a) 0.798; About 79.8% of the variation in proceeds can be explained by the variation in the number of issues, and about 20.2% of the variation is unexplained.

   (b) 8064.633; The standard error of estimate of the proceeds for a specific number of issues is about $8,064,633,000.

■ BUILDING BASIC SKILLS AND VOCABULARY

Graphical Analysis   *In Exercises 1–3, use the graph to answer the question.*

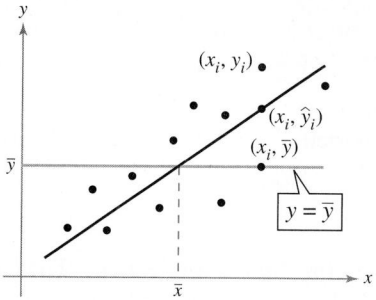

1. Describe the total variation about a regression line in words and in symbols.

2. Describe the explained variation about a regression line in words and in symbols.

3. Describe the unexplained variation about a regression line in words and in symbols.

4. The coefficient of determination r^2 is the ratio of which two types of variations? What does r^2 measure? What does $1 - r^2$ measure?

5. What is the coefficient of determination for two variables that have perfect positive linear correlation or perfect negative linear correlation? Interpret your answer.

6. Two variables have a bivariate normal distribution. Explain what this means.

In Exercises 7–10, use the value of the linear correlation coefficient to calculate the coefficient of determination. What does this tell you about the explained variation of the data about the regression line? About the unexplained variation?

7. $r = 0.465$

8. $r = -0.328$

9. $r = -0.957$

10. $r = 0.881$

■ USING AND INTERPRETING CONCEPTS

Finding Types of Variation and the Coefficient of Determination
In Exercises 11–18, use the data to find (a) the coefficient of determination and interpret the result, and (b) the standard error of estimate s_e and interpret the result.

11. **Stock Offerings**   The number of initial public offerings of stock issued in a recent 12-year period and the total proceeds of these offerings (in millions of U.S. dollars) are shown in the table. The equation of the regression line is $\hat{y} = 104.982x + 14{,}128.671$. *(Source: University of Florida)*

Number of issues, x	318	486	382	79	70	67
Proceeds, y	34,614	64,927	65,088	34,241	22,136	10,068

Number of issues, x	184	168	162	162	21	43
Proceeds, y	32,269	28,593	30,648	35,762	22,762	13,307

12. (a) 0.909; About 90.9% of the variation in the amount of crude oil imported can be explained by the variation in the amount of crude oil produced, and about 9.1% of the variation is unexplained.

(b) 301.658; The standard error of estimate of the amount of crude oil imported for a specific amount of crude oil produced is about 301,658 barrels per day.

13. (a) 0.981; About 98.1% of the variation in sales can be explained by the variation in the total square footage, and about 1.9% of the variation is unexplained.

(b) 30.576; The standard error of estimate of the sales for a specific total square footage is about $30,576,000,000.

14. (a) 0.578; About 57.8% of the variation in the median number of leisure hours per week can be explained by the variation in the median number of work hours per week, and about 42.2% of the variation is unexplained.

(b) 2.013; The standard error of estimate of the median number of leisure hours per week for a specific median number of work hours per week is about 2.013 hours.

15. (a) 0.963; About 96.3% of the variation in wages for federal government employees can be explained by the variation in wages for state government employees, and about 3.7% of the variation is unexplained.

(b) 20.090; The standard error of estimate of the average weekly wages for federal government employees for a specific average weekly wage for state government employees is about $20.09.

16. See Selected Answers, page A130.

12. Crude Oil The table shows the amounts of crude oil (in thousands of barrels per day) produced by the United States and the amounts of crude oil (in thousands of barrels per day) imported by the United States for seven years. The equation of the regression line is $\hat{y} = -2.735x + 27{,}657.823$. *(Source: Energy Information Administration)*

Crude oil produced by U.S., x	5801	5746	5681	5419	5178	5102	5064
Crude oil imported by U.S., y	11,871	11,530	12,264	13,145	13,714	13,707	13,468

13. Retail Space and Sales The table shows the total square footage (in billions) of retailing space at shopping centers and their sales (in billions of U.S. dollars) for 11 years. The equation of the regression line is $\hat{y} = 549.448x - 1881.694$. *(Adapted from International Council of Shopping Centers)*

Total square footage, x	5.0	5.1	5.2	5.3	5.5	5.6
Sales, y	893.8	933.9	980.0	1032.4	1105.3	1181.1

Total square footage, x	5.7	5.8	5.9	6.0	6.1
Sales, y	1221.7	1277.2	1339.2	1432.6	1530.4

14. Work and Leisure Time The median number of work hours per week and the median number of leisure hours per week for people in the United States for 10 recent years are shown in the table. The equation of the regression line is $\hat{y} = -0.646x + 50.734$. *(Source: Louis Harris & Associates)*

Median number of work hours per week, x	40.6	43.1	46.9	47.3	46.8
Median number of leisure hours per week, y	26.2	24.3	19.2	18.1	16.6

Median number of work hours per week, x	48.7	50.0	50.7	50.6	50.8
Median number of leisure hours per week, y	18.8	18.8	19.5	19.2	19.5

15. State and Federal Government Wages The table shows the average weekly wages for state government employees and federal government employees for six years. The equation of the regression line is $\hat{y} = 1.900x - 411.976$. *(Source: U.S. Bureau of Labor Statistics)*

Average weekly wages (state), x	754	770	791	812	844	883
Average weekly wages (federal), y	1001	1043	1111	1151	1198	1248

17. (a) 0.790; About 79.0% of the variation in the gross collections of corporate income taxes can be explained by the variation in the gross collections of individual income taxes, and about 21.0% of the variation is unexplained.

   (b) 42.386; The standard error of estimate of the gross collections of corporate income taxes for a specific gross collection of individual income taxes is about $42,386,000,000.

18. See Selected Answers, page A130.

19. $40{,}116.824 < y < 82{,}624.318$

   You can be 95% confident that the proceeds will be between $40,116,824,000 and $82,624,318,000 when the number of initial offerings is 450 issues.

20. $11{,}783.051 < y < 13{,}447.595$

   You can be 95% confident that the amount of crude oil imported by the United States will be between 11,783,051 and 13,447,595 barrels per day when the amount of crude oil produced by the United States is 5,500,000 barrels per day.

21. $1218.435 < y < 1336.829$

   You can be 90% confident that the shopping center sales will be between $1,218,435,000 and $1,336,829,000 when the total square footage of shopping centers is 5,750,000,000.

22. $17.573 < y < 25.625$

   You can be 90% confident that the median number of leisure hours per week will be between 17.573 and 25.625 hours when the median number of work hours per week is 45.1.

23. $1007.82 < y < 1208.228$

   You can be 99% confident that the average weekly wages of federal government employees will be between $1007.82 and $1208.23 when the average weekly wages of state government employees is $800.

16. **Voter Turnout**   The U.S. voting age population (in millions) and the turnout of the voting age population (in millions) for federal elections for eight nonpresidential election years are shown in the table. The equation of the regression line is $\hat{y} = 0.333x + 7.580$. *(Adapted from Federal Election Commission)*

Voting age population, x	158.4	169.9	178.6	185.8
Turnout in federal elections, y	58.9	67.6	65.0	67.9

Voting age population, x	193.7	200.9	215.5	220.6
Turnout in federal elections, y	75.1	73.1	79.8	80.6

17. **Taxes**   The table shows the gross collections (in billions of dollars) of individual income taxes and corporate income taxes by the U.S. Internal Revenue Service for seven years. The equation of the regression line is $\hat{y} = 0.415x - 186.626$. *(Source: Internal Revenue Service)*

Individual income taxes, x	1038	987	990	1108	1236	1366	1426
Corporate income taxes, y	211	194	231	307	381	396	354

18. **Fund Assets**   The table shows the total assets (in billions of U.S. dollars) of individual retirement accounts (IRAs) and federal pension plans for nine years. The equation of the regression line is $\hat{y} = 0.174x + 432.225$. *(Source: Investment Company Institute)*

IRAs, x	2629	2619	2533	2993	3299
Federal pension plans, y	797	860	894	958	1023

IRAs, x	3652	4220	4736	3572
Federal pension plans, y	1072	1141	1197	1221

Constructing and Interpreting Prediction Intervals   *In Exercises 19–26, construct the indicated prediction interval and interpret the results.*

19. **Proceeds**   Construct a 95% prediction interval for the proceeds from initial public offerings in Exercise 11 when the number of issues is 450.

20. **Crude Oil**   Construct a 95% prediction interval for the amount of crude oil imported by the United States in Exercise 12 when the amount of crude oil (in thousands of barrels per day) produced by the United States is 5500.

21. **Retail Sales**   Using the results of Exercise 13, construct a 90% prediction interval for shopping center sales when the total square footage of shopping centers is 5.75 billion.

22. **Leisure Hours**   Using the results of Exercise 14, construct a 90% prediction interval for the median number of leisure hours per week when the median number of work hours per week is 45.1.

23. **Federal Government Wages**   When the average weekly wages of state government employees is $800, find a 99% prediction interval for the average weekly wages of federal government employees. Use the results of Exercise 15.

24. $67.978 < y < 87.042$

You can be 99% confident that the voter turnout in federal elections will be between 67,978,000 and 87,042,000 when the voting age population is 210 million.

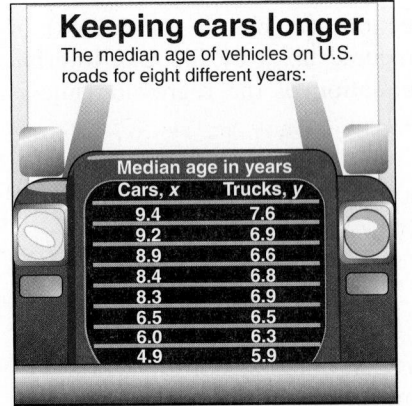

Keeping cars longer
The median age of vehicles on U.S. roads for eight different years:

Median age in years	
Cars, x	Trucks, y
9.4	7.6
9.2	6.9
8.9	6.6
8.4	6.8
8.3	6.9
6.5	6.5
6.0	6.3
4.9	5.9

(Source: Polk Co.)

FIGURE FOR EXERCISES 27–33

25. $213.729 < y < 450.519$

You can be 95% confident that the corporate income taxes collected by the U.S. Internal Revenue Service for a given year will be between $213,729,000,000 and $450,519,000,000 when the U.S. Internal Revenue Service collects $1,250,000,000 in individual income taxes that year.

26. $935.348 < y < 1251.502$

You can be 90% confident that the total assets in federal pension plans will be between $935,348,000,000 and $1,251,502,000,000 when the total assets in IRAs is $3,800,000,000,000.

27.

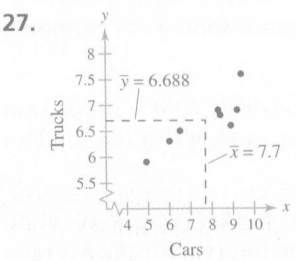

24. Predicting Voter Turnout   When the voting age population is 210 million, construct a 99% prediction interval for the voter turnout in federal elections. Use the results of Exercise 16.

25. Taxes   The U.S. Internal Revenue Service collects $1250 billion in individual income taxes for a given year. Construct a prediction interval for the corporate income taxes collected by the U.S. Internal Revenue Service. Use the results of Exercise 17 and $c = 0.95$.

26. Total Assets   The total assets in IRAs is $3800 billion. Construct a prediction interval for the total assets in federal pension plans. Use the results of Exercise 18 and $c = 0.90$.

Old Vehicles   *In Exercises 27–33, use the information shown at the left.*

27. Scatter Plot   Construct a scatter plot of the data. Show $\bar{y}$ and $\bar{x}$ on the graph.

28. Regression Line   Find and graph the regression line.

29. Deviation   Calculate the explained deviation, the unexplained deviation, and the total deviation for each data point.

30. Variation   Find (a) the explained variation, (b) the unexplained variation, and (c) the total variation.

31. Coefficient of Determination   Find the coefficient of determination. What can you conclude?

32. Error of Estimate   Find the standard error of estimate s_e and interpret the results.

33. Prediction Interval   Construct and interpret a 95% prediction interval for the median age of trucks in use when the median age of cars in use is 7.0 years.

34. Correlation Coefficient and Slope   Recall the formula for the correlation coefficient r and the formula for the slope m of a regression line. Given a set of data, why must the slope m of the data's regression line always have the same sign as the data's correlation coefficient r?

SC   *In Exercises 35 and 36, use StatCrunch and the given data to (a) find the coefficient of determination, (b) find the standard error of estimate s_e, and (c) construct a 95% prediction interval for y using the given value of x.*

35. Trees   The table shows the heights (in feet) and trunk diameters (in inches) of eight trees. The equation of the regression line is $\hat{y} = 0.479x - 24.086$.

$x = 80$ feet

Height, x	70	72	75	76	85	78	77	82
Trunk diameter, y	8.3	10.5	11.0	11.4	14.9	14.0	16.3	15.8

36. Motor Vehicles   The table shows the number of motor vehicle registrations (in millions) and the number of motor vehicle accidents (in millions) in the United States for six years. The equation of the regression line is $\hat{y} = -0.314x + 87.116$.

$x = 235$ million registrations

Registrations, x	229.6	231.4	237.2	241.2	244.2	247.3
Accidents, y	18.3	11.8	10.9	10.7	10.4	10.6

(Adapted from U.S. Federal Highway Administration and National Safety Council)

28. See Selected Answers, page A130.

29. See Odd Answers, page A96.

30. (a) 1.305

(b) 0.442

(c) 1.749

31. 0.746; About 74.6% of the variation in the median ages of trucks in use can be explained by the variation in the median ages of cars in use, and about 25.4% of the variation is unexplained.

32. 0.272; The standard error of estimate of the median age of trucks in use for a specific median age of cars is about 0.272 year.

33. $5.792 < y < 7.22$

You can be 95% confident that the median age of trucks in use will be between 5.792 and 7.22 years when the median age of cars in use is 7.0 years.

34. The slope and correlation coefficient will have the same sign because the denominators of both formulas are always positive while the numerators are always equal.

35. (a) 0.671

(b) 1.780

(c) $9.537 < y < 19.010$

36. (a) 0.522

(b) 2.371

(c) $5.954 < y < 20.470$

37. Fail to reject H_0. There is not enough evidence at the 1% level of significance to support the claim that there is a linear relationship between weight and number of hours slept.

38. Reject H_0. There is enough evidence at the 5% level of significance to support the claim that there is a linear relationship between age and salary.

39. $-118.927 < B < 323.505$

$110.911 < M < 281.393$

40. $-219.558 < B < 424.136$

$72.135 < M < 320.169$

■ EXTENDING CONCEPTS

Hypothesis Testing for Slope   *In Exercises 37 and 38, use the following information.*

When testing the slope M of the regression line for the population, you usually test that the slope is 0, or $H_0: M = 0$. A slope of 0 indicates that there is no linear relationship between x and y. To perform the t-test for the slope M, use the standardized test statistic

$$t = \frac{m}{s_e} \sqrt{\sum x^2 - \frac{(\sum x)^2}{n}}$$

with $n - 2$ degrees of freedom. Then, using the critical values found in Table 5 in Appendix B, make a decision whether to reject or fail to reject the null hypothesis. You can also use the LinRegTTest feature on a TI-83/84 Plus to calculate the standardized test statistic as well as the corresponding P-value. If $P \leq \alpha$, then reject the null hypothesis. If $P > \alpha$, then do not reject H_0.

37. The following table shows the weights (in pounds) and the number of hours slept in a day by a random sample of infants. Test the claim that $M \neq 0$. Use $\alpha = 0.01$. Then interpret the results in the context of the problem. If convenient, use technology to solve the problem.

Weight, x	8.1	10.2	9.9	7.2	6.9	11.2	11	15
Hours slept, y	14.8	14.6	14.1	14.2	13.8	13.2	13.9	12.5

38. The following table shows the ages (in years) and salaries (in thousands of dollars) of a random sample of engineers at a company. Test the claim that $M \neq 0$. Use $\alpha = 0.05$. Then interpret the results in the context of the problem. If convenient, use technology to solve the problem.

Age, x	25	34	29	30	42	38	49	52	35	40
Salary, y	57.5	61.2	59.9	58.7	87.5	67.4	89.2	85.3	69.5	75.1

Confidence Intervals for *y*-Intercept and Slope   *You can construct confidence intervals for the y-intercept B and slope M of the regression line $y = Mx + B$ for the population by using the following inequalities.*

***y*-intercept B:** $b - E < B < b + E$

$$\text{where } E = t_c s_e \sqrt{\frac{1}{n} + \frac{\overline{x}^2}{\sum x^2 - \frac{(\sum x)^2}{n}}} \quad \text{and}$$

slope M: $m - E < M < m + E$

$$\text{where } E = \frac{t_c s_e}{\sqrt{\sum x^2 - \frac{(\sum x)^2}{n}}}$$

The values of m and b are obtained from the sample data, and the critical value t_c is found using Table 5 in Appendix B with $n - 2$ degrees of freedom.

In Exercises 39 and 40, construct the indicated confidence interval for B and M using the gross domestic products and carbon dioxide emissions data found in Example 2.

39. 95% confidence interval

40. 99% confidence interval

9.4 Multiple Regression

Finding a Multiple Regression Equation ▸ Predicting y-Values

WHAT YOU SHOULD LEARN

▸ How to use technology to find a multiple regression equation, the standard error of estimate, and the coefficient of determination

▸ How to use a multiple regression equation to predict y-values

▸ FINDING A MULTIPLE REGRESSION EQUATION

In many instances, a better prediction model can be found for a dependent (response) variable by using more than one independent (explanatory) variable. For instance, a more accurate prediction for the carbon dioxide emissions discussed in previous sections might be made by considering the number of cars as well as the gross domestic product. Models that contain more than one independent variable are multiple regression models.

DEFINITION

A **multiple regression equation** has the form

$$\hat{y} = b + m_1x_1 + m_2x_2 + m_3x_3 + \cdots + m_kx_k$$

where $x_1, x_2, x_3, \ldots, x_k$ are the independent variables, b is the y-intercept, and y is the dependent variable.

The y-intercept b is the value of y when all x_i are 0. Each coefficient m_i is the amount of change in y when the independent variable x_i is changed by one unit and all other independent variables are held constant.

INSIGHT

Because the mathematics associated with multiple regression is complicated, this section focuses on how to use technology to find a multiple regression equation and how to interpret the results.

EXAMPLE 1 Report 44

▸ Finding a Multiple Regression Equation

A researcher wants to determine how employee salaries at a certain company are related to the length of employment, previous experience, and education. The researcher selects eight employees from the company and obtains the following data.

Employee	Salary, y	Employment (in years), x_1	Experience (in years), x_2	Education (in years), x_3
A	57,310	10	2	16
B	57,380	5	6	16
C	54,135	3	1	12
D	56,985	6	5	14
E	58,715	8	8	16
F	60,620	20	0	12
G	59,200	8	4	18
H	60,320	14	6	17

Use MINITAB to find a multiple regression equation that models the data.

▶ **Solution**

Enter the y-values in C1 and the x_1-, x_2-, and x_3-values in C2, C3, and C4, respectively. Select "Regression▶Regression ... " from the *Stat* menu. Using the salaries as the response variable and the remaining data as the predictors, you should obtain results similar to the following.

MINITAB

Regression Analysis: Salary, y versus x1, x2, x3

The regression equation is
Salary, y = 49764 + 364 x1 + 228 x2 + 267 x3

Predictor	Coef	SE Coef	T	P
Constant	49764 b	1981	25.12	0.000
x1	364.41 m_1	48.32	7.54	0.002
x2	227.6 m_2	123.8	1.84	0.140
x3	266.9 m_3	147.4	1.81	0.144

S = 659.490   R-Sq = 94.4%   R-Sq(adj) = 90.2%

The regression equation is $\hat{y} = 49{,}764 + 364x_1 + 228x_2 + 267x_3$.

▶ **Try It Yourself 1**

A statistics professor wants to determine how students' final grades are related to the midterm exam grades and number of classes missed. The professor selects 10 students from her class and obtains the following data.

Student	Final grade, y	Midterm exam, x_1	Classes missed, x_2
1	81	75	1
2	90	80	0
3	86	91	2
4	76	80	3
5	51	62	6
6	75	90	4
7	44	60	7
8	81	82	2
9	94	88	0
10	93	96	1

Use technology to find a multiple regression equation that models the data.

a. *Enter* the data.
b. *Calculate* the regression line.   *Answer: Page A45*

MINITAB displays much more than the regression equation and the coefficients of the independent variables. For instance, it also displays the standard error of estimate, denoted by S, and the coefficient of determination, denoted by R-Sq. In Example 1, $S = 659.490$ and R-$Sq = 94.4\%$. So, the standard error of estimate is \$659.49. The coefficient of determination tells you that 94.4% of the variation in y can be explained by the multiple regression model. The remaining 5.6% is unexplained and is due to other factors or chance.

▶ PREDICTING *y*-VALUES

After finding the equation of the multiple regression line, you can use the equation to predict *y*-values over the range of the data. To predict *y*-values, substitute the given value for each independent variable into the equation, then calculate $\hat{y}$.

PICTURING THE WORLD

In a lake in Finland, 159 fish of 7 species were caught and measured for weight *G* (in grams), length *L* (in centimeters), height *H*, and width *W* (*H* and *W* are percents of *L*). The regression equation for *G* and *L* is

$$G = -491 + 28.5L,$$
$$r \approx 0.925, r^2 \approx 0.855.$$

When all four variables are used, the regression equation is

$$G = -712 + 28.3L + 1.46H + 13.3W,$$
$$r \approx 0.930, r^2 \approx 0.865.$$

(Source: Journal of Statistics Education)

Predict the weight of a fish with the following measurements: L = 40, H = 17, and W = 11. How do your predictions vary when you use a single variable versus many variables? Which do you think is more accurate?

The predicted weight of the fish using the single variable model is 649 grams and the predicted weight of the fish using the model with many variables is 591.12 grams. These predictions vary greatly. Because the coefficient of determination is greater for the model with many variables, this model is most likely more accurate.

EXAMPLE 2

▶ Predicting *y*-Values Using Multiple Regression Equations

Use the regression equation found in Example 1 to predict an employee's salary given the following conditions.

1. 12 years of current employment, 5 years of experience, and 16 years of education

2. 4 years of current employment, 2 years of experience, and 12 years of education

3. 8 years of current employment, 7 years of experience, and 17 years of education

▶ Solution

To predict each employee's salary, substitute the values for x_1, x_2, and x_3 into the regression equation. Then calculate $\hat{y}$.

1. $\hat{y} = 49{,}764 + 364x_1 + 228x_2 + 267x_3$

$\qquad = 49{,}764 + 364(12) + 228(5) + 267(16)$

$\qquad = 59{,}544$

The employee's predicted salary is $59,544.

2. $\hat{y} = 49{,}764 + 364x_1 + 228x_2 + 267x_3$

$\qquad = 49{,}764 + 364(4) + 228(2) + 267(12)$

$\qquad = 54{,}880$

The employee's predicted salary is $54,880.

3. $\hat{y} = 49{,}764 + 364x_1 + 228x_2 + 267x_3$

$\qquad = 49{,}764 + 364(8) + 228(7) + 267(17)$

$\qquad = 58{,}811$

The employee's predicted salary is $58,811.

▶ Try It Yourself 2

Use the regression equation found in Try It Yourself 1 to predict a student's final grade given the following conditions.

1. A student has a midterm exam score of 89 and misses 1 class.
2. A student has a midterm exam score of 78 and misses 3 classes.
3. A student has a midterm exam score of 83 and misses 2 classes.

a. *Substitute* the midterm score for x_1 into the regression equation.
b. *Substitute* the corresponding number of missed classes for x_2 into the regression equation.
c. *Calculate* $\hat{y}$.
d. What is each student's final grade?

Answer: Page A45

9.4 EXERCISES

1. (a) 39,103.5 pounds per acre
   (b) 39,939.1 pounds per acre
   (c) 38,063.5 pounds per acre
   (d) 39,052.4 pounds per acre
2. (a) 27.25 bushels per acre
   (b) 27.1 bushels per acre
   (c) 30.325 bushels per acre
   (d) 25.9 bushels per acre
3. (a) 7.5 cubic feet
   (b) 16.8 cubic feet
   (c) 51.9 cubic feet
   (d) 62.1 cubic feet
4. (a) 4316.7 kilograms
   (b) 2126.55 kilograms
   (c) 3564.3 kilograms
   (d) 734.75 kilograms

BUILDING BASIC SKILLS AND VOCABULARY

Predicting y-Values *In Exercises 1–4, use the multiple regression equation to predict the y-values for the given values of the independent variables.*

1. **Potato Yield** The equation used to predict the annual potato yield (in pounds per acre) is

$$\hat{y} = 61{,}298 + 57.56x_1 - 78.45x_2$$

where x_1 is the number of acres planted (in thousands) and x_2 is the number of acres harvested (in thousands). *(Adapted from United States Department of Agriculture)*

(a) $x_1 = 1100$, $x_2 = 1090$ (b) $x_1 = 1060$, $x_2 = 1050$
(c) $x_1 = 1300$, $x_2 = 1250$ (d) $x_1 = 1140$, $x_2 = 1120$

2. **Rye Yield** The equation used to predict the annual rye yield (in bushels per acre) is

$$\hat{y} = 22 - 0.027x_1 + 0.156x_2$$

where x_1 is the number of acres planted (in thousands) and x_2 is the number of acres harvested (in thousands). *(Source: United States Department of Agriculture)*

(a) $x_1 = 1250$, $x_2 = 250$ (b) $x_1 = 1400$, $x_2 = 275$
(c) $x_1 = 1425$, $x_2 = 300$ (d) $x_1 = 1300$, $x_2 = 250$

3. **Black Cherry Tree Volume** The volume (in cubic feet) of a black cherry tree can be modeled by the equation

$$\hat{y} = -52.2 + 0.3x_1 + 4.5x_2$$

where x_1 is the tree's height (in feet) and x_2 is the tree's diameter (in inches). *(Source: Journal of the Royal Statistical Society)*

(a) $x_1 = 70$, $x_2 = 8.6$ (b) $x_1 = 65$, $x_2 = 11.0$
(c) $x_1 = 83$, $x_2 = 17.6$ (d) $x_1 = 87$, $x_2 = 19.6$

4. **Elephant Weight** The equation used to predict the weight of an elephant (in kilograms) is

$$\hat{y} = -4016 + 11.5x_1 + 7.55x_2 + 12.5x_3$$

where x_1 represents the girth of the elephant (in centimeters), x_2 represents the length of the elephant (in centimeters), and x_3 represents the circumference of a footpad (in centimeters). *(Source: Field Trip Earth)*

(a) $x_1 = 421$, $x_2 = 224$, $x_3 = 144$ (b) $x_1 = 311$, $x_2 = 171$, $x_3 = 102$
(c) $x_1 = 376$, $x_2 = 226$, $x_3 = 124$ (d) $x_1 = 231$, $x_2 = 135$, $x_3 = 86$

USING AND INTERPRETING CONCEPTS

Finding a Multiple Regression Equation *In Exercises 5 and 6, use technology to find the multiple regression equation for the data shown in the table. Then answer the following and interpret the results.*

(a) *What is the standard error of estimate?*

(b) *What is the coefficient of determination?*

5. See Odd Answers, page A97.

   (a) 28.489; The standard error of estimate of the predicted sales given specific total square footage and number of shopping centers is about $28.489 billion.

   (b) 0.985; The multiple regression model explains about 98.5% of the variation in y.

6. $\hat{y} = -16.984 - 0.007x_1 + 0.524x_2$

   (a) 1.756; The standard error of estimate of the predicted shareholder's equity given specific net sales and total assets is about $1.756 billion.

   (b) 0.963; The multiple regression model explains about 96.3% of the variation in y.

7. See Odd Answers, page A97. The equation is the same.

8. $\hat{y} = -16.984 - 0.007x_1 + 0.524x_2$; The equation is the same.

9. 0.981; About 98.1% of the variation in y can be explained by the relationship between variables; $r^2_{adj} < r^2$.

10. 0.925; About 92.5% of the variation in y can be explained by the relationship between variables; $r^2_{adj} < r^2$.

---

5. Sales The total square footage (in billions) of retailing space at shopping centers, the number (in thousands) of shopping centers, and the sales (in billions of dollars) for shopping centers for a recent 11-year period are shown in the table. *(Adapted from International Council of Shopping Centers)*

Sales, y	Total square footage, x_1	Number of shopping centers, x_2
893.8	5.0	41.2
933.9	5.1	42.1
980.0	5.2	43.0
1032.4	5.3	43.7
1105.3	5.5	44.4
1181.1	5.6	45.1
1221.7	5.7	45.8
1277.2	5.8	46.4
1339.2	5.9	47.1
1432.6	6.0	47.8
1530.4	6.1	48.7

6. Shareholder's Equity The following table shows the net sales (in billions of dollars), total assets (in billions of dollars), and shareholder's equity (in billions of dollars) for Wal-Mart for a recent five-year period. *(Adapted from Wal-Mart Stores, Inc.)*

Shareholder's equity, y	Net sales, x_1	Total assets, x_2
53.2	308.9	138.8
61.6	344.8	151.6
64.6	373.8	163.5
65.3	401.1	163.4
70.7	405.0	170.7

7. **SC** Use StatCrunch to find the multiple regression equation for the data in Exercise 5. Compare this result with the equation found in Exercise 5.

8. **SC** Use StatCrunch to find the multiple regression equation for the data in Exercise 6. Compare this result with the equation found in Exercise 6.

■ EXTENDING CONCEPTS

Adjusted r^2 *The calculation of r^2, the coefficient of determination, depends on the number of data pairs and the number of independent variables. An adjusted value of r^2 can be calculated, based on the number of degrees of freedom, as follows.*

$$r^2_{adj} = 1 - \left[\frac{(1 - r^2)(n - 1)}{n - k - 1} \right]$$

where n is the number of data pairs and k is the number of independent variables.

In Exercises 9 and 10, after calculating r^2_{adj}, determine the percentage of the variation in y that can be explained by the relationships between variables according to r^2_{adj}. Compare this result with the one obtained using r^2.

9. Calculate r^2_{adj} for the data in Exercise 5.

10. Calculate r^2_{adj} for the data in Exercise 6.

USES AND ABUSES

Uses

Correlation and Regression   Correlation and regression analysis can be used to determine whether there is a significant relationship between two variables. If there is, you can use one of the variables to predict the value of the other variable. For instance, educators have used correlation and regression analysis to determine that there is a significant correlation between a student's SAT score and the grade point average from a student's freshman year at college. Consequently, many colleges and universities use SAT scores of high school applicants as a predictor of the applicant's initial success at college.

Abuses

Confusing Correlation and Causation   The most common abuse of correlation in studies is to confuse the concepts of correlation with those of causation (see page 494). Good SAT scores do not cause good college grades. Rather, there are other variables, such as good study habits and motivation, that contribute to both. When a strong correlation is found between two variables, look for other variables that are correlated with both.

Considering Only Linear Correlation   The correlation studied in this chapter is linear correlation. When the correlation coefficient is close to 1 or close to -1, the data points can be modeled by a straight line. It is possible that a correlation coefficient is close to 0 but there is still a strong correlation of a different type. Consider the data listed in the table at the left. The value of the correlation coefficient is 0; however, the data are perfectly correlated with the equation $x^2 + y^2 = 1$, as shown in the graph.

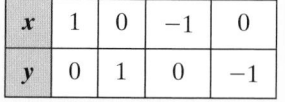

x	1	0	-1	0
y	0	1	0	-1

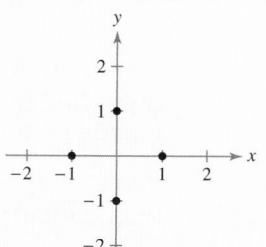

Ethics

When data are collected, all of the data should be used when calculating statistics. In this chapter, you learned that before finding the equation of a regression line, it is helpful to construct a scatter plot of the data to check for outliers, gaps, and clusters in the data. Researchers cannot use only those data points that fit their hypotheses or those that show a significant correlation. Although eliminating outliers may help a data set coincide with predicted patterns or fit a regression line, it is unethical to amend data in such a way. An outlier or any other point that influences a regression model can be removed only if it is properly justified.

    In most cases, the best and sometimes safest approach for presenting statistical measurements is with and without an outlier being included. By doing this, the decision as to whether or not to recognize the outlier is left to the reader.

■ EXERCISES

1. ***Confusing Correlation and Causation***   Find an example of an article that confuses correlation and causation. Discuss other variables that could contribute to the relationship between the variables.

2. ***Considering Only Linear Correlation***   Find an example of two real-life variables that have a nonlinear correlation.

9 CHAPTER SUMMARY

What did you **learn?**	EXAMPLE(S)	REVIEW EXERCISES

Section 9.1

■ How to construct a scatter plot	1–3	1–4
■ How to find a correlation coefficient	4, 5	1–4

$$r = \frac{n\Sigma xy - (\Sigma x)(\Sigma y)}{\sqrt{n\Sigma x^2 - (\Sigma x)^2}\sqrt{n\Sigma y^2 - (\Sigma y)^2}}$$

■ How to perform a hypothesis test for a population correlation coefficient ρ	7	5–10

$$t = \frac{r}{\sqrt{\dfrac{1 - r^2}{n - 2}}}$$

Section 9.2

■ How to find the equation of a regression line, $\hat{y} = mx + b$	1, 2	11–14

$$m = \frac{n\Sigma xy - (\Sigma x)(\Sigma y)}{n\Sigma x^2 - (\Sigma x)^2}$$
$$b = \bar{y} - m\bar{x}$$
$$= \frac{\Sigma y}{n} - m\frac{\Sigma x}{n}$$

■ How to predict y-values using a regression equation	3	15–18

Section 9.3

■ How to find and interpret the coefficient of determination r^2	1	19–24
■ How to find and interpret the standard error of estimate for a regression line	2	23, 24

$$s_e = \sqrt{\frac{\Sigma(y_i - \hat{y}_i)^2}{n - 2}} = \sqrt{\frac{\Sigma y^2 - b\Sigma y - m\Sigma xy}{n - 2}}$$

■ How to construct and interpret a prediction interval for y, $\hat{y} - E < y < \hat{y} + E$	3	25–30

$$E = t_c s_e \sqrt{1 + \frac{1}{n} + \frac{n(x_0 - \bar{x})^2}{n\Sigma x^2 - (\Sigma x)^2}}$$

Section 9.4

■ How to use technology to find a multiple regression equation, the standard error of estimate, and the coefficient of determination	1	31, 32
■ How to use a multiple regression equation to predict y-values	2	33, 34

$$\hat{y} = b + m_1 x_1 + m_2 x_2 + m_3 x_3 + \cdots + m_k x_k$$

9 REVIEW EXERCISES

1.

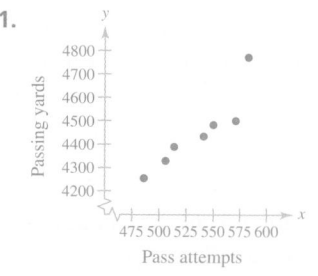

0.912; strong positive linear correlation; the number of passing yards increases as the number of pass attempts increases.

2. See Selected Answers, page A130.

0.297; weak positive linear correlation; the number of wildland acres burned increases as the number of wildland fires increases.

3. See Odd Answers, page A97.

0.338; weak positive linear correlation; brain size increases as IQ increases.

4. See Selected Answers, page A130.

0.979; strong positive linear correlation; the number of cavities increases as sugar consumption increases.

5. There is not enough evidence at the 1% level of significance to conclude that there is a significant linear correlation.

6. There is enough evidence at the 5% level of significance to conclude that there is a significant linear correlation.

7. There is enough evidence at the 5% level of significance to conclude that there is a significant linear correlation between a quarterback's pass attempts and passing yards.

8. There is not enough evidence at the 5% level of significance to conclude that there is a significant linear correlation between the number of wildland fires and the number of acres burned.

■ SECTION 9.1

In Exercises 1–4, display the data in a scatter plot. Then calculate the sample correlation coefficient r. Determine whether there is a positive linear correlation, a negative linear correlation, or no linear correlation between the variables. What can you conclude?

1. The number of pass attempts and passing yards for seven professional quarterbacks for a recent regular season *(Source: National Football League)*

Pass attempts, x	583	571	550	541	506	514	486
Passing yards, y	4770	4500	4483	4434	4328	4388	4254

2. The number of wildland fires (in thousands) and the number of wildland acres burned (in millions) in the United States for eight years *(Source: National Interagency Coordinate Center)*

Fires, x	84.1	73.5	63.6	65.5	66.8	96.4	85.7	79.0
Acres, y	3.6	7.2	4.0	8.1	8.7	9.9	9.3	5.3

3. The IQ and brain size, as measured by the total pixel count (in thousands) from an MRI scan, for nine female college students *(Adapted from Intelligence)*

IQ, x	138	140	96	83	101	135	85	77	88
Pixel count, y	991	856	879	865	808	791	799	794	894

4. The annual per capita sugar consumption (in kilograms) and the average number of cavities of 11- and 12-year-old children in seven countries

Sugar consumption, x	2.1	5.0	6.3	6.5	7.7	8.7	11.6
Cavities, y	0.59	1.51	1.55	1.70	2.18	2.10	2.73

In Exercises 5 and 6, use the given sample statistics to test the claim about the population correlation coefficient ρ at the indicated level of significance α.

5. Claim: $\rho \neq 0$; $\alpha = 0.01$. Sample statistics: $r = 0.24$, $n = 26$

6. Claim: $\rho \neq 0$; $\alpha = 0.05$. Sample statistics: $r = -0.55$, $n = 22$

In Exercises 7–10, test the claim about the population correlation coefficient ρ at the indicated level of significance α. Then interpret the decision in the context of the original claim.

7. Refer to the data in Exercise 1. At $\alpha = 0.05$, test the claim that there is a significant linear correlation between a quarterback's pass attempts and passing yards.

8. Refer to the data in Exercise 2. At $\alpha = 0.05$, is there enough evidence to conclude that there is a significant linear correlation between the number of wildland fires and the number of acres burned?

9. There is not enough evidence at the 1% level of significance to conclude that there is a significant linear correlation between IQ and brain size.

10. There is enough evidence at the 1% level of significance to conclude that there is a significant linear correlation between sugar consumption and tooth decay.

11. $\hat{y} = 0.038x - 3.529$

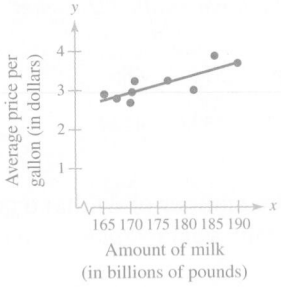

Amount of milk
(in billions of pounds)

$r \approx 0.821$

12. See Selected Answers, page A130.

13. $\hat{y} = -0.086x + 10.450$

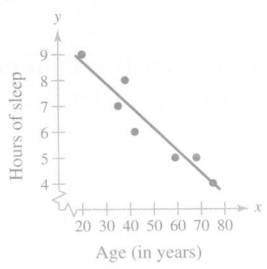

Age (in years)

$r \approx -0.949$

14. See Selected Answers, page A131.

15. (a) It is not meaningful to predict the value of *y* for *x* = 160 because *x* = 160 is outside the range of the original data.

(b) $3.12

(c) $3.31

(d) It is not meaningful to predict the value of *y* for *x* = 200 because *x* = 200 is outside the range of the original data.

16. (a) 4.818 hours

(b) 5.141 hours

(c) 5.41 hours

(d) It is not meaningful to predict the value of *y* for *x* = 5 because *x* = 5 is outside the range of the original data.

9. Refer to the data in Exercise 3. At $\alpha = 0.01$, test the claim that there is a significant linear correlation between a female college student's IQ and brain size.

10. Refer to the data in Exercise 4. At $\alpha = 0.01$, is there enough evidence to conclude that there is a significant linear correlation between sugar consumption and tooth decay?

■ SECTION 9.2

In Exercises 11–14, find the equation of the regression line for the given data. Then construct a scatter plot of the data and draw the regression line. Can you make a guess about the sign and magnitude of r? Calculate r and check your guess. If convenient, use technology to solve the problem.

11. The amounts of milk (in billions of pounds) produced in the United States and the average prices per gallon of milk for nine years *(Adapted from U.S. Department of Agriculture and U.S. Bureau of Labor Statistics)*

Milk produced, *x*	167.6	165.3	170.1	170.4	170.9
Price per gallon, *y*	2.79	2.90	2.68	2.95	3.23

Milk produced, *x*	177.0	181.8	185.7	190.0
Price per gallon, *y*	3.24	3.00	3.87	3.68

12. The average times (in hours) per day spent watching television for men and women for the last 10 years *(Adapted from The Nielsen Company)*

Men, *x*	4.03	4.18	4.32	4.37	4.48	4.43	4.52	4.58	4.65	4.82
Women, *y*	4.67	4.77	4.85	4.97	5.08	5.12	5.28	5.28	5.32	5.42

13. The ages (in years) and the number of hours of sleep in one night for seven adults

Age, *x*	35	20	59	42	68	38	75
Hours of sleep, *y*	7	9	5	6	5	8	4

14. The engine displacements (in cubic inches) and the fuel efficiencies (in miles per gallon) of seven automobiles

Displacement, *x*	170	134	220	305	109	256	322
Fuel efficiency, *y*	29.5	34.5	23.0	17.0	33.5	23.0	15.5

In Exercises 15–18, use the regression equations found in Exercises 11–14 to predict the value of y for each value of x, if meaningful. If not, explain why not. (Each pair of variables has a significant correlation.)

15. Refer to Exercise 11. What price per gallon would you predict for a milk production of (a) 160 billion pounds? (b) 175 billion pounds? (c) 180 billion pounds? (d) 200 billion pounds?

16. Refer to Exercise 12. What average time per day spent watching television for women would you predict when the average time per day for men is (a) 4.2 hours? (b) 4.5 hours? (c) 4.75 hours? (d) 5 hours?

17. (a) It is not meaningful to predict the value of y for $x = 18$ because $x = 18$ is outside the range of the original data.

(b) 8.3 hours

(c) It is not meaningful to predict the value of y for $x = 85$ because $x = 85$ is outside the range of the original data.

(d) 6.15 hours

18. See Selected Answers, page A131.

19. 0.203; About 20.3% of the variation is explained. About 79.7% of the variation is unexplained.

20. 0.878; About 87.8% of the variation is explained. About 12.2% of the variation is unexplained.

21. 0.412; About 41.2% of the variation is explained. About 58.8% of the variation is unexplained.

22. 0.632; About 63.2% of the variation is explained. About 36.8% of the variation is unexplained.

23. (a) 0.679; About 67.9% of the variation in the fuel efficiency of the compact sports sedans can be explained by the variation in their prices, and about 32.1% of the variation is unexplained.

(b) 1.138; The standard error of estimate of the fuel efficiency of the compact sports sedans for a specific price of the compact sports sedans is about 1.138 miles per gallon.

24. (a) 0.446; About 44.6% of the variation in the price of gas grills can be explained by the variation in the cooking areas, and about 55.4% of the variation is unexplained.

(b) 235.079; The standard error of estimate of the price of a gas grill for a specific cooking area is about $235.08.

25. $2.997 < y < 4.025$

You can be 90% confident that the price per gallon of milk will be between $3.00 and $4.03 when 185 billion pounds of milk is produced.

26. See Selected Answers, page A131.

27. See Odd Answers, page A98.

28. See Selected Answers, page A131.

17. Refer to Exercise 13. How many hours of sleep would you predict for an adult of age (a) 18 years? (b) 25 years? (c) 85 years? (d) 50 years?

18. Refer to Exercise 14. What fuel efficiency rating would you predict for a car with an engine displacement of (a) 86 cubic inches? (b) 198 cubic inches? (c) 289 cubic inches? (d) 407 cubic inches?

■ SECTION 9.3

In Exercises 19–22, use the value of the linear correlation coefficient to calculate the coefficient of determination. What does this tell you about the explained variation of the data about the regression line? About the unexplained variation?

19. $r = -0.450$ **20.** $r = -0.937$

21. $r = 0.642$ **22.** $r = 0.795$

In Exercises 23 and 24, use the data to find the (a) coefficient of determination r^2 and interpret the result, and (b) standard error of estimate s_e and interpret the result.

23. The table shows the prices (in thousands of dollars) and fuel efficiencies (in miles per gallon) for nine compact sports sedans. The regression equation is $\hat{y} = -0.414x + 37.147$. *(Adapted from Consumer Reports)*

Price, x	37.2	40.8	29.7	33.7	37.5	32.7	39.2	37.3	31.6
Fuel efficiency, x	21	19	25	24	22	24	23	21	23

24. The table shows the cooking areas (in square inches) of 18 gas grills and their prices (in dollars). The regression equation is $\hat{y} = 1.454x - 532.053$. *(Source: Lowe's)*

Area, x	780	530	942	660	600	732	660	640	869
Price, y	359	98	547	299	449	799	699	199	1049

Area, x	860	700	942	890	733	732	464	869	600
Price, y	499	248	597	999	428	849	99	999	399

In Exercises 25–30, construct the indicated prediction interval and interpret the results.

25. Construct a 90% prediction interval for the price per gallon of milk in Exercise 11 when 185 billion pounds of milk is produced.

26. Construct a 90% prediction interval for the average time women spend per day watching television in Exercise 12 when the average time men spend per day watching television is 4.25 hours.

27. Construct a 95% prediction interval for the number of hours of sleep for an adult in Exercise 13 who is 45 years old.

28. Construct a 95% prediction interval for the fuel efficiency of an automobile in Exercise 14 that has an engine displacement of 265 cubic inches.

29. $16.119 < y < 25.137$

You can be 99% confident that the fuel efficiency of the compact sports sedan that costs $39,900 will be between 16.119 and 25.137 miles per gallon.

30. $44.066 < y < 1509.028$

You can be 99% confident that the price will be between $44.07 and $1509.03 when the cooking area is 900 square inches.

31. $\hat{y} = 3.674 + 1.287x_1 - 7.531x_2$

32. 0.710; 0.943; About 94.3% of the variation of y can be explained by the model.

33. (a) 21.705

(b) 25.21

(c) 30.1

(d) 25.86

34. (a) 11.272

(b) 14.695

(c) 14.380

(d) 9.232

---

29. Construct a 99% prediction interval for the fuel efficiency of a compact sports sedan in Exercise 23 that costs $39,900.

30. Construct a 99% prediction interval for the price of a gas grill in Exercise 24 with a usable cooking area of 900 square inches.

■ SECTION 9.4

In Exercises 31 and 32, use the data in the table, which shows the carbon monoxide, tar, and nicotine content, all in milligrams, of 14 brands of U.S. cigarettes. (Source: Federal Trade Commission)

Carbon monoxide, y	Tar, x_1	Nicotine, x_2
15	16	1.1
17	16	1.0
11	10	0.8
12	11	0.9
14	13	0.8
16	14	0.8
14	16	1.2
16	16	1.2
10	10	0.8
18	19	1.4
17	17	1.2
11	12	1.0
10	9	0.7
14	15	1.2

31. Use technology to find the multiple regression equation for the data.

32. Find the standard error of estimate s_e and the coefficient of determination r^2. What percentage of the variation of y can be explained by the regression equation?

In Exercises 33 and 34, use the multiple regression equation to predict the y-values for the given values of the independent variables.

33. An equation that can be used to predict fuel economy (in miles per gallon) for automobiles is $\hat{y} = 41.3 - 0.004x_1 - 0.0049x_2$, where x_1 is the engine displacement (in cubic inches) and x_2 is the vehicle weight (in pounds).

(a) $x_1 = 305, x_2 = 3750$

(b) $x_1 = 225, x_2 = 3100$

(c) $x_1 = 105, x_2 = 2200$

(d) $x_1 = 185, x_2 = 3000$

34. Use the regression equation found in Exercise 31.

(a) $x_1 = 10, x_2 = 0.7$

(b) $x_1 = 15, x_2 = 1.1$

(c) $x_1 = 13, x_2 = 0.8$

(d) $x_1 = 9, x_2 = 0.8$

9 CHAPTER QUIZ

1.

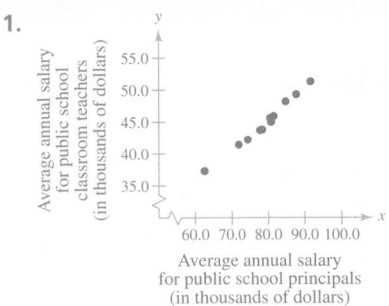

Average annual salary for public school classroom teachers (in thousands of dollars) vs. Average annual salary for public school principals (in thousands of dollars)

The data appear to have a positive linear correlation. As x increases, y tends to increase.

2. 0.993; strong positive linear correlation; public school classroom teachers' salaries increase as public school principals' salaries increase.

3. Reject H_0. There is enough evidence at the 5% level of significance to conclude that there is a significant linear correlation between public school principals' salaries and public school classroom teachers' salaries.

4. $\hat{y} = 0.491x + 5.977$

See Odd Answers, page A98.

5. $50,412.50

6. 0.986; About 98.6% of the variation in the average annual salaries of public school classroom teachers can be explained by the variation in the average annual salaries of public school principals, and about 1.4% of the variation is unexplained.

7. 0.490; The standard error of estimate of the average annual salary of public school classroom teachers for a specific average annual salary of public school principals is about $490.

8. $46.887 < y < 49.273$

You can be 95% confident that the average annual salary of public school classroom teachers will be between $46,887 and $49,273 when the average annual salary of public school principals is $85,750.

9. (a) $59.30
  (b) $30.53
  (c) $45.67
  (d) $35.83

Take this quiz as you would take a quiz in class. After you are done, check your work against the answers given in the back of the book.

For Exercises 1–8, use the data in the table, which shows the average annual salaries (both in thousands of dollars) for public school principals and public school classroom teachers in the United States for 11 years. (Adapted from Educational Research Service)

Principals, x	Classroom teachers, y
62.5	37.3
71.9	41.4
74.4	42.2
77.8	43.7
78.4	43.8
80.8	45.0
80.5	45.6
81.5	45.9
84.8	48.2
87.7	49.3
91.6	51.3

1. Construct a scatter plot for the data. Do the data appear to have a positive linear correlation, a negative linear correlation, or no linear correlation? Explain.

2. Calculate the correlation coefficient r. What can you conclude?

3. Test the level of significance of the correlation coefficient r. Use $\alpha = 0.05$.

4. Find the equation of the regression line for the data. Draw the regression line on the scatter plot.

5. Use the regression equation to predict the average annual salary of public school classroom teachers when the average annual salary of public school principals is $90,500.

6. Find the coefficient of determination r^2 and interpret the result.

7. Find the standard error of estimate s_e and interpret the result.

8. Construct a 95% prediction interval for the average annual salary of public school classroom teachers when the average annual salary of public school principals is $85,750. Interpret the results.

9. Stock Price The equation used to predict the stock price (in dollars) at the end of the year for McDonald's Corporation is

$$\hat{y} = -47 + 5.91x_1 - 1.99x_2$$

where x_1 is the total revenue (in billions of dollars) and x_2 is the shareholders' equity (in billions of dollars). Use the multiple regression equation to predict the y-values for the given values of the independent variables. *(Adapted from McDonald's Corporation)*

  (a) $x_1 = 22.7$, $x_2 = 14.0$
  (b) $x_1 = 17.9$, $x_2 = 14.2$
  (c) $x_1 = 20.9$, $x_2 = 15.5$
  (d) $x_1 = 19.1$, $x_2 = 15.1$

PUTTING IT ALL TOGETHER

Real Statistics — Real Decisions

Acid rain affects the environment by increasing the acidity of lakes and streams to dangerous levels, damaging trees and soil, accelerating the decay of building materials and paint, and destroying national monuments. The goal of the Environmental Protection Agency's (EPA) Acid Rain Program is to achieve environmental health benefits by reducing the emissions of the primary causes of acid rain: sulfur dioxide and nitrogen oxides.

You work for the EPA and you want to determine if there is a significant correlation between sulfur dioxide emissions and nitrogen oxides emissions.

■ EXERCISES

1. Analyzing the Data

(a) The data in the table show the sulfur dioxide emissions (in millions of tons) and the nitrogen oxides emissions (in millions of tons) for 14 years. Construct a scatter plot of the data and make a conclusion about the type of correlation between sulfur dioxide emissions and nitrogen oxides emissions.

(b) Calculate the correlation coefficient r and verify your conclusion in part (a).

(c) Test the significance of the correlation coefficient found in part (b). Use $\alpha = 0.05$.

(d) Find the equation of the regression line for sulfur dioxide emissions and nitrogen oxides emissions. Add the graph of the regression line to your scatter plot in part (a). Does the regression line appear to be a good fit?

(e) Can you use the equation of the regression line to predict the nitrogen oxides emission given the sulfur dioxide emission? Why or why not?

(f) Find the coefficient of determination r^2 and the standard error of estimate s_e. Interpret your results.

2. Making Predictions

The EPA set a goal of reducing sulfur dioxide emissions levels by 10 million tons from 1980 levels of 17.3 million tons. Construct a 95% prediction interval for the nitrogen oxides emissions for this sulfur dioxide emissions goal level. Interpret the results.

Sulfur dioxide emissions, x	Nitrogen oxides emissions, y
11.8	5.8
12.5	6.0
12.9	6.0
13.1	6.0
12.5	5.5
11.2	5.1
10.6	4.7
10.2	4.5
10.6	4.2
10.3	3.8
10.2	3.6
9.4	3.4
8.9	3.3
7.6	3.0

Source: Environmental Protection Agency

TECHNOLOGY

 **U.S. Food and Drug Administration**

NUTRIENTS IN BREAKFAST CEREALS

The U.S. Food and Drug Administration (FDA) requires nutrition labeling for most foods. Under FDA regulations, manufacturers are required to list the amounts of certain nutrients in their foods, such as calories, sugar, fat, and carbohydrates. This nutritional information is displayed in the "Nutrition Facts" panel on the food's package.

The table shows the following nutritional content for one cup of each of 21 different breakfast cereals.

C = calories

S = sugar in grams

F = fat in grams

R = carbohydrates in grams

Cereal	C	S	F	R
Apple Jacks®	100	12	0.5	25
Berry Burst Cheerios®	130	11	1.5	29
Cheerios®	100	1	2	20
Cocoa Puffs®	130	15	2	31
Cookie Crisp®	130	13	1.5	29
Corn Chex®	120	3	0.5	26
Corn Flakes®	100	2	0	24
Corn Pops®	120	10	0	29
Count Chocula®	150	16	1.5	31
Crispix®	110	4	0	25
Froot Loops®	110	12	1	25
Frosted Flakes®	150	15	0	36
Golden Grahams®	160	15	1.5	35
Honey Nut Cheerios®	150	12	2	29
Lucky Charms®	150	15	1.5	29
Multi Grain Cheerios®	110	6	1	23
Raisin Bran®	190	19	1.5	45
Rice Krispies®	100	3	0	23
Special K®	120	4	0.5	23
Trix®	120	11	1.5	28
Wheaties®	130	5	0.5	29

▨ EXERCISES

1. Use a technology tool to draw a scatter plot of the following (x, y) pairs in the data set.

(a) (calories, sugar)

(b) (calories, fat)

(c) (calories, carbohydrates)

(d) (sugar, fat)

(e) (sugar, carbohydrates)

(f) (fat, carbohydrates)

2. From the scatter plots in Exercise 1, which pairs of variables appear to have a strong linear correlation?

3. Use a technology tool to find the correlation coefficient for each pair of variables in Exercise 1. Which has the strongest linear correlation?

4. Use a technology tool to find an equation of a regression line for the following variables.

(a) (calories, sugar)

(b) (calories, carbohydrates)

5. Use the results of Exercise 4 to predict the following.

(a) The sugar content of one cup of cereal that has a caloric content of 120 calories

(b) The carbohydrate content of one cup of cereal that has a caloric content of 120 calories

6. Use a technology tool to find the multiple regression equations of the following forms.

(a) $C = b + m_1 S + m_2 F + m_3 R$

(b) $C = b + m_1 S + m_2 R$

7. Use the equations from Exercise 6 to predict the caloric content of 1 cup of cereal that has 7 grams of sugar, 0.5 gram of fat, and 31 grams of carbohydrates.

Extended solutions are given in the *Technology Supplement*.
Technical instruction is provided for MINITAB, Excel, and the TI-83/84 Plus.

CHI-SQUARE TESTS AND THE *F*-DISTRIBUTION

10 CHAPTER

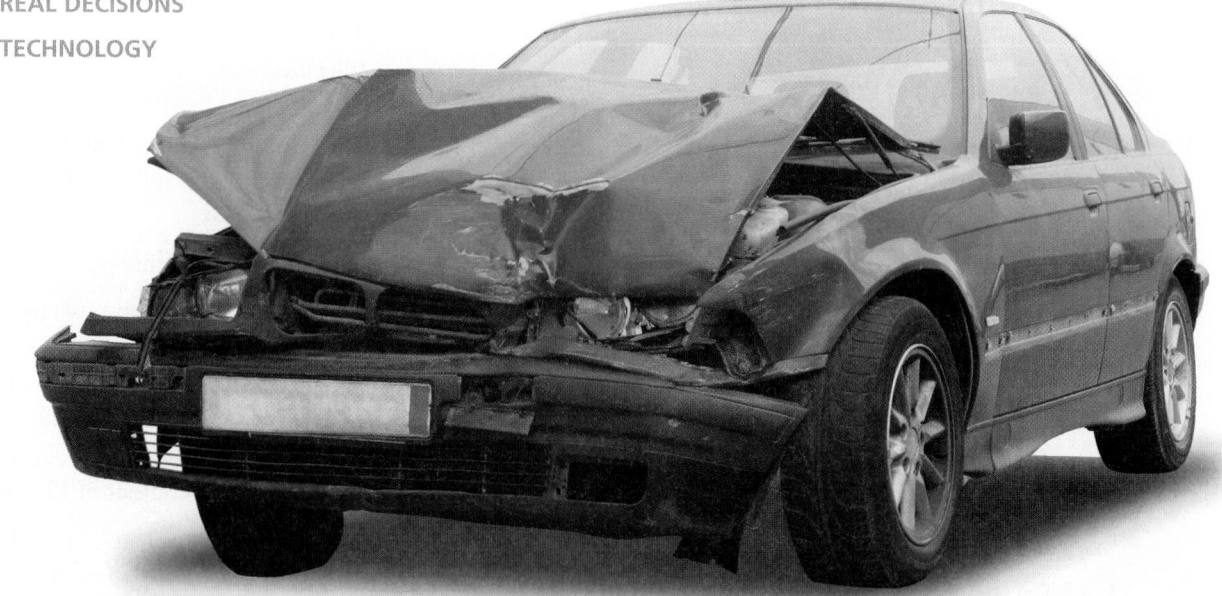

Crash tests performed by the Insurance Institute for Highway Safety demonstrate how a vehicle will react when in a realistic collision. Tests are performed on the front, side, and rear of the vehicles. Results of these tests are classified using the ratings *good*, *acceptable*, *marginal*, and *poor*.

The Insurance Institute for Highway Safety buys new vehicles each year and crashes them into a barrier at 40 miles per hour to compare how different vehicles protect drivers in a frontal offset crash. In this test, 40% of the total width of the vehicle strikes the barrier on the driver side. The forces and impacts that occur during a crash test are measured by equipping dummies with special instruments and placing them in the car. The crash test results include data on head, chest, and leg injuries. For a low crash test number, the injury potential is low. If the crash test number is high, then the injury potential is high. Using the techniques of Chapter 8, you can determine if the mean chest injury potential is the same for pickups and minivans. (Assume the population variances are equal.) The sample statistics are as follows. *(Adapted from Insurance Institute for Highway Safety)*

Vehicle	Number	Mean chest injury	Standard deviation
Minivans	$n_1 = 9$	$\overline{x}_1 = 29.9$	$s_1 = 3.33$
Pickups	$n_2 = 19$	$\overline{x}_2 = 30.4$	$s_2 = 4.21$

For the means of chest injury, the *P*-value for the hypothesis that $\mu_1 = \mu_2$ is about 0.7575. At $\alpha = 0.05$, you fail to reject the null hypothesis. So, you do not have enough evidence to conclude that there is a significant difference in the means of the chest injury potential in a frontal offset crash at 40 miles per hour for minivans and pickups.

In Chapter 8, you learned how to test a hypothesis that compares two populations by basing your decisions on sample statistics and their distributions. In this chapter, you will learn how to test a hypothesis that compares three or more populations.

For instance, in addition to the crash tests for minivans and pickups, a third group of vehicles was also tested. The results for all three types of vehicles are as follows.

Vehicle	Number	Mean chest injury	Standard deviation
Minivans	$n_1 = 9$	$\overline{x}_1 = 29.9$	$s_1 = 3.33$
Pickups	$n_2 = 19$	$\overline{x}_2 = 30.4$	$s_2 = 4.21$
Midsize SUVs	$n_3 = 32$	$\overline{x}_3 = 34.1$	$s_3 = 5.22$

From these three samples, is there evidence of a difference in chest injury potential among minivans, pickups, and midsize SUVs in a frontal offset crash at 40 miles per hour?

In this chapter, you will learn that you can answer this question by testing the hypothesis that the three means are equal. For the means of chest injury, the *P*-value for the hypothesis that $\mu_1 = \mu_2 = \mu_3$ is about 0.0088. At $\alpha = 0.05$, you can reject the null hypothesis. So, you can conclude that for the three types of vehicles tested, at least one of the means of the chest injury potential in a frontal offset crash at 40 miles per hour is different from the others.

10.1 Goodness-of-Fit Test

WHAT YOU SHOULD LEARN

▸ How to use the chi-square distribution to test whether a frequency distribution fits a claimed distribution

The Chi-Square Goodness-of-Fit Test

▸ THE CHI-SQUARE GOODNESS-OF-FIT TEST

Suppose a tax preparation company wants to determine the proportions of people who used different methods to prepare their taxes. To determine these proportions, the company can perform a multinomial experiment. A **multinomial experiment** is a probability experiment consisting of a fixed number of independent trials in which there are more than two possible outcomes for each trial. The probability of each outcome is fixed, and each outcome is classified into **categories.** (Remember from Section 4.2 that a binomial experiment has only two possible outcomes.)

Now, suppose the company wants to test a previous survey's claim concerning the distribution of proportions of people who used different methods to prepare their taxes. To do so, the company could compare the distribution of proportions obtained in the multinomial experiment with the previous survey's specified distribution. How can the company compare the distributions? The answer is, perform a *chi-square goodness-of-fit test.*

> ### DEFINITION
>
> A **chi-square goodness-of-fit test** is used to test whether a frequency distribution fits an expected distribution.

INSIGHT

The hypothesis tests described in Sections 10.1 and 10.2 can be used for qualitative data.

To begin a goodness-of-fit test, you must first state a null and an alternative hypothesis. Generally, the null hypothesis states that the frequency distribution fits the specified distribution and the alternative hypothesis states that the frequency distribution does not fit the specified distribution.

For instance, suppose the previous survey claims that the distribution of people who used different methods to prepare their taxes is as shown below.

Distribution of tax preparation methods	
Accountant	25%
By hand	20%
Computer software	35%
Friend/family	5%
Tax preparation service	15%

To test the previous survey's claim, the company can perform a chi-square goodness-of-fit test using the following null and alternative hypotheses.

H_0: The distribution of tax preparation methods is 25% by accountant, 20% by hand, 35% by computer software, 5% by friend or family, and 15% by tax preparation service. (Claim)

H_a: The distribution of tax preparation methods differs from the claimed or expected distribution.

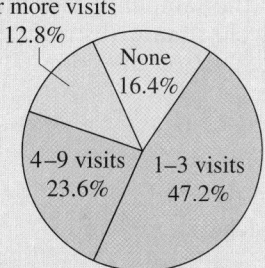

To calculate the test statistic for the chi-square goodness-of-fit test, you can use *observed frequencies* and *expected frequencies*. To calculate the expected frequencies, you must assume the null hypothesis is true.

DEFINITION

The **observed frequency** *O* of a category is the frequency for the category observed in the sample data.

The **expected frequency** *E* of a category is the *calculated* frequency for the category. Expected frequencies are obtained assuming the specified (or hypothesized) distribution. The expected frequency for the *i*th category is

$$E_i = np_i$$

where *n* is the number of trials (the sample size) and p_i is the assumed probability of the *i*th category.

EXAMPLE 1

▶ Finding Observed Frequencies and Expected Frequencies

A tax preparation company randomly selects 300 adults and asks them how they prepare their taxes. The results are shown at the right. Find the observed frequency and the expected frequency for each tax preparation method. (*Adapted from National Retail Federation*)

Survey results ($n = 300$)	
Accountant	71
By hand	40
Computer software	101
Friend/family	35
Tax preparation service	53

▶ Solution

The observed frequency for each tax preparation method is the number of adults in the survey naming a particular tax preparation method. The expected frequency for each tax preparation method is the product of the number of adults in the survey and the probability that an adult will name a particular tax preparation method. The observed frequencies and expected frequencies are shown in the following table.

Tax preparation method	% of people	Observed frequency	Expected frequency
Accountant	25%	71	$300(0.25) = 75$
By hand	20%	40	$300(0.20) = 60$
Computer software	35%	101	$300(0.35) = 105$
Friend/family	5%	35	$300(0.05) = 15$
Tax preparation service	15%	53	$300(0.15) = 45$

▶ Try It Yourself 1

Suppose the tax preparation company randomly selects 500 adults. Find the expected frequency for each tax preparation method.

Multiply 500 by the probability that an adult will name each particular tax preparation method to find the *expected frequencies*. *Answer: Page A45*

For the chi-square goodness-of-fit test to be used, the following must be true.

1. The observed frequencies must be obtained using a random sample.
2. Each expected frequency must be greater than or equal to 5.

If the expected frequency of a category is less than 5, it may be possible to combine it with another category to meet the requirements.

STUDY TIP

Remember that a chi-square distribution is positively skewed and its shape is determined by the degrees of freedom. The graph is not symmetric, but it appears to become more symmetric as the degrees of freedom increase, as shown in Section 6.4.

THE CHI-SQUARE GOODNESS-OF-FIT TEST

If the conditions listed above are satisfied, then the sampling distribution for the goodness-of-fit test is approximated by a chi-square distribution with $k - 1$ degrees of freedom, where k is the number of categories. The **test statistic** for the chi-square goodness-of-fit test is

$$\chi^2 = \Sigma \frac{(O - E)^2}{E}$$

where O represents the observed frequency of each category and E represents the expected frequency of each category.

When the observed frequencies closely match the expected frequencies, the differences between O and E will be small and the chi-square test statistic will be close to 0. As such, the null hypothesis is unlikely to be rejected. However, when there are large discrepancies between the observed frequencies and the expected frequencies, the differences between O and E will be large, resulting in a large chi-square test statistic. A large chi-square test statistic is evidence for rejecting the null hypothesis. So, the chi-square goodness-of-fit test is always a right-tailed test.

GUIDELINES

Performing a Chi-Square Goodness-of-Fit Test

Verify that the expected frequency is at least 5 for each category.

IN WORDS	IN SYMBOLS
1. Identify the claim. State the null and alternative hypotheses.	State H_0 and H_a.
2. Specify the level of significance.	Identify α.
3. Determine the degrees of freedom.	d.f. $= k - 1$
4. Determine the critical value.	Use Table 6 in Appendix B.
5. Determine the rejection region.	
6. Find the test statistic and sketch the sampling distribution.	$\chi^2 = \Sigma \dfrac{(O - E)^2}{E}$
7. Make a decision to reject or fail to reject the null hypothesis.	If χ^2 is in the rejection region, reject H_0. Otherwise, fail to reject H_0.
8. Interpret the decision in the context of the original claim.	

EXAMPLE 2   (SC) Report 45

▶ **Performing a Chi-Square Goodness-of-Fit Test**

The tax preparation methods of adults from a previous survey are distributed as shown in the table at the left below. A tax preparation company randomly selects 300 adults and asks them how they prepare their taxes. The results are shown in the table at the right below. At $\alpha = 0.01$, perform a chi-square goodness-of-fit test to test whether the distributions are different. *(Adapted from National Retail Federation)*

Distribution of tax preparation methods	
Accountant	25%
By hand	20%
Computer software	35%
Friend/family	5%
Tax preparation service	15%

Survey results ($n = 300$)	
Accountant	71
By hand	40
Computer software	101
Friend/family	35
Tax preparation service	53

▶ **Solution**

The observed and expected frequencies are shown in the table at the left. The expected frequencies were calculated in Example 1. Because the observed frequencies were obtained using a random sample and each expected frequency is at least 5, you can use the chi-square goodness-of-fit test to test the proposed distribution. The null and alternative hypotheses are as follows.

H_0: The distribution of tax preparation methods is 25% by accountant, 20% by hand, 35% by computer software, 5% by friend or family, and 15% by tax preparation service.

H_a: The distribution of tax preparation methods differs from the claimed or expected distribution. (Claim)

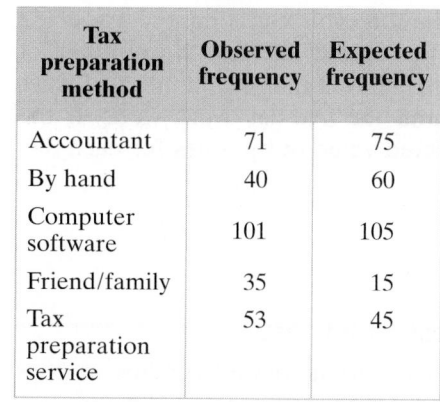

Tax preparation method	Observed frequency	Expected frequency
Accountant	71	75
By hand	40	60
Computer software	101	105
Friend/family	35	15
Tax preparation service	53	45

Because there are 5 categories, the chi-square distribution has $k - 1 = 5 - 1 = 4$ degrees of freedom. With d.f. = 4 and $\alpha = 0.01$, the critical value is $\chi_0^2 = 13.277$. With the observed and expected frequencies, the chi-square test statistic is

$$\chi^2 = \Sigma \frac{(O - E)^2}{E}$$

$$= \frac{(71 - 75)^2}{75} + \frac{(40 - 60)^2}{60} + \frac{(101 - 105)^2}{105}$$

$$+ \frac{(35 - 15)^2}{15} + \frac{(53 - 45)^2}{45}$$

$$\approx 35.121.$$

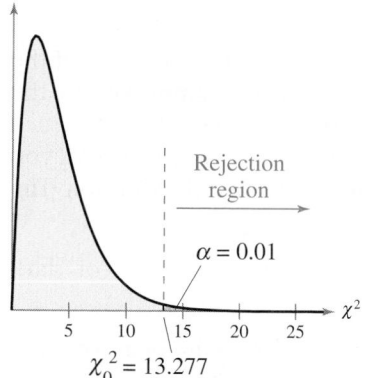

Rejection region

$\alpha = 0.01$

$\chi_0^2 = 13.277$

The graph at the left shows the location of the rejection region. Because χ^2 is in the rejection region, you should reject the null hypothesis.

Interpretation   There is enough evidence at the 1% level of significance to conclude that the distribution of tax preparation methods differs from the previous survey's claimed or expected distribution.

Ages	Previous age distribution	Survey results
0–9	16%	76
10–19	20%	84
20–29	8%	30
30–39	14%	60
40–49	15%	54
50–59	12%	40
60–69	10%	42
70+	5%	14

▸ **Try It Yourself 2**

A sociologist claims that the age distribution for the residents of a certain city is different than it was 10 years ago. The distribution of ages 10 years ago is shown in the table at the left. You randomly select 400 residents and record the age of each. The survey results are shown in the table. At $\alpha = 0.05$, perform a chi-square goodness-of-fit test to test whether the distribution has changed.

a. *Verify* that the expected frequency is at least 5 for each category.
b. Identify the *claimed distribution* and state H_0 and H_a.
c. Specify the *level of significance* α.
d. Determine the *degrees of freedom*.
e. Determine the *critical value* and the *rejection region*.
f. Find the *chi-square test statistic. Sketch* a graph.
g. *Decide* whether to reject the null hypothesis.
h. *Interpret* the decision in the context of the original claim.

Answer: Page A45

The chi-square goodness-of-fit test is often used to determine whether a distribution is uniform. For such tests, the expected frequencies of the categories are equal. When testing a uniform distribution, you can find the expected frequency of each category by dividing the sample size by the number of categories. For instance, suppose a company believes that the number of sales made by its sales force is uniform throughout the five-day work week. If the sample consists of 1000 sales, then the expected value of the sales for each day will be $1000/5 = 200$.

EXAMPLE 3   Report 46

▸ **Performing a Chi-Square Goodness-of-Fit Test**

A researcher claims that the number of different-colored candies in bags of dark chocolate M&M's is uniformly distributed. To test this claim, you randomly select a bag that contains 500 dark chocolate M&M's. The results are shown in the table at the left. At $\alpha = 0.10$, perform a chi-square goodness-of-fit test to test the claimed or expected distribution. *(Adapted from Mars, Incorporated)*

Color	Frequency, f
Brown	80
Yellow	95
Red	88
Blue	83
Orange	76
Green	78

▸ **Solution**

The claim is that the distribution is uniform, so the expected frequencies of the colors are equal. To find each expected frequency, divide the sample size by the number of colors. So, for each color, $E = 500/6 \approx 83.33$. Because each expected frequency is at least 5 and the M&M's were randomly selected, you can use the chi-square goodness-of-fit test to test the claimed distribution. The null and alternative hypotheses are as follows.

H_0: The distribution of the different-colored candies in bags of dark chocolate M&M's is uniform. (Claim)

H_a: The distribution of the different-colored candies in bags of dark chocolate M&M's is not uniform.

Because there are 6 categories, the chi-square distribution has $k - 1 = 6 - 1 = 5$ degrees of freedom. Using d.f. = 5 and $\alpha = 0.10$, the critical value is $\chi_0^2 = 9.236$. With the observed and expected frequencies, the chi-square test statistic is shown in the following table.

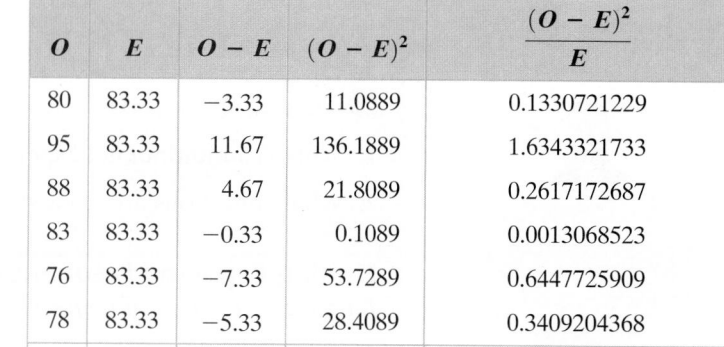

O	E	$O - E$	$(O - E)^2$	$\dfrac{(O - E)^2}{E}$
80	83.33	−3.33	11.0889	0.1330721229
95	83.33	11.67	136.1889	1.6343321733
88	83.33	4.67	21.8089	0.2617172687
83	83.33	−0.33	0.1089	0.0013068523
76	83.33	−7.33	53.7289	0.6447725909
78	83.33	−5.33	28.4089	0.3409204368

$$\chi^2 = \Sigma \frac{(O - E)^2}{E} \approx 3.016$$

The graph shows the location of the rejection region and the chi-square test statistic. Because χ^2 is not in the rejection region, you should fail to reject the null hypothesis.

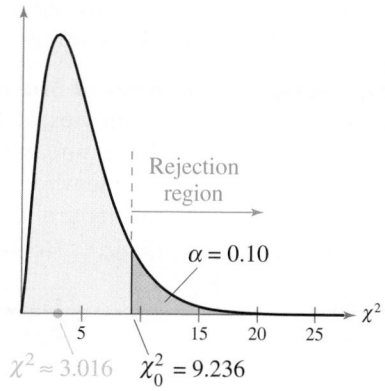

Interpretation   There is not enough evidence at the 10% level of significance to reject the claim that the distribution of the different-colored candies in bags of dark chocolate M&M's is uniform.

▶ **Try It Yourself 3**

A researcher claims that the number of different-colored candies in bags of peanut M&M's is uniformly distributed. To test this claim, you randomly select a bag that contains 180 peanut M&M's. The results are shown in the table at the left. Using $\alpha = 0.05$, perform a chi-square goodness-of-fit test to test the claimed or expected distribution. *(Adapted from Mars, Incorporated)*

Color	Frequency, f
Brown	22
Yellow	27
Red	22
Blue	41
Orange	41
Green	27

a. *Verify* that the expected frequency is at least 5 for each category.
b. Identify the *claimed distribution* and state H_0 and H_a.
c. Specify the *level of significance* α.
d. Determine the *degrees of freedom*.
e. Determine the *critical value* and the *rejection region*.
f. Find the *chi-square test statistic*. *Sketch* a graph.
g. *Decide* whether to reject the null hypothesis.
h. *Interpret* the decision in the context of the original claim.

Answer: Page A46

10.1 EXERCISES

FOR EXTRA HELP:
MyStatLab

1. A multinomial experiment is a probability experiment consisting of a fixed number of independent trials in which there are more than two possible outcomes for each trial. The probability of each outcome is fixed, and each outcome is classified into categories.

2. The observed frequencies must be obtained using a random sample, and each expected frequency must be greater than or equal to 5.

3. 45

4. 450

5. 57.5

6. 33.2

7. (a) H_0: The distribution of the ages of moviegoers is 26.7% ages 2–17, 19.8% ages 18–24, 19.7% ages 25–39, 14% ages 40–49, and 19.8% ages 50+. (claim)

   H_a: The distribution of ages differs from the claimed or expected distribution.

   (b) $\chi_0^2 = 7.779$;

   Rejection region: $\chi^2 > 7.779$

   (c) 7.256

   (d) Fail to reject H_0.

   (e) There is enough evidence at the 10% level of significance to conclude that the distribution of the ages of moviegoers and the claimed or expected distribution are the same.

BUILDING BASIC SKILLS AND VOCABULARY

1. What is a multinomial experiment?

2. What conditions are necessary to use the chi-square goodness-of-fit test?

Finding Expected Frequencies *In Exercises 3–6, find the expected frequency for the given values of n and p_i.*

3. $n = 150$, $p_i = 0.3$

4. $n = 500$, $p_i = 0.9$

5. $n = 230$, $p_i = 0.25$

6. $n = 415$, $p_i = 0.08$

USING AND INTERPRETING CONCEPTS

Performing a Chi-Square Goodness-of-Fit Test *In Exercises 7–16, (a) identify the claimed distribution and state H_0 and H_a, (b) find the critical value and identify the rejection region, (c) find the chi-square test statistic, (d) decide whether to reject or fail to reject the null hypothesis, and (e) interpret the decision in the context of the original claim.*

7. **Ages of Moviegoers** Results from a previous survey asking people who go to movies at least once a month for their ages are shown in the graph. To determine whether this distribution is still the same, you randomly select 1000 people who go to movies at least once a month and record the age of each. The results are shown in the table. At $\alpha = 0.10$, are the distributions the same? *(Source: Motion Picture Association of America)*

Survey results	
Age	Frequency, *f*
2–17	240
18–24	214
25–39	183
40–49	156
50+	207

8. **Coffee** Results from a previous survey asking coffee drinkers how much coffee they drink are shown in the graph. To determine whether this distribution is still the same, you randomly select 1600 coffee drinkers and ask them how much coffee they drink. The results are shown in the table. At $\alpha = 0.05$, are the distributions the same? *(Source: Braun Research)*

Survey results	
Response	Frequency, *f*
2 cups a week	206
1 cup a week	193
1 cup a day	462
2 or more cups a day	739

8. (a) H_0: The distribution of the amounts of coffee people drink is 15% 2 cups a week, 13% 1 cup a week, 27% 1 cup a day, and 45% 2 or more cups a day. (claim)

   H_a: The distribution of amounts differs from the claimed or expected distribution.

   (b) $\chi_0^2 = 7.815$;

   Rejection region: $\chi^2 > 7.815$

   (c) 8.483

   (d) Reject H_0.

   (e) See Selected Answers, page A131.

9. (a) H_0: The distribution of the days people order food for delivery is 7% Sunday, 4% Monday, 6% Tuesday, 13% Wednesday, 10% Thursday, 36% Friday, and 24% Saturday.

   H_a: The distribution of days differs from the claimed or expected distribution. (claim)

   (b) $\chi_0^2 = 16.812$;

   Rejection region: $\chi^2 > 16.812$

   (c) 17.595

   (d) Reject H_0.

   (e) There is enough evidence at the 1% level of significance to conclude that there has been a change in the claimed or expected distribution.

10. See Selected Answers, page A131.

11. (a) H_0: The distribution of the number of homicide crimes in California by season is uniform. (claim)

    H_a: The distribution of homicides by season is not uniform.

    (b) $\chi_0^2 = 7.815$;

    Rejection region: $\chi^2 > 7.815$

    (c) 0.727

    (d) Fail to reject H_0.

    (e) There is not enough evidence at the 5% level of significance to reject the claim that distribution of the number of homicide crimes in California by season is uniform.

9. **Ordering Delivery**   Results from a previous survey asking people which day of the week they are most likely to order food for delivery are shown in the graph. To determine whether this distribution has changed, you randomly select 500 people and record which day of the week each is most likely to order food for delivery. The results are shown in the table. At $\alpha = 0.01$, can you conclude that there has been a change in the claimed or expected distribution? *(Source: Technomic, Inc.)*

Survey results	
Day	**Frequency, f**
Sunday	43
Monday	16
Tuesday	25
Wednesday	49
Thursday	46
Friday	168
Saturday	153

10. **Reasons Workers Leave**   A personnel director believes that the distribution of the reasons workers leave their jobs is different from the one shown in the graph. The director randomly selects 200 workers who recently left their jobs and records each worker's reason for doing so. The results are shown in the table. At $\alpha = 0.01$, are the distributions different? *(Source: Robert Half International, Inc.)*

Survey results	
Response	**Frequency, f**
Limited advancement potential	78
Lack of recognition	52
Low salary/benefits	30
Unhappy with mgmt.	25
Bored/don't know	15

11. **Homicides by Season**   A researcher believes that the number of homicide crimes in California by season is uniformly distributed. To test this claim, you randomly select 1200 homicides from a recent year and record the season when each happened. The results are shown in the table. At $\alpha = 0.05$, can you reject the claim that the distribution is uniform? *(Adapted from California Department of Justice)*

Season	Frequency, f
Spring	312
Summer	299
Fall	297
Winter	292

12. (a) H_0: The distribution of the number of homicide crimes in California by month is uniform. (claim)

H_a: The distribution of homicides by month is not uniform.

(b) $\chi_0^2 = 17.275$;

Rejection region: $\chi^2 > 17.275$

(c) 10.320

(d) Fail to reject H_0.

(e) There is not enough evidence at the 10% level of significance to reject the claim that distribution of the number of homicide crimes in California by month is uniform.

13. (a) H_0: The distribution of the opinions of U.S. parents on whether a college education is worth the expense is 55% strongly agree, 30% somewhat agree, 5% neither agree nor disagree, 6% somewhat disagree, and 4% strongly disagree.

H_a: The distribution of opinions differs from the claimed or expected distribution. (claim)

(b) $\chi_0^2 = 9.488$;

Rejection region: $\chi^2 > 9.488$

(c) 65.236

(d) Reject H_0.

(e) There is enough evidence at the 5% level of significance to conclude that the distribution of the opinions of U.S. parents on whether a college education is worth the expense differs from the claimed or expected distribution.

14. (a) H_0: The distribution of the opinions of U.S. female adults on which is more important to save for, your child's college education or your own retirement, is 50% child's college education, 37% retirement, and 13% not sure. (claim)

H_a: The distribution of the opinions of U.S. female adults differs from the distribution of the opinions of U.S. male adults.

12. Homicides by Month A researcher believes that the number of homicide crimes in California by month is uniformly distributed. To test this claim, you randomly select 1200 homicides from a recent year and record the month when each happened. The results are shown in the table. At $\alpha = 0.10$, can you reject the claim that the distribution is uniform? *(Adapted from California Department of Justice)*

Month	Frequency, f	Month	Frequency, f
January	98	July	84
February	103	August	109
March	114	September	112
April	92	October	95
May	106	November	91
June	106	December	90

13. College Education The pie chart shows the distribution of the opinions of U.S. parents on whether a college education is worth the expense. An economist believes that the distribution of the opinions of U.S. teenagers is different from the distribution for U.S. parents. The economist randomly selects 200 U.S. teenagers and asks each whether a college education is worth the expense. The results are shown in the table. At $\alpha = 0.05$, are the distributions different? *(Adapted from Upromise, Inc.)*

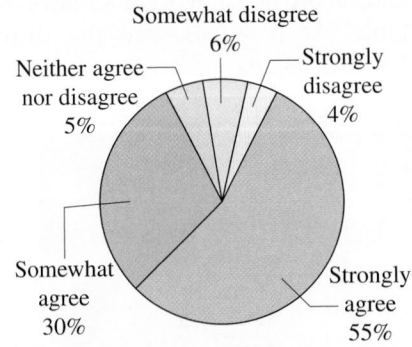

Survey results	
Response	**Frequency, f**
Strongly agree	86
Somewhat agree	62
Neither agree nor disagree	34
Somewhat disagree	14
Strongly disagree	4

14. Saving for the Future The pie chart shows the distribution of the opinions of U.S. male adults on which is more important to save for, your child's college education or your own retirement. A financial services company believes that the distribution of the opinions of U.S. female adults is the same as the distribution for U.S. male adults. The company randomly selects 400 U.S. female adults and asks each which is more important—saving for your child's college education or saving for your own retirement. The results are shown in the table. At $\alpha = 0.10$, are the distributions the same? *(Adapted from Country Financial)*

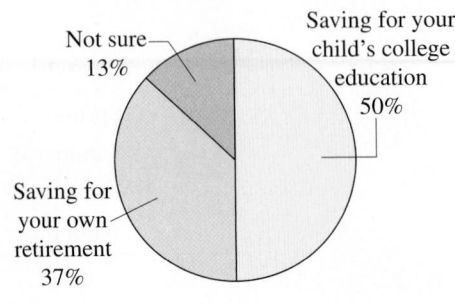

Survey results	
Response	**Frequency, f**
Saving for your child's college education	180
Saving for your own retirement	172
Not sure	48

(b) $\chi_0^2 = 4.605$;
   Rejection region: $\chi^2 > 4.605$

(c) 6.200

(d) Reject H_0.

(e) See Selected Answers,
   page A131.

15. (a) H_0: The distribution of
   prospective home buyers by
   the size they want their next
   house to be is uniform. (claim)

   H_a: The distribution of
   prospective home buyers by
   the size they want their next
   house to be is not uniform.

(b) $\chi_0^2 = 5.991$;
   Rejection region: $\chi^2 > 5.991$

(c) 10.308

(d) Reject H_0.

(e) There is enough evidence at
   the 5% level of significance
   to reject the claim that the
   distribution of prospective
   home buyers by the size they
   want their next house to be
   is uniform.

16. (a) H_0: The distribution of the
   number of births by day of
   the week is uniform. (claim)

   H_a: The distribution of the
   number of births is not
   uniform.

(b) $\chi_0^2 = 16.812$;
   Rejection region:
   $\chi^2 > 16.812$

(c) 26.800

(d) Reject H_0.

(e) There is enough evidence at
   the 1% level of significance
   to reject the claim that the
   distribution of the number
   of births by day of the week
   is uniform.

17. Chi-Square goodness-of-fit
   results:

   Observed: Recent survey

   Expected: Previous survey

   See Odd Answers, page A99.

   $P = 0.0285 < 0.10$, so reject H_0.
   There is enough evidence at
   the 10% level of significance
   to conclude that there has
   been a change in the claimed
   or expected distribution of
   U.S. adults' favorite sports.

15. Home Sizes   An organization believes that the number of prospective home buyers who want their next house to be larger, smaller, or the same size as their current house is uniformly distributed. To test this claim, you randomly select 800 prospective home buyers and ask them what size they want their next house to be. The results are shown in the table. At $\alpha = 0.05$, can you reject the claim that the distribution is uniform? *(Adapted from Better Homes and Gardens)*

Response	Frequency, f
Larger	285
Same size	224
Smaller	291

16. Births by Day of the Week   A doctor believes that the number of births by day of the week is uniformly distributed. To test this claim, you randomly select 700 births from a recent year and record the day of the week on which each takes place. The results are shown below. At $\alpha = 0.01$, can you reject the claim that the distribution is uniform? *(Adapted from National Center for Health Statistics)*

Day	Frequency, f
Sunday	65
Monday	103
Tuesday	114
Wednesday	116
Thursday	115
Friday	112
Saturday	75

SC *In Exercises 17 and 18, use StatCrunch to perform a chi-square goodness-of-fit test. Decide whether to reject the null hypothesis. Then, interpret the decision in the context of the original claim.*

17. Favorite Sport   Results from a survey five years ago asking U.S. adults their favorite sport are shown in the pie chart. To determine whether this distribution has changed, a research organization randomly selects 400 U.S. adults and records each adult's favorite sport. The results are shown in the table. At $\alpha = 0.10$, can you conclude that there has been a change in the claimed or expected distribution? *(Adapted from Harris Interactive)*

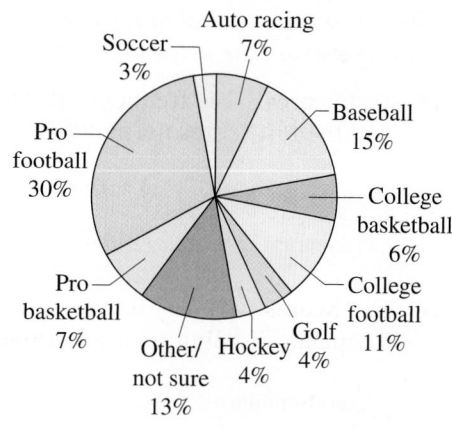

Survey results	
Sport	Frequency, f
Auto racing	36
Baseball	64
College basketball	12
College football	48
Golf	16
Hockey	16
Other/not sure	40
Pro basketball	20
Pro football	140
Soccer	8

18. Chi-Square goodness-of-fit results:

Observed: Not married

Expected: Married

See Selected Answers, page A131.

$P = 0.2172 > 0.01$, so fail to reject H_0. There is not enough evidence at the 1% level of significance to conclude that the distribution of the opinions of U.S. adults who are not married differs from the opinions of U.S. adults who are married.

19. (a) The expected frequencies are 17, 63, 79, 34, and 5.

(b) $\chi_0^2 = 13.277$;

Rejection region:
$\chi^2 > 13.277$

(c) 0.613

(d) Fail to reject H_0.

(e) There is not enough evidence at the 1% level of significance to reject the claim that the test scores are normally distributed.

20. (a) The expected frequencies are 27, 103, 156, 90, and 20.

(b) $\chi_0^2 = 9.488$;

Rejection region: $\chi^2 > 9.488$

(c) 1.029

(d) Fail to reject H_0.

(e) There is not enough evidence at the 5% level of significance to reject the claim that the test scores are normally distributed.

18. Paying Bills The pie chart shows the distribution of the opinions of U.S. adults who are married on how long they could go between jobs without any income and still be able to pay all of their bills on time. A researcher believes that the distribution of the opinions of U.S. adults who are not married is different from the distribution for U.S. adults who are married. The researcher randomly selects 250 U.S. adults who are not married and asks them how long they could go between jobs without any income and still be able to pay all of their bills on time. The results are shown in the table. At $\alpha = 0.01$, are the distributions different? *(Adapted from Country Financial)*

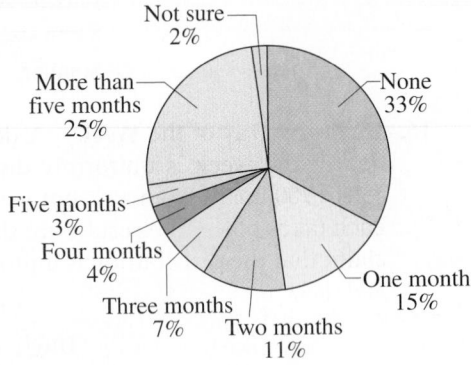

Survey results	
Response	**Frequency, f**
None	83
One month	46
Two months	37
Three months	11
Four months	9
Five months	7
More than five months	52
Not sure	5

▪ EXTENDING CONCEPTS

Testing for Normality *Using a chi-square goodness-of-fit test, you can decide, with some degree of certainty, whether a variable is normally distributed. In all chi-square tests for normality, the null and alternative hypotheses are as follows.*

> H_0: *The variable has a normal distribution.*
>
> H_a: *The variable does not have a normal distribution.*

To determine the expected frequencies when performing a chi-square test for normality, first find the mean and standard deviation of the frequency distribution. Then, use the mean and standard deviation to compute the z-score for each class boundary. Then, use the z-scores to calculate the area under the standard normal curve for each class. Multiplying the resulting class areas by the sample size yields the expected frequency for each class.

In Exercises 19 and 20, (a) find the expected frequencies, (b) find the critical value and identify the rejection region, (c) find the chi-square test statistic, (d) decide whether to reject or fail to reject the null hypothesis, and (e) interpret the decision in the context of the original claim.

19. Test Scores The frequency distribution shows the results of 200 test scores. Are the test scores normally distributed? Use $\alpha = 0.01$.

Class boundaries	49.5–58.5	58.5–67.5	67.5–76.5	76.5–85.5	85.5–94.5
Frequency, f	19	61	82	34	4

20. Test Scores At $\alpha = 0.05$, test the claim that the 400 test scores shown in the frequency distribution are normally distributed.

Class boundaries	50.5–60.5	60.5–70.5	70.5–80.5	80.5–90.5	90.5–100.5
Frequency, f	28	106	151	97	18

10.2   Independence

WHAT YOU SHOULD LEARN

▸ How to use a contingency table to find expected frequencies

▸ How to use a chi-square distribution to test whether two variables are independent

Contingency Tables ▸ The Chi-Square Test for Independence

▸ CONTINGENCY TABLES

In Section 3.2, you learned that two events are *independent* if the occurrence of one event does not affect the probability of the occurrence of the other event. For instance, the outcomes of a roll of a die and a toss of a coin are independent. But, suppose a medical researcher wants to determine if there is a relationship between caffeine consumption and heart attack risk. Are these variables independent or are they dependent? In this section, you will learn how to use the chi-square test for independence to answer such a question. To perform a chi-square test for independence, you will use sample data that are organized in a *contingency table*.

DEFINITION

An *r* × *c* **contingency table** shows the observed frequencies for two variables. The observed frequencies are arranged in *r* rows and *c* columns. The intersection of a row and a column is called a **cell.**

For instance, the following table is a 2 × 5 contingency table. It has two rows and five columns and shows the results of a random sample of 2200 adults classified by their favorite way to eat ice cream and gender. From the table, you can see that 204 of the adults who prefer ice cream in a sundae are males, and 180 of the adults who prefer ice cream in a sundae are females.

Gender	Favorite way to eat ice cream				
	Cup	Cone	Sundae	Sandwich	Other
Male	600	288	204	24	84
Female	410	340	180	20	50

(Adapted from Harris Interactive)

Assuming the two variables of study in a contingency table are independent, you can use the contingency table to find the expected frequency for each cell. The formula for calculating the expected frequency for each cell is given below.

STUDY TIP

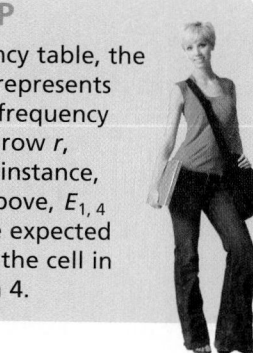

In a contingency table, the notation $E_{r,c}$ represents the expected frequency for the cell in row *r*, column *c*. For instance, in the table above, $E_{1,4}$ represents the expected frequency for the cell in row 1, column 4.

FINDING THE EXPECTED FREQUENCY FOR CONTINGENCY TABLE CELLS

The expected frequency for a cell $E_{r,c}$ in a contingency table is

$$\text{Expected frequency } E_{r,c} = \frac{(\text{Sum of row } r) \times (\text{Sum of column } c)}{\text{Sample size}}.$$

When you find the sum of each row and column in a contingency table, you are calculating the **marginal frequencies.** A marginal frequency is the frequency that an entire category of one of the variables occurs. For instance, in the table above, the marginal frequency for adults who prefer ice cream in a cone is 288 + 340 = 628. The observed frequencies in the interior of a contingency table are called **joint frequencies.** The marginal frequencies for the contingency table in Example 1 have already been calculated.

EXAMPLE 1

▶ **Finding Expected Frequencies**

Find the expected frequency for each cell in the contingency table. Assume that the variables, favorite way to eat ice cream and gender, are independent.

Gender	Favorite way to eat ice cream					Total
	Cup	Cone	Sundae	Sandwich	Other	
Male	600	288	204	24	84	1200
Female	410	340	180	20	50	1000
Total	1010	628	384	44	134	2200

▶ **Solution**

After calculating the marginal frequencies, you can use the formula

$$\text{Expected frequency } E_{r,c} = \frac{(\text{Sum of row } r) \times (\text{Sum of column } c)}{\text{Sample size}}$$

to find each expected frequency as shown.

$$E_{1,1} = \frac{1200 \cdot 1010}{2200} \approx 550.91 \qquad E_{1,2} = \frac{1200 \cdot 628}{2200} \approx 342.55$$

$$E_{1,3} = \frac{1200 \cdot 384}{2200} \approx 209.45 \qquad E_{1,4} = \frac{1200 \cdot 44}{2200} = 24$$

$$E_{1,5} = \frac{1200 \cdot 134}{2200} \approx 73.09 \qquad E_{2,1} = \frac{1000 \cdot 1010}{2200} \approx 459.09$$

$$E_{2,2} = \frac{1000 \cdot 628}{2200} \approx 285.45 \qquad E_{2,3} = \frac{1000 \cdot 384}{2200} \approx 174.55$$

$$E_{2,4} = \frac{1000 \cdot 44}{2200} = 20 \qquad E_{2,5} = \frac{1000 \cdot 134}{2200} \approx 60.91$$

INSIGHT

In Example 1, once the expected frequency for $E_{1,1}$ has been calculated to be 550.91, you can determine the expected frequency for $E_{2,1}$ to be 1010 − 550.91 = 459.09. That is, the expected frequency for the last cell in each row or column can be found by subtracting from the total.

▶ **Try It Yourself 1**

The marketing consultant for a travel agency wants to determine whether certain travel concerns are related to travel purpose. The contingency table shows the results of a random sample of 300 travelers classified by their primary travel concern and travel purpose. Assume that the variables travel concern and travel purpose are independent. Find the expected frequency for each cell. *(Adapted from NPD Group for Embassy Suites)*

Travel purpose	Travel concern			
	Hotel room	Leg room on plane	Rental car size	Other
Business	36	108	14	22
Leisure	38	54	14	14

a. Calculate the *marginal frequencies*.
b. Determine the *sample size*.
c. Use the *formula* to find the *expected frequency* for each cell.

Answer: Page A46

▶ THE CHI-SQUARE TEST FOR INDEPENDENCE

After finding the expected frequencies, you can test whether the variables are independent using a *chi-square independence test*.

DEFINITION

A **chi-square independence test** is used to test the independence of two variables. Using a chi-square test, you can determine whether the occurrence of one variable affects the probability of the occurrence of the other variable.

For the chi-square independence test to be used, the following conditions must be true.

1. The observed frequencies must be obtained using a random sample.
2. Each expected frequency must be greater than or equal to 5.

THE CHI-SQUARE INDEPENDENCE TEST

If the conditions listed above are satisfied, then the sampling distribution for the chi-square independence test is approximated by a chi-square distribution with

$$(r - 1)(c - 1)$$

degrees of freedom, where r and c are the number of rows and columns, respectively, of a contingency table. The **test statistic** for the chi-square independence test is

$$\chi^2 = \Sigma \frac{(O - E)^2}{E}$$

where O represents the observed frequencies and E represents the expected frequencies.

To begin the independence test, you must first state a null hypothesis and an alternative hypothesis. For a chi-square independence test, the null and alternative hypotheses are always some variation of the following statements.

H_0: The variables are independent.

H_a: The variables are dependent.

The expected frequencies are calculated on the assumption that the two variables are independent. If the variables are independent, then you can expect little difference between the observed frequencies and the expected frequencies. When the observed frequencies closely match the expected frequencies, the differences between O and E will be small and the chi-square test statistic will be close to 0. As such, the null hypothesis is unlikely to be rejected.

However, if the variables are dependent, there will be large discrepancies between the observed frequencies and the expected frequencies. When the differences between O and E are large, the chi-square test statistic is also large. A large chi-square test statistic is evidence for rejecting the null hypothesis. So, the chi-square independence test is always a right-tailed test.

GUIDELINES

Performing a Chi-Square Test for Independence

IN WORDS	IN SYMBOLS
1. Identify the claim. State the null and alternative hypotheses.	State H_0 and H_a.
2. Specify the level of significance.	Identify α.
3. Determine the degrees of freedom.	d.f. $= (r − 1)(c − 1)$
4. Determine the critical value.	Use Table 6 in Appendix B.
5. Determine the rejection region.	
6. Find the test statistic and sketch the sampling distribution.	$\chi^2 = \Sigma \dfrac{(O − E)^2}{E}$
7. Make a decision to reject or fail to reject the null hypothesis.	If χ^2 is in the rejection region, reject H_0. Otherwise, fail to reject H_0.
8. Interpret the decision in the context of the original claim.	

EXAMPLE 2   Report 47

▶ Performing a Chi-Square Independence Test

The contingency table shows the results of a random sample of 2200 adults classified by their favorite way to eat ice cream and gender. The expected frequencies are displayed in parentheses. At $\alpha = 0.01$, can you conclude that the adults' favorite ways to eat ice cream are related to gender?

	Favorite way to eat ice cream					
Gender	**Cup**	**Cone**	**Sundae**	**Sandwich**	**Other**	**Total**
Male	600 (550.91)	288 (342.55)	204 (209.45)	24 (24)	84 (73.09)	1200
Female	410 (459.09)	340 (285.45)	180 (174.55)	20 (20)	50 (60.91)	1000
Total	1010	628	384	44	134	2200

▶ Solution

The expected frequencies were calculated in Example 1. Because each expected frequency is at least 5 and the adults were randomly selected, you can use the chi-square independence test to test whether the variables are independent. The null and alternative hypotheses are as follows.

H_0: The adults' favorite ways to eat ice cream are independent of gender.

H_a: The adults' favorite ways to eat ice cream are dependent on gender. (Claim)

Because the contingency table has two rows and five columns, the chi-square distribution has $(r - 1)(c - 1) = (2 - 1)(5 - 1) = 4$ degrees of freedom. Because d.f. = 4 and $\alpha = 0.01$, the critical value is $\chi_0^2 = 13.277$. With the observed and expected frequencies, the chi-square test statistic is as shown.

O	E	$O - E$	$(O - E)^2$	$\dfrac{(O - E)^2}{E}$
600	550.91	49.09	2409.8281	4.3743
288	342.55	−54.55	2975.7025	8.6869
204	209.45	−5.45	29.7025	0.1418
24	24	0	0	0
84	73.09	10.91	119.0281	1.6285
410	459.09	−49.09	2409.8281	5.2491
340	285.45	54.55	2975.7025	10.4246
180	174.55	5.45	29.7025	0.1702
20	20	0	0	0
50	60.91	−10.91	119.0281	1.9542

$$\chi^2 = \Sigma \frac{(O - E)^2}{E} \approx 32.630$$

The graph at the left shows the location of the rejection region. Because $\chi^2 \approx 32.630$ is in the rejection region, you should decide to reject the null hypothesis.

Interpretation There is enough evidence at the 1% level of significance to conclude that the adults' favorite ways to eat ice cream and gender are dependent.

Rejection region

$\alpha = 0.01$

$\chi_0^2 = 13.277$

▶ Try It Yourself 2

The marketing consultant for a travel agency wants to determine whether travel concerns are related to travel purpose. The contingency table shows the results of a random sample of 300 travelers classified by their primary travel concern and travel purpose. At $\alpha = 0.01$, can the consultant conclude that the travel concerns depend on the purpose of travel? (The expected frequencies are displayed in parentheses.) *(Adapted from NPD Group for Embassy Suites)*

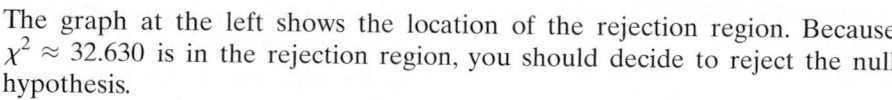

Travel purpose	Travel concern				
	Hotel room	Leg room on plane	Rental car size	Other	Total
Business	36 (44.4)	108 (97.2)	14 (16.8)	22 (21.6)	180
Leisure	38 (29.6)	54 (64.8)	14 (11.2)	14 (14.4)	120
Total	74	162	28	36	300

a. Identify the *claim* and state H_0 and H_a.
b. Specify the *level of significance* α.
c. Determine the *degrees of freedom*.
d. Determine the *critical value* and the *rejection region*.
e. Use the *observed* and *expected frequencies* to find the *chi-square test statistic*. *Sketch* a graph.
f. *Decide* whether to reject the null hypothesis.
g. *Interpret* the decision in the context of the original claim.

Answer: Page A46

EXAMPLE 3

▶ Using Technology for a Chi-Square Independence Test

A health club manager wants to determine whether the number of days per week that college students spend exercising is related to gender. A random sample of 275 college students is selected and the results are classified as shown in the table. At $\alpha = 0.05$, is there enough evidence to conclude that the number of days spent exercising per week is related to gender?

| | Days per week spent exercising | | | | |
Gender	0–1	2–3	4–5	6–7	Total
Male	40	53	26	6	125
Female	34	68	37	11	150
Total	74	121	63	17	275

▶ **Solution**   The null and alternative hypotheses can be stated as follows.

H_0: The number of days spent exercising per week is independent of gender.

H_a: The number of days spent exercising per week depends on gender. (Claim)

Using a TI-83/84 Plus, enter the observed frequencies into Matrix A and the expected frequencies into Matrix B, making sure that each expected frequency is greater than or equal to 5. To perform a chi-square independence test, begin with the STAT keystroke and choose the TESTS menu and select $C: \chi^2 - Test$. Then set up the chi-square test as shown in the top-left screen.

The other displays at the left show the results of selecting *Calculate* or *Draw*. Because d.f. = 3 and $\alpha = 0.05$, the critical value is $\chi_0^2 = 7.815$. So, the rejection region is $\chi^2 > 7.815$. The test statistic $\chi^2 \approx 3.493$ is not in the rejection region, so you should fail to reject the null hypothesis.

Interpretation   There is not enough evidence to conclude that the number of days spent exercising per week is related to gender.

▶ Try It Yourself 3

A researcher wants to determine if age is related to whether or not a tax cut would influence an adult to purchase a hybrid vehicle. A random sample of 1250 adults is selected and the results are classified as shown in the table. At $\alpha = 0.01$, is there enough evidence to conclude that the adults' ages are related to the response? *(Adapted from HNTB)*

| | Age | | | |
Response	18–34	35–54	55 and older	Total
Yes	257	189	143	589
No	218	261	182	661
Total	475	450	325	1250

a. Identify the *claim* and state H_0 and H_a.
b. Use a *technology tool* to enter the observed and expected frequencies into matrices.
c. Determine the *critical value* and the *rejection region*.
d. Use the technology tool to find the *chi-square test statistic*.
e. *Decide* whether to reject the null hypothesis. Use a graph if necessary.
f. *Interpret* the decision in the context of the original claim.

Answer: Page A46

TI-83/84 PLUS

χ^2–Test
  Observed: [A]
  Expected: [B]
  Calculate Draw

TI-83/84 PLUS

χ^2–Test
  χ^2=3.493357223
  p=.321624691
  df=3

TI-83/84 PLUS

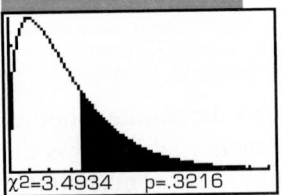

χ2=3.4934   p=.3216

STUDY TIP

You can also use a *P*-value to perform a chi-square test for independence. For instance, in Example 3, note that the TI-83/84 Plus displays $P = .321624691$. Because $P > \alpha$, you should fail to reject the null hypothesis.

10.2 EXERCISES

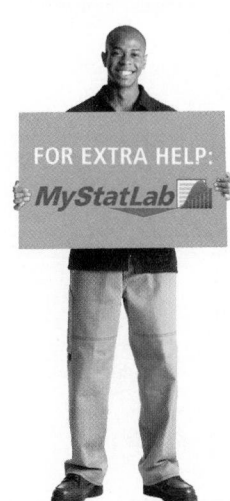

FOR EXTRA HELP:
MyStatLab

1. Find the sum of the row and the sum of the column in which the cell is located. Find the product of these sums. Divide the product by the sample size.

2. In a contingency table, a marginal frequency is the frequency that an entire category of a variable occurs, whereas a joint frequency is a frequency from a cell in the interior of a contingency table.

3. See Odd Answers, page A99.

4. A chi-square independence test is always a right-tailed test because if the variables are dependent, then the chi-square test statistic will be large, which is evidence for rejecting the null hypothesis.

5. False. If the two variables of a chi-square test for independence are dependent, then you can expect a large difference between the observed frequencies and the expected frequencies.

6. True

7. See Odd Answers, page A100.

8. See Selected Answers, page A131.

9. See Odd Answers, page A100.

10. See Selected Answers, page A131.

■ BUILDING BASIC SKILLS AND VOCABULARY

1. Explain how to find the expected frequency for a cell in a contingency table.

2. Explain the difference between marginal frequencies and joint frequencies in a contingency table.

3. Explain how the chi-square test for independence and the chi-square goodness-of-fit test are similar. How are they different?

4. Explain why the chi-square independence test is always a right-tailed test.

True or False?   *In Exercises 5 and 6, determine whether the statement is true or false. If it is false, rewrite it as a true statement.*

5. If the two variables of the chi-square test for independence are dependent, then you can expect little difference between the observed frequencies and the expected frequencies.

6. If the test statistic for the chi-square independence test is large, you will, in most cases, reject the null hypothesis.

Finding Expected Frequencies   *In Exercises 7–12, (a) calculate the marginal frequencies, and (b) find the expected frequency for each cell in the contingency table. Assume that the variables are independent.*

7.

	Athlete has	
Result	Stretched	Not stretched
Injury	18	22
No injury	211	189

8.

	Treatment	
Result	Drug	Placebo
Nausea	36	13
No nausea	254	262

9.

	Preference		
Bank employee	New procedure	Old procedure	No preference
Teller	92	351	50
Customer service representative	76	42	8

10.

	Rating		
Size of restaurant	Excellent	Fair	Poor
Seats 100 or fewer	182	203	165
Seats over 100	180	311	159

11. See Odd Answers, page A100.

12. See Selected Answers, page A131.

13. (a) H_0: Skill level in a subject is independent of location. (claim)

   H_a: Skill level in a subject is dependent on location.

   (b) d.f. = 2; $\chi_0^2 = 9.210$;

   Rejection region: $\chi^2 > 9.210$

   (c) 0.297

   (d) Fail to reject H_0. There is not enough evidence at the 1% level of significance to reject the claim that skill level in a subject is independent of location.

14. (a) H_0: Attitudes about safety are independent of the type of school.

   H_a: Attitudes about safety are dependent on the type of school. (claim)

   (b) d.f. = 1; $\chi_0^2 = 6.635$;

   Rejection region: $\chi^2 > 6.635$

   (c) 8.691

   (d) Reject H_0. There is enough evidence at the 1% level of significance to conclude that attitudes about the safety steps taken by the school staff are dependent on the type of school.

15. (a) H_0: The number of times former smokers tried to quit is independent of gender.

   H_a: The number of times former smokers tried to quit is dependent on gender. (claim)

   (b) d.f. = 2; $\chi_0^2 = 5.991$;

   Rejection region: $\chi^2 > 5.991$

   (c) 0.002

   (d) Fail to reject H_0. There is not enough evidence at the 5% level of significance to conclude that the number of times former smokers tried to quit is dependent on gender.

11.

	Type of car			
Gender	Compact	Full-size	SUV	Truck/van
Male	28	39	21	22
Female	24	32	20	14

12.

	Age				
Type of movie rented	18–24	25–34	35–44	45–64	65 and older
Comedy	38	30	24	10	8
Action	15	17	16	9	5
Drama	12	11	19	25	13

■ USING AND INTERPRETING CONCEPTS

Performing a Chi-Square Test for Independence *In Exercises 13–22, perform the indicated chi-square test for independence by doing the following.*

(a) *Identify the claim and state the null and alternative hypotheses.*

(b) *Determine the degrees of freedom, find the critical value, and identify the rejection region.*

(c) *Calculate the test statistic. If convenient, use technology.*

(d) *Decide to reject or fail to reject the null hypothesis. Then interpret the decision in the context of the original claim.*

13. Achievement and School Location Is achieving a basic skill level in a subject related to the location of the school? The results of a random sample of students by the location of school and the number of those students achieving basic skill levels in three subjects is shown in the contingency table. At $\alpha = 0.01$, test the hypothesis that the variables are independent. *(Adapted from HUD State of the Cities Report)*

	Subject		
Location of school	Reading	Math	Science
Urban	43	42	38
Suburban	63	66	65

14. Attitudes about Safety The results of a random sample of students by type of school and their attitudes on safety steps taken by the school staff are shown in the contingency table. At $\alpha = 0.01$, can you conclude that attitudes about the safety steps taken by the school staff are related to the type of school? *(Adapted from Horatio Alger Association)*

	School staff has	
Type of school	Taken all steps necessary for student safety	Taken some steps toward student safety
Public	40	51
Private	64	34

16. (a) H_0: The adults' ratings of the movie are independent of gender.

H_a: The adults' ratings of the movie are dependent on gender. (claim)

(b) d.f. = 3; $\chi_0^2 = 7.815$;
Rejection region: $\chi^2 > 7.815$

(c) 3.430

(d) Fail to reject H_0. There is not enough evidence at the 5% level of significance to conclude that the adults' ratings of the movie are dependent on gender.

17. (a) H_0: Results are independent of the type of treatment.

H_a: Results are dependent on the type of treatment. (claim)

(b) d.f. = 1; $\chi_0^2 = 2.706$;
Rejection region: $\chi^2 > 2.706$

(c) 5.106

(d) Reject H_0. There is enough evidence at the 10% level of significance to conclude that results are dependent on the type of treatment. Answers will vary.

18. (a) H_0: The result is independent of the type of treatment.

H_a: The result is dependent on the type of treatment. (claim)

(b) d.f. = 2; $\chi_0^2 = 4.605$;
Rejection region: $\chi^2 > 4.605$

(c) 14.070

(d) Reject H_0. There is enough evidence at the 10% level of significance to conclude that the result is dependent on the type of treatment.

15. Trying to Quit Smoking The contingency table shows the number of times a random sample of former smokers tried to quit smoking before they were habit-free and gender. At $\alpha = 0.05$, can you conclude that the number of times they tried to quit before they were habit-free is related to gender? *(Adapted from Porter Novelli Health Styles for the American Lung Association)*

Gender	Number of times tried to quit before habit-free		
	1	2–3	4 or more
Male	271	257	149
Female	146	139	80

16. Reviewing a Movie The contingency table shows how a random sample of adults rated a newly released movie and gender. At $\alpha = 0.05$, can you conclude that the adults' ratings are related to gender?

Gender	Rating			
	Excellent	Good	Fair	Poor
Male	97	42	26	5
Female	101	33	25	11

17. Obsessive-Compulsive Disorder The results of a random sample of patients with obsessive-compulsive disorder treated with a drug or with a placebo are shown in the contingency table. At $\alpha = 0.10$, can you conclude that the treatment is related to the result? On the basis of these results, would you recommend using the drug as part of a treatment for obsessive-compulsive disorder? *(Adapted from The Journal of the American Medical Association)*

Result	Treatment	
	Drug	Placebo
Improvement	39	25
No change	54	70

18. Musculoskeletal Injury The results of a random sample of children with pain from musculoskeletal injuries treated with acetaminophen, ibuprofen, or codeine are shown in the contingency table. At $\alpha = 0.10$, can you conclude that the treatment is related to the result? *(Adapted from American Academy of Pediatrics)*

Result	Treatment		
	Acetaminophen	Ibuprofen	Codeine
Significant improvement	58	81	61
Slight improvement	42	19	39

19. (a) H_0: Reasons are independent of the type of worker.

H_a: Reasons are dependent on the type of worker. (claim)

(b) d.f. = 2; $\chi_0^2 = 9.210$;

Rejection region: $\chi^2 > 9.210$

(c) 7.326

(d) Fail to reject H_0. There is not enough evidence at the 1% level of significance to conclude that reasons for continuing education are dependent on the type of worker. On the basis of these results, marketing strategies should not differ between technical and nontechnical audiences in regard to reasons for continuing education.

20. (a) H_0: The aspect of career development that is considered to be most important is independent of age.

H_a: The aspect of career development that is considered to be most important is dependent on age. (claim)

(b) d.f. = 4; $\chi_0^2 = 13.277$;

Rejection region: $\chi^2 > 13.277$

(c) 5.757

(d) Fail to reject H_0. There is not enough evidence at the 1% level of significance to conclude that age is related to which aspect of career development is considered to be most important.

21. (a) H_0: Type of crash is independent of the type of vehicle.

H_a: Type of crash is dependent on the type of vehicle. (claim)

(b) d.f. = 2; $\chi_0^2 = 5.991$;

Rejection region: $\chi^2 > 5.991$

(c) 144.099

(d) Reject H_0. There is enough evidence at the 5% level of significance to conclude that the type of crash is dependent on the type of vehicle.

19. **Continuing Education** You work for a college's continuing education department and want to determine whether the reasons given by workers for continuing their education are related to job type. In your study, you randomly collect the data shown in the contingency table. At $\alpha = 0.01$, can you conclude that the reason and the type of worker are dependent? How could you use this information in your marketing efforts? *(Adapted from Market Research Institute for George Mason University)*

	Reason		
Type of worker	**Professional**	**Personal**	**Professional and personal**
Technical	30	36	41
Other	47	25	30

20. **Ages and Goals** You are investigating the relationship between the ages of U.S. adults and what aspect of career development they consider to be the most important. You randomly collect the data shown in the contingency table. At $\alpha = 0.01$, is there enough evidence to conclude that age is related to which aspect of career development is considered to be most important? *(Adapted from Harris Interactive)*

	Career development aspect		
Age	**Learning new skills**	**Pay increases**	**Career path**
18–26 years	31	22	21
27–41 years	27	31	33
42–61 years	19	14	8

21. **Vehicles and Crashes** You work for an insurance company and are studying the relationship between types of crashes and the vehicles involved in passenger vehicle occupant deaths. As part of your study, you randomly select 4270 vehicle crashes and organize the resulting data as shown in the contingency table. At $\alpha = 0.05$, can you conclude that the type of crash depends on the type of vehicle? *(Adapted from Insurance Institute for Highway Safety)*

	Vehicle		
Type of crash	**Car**	**Pickup**	**Sport utility**
Single-vehicle	1237	547	479
Multiple-vehicle	1453	307	247

22. **Library Internet Access Speed** The contingency table shows a random sample of urban, suburban, and rural libraries and the speed of their Internet access. In the table, mbps represents megabits per second. At $\alpha = 0.01$, can you conclude that the metropolitan status of libraries and Internet access speed are related? *(Adapted from Center for Library and Information Innovation)*

	Metropolitan status		
Access speed	**Urban**	**Suburban**	**Rural**
1.4 mbps or less	5	20	58
1.5 mbps – 3.0 mbps	24	46	65
Greater than 3.0 mbps	37	59	64

22. (a) H_0: Metropolitan status of libraries and Internet access speed are independent.

H_a: Metropolitan status of libraries and Internet access speed are dependent. (claim)

(b) d.f. = 4; χ_0^2 = 13.277;

Rejection region: $\chi^2 > 13.277$

(c) 21.865

(d) Reject H_0. There is enough evidence at the 1% level of significance to conclude that metropolitan status of libraries and Internet access speed are related.

23. (a)–(b) See Odd Answers, page A100.

(c) Reject H_0. There is enough evidence at the 1% level of significance to conclude that the decision to borrow money is dependent on the child's expected income after graduation.

24. (a)–(b) See Selected Answers, page A132.

(c) Fail to reject H_0. There is not enough evidence at the 5% level of significance to conclude that age is related to gender in fatal alcohol-related accidents of passenger vehicle drivers with blood alcohol concentrations greater than or equal to 0.08.

SC *In Exercises 23 and 24, use StatCrunch to (a) find the marginal frequencies, (b) find the expected frequencies for each cell in the contingency table, and (c) perform the indicated chi-square test for independence.*

23. Financing and Education A financial aid officer is studying the relationship between family decisions to borrow money to finance their child's education and their child's expected income after graduation. As part of the study, 440 families are randomly selected and the resulting data are organized as shown in the contingency table. At $\alpha = 0.01$, can you conclude that the decision to borrow money is related to the child's expected income after graduation? *(Adapted from Sallie Mae, Inc.)*

Expected income	Impact on decision to borrow money			
	More likely	Less likely	Did not make a difference	Did not consider it
Less than $35,000	37	10	22	25
$35,000–$50,000	28	12	15	16
$50,000–$100,000	55	9	65	48
Greater than $100,000	36	1	29	32

24. Alcohol-Related Accidents The contingency table shows the results of a random sample of fatally injured passenger vehicle drivers (with blood alcohol concentrations greater than or equal to 0.08) by age and gender. At $\alpha = 0.05$, can you conclude that age is related to gender in such alcohol-related accidents? *(Adapted from Insurance Institute for Highway Safety)*

Gender	Age					
	16–20	21–30	31–40	41–50	51–60	61 and older
Male	45	170	90	72	45	26
Female	9	30	21	17	10	5

■ EXTENDING CONCEPTS

Homogeneity of Proportions Test *In Exercises 25–28, use the following information. Another chi-square test that involves a contingency table is the* **homogeneity of proportions test**. *This test is used to determine if several proportions are equal when samples are taken from different populations. Before the populations are sampled and the contingency table is made, the sample sizes are determined. After randomly sampling different populations, you can test whether the proportion of elements in a category is the same for each population using the same guidelines in performing a chi-square independence test. The null and alternative hypotheses are always some variation of the following statements.*

H_0: *The proportions are equal.*

H_a: *At least one of the proportions is different from the others.*

Performing a homogeneity of proportions test requires that the observed frequencies be obtained using a random sample, and each expected frequency must be greater than or equal to 5.

25. Fail to reject H_0. There is not enough evidence at the 5% level of significance to reject the claim that the proportions of motor vehicle crash deaths involving males or females are the same for each age group.

26. Reject H_0. There is enough evidence at the 10% level of significance to reject the claim that the proportions of the results for drug and placebo treatments are the same.

27. Right-tailed

28. Answers will vary. *Sample answer:* Both tests are very similar, but the chi-square test for independence tests whether the occurrence of one variable affects the probability of the occurrence of another variable, while the chi-square homogeneity of proportions test determines whether the proportions for categories from a population follow the same distribution as another population.

25. Motor Vehicle Crash Deaths The contingency table shows the results of a random sample of motor vehicle crash deaths by age and gender. At $\alpha = 0.05$, perform a homogeneity of proportions test on the claim that the proportions of motor vehicle crash deaths involving males or females are the same for each age group. *(Adapted from Insurance Institute for Highway Safety)*

	Age			
Gender	**16–24**	**25–34**	**35–44**	**45–54**
Male	123	97	82	82
Female	46	28	28	32

	Age			
Gender	**55–64**	**65–74**	**75–84**	**85 and older**
Male	56	31	26	14
Female	22	18	18	7

26. Obsessive-Compulsive Disorder The contingency table shows the results of a random sample of patients with obsessive-compulsive disorder after being treated with a drug or with a placebo. At $\alpha = 0.10$, perform a homogeneity of proportions test on the claim that the proportions of the results for drug and placebo treatments are the same. *(Adapted from The Journal of the American Medical Association)*

	Treatment	
Result	**Drug**	**Placebo**
Improvement	39	25
No change	54	70

27. Is the chi-square homogeneity of proportions test a left-tailed, right-tailed, or two-tailed test?

28. Explain how the chi-square test for independence is different from the chi-square homogeneity of proportions test.

Contingency Tables and Relative Frequencies *In Exercises 29–31, use the following information.*

The frequencies in a contingency table can be written as relative frequencies by dividing each frequency by the sample size. The contingency table below shows the number of U.S. adults (in millions) ages 25 and over by employment status and educational attainment. *(Adapted from U.S. Census Bureau)*

	Educational attainment			
Status	**Not a high school graduate**	**High school graduate**	**Some college, no degree**	**Associate's, bachelor's, or advanced degree**
Employed	10.8	35.9	22.3	56.9
Unemployed	1.2	2.2	1.0	1.4
Not in the labor force	14.3	23.1	10.5	16.7

29. See Odd Answers, page A101.

30. (a) 0.7%

    (b) 5.3%

    (c) 18.3%

    (d) 32.9%

    (e) 31.2%

31. Several of the expected frequencies are less than 5.

32. See Selected Answers, page A132.

33. 45.2%

34. 16.3%

35. See Odd Answers, page A101.

36. 22.3%

37. 4.6%

38. See Selected Answers, page A132.

39. Answers will vary. *Sample answer:* As educational attainment increases, employment increases.

---

29. Rewrite the contingency table using relative frequencies.

30. What percent of U.S. adults ages 25 and over

    (a) have a degree and are unemployed?

    (b) have some college education, but no degree, and are not in the labor force?

    (c) are employed and high school graduates?

    (d) are not in the labor force?

    (e) are high school graduates?

31. Explain why you cannot perform the chi-square independence test on these data.

Conditional Relative Frequencies *In Exercises 32–39, use the contingency table from Exercises 29–31, and the following information.*

Relative frequencies can also be calculated based on the row totals (by dividing each row entry by the row's total) or the column totals (by dividing each column entry by the column's total). These frequencies are **conditional relative frequencies** and can be used to determine if an association exists between two categories in a contingency table.

32. Calculate the conditional relative frequencies in the contingency table based on the row totals.

33. What percent of U.S. adults ages 25 and over who are employed have a degree?

34. What percent of U.S. adults ages 25 and over who are not in the labor force have some college education, but no degree?

35. Calculate the conditional relative frequencies in the contingency table based on the column totals.

36. What percent of U.S. adults ages 25 and over who have a degree are not in the labor force?

37. What percent of U.S. adults ages 25 and over who are not high school graduates are unemployed?

38. Use your results from Exercise 35 to construct a bar graph that shows the percentages of U.S. adults ages 25 and over based on employment status. Each category of employment status will have four bars, representing the four levels of educational attainment mentioned in the contingency table.

39. What conclusions can you make from the bar graph you constructed in Exercise 38?

CASE STUDY

Fast Food Survey

With the growing trend toward healthier eating, fast food chains are revising their menus. Some chains have added healthier options, such as salads, while other chains are grilling foods instead of frying them. *QSR Magazine* conducted a recent survey of 673 U.S. consumers regarding their attitudes and preferences about fast food.

One question in the survey asks:

Do you agree that, on the whole, fast food menus have gotten healthier over the past 3 years?

The pie chart shows the response to the question on a national level. The contingency table shows the results classified by gender and response.

Are Fast Food Menus Healthier?

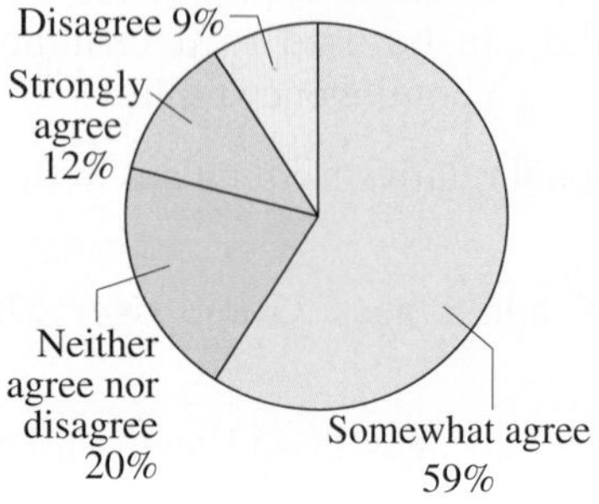

Disagree 9%
Strongly agree 12%
Neither agree nor disagree 20%
Somewhat agree 59%

	Gender	
Response	**Female**	**Male**
Somewhat agree	286	114
Neither agree nor disagree	76	58
Strongly agree	62	19
Disagree	34	24

■ EXERCISES

1. Assuming the variables gender and response are independent, did female respondents or male respondents exceed the expected number of "somewhat agree" responses?

2. Assuming the variables gender and response are independent, did female respondents or male respondents exceed the expected number of "neither agree nor disagree" responses?

3. At $\alpha = 0.01$, perform a chi-square independence test to determine whether the variables response and gender are independent. What can you conclude?

In Exercises 4 and 5, perform a chi-square goodness-of-fit test to compare the national distribution of responses with the distribution of each gender. Use the national distribution as the claimed distribution. Use $\alpha = 0.05$.

4. Compare the distribution of responses by females with the national distribution. What can you conclude?

5. Compare the distribution of responses by males with the national distribution. What can you conclude?

6. In addition to the variables used in the Case Study, what other variables do you think are important to consider when studying the distribution of U.S. consumers' attitudes about healthy fast food?

10.3 Comparing Two Variances

WHAT YOU SHOULD LEARN

▸ How to interpret the F-distribution and use an F-table to find critical values

▸ How to perform a two-sample F-test to compare two variances

The *F*-Distribution ▸ The Two-Sample *F*-Test for Variances

▸ THE *F*-DISTRIBUTION

In Chapter 8, you learned how to perform hypothesis tests to compare population means and population proportions. Recall from Section 8.2 that the *t*-test for the difference between two population means depends on whether the population variances are equal. To determine whether the population variances are equal, you can perform a two-sample *F*-test.

In this section, you will learn about the *F-distribution* and how it can be used to compare two variances.

Note to Instructor

If you prefer, you can cover this section before Section 8.2 so students can test whether variances of two populations are equal.

DEFINITION

Let s_1^2 and s_2^2 represent the sample variances of two different populations. If both populations are normal and the population variances σ_1^2 and σ_2^2 are equal, then the sampling distribution of

$$F = \frac{s_1^2}{s_2^2}$$

is called an **F-distribution.** Several properties of the *F*-distribution are as follows.

1. The *F*-distribution is a family of curves each of which is determined by two types of degrees of freedom: the degrees of freedom corresponding to the variance in the numerator, denoted by **d.f.$_N$**, and the degrees of freedom corresponding to the variance in the denominator, denoted by **d.f.$_D$**.
2. *F*-distributions are positively skewed.
3. The total area under each curve of an *F*-distribution is equal to 1.
4. *F*-values are always greater than or equal to 0.
5. For all *F*-distributions, the mean value of *F* is approximately equal to 1.

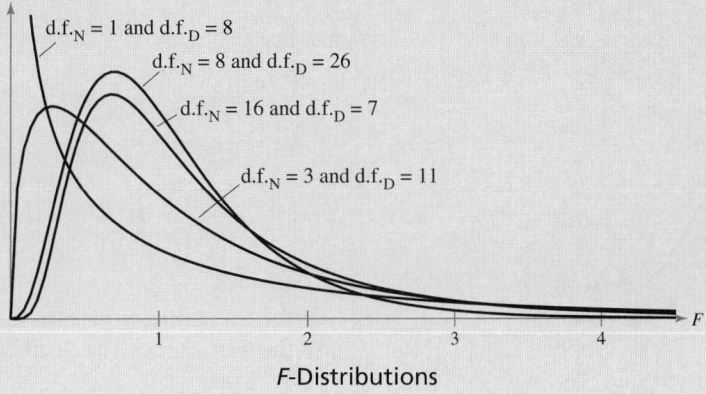

d.f.$_N$ = 1 and d.f.$_D$ = 8
d.f.$_N$ = 8 and d.f.$_D$ = 26
d.f.$_N$ = 16 and d.f.$_D$ = 7
d.f.$_N$ = 3 and d.f.$_D$ = 11

F-Distributions

Table 7 in Appendix B lists the critical values for the *F*-distribution for selected levels of significance α and degrees of freedom d.f.$_N$ and d.f.$_D$.

GUIDELINES

Finding Critical Values for the *F*-Distribution

1. Specify the level of significance α.
2. Determine the degrees of freedom for the numerator d.f.$_N$.
3. Determine the degrees of freedom for the denominator d.f.$_D$.
4. Use Table 7 in Appendix B to find the critical value. If the hypothesis test is

   a. one-tailed, use the α *F*-table.

   b. two-tailed, use the $\frac{1}{2}\alpha$ *F*-table.

EXAMPLE 1

▸ **Finding Critical *F*-Values for a Right-Tailed Test**

Find the critical *F*-value for a right-tailed test when $\alpha = 0.10$, d.f.$_N = 5$, and d.f.$_D = 28$.

▸ **Solution**

A portion of Table 7 is shown below. Using the $\alpha = 0.10$ *F*-table with d.f.$_N = 5$ and d.f.$_D = 28$, you can find the critical value, as shown by the highlighted areas in the table.

d.f.$_D$: Degrees of freedom, denominator	$\alpha = 0.10$							
	d.f.$_N$: Degrees of freedom, numerator							
	1	2	3	4	5	6	7	8
1	39.86	49.50	53.59	55.83	57.24	58.20	58.91	59.44
2	8.53	9.00	9.16	9.24	9.29	9.33	9.35	9.37
26	2.91	2.52	2.31	2.17	2.08	2.01	1.96	1.92
27	2.90	2.51	2.30	2.17	2.07	2.00	1.95	1.91
28	2.89	2.50	2.29	2.16	2.06	2.00	1.94	1.90
29	2.89	2.50	2.28	2.15	2.06	1.99	1.93	1.89
30	2.88	2.49	2.28	2.14	2.05	1.98	1.93	1.88

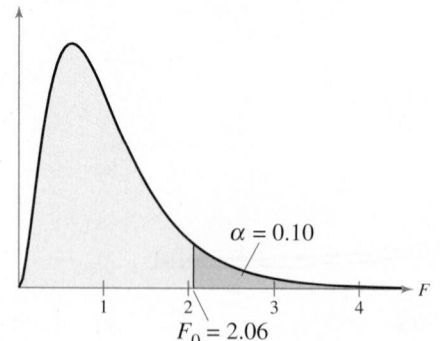

From the table, you can see that the critical value is $F_0 = 2.06$. The graph at the left shows the *F*-distribution for $\alpha = 0.10$, d.f.$_N = 5$, d.f.$_D = 28$, and $F_0 = 2.06$.

▸ **Try It Yourself 1**

Find the critical *F*-value for a right-tailed test when $\alpha = 0.05$, d.f.$_N = 8$, and d.f.$_D = 20$.

a. Specify the *level of significance* α.
b. *Use* Table 7 in Appendix B to find the critical value. *Answer: Page A46*

When performing a two-tailed hypothesis test using the *F*-distribution, you need only to find the right-tailed critical value. You must, however, remember to use the $\frac{1}{2}\alpha$ *F*-table.

EXAMPLE 2

▸ **Finding Critical *F*-Values for a Two-Tailed Test**

Find the critical *F*-value for a two-tailed test when $\alpha = 0.05$, d.f.$_N = 4$, and d.f.$_D = 8$.

▸ **Solution**

A portion of Table 7 is shown below. Using the

$$\frac{1}{2}\alpha = \frac{1}{2}(0.05) = 0.025$$

F-table with d.f.$_N = 4$, and d.f.$_D = 8$, you can find the critical value, as shown by the highlighted areas in the table.

d.f.$_D$: Degrees of freedom, denominator	$\alpha = 0.025$							
	d.f.$_N$: Degrees of freedom, numerator							
	1	2	3	4	5	6	7	8
1	647.8	799.5	864.2	899.6	921.8	937.1	948.2	956.7
2	38.51	39.00	39.17	39.25	39.30	39.33	39.36	39.37
3	17.44	16.04	15.44	15.10	14.88	14.73	14.62	14.54
4	12.22	10.65	9.98	9.60	9.36	9.20	9.07	8.98
5	10.01	8.43	7.76	7.39	7.15	6.98	6.85	6.76
6	8.81	7.26	6.60	6.23	5.99	5.82	5.70	5.60
7	8.07	6.54	5.89	5.52	5.29	5.12	4.99	4.90
8	7.57	6.06	5.42	5.05	4.82	4.65	4.53	4.43
9	7.21	5.71	5.08	4.72	4.48	4.32	4.20	4.10

From the table, the critical value is $F_0 = 5.05$. The graph shows the *F*-distribution for $\frac{1}{2}\alpha = 0.025$, d.f.$_N = 4$, d.f.$_D = 8$, and $F_0 = 5.05$.

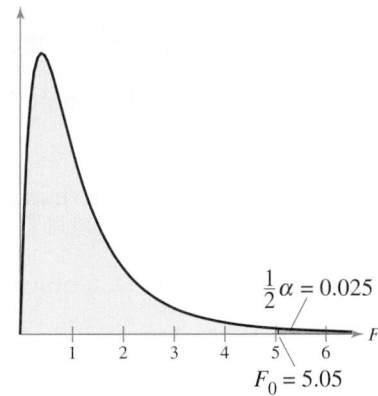

▸ **Try It Yourself 2**

Find the critical *F*-value for a two-tailed test when $\alpha = 0.01$, d.f.$_N = 2$, and d.f.$_D = 5$.

a. Specify the *level of significance* α.
b. *Use* Table 7 in Appendix B with $\frac{1}{2}\alpha$ to find the critical value.

Answer: Page A46

▸ **THE TWO-SAMPLE *F*-TEST FOR VARIANCES**

In the remainder of this section, you will learn how to perform a two-sample *F*-test for comparing two population variances using a sample from each population. Such a test has three conditions that must be met.

1. The samples must be randomly selected.

2. The samples must be independent.

3. Each population must have a normal distribution.

If these requirements are met, you can use the *F*-test to compare the population variances σ_1^2 and σ_2^2.

TWO-SAMPLE *F*-TEST FOR VARIANCES

A **two-sample *F*-test** is used to compare two population variances σ_1^2 and σ_2^2 when a sample is randomly selected from each population. The populations must be independent and normally distributed. The **test statistic** is

$$F = \frac{s_1^2}{s_2^2}$$

where s_1^2 and s_2^2 represent the sample variances with $s_1^2 \geq s_2^2$. The numerator has d.f.$_N$ $= n_1 - 1$ degrees of freedom and the denominator has d.f.$_D$ $= n_2 - 1$ degrees of freedom, where n_1 is the size of the sample having variance s_1^2 and n_2 is the size of the sample having variance s_2^2.

GUIDELINES

Using a Two-Sample *F*-Test to Compare σ_1^2 and σ_2^2

IN WORDS	IN SYMBOLS
1. Identify the claim. State the null and alternative hypotheses.	State H_0 and H_a.
2. Specify the level of significance.	Identify α.
3. Determine the degrees of freedom.	d.f.$_N$ $= n_1 - 1$ d.f.$_D$ $= n_2 - 1$
4. Determine the critical value.	Use Table 7 in Appendix B.
5. Determine the rejection region.	
6. Find the test statistic and sketch the sampling distribution.	$F = \dfrac{s_1^2}{s_2^2}$
7. Make a decision to reject or fail to reject the null hypothesis.	If F is in the rejection region, reject H_0. Otherwise, fail to reject H_0.
8. Interpret the decision in the context of the original claim.	

Miami	Fort Lauderdale
139.0	85.5
138.8	80.9
135.5	91.2
150.9	75.5
155.0	78.0
154.7	69.9
149.9	70.5
150.5	73.6
134.5	105.9
125.0	70.0

Assuming each population of selling prices is normally distributed, is it possible to use a two-sample F-test to compare the population variances?

Yes, because the samples are randomly selected and independent, and the population is normally distributed.

Normal solution	Treated solution
$n = 25$	$n = 20$
$s^2 = 180$	$s^2 = 56$

EXAMPLE 3

▸ Performing a Two-Sample *F*-Test

A restaurant manager is designing a system that is intended to decrease the variance of the time customers wait before their meals are served. Under the old system, a random sample of 10 customers had a variance of 400. Under the new system, a random sample of 21 customers had a variance of 256. At $\alpha = 0.10$, is there enough evidence to convince the manager to switch to the new system? Assume both populations are normally distributed.

▸ **Solution**   Because $400 > 256$, $s_1^2 = 400$, and $s_2^2 = 256$. Therefore, s_1^2 and σ_1^2 represent the sample and population variances for the old system, respectively. With the claim "the variance of the waiting times under the new system is less than the variance of the waiting times under the old system," the null and alternative hypotheses are

$$H_0: \sigma_1^2 \leq \sigma_2^2 \quad \text{and} \quad H_a: \sigma_1^2 > \sigma_2^2. \text{ (Claim)}$$

Because the test is a right-tailed test with $\alpha = 0.10$, $\text{d.f.}_N = n_1 - 1 = 10 - 1 = 9$, and $\text{d.f.}_D = n_2 - 1 = 21 - 1 = 20$, the critical value is $F_0 = 1.96$. So, the rejection region is $F > 1.96$. With the F-test, the test statistic is

$$F = \frac{s_1^2}{s_2^2} = \frac{400}{256} \approx 1.56.$$

The graph shows the location of the rejection region and the test statistic. Because F is not in the rejection region, you should fail to reject the null hypothesis.

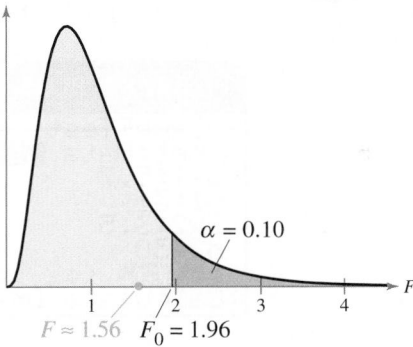

Interpretation   There is not enough evidence at the 10% level of significance to convince the manager to switch to the new system.

▸ Try It Yourself 3

A medical researcher claims that a specially treated intravenous solution decreases the variance of the time required for nutrients to enter the bloodstream. Independent samples from each type of solution are randomly selected, and the results are shown in the table at the left. At $\alpha = 0.01$, is there enough evidence to support the researcher's claim? Assume the populations are normally distributed.

a. Identify the *claim* and state H_0 and H_a.
b. Specify the *level of significance* α.
c. Determine the *degrees of freedom* for the numerator and for the denominator.
d. Determine the *critical value* and the *rejection region*.
e. Use the *F-test* to find the *test statistic F. Sketch* a graph.
f. *Decide* whether to reject the null hypothesis.
g. *Interpret* the decision in the context of the original claim.

Answer: Page A46

Stock A	Stock B
$n_2 = 30$	$n_1 = 31$
$s_2 = 3.5$	$s_1 = 5.7$

EXAMPLE 4   SC Report 48

▶ Using Technology for a Two-Sample *F*-Test

You want to purchase stock in a company and are deciding between two different stocks. Because a stock's risk can be associated with the standard deviation of its daily closing prices, you randomly select samples of the daily closing prices for each stock to obtain the results shown at the left. At $\alpha = 0.05$, can you conclude that one of the two stocks is a riskier investment? Assume the stock closing prices are normally distributed.

▶ Solution

Because $5.7^2 > 3.5^2$, $s_1^2 = 5.7^2$, and $s_2^2 = 3.5^2$. Therefore, s_1^2 and σ_1^2 represent the sample and population variances for Stock B, respectively. With the claim "one of the two stocks is a riskier investment," the null and alternative hypotheses are

$$H_0: \sigma_1^2 = \sigma_2^2 \quad \text{and} \quad H_a: \sigma_1^2 \neq \sigma_2^2. \text{ (Claim)}$$

Because the test is a two-tailed test with $\frac{1}{2}\alpha = \frac{1}{2}(0.05) = 0.025$, d.f.$_N$ = $n_1 - 1 = 31 - 1 = 30$, and d.f.$_D$ = $n_2 - 1 = 30 - 1 = 29$, the critical value is $F_0 = 2.09$. So, the rejection region is $F > 2.09$.

To perform a two-sample *F*-test using a TI-83/84 Plus, begin with the STAT keystroke. Choose the TESTS menu and select *D:2–SampFTest*. Then set up the two-sample *F*-test as shown in the first screen below. Because you are entering the descriptive statistics, select the *Stats* input option. When entering the original data, select the *Data* input option. The other displays below show the results of selecting *Calculate* or *Draw*.

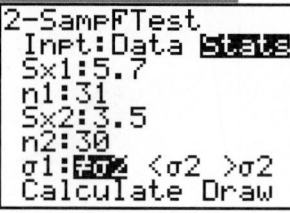

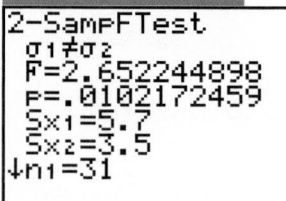

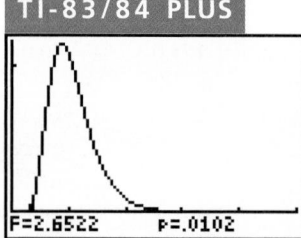

The test statistic $F \approx 2.652$ is in the rejection region, so you should reject the null hypothesis.

Interpretation There is enough evidence at the 5% level of significance to support the claim that one of the two stocks is a riskier investment.

▶ Try It Yourself 4

Location A	Location B
$n = 16$	$n = 22$
$s = 0.95$	$s = 0.78$

A biologist claims that the pH levels of the soil in two geographic locations have equal standard deviations. Independent samples from each location are randomly selected, and the results are shown at the left. At $\alpha = 0.01$, is there enough evidence to reject the biologist's claim? Assume the pH levels are normally distributed.

a. Identify the *claim* and state H_0 and H_a.
b. Specify the *level of significance* α.
c. Determine the *degrees of freedom* for the numerator and for the denominator.
d. Determine the *critical value* and the *rejection region*.
e. Use a technology tool to find the *test statistic F*.
f. *Decide* whether to reject the null hypothesis.
g. *Interpret* the decision in the context of the original claim.

Answer: Page A46

10.3 EXERCISES

BUILDING BASIC SKILLS AND VOCABULARY

1. Explain how to find the critical value for an F-test.
2. List five properties of the F-distribution.
3. List the three conditions that must be met in order to use a two-sample F-test.
4. Explain how to determine the values of d.f.$_N$ and d.f.$_D$ when performing a two-sample F-test.

In Exercises 5–8, find the critical F-value for a right-tailed test using the indicated level of significance α and degrees of freedom d.f.$_N$ and d.f.$_D$.

5. $\alpha = 0.05$, d.f.$_N = 9$, d.f.$_D = 16$
6. $\alpha = 0.01$, d.f.$_N = 2$, d.f.$_D = 11$
7. $\alpha = 0.10$, d.f.$_N = 10$, d.f.$_D = 15$
8. $\alpha = 0.025$, d.f.$_N = 7$, d.f.$_D = 3$

In Exercises 9–12, find the critical F-value for a two-tailed test using the indicated level of significance α and degrees of freedom d.f.$_N$ and d.f.$_D$.

9. $\alpha = 0.01$, d.f.$_N = 6$, d.f.$_D = 7$
10. $\alpha = 0.10$, d.f.$_N = 24$, d.f.$_D = 28$
11. $\alpha = 0.05$, d.f.$_N = 60$, d.f.$_D = 40$
12. $\alpha = 0.05$, d.f.$_N = 27$, d.f.$_D = 19$

In Exercises 13–18, test the claim about the difference between two population variances σ_1^2 and σ_2^2 at the given level of significance α using the given sample statistics. Assume the sample statistics are from independent samples that are randomly selected and each population has a normal distribution.

13. Claim: $\sigma_1^2 > \sigma_2^2$; $\alpha = 0.10$.
    Sample statistics: $s_1^2 = 773$, $n_1 = 5$; $s_2^2 = 765$, $n_2 = 6$

14. Claim: $\sigma_1^2 = \sigma_2^2$; $\alpha = 0.05$.
    Sample statistics: $s_1^2 = 310$, $n_1 = 7$; $s_2^2 = 297$, $n_2 = 8$

15. Claim: $\sigma_1^2 \le \sigma_2^2$; $\alpha = 0.01$.
    Sample statistics: $s_1^2 = 842$, $n_1 = 11$; $s_2^2 = 836$, $n_2 = 10$

16. Claim: $\sigma_1^2 \ne \sigma_2^2$; $\alpha = 0.05$.
    Sample statistics: $s_1^2 = 245$, $n_1 = 31$; $s_2^2 = 112$, $n_2 = 28$

17. Claim: $\sigma_1^2 = \sigma_2^2$; $\alpha = 0.01$.
    Sample statistics: $s_1^2 = 9.8$, $n_1 = 13$; $s_2^2 = 2.5$, $n_2 = 20$

18. Claim: $\sigma_1^2 > \sigma_2^2$; $\alpha = 0.05$.
    Sample statistics: $s_1^2 = 44.6$, $n_1 = 16$; $s_2^2 = 39.3$, $n_2 = 12$

USING AND INTERPRETING CONCEPTS

Comparing Two Variances *In Exercises 19–26, (a) identify the claim and state H_0 and H_a, (b) determine the critical value and the rejection region, (c) find the test statistic F, (d) decide whether to reject or fail to reject the null hypothesis, and (e) interpret the decision in the context of the original claim. If convenient, use technology to solve the problem. In each exercise, assume the samples are independent and each population has a normal distribution.*

19. **Life of Appliances** Company A claims that the variance of the life of its appliances is less than the variance of the life of Company B's appliances. A random sample of the lives of 20 of Company A's appliances has a variance of 2.6. A random sample of the lives of 25 of Company B's appliances has a variance of 2.8. At $\alpha = 0.05$, can you support Company A's claim?

Answers (side column)

1. Specify the level of significance α. Determine the degrees of freedom for the numerator and denominator. Use Table 7 in Appendix B to find the critical value F.

2. See Selected Answers, page A132.

3. (1) The samples must be randomly selected, (2) the samples must be independent, and (3) each population must have a normal distribution.

4. See Selected Answers, page A132.

5. 2.54

6. 7.21

7. 2.06

8. 14.62

9. 9.16

10. 1.91

11. 1.80

12. 2.42

13. Fail to reject H_0. There is not enough evidence at the 10% level of significance to support the claim.

14. See Selected Answers, page A132.

15. Fail to reject H_0. There is not enough evidence at the 1% level of significance to reject the claim.

16. See Selected Answers, page A132.

17. Reject H_0. There is enough evidence at the 1% level of significance to reject the claim.

18. See Selected Answers, page A132.

19. See Odd Answers, page A101.

Company A		Company B		
250	350	450	400	350
650	550	250	350	190
285		550		

TABLE FOR EXERCISE 21

Brand A			Brand B		
150	170	140	270	250	300
160	140	130	250	290	280
160	160	130	240	310	240
180			330		

TABLE FOR EXERCISE 22

20. See Selected Answers, page A132.

21. (a) H_0: $\sigma_1^2 = \sigma_2^2$;
   H_a: $\sigma_1^2 \neq \sigma_2^2$ (claim)

(b) $F_0 = 6.23$;
   Rejection region: $F > 6.23$

(c) 2.10

(d) Fail to reject H_0.

(e) There is not enough evidence at the 5% level of significance to conclude that the variances of the prices differ between the two companies.

22. See Selected Answers, page A132.

23. (a) H_0: $\sigma_1^2 = \sigma_2^2$ (claim);
   H_a: $\sigma_1^2 \neq \sigma_2^2$

(b) $F_0 = 2.635$;
   Rejection region: $F > 2.635$

(c) 1.282

(d) Fail to reject H_0.

(e) There is not enough evidence at the 10% level of significance to reject the administrator's claim that the standard deviations of science assessment test scores for eighth grade students are the same in Districts 1 and 2.

24. (a) H_0: $\sigma_1^2 = \sigma_2^2$ (claim);
   H_a: $\sigma_1^2 \neq \sigma_2^2$

(b) $F_0 = 5.20$;
   Rejection region: $F > 5.20$

(c) 1.26

(d) Fail to reject H_0.

20. Fuel Consumption An automobile manufacturer claims that the variance of the fuel consumption for its hybrid vehicles is less than the variance of the fuel consumption for the hybrid vehicles of a top competitor. A random sample of the fuel consumption of 19 of the manufacturer's hybrids has a variance of 0.24. A random sample of the fuel consumption of 21 of its competitor's hybrids has a variance of 0.77. At $\alpha = 0.01$, can you support the manufacturer's claim? *(Adapted from GreenHybrid)*

21. Home Theater Prices The table shows the prices (in dollars) for a random sample of home theater systems. At $\alpha = 0.05$, can you conclude that the variances of the prices differ between the two companies? *(Adapted from Best Buy)*

22. Ice Cream and Calories The table shows the numbers of calories in a serving for a random sample of ice cream flavors for two brands. At $\alpha = 0.10$, can you conclude that the variances of the numbers of calories differ between the two brands? *(Source: Perry's Ice Cream and Ben & Jerry's Homemade, Inc.)*

23. Science Assessment Tests In a recent interview, a state school administrator stated that the standard deviations of science assessment test scores for eighth grade students are the same in Districts 1 and 2. A random sample of 12 test scores from District 1 has a standard deviation of 36.8 points, and a random sample of 14 test scores from District 2 has a standard deviation of 32.5 points. At $\alpha = 0.10$, can you reject the administrator's claim? *(Adapted from National Center for Educational Statistics)*

24. U.S. History Assessment Tests A school administrator reports that the standard deviations of U.S. history assessment test scores for eighth grade students are the same in Districts 1 and 2. As proof, the administrator gives the results of a study of test scores in each district. The study shows that a random sample of 10 test scores from District 1 has a standard deviation of 33.9 points, and a random sample of 13 test scores from District 2 has a standard deviation of 30.2 points. At $\alpha = 0.01$, can you reject the administrator's claim? *(Adapted from National Center for Educational Statistics)*

25. Annual Salaries The annual salaries for a random sample of 16 actuaries working in New York have a standard deviation of $14,900. The annual salaries for a random sample of 17 actuaries working in California have a standard deviation of $9600. At $\alpha = 0.05$, can you conclude that the standard deviation of the annual salaries for actuaries is greater in New York than in California? *(Adapted from America's Career InfoNet)*

26. Annual Salaries An employment information service claims the standard deviation of the annual salaries for public relations managers is greater in Florida than in Louisiana. The annual salaries for a random sample of 28 public relations managers in Florida have a standard deviation of $10,100. The annual salaries for a random sample of 24 public relations managers in Louisiana have a standard deviation of $6400. At $\alpha = 0.05$, can you support the service's claim? *(Adapted from America's Career InfoNet)*

SC *In Exercises 27–30, use StatCrunch to test the claim about the difference between two population variances σ_1^2 and σ_2^2 at the given level of significance α using the given sample statistics. Assume the sample statistics are from independent samples that are randomly selected and each population has a normal distribution.*

27. Claim: $\sigma_1^2 = \sigma_2^2$; $\alpha = 0.10$.
Sample statistics: $s_1^2 = 156.25$,
$n_1 = 15$; $s_2^2 = 295.84$, $n_2 = 18$

28. Claim: $\sigma_1^2 \neq \sigma_2^2$; $\alpha = 0.05$.
Sample statistics: $s_1^2 = 31.36$,
$n_1 = 24$; $s_2^2 = 11.56$, $n_2 = 20$

(e) There is not enough evidence at the 1% level of significance to reject the administrator's claim that the standard deviations of U.S. history assessment test scores for eighth grade students are the same in Districts 1 and 2.

25. (a) H_0: $\sigma_1^2 \le \sigma_2^2$;
H_a: $\sigma_1^2 > \sigma_2^2$ (claim)

(b) $F_0 = 2.35$;
Rejection region: $F > 2.35$

(c) 2.41

(d) Reject H_0.

(e) There is enough evidence at the 5% level of significance to conclude that the standard deviation of the annual salaries for actuaries is greater in New York than in California.

26. See Selected Answers, page A132.

27. Hypothesis test results:

σ_1^2 : variance of population 1
σ_2^2 : variance of population 2
σ_1^2/σ_2^2 : variance ratio
H_0 : $\sigma_1^2/\sigma_2^2 = 1$
H_A : $\sigma_1^2/\sigma_2^2 \ne 1$

See Odd Answers, page A101.

$P = 0.2333 > 0.10$, so fail to reject H_0. There is not enough evidence at the 10% level of significance to reject the claim.

28. See Selected Answers, page A132.

29. Hypothesis test results:

σ_1^2 : variance of population 1
σ_2^2 : variance of population 2
σ_1^2/σ_2^2 : variance ratio
H_0 : $\sigma_1^2/\sigma_2^2 = 1$
H_A : $\sigma_1^2/\sigma_2^2 > 1$

See Odd Answers, page A101.

$P = 0.0293 < 0.05$, so reject H_0. There is enough evidence at the 5% level of significance to reject the claim.

30. See Selected Answers, page A133.

31. Right-tailed: 14.73
Left-tailed: 0.15

32. Right-tailed: 2.33
Left-tailed: 0.45

33. (0.375, 3.774)

34. (0.485, 4.880)

29. Claim: $\sigma_1^2 \le \sigma_2^2$; $\alpha = 0.05$.
Sample statistics: $s_1^2 = 416.16$, $n_1 = 22$; $s_2^2 = 193.21$, $n_2 = 29$

30. Claim: $\sigma_1^2 > \sigma_2^2$; $\alpha = 0.01$.
Sample statistics: $s_1^2 = 828$, $n_1 = 7$; $s_2^2 = 697$, $n_2 = 13$

■ EXTENDING CONCEPTS

Finding Left-Tailed Critical F-Values   *In this section you learned that if s_1^2 is larger than s_2^2, then you only need to calculate the right-tailed critical F-value for a two-tailed test. For other applications of the F-distribution, you will need to calculate the left-tailed critical F-value. To calculate the left-tailed critical F-value, do the following.*

(1) Interchange the values for d.f.$_N$ and d.f.$_D$.

(2) Find the corresponding F-value in Table 7.

(3) Calculate the reciprocal of the F-value to obtain the left-tailed critical F-value.

In Exercises 31 and 32, find the right- and left-tailed critical F-values for a two-tailed test with the given values of α, d.f.$_N$, and d.f.$_D$.

31. $\alpha = 0.05$, d.f.$_N = 6$, d.f.$_D = 3$ **32.** $\alpha = 0.10$, d.f.$_N = 20$, d.f.$_D = 15$

Confidence Interval for σ_1^2/σ_2^2   *When s_1^2 and s_2^2 are the variances of randomly selected, independent samples from normally distributed populations, then a confidence interval for σ_1^2/σ_2^2 is*

$$\frac{s_1^2}{s_2^2}F_L < \frac{\sigma_1^2}{\sigma_2^2} < \frac{s_1^2}{s_2^2}F_R$$

where F_L is the left-tailed critical F-value and F_R is the right-tailed critical F-value.

In Exercises 33 and 34, construct the indicated confidence interval for σ_1^2/σ_2^2. Assume the samples are independent and each population has a normal distribution.

33. Cholesterol Contents   In a recent study of the cholesterol contents of grilled chicken sandwiches served at fast food restaurants, a nutritionist found that random samples of sandwiches from Burger King and from McDonald's had the

Cholesterol contents of grilled chicken sandwiches		
Restaurant	Burger King	McDonald's
Sample variance	$s_1^2 = 10.89$	$s_2^2 = 9.61$
Sample size	$n_1 = 16$	$n_2 = 12$

sample statistics shown in the table. Construct a 95% confidence interval for σ_1^2/σ_2^2, where σ_1^2 and σ_2^2 are the variances of the cholesterol contents of grilled chicken sandwiches from Burger King and McDonald's, respectively. *(Adapted from Burger King Brands, Inc. and McDonald's Corporation)*

34. Carbohydrate Contents   A fast food study found that the carbohydrate contents of 16 randomly selected grilled chicken sandwiches from Burger King had a variance of 5.29. The study also found that the carbohydrate contents of 12 randomly selected grilled chicken sandwiches from McDonald's had a variance of 3.61. Construct a 95% confidence interval for σ_1^2/σ_2^2, where σ_1^2 and σ_2^2 are the variances of the carbohydrate contents of grilled chicken sandwiches from Burger King and McDonald's, respectively. *(Adapted from Burger King Brands, Inc. and McDonald's Corporation)*

10.4 Analysis of Variance

One-Way ANOVA ▸ Two-Way ANOVA

▸ ONE-WAY ANOVA

Suppose a medical researcher is analyzing the effectiveness of three types of pain relievers and wants to determine whether there is a difference in the mean lengths of time it takes the three medications to provide relief. To determine whether such a difference exists, the researcher can use the *F*-distribution together with a technique called *analysis of variance*. Because one independent variable is being studied, the process is called *one-way analysis of variance*.

> **DEFINITION**
>
> **One-way analysis of variance** is a hypothesis-testing technique that is used to compare the means of three or more populations. Analysis of variance is usually abbreviated as **ANOVA.**

To begin a one-way analysis of variance test, you should first state the null and alternative hypotheses. For a one-way ANOVA test, the null and alternative hypotheses are always similar to the following statements.

H_0: $\mu_1 = \mu_2 = \mu_3 = \cdots = \mu_k$ (All population means are equal.)

H_a: At least one mean is different from the others.

When you reject the null hypothesis in an ANOVA test, you can conclude that at least one of the means is different from the others. Without performing more statistical tests, however, you cannot determine which of the means is different.

In a one-way ANOVA test, the following conditions must be true.

1. Each sample must be randomly selected from a normal, or approximately normal, population.
2. The samples must be independent of each other.
3. Each population must have the same variance.

The test statistic for a one-way ANOVA test is the ratio of two variances: the variance between samples and the variance within samples.

$$\text{Test statistic} = \frac{\text{Variance between samples}}{\text{Variance within samples}}$$

1. The variance between samples MS_B measures the differences related to the treatment given to each sample and is sometimes called the **mean square between.**
2. The variance within samples MS_W measures the differences related to entries within the same sample. This variance, sometimes called the **mean square within,** is usually due to sampling error.

ONE-WAY ANALYSIS OF VARIANCE TEST

If the conditions for a one-way analysis of variance test are satisfied, then the sampling distribution for the test is approximated by the F-distribution. The **test statistic** is

$$F = \frac{MS_B}{MS_W}.$$

The degrees of freedom for the F-test are

$$\text{d.f.}_N = k - 1$$

and

$$\text{d.f.}_D = N - k$$

where k is the number of samples and N is the sum of the sample sizes.

If there is little or no difference between the means, then MS_B will be approximately equal to MS_W and the test statistic will be approximately 1. Values of F close to 1 suggest that you should fail to reject the null hypothesis. However, if one of the means differs significantly from the others, MS_B will be greater than MS_W and the test statistic will be greater than 1. Values of F significantly greater than 1 suggest that you should reject the null hypothesis. As such, all one-way ANOVA tests are right-tailed tests. That is, if the test statistic is greater than the critical value, H_0 will be rejected.

STUDY TIP

The notations n_i, $\overline{x}_i$, and s_i^2 represent the sample size, mean, and variance of the ith sample, respectively. $\overline{\overline{x}}$ is sometimes called the **grand mean**.

GUIDELINES

Finding the Test Statistic for a One-Way ANOVA Test

IN WORDS	IN SYMBOLS
1. Find the mean and variance of each sample.	$\overline{x} = \dfrac{\Sigma x}{n}$, $s^2 = \dfrac{\Sigma(x - \overline{x})^2}{n - 1}$
2. Find the mean of all entries in all samples (the grand mean).	$\overline{\overline{x}} = \dfrac{\Sigma x}{N}$
3. Find the sum of squares between the samples.	$SS_B = \Sigma n_i(\overline{x}_i - \overline{\overline{x}})^2$
4. Find the sum of squares within the samples.	$SS_W = \Sigma(n_i - 1)s_i^2$
5. Find the variance between the samples.	$MS_B = \dfrac{SS_B}{\text{d.f.}_N} = \dfrac{\Sigma n_i(\overline{x}_i - \overline{\overline{x}})^2}{k - 1}$
6. Find the variance within the samples.	$MS_W = \dfrac{SS_W}{\text{d.f.}_D} = \dfrac{\Sigma(n_i - 1)s_i^2}{N - k}$
7. Find the test statistic.	$F = \dfrac{MS_B}{MS_W}$

Note that in Step 1, you are summing the values from just one sample. In Step 2, you are summing the values from all of the samples.

The notation SS_B represents the sum of squares between the samples.

$$SS_B = n_1(\overline{x}_1 - \overline{\overline{x}})^2 + n_2(\overline{x}_2 - \overline{\overline{x}})^2 + \cdots + n_k(\overline{x}_k - \overline{\overline{x}})^2$$

$$= \Sigma n_i(\overline{x}_i - \overline{\overline{x}})^2$$

The notation SS_W represents the sum of squares within the samples.

$$SS_W = (n_1 - 1)s_1^2 + (n_2 - 1)s_2^2 + \cdots + (n_k - 1)s_k^2$$

$$= \Sigma(n_i - 1)s_i^2$$

GUIDELINES

Performing a One-Way Analysis of Variance Test

IN WORDS	IN SYMBOLS
1. Identify the claim. State the null and alternative hypotheses.	State H_0 and H_a.
2. Specify the level of significance.	Identify α.
3. Determine the degrees of freedom.	$\text{d.f.}_N = k - 1$ $\text{d.f.}_D = N - k$
4. Determine the critical value.	Use Table 7 in Appendix B.
5. Determine the rejection region.	
6. Find the test statistic and sketch the sampling distribution.	$F = \dfrac{MS_B}{MS_W}$
7. Make a decision to reject or fail to reject the null hypothesis.	If F is in the rejection region, reject H_0. Otherwise, fail to reject H_0.
8. Interpret the decision in the context of the original claim.	

Tables are a convenient way to summarize the results of a one-way analysis of variance test. ANOVA summary tables are set up as shown below.

ANOVA Summary Table

Variation	Sum of squares	Degrees of freedom	Mean squares	*F*
Between	SS_B	d.f._N	$MS_B = \dfrac{SS_B}{\text{d.f.}_N}$	$\dfrac{MS_B}{MS_W}$
Within	SS_W	d.f._D	$MS_W = \dfrac{SS_W}{\text{d.f.}_D}$	

EXAMPLE 1

▶ Performing a One-Way ANOVA Test

A medical researcher wants to determine whether there is a difference in the mean lengths of time it takes three types of pain relievers to provide relief from headache pain. Several headache sufferers are randomly selected and given one of the three medications. Each headache sufferer records the time (in minutes) it takes the medication to begin working. The results are shown in the table. At $\alpha = 0.01$, can you conclude that at least one mean time is different from the others? Assume that each population of relief times is normally distributed and that the population variances are equal.

Medication 1	Medication 2	Medication 3
12	16	14
15	14	17
17	21	20
12	15	15
	19	
$n_1 = 4$	$n_2 = 5$	$n_3 = 4$
$\bar{x}_1 = \frac{56}{4} = 14$	$\bar{x}_2 = \frac{85}{5} = 17$	$\bar{x}_3 = \frac{66}{4} = 16.5$
$s_1^2 = 6$	$s_2^2 = 8.5$	$s_3^2 = 7$

▶ Solution
The null and alternative hypotheses are as follows.

H_0: $\mu_1 = \mu_2 = \mu_3$

H_a: At least one mean is different from the others. (Claim)

Because there are $k = 3$ samples, d.f.$_N$ = $k - 1 = 3 - 1 = 2$. The sum of the sample sizes is $N = n_1 + n_2 + n_3 = 4 + 5 + 4 = 13$. So, d.f.$_D$ = $N - k = 13 - 3 = 10$. Using d.f.$_N$ = 2, d.f.$_D$ = 10, and $\alpha = 0.01$, the critical value is $F_0 = 7.56$. To find the test statistic, first calculate $\bar{\bar{x}}$, MS_B, and MS_W.

$$\bar{\bar{x}} = \frac{\Sigma x}{N} = \frac{56 + 85 + 66}{13} \approx 15.92$$

$$MS_B = \frac{SS_B}{\text{d.f.}_N} = \frac{\Sigma n_i (\bar{x}_i - \bar{\bar{x}})^2}{k - 1}$$

$$\approx \frac{4(14 - 15.92)^2 + 5(17 - 15.92)^2 + 4(16.5 - 15.92)^2}{3 - 1}$$

$$= \frac{21.9232}{2} = 10.9616$$

$$MS_W = \frac{SS_W}{\text{d.f.}_D} = \frac{\Sigma (n_i - 1)s_i^2}{N - k}$$

$$= \frac{(4 - 1)(6) + (5 - 1)(8.5) + (4 - 1)(7)}{13 - 3} = \frac{73}{10} = 7.3$$

Using $MS_B \approx 10.9616$ and $MS_W = 7.3$, the test statistic is

$$F = \frac{MS_B}{MS_W} \approx \frac{10.9616}{7.3} \approx 1.50.$$

The graph shows the location of the rejection region and the test statistic. Because F is not in the rejection region, you should fail to reject the null hypothesis.

Interpretation There is not enough evidence at the 1% level of significance to conclude that there is a difference in the mean length of time it takes the three pain relievers to provide relief from headache pain.

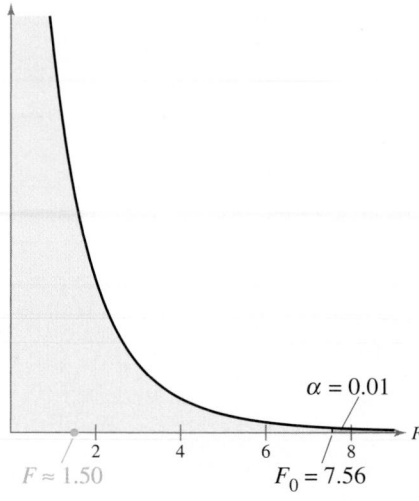

The ANOVA summary table for Example 1 is shown below.

Variation	Sum of squares	Degrees of freedom	Mean squares	*F*
Between	21.9232	2	10.9616	1.50
Within	73	10	7.3	

▸ Try It Yourself 1

A sales analyst wants to determine whether there is a difference in the mean monthly sales of a company's four sales regions. Several salespersons from each region are randomly selected and they provide their sales amounts (in thousands of dollars) for the previous month. The results are shown in the table. At $\alpha = 0.05$, can the analyst conclude that there is a difference in the mean monthly sales among the sales regions? Assume that each population of sales is normally distributed and that the population variances are equal.

North	East	South	West
34	47	40	21
28	36	30	30
18	30	41	24
24	38	29	37
	44		23
$n_1 = 4$	$n_2 = 5$	$n_3 = 4$	$n_4 = 5$
$\bar{x}_1 = 26$	$\bar{x}_2 = 39$	$\bar{x}_3 = 35$	$\bar{x}_4 = 27$
$s_1^2 \approx 45.33$	$s_2^2 = 45$	$s_3^2 \approx 40.67$	$s_4^2 = 42.5$

a. Identify the *claim* and state H_0 and H_a.
b. Specify the *level of significance* α.
c. Determine the *degrees of freedom* for the *numerator* and for the *denominator*.
d. Determine the *critical value* and the *rejection region*.
e. Find the *test statistic F. Sketch* a graph.
f. *Decide* whether to reject the null hypothesis.
g. *Interpret* the decision in the context of the original claim.

Answer: Page A46

Using technology greatly simplifies the one-way ANOVA process. When using a technology tool such as Excel, MINITAB, or the TI-83/84 Plus to perform a one-way analysis of variance test, you can use P-values to decide whether to reject the null hypothesis. If the P-value is less than α, you should reject H_0.

Actor	Athlete	Musician
55	110	110
55	45	87
45	45	70
40	40	65
28	33	52
27	30	35
25	30	35
25	22	30
17	15	25
15	10	18
12		14
10		
8		
7		
6		

EXAMPLE 2   Report 49

▶ Using Technology to Perform ANOVA Tests

A researcher believes that the mean earnings of top-paid actors, athletes, and musicians are the same. The earnings (in millions of dollars) for several randomly selected people from each category are shown in the table at the left. Assume that the populations are normally distributed, the samples are independent, and the population variances are equal. At $\alpha = 0.10$, can you reject the claim that the mean earnings are the same for the three categories? Use a technology tool. *(Source: Forbes.com LLC)*

▶ **Solution**   The null and alternative hypotheses are as follows.

H_0: $\mu_1 = \mu_2 = \mu_3$ (Claim)

H_a: At least one mean is different from the others.

The results obtained by performing the test on a TI-83/84 Plus are shown below. From the results, you can see that $P \approx 0.06$. Because $P < \alpha$, you should reject the null hypothesis.

TI-83/84 PLUS

One-way ANOVA
  F=3.051763393
  p=.0608019049
  Factor
    df=2
    SS=3766.36364
  ↓ MS=1883.18182

TI-83/84 PLUS

One-way ANOVA
  ↑ MS=1883.18182
  Error
    df=33
    SS=20363.6364
    MS=617.07989
  Sxp=24.8410928

Interpretation   There is enough evidence at the 10% level of significance to reject the claim that the mean earnings are the same.

▶ **Try It Yourself 2**

The data shown in the table represent the GPAs of randomly selected freshmen, sophomores, juniors, and seniors. At $\alpha = 0.05$, can you conclude that there is a difference in the means of the GPAs? Assume that the populations of GPAs are normally distributed and that the population variances are equal. Use a technology tool.

Freshmen	2.34	2.38	3.31	2.39	3.40	2.70	2.34			
Sophomores	3.26	2.22	3.26	3.29	2.95	3.01	3.13	3.59	2.84	3.00
Juniors	2.80	2.60	2.49	2.83	2.34	3.23	3.49	3.03	2.87	
Seniors	3.31	2.35	3.27	2.86	2.78	2.75	3.05	3.31		

a. Identify the *claim* and state H_0 and H_a.
b. Use a *technology tool* to enter the data.
c. *Perform* the ANOVA test to find the P-value.
d. *Decide* whether to reject the null hypothesis.
e. *Interpret* the decision in the context of the claim.   *Answer: Page A47*

▶ **TWO-WAY ANOVA**

When you want to test the effect of *two* independent variables, or factors, on one dependent variable, you can use a **two-way analysis of variance test.** For instance, suppose a medical researcher wants to test the effect of gender *and* type of medication on the mean length of time it takes pain relievers to provide relief. To perform such an experiment, the researcher can use the following two-way ANOVA block design.

	Gender	
	M	**F**
I	Males taking type I	Females taking type I
II	Males taking type II	Females taking type II
III	Males taking type III	Females taking type III

Type of medication

A two-way ANOVA test has three null hypotheses—one for each main effect and one for the interaction effect. A **main effect** is the effect of one independent variable on the dependent variable, and the **interaction effect** is the effect of both independent variables on the dependent variable. For instance, the hypotheses for the pain reliever experiment are as follows.

Hypotheses for main effects:

H_0: Gender has no effect on the mean length of time it takes a pain reliever to provide relief.

H_a: Gender has an effect on the mean length of time it takes a pain reliever to provide relief.

H_0: The type of medication has no effect on the mean length of time it takes a pain reliever to provide relief.

H_a: The type of medication has an effect on the mean length of time it takes a pain reliever to provide relief.

Hypotheses for interaction effect:

H_0: There is no interaction effect between gender and type of medication on the mean length of time it takes a pain reliever to provide relief.

H_a: There is an interaction effect between gender and type of medication on the mean length of time it takes a pain reliever to provide relief.

To test these hypotheses, you can perform a two-way ANOVA test. Using the *F*-distribution, a two-way ANOVA test calculates an *F*-test statistic for each hypothesis. As a result, it is possible to reject none, one, two, or all of the null hypotheses. The statistics involved with a two-way ANOVA test is beyond the scope of this course. You can, however, use a technology tool such as MINITAB to perform a two-way ANOVA test.

10.4 EXERCISES

FOR EXTRA HELP:

MyStatLab

1. H_0: $\mu_1 = \mu_2 = \mu_3 = \ldots = \mu_k$
   H_a: At least one of the means is different from the others.

2. Each sample must be randomly selected from a normal, or approximately normal, population. The samples must be independent of each other. Each population must have the same variance.

3. The MS_B measures the differences related to the treatment given to each sample. The MS_W measures the differences related to entries within the same sample.

4. There are three null hypotheses for a two-way ANOVA test. There is one for each main effect and one for the interaction effect.

5. (a) H_0: $\mu_1 = \mu_2 = \mu_3$
   H_a: At least one mean is different from the others. (claim)

   (b) $F_0 = 3.37$;
   Rejection region: $F > 3.37$

   (c) 1.02

   (d) Fail to reject H_0.

   (e) There is not enough evidence at the 5% level of significance to conclude that the mean costs per ounce are different.

6. See Selected Answers, page A133.

▮ BUILDING BASIC SKILLS AND VOCABULARY

1. State the null and alternative hypotheses for a one-way ANOVA test.

2. What conditions are necessary in order to use a one-way ANOVA test?

3. Describe the difference between the variance between samples MS_B and the variance within samples MS_W.

4. Describe the hypotheses for a two-way ANOVA test.

▮ USING AND INTERPRETING CONCEPTS

Performing a One-Way ANOVA Test   *In Exercises 5–14, (a) identify the claim and state H_0 and H_a, (b) find the critical value and identify the rejection region, (c) find the test statistic F, (d) decide whether to reject or fail to reject the null hypothesis, and (e) interpret the decision in the context of the original claim. If convenient, use technology to solve the problem. In each exercise, assume that each population is normally distributed and that the population variances are equal.*

5. Toothpaste   The table shows the cost per ounce (in dollars) for a random sample of toothpastes exhibiting very good stain removal, good stain removal, and fair stain removal. At $\alpha = 0.05$, can you conclude that the mean costs per ounce are different?
(Source: Consumer Reports)

Very good stain removal	Good stain removal	Fair stain removal
0.47	0.60	0.34
0.49	0.64	0.46
0.33	1.05	1.31
1.52	2.73	0.44
0.64	0.58	0.60
0.36	0.75	
0.41	0.22	
0.37	0.33	
0.48	0.42	
0.50	0.46	
0.51	0.98	
0.35	1.16	

6. Automobile Batteries   The prices (in dollars) of 17 randomly selected automobile batteries are shown in the table. The prices are classified according to battery type. At $\alpha = 0.05$, is there enough evidence to conclude that at least one of the mean battery prices is different from the others?
(Source: Consumer Reports)

Group size 35	90	90	75	105	65	
Group size 65	100	75	105	100	100	75
Group size 24/24F	115	75	75	90	110	90

7. (a) H_0: $\mu_1 = \mu_2 = \mu_3$

H_a: At least one mean is different from the others. (claim)

(b) $F_0 = 5.49$; Rejection region: $F > 5.49$

(c) 21.99

(d) Reject H_0.

(e) There is enough evidence at the 1% level of significance to conclude that at least one mean salary is different.

8. (a) H_0: $\mu_1 = \mu_2 = \mu_3 = \mu_4 = \mu_5$

H_a: At least one mean is different from the others. (claim)

(b) $F_0 = 2.53$; Rejection region: $F > 2.53$

(c) 0.27

(d) Fail to reject H_0.

(e) There is not enough evidence at the 5% level of significance to conclude that at least one mean age is different.

9. (a) H_0: $\mu_1 = \mu_2 = \mu_3 = \mu_4 = \mu_5$

H_a: At least one mean is different from the others. (claim)

(b) $F_0 = 4.37$; Rejection region: $F > 4.37$

(c) 12.61

(d) Reject H_0.

(e) There is enough evidence at the 1% level of significance to conclude that at least one mean cost per mile is different.

 7. Government Salaries   The table shows the salaries (in thousands of dollars) of randomly selected individuals from the federal, state, and local levels of government. At $\alpha = 0.01$, can you conclude that at least one mean salary is different? *(Adapted from Bureau of Labor Statistics)*

Federal	State	Local
63.7	50.6	47.1
56.4	34.7	36.6
67.8	51.7	40.9
75.6	52.2	39.3
74.9	54.4	49.9
79.0	59.5	44.0
49.6	37.6	58.6
64.5	48.1	39.1
74.2	51.3	35.5
57.9	45.1	31.7

8. Late Night Hosts   The table shows the ages (in years) of randomly selected viewers for several late night hosts. At $\alpha = 0.05$, can you conclude that at least one mean age is different? *(Adapted from The New York Times)*

Jimmy Fallon	Craig Ferguson	Jimmy Kimmel	Jay Leno	David Letterman
19	22	18	41	27
28	27	21	43	34
37	32	33	44	41
43	37	42	47	43
48	45	48	49	48
48	53	57	51	54
53	55	59	59	57
54	62	61	61	59
54	64	62	62	60
57	67	64	64	63
62	68	68	65	69
67	70	68	67	71
79	72	75	74	73

9. Cost Per Mile   The table shows the cost per mile (in cents) for a random sample of automobiles. At $\alpha = 0.01$, can you conclude that at least one mean cost per mile is different? *(Adapted from American Automobile Association)*

Small sedan	Medium sedan	Large sedan	4WD SUV	Minivan
40	62	59	84	63
38	44	68	63	73
46	58	78	72	56
51	54	70	75	48
43	59	75		67
	47	67		

10. (a) H_0: $\mu_1 = \mu_2 = \mu_3 = \mu_4$ (claim)

H_a: At least one mean is different from the others.

(b) $F_0 = 2.28$;
Rejection region: $F > 2.28$

(c) 2.55

(d) Reject H_0.

(e) There is enough evidence at the 10% level of significance to reject the claim that the mean scores are the same for all regions.

11. (a) H_0: $\mu_1 = \mu_2 = \mu_3 = \mu_4$ (claim)

H_a: At least one mean is different from the others.

(b) $F_0 = 4.54$;
Rejection region: $F > 4.54$

(c) 0.56

(d) Fail to reject H_0.

(e) There is not enough evidence at the 1% level of significance for the company to reject the claim that the mean number of days patients spend at the hospital is the same for all four regions.

12. (a) See Selected Answers, page A133.

H_a: At least one mean is different from the others. (claim)

(b) $F_0 = 4.23$;

Rejection region: $F > 4.23$

(c) 1.63

(d) Fail to reject H_0.

(e) There is not enough evidence at the 5% level of significance to conclude that the mean salary is different in at least one of the areas.

10. Well-Being Index The well-being index is a way to measure how people are faring physically, emotionally, socially, and professionally, as well as to rate the overall quality of their lives and their outlooks for the future. The table shows the well-being index scores for a random sample of states from four regions of the United States. At $\alpha = 0.10$, can you reject the claim that the mean score is the same for all regions? (*Adapted from Gallup and Healthways*)

Northeast	Midwest	South	West
66.9	67.3	67.0	70.2
67.4	67.6	66.2	68.3
65.0	67.8	66.1	67.3
65.4	67.2	66.8	68.3
66.7	66.5	64.8	66.4
64.2	64.9	64.9	67.3
	63.6	65.1	63.8
	63.9	64.2	65.3
		63.9	66.0
		64.0	
		60.5	

11. Days Spent at the Hospital In a recent study, a health insurance company investigated the number of days patients spent at the hospital. In part of the study, the company randomly selected patients from various parts of the United States and recorded the number of days each patient spent at the hospital. The results of the study are shown in the table. At $\alpha = 0.01$, can the company reject the claim that the mean number of days patients spend at the hospital is the same for all four regions? (*Adapted from National Center for Health Statistics*)

Northeast	Midwest	South	West
6	6	3	3
4	6	5	4
7	7	6	6
2	3	6	4
3	5	3	6
4	4	7	6
6	4	4	5
8	3		2
9	2		

12. Personal Income The table shows the salaries of randomly selected individuals from six large metropolitan areas. At $\alpha = 0.05$, can you conclude that the mean salary is different in at least one of the areas? (*Adapted from U.S. Bureau of Economic Analysis*)

Chicago	Dallas	Miami	Denver	San Diego	Seattle
41,950	34,315	43,500	41,400	46,000	56,135
36,100	31,500	47,350	42,580	43,100	46,500
45,200	38,000	34,700	46,600	41,550	44,400
51,400	49,495	46,500	49,900	52,300	51,000
50,920	38,700	39,050	53,175	39,400	55,875
40,500	51,050	42,900			44,000
	45,060	47,700			

13. (a) H_0: $\mu_1 = \mu_2 = \mu_3 = \mu_4$

   H_a: At least one mean is different from the others. (claim)

(b) $F_0 = 2.255$;
   Rejection region: $F > 2.255$

(c) 3.107

(d) Reject H_0.

(e) There is enough evidence at the 10% level of significance to conclude that the mean energy consumption of at least one region is different from the others.

14. (a) H_0: $\mu_1 = \mu_2 = \mu_3 = \mu_4$ (claim)

   H_a: At least one mean is different from the others.

(b) $F_0 = 2.93$;
   Rejection region: $F > 2.93$

(c) 4.29

(d) Reject H_0.

(e) There is enough evidence at the 5% level of significance to reject the claim that the mean amounts spent are equal for all regions.

15. Analysis of Variance results:

Data stored in separate columns.

Column means

Column	n	Mean	Std. Error
Grade 9	8	84.375	9.531784
Grade 10	8	79.25	9.090321
Grade 11	8	76.625	7.648383
Grade 12	8	70.75	6.9224014

ANOVA table

See Odd Answers, page A102.

$P = 0.7129 > 0.01$, so fail to reject H_0. There is not enough evidence at the 1% level of significance to reject the claim that the mean numbers of female students who played on a sports team are equal for all grades.

13. Energy Consumption   The table shows the energy consumed (in millions of Btu) in one year for a random sample of households from four regions of the United States. At $\alpha = 0.10$, can you conclude that the mean energy consumption of at least one region is different from the others? *(Adapted from U.S. Energy Information Administration)*

Northeast	Midwest	South	West
101.5	56.7	68.8	31.0
109.5	174.8	92.8	46.2
153.6	61.6	56.5	127.7
129.0	79.3	96.6	61.4
160.3	160.9	57.0	108.4
114.6	179.9	51.3	69.8
85.2	98.6	63.2	44.5
173.0	132.1	50.6	124.8
73.6	89.5	182.0	98.6
	155.5		58.3
	61.9		

14. Amount Spent on Energy   The table shows the amount spent (in dollars) on energy in one year for a random sample of households from four regions of the United States. At $\alpha = 0.05$, can you reject the claim that the mean amounts spent are equal for all regions? *(Adapted from U.S. Energy Information Administration)*

Northeast	Midwest	South	West
1456	1940	1457	1168
3025	1570	2202	1927
2029	1972	1883	989
1735	1924	1310	2022
1956	1820	1876	1330
3078	2144	1578	1184
3023	1319	1980	1819
1709	1655		
3425	1730		
1684			

SC   *In Exercises 15 and 16, use StatCrunch to perform a one-way ANOVA test. Decide whether to reject the null hypothesis. Then, interpret the decision in the context of the original claim.*

15. Sports Team Involvement   The table shows the number of female students who played on a sports team in grades 9 through 12 for a random sample of 8 high schools in a state. At $\alpha = 0.01$, can you reject the claim that the mean numbers of female students who played on a sports team are equal for all grades?

Grade 9	82	91	53	133	64	112	63	77
Grade 10	77	87	58	125	51	106	58	72
Grade 11	86	81	46	115	56	87	62	80
Grade 12	65	77	42	102	49	84	65	82

16. Analysis of Variance results:

Data stored in separate columns.
See Selected Answers, page A133.

$P = 0.1752 > 0.10$, so fail to reject H_0. There is not enough evidence at the 10% level of significance to conclude that at least one mean sale price is different.

17. Fail to reject all null hypotheses. The interaction between the advertising medium and the length of the ad has no effect on the rating and therefore there is no significant difference in the means of the ratings.

18. Type of vehicle has an effect on the mean number of vehicles sold in a month. There is an interaction effect between type of vehicle and gender on the number of vehicles sold in a month.

16. Housing Prices The table shows the sale prices (in thousands of dollars) of randomly selected one-family houses in three cities. At $\alpha = 0.10$, can you conclude that at least one mean sale price is different? *(Adapted from National Association of Realtors)*

Gainesville	Orlando	Tampa
179.0	253.9	229.9
151.5	211.1	114.9
196.6	195.3	210.2
192.3	197.5	202.7
254.7	217.9	149.1
212.4	244.8	166.0
92.8	263.2	133.3
210.6	154.7	213.4
180.5	173.9	104.7
226.0	183.3	215.4
179.0		

■ EXTENDING CONCEPTS

Using Technology to Perform a Two-Way ANOVA Test *In Exercises 17–20, use a technology tool and the given block design to perform a two-way ANOVA test. Use $\alpha = 0.10$. Interpret the results.*

17. Advertising A study was conducted in which a random sample of 20 adults was asked to rate the effectiveness of advertisements. Each adult rated a radio or television advertisement that lasted 30 or 60 seconds. The block design shows these ratings (on a scale of 1 to 5, with 5 being extremely effective).

Advertising medium

Length of ad	Radio	Television
30 sec	2, 3, 5, 1, 3	3, 5, 4, 1, 2
60 sec	1, 4, 2, 2, 5	2, 5, 3, 4, 4

18. Vehicle Sales The owner of a car dealership wants to determine if the gender of a salesperson and the type of vehicle sold affect the number of vehicles sold in a month. The block design shows the number of vehicles, listed by type, sold in a month by a random sample of eight salespeople.

Type of vehicle

Gender	Car	Truck	Van/SUV
Male	6, 5, 4, 5	2, 2, 1, 3	4, 3, 4, 2
Female	5, 7, 8, 7	1, 0, 1, 2	4, 2, 0, 1

19. Fail to reject all null hypotheses. The interaction between age and gender has no effect on GPA and therefore there is no significant difference in the means of the GPAs.

20. There is an interaction effect between technician and brand on the mean time it takes to repair a disk drive.

21. $CV_{Scheffé} = 10.98$

(1, 2) → 22.396 → Significant difference

(1, 3) → 40.837 → Significant difference

(2, 3) → 2.749 → No difference

22. $CV_{Scheffé} = 17.48$

(1, 2) → 5.237 → No difference

(1, 3) → 32.479 → Significant difference

(1, 4) → 35.270 → Significant difference

(1, 5) → 14.062 → No difference

(2, 3) → 12.795 → No difference

(2, 4) → 16.201 → No difference

(2, 5) → 2.651 → No difference

(3, 4) → 0.682 → No difference

(3, 5) → 3.177 → No difference

(4, 5) → 5.776 → No difference

23. $CV_{Scheffé} = 6.84$

(1, 2) → 0.032 → No difference

(1, 3) → 1.490 → No difference

(1, 4) → 1.345 → No difference

(2, 3) → 2.374 → No difference

(2, 4) → 1.122 → No difference

(3, 4) → 7.499 → Significant difference

24. $CV_{Scheffé} = 8.79$

(1, 2) → 5.416 → No difference

(1, 3) → 5.277 → No difference

(1, 4) → 11.462 → Significant difference

(2, 3) → 0.015 → No difference

(2, 4) → 1.413 → No difference

(3, 4) → 1.007 → No difference

19. Grade Point Average A study was conducted in which a random sample of 24 high school students was asked to give their grade point average (GPA). The block design shows the GPAs of male and female students from four different age groups.

		Age		
	Under 15	15–16	17–18	Over 18
Male	2.5, 2.1, 3.8	4.0, 1.4, 2.0	3.5, 2.2, 2.0	3.1, 0.7, 2.8
Female	4.0, 2.1, 1.9	3.5, 3.0, 2.1	4.0, 2.2, 1.7	1.6, 2.5, 3.6

(Gender)

20. Disk Drive Repairs The manager of a computer repair service wants to determine whether there is a difference in the time it takes four technicians to repair different brands of disk drives. The block design shows the times (in minutes) it took for each technician to repair three disk drives of each brand.

	Technician 1	Technician 2	Technician 3	Technician 4
Brand A	67, 82, 64	42, 56, 39	69, 47, 38	70, 44, 50
Brand B	44, 62, 55	47, 58, 62	55, 45, 66	47, 29, 40
Brand C	47, 36, 68	39, 74, 51	74, 80, 70	45, 62, 59

Technician (column header); Brand (row header)

The Scheffé Test *If the null hypothesis is rejected in a one-way ANOVA test of three or more means, a **Scheffé Test** can be performed to find which means have a significant difference. In a Scheffé Test, the means are compared two at a time. For instance, with three means you would have the following comparisons: $\bar{x}_1$ versus $\bar{x}_2$, $\bar{x}_1$ versus $\bar{x}_3$, and $\bar{x}_2$ versus $\bar{x}_3$. For each comparison, calculate*

$$\frac{(\bar{x}_a - \bar{x}_b)^2}{\frac{SS_W}{\sum(n_i - 1)}[(1/n_a) + (1/n_b)]}$$

where $\bar{x}_a$ and $\bar{x}_b$ are the means being compared and n_a and n_b are the corresponding sample sizes. Calculate the critical value using the same steps as in a one-way ANOVA test and multiply the result by $k - 1$. Then compare the value that is calculated using the formula above with the critical value. The means have a significant difference if the critical value is less than the value calculated using the formula above.

Use the information above to solve Exercises 21–24.

21. Refer to the data in Exercise 7. At $\alpha = 0.01$, perform a Scheffé Test to determine which means have a significant difference.

22. Refer to the data in Exercise 9. At $\alpha = 0.01$, perform a Scheffé Test to determine which means have a significant difference.

23. Refer to the data in Exercise 10. At $\alpha = 0.10$, perform a Scheffé Test to determine which means have a significant difference.

24. Refer to the data in Exercise 14. At $\alpha = 0.05$, perform a Scheffé Test to determine which means have a significant difference.

USES AND ABUSES

Uses

One-Way Analysis of Variance (ANOVA) ANOVA can help you make important decisions about the allocation of resources. For instance, suppose you work for a large manufacturing company and part of your responsibility is to determine the distribution of the company's sales throughout the world and decide where to focus the company's efforts. Because wrong decisions will cost your company money, you want to make sure that you make the right decisions.

Abuses

Preconceived Notions There are several ways that the tests presented in this chapter can be abused. For instance, it is easy to allow preconceived notions to affect the results of a chi-square goodness-of-fit test and a test for independence. When testing to see whether a distribution has changed, do not let the existing distribution "cloud" the study results. Similarly, when determining whether two variables are independent, do not let your intuition "get in the way." As with any hypothesis test, you must properly gather appropriate data and perform the corresponding test before you can reach a logical conclusion.

Incorrect Interpretation of Rejection of Null Hypothesis It is important to remember that when you reject the null hypothesis of an ANOVA test, you are simply stating that you have enough evidence to determine that at least one of the population means is different from the others. You are not finding them all to be different. One way to further test which of the population means differs from the others is explained in Extending Concepts in Section 10.4 Exercises.

◼ EXERCISES

1. ***Preconceived Notions*** ANOVA depends on having independent variables. Describe an abuse that might occur by having dependent variables. Then describe how the abuse could be avoided.

2. ***Incorrect Interpretation of Rejection of Null Hypothesis*** Find an example of the use of ANOVA. In that use, describe what would be meant by "rejection of the null hypothesis." How should rejection of the null hypothesis be correctly interpreted?

10   CHAPTER SUMMARY

What did you learn?

	EXAMPLE(S)	REVIEW EXERCISES
Section 10.1		
■ How to use the chi-square distribution to test whether a frequency distribution fits a claimed distribution $$\chi^2 = \Sigma \frac{(O - E)^2}{E}$$	*1–3*	*1–4*
Section 10.2		
■ How to use a contingency table to find expected frequencies $$E_{r,c} = \frac{(\text{Sum of row } r) \times (\text{Sum of column } c)}{\text{Sample size}}$$	*1*	*5–8*
■ How to use a chi-square distribution to test whether two variables are independent	*2, 3*	*5–8*
Section 10.3		
■ How to interpret the *F*-distribution and use an *F*-table to find critical values $$F = \frac{s_1^2}{s_2^2}$$	*1, 2*	*9–16*
■ How to perform a two-sample *F*-test to compare two variances	*3, 4*	*17–22*
Section 10.4		
■ How to use one-way analysis of variance to test claims involving three or more means $$F = \frac{MS_B}{MS_W}$$	*1, 2*	*23, 24*

10 REVIEW EXERCISES

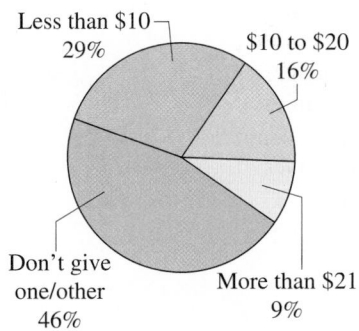

FIGURE FOR EXERCISE 1

1. (a) H_0: The distribution of the allowance amounts is 29% less than \$10, 16% \$10 to \$20, 9% more than \$21, and 46% don't give one/other.

H_a: The distribution of amounts differs from the claimed or expected distribution. (claim)

(b) $\chi_0^2 = 6.251$;
Rejection region: $\chi^2 > 6.251$

(c) 4.886

(d) Fail to reject H_0.

(e) There is not enough evidence at the 10% level of significance to conclude that there has been a change in the claimed or expected distribution.

2. See Selected Answers, page A133.

3. (a) See Odd Answers, page A103.

(b) $\chi_0^2 = 7.815$;
Rejection region: $\chi^2 > 7.815$

(c) 0.503

(d) Fail to reject H_0.

(e) See Odd Answers, page A103.

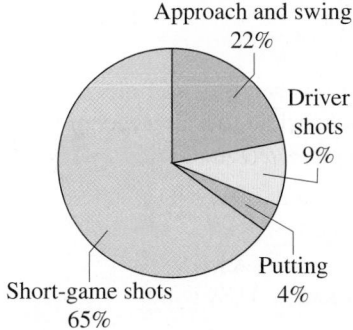

FIGURE FOR EXERCISE 3

■ SECTION 10.1

In Exercises 1–4, (a) identify the claimed distribution and state H_0 and H_a, (b) find the critical value and identify the rejection region, (c) find the chi-square test statistic, (d) decide whether to reject or fail to reject the null hypothesis, and (e) interpret the decision in the context of the original claim.

1. Results from a previous survey asking parents how much they give for an allowance are shown in the pie chart. To determine whether the distribution has changed, you randomly select 1103 parents and ask them how much they give for an allowance. The results are shown in the table. At $\alpha = 0.10$, can you conclude that there has been a change in the claimed or expected distribution? *(Adapted from Echo Research)*

Survey results	
Response	**Frequency, f**
Less than \$10	353
\$10 to \$20	167
More than \$21	94
Don't give one/other	489

2. Results from a survey 10 years ago asking people how long their office visits with a physician were are shown in the pie chart. To determine whether this distribution has changed, a research organization randomly selects 350 people and asks them how long their office visits with a physician were. The results are shown in the table. At $\alpha = 0.01$, can you conclude that there has been a change in the claimed or expected distribution? *(Adapted from National Center for Health Statistics)*

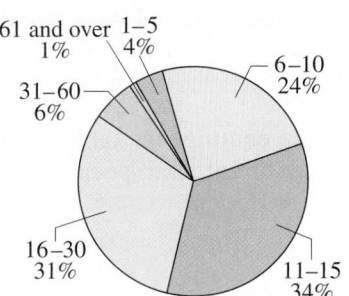

Survey results	
Minutes	**Frequency, f**
1–5	9
6–10	62
11–15	126
16–30	129
31–60	23
61 and over	1

3. Results from a previous survey asking golf students what they need the most help with in golf are shown in the pie chart. To determine whether the distribution is the same, a golf instructor randomly selects 435 golf students and asks them what they need the most help with in golf. The results are shown in the table. At $\alpha = 0.05$, are the distributions the same? *(Adapted from PGA of America)*

Survey results	
Response	**Frequency, f**
Short-game shots	276
Approach and swing	99
Driver shots	42
Putting	18

4. (a) H_0: The distribution of responses about which industry has the most trustworthy advertising is uniform. (claim)

H_a: The distribution of responses is not uniform.

(b) $\chi_0^2 = 9.488$;

Rejection region: $\chi^2 > 9.488$

(c) 47.200

(d) Reject H_0.

(e) There is enough evidence at the 5% level of significance to reject the claim that the distribution of responses about which industry has the most trustworthy advertising is uniform.

5. (a) $E_{1,1} = 63$, $E_{1,2} = 356.4$,
$E_{1,3} = 319.8$, $E_{1,4} = 310.8$,
$E_{2,1} = 147$, $E_{2,2} = 831.6$,
$E_{2,3} = 746.2$, $E_{2,4} = 725.2$

(b) Reject H_0.

(c) There is enough evidence at the 1% level of significance to conclude that public school teachers' gender and years of full-time teaching experience are related.

6. (a) $E_{1,1} = 94.25$, $E_{1,2} = 82.65$,
$E_{1,3} = 50.75$, $E_{1,4} = 4.35$,
$E_{2,1} = 100.75$, $E_{2,2} = 88.35$,
$E_{2,3} = 54.25$, $E_{2,4} = 4.65$

(b) Reject H_0.

(c) There is enough evidence at the 5% level of significance to conclude that the type of vehicle owned is dependent on gender.

7. (a) $E_{1,1} \approx 54.86$, $E_{1,2} \approx 40.38$,
$E_{1,3} \approx 22.10$, $E_{1,4} = 16.00$,
$E_{1,5} \approx 26.67$, $E_{2,1} \approx 17.14$,
$E_{2,2} \approx 12.62$, $E_{2,3} \approx 6.90$,
$E_{2,4} = 5.00$, $E_{2,5} \approx 8.33$

(b) Reject H_0.

(c) There is enough evidence at the 1% level of significance to conclude that a species' status (endangered or threatened) is dependent on vertebrate group.

4. An organization believes that the thoughts of adults ages 55 and over on which industry has the most trustworthy advertising is uniformly distributed. To test this claim, you randomly select 800 adults ages 55 and over and ask each which industry has the most trustworthy advertising. The results are shown in the table. At $\alpha = 0.05$, can you reject the claim that the distribution is uniform? *(Adapted from Harris Interactive)*

Response	Frequency, f
Auto companies	128
Fast food companies	192
Financial services companies	112
Pharmaceutical companies	152
Soft drink companies	216

■ SECTION 10.2

In Exercises 5–8, use the given contingency table to (a) find the expected frequencies of each cell in the table, (b) perform a chi-square test for independence, and (c) comment on the relationship between the two variables. Assume the variables are independent. If convenient, use technology to solve the problem.

5. The contingency table shows the results of a random sample of public elementary and secondary school teachers by gender and years of full-time teaching experience. Use $\alpha = 0.01$. *(Adapted from U.S. National Center for Education Statistics)*

	Years of full-time teaching experience			
Gender	Less than 3 years	3–9 years	10–20 years	20 years or more
Male	58	377	280	335
Female	152	811	786	701

6. The contingency table shows the results of a random sample of individuals by gender and type of vehicle owned. Use $\alpha = 0.05$.

	Type of vehicle owned			
Gender	Car	Truck	SUV	Van
Male	85	96	45	6
Female	110	75	60	3

7. The contingency table shows the results of a random sample of endangered and threatened species by status and vertebrate group. Use $\alpha = 0.01$. *(Adapted from U.S. Fish and Wildlife Service)*

	Vertebrate group				
Status	Mammals	Birds	Reptiles	Amphibians	Fish
Endangered	62	48	17	12	21
Threatened	10	5	12	9	14

8. The contingency table shows the distribution of a random sample of fatal pedestrian motor vehicle collisions by time of day and gender in a recent year. Use $\alpha = 0.10$. *(Adapted from National Highway Traffic Safety Administration)*

	Time of day			
Gender	12 A.M.–5:59 A.M.	6 A.M.–11:59 A.M.	12 P.M.–5:59 P.M.	6 P.M.–11:59 P.M.
Male	654	591	909	928
Female	255	365	601	505

8. (a) $E_{1,1} \approx 582.68$, $E_{1,2} \approx 612.81$, $E_{1,3} \approx 967.93$, $E_{1,4} \approx 918.57$, $E_{2,1} \approx 326.32$, $E_{2,2} \approx 343.19$, $E_{2,3} \approx 542.07$, $E_{2,4} \approx 514.43$

(b) Reject H_0.

(c) There is enough evidence at the 10% level of significance to conclude that the time of day of a collision is dependent on gender.

9. 2.295

10. 4.71

11. 2.39

12. 2.01

13. 2.06

14. 4.36

15. 2.08

16. 4.73

17. Fail to reject H_0. There is not enough evidence at the 1% level of significance to reject the claim.

18. Reject H_0. There is enough evidence at the 10% level of significance to support the claim.

19. Fail to reject H_0. There is not enough evidence at the 10% level of significance to support the claim that the variation in wheat production is greater in Garfield County than in Kay County.

20. Reject H_0. There is enough evidence at the 1% level of significance to reject the claim that the standard deviations of hotel room rates for San Francisco, CA and Sacramento, CA are the same.

21. Fail to reject H_0. There is not enough evidence at the 1% level of significance to support the claim that the test score variance for females is different from that for males.

SECTION 10.3

In Exercises 9–12, find the critical F-value for a right-tailed test using the indicated level of significance α and degrees of freedom d.f.$_N$ and d.f.$_D$.

9. $\alpha = 0.05$, d.f.$_N = 6$, d.f.$_D = 50$

10. $\alpha = 0.01$, d.f.$_N = 12$, d.f.$_D = 10$

11. $\alpha = 0.10$, d.f.$_N = 5$, d.f.$_D = 12$

12. $\alpha = 0.05$, d.f.$_N = 20$, d.f.$_D = 25$

In Exercises 13–16, find the critical F-value for a two-tailed test using the indicated level of significance α and degrees of freedom d.f.$_N$ and d.f.$_D$.

13. $\alpha = 0.10$, d.f.$_N = 15$, d.f.$_D = 27$

14. $\alpha = 0.05$, d.f.$_N = 9$, d.f.$_D = 8$

15. $\alpha = 0.01$, d.f.$_N = 40$, d.f.$_D = 60$

16. $\alpha = 0.01$, d.f.$_N = 11$, d.f.$_D = 13$

In Exercises 17 and 18, test the claim about the difference between two population variances σ_1^2 and σ_2^2 at the indicated level of significance α using the given sample statistics. Assume the sample statistics are from independent samples that are randomly selected and each population has a normal distribution.

17. Claim: $\sigma_1^2 \leq \sigma_2^2$; $\alpha = 0.01$. Sample statistics: $s_1^2 = 653$, $n_1 = 16$; $s_2^2 = 270$, $n_2 = 21$

18. Claim: $\sigma_1^2 \neq \sigma_2^2$; $\alpha = 0.10$. Sample statistics: $s_1^2 = 87.3$, $n_1 = 31$; $s_2^2 = 45.5$, $n_2 = 29$

In Exercises 19–22, test the claim about two population variances at the indicated level of significance α. Interpret the results in the context of the claim. If convenient, use technology to solve the problem. In each exercise, assume the samples are independent and each population has a normal distribution.

19. An agricultural analyst is comparing the wheat production in Oklahoma counties. The analyst claims that the variation in wheat production is greater in Garfield County than in Kay County. A random sample of 21 Garfield County farms yields a standard deviation of 0.76 bushel per acre. A random sample of 16 Kay County farms is found to have a standard deviation of 0.58 bushel per acre. Test the analyst's claim at $\alpha = 0.10$. *(Adapted from Environmental Verification and Analysis Center—University of Oklahoma)*

20. A travel consultant indicates that the standard deviations of hotel room rates for San Francisco, CA and Sacramento, CA are the same. A random sample of 36 hotel room rates in San Francisco has a standard deviation of $75 and a random sample of 31 hotel room rates in Sacramento has a standard deviation of $44. At $\alpha = 0.01$, can you reject the travel consultant's claim? *(Adapted from I-Map Data Systems LLC)*

21. The table shows the SAT verbal test scores for 9 randomly selected female students and 13 randomly selected male students. Assume that SAT verbal test scores are normally distributed. At $\alpha = 0.01$, test the claim that the test score variance for females is different from that for males.

Female	480	610	340	630	520	690	540
Male	560	680	360	530	380	460	630

Female	600	800				
Male	310	730	740	520	560	400

22. Reject H_0. There is enough evidence at the 5% level of significance to support the claim that the new mold produces inserts that are less variable in diameter than those produced with the current mold.

23. Reject H_0. There is enough evidence at the 10% level of significance to conclude that at least one of the mean costs is different from the others.

24. Fail to reject H_0. There is not enough evidence at the 5% level of significance to conclude that at least one of the mean incomes is different from the others.

 22. A plastics company that produces automobile dashboard inserts has just received a new injection mold that is supposedly more consistent than the company's current mold. A quality technician wishes to test whether this new mold will produce inserts that are less variable in diameter than those produced with the company's current mold. The table shows independent random samples (of size 12) of insert diameters (in centimeters) for both the current and new molds. At $\alpha = 0.05$, test the claim that the new mold produces inserts that are less variable in diameter than the inserts the current mold produces.

New	9.611	9.618	9.594	9.580	9.611	9.597
Current	9.571	9.642	9.650	9.651	9.596	9.636

New	9.638	9.568	9.605	9.603	9.647	9.590
Current	9.570	9.537	9.641	9.625	9.626	9.579

■ SECTION 10.4

In Exercises 23 and 24, use the given sample data to perform a one-way ANOVA test using the indicated level of significance α. What can you conclude? Assume that each sample is drawn from a normal, or approximately normal, population, that the samples are independent of each other, and that the populations have the same variances. If convenient, use technology to solve the problem.

23. The table at the right shows the residential electricity cost (in dollars per million Btu) in one year for a random sample of households in four regions of the United States. Use $\alpha = 0.10$ to test for differences among the means for the four regions. *(Adapted from U.S. Energy Information Administration)*

Northeast	Midwest	South	West
40.24	18.40	22.85	35.03
28.18	26.66	29.79	31.51
35.67	28.27	18.93	20.28
34.18	21.38	21.81	28.82
39.03	24.64	25.47	24.07
30.74	20.15	23.64	27.60
32.65	29.77	19.82	29.25
29.98	25.08	28.15	18.57

24. The table at the right shows the annual income (in dollars) for a random sample of families in four regions of the United States. Use $\alpha = 0.05$ to test for differences among the means for the four regions. *(Adapted from U.S. Census Bureau)*

Northeast	Midwest	South	West
74,833	57,116	54,248	71,690
66,098	80,729	77,990	64,098
74,961	78,788	42,293	60,307
51,089	59,543	51,998	58,385
71,920	57,093	40,161	61,862
	72,305	64,698	61,609
	39,019		

10 CHAPTER QUIZ

1. (a) $H_0: \sigma_1^2 = \sigma_2^2$;
   $H_a: \sigma_1^2 \neq \sigma_2^2$ (claim)
   (b) 0.01
   (c) $F_0 = 3.80$
   (d) Rejection region: $F > 3.80$
   (e) 2.12
   (f) Fail to reject H_0.
   (g) There is not enough evidence at the 1% level of significance to conclude that the variances in annual wages for San Francisco, CA and Baltimore, MD are different.

2. (a) $H_0: \mu_1 = \mu_2 = \mu_3$ (claim)
   $H_a:$ At least one mean is different from the others.
   (b) 0.10
   (c) $F_0 = 2.44$
   (d) Rejection region: $F > 2.44$
   (e) 27.48
   (f) Reject H_0.
   (g) There is enough evidence at the 10% level of significance to reject the claim that the mean annual wages are equal for all three cities.

3. (a) $H_0:$ The distribution of educational achievement for people in the United States ages 35–44 is 13.4% not a high school graduate, 31.2% high school graduate, 17.2% some college, no degree; 8.8% associate's degree, 19.1% bachelor's degree, and 10.3% advanced degree.

   $H_a:$ The distribution of educational achievement for people in the United States ages 35–44 differs from the claimed distribution. (claim)
   (b) 0.05
   (c) $\chi_0^2 = 11.071$
   (d) Rejection region: $\chi^2 > 11.071$
   (e) 3.799
   (f) Fail to reject H_0.
   (g) See Odd Answers, page A103.

4. See Odd Answers, page A104.

Take this quiz as you would take a quiz in class. After you are done, check your work against the answers given in the back of the book.

For each exercise,

(a) state H_0 and H_a.

(b) specify the level of significance α.

(c) determine the critical value.

(d) determine the rejection region.

(e) find the test statistic.

(f) make a decision.

(g) interpret the results in the context of the problem.

If convenient, use technology to solve the problem.

For Exercises 1 and 2, use the following data. The data list the annual wages (in thousands of dollars) for randomly selected individuals from three metropolitan areas. Assume the wages are normally distributed and that the samples are independent. (Adapted from U.S. Bureau of Economic Analysis)

San Francisco, CA: 64.5, 75.9, 47.5, 52.3, 45.9, 59.7, 71.2, 74.1, 65.4, 61.9, 60.9, 58.7, 54.6

Baltimore, MD: 45.9, 39.8, 46.2, 44.9, 37.5, 52.9, 57.5, 49.7, 48.1, 47.9, 55.9, 35.5, 39.9, 40.9, 45.4

Jacksonville, FL: 31.3, 29.3, 39.2, 45.7, 34.9, 31.5, 41.5, 49.8, 52.6, 40.1, 42.9, 33.4, 30.5, 50.2, 34.7

1. At $\alpha = 0.01$, is there enough evidence to conclude that the variances in annual wages for San Francisco, CA and Baltimore, MD are different?

2. Are the mean annual wages equal for all three cities? Use $\alpha = 0.10$. Assume that the population variances are equal.

For Exercises 3 and 4, use the following table. The table lists the distribution of educational achievement for people in the United States ages 25 and older. It also lists the results of a random survey for two additional age categories. (Adapted from U.S. Census Bureau)

	25 and older	35–44	65–74
Not a H.S. graduate	13.4%	36	86
High school graduate	31.2%	92	161
Some college, no degree	17.2%	55	72
Associate's degree	8.8%	32	27
Bachelor's degree	19.1%	70	60
Advanced degree	10.3%	36	45

3. Does the distribution for people in the United States ages 25 and older differ from the distribution for people in the United States ages 35–44? Use $\alpha = 0.05$.

4. Does the distribution for people in the United States ages 25 and older differ from the distribution for people in the United States ages 65–74? Use $\alpha = 0.01$.

PUTTING IT ALL TOGETHER

Real Statistics — Real Decisions

The National Fraud Information Center (NFIC) was established in 1992 by the National Consumers League (NCL) to combat the growing problem of telemarketing fraud by improving prevention and enforcement. NCL works to protect and promote social and economic justice for consumers and workers in the United States and abroad.

You work for the NCL's Fraud Center as a statistical analyst. You are studying data on telemarketing fraud. Part of your analysis involves testing the goodness of fit, testing for independence, comparing variances, and performing ANOVA.

■ EXERCISES

1. *Goodness of Fit*

A claimed distribution for the ages of telemarketing fraud victims is shown in the table at the right. The results of a survey of 1000 randomly selected telemarketing fraud victims are also shown in the table. Using $\alpha = 0.01$, perform a chi-square goodness-of-fit test to test the claimed distribution. What can you conclude? Do you think the claimed distribution is valid? Why or why not?

2. *Independence*

The following contingency table shows the results of a random sample of 2000 telemarketing fraud victims classified by age and type of fraud. The frauds were committed using bogus sweepstakes or credit card offers.

(a) Calculate the expected frequency for each cell in the contingency table. Assume the variables age and type of fraud are independent.

(b) Can you conclude that the ages of the victims are related to the type of fraud? Use $\alpha = 0.01$.

Ages	Claimed distribution	Survey results
Under 20	1%	30
20–29	14%	200
30–39	17%	300
40–49	18%	270
50–59	18%	150
60–69	12%	40
70+	20%	10

TABLE FOR EXERCISE 1

Type of fraud	Age								
	Under 20	**20–29**	**30–39**	**40–49**	**50–59**	**60–69**	**70–79**	**80+**	**Total**
Sweepstakes	10	60	70	130	90	160	280	200	1000
Credit cards	20	180	260	240	180	70	30	20	1000
Total	30	240	330	370	270	230	310	220	2000

TECHNOLOGY

MINITAB   EXCEL   TI-83/84 PLUS

TEACHER SALARIES

In 1916, the American Federation of Teachers (AFT) was formed by three teacher groups in Chicago, Illinois and locals from Gary, Indiana; New York, New York; Scranton, Pennsylvania; and Washington, D.C. Today, the AFT represents over 1.4 million teachers, higher education faculty and staff, school support staff, state and municipal employees, and health care professionals.

Each year, the AFT publishes the *Survey and Analysis of Teacher Salary Trends*. This report focuses on national trends in teacher salaries, state comparisons, beginning teacher salaries, and salary data and living costs for the nation's 50 largest cities.

The table at the right shows the salaries of a random sample of teachers from California, Ohio, and Texas.

Teacher salaries		
California	**Ohio**	**Texas**
66,645	49,400	45,300
56,622	50,500	47,800
47,400	46,750	43,400
65,000	62,125	39,605
52,150	68,900	52,425
69,200	45,300	57,200
74,400	49,080	45,000
59,400	54,525	50,150
68,378	51,400	39,500
64,873	49,800	41,680
59,395	48,950	43,075
62,000	69,300	46,450
69,200	41,860	51,970
73,400	56,526	41,795
71,405	57,900	40,100
58,200	54,300	40,822

■ EXERCISES

In Exercises 1–3, refer to the following samples. Use $\alpha = 0.05$.

(a) California teachers

(b) Ohio teachers

(c) Texas teachers

1. Are the samples independent of each other? Explain.

2. Use a technology tool to determine whether each sample is from a normal, or approximately normal, population.

3. Use a technology tool to determine whether the samples were selected from populations having equal variances.

4. Using the results of Exercises 1–3, discuss whether the three conditions for a one-way ANOVA test are satisfied. If so, use a technology tool to test the claim that teachers from California, Ohio, and Texas have the same mean salary. Use $\alpha = 0.05$.

5. Repeat Exercises 1–4 using the data in the table below. The table displays the salaries of random samples of teachers from Alaska, Nevada, and New York.

Teacher salaries		
Alaska	**Nevada**	**New York**
64,350	72,320	78,106
55,425	55,750	87,500
72,100	36,800	89,100
54,900	45,556	42,300
37,160	66,800	47,350
48,868	44,874	39,781
62,585	44,805	37,660
56,185	41,452	54,100
54,232	49,678	93,100
50,252	54,459	38,150
49,269	63,084	90,100
40,160	45,465	40,108
42,585	35,699	45,714
53,495	47,505	83,850
50,262	45,002	38,099
82,870	41,387	48,102

Extended solutions are given in the *Technology Supplement*.
Technical instruction is provided for MINITAB, Excel, and the TI-83/84 Plus.

11 NONPARAMETRIC TESTS

In a recent year, the most common form of reported identity theft was credit card fraud (20%), followed by government documents/benefits fraud (15%), employment fraud (15%), and phone or utilities fraud (13%).

Up to this point in the text, you have studied dozens of different statistical formulas and tests that can help you in a decision-making process. Specific conditions had to be satisfied in order to use these formulas and tests.

Suppose it is believed that as the number of fraud complaints in a state increases, the number of identity theft victims also increases. Can this belief be supported by actual data? The data below show the number of fraud complaints and the number of identify theft victims for 25 randomly selected states in a recent year.

| Fraud complaints | 29,506 | 22,805 | 1535 | 10,556 | 8099 | 106,623 | 2630 | 8978 | |
| Identity theft victims | 8237 | 8363 | 296 | 3292 | 2696 | 51,140 | 759 | 3819 | |

| Fraud complaints | 5895 | 57,472 | 19,585 | 15,159 | 2253 | 12,584 | 4807 | 7101 | |
| Identity theft victims | 1347 | 24,440 | 5412 | 4589 | 490 | 2937 | 2081 | 2005 | |

| Fraud complaints | 8173 | 24,695 | 7345 | 15,515 | 21,730 | 20,610 | 13,259 | 4498 | 30,578 |
| Identity theft victims | 2396 | 6349 | 1775 | 5408 | 5855 | 9683 | 3528 | 2367 | 13,726 |

In this chapter, you will study additional statistical tests that do not require the population distribution to meet any specific conditions. Each of these tests has usefulness in real-life applications.

With the data above, the number of fraud complaints F and the number of identity theft victims V can be related by the regression equation $V = 0.472F - 1802.101$. The correlation coefficient is approximately 0.987, so there is a strong positive correlation. You can determine that the correlation is significant by using Table 11 in Appendix B, but the V-values do not pass the normality requirement.

So, although a simple correlation test might indicate a relationship between the number of fraud complaints and the number of identity theft victims, one might question the results because the data do not fit the requirements

for the test. Similar tests you will study in this chapter, such as Spearman's rank correlation test, will give you additional information. The Spearman's rank correlation coefficient for this data is approximately 0.971. At $\alpha = 0.01$, there is in fact a significant correlation between the number of fraud complaints and the number of identity theft victims for each state.

Number of Fraud Complaints and Identity Theft Victims for 25 States

11.1 The Sign Test

The Sign Test for a Population Median ▸ The Paired-Sample Sign Test

WHAT YOU SHOULD LEARN

▸ How to use the sign test to test a population median

▸ How to use the paired-sample sign test to test the difference between two population medians (dependent samples)

▸ THE SIGN TEST FOR A POPULATION MEDIAN

Many of the hypothesis tests studied so far have imposed one or more requirements for a population distribution. For instance, some tests require that a population must have a normal distribution, and other tests require that population variances be equal. What if, for a given test, such requirements cannot be met? For these cases, statisticians have developed hypothesis tests that are "distribution free." Such tests are called *nonparametric tests*.

DEFINITION

A **nonparametric test** is a hypothesis test that does not require any specific conditions concerning the shapes of population distributions or the values of population parameters.

Nonparametric tests are usually easier to perform than corresponding parametric tests. However, they are usually less efficient than parametric tests. Stronger evidence is required to reject a null hypothesis using the results of a nonparametric test. Consequently, whenever possible, you should use a parametric test. One of the easiest nonparametric tests to perform is the *sign test*.

INSIGHT

For many nonparametric tests, statisticians test the median instead of the mean.

DEFINITION

The **sign test** is a nonparametric test that can be used to test a population median against a hypothesized value k.

The sign test for a population median can be left-tailed, right-tailed, or two-tailed. The null and alternative hypotheses for each type of test are as follows.

Left-tailed test:
H_0: median $\geq k$ and H_a: median $< k$

Right-tailed test:
H_0: median $\leq k$ and H_a: median $> k$

Two-tailed test:
H_0: median $= k$ and H_a: median $\neq k$

To use the sign test, first compare each entry in the sample with the hypothesized median k. If the entry is below the median, assign it a $-$ sign; if the entry is above the median, assign it a $+$ sign; and if the entry is equal to the median, assign it a 0. Then compare the number of $+$ and $-$ signs. (The 0's are ignored.) If there is a large difference between the number of $+$ signs and the number of $-$ signs, it is likely that the median is different from the hypothesized value and the null hypothesis should be rejected.

Table 8 in Appendix B lists the critical values for the sign test for selected levels of significance and sample sizes. When the sign test is used, the sample size n is the total number of $+$ and $-$ signs. If the sample size is greater than 25, you can use the standard normal distribution to find the critical values.

TEST STATISTIC FOR THE SIGN TEST

When $n \leq 25$, the **test statistic** for the sign test is x, the smaller number of $+$ or $-$ signs.

When $n > 25$, the **test statistic** for the sign test is

$$z = \frac{(x + 0.5) - 0.5n}{\dfrac{\sqrt{n}}{2}}$$

where x is the smaller number of $+$ or $-$ signs and n is the sample size, i.e., the total number of $+$ and $-$ signs.

Because x is defined to be the smaller number of $+$ or $-$ signs, the rejection region is always in the left tail. Consequently, the sign test for a population median is always a left-tailed test or a two-tailed test. When the test is two-tailed, use only the left-tailed critical value. (If x is defined to be the larger number of $+$ or $-$ signs, the rejection region is always in the right tail. Right-tailed sign tests are presented in the exercises.)

GUIDELINES

Performing a Sign Test for a Population Median

IN WORDS	IN SYMBOLS
1. Identify the claim. State the null and alternative hypotheses.	State H_0 and H_a.
2. Specify the level of significance.	Identify α.
3. Determine the sample size n by assigning $+$ signs and $-$ signs to the sample data.	$n = $ total number of $+$ and $-$ signs
4. Determine the critical value.	If $n \leq 25$, use Table 8 in Appendix B. If $n > 25$, use Table 4 in Appendix B.
5. Find the test statistic.	If $n \leq 25$, use x. If $n > 25$, use $$z = \frac{(x + 0.5) - 0.5n}{\dfrac{\sqrt{n}}{2}}.$$
6. Make a decision to reject or fail to reject the null hypothesis.	If the test statistic is less than or equal to the critical value, reject H_0. Otherwise, fail to reject H_0.
7. Interpret the decision in the context of the original claim.	

EXAMPLE 1  SC Report 50

▶ Using the Sign Test

A website administrator for a company claims that the median number of visitors per day to the company's website is no more than 1500. An employee doubts the accuracy of this claim. The number of visitors per day for 20 randomly selected days are listed below. At $\alpha = 0.05$, can the employee reject the administrator's claim?

1469	1462	1634	1602	1500
1463	1476	1570	1544	1452
1487	1523	1525	1548	1511
1579	1620	1568	1492	1649

▶ Solution

The claim is "the median number of visitors per day to the company's website is no more than 1500." So, the null and alternative hypotheses are

H_0: median $\leq$ 1500 (Claim)   and H_a: median $>$ 1500.

The results of comparing each data entry with the hypothesized median 1500 are shown.

−	−	+	+	0
−	−	+	+	−
−	+	+	+	+
+	+	+	−	+

You can see that there are 7 − signs and 12 + signs. So, $n = 12 + 7 = 19$. Because $n \leq 25$, use Table 8 to find the critical value. The test is a one-tailed test with $\alpha = 0.05$ and $n = 19$. So, the critical value is 5. Because $n \leq 25$, the test statistic x is the smaller number of + or − signs. So, $x = 7$. Because $x = 7$ is greater than the critical value, the employee should fail to reject the null hypothesis.

Interpretation   There is not enough evidence at the 5% level of significance for the employee to reject the website administrator's claim that the median number of visitors per day to the company's website is no more than 1500.

▶ Try It Yourself 1

A real estate agency claims that the median number of days a home is on the market in its city is greater than 120. A homeowner wants to verify the accuracy of this claim. The number of days on the market for 24 randomly selected homes is shown below. At $\alpha = 0.025$, can the homeowner support the agency's claim?

118	167	72	79	76	106	102	113
73	119	162	114	120	93	135	147
77	157	115	88	152	70	65	91

a. Identify the *claim* and state H_0 and H_a.
b. Specify the *level of significance* α.
c. Determine the *sample size n*.
d. Determine the *critical value*.
e. Find the *test statistic x*.
f. *Decide* whether to reject the null hypothesis.
g. *Interpret* the decision in the context of the original claim.

Answer: Page A47

In 2008, people in the United States spent a total of about $15.5 billion on candy. The U.S. Department of Commerce reported that in 2008, the average person in the United States ate about 22.4 pounds of candy.

Candy Consumption

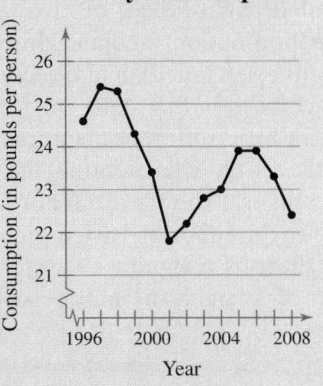

If you were to test the U.S. Department of Commerce's claim concerning per capita candy consumption, would you use a parametric test or a nonparametric test? What factors must you consider?

Nonparametric test; Factors that must be considered include sample size, whether the population is normally distributed, and whether the samples were randomly selected.

EXAMPLE 2

▶ Using the Sign Test

An organization claims that the median annual attendance for museums in the United States is at least 39,000. A random sample of 125 museums reveals that the annual attendances for 79 museums were less than 39,000, the annual attendances for 42 museums were more than 39,000, and the annual attendances for 4 museums were 39,000. At $\alpha = 0.01$, is there enough evidence to reject the organization's claim? *(Adapted from American Association of Museums)*

▶ Solution

The claim is "the median annual attendance for museums in the United States is at least 39,000." So, the null and alternative hypotheses are

$$H_0: \text{median} \geq 39,000 \text{ (Claim)} \quad \text{and} \quad H_a: \text{median} < 39,000.$$

Because $n > 25$, use Table 4, the Standard Normal Table, to find the critical value. Because the test is a left-tailed test with $\alpha = 0.01$, the critical value is $z_0 = -2.33$. Of the 125 museums, there are $79 -$ signs and $42 +$ signs. When the 0s are ignored, the sample size is

$$n = 79 + 42 = 121, \quad \text{and} \quad x = 42.$$

With these values, the test statistic is

$$z = \frac{(42 + 0.5) - 0.5(121)}{\sqrt{121}/2}$$

$$= \frac{-18}{5.5}$$

$$\approx -3.27.$$

The graph at the right shows the location of the rejection region and the test statistic z. Because z is less than the critical value, it is in the rejection region. So, you should reject the null hypothesis.

Interpretation There is enough evidence at the 1% level of significance to reject the organization's claim that the median annual attendance for museums in the United States is at least 39,000.

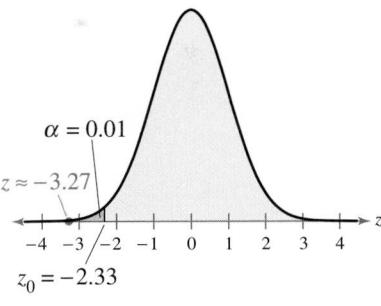

$\alpha = 0.01$

$z \approx -3.27$

$z_0 = -2.33$

▶ Try It Yourself 2

An organization claims that the median age of automobiles in operation in the United States is 9.4 years. A random sample of 95 automobiles reveals that 41 automobiles were less than 9.4 years old and 51 automobiles were more than 9.4 years old. At $\alpha = 0.10$, can you reject the organization's claim? *(Source: Bureau of Transportation Statistics)*

a. Identify the *claim* and state H_0 and H_a.
b. Specify the *level of significance* α.
c. Determine the *sample size n*.
d. Determine the *critical value*.
e. Find the *test statistic z*.
f. *Decide* whether to reject the null hypothesis.
g. *Interpret* the decision in the context of the original claim.

Answer: Page A47

▶ THE PAIRED-SAMPLE SIGN TEST

In Section 8.3, you learned how to use a *t*-test for the difference between means of dependent samples. That test required both populations to be normally distributed. If the parametric condition of normality cannot be satisfied, you can use the paired-sample sign test to test the difference between two population medians. To perform the paired-sample sign test for the difference between two population medians, the following conditions must be met.

1. A sample must be randomly selected from each population.

2. The samples must be dependent (paired).

The paired-sample sign test can be left-tailed, right-tailed, or two-tailed. This test is similar to the sign test for a single population median. However, instead of comparing each data entry with a hypothesized median and recording a +, −, or 0, you find the difference between corresponding data entries and record the sign of the difference. Generally, to find the difference, subtract the entry representing the second variable from the entry representing the first variable. Then compare the number of + and − signs. (The 0's are ignored.) If the number of + signs is approximately equal to the number of − signs, the null hypothesis should not be rejected. If, however, there is a significant difference between the number of + signs and the number of − signs, the null hypothesis should be rejected.

GUIDELINES

Performing a Paired-Sample Sign Test

IN WORDS	IN SYMBOLS
1. Identify the claim. State the null and alternative hypotheses.	State H_0 and H_a.
2. Specify the level of significance.	Identify α.
3. Determine the sample size *n* by finding the difference for each data pair. Assign a + sign for a positive difference, a − sign for a negative difference, and a 0 for no difference.	n = total number of + and − signs
4. Determine the critical value.	Use Table 8 in Appendix B.
5. Find the test statistic.	x = smaller number of + or − signs
6. Make a decision to reject or fail to reject the null hypothesis.	If the test statistic is less than or equal to the critical value, reject H_0. Otherwise, fail to reject H_0.
7. Interpret the decision in the context of the original claim.	

EXAMPLE 3

▸ Using the Paired-Sample Sign Test

A psychologist claims that the number of repeat offenders will decrease if first-time offenders complete a particular rehabilitation course. You randomly select 10 prisons and record the number of repeat offenders during a two-year period. Then, after first-time offenders complete the course, you record the number of repeat offenders at each prison for another two-year period. The results are shown in the following table. At $\alpha = 0.025$, can you support the psychologist's claim?

Prison	1	2	3	4	5	6	7	8	9	10
Before	21	34	9	45	30	54	37	36	33	40
After	19	22	16	31	21	30	22	18	17	21

▸ Solution

To support the psychologist's claim, you could use the following null and alternative hypotheses.

H_0: The number of repeat offenders will not decrease.

H_a: The number of repeat offenders will decrease. (Claim)

The table below shows the sign of the differences between the "before" and "after" data.

Prison	1	2	3	4	5	6	7	8	9	10
Before	21	34	9	45	30	54	37	36	33	40
After	19	22	16	31	21	30	22	18	17	21
Sign	+	+	−	+	+	+	+	+	+	+

You can see that there is $1 -$ sign and there are $9 +$ signs. So, $n = 1 + 9 = 10$. In Table 8 with $\alpha = 0.025$ (one-tailed) and $n = 10$, the critical value is 1. The test statistic x is the smaller number of $+$ or $-$ signs. So, $x = 1$. Because x is equal to the critical value, you should reject the null hypothesis.

Interpretation   There is enough evidence at the 2.5% level of significance to support the psychologist's claim that the number of repeat offenders will decrease.

▸ Try It Yourself 3

A medical researcher claims that a new vaccine will decrease the number of colds in adults. You randomly select 14 adults and record the number of colds each has in a one-year period. After giving the vaccine to each adult, you again record the number of colds each has in a one-year period. The results are shown in the table at the left. At $\alpha = 0.05$, can you support the researcher's claim?

a. Identify the *claim* and state H_0 and H_a.
b. Specify the *level of significance α*.
c. Determine the *sample size n*.
d. Determine the *critical value*.
e. Find the *test statistic x*.
f. *Decide* whether to reject the null hypothesis.
g. *Interpret* the decision in the context of the original claim.

Answer: Page A47

Adult	Before vaccine	After vaccine
1	3	2
2	4	1
3	2	0
4	1	1
5	3	1
6	6	3
7	4	3
8	5	2
9	2	2
10	0	2
11	2	3
12	5	4
13	3	3
14	3	2

11.1 EXERCISES

FOR EXTRA HELP:
MyStatLab

1. A nonparametric test is a
   hypothesis test that does not
   require any specific conditions
   concerning the shapes of
   populations or the values of
   population parameters.

   A nonparametric test is usually
   easier to perform than its
   corresponding parametric test,
   but the nonparametric test is
   usually less efficient.

2. Median

3. See Odd Answers, page A104.

4. See Selected Answers, page A133.

5. Identify the claim and state H_0
   and H_a. Identify the level of
   significance and sample size.
   Find the critical value using
   Table 8 (if $n \leq 25$) or Table 4
   ($n > 25$). Calculate the test
   statistic. Make a decision and
   interpret it in the context of
   the problem.

6. See Selected Answers, page A133.

7. (a) H_0: median $\leq$ \$300;
      H_a: median $>$ \$300 (claim)

   (b) 1   (c) 5

   (d) Fail to reject H_0.

   (e) There is not enough evidence
       at the 1% level of significance
       for the accountant to
       conclude that the median
       amount of new credit card
       charges for the previous
       month was more than \$300.

8. See Selected Answers, page A133.

9. See Odd Answers, page A104.

■ BUILDING BASIC SKILLS AND VOCABULARY

1. What is a nonparametric test? How does a nonparametric test differ from a parametric test? What are the advantages and disadvantages of using a nonparametric test?

2. When the sign test is used, what population parameter is being tested?

3. Describe the test statistic for the sign test when the sample size n is less than or equal to 25 and when n is greater than 25.

4. In your own words, explain why the hypothesis test discussed in this section is called the sign test.

5. Explain how to use the sign test to test a population median.

6. List the two conditions that must be met in order to use the paired-sample sign test.

■ USING AND INTERPRETING CONCEPTS

Performing a Sign Test *In Exercises 7–22, (a) identify the claim and state H_0 and H_a, (b) determine the critical value, (c) find the test statistic, (d) decide whether to reject or fail to reject the null hypothesis, and (e) interpret the decision in the context of the original claim.*

7. **Credit Card Charges** In order to estimate the median amount of new credit card charges for the previous month, a financial service accountant randomly selects 12 credit card accounts and records the amount of new charges for each account for the previous month. The amounts (in dollars) are listed below. At $\alpha = 0.01$, can the accountant conclude that the median amount of new credit card charges for the previous month was more than \$300? *(Adapted from Board of Governors of the Federal Reserve System)*

   | 346.71 | 382.59 | 255.03 | 202.17 | 309.80 | 265.88 |
   | 299.41 | 270.38 | 296.54 | 318.46 | 245.92 | 309.47 |

8. **Temperature** A meteorologist estimates that the median daily high temperature for the month of July in Pittsburgh is 83° Fahrenheit. The high temperatures (in degrees Fahrenheit) for 15 randomly selected July days in Pittsburgh are listed below. At $\alpha = 0.01$, is there enough evidence to reject the meteorologist's claim? *(Adapted from U.S. National Oceanic and Atmospheric Administration)*

   | 74 | 79 | 81 | 86 | 90 | 79 | 81 | 83 | 81 | 74 | 78 | 76 | 84 | 82 | 85 |

9. **Sales Prices of Homes** A real estate agent believes that the median sales price of new privately owned one-family homes sold in the past year is \$198,000 or less. The sales prices (in dollars) of 10 randomly selected homes are listed below. At $\alpha = 0.05$, is there enough evidence to reject the agent's claim? *(Adapted from National Association of Realtors)*

   | 205,800 | 234,500 | 210,900 | 195,700 | 145,200 |
   | 198,900 | 254,000 | 175,900 | 189,500 | 212,500 |

10. See Selected Answers, page A133.

11. (a) H_0: median $\geq$ \$3000 (claim);
H_a: median $<$ \$3000

(b) -2.05 (c) -1.47

(d) Fail to reject H_0.

(e) There is not enough evidence at the 2% level of significance to reject the institution's claim that the median amount of credit card debt for families holding such debts is at least \$3000.

12. See Selected Answers, page A133.

13. (a) H_0: median $\leq$ 30;
H_a: median $>$ 30 (claim)

(b) 4   (c) 10

(d) Fail to reject H_0.

(e) There is not enough evidence at the 1% level of significance to support the research group's claim that the median age of Twitter® users is greater than 30 years old.

14. See Selected Answers, page A133.

15. (a) H_0: median $=$ 4 (claim);
H_a: median $\neq$ 4

(b) -1.96 (c) -1.90

(d) Fail to reject H_0.

(e) There is not enough evidence at the 5% level of significance to reject the organization's claim that the median number of rooms in renter-occupied units is 4.

16. (a) H_0: median $=$ 1350 (claim);
H_a: median $\neq$ 1350

(b) 5   (c) 7

(d) Fail to reject H_0.

(e) There is not enough evidence at the 10% level of significance to reject the organization's claim that the median square footage of renter-occupied units is 1350 square feet.

17. (a) H_0: median $=$ \$37.06 (claim);
H_a: median $\neq$ \$37.06

(b) -2.575 (c) -0.91

(d) Fail to reject H_0.

(e) There is not enough evidence at the 1% level of significance to reject the labor organization's claim that the median hourly wage of computer systems analysts is \$37.06.

10. Temperature During a weather report, a meteorologist states that the median daily high temperature for the month of January in San Diego is 66° Fahrenheit. The high temperatures (in degrees Fahrenheit) for 16 randomly selected January days in San Diego are listed below. At $\alpha = 0.01$, can you reject the meteorologist's claim? *(Adapted from U.S. National Oceanic and Atmospheric Administration)*

78 74 72 72 70 70 72 78 74 71 72 74 77 79 75 73

11. Credit Card Debt A financial services institution reports that the median amount of credit card debt for families holding such debts is at least \$3000. In a random sample of 104 families holding debt, the debts of 60 families were less than \$3000 and the debts of 44 families were greater than \$3000. At $\alpha = 0.02$, can you reject the institution's claim? *(Adapted from Board of Governors of the Federal Reserve System)*

12. Financial Debt A financial services accountant estimates that the median amount of financial debt for families holding such debts is less than \$65,000. In a random sample of 70 families holding debts, the debts of 24 families were less than \$65,000 and the debts of 46 families were greater than \$65,000. At $\alpha = 0.025$, can you support the accountant's estimate? *(Adapted from Board of Governors of the Federal Reserve System)*

13. Twitter® Users A research group claims that the median age of Twitter® users is greater than 30 years old. In a random sample of 24 Twitter® users, 11 are less than 30 years old, 10 are more than 30 years old, and 3 are 30 years old. At $\alpha = 0.01$, can you support the research group's claim? *(Adapted from Pew Research Center)*

14. Facebook® Users A research group claims that the median age of Facebook® users is less than 32 years old. In a random sample of 20 Facebook® users, 5 are less than 32 years old, 13 are more than 32 years old, and 2 are 32 years old. At $\alpha = 0.05$, can you support the research group's claim? *(Adapted from Pew Research Center)*

15. Unit Size A renters' organization claims that the median number of rooms in renter-occupied units is four. You randomly select 120 renter-occupied units and obtain the results shown below. At $\alpha = 0.05$, can you reject the organization's claim? *(Adapted from U.S. Census Bureau)*

Unit size	Number of units
Fewer than 4 rooms	31
4 rooms	40
More than 4 rooms	49

TABLE FOR EXERCISE 15

Unit size	Number of units
Less than 1350	7
1350	3
More than 1350	12

TABLE FOR EXERCISE 16

16. Square Footage A renters' organization believes that the median square footage of renter-occupied units is 1350 square feet. To test this claim, you randomly select 22 renter-occupied units and obtain the results shown above. At $\alpha = 0.10$, can you reject the organization's claim? *(Adapted from U.S. Census Bureau)*

17. Hourly Wages A labor organization estimates that the median hourly wage of computer systems analysts is \$37.06. In a random sample of 45 computer systems analysts, 18 are paid less than \$37.06 per hour, 25 are paid more than \$37.06 per hour, and 2 are paid \$37.06 per hour. At $\alpha = 0.01$, can you reject the labor organization's claim? *(Adapted from U.S. Bureau of Labor Statistics)*

18. (a) H_0: median $\geq$ \$55.89 (claim);
H_a: median $<$ \$55.89

(b) 6 (c) 5 (d) Reject H_0.

(e) There is enough evidence at the 5% level of significance to reject the organization's claim that the median hourly wage of podiatrists is at least \$55.89.

19. (a) H_0: The lower back pain intensity scores have not decreased.

H_a: The lower back pain intensity scores have decreased. (claim)

(b) 1 (c) 0 (d) Reject H_0.

(e) There is enough evidence at the 5% level of significance to conclude that the lower back pain intensity scores were lower after the acupuncture.

20. (a) H_0: The lower back pain intensity scores have not decreased.

H_a: The lower back pain intensity scores have decreased. (claim)

(b) 2 (c) 4

(d) Fail to reject H_0.

(e) There is not enough evidence at the 5% level of significance to conclude that the lower back pain intensity scores decreased after taking the anti-inflammatory drugs.

21. (a) H_0: The SAT scores have not improved.

H_a: The SAT scores have improved. (claim)

(b) 2 (c) 4

(d) Fail to reject H_0.

(e) There is not enough evidence at the 5% level of significance to conclude that the critical reading SAT scores improved.

18. Hourly Wages A labor organization estimates that the median hourly wage of podiatrists is at least \$55.89. In a random sample of 23 podiatrists, 17 are paid less than \$55.89 per hour, 5 are paid more than \$55.89 per hour, and 1 is paid \$55.89 per hour. At $\alpha = 0.05$, can you reject the labor organization's claim? *(Adapted from U.S. Bureau of Labor Statistics)*

19. Lower Back Pain The table shows the lower back pain intensity scores for eight patients before and after receiving acupuncture for eight weeks. At $\alpha = 0.05$, is there enough evidence to conclude that the lower back pain intensity scores decreased after the acupuncture? *(Adapted from Archives of Internal Medicine)*

Patient	1	2	3	4	5	6	7	8
Intensity score (before)	59.2	46.3	65.4	74.0	79.3	81.6	44.4	59.1
Intensity score (after)	12.4	22.5	18.6	59.3	70.1	70.2	13.2	25.9

20. Lower Back Pain The table shows the lower back pain intensity scores for 12 patients before and after taking anti-inflammatory drugs for 8 weeks. At $\alpha = 0.05$, is there enough evidence to conclude that the lower back pain intensity scores decreased after taking anti-inflammatory drugs? *(Adapted from Archives of Internal Medicine)*

Patient	1	2	3	4	5	6
Intensity score (before)	71.0	42.1	79.1	57.5	64.0	60.4
Intensity score (after)	60.1	23.4	86.2	62.1	44.2	49.7

Patient	7	8	9	10	11	12
Intensity score (before)	68.3	95.2	48.1	78.6	65.4	59.9
Intensity score (after)	58.3	72.6	51.8	82.5	63.2	47.9

21. Improving SAT Scores A tutoring agency believes that by completing a special course, students can improve their critical reading SAT scores. In part of a study, 12 students take the critical reading part of the SAT, complete the special course, then take the critical reading part of the SAT again. The students' scores are shown below. At $\alpha = 0.05$, is there enough evidence to conclude that the students' critical reading SAT scores improved?

Student	1	2	3	4	5	6
Score on first SAT	308	456	352	433	306	471
Score on second SAT	300	524	409	419	304	483

Student	7	8	9	10	11	12
Score on first SAT	538	207	205	351	360	251
Score on second SAT	708	253	399	350	480	303

22. See Selected Answers, page A134.

23. (a) Reject H_0.

 (b) There is enough evidence at the 5% level of significance to reject the claim that the proportion of adults who feel older than their real age is equal to the proportion of adults who feel younger than their real age.

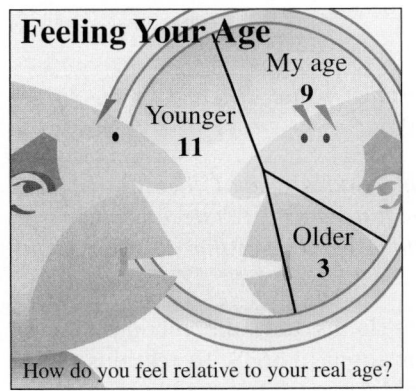

Feeling Your Age

My age **9**

Younger **11**

Older **3**

How do you feel relative to your real age?

Adapted from USA TODAY Snapshot, September 1, 2009.

FIGURE FOR EXERCISE 23

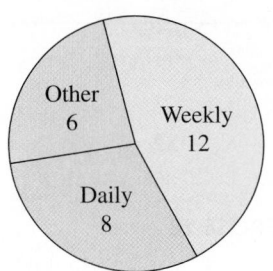

Other **6**

Weekly **12**

Daily **8**

FIGURE FOR EXERCISE 24

24. (a) Fail to reject H_0.

 (b) There is not enough evidence at the 5% level of significance to reject the claim that the proportion of adults who contact their parents by phone weekly is equal to the proportion of adults who contact their parents by phone daily.

25. See Odd Answers, page A105.

26. See Selected Answers, page A134.

22. SAT Scores   Students at a certain school are required to take the SAT twice. The table shows both critical reading SAT scores for 12 students. At $\alpha = 0.01$, can you conclude that the students' critical reading scores improved the second time they took the SAT?

Student	1	2	3	4	5	6
Score on first SAT	445	510	429	452	629	453
Score on second SAT	446	571	517	478	610	453

Student	7	8	9	10	11	12
Score on first SAT	358	477	325	513	636	571
Score on second SAT	378	532	299	501	648	603

23. Feeling Your Age   A research organization conducts a survey by randomly selecting adults and asking them how they feel relative to their real age. The results are shown in the figure. *(Adapted from Pew Research Center)*

 (a) Use a sign test to test the null hypothesis that the proportion of adults who feel older than their real age is equal to the proportion of adults who feel younger than their real age. Assign a + sign to adults who feel older than their real age, assign a − sign to adults who feel younger than their real age, and assign a 0 to adults who feel their age. Use $\alpha = 0.05$.

 (b) What can you conclude?

24. Contacting Parents   A research organization conducts a survey by randomly selecting adults and asking them how frequently they contact their parents by phone. The results are shown in the figure. *(Adapted from Pew Research Center)*

 (a) Use a sign test to test the null hypothesis that the proportion of adults who contact their parents by phone weekly is equal to the proportion of adults who contact their parents by phone daily. Assign a + sign to adults who contact their parents by phone weekly, assign a − sign to adults who contact their parents by phone daily, and assign a 0 to adults who answer "other." Use $\alpha = 0.05$.

 (b) What can you conclude?

SC *In Exercises 25 and 26, use StatCrunch to help you test the claim about the population median.*

25. Hourly Wages   A labor organization claims that the median hourly wage of tool and die makers is $22.55. The hourly wages (in dollars) of 14 randomly selected tool and die makers are listed below. At $\alpha = 0.05$, is there enough evidence to reject the labor organization's claim? *(Adapted from U.S. Bureau of Labor Statistics)*

21.75   23.10   20.50   25.80   29.25   26.35   27.40
22.90   23.50   22.55   32.70   30.05   29.80   34.85

26. Viewing Audience   A television network claims that the median age of viewers for the Masters Golf Tournament is greater than 57 years. The ages of 24 randomly selected viewers are listed below. At $\alpha = 0.01$, is there enough evidence to support the network's claim? *(Adapted from ESPN)*

60   85   70   59   42   21   57   25   65   71   33   40
54   50   57   49   50   30   27   57   17   90   35   46

27. (a) H_0: median $\leq$ \$638 (claim);
H_a: median $>$ \$638

(b) 2.33   (c) 1.46

(d) Fail to reject H_0.

(e) There is not enough evidence at the 1% level of significance to reject the organization's claim that the median weekly earnings of female workers is less than or equal to \$638.

28. (a) H_0: median $\leq$ \$798;
H_a: median $>$ \$798 (claim)

(b) 2.33   (c) 2.55

(d) Reject H_0.

(e) There is enough evidence at the 1% level of significance to support the organization's claim that the median weekly earnings of male workers is greater than \$798.

29. (a) H_0: median $\leq$ 26 (claim);
H_a: median $>$ 26

(b) 1.645   (c) 1.302

(d) Fail to reject H_0.

(e) There is not enough evidence at the 5% level of significance to reject the counselor's claim that the median age of brides at the time of their first marriage is less than or equal to 26 years.

30. (a) H_0: median $\leq$ 28;
H_a: median $>$ 28 (claim)

(b) 1.645   (c) 1.203

(d) Fail to reject H_0.

(e) There is not enough evidence at the 5% level of significance to support the counselor's claim that the median age of grooms at the time of their first marriage is greater than 28 years.

■ EXTENDING CONCEPTS

More on Sign Tests   *When you are using a sign test for $n > 25$ and the test is left-tailed, you know you can reject the null hypothesis if the test statistic*

$$z = \frac{(x + 0.5) - 0.5n}{\frac{\sqrt{n}}{2}}$$

*is less than or equal to the **left-tailed** critical value, where x is the **smaller** number of $+$ or $-$ signs. For a right-tailed test, you can reject the null hypothesis if the test statistic*

$$z = \frac{(x - 0.5) - 0.5n}{\frac{\sqrt{n}}{2}}$$

*is greater than or equal to the **right-tailed** critical value, where x is the **larger** number of $+$ or $-$ signs.*

In Exercises 27–30, (a) write the claim mathematically and identify H_0 and H_a, (b) determine the critical value, (c) find the test statistic, (d) decide whether to reject or fail to reject the null hypothesis, and (e) interpret the decision in the context of the original claim.

27. Weekly Earnings   A labor organization claims that the median weekly earnings of female workers is less than or equal to \$638. To test this claim, you randomly select 50 female workers and ask each to provide her weekly earnings. The results are shown in the table. At $\alpha = 0.01$, can you reject the organization's claim? *(Adapted from U.S. Bureau of Labor Statistics)*

Weekly earnings	Number of workers
Less than \$638	18
\$638	3
More than \$638	29

TABLE FOR EXERCISE 27

Weekly earnings	Number of workers
Less than \$798	23
\$798	2
More than \$798	45

TABLE FOR EXERCISE 28

28. Weekly Earnings   A labor organization states that the median weekly earnings of male workers is greater than \$798. To test this claim, you randomly select 70 male workers and ask each to provide his weekly earnings. The results are shown in the table. At $\alpha = 0.01$, can you support the organization's claim? *(Adapted from U.S. Bureau of Labor Statistics)*

29. Ages of Brides   A marriage counselor estimates that the median age of brides at the time of their first marriage is less than or equal to 26 years. In a random sample of 65 brides, 24 are less than 26 years old, 35 are more than 26 years old, and 6 are 26 years old. At $\alpha = 0.05$, can you reject the counselor's claim? *(Adapted from U.S. Census Bureau)*

30. Ages of Grooms   A marriage counselor estimates that the median age of grooms at the time of their first marriage is greater than 28 years. In a random sample of 56 grooms, 33 are less than 28 years old, 23 are more than 28 years old, and none are 28 years old. At $\alpha = 0.05$, can you support the counselor's claim? *(Adapted from U.S. Census Bureau)*

11.2 The Wilcoxon Tests

The Wilcoxon Signed-Rank Test ▸ The Wilcoxon Rank Sum Test

▸ THE WILCOXON SIGNED-RANK TEST

In this section, you will study the Wilcoxon signed-rank test and the Wilcoxon rank sum test. Unlike the sign test from Section 11.1, the strength of these two nonparametric tests is that each considers the magnitude, or size, of the data entries.

In Section 8.3, you used a *t*-test together with dependent samples to determine whether there was a difference between two populations. To use the *t*-test to test such a difference, you must assume (or know) that the dependent samples are randomly selected from populations having a normal distribution. But, what if this assumption cannot be made? Instead of using the two-sample *t*-test, you can use the *Wilcoxon signed-rank test*.

DEFINITION

The **Wilcoxon signed-rank test** is a nonparametric test that can be used to determine whether two *dependent* samples were selected from populations having the same distribution.

GUIDELINES

Performing a Wilcoxon Signed-Rank Test

IN WORDS	IN SYMBOLS
1. Identify the claim. State the null and alternative hypotheses.	State H_0 and H_a.
2. Specify the level of significance.	Identify α.
3. Determine the sample size n, which is the number of pairs of data for which the difference is not 0.	
4. Determine the critical value.	Use Table 9 in Appendix B.
5. Find the test statistic w_s.	Headers: **Sample 1, Sample 2, Difference, Absolute value, Rank,** and **Signed rank.** Signed rank takes on the same sign as its corresponding difference.
a. Complete a table using the headers listed at the right.	
b. Find the sum of the positive ranks and the sum of the negative ranks.	
c. Select the smaller absolute value of the sums.	
6. Make a decision to reject or fail to reject the null hypothesis.	If w_s is less than or equal to the critical value, reject H_0. Otherwise, fail to reject H_0.
7. Interpret the decision in the context of the original claim.	

EXAMPLE 1

▸ Performing a Wilcoxon Signed-Rank Test

A golf club manufacturer believes that golfers can lower their scores by using the manufacturer's newly designed golf clubs. The scores of 10 golfers while using the old design and while using the new design are shown in the table. At $\alpha = 0.05$, can you support the manufacturer's claim?

Golfer	1	2	3	4	5	6	7	8	9	10
Score (old design)	89	84	96	74	91	85	95	82	92	81
Score (new design)	83	83	92	76	91	80	87	85	90	77

▸ Solution

The claim is "golfers can lower their scores." To test this claim, use the following null and alternative hypotheses.

H_0: The new design does not lower scores.

H_a: The new design lowers scores. (Claim)

This Wilcoxon signed-rank test is a one-tailed test with $\alpha = 0.05$, and because one data pair has a difference of 0, $n = 9$ instead of 10. From Table 9 in Appendix B, the critical value is 8. To find the test statistic w_s, complete a table as shown below.

Score (old design)	Score (new design)	Difference	Absolute value	Rank	Signed rank
89	83	6	6	8	8
84	83	1	1	1	1
96	92	4	4	5.5	5.5
74	76	−2	2	2.5	−2.5
91	91	0	0	—	—
85	80	5	5	7	7
95	87	8	8	9	9
82	85	−3	3	4	−4
92	90	2	2	2.5	2.5
81	77	4	4	5.5	5.5

STUDY TIP

Do not assign a rank to any difference of 0. In the case of a tie between data entries, use the average of the corresponding ranks. For instance, if two data entries are tied for the fifth rank, use the average of 5 and 6, which is 5.5, as the rank for both entries. The next data entry will be assigned a rank of 7, not 6.

If three entries are tied for the fifth rank, use the average of 5, 6, and 7, which is 6, as the rank for all three data entries. The next data entry will be assigned a rank of 8.

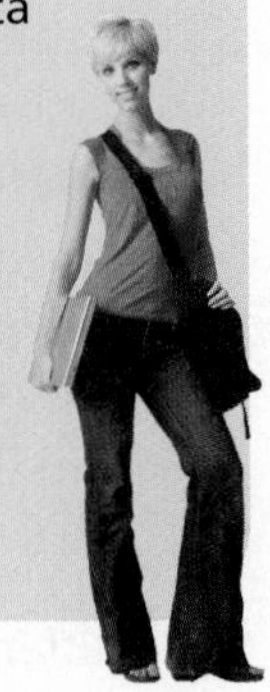

The sum of the negative ranks is

$$-2.5 + (-4) = -6.5.$$

The sum of the positive ranks is

$$8 + 1 + 5.5 + 7 + 9 + 2.5 + 5.5 = 38.5.$$

The test statistic is the smaller absolute value of these two sums. Because $|-6.5| < |38.5|$, the test statistic is $w_s = 6.5$. Because the test statistic is less than the critical value, that is, $6.5 < 8$, you should decide to reject the null hypothesis.

Interpretation There is enough evidence at the 5% level of significance to support the claim that golfers can lower their scores by using the newly designed clubs.

▶ **Try It Yourself 1**

A quality control inspector wants to test the claim that a spray-on water repellent is effective. To test this claim, he selects 12 pieces of fabric, sprays water on each, and measures the amount of water repelled (in milliliters). He then applies the water repellent and repeats the experiment. The results are shown in the table. At $\alpha = 0.01$, can he conclude that the water repellent is effective?

Fabric	1	2	3	4	5	6	7	8	9	10	11	12
No repellent	8	7	7	4	6	10	9	5	9	11	8	4
Repellent applied	15	12	11	6	6	8	8	6	12	8	14	8

a. Identify the *claim* and state H_0 and H_a.
b. Specify the *level of significance* α.
c. Determine the *sample size n*.
d. Determine the *critical value*.
e. Find the *test statistic w_s* by making a table, finding the sum of the positive ranks and the sum of the negative ranks, and finding the absolute value of each.
f. *Decide* whether to reject the null hypothesis.
g. *Interpret* the decision in the context of the original claim.

Answer: Page A47

▶ THE WILCOXON RANK SUM TEST

In Sections 8.1 and 8.2, you used a *z*-test or a *t*-test together with independent samples to determine whether there was a difference between two populations. To use the *z*-test to test such a difference, you must assume (or know) that the independent samples are randomly selected and that either each sample size is at least 30 or each population has a normal distribution with a known standard deviation. To use the *t*-test to test such a difference, you must assume (or know) that the independent samples are randomly selected from populations having a normal distribution. But, what if these assumptions cannot be made? You can still compare the populations using the *Wilcoxon rank sum test*.

DEFINITION

The **Wilcoxon rank sum test** is a nonparametric test that can be used to determine whether two *independent* samples were selected from populations having the same distribution.

A requirement for the Wilcoxon rank sum test is that the sample sizes of both samples must be at least 10. When calculating the test statistic for the Wilcoxon rank sum test, let n_1 represent the sample size of the smaller sample and n_2 represent the sample size of the larger sample. If the two samples have the same size, it does not matter which one is n_1 or n_2.

When calculating the sum of the ranks R, combine both samples and rank the combined data. Then sum the ranks for the smaller of the two samples. If the two samples have the same size, you can use the ranks from either sample, but you must use the ranks from the sample you associate with n_1.

TEST STATISTIC FOR THE WILCOXON RANK SUM TEST

Given two independent samples, the **test statistic z** for the Wilcoxon rank sum test is

$$z = \frac{R - \mu_R}{\sigma_R}$$

where

R = sum of the ranks for the smaller sample,

$$\mu_R = \frac{n_1(n_1 + n_2 + 1)}{2},$$

and

$$\sigma_R = \sqrt{\frac{n_1 n_2 (n_1 + n_2 + 1)}{12}}.$$

GUIDELINES

Performing a Wilcoxon Rank Sum Test

IN WORDS	IN SYMBOLS
1. Identify the claim. State the null and alternative hypotheses.	State H_0 and H_a.
2. Specify the level of significance.	Identify α.
3. Determine the critical value(s) and the rejection region(s).	Use Table 4 in Appendix B.
4. Determine the sample sizes.	$n_1 \leq n_2$
5. Find the sum of the ranks for the smaller sample.	R
a. List the combined data in ascending order.	
b. Rank the combined data.	
c. Add the sum of the ranks for the smaller sample, n_1.	
6. Find the test statistic and sketch the sampling distribution.	$z = \dfrac{R - \mu_R}{\sigma_R}$
7. Make a decision to reject or fail to reject the null hypothesis.	If z is in the rejection region, reject H_0. Otherwise, fail to reject H_0.
8. Interpret the decision in the context of the original claim.	

EXAMPLE 2

▶ Performing a Wilcoxon Rank Sum Test

The table shows the earnings (in thousands of dollars) of a random sample of 10 male and 12 female pharmaceutical sales representatives. At $\alpha = 0.10$, can you conclude that there is a difference between the males' and females' earnings?

Male earnings	78	93	114	101	98	94	86	95	117	99		
Female earnings	86	77	101	93	85	98	91	87	84	97	100	90

▶ Solution

The claim is "there is a difference between the males' and females' earnings." The null and alternative hypotheses for this test are as follows.

H_0: There is no difference between the males' and the females' earnings.

H_a: There is a difference between the males' and the females' earnings.
  (Claim)

Because the test is a two-tailed test with $\alpha = 0.10$, the critical values are $-z_0 = -1.645$ and $z_0 = 1.645$. The rejection regions are $z < -1.645$ and $z > 1.645$.

The sample size for men is 10 and the sample size for women is 12. Because $10 < 12$, $n_1 = 10$ and $n_2 = 12$. Before calculating the test statistic, you must find the values of R, μ_R, and σ_R. The table shows the combined data listed in ascending order and the corresponding ranks.

Ordered data	Sample	Rank	Ordered data	Sample	Rank
77	F	1	94	M	12
78	M	2	95	M	13
84	F	3	97	F	14
85	F	4	98	M	15.5
86	M	5.5	98	F	15.5
86	F	5.5	99	M	17
87	F	7	100	F	18
90	F	8	101	M	19.5
91	F	9	101	F	19.5
93	M	10.5	114	M	21
93	F	10.5	117	M	22

Because the smaller sample is the sample of males, R is the sum of the male rankings.

$$R = 2 + 5.5 + 10.5 + 12 + 13 + 15.5 + 17 + 19.5 + 21 + 22$$
$$= 138$$

Using $n_1 = 10$ and $n_2 = 12$, you can find μ_R and σ_R as follows.

$$\mu_R = \frac{n_1(n_1 + n_2 + 1)}{2} = \frac{10(10 + 12 + 1)}{2} = \frac{230}{2} = 115$$

$$\sigma_R = \sqrt{\frac{n_1 n_2 (n_1 + n_2 + 1)}{12}}$$

$$= \sqrt{\frac{(10)(12)(10 + 12 + 1)}{12}}$$

$$= \sqrt{\frac{2760}{12}}$$

$$= \sqrt{230}$$

$$\approx 15.17$$

When $R = 138$, $\mu_R = 115$, and $\sigma_R \approx 15.17$, the test statistic is

$$z = \frac{R - \mu_R}{\sigma_R}$$

$$\approx \frac{138 - 115}{15.17}$$

$$\approx 1.52.$$

From the graph at the right, you can see that the test statistic z is not in the rejection region. So, you should decide to fail to reject the null hypothesis.

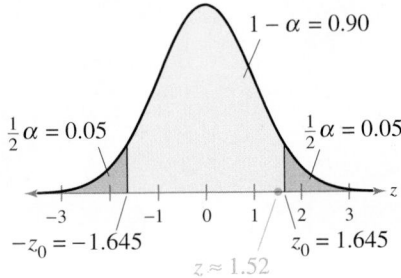

Interpretation There is not enough evidence at the 10% level of significance to conclude that there is a difference between the males' and females' earnings.

▶ **Try It Yourself 2**

You are investigating the automobile insurance claims paid (in thousands of dollars) by two insurance companies. The table shows a random, independent sample of 12 claims paid by the two insurance companies. At $\alpha = 0.05$, can you conclude that there is a difference in the claims paid by the companies?

Company A	6.2	10.6	2.5	4.5	6.5	7.4
Company B	7.3	5.6	3.4	1.8	2.2	4.7

Company A	9.9	3.0	5.8	3.9	6.0	6.3
Company B	10.8	4.1	1.7	3.0	4.4	5.3

a. Identify the *claim* and state H_0 and H_a.
b. Specify the *level of significance* α.
c. Determine the *critical value(s)* and the *rejection region(s)*.
d. Determine the *sample sizes* n_1 and n_2.
e. *List* the combined data in ascending order, *rank* the data, and *find* the sum of the ranks of the smaller sample.
f. Find the *test statistic z. Sketch* a graph.
g. *Decide* whether to reject the null hypothesis.
h. *Interpret* the decision in the context of the original claim.

Answer: Page A47

11.2   EXERCISES

1. If the samples are dependent, use a Wilcoxon signed-rank test. If the samples are independent, use a Wilcoxon rank sum test.

2. The sample size for both samples must be at least 10 to use the Wilcoxon rank sum test.

3. (a) H_0: There is no reduction in diastolic blood pressure. (claim)

   H_a: There is a reduction in diastolic blood pressure.

   (b) Wilcoxon signed-rank test

   (c) 10   (d) 17

   (e) Fail to reject H_0.

   (f) There is not enough evidence at the 1% level of significance to reject the claim that there was no reduction in diastolic blood pressure.

4. (a) H_0: There is no difference in salaries. (claim)

   H_a: There is a difference in salaries.

   (b) Wilcoxon rank sum test

   (c) ±1.645

   (d) −1.663 (or 1.663)

   (e) Reject H_0.

   (f) There is enough evidence at the 10% level of significance to reject the analyst's claim that there is no difference in the salaries earned by workers in the wholesale trade and manufacturing industries.

■ BUILDING BASIC SKILLS AND VOCABULARY

1. How do you know whether to use a Wilcoxon signed-rank test or a Wilcoxon rank sum test?

2. What is the requirement for the sample size of both samples when using the Wilcoxon rank sum test?

■ USING AND INTERPRETING CONCEPTS

Performing a Wilcoxon Test   *In Exercises 3–8,*

(a) identify the claim and state H_0 and H_a.

(b) decide whether to use a Wilcoxon signed-rank test or a Wilcoxon rank sum test.

(c) determine the critical value(s).

(d) find the test statistic.

(e) decide whether to reject or fail to reject the null hypothesis.

(f) interpret the decision in the context of the original claim.

3. **Calcium Supplements and Blood Pressure**   In a study testing the effects of calcium supplements on blood pressure in men, 12 men were randomly chosen and given a calcium supplement for 12 weeks. The measurements shown in the table are for each subject's diastolic blood pressure taken before and after the 12-week treatment period. At $\alpha = 0.01$, can you reject the claim that there was no reduction in diastolic blood pressure? *(Adapted from The Journal of the American Medical Association)*

Patient	1	2	3	4	5	6
Before treatment	108	109	120	129	112	111
After treatment	99	115	105	116	115	117

Patient	7	8	9	10	11	12
Before treatment	117	135	124	118	130	115
After treatment	108	122	120	126	128	106

4. **Wholesale Trade and Manufacturing**   A private industry analyst claims that there is no difference in the salaries earned by workers in the wholesale trade and manufacturing industries. A random sample of 10 wholesale trade and 10 manufacturing workers and their salaries (in thousands of dollars) are shown in the table. At $\alpha = 0.10$, can you reject the analyst's claim? *(Adapted from U.S. Bureau of Economic Analysis)*

Wholesale trade	62	55	56	70	53	59	64	67	65	62
Manufacturing	62	58	47	65	45	56	67	49	55	43

5. (a) H_0: The cost of prescription drugs is not lower in Canada than in the United States.

   H_a: The cost of prescription drugs is lower in Canada than in the United States. (claim)

   (b) Wilcoxon signed-rank test

   (c) 4   (d) 6

   (e) Fail to reject H_0.

   (f) There is not enough evidence at the 5% level of significance for the researcher to conclude that the cost of prescription drugs is lower in Canada than in the United States.

6. (a) H_0: There is no difference in the earnings.

   H_a: There is a difference in the earnings. (claim)

   (b) Wilcoxon rank sum test

   (c) ± 1.96 (d) -3.66

   (e) Reject H_0.

   (f) There is enough evidence at the 5% level of significance to support the administrator's belief that there is a difference in the earnings of people with bachelor's degrees and those with associate's degrees.

7. (a) H_0: There is no difference in salaries.

   H_a: There is a difference in salaries. (claim)

   (b) Wilcoxon rank sum test

   (c) ± 1.96 (d) -1.94

   (e) Fail to reject H_0.

   (f) There is not enough evidence at the 5% level of significance to support the representative's claim that there is a difference in the salaries earned by teachers in Wisconsin and Michigan.

5. **Drug Prices** A researcher wants to determine whether the cost of prescription drugs is lower in Canada than in the United States. The researcher selects seven of the most popular brand-name prescription drugs and records the cost per pill (in U.S. dollars) of each. The results are shown in the table. At $\alpha = 0.05$, can the researcher conclude that the cost of prescription drugs is lower in Canada than in the United States? *(Adapted from Annals of Internal Medicine)*

Drug	1	2	3	4	5	6	7
Cost in U.S.	1.26	1.76	4.19	3.36	1.80	9.91	3.95
Cost in Canada	1.04	0.82	2.22	2.22	1.31	11.47	2.63

6. **Earnings by Degree** A college administrator believes that there is a difference in the earnings of people with bachelor's degrees and those with associate's degrees. The table shows the earnings (in thousands of dollars) of a random sample of 11 people with bachelor's degrees and 10 people with associate's degrees. At $\alpha = 0.05$, is there enough evidence to support the administrator's belief? *(Adapted from U.S. Census Bureau)*

Bachelor's degree	54	50	63	76	70	50	44	56	60	52	54
Associate's degree	36	39	47	33	38	38	45	45	42	34	

7. **Teacher Salaries** A teacher's union representative claims that there is a difference in the salaries earned by teachers in Wisconsin and Michigan. The table shows the salaries (in thousands of dollars) of a random sample of 11 teachers from Wisconsin and 12 teachers from Michigan. At $\alpha = 0.05$, is there enough evidence to support the representative's claim? *(Adapted from National Education Association)*

Wisconsin	51	59	52	46	51	55	53	51	50	50	64	
Michigan	57	61	51	58	53	63	57	63	55	49	54	72

8. **Heart Rate** A physician wants to determine whether an experimental medication affects an individual's heart rate. The physician selects 15 patients and measures the heart rate of each. The subjects then take the medication and have their heart rates measured after one hour. The results are shown in the table. At $\alpha = 0.05$, can the physician conclude that the experimental medication affects an individual's heart rate?

Patient	1	2	3	4	5	6	7	8
Heart rate (before)	72	81	75	76	79	74	65	67
Heart rate (after)	73	80	75	79	74	76	73	67

Patient	9	10	11	12	13	14	15
Heart rate (before)	76	83	66	75	76	78	68
Heart rate (after)	74	77	70	77	76	75	74

8. (a) H_0: The experimental medication does not affect an individual's heart rate.

H_a: The experimental medication affects an individual's heart rate. (claim)

(b) Wilcoxon signed-rank test

(c) 25   (d) 31.5

(e) Fail to reject H_0.

(f) There is not enough evidence at the 5% level of significance for the physician to conclude that the experimental medication affects an individual's heart rate.

9. Reject H_0. There is enough evidence at the 10% level of significance for the engineer to conclude that the gas mileage is improved.

10. Reject H_0. There is enough evidence at the 5% level of significance to support the engineer's claim that the fuel additive improves gas mileage.

■ EXTENDING CONCEPTS

Wilcoxon Signed-Rank Test for $n > 30$ *If you are performing a Wilcoxon signed-rank test and the sample size n is greater than 30, you can use the Standard Normal Table and the following formula to find the test statistic.*

$$z = \frac{w_s - \dfrac{n(n+1)}{4}}{\sqrt{\dfrac{n(n+1)(2n+1)}{24}}}$$

In Exercises 9 and 10, perform the indicated Wilcoxon signed-rank test using the test statistic for $n > 30$.

9. Fuel Additive   A petroleum engineer wants to know whether a certain fuel additive improves a car's gas mileage. To decide, the engineer records the gas mileages (in miles per gallon) of 33 cars with and without the additive. The results are shown in the table. At $\alpha = 0.10$, can the engineer conclude that the gas mileage is improved?

Car	1	2	3	4	5	6	7	8	9	10	11
Without additive	36.4	36.4	36.6	36.6	36.8	36.9	37.0	37.1	37.2	37.2	36.7
With additive	36.7	36.9	37.0	37.5	38.0	38.1	38.4	38.7	38.8	38.9	36.3

Car	12	13	14	15	16	17	18	19	20	21	22
Without additive	37.5	37.6	37.8	37.9	37.9	38.1	38.4	40.2	40.5	40.9	35.0
With additive	38.9	39.0	39.1	39.4	39.4	39.5	39.8	40.0	40.0	40.1	36.3

Car	23	24	25	26	27	28	29	30	31	32	33
Without additive	32.7	33.6	34.2	35.1	35.2	35.3	35.5	35.9	36.0	36.1	37.2
With additive	32.8	34.2	34.7	34.9	34.9	35.3	35.9	36.4	36.6	36.6	38.3

10. Fuel Additive   A petroleum engineer claims that a fuel additive improves gas mileage. The table shows the gas mileages (in miles per gallon) of 32 cars measured with and without the fuel additive. Test the petroleum engineer's claim at $\alpha = 0.05$.

Car	1	2	3	4	5	6	7	8	9	10	11
Without additive	34.0	34.2	34.4	34.4	34.6	34.8	35.6	35.7	30.2	31.6	32.3
With additive	36.6	36.7	37.2	37.2	37.3	37.4	37.6	37.7	34.2	34.9	34.9

Car	12	13	14	15	16	17	18	19	20	21	22
Without additive	33.0	33.1	33.7	33.7	33.8	35.7	36.1	36.1	36.6	36.6	36.8
With additive	34.9	35.7	36.0	36.2	36.5	37.8	38.1	38.2	38.3	38.3	38.7

Car	23	24	25	26	27	28	29	30	31	32
Without additive	37.1	37.1	37.2	37.9	37.9	38.0	38.0	38.4	38.8	42.1
With additive	38.8	38.9	39.1	39.1	39.2	39.4	39.8	40.3	40.8	43.2

College Ranks

Each year, Forbes and the Center for College Affordability and Productivity release a list of the best colleges in America. Six hundred undergraduate colleges and universities are ranked according to quality of education, 4-year graduation rate, post-graduate success, average student debt after 4 years, and number of students or faculty who have won competitive awards, such as Rhodes Scholarships or Nobel Prizes.

The table shows freshman class size by state for randomly selected colleges on the 2009 list.

Freshman Class Size			
CA	**MA**	**NC**	**PA**
236	540	3865	372
1703	1666	1699	327
320	596	1201	366
382	439	2073	588
202	1048	2781	957
202	2167	1291	453
458	643	2492	2400
252	754	3090	601
467	1297	4538	613
574	518	4804	399

■ EXERCISES

1. Construct a side-by-side box-and-whisker plot for the four states. Do any of the median freshman class sizes appear to be the same? Do any appear to be different?

In Exercises 2–5, use the sign test to test the claim. What can you conclude? Use $\alpha = 0.05$.

2. The median freshman class size at a California college is less than or equal to 400.

3. The median freshman class size at a Massachusetts college is greater than or equal to 750.

4. The median freshman class size at a Pennsylvania college is 500.

5. The median freshman class size at a North Carolina college is different from 2400.

In Exercises 6 and 7, use the Wilcoxon rank sum test to test the claim. Use $\alpha = 0.01$.

6. There is no difference between freshman class sizes for Pennsylvania colleges and California colleges.

7. There is a difference between freshman class sizes for Massachusetts colleges and North Carolina colleges.

11.3 The Kruskal-Wallis Test

The Kruskal-Wallis Test

▸ THE KRUSKAL-WALLIS TEST

In Section 10.4, you learned how to use one-way ANOVA techniques to compare the means of three or more populations. When using one-way ANOVA, you should verify that each independent sample is selected from a population that is normally, or approximately normally, distributed. If, however, you cannot verify that the populations are normal, you can still compare the distributions of three or more populations. To do so, you can use the *Kruskal-Wallis test*.

DEFINITION

The **Kruskal-Wallis test** is a nonparametric test that can be used to determine whether three or more independent samples were selected from populations having the same distribution.

The null and alternative hypotheses for the Kruskal-Wallis test are as follows.

H_0: There is no difference in the distribution of the populations.

H_a: There is a difference in the distribution of the populations.

Two conditions for using the Kruskal-Wallis test are that each sample must be randomly selected and the size of each sample must be at least 5. If these conditions are met, then the sampling distribution for the Kruskal-Wallis test is approximated by a chi-square distribution with $k - 1$ degrees of freedom, where k is the number of samples. You can calculate the Kruskal-Wallis test statistic using the following formula.

TEST STATISTIC FOR THE KRUSKAL-WALLIS TEST

Given three or more independent samples, the **test statistic** for the Kruskal-Wallis test is

$$H = \frac{12}{N(N+1)}\left(\frac{R_1^2}{n_1} + \frac{R_2^2}{n_2} + \cdots + \frac{R_k^2}{n_k}\right) - 3(N+1)$$

where

k represents the number of samples,

n_i is the size of the ith sample,

N is the sum of the sample sizes,

and

R_i is the sum of the ranks of the ith sample.

Performing a Kruskal-Wallis test consists of combining and ranking the sample data. The data are then separated according to sample and the sum of the ranks of each sample is calculated.

These sums are then used to calculate the test statistic H, which is an approximation of the variance of the rank sums. If the samples are selected from populations having the same distribution, the sums of the ranks will be approximately equal, H will be small, and the null hypothesis should not be rejected.

If, however, the samples are selected from populations not having the same distribution, the sums of the ranks will be quite different, H will be large, and the null hypothesis should be rejected.

Because the null hypothesis is rejected only when H is significantly large, the Kruskal-Wallis test is always a right-tailed test.

GUIDELINES

Performing a Kruskal-Wallis Test

IN WORDS	IN SYMBOLS
1. Identify the claim. State the null and alternative hypotheses.	State H_0 and H_a.
2. Specify the level of significance.	Identify α.
3. Identify the degrees of freedom.	d.f. $= k - 1$
4. Determine the critical value and the rejection region.	Use Table 6 in Appendix B.
5. Find the sum of the ranks for each sample.	
a. List the combined data in ascending order.	
b. Rank the combined data.	
6. Find the test statistic and sketch the sampling distribution.	$H = \dfrac{12}{N(N+1)} \cdot$ $\left(\dfrac{R_1^2}{n_1} + \dfrac{R_2^2}{n_2} + \cdots + \dfrac{R_k^2}{n_k} \right)$ $- 3(N+1)$
7. Make a decision to reject or fail to reject the null hypothesis.	If H is in the rejection region, reject H_0. Otherwise, fail to reject H_0.
8. Interpret the decision in the context of the original claim.	

EXAMPLE 1    Report 51

▶ Performing a Kruskal-Wallis Test

You want to compare the number of crimes reported in three police precincts in a city. To do so, you randomly select 10 weeks for each precinct and record the number of crimes reported. The results are shown in the table. At $\alpha = 0.01$, can you conclude that the distributions of crimes reported in the three police precincts are different?

Number of Crimes Reported for the Week		
101st Precinct (Sample 1)	106th Precinct (Sample 2)	113th Precinct (Sample 3)
60	65	69
52	55	51
49	64	70
52	66	61
50	53	67
48	58	65
57	50	62
45	54	59
44	70	60
56	62	63

▶ Solution

You want to test the claim that there is a difference in the number of crimes reported in the three precincts. The null and alternative hypotheses are as follows.

H_0: There is no difference in the number of crimes reported in the three precincts.

H_a: There is a difference in the number of crimes reported in the three precincts. (Claim)

The test is a right-tailed test with $\alpha = 0.01$ and d.f. $= k - 1 = 3 - 1 = 2$. From Table 6, the critical value is $\chi_0^2 = 9.210$. Before calculating the test statistic, you must find the sum of the ranks for each sample. The table shows the combined data listed in ascending order and the corresponding ranks.

Ordered data	Sample	Rank	Ordered data	Sample	Rank	Ordered data	Sample	Rank
44	101st	1	54	106th	11	62	113th	20.5
45	101st	2	55	106th	12	63	113th	22
48	101st	3	56	101st	13	64	106th	23
49	101st	4	57	101st	14	65	106th	24.5
50	101st	5.5	58	106th	15	65	113th	24.5
50	106th	5.5	59	113th	16	66	106th	26
51	113th	7	60	101st	17.5	67	113th	27
52	101st	8.5	60	113th	17.5	69	113th	28
52	101st	8.5	61	113th	19	70	106th	29.5
53	106th	10	62	106th	20.5	70	113th	29.5

PICTURING THE WORLD

The following randomly collected data were used to compare the water temperatures (in degrees Fahrenheit) of cities bordering the Gulf of Mexico. (Adapted from National Oceanographic Data Center)

Cedar Key, FL (Sample 1)	Eugene Island, LA (Sample 2)	Dauphin Island, AL (Sample 3)
62	51	63
69	55	51
77	57	54
59	63	60
60	74	75
75	82	80
83	85	70
65	60	78
79	64	82
86	76	84
82	83	
	86	

At $\alpha = 0.05$, can you conclude that the temperature distributions of the three cities are different?

There is not enough evidence at the 5% level of significance to conclude that the temperature distributions of the three cities are different.

The sum of the ranks for each sample is as follows.

$$R_1 = 1 + 2 + 3 + 4 + 5.5 + 8.5 + 8.5 + 13 + 14 + 17.5 = 77$$

$$R_2 = 5.5 + 10 + 11 + 12 + 15 + 20.5 + 23 + 24.5 + 26 + 29.5 = 177$$

$$R_3 = 7 + 16 + 17.5 + 19 + 20.5 + 22 + 24.5 + 27 + 28 + 29.5 = 211$$

Using these sums and the values $n_1 = 10$, $n_2 = 10$, $n_3 = 10$, and $N = 30$, the test statistic is

$$H = \frac{12}{30(30 + 1)} \left(\frac{77^2}{10} + \frac{177^2}{10} + \frac{211^2}{10} \right) - 3(30 + 1) \approx 12.521.$$

From the graph at the right, you can see that the test statistic H is in the rejection region. So, you should decide to reject the null hypothesis.

Interpretation There is enough evidence at the 1% level of significance to support the claim that there is a difference in the number of crimes reported in the three police precincts.

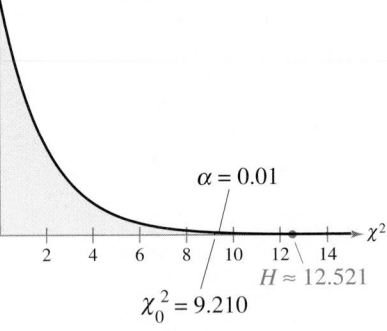

▶ **Try It Yourself 1**

You want to compare the salaries of veterinarians who work in California, New York, and Pennsylvania. To compare the salaries, you randomly select several veterinarians in each state and record their salaries. The salaries (in thousands of dollars) are listed in the table. At $\alpha = 0.05$, can you conclude that the distributions of the veterinarians' salaries in these three states are different? *(Adapted from U.S. Bureau of Labor Statistics)*

Sample Salaries		
CA (Sample 1)	NY (Sample 2)	PA (Sample 3)
99.95	94.40	99.20
97.50	99.75	103.70
98.85	97.50	110.45
100.75	101.97	95.15
101.20	93.10	88.80
96.25	102.35	99.99
99.70	97.89	100.55
88.28	92.50	97.25
113.90	101.55	97.44
103.20		

a. Identify the *claim* and state H_0 and H_a.
b. Specify the *level of significance* α.
c. Identify the *degrees of freedom*.
d. Determine the *critical value* and the *rejection region*.
e. *List* the combined data in ascending order, *rank* the data, and *find* the sum of the ranks of each sample.
f. Find the *test statistic H. Sketch* a graph.
g. *Decide* whether to reject the null hypothesis.
h. *Interpret* the decision in the context of the original claim.

Answer: Page A48

11.3 EXERCISES

■ BUILDING BASIC SKILLS AND VOCABULARY

1. What are the conditions for using a Kruskal-Wallis test?

2. Explain why the Kruskal-Wallis test is always a right-tailed test.

■ USING AND INTERPRETING CONCEPTS

Performing a Kruskal-Wallis Test   *In Exercises 3–6, (a) identify the claim and state H_0 and H_a, (b) determine the critical value, (c) find the sums of the ranks for each sample and calculate the test statistic, (d) decide whether to reject or fail to reject the null hypothesis, and (e) interpret the decision in the context of the original claim.*

3. Home Insurance   The table shows the annual premiums for a random sample of home insurance policies in Connecticut, Massachusetts, and Virginia. At $\alpha = 0.05$, can you conclude that the distributions of the annual premiums in these three states are different? *(Adapted from National Association of Insurance Commissioners)*

State	Annual Premium (in dollars)						
Connecticut	930	725	890	1040	1165	806	947
Massachusetts	1105	1025	980	1295	1110	889	757
Virginia	815	730	546	625	912	618	535

4. Hourly Rates   A researcher wants to determine whether there is a difference in the hourly pay rates for registered nurses in three states: Indiana, Kentucky, and Ohio. The researcher randomly selects several registered nurses in each state and records the hourly pay rate for each in the table shown. At $\alpha = 0.05$, can the researcher conclude that the distributions of the registered nurses' hourly pay rates in these three states are different? *(Adapted from U.S. Bureau of Labor Statistics)*

State	Hourly Pay Rate (in dollars)						
Indiana	27.80	28.25	26.65	27.40	30.24	25.10	29.44
Kentucky	26.95	25.58	28.10	30.20	28.55	31.60	24.60
Ohio	25.75	30.15	31.55	31.82	25.25	27.80	

5. Annual Salaries   The table shows the annual salaries for a random sample of workers in Kentucky, North Carolina, South Carolina, and West Virginia. At $\alpha = 0.10$, can you conclude that the distributions of the annual salaries in these four states are different? *(Adapted from U.S. Bureau of Labor Statistics)*

State	Annual Salary (in thousands of dollars)					
Kentucky	32.5	34.2	43.1	54.7	30.9	25.5
North Carolina	40.5	38.9	33.6	51.3	32.5	36.6
South Carolina	27.8	35.4	41.5	40.9	32.7	34.1
West Virginia	27.1	38.2	28.9	37.4	42.6	30.4

1. The conditions for using a Kruskal-Wallis test are that each sample must be randomly selected and the size of each sample must be at least 5.

2. The Kruskal-Wallis test is always a right-tailed test because the null hypothesis is rejected only when H is significantly large.

3. (a) H_0: There is no difference in the premiums.

H_a: There is a difference in the premiums. (claim)

(b) 5.991   (c) 9.506

(d) Reject H_0.

(e) There is enough evidence at the 5% level of significance to conclude that the distributions of the annual premiums of the three states are different.

4. See Selected Answers, page A134.

5. (a) H_0: There is no difference in the salaries.

H_a: There is a difference in the salaries. (claim)

(b) 6.251   (c) 1.202

(d) Fail to reject H_0.

(e) There is not enough evidence at the 10% level of significance to conclude that the distributions of the annual salaries in the four states are different.

6. (a) H_0: There is no difference in the amounts of caffeine.

H_a: There is a difference in the amounts of caffeine. (claim)

(b) 11.345   (c) 18.197

(d) Reject H_0.

(e) There is enough evidence at the 1% level of significance to conclude that the distributions of the amounts of caffeine in the four beverages are different.

Number of Job Offers			
A	**B**	**C**	**D**
5	8	5	2
4	10	4	3
7	9	3	5
6	7	5	4
5	10	7	2
4	6	8	3

TABLE FOR EXERCISES 7 AND 8

7. Kruskal-Wallis results:

Data stored in separate columns.

Chi Square = 8.0965185 (adjusted for ties)

DF = 2

P-value = 0.0175

Column	n	Median	Ave. Rank
A	6	5	6.75
B	6	8.5	14.5
C	6	5	7.25

$P = 0.0175 > 0.01$, so fail to reject H_0. There is not enough evidence at the 1% level of significance to conclude that the distributions of the number of job offers at Colleges A, B, and C are different.

8. See Selected Answers, page A134.

9. (a) Fail to reject H_0.

(b) Fail to reject H_0.

Both tests come to the same decision, which is that there is not enough evidence to support the claim that there is a difference in the number of days spent in the hospital.

10. See Selected Answers, page A134.

6. Caffeine Content   The table shows the amounts of caffeine (in milligrams) in 16-ounce servings for a random sample of beverages. At $\alpha = 0.01$, can you conclude that the distributions of the amounts of caffeine in these four beverages are different? *(Source: Center for Science in the Public Interest)*

Beverage	Amount of Caffeine in 16-ounce Serving (in milligrams)						
Coffees	320	300	206	150	266		
Soft drinks	95	96	56	51	71	72	47
Energy drinks	200	141	160	152	154	166	
Teas	100	106	42	15	32	10	

SC   *In Exercises 7 and 8, use StatCrunch and the table at the left, which shows the number of job offers received by mechanical engineers who recently graduated from four colleges (A, B, C, D).*

7. At $\alpha = 0.01$, can you conclude that the distributions of the number of job offers at Colleges A, B, and C are different?

8. At $\alpha = 0.01$, can you conclude that the distributions of the number of job offers at all four colleges are different?

■ EXTENDING CONCEPTS

Comparing Two Tests   *In Exercises 9 and 10, perform the indicated test using (a) a Kruskal-Wallis test and (b) a one-way ANOVA test, assuming that each population is normally distributed and the population variances are equal. Compare the results. If convenient, use technology to solve the problem.*

9. Hospital Patient Stays   An insurance underwriter reports that the mean number of days patients spend in a hospital differs according to the region of the United States in which the patient lives. The table shows the number of days randomly selected patients spent in a hospital in four U.S. regions. At $\alpha = 0.01$, can you support the underwriter's claim? *(Adapted from U.S. National Center for Health Statistics)*

Region	Number of Days									
Northeast	8	6	6	3	5	11	3	8	1	6
Midwest	5	4	3	9	1	4	6	3	4	7
South	5	8	1	5	8	7	5	1		
West	2	3	6	6	5	4	3	6	5	

10. Energy Consumption   The table shows the energy consumed (in millions of Btu) in one year for a random sample of households from four U.S. regions. At $\alpha = 0.01$, can you conclude that the mean energy consumptions are different? *(Adapted from U.S. Energy Information Administration)*

Region	Energy Consumed (in millions of Btu)										
Northeast	72	106	151	138	104	108	95	134	100	174	
Midwest	84	183	194	165	120	212	148	129	113	62	97
South	91	40	72	91	147	74	70	67			
West	74	32	78	28	106	39	118	63	70	56	

11.4 Rank Correlation

WHAT YOU SHOULD LEARN

▸ How to use the Spearman rank correlation coefficient to determine whether the correlation between two variables is significant

The Spearman Rank Correlation Coefficient

▸ THE SPEARMAN RANK CORRELATION COEFFICIENT

In Section 9.1, you learned how to measure the strength of the relationship between two variables using the Pearson correlation coefficient r. Two requirements for the Pearson correlation coefficient are that the variables are linearly related and that the population represented by each variable is normally distributed. If these requirements cannot be met, you can examine the relationship between two variables using the nonparametric equivalent to the Pearson correlation coefficient—the *Spearman rank correlation coefficient*.

The Spearman rank correlation coefficient has several advantages over the Pearson correlation coefficient. For instance, the Spearman rank correlation coefficient can be used to describe the relationship between linear or nonlinear data. The Spearman rank correlation coefficient can be used for data at the ordinal level. And, the Spearman rank correlation coefficient is easier to calculate by hand than the Pearson coefficient.

DEFINITION

The **Spearman rank correlation coefficient r_s** is a measure of the strength of the relationship between two variables. The Spearman rank correlation coefficient is calculated using the ranks of paired sample data entries. If there are no ties in the ranks of either variable, then the formula for the Spearman rank correlation coefficient is

$$r_s = 1 - \frac{6\sum d^2}{n(n^2 - 1)}$$

where n is the number of paired data entries and d is the difference between the ranks of a paired data entry. If there are ties in the ranks and the number of ties is small relative to the number of data pairs, then the formula can still be used to approximate r_s.

The values of r_s range from -1 to 1, inclusive. If the ranks of corresponding data pairs are exactly identical, r_s is equal to 1. If the ranks are in "reverse" order, r_s is equal to -1. If the ranks of corresponding data pairs have no relationship, r_s is equal to 0.

After calculating the Spearman rank correlation coefficient, you can determine whether the correlation between the variables is significant. You can make this determination by performing a hypothesis test for the population correlation coefficient ρ_s. The null and alternative hypotheses for this test are as follows.

H_0: $\rho_s = 0$ (There is no correlation between the variables.)

H_a: $\rho_s \neq 0$ (There is a significant correlation between the variables.)

The critical values for the Spearman rank correlation coefficient are listed in Table 10 of Appendix B. Table 10 lists critical values for selected levels of significance and for sample sizes of 30 or less. The test statistic for the hypothesis test is the Spearman rank correlation coefficient r_s.

GUIDELINES

Testing the Significance of the Spearman Rank Correlation Coefficient

IN WORDS	IN SYMBOLS
1. State the null and alternative hypotheses.	State H_0 and H_a.
2. Specify the level of significance.	Identify α.
3. Determine the critical value.	Use Table 10 in Appendix B.
4. Find the test statistic.	$r_s = 1 - \dfrac{6\Sigma d^2}{n(n^2 - 1)}$
5. Make a decision to reject or fail to reject the null hypothesis.	If $|r_s|$ is greater than the critical value, reject H_0. Otherwise, fail to reject H_0.
6. Interpret the decision in the context of the original claim.	

EXAMPLE 1

▶ The Spearman Rank Correlation Coefficient

The table shows the school enrollments (in millions) at all levels of education for males and females from 2000 to 2007. At $\alpha = 0.05$, can you conclude that there is a correlation between the number of males and females enrolled in school? *(Source: U.S. Census Bureau)*

Year	Male	Female
2000	35.8	36.4
2001	36.3	36.9
2002	36.8	37.3
2003	37.3	37.6
2004	37.4	38.0
2005	37.4	38.4
2006	37.2	38.0
2007	37.6	38.4

▶ Solution

The null and alternative hypotheses are as follows.

H_0: $\rho_s = 0$ (There is no correlation between the number of males and females enrolled in school.)

H_a: $\rho_s \neq 0$ (There is a correlation between the number of males and females enrolled in school.) (Claim)

 PICTURING THE WORLD

The table shows the retail prices (in dollars per pound) for 100% ground beef and fresh whole chicken in the United States from 2000 to 2008. (Source: U.S. Bureau of Labor Statistics)

Year	Beef	Chicken
2000	1.63	1.08
2001	1.71	1.11
2002	1.69	1.05
2003	2.23	1.05
2004	2.14	1.03
2005	2.30	1.06
2006	2.26	1.06
2007	2.23	1.17
2008	2.41	1.31

Does a correlation exist between ground beef and chicken prices in the United States from 2000 to 2008? Use α = 0.10.

There is not enough evidence at the 10% level of significance to conclude that there is a correlation between ground beef and chicken prices from 2000 to 2008.

Each data set has eight entries. From Table 10 with $\alpha = 0.05$ and $n = 8$, the critical value is 0.738. Before calculating the test statistic, you must find Σd^2, the sum of the squares of the differences of the ranks of the data sets. You can use a table to calculate d^2, as shown below.

Male	Rank	Female	Rank	d	d^2
35.8	1	36.4	1	0	0
36.3	2	36.9	2	0	0
36.8	3	37.3	3	0	0
37.3	5	37.6	4	1	1
37.4	6.5	38.0	5.5	1	1
37.4	6.5	38.4	7.5	−1	1
37.2	4	38.0	5.5	−1.5	2.25
37.6	8	38.4	7.5	0.5	0.25
					$\Sigma d^2 = 5.5$

When $n = 8$ and $\Sigma d^2 = 5.5$, the test statistic is

$$r_s \approx 1 - \frac{6\Sigma d^2}{n(n^2 - 1)}$$
$$= 1 - \frac{6(5.5)}{8(8^2 - 1)}$$
$$\approx 0.935.$$

Because $|0.935| > 0.738$, you should reject the null hypothesis.

Interpretation There is enough evidence at the 5% level of significance to conclude that there is a correlation between the number of males and females enrolled in school.

▶ **Try It Yourself 1**

The table shows the number of males and females (in thousands) who received their doctoral degrees from 2001 to 2007. At $\alpha = 0.01$, can you conclude that there is a correlation between the number of males and females who received doctoral degrees? *(Source: U.S. National Center for Education Statistics)*

Year	2001	2002	2003	2004	2005	2006	2007
Male	25	24	24	25	27	29	30
Female	20	20	22	23	26	27	30

a. State the *null* and *alternative hypotheses*.
b. Specify the *level of significance* α.
c. Determine the *critical value*.
d. *Use a table* to calculate Σd^2.
e. Find the *test statistic* r_s.
f. *Decide* whether to reject the null hypothesis.
g. *Interpret* the decision in the context of the original claim.

Answer: Page A48

11.4 EXERCISES

1. The Spearman rank correlation coefficient can be used to describe the relationship between linear or nonlinear data. Also, it can be used for data at the ordinal level and it is easier to calculate by hand than the Pearson correlation coefficient.

2. Both the Spearman rank correlation coefficient and the Pearson correlation coefficient range from -1 to 1.

3. The ranks of the corresponding data are identical when r_s is equal to 1. The ranks are in "reverse" order when r_s is equal to -1. The ranks have no relationship when r_s is equal to 0.

4. The value of r_s represents the correlation coefficient of the sample data, whereas ρ_s represents the correlation coefficient of the entire population.

5. (a) $H_0\colon \rho_s = 0$; $H_a\colon \rho_s \neq 0$ (claim)

   (b) 0.929   (c) 0.857

   (d) Fail to reject H_0.

   (e) There is not enough evidence at the 1% level of significance to support the claim that there is a correlation between debt and income in the farming business.

■ BUILDING BASIC SKILLS AND VOCABULARY

1. What are some advantages of the Spearman rank correlation coefficient over the Pearson correlation coefficient?

2. Describe the ranges of the Spearman rank correlation coefficient and the Pearson correlation coefficient.

3. What does it mean when r_s is equal to 1? What does it mean when r_s is equal to -1? What does it mean when r_s is equal to 0?

4. Explain, in your own words, what r_s and ρ_s represent in Example 1.

■ USING AND INTERPRETING CONCEPTS

Testing a Claim *In Exercises 5–8, (a) identify the claim and state H_0 and H_a, (b) determine the critical value using Table 10 in Appendix B, (c) find the test statistic (d) decide whether to reject or fail to reject the null hypothesis, and (e) interpret the decision in the context of the original claim.*

5. **Farming: Debt and Income** In an agricultural report, a commodities analyst suggests that there is a correlation between debt and income in the farming business. The table shows the total debts and total incomes for farms in seven states for a recent year. At $\alpha = 0.01$, is there enough evidence to support the analyst's claim? *(Adapted from U.S. Department of Agriculture)*

State	Debt (in millions of dollars)	Income (in millions of dollars)
California	19,955	28,926
Illinois	10,480	8,630
Iowa	14,434	12,942
Minnesota	9,982	8,807
Nebraska	10,085	11,028
North Carolina	4,235	7,008
Texas	13,286	15,268

6. **Exercise Machines** The table shows the overall scores and the prices for 11 different models of elliptical exercise machines. The overall score represents the ergonomics, exercise range, ease of use, construction, heart-rate monitoring, and safety. At $\alpha = 0.05$, can you conclude that there is a correlation between the overall score and the price? *(Source: Consumer Reports)*

Overall score	85	78	77	75	73	71
Price (in dollars)	2600	2800	3700	1700	1300	900

Overall score	66	66	64	62	58
Price (in dollars)	1000	1400	1800	1000	700

6. (a) H_0: $\rho_s = 0$; H_a: $\rho_s \neq 0$ (claim)

(b) 0.618   (c) 0.730

(d) Reject H_0.

(e) There is enough evidence at the 5% level of significance to conclude that there is a correlation between the overall score and the price.

7. (a) H_0: $\rho_s = 0$; H_a: $\rho_s \neq 0$ (claim)

(b) 0.833   (c) 0.950

(d) Reject H_0.

(e) There is enough evidence at the 1% level of significance to conclude that there is a correlation between the oat and wheat prices.

8. (a) H_0: $\rho_s = 0$; H_a: $\rho_s \neq 0$ (claim)

(b) 0.497   (c) −0.154

(d) Fail to reject H_0.

(e) There is not enough evidence at the 10% level of significance to conclude that there is a correlation between the overall score and the price.

9. Fail to reject H_0. There is not enough evidence at the 5% level of significance to conclude that there is a correlation between science achievement scores and GNI.

10. Fail to reject H_0. There is not enough evidence at the 5% level of significance to conclude that there is a correlation between mathematics achievement scores and GNI.

11. Reject H_0. There is enough evidence at the 5% level of significance to conclude that there is a correlation between science and mathematics achievement scores.

12. Answers will vary. *Sample answer:* Although there is a strong correlation between math and science scores, there is not a significant correlation between the scores and GNI.

7. Crop Prices   The table shows the prices (in dollars per bushel) received by U.S. farmers for oat and wheat from 2000 to 2008. At $\alpha = 0.01$, can you conclude that there is a correlation between the oat and wheat prices? *(Source: U.S. Department of Agriculture)*

Year	2000	2001	2002	2003	2004	2005	2006	2007	2008
Oat	1.10	1.59	1.81	1.48	1.48	1.63	1.87	2.63	3.10
Wheat	2.62	2.78	3.56	3.40	3.40	3.42	4.26	6.48	6.80

8. Vacuum Cleaners   The table shows the overall scores and the prices for 12 different models of vacuum cleaners. The overall score represents carpet and bare-floor cleaning, airflow, handling, noise, and emissions. At $\alpha = 0.10$, can you conclude that there is a correlation between the overall score and the price? *(Source: Consumer Reports)*

Overall score	73	65	60	71	62	39
Price (in dollars)	230	400	600	350	100	300

Overall score	67	64	68	60	70	55
Price (in dollars)	600	700	140	200	80	300

Test Scores and GNI   *In Exercises 9–12, use the following table. The table shows the average achievement scores of 15-year-olds in science and mathematics along with the gross national income (GNI) of nine countries for a recent year. (The GNI is a measure of the total value of goods and services produced by the economy of a country.)* *(Adapted from Organization for Economic Cooperation and Development; The World Bank)*

Country	Science average	Mathematics average	GNI (in billions of dollars)
Canada	534	527	1307
France	495	496	2467
Germany	516	504	3207
Italy	475	462	1988
Japan	531	523	4829
Mexico	410	406	989
Spain	488	480	1314
Sweden	503	502	438
United States	489	474	13,886

9. Science and GNI   At $\alpha = 0.05$, can you conclude that there is a correlation between science achievement scores and GNI?

10. Math and GNI   At $\alpha = 0.05$, can you conclude that there is a correlation between mathematics achievement scores and GNI?

11. Science and Math   At $\alpha = 0.05$, can you conclude that there is a correlation between science and mathematics achievement scores?

12. Writing a Summary   Use the results from Exercises 9–11 to write a summary about the correlation (or lack of correlation) between test scores and GNI.

13. Fail to reject H_0. There is not enough evidence at the 5% level of significance to conclude that there is a correlation between average hours worked and the number of on-the-job injuries.

14. Fail to reject H_0. There is not enough evidence at the 5% level of significance to conclude that there is a correlation between average hours worked and the number of on-the-job injuries.

■ EXTENDING CONCEPTS

Testing the Rank Correlation Coefficient for $n > 30$
If you are testing the significance of the Spearman rank correlation coefficient and the sample size n is greater than 30, you can use the following expression to find the critical value.

$$\frac{\pm z}{\sqrt{n-1}}, \; z \text{ corresponds to the level of significance}$$

In Exercises 13 and 14, perform the indicated test.

13. Work Injuries The table shows the average hours worked per week and the number of on-the-job injuries for a random sample of U.S. industries in a recent year. At $\alpha = 0.05$, can you conclude that there is a correlation between average hours worked and the number of on-the-job injuries? *(Adapted from U.S. Bureau of Labor Statistics; National Safety Council)*

Hours worked	47.6	44.1	45.6	45.5	44.5	47.3	44.6	45.9	45.5	43.7	44.8	42.5
Injuries	16	33	25	33	18	20	21	18	21	28	15	26

Hours worked	46.5	42.3	45.5	41.8	43.1	44.4	44.5	43.7	44.9	47.8	46.6	45.5
Injuries	34	32	26	28	22	19	23	20	28	24	26	29

Hours worked	43.5	42.8	44.8	43.5	47.0	44.5	50.1	46.7	43.1
Injuries	21	28	23	26	24	20	28	26	25

14. Work Injuries in Construction The table shows the average hours worked per week and the number of on-the-job injuries for a random sample of U.S. construction companies in a recent year. At $\alpha = 0.05$, can you conclude that there is a correlation between average hours worked and the number of on-the-job injuries? *(Adapted from U.S. Bureau of Labor Statistics; National Safety Council)*

Hours worked	40.5	38.3	37.8	38.2	38.6	41.2	39.0	41.0	40.6	44.1	39.7	41.2
Injuries	12	13	19	18	22	22	17	13	15	10	18	19

Hours worked	41.1	38.2	42.3	39.2	36.1	36.2	38.7	36.0	37.3	36.5	37.9	38.0
Injuries	13	24	12	12	13	15	18	11	24	16	13	23

Hours worked	36.7	40.1	35.5	38.2	42.3	39.0	39.6	39.1	39.6	39.1
Injuries	14	10	5	14	13	18	15	23	15	23

11.5 The Runs Test

WHAT YOU SHOULD LEARN

▸ How to use the runs test to determine whether a data set is random

The Runs Test for Randomness

▸ THE RUNS TEST FOR RANDOMNESS

In obtaining a sample of data, it is important for the data to be selected randomly. But how do you know if the sample data are truly random? One way to test for randomness in a data set is to use a *runs test for randomness*.

Before using a runs test for randomness, you must first know how to determine the number of runs in a data set.

DEFINITION

A **run** is a sequence of data having the same characteristic. Each run is preceded by and followed by data with a different characteristic or by no data at all. The number of data in a run is called the **length** of the run.

EXAMPLE 1

▸ **Finding the Number of Runs**

A liquid-dispensing machine has been designed to fill one-liter bottles. A quality control inspector decides whether each bottle is filled to an acceptable level and passes inspection (P) or fails inspection (F). Determine the number of runs for each sequence and find the length of each run.

1. $P\,P\,P\,P\,P\,P\,P\,P\,F\,F\,F\,F\,F\,F\,F\,F$

2. $P\,F\,P\,F\,P\,F\,P\,F\,P\,F\,P\,F\,P\,F\,P\,F$

3. $P\,P\,F\,F\,F\,F\,P\,F\,F\,F\,P\,P\,P\,P\,P$

▸ **Solution**

1. There are two runs. The first 8 P's form a run of length 8 and the first 8 F's form another run of length 8, as shown below.

$$\underbrace{P\,P\,P\,P\,P\,P\,P\,P}_{\text{1st run}}\quad\underbrace{F\,F\,F\,F\,F\,F\,F\,F}_{\text{2nd run}}$$

2. There are 16 runs each of length 1, as shown below.

$$\underbrace{P}\ \underbrace{F}\ P\ F\ P\ F\ P\ F\ P\ F\ P\ F\ P\ F\ P\ \underbrace{F}$$
1st run 2nd run...   ...16th run

3. There are 5 runs, the first of length 2, the second of length 4, the third of length 1, the fourth of length 3, and the fifth of length 6, as shown below.

$$\underbrace{P\,P}_{\text{1st run}}\ \underbrace{F\,F\,F\,F}_{\text{2nd run}}\ \underbrace{P}_{\text{3rd run}}\ \underbrace{F\,F\,F}_{\text{4th run}}\ \underbrace{P\,P\,P\,P\,P}_{\text{5th run}}$$

▶ **Try It Yourself 1**

A machine produces engine parts. An inspector measures the diameter of each engine part and determines if the part passes inspection (P) or fails inspection (F). The results are shown below. Determine the number of runs in the sequence and find the length of each run.

$$P\ P\ P\ F\ P\ F\ P\ P\ P\ P\ F\ F\ P\ F\ P\ P\ F\ F\ F\ P\ P\ P\ F\ P\ P\ P$$

a. *Separate the data* each time there is a change in the characteristic of the data.
b. *Count the number of groups* to determine the number of runs.
c. *Count the number of data* within each run to determine the length.

Answer: Page A48

When each value in a set of data can be categorized into one of two separate categories, you can use the runs test for randomness to determine whether the data are random.

DEFINITION

The **runs test for randomness** is a nonparametric test that can be used to determine whether a sequence of sample data is random.

The runs test for randomness considers the number of runs in a sequence of sample data in order to test whether a sequence is random. If a sequence has too few or too many runs, it is usually not random. For instance, the sequence

$$P\ P\ P\ P\ P\ P\ P\ P\ F\ F\ F\ F\ F\ F\ F\ F$$

from Example 1, Part 1, has too few runs (only 2 runs). The sequence

$$P\ F\ P\ F\ P\ F\ P\ F\ P\ F\ P\ F\ P\ F\ P\ F$$

from Example 1, Part 2, has too many runs (16 runs). So, these sample data are probably not random.

You can use a hypothesis test to determine whether the number of runs in a sequence of sample data is too high or too low. The runs test is a two-tailed test, and the null and alternative hypotheses are as follows.

H_0: The sequence of data is random.

H_a: The sequence of data is not random.

When using the runs test, let n_1 represent the number of data that have one characteristic and let n_2 represent the number of data that have the second characteristic. It does not matter which characteristic you choose to be represented by n_1. Let G represent the number of runs.

n_1 = number of data with one characteristic

n_2 = number of data with the other characteristic

G = number of runs

Table 12 in Appendix B lists the critical values for the runs test for selected values of n_1 and n_2 at the $\alpha = 0.05$ level of significance. (In this text, you will use only the $\alpha = 0.05$ level of significance when performing runs tests.) If n_1 or n_2 is greater than 20, you can use the standard normal distribution to find the critical values.

You can calculate the test statistic for the runs test as follows.

TEST STATISTIC FOR THE RUNS TEST

When $n_1 \leq 20$ and $n_2 \leq 20$, the **test statistic** for the runs test is G, the number of runs.

When $n_1 > 20$ or $n_2 > 20$, the **test statistic** for the runs test is

$$z = \frac{G - \mu_G}{\sigma_G}$$

where

$$\mu_G = \frac{2n_1 n_2}{n_1 + n_2} + 1 \quad \text{and} \quad \sigma_G = \sqrt{\frac{2n_1 n_2 (2n_1 n_2 - n_1 - n_2)}{(n_1 + n_2)^2 (n_1 + n_2 - 1)}}.$$

GUIDELINES

Performing a Runs Test for Randomness

IN WORDS	IN SYMBOLS
1. Identify the claim. State the null and alternative hypotheses.	State H_0 and H_a.
2. Specify the level of significance. (Use $\alpha = 0.05$ for the runs test.)	Identify α.
3. Determine the number of data that have each characteristic and the number of runs.	Determine n_1, n_2, and G.
4. Determine the critical values.	If $n_1 \leq 20$ and $n_2 \leq 20$, use Table 12 in Appendix B.
	If $n_1 > 20$ or $n_2 > 20$, use Table 4 in Appendix B.
5. Find the test statistic.	If $n_1 \leq 20$ and $n_2 \leq 20$, use G.
	If $n_1 > 20$ or $n_2 > 20$, use $z = \dfrac{G - \mu_G}{\sigma_G}.$
6. Make a decision to reject or fail to reject the null hypothesis.	If G is less than or equal to the lower critical value or greater than or equal to the upper critical value, reject H_0. Otherwise, fail to reject H_0.
	Or, if z is in the rejection region, reject H_0. Otherwise, fail to reject H_0.
7. Interpret the decision in the context of the original claim.	

EXAMPLE 2

▶ **Using the Runs Test**

As people enter a concert, an usher records where they are sitting. The results for 13 people are shown, where L represents a lawn seat and P represents a pavilion seat. At $\alpha = 0.05$, can you conclude that the sequence of seat locations is not random?

$$L\ L\ L\ P\ P\ L\ P\ P\ P\ L\ L\ P\ L$$

▶ **Solution**

The claim is "the sequence of seat locations is not random." To test this claim, use the following null and alternative hypotheses.

H_0: The sequence of seat locations is random.

H_a: The sequence of seat locations is not random. (Claim)

To find the critical values, first determine n_1, the number of L's; n_2, the number of P's; and G, the number of runs.

$L\ L\ L$	$P\ P$	L	$P\ P\ P$	$L\ L$	P	L
1st run	2nd run	3rd run	4th run	5th run	6th run	7th run

$n_1 = $ number of L's $= 7$

$n_2 = $ number of P's $= 6$

$G = $ number of runs $= 7$

Because $n_1 \leq 20$, $n_2 \leq 20$, and $\alpha = 0.05$, use Table 12 to find the lower critical value 3 and the upper critical value 12. The test statistic is the number of runs $G = 7$. Because the test statistic G is between the critical values 3 and 12, you should fail to reject the null hypothesis.

Interpretation   There is not enough evidence at the 5% level of significance to support the claim that the sequence of seat locations is not random. So, it appears that the sequence of seat locations is random.

▶ **Try It Yourself 2**

The genders of 15 students as they enter a classroom are shown below, where F represents a female and M represents a male. At $\alpha = 0.05$, can you conclude that the sequence of genders is not random?

$$M\ F\ F\ F\ M\ M\ F\ F\ M\ F\ M\ M\ F\ F\ F$$

a. Identify the *claim* and state H_0 and H_a.
b. Specify the *level of significance* α.
c. Determine n_1, n_2, and G.
d. Determine the *critical values*.
e. Find the *test statistic* G.
f. *Decide* whether to reject the null hypothesis.
g. *Interpret* the decision in the context of the original claim.

Answer: Page A48

EXAMPLE 3

▶ **Using the Runs Test**

You want to determine whether the selection of recently hired employees in a large company is random with respect to gender. The genders of 36 recently hired employees are shown below, where F represents a female and M represents a male. At $\alpha = 0.05$, can you conclude that the sequence of employees is not random?

M M F F F F M M M M M M
F F F F F M M M M M M M
F F F M M M M F M M F M

▶ **Solution**

The claim is "the sequence of employees is not random." To test this claim, use the following null and alternative hypotheses.

H_0: The sequence of employees is random.

H_a: The sequence of employees is not random. (Claim)

To find the critical values, first determine n_1, the number of F's; n_2, the number of M's; and G, the number of runs.

$\underbrace{M\ M}_{\text{1st run}}\quad \underbrace{F\ F\ F\ F}_{\text{2nd run}}\quad \underbrace{M\ M\ M\ M\ M}_{\text{3rd run}}$

$\underbrace{F\ F\ F\ F\ F}_{\text{4th run}}\quad \underbrace{M\ M\ M\ M\ M\ M\ M}_{\text{5th run}}$

$\underbrace{F\ F\ F}_{\text{6th run}}\quad \underbrace{M\ M\ M\ M}_{\text{7th run}}\quad \underbrace{F}_{\substack{\text{8th}\\\text{run}}}\quad \underbrace{M\ M}_{\substack{\text{9th}\\\text{run}}}\quad \underbrace{F}_{\substack{\text{10th}\\\text{run}}}\quad \underbrace{M}_{\substack{\text{11th}\\\text{run}}}$

$n_1 = \text{number of } F\text{'s} = 14$

$n_2 = \text{number of } M\text{'s} = 22$

$G = \text{number of runs} = 11$

Because $n_2 > 20$, use Table 4 in Appendix B to find the critical values. Because the test is a two-tailed test with $\alpha = 0.05$, the critical values are

$-z_0 = -1.96$

and

$z_0 = 1.96$.

Before calculating the test statistic, find the values of μ_G and σ_G, as follows.

$$\mu_G = \frac{2n_1n_2}{n_1 + n_2} + 1$$

$$= \frac{2(14)(22)}{14 + 22} + 1$$

$$= \frac{616}{36} + 1$$

$$\approx 18.11$$

The table shows the National Football League conference of each winning team from Super Bowl I to Super Bowl XLIV, where A represents the American Football Conference and N represents the National Football Conference. (Source: National Football League)

Year	Confer-ence	Year	Confer-ence
1967	N	1989	N
1968	N	1990	N
1969	A	1991	N
1970	A	1992	N
1971	A	1993	N
1972	N	1994	N
1973	A	1995	N
1974	A	1996	N
1975	A	1997	N
1976	A	1998	A
1977	A	1999	A
1978	N	2000	N
1979	A	2001	A
1980	A	2002	A
1981	A	2003	N
1982	N	2004	A
1983	N	2005	A
1984	A	2006	A
1985	N	2007	A
1986	N	2008	N
1987	N	2009	A
1988	N	2010	N

At $\alpha = 0.05$, can you conclude that the sequence of conferences of Super Bowl winning teams is not random?

There is not enough evidence at the 5% level of significance to conclude that the sequence of conferences of Super Bowl winning teams is not random.

$$\sigma_G = \sqrt{\frac{2n_1 n_2 (2n_1 n_2 - n_1 - n_2)}{(n_1 + n_2)^2 (n_1 + n_2 - 1)}}$$

$$= \sqrt{\frac{2(14)(22)[2(14)(22) - 14 - 22]}{(14 + 22)^2 (14 + 22 - 1)}}$$

$$\approx 2.81$$

You can find the test statistic as follows.

$$z = \frac{G - \mu_G}{\sigma_G}$$

$$\approx \frac{11 - 18.11}{2.81}$$

$$\approx -2.53$$

From the graph below, you can see that the test statistic z is in the rejection region. So, you should decide to reject the null hypothesis.

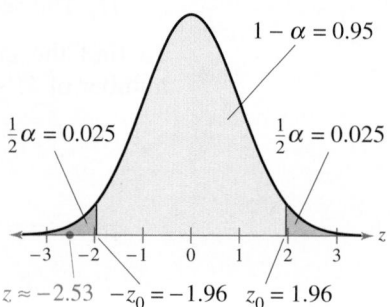

Interpretation You have enough evidence at the 5% level of significance to support the claim that the sequence of employees with respect to gender is not random.

▶ **Try It Yourself 3**

Let S represent a day in a small town in which it snowed and let N represent a day in the same town in which it did not snow. The following are the snowfall results for the entire month of January. At $\alpha = 0.05$, can you conclude that the sequence is not random?

$$N \; N \; N \; S \; S \; N \; N \; S \; N \; S \; N \; N \; N \; N \; S$$
$$N \; S \; N \; S \; N \; N \; S \; N \; S \; S \; N \; N \; N \; N \; N$$

a. Identify the *claim* and state H_0 and H_a.
b. Specify the *level of significance* α.
c. Determine n_1, n_2, and G.
d. Determine the *critical values*.
e. Find the *test statistic* z.
f. *Decide* whether to reject the null hypothesis.
g. *Interpret* the decision in the context of the original claim.

Answer: Page A48

When n_1 or n_2 is greater than 20, you can also use a P-value to perform a hypothesis test for the randomness of the data. In Example 3, you can calculate the P-value to be 0.0114. Because $P < \alpha$, you should reject the null hypothesis.

11.5   EXERCISES

■ BUILDING BASIC SKILLS AND VOCABULARY

1. In your own words, explain why the hypothesis test discussed in this section is called the runs test.

2. Describe the test statistic for the runs test when the sample sizes n_1 and n_2 are less than or equal to 20 and when either n_1 or n_2 is greater than 20.

■ USING AND INTERPRETING CONCEPTS

Finding the Number of Runs   *In Exercises 3–6, determine the number of runs in the given sequence. Then find the length of each run.*

3. *T F T F T T T F F F T F*

4. *U U D D U D U U D D U D U U*

5. *M F M F M F F F F F F M M M F F M M M M*

6. *A A A B B B A B B A A A A A A A B A A B A B B*

7. Find the values of n_1 and n_2 in Exercise 3.

8. Find the values of n_1 and n_2 in Exercise 4.

9. Find the values of n_1 and n_2 in Exercise 5.

10. Find the values of n_1 and n_2 in Exercise 6.

Finding Critical Values   *In Exercises 11–14, use the given sequence and Table 12 in Appendix B to determine the number of runs that are considered too high and the number of runs that are considered too low for the data to be in random order.*

11. *T F T F T F T F T F T F*

12. *M F M M M M M M M F F M M*

13. *N S S S N N N N S N S N S S N N N*

14. *X X X X X X X Y Y Y Y Y Y Y Y Y Y Y Y Y Y Y*

Performing a Runs Test   *In Exercises 15–20, use the runs test to (a) identify the claim and state H_0 and H_a, (b) determine the critical values using Table 4 or Table 12 in Appendix B, (c) find the test statistic, (d) decide whether to reject or fail to reject the null hypothesis, and (e) interpret the decision in the context of the original claim. Use $\alpha = 0.05$.*

15. Coin Toss   A coach records the results of the coin toss at the beginning of each football game for a season. The results are shown, where *H* represents heads and *T* represents tails. The coach claimed the tosses were not random. Use the runs test to test the coach's claim.

H T T T H T H H T T T T H T H H

16. Senate   The sequence shows the majority party of the U.S. Senate after each election for a recent group of years, where *R* represents the Republican party and *D* represents the Democratic party. Can you conclude that the sequence is not random? *(Source: United States Senate)*

R D D D R R R R R R D D D D D D
R D D R D D D D D D D D D D D D
R R R D D D D R R R D R R D D

---

1. Answers will vary. *Sample answer:* It is called the runs test because it considers the number of runs of data in a sample to determine whether the sequence of data was randomly selected.

2. See Selected Answers, page A134.

3. Number of runs: 8
Run lengths: 1, 1, 1, 1, 3, 3, 1, 1

4. See Selected Answers, page A134.

5. Number of runs: 9
Run lengths: 1, 1, 1, 1, 1, 6, 3, 2, 4

6. See Selected Answers, page A134.

7. n_1 = number of *T*'s = 6
n_2 = number of *F*'s = 6

8. See Selected Answers, page A134.

9. n_1 = number of *M*'s = 10
n_2 = number of *F*'s = 10

10. See Selected Answers, page A134.

11. too high: 11; too low: 3

12. too high: 8; too low: 2

13. too high: 14; too low: 5

14. too high: 15; too low: 5

15. (a) H_0: The coin tosses were random.

H_a: The coin tosses were not random. (claim)

(b) lower critical value = 4
upper critical value = 14

(c) 9   (d) Fail to reject H_0.

(e) There is not enough evidence at the 5% level of significance to support the claim that the coin tosses were not random.

16. See Selected Answers, page A134.

17. (a) H_0: The sequence of leagues of winning teams is random.

   H_a: The sequence of leagues of winning teams is not random. (claim)

   (b) ± 1.96 (c) 1.79

   (d) Fail to reject H_0.

   (e) There is not enough evidence at the 5% level of significance to conclude that the sequence of leagues of World Series winning teams is not random.

18. (a) H_0: The sequence of digits was randomly generated.

   H_a: The sequence of digits was not randomly generated. (claim)

   (b) lower critical value = 11 upper critical value = 23

   (c) 9   (d) Reject H_0.

   (e) There is enough evidence at the 5% level of significance to support the claim that the sequence of digits was not randomly generated.

19. (a) H_0: The microchips are random by gender. (claim)

   H_a: The microchips are not random by gender.

   (b) lower critical value = 8 upper critical value = 18

   (c) 12   (d) Fail to reject H_0.

   (e) There is not enough evidence at the 5% level of significance to reject the claim that the microchips are random by gender.

20. See Selected Answers, page A134.

21. Fail to reject H_0. There is not enough evidence at the 5% level of significance to support the claim that the daily high temperatures do not occur randomly.

22. Fail to reject H_0. There is not enough evidence at the 5% level of significance to conclude that the scores do not occur randomly.

23. Answers will vary.

---

17. **Baseball**   The sequence shows the Major League Baseball league of each World Series winning team from 1969 to 2009, where N represents the National League and A represents the American League. Can you conclude that the sequence of leagues of World Series winning teams is not random? *(Source: Major League Baseball)*

   N A N A A A N N A A N N N N A A A N A N
   A N A A A N A N A A A N A N A A N A N A

18. **Number Generator**   A number generator outputs the sequence of digits shown, where O represents an odd digit and E represents an even digit. Test the claim that the digits were not randomly generated.

   O O O E E E E O O O O O E E E
   O O E E E E O O O O E E E E O O

19. **Dog Identifications**   A team of veterinarians record, in order, the genders of every dog that is microchipped at their pet hospital in one month. The genders of recently microchipped dogs are shown, where F represents a female and M represents a male. A veterinarian claims that the microchips are random by gender. Do you have enough evidence to reject the doctor's claim?

   M M F M F F F F F M M M F F F
   M F F F F M F F F M F F F

20. **Golf Tournament**   A golf tournament official records whether each past winner is American-born (A) or foreign-born (F). The results are shown for every year the tournament has existed. Can you conclude that the sequence is not random?

   F F A F F A F F A F F A F F A F F A F F F F F
   A F F A F F A F F A F F A F A F A F F A F F F F A
   F F F F A F F F A

■ EXTENDING CONCEPTS

Runs Test with Quantitative Data   *In Exercises 21–23, use the following information to perform a runs test. You can also use the runs test for randomness with quantitative data. First, calculate the median. Then assign + to those values above the median and − to those values below the median. Ignore any values that are equal to the median. Use $\alpha = 0.05$.*

21. **Daily High Temperatures**   The sequence shows the daily high temperatures (in degrees Fahrenheit) for a city during the month of July. Test the claim that the daily high temperatures do not occur randomly.

   84 87 92 93 95 84 82 83 81 87 92 98 99 93 84 85
   86 92 91 95 84 92 83 81 87 92 98 89 93 84 85

22. **Exam Scores**   The sequence shows the exam scores of a class based on the order in which the students finished the test. Test the claim that the scores occur randomly.

   83 94 80 76 92 89 65 75 82 87 90 91 81 99 97 72
   72 89 90 92 87 76 74 66 88 81 90 92 89 76 80

23. Use a technology tool to generate a sequence of 30 numbers from 1 to 99, inclusive. Test the claim that the sequence of numbers is not random.

USES AND ABUSES

Uses

Nonparametric Tests   Before you could perform many of the hypothesis tests you learned about in previous chapters, you had to ensure that certain conditions about the population were satisfied. For instance, before you could run a *t*-test, you had to verify that the population was normally distributed. One advantage of the nonparametric tests shown in this chapter is that they are distribution free. That is, they do not require any particular information about the population or populations being tested. Another advantage of nonparametric tests is that they are easier to perform than their parametric counterparts. This means that they are easier to understand and quicker to use. Nonparametric tests can often be used when data are at the nominal or ordinal level.

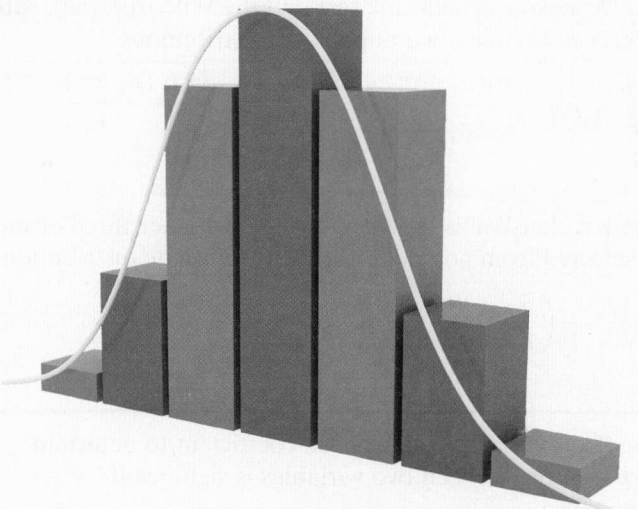

Abuses

Insufficient Evidence   Stronger evidence is needed to reject a null hypothesis in a nonparametric test than in a corresponding parametric test. That is, when you are trying to support a claim represented by the alternative hypothesis, you might need a larger sample when performing a nonparametric test. If the outcome of a nonparametric test results in failure to reject the null hypothesis, you should investigate the sample size used. It may be that a larger sample will produce different results.

Using an Inappropriate Test   In general, when information about the population (such as the condition of normality) is known, it is more efficient to use a parametric test. However, if information about the population is not known, nonparametric tests can be helpful.

■ EXERCISES

1. ***Insufficient Evidence***   Give an example of a nonparametric test in which there is not enough evidence to reject the null hypothesis.

2. ***Using an Inappropriate Test***   Discuss the nonparametric tests described in this chapter and match each test with its parametric counterpart, which you studied in earlier chapters.

11 CHAPTER SUMMARY

What did you **learn?**	EXAMPLE(S)	REVIEW EXERCISES
Section 11.1		
■ How to use the sign test to test a population median $$z = \frac{(x + 0.5) - 0.5n}{\frac{\sqrt{n}}{2}}$$	1, 2	1–3, 6
■ How to use the paired-sample sign test to test the difference between two population medians (dependent samples)	3	4, 5
Section 11.2		
■ How to use the Wilcoxon signed-rank test and the Wilcoxon rank sum test to test the difference between two population distributions $$z = \frac{R - \mu_R}{\sigma_R}, \quad \mu_R = \frac{n_1(n_1 + n_2 + 1)}{2}, \quad \sigma_R = \sqrt{\frac{n_1 n_2(n_1 + n_2 + 1)}{12}}$$	1, 2	7, 8
Section 11.3		
■ How to use the Kruskal-Wallis test to determine whether three or more samples were selected from populations having the same distribution $$H = \frac{12}{N(N + 1)}\left(\frac{R_1^2}{n_1} + \frac{R_2^2}{n_2} + \cdots + \frac{R_k^2}{n_k}\right) - 3(N + 1)$$	1	9, 10
Section 11.4		
■ How to use the Spearman rank correlation coefficient to determine whether the correlation between two variables is significant $$r_s = 1 - \frac{6\sum d^2}{n(n^2 - 1)}$$	1	11, 12
Section 11.5		
■ How to use the runs test to determine whether a data set is random $$G = \text{number of runs}, \quad z = \frac{G - \mu_G}{\sigma_G}, \quad \mu_G = \frac{2n_1 n_2}{n_1 + n_2} + 1, \quad \sigma_G = \sqrt{\frac{2n_1 n_2(2n_1 n_2 - n_1 - n_2)}{(n_1 + n_2)^2(n_1 + n_2 - 1)}}$$	1–3	13, 14

The table summarizes parametric and nonparametric tests. Always use the parametric test if the conditions for that test are satisfied.

Test application	Parametric test	Nonparametric test
One-sample tests	z-test for a population mean t-test for a population mean	Sign test for a population median
Two-sample tests Dependent samples	t-test for the difference between means	Paired-sample sign test Wilcoxon signed-rank test
Independent samples	z-test for the difference between means t-test for the difference between means	Wilcoxon rank sum test
Tests involving three or more samples	One-way ANOVA	Kruskal-Wallis test
Correlation	Pearson correlation coefficient	Spearman rank correlation coefficient
Randomness	(No parametric test)	Runs test

11 REVIEW EXERCISES

1. (a) H_0: median ≤ 650 (claim);
H_a: median > 650

(b) 2   (c) 7

(d) Fail to reject H_0.

(e) There is not enough evidence at the 1% level of significance to reject the bank manager's claim that the median number of customers per day is no more than 650.

2. (a) H_0: median ≥ 710 (claim);
H_a: median < 710

(b) 2   (c) 5

(d) Fail to reject H_0.

(e) There is not enough evidence at the 5% level of significance to reject the company's claim that the median credit score for U.S. adults is at least 710.

3. (a) H_0: median $= 2$ (claim);
H_a: median $\neq 2$

(b) -1.645 (c) -3.26

(d) Reject H_0.

(e) There is enough evidence at the 10% level of significance to reject the agency's claim that the median sentence length for all federal prisoners is 2 years.

4. (a) H_0: There was no reduction in diastolic blood pressure. (claim)

H_a: There was a reduction in diastolic blood pressure.

(b) 1   (c) 4

(d) Fail to reject H_0.

(e) There is not enough evidence at the 5% level of significance to reject the claim that there was no reduction in diastolic blood pressure.

■ SECTION 11.1

In Exercises 1–6, use a sign test to test the claim by doing the following.

(a) Identify the claim and state H_0 and H_a.

(b) Determine the critical value.

(c) Find the test statistic.

(d) Decide whether to reject or fail to reject the null hypothesis.

(e) Interpret the decision in the context of the original claim.

1. A bank manager claims that the median number of customers per day is no more than 650. The number of bank customers per day for 17 randomly selected days are listed below. At $\alpha = 0.01$, can you reject the bank manager's claim?

675	665	601	642	554	653	639	650	645
550	677	569	650	660	682	689	590	

2. A company claims that the median credit score for U.S. adults is at least 710. The credit scores for 13 randomly selected U.S. adults are listed below. At $\alpha = 0.05$, can you reject the company's claim? *(Adapted from Fair Isaac Corporation)*

750	782	805	695	700	706	625
589	690	772	745	704	710	

3. A government agency estimates that the median sentence length for all federal prisoners is 2 years. In a random sample of 180 federal prisoners, 65 have sentence lengths that are less than 2 years, 109 have sentence lengths that are more than 2 years, and 6 have sentence lengths that are 2 years. At $\alpha = 0.10$, can you reject the agency's claim? *(Adapted from United States Sentencing Commission)*

4. In a study testing the effects of calcium supplements on blood pressure in men, 10 randomly selected men were given a calcium supplement for 12 weeks. The following measurements are for each subject's diastolic blood pressure taken before and after the 12-week treatment period. At $\alpha = 0.05$, can you reject the claim that there was no reduction in diastolic blood pressure? *(Adapted from the American Medical Association)*

Patient	1	2	3	4	5	6	7
Before treatment	107	110	123	129	112	111	107
After treatment	100	114	105	112	115	116	106

Patient	8	9	10
Before treatment	112	136	102
After treatment	102	125	104

5. (a) H_0: There is no reduction in diastolic blood pressure. (claim)

H_a: There is a reduction in diastolic blood pressure.

(b) 2 (c) 3

(d) Fail to reject H_0.

(e) There is not enough evidence at the 5% level of significance to reject the claim that there was no reduction in diastolic blood pressure.

6. (a) H_0: median = $68,500 (claim)

H_a: median ≠ $68,500

(b) −1.96 (c) −2.33

(d) Reject H_0.

(e) There is enough evidence at the 5% level of significance to reject the claim that the median starting salary is $68,500.

7. (a) Independent; Wilcoxon rank sum test

(b) H_0: There is no difference in the total times to earn a doctorate degree by female and male graduate students.

H_a: There is a difference in the total times to earn a doctorate degree by female and male graduate students. (claim)

(c) ±2.575

(d) −1.357 (or 1.357)

(e) Fail to reject H_0.

(f) There is not enough evidence at the 1% level of significance to support the claim that there is a difference in the total times to earn a doctorate degree by female and male graduate students.

5. In a study testing the effects of an herbal supplement on blood pressure in men, 11 randomly selected men were given an herbal supplement for 12 weeks. The following measurements are for each subject's diastolic blood pressure taken before and after the 12-week treatment period. At $\alpha = 0.05$, can you reject the claim that there was no reduction in diastolic blood pressure? *(Adapted from The Journal of the American Medical Association)*

Patient	1	2	3	4	5	6	7
Before treatment	123	109	112	102	98	114	119
After treatment	124	97	113	105	95	119	114

Patient	8	9	10	11
Before treatment	112	110	117	130
After treatment	114	121	118	133

6. An association claims that the median annual salary of lawyers 9 months after graduation from law school is $68,500. In a random sample of 125 lawyers 9 months after graduation from law school, 76 were paid less than $68,500, and 49 were paid more than $68,500. At $\alpha = 0.05$, can you reject the association's claim? *(Adapted from National Association of Law Placement)*

■ SECTION 11.2

In Exercises 7 and 8, use a Wilcoxon test to test the claim by doing the following.

(a) *Decide whether the samples are dependent or independent; then choose the appropriate Wilcoxon test.*

(b) *Identify the claim and state H_0 and H_a.*

(c) *Determine the critical values.*

(d) *Find the test statistic.*

(e) *Decide whether to reject or fail to reject the null hypothesis.*

(f) *Interpret the decision in the context of the original claim.*

7. A career placement advisor estimates that there is a difference in the total times required to earn a doctorate degree by female and male graduate students. The table shows the total times to earn a doctorate for a random sample of 12 female and 12 male graduate students. At $\alpha = 0.01$, can you support the advisor's claim? *(Adapted from National Opinion Research Council)*

Gender	Total Time (in years)											
Female	13	12	10	13	12	9	11	14	7	7	9	10
Male	11	8	9	11	10	8	8	10	11	9	10	8

8. (a) Dependent; Wilcoxon signed-rank test

(b) H_0: The new drug does not affect the number of headache hours experienced.

H_a: The new drug does affect the number of headache hours experienced. (claim)

(c) 4   (d) 1.5   (e) Reject H_0.

(f) There is enough evidence at the 5% level of significance to support the claim that the new drug does affect the number of headache hours experienced.

9. (a) H_0: There is no difference in the ages of doctorate recipients among the fields of study.

H_a: There is a difference in the ages of doctorate recipients among the fields of study. (claim)

(b) 9.210   (c) 6.741

(d) Fail to reject H_0.

(e) There is not enough evidence at the 1% level of significance to conclude that the distributions of ages of the doctorate recipients in these three fields of study are different.

10. (a) H_0: There is no difference in starting salaries among the fields of engineering.

H_a: There is a difference in starting salaries among the fields of engineering. (claim)

(b) 7.815   (c) 24.833

(d) Reject H_0.

(e) There is enough evidence at the 5% level of significance to conclude that the distributions of the starting salaries in the four fields of engineering are different.

8. A medical researcher claims that a new drug affects the number of headache hours experienced by headache sufferers. The number of headache hours (per day) experienced by eight randomly selected patients before and after taking the drug are shown in the table. At $\alpha = 0.05$, can you support the researcher's claim?

Patient	1	2	3	4	5	6	7	8
Headache hours (before)	0.9	2.3	2.7	2.4	2.9	1.9	1.2	3.1
Headache hours (after)	1.4	1.5	1.4	1.8	1.3	0.6	0.7	1.9

■ SECTION 11.3

In Exercises 9 and 10, use the Kruskal-Wallis test to test the claim by doing the following.

(a) Identify the claim and state H_0 and H_a.

(b) Determine the critical value.

(c) Find the sums of the ranks for each sample and calculate the test statistic.

(d) Decide whether to reject or fail to reject the null hypothesis.

(e) Interpret the decision in the context of the original claim.

9. The table shows the ages for a random sample of doctorate recipients in three fields of study. At $\alpha = 0.01$, can you conclude that the distributions of the ages of the doctorate recipients in these three fields of study are different? *(Adapted from Survey of Earned Doctorates)*

Field of Study	Age										
Life sciences	31	32	34	31	30	32	35	31	32	34	29
Physical sciences	30	31	32	31	30	29	31	30	32	33	30
Social sciences	32	35	31	33	34	31	35	36	32	30	33

10. The table shows the starting salary offers for a random sample of college graduates in four fields of engineering. At $\alpha = 0.05$, can you conclude that the distributions of the starting salaries in these four fields of engineering are different? *(Adapted from National Association of Colleges and Employers)*

Field of Engineering	Starting Salary (in thousands of dollars)									
Chemical engineering	66.4	63.9	69.7	68.5	62.3	67.9	65.5	63.7	67.4	69.1
Computer engineering	61.1	60.5	58.7	59.3	62.4	65.5	59.9	63.1	61.4	59.3
Electrical engineering	59.3	57.9	58.5	56.8	60.0	59.7	61.3	60.5	59.5	59.8
Mechanical engineering	58.9	58.2	59.0	57.1	59.0	58.7	61.5	62.0	58.3	56.1

11. (a) H_0: $\rho_s = 0$; H_a: $\rho_s \neq 0$ (claim)

(b) 0.786   (c) 0.830

(d) Reject H_0.

(e) There is enough evidence at the 5% level of significance to conclude that there is a correlation between overall score and price.

12. (a) H_0: $\rho_s = 0$; H_a: $\rho_s \neq 0$ (claim)

(b) 0.600   (c) 0.213

(d) Fail to reject H_0.

(e) There is not enough evidence at the 10% level of significance to conclude that there is a correlation between overall score and price.

13. (a) H_0: The traffic stops were random by gender.

H_a: The traffic stops were not random by gender. (claim)

(b) lower critical value = 8 upper critical value = 19

(c) 14   (d) Fail to reject H_0.

(e) There is not enough evidence at the 5% level of significance to support the claim that the stops were not random by gender.

14. (a) H_0: The departure status of buses is random.

H_a: The departure status of buses is not random. (claim)

(b) lower critical value = 5 upper critical value = 14

(c) 5   (d) Reject H_0.

(e) There is enough evidence at the 5% level of significance to support the claim that the departure status of the buses is not random.

■ SECTION 11.4

In Exercises 11 and 12, use the Spearman rank correlation coefficient to test the claim by doing the following.

(a) Identify the claim mathematically and state H_0 and H_a.

(b) Determine the critical value using Table 10 in Appendix B.

(c) Find the test statistic.

(d) Decide whether to reject the null hypothesis.

(e) Interpret the decision in the context of the original claim.

11. The table shows the overall scores and the prices for seven randomly selected Blu-ray™ players. The overall score is based mainly on picture quality. At $\alpha = 0.05$, can you conclude that there is a correlation between overall score and price? *(Source: Consumer Reports)*

Overall score	93	91	90	87	85	74	69
Price (in dollars)	500	300	500	150	250	200	130

12. The table shows the overall scores and the prices per gallon for nine randomly selected interior paints. The overall score represents hiding, surface smoothness, and resistance to staining, scrubbing, gloss change, sticking, mildew, and fading. At $\alpha = 0.10$, can you conclude that there is a correlation between overall score and price? *(Adapted from Consumer Reports)*

Overall score	86	84	82	81	75	74	71	69	62
Price per gallon (in dollars)	33	32	20	45	19	25	25	18	37

■ SECTION 11.5

In Exercises 13 and 14, use the runs test to (a) identify the claim and state H_0 and H_a, (b) determine the critical values using Table 4 or Table 12 in Appendix B, (c) calculate the test statistic, (d) decide whether to reject or fail to reject the null hypothesis, and (e) interpret the decision in the context of the original claim. Use $a = 0.05$.

13. A highway patrol officer stops speeding vehicles on an interstate highway. The following shows the genders of the last 25 drivers who were stopped, where F represents a female driver and M represents a male driver. Can you conclude that the stops were not random by gender?

F M M M F M F M F F F M M
F F F M M M F M M M F F M

14. The following data represent the departure status of the last 18 buses to leave a bus station, where T represents a bus that departed on time and L represents a bus that departed late. Can you conclude that the departure status of the buses is not random?

T T T T L L L L T
L L L T T T T T T

11 CHAPTER QUIZ

Take this quiz as you would take a quiz in class. After you are done, check your work against the answers given in the back of the book.

For each exercise, (a) identify the claim and state H_0 and H_a, (b) decide which test to use, (c) determine the critical value(s), (d) find the test statistic, (e) decide whether to reject or fail to reject the null hypothesis, and (f) interpret the decision in the context of the original claim.

1. A labor organization claims that there is a difference in the hourly earnings of union and nonunion workers in state and local governments. A random sample of 10 union and 10 nonunion workers in state and local governments and their hourly earnings are listed in the tables. At $\alpha = 0.10$, can you support the organization's claim? *(Adapted from U.S. Bureau of Labor Statistics)*

Union					Nonunion				
27.20	25.60	29.75	32.97	30.33	24.80	21.75	19.85	25.60	20.70
25.30	24.80	26.50	25.05	24.20	23.40	21.15	20.90	20.05	19.10

2. An organization claims that the median number of annual volunteer hours is 52. In a random sample of 75 people who volunteered last year, 47 volunteered for less than 52 hours, 23 volunteered for more than 52 hours, and 5 volunteered for 52 hours. At $\alpha = 0.05$, can you reject the organization's claim? *(Adapted from U.S. Bureau of Labor Statistics)*

3. The table shows the sales prices for a random sample of apartment condominiums and cooperatives in four U.S. regions. At $\alpha = 0.01$, can you conclude that the distributions of the sales prices in these four regions are different? *(Adapted from National Association of Realtors)*

Region	Sales Price (in thousands of dollars)							
Northeast	252.5	245.5	237.9	270.2	265.9	250.0	259.4	238.6
Midwest	188.9	205.1	200.9	175.9	170.5	191.9	185.3	187.1
South	175.5	150.9	149.8	164.6	169.5	190.5	172.6	161.0
West	218.5	201.9	255.7	230.0	189.9	225.7	220.0	206.3

4. A meteorologist wants to determine whether days with rain occur randomly in April in his home town. To do so, the meteorologist records whether it rains for each day in April. The results are shown, where R represents a day with rain and N represents a day with no rain. At $\alpha = 0.05$, can the meteorologist conclude that days with rain are not random?

    N R R N N N N R N R R N R R R
    N R R R R N N N N R N R N N R

5. The table shows the number of larceny-thefts (per 100,000 population) and the number of motor vehicle thefts (per 100,000 population) in six randomly selected large U.S. cities. At $\alpha = 0.10$, can you conclude that there is a correlation between the number of larceny-thefts and the number of motor vehicle thefts? *(Source: U.S. Department of Justice)*

Larceny-thefts	1403	1506	2937	3449	2728	3042
Motor vehicle thefts	161	608	659	897	774	945

---

1. (a) H_0: There is no difference in the hourly earnings.

    H_a: There is a difference in the hourly earnings. (claim)

    (b) Wilcoxon rank sum test

    (c) ± 1.645

    (d) -3.326 (*or* 3.326)

    (e) Reject H_0.

    (f) There is enough evidence at the 10% level of significance to support the organization's claim that there is a difference in the hourly earnings of union and nonunion workers in state and local governments.

2. (a) H_0: median $= 52$ (claim); H_a: median $\neq 52$

    (b) Sign test   (c) ± 1.96

    (d) -2.75 (e) Reject H_0.

    (f) There is enough evidence at the 5% level of significance to reject the organization's claim that the median number of annual volunteer hours is 52 hours.

3. (a) H_0: There is no difference in the sales prices among the regions.

    H_a: There is a difference in the sales prices among the regions. (claim)

    (b) Kruskal-Wallis test

    (c) 11.345   (d) 25.957

    (e) Reject H_0.

    (f) There is enough evidence at the 1% level of significance to conclude that the distributions of the sales prices in these regions are different.

4. See Odd Answers, page A107.

5. (a) H_0: $\rho_s = 0$; H_a: $\rho_s \neq 0$ (claim)

    (b) Spearman rank correlation coefficient

    (c) 0.829   (d) 0.886

    (e) Reject H_0.

    (f) There is enough evidence at the 10% level of significance to conclude that there is a correlation between the number of larceny-thefts and the number of motor vehicle thefts.

PUTTING IT ALL TOGETHER

Real Statistics — Real Decisions

In a recent year, according to the Bureau of Labor Statistics, the median number of years that wage and salary workers had been with their current employer (called employee tenure) was 4.1 years. Information on employee tenure has been gathered since 1996 using the *Current Population Survey* (*CPS*), a monthly survey of about 60,000 households that provides information on employment, unemployment, earnings, demographics, and other characteristics of the U.S. population ages 16 and over. With respect to employee tenure, the questions measure how long workers have been with their current employers, not how long they plan to stay with their employers.

www.bls.gov

Employee Tenure of 20 Workers		
4.6	2.6	3.3
2.8	1.5	1.9
4.0	5.0	3.9
5.1	3.7	5.4
3.6	3.9	6.2
1.7	4.6	3.1
4.4	3.6	

TABLE FOR EXERCISE 2

 EXERCISES

1. *How Would You Do It?*

(a) What sampling technique would you use to select the sample for the *CPS*?

(b) Do you think the technique in part (a) will give you a sample that is representative of the U.S. population? Why or why not?

(c) Identify possible flaws or biases in the survey on the basis of the technique you chose in part (a).

2. *Is There a Difference?*

A congressional representative claims that the median tenure for workers from the representative's district is less than the national median tenure of 4.1 years. The claim is based on the representative's data and is shown in the table at the right above. (Assume that the employees were randomly selected.)

(a) Is it possible that the claim is true? What questions should you ask about how the data were collected?

(b) How would you test the representative's claim? Can you use a parametric test, or do you need to use a nonparametric test?

(c) State the null hypothesis and the alternative hypothesis.

(d) Test the claim using $\alpha = 0.05$. What can you conclude?

3. *Comparing Male and Female Employee Tenures*

A congressional representative claims that there is a difference between the median tenures for male workers and female workers. The claim is based on the representative's data and is shown in the table at the right. (Assume that the employees were randomly selected from the representative's district.)

(a) How would you test the representative's claim? Can you use a parametric test, or do you need to use a nonparametric test?

(b) State the null hypothesis and the alternative hypothesis.

(c) Test the claim using $\alpha = 0.05$. What can you conclude?

Employee tenure for a sample of male workers	Employee tenure for a sample of female workers
3.9	4.4
4.4	4.9
4.7	5.4
4.3	4.3
4.9	4.0
3.8	1.8
3.6	5.1
4.7	5.1
2.3	3.3
6.5	2.2
0.9	5.2
5.1	3.0
	1.3
	4.0

TABLE FOR EXERCISE 3

TECHNOLOGY

U.S. INCOME AND ECONOMIC RESEARCH

The National Bureau of Economic Research (NBER) is a private, nonprofit, nonpartisan research organization. The NBER provides information for better understanding of how the U.S. economy works. Researchers at the NBER concentrate on four types of empirical research: developing new statistical measurements, estimating quantitative models of economic behavior, assessing the effects of public policies on the U.S. economy, and projecting the effects of alternative policy proposals.

One of the NBER's interests is the median income of people in different regions of the United States. The table at the right shows the annual incomes (in dollars) of a random sample of people (15 years and over) in a recent year in four U.S. regions: Northeast, Midwest, South, and West.

Annual Income of People (in dollars)			
Northeast	**Midwest**	**South**	**West**
47,000	30,035	24,030	41,180
35,145	32,235	37,943	35,298
31,497	31,010	36,280	29,114
27,500	37,660	38,738	36,180
28,500	36,224	22,275	38,558
35,400	35,510	27,975	27,680
33,810	34,535	28,275	33,080
32,500	46,035	35,073	44,930
29,950	23,331	39,730	29,408
25,100	44,213	36,775	26,180
42,700	29,405	25,675	32,956
49,950	31,695	29,875	40,744

■ EXERCISES

In Exercises 1–5, refer to the annual income of people in the table. Use $\alpha = 0.05$ for all tests.

1. Construct a box-and-whisker plot for each region. Do the median annual incomes appear to differ between regions?

2. Use a technology tool to perform a sign test to test the claim that the median annual income in the Midwest is greater than $30,000.

3. Use a technology tool to perform a Wilcoxon rank sum test to test the claim that the median annual incomes in the Northeast and South are the same.

4. Use a technology tool to perform a Kruskal-Wallis test to test the claim that the distributions of annual incomes for all four regions are the same.

5. Use a technology tool to perform a one-way ANOVA to test the claim that the average annual incomes for all four regions are the same. Assume that the populations of incomes are normally distributed, the samples are independent, and the population variances are equal. How do your results compare with those in Exercise 4?

6. Repeat Exercises 1, 3, 4, and 5 using the data in the following table. The table shows the annual incomes (in dollars) of a random sample of families in a recent year in four U.S. regions: Northeast, Midwest, South, and West.

Annual Income of Families (in dollars)			
Northeast	**Midwest**	**South**	**West**
58,010	57,680	55,200	61,808
107,465	83,260	57,787	70,125
75,800	39,060	49,400	51,982
106,770	55,260	90,200	67,330
65,780	72,216	55,209	53,830
51,500	52,048	35,200	79,220
46,366	67,760	60,300	61,108
66,750	61,860	38,756	86,130
48,800	64,920	64,621	86,955
73,520	57,260	52,800	42,130
44,795	62,260	77,650	66,650
65,650	59,596	51,085	47,910
58,500	70,510	59,200	62,364
72,800	61,460	45,100	66,880
57,799	77,960	55,562	46,923

Extended solutions are given in the *Technology Supplement*.
Technical instruction is provided for MINITAB, Excel, and the TI-83/84 Plus.

CUMULATIVE REVIEW
Chapters 9 – 11

Men, x	Women, y
10.80	12.20
10.38	11.90
10.30	11.50
10.30	12.20
10.79	11.67
10.62	11.82
10.32	11.18
10.06	11.49
9.95	11.08
10.14	11.07
10.06	11.08
10.25	11.06
9.99	10.97
9.92	10.54
9.96	10.82
9.84	10.94
9.87	10.75
9.85	10.93
9.69	10.78

TABLE FOR EXERCISE 1

Sections 9.1 and 9.2

1. The table at the left shows the winning times (in seconds) for the men's and women's 100-meter runs in the Summer Olympics from 1928 to 2008. *(Source: The International Association of Athletics Federations)*

(a) Display the data in a scatter plot, calculate the correlation coefficient r, and make a conclusion about the type of correlation.

(b) Test the level of significance of the correlation coefficient r found in part (a). Use $\alpha = 0.05$.

(c) Find the equation of the regression line for the data. Draw the regression line on the scatter plot.

(d) Use the regression line to predict the women's 100-meter time when the men's 100-meter time is 9.90 seconds.

Section 11.2

2. An employment agency claims that there is a difference in the weekly earnings of workers who are union members and workers who are not union members. A random sample of nine union members and eight nonunion members and their weekly earnings (in dollars) are shown in the table. At $\alpha = 0.05$, can you support the agency's claim? *(Adapted from U.S. Bureau of Labor Statistics)*

Union member	855	994	692	800	884	991	1040	904	930
Not a union member	758	691	862	557	655	814	803	638	

Section 11.1

3. An investment company claims that the median age of people with mutual funds is 50 years. The ages (in years) of 20 randomly selected mutual fund owners are listed below. At $\alpha = 0.01$, is there enough evidence to reject the company's claim? *(Adapted from Investment Company Institute)*

46	34	33	27	58	64	54	36	38	42
26	51	49	44	46	50	39	34	51	63

Section 10.4

4. The table at the right shows the residential natural gas expenditures (in dollars) in one year for a random sample of households in four regions of the United States. Assume that the populations are normally distributed and the population variances are equal. At $\alpha = 0.10$, can you reject the claim that the mean expenditures are equal for all four regions? *(Adapted from U.S. Energy Information Administration)*

Northeast	Midwest	South	West
1478	393	434	625
649	980	319	538
834	609	694	1045
1173	1157	678	497
1013	865	856	305
1565	1337	499	358
655	870	451	549
648	810	1021	633

Section 9.4

5. The equation used to predict sweet potato yield (in pounds per acre) is $\hat{y} = 11{,}182 + 174.53x_1 - 104.41x_2$, where x_1 is the number of acres planted (in thousands) and x_2 is the number of acres harvested (in thousands). Use the multiple regression equation to predict the y-values for the given values of the independent variables. *(Adapted from United States Department of Agriculture)*

(a) $x_1 = 91$, $x_2 = 88$

(b) $x_1 = 110$, $x_2 = 98$

Section 10.3

6. A school administrator reports that the standard deviations of reading test scores for eighth grade students are the same in Colorado and Utah. A random sample of 16 test scores from Colorado has a standard deviation of 34.6 points and a random sample of 15 test scores from Utah has a standard deviation of 33.2 points. At $\alpha = 0.10$, can you reject the administrator's claim? Assume the samples are independent and each population has a normal distribution. *(Adapted from National Center for Education Statistics)*

Section 11.3

7. An employment agency representative wants to determine whether there is a difference in the annual household incomes in four regions of the United States. To do so, the representative randomly selects several households in each region and records the annual household income for each in the table. At $\alpha = 0.01$, can the representative conclude that the distributions of the annual household incomes in these regions are different? *(Adapted from U.S. Census Bureau)*

Region	Household Income (in thousands of dollars)						
Northeast	54.3	47.1	55.7	54.8	50.0	52.5	51.6
Midwest	49.3	54.4	45.2	48.5	50.7	51.8	52.0
South	44.4	45.6	49.2	41.5	46.4	49.2	47.0
West	56.8	54.7	51.4	53.5	52.4	54.0	55.9

Section 10.1

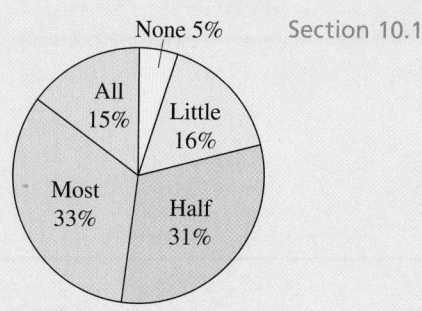

FIGURE FOR EXERCISE 8

8. Results from a previous survey asking U.S. parents how much they intend to contribute to the college costs of their children are shown in the pie chart. To determine whether this distribution is still the same, you randomly select 900 U.S. parents and ask them how much they intend to contribute to the college costs of their children. The results are shown in the table. At $\alpha = 0.05$, are the distributions different? *(Adapted from Sallie Mae, Inc.)*

Survey results	
Response	Frequency, f
None	31
Little	164
Half	277
Most	305
All	123

Section 9.3

9. The table shows the metacarpal bone lengths (in centimeters) and the heights (in centimeters) of nine adults. The equation of the regression line is $\hat{y} = 1.700x + 94.428$. *(Adapted from the American Journal of Physical Anthropology)*

Metacarpal bone length, x	45	51	39	41	48	49	46	43	47
Height, y	171	178	157	163	172	183	173	175	173

(a) Find the coefficient of determination and interpret the results.

(b) Find the standard error of estimate s_e and interpret the results.

(c) Construct a 95% prediction interval for the height of an adult when his or her metacarpal bone length is 50 centimeters. Interpret the results.

Section 11.4

10. The table shows the overall scores and the prices of eight all-season tires. The overall score represents safety-related tests, such as braking, handling, and resistance to hydroplaning. At $\alpha = 0.10$, can you conclude that there is a correlation between the overall score and the price? Use the Spearman rank correlation coefficient. *(Source: Consumer Reports)*

Overall score	74	82	78	84	80	64	70	74
Price (in dollars)	77	96	77	116	98	67	70	81

In this appendix, we use a 0-to-z table as an alternative development of the standard normal distribution. It is intended that this appendix be used after completion of the "Properties of a Normal Distribution" subsection of Section 5.1 in the text. If used, this appendix should replace the material in the "Standard Normal Distribution" subsection of Section 5.1 except for the exercises.

Standard Normal Distribution (0-to-z)

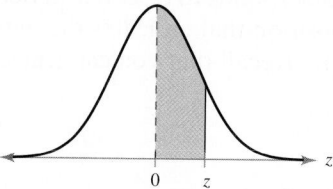

z	.00	.01	.02	.03	.04	.05	.06	.07	.08	.09
0.0	.0000	.0040	.0080	.0120	.0160	.0199	.0239	.0279	.0319	.0359
0.1	.0398	.0438	.0478	.0517	.0557	.0596	.0636	.0675	.0714	.0753
0.2	.0793	.0832	.0871	.0910	.0948	.0987	.1026	.1064	.1103	.1141
0.3	.1179	.1217	.1255	.1293	.1331	.1368	.1406	.1443	.1480	.1517
0.4	.1554	.1591	.1628	.1664	.1700	.1736	.1772	.1808	.1844	.1879
0.5	.1915	.1950	.1985	.2019	.2054	.2088	.2123	.2157	.2190	.2224
0.6	.2257	.2291	.2324	.2357	.2389	.2422	.2454	.2486	.2517	.2549
0.7	.2580	.2611	.2642	.2673	.2704	.2734	.2764	.2794	.2823	.2852
0.8	.2881	.2910	.2939	.2967	.2995	.3023	.3051	.3078	.3106	.3133
0.9	.3159	.3186	.3212	.3238	.3264	.3289	.3315	.3340	.3365	.3389
1.0	.3413	.3438	.3461	.3485	.3508	.3531	.3554	.3577	.3599	.3621
1.1	.3643	.3665	.3686	.3708	.3729	.3749	.3770	.3790	.3810	.3830
1.2	.3849	.3869	.3888	.3907	.3925	.3944	.3962	.3980	.3997	.4015
1.3	.4032	.4049	.4066	.4082	.4099	.4115	.4131	.4147	.4162	.4177
1.4	.4192	.4207	.4222	.4236	.4251	.4265	.4279	.4292	.4306	.4319
1.5	.4332	.4345	.4357	.4370	.4382	.4394	.4406	.4418	.4429	.4441
1.6	.4452	.4463	.4474	.4484	.4495	.4505	.4515	.4525	.4535	.4545
1.7	.4554	.4564	.4573	.4582	.4591	.4599	.4608	.4616	.4625	.4633
1.8	.4641	.4649	.4656	.4664	.4671	.4678	.4686	.4693	.4699	.4706
1.9	.4713	.4719	.4726	.4732	.4738	.4744	.4750	.4756	.4761	.4767
2.0	.4772	.4778	.4783	.4788	.4793	.4798	.4803	.4808	.4812	.4817
2.1	.4821	.4826	.4830	.4834	.4838	.4842	.4846	.4850	.4854	.4857
2.2	.4861	.4864	.4868	.4871	.4875	.4878	.4881	.4884	.4887	.4890
2.3	.4893	.4896	.4898	.4901	.4904	.4906	.4909	.4911	.4913	.4916
2.4	.4918	.4920	.4922	.4925	.4927	.4929	.4931	.4932	.4934	.4936
2.5	.4938	.4940	.4941	.4943	.4945	.4946	.4948	.4949	.4951	.4952
2.6	.4953	.4955	.4956	.4957	.4959	.4960	.4961	.4962	.4963	.4964
2.7	.4965	.4966	.4967	.4968	.4969	.4970	.4971	.4972	.4973	.4974
2.8	.4974	.4975	.4976	.4977	.4977	.4978	.4979	.4979	.4980	.4981
2.9	.4981	.4982	.4982	.4983	.4984	.4984	.4985	.4985	.4986	.4986
3.0	.4987	.4987	.4987	.4988	.4988	.4989	.4989	.4989	.4990	.4990
3.1	.4990	.4991	.4991	.4991	.4992	.4992	.4992	.4992	.4993	.4993
3.2	.4993	.4993	.4994	.4994	.4994	.4994	.4994	.4995	.4995	.4995
3.3	.4995	.4995	.4995	.4996	.4996	.4996	.4996	.4996	.4996	.4997
3.4	.4997	.4997	.4997	.4997	.4997	.4997	.4997	.4997	.4997	.4998

Reprinted with permission of Gale Mosteller, executor of estate of Frederick Mosteller, 3830 13th Street North, Arlington, VA 22201   mosteller.g@ei.com.

A Alternative Presentation of the Standard Normal Distribution

WHAT YOU SHOULD LEARN

▸ How to find areas under the standard normal curve

▸ **THE STANDARD NORMAL DISTRIBUTION**

There are infinitely many normal distributions, each with its own mean and standard deviation. The normal distribution with a mean of 0 and a standard deviation of 1 is called the *standard normal distribution*. The horizontal scale of the graph of the standard normal distribution corresponds to *z*-scores. In Section 2.5, you learned that a *z*-score is a measure of position that indicates the number of standard deviations a value lies from the mean. Recall that you can transform an *x*-value to a *z*-score using the formula

$$z = \frac{\text{Value} - \text{Mean}}{\text{Standard deviation}} = \frac{x - \mu}{\sigma}.$$

INSIGHT

Because every normal distribution can be transformed to the standard normal distribution, you can use *z*-scores and the standard normal curve to find areas (and therefore probabilities) under any normal curve.

DEFINITION

The **standard normal distribution** is a normal distribution with a mean of 0 and a standard deviation of 1.

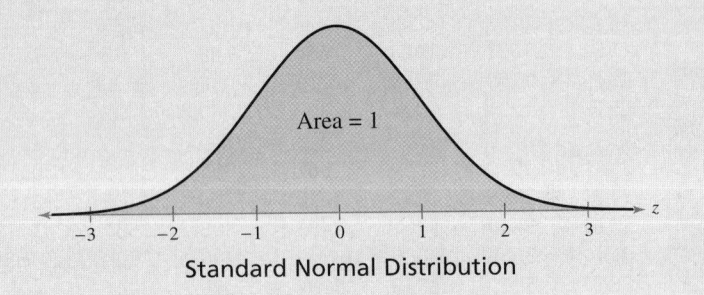

Area = 1

-3 -2 -1 0 1 2 3 z

Standard Normal Distribution

Note to Instructor

Mention that the formula for a normal probability density function on page 236 is greatly simplified when $\mu = 0$ and $\sigma = 1$.

$$y = \frac{e^{-x^2/2}}{\sqrt{2\pi}}$$

If each data value of a normally distributed random variable *x* is transformed into a *z*-score, the result will be the standard normal distribution. When this transformation takes place, the area that falls in the interval under the nonstandard normal curve is the *same* as that under the standard normal curve within the corresponding *z*-boundaries.

In Section 2.4, you learned to use the Empirical Rule to approximate areas under a normal curve when values of the random variable *x* corresponded to $-3, -2, -1, 0, 1, 2,$ or 3 standard deviations from the mean. Now, you will learn to calculate areas corresponding to other *x*-values. After you transform an *x*-value to a *z*-score, you can use the Standard Normal Table (0-to-*z*) on page A1. The table lists the area under the standard normal curve between 0 and the given *z*-score. As you examine the table, notice the following.

STUDY TIP

It is important that you know the difference between *x* and *z*. The random variable *x* is sometimes called a raw score and represents values in a nonstandard normal distribution, whereas *z* represents values in the standard normal distribution.

PROPERTIES OF THE STANDARD NORMAL DISTRIBUTION

1. The distribution is symmetric about the mean $(z = 0)$.
2. The area under the standard normal curve to the left of $z = 0$ is 0.5 and the area to the right of $z = 0$ is 0.5.
3. The area under the standard normal curve increases as the distance between 0 and *z* increases.

At first glance, the table on page A1 appears to give areas for positive z-scores only. However, because of the symmetry of the standard normal curve, the table also gives areas for negative z-scores (see Example 1).

EXAMPLE 1

▸ **Using the Standard Normal Table (0-to-z)**

1. Find the area under the standard normal curve between $z = 0$ and $z = 1.15$.
2. Find the z-scores that correspond to an area of 0.0948.

▸ **Solution**

1. Find the area that corresponds to $z = 1.15$ by finding 1.1 in the left column and then moving across the row to the column under 0.05. The number in that row and column is 0.3749. So, the area between $z = 0$ and $z = 1.15$ is 0.3749.

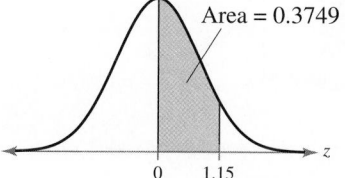

Area = 0.3749

z	.00	.01	.02	.03	.04	.05	.06
0.0	.0000	.0040	.0080	.0120	.0160	.0199	.0239
0.1	.0398	.0438	.0478	.0517	.0557	.0596	.0636
0.2	.0793	.0832	.0871	.0910	.0948	.0987	.1026
0.3	.1179	.1217	.1255	.1293	.1331	.1368	.1406

z	.00	.01	.02	.03	.04	.05	.06
0.9	.3159	.3186	.3212	.3238	.3264	.3289	.3315
1.0	.3413	.3438	.3461	.3485	.3508	.3531	.3554
1.1	.3643	.3665	.3686	.3708	.3729	.3749	.3770
1.2	.3849	.3869	.3888	.3907	.3925	.3944	.3962
1.3	.4032	.4049	.4066	.4082	.4099	.4115	.4131
1.4	.4192	.4207	.4222	.4236	.4251	.4265	.4279

2. Find the z-scores that correspond to an area of 0.0948 by locating 0.0948 in the table. The values at the beginning of the corresponding row and at the top of the corresponding column give the z-score. For an area of 0.0948, the row value is 0.2 and the column value is 0.04. So, the z-scores are $z = -0.24$ and $z = 0.24$.

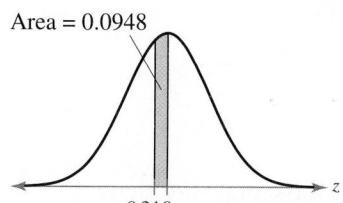

Area = 0.0948

z	.00	.01	.02	.03	.04	.05	.06
0.0	.0000	.0040	.0080	.0120	.0160	.0199	.0239
0.1	.0398	.0438	.0478	.0517	.0557	.0596	.0636
0.2	.0793	.0832	.0871	.0910	.0948	.0987	.1026
0.3	.1179	.1217	.1255	.1293	.1331	.1368	.1406
0.4	.1554	.1591	.1628	.1664	.1700	.1736	.1772
0.5	.1915	.1950	.1985	.2019	.2054	.2088	.2123

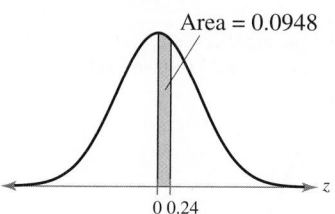

Area = 0.0948

▸ **Try It Yourself 1**

1. Find the area under the standard normal curve between $z = 0$ and $z = 2.19$.

   Locate the given z-score and *find the corresponding area* in the Standard Normal Table (0-to-z) on page A1.

2. Find the z-scores that correspond to an area of 0.4850.

   Locate the given area in the Standard Normal Table (0-to-z) on page A1 and *find the corresponding z-score.*   *Answer: Page A49*

Use the following guidelines to find various types of areas under the standard normal curve.

Finding Areas Under the Standard Normal Curve

1. Sketch the standard normal curve and shade the appropriate area under the curve.
2. Use the Standard Normal Table (0-to-z) on page A1 to find the area that corresponds to the given z-score(s).
3. Find the desired area by following the directions for each case shown.

a. Area to the left of z

i. When $z < 0$, *subtract* the area from 0.5.

2. Subtract to find the area to the left of $z = -1.23$; $0.5 - 0.3907 = 0.1093$.

1. The area between $z = 0$ and $z = -1.23$ is 0.3907.

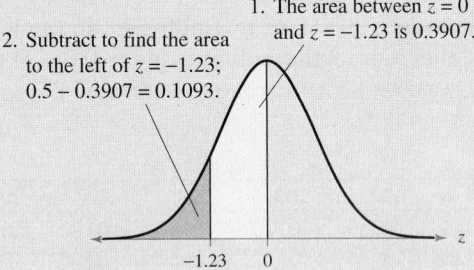

ii. When $z > 0$, *add* 0.5 to the area.

2. Add to find the area to the left of $z = 1.23$; $0.5 + 0.3907 = 0.8907$.

1. The area between $z = 0$ and $z = 1.23$ is 0.3907.

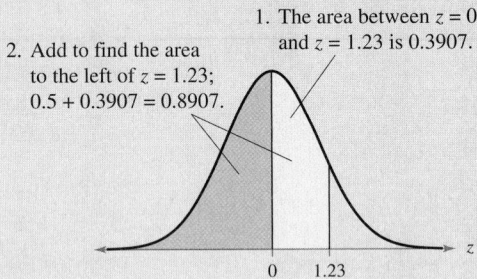

b. Area to the right of z

i. When $z < 0$, *add* 0.5 to the area.

1. The area between $z = 0$ and $z = -1.23$ is 0.3907.
2. Add to find the area to the right of $z = -1.23$; $0.5 + 0.3907 = 0.8907$.

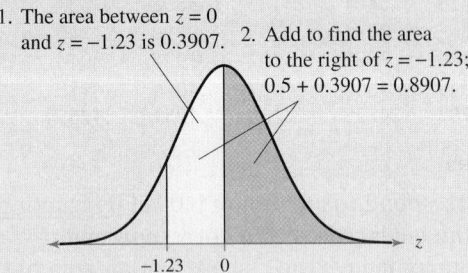

ii. When $z > 0$, *subtract* the area from 0.5.

1. The area between $z = 0$ and $z = 1.23$ is 0.3907.
2. Subtract to find the area to the right of $z = 1.23$; $0.5 - 0.3907 = 0.1093$.

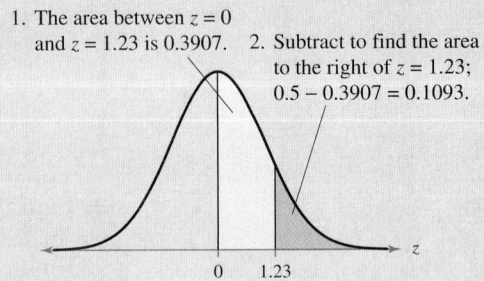

c. Area between two z-scores

i. When the two z-scores have the same sign (both positive or both negative), *subtract* the smaller area from the larger area.

1. The area between $z = 0$ and $z_1 = 1.23$ is 0.3907.
2. The area between $z = 0$ and $z_2 = 2.5$ is 0.4938.

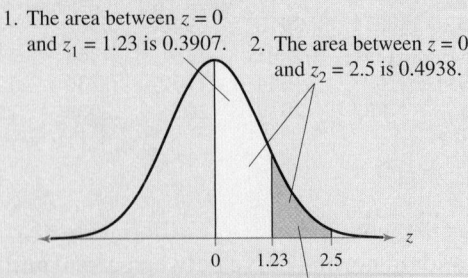

3. Subtract to find the area between $z_1 = 1.23$ and $z_2 = 2.5$; $0.4938 - 0.3907 = 0.1031$.

ii. When the two z-scores have opposite signs (one negative and one positive), *add* the areas.

2. The area between $z = 0$ and $z_2 = -0.5$ is 0.1915.

1. The area between $z = 0$ and $z_1 = 1.23$ is 0.3907.

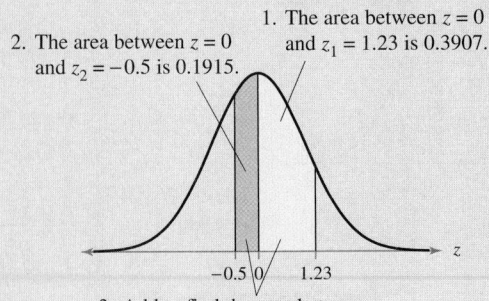

3. Add to find the area between $z_1 = 1.23$ and $z_2 = -0.5$; $0.3907 + 0.1915 = 0.5822$.

EXAMPLE 2

▸ **Finding Area Under the Standard Normal Curve**

Find the area under the standard normal curve to the left of $z = -0.99$.

▸ **Solution**

The area under the standard normal curve to the left of $z = -0.99$ is shown.

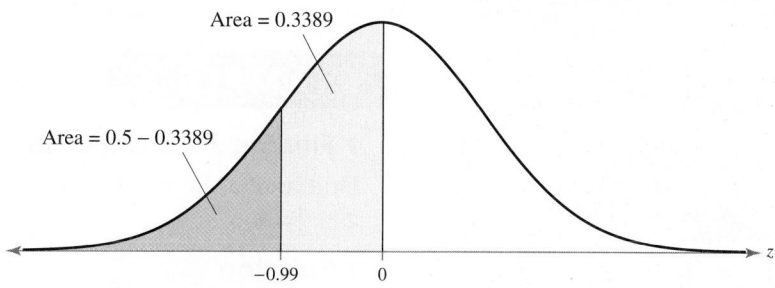

From the Standard Normal Table (0-to-z), the area corresponding to $z = -0.99$ is 0.3389. Because the area to the left of $z = 0$ is 0.5, the area to the left of $z = -0.99$ is $0.5 - 0.3389 = 0.1611$.

▸ **Try It Yourself 2**

Find the area under the standard normal curve to the left of $z = 2.13$.

a. *Draw* the standard normal curve and shade the area under the curve and to the left of $z = 2.13$.
b. Use the Standard Normal Table (0-to-z) on page A1 to *find the area* that corresponds to $z = 2.13$.
c. *Add* 0.5 to the resulting area.   *Answer: Page A49*

EXAMPLE 3

▸ **Finding Area Under the Standard Normal Curve**

Find the area under the standard normal curve to the right of $z = 1.06$.

▸ **Solution**

The area under the standard normal curve to the right of $z = 1.06$ is shown.

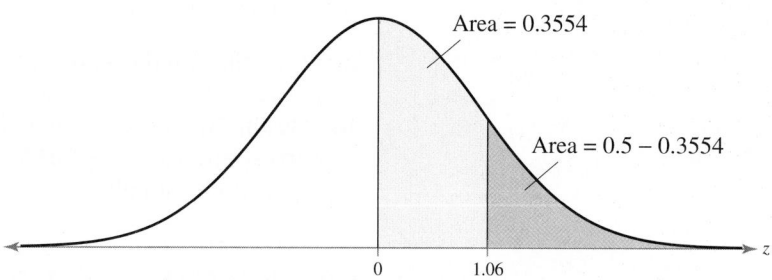

From the Standard Normal Table (0-to-z), the area corresponding to $z = 1.06$ is 0.3554. Because the area to the right of $z = 0$ is 0.5, the area to the right of $z = 1.06$ is $0.5 - 0.3554 = 0.1446$.

PICTURING THE WORLD

According to one publication, the number of births in a recent year was 4,317,000. The weights of the newborns can be approximated by a normal distribution, as shown by the following graph. (Adapted from National Center for Health Statistics)

Weights of Newborns

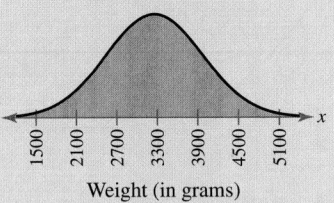

Weight (in grams)

Find the z-scores that correspond to weights of 2000, 3000, and 4000 grams. Are any of these unusually heavy or light?

−2.17, −0.50, 1.17; A weight of 2000 grams is unusually light because it is more than 2 standard deviations from the mean.

▶ **Try It Yourself 3**

Find the area under the standard normal curve to the right of $z = -2.16$.

a. *Draw* the standard normal curve and shade the area below the curve and to the right of $z = -2.16$.
b. Use the Standard Normal Table (0-to-z) on page A1 to *find the area* that corresponds to $z = -2.16$.
c. *Add* 0.5 to the resulting area.   *Answer: Page A49*

EXAMPLE 4

▶ **Finding Area Under the Standard Normal Curve**

Find the area under the standard normal curve between $z = -1.5$ and $z = 1.25$.

▶ **Solution**

The area under the standard normal curve between $z = -1.5$ and $z = 1.25$ is shown.

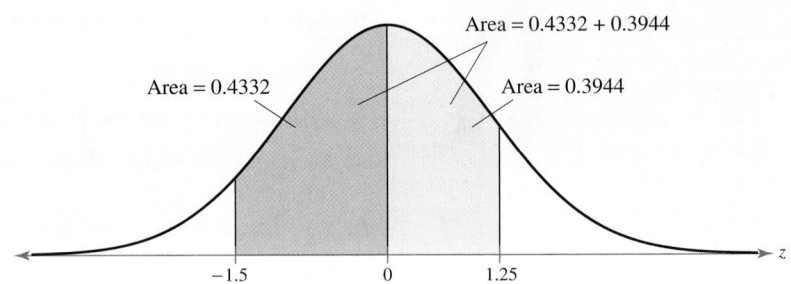

From the Standard Normal Table, the area corresponding to $z = -1.5$ is 0.4332 and the area corresponding to $z = 1.25$ is 0.3944. To find the area between these two z-scores, add the resulting areas.

Area $= 0.4332 + 0.3944 = 0.8276$

Interpretation   So, 82.76% of the area under the curve falls between $z = -1.5$ and $z = 1.25$.

▶ **Try It Yourself 4**

Find the area under the standard normal curve between $z = -2.165$ and $z = -1.35$.

a. *Draw* the standard normal curve and shade the area below the curve that is between $z = -2.165$ and $z = -1.35$.
b. Use the Standard Normal Table (0-to-z) on page A1 to *find the areas* that correspond to $z = -2.165$ and to $z = -1.35$.
c. *Subtract* the smaller area from the larger area.   *Answer: Page A49*

Recall that in Section 2.5 you learned, using the Empirical Rule, that values lying more than two standard deviations from the mean are considered unusual. Values lying more than three standard deviations from the mean are considered *very* unusual. So, if a z-score is greater than 2 or less than −2, it is unusual. If a z-score is greater than 3 or less than −3, it is *very* unusual.

Table 1— Random Numbers

92630	78240	19267	95457	53497	23894	37708	79862	76471	66418
79445	78735	71549	44843	26104	67318	00701	34986	66751	99723
59654	71966	27386	50004	05358	94031	29281	18544	52429	06080
31524	49587	76612	39789	13537	48086	59483	60680	84675	53014
06348	76938	90379	51392	55887	71015	09209	79157	24440	30244
28703	51709	94456	48396	73780	06436	86641	69239	57662	80181
68108	89266	94730	95761	75023	48464	65544	96583	18911	16391
99938	90704	93621	66330	33393	95261	95349	51769	91616	33238
91543	73196	34449	63513	83834	99411	58826	40456	69268	48562
42103	02781	73920	56297	72678	12249	25270	36678	21313	75767
17138	27584	25296	28387	51350	61664	37893	05363	44143	42677
28297	14280	54524	21618	95320	38174	60579	08089	94999	78460
09331	56712	51333	06289	75345	08811	82711	57392	25252	30333
31295	04204	93712	51287	05754	79396	87399	51773	33075	97061
36146	15560	27592	42089	99281	59640	15221	96079	09961	05371
29553	18432	13630	05529	02791	81017	49027	79031	50912	09399
23501	22642	63081	08191	89420	67800	55137	54707	32945	64522
57888	85846	67967	07835	11314	01545	48535	17142	08552	67457
55336	71264	88472	04334	63919	36394	11196	92470	70543	29776
10087	10072	55980	64688	68239	20461	89381	93809	00796	95945
34101	81277	66090	88872	37818	72142	67140	50785	21380	16703
53362	44940	60430	22834	14130	96593	23298	56203	92671	15925
82975	66158	84731	19436	55790	69229	28661	13675	99318	76873
54827	84673	22898	08094	14326	87038	42892	21127	30712	48489
25464	59098	27436	89421	80754	89924	19097	67737	80368	08795
67609	60214	41475	84950	40133	02546	09570	45682	50165	15609
44921	70924	61295	51137	47596	86735	35561	76649	18217	63446
33170	30972	98130	95828	49786	13301	36081	80761	33985	68621
84687	85445	06208	17654	51333	02878	35010	67578	61574	20749
71886	56450	36567	09395	96951	35507	17555	35212	69106	01679
00475	02224	74722	14721	40215	21351	08596	45625	83981	63748
25993	38881	68361	59560	41274	69742	40703	37993	03435	18873
92882	53178	99195	93803	56985	53089	15305	50522	55900	43026
25138	26810	07093	15677	60688	04410	24505	37890	67186	62829
84631	71882	12991	83028	82484	90339	91950	74579	03539	90122
34003	92326	12793	61453	48121	74271	28363	66561	75220	35908
53775	45749	05734	86169	42762	70175	97310	73894	88606	19994
59316	97885	72807	54966	60859	11932	35265	71601	55577	67715
20479	66557	50705	26999	09854	52591	14063	30214	19890	19292
86180	84931	25455	26044	02227	52015	21820	50599	51671	65411
21451	68001	72710	40261	61281	13172	63819	48970	51732	54113
98062	68375	80089	24135	72355	95428	11808	29740	81644	86610
01788	64429	14430	94575	75153	94576	61393	96192	03227	32258
62465	04841	43272	68702	01274	05437	22953	18946	99053	41690
94324	31089	84159	92933	99989	89500	91586	02802	69471	68274
05797	43984	21575	09908	70221	19791	51578	36432	33494	79888
10395	14289	52185	09721	25789	38562	54794	04897	59012	89251
35177	56986	25549	59730	64718	52630	31100	62384	49483	11409
25633	89619	75882	98256	02126	72099	57183	55887	09320	73463
16464	48280	94254	45777	45150	68865	11382	11782	22695	41988

Table 2— Binomial Distribution

This table shows the probability of x successes in n independent trials, each with probability of success p.

											p										
n	x	.01	.05	.10	.15	.20	.25	.30	.35	.40	.45	.50	.55	.60	.65	.70	.75	.80	.85	.90	.95
2	0	.980	.902	.810	.723	.640	.563	.490	.423	.360	.303	.250	.203	.160	.123	.090	.063	.040	.023	.010	.002
	1	.020	.095	.180	.255	.320	.375	.420	.455	.480	.495	.500	.495	.480	.455	.420	.375	.320	.255	.180	.095
	2	.000	.002	.010	.023	.040	.063	.090	.123	.160	.203	.250	.303	.360	.423	.490	.563	.640	.723	.810	.902
3	0	.970	.857	.729	.614	.512	.422	.343	.275	.216	.166	.125	.091	.064	.043	.027	.016	.008	.003	.001	.000
	1	.029	.135	.243	.325	.384	.422	.441	.444	.432	.408	.375	.334	.288	.239	.189	.141	.096	.057	.027	.007
	2	.000	.007	.027	.057	.096	.141	.189	.239	.288	.334	.375	.408	.432	.444	.441	.422	.384	.325	.243	.135
	3	.000	.000	.001	.003	.008	.016	.027	.043	.064	.091	.125	.166	.216	.275	.343	.422	.512	.614	.729	.857
4	0	.961	.815	.656	.522	.410	.316	.240	.179	.130	.092	.062	.041	.026	.015	.008	.004	.002	.001	.000	.000
	1	.039	.171	.292	.368	.410	.422	.412	.384	.346	.300	.250	.200	.154	.112	.076	.047	.026	.011	.004	.000
	2	.001	.014	.049	.098	.154	.211	.265	.311	.346	.368	.375	.368	.346	.311	.265	.211	.154	.098	.049	.014
	3	.000	.000	.004	.011	.026	.047	.076	.112	.154	.200	.250	.300	.346	.384	.412	.422	.410	.368	.292	.171
	4	.000	.000	.000	.001	.002	.004	.008	.015	.026	.041	.062	.092	.130	.179	.240	.316	.410	.522	.656	.815
5	0	.951	.774	.590	.444	.328	.237	.168	.116	.078	.050	.031	.019	.010	.005	.002	.001	.000	.000	.000	.000
	1	.048	.204	.328	.392	.410	.396	.360	.312	.259	.206	.156	.113	.077	.049	.028	.015	.006	.002	.000	.000
	2	.001	.021	.073	.138	.205	.264	.309	.336	.346	.337	.312	.276	.230	.181	.132	.088	.051	.024	.008	.001
	3	.000	.001	.008	.024	.051	.088	.132	.181	.230	.276	.312	.337	.346	.336	.309	.264	.205	.138	.073	.021
	4	.000	.000	.000	.002	.006	.015	.028	.049	.077	.113	.156	.206	.259	.312	.360	.396	.410	.392	.328	.204
	5	.000	.000	.000	.000	.000	.001	.002	.005	.010	.019	.031	.050	.078	.116	.168	.237	.328	.444	.590	.774
6	0	.941	.735	.531	.377	.262	.178	.118	.075	.047	.028	.016	.008	.004	.002	.001	.000	.000	.000	.000	.000
	1	.057	.232	.354	.399	.393	.356	.303	.244	.187	.136	.094	.061	.037	.020	.010	.004	.002	.000	.000	.000
	2	.001	.031	.098	.176	.246	.297	.324	.328	.311	.278	.234	.186	.138	.095	.060	.033	.015	.006	.001	.000
	3	.000	.002	.015	.042	.082	.132	.185	.236	.276	.303	.312	.303	.276	.236	.185	.132	.082	.042	.015	.002
	4	.000	.000	.001	.006	.015	.033	.060	.095	.138	.186	.234	.278	.311	.328	.324	.297	.246	.176	.098	.031
	5	.000	.000	.000	.000	.002	.004	.010	.020	.037	.061	.094	.136	.187	.244	.303	.356	.393	.399	.354	.232
	6	.000	.000	.000	.000	.000	.000	.001	.002	.004	.008	.016	.028	.047	.075	.118	.178	.262	.377	.531	.735
7	0	.932	.698	.478	.321	.210	.133	.082	.049	.028	.015	.008	.004	.002	.001	.000	.000	.000	.000	.000	.000
	1	.066	.257	.372	.396	.367	.311	.247	.185	.131	.087	.055	.032	.017	.008	.004	.001	.000	.000	.000	.000
	2	.002	.041	.124	.210	.275	.311	.318	.299	.261	.214	.164	.117	.077	.047	.025	.012	.004	.001	.000	.000
	3	.000	.004	.023	.062	.115	.173	.227	.268	.290	.292	.273	.239	.194	.144	.097	.058	.029	.011	.003	.000
	4	.000	.000	.003	.011	.029	.058	.097	.144	.194	.239	.273	.292	.290	.268	.227	.173	.115	.062	.023	.004
	5	.000	.000	.000	.001	.004	.012	.025	.047	.077	.117	.164	.214	.261	.299	.318	.311	.275	.210	.124	.041
	6	.000	.000	.000	.000	.001	.004	.008	.017	.032	.055	.087	.131	.185	.247	.311	.367	.396	.372	.257
	7	.000	.000	.000	.000	.000	.000	.000	.001	.002	.004	.008	.015	.028	.049	.082	.133	.210	.321	.478	.698
8	0	.923	.663	.430	.272	.168	.100	.058	.032	.017	.008	.004	.002	.001	.000	.000	.000	.000	.000	.000	.000
	1	.075	.279	.383	.385	.336	.267	.198	.137	.090	.055	.031	.016	.008	.003	.001	.000	.000	.000	.000	.000
	2	.003	.051	.149	.238	.294	.311	.296	.259	.209	.157	.109	.070	.041	.022	.010	.004	.001	.000	.000	.000
	3	.000	.005	.033	.084	.147	.208	.254	.279	.279	.257	.219	.172	.124	.081	.047	.023	.009	.003	.000	.000
	4	.000	.000	.005	.018	.046	.087	.136	.188	.232	.263	.273	.263	.232	.188	.136	.087	.046	.018	.005	.000
	5	.000	.000	.000	.003	.009	.023	.047	.081	.124	.172	.219	.257	.279	.279	.254	.208	.147	.084	.033	.005
	6	.000	.000	.000	.000	.001	.004	.010	.022	.041	.070	.109	.157	.209	.259	.296	.311	.294	.238	.149	.051
	7	.000	.000	.000	.000	.000	.000	.001	.003	.008	.016	.031	.055	.090	.137	.198	.267	.336	.385	.383	.279
	8	.000	.000	.000	.000	.000	.000	.000	.000	.001	.002	.004	.008	.017	.032	.058	.100	.168	.272	.430	.663
9	0	.914	.630	.387	.232	.134	.075	.040	.021	.010	.005	.002	.001	.000	.000	.000	.000	.000	.000	.000	.000
	1	.083	.299	.387	.368	.302	.225	.156	.100	.060	.034	.018	.008	.004	.001	.000	.000	.000	.000	.000	.000
	2	.003	.063	.172	.260	.302	.300	.267	.216	.161	.111	.070	.041	.021	.010	.004	.001	.000	.000	.000	.000
	3	.000	.008	.045	.107	.176	.234	.267	.272	.251	.212	.164	.116	.074	.042	.021	.009	.003	.001	.000	.000
	4	.000	.001	.007	.028	.066	.117	.172	.219	.251	.260	.246	.213	.167	.118	.074	.039	.017	.005	.001	.000
	5	.000	.000	.001	.005	.017	.039	.074	.118	.167	.213	.246	.260	.251	.219	.172	.117	.066	.028	.007	.001
	6	.000	.000	.000	.001	.003	.009	.021	.042	.074	.116	.164	.212	.251	.272	.267	.234	.176	.107	.045	.008
	7	.000	.000	.000	.000	.000	.001	.004	.010	.021	.041	.070	.111	.161	.216	.267	.300	.302	.260	.172	.063
	8	.000	.000	.000	.000	.000	.000	.000	.001	.004	.008	.018	.034	.060	.100	.156	.225	.302	.368	.387	.299
	9	.000	.000	.000	.000	.000	.000	.000	.000	.000	.001	.002	.005	.010	.021	.040	.075	.134	.232	.387	.630

Table 2—Binomial Distribution *(continued)*

| | | | | | | | | | | | *p* | | | | | | | | | | |
n	x	.01	.05	.10	.15	.20	.25	.30	.35	.40	.45	.50	.55	.60	.65	.70	.75	.80	.85	.90	.95
10	0	.904	.599	.349	.197	.107	.056	.028	.014	.006	.003	.001	.000	.000	.000	.000	.000	.000	.000	.000	.000
	1	.091	.315	.387	.347	.268	.188	.121	.072	.040	.021	.010	.004	.002	.000	.000	.000	.000	.000	.000	.000
	2	.004	.075	.194	.276	.302	.282	.233	.176	.121	.076	.044	.023	.011	.004	.001	.000	.000	.000	.000	.000
	3	.000	.010	.057	.130	.201	.250	.267	.252	.215	.166	.117	.075	.042	.021	.009	.003	.001	.000	.000	.000
	4	.000	.001	.011	.040	.088	.146	.200	.238	.251	.238	.205	.160	.111	.069	.037	.016	.006	.001	.000	.000
	5	.000	.000	.001	.008	.026	.058	.103	.154	.201	.234	.246	.234	.201	.154	.103	.058	.026	.008	.001	.000
	6	.000	.000	.000	.001	.006	.016	.037	.069	.111	.160	.205	.238	.251	.238	.200	.146	.088	.040	.011	.001
	7	.000	.000	.000	.000	.001	.003	.009	.021	.042	.075	.117	.166	.215	.252	.267	.250	.201	.130	.057	.010
	8	.000	.000	.000	.000	.000	.000	.001	.004	.011	.023	.044	.076	.121	.176	.233	.282	.302	.276	.194	.075
	9	.000	.000	.000	.000	.000	.000	.000	.000	.002	.004	.010	.021	.040	.072	.121	.188	.268	.347	.387	.315
	10	.000	.000	.000	.000	.000	.000	.000	.000	.000	.000	.001	.003	.006	.014	.028	.056	.107	.197	.349	.599
11	0	.895	.569	.314	.167	.086	.042	.020	.009	.004	.001	.000	.000	.000	.000	.000	.000	.000	.000	.000	.000
	1	.099	.329	.384	.325	.236	.155	.093	.052	.027	.013	.005	.002	.001	.000	.000	.000	.000	.000	.000	.000
	2	.005	.087	.213	.287	.295	.258	.200	.140	.089	.051	.027	.013	.005	.002	.001	.000	.000	.000	.000	.000
	3	.000	.014	.071	.152	.221	.258	.257	.225	.177	.126	.081	.046	.023	.010	.004	.001	.000	.000	.000	.000
	4	.000	.001	.016	.054	.111	.172	.220	.243	.236	.206	.161	.113	.070	.038	.017	.006	.002	.000	.000	.000
	5	.000	.000	.002	.013	.039	.080	.132	.183	.221	.236	.226	.193	.147	.099	.057	.027	.010	.002	.000	.000
	6	.000	.000	.000	.002	.010	.027	.057	.099	.147	.193	.226	.236	.221	.183	.132	.080	.039	.013	.002	.000
	7	.000	.000	.000	.000	.002	.006	.017	.038	.070	.113	.161	.206	.236	.243	.220	.172	.111	.054	.016	.001
	8	.000	.000	.000	.000	.000	.001	.004	.010	.023	.046	.081	.126	.177	.225	.257	.258	.221	.152	.071	.014
	9	.000	.000	.000	.000	.000	.000	.001	.002	.005	.013	.027	.051	.089	.140	.200	.258	.295	.287	.213	.087
	10	.000	.000	.000	.000	.000	.000	.000	.000	.001	.002	.005	.013	.027	.052	.093	.155	.236	.325	.384	.329
	11	.000	.000	.000	.000	.000	.000	.000	.000	.000	.000	.000	.001	.004	.009	.020	.042	.086	.167	.314	.569
12	0	.886	.540	.282	.142	.069	.032	.014	.006	.002	.001	.000	.000	.000	.000	.000	.000	.000	.000	.000	.000
	1	.107	.341	.377	.301	.206	.127	.071	.037	.017	.008	.003	.001	.000	.000	.000	.000	.000	.000	.000	.000
	2	.006	.099	.230	.292	.283	.232	.168	.109	.064	.034	.016	.007	.002	.001	.000	.000	.000	.000	.000	.000
	3	.000	.017	.085	.172	.236	.258	.240	.195	.142	.092	.054	.028	.012	.005	.001	.000	.000	.000	.000	.000
	4	.000	.002	.021	.068	.133	.194	.231	.237	.213	.170	.121	.076	.042	.020	.008	.002	.001	.000	.000	.000
	5	.000	.000	.004	.019	.053	.103	.158	.204	.227	.223	.193	.149	.101	.059	.029	.011	.003	.001	.000	.000
	6	.000	.000	.000	.004	.016	.040	.079	.128	.177	.212	.226	.212	.177	.128	.079	.040	.016	.004	.000	.000
	7	.000	.000	.000	.001	.003	.011	.029	.059	.101	.149	.193	.223	.227	.204	.158	.103	.053	.019	.004	.000
	8	.000	.000	.000	.000	.001	.002	.008	.020	.042	.076	.121	.170	.213	.237	.231	.194	.133	.068	.021	.002
	9	.000	.000	.000	.000	.000	.000	.001	.005	.012	.028	.054	.092	.142	.195	.240	.258	.236	.172	.085	.017
	10	.000	.000	.000	.000	.000	.000	.000	.001	.002	.007	.016	.034	.064	.109	.168	.232	.283	.292	.230	.099
	11	.000	.000	.000	.000	.000	.000	.000	.000	.000	.001	.003	.008	.017	.037	.071	.127	.206	.301	.377	.341
	12	.000	.000	.000	.000	.000	.000	.000	.000	.000	.000	.000	.001	.002	.006	.014	.032	.069	.142	.282	.540
15	0	.860	.463	.206	.087	.035	.013	.005	.002	.000	.000	.000	.000	.000	.000	.000	.000	.000	.000	.000	.000
	1	.130	.366	.343	.231	.132	.067	.031	.013	.005	.002	.000	.000	.000	.000	.000	.000	.000	.000	.000	.000
	2	.009	.135	.267	.286	.231	.156	.092	.048	.022	.009	.003	.001	.000	.000	.000	.000	.000	.000	.000	.000
	3	.000	.031	.129	.218	.250	.225	.170	.111	.063	.032	.014	.005	.002	.000	.000	.000	.000	.000	.000	.000
	4	.000	.005	.043	.116	.188	.225	.219	.179	.127	.078	.042	.019	.007	.002	.001	.000	.000	.000	.000	.000
	5	.000	.001	.010	.045	.103	.165	.206	.212	.186	.140	.092	.051	.024	.010	.003	.001	.000	.000	.000	.000
	6	.000	.000	.002	.013	.043	.092	.147	.191	.207	.191	.153	.105	.061	.030	.012	.003	.001	.000	.000	.000
	7	.000	.000	.000	.003	.014	.039	.081	.132	.177	.201	.196	.165	.118	.071	.035	.013	.003	.001	.000	.000
	8	.000	.000	.000	.001	.003	.013	.035	.071	.118	.165	.196	.201	.177	.132	.081	.039	.014	.003	.000	.000
	9	.000	.000	.000	.000	.001	.003	.012	.030	.061	.105	.153	.191	.207	.191	.147	.092	.043	.013	.002	.000
	10	.000	.000	.000	.000	.000	.001	.003	.010	.024	.051	.092	.140	.186	.212	.206	.165	.103	.045	.010	.001
	11	.000	.000	.000	.000	.000	.000	.001	.002	.007	.019	.042	.078	.127	.179	.219	.225	.188	.116	.043	.005
	12	.000	.000	.000	.000	.000	.000	.000	.000	.002	.005	.014	.032	.063	.111	.170	.225	.250	.218	.129	.031
	13	.000	.000	.000	.000	.000	.000	.000	.000	.000	.001	.003	.009	.022	.048	.092	.156	.231	.286	.267	.135
	14	.000	.000	.000	.000	.000	.000	.000	.000	.000	.000	.000	.002	.005	.013	.031	.067	.132	.231	.343	.366
	15	.000	.000	.000	.000	.000	.000	.000	.000	.000	.000	.000	.000	.000	.002	.005	.013	.035	.087	.206	.463

Table 2 — Binomial Distribution *(continued)*

												p									
n	*x*	.01	.05	.10	.15	.20	.25	.30	.35	.40	.45	.50	.55	.60	.65	.70	.75	.80	.85	.90	.95
16	0	.851	.440	.185	.074	.028	.010	.003	.001	.000	.000	.000	.000	.000	.000	.000	.000	.000	.000	.000	.000
	1	.138	.371	.329	.210	.113	.053	.023	.009	.003	.001	.000	.000	.000	.000	.000	.000	.000	.000	.000	.000
	2	.010	.146	.275	.277	.211	.134	.073	.035	.015	.006	.002	.001	.000	.000	.000	.000	.000	.000	.000	.000
	3	.000	.036	.142	.229	.246	.208	.146	.089	.047	.022	.009	.003	.001	.000	.000	.000	.000	.000	.000	.000
	4	.000	.006	.051	.131	.200	.225	.204	.155	.101	.057	.028	.011	.004	.001	.000	.000	.000	.000	.000	.000
	5	.000	.001	.014	.056	.120	.180	.210	.201	.162	.112	.067	.034	.014	.005	.001	.000	.000	.000	.000	.000
	6	.000	.000	.003	.018	.055	.110	.165	.198	.198	.168	.122	.075	.039	.017	.006	.001	.000	.000	.000	.000
	7	.000	.000	.000	.005	.020	.052	.101	.152	.189	.197	.175	.132	.084	.044	.019	.006	.001	.000	.000	.000
	8	.000	.000	.000	.001	.006	.020	.049	.092	.142	.181	.196	.181	.142	.092	.049	.020	.006	.001	.000	.000
	9	.000	.000	.000	.000	.001	.006	.019	.044	.084	.132	.175	.197	.189	.152	.101	.052	.020	.005	.000	.000
	10	.000	.000	.000	.000	.000	.001	.006	.017	.039	.075	.122	.168	.198	.198	.165	.110	.055	.018	.003	.000
	11	.000	.000	.000	.000	.000	.000	.001	.005	.014	.034	.067	.112	.162	.201	.210	.180	.120	.056	.014	.001
	12	.000	.000	.000	.000	.000	.000	.000	.001	.004	.011	.028	.057	.101	.155	.204	.225	.200	.131	.051	.006
	13	.000	.000	.000	.000	.000	.000	.000	.000	.001	.003	.009	.022	.047	.089	.146	.208	.246	.229	.142	.036
	14	.000	.000	.000	.000	.000	.000	.000	.000	.000	.001	.002	.006	.015	.035	.073	.134	.211	.277	.275	.146
	15	.000	.000	.000	.000	.000	.000	.000	.000	.000	.000	.000	.001	.003	.009	.023	.053	.113	.210	.329	.371
	16	.000	.000	.000	.000	.000	.000	.000	.000	.000	.000	.000	.000	.001	.003	.010	.028	.074	.185	.440	
20	0	.818	.358	.122	.039	.012	.003	.001	.000	.000	.000	.000	.000	.000	.000	.000	.000	.000	.000	.000	.000
	1	.165	.377	.270	.137	.058	.021	.007	.002	.000	.000	.000	.000	.000	.000	.000	.000	.000	.000	.000	.000
	2	.016	.189	.285	.229	.137	.067	.028	.010	.003	.001	.000	.000	.000	.000	.000	.000	.000	.000	.000	.000
	3	.001	.060	.190	.243	.205	.134	.072	.032	.012	.004	.001	.000	.000	.000	.000	.000	.000	.000	.000	.000
	4	.000	.013	.090	.182	.218	.190	.130	.074	.035	.014	.005	.001	.000	.000	.000	.000	.000	.000	.000	.000
	5	.000	.002	.032	.103	.175	.202	.179	.127	.075	.036	.015	.005	.001	.000	.000	.000	.000	.000	.000	.000
	6	.000	.000	.009	.045	.109	.169	.192	.171	.124	.075	.036	.015	.005	.001	.000	.000	.000	.000	.000	.000
	7	.000	.000	.002	.016	.055	.112	.164	.184	.166	.122	.074	.037	.015	.005	.001	.000	.000	.000	.000	.000
	8	.000	.000	.000	.005	.022	.061	.114	.161	.180	.162	.120	.073	.035	.014	.004	.001	.000	.000	.000	.000
	9	.000	.000	.000	.001	.007	.027	.065	.116	.160	.177	.160	.119	.071	.034	.012	.003	.000	.000	.000	.000
	10	.000	.000	.000	.000	.002	.010	.031	.069	.117	.159	.176	.159	.117	.069	.031	.010	.002	.000	.000	.000
	11	.000	.000	.000	.000	.000	.003	.012	.034	.071	.119	.160	.177	.160	.116	.065	.027	.007	.001	.000	.000
	12	.000	.000	.000	.000	.000	.001	.004	.014	.035	.073	.120	.162	.180	.161	.114	.061	.022	.005	.000	.000
	13	.000	.000	.000	.000	.000	.000	.001	.005	.015	.037	.074	.122	.166	.184	.164	.112	.055	.016	.002	.000
	14	.000	.000	.000	.000	.000	.000	.000	.001	.005	.015	.037	.075	.124	.171	.192	.169	.109	.045	.009	.000
	15	.000	.000	.000	.000	.000	.000	.000	.000	.001	.005	.015	.036	.075	.127	.179	.202	.175	.103	.032	.002
	16	.000	.000	.000	.000	.000	.000	.000	.000	.000	.001	.005	.014	.035	.074	.130	.190	.218	.182	.090	.013
	17	.000	.000	.000	.000	.000	.000	.000	.000	.000	.000	.001	.004	.012	.032	.072	.134	.205	.243	.190	.060
	18	.000	.000	.000	.000	.000	.000	.000	.000	.000	.000	.000	.001	.003	.010	.028	.067	.137	.229	.285	.189
	19	.000	.000	.000	.000	.000	.000	.000	.000	.000	.000	.000	.000	.002	.007	.021	.058	.137	.270	.377	
	20	.000	.000	.000	.000	.000	.000	.000	.000	.000	.000	.000	.000	.000	.001	.003	.012	.039	.122	.358	

Table 3 — Poisson Distribution

x	μ									
	0.1	**0.2**	**0.3**	**0.4**	**0.5**	**0.6**	**0.7**	**0.8**	**0.9**	**1.0**
0	.9048	.8187	.7408	.6703	.6065	.5488	.4966	.4493	.4066	.3679
1	.0905	.1637	.2222	.2681	.3033	.3293	.3476	.3595	.3659	.3679
2	.0045	.0164	.0333	.0536	.0758	.0988	.1217	.1438	.1647	.1839
3	.0002	.0011	.0033	.0072	.0126	.0198	.0284	.0383	.0494	.0613
4	.0000	.0001	.0003	.0007	.0016	.0030	.0050	.0077	.0111	.0153
5	.0000	.0000	.0000	.0001	.0002	.0004	.0007	.0012	.0020	.0031
6	.0000	.0000	.0000	.0000	.0000	.0000	.0001	.0002	.0003	.0005
7	.0000	.0000	.0000	.0000	.0000	.0000	.0000	.0000	.0000	.0001

x	μ									
	1.1	**1.2**	**1.3**	**1.4**	**1.5**	**1.6**	**1.7**	**1.8**	**1.9**	**2.0**
0	.3329	.3012	.2725	.2466	.2231	.2019	.1827	.1653	.1496	.1353
1	.3662	.3614	.3543	.3452	.3347	.3230	.3106	.2975	.2842	.2707
2	.2014	.2169	.2303	.2417	.2510	.2584	.2640	.2678	.2700	.2707
3	.0738	.0867	.0998	.1128	.1255	.1378	.1496	.1607	.1710	.1804
4	.0203	.0260	.0324	.0395	.0471	.0551	.0636	.0723	.0812	.0902
5	.0045	.0062	.0084	.0111	.0141	.0176	.0216	.0260	.0309	.0361
6	.0008	.0012	.0018	.0026	.0035	.0047	.0061	.0078	.0098	.0120
7	.0001	.0002	.0003	.0005	.0008	.0011	.0015	.0020	.0027	.0034
8	.0000	.0000	.0001	.0001	.0001	.0002	.0003	.0005	.0006	.0009
9	.0000	.0000	.0000	.0000	.0000	.0000	.0001	.0001	.0001	.0002

x	μ									
	2.1	**2.2**	**2.3**	**2.4**	**2.5**	**2.6**	**2.7**	**2.8**	**2.9**	**3.0**
0	.1225	.1108	.1003	.0907	.0821	.0743	.0672	.0608	.0550	.0498
1	.2572	.2438	.2306	.2177	.2052	.1931	.1815	.1703	.1596	.1494
2	.2700	.2681	.2652	.2613	.2565	.2510	.2450	.2384	.2314	.2240
3	.1890	.1966	.2033	.2090	.2138	.2176	.2205	.2225	.2237	.2240
4	.0992	.1082	.1169	.1254	.1336	.1414	.1488	.1557	.1622	.1680
5	.0417	.0476	.0538	.0602	.0668	.0735	.0804	.0872	.0940	.1008
6	.0146	.0174	.0206	.0241	.0278	.0319	.0362	.0407	.0455	.0504
7	.0044	.0055	.0068	.0083	.0099	.0118	.0139	.0163	.0188	.0216
8	.0011	.0015	.0019	.0025	.0031	.0038	.0047	.0057	.0068	.0081
9	.0003	.0004	.0005	.0007	.0009	.0011	.0014	.0018	.0022	.0027
10	.0001	.0001	.0001	.0002	.0002	.0003	.0004	.0005	.0006	.0008
11	.0000	.0000	.0000	.0000	.0000	.0001	.0001	.0001	.0002	.0002
12	.0000	.0000	.0000	.0000	.0000	.0000	.0000	.0000	.0000	.0001

x	μ									
	3.1	**3.2**	**3.3**	**3.4**	**3.5**	**3.6**	**3.7**	**3.8**	**3.9**	**4.0**
0	.0450	.0408	.0369	.0334	.0302	.0273	.0247	.0224	.0202	.0183
1	.1397	.1304	.1217	.1135	.1057	.0984	.0915	.0850	.0789	.0733
2	.2165	.2087	.2008	.1929	.1850	.1771	.1692	.1615	.1539	.1465
3	.2237	.2226	.2209	.2186	.2158	.2125	.2087	.2046	.2001	.1954
4	.1734	.1781	.1823	.1858	.1888	.1912	.1931	.1944	.1951	.1954
5	.1075	.1140	.1203	.1264	.1322	.1377	.1429	.1477	.1522	.1563
6	.0555	.0608	.0662	.0716	.0771	.0826	.0881	.0936	.0989	.1042
7	.0246	.0278	.0312	.0348	.0385	.0425	.0466	.0508	.0551	.0595
8	.0095	.0111	.0129	.0148	.0169	.0191	.0215	.0241	.0269	.0298
9	.0033	.0040	.0047	.0056	.0066	.0076	.0089	.0102	.0116	.0132
10	.0010	.0013	.0016	.0019	.0023	.0028	.0033	.0039	.0045	.0053
11	.0003	.0004	.0005	.0006	.0007	.0009	.0011	.0013	.0016	.0019
12	.0001	.0001	.0001	.0002	.0002	.0003	.0003	.0004	.0005	.0006
13	.0000	.0000	.0000	.0000	.0001	.0001	.0001	.0001	.0002	.0002
14	.0000	.0000	.0000	.0000	.0000	.0000	.0000	.0000	.0000	.0001

Reprinted with permission from W. H. Beyer, *Handbook of Tables for Probability and Statistics*, 2e, CRC Press, Boca Raton, Florida, 1986.

Table 3— Poisson Distribution *(continued)*

x	μ									
	4.1	**4.2**	**4.3**	**4.4**	**4.5**	**4.6**	**4.7**	**4.8**	**4.9**	**5.0**
0	.0166	.0150	.0136	.0123	.0111	.0101	.0091	.0082	.0074	.0067
1	.0679	.0630	.0583	.0540	.0500	.0462	.0427	.0395	.0365	.0337
2	.1393	.1323	.1254	.1188	.1125	.1063	.1005	.0948	.0894	.0842
3	.1904	.1852	.1798	.1743	.1687	.1631	.1574	.1517	.1460	.1404
4	.1951	.1944	.1933	.1917	.1898	.1875	.1849	.1820	.1789	.1755
5	.1600	.1633	.1662	.1687	.1708	.1725	.1738	.1747	.1753	.1755
6	.1093	.1143	.1191	.1237	.1281	.1323	.1362	.1398	.1432	.1462
7	.0640	.0686	.0732	.0778	.0824	.0869	.0914	.0959	.1002	.1044
8	.0328	.0360	.0393	.0428	.0463	.0500	.0537	.0575	.0614	.0653
9	.0150	.0168	.0188	.0209	.0232	.0255	.0280	.0307	.0334	.0363
10	.0061	.0071	.0081	.0092	.0104	.0118	.0132	.0147	.0164	.0181
11	.0023	.0027	.0032	.0037	.0043	.0049	.0056	.0064	.0073	.0082
12	.0008	.0009	.0011	.0014	.0016	.0019	.0022	.0026	.0030	.0034
13	.0002	.0003	.0004	.0005	.0006	.0007	.0008	.0009	.0011	.0013
14	.0001	.0001	.0001	.0001	.0002	.0002	.0003	.0003	.0004	.0005
15	.0000	.0000	.0000	.0000	.0001	.0001	.0001	.0001	.0001	.0002

x	μ									
	5.1	**5.2**	**5.3**	**5.4**	**5.5**	**5.6**	**5.7**	**5.8**	**5.9**	**6.0**
0	.0061	.0055	.0050	.0045	.0041	.0037	.0033	.0030	.0027	.0025
1	.0311	.0287	.0265	.0244	.0225	.0207	.0191	.0176	.0162	.0149
2	.0793	.0746	.0701	.0659	.0618	.0580	.0544	.0509	.0477	.0446
3	.1348	.1293	.1239	.1185	.1133	.1082	.1033	.0985	.0938	.0892
4	.1719	.1681	.1641	.1600	.1558	.1515	.1472	.1428	.1383	.1339
5	.1753	.1748	.1740	.1728	.1714	.1697	.1678	.1656	.1632	.1606
6	.1490	.1515	.1537	.1555	.1571	.1584	.1594	.1601	.1605	.1606
7	.1086	.1125	.1163	.1200	.1234	.1267	.1298	.1326	.1353	.1377
8	.0692	.0731	.0771	.0810	.0849	.0887	.0925	.0962	.0998	.1033
9	.0392	.0423	.0454	.0486	.0519	.0552	.0586	.0620	.0654	.0688
10	.0200	.0220	.0241	.0262	.0285	.0309	.0334	.0359	.0386	.0413
11	.0093	.0104	.0116	.0129	.0143	.0157	.0173	.0190	.0207	.0225
12	.0039	.0045	.0051	.0058	.0065	.0073	.0082	.0092	.0102	.0113
13	.0015	.0018	.0021	.0024	.0028	.0032	.0036	.0041	.0046	.0052
14	.0006	.0007	.0008	.0009	.0011	.0013	.0015	.0017	.0019	.0022
15	.0002	.0002	.0003	.0003	.0004	.0005	.0006	.0007	.0008	.0009
16	.0001	.0001	.0001	.0001	.0001	.0002	.0002	.0002	.0003	.0003
17	.0000	.0000	.0000	.0000	.0000	.0000	.0001	.0001	.0001	.0001

Table 3 — Poisson Distribution *(continued)*

x	6.1	6.2	6.3	6.4	6.5	6.6	6.7	6.8	6.9	7.0
0	.0022	.0020	.0018	.0017	.0015	.0014	.0012	.0011	.0010	.0009
1	.0137	.0126	.0116	.0106	.0098	.0090	.0082	.0076	.0070	.0064
2	.0417	.0390	.0364	.0340	.0318	.0296	.0276	.0258	.0240	.0223
3	.0848	.0806	.0765	.0726	.0688	.0652	.0617	.0584	.0552	.0521
4	.1294	.1249	.1205	.1162	.1118	.1076	.1034	.0992	.0952	.0912
5	.1579	.1549	.1519	.1487	.1454	.1420	.1385	.1349	.1314	.1277
6	.1605	.1601	.1595	.1586	.1575	.1562	.1546	.1529	.1511	.1490
7	.1399	.1418	.1435	.1450	.1462	.1472	.1480	.1486	.1489	.1490
8	.1066	.1099	.1130	.1160	.1188	.1215	.1240	.1263	.1284	.1304
9	.0723	.0757	.0791	.0825	.0858	.0891	.0923	.0954	.0985	.1014
10	.0441	.0469	.0498	.0528	.0558	.0588	.0618	.0649	.0679	.0710
11	.0245	.0265	.0285	.0307	.0330	.0353	.0377	.0401	.0426	.0452
12	.0124	.0137	.0150	.0164	.0179	.0194	.0210	.0227	.0245	.0264
13	.0058	.0065	.0073	.0081	.0089	.0098	.0108	.0119	.0130	.0142
14	.0025	.0029	.0033	.0037	.0041	.0046	.0052	.0058	.0064	.0071
15	.0010	.0012	.0014	.0016	.0018	.0020	.0023	.0026	.0029	.0033
16	.0004	.0005	.0005	.0006	.0007	.0008	.0010	.0011	.0013	.0014
17	.0001	.0002	.0002	.0002	.0003	.0003	.0004	.0004	.0005	.0006
18	.0000	.0001	.0001	.0001	.0001	.0001	.0001	.0002	.0002	.0002
19	.0000	.0000	.0000	.0000	.0000	.0000	.0000	.0001	.0001	.0001

x	7.1	7.2	7.3	7.4	7.5	7.6	7.7	7.8	7.9	8.0
0	.0008	.0007	.0007	.0006	.0006	.0005	.0005	.0004	.0004	.0003
1	.0059	.0054	.0049	.0045	.0041	.0038	.0035	.0032	.0029	.0027
2	.0208	.0194	.0180	.0167	.0156	.0145	.0134	.0125	.0116	.0107
3	.0492	.0464	.0438	.0413	.0389	.0366	.0345	.0324	.0305	.0286
4	.0874	.0836	.0799	.0764	.0729	.0696	.0663	.0632	.0602	.0573
5	.1241	.1204	.1167	.1130	.1094	.1057	.1021	.0986	.0951	.0916
6	.1468	.1445	.1420	.1394	.1367	.1339	.1311	.1282	.1252	.1221
7	.1489	.1486	.1481	.1474	.1465	.1454	.1442	.1428	.1413	.1396
8	.1321	.1337	.1351	.1363	.1373	.1382	.1388	.1392	.1395	.1396
9	.1042	.1070	.1096	.1121	.1144	.1167	.1187	.1207	.1224	.1241
10	.0740	.0770	.0800	.0829	.0858	.0887	.0914	.0941	.0967	.0993
11	.0478	.0504	.0531	.0558	.0585	.0613	.0640	.0667	.0695	.0722
12	.0283	.0303	.0323	.0344	.0366	.0388	.0411	.0434	.0457	.0481
13	.0154	.0168	.0181	.0196	.0211	.0227	.0243	.0260	.0278	.0296
14	.0078	.0086	.0095	.0104	.0113	.0123	.0134	.0145	.0157	.0169
15	.0037	.0041	.0046	.0051	.0057	.0062	.0069	.0075	.0083	.0090
16	.0016	.0019	.0021	.0024	.0026	.0030	.0033	.0037	.0041	.0045
17	.0007	.0008	.0009	.0010	.0012	.0013	.0015	.0017	.0019	.0021
18	.0003	.0003	.0004	.0004	.0005	.0006	.0006	.0007	.0008	.0009
19	.0001	.0001	.0001	.0002	.0002	.0002	.0003	.0003	.0003	.0004
20	.0000	.0000	.0001	.0001	.0001	.0001	.0001	.0001	.0001	.0002
21	.0000	.0000	.0000	.0000	.0000	.0000	.0000	.0000	.0001	.0001

Table 3— Poisson Distribution *(continued)*

	μ									
x	8.1	8.2	8.3	8.4	8.5	8.6	8.7	8.8	8.9	9.0
0	.0003	.0003	.0002	.0002	.0002	.0002	.0002	.0002	.0001	.0001
1	.0025	.0023	.0021	.0019	.0017	.0016	.0014	.0013	.0012	.0011
2	.0100	.0092	.0086	.0079	.0074	.0068	.0063	.0058	.0054	.0050
3	.0269	.0252	.0237	.0222	.0208	.0195	.0183	.0171	.0160	.0150
4	.0544	.0517	.0491	.0466	.0443	.0420	.0398	.0377	.0357	.0337
5	.0882	.0849	.0816	.0784	.0752	.0722	.0692	.0663	.0635	.0607
6	.1191	.1160	.1128	.1097	.1066	.1034	.1003	.0972	.0941	.0911
7	.1378	.1358	.1338	.1317	.1294	.1271	.1247	.1222	.1197	.1171
8	.1395	.1392	.1388	.1382	.1375	.1366	.1356	.1344	.1332	.1318
9	.1256	.1269	.1280	.1290	.1299	.1306	.1311	.1315	.1317	.1318
10	.1017	.1040	.1063	.1084	.1104	.1123	.1140	.1157	.1172	.1186
11	.0749	.0776	.0802	.0828	.0853	.0878	.0902	.0925	.0948	.0970
12	.0505	.0530	.0555	.0579	.0604	.0629	.0654	.0679	.0703	.0728
13	.0315	.0334	.0354	.0374	.0395	.0416	.0438	.0459	.0481	.0504
14	.0182	.0196	.0210	.0225	.0240	.0256	.0272	.0289	.0306	.0324
15	.0098	.0107	.0116	.0126	.0136	.0147	.0158	.0169	.0182	.0194
16	.0050	.0055	.0060	.0066	.0072	.0079	.0086	.0093	.0101	.0109
17	.0024	.0026	.0029	.0033	.0036	.0040	.0044	.0048	.0053	.0058
18	.0011	.0012	.0014	.0015	.0017	.0019	.0021	.0024	.0026	.0029
19	.0005	.0005	.0006	.0007	.0008	.0009	.0010	.0011	.0012	.0014
20	.0002	.0002	.0002	.0003	.0003	.0004	.0004	.0005	.0005	.0006
21	.0001	.0001	.0001	.0001	.0001	.0002	.0002	.0002	.0002	.0003
22	.0000	.0000	.0000	.0000	.0001	.0001	.0001	.0001	.0001	.0001

	μ									
x	9.1	9.2	9.3	9.4	9.5	9.6	9.7	9.8	9.9	10.0
0	.0001	.0001	.0001	.0001	.0001	.0001	.0001	.0001	.0001	.0000
1	.0010	.0009	.0009	.0008	.0007	.0007	.0006	.0005	.0005	.0005
2	.0046	.0043	.0040	.0037	.0034	.0031	.0029	.0027	.0025	.0023
3	.0140	.0131	.0123	.0115	.0107	.0100	.0093	.0087	.0081	.0076
4	.0319	.0302	.0285	.0269	.0254	.0240	.0226	.0213	.0201	.0189
5	.0581	.0555	.0530	.0506	.0483	.0460	.0439	.0418	.0398	.0378
6	.0881	.0851	.0822	.0793	.0764	.0736	.0709	.0682	.0656	.0631
7	.1145	.1118	.1091	.1064	.1037	.1010	.0982	.0955	.0928	.0901
8	.1302	.1286	.1269	.1251	.1232	.1212	.1191	.1170	.1148	.1126
9	.1317	.1315	.1311	.1306	.1300	.1293	.1284	.1274	.1263	.1251
10	.1198	.1210	.1219	.1228	.1235	.1241	.1245	.1249	.1250	.1251
11	.0991	.1012	.1031	.1049	.1067	.1083	.1098	.1112	.1125	.1137
12	.0752	.0776	.0799	.0822	.0844	.0866	.0888	.0908	.0928	.0948
13	.0526	.0549	.0572	.0594	.0617	.0640	.0662	.0685	.0707	.0729
14	.0342	.0361	.0380	.0399	.0419	.0439	.0459	.0479	.0500	.0521
15	.0208	.0221	.0235	.0250	.0265	.0281	.0297	.0313	.0330	.0347
16	.0118	.0127	.0137	.0147	.0157	.0168	.0180	.0192	.0204	.0217
17	.0063	.0069	.0075	.0081	.0088	.0095	.0103	.0111	.0119	.0128
18	.0032	.0035	.0039	.0042	.0046	.0051	.0055	.0060	.0065	.0071
19	.0015	.0017	.0019	.0021	.0023	.0026	.0028	.0031	.0034	.0037
20	.0007	.0008	.0009	.0010	.0011	.0012	.0014	.0015	.0017	.0019
21	.0003	.0003	.0004	.0004	.0005	.0006	.0006	.0007	.0008	.0009
22	.0001	.0001	.0002	.0002	.0002	.0002	.0003	.0003	.0004	.0004
23	.0000	.0001	.0001	.0001	.0001	.0001	.0001	.0001	.0002	.0002
24	.0000	.0000	.0000	.0000	.0000	.0000	.0000	.0001	.0001	.0001

Table 3— Poisson Distribution *(continued)*

	μ									
x	11	12	13	14	15	16	17	18	19	20
0	.0000	.0000	.0000	.0000	.0000	.0000	.0000	.0000	.0000	.0000
1	.0002	.0001	.0000	.0000	.0000	.0000	.0000	.0000	.0000	.0000
2	.0010	.0004	.0002	.0001	.0000	.0000	.0000	.0000	.0000	.0000
3	.0037	.0018	.0008	.0004	.0002	.0001	.0000	.0000	.0000	.0000
4	.0102	.0053	.0027	.0013	.0006	.0003	.0001	.0001	.0000	.0000
5	.0224	.0127	.0070	.0037	.0019	.0010	.0005	.0002	.0001	.0001
6	.0411	.0255	.0152	.0087	.0048	.0026	.0014	.0007	.0004	.0002
7	.0646	.0437	.0281	.0174	.0104	.0060	.0034	.0018	.0010	.0005
8	.0888	.0655	.0457	.0304	.0194	.0120	.0072	.0042	.0024	.0013
9	.1085	.0874	.0661	.0473	.0324	.0213	.0135	.0083	.0050	.0029
10	.1194	.1048	.0859	.0663	.0486	.0341	.0230	.0150	.0095	.0058
11	.1194	.1144	.1015	.0844	.0663	.0496	.0355	.0245	.0164	.0106
12	.1094	.1144	.1099	.0984	.0829	.0661	.0504	.0368	.0259	.0176
13	.0926	.1056	.1099	.1060	.0956	.0814	.0658	.0509	.0378	.0271
14	.0728	.0905	.1021	.1060	.1024	.0930	.0800	.0655	.0514	.0387
15	.0534	.0724	.0885	.0989	.1024	.0992	.0906	.0786	.0650	.0516
16	.0367	.0543	.0719	.0866	.0960	.0992	.0963	.0884	.0772	.0646
17	.0237	.0383	.0550	.0713	.0847	.0934	.0963	.0936	.0863	.0760
18	.0145	.0256	.0397	.0554	.0706	.0830	.0909	.0936	.0911	.0844
19	.0084	.0161	.0272	.0409	.0557	.0699	.0814	.0887	.0911	.0888
20	.0046	.0097	.0177	.0286	.0418	.0559	.0692	.0798	.0866	.0888

	μ									
x	11	12	13	14	15	16	17	18	19	20
21	.0024	.0055	.0109	.0191	.0299	.0426	.0560	.0684	.0783	.0846
22	.0012	.0030	.0065	.0121	.0204	.0310	.0433	.0560	.0676	.0769
23	.0006	.0016	.0037	.0074	.0133	.0216	.0320	.0438	.0559	.0669
24	.0003	.0008	.0020	.0043	.0083	.0144	.0226	.0328	.0442	.0557
25	.0001	.0004	.0010	.0024	.0050	.0092	.0154	.0237	.0336	.0446
26	.0000	.0002	.0005	.0013	.0029	.0057	.0101	.0164	.0246	.0343
27	.0000	.0001	.0002	.0007	.0016	.0034	.0063	.0109	.0173	.0254
28	.0000	.0000	.0001	.0003	.0009	.0019	.0038	.0070	.0117	.0181
29	.0000	.0000	.0001	.0002	.0004	.0011	.0023	.0044	.0077	.0125
30	.0000	.0000	.0000	.0001	.0002	.0006	.0013	.0026	.0049	.0083
31	.0000	.0000	.0000	.0000	.0001	.0003	.0007	.0015	.0030	.0054
32	.0000	.0000	.0000	.0000	.0001	.0001	.0004	.0009	.0018	.0034
33	.0000	.0000	.0000	.0000	.0000	.0001	.0002	.0005	.0010	.0020
34	.0000	.0000	.0000	.0000	.0000	.0000	.0001	.0002	.0006	.0012
35	.0000	.0000	.0000	.0000	.0000	.0000	.0000	.0001	.0003	.0007
36	.0000	.0000	.0000	.0000	.0000	.0000	.0000	.0001	.0002	.0004
37	.0000	.0000	.0000	.0000	.0000	.0000	.0000	.0000	.0001	.0002
38	.0000	.0000	.0000	.0000	.0000	.0000	.0000	.0000	.0000	.0001
39	.0000	.0000	.0000	.0000	.0000	.0000	.0000	.0000	.0000	.0001

Table 4 — Standard Normal Distribution

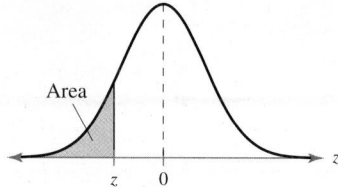

z	.09	.08	.07	.06	.05	.04	.03	.02	.01	.00
−3.4	.0002	.0003	.0003	.0003	.0003	.0003	.0003	.0003	.0003	.0003
−3.3	.0003	.0004	.0004	.0004	.0004	.0004	.0004	.0005	.0005	.0005
−3.2	.0005	.0005	.0005	.0006	.0006	.0006	.0006	.0006	.0007	.0007
−3.1	.0007	.0007	.0008	.0008	.0008	.0008	.0009	.0009	.0009	.0010
−3.0	.0010	.0010	.0011	.0011	.0011	.0012	.0012	.0013	.0013	.0013
−2.9	.0014	.0014	.0015	.0015	.0016	.0016	.0017	.0018	.0018	.0019
−2.8	.0019	.0020	.0021	.0021	.0022	.0023	.0023	.0024	.0025	.0026
−2.7	.0026	.0027	.0028	.0029	.0030	.0031	.0032	.0033	.0034	.0035
−2.6	.0036	.0037	.0038	.0039	.0040	.0041	.0043	.0044	.0045	.0047
−2.5	.0048	.0049	.0051	.0052	.0054	.0055	.0057	.0059	.0060	.0062
−2.4	.0064	.0066	.0068	.0069	.0071	.0073	.0075	.0078	.0080	.0082
−2.3	.0084	.0087	.0089	.0091	.0094	.0096	.0099	.0102	.0104	.0107
−2.2	.0110	.0113	.0116	.0119	.0122	.0125	.0129	.0132	.0136	.0139
−2.1	.0143	.0146	.0150	.0154	.0158	.0162	.0166	.0170	.0174	.0179
−2.0	.0183	.0188	.0192	.0197	.0202	.0207	.0212	.0217	.0222	.0228
−1.9	.0233	.0239	.0244	.0250	.0256	.0262	.0268	.0274	.0281	.0287
−1.8	.0294	.0301	.0307	.0314	.0322	.0329	.0336	.0344	.0351	.0359
−1.7	.0367	.0375	.0384	.0392	.0401	.0409	.0418	.0427	.0436	.0446
−1.6	.0455	.0465	.0475	.0485	.0495	.0505	.0516	.0526	.0537	.0548
−1.5	.0559	.0571	.0582	.0594	.0606	.0618	.0630	.0643	.0655	.0668
−1.4	.0681	.0694	.0708	.0721	.0735	.0749	.0764	.0778	.0793	.0808
−1.3	.0823	.0838	.0853	.0869	.0885	.0901	.0918	.0934	.0951	.0968
−1.2	.0985	.1003	.1020	.1038	.1056	.1075	.1093	.1112	.1131	.1151
−1.1	.1170	.1190	.1210	.1230	.1251	.1271	.1292	.1314	.1335	.1357
−1.0	.1379	.1401	.1423	.1446	.1469	.1492	.1515	.1539	.1562	.1587
−0.9	.1611	.1635	.1660	.1685	.1711	.1736	.1762	.1788	.1814	.1841
−0.8	.1867	.1894	.1922	.1949	.1977	.2005	.2033	.2061	.2090	.2119
−0.7	.2148	.2177	.2206	.2236	.2266	.2296	.2327	.2358	.2389	.2420
−0.6	.2451	.2483	.2514	.2546	.2578	.2611	.2643	.2676	.2709	.2743
−0.5	.2776	.2810	.2843	.2877	.2912	.2946	.2981	.3015	.3050	.3085
−0.4	.3121	.3156	.3192	.3228	.3264	.3300	.3336	.3372	.3409	.3446
−0.3	.3483	.3520	.3557	.3594	.3632	.3669	.3707	.3745	.3783	.3821
−0.2	.3859	.3897	.3936	.3974	.4013	.4052	.4090	.4129	.4168	.4207
−0.1	.4247	.4286	.4325	.4364	.4404	.4443	.4483	.4522	.4562	.4602
−0.0	.4641	.4681	.4721	.4761	.4801	.4840	.4880	.4920	.4960	.5000

Critical Values

Level of Confidence c	z_c
0.80	1.28
0.90	1.645
0.95	1.96
0.99	2.575

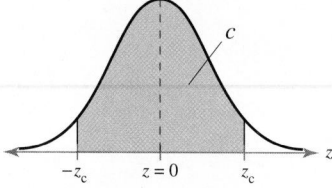

Table A-3, pp. 681–682 from *Probability and Statistics for Engineers and Scientists*, 6e by Walpole, Meyers, and Myers. Copyright 1997. Reprinted by permission of Pearson Prentice Hall, Upper Saddle River, N.J.

Table 4 — Standard Normal Distribution *(continued)*

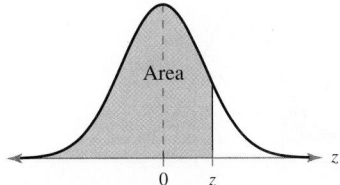

z	.00	.01	.02	.03	.04	.05	.06	.07	.08	.09
0.0	.5000	.5040	.5080	.5120	.5160	.5199	.5239	.5279	.5319	.5359
0.1	.5398	.5438	.5478	.5517	.5557	.5596	.5636	.5675	.5714	.5753
0.2	.5793	.5832	.5871	.5910	.5948	.5987	.6026	.6064	.6103	.6141
0.3	.6179	.6217	.6255	.6293	.6331	.6368	.6406	.6443	.6480	.6517
0.4	.6554	.6591	.6628	.6664	.6700	.6736	.6772	.6808	.6844	.6879
0.5	.6915	.6950	.6985	.7019	.7054	.7088	.7123	.7157	.7190	.7224
0.6	.7257	.7291	.7324	.7357	.7389	.7422	.7454	.7486	.7517	.7549
0.7	.7580	.7611	.7642	.7673	.7704	.7734	.7764	.7794	.7823	.7852
0.8	.7881	.7910	.7939	.7967	.7995	.8023	.8051	.8078	.8106	.8133
0.9	.8159	.8186	.8212	.8238	.8264	.8289	.8315	.8340	.8365	.8389
1.0	.8413	.8438	.8461	.8485	.8508	.8531	.8554	.8577	.8599	.8621
1.1	.8643	.8665	.8686	.8708	.8729	.8749	.8770	.8790	.8810	.8830
1.2	.8849	.8869	.8888	.8907	.8925	.8944	.8962	.8980	.8997	.9015
1.3	.9032	.9049	.9066	.9082	.9099	.9115	.9131	.9147	.9162	.9177
1.4	.9192	.9207	.9222	.9236	.9251	.9265	.9279	.9292	.9306	.9319
1.5	.9332	.9345	.9357	.9370	.9382	.9394	.9406	.9418	.9429	.9441
1.6	.9452	.9463	.9474	.9484	.9495	.9505	.9515	.9525	.9535	.9545
1.7	.9554	.9564	.9573	.9582	.9591	.9599	.9608	.9616	.9625	.9633
1.8	.9641	.9649	.9656	.9664	.9671	.9678	.9686	.9693	.9699	.9706
1.9	.9713	.9719	.9726	.9732	.9738	.9744	.9750	.9756	.9761	.9767
2.0	.9772	.9778	.9783	.9788	.9793	.9798	.9803	.9808	.9812	.9817
2.1	.9821	.9826	.9830	.9834	.9838	.9842	.9846	.9850	.9854	.9857
2.2	.9861	.9864	.9868	.9871	.9875	.9878	.9881	.9884	.9887	.9890
2.3	.9893	.9896	.9898	.9901	.9904	.9906	.9909	.9911	.9913	.9916
2.4	.9918	.9920	.9922	.9925	.9927	.9929	.9931	.9932	.9934	.9936
2.5	.9938	.9940	.9941	.9943	.9945	.9946	.9948	.9949	.9951	.9952
2.6	.9953	.9955	.9956	.9957	.9959	.9960	.9961	.9962	.9963	.9964
2.7	.9965	.9966	.9967	.9968	.9969	.9970	.9971	.9972	.9973	.9974
2.8	.9974	.9975	.9976	.9977	.9977	.9978	.9979	.9979	.9980	.9981
2.9	.9981	.9982	.9982	.9983	.9984	.9984	.9985	.9985	.9986	.9986
3.0	.9987	.9987	.9987	.9988	.9988	.9989	.9989	.9989	.9990	.9990
3.1	.9990	.9991	.9991	.9991	.9992	.9992	.9992	.9992	.9993	.9993
3.2	.9993	.9993	.9994	.9994	.9994	.9994	.9994	.9995	.9995	.9995
3.3	.9995	.9995	.9995	.9996	.9996	.9996	.9996	.9996	.9996	.9997
3.4	.9997	.9997	.9997	.9997	.9997	.9997	.9997	.9997	.9997	.9998

Table 5 — *t*-Distribution

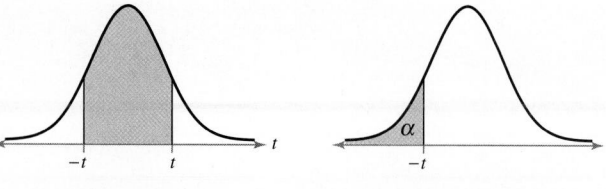

c-confidence interval   Left-tailed test

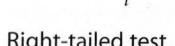

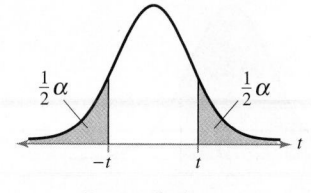

Right-tailed test   Two-tailed test

d.f.	Level of confidence, c	0.50	0.80	0.90	0.95	0.98	0.99
	One tail, α	0.25	0.10	0.05	0.025	0.01	0.005
	Two tails, α	0.50	0.20	0.10	0.05	0.02	0.01
1		1.000	3.078	6.314	12.706	31.821	63.657
2		.816	1.886	2.920	4.303	6.965	9.925
3		.765	1.638	2.353	3.182	4.541	5.841
4		.741	1.533	2.132	2.776	3.747	4.604
5		.727	1.476	2.015	2.571	3.365	4.032
6		.718	1.440	1.943	2.447	3.143	3.707
7		.711	1.415	1.895	2.365	2.998	3.499
8		.706	1.397	1.860	2.306	2.896	3.355
9		.703	1.383	1.833	2.262	2.821	3.250
10		.700	1.372	1.812	2.228	2.764	3.169
11		.697	1.363	1.796	2.201	2.718	3.106
12		.695	1.356	1.782	2.179	2.681	3.055
13		.694	1.350	1.771	2.160	2.650	3.012
14		.692	1.345	1.761	2.145	2.624	2.977
15		.691	1.341	1.753	2.131	2.602	2.947
16		.690	1.337	1.746	2.120	2.583	2.921
17		.689	1.333	1.740	2.110	2.567	2.898
18		.688	1.330	1.734	2.101	2.552	2.878
19		.688	1.328	1.729	2.093	2.539	2.861
20		.687	1.325	1.725	2.086	2.528	2.845
21		.686	1.323	1.721	2.080	2.518	2.831
22		.686	1.321	1.717	2.074	2.508	2.819
23		.685	1.319	1.714	2.069	2.500	2.807
24		.685	1.318	1.711	2.064	2.492	2.797
25		.684	1.316	1.708	2.060	2.485	2.787
26		.684	1.315	1.706	2.056	2.479	2.779
27		.684	1.314	1.703	2.052	2.473	2.771
28		.683	1.313	1.701	2.048	2.467	2.763
29		.683	1.311	1.699	2.045	2.462	2.756
∞		.674	1.282	1.645	1.960	2.326	2.576

Adapted from W. H. Beyer, *Handbook of Tables of Probability and Statistics*, 2e, CRC Press, Boca Raton, Florida, 1986. Reprinted with permission.

Table 6 — Chi-Square Distribution

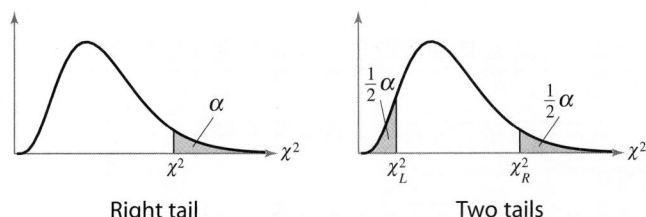

Right tail

Two tails

Degrees of freedom	α									
	0.995	**0.99**	**0.975**	**0.95**	**0.90**	**0.10**	**0.05**	**0.025**	**0.01**	**0.005**
1	—	—	0.001	0.004	0.016	2.706	3.841	5.024	6.635	7.879
2	0.010	0.020	0.051	0.103	0.211	4.605	5.991	7.378	9.210	10.597
3	0.072	0.115	0.216	0.352	0.584	6.251	7.815	9.348	11.345	12.838
4	0.207	0.297	0.484	0.711	1.064	7.779	9.488	11.143	13.277	14.860
5	0.412	0.554	0.831	1.145	1.610	9.236	11.071	12.833	15.086	16.750
6	0.676	0.872	1.237	1.635	2.204	10.645	12.592	14.449	16.812	18.548
7	0.989	1.239	1.690	2.167	2.833	12.017	14.067	16.013	18.475	20.278
8	1.344	1.646	2.180	2.733	3.490	13.362	15.507	17.535	20.090	21.955
9	1.735	2.088	2.700	3.325	4.168	14.684	16.919	19.023	21.666	23.589
10	2.156	2.558	3.247	3.940	4.865	15.987	18.307	20.483	23.209	25.188
11	2.603	3.053	3.816	4.575	5.578	17.275	19.675	21.920	24.725	26.757
12	3.074	3.571	4.404	5.226	6.304	18.549	21.026	23.337	26.217	28.299
13	3.565	4.107	5.009	5.892	7.042	19.812	22.362	24.736	27.688	29.819
14	4.075	4.660	5.629	6.571	7.790	21.064	23.685	26.119	29.141	31.319
15	4.601	5.229	6.262	7.261	8.547	22.307	24.996	27.488	30.578	32.801
16	5.142	5.812	6.908	7.962	9.312	23.542	26.296	28.845	32.000	34.267
17	5.697	6.408	7.564	8.672	10.085	24.769	27.587	30.191	33.409	35.718
18	6.265	7.015	8.231	9.390	10.865	25.989	28.869	31.526	34.805	37.156
19	6.844	7.633	8.907	10.117	11.651	27.204	30.144	32.852	36.191	38.582
20	7.434	8.260	9.591	10.851	12.443	28.412	31.410	34.170	37.566	39.997
21	8.034	8.897	10.283	11.591	13.240	29.615	32.671	35.479	38.932	41.401
22	8.643	9.542	10.982	12.338	14.042	30.813	33.924	36.781	40.289	42.796
23	9.260	10.196	11.689	13.091	14.848	32.007	35.172	38.076	41.638	44.181
24	9.886	10.856	12.401	13.848	15.659	33.196	36.415	39.364	42.980	45.559
25	10.520	11.524	13.120	14.611	16.473	34.382	37.652	40.646	44.314	46.928
26	11.160	12.198	13.844	15.379	17.292	35.563	38.885	41.923	45.642	48.290
27	11.808	12.879	14.573	16.151	18.114	36.741	40.113	43.194	46.963	49.645
28	12.461	13.565	15.308	16.928	18.939	37.916	41.337	44.461	48.278	50.993
29	13.121	14.257	16.047	17.708	19.768	39.087	42.557	45.722	49.588	52.336
30	13.787	14.954	16.791	18.493	20.599	40.256	43.773	46.979	50.892	53.672
40	20.707	22.164	24.433	26.509	29.051	51.805	55.758	59.342	63.691	66.766
50	27.991	29.707	32.357	34.764	37.689	63.167	67.505	71.420	76.154	79.490
60	35.534	37.485	40.482	43.188	46.459	74.397	79.082	83.298	88.379	91.952
70	43.275	45.442	48.758	51.739	55.329	85.527	90.531	95.023	100.425	104.215
80	51.172	53.540	57.153	60.391	64.278	96.578	101.879	106.629	112.329	116.321
90	59.196	61.754	65.647	69.126	73.291	107.565	113.145	118.136	124.116	128.299
100	67.328	70.065	74.222	77.929	82.358	118.498	124.342	129.561	135.807	140.169

Table 7— F-Distribution

α = 0.005

d.f.N: Degrees of freedom, numerator

d.f.D: Degrees of freedom, denominator	1	2	3	4	5	6	7	8	9	10	12	15	20	24	30	40	60	120	∞
1	16211	20000	21615	22500	23056	23437	23715	23925	24091	24224	24426	24630	24836	24940	25044	25148	25253	25359	25465
2	198.5	199.0	199.2	199.2	199.3	199.3	199.4	199.4	199.4	199.4	199.4	199.4	199.4	199.5	199.5	199.5	199.5	199.5	199.5
3	55.55	49.80	47.47	46.19	45.39	44.84	44.43	44.13	43.88	43.69	43.39	43.08	42.78	42.62	42.47	42.31	42.15	41.99	41.83
4	31.33	26.28	24.26	23.15	22.46	21.97	21.62	21.35	21.14	20.97	20.70	20.44	20.17	20.03	19.89	19.75	19.61	19.47	19.32
5	22.78	18.31	16.53	15.56	14.94	14.51	14.20	13.96	13.77	13.62	13.38	13.15	12.90	12.78	12.66	12.53	12.40	12.27	12.14
6	18.63	14.54	12.92	12.03	11.46	11.07	10.79	10.57	10.39	10.25	10.03	9.81	9.59	9.47	9.36	9.24	9.12	9.00	8.88
7	16.24	12.40	10.88	10.05	9.52	9.16	8.89	8.68	8.51	8.38	8.18	7.97	7.75	7.65	7.53	7.42	7.31	7.19	7.08
8	14.69	11.04	9.60	8.81	8.30	7.95	7.69	7.50	7.34	7.21	7.01	6.81	6.61	6.50	6.40	6.29	6.18	6.06	5.95
9	13.61	10.11	8.72	7.96	7.47	7.13	6.88	6.69	6.54	6.42	6.23	6.03	5.83	5.73	5.62	5.52	5.41	5.30	5.19
10	12.83	9.43	8.08	7.34	6.87	6.54	6.30	6.12	5.97	5.85	5.66	5.47	5.27	5.17	5.07	4.97	4.86	4.75	4.64
11	12.23	8.91	7.60	6.88	6.42	6.10	5.86	5.68	5.54	5.42	5.24	5.05	4.86	4.76	4.65	4.55	4.44	4.34	4.23
12	11.75	8.51	7.23	6.52	6.07	5.76	5.52	5.35	5.20	5.09	4.91	4.72	4.53	4.43	4.33	4.23	4.12	4.01	3.90
13	11.37	8.19	6.93	6.23	5.79	5.48	5.25	5.08	4.94	4.82	4.64	4.46	4.27	4.17	4.07	3.97	3.87	3.76	3.65
14	11.06	7.92	6.68	6.00	5.56	5.26	5.03	4.86	4.72	4.60	4.43	4.25	4.06	3.96	3.86	3.76	3.66	3.55	3.44
15	10.80	7.70	6.48	5.80	5.37	5.07	4.85	4.67	4.54	4.42	4.25	4.07	3.88	3.79	3.69	3.58	3.48	3.37	3.26
16	10.58	7.51	6.30	5.64	5.21	4.91	4.69	4.52	4.38	4.27	4.10	3.92	3.73	3.64	3.54	3.44	3.33	3.22	3.11
17	10.38	7.35	6.16	5.50	5.07	4.78	4.56	4.39	4.25	4.14	3.97	3.79	3.61	3.51	3.41	3.31	3.21	3.10	2.98
18	10.22	7.21	6.03	5.37	4.96	4.66	4.44	4.28	4.14	4.03	3.86	3.68	3.50	3.40	3.30	3.20	3.10	2.99	2.87
19	10.07	7.09	5.92	5.27	4.85	4.56	4.34	4.18	4.04	3.93	3.76	3.59	3.40	3.31	3.21	3.11	3.00	2.89	2.78
20	9.94	6.99	5.82	5.17	4.76	4.47	4.26	4.09	3.96	3.85	3.68	3.50	3.32	3.22	3.12	3.02	2.92	2.81	2.69
21	9.83	6.89	5.73	5.09	4.68	4.39	4.18	4.01	3.88	3.77	3.60	3.43	3.24	3.15	3.05	2.95	2.84	2.73	2.61
22	9.73	6.81	5.65	5.02	4.61	4.32	4.11	3.94	3.81	3.70	3.54	3.36	3.18	3.08	2.98	2.88	2.77	2.66	2.55
23	9.63	6.73	5.58	4.95	4.54	4.26	4.05	3.88	3.75	3.64	3.47	3.30	3.12	3.02	2.92	2.82	2.71	2.60	2.48
24	9.55	6.66	5.52	4.89	4.49	4.20	3.99	3.83	3.69	3.59	3.42	3.25	3.06	2.97	2.87	2.77	2.66	2.55	2.43
25	9.48	6.60	5.46	4.84	4.43	4.15	3.94	3.78	3.64	3.54	3.37	3.20	3.01	2.92	2.82	2.72	2.61	2.50	2.38
26	9.41	6.54	5.41	4.79	4.38	4.10	3.89	3.73	3.60	3.49	3.33	3.15	2.97	2.87	2.77	2.67	2.56	2.45	2.33
27	9.34	6.49	5.36	4.74	4.34	4.06	3.85	3.69	3.56	3.45	3.28	3.11	2.93	2.83	2.73	2.63	2.52	2.41	2.29
28	9.28	6.44	5.32	4.70	4.30	4.02	3.81	3.65	3.52	3.41	3.25	3.07	2.89	2.79	2.69	2.59	2.48	2.37	2.25
29	9.23	6.40	5.28	4.66	4.26	3.98	3.77	3.61	3.48	3.38	3.21	3.04	2.86	2.76	2.66	2.56	2.45	2.33	2.24
30	9.18	6.35	5.24	4.62	4.23	3.95	3.74	3.58	3.45	3.34	3.18	3.01	2.82	2.73	2.63	2.52	2.42	2.30	2.18
40	8.83	6.07	4.98	4.37	3.99	3.71	3.51	3.35	3.22	3.12	2.95	2.78	2.60	2.50	2.40	2.30	2.18	2.06	1.93
60	8.49	5.79	4.73	4.14	3.76	3.49	3.29	3.13	3.01	2.90	2.74	2.57	2.39	2.29	2.19	2.08	1.96	1.83	1.69
120	8.18	5.54	4.50	3.92	3.55	3.28	3.09	2.93	2.81	2.71	2.54	2.37	2.19	2.09	1.98	1.87	1.75	1.61	1.43
∞	7.88	5.30	4.28	3.72	3.35	3.09	2.90	2.74	2.62	2.52	2.36	2.19	2.00	1.90	1.79	1.67	1.53	1.36	1.00

Table 7— *F*-Distribution (continued)

$\alpha = 0.01$

d.f._D: Degrees of freedom, denominator	d.f._N: Degrees of freedom, numerator																		
	1	2	3	4	5	6	7	8	9	10	12	15	20	24	30	40	60	120	∞
1	4052	4999.5	5403	5625	5764	5859	5928	5982	6022	6056	6106	6157	6209	6235	6261	6287	6313	6339	6366
2	98.50	99.00	99.17	99.25	99.30	99.33	99.36	99.37	99.39	99.40	99.42	99.43	99.45	99.46	99.47	99.47	99.48	99.49	99.50
3	34.12	30.82	29.46	28.71	28.24	27.91	27.67	27.49	27.35	27.23	27.05	26.87	26.69	26.60	26.50	26.41	26.32	26.22	26.13
4	21.20	18.00	16.69	15.98	15.52	15.21	14.98	14.80	14.66	14.55	14.37	14.20	14.02	13.93	13.84	13.75	13.65	13.56	13.46
5	16.26	13.27	12.06	11.39	10.97	10.67	10.46	10.29	10.16	10.05	9.89	9.72	9.55	9.47	9.38	9.29	9.20	9.11	9.02
6	13.75	10.92	9.78	9.15	8.75	8.47	8.26	8.10	7.98	7.87	7.72	7.56	7.40	7.31	7.23	7.14	7.06	6.97	6.88
7	12.25	9.55	8.45	7.85	7.46	7.19	6.99	6.84	6.72	6.62	6.47	6.31	6.16	6.07	5.99	5.91	5.82	5.74	5.65
8	11.26	8.65	7.59	7.01	6.63	6.37	6.18	6.03	5.91	5.81	5.67	5.52	5.36	5.28	5.20	5.12	5.03	4.95	4.86
9	10.56	8.02	6.99	6.42	6.06	5.80	5.61	5.47	5.35	5.26	5.11	4.96	4.81	4.73	4.65	4.57	4.48	4.40	4.31
10	10.04	7.56	6.55	5.99	5.64	5.39	5.20	5.06	4.94	4.85	4.71	4.56	4.41	4.33	4.25	4.17	4.08	4.00	3.91
11	9.65	7.21	6.22	5.67	5.32	5.07	4.89	4.74	4.63	4.54	4.40	4.25	4.10	4.02	3.94	3.86	3.78	3.69	3.60
12	9.33	6.93	5.95	5.41	5.06	4.82	4.64	4.50	4.39	4.30	4.16	4.01	3.86	3.78	3.70	3.62	3.54	3.45	3.36
13	9.07	6.70	5.74	5.21	4.86	4.62	4.44	4.30	4.19	4.10	3.96	3.82	3.66	3.59	3.51	3.43	3.34	3.25	3.17
14	8.86	6.51	5.56	5.04	4.69	4.46	4.28	4.14	4.03	3.94	3.80	3.66	3.51	3.43	3.35	3.27	3.18	3.09	3.00
15	8.68	6.36	5.42	4.89	4.56	4.32	4.14	4.00	3.89	3.80	3.67	3.52	3.37	3.29	3.21	3.13	3.05	2.96	2.87
16	8.53	6.23	5.29	4.77	4.44	4.20	4.03	3.89	3.78	3.69	3.55	3.41	3.26	3.18	3.10	3.02	2.93	2.84	2.75
17	8.40	6.11	5.18	4.67	4.34	4.10	3.93	3.79	3.68	3.59	3.46	3.31	3.16	3.08	3.00	2.92	2.83	2.75	2.65
18	8.29	6.01	5.09	4.58	4.25	4.01	3.84	3.71	3.60	3.51	3.37	3.23	3.08	3.00	2.92	2.84	2.75	2.66	2.57
19	8.18	5.93	5.01	4.50	4.17	3.94	3.77	3.63	3.52	3.43	3.30	3.15	3.00	2.92	2.84	2.76	2.67	2.58	2.49
20	8.10	5.85	4.94	4.43	4.10	3.87	3.70	3.56	3.46	3.37	3.23	3.09	2.94	2.86	2.78	2.69	2.61	2.52	2.42
21	8.02	5.78	4.87	4.37	4.04	3.81	3.64	3.51	3.40	3.31	3.17	3.03	2.88	2.80	2.72	2.64	2.55	2.46	2.36
22	7.95	5.72	4.82	4.31	3.99	3.76	3.59	3.45	3.35	3.26	3.12	2.98	2.83	2.75	2.67	2.58	2.50	2.40	2.31
23	7.88	5.66	4.76	4.26	3.94	3.71	3.54	3.41	3.30	3.21	3.07	2.93	2.78	2.70	2.62	2.54	2.45	2.35	2.26
24	7.82	5.61	4.72	4.22	3.90	3.67	3.50	3.36	3.26	3.17	3.03	2.89	2.74	2.66	2.58	2.49	2.40	2.31	2.21
25	7.77	5.57	4.68	4.18	3.85	3.63	3.46	3.32	3.22	3.13	2.99	2.85	2.70	2.62	2.54	2.45	2.36	2.27	2.17
26	7.72	5.53	4.64	4.14	3.82	3.59	3.42	3.29	3.18	3.09	2.96	2.81	2.66	2.58	2.50	2.42	2.33	2.23	2.13
27	7.68	5.49	4.60	4.11	3.78	3.56	3.39	3.26	3.15	3.06	2.93	2.78	2.63	2.55	2.47	2.38	2.29	2.20	2.10
28	7.64	5.45	4.57	4.07	3.75	3.53	3.36	3.23	3.12	3.03	2.90	2.75	2.60	2.52	2.44	2.35	2.26	2.17	2.06
29	7.60	5.42	4.54	4.04	3.73	3.50	3.33	3.20	3.09	3.00	2.87	2.73	2.57	2.49	2.41	2.33	2.23	2.14	2.03
30	7.56	5.39	4.51	4.02	3.70	3.47	3.30	3.17	3.07	2.98	2.84	2.70	2.55	2.47	2.39	2.30	2.21	2.11	2.01
40	7.31	5.18	4.31	3.83	3.51	3.29	3.12	2.99	2.89	2.80	2.66	2.52	2.37	2.29	2.20	2.11	2.02	1.92	1.80
60	7.08	4.98	4.13	3.65	3.34	3.12	2.95	2.82	2.72	2.63	2.50	2.35	2.20	2.12	2.03	1.94	1.84	1.73	1.60
120	6.85	4.79	3.95	3.48	3.17	2.96	2.79	2.66	2.56	2.47	2.34	2.19	2.03	1.95	1.86	1.76	1.66	1.53	1.38
∞	6.63	4.61	3.78	3.32	3.02	2.80	2.64	2.51	2.41	2.32	2.18	2.04	1.88	1.79	1.70	1.59	1.47	1.32	1.00

Table 7— F-Distribution (continued)

$\alpha = 0.025$

d.f._D: Degrees of freedom, denominator	d.f._N: Degrees of freedom, numerator																		
	1	2	3	4	5	6	7	8	9	10	12	15	20	24	30	40	60	120	∞
1	647.8	799.5	864.2	899.6	921.8	937.1	948.2	956.7	963.3	968.6	976.7	984.9	993.1	997.2	1001	1006	1010	1014	1018
2	38.51	39.00	39.17	39.25	39.30	39.33	39.36	39.37	39.39	39.40	39.41	39.43	39.45	39.46	39.46	39.47	39.48	39.49	39.50
3	17.44	16.04	15.44	15.10	14.88	14.73	14.62	14.54	14.47	14.42	14.34	14.25	14.17	14.12	14.08	14.04	13.99	13.95	13.90
4	12.22	10.65	9.98	9.60	9.36	9.20	9.07	8.98	8.90	8.84	8.75	8.66	8.56	8.51	8.46	8.41	8.36	8.31	8.26
5	10.01	8.43	7.76	7.39	7.15	6.98	6.85	6.76	6.68	6.62	6.52	6.43	6.33	6.28	6.23	6.18	6.12	6.07	6.02
6	8.81	7.26	6.60	6.23	5.99	5.82	5.70	5.60	5.52	5.46	5.37	5.27	5.17	5.12	5.07	5.01	4.96	4.90	4.85
7	8.07	6.54	5.89	5.52	5.29	5.12	4.99	4.90	4.82	4.76	4.67	4.57	4.47	4.42	4.36	4.31	4.25	4.20	4.14
8	7.57	6.06	5.42	5.05	4.82	4.65	4.53	4.43	4.36	4.30	4.20	4.10	4.00	3.95	3.89	3.84	3.78	3.73	3.67
9	7.21	5.71	5.08	4.72	4.48	4.32	4.20	4.10	4.03	3.96	3.87	3.77	3.67	3.61	3.56	3.51	3.45	3.39	3.33
10	6.94	5.46	4.83	4.47	4.24	4.07	3.95	3.85	3.78	3.72	3.62	3.52	3.42	3.37	3.31	3.26	3.20	3.14	3.08
11	6.72	5.26	4.63	4.28	4.04	3.88	3.76	3.66	3.59	3.53	3.43	3.33	3.23	3.17	3.12	3.06	3.00	2.94	2.88
12	6.55	5.10	4.47	4.12	3.89	3.73	3.61	3.51	3.44	3.37	3.28	3.18	3.07	3.02	2.96	2.91	2.85	2.79	2.72
13	6.41	4.97	4.35	4.00	3.77	3.60	3.48	3.39	3.31	3.25	3.15	3.05	2.95	2.89	2.84	2.78	2.72	2.66	2.60
14	6.30	4.86	4.24	3.89	3.66	3.50	3.38	3.29	3.21	3.15	3.05	2.95	2.84	2.79	2.73	2.67	2.61	2.55	2.49
15	6.20	4.77	4.15	3.80	3.58	3.41	3.29	3.20	3.12	3.06	2.96	2.86	2.76	2.70	2.64	2.59	2.52	2.46	2.40
16	6.12	4.69	4.08	3.73	3.50	3.34	3.22	3.12	3.05	2.99	2.89	2.79	2.68	2.63	2.57	2.51	2.45	2.38	2.32
17	6.04	4.62	4.01	3.66	3.44	3.28	3.16	3.06	2.98	2.92	2.82	2.72	2.62	2.56	2.50	2.44	2.38	2.32	2.25
18	5.98	4.56	3.95	3.61	3.38	3.22	3.10	3.01	2.93	2.87	2.77	2.67	2.56	2.50	2.44	2.38	2.32	2.26	2.19
19	5.92	4.51	3.90	3.56	3.33	3.17	3.05	2.96	2.88	2.82	2.72	2.62	2.51	2.45	2.39	2.33	2.27	2.20	2.13
20	5.87	4.46	3.86	3.51	3.29	3.13	3.01	2.91	2.84	2.77	2.68	2.57	2.46	2.41	2.35	2.29	2.22	2.16	2.09
21	5.83	4.42	3.82	3.48	3.25	3.09	2.97	2.87	2.80	2.73	2.64	2.53	2.42	2.37	2.31	2.25	2.18	2.11	2.04
22	5.79	4.38	3.78	3.44	3.22	3.05	2.93	2.84	2.76	2.70	2.60	2.50	2.39	2.33	2.27	2.21	2.14	2.08	2.00
23	5.75	4.35	3.75	3.41	3.18	3.02	2.90	2.81	2.73	2.67	2.57	2.47	2.36	2.30	2.24	2.18	2.11	2.04	1.97
24	5.72	4.32	3.72	3.38	3.15	2.99	2.87	2.78	2.70	2.64	2.54	2.44	2.33	2.27	2.21	2.15	2.08	2.01	1.94
25	5.69	4.29	3.69	3.35	3.13	2.97	2.85	2.75	2.68	2.61	2.51	2.41	2.30	2.24	2.18	2.12	2.05	1.98	1.91
26	5.66	4.27	3.67	3.33	3.10	2.94	2.82	2.73	2.65	2.59	2.49	2.39	2.28	2.22	2.16	2.09	2.03	1.95	1.88
27	5.63	4.24	3.65	3.31	3.08	2.92	2.80	2.71	2.63	2.57	2.47	2.36	2.25	2.19	2.13	2.07	2.00	1.93	1.85
28	5.61	4.22	3.63	3.29	3.06	2.90	2.78	2.69	2.61	2.55	2.45	2.34	2.23	2.17	2.11	2.05	1.98	1.91	1.83
29	5.59	4.20	3.61	3.27	3.04	2.88	2.76	2.67	2.59	2.53	2.43	2.32	2.21	2.15	2.09	2.03	1.96	1.89	1.81
30	5.57	4.18	3.59	3.25	3.03	2.87	2.75	2.65	2.57	2.51	2.41	2.31	2.20	2.14	2.07	2.01	1.94	1.87	1.79
40	5.42	4.05	3.46	3.13	2.90	2.74	2.62	2.53	2.45	2.39	2.29	2.18	2.07	2.01	1.94	1.88	1.80	1.72	1.64
60	5.29	3.93	3.34	3.01	2.79	2.63	2.51	2.41	2.33	2.27	2.17	2.06	1.94	1.88	1.82	1.74	1.67	1.58	1.48
120	5.15	3.80	3.23	2.89	2.67	2.52	2.39	2.30	2.22	2.16	2.05	1.94	1.82	1.76	1.69	1.61	1.53	1.43	1.31
∞	5.02	3.69	3.12	2.79	2.57	2.41	2.29	2.19	2.11	2.05	1.94	1.83	1.71	1.64	1.57	1.48	1.39	1.27	1.00

Table 7— F-Distribution (continued)

$\alpha = 0.05$

d.f.N: Degrees of freedom, numerator

d.f.D: Degrees of freedom, denominator	1	2	3	4	5	6	7	8	9	10	12	15	20	24	30	40	60	120	∞
1	161.4	199.5	215.7	224.6	230.2	234.0	236.8	238.9	240.5	241.9	243.9	245.9	248.0	249.1	250.1	251.1	252.2	253.3	254.3
2	18.51	19.00	19.16	19.25	19.30	19.33	19.35	19.37	19.38	19.40	19.41	19.43	19.45	19.45	19.46	19.47	19.48	19.49	19.50
3	10.13	9.55	9.28	9.12	9.01	8.94	8.89	8.85	8.81	8.79	8.74	8.70	8.66	8.64	8.62	8.59	8.57	8.55	8.53
4	7.71	6.94	6.59	6.39	6.26	6.16	6.09	6.04	6.00	5.96	5.91	5.86	5.80	5.77	5.75	5.72	5.69	5.66	5.63
5	6.61	5.79	5.41	5.19	5.05	4.95	4.88	4.82	4.77	4.74	4.68	4.62	4.56	4.53	4.50	4.46	4.43	4.40	4.36
6	5.99	5.14	4.76	4.53	4.39	4.28	4.21	4.15	4.10	4.06	4.00	3.94	3.87	3.84	3.81	3.77	3.74	3.70	3.67
7	5.59	4.74	4.35	4.12	3.97	3.87	3.79	3.73	3.68	3.64	3.57	3.51	3.44	3.41	3.38	3.34	3.30	3.27	3.23
8	5.32	4.46	4.07	3.84	3.69	3.58	3.50	3.44	3.39	3.35	3.28	3.22	3.15	3.12	3.08	3.04	3.01	2.97	2.93
9	5.12	4.26	3.86	3.63	3.48	3.37	3.29	3.23	3.18	3.14	3.07	3.01	2.94	2.90	2.86	2.83	2.79	2.75	2.71
10	4.96	4.10	3.71	3.48	3.33	3.22	3.14	3.07	3.02	2.98	2.91	2.85	2.77	2.74	2.70	2.66	2.62	2.58	2.54
11	4.84	3.98	3.59	3.36	3.20	3.09	3.01	2.95	2.90	2.85	2.79	2.72	2.65	2.61	2.57	2.53	2.49	2.45	2.40
12	4.75	3.89	3.49	3.26	3.11	3.00	2.91	2.85	2.80	2.75	2.69	2.62	2.54	2.51	2.47	2.43	2.38	2.34	2.30
13	4.67	3.81	3.41	3.18	3.03	2.92	2.83	2.77	2.71	2.67	2.60	2.53	2.46	2.42	2.38	2.34	2.30	2.25	2.21
14	4.60	3.74	3.34	3.11	2.96	2.85	2.76	2.70	2.65	2.60	2.53	2.46	2.39	2.35	2.31	2.27	2.22	2.18	2.13
15	4.54	3.68	3.29	3.06	2.90	2.79	2.71	2.64	2.59	2.54	2.48	2.40	2.33	2.29	2.25	2.20	2.16	2.11	2.07
16	4.49	3.63	3.24	3.01	2.85	2.74	2.66	2.59	2.54	2.49	2.42	2.35	2.28	2.24	2.19	2.15	2.11	2.06	2.01
17	4.45	3.59	3.20	2.96	2.81	2.70	2.61	2.55	2.49	2.45	2.38	2.31	2.23	2.19	2.15	2.10	2.06	2.01	1.96
18	4.41	3.55	3.16	2.93	2.77	2.66	2.58	2.51	2.46	2.41	2.34	2.27	2.19	2.15	2.11	2.06	2.02	1.97	1.92
19	4.38	3.52	3.13	2.90	2.74	2.63	2.54	2.48	2.42	2.38	2.31	2.23	2.16	2.11	2.07	2.03	1.98	1.93	1.88
20	4.35	3.49	3.10	2.87	2.71	2.60	2.51	2.45	2.39	2.35	2.28	2.20	2.12	2.08	2.04	1.99	1.95	1.90	1.84
21	4.32	3.47	3.07	2.84	2.68	2.57	2.49	2.42	2.37	2.32	2.25	2.18	2.10	2.05	2.01	1.96	1.92	1.87	1.81
22	4.30	3.44	3.05	2.82	2.66	2.55	2.46	2.40	2.34	2.30	2.23	2.15	2.07	2.03	1.98	1.94	1.89	1.84	1.78
23	4.28	3.42	3.03	2.80	2.64	2.53	2.44	2.37	2.32	2.27	2.20	2.13	2.05	2.01	1.96	1.91	1.86	1.81	1.76
24	4.26	3.40	3.01	2.78	2.62	2.51	2.42	2.36	2.30	2.25	2.18	2.11	2.03	1.98	1.94	1.89	1.84	1.79	1.73
25	4.24	3.39	2.99	2.76	2.60	2.49	2.40	2.34	2.28	2.24	2.16	2.09	2.01	1.96	1.92	1.87	1.82	1.77	1.71
26	4.23	3.37	2.98	2.74	2.59	2.47	2.39	2.32	2.27	2.22	2.15	2.07	1.99	1.95	1.90	1.85	1.80	1.75	1.69
27	4.21	3.35	2.96	2.73	2.57	2.46	2.37	2.31	2.25	2.20	2.13	2.06	1.97	1.93	1.88	1.84	1.79	1.73	1.67
28	4.20	3.34	2.95	2.71	2.56	2.45	2.36	2.29	2.24	2.19	2.12	2.04	1.96	1.91	1.87	1.82	1.77	1.71	1.65
29	4.18	3.33	2.93	2.70	2.55	2.43	2.35	2.28	2.22	2.18	2.10	2.03	1.94	1.90	1.85	1.81	1.75	1.70	1.64
30	4.17	3.32	2.92	2.69	2.53	2.42	2.33	2.27	2.21	2.16	2.09	2.01	1.93	1.89	1.84	1.79	1.74	1.68	1.62
40	4.08	3.23	2.84	2.61	2.45	2.34	2.25	2.18	2.12	2.08	2.00	1.92	1.84	1.79	1.74	1.69	1.64	1.58	1.51
60	4.00	3.15	2.76	2.53	2.37	2.25	2.17	2.10	2.04	1.99	1.92	1.84	1.75	1.70	1.65	1.59	1.53	1.47	1.39
120	3.92	3.07	2.68	2.45	2.29	2.17	2.09	2.02	1.96	1.91	1.83	1.75	1.66	1.61	1.55	1.50	1.43	1.35	1.25
∞	3.84	3.00	2.60	2.37	2.21	2.10	2.01	1.94	1.88	1.83	1.75	1.67	1.57	1.52	1.46	1.39	1.32	1.22	1.00

Table 7—F-Distribution (continued)

$\alpha = 0.10$

d.f.$_D$: Degrees of freedom, denominator	\multicolumn{19}{c}{d.f.$_N$: Degrees of freedom, numerator}

d.f.$_D$	1	2	3	4	5	6	7	8	9	10	12	15	20	24	30	40	60	120	∞
1	39.86	49.50	53.59	55.83	57.24	58.20	58.91	59.44	59.86	60.19	60.71	61.22	61.74	62.00	62.26	62.53	62.79	63.06	63.33
2	8.53	9.00	9.16	9.24	9.29	9.33	9.35	9.37	9.38	9.39	9.41	9.42	9.44	9.45	9.46	9.47	9.47	9.48	9.49
3	5.54	5.46	5.39	5.34	5.31	5.28	5.27	5.25	5.24	5.23	5.22	5.20	5.18	5.18	5.17	5.16	5.15	5.14	5.13
4	4.54	4.32	4.19	4.11	4.05	4.01	3.98	3.95	3.94	3.92	3.90	3.87	3.84	3.83	3.82	3.80	3.79	3.78	3.76
5	4.06	3.78	3.62	3.52	3.45	3.40	3.37	3.34	3.32	3.30	3.27	3.24	3.21	3.19	3.17	3.16	3.14	3.12	3.10
6	3.78	3.46	3.29	3.18	3.11	3.05	3.01	2.98	2.96	2.94	2.90	2.87	2.84	2.82	2.80	2.78	2.76	2.74	2.72
7	3.59	3.26	3.07	2.96	2.88	2.83	2.78	2.75	2.72	2.70	2.67	2.63	2.59	2.58	2.56	2.54	2.51	2.49	2.47
8	3.46	3.11	2.92	2.81	2.73	2.67	2.62	2.59	2.56	2.54	2.50	2.46	2.42	2.40	2.38	2.36	2.34	2.32	2.29
9	3.36	3.01	2.81	2.69	2.61	2.55	2.51	2.47	2.44	2.42	2.38	2.34	2.30	2.28	2.25	2.23	2.21	2.18	2.16
10	3.29	2.92	2.73	2.61	2.52	2.46	2.41	2.38	2.35	2.32	2.28	2.24	2.20	2.18	2.16	2.13	2.11	2.08	2.06
11	3.23	2.86	2.66	2.54	2.45	2.39	2.34	2.30	2.27	2.25	2.21	2.17	2.12	2.10	2.08	2.05	2.03	2.00	1.97
12	3.18	2.81	2.61	2.48	2.39	2.33	2.28	2.24	2.21	2.19	2.15	2.10	2.06	2.04	2.01	1.99	1.96	1.93	1.90
13	3.14	2.76	2.56	2.43	2.35	2.28	2.23	2.20	2.16	2.14	2.10	2.05	2.01	1.98	1.96	1.93	1.90	1.88	1.85
14	3.10	2.73	2.52	2.39	2.31	2.24	2.19	2.15	2.12	2.10	2.05	2.01	1.96	1.94	1.91	1.89	1.86	1.83	1.80
15	3.07	2.70	2.49	2.36	2.27	2.21	2.16	2.12	2.09	2.06	2.02	1.97	1.92	1.90	1.87	1.85	1.82	1.79	1.76
16	3.05	2.67	2.46	2.33	2.24	2.18	2.13	2.09	2.06	2.03	1.99	1.94	1.89	1.87	1.84	1.81	1.78	1.75	1.72
17	3.03	2.64	2.44	2.31	2.22	2.15	2.10	2.06	2.03	2.00	1.96	1.91	1.86	1.84	1.81	1.78	1.75	1.72	1.69
18	3.01	2.62	2.42	2.29	2.20	2.13	2.08	2.04	2.00	1.98	1.93	1.89	1.84	1.81	1.78	1.75	1.72	1.69	1.66
19	2.99	2.61	2.40	2.27	2.18	2.11	2.06	2.02	1.98	1.96	1.91	1.86	1.81	1.79	1.76	1.73	1.70	1.67	1.63
20	2.97	2.59	2.38	2.25	2.16	2.09	2.04	2.00	1.96	1.94	1.89	1.84	1.79	1.77	1.74	1.71	1.68	1.64	1.61
21	2.96	2.57	2.36	2.23	2.14	2.08	2.02	1.98	1.95	1.92	1.87	1.83	1.78	1.75	1.72	1.69	1.66	1.62	1.59
22	2.95	2.56	2.35	2.22	2.13	2.06	2.01	1.97	1.93	1.90	1.86	1.81	1.76	1.73	1.70	1.67	1.64	1.60	1.57
23	2.94	2.55	2.34	2.21	2.11	2.05	1.99	1.95	1.92	1.89	1.84	1.80	1.74	1.72	1.69	1.66	1.62	1.59	1.55
24	2.93	2.54	2.33	2.19	2.10	2.04	1.98	1.94	1.91	1.88	1.83	1.78	1.73	1.70	1.67	1.64	1.61	1.57	1.53
25	2.92	2.53	2.32	2.18	2.09	2.02	1.97	1.93	1.89	1.87	1.82	1.77	1.72	1.69	1.66	1.63	1.59	1.56	1.52
26	2.91	2.52	2.31	2.17	2.08	2.01	1.96	1.92	1.88	1.86	1.81	1.76	1.71	1.68	1.65	1.61	1.58	1.54	1.50
27	2.90	2.51	2.30	2.17	2.07	2.00	1.95	1.91	1.87	1.85	1.80	1.75	1.70	1.67	1.64	1.60	1.57	1.53	1.49
28	2.89	2.50	2.29	2.16	2.06	2.00	1.94	1.90	1.87	1.84	1.79	1.74	1.69	1.66	1.63	1.59	1.56	1.52	1.48
29	2.89	2.50	2.28	2.15	2.06	1.99	1.93	1.89	1.86	1.83	1.78	1.73	1.68	1.65	1.62	1.58	1.55	1.51	1.47
30	2.88	2.49	2.28	2.14	2.05	1.98	1.93	1.88	1.85	1.82	1.77	1.72	1.67	1.64	1.61	1.57	1.54	1.50	1.46
40	2.84	2.44	2.23	2.09	2.00	1.93	1.87	1.83	1.79	1.76	1.71	1.66	1.61	1.57	1.54	1.51	1.47	1.42	1.38
60	2.79	2.39	2.18	2.04	1.95	1.87	1.82	1.77	1.74	1.71	1.66	1.60	1.54	1.51	1.48	1.44	1.40	1.35	1.29
120	2.75	2.35	2.13	1.99	1.90	1.82	1.77	1.72	1.68	1.65	1.60	1.55	1.48	1.45	1.41	1.37	1.32	1.26	1.19
∞	2.71	2.30	2.08	1.94	1.85	1.77	1.72	1.67	1.63	1.60	1.55	1.49	1.42	1.38	1.34	1.30	1.24	1.19	1.00

From M. Merrington and C.M. Thompson, "Table of Percentage Points of the Inverted Beta (F) Distribution", *Biometrika* 33 (1943), pp. 74-87, by permission of Oxford University Press.

Table 8 — Critical Values for the Sign Test

Reject the null hypothesis if the test statistic is less than or equal to the value in the table.

	One-tailed, $\alpha = 0.005$	$\alpha = 0.01$	$\alpha = 0.025$	$\alpha = 0.05$
n	Two-tailed, $\alpha = 0.01$	$\alpha = 0.02$	$\alpha = 0.05$	$\alpha = 0.10$
8	0	0	0	1
9	0	0	1	1
10	0	0	1	1
11	0	1	1	2
12	1	1	2	2
13	1	1	2	3
14	1	2	3	3
15	2	2	3	3
16	2	2	3	4
17	2	3	4	4
18	3	3	4	5
19	3	4	4	5
20	3	4	5	5
21	4	4	5	6
22	4	5	5	6
23	4	5	6	7
24	5	5	6	7
25	5	6	6	7

Note: Table 8 is for one-tailed or two-tailed tests. The sample size n represents the total number of + and − signs. The test value is the smaller number of + or − signs.

From *Journal of American Statistical Association* Vol. 41 (1946), pp. 557–66. W. J. Dixon and A. M. Mood. Reprinted with permission.

Table 9 — Critical Values for the Wilcoxon Signed-Rank Test

Reject the null hypothesis if the value of the test statistic w_s is less than or equal to the value given in the table.

	One-tailed, $\alpha = 0.05$	$\alpha = 0.025$	$\alpha = 0.01$	$\alpha = 0.005$
n	Two-tailed, $\alpha = 0.10$	$\alpha = 0.05$	$\alpha = 0.02$	$\alpha = 0.01$
5	1	—	—	—
6	2	1	—	—
7	4	2	0	—
8	6	4	2	0
9	8	6	3	2
10	11	8	5	3
11	14	11	7	5
12	17	14	10	7
13	21	17	13	10
14	26	21	16	13
15	30	25	20	16
16	36	30	24	19
17	41	35	28	23
18	47	40	33	28
19	54	46	38	32
20	60	52	43	37
21	68	59	49	43
22	75	66	56	49
23	83	73	62	55
24	92	81	69	61
25	101	90	77	68
26	110	98	85	76
27	120	107	93	84
28	130	117	102	92
29	141	127	111	100
30	152	137	120	109

From *Some Rapid Approximate Statistical Procedures.* Copyright 1949, 1964 Lederle Laboratories, American Cyanamid Co., Wayne, N.J. Reprinted with permission.

Table 10— Critical Values for the Spearman Rank Correlation

Reject H_0: $\rho_s = 0$ if the absolute value of r_s is greater than the value given in the table.

n	$\alpha = 0.10$	$\alpha = 0.05$	$\alpha = 0.01$
5	0.900	—	—
6	0.829	0.886	—
7	0.714	0.786	0.929
8	0.643	0.738	0.881
9	0.600	0.700	0.833
10	0.564	0.648	0.794
11	0.536	0.618	0.818
12	0.497	0.591	0.780
13	0.475	0.566	0.745
14	0.457	0.545	0.716
15	0.441	0.525	0.689
16	0.425	0.507	0.666
17	0.412	0.490	0.645
18	0.399	0.476	0.625
19	0.388	0.462	0.608
20	0.377	0.450	0.591
21	0.368	0.438	0.576
22	0.359	0.428	0.562
23	0.351	0.418	0.549
24	0.343	0.409	0.537
25	0.336	0.400	0.526
26	0.329	0.392	0.515
27	0.323	0.385	0.505
28	0.317	0.377	0.496
29	0.311	0.370	0.487
30	0.305	0.364	0.478

Reprinted with permission from the Institute of Mathematical Statistics.

Table 11— Critical Values for the Pearson Correlation Coefficient

Reject H_0: $\rho = 0$ if the absolute value of r is greater than the value given in the table.

n	$\alpha = 0.05$	$\alpha = 0.01$
4	0.950	0.990
5	0.878	0.959
6	0.811	0.917
7	0.754	0.875
8	0.707	0.834
9	0.666	0.798
10	0.632	0.765
11	0.602	0.735
12	0.576	0.708
13	0.553	0.684
14	0.532	0.661
15	0.514	0.641
16	0.497	0.623
17	0.482	0.606
18	0.468	0.590
19	0.456	0.575
20	0.444	0.561
21	0.433	0.549
22	0.423	0.537
23	0.413	0.526
24	0.404	0.515
25	0.396	0.505
26	0.388	0.496
27	0.381	0.487
28	0.374	0.479
29	0.367	0.471
30	0.361	0.463
35	0.334	0.430
40	0.312	0.403
45	0.294	0.380
50	0.279	0.361
55	0.266	0.345
60	0.254	0.330
65	0.244	0.317
70	0.235	0.306
75	0.227	0.296
80	0.220	0.286
85	0.213	0.278
90	0.207	0.270
95	0.202	0.263
100	0.197	0.256

The critical values in Table 11 were generated using Excel.

Table 12 — Critical Values for the Number of Runs

Reject the null hypothesis if the test statistic G is less than or equal to the smaller entry or greater than or equal to the larger entry.

Value of n_1		2	3	4	5	6	7	8	9	10	11	12	13	14	15	16	17	18	19	20
										Value of n_2										
2		1	1	1	1	1	1	1	1	1	1	2	2	2	2	2	2	2	2	2
		6	6	6	6	6	6	6	6	6	6	6	6	6	6	6	6	6	6	6
3		1	1	1	1	2	2	2	2	2	2	2	2	2	3	3	3	3	3	3
		6	8	8	8	8	8	8	8	8	8	8	8	8	8	8	8	8	8	8
4		1	1	1	2	2	2	3	3	3	3	3	3	3	4	4	4	4	4	4
		6	8	9	9	9	10	10	10	10	10	10	10	10	10	10	10	10	10	10
5		1	1	2	2	3	3	3	3	3	4	4	4	4	4	4	4	5	5	5
		6	8	9	10	10	11	11	12	12	12	12	12	12	12	12	12	12	12	12
6		1	2	2	3	3	3	3	4	4	4	4	5	5	5	5	5	5	6	6
		6	8	9	10	11	12	12	13	13	13	13	14	14	14	14	14	14	14	14
7		1	2	2	3	3	3	4	4	5	5	5	5	5	6	6	6	6	6	6
		6	8	10	11	12	13	13	14	14	14	14	15	15	15	16	16	16	16	16
8		1	2	3	3	3	4	4	5	5	5	6	6	6	6	6	7	7	7	7
		6	8	10	11	12	13	14	14	15	15	16	16	16	16	17	17	17	17	17
9		1	2	3	3	4	4	5	5	5	6	6	6	7	7	7	7	8	8	8
		6	8	10	12	13	14	14	15	16	16	16	17	17	18	18	18	18	18	18
10		1	2	3	3	4	5	5	5	6	6	7	7	7	7	8	8	8	8	9
		6	8	10	12	13	14	15	16	16	17	17	18	18	18	19	19	19	20	20
11		1	2	3	4	4	5	5	6	6	7	7	7	8	8	8	9	9	9	9
		6	8	10	12	13	14	15	16	17	17	18	19	19	19	20	20	20	21	21
12		2	2	3	4	4	5	6	6	7	7	7	8	8	8	9	9	9	10	10
		6	8	10	12	13	14	16	16	17	18	19	19	20	20	21	21	21	22	22
13		2	2	3	4	5	5	6	6	7	7	8	8	9	9	9	10	10	10	10
		6	8	10	12	14	15	16	17	18	19	19	20	20	21	21	22	22	23	23
14		2	2	3	4	5	5	6	7	7	8	8	9	9	9	10	10	10	11	11
		6	8	10	12	14	15	16	17	18	19	20	20	21	22	22	23	23	23	24
15		2	3	3	4	5	6	6	7	7	8	8	9	9	10	10	11	11	11	12
		6	8	10	12	14	15	16	18	18	19	20	21	22	22	23	23	24	24	25
16		2	3	4	4	5	6	6	7	8	8	9	9	10	10	11	11	11	12	12
		6	8	10	12	14	16	17	18	19	20	21	21	22	23	23	24	25	25	25
17		2	3	4	4	5	6	7	7	8	9	9	10	10	11	11	11	12	12	13
		6	8	10	12	14	16	17	18	19	20	21	22	23	23	24	25	25	26	26
18		2	3	4	5	5	6	7	8	8	9	9	10	10	11	11	12	12	13	13
		6	8	10	12	14	16	17	18	19	20	21	22	23	24	25	25	26	26	27
19		2	3	4	5	6	6	7	8	8	9	10	10	11	11	12	12	13	13	13
		6	8	10	12	14	16	17	18	20	21	22	23	23	24	25	26	26	27	27
20		2	3	4	5	6	6	7	8	9	9	10	10	11	12	12	13	13	13	14
		6	8	10	12	14	16	17	18	20	21	22	23	24	25	25	26	27	27	28

Note: Table 12 is for a two-tailed test with $\alpha = 0.05$.

Reprinted with permission from the Institute of Mathematical Statistics.

APPENDIX C

C   Normal Probability Plots and Their Graphs

WHAT YOU SHOULD LEARN

▸ How to construct and interpret a normal probability plot

Normal Probability Plots

▸ NORMAL PROBABILITY PLOTS

For the majority of problems throughout this book, it has been assumed that a random sample of data is selected from a population that has a normal distribution. Suppose you select a random sample from a population with an unknown distribution. How can you determine if the sample was selected from a population that has a normal distribution?

You have already learned that a histogram or stem-and-leaf plot can reveal the shape of a distribution and any outliers, clusters, or gaps in a distribution. These data displays are useful for assessing large sets of data, but assessing small data sets in this manner can be difficult and unreliable. A reliable method for assessing normality in small data sets is to use a graph called a *normal probability plot*.

DEFINITION

A **normal probability plot** is a graph that plots each observed value from the data set along with its corresponding z-score. The observed values are usually plotted along the horizontal axis while the corresponding z-scores are plotted along the vertical axis.

INSIGHT

A normal probability plot is also called a **normal quantile plot**.

If the plotted points in a normal probability plot are approximately linear, then you can conclude that the data come from a normal distribution. If the plotted points are not approximately linear or follow some type of pattern that is not linear, you can conclude that the data come from a distribution that is not normal. When examining a normal probability plot, look for deviations or clusters of points that stray from the line, which indicate a distribution that is not normal. Individual points that stray from the line in a normal probability plot may be outliers.

Constructing a normal probability plot by hand can be rather tedious. Technology tools such as MINITAB or a TI-83/84 Plus can be used to construct normal probability plots, as shown in Example 1.

EXAMPLE 1

▸ Constructing a Normal Probability Plot

The heights (in inches) of 12 current National Basketball Association players are listed. Use a technology tool to construct a normal probability plot to determine if the data come from a population that has a normal distribution. Identify any possible outliers.

74, 69, 78, 75, 73, 71, 80, 82, 81, 76, 86, 77

Answer: Page A49

STUDY TIP

Here are instructions for constructing a normal probability plot using MINITAB. First, enter the heights into column C1. Then click *Graph*, and select *Probability Plot*. Make sure single probability plot is chosen and click OK. Then double-click C1 to select the data to be graphed. Click *Distribution* and make sure normal is chosen. Click the *Data Display* menu. Select *Symbols only*. Then click OK. Click *Scale* and in the *Y-Scale Type* menu, select *Score*, and click OK. Click *Labels* and title the graph. Then click OK twice.

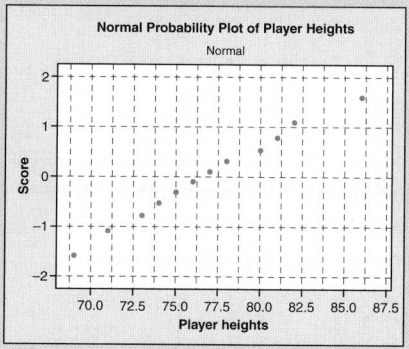

▶ Solution

Using a TI-83/84 Plus, begin by entering the data into List 1. Then use *Stat Plot* to construct the normal probability plot. The plot should look similar to the one shown below. From the scatter plot, it appears that there are no outliers and the points are approximately linear. To construct a normal probability plot using MINITAB, follow the instructions in the margin.

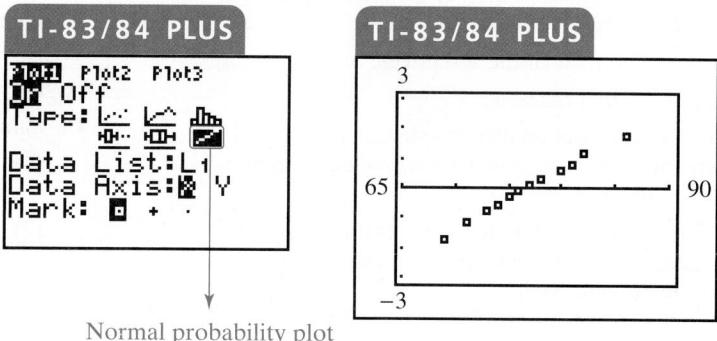

Normal probability plot

Interpretation   Because the points are approximately linear, you can conclude that the sample data come from a population that has a normal distribution.

▶ Try It Yourself 1

The balances (in dollars) on student loans for 18 randomly selected college seniors are listed.

29,150	16,980	12,470	19,235	15,875	8,960
16,105	14,575	39,860	20,170	9,710	19,650
21,590	8,200	18,100	25,530	9,285	10,075

a. *Use* a technology tool to construct a normal probability plot. Are the points approximately linear?
b. *Identify* any possible outliers.
c. *Interpret* your answer.

To see that the points are approximately linear, you can graph the regression line for the original data values and their corresponding z-scores. The regression line for the heights and z-scores from Example 1 is shown in the graph. From the graph, you can see that the points lie along the regression line. You can also approximate the mean of the data set by determining where the line crosses the x-axis.

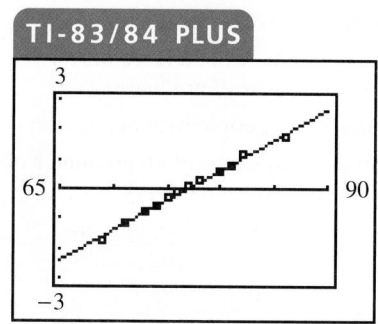

Try It Yourself Answers

CHAPTER 1

Section 1.1

1a. The population consists of the prices per gallon of regular gasoline at all gasoline stations in the United States. The sample consists of the prices per gallon of regular gasoline at the 900 surveyed stations.

b. The data set consists of the 900 prices.

2a. Population **b.** Parameter

3a. Descriptive statistics involve the statement "76% of women and 60% of men had a physical examination within the previous year."

b. An inference drawn from the study is that a higher percentage of women had a physical examination within the previous year.

Section 1.2

1a. City names and city populations

b. City names: Nonnumerical
City populations: Numerical

c. City names: Qualitative
City populations: Quantitative

2a. (1) The final standings represent a ranking of basketball teams.

(2) The collection of phone numbers represents labels.

b. (1) Ordinal, because the data can be put in order.

(2) Nominal, because no mathematical computations can be made.

3a. (1) The data set is the collection of body temperatures.

(2) The data set is the collection of heart rates.

b. (1) Interval, because the data can be ordered and meaningful differences can be calculated, but it does not make sense to write a ratio using the temperatures.

(2) Ratio, because the data can be ordered, meaningful differences can be calculated, the data can be written as a ratio, and the data set contains an inherent zero.

Section 1.3

1a. (1) Focus: Effect of exercise on relieving depression

(2) Focus: Success rates of graduates of a large university in finding a job within one year of graduation

b. (1) Population: Collection of all people with depression

(2) Population: The employment status of all graduates of a large university one year after graduation

c. (1) Experiment

(2) Survey

2a. There is no way to tell why the people quit smoking. They could have quit smoking as a result of either chewing the gum or watching the DVD.

b. Two experiments could be done; one using the gum and the other using the DVD.

3a. Answers will vary. *Sample answer:* Start with the first digits 92630782 . . . .

b. 92|63|07|82|40|19|26

c. 63, 7, 40, 19, 26

4a. Sample selection:

(1) The sample was selected by using the students in a randomly chosen class.

(2) The sample was selected by numbering each student in the school, randomly choosing a starting number, and selecting students at regular intervals from the starting number.

Sampling technique:

(1) Cluster sampling

(2) Systematic sampling

b. (1) The sample may be biased because some classes may be more familiar with stem cell research than other classes and have stronger opinions.

(2) The sample may be biased if there is any regularly occurring pattern in the data.

CHAPTER 2

Section 2.1

1a. 8 classes

b. Min = 35; Max = 89; Class width = 7

c.

Lower limit	Upper limit
35	41
42	48
49	55
56	62
63	69
70	76
77	83
84	90

d. See part (e).

e.

Class	Frequency, f
35–41	2
42–48	5
49–55	7
56–62	7
63–69	10
70–76	5
77–83	8
84–90	6

2 a. See part (b).

b.

Class	Frequency, f	Midpoint	Relative frequency	Cumulative frequency
35–41	2	38	0.04	2
42–48	5	45	0.10	7
49–55	7	52	0.14	14
56–62	7	59	0.14	21
63–69	10	66	0.20	31
70–76	5	73	0.10	36
77–83	8	80	0.16	44
84–90	6	87	0.12	50
	$\Sigma f = 50$		$\Sigma \dfrac{f}{n} = 1$	

c. The most common age bracket for the 50 richest people is 63–69. 72% of the 50 richest people are older than 55. 4% of the 50 richest people are younger than 42. (Answers will vary.)

3 a.

Class boundaries
34.5–41.5
41.5–48.5
48.5–55.5
55.5–62.5
62.5–69.5
69.5–76.5
76.5–83.5
83.5–90.5

b. Use class midpoints for the horizontal scale and frequency for the vertical scale. (Class boundaries can also be used for the horizontal scale.)

c.

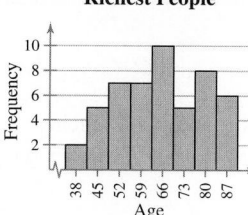

Ages of the 50 Richest People

d. Same as 2(c).

4 a. Same as 3(b).

b. See part (c).

c.

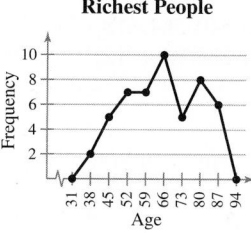

Ages of the 50 Richest People

d. The frequency of ages increases up to 66 and then decreases.

5 abc.

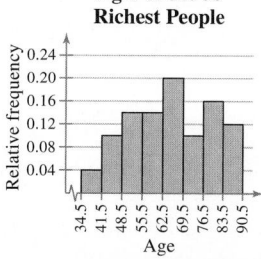

Ages of the 50 Richest People

6 a. Use upper class boundaries for the horizontal scale and cumulative frequency for the vertical scale.

b. See part (c).

c.

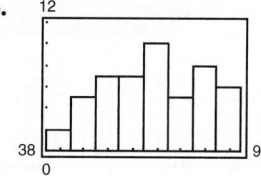

Ages of the 50 Richest People

d. Approximately 40 of the 50 richest people are 80 years old or younger.

e. Answers will vary.

7 a. Enter data.

b.

```
12
 .
 .
 .
38          94
 0
```

Section 2.2

1 a.
```
3
4
5
6
7
8
```

b.
```
3 | 6 5                    Key: 3|6 = 36
4 | 9 7 6 4 3 2
5 | 9 8 7 6 4 4 3 3 1 1
6 | 9 9 8 7 6 6 5 5 4 3 1 1 0
7 | 8 8 7 6 3 3 3 2
8 | 9 9 7 6 6 5 3 3 2 1 0
```

c.
```
3 | 5 6                    Key: 3|5 = 35
4 | 2 3 4 6 7 9
5 | 1 1 3 3 4 4 6 7 8 9
6 | 0 1 1 3 4 5 5 6 6 7 8 9 9
7 | 2 3 3 3 6 7 8 8
8 | 0 1 2 3 3 5 6 6 7 9 9
```

d. More than 50% of the 50 richest people are older than 60. (Answers will vary.)

2 ab.

```
3 |              Key: 3|5 = 35
3 | 5 6
4 | 2 3 4
4 | 6 7 9
5 | 1 1 3 3 4 4
5 | 6 7 8 9
6 | 0 1 1 3 4
6 | 5 5 6 6 7 8 9 9
7 | 2 3 3 3
7 | 6 7 8 8
8 | 0 1 2 3 3
8 | 5 6 6 7 9 9
```

c. Most of the 50 richest people are older than 60. (Answers will vary.)

3 a. Use age for the horizontal axis.

b.

Ages of the 50 Richest People

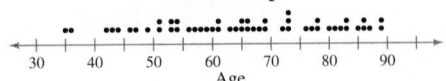

c. A large percentage of the ages are over 60. (Answers will vary.)

4 a.

Type of degree	f	Relative frequency	Angle
Associate's	455	0.23	82.8°
Bachelor's	1052	0.54	194.4°
Master's	325	0.17	61.2°
First professional	71	0.04	14.4°
Doctoral	38	0.02	7.2°
	$\Sigma f = 1941$	$\Sigma \dfrac{f}{n} = 1$	$\Sigma = 360°$

b.

Earned Degrees Conferred in 1990

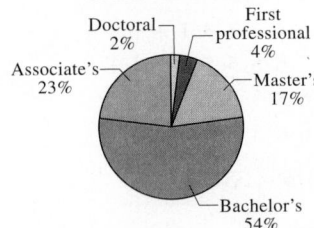

Doctoral 2%
First professional 4%
Associate's 23%
Master's 17%
Bachelor's 54%

c. From 1990 to 2007, as percentages of the total degrees conferred, associate's degrees increased by 1%, bachelor's degrees decreased by 3%, master's degrees increased by 3%, first professional degrees decreased by 1%, and doctoral degrees remained unchanged.

5 a.

Cause	Frequency, f
Auto Dealers	14,668
Auto Repair	9728
Home Furnishing	7792
Computer Sales	5733
Dry Cleaning	4649

b.

Causes of BBB Complaints

c. It appears that the auto industry (dealers and repair shops) account for the largest portion of complaints filed at the BBB. (Answers will vary.)

6 ab.

Salaries

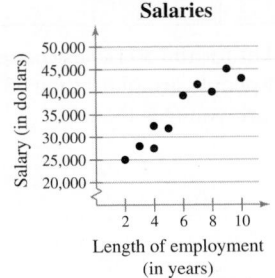

c. It appears that the longer an employee is with the company, the larger the employee's salary will be.

7 ab.

Cellular Phone Bills

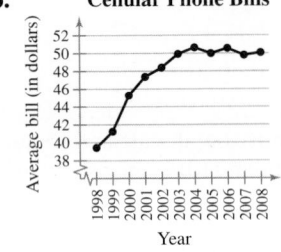

c. The average bill increased from 1998 to 2004, then it hovered around $50.00 from 2004 to 2008.

Section 2.3

1 a. 1193　**b.** 79.5

c. The mean height of the players is about 79.5 inches.

2 a. 18, 18, 19, 19, 19, 20, 21, 21, 21, 21, 23, 24, 24, 26, 27, 27, 29, 30, 30, 30, 33, 33, 34, 35, 38

b. 24

c. The median age of the sample of fans at the concert is 24.

3 a. 25, 60, 80, 97, 100, 130, 140, 200, 220, 250　**b.** 115

c. The median price of the sample of digital photo frames is $115.

4 a. 324, 385, 450, 450, 462, 475, 540, 540, 564, 618, 624, 638, 670, 670, 670, 705, 720, 723, 750, 750, 825, 830, 912, 975, 980, 980, 1100, 1260, 1420, 1650

b. 670

c. The mode of the prices for the sample of South Beach, FL condominiums is $670.

5 a. Yes

b. In this sample, there were more people who thought public cell phone conversations were rude than people who did not or had no opinion.

6 a. 21.6; 21; 20

b. The mean in Example 6 ($\bar{x} \approx 23.8$) was heavily influenced by the entry 65. Neither the median nor the mode was affected as much by the entry 65.

7 ab.

Source	Score, x	Weight, w	xw
Test mean	86	0.50	43.0
Midterm	96	0.15	14.4
Final exam	98	0.20	19.6
Computer lab	98	0.10	9.8
Homework	100	0.05	5.0
		$\sum w = 1.00$	$\sum(x \cdot w) = 91.8$

c. 91.8

d. The weighted mean for the course is 91.8. So you did get an A.

8 abc.

Class	Midpoint, x	Frequency, f	xf
35–41	38	2	76
42–48	45	5	225
49–55	52	7	364
56–62	59	7	413
63–69	66	10	660
70–76	73	5	365
77–83	80	8	640
84–90	87	6	522
		$N = 50$	$\sum(x \cdot f) = 3265$

d. 65.3

Section 2.4

1 a. Min = 23, or $23,000; Max = 58, or $58,000

b. 35, or $35,000

c. The range of the starting salaries for Corporation B, which is 35, or $35,000, is much larger than the range of Corporation A.

2 a. 41.5, or $41,500

b.

Salary, x (1000s of dollars)	Deviation, $x - \mu$ (1000s of dollars)
23	−18.5
29	−12.5
32	−9.5
40	−1.5
41	−0.5
41	−0.5
49	7.5
50	8.5
52	10.5
58	16.5
$\sum x = 415$	$\sum(x - \mu) = 0$

3 ab. $\mu = 41.5$, or $41,500

Salary, x	$x - \mu$	$(x - \mu)^2$
23	−18.5	342.25
29	−12.5	156.25
32	−9.5	90.25
40	−1.5	2.25
41	−0.5	0.25
41	−0.5	0.25
49	7.5	56.25
50	8.5	72.25
52	10.5	110.25
58	16.5	272.25
$\sum x = 415$	$\sum(x - \mu) = 0$	$\sum(x - \mu)^2 = 1102.5$

c. 110.3 **d.** 10.5, or $10,500

e. The population standard deviation is 10.5, or $10,500.

4 a. See 3ab. **b.** 122.5 **c.** 11.1, or $11,100

d. The population standard deviation is 11.1, or $11,100.

5 a. Enter data. **b.** 37.89; 3.98

6 a. 7, 7, 7, 7, 7, 13, 13, 13, 13, 13 **b.** 3

7 a. 1 standard deviation **b.** 34%

c. Approximately 34% of women ages 20–29 are between 64.3 and 66.92 inches tall.

8 a. 0 **b.** 70.6

c. At least 75% of the data lie within 2 standard deviations of the mean. At least 75% of the population of Alaska is between 0 and 70.6 years old.

9 a.

x	f	xf
0	10	0
1	19	19
2	7	14
3	7	21
4	5	20
5	1	5
6	1	6
	$n = 50$	$\sum xf = 85$

b. 1.7

c.

$x - \bar{x}$	$(x - \bar{x})^2$	$(x - \bar{x})^2 f$
−1.7	2.89	28.90
−0.7	0.49	9.31
0.3	0.09	0.63
1.3	1.69	11.83
2.3	5.29	26.45
3.3	10.89	10.89
4.3	18.49	18.49
		$\sum(x - \bar{x})^2 f = 106.5$

d. 1.5

10 a.

Class	x	f	xf
0–99	49.5	380	18,810
100–199	149.5	230	34,385
200–299	249.5	210	52,395
300–399	349.5	50	17,475
400–499	449.5	60	26,970
500+	650.0	70	45,500
		$n = 1000$	$\Sigma xf = 195{,}535$

b. 195.5

c.

$x - \bar{x}$	$(x - \bar{x})^2$	$(x - \bar{x})^2 f$
−146.0	21,316	8,100,080
−46.0	2116	486,680
54.0	2916	612,360
154.0	23,716	1,185,800
254.0	64,516	3,870,960
454.5	206,570.25	14,459,917.5
		$\Sigma(x - \bar{x})^2 f = 28{,}715{,}797.5$

d. 169.5

Section 2.5

1 a. 35, 36, 42, 43, 44, 46, 47, 49, 51, 51, 53, 53, 54, 54, 56, 57, 58, 59, 60, 61, 61, 63, 64, 65, 65, 66, 66, 67, 68, 69, 69, 72, 73, 73, 73, 76, 77, 78, 78, 80, 81, 82, 83, 83, 85, 86, 86, 87, 89, 89

b. 65.5   **c.** 54, 78

d. About one fourth of the 50 richest people are 54 years old or younger; one half are 65.5 years old or younger; and about three fourths of the 50 richest people are 78 years old or younger.

2 a. Enter data.   **b.** 17, 23, 28.5

c. One quarter of the tuition costs is $17,000 or less, one half is $23,000 or less, and three quarters is $28,500 or less.

3 a. 54, 78   **b.** 24

c. The ages of the 50 richest people in the middle portion of the data set vary by at most 24 years.

4 a. Min $= 35, Q_1 = 54, Q_2 = 65.5, Q_3 = 78$, Max $= 89$

bc. Ages of the 50 Richest People

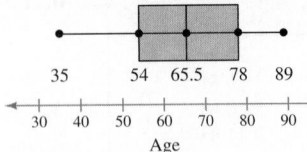

d. It appears that half of the ages are between 54 and 78.

5 a. 50th percentile

b. 50% of the 50 richest people are younger than 66.

6 a. $\mu = 70, \sigma = 8$

$$z_1 = \frac{60 - 70}{8} = -1.25$$

$$z_2 = \frac{71 - 70}{8} = 0.125$$

$$z_3 = \frac{92 - 70}{8} = 2.75$$

b. From the z-scores, $60 is 1.25 standard deviations below the mean, $71 is 0.125 standard deviation above the mean, and $92 is 2.75 standard deviations above the mean.

7 a. Best Actor: $\mu = 43.7, \sigma = 8.7$

Best Actress: $\mu = 35.9, \sigma = 11.4$

b. Sean Penn: $z = 0.49$

Kate Winslet: $z = -0.25$

c. The age of Sean Penn is 0.49 standard deviation above the mean and the age of Kate Winslet is 0.25 standard deviation below the mean. Both z-scores fall between -2 and 2, so neither would be considered unusual. Comparing the two measures indicates that Sean Penn is further above the average age of actors than Kate Winslet is below the average age of actresses. (Answers will vary.)

CHAPTER 3

Section 3.1

1ab. (1)

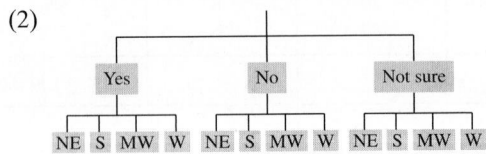

(2)

c. (1) 6   (2) 12

d. (1) Let $Y =$ Yes, $N =$ No, $NS =$ Not sure, $M =$ Male, $F =$ Female.

Sample space $=$

$\{YM, YF, NM, NF, NSM, NSF\}$

(2) Let $Y =$ Yes, $N =$ No, $NS =$ Not sure, $NE =$ Northeast, $S =$ South, $MW =$ Midwest, $W =$ West.

Sample space $=$

$\{YNE, YS, YMW, YW, NNE, NS, NMW, NW,$ $NSNE, NSS, NSMW, NSW\}$

2 a. (1) 6   (2) 1

b. (1) Not a simple event because it is an event that consists of more than a single outcome.

(2) Simple event because it is an event that consists of a single outcome.

3 a. Manufacturer: 4, Size: 2, Color: 5 **b.** 40

c.

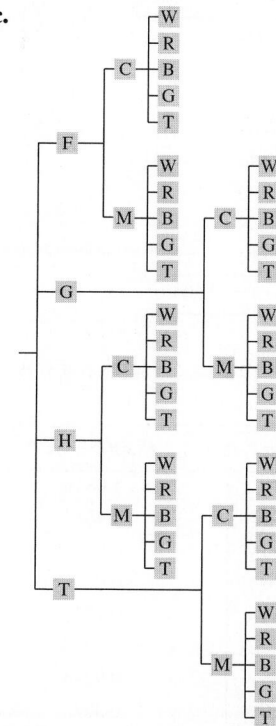

4 a. (1) Each letter is an event (26 choices for each).

(2) Each letter is an event (26, 25, 24, 23, 22, and 21 choices).

(3) Each letter is an event (22, 26, 26, 26, 26, and 26 choices).

b. (1) 308,915,776 (2) 165,765,600 (3) 261,390,272

5 a. (1) 52 (2) 52 (3) 52

b. (1) 1 (2) 13 (3) 52

c. (1) 0.019 (2) 0.25 (3) 1

6 a. The event is "the next claim processed is fraudulent." The frequency is 4.

b. 100 **c.** 0.04

7 a. 54 **b.** 1000 **c.** 0.054

8 a. The event is "salmon successfully passing through a dam on the Columbia River."

b. Estimated **c.** Empirical probability

9 a. 0.18 **b.** 0.82 **c.** $\frac{41}{50}$ or 0.82

10 a. 5 **b.** 0.313

11 a. 10,000,000 **b.** $\dfrac{1}{10,000,000}$

Section 3.2

1 a. (1) 30 and 102 (2) 11 and 50

b. (1) 0.294 (2) 0.22

2 a. (1) Yes (2) No

b. (1) Dependent (2) Independent

3 a. (1) Independent (2) Dependent

b. (1) 0.723 (2) 0.059

4 a. (1) Event (2) Event (3) Complement

b. (1) 0.729 (2) 0.001 (3) 0.999

c. (1) The event cannot be considered unusual because its probability is not less than or equal to 0.05.

(2) The event can be considered unusual because its probability is less than or equal to 0.05.

(3) The event cannot be considered unusual because its probability is not less than or equal to 0.05.

5 a. (1) and (2) $A = \{\text{is female}\}$,

$B = \{\text{works in health field}\}$

b. (1) $P(A \text{ and } B) = P(A) \cdot P(B|A) = (0.65) \cdot (0.25)$

(2) $P(A \text{ and } B') = P(A) \cdot (1 - P(B|A))$

$= (0.65) \cdot (0.75)$

c. (1) 0.163 (2) 0.488

Section 3.3

1 a. (1) None are true. (2) None are true.

(3) All are true.

b. (1) Not mutually exclusive (2) Not mutually exclusive

(3) Mutually exclusive

2 a. (1) Mutually exclusive (2) Not mutually exclusive

b. (1) $\frac{1}{6}, \frac{1}{2}$ (2) $\frac{12}{52}, \frac{13}{52}, \frac{3}{52}$ **c.** (1) 0.667 (2) 0.423

3 a. $A = \{\text{sales between \$0 and \$24,999}\}$

$B = \{\text{sales between \$25,000 and \$49,999}\}$

b. A and B cannot occur at the same time.

A and B are mutually exclusive.

c. $\frac{3}{36}, \frac{5}{36}$ **d.** 0.222

4 a. (1) $A = \{\text{type B}\}$

$B = \{\text{type AB}\}$

(2) $A = \{\text{type O}\}$

$B = \{\text{Rh-positive}\}$

b. (1) A and B cannot occur at the same time.

A and B are mutually exclusive.

(2) A and B can occur at the same time.

A and B are not mutually exclusive.

c. (1) $\frac{45}{409}, \frac{16}{409}$ (2) $\frac{184}{409}, \frac{344}{409}$

d. (1) 0.149 (2) 0.910

5 a. 0.141 **b.** 0.859

Section 3.4

1 a. 8 **b.** 40,320

2 a. 336

b. There are 336 possible ways that the subject can pick a first, second, and third activity.

3 a. $n = 12, r = 4$ **b.** 11,880

4 a. $n = 20, n_1 = 6, n_2 = 9, n_3 = 5$ **b.** 77,597,520

5 a. $n = 20, r = 3$ **b.** 1140

c. There are 1140 different possible three-person committees that can be selected from 20 employees.

6 a. 380   **b.** 0.003

7 a. 1 outcome and 180 distinguishable permutations

b. 0.006

8 a. 3003   **b.** 3,162,510   **c.** 0.0009

9 a. 10   **b.** 220   **c.** 0.045

CHAPTER 4

Section 4.1

1 a. (1) Measured   (2) Counted

b. (1) The random variable is continuous because x can be any speed up to the maximum speed of a Space Shuttle.

(2) The random variable is discrete because the number of calves born on a farm in one year is countable.

2ab.

x	f	$P(x)$
0	16	0.16
1	19	0.19
2	15	0.15
3	21	0.21
4	9	0.09
5	10	0.10
6	8	0.08
7	2	0.02
	$n = 100$	$\Sigma P(x) = 1$

c.

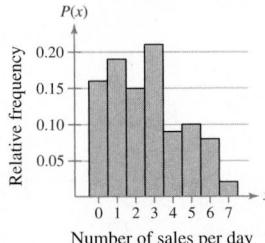

New Employee Sales

3 a. Each $P(x)$ is between 0 and 1.   **b.** $\Sigma P(x) = 1$

c. Because both conditions are met, the distribution is a probability distribution.

4 a. (1) Yes, each outcome is between 0 and 1.

(2) Yes, each outcome is between 0 and 1.

b. (1) Yes   (2) Yes

c. (1) A probability distribution

(2) A probability distribution

5ab.

x	$P(x)$	$xP(x)$
0	0.16	0.00
1	0.19	0.19
2	0.15	0.30
3	0.21	0.63
4	0.09	0.36
5	0.10	0.50
6	0.08	0.48
7	0.02	0.14
	$\Sigma P(x) = 1$	$\Sigma xP(x) = 2.60$

c. $\mu = 2.6$

On average, a new employee makes 2.6 sales per day.

6ab.

x	$P(x)$	$x - \mu$	$(x - \mu)^2$	$P(x)(x - \mu)^2$
0	0.16	-2.6	6.76	1.0816
1	0.19	-1.6	2.56	0.4864
2	0.15	-0.6	0.36	0.0540
3	0.21	0.4	0.16	0.0336
4	0.09	1.4	1.96	0.1764
5	0.10	2.4	5.76	0.5760
6	0.08	3.4	11.56	0.9248
7	0.02	4.4	19.36	0.3872
	$\Sigma P(x) = 1$			$\Sigma P(x)(x - \mu)^2 = 3.72$

c. 1.9

d. Most of the data values differ from the mean by no more than 1.9 sales per day.

7ab.

Gain, x	$1995	$995	$495	$245	$95	$-\$5$
Probability, $P(x)$	$\frac{1}{2000}$	$\frac{1}{2000}$	$\frac{1}{2000}$	$\frac{1}{2000}$	$\frac{1}{2000}$	$\frac{1995}{2000}$

c. $-\$3.08$

d. Because the expected value is negative, you can expect to lose an average of $3.08 for each ticket you buy.

Section 4.2

1 a. Trial: answering a question

Success: question answered correctly

b. Yes

c. It is a binomial experiment;
$n = 10, p = 0.25, q = 0.75, x = 0, 1, 2, 3, 4, 5, 6, 7, 8, 9, 10$

2 a. Trial: drawing a card with replacement

Success: card drawn is a club

Failure: card drawn is not a club

b. $n = 5, p = 0.25, q = 0.75, x = 3$

c. $P(3) = \dfrac{5!}{2! \, 3!} (0.25)^3 (0.75)^2 \approx 0.088$

3a. Trial: selecting an adult and asking a question

Success: selecting an adult who likes texting because it works where talking won't do

Failure: selecting an adult who does not like texting because it works where talking won't do

b. $n = 7, p = 0.75, q = 0.25, x = 0, 1, 2, 3, 4, 5, 6, 7$

c. $P(0) = {}_7C_0 (0.75)^0 (0.25)^7 \approx 0.00006$

$P(1) = {}_7C_1 (0.75)^1 (0.25)^6 \approx 0.00128$

$P(2) = {}_7C_2 (0.75)^2 (0.25)^5 \approx 0.01154$

$P(3) = {}_7C_3 (0.75)^3 (0.25)^4 \approx 0.05768$

$P(4) = {}_7C_4 (0.75)^4 (0.25)^3 \approx 0.17303$

$P(5) = {}_7C_5 (0.75)^5 (0.25)^2 \approx 0.31146$

$P(6) = {}_7C_6 (0.75)^6 (0.25)^1 \approx 0.31146$

$P(7) = {}_7C_7 (0.75)^7 (0.25)^0 \approx 0.13348$

d.

x	$P(x)$
0	0.00006
1	0.00128
2	0.01154
3	0.05768
4	0.17303
5	0.31146
6	0.31146
7	0.13348
	$\Sigma P(x) \approx 1$

4a. $n = 250, p = 0.71, x = 178$ **b.** 0.056

c. The probability that exactly 178 people from a random sample of 250 people in the United States will use more than one topping on their hotdogs is about 0.056.

d. Because 0.056 is not less than or equal to 0.05, this event is not unusual.

5a. (1) $x = 2$ (2) $x = 2, 3, 4,$ or 5   (3) $x = 0$ or 1

b. (1) 0.217   (2) 0.217, 0.058, 0.008, 0.0004; 0.283

(3) 0.308, 0.409; 0.717

c. (1) The probability that exactly two of the five men consider fishing their favorite leisure-time activity is about 0.217.

(2) The probability that at least two of the five men consider fishing their favorite leisure-time activity is about 0.283.

(3) The probability that fewer than two of the five men consider fishing their favorite leisure-time activity is about 0.717.

6a. Trial: selecting a business and asking if it has a website

Success: selecting a business with a website

Failure: selecting a business without a website

b. $n = 10, p = 0.55, x = 4$ **c.** 0.160

d. The probability that exactly 4 of the 10 small businesses have websites is 0.160.

e. Because 0.160 is greater than 0.05, this event is not unusual.

7a. 0.001, 0.022, 0.142, 0.404, 0.430

b.

x	$P(x)$
0	0.001
1	0.022
2	0.142
3	0.404
4	0.430

c.

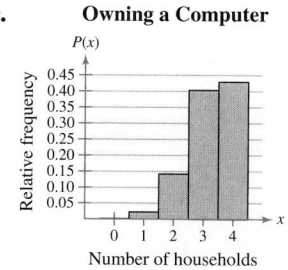

Owning a Computer

Skewed left

d. Yes, it would be unusual if exactly zero or exactly one of the four households owned a computer, because each of these events has a probability that is less than 0.05.

8a. Success: selecting a clear day

$n = 31, p = 0.44, q = 0.56$

b. 13.6   **c.** 7.6   **d.** 2.8

e. On average, there are about 14 clear days during the month of May.

f. A May with fewer than 8 clear days or more than 19 clear days would be unusual.

Section 4.3

1a. 0.74; 0.192   **b.** 0.932

c. The probability that LeBron makes his first free throw shot before his third attempt is 0.932.

2a. $P(0) \approx 0.050$

$P(1) \approx 0.149$

$P(2) \approx 0.224$

$P(3) \approx 0.224$

$P(4) \approx 0.168$

b. 0.815   **c.** 0.185

d. The probability that more than four accidents will occur in any given month at the intersection is 0.185.

3a. 0.10   **b.** 0.10, 3   **c.** 0.0002

d. The probability of finding three brown trout in any given cubic meter of the lake is 0.0002.

e. Because 0.0002 is less than 0.05, this can be considered an unusual event.

CHAPTER 5

Section 5.1

1a. $A: x = 45, B: x = 60, C: x = 45$; B has the greatest mean.

b. Curve C is more spread out, so curve C has the greatest standard deviation.

2a. $x = 660$ **b.** 630, 690; 30

3. (1) 0.0143   (2) 0.9850

4 a. 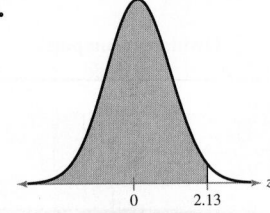 **b.** 0.9834

5 a. 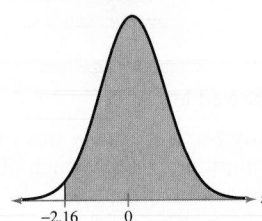 **b.** 0.0154 **c.** 0.9846

6 a. 0.0885 **b.** 0.0152 **c.** 0.0733

d. 7.33% of the area under the curve falls between $z = -2.165$ and $z = -1.35$.

Section 5.2

1 a.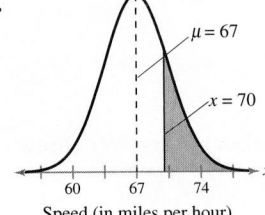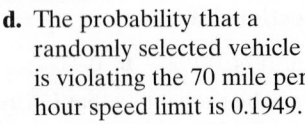

Speed (in miles per hour)

b. 0.86
c. 0.1949
d. The probability that a randomly selected vehicle is violating the 70 mile per hour speed limit is 0.1949.

2 a.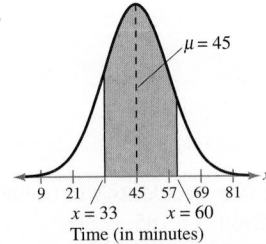

$x = 33$ $x = 60$
Time (in minutes)

b. -1, 1.25
c. 0.1587; 0.8944; 0.7357
d. If 150 shoppers enter the store, then you would expect $150(0.7357) = 110.355$, or about 110, shoppers to be in the store between 33 and 60 minutes.

3 a. Read user's guide for the technology tool.

b. 0.5105

c. The probability that a randomly selected U.S. person's triglyceride level is between 100 and 150 is 0.5105.

Section 5.3

1 a. (1) 0.0384 (2) 0.0250 and 0.9750
bc. (1) -1.77 (2) ± 1.96
2 a. (1) Area = 0.10 (2) Area = 0.20
(3) Area = 0.99
bc. (1) -1.28 (2) -0.84 (3) 2.33
3 a. $\mu = 52$, $\sigma = 15$ **b.** 17.05; 98.5; 60.7
c. 17.05 pounds is below the mean, 60.7 pounds and 98.5 pounds are above the mean.

4ab.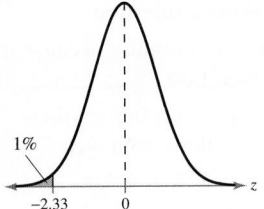

1%

-2.33 0

c. 116.93

d. So, the longest braking distance a Nissan Altima could have and still be in the bottom 1% is about 117 feet.

5ab.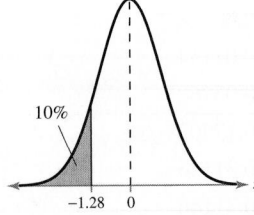

10%

-1.28 0

c. 8.512

d. So, the maximum length of time an employee could have worked and still be laid off is about 8.5 years.

Section 5.4

1a.

Sample	Mean	Sample	Mean	Sample	Mean
1, 1, 1	1	3, 3, 5	3.67	5, 7, 1	4.33
1, 1 ,3	1.67	3, 3, 7	4.33	5, 7, 3	5
1, 1, 5	2.33	3, 5, 1	3	5, 7, 5	5.67
1, 1, 7	3	3, 5, 3	3.67	5, 7, 7	6.33
1, 3, 1	1.67	3, 5, 5	4.33	7, 1, 1	3
1, 3, 3	2.33	3, 5, 7	5	7, 1, 3	3.67
1, 3, 5	3	3, 7, 1	3.67	7, 1, 5	4.33
1, 3, 7	3.67	3, 7, 3	4.33	7 ,1, 7	5
1, 5, 1	2.33	3, 7, 5	5	7, 3, 1	3.67
1, 5, 3	3	3, 7, 7	5.67	7, 3, 3	4.33
1, 5, 5	3.67	5, 1, 1	2.33	7, 3, 5	5
1, 5, 7	4.33	5, 1, 3	3	7, 3, 7	5.67
1, 7, 1	3	5, 1, 5	3.67	7, 5, 1	4.33
1, 7, 3	3.67	5, 1, 7	4.33	7, 5, 3	5
1, 7, 5	4.33	5, 3, 1	3	7, 5, 5	5.67
1, 7, 7	5	5, 3, 3	3.67	7, 5, 7	6.33
3, 1, 1	1.67	5, 3, 5	4.33	7, 7, 1	5
3, 1, 3	2.33	5, 3, 7	5	7, 7, 3	5.67
3, 1, 5	3	5, 5, 1	3.67	7, 7, 5	6.33
3, 1, 7	3.67	5, 5, 3	4.33	7, 7, 7	7
3, 3, 1	2.33	5, 5, 5	5		
3, 3, 3	3	5, 5, 7	5.67		

b.

$\bar{x}$	f	Probability
1	1	0.0156
1.67	3	0.0469
2.33	6	0.0938
3	10	0.1563
3.67	12	0.1875
4.33	12	0.1875
5	10	0.1563
5.67	6	0.0938
6.33	3	0.0469
7	1	0.0156

$\mu_{\bar{x}} = 4$
$(\sigma_{\bar{x}})^2 \approx 1.667$
$\sigma_{\bar{x}} \approx 1.291$

c. $\mu_{\bar{x}} = \mu = 4$

$(\sigma_{\bar{x}})^2 = \dfrac{\sigma^2}{n} = \dfrac{5}{3} \approx 1.667$; $\sigma_{\bar{x}} = \dfrac{\sigma}{\sqrt{n}} = \dfrac{\sqrt{5}}{\sqrt{3}} \approx 1.291$

2 a. 63, 1.4

  b. $n = 64$

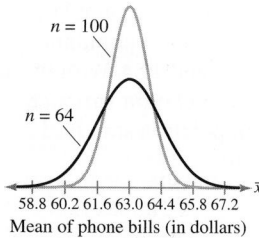

Mean of phone bills (in dollars)

  c. With a smaller sample size, the mean stays the same but the standard deviation increases.

3 a. 3.5, 0.05

  b.

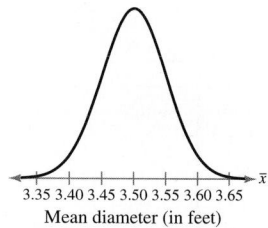

Mean diameter (in feet)

4 a. 25; 0.15   **b.** $-2, 3.33$ **c.** 0.0228; 0.996; 0.9768

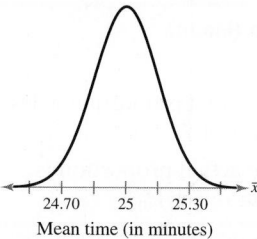

Mean time (in minutes)

  d. Of the samples of 100 drivers ages 15 to 19, 97.68% will have a mean driving time between 24.7 and 25.5 minutes.

5 a. 290,600; 10,392.30

Mean sales price (in dollars)

  b. -2.46 **c.** 0.0069; 0.9931

  d. 99.31% of samples of 12 single-family houses will have a mean sales price greater than $265,000.

6 a. 0.21; 0.66   **b.** 0.5832; 0.7454

  c. There is about a 58% chance that an LCD computer monitor will cost less than $200. There is about a 75% chance that the mean of a sample of 10 LCD computer monitors is less than $200.

Section 5.5

1 a. $n = 125$, $p = 0.05$, $q = 0.95$ **b.** 6.25, 118.75

  c. Normal distribution can be used.   **d.** 6.25, 2.44

2 a. (1) 57, 58, ..., 83   (2) ..., 52, 53, 54

  b. (1) $56.5 < x < 83.5$ (2) $x < 54.5$

3 a. Normal distribution can be used.   **b.** 6.25, 2.44

  c.   **d.** 1.33   **e.** 0.9082; 0.0918

4 a. Normal distribution can be used.   **b.** 116, 6.98

  c.   **d.** -2.22 **e.** 0.0132

5 a. Normal distribution can be used.   **b.** 36, 5.23

  c.   **d.** $-1.82, -1.625$ **e.** 0.0177

CHAPTER 6

Section 6.1

1 a. $\bar{x} = 138.5$

  b. A point estimate for the population mean number of friends is 138.5.

2 a. $z_c = 1.96$, $n = 30$, $s \approx 51.0$ **b.** $E \approx 18.3$

  c. You are 95% confident that the maximum error of the estimate is about 18.3 friends.

3 a. $\bar{x} = 138.5$, $E \approx 18.3$ **b.** 120.2, 156.8

  c. With 95% confidence, you can say that the population mean number of friends is between 120.2 and 156.8. This confidence interval is wider than the one found in Example 3.

4 a. Enter the data.

  b. (121.2, 140.4); (118.7, 142.9); (109.2, 152.4)

  c. As the confidence level increases, so does the width of the interval.

5 a. $n = 30$, $\bar{x} = 22.9$, $\sigma = 1.5$, $z_c = 1.645$, $E \approx 0.5$

  b. (22.4, 23.4) [*Tech:* (22.5, 23.4)]

  c. With 90% confidence, you can say that the mean age of the students is between 22.4 (*Tech:* 22.5) and 23.4 years. Because of the larger sample size, the confidence interval is slightly narrower.

6 a. $z_c = 1.96, E = 10, s \approx 53.0$ **b.** $n = 108$

c. You should have at least 108 users in your sample. Because of the larger margin of error, the sample size needed is much smaller.

Section 6.2

1 a. d.f. $= 21$ **b.** $c = 0.90$ **c.** $t_c = 1.721$

2 a. $t_c = 1.753, E \approx 4.4; t_c = 2.947, E \approx 7.4$

b. $(157.6, 166.4); (154.6, 169.4)$

c. With 90% confidence, you can say that the population mean temperature of coffee sold is between 157.6°F and 166.4°F.

With 99% confidence, you can say that the population mean temperature of coffee sold is between 154.6°F and 169.4°F.

3 a. $t_c = 1.729, E \approx 0.92; t_c = 2.093, E \approx 1.12$

b. $(8.83, 10.67); (8.63, 10.87)$

c. With 90% confidence, you can say that the population mean number of days the car model sits on the lot is between 8.83 and 10.67; with 95% confidence, you can say that the population mean number of days the car model sits on the lot is between 8.63 and 10.87. The 90% confidence interval is slightly narrower.

4. Use a t-distribution because the sample size is small $(n < 30)$, the population is normally distributed, and the population standard deviation is unknown.

Section 6.3

1 a. $x = 181, n = 1006$ **b.** $\hat{p} \approx 0.180$

2 a. $\hat{p} \approx 0.180, \ \hat{q} \approx 0.820$

b. $n\hat{p} \approx 181 > 5$ and $n\hat{q} \approx 825 > 5$

c. $z_c = 1.645, E = 0.020$ **d.** $(0.160, 0.200)$

e. With 90% confidence, you can say that the proportion of adults who think Abraham Lincoln was the greatest president is between 16.0% and 20.0%.

3 a. $\hat{p} = 0.25, \hat{q} = 0.75$

b. $n\hat{p} = 124.5 > 5$ and $n\hat{q} = 373.5 > 5$

c. $z_c = 2.575, E \approx 0.050$ **d.** $(0.20, 0.30)$

e. With 99% confidence, you can say that the proportion of U.S. adults who think that people over 65 are the more dangerous drivers is between 20% and 30%.

4 a. (1) $\hat{p} = 0.5, \hat{q} = 0.5, z_c = 1.645, E = 0.02$

(2) $\hat{p} = 0.11, \hat{q} = 0.89, z_c = 1.645, E = 0.02$

b. (1) 1691.27   (2) 662.30

c. (1) 1692 females   (2) 663 females

Section 6.4

1 a. d.f. $= 29, c = 0.90$ **b.** $0.05, 0.95$ **c.** $42.557, 17.708$

d. 90% of the area under the curve lies between 17.708 and 42.557.

2 a. 42.557, 17.708; 45.722, 16.047

b. $(0.98, 2.36); (0.91, 2.60)$ **c.** $(0.99, 1.54); (0.96, 1.61)$

d. With 90% confidence, you can say that the population variance is between 0.98 and 2.36 and the population standard deviation is between 0.99 and 1.54. With 95% confidence, you can say that the population variance is between 0.91 and 2.60 and the population standard deviation is between 0.96 and 1.61.

CHAPTER 7

Section 7.1

1 a. (1) The mean is not 74 months.

$\mu \neq 74$

(2) The variance is less than or equal to 2.7.

$\sigma^2 \leq 2.7$

(3) The proportion is more than 24%.

$p > 0.24$

b. (1) $\mu = 74$ (2) $\sigma^2 > 2.7$ (3) $p \leq 0.24$

c. (1) $H_0: \mu = 74; H_a: \mu \neq 74$ (claim)

(2) $H_0: \sigma^2 \leq 2.7$ (claim); $H_a: \sigma^2 > 2.7$

(3) $H_0: p \leq 0.24; H_a: p > 0.24$ (claim)

2 a. $H_0: p \leq 0.01; H_a: p > 0.01$

b. A type I error will occur if the actual proportion is less than or equal to 0.01, but you reject H_0.

A type II error will occur if the actual proportion is greater than 0.01, but you fail to reject H_0.

c. A type II error is more serious because you would be misleading the consumer, possibly causing serious injury or death.

3 a. (1) H_0: The mean life of a certain type of automobile battery is 74 months.

H_a: The mean life of a certain type of automobile battery is not 74 months.

$H_0: \mu = 74; H_a: \mu \neq 74$

(2) H_0: The variance of the life of the home theater systems is less than or equal to 2.7.

H_a: The variance of the life of the home theater systems is greater than 2.7.

$H_0: \sigma^2 \leq 2.7; H_a: \sigma^2 > 2.7$

(3) H_0: The proportion of homeowners who feel their house is too small for their family is less than or equal to 24%.

H_a: The proportion of homeowners who feel their house is too small for their family is greater than 24%.

$H_0: p \leq 0.24; H_a: p > 0.24$

b. (1) Two-tailed   (2) Right-tailed   (3) Right-tailed

c. (1)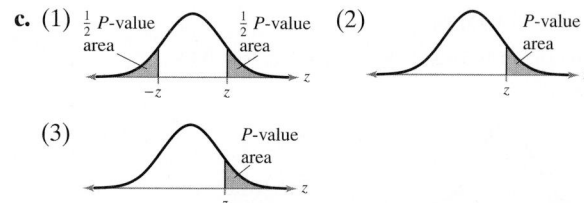

(2)

(3)

4a. There is enough evidence to support the realtor's claim that the proportion of homeowners who feel their house is too small for their family is more than 24%.

b. There is not enough evidence to support the realtor's claim that the proportion of homeowners who feel their house is too small for their family is more than 24%.

5a. (1) Support claim. (2) Reject claim.

b. (1) $H_0: \mu \geq 650$; $H_a: \mu < 650$ (claim)

(2) $H_0: \mu = 98.6$ (claim); $H_a: \mu \neq 98.6$

Section 7.2

1a. (1) $0.0347 > 0.01$ (2) $0.0347 < 0.05$

b. (1) Fail to reject H_0. (2) Reject H_0.

2ab. 0.0436

c. Reject H_0 because $0.0436 < 0.05$.

3a. 0.9495 **b.** 0.1010

c. Fail to reject H_0 because $0.1010 > 0.01$.

4a. The claim is "the mean speed is greater than 35 miles per hour."

$H_0: \mu \leq 35$; $H_a: \mu > 35$ (claim)

b. $\alpha = 0.05$ **c.** 2.5 **d.** 0.0062 **e.** Reject H_0.

f. There is enough evidence at the 5% level of significance to support the claim that the average speed is greater than 35 miles per hour.

5a. The claim is "one of your distributors reports an average of 150 sales per day."

$H_0: \mu = 150$ (claim); $H_a: \mu \neq 150$

b. $\alpha = 0.01$ **c.** −2.76 **d.** 0.0058

e. Reject H_0 because $0.0058 < 0.01$.

f. There is enough evidence at the 1% level of significance to reject the claim that the distributorship averages 150 sales per day.

6a. $0.0440 > 0.01$ **b.** Fail to reject H_0.

7a.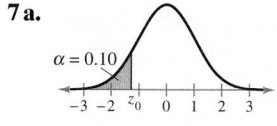

b. 0.1003

c. $z_0 = -1.28$

d. Rejection region: $z < -1.28$

8a.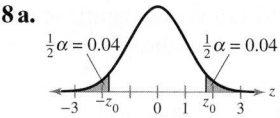

b. 0.0401, 0.9599

c. $-z_0 = -1.75$, $z_0 = 1.75$

d. Rejection regions: $z < -1.75$, $z > 1.75$

9a. The claim is "the mean work day of the company's mechanical engineers is less than 8.5 hours."

$H_0: \mu \geq 8.5$; $H_a: \mu < 8.5$ (claim)

b. $\alpha = 0.01$

c. $z_0 = -2.33$; Rejection region: $z < -2.33$

d. −3.55

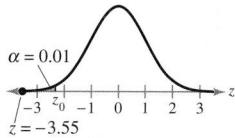

e. Because $-3.55 < -2.33$, reject H_0.

f. There is enough evidence at the 1% level of significance to support the claim that the mean work day is less than 8.5 hours.

10a. $\alpha = 0.01$

b. $-z_0 = -2.575$, $z_0 = 2.575$;

Rejection regions: $z < -2.575$, $z > 2.575$

c.

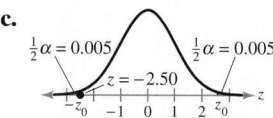

Fail to reject H_0.

d. There is not enough evidence at the 1% level of significance to reject the claim that the mean cost of raising a child from birth to age 2 by husband-wife families in the United States is $13,120.

Section 7.3

1a. 13 **b.** −2.650

2a. 8 **b.** 1.397

3a. 15 **b.** −2.131, 2.131

4a. The claim is "the mean cost of insuring a 2008 Honda CR-V is less than $1200."

$H_0: \mu \geq \$1200$; $H_a: \mu < \$1200$ (claim)

b. $\alpha = 0.10$, d.f. = 6

c. $t_0 = -1.440$; Rejection region: $t < -1.440$

d. −3.61

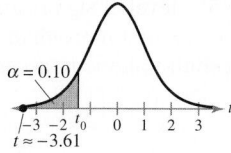

e. Reject H_0.

f. There is enough evidence at the 10% level of significance to support the insurance agent's claim that the mean cost of insuring a 2008 Honda CR-V is less than $1200.

5 a. The claim is "the mean conductivity of the river is 1890 milligrams per liter."

$H_0: \mu = 1890$ (claim); $H_a: \mu \neq 1890$

b. $\alpha = 0.01$, d.f. $= 18$

c. $-t_0 = -2.878$, $t_0 = 2.878$

Rejection regions: $t < -2.878$, $t > 2.878$

d. 3.798

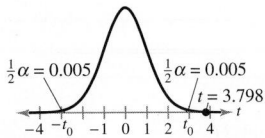

e. Reject H_0.

f. There is enough evidence at the 1% level of significance to reject the company's claim that the mean conductivity of the river is 1890 milligrams per liter.

6 a. The claim is "the mean wait time is at most 18 minutes."

$H_0: \mu \leq 18$ minutes (claim); $H_a: \mu > 18$ minutes

b. 0.9997

c. $0.9997 > 0.05$; Fail to reject H_0.

d. There is not enough evidence at the 5% level of significance to reject the office's claim that the mean wait time is at most 18 minutes..

Section 7.4

1 a. $np = 31.25 > 5$, $nq = 93.75 > 5$

b. The claim is "more than 25% of U.S. adults have used a cellular phone to access the Internet."

$H_0: p \leq 0.25$; $H_a: p > 0.25$ (claim)

c. $\alpha = 0.05$

d. $z_0 = 1.645$; Rejection region: $z > 1.645$

e. 1.81

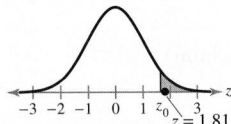

f. Reject H_0.

g. There is enough evidence at the 5% level of significance to support the research center's claim that more than 25% of U.S. adults have used a cellular phone to access the Internet.

2 a. $np = 75 > 5$, $nq = 175 > 5$

b. The claim is "30% of U.S. adults have not purchased a certain brand because they found the advertisements distasteful."

$H_0: p = 0.30$ (claim); $H_a: p \neq 0.30$

c. $\alpha = 0.10$

d. $-z_0 = -1.645$, $z_0 = 1.645$;

Rejection regions: $z < -1.645$, $z > 1.645$

e. 2.07

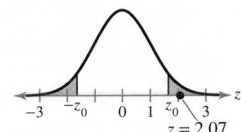

f. Reject H_0.

g. There is enough evidence at the 10% level of significance to reject the claim that 30% of U.S. adults have not purchased a certain brand because they found the advertisements distasteful.

Section 7.5

1 a. d.f. $= 17$, $\alpha = 0.01$ **b.** 33.409

2 a. d.f. $= 29$, $\alpha = 0.05$ **b.** 17.708

3 a. d.f. $= 50$, $\alpha = 0.01$ **b.** 79.490   **c.** 27.991

4 a. The claim is "the variance of the amount of sports drink in a 12-ounce bottle is no more than 0.40."

$H_0: \sigma^2 \leq 0.40$ (claim); $H_a: \sigma^2 > 0.40$

b. $\alpha = 0.01$, d.f. $= 30$

c. $\chi_0^2 = 50.892$; Rejection region: $\chi^2 > 50.892$

d. 56.250   **e.** Reject H_0.

f. There is enough evidence at the 1% level of significance to reject the bottling company's claim that the variance of the amount of sports drink in a 12-ounce bottle is no more than 0.40.

5 a. The claim is "the standard deviation of the lengths of response times is less than 3.7 minutes."

$H_0: \sigma \geq 3.7$; $H_a: \sigma < 3.7$ (claim)

b. $\alpha = 0.05$, d.f. $= 8$

c. $\chi_0^2 = 2.733$; Rejection region: $\chi^2 < 2.733$

d. 5.259   **e.** Fail to reject H_0.

f. There is not enough evidence at the 5% level of significance to support the police chief's claim that the standard deviation of the lengths of response times is less than 3.7 minutes.

6 a. The claim is "the variance of the weight losses is 25.5."

$H_0: \sigma^2 = 25.5$ (claim); $H_a: \sigma^2 \neq 25.5$

b. $\alpha = 0.10$, d.f. $= 12$

c. $\chi_L^2 = 5.226$, $\chi_R^2 = 21.026$;

Rejection regions: $\chi^2 < 5.226$, $\chi^2 > 21.026$

d. 5.082   **e.** Reject H_0.

f. There is enough evidence at the 10% level of significance to reject the company's claim that the variance of the weight losses of the users is 25.5.

CHAPTER 8

Section 8.1

1a. (1) Independent   (2) Dependent

b. (1) Because each sample represents blood pressures of different individuals, and it is not possible to form a pairing between the members of the samples.

(2) Because the samples represent exam scores of the same students, the samples can be paired with respect to each student.

2a. The claim is "there is a difference in the mean annual wages for forensic science technicians working for local and state governments."

$H_0: \mu_1 = \mu_2$; $H_a: \mu_1 \neq \mu_2$ (claim)

b. $\alpha = 0.10$

c. $-z_0 = -1.645$, $z_0 = 1.645$;

Rejection regions: $z < -1.645$, $z > 1.645$

d. 1.667

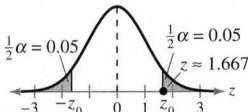

e. Reject H_0.

f. There is enough evidence at the 10% level of significance to support the claim that there is a difference in the mean annual wages for forensic science technicians working for local and state governments.

3a. $z \approx 1.36$; $P \approx 0.0865$

b. Fail to reject H_0.

c. There is not enough evidence at the 5% level of significance to support the travel agency's claim that the average daily cost of meals and lodging for vacationing in Alaska is greater than the same average cost for vacationing in Colorado.

Section 8.2

1a. The claim is "there is a difference in the mean annual earnings based on level of education."

$H_0: \mu_1 = \mu_2$; $H_a: \mu_1 \neq \mu_2$ (claim)

b. $\alpha = 0.01$; d.f. = 11

c. $-t_0 = -3.106$, $t_0 = 3.106$; Rejection regions: $t < -3.106$, $t > 3.106$

d. -4.63

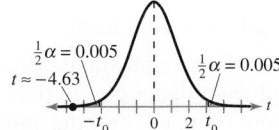

e. Reject H_0.

f. There is enough evidence at the 1% level of significance to support the claim that there is a difference in the mean annual earnings based on level of education.

2a. The claim is "the watt usage of a manufacturer's 17-inch flat panel monitors is less than that of its leading competitor."

$H_0: \mu_1 \geq \mu_2$; $H_a: \mu_1 < \mu_2$ (claim)

b. $\alpha = 0.10$; d.f. = 25

c. $t_0 = -1.316$; Rejection region: $t < -1.316$

d. -3.997

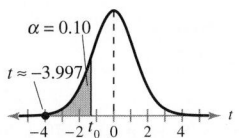

e. Reject H_0.

f. There is enough evidence at the 10% level of significance to support the manufacturer's claim that the watt usage of its monitors is less than that of its leading competitor.

Section 8.3

1a. The claim is "athletes can decrease their times in the 40-yard dash."

$H_0: \mu_d \leq 0$; $H_a: \mu_d > 0$ (claim)

b. $\alpha = 0.05$; d.f. = 11

c. $t_0 = 1.796$; Rejection region: $t > 1.796$

d. $\bar{d} \approx 0.0233$; $s_d \approx 0.0607$

e. 1.333

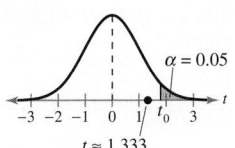

f. Fail to reject H_0.

g. There is not enough evidence at the 5% level of significance to support the claim that athletes can decrease their times in the 40-yard dash.

2a. The claim is "the drug changes the body's temperature."

$H_0: \mu_d = 0$; $H_a: \mu_d \neq 0$ (claim)

b. $\alpha = 0.05$; d.f. = 6

c. $-t_0 = -2.447$, $t_0 = 2.447$; Rejection regions: $t < -2.447$, $t > 2.447$

d. $\bar{d} \approx 0.5571$; $s_d \approx 0.9235$

e. 1.596

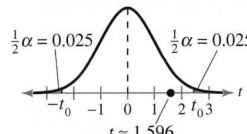

f. Fail to reject H_0.

g. There is not enough evidence at the 5% level of significance to support the claim that the drug changes the body's temperature.

Section 8.4

1a. The claim is "there is a difference between the proportion of male high school students who smoke cigarettes and the proportion of female high school students who smoke cigarettes."

$H_0: p_1 = p_2$; $H_a: p_1 \neq p_2$ (claim)

b. $\alpha = 0.05$

c. $-z_0 = -1.96$, $z_0 = 1.96$;

Rejection regions: $z < -1.96$, $z > 1.96$

d. $\overline{p} \approx 0.1975$; $\overline{q} \approx 0.8025$

e. $n_1\overline{p} \approx 1382.5 > 5$, $n_1\overline{q} \approx 5617.5 > 5$, $n_2\overline{p} \approx 1479.1 > 5$, and $n_2\overline{q} \approx 6009.9 > 5$.

f. 4.23

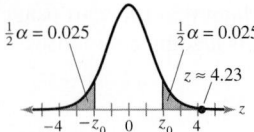

g. Reject H_0.

h. There is enough evidence at the 5% level of significance to support the claim that there is a difference between the proportion of male high school students who smoke cigarettes and the proportion of female high school students who smoke cigarettes.

2a. The claim is "the proportion of male high school students who smoke cigars is greater than the proportion of female high school students who smoke cigars."

$H_0: p_1 \leq p_2$; $H_a: p_1 > p_2$ (claim)

b. $\alpha = 0.05$ **c.** $z_0 = 1.645$; Rejection region: $z > 1.645$

d. $\overline{p} \approx 0.1174$; $\overline{q} \approx 0.8826$

e. $n_1\overline{p} \approx 821.8 > 5$, $n_1\overline{q} \approx 6178.2 > 5$, $n_2\overline{p} \approx 879.2 > 5$, and $n_2\overline{q} = 6609.8 > 5$

f. 17.565

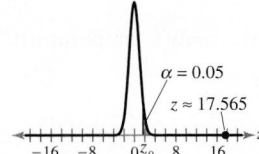

g. Reject H_0.

h. There is enough evidence at the 5% level of significance to support the claim that the proportion of male high school students who smoke cigars is greater than the proportion of female high school students who smoke cigars.

CHAPTER 9

Section 9.1

1ab.

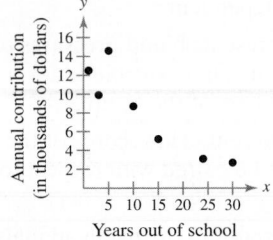

c. Yes, it appears that there is a negative linear correlation. As the number of years out of school increases, the annual contribution decreases.

2ab.

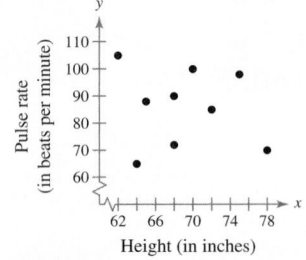

c. No, it appears that there is no linear correlation between height and pulse rate.

3ab.

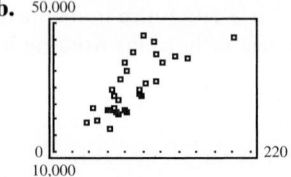

c. Yes, it appears that there is a positive linear correlation. As the team salary increases, the average attendance per home game increases.

4a. 7; $\sum x = 88$, $\sum y = 56.7$, $\sum xy = 435.6$, $\sum x^2 = 1836$, $\sum y^2 = 587.05$

b. -0.908

c. Because r is close to -1, this suggests a strong negative linear correlation between years out of school and annual contribution.

5ab. 0.750

c. Because r is close to 1, this suggests a strong positive linear correlation between the salaries and the average attendances at home games.

6a. 7   **b.** 0.01   **c.** 0.875

d. $|r| \approx 0.908 > 0.875$; The correlation is significant.

e. There is enough evidence at the 1% level of significance to conclude that there is a significant linear correlation between the number of years out of school and the annual contribution.

7a. $H_0: \rho = 0$; $H_a: \rho \neq 0$ **b.** 0.01   **c.** 28

d. $-t_0 = -2.763$, $t_0 = 2.763$;

   Rejection regions: $t < -2.763$, $t > 2.763$

e. 5.995   **f.** Reject H_0.

g. There is enough evidence at the 1% level of significance to conclude that there is a significant linear correlation between the salaries and average attendances at home games for the teams in Major League Baseball.

Section 9.2

1a. $n = 7$, $\Sigma x = 88$, $\Sigma y = 56.7$, $\Sigma xy = 435.6$, $\Sigma x^2 = 1836$

b. $m \approx -0.379875$; $b \approx 12.8756$

c. $\hat{y} = -0.380x + 12.876$

2a. Enter the data.   **b.** $m \approx 189.038015$; $b \approx 13{,}497.9583$

c. $\hat{y} = 189.038x + 13{,}497.958$

3a. (1) $\hat{y} = 12.481(2) + 33.683$

   (2) $\hat{y} = 12.481(3.32) + 33.683$

b. (1) 58.645   (2) 75.120

c. (1) 58.645 minutes   (2) 75.120 minutes

Section 9.3

1a. 0.979   **b.** 0.958

c. About 95.8% of the variation in the times is explained. About 4.2% of the variation is unexplained.

2a.

x_i	y_i	$\hat{y}_i$	$y_i - \hat{y}_i$	$(y_i - \hat{y}_i)^2$
15	26	28.386	−2.386	5.692996
20	32	35.411	−3.411	11.634921
20	38	35.411	2.589	6.702921
30	56	49.461	6.539	42.758521
40	54	63.511	−9.511	90.459121
45	78	70.536	7.464	55.711296
50	80	77.561	2.439	5.948721
60	88	91.611	−3.611	13.039321
				$\Sigma = 231.947818$

b. 8   **c.** 6.218

d. The standard error of estimate of the weekly sales for a specific radio ad time is about $621.80.

3a. $n = 10$, d.f. $= 8$, $t_c = 2.306$, $s_e \approx 138.255$

b. 886.897   **c.** 364.088

d. $522.809 < y < 1250.985$

e. You can be 95% confident that when the gross domestic product is $4 trillion, the carbon dioxide emissions will be between 522.809 and 1250.985 million metric tons.

Section 9.4

1a. Enter the data.

b. $\hat{y} = 46.385 + 0.540x_1 - 4.897x_2$

2ab. (1) $\hat{y} = 46.385 + 0.540(89) - 4.897(1)$

   (2) $\hat{y} = 46.385 + 0.540(78) - 4.897(3)$

   (3) $\hat{y} = 46.385 + 0.540(83) - 4.897(2)$

c. (1) $\hat{y} = 89.548$ (2) $\hat{y} = 73.814$ (3) $\hat{y} = 81.411$

d. (1) 90   (2) 74   (3) 81

CHAPTER 10

Section 10.1

1.

Tax preparation method	% of people	Expected frequency
Accountant	25%	125
By hand	20%	100
Computer software	35%	175
Friend/family	5%	25
Tax preparation service	15%	75

2a. The expected frequencies are 64, 80, 32, 56, 60, 48, 40, and 20, all of which are at least 5.

b. Claimed distribution:

Ages	Distribution
0–9	16%
10–19	20%
20–29	8%
30–39	14%
40–49	15%
50–59	12%
60–69	10%
70+	5%

H_0: The distribution of ages is as shown in table above.

H_a: The distribution of ages differs from the claimed distribution. (claim)

c. 0.05   **d.** 7

e. $\chi_0^2 = 14.067$; Rejection region: $\chi^2 > 14.067$

f. 6.694

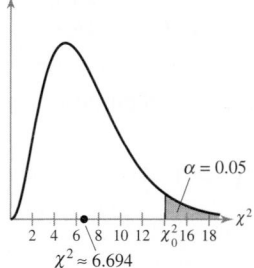

$\chi^2 \approx 6.694$

g. Fail to reject H_0.

h. There is not enough evidence at the 5% level of significance to support the sociologist's claim that the age distribution differs from the age distribution 10 years ag

3 a. The expected frequency for each category is 30, which is at least 5.

b. Claimed distribution:

Color	Distribution
Brown	$16.\overline{6}\%$
Yellow	$16.\overline{6}\%$
Red	$16.\overline{6}\%$
Blue	$16.\overline{6}\%$
Orange	$16.\overline{6}\%$
Green	$16.\overline{6}\%$

H_0: The distribution of colors is uniform, as shown in the table above. (claim)

H_a: The distribution of colors is not uniform.

c. 0.05   **d.** 5

e. $\chi_0^2 = 11.071$; Rejection region: $\chi^2 > 11.071$

f. 12.933

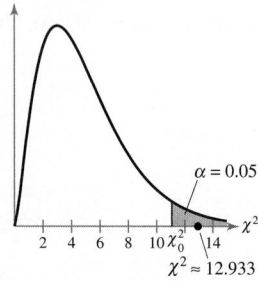

$\alpha = 0.05$

$\chi^2 \approx 12.933$

g. Reject H_0.

h. There is enough evidence at the 5% level of significance to reject the claim that the distribution of different-colored candies in bags of peanut M&M's is uniform.

Section 10.2

1 a. Marginal frequencies: Row 1: 180; Row 2: 120; Column 1: 74; Column 2: 162; Column 3: 28; Column 4: 36

b. 300

c. $E_{1,1} = 44.4$, $E_{1,2} = 97.2$, $E_{1,3} = 16.8$, $E_{1,4} = 21.6$, $E_{2,1} = 29.6$, $E_{2,2} = 64.8$, $E_{2,3} = 11.2$, $E_{2,4} = 14.4$

2 a. H_0: Travel concern is independent of travel purpose.

H_a: Travel concern is dependent on travel purpose. (claim)

b. 0.01   **c.** 3

$\chi_0^2 = 11.345$; Rejection region: $\chi^2 > 11.345$

‸58

$\alpha = 0.01$

g. There is not enough evidence at the 1% level of significance for the consultant to conclude that travel concern is dependent on travel purpose.

3 a. H_0: Whether or not a tax cut would influence an adult to purchase a hybrid vehicle is independent of age.

H_a: Whether or not a tax cut would influence an adult to purchase a hybrid vehicle is dependent on age. (claim)

b. Enter the data.

c. $\chi_0^2 = 9.210$; Rejection region: $\chi^2 > 9.210$

d. 15.306   **e.** Reject H_0.

f. There is enough evidence at the 1% level of significance to conclude that whether or not a tax cut would influence an adult to purchase a hybrid vehicle is dependent on age.

Section 10.3

1 a. 0.05   **b.** 2.45

2 a. 0.01   **b.** 18.31

3 a. H_0: $\sigma_1^2 \le \sigma_2^2$; H_a: $\sigma_1^2 > \sigma_2^2$ (claim)

b. 0.01   **c.** d.f.$_N$ = 24, d.f.$_D$ = 19

d. $F_0 = 2.92$; Rejection region: $F > 2.92$

e. 3.21

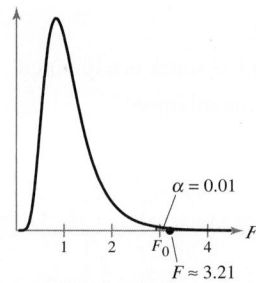

$\alpha = 0.01$

$F \approx 3.21$

f. Reject H_0.

g. There is enough evidence at the 1% level of significance to support the researcher's claim that a specially treated intravenous solution decreases the variance of the time required for nutrients to enter the bloodstream.

4 a. H_0: $\sigma_1 = \sigma_2$ (claim); H_a: $\sigma_1 \ne \sigma_2$

b. 0.01   **c.** d.f.$_N$ = 15, d.f.$_D$ = 21

d. $F_0 = 3.43$; Rejection region: $F > 3.43$

e. 1.48   **f.** Fail to reject H_0.

g. There is not enough evidence at the 1% level of significance to reject the biologist's claim that the pH levels of the soil in the two geographic locations have equal standard deviations.

Section 10.4

1 a. H_0: $\mu_1 = \mu_2 = \mu_3 = \mu_4$

H_a: At least one mean is different from the others. (claim)

b. 0.05   **c.** d.f.$_N$ = 3, d.f.$_D$ = 14

d. $F_0 = 3.34$; Rejection region: $F > 3.34$

e. 4.22

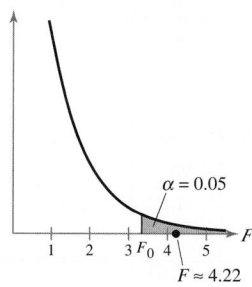

$\alpha = 0.05$

$F \approx 4.22$

f. Reject H_0.

g. There is enough evidence at the 5% level of significance for the analyst to conclude that there is a difference in the mean monthly sales among the sales regions.

2 a. $H_0: \mu_1 = \mu_2 = \mu_3 = \mu_4$

H_a: At least one mean is different from the others. (claim)

b. Enter the data.

c. $F \approx 1.34$; P-value ≈ 0.280

d. Fail to reject H_0.

e. There is not enough evidence at the 5% level of significance to conclude that there is a difference in the means of the GPAs.

CHAPTER 11

Section 11.1

1 a. H_0: median ≤ 120; H_a: median > 120 (claim)

b. 0.025   **c.** 23   **d.** 6   **e.** 6   **f.** Reject H_0.

g. There is enough evidence at the 2.5% level of significance to support the agency's claim that the median number of days a home is on the market in its city is greater than 120.

2 a. H_0: median $= 9.4$ (claim); H_a: median $\neq 9.4$

b. 0.10   **c.** 92   **d.** -1.645 **e.** -0.94

f. Fail to reject H_0.

g. There is not enough evidence at the 10% level of significance to reject the organization's claim that the median age of automobiles in operation in the United States is 9.4 years.

3 a. H_0: The number of colds will not decrease.

H_a: The number of colds will decrease. (claim)

b. 0.05   **c.** 11   **d.** 2   **e.** 2   **f.** Reject H_0.

g. There is enough evidence at the 5% level of significance to support the researcher's claim that a new vaccine will decrease the number of colds in adults.

Section 11.2

1 a. H_0: There is no difference in the amounts of water repelled.

H_a: There is a difference in the amounts of water repelled. (claim)

b. 0.01   **c.** 11   **d.** 5

e.

No repellent	Repellent applied	Difference	Absolute value	Rank	Signed rank
8	15	-7	7	11	-11
7	12	-5	5	9	-9
7	11	-4	4	7.5	-7.5
4	6	-2	2	3.5	-3.5
6	6	0	0		
10	8	2	2	3.5	3.5
9	8	1	1	1.5	1.5
5	6	-1	1	1.5	-1.5
9	12	-3	3	5.5	-5.5
11	8	3	3	5.5	5.5
8	14	-6	6	10	-10
4	8	-4	4	7.5	-7.5

$w_s = 10.5$

f. Fail to reject H_0.

g. There is not enough evidence at the 1% level of significance for the quality control inspector to conclude that the spray-on water repellent is effective.

2 a. H_0: There is no difference in the claims paid by the companies.

H_a: There is a difference in the claims paid by the companies. (claim)

b. 0.05

c. $-z_0 = -1.96$, $z_0 = 1.96$;

Rejection regions: $z < -1.96$, $z > 1.96$

d. $n_1 = 12$ and $n_2 = 12$

e.

Ordered data	Sample	Rank
1.7	B	1
1.8	B	2
2.2	B	3
2.5	A	4
3.0	A	5.5
3.0	B	5.5
3.4	B	7
3.9	A	8
4.1	B	9
4.4	B	10
4.5	A	11
4.7	B	12

Ordered data	Sample	Rank
5.3	B	13
5.6	B	14
5.8	A	15
6.0	A	16
6.2	A	17
6.3	A	18
6.5	A	19
7.3	B	20
7.4	A	21
9.9	A	22
10.6	A	23
10.8	B	24

$R = 120.5$ (or $R = 179.5$)

f. -1.703 (or 1.703)

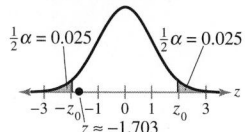

$\frac{1}{2}\alpha = 0.025$ $\frac{1}{2}\alpha = 0.025$

$z \approx -1.703$

g. Fail to reject H_0.

h. There is not enough evidence at the 5% level of significance to conclude that there is a difference in the claims paid by the companies.

Section 11.3

1 a. H_0: There is no difference in the salaries in the three states.

H_a: There is a difference in the salaries in the three states. (claim)

b. 0.05   **c.** 2

d. $\chi_0^2 = 5.991$; Rejection region: $\chi^2 > 5.991$

e.

Ordered data	State	Rank
88.28	CA	1
88.80	PA	2
92.50	NY	3
93.10	NY	4
94.40	NY	5
95.15	PA	6
96.25	CA	7
97.25	PA	8
97.44	PA	9
97.50	CA	10.5
97.50	NY	10.5
97.89	NY	12
98.85	CA	13
99.20	PA	14

Ordered data	State	Rank
99.70	CA	15
99.75	NY	16
99.95	CA	17
99.99	PA	18
100.55	PA	19
100.75	CA	20
101.20	CA	21
101.55	NY	22
101.97	NY	23
102.35	NY	24
103.20	CA	25
103.70	PA	26
110.45	PA	27
113.90	CA	28

$R_1 = 157.5$

$R_2 = 129$

$R_3 = 119.5$

f. 0.433

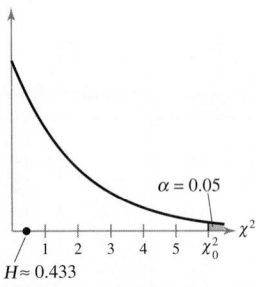

$\alpha = 0.05$

$H \approx 0.433$

g. Fail to reject H_0.

h. There is not enough evidence at the 5% level of significance to conclude that the distributions of the veterinarians' salaries in these three states are different.

Section 11.4

: $\rho_s = 0$; H_a: $\rho_s \neq 0$ (claim)

  c. 0.929

d.

Male	Rank	Female	Rank	d	d^2
25	3.5	20	1.5	2	4
24	1.5	20	1.5	0	0
24	1.5	22	3.0	−1.5	2.25
25	3.5	23	4.0	−0.5	0.25
27	5.0	26	5.0	0	0
29	6.0	27	6.0	0	0
30	7.0	30	7.0	0	0
					$\Sigma d^2 = 6.5$

$\Sigma d^2 = 6.5$

e. 0.884   **f.** Fail to reject H_0.

g. There is not enough evidence at the 1% level of significance to conclude that a significant correlation exists between the number of males and females who received doctoral degrees.

Section 11.5

1 a. *P P P   F   P   F   P P P P   F F   P   F   P P*
F F F   P P P   F   P P P

b. 13

c. 3, 1, 1, 1, 4, 2, 1, 1, 2, 3, 3, 1, 3

2 a. H_0: The sequence of genders is random.

H_a: The sequence of genders is not random. (claim)

b. 0.05

c. n_1 = number of F's = 9

n_2 = number of M's = 6

G = number of runs = 8

d. lower critical value = 4

upper critical value = 13

e. 8   **f.** Fail to reject H_0.

g. There is not enough evidence at the 5% level of significance to support the claim that the sequence of genders is not random.

3 a. H_0: The sequence of weather conditions is random.

H_a: The sequence of weather conditions is not random. (claim)

b. 0.05

c. n_1 = number of N's = 21

n_2 = number of S's = 10

G = number of runs = 17

d. ±1.96   **e.** 1.03   **f.** Fail to reject H_0.

g. There is not enough evidence at the 5% level of significance to support the claim that the sequence of weather conditions is not random.

APPENDIX A

1. (1) 0.4857

    (2) $z = \pm 2.17$

2 a.

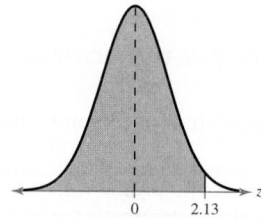

b. 0.4834   **c.** 0.9834

3 a.

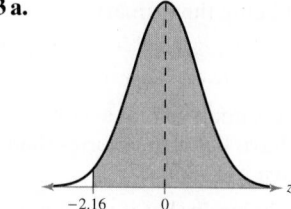

b. 0.4846   **c.** 0.9846

4 a.

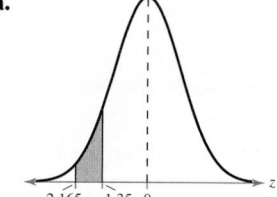

b. 0.4848; 0.4115   **c.** 0.0733

APPENDIX C

1 a.

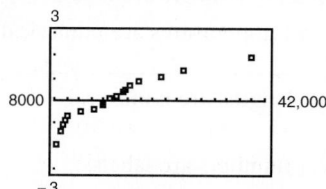

The points do not appear to be approximately linear.

b. 39,860 is a possible outlier because it is far removed from the other entries in the data set.

c. Because the points do not appear to be approximately linear and there is an outlier, you can conclude that the sample data do not come from a population that has a normal distribution.

Odd Answers

CHAPTER 1

Section 1.1 *(page 6)*

1. A sample is a subset of a population.

3. A parameter is a numerical description of a population characteristic. A statistic is a numerical description of a sample characteristic.

5. False. A statistic is a numerical measure that describes a sample characteristic.

7. True

9. False. A population is the collection of *all* outcomes, responses, measurements, or counts that are of interest.

11. Population, because it is a collection of the heights of all the players on the school's basketball team.

13. Sample, because the collection of the 500 spectators is a subset of the population.

15. Sample, because the collection of the 20 patients is a subset of the population.

17. Population, because it is a collection of all the golfers' scores in the tournament.

19. Population, because it is a collection of all the U.S. presidents' political parties.

21. Population: Parties of registered voters in Warren County

    Sample: Parties of Warren County voters who respond to online survey

23. Population: Ages of adults in the United States who own cellular phones

    Sample: Ages of adults in the United States who own Samsung cellular phones

25. Population: Collection of the responses of all adults in the United States

    Sample: Collection of the responses of the 1000 adults surveyed

27. Population: Collection of the immunization status of all adults in the United States

    Sample: Collection of the immunization status of the 1442 adults surveyed

    Population: Collection of the opinions of all registered voters

    ple: Collection of the opinions of the 800 registered surveyed

    on: Collection of the investor role of all women in d States

    lection of the investor role of the 546 U.S. red

    ection of the responses of all Fortune 0 best companies to work for

    of the responses of the 85 companies questionnaire

    000 is a numerical description of a

37. Parameter. The 62 surviving passengers out of 97 total passengers is a numerical description of all of the passengers of the Hindenburg that survived.

39. Statistic. 8% is a numerical description of a sample of computer users.

41. Statistic. 44% is a numerical description of a sample of all people.

43. The statement "more than 56% are the primary investors in their households" is an example of descriptive statistics.

    An inference drawn from the sample is that an association exists between U.S. women and being the primary investors in their households.

45. Answers will vary.

47. (a) An inference drawn from the sample is that senior citizens who live in Florida have better memories than senior citizens who do not live in Florida.

    (b) This inference may incorrectly imply that if you live in Florida you will have a better memory.

49. Answers will vary.

Section 1.2 *(page 13)*

1. Nominal and ordinal

3. False. Data at the ordinal level can be qualitative or quantitative.

5. False. More types of calculations can be performed with data at the interval level than with data at the nominal level.

7. Qualitative, because telephone numbers are labels.

9. Quantitative, because body temperatures are numerical measurements.

11. Quantitative, because song lengths are numerical measurements.

13. Qualitative, because player numbers are labels.

15. Quantitative, because infant weights are numerical measurements.

17. Qualitative, because the poll responses are attributes.

19. Qualitative. Ordinal. Data can be arranged in order, but the differences between data entries make no sense.

21. Qualitative. Nominal. No mathematical computations can be made and data are categorized by region.

23. Qualitative. Ordinal. Data can be arranged in order, but the differences between data entries are not meaningful.

25. Ordinal   27. Nominal

29. (a) Interval   (b) Nominal   (c) Ratio   (d) Ordinal

31. An inherent zero is a zero that implies "none." Answers will vary.

Section 1.3   *(page 23)*

1. In an experiment, a treatment is applied to part of a population and responses are observed. In an observational study, a researcher measures characteristics of interest of a part of a population but does not change existing conditions.

3. In a random sample, every member of the population has an equal chance of being selected. In a simple random sample, every possible sample of the same size has an equal chance of being selected.

5. True

7. False. Using stratified sampling guarantees that members of each group within a population will be sampled.

9. False. A systematic sample is selected by ordering a population in some way and then selecting members of the population at regular intervals.

11. Use a census because all the patients are accessible and the number of patients is not too large.

13. Perform an experiment because you want to measure the effect of a treatment on the human digestive system.

15. Use a simulation because the situation is impractical and dangerous to create in real life.

17. (a) The experimental units are the 30- to 35-year-old females being given the treatment. The treatment is the new allergy drug.

    (b) A problem with the design is that there may be some bias on the part of the researcher if the researcher knows which patients were given the real drug. A way to eliminate this problem would be to make the study into a double-blind experiment.

    (c) The study would be a double-blind study if the researcher did not know which patients received the real drug or the placebo.

19. Simple random sampling is used because each telephone number has an equal chance of being dialed, and all samples of 1400 phone numbers have an equal chance of being selected. The sample may be biased because only homes with telephones will be sampled.

21. Convenience sampling is used because the students are chosen due to their convenience of location. Bias may enter into the sample because the students sampled may not be representative of the population of students.

23. Simple random sampling is used because each customer has an equal chance of being contacted, and all samples of 580 customers have an equal chance of being selected.

25. Stratified sampling is used because a sample is taken from each one-acre subplot.

27. Answers will vary.

29. Answers will vary. *Sample answer:* Treatment group: Jake, Maria, Lucy, Adam, Bridget, Vanessa, Rick, Dan, and Mary. Control group: Mike, Ron, Carlos, Steve, Susan, Kate, Pete, Judy, and Connie. A random number table is used.

31. Census, because it is relatively easy to obtain the ages of the 115 residents.

33. The question is biased because it already suggests that eating whole-grain foods improves your health. The question might be rewritten as "How does eating whole-grain foods affect your health?"

35. The survey question is unbiased.

37. The households sampled represent various locations, ethnic groups, and income brackets. Each of these variables is considered a stratum. Stratified sampling ensures that each segment of the population is represented.

39. Observational studies may be referred to as natural experiments because they involve observing naturally occurring events that are not influenced by the study.

41. (a) Advantage: Usually results in a savings in the survey cost.

    Disadvantage: There tends to be a lower response rate and this can introduce a bias into the sample. Only a certain segment of the population might respond.

    (b) Sampling technique: Convenience sampling

43. If blinding is not used, then the placebo effect is more likely to occur.

45. Both a randomized block design and a stratified sample split their members into groups based on similar characteristics.

Section 1.3 Activity   *(page 26)*

1. Answers will vary. The list contains one number at least twice.

2. The minimum is 1, the maximum is 731, and the number of samples is 8. Answers will vary.

Uses and Abuses for Chapter 1   *(page 27)*

1. Answers will vary.   2. Answers will vary.

Review Answers for Chapter 1   *(page 29)*

1. Population: Collection of the opinions of all U.S. adults about credit cards

    Sample: Collection of the opinions of the 1000 U.S. adults surveyed about credit cards

3. Population: Collection of the average annual percentage rates of all credit cards

    Sample: Collection of the average annual percentage rates of the 39 credit cards sampled

5. Parameter   7. Parameter

9. The statement "the average annual percentage rate (APR) [charged by credit cards] is 12.83%" is an example of descriptive statistics.

    An inference drawn from the sample is that all credit cards have an annual percentage rate of 12.83%.

11. Quantitative, because monthly salaries are numerical measurements.

13. Quantitative, because ages are numerical measurements.

15. Quantitative, because revenues are numerical measurements.

17. Interval. The data can be ordered and meaningful differences can be calculated, but it makes no sense to say that 100 degrees is twice as hot as 50 degrees.

19. Nominal. The data are qualitative and cannot be arranged in a meaningful order.

21. Take a census because CEOs keep accurate records of charitable donations.

23. Perform an experiment because you want to measure the effect of training dogs from animal shelters on inmates.

25. The subjects could be split into male and female and then be randomly assigned to each of the five treatment groups.

27. Answers will vary.

29. Simple random sampling is used because random telephone numbers were generated and called.

31. Cluster sampling is used because each community is considered a cluster and every pregnant woman in a selected community is surveyed.

33. Stratified sampling is used because 25 students are randomly selected from each grade level.

35. Telephone sampling samples only individuals who have telephones, who are available, and who are willing to respond.

37. The selected communities may not be representative of the entire area.

Chapter Quiz for Chapter 1 *(page 31)*

1. Population: Collection of the prostate conditions of all men

Sample: Collection of the prostate conditions of 20,000 men in study

2. (a) Statistic (b) Parameter (c) Statistic

3. (a) Qualitative (b) Quantitative

4. (a) Ordinal, because badge numbers can be ordered and often indicate seniority of service, but no meaningful mathematical computation can be performed.

  (b) Ratio, because one data value can be expressed as a multiple of another.

  (c) Ordinal, because data can be arranged in order, but the differences between data entries make no sense.

  (d) Interval, because meaningful differences between entries can be calculated but a zero entry is not an inherent zero.

5. (a) Perform an experiment because you want to measure the effect of a treatment on lead levels in adults.

  (b) Use a survey because it would be impossible to question everyone in the population.

6. Randomized block design

7. (a) Convenience sampling, because all of the people sampled are in one convenient location.

  (b) Systematic sampling, because every tenth machine part is sampled.

  (c) Stratified sampling, because the population is first stratified and then a sample is collected from each stratum.

8. Convenience sampling

Real Statistics—Real Decisions for Chapter 1 *(page 32)*

1. (a) Answers will vary. (b) Yes (c) Use surveys.

  (d) You may take too large a percentage of your sample from a subgroup of the population that is relatively small.

2. (a) Both, because questions will ask for demographics (qualitative) as well as cost (quantitative).

  (b) Gender, business/recreational: nominal
    Cost of ticket: ratio
    Comfort, safety: ordinal

  (c) Sample (d) Statistics

3. (a) Answers will vary. *Sample answer:* Sample includes only members of the population with access to the Internet.

  (b) Answers will vary.

CHAPTER 2

Section 2.1 *(page 47)*

1. Organizing the data into a frequency distribution may make patterns within the data more evident. Sometimes it is easier to identify patterns of a data set by looking at a graph of the frequency distribution.

3. Class limits determine which numbers can belong to each class.

Class boundaries are the numbers that separate classes without forming gaps between them.

5. The sum of the relative frequencies must be 1 or 100% because it is the sum of all portions or percentages of the data.

7. False. Class width is the difference between lower or upper limits of consecutive classes.

9. False. An ogive is a graph that displays cumulative frequencies.

11. Class width = 8; Lower class limits: 9, 17, 25, 33, 41, 49, 57; Upper class limits: 16, 24, 32, 40, 48, 56, 64

13. Class width = 15; Lower class limits: 17, 32, 47, 62, 77, 92, 107, 122; Upper class limits: 31, 46, 61, 76, 91, 106, 121, 136

15. (a) Class width = 11

  (b) and (c)

Class	Midpoint	Class boundaries
20–30	25	19.5–30.5
31–41	36	30.5–41.5
42–52	47	41.5–52.5
53–63	58	52.5–63.5
64–74	69	63.5–74.5
75–85	80	74.5–85.5
86–96	91	85.5–96.5

17.

Class	Frequency, f	Midpoint	Relative frequency	Cumulative frequency
20–30	19	25	0.05	19
31–41	43	36	0.12	62
42–52	68	47	0.19	130
53–63	69	58	0.19	199
64–74	74	69	0.20	273
75–85	68	80	0.19	341
86–96	24	91	0.07	365
	$\Sigma f = 365$		$\Sigma \dfrac{f}{n} \approx 1$	

19. (a) Number of classes = 7   (b) Least frequency ≈ 10

(c) Greatest frequency ≈ 300   (d) Class width = 10

21. (a) 50   (b) 22.5–23.5 pounds

23. (a) 42   (b) 29.5 pounds   (c) 35   (d) 2

25. (a) Class with greatest relative frequency: 8–9 inches

Class with least relative frequency: 17–18 inches

(b) Greatest relative frequency ≈ 0.195

Least relative frequency ≈ 0.005

(c) Approximately 0.01

27. Class with greatest frequency: 29.5–32.5

Classes with least frequency: 11.5–14.5 and 38.5–41.5

29.

Class	Frequency, f	Midpoint	Relative frequency	Cumulative frequency
0–7	8	3.5	0.32	8
8–15	8	11.5	0.32	16
16–23	3	19.5	0.12	19
24–31	3	27.5	0.12	22
32–39	3	35.5	0.12	25
	$\Sigma f = 25$		$\Sigma \dfrac{f}{n} = 1$	

Classes with greatest frequency: 0–7, 8–15

Classes with least frequency: 16–23, 24–31, 32–39

31.

Class	Frequency, f	Mid-point	Relative frequency	Cumulative frequency
1000–2019	12	1509.5	0.5455	12
2020–3039	3	2529.5	0.1364	15
3040–4059	2	3549.5	0.0909	17
4060–5079	3	4569.5	0.1364	20
5080–6099	1	5589.5	0.0455	21
6100–7119	1	6609.5	0.0455	22
	$\Sigma f = 22$		$\Sigma \dfrac{f}{N} \approx 1$	

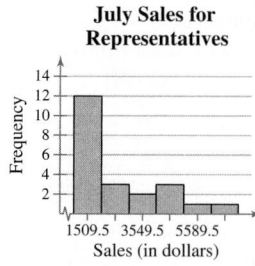

July Sales for Representatives

Frequency / Sales (in dollars)

The graph shows that most of the sales representatives at the company sold between $1000 and $2019. (Answers will vary.)

33.

Class	Frequency, f	Mid-point	Relative frequency	Cumulative frequency
291–318	5	304.5	0.1667	5
319–346	4	332.5	0.1333	9
347–374	3	360.5	0.1000	12
375–402	5	388.5	0.1667	17
403–430	6	416.5	0.2000	23
431–458	4	444.5	0.1333	27
459–486	1	472.5	0.0333	28
487–514	2	500.5	0.0667	30
	$\Sigma f = 30$		$\Sigma \dfrac{f}{n} = 1$	

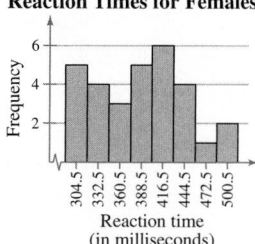

Reaction Times for Females

Frequency / Reaction time (in milliseconds)

The graph shows that the most frequent reaction times were between 403 and 430 milliseconds. (Answers will vary.)

35.

Class	Frequency, f	Mid-point	Relative frequency	Cumulative frequency
24–30	9	27	0.30	9
31–37	8	34	0.27	17
38–44	10	41	0.33	27
45–51	2	48	0.07	29
52–58	1	55	0.03	30
	$\Sigma f = 30$		$\Sigma \dfrac{f}{n} = 1$	

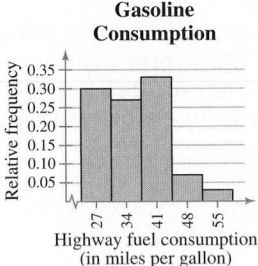

Gasoline Consumption

Highway fuel consumption (in miles per gallon)

Class with greatest relative frequency: 38–44
Class with least relative frequency: 52–58

37.

Class	Frequency, f	Mid-point	Relative frequency	Cumulative frequency
138–202	12	170	0.46	12
203–267	6	235	0.23	18
268–332	4	300	0.15	22
333–397	1	365	0.04	23
398–462	3	430	0.12	26
	$\Sigma f = 26$		$\Sigma \dfrac{f}{n} = 1$	

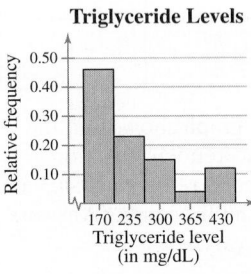

Triglyceride Levels

Triglyceride level (in mg/dL)

Class with greatest relative frequency: 138–202
Class with least relative frequency: 333–397

39.

Class	Frequency, f	Relative frequency	Cumulative frequency
52–55	3	0.125	3
56–59	3	0.125	6
60–63	9	0.375	15
64–67	4	0.167	19
68–71	4	0.167	23
72–75	1	0.042	24
	$\Sigma f = 24$	$\Sigma \dfrac{f}{n} \approx 1$	

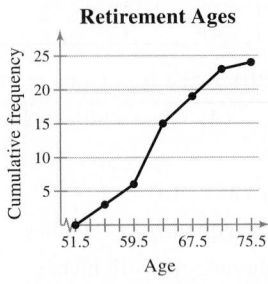

Retirement Ages

Age

Location of the greatest increase in frequency: 60–63

41.

Class	Frequency, f	Mid-point	Relative frequency	Cumulative frequency
47–57	1	52	0.05	1
58–68	1	63	0.05	2
69–79	5	74	0.25	7
80–90	8	85	0.40	15
91–101	5	96	0.25	20
	$\Sigma f = 20$		$\Sigma \dfrac{f}{N} = 1$	

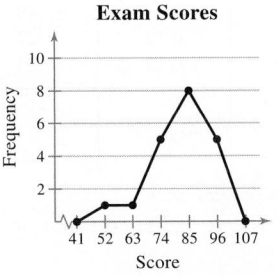

Exam Scores

Score

The graph shows that the most frequent exam scores were between 80 and 90. (Answers will vary.)

43. (a)

Class	Frequency, f	Midpoint	Relative frequency	Cumulative frequency
65–74	4	69.5	0.17	4
75–84	7	79.5	0.29	11
85–94	4	89.5	0.17	15
95–104	5	99.5	0.21	20
105–114	3	109.5	0.13	23
115–124	1	119.5	0.04	24
	$\Sigma f = 24$		$\Sigma \dfrac{f}{n} \approx 1$	

(b)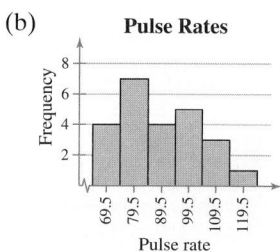
Pulse Rates

(c)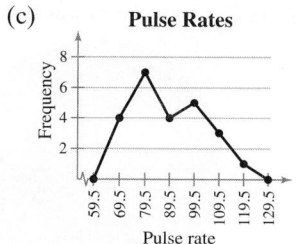
Pulse Rates

(d)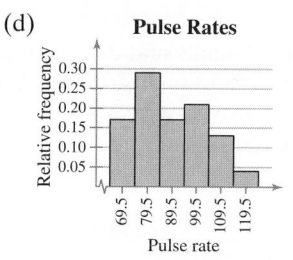
Pulse Rates

(e)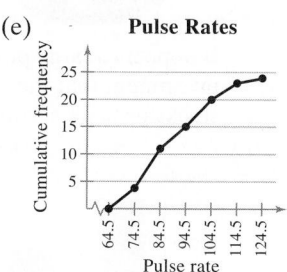
Pulse Rates

45.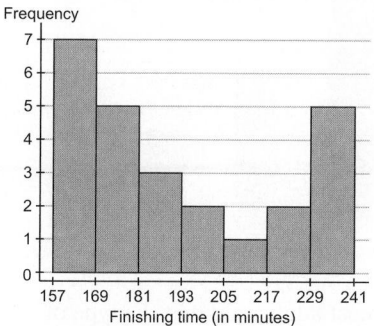
Finishing Times of Marathon Runners

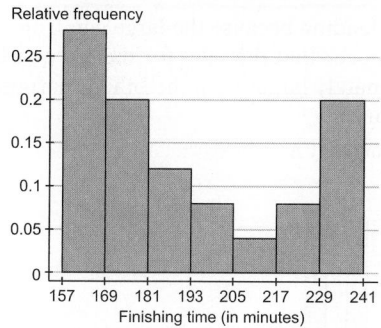
Finishing Times of Marathon Runners

47. (a)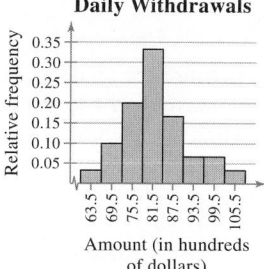
Daily Withdrawals

(b) 16.7%, because the sum of the relative frequencies for the last three classes is 0.167.

(c) $9600, because the sum of the relative frequencies for the last two classes is 0.10.

Section 2.2   *(page 60)*

1. Quantitative: stem-and-leaf plot, dot plot, histogram, scatter plot, time series chart

Qualitative: pie chart, Pareto chart

3. Both the stem-and-leaf plot and the dot plot allow you to see how data are distributed, to determine specific data entries, and to identify unusual data values.

5. b   **6.** d   **7.** a   **8.** c

9. 27, 32, 41, 43, 43, 44, 47, 47, 48, 50, 51, 51, 52, 53, 53, 53, 54, 54, 54, 54, 55, 56, 56, 58, 59, 68, 68, 68, 73, 78, 78, 85

Max: 85; Min: 27

11. 13, 13, 14, 14, 14, 15, 15, 15, 15, 15, 16, 17, 17, 18, 19

Max: 19; Min: 13

13. Answers will vary. *Sample answer:* Users spend the most amount of time on MySpace and the least amount of time on Twitter.

15. Answers will vary. *Sample answer:* Tailgaters irk drivers the most, and too-cautious drivers irk drivers the least.

17. Key: $6|7 = 67$

```
6 | 7 8
7 | 3 5 5 6 9
8 | 0 0 2 3 5 5 7 7 8
9 | 0 1 1 1 2 4 5 5
```

It appears that most grades for the biology midterm were in the 80s or 90s. (Answers will vary.)

19. Key: $4|3 = 4.3$

```
4 | 3 9
5 | 1 8 8 8 9
6 | 4 8 9 9 9
7 | 0 0 2 2 2 5
8 | 0 1
```

It appears that most ice had a thickness of 5.8 centimeters to 7.2 centimeters. (Answers will vary.)

21.
Systolic Blood Pressures

100 110 120 130 140 150 160 170 180 190 200
Systolic blood pressure (in mmHg)

It appears that systolic blood pressure tends to be between 120 and 150 millimeters of mercury. (Answers will vary.)

23.
Marathon Winners' Countries of Origin

Kenya 20%
United States 37.5%
Mexico 10%
Great Britain 2.5%
Italy 10%
New Zealand 2.5%
Brazil 5%
Ethiopia 2.5%
South Africa 5%
Tanzania 2.5%   Morocco 2.5%

Most of the New York City Marathon winners are from the United States and Kenya. (Answers will vary.)

25.

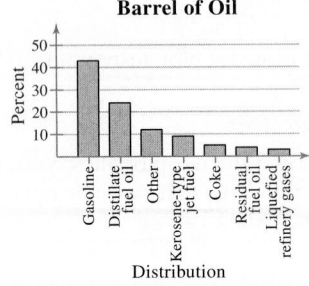

It appears that the largest portion of a 42-gallon barrel of crude oil is used for making gasoline. (Answers will vary.)

27.

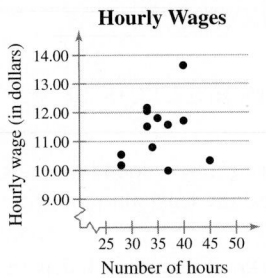

It appears that there is no relation between wages and hours worked. (Answers will vary.)

29.

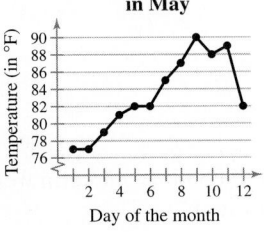

It appears that it was hottest from May 7th to May 11th. (Answers will vary.)

31. Variable: Scores

Decimal point is 1 digit(s) to the right of the colon.

5 : 5
6 : 2
6 : 8
7 : 0 1
7 : 5 6
8 : 0 2 3
8 : 5 6 7 8 8 9
9 : 0 3 3
9 : 5 5 8 9
10 : 0

It appears that most scores on the final exam in economics were in the 80s and 90s. (Answers will vary.)

33. (a)

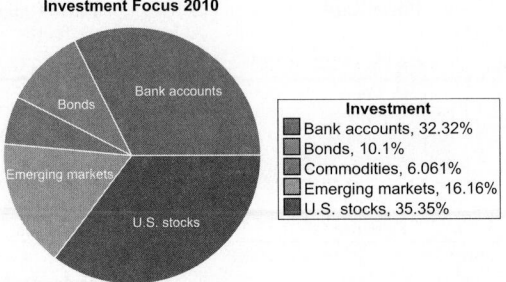

It appears a large portion of adults said that the type of investment that they would focus on in 2010 was U.S. stocks or bank accounts. (Answers will vary.)

(b)

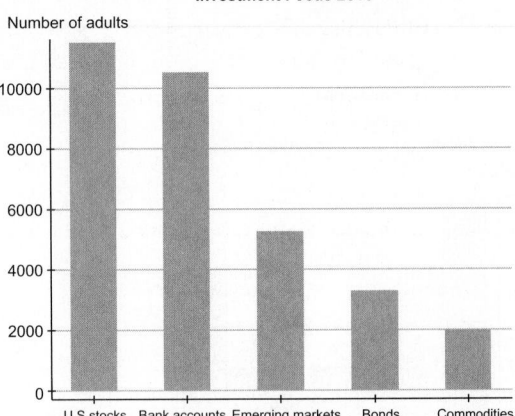

It appears that most adults said that the type of investment that they would focus on in 2010 was U.S. stocks or bank accounts. (Answers will vary.)

35. (a) The graph is misleading because the large gap from 0 to 90 makes it appear that the sales for the 3rd quarter are disproportionately larger than the other quarters. (Answers will vary.)

(b)

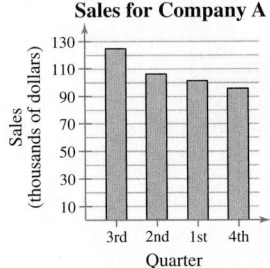

37. (a) The graph is misleading because the angle makes it appear as though the 3rd quarter had a larger percent of sales than the others, when the 1st and 3rd quarters have the same percent.

(b)

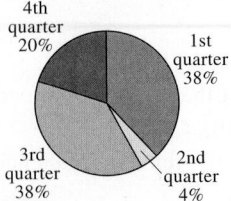

39. (a) At Law Firm A, the lowest salary was $90,000 and the highest salary was $203,000; at Law Firm B, the lowest salary was $90,000 and the highest salary was $190,000.

  (b) There are 30 lawyers at Law Firm A and 32 lawyers at Law Firm B.

  (c) At Law Firm A, the salaries tend to be clustered at the far ends of the distribution range and at Law Firm B, the salaries tend to fall in the middle of the distribution range.

Section 2.3 (page 72)

1. True   **3.** True   **5.** 1, 2, 2, 2, 3 (Answers will vary.)

7. 2, 5, 7, 9, 35 (Answers will vary.)

9. The shape of the distribution is skewed right because the bars have a "tail" to the right.

11. The shape of the distribution is uniform because the bars are approximately the same height.

13. (11), because the distribution of values ranges from 1 to 12 and has (approximately) equal frequencies.

15. (12), because the distribution has a maximum value of 90 and is skewed left due to a few students scoring much lower than the majority of the students.

17. $\bar{x} \approx 4.9$; median = 5; mode = 4

19. $\bar{x} \approx 11.0$; median = 11.0; mode = 11.7; The mode does not represent the center of the data because 11.7 is the largest number in the data set.

21. $\bar{x} \approx 21.46$; median = 21.95; mode = 20.4

23. $\bar{x}$ = not possible; median = not possible; mode = "Eyeglasses"; The mean and median cannot be found because the data are at the nominal level of measurement.

25. $\bar{x} \approx 170.63$; median = 169.3; mode = none; There is no mode because no data point is repeated.

27. $\bar{x}$ = 168.7; median = 162.5; mode = 125; The mode does not represent the center of the data because 125 is the smallest number in the data set.

29. $\bar{x} \approx 14.11$; median = 14.25; mode = 2.5; The mode does not represent the center of the data because 2.5 is much smaller than most of the data in the set.

31. $\bar{x} \approx 29.82$; median = 32; mode = 24, 35

33. $\bar{x} \approx 19.5$; median = 20; mode = 15

35. The data are skewed right.

  A = mode, because it is the data entry that occurred most often.

  B = median, because the median is to the left of the mean in a skewed right distribution.

  C = mean, because the mean is to the right of the median in a skewed right distribution.

37. Mode, because the data are at the nominal level of measurement.

39. Mean, because there are no outliers.

41. 89   **43.** $612.73   **45.** 2.8   **47.** 87

49. 36.2 miles per gallon   **51.** 35.8 years old

53.

Class	Frequency, f	Midpoint
127–161	9	144
162–196	8	179
197–231	3	214
232–266	3	249
267–301	1	284
	$\Sigma f = 24$	

Hospital Beds   Positively skewed

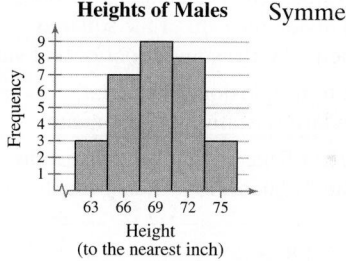

55.

Class	Frequency, f	Midpoint
62–64	3	63
65–67	7	66
68–70	9	69
71–73	8	72
74–76	3	75
	$\Sigma f = 30$	

Heights of Males   Symmetric

57. (a) $\bar{x}$ = 6.005   (b) $\bar{x}$ = 5.945
    median = 6.01   median = 6.01

  (c) Mean

59. Summary statistics:

Column	n	Mean	Median	Min	Max
Amount (in dollars)	11	112.11364	105.25	79	151.5

61. (a) $\bar{x}$ = 358, median = 375

  (b) $\bar{x}$ = 1074, median = 1125

  (c) The mean and median in part (b) are three times the mean and median in part (a).

  (d) If you multiply the mean and median from part (b) by 12, you will get the mean and median of the data set in inches.

63. Car A, because the midrange is the largest.

65. (a) 49.2

  (b) $\bar{x} \approx 49.2$; median = 46.5; mode = 36, 37, 51; midrange = 50.5

(c) Using the trimmed mean eliminates potential outliers that could affect the mean of the entries.

Section 2.3 Activity (page 79)

1. The distribution is symmetric. The mean and median both decrease slightly. Over time, the median will decrease dramatically and the mean will also decrease, but to a lesser degree.

2. Neither the mean nor the median can be any of the points that were plotted. Because there are 10 points in each output region, the mean will fall somewhere between the two regions. By the same logic, the median will be the average of the greatest point between 0 and 0.75 and the least point between 20 and 25.

Section 2.4 (page 90)

1. The range is the difference between the maximum and minimum values of a data set. The advantage of the range is that it is easy to calculate. The disadvantage is that it uses only two entries from the data set.

3. The units of variance are squared. Its units are meaningless. (Example: dollars^2)

5. $\{9, 9, 9, 9, 9, 9, 9\}$

7. When calculating the population standard deviation, you divide the sum of the squared deviations by N, then take the square root of that value. When calculating the sample standard deviation, you divide the sum of the squared deviations by $n - 1$, then take the square root of that value.

9. Similarity: Both estimate proportions of the data contained within k standard deviations of the mean.

   Difference: The Empirical Rule assumes the distribution is bell-shaped; Chebychev's Theorem makes no such assumption.

11. Range = 7, $\mu = 9$, $\sigma^2 = 4.8$, $\sigma \approx 2.2$

13. Range = 15, $\bar{x} = 12$, $s^2 \approx 21$, $s \approx 4.6$

15. 73   17. 24

19. (a) Range = 17.8   (b) Range = 39.8

21. The data set in (a) has a standard deviation of 24 and the data set in (b) has a standard deviation of 16, because the data in (a) have more variability.

23. Company B; An offer of $33,000 is two standard deviations from the mean of Company A's starting salaries, which makes it unlikely. The same offer is within one standard deviation of the mean of Company B's starting salaries, which makes the offer likely.

25. (a) Dallas: $\bar{x} \approx 44.28$; median = 44.7; range = 11.3; $s^2 \approx 18.33$; $s \approx 4.28$

   New York City: $\bar{x} \approx 50.91$; median = 50.6; range = 17.8; $s^2 \approx 50.36$; $s \approx 7.10$

   (b) It appears from the data that the annual salaries in New York City are more variable than the annual salaries in Dallas. The annual salaries in Dallas have a lower mean and a lower median than the annual salaries in New York City.

27. (a) Males: $\bar{x} \approx 1643$; median = 1679.5; range = 1087; $s^2 \approx 116,477.4$; $s \approx 341.3$

   Females: $\bar{x} \approx 1709.1$; median = 1686.5; range = 947; $s^2 \approx 91,625.0$; $s \approx 302.7$

   (b) It appears from the data that the SAT scores for males are more variable than the SAT scores for females. The SAT scores for males have a lower mean and median than the SAT scores for females.

29. (a) Greatest sample standard deviation: (ii)

   Data set (ii) has more entries that are farther away from the mean.

   Least sample standard deviation: (iii)

   Data set (iii) has more entries that are close to the mean.

   (b) The three data sets have the same mean but have different standard deviations.

31. (a) Greatest sample standard deviation: (ii)

   Data set (ii) has more entries that are farther away from the mean.

   Least sample standard deviation: (iii)

   Data set (iii) has more entries that are close to the mean.

   (b) The three data sets have the same mean, median, and mode but have different standard deviations.

33. 68%   35. (a) 51   (b) 17

37. $2180, $1000, $2000, $950; $2180 is very unusual because it is more than 3 standard deviations from the mean.

39. 24   41. $\bar{x} \approx 2.1$, $s \approx 1.3$

43.

Class	Midpoint, x	f	xf
1–3	2	3	6
4–6	5	6	30
7–9	8	13	104
10–12	11	7	77
13–15	14	3	42
		$N = 32$	$\Sigma xf = 259$

$x - \mu$	$(x - \mu)^2$	$(x - \mu)^2 f$
−6.1	37.21	111.63
−3.1	9.61	57.66
−0.1	0.01	0.13
2.9	8.41	58.87
5.9	34.81	104.43
		$\Sigma (x - \mu)^2 f = 332.72$

$\mu \approx 8.1$

$\sigma \approx 3.2$

45.

Midpoint, x	f	xf	$x - \bar{x}$	$(x - \bar{x})^2$	$(x - \bar{x})^2 f$
70.5	1	70.5	−44	1936	1936
92.5	12	1110.0	−22	484	5808
114.5	25	2862.5	0	0	0
136.5	10	1365.0	22	484	4840
158.5	2	317.0	44	1936	3872
	$n = 50$	$\sum xf = 5725$			$\sum (x - \bar{x})^2 f = 16{,}456$

$\bar{x} = 114.5$

$s \approx 18.33$

47.

Class	Midpoint, x	f	xf
0–4	2.0	22.1	44.20
5–14	9.5	43.4	412.30
15–19	17.0	21.2	360.40
20–24	22.0	22.3	490.60
25–34	29.5	44.5	1312.75
35–44	39.5	41.3	1631.35
45–64	54.5	83.9	4572.55
65+	70.0	46.8	3276.00
		$n = 325.5$	$\sum xf = 12{,}100.15$

$x - \bar{x}$	$(x - \bar{x})^2$	$(x - \bar{x})^2 f$
−35.17	1236.93	27,336.15
−27.67	765.63	33,228.34
−20.17	406.83	8624.80
−15.17	230.13	5131.90
−7.67	58.83	2617.94
2.33	5.43	224.26
17.33	300.33	25,197.69
32.83	1077.81	50,441.51
	$\sum (x - \bar{x})^2 f = 152{,}802.59$	

$\bar{x} \approx 37.17$

$s \approx 21.70$

49. Summary statistics:

Column	n	Mean	Variance
Amount (in dollars)	15	58.8	239.74286

Std. Dev.	Median	Range	Min	Max
15.483632	60	59	30	89

51. $CV_{\text{heights}} = \dfrac{3.29}{72.75} \cdot 100\% \approx 4.5\%$

$CV_{\text{weights}} = \dfrac{17.69}{187.83} \cdot 100\% \approx 9.4\%$

It appears that weight is more variable than height.

53. (a) $\bar{x} \approx 41.5$, $s \approx 5.3$

(b) $\bar{x} \approx 43.6$, $s \approx 5.6$

(c) $\bar{x} \approx 3.5$, $s \approx 0.4$

(d) When each entry is multiplied by a constant k, the new sample mean is $k \cdot \bar{x}$, and the new sample standard deviation is $k \cdot s$.

55. (a) Males: 249, Females: 245.4; The mean absolute deviation is less than the sample standard deviation.

(b) Team A: 0.0315, Team B: 0.0199; The mean absolute deviation is less than the sample standard deviation.

57. (a) $P \approx -2.61$

The data are skewed left.

(b) $P \approx 4.12$

The data are skewed right.

(c) $P = 0$

The data are symmetric.

(d) $P = 1$

The data are skewed right.

Section 2.4 Activity *(page 98)*

1. When a point with a value of 15 is added, the mean remains constant and the standard deviation decreases; When a point with a value of 20 is added, the mean is raised and the standard deviation increases. (Answers will vary.)

2. To get the largest standard deviation, plot four of the points at 30 and four of the points at 40; To get the smallest standard deviation, plot all of the points at the same number.

Section 2.5 *(page 107)*

1. The soccer team scored fewer points per game than 75% of the teams in the league.

3. The student scored higher than 78% of the students who took the actuarial exam.

5. The interquartile range of a data set can be used to identify outliers because data values that are greater than $Q_3 + 1.5(\text{IQR})$ or less than $Q_1 - 1.5(\text{IQR})$ are considered outliers.

7. False. The median of a data set is a fractile, but the mean may or may not be a fractile depending on the distribution of the data.

9. True

11. False. The 50th percentile is equivalent to Q_2.

13. False. A z-score of -2.5 is considered unusual.

15. (a) Min = 10, $Q_1 = 13$, $Q_2 = 15$, $Q_3 = 17$, Max = 20

(b) IQR = 4

17. (a) Min = 900, $Q_1 = 1250$, $Q_2 = 1500$, $Q_3 = 1950$, Max = 2100

(b) IQR = 700

19. (a) Min = −1.9, $Q_1 = -0.5$, $Q_2 = 0.1$, $Q_3 = 0.7$, Max = 2.1

(b) IQR = 1.2

21. (a) Min = 24, Q_1 = 28, Q_2 = 35, Q_3 = 41, Max = 60

(b)

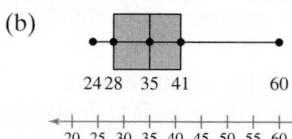

23. (a) Min = 1, Q_1 = 4.5, Q_2 = 6, Q_3 = 7.5, Max = 9

(b)

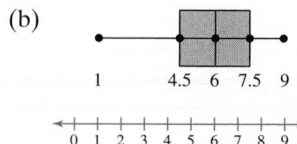

25. None. The data are not skewed or symmetric.

27. Skewed left. Most of the data lie to the right on the box plot.

29. Q_1 = B, Q_2 = A, Q_3 = C, because about one quarter of the data fall on or below 17, 18.5 is the median of the entire data set, and about three quarters of the data fall on or below 20.

31. (a) Q_1 = 2, Q_2 = 4, Q_3 = 5

(b) **Watching Television**

Number of hours

33. (a) Q_1 = 3, Q_2 = 3.85, Q_3 = 5.2

(b) **Airplane Distances**

Distance (in miles)

35. (a) 5 (b) 50% (c) 25%

37. A → z = −1.43

B → z = 0

C → z = 2.14

A z-score of 2.14 would be unusual.

39. (a) Statistics: $z = \dfrac{75 - 63}{7} \approx 1.71$

Biology: $z = \dfrac{25 - 23}{3.9} \approx 0.51$

(b) The student did better on the statistics test.

41. (a) Statistics: $z = \dfrac{78 - 63}{7} \approx 2.14$

Biology: $z = \dfrac{29 - 23}{3.9} \approx 1.54$

(b) The student did better on the statistics test.

43. (a) $z_1 = \dfrac{34{,}000 - 35{,}000}{2250} \approx -0.44$

$z_2 = \dfrac{37{,}000 - 35{,}000}{2250} \approx 0.89$

$z_3 = \dfrac{30{,}000 - 35{,}000}{2250} \approx -2.22$

The tire with a life span of 30,000 miles has an unusually short life span.

(b) For 30,500, 2.5th percentile

For 37,250, 84th percentile

For 35,000, 50th percentile

45. 72 inches; 60% of the heights are below 72 inches.

47. $z_1 = \dfrac{74 - 69.9}{3.0} \approx 1.37$

$z_2 = \dfrac{62 - 69.9}{3.0} \approx -2.63$

$z_3 = \dfrac{80 - 69.9}{3.0} \approx 3.37$

The heights of 62 and 80 inches are unusual.

49. $z = \dfrac{71.1 - 69.9}{3.0} = 0.4$

About the 50th percentile

51. (a) Min = 27, Q_1 = 42, Q_2 = 49, Q_3 = 56, Max = 82

(b) **Ages of Executives**

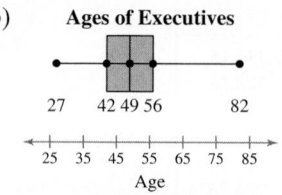

Age

(c) Half of the executives are between 42 and 56 years old.

(d) 49, because half of the executives are older and half are younger.

(e) The age groups 20–29, 70–79, and 80–89 would all be considered unusual because they lie more than two standard deviations from the mean.

53. 33.75 **55.** 19.8

57. **Credit Card Purchases**

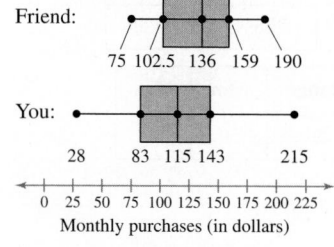

Monthly purchases (in dollars)

The shape of your bill is symmetric, and the shape of your friend's bill is uniform.

59. 40th percentile

61. (a) 62, 95

(b)

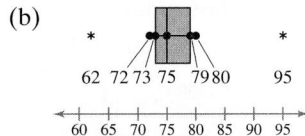

63. (a) **Summary statistics:**

Column	Min	Q1	Median	Q3	Max
Weight (in pounds)	165	230	262.5	294	395

(b) **Weights of Professional Football Players**

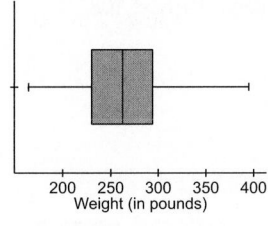

(c) **Weights of Professional Football Players**

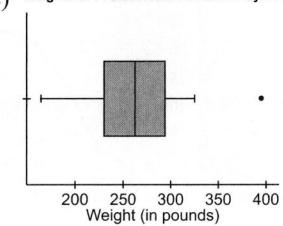

Uses and Abuses for Chapter 2   *(page 113)*

1. Answers will vary.

2. No, it is not ethical because it misleads the consumer to believe that oatmeal is more effective at lowering cholesterol than it may actually be.

Review Answers for Chapter 2   *(page 115)*

1.

Class	Mid-point	Boundaries	Frequency, f	Relative frequency	Cumulative frequency
8–12	10	7.5–12.5	2	0.10	2
13–17	15	12.5–17.5	10	0.50	12
18–22	20	17.5–22.5	5	0.25	17
23–27	25	22.5–27.5	1	0.05	18
28–32	30	27.5–32.5	2	0.10	20
			$\sum f = 20$	$\sum \dfrac{f}{n} = 1$	

3.   **Liquid Volume 12-oz Cans**

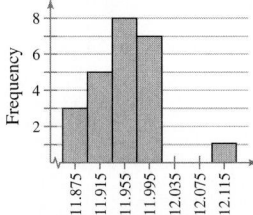

Actual volume (in ounces)

5.

Class	Midpoint	Frequency, f
79–93	86	9
94–108	101	12
109–123	116	5
124–138	131	3
139–153	146	2
154–168	161	1
		$\sum f = 32$

Rooms Reserved

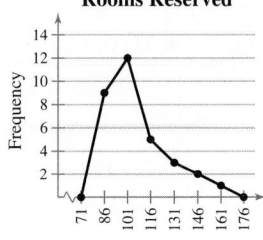

Number of rooms

7.

```
1 | 0 0              Key: 1|0 = 10
2 | 0 0 2 5 5
3 | 0 3 4 5 5 8
4 | 1 2 4 4 7 8
5 | 2 3 3 7 9
6 | 1 1 5
7 | 1 5
8 | 9
```

9.   **Heights of Buildings**

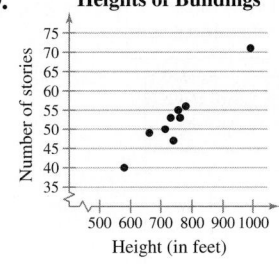

Height (in feet)

The number of stories appears to increase with height.

11.

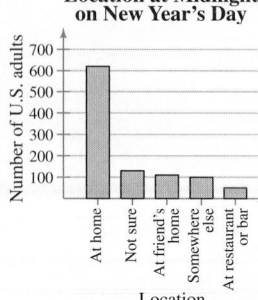

Location at Midnight on New Year's Day

13. $\bar{x} = 29.15$; median $= 29.5$; mode $= 29.5$

15. 17.8   **17.** 82.1   **19.** Skewed   **21.** Skewed left

23. Median; When a distribution is skewed left, the mean is to the left of the median.

25. \$2.80   **27.** $\mu \approx 6.9, \sigma \approx 4.6$

29. $\bar{x} = 2453.4, s \approx 306.1$

31. Between \$41.50 and \$56.50   **33.** 30 customers

35. $\bar{x} \approx 2.5, s \approx 1.2$

37. Min $= 42, Q_1 = 47.5, Q_2 = 53, Q_3 = 54$, Max $= 60$

39.

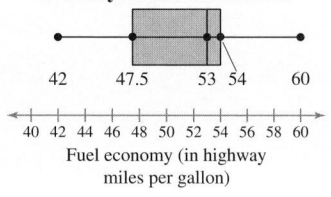

Motorcycle Fuel Economies

Fuel economy (in highway miles per gallon)

41. 4.5   **43.** 35% scored higher than 75.

45. Not unusual   **47.** Unusual

Chapter Quiz for Chapter 2 *(page 119)*

1. (a)

Class	Midpoint	Class boundaries
101–112	106.5	100.5–112.5
113–124	118.5	112.5–124.5
125–136	130.5	124.5–136.5
137–148	142.5	136.5–148.5
149–160	154.5	148.5–160.5

Frequency, f	Relative frequency	Cumulative frequency
3	0.12	3
11	0.44	14
7	0.28	21
2	0.08	23
2	0.08	25

(b) Frequency histogram and polygon

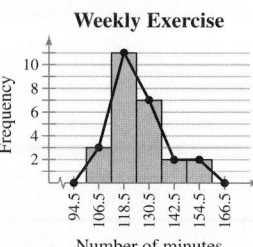

Weekly Exercise

(c) Relative frequency histogram

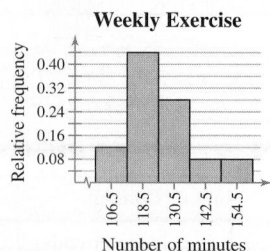

Weekly Exercise

(d) Skewed

(e)

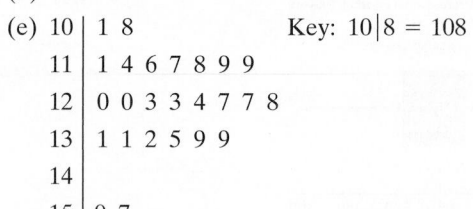

```
10 | 1 8              Key: 10|8 = 108
11 | 1 4 6 7 8 9 9
12 | 0 0 3 3 4 7 7 8
13 | 1 1 2 5 9 9
14 |
15 | 0 7
```

(f)

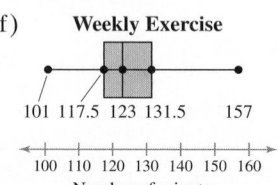

Weekly Exercise

(g)

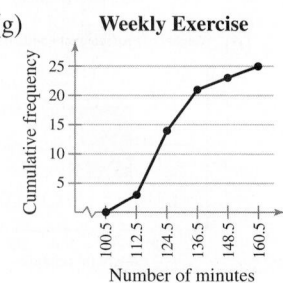

Weekly Exercise

2. 125.2, 13.0

3. (a)

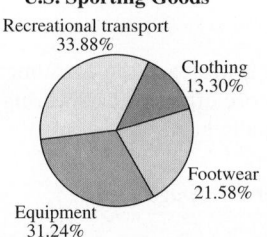

U.S. Sporting Goods

Recreational transport 33.88%
Clothing 13.30%
Footwear 21.58%
Equipment 31.24%

(b)

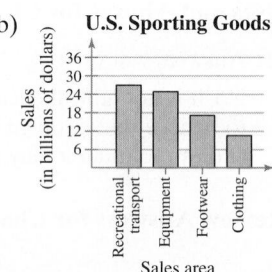

U.S. Sporting Goods

4. (a) $\bar{x} \approx 751.6$; median $= 784.5$; mode $=$ none

The mean best describes a typical salary because there are no outliers.

(b) Range $= 575; s^2 \approx 48,135.1; s \approx 219.4$

5. Between \$125,000 and \$185,000

6. (a) $z = 3.0$, unusual   (b) $z \approx -6.67$, very unusual
(c) $z \approx 1.33$ (d) $z = -2.2$, unusual

7. (a) Min $= 59, Q_1 = 74, Q_2 = 83.5, Q_3 = 88$, Max $= 103$
(b) 14
(c)

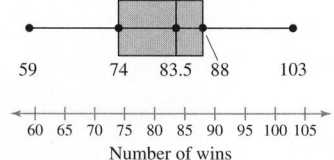

Wins for Each Team

Number of wins

Real Statistics–Real Decisions for Chapter 2   *(page 120)*

1. (a) Find the average cost of renting an apartment for each area and do a comparison.

  (b) The mean would best represent the data sets for the four areas of the city.

  (c) Area A: $\bar{x} = \$1005.50$

   Area B: $\bar{x} = \$887.00$

   Area C: $\bar{x} = \$881.00$

   Area D: $\bar{x} = \$945.50$

2. (a) Construct a Pareto chart, because the data are quantitative and a Pareto chart positions data in order of decreasing height, with the tallest bar positioned at the left.

  (b)

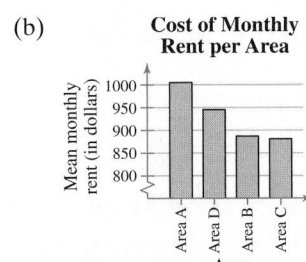

Cost of Monthly Rent per Area

  (c) Yes. From the Pareto chart you can see that Area A has the highest average cost of monthly rent, followed by Area D, Area B, and Area C.

3. (a) You could use the range and sample standard deviation for each area.

  (b)

Area A	*Area B*
$s \approx \$123.07$	$s \approx \$144.91$
range $= \$415.00$	range $= \$421.00$

Area C	*Area D*
$s \approx \$146.21$	$s \approx \$138.70$
range $= \$460.00$	range $= \$497.00$

  (c) No. Area A has the lowest range and standard deviation, so the rents in Areas B–D are more spread out. There could be one or two inexpensive rents that lower the means for these areas. It is possible that the population means of Areas B–D are close to the populations mean of Area A.

4. (a) Answers will vary.

  (b) Location, weather, population

Cumulative Review Answers for Chapters 1–2   *(page 124)*

1. Systematic sampling. A bias may enter this study if the machine makes a consistent error.

2. Random sampling. A bias of this type of study is that the researchers did not include people without telephones.

3.

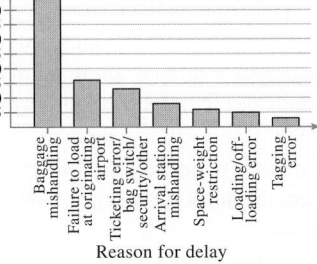

Reason for Baggage Delay

4. Parameter. All Major League Baseball players are included.

5. Statistic. The 19% is a numerical description of the 1000 voters surveyed in the United States.

6. (a) 95%   (b) 38

  (c) $z_1 = \dfrac{90{,}500 - 83{,}500}{1500} \approx 4.67$

   $z_2 = \dfrac{79{,}750 - 83{,}500}{1500} = -2.5$

   $z_3 = \dfrac{82{,}600 - 83{,}500}{1500} = -0.6$

   The salaries of \$90,500 and \$79,750 are unusual.

7. Population: Collection of the career interests of all college and university students

   Sample: Collection of the career interests of the 195 college and university students whose career counselors were surveyed

8. Population: Collection of the life spans of all people

   Sample: Collection of the life spans of the 232,606 people in the study

9. Census. There are only 100 members in the Senate.

10. Experiment. An experiment could compare a control group that has recess and a treatment group that has recess removed.

11. Quantitative. The data are at the ratio level.

12. Qualitative. The data are at the nominal level.

13. (a) Min $= 0$, $Q_1 = 2$, $Q_2 = 12.5$, $Q_3 = 39$, Max $= 136$

  (b) **Number of Tornadoes by State**

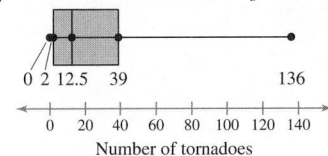

  (c) The distribution of the number of tornadoes is skewed right.

14. 88.9

15. (a) $\bar{x} \approx 5.49$; median = 5.4; mode = none; Both the mean and the median accurately describe a typical American alligator tail length. (Answers will vary.)

(b) Range = 4.1; $s^2 = 2.34$; $s = 1.53$; The maximum difference in alligator tail lengths is about 4.1 feet, and about 68% of alligator tail lengths will fall between 3.96 feet and 7.02 feet.

16. (a) An inference drawn from the sample is that the number of deaths due to heart disease for women will continue to decrease.

(b) This inference may incorrectly imply that women will have less of a chance of dying of heart disease in the future.

17.

Class	Class boundaries	Midpoint
0–8	−0.5–8.5	4
9–17	8.5–17.5	13
18–26	17.5–26.5	22
27–35	26.5–35.5	31
36–44	35.5–44.5	40
45–53	44.5–53.5	49
54–62	53.5–62.5	58
63–71	62.5–71.5	67

Frequency, f	Relative frequency	Cumulative frequency
8	0.27	8
5	0.17	13
7	0.23	20
3	0.10	23
4	0.13	27
1	0.03	28
0	0.00	28
2	0.07	30
$\Sigma f = 30$	$\Sigma \dfrac{f}{n} = 1$	

18. The distribution is skewed right.

19.

Montreal Canadiens Points Scored

Relative frequency vs. Number of points scored (per player)

Class with greatest frequency: 0–8

Class with least frequency: 54–62

CHAPTER 3

Section 3.1 (page 138)

1. An outcome is the result of a single trial in a probability experiment, whereas an event is a set of one or more outcomes.

3. The probability of an event cannot exceed 100%.

5. The law of large numbers states that as an experiment is repeated over and over, the probabilities found in the experiment will approach the actual probabilities of the event. Examples will vary.

7. False. If you roll a six-sided die six times, the probability of rolling an even number at least once is approximately 0.984.

9. False. A probability of less than 0.05 indicates an unusual event.

11. b   **12.** d   **13.** c   **14.** a

15. {A, B, C, D, E, F, G, H, I, J, K, L, M, N, O, P, Q, R, S, T, U, V, W, X, Y, Z}; 26

17. {A♥, K♥, Q♥, J♥, 10♥, 9♥, 8♥, 7♥, 6♥, 5♥, 4♥, 3♥, 2♥, A♦, K♦, Q♦, J♦, 10♦, 9♦, 8♦, 7♦, 6♦, 5♦, 4♦, 3♦, 2♦, A♠, K♠, Q♠, J♠, 10♠, 9♠, 8♠, 7♠, 6♠, 5♠, 4♠, 3♠, 2♠, A♣, K♣, Q♣, J♣, 10♣, 9♣, 8♣, 7♣, 6♣, 5♣, 4♣, 3♣, 2♣}; 52

19.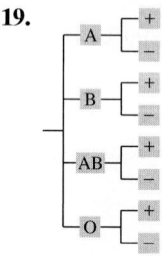

{(A, +), (A, −), (B, +), (B, −), (AB, +), (AB, −), (O, +), (O, −)}, where (A, +) represents positive Rh-factor with blood type A and (A, −) represents negative Rh-factor with blood type A; 8.

21. 1; Simple event because it is an event that consists of a single outcome.

23. 4; Not a simple event because it is an event that consists of more than a single outcome.

25. 204   **27.** 4500   **29.** 0.083   **31.** 0.667   **33.** 0.417

35. Empirical probability because company records were used to calculate the frequency of a washing machine breaking down.

37. 0.159   **39.** 0.000953   **41.** 0.042; Yes   **43.** 0.208; No

45. (a) 1000   (b) 0.001   (c) 0.999

47. {(SSS), (SSR), (SRS), (SRR), (RSS), (RSR), (RRS), (RRR)}

49. {(SSR), (SRS), (RSS)}

51. (a)

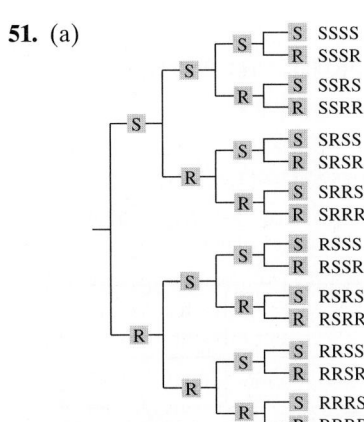

(b) {(SSSS), (SSSR), (SSRS), (SSRR), (SRSS), (SRSR), (SRRS), (SRRR), (RSSS), (RSSR), (RSRS), (RSRR), (RRSS), (RRSR), (RRRS), (RRRR)}

(c) {(SSSR), (SSRS), (SRSS), (RSSS)}

53. 0.399 **55.** 0.040 **57.** 0.936 **59.** 0.033 **61.** 0.275

63. Yes; The event in Exercise 55 can be considered unusual because its probability is 0.05 or less.

65. (a) 0.5 (b) 0.25 (c) 0.25

67. 0.795 **69.** 0.205

71. (a) 0.225 (b) 0.133

(c) 0.017; This event is unusual because its probability is 0.05 or less.

73. The probability of randomly choosing a tea drinker who does not have a college degree

75. (a)

Sum	Probability
2	0.028
3	0.056
4	0.083
5	0.111
6	0.139
7	0.167
8	0.139
9	0.111
10	0.083
11	0.056
12	0.028

(b) Answers will vary.

(c) Answers will vary.

77. The first game; The probability of winning the second game is $\frac{1}{11} \approx 0.091$, which is less than $\frac{1}{10}$.

79. $13 : 39 = 1 : 3$

81. p = number of successful outcomes

q = number of unsuccessful outcomes

$$P(A) = \frac{\text{number of successful outcomes}}{\text{total number of outcomes}} = \frac{p}{p + q}$$

Section 3.1 Activity   *(page 144)*

1–2. Answers will vary.

Section 3.2   *(page 150)*

1. Two events are independent if the occurrence of one of the events does not affect the probability of the occurrence of the other event, whereas two events are dependent if the occurrence of one of the events does affect the probability of the occurrence of the other event.

3. The notation $P(B|A)$ means the probability of B, given A.

5. False. If two events are independent, then $P(A|B) = P(A)$.

7. Independent. The outcome of the first draw does not affect the outcome of the second draw.

9. Dependent. The outcome of a father having hazel eyes affects the outcome of a daughter having hazel eyes.

11. Dependent. The sum of the rolls depends on which numbers came up on the first and second rolls.

13. Events: moderate to severe sleep apnea, high blood pressure; Dependent. People with moderate to severe sleep apnea are more likely to have high blood pressure.

15. Events: exposure to aluminum, Alzheimer's disease; Independent. Exposure to everyday sources of aluminum does not cause Alzheimer's disease.

17. (a) 0.6 (b) 0.001

(c) Dependent.

P(developing breast cancer|gene)

$\neq$ (developing breast cancer)

19. (a) 0.308 (b) 0.788 (c) 0.757 (d) 0.596

(e) Dependent.

P(taking a summer vacation|family owns a computer)

$\neq P$(taking a summer vacation)

21. (a) 0.093 (b) 0.75

(c) No, the probability is not unusual because it is not less than or equal to 0.05.

23. 0.745

25. (a) 0.017 (b) 0.757 (c) 0.243

(d) The event in part (a) is unusual because its probability is less than or equal to 0.05.

27. (a) 0.481 (b) 0.465 (c) 0.449

(d) Dependent.

P(having less than one month's income saved|being male)

$\neq P$(having less than one month's income saved)

29. (a) 0.00000590 (b) 0.624 (c) 0.376

31. (a) 0.25 (b) 0.063 (c) 0.000977

(d) 0.237 (e) 0.763

33. (a) 0.011 (b) 0.458 **35.** 0.444

37. 0.167 **39.** (a) 0.074 (b) 0.999 **41.** 0.954

Section 3.3   *(page 161)*

1. $P(A \text{ and } B) = 0$ because A and B cannot occur at the same time.

3. True

5. False. The probability that event *A* or event *B* will occur is $P(A \text{ or } B) = P(A) + P(B) - P(A \text{ and } B)$.

7. Not mutually exclusive. A student can be an athlete and on the Dean's list.

9. Not mutually exclusive. A public school teacher can be female and 25 years old.

11. Mutually exclusive. A student cannot have a birthday in both months.

13. (a) Not mutually exclusive. For five weeks the events overlapped.
   (b) 0.423

15. (a) Not mutually exclusive. A carton can have a puncture and a smashed corner.
   (b) 0.126

17. (a) 0.308   (b) 0.538   (c) 0.308

19. (a) 0.067   (b) 0.839   (c) 0.199

21. (a) 0.949   (b) 0.388

23. (a) 0.573   (b) 0.962   (c) 0.573
   (d) Not mutually exclusive. A male can be a nursing major.

25. (a) 0.461   (b) 0.762   (c) 0.589   (d) 0.922
   (e) Not mutually exclusive. A female can be frequently involved in charity work.

27. Answers will vary.   **29.** 0.55

Section 3.3 Activity   *(page 166)*

1. 0.333   **2.** Answers will vary.

3. The theoretical probability is 0.5, so the green line should be placed there.

Section 3.4   *(page 174)*

1. The number of ordered arrangements of *n* objects taken *r* at a time. An example of a permutation is the number of seating arrangements of you and three of your friends.

3. False. A permutation is an ordered arrangement of objects.

5. True   **7.** 15,120   **9.** 56   **11.** 203,490   **13.** 0.030

15. Permutation. The order of the eight cars in line matters.

17. Combination. The order does not matter because the position of one captain is the same as the other.

19. 5040   **21.** 720   **23.** 20,358,520   **25.** 320,089,770

27. 50,400   **29.** 6240   **31.** 86,296,950

33. (a) 720   (b) sample
   (c) 0.0014; Yes, the event can be considered unusual because its probability is less than or equal to 0.05.

35. (a) 12   (b) tree
   (c) 0.083; No, the event cannot be considered unusual because its probability is not less than or equal to 0.05.

37. (a) 907,200   (b) population
   (c) 0.000001; Yes, the event can be considered unusual because its probability is less than or equal to 0.05.

39. 0.005   **41.** (a) 0.016   (b) 0.385

43. (a) 70   (b) 16   (c) 0.086

45. (a) 67,600,000   (b) 19,656,000   (c) 0.000000015

47. (a) 120   (b) 12   (c) 12   (d) 0.4

49. 0.000022   **51.** 6.00×10^{-20}

53. (a) 658,008   (b) 0.00000152

55. (a) 0.0002   (b) 0.0014   (c) 0.0211   (d) 0.0659

57. 1001; 1000

59.

Team (worst team first)	1	2	3	4	5
Probability	0.250	0.199	0.156	0.119	0.088

Team (worst team first)	6	7	8	9	10
Probability	0.063	0.043	0.028	0.017	0.011

Team (worst team first)	11	12	13	14
Probability	0.008	0.007	0.006	0.005

Events in which any of Teams 7–14 win the first pick would be considered unusual because the probabilities are all less than or equal to 0.05.

61. 0.314

Uses and Abuses for Chapter 3   *(page 179)*

1. (a) 0.000001   (b) 0.001   (c) 0.001

2. The probability that a randomly chosen person owns a pickup or an SUV can equal 0.55 if no one in the town owns both a pickup and an SUV. The probability cannot equal 0.60 because 0.60 > 0.25 + 0.30. (Answers will vary.)

Review Exercises for Chapter 3   *(page 181)*

1. Sample space:

{HHHH, HHHT, HHTH, HHTT, HTHH, HTHT, HTTH, HTTT, THHH, THHT, THTH, THTT, TTHH, TTHT, TTTH, TTTT}; 4

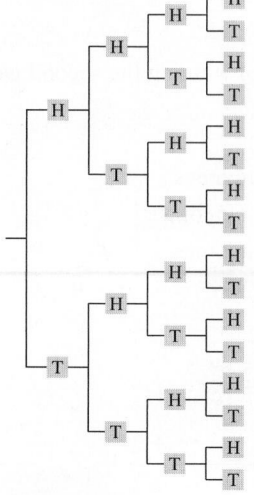

3. Sample space:
{January, February, March, April, May, June, July, August, September, October, November, December}; 3

5. 84

7. Empirical probability because it is based on observations obtained from probability experiments.

9. Subjective probability because it is based on opinion.

11. Classical probability because all of the outcomes in the event and the sample space can be counted.

13. 0.215   **15.** 1.25×10^{-7} **17.** 0.92

19. Independent. The outcomes of the first four coin tosses do not affect the outcome of the fifth coin toss.

21. Dependent. The outcome of getting high grades affects the outcome of being awarded an academic scholarship.

23. 0.025; Yes, the event is unusual because its probability is less than or equal to 0.05.

25. Mutually exclusive. A jelly bean cannot be both completely red and completely yellow.

27. Mutually exclusive. A person cannot be registered to vote in more than one state.

29. 0.60   **31.** 0.538   **33.** 0.583   **35.** 0.291

37. 0.188   **39.** 0.703   **41.** 110   **43.** 35

45. 254,251,200   **47.** 2730   **49.** 2380

51. 0.00000923; unusual

53. (a) 0.955; not unusual   (b) 0.000000761; unusual
(c) 0.045; unusual   (d) 0.999999239; not unusual

55. (a) 0.071; not unusual   (b) 0.005; unusual
(c) 0.429; not unusual   (d) 0.114; not unusual

Chapter Quiz for Chapter 3   *(page 185)*

1. (a) 0.523   (b) 0.508   (c) 0.545   (d) 0.772
(e) 0.025   (f) 0.673   (g) 0.094   (h) 0.574

2. The event in part (e) is unusual because its probability is less than or equal to 0.05.

3. Not mutually exclusive. A golfer can score the best round in a four-round tournament and still lose the tournament.

Dependent. One event can affect the occurrence of the second event.

4. (a) 2,481,115   (b) 1   (c) 2,572,999

5. (a) 0.964   (b) 0.000000389   (c) 0.9999996

6. 450,000   **7.** 657,720

Real Statistics–Real Decisions for Chapter 3   *(page 186)*

1. (a) Answers will vary.
(b) Use the Multiplication Rule, Fundamental Counting Principle, and combinations.

2. If you played only the red ball, the probability of matching it is $\frac{1}{39}$. However, because you must pick five white balls, you must get the white balls wrong. So, using the Multiplication Rule, you get

P(matching only the red
ball and not matching any $= \frac{1}{39} \cdot \frac{54}{59} \cdot \frac{53}{58} \cdot \frac{52}{57} \cdot \frac{51}{56} \cdot \frac{50}{55}$
of the five white balls)

≈ 0.016

$\approx \frac{1}{62}.$

3. The overall probability of winning a prize is determined by calculating the number of ways to win and dividing by the total number of outcomes.

To calculate the number of ways to win something, you must use combinations.

CHAPTER 4

Section 4.1   *(page 197)*

1. A random variable represents a numerical value associated with each outcome of a probability experiment.
Examples: Answers will vary.

3. No; Expected value may not be a possible value of x for one trial, but it represents the average value of x over a large number of trials.

5. False. In most applications, discrete random variables represent counted data, while continuous random variables represent measured data.

7. True

9. Discrete; Attendance is a random variable that is countable.

11. Continuous; Distance traveled is a random variable that must be measured.

13. Discrete; The number of books in a library is a random variable that is countable.

15. Continuous; The volume of blood drawn for a blood test is a random variable that must be measured.

17. Discrete; The number of messages posted each month on a social networking site is a random variable that is countable.

19. Continuous; The amount of snow that fell in Nome, Alaska last winter is a random variable that cannot be counted.

21. (a) 0.35   (b) 0.90   **23.** 0.22   **25.** Yes

27. (a)

x	$P(x)$
0	0.686
1	0.195
2	0.077
3	0.022
4	0.013
5	0.006
	$\sum P(x) \approx 1$

(b)

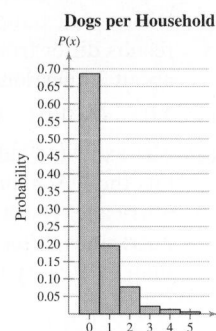

Dogs per Household

Skewed right

(c) 0.5, 0.8, 0.9

(d) The mean is 0.5, so the average number of dogs per household is about 0 or 1 dog. The standard deviation is 0.9, so most of the households differ from the mean by no more than about 1 dog.

29. (a)

x	P(x)
0	0.01
1	0.17
2	0.28
3	0.54

(b)

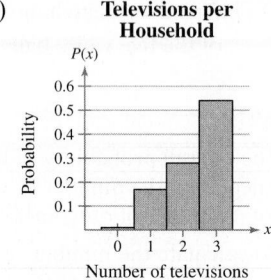

Televisions per Household

Skewed left

(c) 2.4, 0.6, 0.8

(d) The mean is 2.4, so the average household in the town has about 2 televisions. The standard deviation is 0.8, so most of the households differ from the mean by no more than about 1 television.

31. (a)

x	P(x)
0	0.031
1	0.063
2	0.151
3	0.297
4	0.219
5	0.156
6	0.083
	$\Sigma P(x) = 1$

(b)

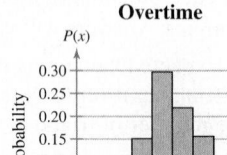

Overtime

Approximately symmetric

(c) 3.4, 2.1, 1.5

(d) The mean is 3.4, so the average employee worked 3.4 hours of overtime. The standard deviation is 1.5, so the overtime worked by most of the employees differed from the mean by no more than 1.5 hours.

33. An expected value of 0 means that the money gained is equal to the money spent, representing the break-even point.

35. (a) 5.3 (b) 3.3 (c) 1.8 (d) 5.3

(e) The expected value is 5.3, so an average student is expected to answer about 5 questions correctly. The standard deviation is 1.8, so most of the students' quiz results differ from the expected value by no more than about 2 questions.

37. (a) 2.0 (b) 1.0 (c) 1.0 (d) 2.0

(e) The expected value is 2.0, so an average hurricane that hits the U.S. mainland is expected to be a category 2 hurricane. The standard deviation is 1.0, so most of the hurricanes differ from the expected value by no more than 1 category level.

39. (a) 2.5 (b) 1.9 (c) 1.4 (d) 2.5

(e) The expected value is 2.5, so an average household is expected to have either 2 or 3 people. The standard deviation is 1.4, so most of the household sizes differ from the expected value by no more than 1 or 2 people.

41. (a) 0.881 (b) 0.314 (c) 0.294

43. A household with three dogs is unusual because the probability of this event is 0.022, which is less than 0.05.

45. −$0.05

47. (a)

x	P(x)
0	0.432
1	0.403
2	0.137
3	0.029
	$\Sigma P(x) \approx 1$

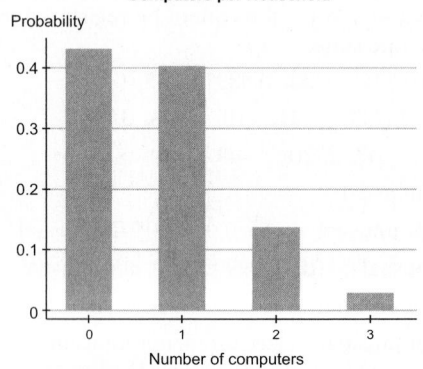

Computers per Household

(b) Skewed right

49. $38,800 **51.** 3020; 28

Section 4.2 *(page 211)*

1. Each trial is independent of the other trials if the outcome of one trial does not affect the outcome of any of the other trials.

3. (a) $p = 0.50$ (b) $p = 0.20$ (c) $p = 0.80$

5. (a) $n = 12$ (b) $n = 4$ (c) $n = 8$

As n increases, the distribution becomes more symmetric.

7. (a) $x = 0, 1, 2, 3, 4, 11, 12$

(b) $x = 0$

(c) $x = 0, 1, 2, 8$

9. Binomial experiment

Success: baby recovers

$n = 5, p = 0.80, q = 0.20, x = 0, 1, 2, 3, 4, 5$

11. Binomial experiment

Success: selecting an officer who is postponing or reducing the amount of vacation

$n = 20, p = 0.31, q = 0.69, x = 0, 1, 2, \ldots, 20$

13. 20, 12, 3.5 **15.** 32.2, 23.9, 4.9

17. (a) 0.088 (b) 0.104 (c) 0.896

19. (a) 0.111   (b) 0.152   (c) 0.848

21. (a) 0.257   (b) 0.220   (c) 0.780

23. (a) 0.187   (b) 0.605   (c) 0.084

25. (a) 0.255   (b) 0.562   (c) 0.783

27. (a) $n = 6$, $p = 0.63$ (b)

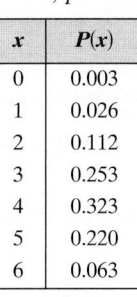

x	$P(x)$
0	0.003
1	0.026
2	0.112
3	0.253
4	0.323
5	0.220
6	0.063

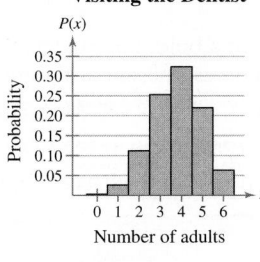

Visiting the Dentist

Skewed left

(c) 3.8, 1.4, 1.2

(d) On average, 3.8 out of 6 adults are visiting the dentist less because of the economy. The standard deviation is 1.2, so most samples of 6 adults would differ from the mean by no more than 1.2 people. The values $x = 0$ and $x = 1$ would be unusual because their probabilities are less than 0.05.

29. (a) $n = 4$, $p = 0.05$ (b)

x	$P(x)$
0	0.814506
1	0.171475
2	0.013538
3	0.000475
4	0.000006

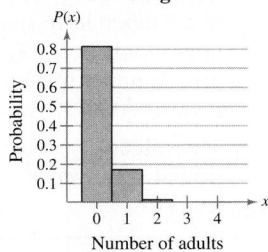

Donating Blood

Skewed right

(c) 0.2, 0.2, 0.4

(d) On average, 0.2 eligible adult out of every 4 gives blood. The standard deviation is 0.4, so most samples of four eligible adults would differ from the mean by at most 0.4 adult.

$x = 2, 3,$ and 4 would be unusual because their probabilities are less than 0.05.

31. (a) $n = 6$, $p = 0.37$ (b) 0.323   (c) 0.029

x	$P(x)$
0	0.063
1	0.220
2	0.323
3	0.253
4	0.112
5	0.026
6	0.003

33. 2.2, 1.2

On average, 2.2 out of 6 travelers would name "crying kids" as the most annoying. The standard deviation is 1.2, so most samples of 6 travelers would differ from the mean by at most 1.2 travelers. The values $x = 5$ and $x = 6$ would be unusual because their probabilities are less than 0.05.

35. (a) 0.081   (b) 0.541

(c) 0.022; This event is unusual because its probability is less than 0.05.

37. 0.033

4.2 Activity   *(page 216)*

1–3. Answers will vary.

Section 4.3   *(page 222)*

1. 0.080   **3.** 0.062   **5.** 0.175   **7.** 0.251

9. In a binomial distribution, the value of x represents the number of successes in n trials, and in a geometric distribution the value of x represents the first trial that results in a success.

11. Geometric. You are interested in counting the number of trials until the first success.

13. Binomial. You are interested in counting the number of successes out of n trials.

15. (a) 0.082   (b) 0.469   (c) 0.531

17. (a) 0.195   (b) 0.434 (*Tech:* 0.433)

(c) 0.566 (*Tech:* 0.567)

19. (a) 0.329   (b) 0.878   (c) 0.122

21. (a) 0.105   (b) 0.578   (c) 0.316

23. (a) 0.140

(b) 0.042; This event is unusual because its probability is less than 0.05.

(c) 0.064

25. (a) 0.1254235482

(b) 0.1254084986; The results are approximately the same.

27. (a) 1000, 999,000, 999.5

On average you would have to play 1000 times in order to win the lottery. The standard deviation is 999.5 times.

(b) 1000 times

Lose money. On average you would win $500 once in every 1000 times you play the lottery. So, the net gain would be −$500.

29. (a) 3.9, 2.0; The standard deviation is 2.0 strokes, so most of Phil's scores per hole differ from the mean by no more than 2.0 strokes.

(b) 0.385

Uses and Abuses for Chapter 4   *(page 225)*

1. 40, 0.081   **2.** 0.739; Answers will vary.

3. The probability of finding 36 adults out of 100 who prefer Brand A is 0.059. So, the manufacturer's claim is believable because 0.059 > 0.05.

4. The probability of finding 25 adults out of 100 who prefer Brand A is 0.000627. So, the manufacturer's claim is not believable.

Review Answers for Chapter 4   *(page 227)*

1. Continuous; The length of time spent sleeping is a random variable that cannot be counted.

3. Discrete   5. Continuous   7. No, $\sum P(x) \neq 1$.   9. Yes

11. (a)

x	f	$P(x)$
2	3	0.005
3	12	0.018
4	72	0.111
5	115	0.177
6	169	0.260
7	120	0.185
8	83	0.128
9	48	0.074
10	22	0.034
11	6	0.009
	$n = 650$	$\sum P(x) \approx 1$

(b)

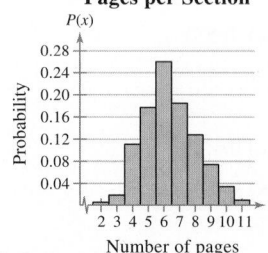

Pages per Section

Approximately symmetric

(c) 6.4, 2.9, 1.7

(d) The mean is 6.4, so the average number of pages per section is about 6 pages. The standard deviation is 1.7, so most of the sections differ from the mean by no more than about 2 pages.

13. (a)

x	f	$P(x)$
0	5	0.020
1	35	0.140
2	68	0.272
3	73	0.292
4	42	0.168
5	19	0.076
6	8	0.032
	$n = 250$	$\sum P(x) = 1$

(b)

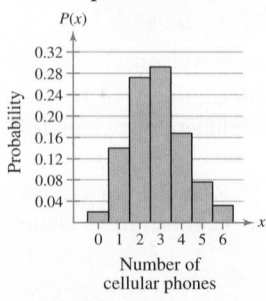

Cellular Phones per Household

Approximately symmetric

(c) 2.8, 1.7, 1.3

(d) The mean is 2.8, so the average number of cellular phones per household is about 3. The standard deviation is 1.3, so most of the households differ from the mean by no more than about 1 cellular phone.

15. 3.4

17. No; In a binomial experiment, there are only two possible outcomes: success or failure.

19. Yes; $n = 12$, $p = 0.24$, $q = 0.76$, $x = 0, 1, \ldots, 12$

21. (a) 0.208   (b) 0.322 (*Tech:* 0.321)   (c) 0.114

23. (a) 0.196   (b) 0.332   (c) 0.137

25. (a)

x	$P(x)$
0	0.125
1	0.323
2	0.332
3	0.171
4	0.044
5	0.005

(b)

Help with Chores

Skewed right

(c) 1.7, 1.1, 1.1; The mean is 1.7, so an average of 1.7 out of 5 women have spouses who never help with household chores. The standard deviation is 1.1, so most samples of 5 women differ from the mean by no more than 1.1 women.

(d) The values $x = 4$ and $x = 5$ are unusual because their probabilities are less than 0.05.

27. (a)

x	$P(x)$
0	0.130
1	0.346
2	0.346
3	0.154
4	0.026

(b)

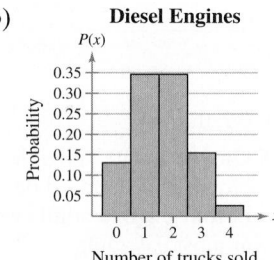

Diesel Engines

Skewed right

(c) 1.6, 1.0, 1.0; The mean is 1.6, so an average of 1.6 out of 4 trucks have diesel engines. The standard deviation is 1.0, so most samples of 4 trucks differ from the mean by no more than 1 truck.

(d) The value $x = 4$ is unusual because its probability is less than 0.05.

29. (a) 0.134   (b) 0.186   (c) 0.176

31. (a) 0.765   (b) 0.205   (c) 0.997   (d) 0.030; unusual

33. The probability increases as the rate increases, and decreases as the rate decreases.

Chapter Quiz for Chapter 4   *(page 231)*

1. (a) Discrete; The number of lightning strikes that occur in Wyoming during the month of June is a random variable that is countable.

(b) Continuous; The fuel (in gallons) used by the Space Shuttle during takeoff is a random variable that has an infinite number of possible outcomes and cannot be counted.

2. (a)

x	f	P(x)
1	114	0.400
2	74	0.260
3	76	0.267
4	18	0.063
5	3	0.011
	n = 285	ΣP(x) ≈ 1

(b)

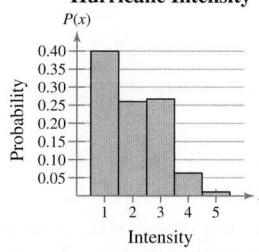

Hurricane Intensity

Skewed right

(c) 2.0, 1.0, 1.0

On average, the intensity of a hurricane will be 2.0. The standard deviation is 1.0, so most hurricane intensities will differ from the mean by no more than 1.0.

(d) 0.074

3. (a)

x	P(x)
0	0.00001
1	0.00039
2	0.00549
3	0.04145
4	0.17618
5	0.39933
6	0.37715

(b)

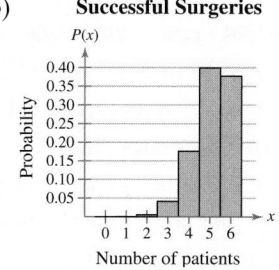

Successful Surgeries

Skewed left

(c) 5.1, 0.8, 0.9; The average number of successful surgeries is 5.1 out of 6. The standard deviation is 0.9, so most samples of 6 surgeries differ from the mean by no more than 0.9 surgery.

(d) 0.041; Yes, this event is unusual because 0.041 < 0.05.

(e) 0.047; Yes, this event is unusual because 0.047 < 0.05.

4. (a) 0.175   (b) 0.440   (c) 0.007

5. 0.038; Yes, this event is unusual because 0.038 < 0.05.

6. 0.335; No, this event is not unusual because 0.335 > 0.05.

Real Statistics–Real Decisions for Chapter 4   *(page 232)*

1. (a) Answers will vary. For instance, calculate the probability of obtaining 0 clinical pregnancies out of 10 randomly selected ART cycles.

(b) Binomial. The distribution is discrete because the number of clinical pregnancies is countable.

2. $n = 10$, $p = 0.349$, $P(0) = 0.014$

x	P(x)
0	0.01367
1	0.07329
2	0.17681
3	0.25277
4	0.23714
5	0.15256
6	0.06815
7	0.02088
8	0.00420
9	0.00050
10	0.00003

Answers will vary. *Sample answer:* Because $P(0) = 0.014$, this event is unusual but not impossible.

3. (a) Suspicious, because the probability is very small.

(b) Not suspicious, because the probability is not that small.

CHAPTER 5

Section 5.1   *(page 244)*

1. Answers will vary.

3. 1

5. Answers will vary.

Similarities: The two curves will have the same line of symmetry.

Differences: The curve with the larger standard deviation will be more spread out than the curve with the smaller standard deviation.

7. $\mu = 0$, $\sigma = 1$

9. "The" standard normal distribution is used to describe one specific normal distribution ($\mu = 0$, $\sigma = 1$). "A" normal distribution is used to describe a normal distribution with any mean and standard deviation.

11. No, the graph crosses the x-axis.

13. Yes, the graph fulfills the properties of the normal distribution.

15. No, the graph is skewed right.

17. It is normal because it is bell-shaped and nearly symmetric.

19. 0.0968   **21.** 0.0228   **23.** 0.4878   **25.** 0.5319

27. 0.005   **29.** 0.7422   **31.** 0.6387   **33.** 0.4979

35. 0.95   **37.** 0.2006 (*Tech:* 0.2005)

39. (a)

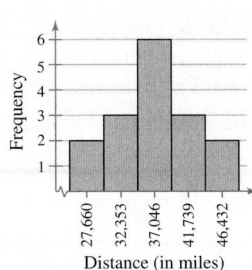

Life Spans of Tires

It is reasonable to assume that the life spans are normally distributed because the histogram is symmetric and bell-shaped.

(b) 37,234.7, 6259.2

(c) The sample mean of 37,234.7 hours is less than the claimed mean, so, on average, the tires in the sample lasted for a shorter time. The sample standard deviation of 6259.2 is greater than the claimed standard deviation, so the tires in the sample had a greater variation in life span than the manufacturer's claim.

41. (a) A = 105; B = 113; C = 121; D = 127

(b) −2.78; −0.56; 1.67; 3.33

(c) $x = 105$ is unusual because its corresponding z-score (−2.78) lies more than 2 standard deviations from the mean, and $x = 127$ is very unusual because its corresponding z-score (3.33) lies more than 3 standard deviations from the mean.

43. (a) A = 1241; B = 1392; C = 1924; D = 2202

(b) −0.86; −0.375; 1.33; 2.22

(c) $x = 2202$ is unusual because its corresponding z-score (2.22) lies more than 2 standard deviations from the mean.

45. 0.9750   **47.** 0.9775   **49.** 0.84   **51.** 0.9265

53. 0.0148   **55.** 0.3133   **57.** 0.901 (*Tech:* 0.9011)

59. 0.0098 (*Tech:* 0.0099)

61.

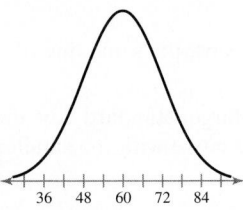

The normal distribution curve is centered at its mean (60) and has 2 points of inflection (48 and 72) representing $\mu \pm \sigma$.

63. (a) Area under curve = area of square = (1)(1) = 1

(b) 0.25   (c) 0.4

Section 5.2   (page 252)

1. 0.4207   **3.** 0.3446   **5.** 0.1787 (*Tech:* 0.1788)

7. 0.3442 (*Tech:* 0.3451)   **9.** 0.2747 (*Tech:* 0.2737)

11. 0.3387 (*Tech:* 0.3385)

13. (a) 0.0968   (b) 0.6612   (c) 0.2420

(d) No, none of the events are unusual because their probabilities are greater than 0.05.

15. (a) 0.1867 (*Tech:* 0.1870)   (b) 0.4171 (*Tech:* 0.4176)

(c) 0.0166 (*Tech:* 0.0167)

(d) Yes, the event in part (c) is unusual because its probability is less than 0.05.

17. (a) 0.0228   (b) 0.927   (c) 0.0013

19. (a) 0.0073   (b) 0.7215 (*Tech:* 0.7218)   (c) 0.0228

21. (a) 83.15% (*Tech:* 83.25%)

(b) 305 scores (*Tech:* 304 scores)

23. (a) 66.28% (*Tech:* 66.4%)   (b) 22 men

25. (a) 99.87%   (b) 1 adult

27. 1.5% (*Tech:* 1.51%); It is unusual for a battery to have a life span that is more than 2065 hours because the probability is less than 0.05.

29. (a) 0.3085   (b) 0.1499

(c) 0.0668; No, because 0.0668 > 0.05, this event is not unusual.

31. Out of control, because there is a point more than three standard deviations beyond the mean.

33. Out of control, because there are nine consecutive points below the mean, and two out of three consecutive points lie more than two standard deviations from the mean.

Section 5.3   (page 262)

1. −0.81   **3.** 2.39   **5.** −1.645   **7.** 1.555   **9.** −1.04

11. 1.175   **13.** −0.67   **15.** 0.67   **17.** −0.38

19. −0.58   **21.** −1.645, 1.645   **23.** −1.18   **25.** 1.18

27. −1.28, 1.28   **29.** −0.06, 0.06

31. (a) 68.58 inches   (b) 62.56 inches (*Tech:* 62.55 inches)

33. (a) 161.72 days (*Tech:* 161.73 days)

(b) 221.22 days (*Tech:* 221.33 days)

35. (a) 7.75 hours (*Tech:* 7.74 hours)

(b) 5.43 hours and 6.77 hours

37. 32.61 ounces

39. (a) 18.88 pounds (*Tech:* 18.90 pounds)

(b) 12.04 pounds (*Tech:* 12.05 pounds)

41. Tires that wear out by 26,800 miles (*Tech:* 26,796 miles) will be replaced free of charge.

Section 5.4   (page 274)

1. 150, 3.536   **3.** 150, 1.581

5. False. As the size of a sample increases, the mean of the distribution of sample means does not change.

7. False. A sampling distribution is normal if either $n \geq 30$ or the population is normal.

9. (c), because $\mu_{\bar{x}} = 16.5$, $\sigma_{\bar{x}} = 1.19$, and the graph approximates a normal curve.

11.

Sample	Mean	Sample	Mean	Sample	Mean
2, 2, 2	2	4, 4, 8	5.33	8, 16, 2	8.67
2, 2, 4	2.67	4, 4, 16	8	8, 16, 4	9.33
2, 2, 8	4	4, 8, 2	4.67	8, 16, 8	10.67
2, 2, 16	6.67	4, 8, 4	5.33	8, 16, 16	13.33
2, 4, 2	2.67	4, 8, 8	6.67	16, 2, 2	6.67
2, 4, 4	3.33	4, 8, 16	9.33	16, 2, 4	7.33
2, 4, 8	4.67	4, 16, 2	7.33	16, 2, 8	8.67
2, 4, 16	7.33	4, 16, 4	8	16, 2, 16	11.33
2, 8, 2	4	4, 16, 8	9.33	16, 4, 2	7.33
2, 8, 4	4.67	4, 16, 16	12	16, 4, 4	8
2, 8, 8	6	8, 2, 2	4	16, 4, 8	9.33
2, 8, 16	8.67	8, 2, 4	4.67	16, 4, 16	12
2, 16, 2	6.67	8, 2, 8	6	16, 8, 2	8.67
2, 16, 4	7.33	8, 2, 16	8.67	16, 8, 4	9.33
2, 16, 8	8.67	8, 4, 2	4.67	16, 8, 8	10.67
2, 16, 16	11.33	8, 4, 4	5.33	16, 8, 16	13.33
4, 2, 2	2.67	8, 4, 8	6.67	16, 16, 2	11.33
4, 2, 4	3.33	8, 4, 16	9.33	16, 16, 4	12
4, 2, 8	4.67	8, 8, 2	6	16, 16, 8	13.33
4, 2, 16	7.33	8, 8, 4	6.67	16, 16, 16	16
4, 4, 2	3.33	8, 8, 8	8		
4, 4, 4	4	8, 8, 16	10.67		

$\mu = 7.5$, $\sigma \approx 5.36$

$\mu_{\bar{x}} = 7.5$, $\sigma_{\bar{x}} \approx 3.09$

The means are equal but the standard deviation of the sampling distribution is smaller.

13. 0.9726; not unusual   **15.** 0.0351 (*Tech:* 0.0349); unusual

17. 7.6, 0.101   **19.** 235, 13.864

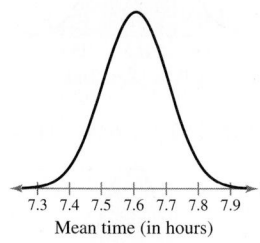

7.3 7.4 7.5 7.6 7.7 7.8 7.9 $\bar{x}$
Mean time (in hours)

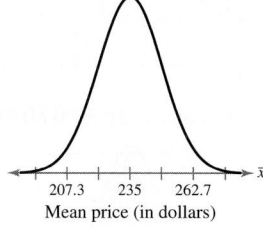

207.3   235   262.7 $\bar{x}$
Mean price (in dollars)

21. 188.4, 10.9

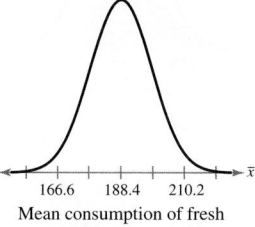

166.6   188.4   210.2 $\bar{x}$
Mean consumption of fresh
vegetables (in pounds)

23. $n = 24$: 7.6, 0.07; $n = 36$: 7.6, 0.06

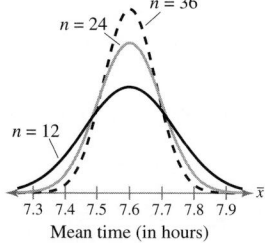

$n = 36$
$n = 24$
$n = 12$

7.3 7.4 7.5 7.6 7.7 7.8 7.9 $\bar{x}$
Mean time (in hours)

As the sample size increases, the standard error decreases, while the mean of the sample means remains constant.

25. 0.0003; Only 0.03% of samples of 35 specialists will have a mean salary less than $60,000. This is an extremely unusual event.

27. 0.9078 (*Tech:* 0.9083); About 91% of samples of 32 gas stations that week will have a mean price between $2.695 and $2.725.

29. ≈ 0 (*Tech:* 0.0000002); There is almost no chance that a random sample of 60 women will have a mean height greater than 66 inches. This event is almost impossible.

31. It is more likely to select a sample of 20 women with a mean height less than 70 inches because the sample of 20 has a higher probability.

33. Yes, it is very unlikely that you would have randomly sampled 40 cans with a mean equal to 127.9 ounces because it is more than 2 standard deviations from the mean of the sample means.

35. (a) 0.0008   (b) Claim is inaccurate.

(c) No, assuming the manufacturer's claim is true, because 96.25 is within 1 standard deviation of the mean for an individual board.

37. (a) 0.0002   (b) Claim is inaccurate.

(c) No, assuming the manufacturer's claim is true, because 49,721 is within 1 standard deviation of the mean for an individual tire.

39. No, because the *z*-score (0.88) is not unusual.

41. Yes, the finite correction factor should be used; 0.0003

43.

Sample	Number of boys from 3 births	Proportion of boys from 3 births
bbb	3	1
bbg	2	$\frac{2}{3}$
bgb	2	$\frac{2}{3}$
gbb	2	$\frac{2}{3}$
bgg	1	$\frac{1}{3}$
gbg	1	$\frac{1}{3}$
ggb	1	$\frac{1}{3}$
ggg	0	0

45.

Sample	Numerical representation	Sample mean
bbb	111	1
bbg	110	$\frac{2}{3}$
bgb	101	$\frac{2}{3}$
gbb	011	$\frac{2}{3}$
bgg	100	$\frac{1}{3}$
gbg	010	$\frac{1}{3}$
ggb	001	$\frac{1}{3}$
ggg	000	0

The sample means are equal to the proportions.

47. 0.0446 (*Tech:* 0.0441); About 4.5% (*Tech:* 4.4%) of samples of 105 female heart transplant patients will have a mean 3-year survival rate of less than 70%. Because the probability is less than 0.05, this is an unusual event.

Section 5.4 Activity *(page 280)*

1–2. Answers will vary.

Section 5.5 *(page 287)*

1. Properties of a binomial experiment:
 (1) The experiment is repeated for a fixed number of independent trials.
 (2) There are two possible outcomes: success or failure.
 (3) The probability of success is the same for each trial.
 (4) The random variable x counts the number of successful trials.

3. Cannot use normal distribution.

5. Cannot use normal distribution.

7. Cannot use normal distribution because $nq < 5$.

9. Can use normal distribution; $\mu = 27.5$, $\sigma \approx 3.52$

11. Cannot use normal distribution because $nq < 5$.

13. a   **14.** d   **15.** c   **16.** b

17. The probability of getting fewer than 25 successes; $P(x < 24.5)$

19. The probability of getting exactly 33 successes; $P(32.5 < x < 33.5)$

21. The probability of getting at most 150 successes; $P(x < 150.5)$

23. Can use normal distribution.
 (a) 0.0782 (*Tech:* 0.0785)   (b) 0.9147 (*Tech:* 0.9151)

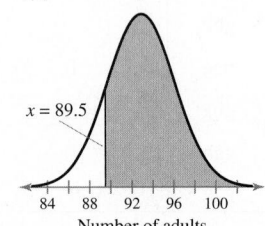

 (c) 0.0853 (*Tech:* 0.0849)

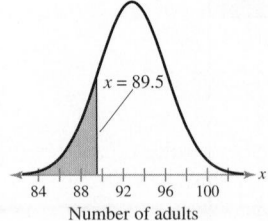

 (d) No, none of the probabilities are less than 0.05.

25. Can use normal distribution.
 (a) ≈ 1 (b) 0.9798 (*Tech:* 0.9801)

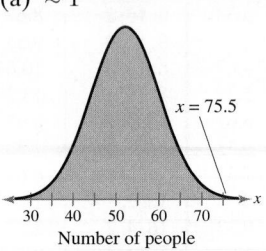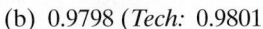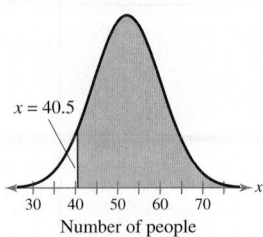

 (c) 0.6097 (*Tech:* 0.6109)

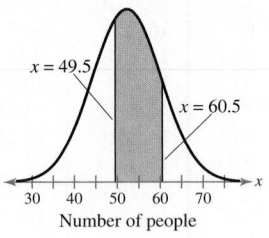

 (d) No, none of the probabilities are less than 0.05.

27. Can use normal distribution.
 (a) 0.0692 (*Tech:* 0.0691)   (b) 0.8770 (*Tech:* 0.8771)

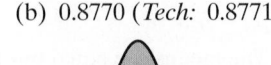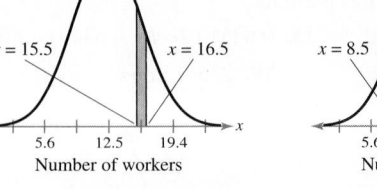

 (c) 0.8078 (*Tech:* 0.8080)   (d) 0.8212 (*Tech:* 0.8221)

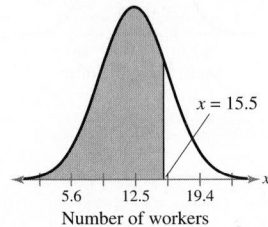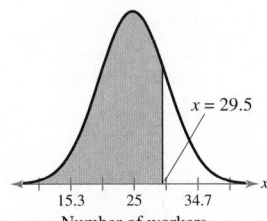

29. (a) Can use normal distribution.
  0.0069

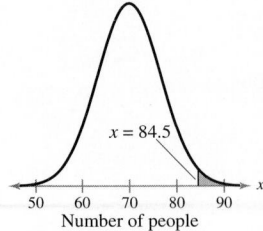

(b) Can use normal distribution.
0.3557 (*Tech:* 0.3545)

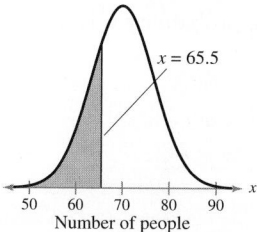

(c) Can use normal distribution.
0.0558 (*Tech:* 0.0595)

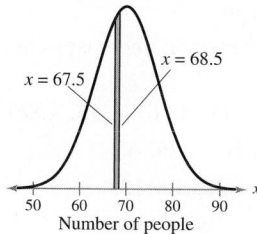

(d) Cannot use normal distribution because $np < 5$ and $nq < 5$; 0.002

31. Binomial: 0.549; Normal: 0.5463 (*Tech:* 0.5466); The results are about the same.

33. Highly unlikely. Answers will vary.   **35.** 0.1020

Uses and Abuses for Chapter 5   *(page 291)*

1. (a) Not unusual; A sample mean of 115 is less than 2 standard deviations from the population mean.

(b) Not unusual; A sample mean of 105 lies within 2 standard deviations of the population mean.

2. The ages of students at a high school may not be normally distributed.

3. Answers will vary.

Review Answers for Chapter 5   *(page 293)*

1. $\mu = 15, \sigma = 3$

3. Curve B has the greatest mean because its line of symmetry occurs the farthest to the right.

5. $-2.25 ; 0.5; 2; 3.5$ **7.** 0.6772   **9.** 0.6293   **11.** 0.7157

13. 0.00235 (*Tech:* 0.00236)   **15.** 0.4495

17. 0.4365 (*Tech:* 0.4364)   **19.** 0.1336

21. $A = 8; B = 17; C = 23; D = 29$ **23.** 0.8997

25. 0.9236 (*Tech:* 0.9237)   **27.** 0.0124   **29.** 0.8944

31. 0.2266   **33.** 0.2684 (*Tech:* 0.2685)

35. (a) 0.3156   (b) 0.3099   (c) 0.3446

37. No, none of the events are unusual because their probabilities are greater than 0.05.

39. -0.07 **41.** 1.13   **43.** 1.04   **45.** 0.51

47. 42.5 meters   **49.** 51.6 meters   **51.** 50.8 meters

53. $\mu = 145, \sigma = 45$

$\mu_{\bar{x}} = 145, \sigma_{\bar{x}} \approx 25.98$

The means are the same, but $\sigma_{\bar{x}}$ is less than σ.

55. 76, 3.465

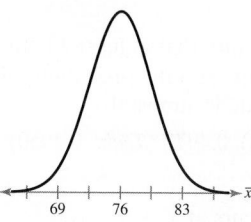

57. (a) 0.0485 (*Tech:* 0.0482)

(b) 0.8180

(c) 0.0823 (*Tech:* 0.0829)

(a) and (c) are smaller, (b) is larger. This is to be expected because the standard error of the sample means is smaller.

59. (a) 0.1867 (*Tech:* 0.1855)   (b) ≈ 0

61. 0.0019 (*Tech:* 0.0018)

63. Cannot use normal distribution because $nq < 5$.

65. $P(x > 24.5)$ **67.** $P(44.5 < x < 45.5)$

69. Can use normal distribution.

≈ 0 (*Tech:* 0.0002)

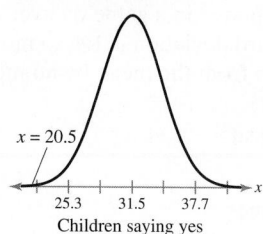

Chapter Quiz for Chapter 5   *(page 297)*

1. (a) 0.9945   (b) 0.9990

(c) 0.6212   (d) 0.83685 (*Tech:* 0.83692)

2. (a) 0.9198 (*Tech:* 0.9199)   (b) 0.1940 (*Tech:* 0.1938)

(c) 0.0456 (*Tech:* 0.0455)

3. 0.0475 (*Tech:* 0.0478); Yes, the event is unusual because its probability is less than 0.05.

4. 0.2586 (*Tech:* 0.2611); No, the event is not unusual because its probability is greater than 0.05.

5. 21.19%   **6.** 503 students (*Tech:* 505 students)

7. 125   **8.** 80

9. 0.0049; About 0.5% of samples of 60 students will have a mean IQ score greater than 105. This is a very unusual event.

10. More likely to select one student with an IQ score greater than 105 because the standard error of the mean is less than the standard deviation.

11. Can use normal distribution.

$\mu = 28.35, \sigma \approx 2.32$

12. 0.0004; This event is extremely unusual because its probability is much less than 0.05.

Real Statistics–Real Decisions for Chapter 5   (page 298)

1. (a) 0.0014   (b) 0.9495   (c) ≈ 0

   (d) There is a very high probability that at least 40 out of 60 employees will participate, and the probability that fewer than 20 will participate is almost 0.

2. (a) 0.2514 (*Tech:* 0.2525)   (b) 0.4972 (*Tech:* 0.4950)

   (c) 0.2514 (*Tech:* 0.2525)

3. (a) 3; The line of symmetry occurs at $x = 3$.

   (b) Yes   (c) Answers will vary.

Cumulative Review Answers for Chapters 3–5   (page 300)

1. (a) $np = 7.5 \geq 5$, $nq = 42.5 \geq 5$

   (b) 0.9973

   (c) Yes, because the probability is less than 0.05.

2. (a) 3.1   (b) 1.6   (c) 1.3   (d) 3.1

   (e) The size of a family household on average is about 3 persons. The standard deviation is 1.3, so most households differ from the mean by no more than about 1 person.

3. (a) 3.6   (b) 1.9   (c) 1.4   (d) 3.6

   (e) The number of fouls for a player in a game on average is about 4 fouls. The standard deviation is 1.4, so most of the player's games differ from the mean by no more than about 1 or 2 fouls.

4. (a) 0.476   (b) 0.78   (c) 0.659

5. (a) 43,680   (b) 0.0192

6. 0.7642   **7.** 0.0010   **8.** 0.7995

9. 0.4984   **10.** 0.2862   **11.** 0.5905

12. (a) 0.0462; unusual, because the probability is less than 0.05

   (b) 0.6029

   (c) 0.0139; unusual, because the probability is less than 0.05

13. (a) 0.0048   (b) 0.0149   (c) 0.9511

14. (a) 0.2777   (b) 0.8657

   (c) Dependent. P(being a public school teacher | having 20 years or more of full-time teaching experience) ≠ P(being a public school teacher)

   (d) 0.8881   (e) 0.4177

15. (a) 70, 0.1897   (b) 0.0006

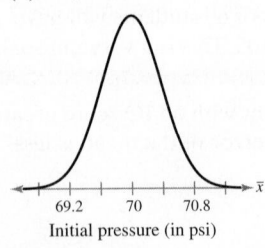

Initial pressure (in psi)

16. (a) 0.0548   (b) 0.6547   (c) 52.2 months

17. (a) 495   (b) 0.0020

18. (a) 0.0278; unusual, because the probability is less than 0.05

   (b) 0.2272   (c) 0.5982

CHAPTER 6

Section 6.1   (page 311)

1. You are more likely to be correct using an interval estimate because it is unlikely that a point estimate will exactly equal the population mean.

3. d; As the level of confidence increases, z_c increases, causing wider intervals.

5. 1.28   **7.** 1.15   **9.** −0.47   **11.** 1.76   **13.** 1.861

15. 0.192   **17.** c   **18.** d   **19.** b   **20.** a

21. (12.0, 12.6)   **23.** (9.7, 11.3)   **25.** 1.4, 13.4

27. 0.17, 1.88   **29.** 126   **31.** 7   **33.** 1.95, 28.15

35. (428.68, 476.92); (424.06, 481.54)

   With 90% confidence, you can say that the population mean price is between $428.68 and $476.92; with 95% confidence, you can say that the population mean price is between $424.06 and $481.54. The 95% CI is wider.

37. (87.0, 111.6); (84.7, 113.9)

   With 90% confidence, you can say that the population mean is between 87.0 and 111.6 calories; with 95% confidence, you can say that the population mean is between 84.7 and 113.9 calories. The 95% CI is wider.

39. (2532.20, 2767.80)

   With 95% confidence, you can say that the population mean cost is between $2532.20 and $2767.80.

41. (2556.87, 2743.13) [*Tech:* (2556.90, 2743.10)]

   The $n = 50$ CI is wider because a smaller sample is taken, giving less information about the population.

43. (3.09, 3.15)

   With 95% confidence, you can say that the population mean time is between 3.09 and 3.15 minutes.

45. (3.10, 3.14)

   The $s = 0.09$ CI is wider because of the increased variability within the sample.

47. (a) An increase in the level of confidence will widen the confidence interval.

   (b) An increase in the sample size will narrow the confidence interval.

   (c) An increase in the standard deviation will widen the confidence interval.

49. (14.6, 15.6); (14.4, 15.8)

   With 90% confidence, you can say that the population mean length of time is between 14.6 and 15.6 minutes; with 99% confidence, you can say that the population mean length of time is between 14.4 and 15.8 minutes. The 99% CI is wider.

51. 89

53. (a) 121 servings   (b) 208 servings

  (c) The 99% CI requires a larger sample because more information is needed from the population to be 99% confident.

55. (a) 32 cans   (b) 87 cans

  $E = 0.15$ requires a larger sample size. As the error size decreases, a larger sample must be taken to obtain enough information from the population to ensure the desired accuracy.

57. (a) 16 sheets   (b) 62 sheets

  $E = 0.0625$ requires a larger sample size. As the error size decreases, a larger sample must be taken to obtain enough information from the population to ensure the desired accuracy.

59. (a) 42 soccer balls   (b) 60 soccer balls

  $\sigma = 0.3$ requires a larger sample size. Due to the increased variability in the population, a larger sample size is needed to ensure the desired accuracy.

61. (a) An increase in the level of confidence will increase the minimum sample size required.

  (b) An increase (larger E) in the error tolerance will decrease the minimum sample size required.

  (c) An increase in the population standard deviation will increase the minimum sample size required.

63. (212.8, 221.4)

  With 95% confidence, you can say that the population mean airfare price is between \$212.8 and \$221.4.

65. **80% confidence interval results:**

  μ : population mean

  Standard deviation = 344.9

Mean	n	Sample Mean	Std. Err.	L. Limit	U. Limit
μ	30	1042.7	62.969837	962.0009	1123.399

  90% confidence interval results:

  μ : population mean

  Standard deviation = 344.9

Mean	n	Sample Mean	Std. Err.	L. Limit	U. Limit
μ	30	1042.7	62.969837	939.12384	1146.2761

  95% confidence interval results:

  μ : population mean

  Standard deviation = 344.9

Mean	n	Sample Mean	Std. Err.	L. Limit	U. Limit
μ	30	1042.7	62.969837	919.2814	1166.1187

With 80% confidence, you can say that the population mean sodium content is between 962.0 and 1123.4 milligrams; with 90% confidence, you can say it is between 939.1 and 1146.3 milligrams; with 95% confidence, you can say it is between 919.3 and 1166.1 milligrams.

67. (a) 0.707   (b) 0.949   (c) 0.962   (d) 0.975

  (e) The finite population correction factor approaches 1 as the sample size decreases and the population size remains the same.

69. *Sample answer:*

$$E = \frac{z_c \sigma}{\sqrt{n}} \qquad \text{Write original equation}$$

$$E\sqrt{n} = z_c \sigma \qquad \text{Multiply each side by } \sqrt{n}.$$

$$\sqrt{n} = \frac{z_c \sigma}{E} \qquad \text{Divide each side by } E.$$

$$n = \left(\frac{z_c \sigma}{E}\right)^2 \qquad \text{Square each side.}$$

Section 6.2   *(page 323)*

1. 1.833   **3.** 2.947   **5.** 2.7   **7.** 1.2

9. (a) (10.9, 14.1)   (b) The t-CI is wider.

11. (a) (4.1, 4.5)

  (b) When rounded to the nearest tenth, the normal CI and the t-CI have the same width.

13. 3.7, 18.4

15. 9.5, 74.1

17. 6.0; (29.5, 41.5); With 95% confidence, you can say that the population mean commute time is between 29.5 and 41.5 minutes.

19. 6.4; (29.1, 41.9); With 95% confidence, you can say that the population mean commute time is between 29.1 and 41.9 minutes. This confidence interval is slightly wider than the one found in Exercise 17.

21. (a) (3.80, 5.20)

  (b) (4.41, 4.59); The t-CI in part (a) is wider.

23. (a) 90,182.9   (b) 3724.9   (c) (87,438.6, 92,927.2)

25. (a) 1767.7   (b) 252.2   (c) (1541.5, 1993.8)

27. Use a normal distribution because $n \geq 30$.

  (26.0, 29.4); With 95% confidence, you can say that the population mean BMI is between 26.0 and 29.4.

29. Use a t-distribution because $n < 30$, the miles per gallon are normally distributed, and σ is unknown.

  (20.5, 23.3) [*Tech:* (20.5, 23.4)]; With 95% confidence, you can say that the population mean is between 20.5 and 23.3 miles per gallon.

31. Cannot use normal or t-distribution because $n < 30$ and the times are not normally distributed.

33. **90% confidence interval results:**

μ : mean of Variable

Variable	Sample Mean	Std. Err.
Time (in hours)	12.194445	0.4136141

DF	L. Limit	U. Limit
17	11.474918	12.91397

95% confidence interval results:

μ : mean of Variable

Variable	Sample Mean	Std. Err.
Time (in hours)	12.194445	0.4136141

DF	L. Limit	U. Limit
17	11.321795	13.067094

99% confidence interval results:

μ : mean of Variable

Variable	Sample Mean	Std. Err.
Time (in hours)	12.194445	0.4136141

DF	L. Limit	U. Limit
17	10.995695	13.393193

With 90% confidence, you can say that the population mean time spent on homework is between 11.5 and 12.9 hours; with 95% confidence, you can say it is between 11.3 and 13.1 hours; and with 99% confidence, you can say it is between 11.0 and 13.4 hours. As the level of confidence increases, the intervals get wider.

35. No; They are not making good tennis balls because the desired bounce height of 55.5 inches is not between 55.9 and 56.1 inches.

Activity 6.2 *(page 326)*

1–2. Answers will vary.

Section 6.3 *(page 332)*

1. False. To estimate the value of p, the population proportion of successes, use the point estimate $\hat{p} = x/n$.

3. 0.750, 0.250   **5.** 0.423, 0.577

7. $E = 0.014$, $\hat{p} = 0.919$ **9.** $E = 0.042$, $\hat{p} = 0.554$

11. (0.557, 0.619) [*Tech:* (0.556, 0.619)];

(0.551, 0.625) [*Tech:* (0.550, 0.625)];

With 90% confidence, you can say that the population proportion of U.S. males ages 18–64 who say they have gone to the dentist in the past year is between 55.7% (*Tech:* 55.6%) and 61.9%; with 95% confidence, you can say it is between 55.1% (*Tech:* 55.0%) and 62.5%. The 95% confidence interval is slightly wider.

13. (0.438, 0.484);

With 99% confidence, you can say that the population proportion of U.S. adults who say they have started paying bills online in the last year is between 43.8% and 48.4%.

15. (0.622, 0.644)

17. (a) 601 adults   (b) 413 adults

(c) Having an estimate of the population proportion reduces the minimum sample size needed.

19. (a) 752 adults   (b) 483 adults

(c) Having an estimate of the population proportion reduces the minimum sample size needed.

21. (a) (0.234, 0.306)

(b) (0.450, 0.530)

(c) (0.275, 0.345)

23. (a) (0.274, 0.366)   (b) (0.511, 0.609)

25. No, it is unlikely that the two proportions are equal because the confidence intervals estimating the proportions do not overlap. The 99% confidence intervals are (0.260, 0.380) and (0.496, 0.624). Although these intervals are wider, they still do not overlap.

27. **90% confidence interval results:**

p : proportion of successes for population

Method: Standard-Wald

Proportion	Count	Total	Sample Prop.
p	802	1025	0.78243905

Std. Err.	L. Limit	U. Limit
0.012887059	0.7612417	0.8036364

95% confidence interval results:

p : proportion of successes for population

Method: Standard-Wald

Proportion	Count	Total	Sample Prop.
p	802	1025	0.78243905

Std. Err.	L. Limit	U. Limit
0.012887059	0.75718087	0.8076972

99% confidence interval results:

p : proportion of successes for population

Method: Standard-Wald

Proportion	Count	Total	Sample Prop.
p	802	1025	0.78243905

Std. Err.	L. Limit	U. Limit
0.012887059	0.74924415	0.8156339

With 90% confidence, you can say that the population proportion of U.S. adults who disapprove of the job Congress is doing is between 76.1% and 80.4%; with 95% confidence, you can say it is between 75.7% and 80.8%; and with 99% confidence, you can say it is between 74.9% and 81.6%. As the level of confidence increases, the intervals get wider.

29. (0.304, 0.324) is approximately a 97.6% CI.

31. If $n\hat{p} < 5$ or $n\hat{q} < 5$, the sampling distribution of $\hat{p}$ may not be normally distributed, so z_c cannot be used to calculate the confidence interval.

33.

$\hat{p}$	$\hat{q} = 1 - \hat{p}$	$\hat{p}\hat{q}$	$\hat{p}$	$\hat{q} = 1 - \hat{p}$	$\hat{p}\hat{q}$
0.0	1.0	0.00	0.45	0.55	0.2475
0.1	0.9	0.09	0.46	0.54	0.2484
0.2	0.8	0.16	0.47	0.53	0.2491
0.3	0.7	0.21	0.48	0.52	0.2496
0.4	0.6	0.24	0.49	0.51	0.2499
0.5	0.5	0.25	0.50	0.50	0.2500
0.6	0.4	0.24	0.51	0.49	0.2499
0.7	0.3	0.21	0.52	0.48	0.2496
0.8	0.2	0.16	0.53	0.47	0.2491
0.9	0.1	0.09	0.54	0.46	0.2484
1.0	0.0	0.00	0.55	0.45	0.2475

$\hat{p} = 0.5$ gives the maximum value of $\hat{p}\hat{q}$.

Activity 6.3 *(page 336)*

1–2. Answers will vary.

Section 6.4 *(page 341)*

1. Yes

3. 14.067, 2.167   **5.** 32.852, 8.907   **7.** 52.336, 13.121

9. (a) (0.0000413, 0.000157)   (b) (0.00643, 0.0125)

With 90% confidence, you can say that the population variance is between 0.0000413 and 0.000157, and the population standard deviation is between 0.00643 and 0.0125 milligram.

11. (a) (0.0305, 0.191)   (b) (0.175, 0.438)

With 99% confidence, you can say that the population variance is between 0.0305 and 0.191, and the population standard deviation is between 0.175 and 0.438 hour.

13. (a) (6.63, 55.46)   (b) (2.58, 7.45)

With 99% confidence, you can say that the population variance is between 6.63 and 55.46, and the population standard deviation is between 2.58 and 7.45 dollars per year.

15. (a) (380.0, 3942.6)   (b) (19.5, 62.8)

With 98% confidence, you can say that the population variance is between 380.0 and 3942.6, and the population standard deviation is between $19.5 and $62.8.

17. (a) (22.5, 98.7)   (b) (4.7, 9.9)

With 95% confidence, you can say that the population variance is between 22.5 and 98.7, and the population standard deviation is between 4.7 and 9.9 beats per minute.

19. (a) (128, 492)   (b) (11, 22)

With 95% confidence, you can say that the population variance is between 128 and 492, and the population standard deviation is between 11 and 22 grains per gallon.

21. (a) (9,104,741, 25,615,326)   (b) (3017, 5061)

With 80% confidence, you can say that the population variance is between 9,104,741 and 25,615,326, and the population standard deviation is between $3017 and $5061.

23. (a) (7.0, 30.6)   (b) (2.6, 5.5)

With 98% confidence, you can say that the population variance is between 7.0 and 30.6, and the population standard deviation is between 2.6 and 5.5 minutes.

25. 95% confidence interval results:

σ^2 : variance of Variable

Variance	Sample Var.	DF	L. Limit	U. Limit
σ^2	11.56	29	7.332092	20.891039

(2.71, 4.57)

27. 90% confidence interval results:

σ^2 : variance of Variable

Variance	Sample Var.	DF	L. Limit	U. Limit
σ^2	1225	17	754.8815	2401.4731

(27, 49)

29. Yes, because all of the values in the confidence interval are less than 0.015.

31. Answers will vary. *Sample answer:* Unlike a confidence interval for a population mean or proportion, a confidence interval for a population variance does not have a margin of error. The left and right endpoints must be calculated separately.

Uses and Abuses for Chapter 6 *(page 344)*

1–2. Answers will vary.

Review Answers for Chapter 6 *(page 346)*

1. (a) 103.5   (b) 9.0   **3.** (15.6, 16.0)   **5.** 1.675, 22.425

7. 47 people   **9.** 49 people   **11.** 1.383   **13.** 2.624

15. $n = 20$ **17.** 11.2   **19.** 0.7   **21.** (60.9, 83.3)

23. (6.1, 7.5)   **25.** (2050, 2386)   **27.** 0.81, 0.19

29. 0.540, 0.460   **31.** 0.140, 0.860   **33.** 0.490, 0.510

35. (0.790, 0.830)

With 95% confidence, you can say that the population proportion of U.S. adults who say they will participate in the 2010 Census is between 79.0% and 83.0%.

37. (0.514, 0.566) [*Tech:* (0.514, 0.565)]

With 90% confidence, you can say that the population proportion of U.S. adults who say they have worked the night shift at some point in their lives is between 51.4% and 56.6% (*Tech:* 56.5%).

39. (0.112, 0.168)

With 99% confidence, you can say that the population proportion of U.S. adults who say that the cost of healthcare is the most important financial problem facing their family today is between 11.2% and 16.8%.

41. (0.466, 0.514)

With 80% confidence, you can say that the population proportion of parents with kids 4 to 8 years old who say they know their state booster seat law is between 46.6% and 51.4%.

43. (a) 385 adults (b) 359 adults

(c) Having an estimate of the population proportion reduces the minimum sample size needed.

45. 23.337, 4.404 **47.** 14.067, 2.167

49. (27.2, 113.5); (5.2, 10.7) **51.** (0.80, 3.07); (0.89, 1.75)

Chapter Quiz for Chapter 6 *(page 349)*

1. (a) 6.85

(b) 0.65; You are 95% confident that the margin of error for the population mean is about 0.65 minute.

(c) (6.20, 7.50)

With 95% confidence, you can say that the population mean amount of time is between 6.20 and 7.50 minutes.

2. 39 college students

3. (a) 33.11; 2.38

(b) (31.73, 34.49)

With 90% confidence, you can say that the population mean time played in the season is between 31.73 and 34.49 minutes.

(c) (30.38, 35.84)

With 90% confidence, you can say that the population mean time played in the season is between 30.38 and 35.84 minutes. This confidence interval is wider than the one found in part (b).

4. (6510, 7138)

5. (a) 0.780 (b) (0.762, 0.798) [*Tech:* (0.762, 0.799)]

(c) 712 adults

6. (a) (2.10, 5.99) (b) (1.45, 2.45)

Real Statistics–Real Decisions for Chapter 6 *(page 350)*

1. (a) Yes, there has been a change in the mean concentration level because the confidence interval for Year 1 does not overlap the confidence interval for Year 2.

(b) No, there has not been a change in the mean concentration level because the confidence interval for Year 2 overlaps the confidence interval for Year 3.

(c) Yes, there has been a change in the mean concentration level because the confidence interval for Year 1 does not overlap the confidence interval for Year 3.

2. The concentrations of cyanide in the drinking water have increased over the three-year period.

3. The width of the confidence interval for Year 2 may have been caused by greater variation in the levels of cyanide than in the other years, which may be the result of outliers.

4. (a) The sampling distribution of the sample means was used because the "mean concentration" was used. The sample mean is the most unbiased point estimate of the population mean.

(b) No, because typically σ is unknown. They could have used the sample standard deviation.

CHAPTER 7

Section 7.1 *(page 367)*

1. The two types of hypotheses used in a hypothesis test are the null hypothesis and the alternative hypothesis.

The alternative hypothesis is the complement of the null hypothesis.

3. You can reject the null hypothesis, or you can fail to reject the null hypothesis.

5. False. In a hypothesis test, you assume the null hypothesis is true.

7. True

9. False. A small P-value in a test will favor rejection of the null hypothesis.

11. $H_0: \mu \leq 645$ (claim); $H_a: \mu > 645$

13. $H_0: \sigma = 5$; $H_a: \sigma \neq 5$ (claim)

15. $H_0: p \geq 0.45$; $H_a: p < 0.45$ (claim)

17. c; $H_0: \mu \leq 3$ **18.** d; $H_0: \mu \geq 3$

19. b; $H_0: \mu = 3$ **20.** a; $H_0: \mu \leq 2$

21. Right-tailed **23.** Two-tailed

25. $\mu > 750$

$H_0: \mu \leq 750$; $H_a: \mu > 750$ (claim)

27. $\sigma \leq 320$

$H_0: \sigma \leq 320$ (claim); $H_a: \sigma > 320$

29. $\mu < 45$

$H_0: \mu \geq 45$; $H_a: \mu < 45$ (claim)

31. A type I error will occur if the actual proportion of new customers who return to buy their next piece of furniture is at least 0.60, but you reject $H_0: p \geq 0.60$.

A type II error will occur if the actual proportion of new customers who return to buy their next piece of furniture is less than 0.60, but you fail to reject $H_0: p \geq 0.60$.

33. A type I error will occur if the actual standard deviation of the length of time to play a game is less than or equal to 12 minutes, but you reject $H_0: \sigma \leq 12$.

A type II error will occur if the actual standard deviation of the length of time to play a game is greater than 12 minutes, but you fail to reject $H_0: \sigma \leq 12$.

35. A type I error will occur if the actual proportion of applicants who become police officers is at most 0.20, but you reject H_0: $p \le 0.20$.

A type II error will occur if the actual proportion of applicants who become police officers is greater than 0.20, but you fail to reject H_0: $p \le 0.20$.

37. H_0: The proportion of homeowners who have a home security alarm is greater than or equal to 14%.

H_a: The proportion of homeowners who have a home security alarm is less than 14%.

H_0: $p \ge 0.14$; H_a: $p < 0.14$

Left-tailed because the alternative hypothesis contains $<$.

39. H_0: The standard deviation of the 18-hole scores for a golfer is greater than or equal to 2.1 strokes.

H_a: The standard deviation of the 18-hole scores for a golfer is less than 2.1 strokes.

H_0: $\sigma \ge 2.1$; H_a: $\sigma < 2.1$

Left-tailed because the alternative hypothesis contains $<$.

41. H_0: The mean length of the baseball team's games is greater than or equal to 2.5 hours.

H_a: The mean length of the baseball team's games is less than 2.5 hours.

H_0: $\mu \ge 2.5$; H_a: $\mu < 2.5$

Left-tailed because the alternative hypothesis contains $<$.

43. (a) There is enough evidence to support the scientist's claim that the mean incubation period for swan eggs is less than 40 days.

(b) There is not enough evidence to support the scientist's claim that the mean incubation period for swan eggs is less than 40 days.

45. (a) There is enough evidence to support the U.S. Department of Labor's claim that the proportion of full-time workers earning over $450 per week is greater than 75%.

(b) There is not enough evidence to support the U.S. Department of Labor's claim that the proportion of full-time workers earning over $450 per week is greater than 75%.

47. (a) There is enough evidence to support the researcher's claim that the proportion of people who have had no health care visits in the past year is less than 17%.

(b) There is not enough evidence to support the researcher's claim that the proportion of people who have had no health care visits in the past year is less than 17%.

49. H_0: $\mu \ge 60$; H_a: $\mu < 60$

51. (a) H_0: $\mu \ge 15$; H_a: $\mu < 15$

(b) H_0: $\mu \le 15$; H_a: $\mu > 15$

53. If you decrease α, you are decreasing the probability that you will reject H_0. Therefore, you are increasing the probability of failing to reject H_0. This could increase β, the probability of failing to reject H_0 when H_0 is false.

55. Yes; If the P-value is less than $\alpha = 0.05$, it is also less than $\alpha = 0.10$.

57. (a) Fail to reject H_0 because the confidence interval includes values greater than 70.

(b) Reject H_0 because the confidence interval is located entirely to the left of 70.

(c) Fail to reject H_0 because the confidence interval includes values greater than 70.

59. (a) Reject H_0 because the confidence interval is located entirely to the right of 0.20.

(b) Fail to reject H_0 because the confidence interval includes values less than 0.20.

(c) Fail to reject H_0 because the confidence interval includes values less than 0.20.

Section 7.2   *(page 381)*

1. In the z-test using rejection region(s), the test statistic is compared with critical values. The z-test using a P-value compares the P-value with the level of significance α.

3. $P = 0.0934$; Reject H_0.   **5.** $P = 0.0069$; Reject H_0.

7. $P = 0.0930$; Fail to reject H_0.

9. b   **10.** d   **11.** c   **12.** a

13. (a) Fail to reject H_0.

(b) Reject H_0.

15. Fail to reject H_0.

17. 1.645

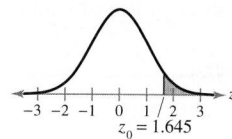

19. -1.88

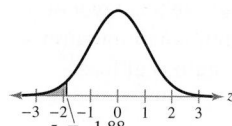

21. -2.33, 2.33

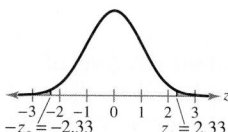

23. (a) Fail to reject H_0 because $z < 1.285$.

(b) Fail to reject H_0 because $z < 1.285$.

(c) Fail to reject H_0 because $z < 1.285$.

(d) Reject H_0 because $z > 1.285$.

25. Reject H_0. There is enough evidence at the 5% level of significance to reject the claim.

27. Reject H_0. There is enough evidence at the 2% level of significance to support the claim.

29. (a) $H_0: \mu \le 30$; $H_a: \mu > 30$ (claim) (b) 2.83; 0.9977
(c) 0.0023 (d) Reject H_0.
(e) There is enough evidence at the 1% level of significance to support the student's claim that the mean raw score for the school's applicants is more than 30.

31. (a) $H_0: \mu = 28.5$ (claim); $H_a: \mu \ne 28.5$
(b) -1.71; 0.0436 (c) 0.0872 (*Tech:* 0.0878)
(d) Fail to reject H_0.
(e) There is not enough evidence at the 8% level of significance to reject the U.S. Department of Agriculture's claim that the mean consumption of bottled water by a person in the United States is 28.5 gallons per year.

33. (a) $H_0: \mu = 15$ (claim); $H_a: \mu \ne 15$
(b) -0.22; 0.4129 (*Tech:* 0.4135)
(c) 0.8258 (*Tech:* 0.8270) (d) Fail to reject H_0.
(e) There is not enough evidence at the 5% level of significance to reject the claim that the mean time it takes smokers to quit smoking permanently is 15 years.

35. (a) $H_0: \mu = 40$ (claim); $H_a: \mu \ne 40$
(b) $-z_0 = -2.575$, $z_0 = 2.575$;
Rejection regions: $z < -2.575$, $z > 2.575$
(c) -0.584 (d) Fail to reject H_0.
(e) There is not enough evidence at the 1% level of significance to reject the company's claim that the mean caffeine content per 12-ounce bottle of cola is 40 milligrams.

37. (a) $H_0: \mu \ge 750$ (claim); $H_a: \mu < 750$
(b) $z_0 = -2.05$; Rejection region: $z < -2.05$
(c) -0.5 (d) Fail to reject H_0.
(e) There is not enough evidence at the 2% level of significance to reject the light bulb manufacturer's claim that the mean life of the bulb is at least 750 hours.

39. (a) $H_0: \mu \le 32$; $H_a: \mu > 32$ (claim)
(b) $z_0 = 1.555$; Rejection region: $z > 1.555$
(c) -1.478 (d) Fail to reject H_0.
(e) There is not enough evidence at the 6% level of significance to support the scientist's claim that the mean nitrogen dioxide level in Calgary is greater than 32 parts per billion.

41. (a) $H_0: \mu \ge 10$ (claim); $H_a: \mu < 10$
(b) $z_0 = -1.88$; Rejection region: $z < -1.88$
(c) -0.51 (d) Fail to reject H_0.
(e) There is not enough evidence at the 3% level of significance to reject the weight loss program's claim that the mean weight loss after one month is at least 10 pounds.

43. Hypothesis test results:

μ : population mean

$H_0: \mu = 58$

$H_A: \mu \ne 58$

Standard deviation = 2.35

Mean	n	Sample Mean	Std. Err.	Z-Stat	P-value
μ	80	57.6	0.262738	-1.5224292	0.1279

$P = 0.1279 > 0.10$, so fail to reject H_0. There is not enough evidence at the 10% level of significance to reject the claim.

45. Hypothesis test results:

μ : population mean

$H_0: \mu = 1210$

$H_A: \mu > 1210$

Standard deviation = 205.87

Mean	n	Sample Mean	Std. Err.	Z-Stat	P-value
μ	250	1234.21	13.020362	1.8593953	0.0315

$P = 0.0315 < 0.08$, so reject H_0. There is enough evidence at the 8% level of significance to reject the claim.

47. Fail to reject H_0 because the standardized test statistic $z = -1.86$ is not in the rejection region ($z < -2.33$).

49. b, d; If $\alpha = 0.05$, the rejection region is $z < -1.645$; because $z = -1.86$ is in the rejection region, you can reject H_0. If $n = 50$, the standardized test statistic is $z = -2.40$; because $z = -2.40$ is in the rejection region ($z < -2.33$), you can reject H_0.

Section 7.3 *(page 393)*

1. Identify the level of significance α and the degrees of freedom, d.f. $= n - 1$. Find the critical value(s) using the t-distribution table in the row with $n - 1$ d.f. If the hypothesis test is
(1) left-tailed, use the "One Tail, α" column with a negative sign.
(2) right-tailed, use the "One Tail, α" column with a positive sign.
(3) two-tailed, use the "Two Tails, α" column with a negative and a positive sign.

3. 1.717 **5.** -1.328 **7.** -2.056, 2.056

9. (a) Fail to reject H_0 because $t > -2.086$.
(b) Fail to reject H_0 because $t > -2.086$.
(c) Fail to reject H_0 because $t > -2.086$.
(d) Reject H_0 because $t < -2.086$.

11. (a) Fail to reject H_0 because $-2.602 < t < 2.602$.
(b) Fail to reject H_0 because $-2.602 < t < 2.602$.
(c) Reject H_0 because $t > 2.602$.
(d) Reject H_0 because $t < -2.602$.

13. Fail to reject H_0. There is not enough evidence at the 1% level of significance to reject the claim.

15. Reject H_0. There is enough evidence at the 1% level of significance to reject the claim.

17. (a) $H_0: \mu = 18{,}000$ (claim); $H_a: \mu \neq 18{,}000$

(b) $-t_0 = -2.145$, $t_0 = 2.145$;
Rejection regions: $t < -2.145$, $t > 2.145$

(c) 1.21   (d) Fail to reject H_0.

(e) There is not enough evidence at the 5% level of significance to reject the dealer's claim that the mean price of a 2008 Subaru Forester is $18,000.

19. (a) $H_0: \mu \leq 60$; $H_a: \mu > 60$ (claim)

(b) $t_0 = 1.943$; Rejection region: $t > 1.943$

(c) 2.12   (d) Reject H_0.

(e) There is enough evidence at the 5% level of significance to support the board's claim that the mean number of hours worked per week by surgical faculty who teach at an academic institution is more than 60 hours.

21. (a) $H_0: \mu \leq 1$; $H_a: \mu > 1$ (claim)

(b) $t_0 = 1.356$; Rejection region: $t > 1.356$

(c) 6.44   (d) Reject H_0.

(e) There is enough evidence at the 10% level of significance to support the environmentalist's claim that the mean amount of waste recycled by adults in the United States is more than 1 pound per person per day.

23. (a) $H_0: \mu = \$26{,}000$ (claim); $H_a: \mu \neq \$26{,}000$

(b) $-t_0 = -2.262$, $t_0 = 2.262$;
Rejection regions: $t < -2.262$, $t > 2.262$

(c) -0.15 (d) Fail to reject H_0.

(e) There is not enough evidence at the 5% level of significance to reject the employment information service's claim that the mean salary for full-time male workers over age 25 without a high school diploma is $26,000.

25. (a) $H_0: \mu \leq 45$; $H_a: \mu > 45$ (claim)

(b) 0.0052   (c) Reject H_0.

(d) There is enough evidence at the 10% level of significance to support the county's claim that the mean speed of the vehicles is greater than 45 miles per hour.

27. (a) $H_0: \mu = \$105$ (claim); $H_a: \mu \neq \$105$

(b) 0.0165   (c) Fail to reject H_0.

(d) There is not enough evidence at the 1% level of significance to reject the travel association's claim that the mean daily meal cost for two adults traveling together on vacation in San Francisco is $105.

29. (a) $H_0: \mu \geq 32$; $H_a: \mu < 32$ (claim)

(b) 0.0344   (c) Reject H_0.

(d) There is enough evidence at the 5% level of significance to support the brochure's claim that the mean class size for full-time faculty is fewer than 32 students.

31. Hypothesis test results:

μ : population mean

$H_0 : \mu = 75$

$H_A : \mu > 75$

Mean	Sample Mean	Std. Err.	DF	T-Stat	P-value
μ	73.6	0.62757164	25	-2.2308211	0.9825

$P = 0.9825 > 0.05$, so fail to reject H_0. There is not enough evidence at the 5% level of significance to reject the claim.

33. Hypothesis test results:

μ : population mean

$H_0 : \mu = 188$

$H_A : \mu < 188$

Mean	Sample Mean	Std. Err.	DF	T-Stat	P-value
μ	186	4	8	-0.5	0.3153

$P = 0.3153 > 0.05$, so fail to reject H_0. There is not enough evidence at the 5% level of significance to support the claim.

35. Because the P-value $= 0.0748 > 0.05$, fail to reject H_0.

37. Use the t-distribution because the population is normal, $n < 30$, and σ is unknown.

Fail to reject H_0. There is not enough evidence at the 5% level of significance to reject the car company's claim that the mean gas mileage for the luxury sedan is at least 23 miles per gallon.

39. More likely; For degrees of freedom less than 30, the tails of a t-distribution curve are thicker than those of a standard normal distribution curve. So, if you incorrectly use a standard normal sampling distribution instead of a t-sampling distribution, the area under the curve at the tails will be smaller than what it would be for the t-test, meaning the critical value(s) will lie closer to the mean. This makes it more likely for the test statistic to be in the rejection region(s). This result is the same regardless of whether the test is left-tailed, right-tailed, or two-tailed; in each case, the tail thickness affects the location of the critical value(s).

Section 7.3 Activity   *(page 397)*

1–3. Answers will vary.

Section 7.4   *(page 401)*

1. If $np \geq 5$ and $nq \geq 5$, the normal distribution can be used.

3. Cannot use normal distribution.

5. Can use normal distribution.

Fail to reject H_0. There is not enough evidence at the 5% level of significance to support the claim.

7. Can use normal distribution.

Fail to reject H_0. There is not enough evidence at the 5% level of significance to reject the claim.

9. (a) H_0: $p \geq 0.25$; H_a: $p < 0.25$ (claim)

(b) $z_0 = -1.645$; Rejection region: $z < -1.645$

(c) -2.12 (d) Reject H_0.

(e) There is enough evidence at the 5% level of significance to support the researcher's claim that less than 25% of U.S. adults are smokers.

11. (a) H_0: $p \leq 0.50$ (claim); H_a: $p > 0.50$

(b) $z_0 = 2.33$; Rejection region: $z > 2.33$

(c) 1.96   (d) Fail to reject H_0.

(e) There is not enough evidence at the 1% level of significance to reject the research center's claim that at most 50% of people believe that drivers should be allowed to use cellular phones with hands-free devices while driving.

13. (a) H_0: $p \leq 0.75$; H_a: $p > 0.75$ (claim)

(b) $z_0 = 1.28$; Rejection region: $z > 1.28$

(c) 1.98   (d) Reject H_0.

(e) There is enough evidence at the 10% level of significance to support the research center's claim that more than 75% of females ages 20–29 are taller than 62 inches.

15. (a) H_0: $p \geq 0.35$; H_a: $p < 0.35$ (claim)

(b) $z_0 = -1.28$; Rejection region: $z < -1.28$

(c) 1.68   (d) Fail to reject H_0.

(e) There is not enough evidence at the 10% level of significance to support the humane society's claim that less than 35% of U.S. households own a dog.

17. Fail to reject H_0. There is not enough evidence at the 5% level of significance to reject the claim that at least 52% of adults are more likely to buy a product when there are free samples.

19. (a) H_0: $p \geq 0.35$; H_a: $p < 0.35$ (claim)

(b) $z_0 = -1.28$; Rejection region: $z < -1.28$

(c) 1.68   (d) Fail to reject H_0.

(e) There is not enough evidence at the 10% level of significance to support the humane society's claim that less than 35% of U.S. households own a dog.

The results are the same.

Section 7.4 Activity *(page 403)*

1–2. Answers will vary.

Section 7.5 *(page 410)*

1. Specify the level of significance α. Determine the degrees of freedom. Determine the critical values using the χ^2- distribution. For a right-tailed test, use the value that corresponds to d.f. and α; for a left-tailed test, use the value that corresponds to d.f. and $1 - \alpha$; for a two-tailed test, use the values that correspond to d.f. and $\frac{1}{2}\alpha$, and d.f. and $1 - \frac{1}{2}\alpha$.

3. The requirement of a normal distribution is more important when testing a standard deviation than when testing a mean. If the population is not normal, the results of a χ^2-test can be misleading because the χ^2-test is not as robust as the tests for the population mean.

5. 38.885   **7.** 0.872   **9.** 60.391, 101.879

11. (a) Fail to reject H_0.   (b) Fail to reject H_0.

(c) Fail to reject H_0.   (d) Reject H_0.

13. (a) Fail to reject H_0.   (b) Reject H_0.

(c) Reject H_0.   (d) Fail to reject H_0.

15. Fail to reject H_0. There is not enough evidence at the 5% level of significance to reject the claim.

17. Reject H_0. There is enough evidence at the 10% level of significance to reject the claim.

19. (a) H_0: $\sigma^2 = 1.25$ (claim); H_a: $\sigma^2 \neq 1.25$

(b) $\chi_L^2 = 10.283$, $\chi_R^2 = 35.479$;

Rejection regions: $\chi^2 < 10.283$, $\chi^2 > 35.479$

(c) 22.68   (d) Fail to reject H_0.

(e) There is not enough evidence at the 5% level of significance to reject the manufacturer's claim that the variance of the number of grams of carbohydrates in servings of its tortilla chips is 1.25.

21. (a) H_0: $\sigma \geq 36$; H_a: $\sigma < 36$ (claim)

(b) $\chi_0^2 = 13.240$; Rejection region: $\chi^2 < 13.240$

(c) 18.076   (d) Fail to reject H_0.

(e) There is not enough evidence at the 10% level of significance to support the test administrator's claim that the standard deviation for eighth graders on the examination is less than 36 points.

23. (a) H_0: $\sigma \leq 25$ (claim); H_a: $\sigma > 25$

(b) $\chi_0^2 = 36.741$; Rejection region: $\chi^2 > 36.741$

(c) 41.515   (d) Reject H_0.

(e) There is enough evidence at the 10% level of significance to reject the weather service's claim that the standard deviation of the number of fatalities per year from tornadoes is no more than 25.

25. (a) H_0: $\sigma \geq \$3500$; H_a: $\sigma < \$3500$ (claim)

(b) $\chi_0^2 = 18.114$; Rejection region: $\chi^2 < 18.114$

(c) 37.051   (d) Fail to reject H_0.

(e) There is not enough evidence at the 10% level of significance to support the insurance agent's claim that the standard deviation of the total charges for patients involved in a crash in which the vehicle struck a construction barricade is less than $3500.

27. (a) H_0: $\sigma \leq 6100$; H_a: $\sigma > 6100$ (claim)

(b) $\chi_0^2 = 27.587$; Rejection region: $\chi^2 > 27.587$

(c) 27.897   (d) Reject H_0.

(e) There is enough evidence at the 5% level of significance to support the claim that the standard deviation of the annual salaries of environmental engineers is greater than $6100.

29. Hypothesis test results:

σ^2 : variance of Variable

$H_0 : \sigma^2 = 9$

$H_A : \sigma^2 < 9$

Variance	Sample Var.	DF	Chi-Square Stat	P-value
σ^2	2.03	9	2.03	0.009

Reject H_0. There is enough evidence at the 1% level of significance to reject the claim.

31. $\sigma^2 = 4.5^2 = 20.25$

$s^2 = 5.8^2 = 33.64$

Hypothesis test results:

σ^2 : variance of Variable

$H_0 : \sigma^2 = 20.25$

$H_A : \sigma^2 > 20.25$

Variance	Sample Var.	DF	Chi-Square Stat	P-value
σ^2	33.64	14	23.257284	0.0562

Fail to reject H_0. There is not enough evidence at the 5% level of significance to support the claim.

33. P-value = 0.9059   **35.** P-value = 0.0462

Fail to reject H_0.   Reject H_0.

Uses and Abuses for Chapter 7   *(page 413)*

1. Answers will vary.

2. H_0: $p = 0.73$; Answers will vary.

3. Answers will vary.

4. Answers will vary.

Review Answers for Chapter 7   *(page 417)*

1. H_0: $\mu \le 375$ (claim); H_a: $\mu > 375$

3. H_0: $p \ge 0.205$; H_a: $p < 0.205$ (claim)

5. H_0: $\sigma \le 1.9$; H_a: $\sigma > 1.9$ (claim)

7. (a) H_0: $p = 0.71$ (claim); H_a: $p \ne 0.71$

(b) A type I error will occur if the actual proportion of Americans who support plans to order deep cuts in executive compensation at companies that have received federal bailout funds is 71%, but you reject H_0: $p = 0.71$.

A type II error will occur if the actual proportion is not 71%, but you fail to reject H_0: $p = 0.71$.

(c) Two-tailed because the alternative hypothesis contains $\ne$.

(d) There is enough evidence to reject the news outlet's claim that the proportion of Americans who support plans to order deep cuts in executive compensation at companies that have received federal bailout funds is 71%.

(e) There is not enough evidence to reject the news outlet's claim that the proportion of Americans who support plans to order deep cuts in executive compensation at companies that have received federal bailout funds is 71%.

9. (a) H_0: $\sigma \le 50$ (claim); H_a: $\sigma > 50$

(b) A type I error will occur if the actual standard deviation of the sodium content in one serving of a certain soup is no more than 50 milligrams, but you reject H_0: $\sigma \le 50$.

A type II error will occur if the actual standard deviation of the sodium content in one serving of a certain soup is more than 50 milligrams, but you fail to reject H_0: $\sigma \le 50$.

(c) Right-tailed because the alternative hypothesis contains $>$.

(d) There is enough evidence to reject the soup maker's claim that the standard deviation of the sodium content in one serving of a certain soup is no more than 50 milligrams.

(e) There is not enough evidence to reject the soup maker's claim that the standard deviation of the sodium content in one serving of a certain soup is no more than 50 milligrams.

11. 0.1736; Fail to reject H_0.

13. H_0: $\mu \le 0.05$ (claim); H_a: $\mu > 0.05$

$z = 2.20$; P-value = 0.0139

$\alpha = 0.10 \Rightarrow$ Reject H_0.

$\alpha = 0.05 \Rightarrow$ Reject H_0.

$\alpha = 0.01 \Rightarrow$ Fail to reject H_0.

15. -2.05

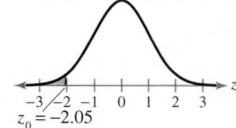

17. 1.96

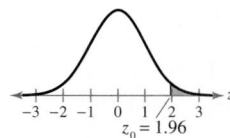

19. Fail to reject H_0 because $-1.645 < z < 1.645$.

21. Fail to reject H_0 because $-1.645 < z < 1.645$.

23. Reject H_0. There is enough evidence at the 5% level of significance to reject the claim.

25. Fail to reject H_0. There is not enough evidence at the 1% level of significance to support the claim.

27. Fail to reject H_0. There is not enough evidence at the 1% level of significance to reject the U.S. Department of Agriculture's claim that the mean cost of raising a child from birth to age 2 by husband-wife families in rural areas is $10,380.

29. $-2.093, 2.093$ **31.** -2.977

33. Fail to reject H_0. There is not enough evidence at the 5% level of significance to support the claim.

35. Fail to reject H_0. There is not enough evidence at the 10% level of significance to reject the claim.

37. Reject H_0. There is enough evidence at the 1% level of significance to reject the claim.

39. Fail to reject H_0. There is not enough evidence at the 10% level of significance to reject the advertisement's claim that the mean monthly cost of joining a health club is $25.

41. There is not enough evidence at the 1% level of significance to reject the education publication's claim that the mean expenditure per student in public elementary and secondary schools is at least $10,200.

43. Can use normal distribution.

Fail to reject H_0. There is not enough evidence at the 5% level of significance to reject the claim.

45. Can use normal distribution.

Fail to reject H_0. There is not enough evidence at the 8% level of significance to support the claim.

47. Cannot use normal distribution.

49. Can use normal distribution.

Fail to reject H_0. There is not enough evidence at the 2% level of significance to support the claim.

51. Reject H_0. There is enough evidence at the 2% level of significance to support the polling agency's claim that over 16% of U.S. adults are without health care coverage.

53. 30.144   **55.** 63.167

57. Reject H_0. There is enough evidence at the 10% level of significance to support the claim.

59. Fail to reject H_0. There is not enough evidence at the 5% level of significance to reject the claim.

61. Reject H_0. There is enough evidence at the 0.5% level of significance to reject the bolt manufacturer's claim that the variance is at most 0.01.

63. You can reject H_0 at the 5% level of significance because $\chi^2 = 43.94 > 41.923$.

Chapter Quiz for Chapter 7   *(page 421)*

1. (a) H_0: $\mu \geq 170$ (claim); H_a: $\mu < 170$

(b) One-tailed because the alternative hypothesis contains $<$; z-test because $n \geq 30$

(c) $z_0 = -1.88$; Rejection region: $z < -1.88$

(d) -2.59

(e) Reject H_0.

(f) There is enough evidence at the 3% level of significance to reject the service's claim that the mean consumption of vegetables and melons by people in the United States is at least 170 pounds per person.

2. (a) H_0: $\mu \geq 7.25$ (claim); H_a: $\mu < 7.25$

(b) One-tailed because the alternative hypothesis contains $<$; t-test because $n < 30$, σ is unknown, and the population is normally distributed

(c) $t_0 = -1.796$; Rejection region: $t < -1.796$

(d) -1.283

(e) Fail to reject H_0.

(f) There is not enough evidence at the 5% level of significance to reject the company's claim that the mean hat size for a male is at least 7.25.

3. (a) H_0: $p \leq 0.10$ (claim); H_a: $p > 0.10$

(b) One-tailed because the alternative hypothesis contains $>$; z-test because $np > 5$ and $nq > 5$

(c) $z_0 = 1.75$; Rejection region: $z > 1.75$

(d) 0.75

(e) Fail to reject H_0.

(f) There is not enough evidence at the 4% level of significance to reject the microwave oven maker's claim that no more than 10% of its microwaves need repair during the first 5 years of use.

4. (a) H_0: $\sigma = 112$ (claim); H_a: $\sigma \neq 112$

(b) Two-tailed because the alternative hypothesis contains $\neq$; χ^2-test because the test is for a standard deviation and the population is normally distributed

(c) $\chi_L^2 = 9.390$, $\chi_R^2 = 28.869$;
Rejection regions: $\chi^2 < 9.390$, $\chi^2 > 28.869$

(d) 29.343

(e) Reject H_0.

(f) There is enough evidence at the 10% level of significance to reject the state school administrator's claim that the standard deviation of SAT critical reading test scores is 112.

5. (a) H_0: $\mu = \$62,569$ (claim); H_a: $\mu \neq \$62,569$

(b) Two-tailed because the alternative hypothesis contains $\neq$; t-test because $n < 30$, σ is unknown, and the population is normally distributed

(c) Not necessary   (d) $-2.175; 0.0473$

(e) Reject H_0.

(f) There is enough evidence at the 5% level of significance to reject the agency's claim that the mean income for full-time workers ages 25 to 34 with a master's degree is $62,569.

6. (a) H_0: $\mu = \$201$ (claim); H_a: $\mu \neq \$201$

(b) Two-tailed because the alternative hypothesis contains $\neq$; z-test because $n \geq 30$

(c) Not necessary

(d) 0.0030 (*Tech:* 0.0031)

(e) Reject H_0.

(f) There is enough evidence at the 5% level of significance to reject the tourist agency's claim that the mean daily cost of meals and lodging for a family of 4 traveling in the state of Kansas is $201.

1. (a)–(c) Answers will vary.

2. Fail to reject H_0. There is not enough evidence at the 5% level of significance to support PepsiCo's claim that more than 50% of cola drinkers prefer Pepsi® over Coca-Cola®.

3. Knowing the brand may influence participants' decisions.

4. (a)–(c) Answers will vary.

CHAPTER 8

Section 8.1 *(page 434)*

1. Two samples are dependent if each member of one sample corresponds to a member of the other sample. Example: The weights of 22 people before starting an exercise program and the weights of the same 22 people 6 weeks after starting the exercise program.

Two samples are independent if the sample selected from one population is not related to the sample selected from the other population. Example: The weights of 25 cats and the weights of 20 dogs.

3. Use *P*-values.

5. Independent because different students were sampled.

7. Dependent because the same football players were sampled.

9. Independent because different boats were sampled.

11. Dependent because the same tire sets were sampled.

13. (a) 2   (b) 2.95   (c) In the rejection region.

(d) Reject H_0. There is enough evidence at the 1% level of significance to reject the claim.

15. (a) 3   (b) 0.18   (c) Not in the rejection region.

(d) Fail to reject H_0. There is not enough evidence at the 5% level of significance to support the claim.

17. Fail to reject H_0. There is not enough evidence at the 1% level of significance to support the claim.

19. Reject H_0.

21. (a) The claim is "the mean braking distances are different for the two types of tires."

$H_0: \mu_1 = \mu_2$; $H_a: \mu_1 \neq \mu_2$ (claim)

(b) $-z_0 = -1.645$, $z_0 = 1.645$;

Rejection regions: $z < -1.645$, $z > 1.645$

(c) -2.786 (d) Reject H_0.

(e) There is enough evidence at the 10% level of significance to support the safety engineer's claim that the mean braking distances are different for the two types of tires.

23. (a) The claim is "Region A's average wind speed is greater than Region B's."

$H_0: \mu_1 \leq \mu_2$; $H_a: \mu_1 > \mu_2$ (claim)

(b) $z_0 = 1.645$; Rejection region: $z > 1.645$

(c) 1.53   (d) Fail to reject H_0.

(e) There is not enough evidence at the 5% level of significance to conclude that Region A's average wind speed is greater than Region B's.

25. (a) The claim is "male and female high school students have equal ACT scores."

$H_0: \mu_1 = \mu_2$ (claim); $H_a: \mu_1 \neq \mu_2$

(b) $-z_0 = -2.575$, $z_0 = 2.575$;

Rejection regions: $z < -2.575$, $z > 2.575$

(c) 0.202   (d) Fail to reject H_0.

(e) There is not enough evidence at the 1% level of significance to reject the claim that male and female high school students have equal ACT scores.

27. (a) The claim is "the average home sales price in Dallas, Texas is the same as in Austin, Texas."

$H_0: \mu_1 = \mu_2$ (claim); $H_a: \mu_1 \neq \mu_2$

(b) $-z_0 = -1.645$, $z_0 = 1.645$;

Rejection regions: $z < -1.645$, $z > 1.645$

(c) -1.30 (d) Fail to reject H_0.

(e) There is not enough evidence at the 10% level of significance to reject the real estate agency's claim that the average home sales price in Dallas, Texas is the same as in Austin, Texas.

29. (a) The claim is "the average home sales price in Dallas, Texas is the same as in Austin, Texas."

$H_0: \mu_1 = \mu_2$ (claim); $H_a: \mu_1 \neq \mu_2$

(b) $-z_0 = -1.645$, $z_0 = 1.645$;

Rejection regions: $z < -1.645$, $z > 1.645$

(c) 1.86   (d) Reject H_0.

(e) There is enough evidence at the 10% level of significance to reject the real estate agency's claim that the average home sales price in Dallas, Texas is the same as in Austin, Texas.

The new samples do lead to a different conclusion.

31. (a) The claim is "children ages 6–17 spent more time watching television in 1981 than children ages 6–17 do today."

$H_0: \mu_1 \leq \mu_2$; $H_a: \mu_1 > \mu_2$ (claim)

(b) $z_0 = 1.96$; Rejection region: $z > 1.96$

(c) 3.01   (d) Reject H_0.

(e) There is enough evidence at the 2.5% level of significance to support the sociologist's claim that children ages 6–17 spent more time watching television in 1981 than children ages 6–17 do today.

33. (a) The claim is "there is no difference in the mean washer diameter manufactured by two different methods."

$H_0: \mu_1 = \mu_2$ (claim); $H_a: \mu_1 \neq \mu_2$

(b) $-z_0 = -2.575$, $z_0 = 2.575$;

Rejection regions: $z < -2.575$, $z > 2.575$

(c) 64.978   (d) Reject H_0.

(e) There is enough evidence at the 1% level of significance to reject the production engineer's claim that there is no difference in the mean washer diameter manufactured by two different methods.

35. They are equivalent through algebraic manipulation of the equation.

$$\mu_1 = \mu_2 \Rightarrow \mu_1 - \mu_2 = 0$$

37. Hypothesis test results:

μ_1 : mean of population 1 (Std. Dev. = 5.4)

μ_2 : mean of population 2 (Std. Dev. = 7.5)

$\mu_1 - \mu_2$: mean difference

$H_0 : \mu_1 - \mu_2 = 0$

$H_A : \mu_1 - \mu_2 \neq 0$

Difference	n_1	n_2	Sample Mean
$\mu_1 - \mu_2$	50	45	4

Std. Err.	Z-Stat	P-value
1.3539572	2.9543033	0.0031

$P = 0.0031 < 0.01$, so reject H_0.

There is enough evidence at the 1% level of significance to support the claim.

39. Hypothesis test results:

μ_1 : mean of population 1 (Std. Dev. = 0.92)

μ_2 : mean of population 2 (Std. Dev. = 0.73)

$\mu_1 - \mu_2$: mean difference

$H_0 : \mu_1 - \mu_2 = 0$

$H_A : \mu_1 - \mu_2 < 0$

Difference	n_1	n_2	Sample Mean
$\mu_1 - \mu_2$	35	40	-0.32

Std. Err.	Z-Stat	P-value
0.193663	-1.6523548	0.0492

$P = 0.0492 < 0.05$, so reject H_0.

There is enough evidence at the 5% level of significance to reject the claim.

41. $H_0: \mu_1 - \mu_2 = -9$ (claim); $H_a: \mu_1 - \mu_2 \neq -9$

Fail to reject H_0.

There is not enough evidence at the 1% level of significance to reject the claim that children spend 9 hours a week more in day care or preschool today than in 1981.

43. $H_0: \mu_1 - \mu_2 \leq 10,000$; $H_a: \mu_1 - \mu_2 > 10,000$ (claim)

Reject H_0. There is enough evidence at the 5% level of significance to support the claim that the difference in mean annual salaries of microbiologists in Maryland and California is more than $10,000.

45. $-3.6 < \mu_1 - \mu_2 < -0.2$

47. $H_0: \mu_1 \geq \mu_2$; $H_a: \mu_1 < \mu_2$ (claim)

Reject H_0. There is enough evidence at the 5% level of significance to support the claim. You should recommend the DASH diet and exercise program over the traditional diet and exercise program because the mean systolic blood pressure was significantly lower in the DASH program.

49. The 95% CI for $\mu_1 - \mu_2$ in Exercise 45 contained only values less than 0 and, as found in Exercise 47, there was enough evidence at the 5% level of significance to support the claim.

If the CI for $\mu_1 - \mu_2$ contains only negative numbers, you reject H_0 because the null hypothesis states that $\mu_1 - \mu_2$ is greater than or equal to 0.

Section 8.2 *(page 446)*

1. (1) The samples must be randomly selected.

(2) The samples must be independent.

(3) Each population must have a normal distribution.

3. (a) $-t_0 = -1.714$, $t_0 = 1.714$

(b) $-t_0 = -1.812$, $t_0 = 1.812$

5. (a) $t_0 = -1.746$

(b) $t_0 = -1.943$

7. (a) $t_0 = 1.729$

(b) $t_0 = 1.895$

9. (a) -1.8 (b) -1.70

(c) Not in the rejection region.

(d) Fail to reject H_0.

11. (a) 105   (b) 2.05

(c) In the rejection region.

(d) Reject H_0.

13. (a) The claim is "the mean annual costs of routine veterinarian visits for dogs and cats are the same."

$H_0: \mu_1 = \mu_2$ (claim); $H_a: \mu_1 \neq \mu_2$

(b) $-t_0 = -1.943$, $t_0 = 1.943$;

Rejection regions: $t < -1.943$, $t > 1.943$

(c) 1.90   (d) Fail to reject H_0.

(e) There is not enough evidence at the 10% level of significance to reject the pet association's claim that the mean annual costs of routine veterinarian visits for dogs and cats are the same.

15. (a) The claim is "the mean bumper repair cost is less for mini cars than for midsize cars."

$H_0: \mu_1 \geq \mu_2$; $H_a: \mu_1 < \mu_2$ (claim)

(b) $t_0 = -1.325$; Rejection region: $t < -1.325$

(c) -0.93 (d) Fail to reject H_0.

(e) There is not enough evidence at the 10% level of significance to support the claim that the mean bumper repair cost is less for mini cars than for midsize cars.

17. (a) The claim is "the mean household income is greater in Allegheny County than it is in Erie County."

$H_0: \mu_1 \leq \mu_2$; $H_a: \mu_1 > \mu_2$ (claim)

(b) $t_0 = 1.761$; Rejection region: $t > 1.761$

(c) 1.99   (d) Reject H_0.

(e) There is enough evidence at the 5% level of significance to support the personnel director's claim that the mean household income is greater in Allegheny County than it is in Erie County.

19. (a) The claim is "the new treatment makes a difference in the tensile strength of steel bars."

$H_0: \mu_1 = \mu_2$; $H_a: \mu_1 \neq \mu_2$ (claim)

(b) $-t_0 = -2.831$, $t_0 = 2.831$;

Rejection regions: $t < -2.831$, $t > 2.831$

(c) -2.76 (d) Fail to reject H_0.

(e) There is not enough evidence at the 1% level of significance to support the claim that the new treatment makes a difference in the tensile strength of steel bars.

21. (a) The claim is "the new method of teaching reading produces higher reading test scores than the old method."

$H_0: \mu_1 \geq \mu_2$; $H_a: \mu_1 < \mu_2$ (claim)

(b) $t_0 = -1.282$; Rejection region: $t < -1.282$

(c) -4.295 (d) Reject H_0.

(e) There is enough evidence at the 10% level of significance to support the claim that the new method of teaching reading produces higher reading test scores than the old method and to recommend changing to the new method.

23. Hypothesis test results:

μ_1 : mean of population 1

μ_2 : mean of population 2

$\mu_1 - \mu_2$: mean difference

$H_0: \mu_1 - \mu_2 = 0$

$H_A: \mu_1 - \mu_2 > 0$

(with pooled variances)

Difference	Sample Mean	Std. Err.
$\mu_1 - \mu_2$	-8	16.985794

DF	T-Stat	P-value
22	-0.47098184	0.6789

$P = 0.6789 > 0.10$, so fail to reject H_0.

There is not enough evidence at the 10% level of significance to support the claim.

25. Hypothesis test results:

μ_1 : mean of population 1

μ_2 : mean of population 2

$\mu_1 - \mu_2$: mean difference

$H_0: \mu_1 - \mu_2 = 0$

$H_A: \mu_1 - \mu_2 < 0$

(without pooled variances)

Difference	Sample Mean	Std. Err.
$\mu_1 - \mu_2$	-43	28.12301

DF	T-Stat	P-value
18.990595	-1.5289971	0.0714

$P = 0.0714 > 0.05$, so fail to reject H_0.

There is not enough evidence at the 5% level of significance to reject the claim.

27. $45 < \mu_1 - \mu_2 < 307$ **29.** $11 < \mu_1 - \mu_2 < 35$

Section 8.3   *(page 456)*

1. (1) Each sample must be randomly selected.

(2) Each member of the first sample must be paired with a member of the second sample.

(3) Both populations must be normally distributed.

3. Left-tailed test; Fail to reject H_0.

5. Right-tailed test; Reject H_0.

7. Left-tailed test; Reject H_0.

9. (a) The claim is "a grammar seminar will help students reduce the number of grammatical errors."

$H_0: \mu_d \leq 0$; $H_a: \mu_d > 0$ (claim)

(b) $t_0 = 3.143$; Rejection region: $t > 3.143$

(c) $\overline{d} \approx 3.143$; $s_d \approx 2.035$

(d) 4.085   (e) Reject H_0.

(f) There is enough evidence at the 1% level of significance to support the teacher's claim that a grammar seminar will help students reduce the number of grammatical errors.

11. (a) The claim is "a particular exercise program will help participants lose weight after one month."

$H_0: \mu_d \leq 0$; $H_a: \mu_d > 0$ (claim)

(b) $t_0 = 1.363$; Rejection region: $t > 1.363$

(c) $\overline{d} = 3.75$; $s_d \approx 7.841$

(d) 1.657   (e) Reject H_0.

(f) There is enough evidence at the 10% level of significance to support the nutritionist's claim that the exercise program helps participants lose weight after one month.

13. (a) The claim is "soft tissue therapy and spinal manipulation help to reduce the length of time patients suffer from headaches."

$H_0: \mu_d \leq 0$; $H_a: \mu_d > 0$ (claim)

(b) $t_0 = 2.764$; Rejection region: $t > 2.764$

(c) $\overline{d} \approx 1.255$; $s_d \approx 0.441$

(d) 9.429   (e) Reject H_0.

(f) There is enough evidence at the 1% level of significance to support the physical therapist's claim that soft tissue therapy and spinal manipulation help reduce the length of time patients suffer from headaches.

15. (a) The claim is "the new drug reduces systolic blood pressure."

$H_0: \mu_d \leq 0$; $H_a: \mu_d > 0$ (claim)

(b) $t_0 = 1.895$; Rejection region: $t > 1.895$

(c) $\overline{d} = 14.75$; $s_d \approx 6.861$

(d) 6.081   (e) Reject H_0.

(f) There is enough evidence at the 5% level of significance to support the pharmaceutical company's claim that its new drug reduces systolic blood pressure.

17. (a) The claim is "the product ratings have changed from last year to this year."

H_0: $\mu_d = 0$; H_a: $\mu_d \neq 0$ (claim)

(b) $-t_0 = -2.365$, $t_0 = 2.365$

Rejection regions: $t < -2.365$, $t > 2.365$

(c) $\bar{d} = -1$; $s_d \approx 1.309$

(d) -2.160 (e) Fail to reject H_0.

(f) There is not enough evidence at the 5% level of significance to support the claim that the product ratings have changed from last year to this year.

19. Hypothesis test results:

$\mu_1 - \mu_2$: mean of the paired difference between Cholesterol (before) and Cholesterol (after)

$H_0 : \mu_1 - \mu_2 = 0$

$H_A : \mu_1 - \mu_2 > 0$

Difference	Sample Diff.
Cholesterol (before) − Cholesterol (after)	2.857143

Std. Err.	DF	T-Stat	P-value
1.6822401	6	1.6984155	0.0702

$P = 0.0702 > 0.05$, so fail to reject H_0.

There is not enough evidence at the 5% level of significance to support the claim that the new cereal lowers total blood cholesterol levels.

21. Yes; $P \approx 0.0003 < 0.05$, so you reject H_0.

23. $-1.76 < \mu_d < -1.29$

Section 8.4 (page 465)

1. (1) The samples must be randomly selected.

(2) The samples must be independent.

(3) $n_1\bar{p} \geq 5$, $n_1\bar{q} \geq 5$, $n_2\bar{p} \geq 5$, and $n_2\bar{q} \geq 5$

3. Can use normal sampling distribution; Fail to reject H_0.

5. Can use normal sampling distribution; Reject H_0.

7. Can use normal sampling distribution; Fail to reject H_0.

9. (a) The claim is "there is a difference in the proportion of subjects who feel all or mostly better after 4 weeks between subjects who used magnetic insoles and subjects who used nonmagnetic insoles."

H_0: $p_1 = p_2$; H_a: $p_1 \neq p_2$ (claim)

(b) $-z_0 = -2.575$, $z_0 = 2.575$;

Rejection regions: $z < -2.575$, $z > 2.575$

(c) -1.24 (d) Fail to reject H_0.

(e) There is not enough evidence at the 1% level of significance to support the claim that there is a difference in the proportion of subjects who feel all or mostly better after 4 weeks between subjects who used magnetic insoles and subjects who used nonmagnetic insoles.

11. (a) The claim is "the proportion of males who enrolled in college is less than the proportion of females who enrolled in college."

H_0: $p_1 \geq p_2$; H_a: $p_1 < p_2$ (claim)

(b) $z_0 = -1.645$; Rejection region: $z < -1.645$

(c) -4.22 (d) Reject H_0.

(e) There is enough evidence at the 5% level of significance to support the claim that the proportion of males who enrolled in college is less than the proportion of females who enrolled in college.

13. (a) The claim is "the proportion of subjects who are pain-free is the same for the two groups."

H_0: $p_1 = p_2$ (claim); H_a: $p_1 \neq p_2$

(b) $-z_0 = -1.96$, $z_0 = 1.96$;

Rejection regions: $z < -1.96$, $z > 1.96$

(c) 5.62 (*Tech:* 5.58) (d) Reject H_0.

(e) There is enough evidence at the 5% level of significance to reject the claim that the proportion of subjects who are pain-free is the same for the two groups.

15. (a) The claim is "the proportion of motorcyclists who wear a helmet is now greater."

H_0: $p_1 \leq p_2$; H_a: $p_1 > p_2$ (claim)

(b) $z_0 = 1.645$; Rejection region: $z > 1.645$

(c) 1.37 (d) Fail to reject H_0.

(e) There is not enough evidence at the 5% level of significance to support the claim that the proportion of motorcyclists who wear a helmet is now greater.

17. (a) The claim is "the proportion of Internet users is the same for the two age groups."

H_0: $p_1 = p_2$ (claim); H_a: $p_1 \neq p_2$

(b) $-z_0 = -2.575$, $z_0 = 2.575$;

Rejection regions: $z < -2.575$, $z > 2.575$

(c) 5.31 (d) Reject H_0.

(e) There is enough evidence at the 1% level of significance to reject the claim that the proportion of Internet users is the same for the two age groups.

19. There is enough evidence at the 5% level of significance to reject the claim that the proportion of customers who wait 20 minutes or less is the same at the Fairfax North and Fairfax South offices.

21. There is enough evidence at the 10% level of significance to support the claim that the proportion of customers who wait 20 minutes or less at the Roanoke office is less than the proportion of customers who wait 20 minutes or less at the Staunton office.

23. No; When $\alpha = 0.01$, the rejection region becomes $z < -2.33$. Because $-2.02 > -2.33$, you fail to reject H_0. There is not enough evidence at the 1% level of significance to support the claim that the proportion of customers who wait 20 minutes or less at the Roanoke office is less than the proportion of customers who wait 20 minutes or less at the Staunton office.

25. Hypothesis test results:

p_1 : proportion of successes for population 1

p_2 : proportion of successes for population 2

$p_1 - p_2$: difference in proportions

$H_0 : p_1 - p_2 = 0$

$H_A : p_1 - p_2 > 0$

Difference	Count1	Total1	Count2	Total2
$p_1 - p_2$	7501	13300	8120	14500

Sample Diff.	Std. Err.	Z-Stat	P-value
0.0039849626	0.0059570055	0.66895396	0.2518

$P = 0.2518 > 0.05$, so fail to reject H_0.

There is not enough evidence at the 5% level of significance to support the claim that the proportion of men ages 18 to 24 living in their parents' homes was greater in 2000 than in 2009.

27. Hypothesis test results:

p_1 : proportion of successes for population 1

p_2 : proportion of successes for population 2

$p_1 - p_2$: difference in proportions

$H_0 : p_1 - p_2 = 0$

$H_A : p_1 - p_2 \neq 0$

Difference	Count1	Total1	Count2	Total2
$p_1 - p_2$	7501	13300	5610	13200

Sample Diff.	Std. Err.	Z-Stat	P-value
0.13898496	0.006142657	22.626196	<0.0001

$P < 0.0001 < 0.01$, so reject H_0.

There is enough evidence at the 1% level of significance to reject the claim that the proportion of 18- to 24-year-olds living in their parents' homes in 2000 was the same for men and women.

29. $-0.028 < p_1 - p_2 < -0.012$

Uses and Abuses for Chapter 8 *(page 469)*

1. Answers will vary.

2. Blind: The patients do not know which group (medicine or placebo) they belong to.

Double Blind: Both the researcher and patient do not know which group (medicine or placebo) that the patient belongs to.

Review Answers for Chapter 8 *(page 471)*

1. Dependent because the same cities were sampled.

3. Fail to reject H_0. There is not enough evidence at the 5% level of significance to reject the claim.

5. Reject H_0. There is enough evidence at the 10% level of significance to support the claim.

7. (a) The claim is "the Wendy's fish sandwich has less sodium than the Long John Silver's fish sandwich."

$H_0: \mu_1 \geq \mu_2$; $H_a: \mu_1 < \mu_2$ (claim)

(b) $-z_0 = -1.645$; Rejection region: $z < -1.645$

(c) -9.20 (d) Reject H_0.

(e) There is enough evidence at the 5% level of significance to support the claim that the Wendy's fish sandwich has less sodium than the Long John Silver's fish sandwich.

9. Yes; The new rejection region is $z < -2.33$, which contains $z = -9.20$, so you still reject H_0.

11. Reject H_0. There is enough evidence at the 5% level of significance to reject the claim.

13. Fail to reject H_0. There is not enough evidence at the 5% level of significance to reject the claim.

15. Reject H_0. There is enough evidence at the 1% level of significance to support the claim.

17. (a) The claim is "third graders taught with the directed reading activities scored higher than those taught without the activities."

$H_0: \mu_1 \leq \mu_2$; $H_a: \mu_1 > \mu_2$ (claim)

(b) $t_0 = 1.645$; Rejection region: $t > 1.645$

(c) 2.267 (d) Reject H_0.

(e) There is enough evidence at the 5% level of significance to support the claim that third graders taught with the directed reading activities scored higher than those taught without the activities.

19. Two-tailed test; Reject H_0.

21. Right-tailed test; Reject H_0.

23. (a) The claim is "the men's systolic blood pressure decreased."

$H_0: \mu_d \leq 0$; $H_a: \mu_d > 0$ (claim)

(b) $t_0 = 1.383$; Rejection region: $t > 1.383$

(c) $\bar{d} = 5$; $s_d \approx 8.743$ (d) 1.808 (e) Reject H_0.

(f) There is enough evidence at the 10% level of significance to support the claim that the men's systolic blood pressure decreased.

25. Can use normal sampling distribution; Fail to reject H_0.

27. Can use normal sampling distribution; Reject H_0.

29. (a) The claim is "the proportions of U.S. adults who considered the amount of federal income tax they had to pay to be too high were the same for the two years."

$H_0: p_1 = p_2$ (claim); $H_a: p_1 \neq p_2$

(b) $-z_0 = -2.575$, $z_0 = 2.575$;

Rejection regions: $z < -2.575$, $z > 2.575$

(c) 2.65 (d) Reject H_0.

(e) There is enough evidence at the 1% level of significance to reject the claim that the proportions of U.S. adults who considered the amount of federal income tax they had to pay to be too high were the same for the two years.

31. Yes; When $\alpha = 0.05$, the rejection regions become $z < -1.96$ and $z > 1.96$. Because $2.65 > 1.96$, you still reject H_0. There is enough evidence at the 5% level of significance to reject the claim that the proportions of U.S. adults who considered the amount of federal income tax they had to pay to be too high were the same for the two years.

Chapter Quiz for Chapter 8 *(page 475)*

1. (a) $H_0: \mu_1 \le \mu_2$; $H_a: \mu_1 > \mu_2$ (claim)

(b) One-tailed because H_a contains $>$; z-test because n_1 and n_2 are each greater than 30.

(c) $z_0 = 1.645$; Rejection region: $z > 1.645$

(d) 0.585   (e) Fail to reject H_0.

(f) There is not enough evidence at the 5% level of significance to support the claim that the mean score on the science assessment for the male high school students was higher than for the female high school students.

2. (a) $H_0: \mu_1 = \mu_2$ (claim); $H_a: \mu_1 \ne \mu_2$

(b) Two-tailed because H_a contains $\ne$; t-test because n_1 and n_2 are less than 30, the samples are independent, and the populations are normally distributed.

(c) $-t_0 = -2.779$, $t_0 = 2.779$;

Rejection regions: $t < -2.779$, $t > 2.779$

(d) 0.341   (e) Fail to reject H_0.

(f) There is not enough evidence at the 1% level of significance to reject the teacher's claim that the mean scores on the science assessment test are the same for fourth grade boys and girls.

3. (a) $H_0: p_1 = p_2$ (claim); $H_a: p_1 \ne p_2$

(b) Two-tailed because H_a contains $\ne$; z-test because you are testing proportions and $n_1 \bar{p}$, $n_1 \bar{q}$, $n_2 \bar{p}$, and $n_2 \bar{q} \ge 5$.

(c) $-z_0 = 1.645$, $z_0 = 1.645$;

Rejection regions: $z < -1.645$, $z > 1.645$

(d) 1.32 (e) Fail to reject H_0.

(f) There is not enough evidence at the 10% level of significance to reject the claim that the proportion of U.S. adults who are worried that they or someone in their family will become a victim of terrorism has not changed.

4. (a) $H_0: \mu_d \ge 0$; $H_a: \mu_d < 0$ (claim)

(b) One-tailed because H_a contains $<$; t-test because both populations are normally distributed and the samples are dependent.

(c) $t_0 = -2.718$; Rejection region: $t < -2.718$

(d) -5.07 (e) Reject H_0.

(f) There is enough evidence at the 1% level of significance to support the claim that the seminar helps adults increase their credit scores.

Real Statistics–Real Decisions for Chapter 8 *(page 476)*

1. (a) Answers will vary. *Sample answer:* Divide the records into groups according to the inpatients' ages, and then randomly select records from each group.

(b) Answers will vary. *Sample answer:* Divide the records into groups according to geographic regions, and then randomly select records from each group.

(c) Answers will vary. *Sample answer:* Assign a different number to each record, randomly choose a starting number, and then select every 50th record.

(d) Answers will vary. *Sample answer:* Assign a different number to each record, and then use a table of random numbers to generate a sample of numbers.

2. (a) Answers will vary.   (b) Answers will vary.

3. Use a t-test; independent; yes, you need to know if the population distributions are normal or not; yes. you need to know if the population variances are equal or not.

4. There is not enough evidence at the 10% level of significance to support the claim that there is a difference in the mean length of hospital stays for inpatients.

This decision does not support the calim.

Cumulative Review Chapters 6–8 *(page 480)*

1. (a) $(0.109, 0.151)$

(b) There is enough evidence at the 5% level of significance to support the researcher's claim that more than 10% of people who attend community college are age 40 or older.

2. There is enough evidence at the 10% level of significance to support the claim that the fuel additive improved gas mileage.

3. $(25.94, 28.00)$; z-distribution

4. $(2.75, 4.17)$; t-distribution

5. $(10.7, 13.5)$; t-distribution

6. $(7.69, 8.73)$; t-distribution

7. There is enough evidence at the 10% level of significance to support the pediatrician's claim that the mean birth weight of a single-birth baby is greater than the mean birth weight of a baby that has a twin.

8. $H_0: \mu \ge 33$; $H_a: \mu < 33$ (claim)

9. $H_0: p \ge 0.19$ (claim); $H_a: p < 0.19$

10. $H_0: \sigma = 0.63$ (claim); $H_a: \sigma \ne 0.63$

11. $H_0: \mu = 2.28$; $H_a: \mu \ne 2.28$ (claim)

12. (a) $(5.1, 22.8)$ (b) $(2.3, 4.8)$

(c) There is not enough evidence at the 1% level of significance to support the pharmacist's claim that the standard deviation of the mean number of chronic medications taken by elderly adults in the community is less than 2.5 medications.

13. There is enough evidence at the 5% level of significance to support the organization's claim that the mean SAT scores for male athletes and male non-athletes at a college are different.

14. (a) (37,732.2, 40,060.7)

(b) There is not enough evidence at the 5% level of significance to reject the claim that the mean annual earnings for translators is $40,000.

15. There is not enough evidence at the 10% level of significance to reject the claim that the proportions of players sustaining head and neck injuries are the same for the two groups,

16. (a) (41.5, 42.5)

(b) There is enough evidence at the 5% level of significance to reject the zoologist's claim that the mean incubation period for ostriches is at least 45 days.

CHAPTER 9

Section 9.1   *(page 495)*

1. Increase

3. The range of values for the correlation coefficient is -1 to 1, inclusive.

5. Answers will vary. *Sample answer:*

Perfect positive linear correlation: price per gallon of gasoline and total cost of gasoline

Perfect negative linear correlation: distance from door and height of wheelchair ramp

7. r is the sample correlation coefficient, while ρ is the population correlation coefficient.

9. Negative linear correlation

11. Perfect negative linear correlation

13. Positive linear correlation

15. c; You would expect a positive linear correlation between age and income.

16. d; You would not expect age and height to be correlated.

17. b; You would expect a negative linear correlation between age and balance on student loans.

18. a; You would expect the relationship between age and body temperature to be fairly constant.

19. Explanatory variable: Amount of water consumed

Response variable: Weight loss

21. (a)

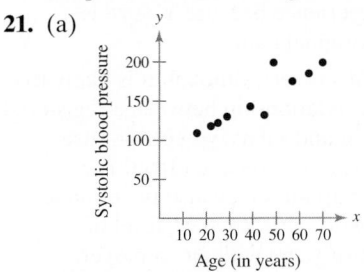

(b) 0.908

(c) Strong positive linear correlation

23. (a)

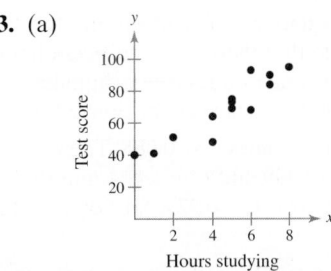

(b) 0.923   (c) Strong positive linear correlation

25. (a)

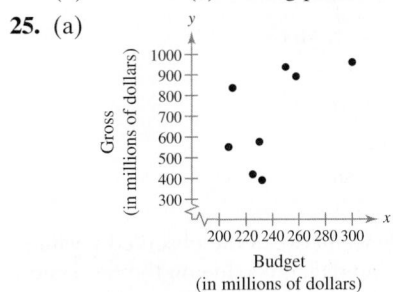

(b) 0.604   (c) Weak positive linear correlation

27. (a)

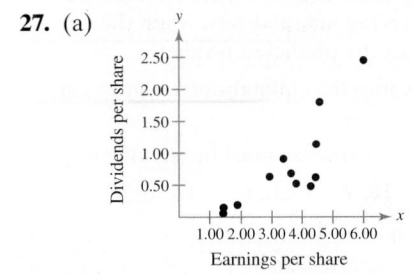

(b) 0.828   (c) Strong positive linear correlation

29. The correlation coefficient becomes $r \approx 0.621$. The new data entry is an outlier, so the linear correlation is weaker.

31. There is not enough evidence at the 1% level of significance to conclude that there is a significant linear correlation between vehicle weight and the variability in braking distance.

33. There is enough evidence at the 1% level of significance to conclude that there is a significant linear correlation between the number of hours spent studying for a test and the score received on the test.

35. There is enough evidence at the 1% level of significance to conclude that there is a significant linear correlation between earnings per share and dividends per share.

37. (a)

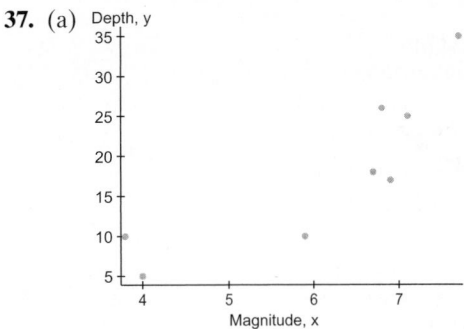

(b) 0.848

(c) Reject H_0. There is enough evidence at the 1% level of significance to conclude that there is a significant linear correlation between the magnitudes of earthquakes and their depths below the surface at the epicenter.

39. The correlation coefficient becomes $r \approx 0.085$. The new rejection regions are $t < -3.499$ and $t > 3.499$, and the new standardized test statistic is $t \approx 0.227$. So, you now fail to reject H_0.

41. 0.883; 0.883; The correlation coefficient remains unchanged when the x-values and y-values are switched.

43. Answers will vary.

Activity 9.1 *(page 500)*

1–4. Answers will vary.

Section 9.2 *(page 505)*

1. A residual is the difference between the observed y-value of a data point and the predicted y-value on the regression line for the x-coordinate of the data point. A residual is positive when the data point is above the line, negative when the point is below the line, and zero when the observed y-value equals the predicted y-value.

3. Substitute a value of x into the equation of a regression line and solve for y.

5. The correlation between variables must be significant.

7. b   **8.** a   **9.** e   **10.** c   **11.** f   **12.** d

13. c   **14.** b   **15.** a   **16.** d

17. $\hat{y} = 0.065x + 0.465$

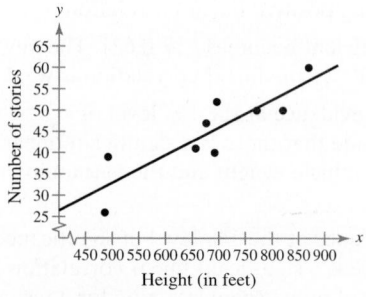

(a) 52 stories   (b) 49 stories

(c) It is not meaningful to predict the value of y for $x = 400$ because $x = 400$ is outside the range of the original data.

(d) 41 stories

19. $\hat{y} = 7.350x + 34.617$

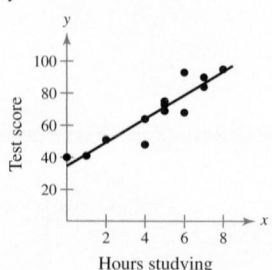

(a) 57   (b) 82

(c) It is not meaningful to predict the value of y for $x = 13$ because $x = 13$ is outside the range of the original data.

(d) 68

21. $\hat{y} = 2.472x + 80.813$

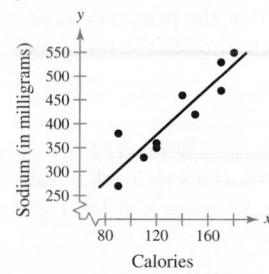

(a) 501.053 milligrams

(b) 328.013 milligrams

(c) 426.893 milligrams

(d) It is not meaningful to predict the value of y for $x = 210$ because $x = 210$ is outside the range of the original data.

23. $\hat{y} = 1.870x + 51.360$

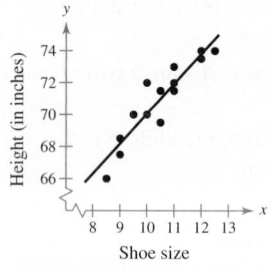

(a) 72.865 inches

(b) 66.32 inches

(c) It is not meaningful to predict the value of y for $x = 15.5$ because $x = 15.5$ is outside the range of the original data.

(d) 70.06 inches

25. Strong positive linear correlation; As the years of experience of the registered nurses increase, their salaries tend to increase.

27. No, it is not meaningful to predict a salary for a registered nurse with 28 years of experience because $x = 28$ is outside the range of the original data.

29. Answers will vary. *Sample answer:* Although it is likely that there is a cause-and-effect relationship between a registered nurse's years of experience and salary, you cannot use significant correlation to claim cause and effect. The relationship between the variables may also be influenced by other factors, such as work performance, level of education, or the number of years with an employer.

31. (a) $\hat{y} = -0.159x + 5.827$

(b) -0.852

(c)

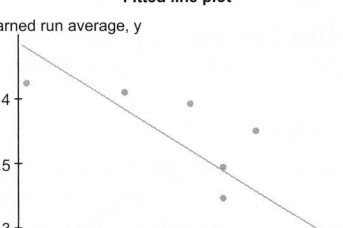

33. (a) $\hat{y} = -4.297x + 94.200$

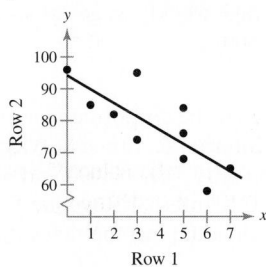

(b) $\hat{y} = -0.141x + 14.763$

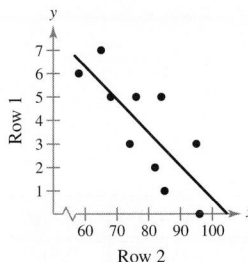

(c) The slope of the line keeps the same sign, but the values of m and b change.

35. (a) $\hat{y} = 0.139x + 21.024$

(b)

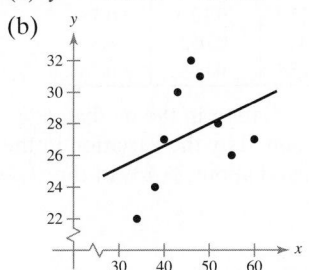

(c)

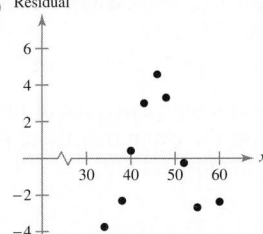

(d) The residual plot shows a pattern because the residuals do not fluctuate about 0. This implies that the regression line is not a good representation of the relationship between the two variables.

37. (a)

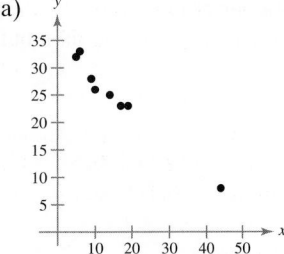

(b) The point $(44, 8)$ may be an outlier.

(c) The point $(44, 8)$ is not an influential point because the slopes and y-intercepts of the regression lines with the point included and without the point included are not significantly different.

39. $\hat{y} = 654.536x - 1214.857$

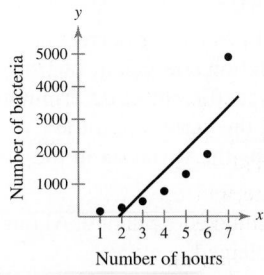

41. $y = 93.028(1.712)^x$

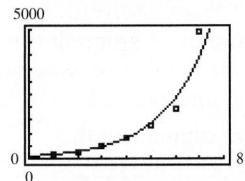

43. $\hat{y} = -78.929x + 576.179$

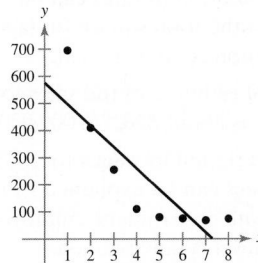

45. $y = 782.300x^{-1.251}$

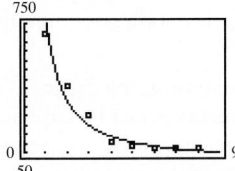

47. $y = 25.035 + 19.599 \ln x$

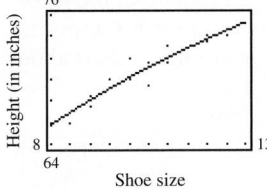

49. The logarithmic equation is a better model for the data. The graph of the logarithmic equation fits the data better than the regression line.

Activity 9.2 *(page 511)*

1–4. Answers will vary.

Section 9.3 *(page 519)*

1. The total variation is the sum of the squares of the differences between the y-values of each ordered pair and the mean of the y-values of the ordered pairs, or $\Sigma(y_i - \bar{y})^2$.

3. The unexplained variation is the sum of the squares of the differences between the observed y-values and the predicted y-values, or $\Sigma(y_i - \hat{y}_i)^2$.

5. Two variables that have perfect positive or perfect negative linear correlation have a correlation coefficient of 1 or -1, respectively. In either case, the coefficient of determination is 1, which means that 100% of the variation in the response variable is explained by the variation in the explanatory variable.

7. 0.216; About 21.6% of the variation is explained. About 78.4% of the variation is unexplained.

9. 0.916; About 91.6% of the variation is explained. About 8.4% of the variation is unexplained.

11. (a) 0.798; About 79.8% of the variation in proceeds can be explained by the variation in the number of issues, and about 20.2% of the variation is unexplained.

(b) 8064.633; The standard error of estimate of the proceeds for a specific number of issues is about $8,064,633,000.

13. (a) 0.981; About 98.1% of the variation in sales can be explained by the variation in the total square footage, and about 1.9% of the variation is unexplained.

(b) 30.576; The standard error of estimate of the sales for a specific total square footage is about $30,576,000,000.

15. (a) 0.963; About 96.3% of the variation in wages for federal government employees can be explained by the variation in wages for state government employees, and about 3.7% of the variation is unexplained.

(b) 20.090; The standard error of estimate of the average weekly wages for federal government employees for a specific average weekly wage for state government employees is about $20.09.

17. (a) 0.790; About 79.0% of the variation in the gross collections of corporate income taxes can be explained by the variation in the gross collections of individual income taxes, and about 21.0% of the variation is unexplained.

(b) 42.386; The standard error of estimate of the gross collections of corporate income taxes for a specific gross collection of individual income taxes is about $42,386,000,000.

19. $40,116.824 < y < 82,624.318$

You can be 95% confident that the proceeds will be between $40,116,824,000 and $82,624,318,000 when the number of initial offerings is 450 issues.

21. $1218.435 < y < 1336.829$

You can be 90% confident that the shopping center sales will be between $1,218,435,000 and $1,336,829,000 when the total square footage of shopping centers is 5,750,000,000.

23. $1007.82 < y < 1208.228$

You can be 99% confident that the average weekly wages of federal government employees will be between $1007.82 and $1208.23 when the average weekly wages of state government employees is $800.

25. $213.729 < y < 450.519$

You can be 95% confident that the corporate income taxes collected by the U.S. Internal Revenue Service for a given year will be between $213,729,000,000 and $450,519,000,000 when the U.S. Internal Revenue Service collects $1,250,000,000 in individual income taxes that year.

27.

29.

x_i	y_i	$\hat{y}_i$	$\hat{y}_i - \bar{y}$	$y_i - \hat{y}_i$	$y_i - \bar{y}$
9.4	7.6	7.1252	0.4372	0.4748	0.912
9.2	6.9	7.0736	0.3856	-0.1736	0.212
8.9	6.6	6.9962	0.3082	-0.3962	-0.088
8.4	6.8	6.8672	0.1792	-0.0672	0.112
8.3	6.9	6.8414	0.1534	0.0586	0.212
6.5	6.5	6.377	-0.311	0.123	-0.188
6	6.3	6.248	-0.44	0.052	-0.388
4.9	5.9	5.9642	-0.7238	-0.0642	-0.788

31. 0.746; About 74.6% of the variation in the median ages of trucks in use can be explained by the variation in the median ages of cars in use, and about 25.4% of the variation is unexplained.

33. $5.792 < y < 7.22$

You can be 95% confident that the median age of trucks in use will be between 5.792 and 7.22 years when the median age of cars in use is 7.0 years.

35. (a) 0.671   (b) 1.780   (c) $9.537 < y < 19.010$

37. Fail to reject H_0. There is not enough evidence at the 1% level of significance to support the claim that there is a linear relationship between weight and number of hours slept.

39. $-118.927 < B < 323.505$

$110.911 < M < 281.393$

Section 9.4 *(page 527)*

1. (a) 39,103.5 pounds per acre
   (b) 39,939.1 pounds per acre
   (c) 38,063.5 pounds per acre
   (d) 39,052.4 pounds per acre

3. (a) 7.5 cubic feet   (b) 16.8 cubic feet
   (c) 51.9 cubic feet   (d) 62.1 cubic feet

5. $\hat{y} = -2518.364 + 126.822x_1 + 66.360x_2$
   (a) 28.489; The standard error of estimate of the predicted sales given specific total square footage and number of shopping centers is about \$28.489 billion.
   (b) 0.985; The multiple regression model explains about 98.5% of the variation in y.

7. $\hat{y} = -2518.364 + 126.822x_1 + 66.360x_2$; The equation is the same.

9. 0.981; About 98.1% of the variation in y can be explained by the relationship between variables; $r^2_{adj} < r^2$.

Uses and Abuses for Chapter 9 *(page 529)*

1. Answers will vary.   2. Answers will vary.

Review Answers for Chapter 9 *(page 531)*

1.
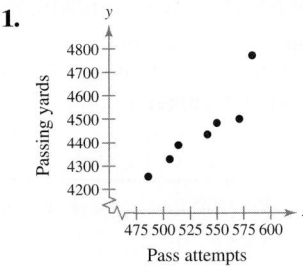
0.912; strong positive linear correlation; the number of passing yards increases as the number of pass attempts increases.

3.
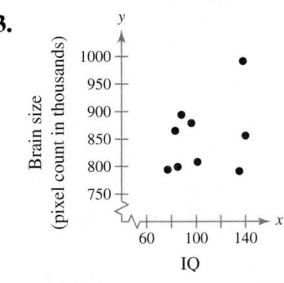
0.338; weak positive linear correlation; brain size increases as IQ increases.

5. There is not enough evidence at the 1% level of significance to conclude that there is a significant linear correlation.

7. There is enough evidence at the 5% level of significance to conclude that there is a significant linear correlation between a quarterback's pass attempts and passing yards.

9. There is not enough evidence at the 1% level of significance to conclude that there is a significant linear correlation between IQ and brain size.

11. $\hat{y} = 0.038x - 3.529$

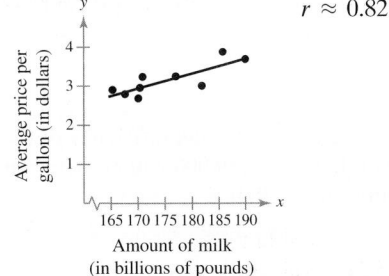
$r \approx 0.821$

13. $\hat{y} = -0.086x + 10.450$

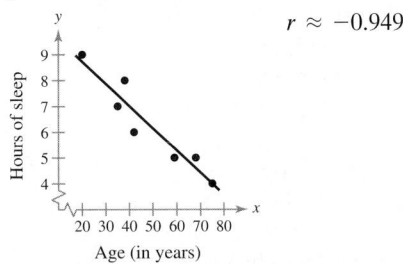
$r \approx -0.949$

15. (a) It is not meaningful to predict the value of y for $x = 160$ because $x = 160$ is outside the range of the original data.
   (b) \$3.12   (c) \$3.31
   (d) It is not meaningful to predict the value of y for $x = 200$ because $x = 200$ is outside the range of the original data.

17. (a) It is not meaningful to predict the value of y for $x = 18$ because $x = 18$ is outside the range of the original data.
   (b) 8.3 hours
   (c) It is not meaningful to predict the value of y for $x = 85$ because $x = 85$ is outside the range of the original data.
   (d) 6.15 hours

19. 0.203; About 20.3% of the variation is explained. About 79.7% of the variation is unexplained.

21. 0.412; About 41.2% of the variation is explained. About 58.8% of the variation is unexplained.

23. (a) 0.679; About 67.9% of the variation in the fuel efficiency of the compact sports sedans can be explained by the variation in their prices, and about 32.1% of the variation is unexplained.
   (b) 1.138; The standard error of estimate of the fuel efficiency of the compact sports sedans for a specific price of the compact sports sedans is about 1.138 miles per gallon.

25. $2.997 < y < 4.025$

   You can be 90% confident that the price per gallon of milk will be between \$3.00 and \$4.03 when 185 billion pounds of milk is produced.

27. $4.865 < y < 8.295$

You can be 95% confident that the hours slept will be between 4.865 and 8.295 hours for a person who is 45 years old.

29. $16.119 < y < 25.137$

You can be 99% confident that the fuel efficiency of the compact sports sedan that costs \$39,900 will be between 16.119 and 25.137 miles per gallon.

31. $\hat{y} = 3.674 + 1.287x_1 - 7.531x_2$

33. (a) 21.705   (b) 25.21   (c) 30.1   (d) 25.86

Chapter Quiz for Chapter 9   *(page 535)*

1.

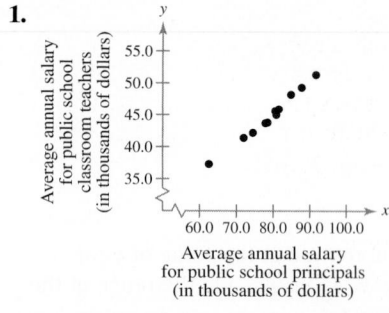

The data appear to have a positive linear correlation. As x increases, y tends to increase.

2. 0.993; strong positive linear correlation; public school classroom teachers' salaries increase as public school principals' salaries increase.

3. Reject H_0. There is enough evidence at the 5% level of significance to conclude that there is a significant linear correlation between public school principals' salaries and public school classroom teachers' salaries.

4. $\hat{y} = 0.491x + 5.977$

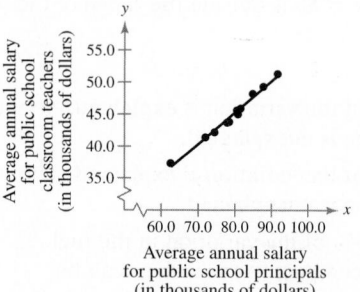

5. \$50,412.50

6. 0.986; About 98.6% of the variation in the average annual salaries of public school classroom teachers can be explained by the variation in the average annual salaries of public school principals, and about 1.4% of the variation is unexplained.

7. 0.490; The standard error of estimate of the average annual salary of public school classroom teachers for a specific average annual salary of public school principals is about \$490.

8. $46.887 < y < 49.273$

You can be 95% confident that the average annual salary of public school classroom teachers will be between \$46,887 and \$49,273 when the average annual salary of public school principals is \$85,750.

9. (a) \$59.30

(b) \$30.53

(c) \$45.67

(d) \$35.83

Real Statistics–Real Decisions for Chapter 9   *(page 536)*

1. (a)

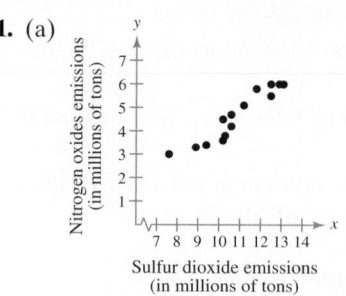

It appears that there is a positive linear correlation. As the sulfur dioxide emissions increase, the nitrogen oxides emissions increase.

(b) $r \approx 0.947$; There is a strong positive linear correlation.

(c) There is enough evidence at the 5% level of significance to conclude that there is a significant linear correlation between sulfur dioxide emissions and nitrogen oxides emissions.

(d) $\hat{y} = 0.652x - 2.438$

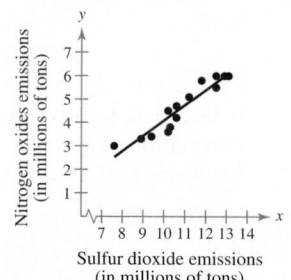

Yes, the line appears to be a good fit.

(e) Yes, for x-values that are within the range of the data set.

(f) $r^2 \approx 0.898$; About 89.8% of the variation in nitrogen oxides emissions can be explained by the variation in sulfur dioxide emissions, and about 10.2% of the variation is unexplained.

$s_e \approx 0.368$; The standard error of estimate of nitrogen oxides emissions for a specific sulfur dioxide emission is about 368,000 tons.

2. $1.358 < y < 3.286$

You can be 95% confident that the nitrogen oxide emissions will be between 1.358 and 3.286 million tons when the sulfur dioxide emissions are $17.3 - 10 = 7.3$ million tons.

CHAPTER 10

Section 10.1 (page 546)

1. A multinomial experiment is a probability experiment consisting of a fixed number of independent trials in which there are more than two possible outcomes for each trial. The probability of each outcome is fixed, and each outcome is classified into categories.

3. 45   5. 57.5

7. (a) H_0: The distribution of the ages of moviegoers is 26.7% ages 2–17, 19.8% ages 18–24, 19.7% ages 25–39, 14% ages 40–49, and 19.8% ages 50+. (claim)

   H_a: The distribution of ages differs from the claimed or expected distribution.

   (b) $\chi_0^2 = 7.779$; Rejection region: $\chi^2 > 7.779$

   (c) 7.256

   (d) Fail to reject H_0.

   (e) There is enough evidence at the 10% level of significance to conclude that the distribution of the ages of moviegoers and the claimed or expected distribution are the same.

9. (a) H_0: The distribution of the days people order food for delivery is 7% Sunday, 4% Monday, 6% Tuesday, 13% Wednesday, 10% Thursday, 36% Friday, and 24% Saturday.

   H_a: The distribution of days differs from the claimed or expected distribution. (claim)

   (b) $\chi_0^2 = 16.812$;

   Rejection region: $\chi^2 > 16.812$

   (c) 17.595

   (d) Reject H_0.

   (e) There is enough evidence at the 1% level of significance to conclude that there has been a change in the claimed or expected distribution.

11. (a) H_0: The distribution of the number of homicide crimes in California by season is uniform. (claim)

    H_a: The distribution of homicides by season is not uniform.

    (b) $\chi_0^2 = 7.815$; Rejection region: $\chi^2 > 7.815$

    (c) 0.727

    (d) Fail to reject H_0.

    (e) There is not enough evidence at the 5% level of significance to reject the claim that distribution of the number of homicide crimes in California by season is uniform.

13. (a) H_0: The distribution of the opinions of U.S. parents on whether a college education is worth the expense is 55% strongly agree, 30% somewhat agree, 5% neither agree nor disagree, 6% somewhat disagree, and 4% strongly disagree.

    H_a: The distribution of opinions differs from the claimed or expected distribution. (claim)

    (b) $\chi_0^2 = 9.488$; Rejection region: $\chi^2 > 9.488$

(c) 65.236

(d) Reject H_0.

(e) There is enough evidence at the 5% level of significance to conclude that the distribution of the opinions of U.S. parents on whether a college education is worth the expense differs from the claimed or expected distribution.

15. (a) H_0: The distribution of prospective home buyers by the size they want their next house to be is uniform. (claim)

    H_a: The distribution of prospective home buyers by the size they want their next house to be is not uniform.

    (b) $\chi_0^2 = 5.991$; Rejection region: $\chi^2 > 5.991$ (c) 10.308

    (d) Reject H_0.

    (e) There is enough evidence at the 5% level of significance to reject the claim that the distribution of prospective home buyers by the size they want their next house to be is uniform.

17. **Chi-Square goodness-of-fit results:**

    Observed: Recent survey

    Expected: Previous survey

N	DF	Chi-Square	P-Value
400	9	18.637629	0.0285

    $P = 0.0285$, so reject H_0. There is enough evidence at the 10% level of significance to conclude that there has been a change in the claimed or expected distribution of U.S. adults' favorite sports.

19. (a) The expected frequencies are 17, 63, 79, 34, and 5.

    (b) $\chi_0^2 = 13.277$; Rejection region: $\chi^2 > 13.277$

    (c) 0.613   (d) Fail to reject H_0.

    (e) There is not enough evidence at the 1% level of significance to reject the claim that the test scores are normally distributed.

Section 10.2 (page 557)

1. Find the sum of the row and the sum of the column in which the cell is located. Find the product of these sums. Divide the product by the sample size.

3. Answers will vary. *Sample answer:* For both the chi-square test for independence and the chi-square goodness-of-fit test, you are testing a claim about data that are in categories. However, the chi-square goodness-of-fit test has only one data value per category, while the chi-square test for independence has multiple data values per category.

   Both tests compare observed and expected frequencies. However, the chi-square goodness-of-fit test simply compares the distributions, whereas the chi-square test for independence compares them and then draws a conclusion about the dependence or independence of the variables.

5. False. If the two variables of a chi-square test for independence are dependent, then you can expect a large difference between the observed frequencies and the expected frequencies.

7. (a)–(b)

	Athlete has		
Result	Stretched	Not stretched	Total
Injury	18 (20.82)	22 (19.18)	40
No injury	211 (208.18)	189 (191.82)	400
Total	229	211	440

9. (a)–(b)

	Preference			
Bank employee	New procedure	Old procedure	No preference	Total
Teller	92 (133.80)	351 (313.00)	50 (46.19)	493
Customer service	76 (34.20)	42 (80.00)	8 (11.81)	126
Total	168	393	58	619

11. (a)–(b)

	Type of car				
Gender	Compact	Full-size	SUV	Truck/van	Total
Male	28 (28.6)	39 (39.05)	21 (22.55)	22 (19.8)	110
Female	24 (23.4)	32 (31.95)	20 (18.45)	14 (16.2)	90
Total	52	71	41	36	200

13. (a) H_0: Skill level in a subject is independent of location. (claim)

H_a: Skill level in a subject is dependent on location.

(b) d.f. $= 2$; $\chi_0^2 = 9.210$; Rejection region: $\chi^2 > 9.210$

(c) 0.297

(d) Fail to reject H_0. There is not enough evidence at the 1% level of significance to reject the claim that skill level in a subject is independent of location.

15. (a) H_0: The number of times former smokers tried to quit is independent of gender.

H_a: The number of times former smokers tried to quit is dependent on gender. (claim)

(b) d.f. $= 2$; $\chi_0^2 = 5.991$; Rejection region: $\chi^2 > 5.991$

(c) 0.002

(d) Fail to reject H_0. There is not enough evidence at the 5% level of significance to conclude that the number of times former smokers tried to quit is dependent on gender.

17. (a) H_0: Results are independent of the type of treatment.

H_a: Results are dependent on the type of treatment. (claim)

(b) d.f. $= 1$; $\chi_0^2 = 2.706$; Rejection region: $\chi^2 > 2.706$

(c) 5.106

(d) Reject H_0. There is enough evidence at the 10% level of significance to conclude that results are dependent on the type of treatment. Answers will vary.

19. (a) H_0: Reasons are independent of the type of worker.

H_a: Reasons are dependent on the type of worker. (claim)

(b) d.f. $= 2$; $\chi_0^2 = 9.210$; Rejection region: $\chi^2 > 9.210$

(c) 7.326

(d) Fail to reject H_0. There is not enough evidence at the 1% level of significance to conclude that reasons for continuing education are dependent on the type of worker. On the basis of these results, marketing strategies should not differ between technical and nontechnical audiences in regard to reasons for continuing education.

21. (a) H_0: Type of crash is independent of the type of vehicle.

H_a: Type of crash is dependent on the type of vehicle. (claim)

(b) d.f. $= 2$; $\chi_0^2 = 5.991$; Rejection region: $\chi^2 > 5.991$

(c) 144.099

(d) Reject H_0. There is enough evidence at the 5% level of significance to conclude that the type of crash is dependent on the type of vehicle.

23. (a)–(b) Contingency table results:

Rows: Expected income

Columns: None

Cell format
Count
Expected count

	More likely	Less likely	Did not make a difference
Less than $35,000	37 / 33.33	10 / 6.836	22 / 27.99
$35,000 to $50,000	28 / 25.17	12 / 5.164	15 / 21.14
$50,000 to $100,000	55 / 62.75	9 / 12.87	65 / 52.7
Greater than $100,000	36 / 34.75	1 / 7.127	29 / 29.18
Total	156	32	131

	Did not consider it	Total
Less than $35,000	25 / 25.85	94
$35,000 to $50,000	16 / 19.52	71
$50,000 to $100,000	48 / 48.68	177
Greater than $100,000	32 / 26.95	98
Total	121	440

Statistic	DF	Value	P-value
Chi-square	9	26.22966	0.0019

(c) Reject H_0. There is enough evidence at the 1% level of significance to conclude that the decision to borrow money is dependent on the child's expected income after graduation.

25. Fail to reject H_0. There is not enough evidence at the 5% level of significance to reject the claim that the proportions of motor vehicle crash deaths involving males or females are the same for each age group.

27. Right-tailed

29.

Status	Educational attainment			
	Not a high school graduate	High school graduate	Some college, no degree	Associate's, bachelor's, or advanced degree
Employed	0.055	0.183	0.114	0.290
Unemployed	0.006	0.011	0.005	0.007
Not in the labor force	0.073	0.118	0.053	0.085

31. Several of the expected frequencies are less than 5.

33. 45.2%

35.

Status	Educational attainment			
	Not a high school graduate	High school graduate	Some college, no degree	Associate's, bachelor's, or advanced degree
Employed	0.411	0.587	0.660	0.759
Unemployed	0.046	0.036	0.030	0.019
Not in the labor force	0.544	0.377	0.311	0.223

37. 4.6%

39. Answers will vary. *Sample answer:* As educational attainment increases, employment increases.

Section 10.3 *(page 571)*

1. Specify the level of significance α. Determine the degrees of freedom for the numerator and denominator. Use Table 7 in Appendix B to find the critical value F.

3. (1) The samples must be randomly selected, (2) the samples must be independent, and (3) each population must have a normal distribution.

5. 2.54   7. 2.06   9. 9.16   11. 1.80

13. Fail to reject H_0. There is not enough evidence at the 10% level of significance to support the claim.

15. Fail to reject H_0. There is not enough evidence at the 1% level of significance to reject the claim.

17. Reject H_0. There is enough evidence at the 1% level of significance to reject the claim.

19. (a) H_0: $\sigma_1^2 \le \sigma_2^2$; H_a: $\sigma_1^2 > \sigma_2^2$ (claim)
 (b) $F_0 = 2.11$; Rejection region: $F > 2.11$
 (c) 1.08
 (d) Fail to reject H_0.
 (e) There is not enough evidence at the 5% level of significance to support Company A's claim that the variance of the life of its appliances is less than the variance of the life of Company B's appliances.

21. (a) H_0: $\sigma_1^2 = \sigma_2^2$; H_a: $\sigma_1^2 \ne \sigma_2^2$ (claim)
 (b) $F_0 = 6.23$; Rejection region: $F > 6.23$
 (c) 2.10
 (d) Fail to reject H_0.
 (e) There is not enough evidence at the 5% level of significance to conclude that the variances of the prices differ between the two companies.

23. (a) H_0: $\sigma_1^2 = \sigma_2^2$ (claim); H_a: $\sigma_1^2 \ne \sigma_2^2$
 (b) $F_0 = 2.635$; Rejection region: $F > 2.635$
 (c) 1.282
 (d) Fail to reject H_0.
 (e) There is not enough evidence at the 10% level of significance to reject the administrator's claim that the standard deviations of science assessment test scores for eighth grade students are the same in Districts 1 and 2.

25. (a) H_0: $\sigma_1^2 \le \sigma_2^2$; H_a: $\sigma_1^2 > \sigma_2^2$ (claim)
 (b) $F_0 = 2.35$; Rejection region: $F > 2.35$
 (c) 2.41
 (d) Reject H_0.
 (e) There is enough evidence at the 5% level of significance to conclude that the standard deviation of the annual salaries for actuaries is greater in New York than in California.

27. **Hypothesis test results:**
 σ_1^2: variance of population 1
 σ_2^2: variance of population 2
 σ_1^2/σ_2^2: variance ratio
 H_0: $\sigma_1^2/\sigma_2^2 = 1$
 H_A: $\sigma_1^2/\sigma_2^2 \ne 1$

Ratio	n1	n2	Sample Ratio	F-Stat	P-value
σ_1^2/σ_2^2	15	18	0.5281571	0.5281571	0.2333

$P = 0.2333 > 0.10$, so fail to reject H_0. There is not enough evidence at the 10% level of significance to reject the claim.

29. Hypothesis test results:

σ_1^2: variance of population 1

σ_2^2: variance of population 2

σ_1^2/σ_2^2: variance ratio

$H_0: \sigma_1^2/\sigma_2^2 = 1$

$H_A: \sigma_1^2/\sigma_2^2 > 1$

Ratio	n1	n2	Sample Ratio	F-Stat	P-value
σ_1^2/σ_2^2	22	29	2.153926	2.153926	0.0293

$P = 0.0293 < 0.05$, so reject H_0. There is enough evidence at the 5% level of significance to reject the claim.

31. Right-tailed: 14.73 **33.** (0.375, 3.774)

Left-tailed: 0.15

Section 10.4 *(page 581)*

1. $H_0: \mu_1 = \mu_2 = \mu_3 = \ldots = \mu_k$

H_a: At least one of the means is different from the others.

3. The MS_B measures the differences related to the treatment given to each sample. The MS_W measures the differences related to entries within the same sample.

5. (a) $H_0: \mu_1 = \mu_2 = \mu_3$

H_a: At least one mean is different from the others. (claim)

(b) $F_0 = 3.37$;

Rejection region: $F > 3.37$

(c) 1.02

(d) Fail to reject H_0.

(e) There is not enough evidence at the 5% level of significance to conclude that the mean costs per ounce are different.

7. (a) $H_0: \mu_1 = \mu_2 = \mu_3$

H_a: At least one mean is different from the others. (claim)

(b) $F_0 = 5.49$; Rejection region: $F > 5.49$

(c) 21.99

(d) Reject H_0.

(e) There is enough evidence at the 1% level of significance to conclude that at least one mean salary is different.

9. (a) $H_0: \mu_1 = \mu_2 = \mu_3 = \mu_4 = \mu_5$

H_a: At least one mean is different from the others. (claim)

(b) $F_0 = 4.37$; Rejection region: $F > 4.37$

(c) 12.61

(d) Reject H_0.

(e) There is enough evidence at the 1% level of significance to conclude that at least one mean cost per mile is different.

11. (a) $H_0: \mu_1 = \mu_2 = \mu_3 = \mu_4$ (claim)

H_a: At least one mean is different from the others.

(b) $F_0 = 4.54$; Rejection region: $F > 4.54$

(c) 0.56

(d) Fail to reject H_0.

(e) There is not enough evidence at the 1% level of significance for the company to reject the claim that the mean number of days patients spend at the hospital is the same for all four regions.

13. (a) $H_0: \mu_1 = \mu_2 = \mu_3 = \mu_4$

H_a: At least one mean is different from the others. (claim)

(b) $F_0 = 2.255$; Rejection region: $F > 2.255$

(c) 3.107

(d) Reject H_0.

(e) There is enough evidence at the 10% level of significance to conclude that the mean energy consumption of at least one region is different from the others.

15. Analysis of Variance results:

Data stored in separate columns.

Column means

Column	n	Mean	Std. Error
Grade 9	8	84.375	9.531784
Grade 10	8	79.25	9.090321
Grade 11	8	76.625	7.648383
Grade 12	8	70.75	6.9224014

ANOVA table

Source	df	SS	MS
Treatments	3	771.25	257.08334
Error	28	15674.75	559.8125
Total	31	16446	

F-Stat	P-Value
0.45923114	0.7129

$P = 0.7129 > 0.01$, so fail to reject H_0. There is not enough evidence at the 1% level of significance to reject the claim that the mean numbers of female students who played on a sports team are equal for all grades.

17. Fail to reject all null hypotheses. The interaction between the advertising medium and the length of the ad has no effect on the rating and therefore there is no significant difference in the means of the ratings.

19. Fail to reject all null hypotheses. The interaction between age and gender has no effect on GPA and therefore there is no significant difference in the means of the GPAs.

21. $CV_{Scheffé} = 10.98$

$(1, 2) \rightarrow 22.396 \rightarrow$ Significant difference

$(1, 3) \rightarrow 40.837 \rightarrow$ Significant difference

$(2, 3) \rightarrow 2.749 \rightarrow$ No difference

23. $CV_{Scheffé} = 6.84$

$(1, 2) \rightarrow 0.032 \rightarrow$ No difference

$(1, 3) \rightarrow 1.490 \rightarrow$ No difference

$(1, 4) \rightarrow 1.345 \rightarrow$ No difference

$(2, 3) \rightarrow 2.374 \rightarrow$ No difference

$(2, 4) \rightarrow 1.122 \rightarrow$ No difference

$(3, 4) \rightarrow 7.499 \rightarrow$ Significant difference

Uses and Abuses for Chapter 10   *(page 587)*

1–2. Answers will vary.

Review Answers for Chapter 10   *(page 589)*

1. (a) H_0: The distribution of the allowance amounts is 29% less than \$10, 16% \$10 to \$20, 9% more than \$21, and 46% don't give one/other.

H_a: The distribution of amounts differs from the claimed or expected distribution. (claim)

(b) $\chi_0^2 = 6.251$; Rejection region: $\chi^2 > 6.251$

(c) 4.886   (d) Fail to reject H_0.

(e) There is not enough evidence at the 10% level of significance to conclude that there has been a change in the claimed or expected distribution.

3. (a) H_0: The distribution of responses from golf students about what they need the most help with is 22% approach and swing, 9% driver shots, 4% putting, and 65% short-game shots. (claim)

H_a: The distribution of responses differs from the claimed or expected distribution.

(b) $\chi_0^2 = 7.815$; Rejection region: $\chi^2 > 7.815$

(c) 0.503   (d) Fail to reject H_0.

(e) There is enough evidence at the 5% level of significance to conclude that the distribution of golf students' responses is the same as the claimed or expected distribution.

5. (a) $E_{1,1} = 63$, $E_{1,2} = 356.4$, $E_{1,3} = 319.8$, $E_{1,4} = 310.8$, $E_{2,1} = 147$, $E_{2,2} = 831.6$, $E_{2,3} = 746.2$, $E_{2,4} = 725.2$

(b) Reject H_0.

(c) There is enough evidence at the 1% level of significance to conclude that public school teachers' gender and years of full-time teaching experience are related.

7. (a) $E_{1,1} \approx 54.86$, $E_{1,2} \approx 40.38$, $E_{1,3} \approx 22.10$, $E_{1,4} = 16.00$, $E_{1,5} \approx 26.67$, $E_{2,1} \approx 17.14$, $E_{2,2} \approx 12.62$, $E_{2,3} \approx 6.90$, $E_{2,4} = 5.00$, $E_{2,5} \approx 8.33$

(b) Reject H_0.

(c) There is enough evidence at the 1% level of significance to conclude that a species' status (endangered or threatened) is dependent on vertebrate group.

9. 2.295   **11.** 2.39   **13.** 2.06   **15.** 2.08

17. Fail to reject H_0. There is not enough evidence at the 1% level of significance to reject the claim.

19. Fail to reject H_0. There is not enough evidence at the 10% level of significance to support the claim that the variation in wheat production is greater in Garfield County than in Kay County.

21. Fail to reject H_0. There is not enough evidence at the 1% level of significance to support the claim that the test score variance for females is different from that for males.

23. Reject H_0. There is enough evidence at the 10% level of significance to conclude that at least one of the mean costs is different from the others.

Chapter Quiz for Chapter 10   *(page 593)*

1. (a) H_0: $\sigma_1^2 = \sigma_2^2$; H_a: $\sigma_1^2 \neq \sigma_2^2$ (claim)

(b) 0.01   (c) $F_0 = 3.80$

(d) Rejection region: $F > 3.80$

(e) 2.12   (f) Fail to reject H_0.

(g) There is not enough evidence at the 1% level of significance to conclude that the variances in annual wages for San Francisco, CA and Baltimore, MD are different.

2. (a) H_0: $\mu_1 = \mu_2 = \mu_3$ (claim)

H_a: At least one mean is different from the others.

(b) 0.10   (c) $F_0 = 2.44$

(d) Rejection region: $F > 2.44$

(e) 27.48   (f) Reject H_0.

(g) There is enough evidence at the 10% level of significance to reject the claim that the mean annual wages are equal for all three cities.

3. (a) H_0: The distribution of educational achievement for people in the United States ages 35–44 is 13.4% not a high school graduate, 31.2% high school graduate, 17.2% some college, no degree; 8.8% associate's degree, 19.1% bachelor's degree, and 10.3% advanced degree.

H_a: The distribution of educational achievement for people in the United States ages 35–44 differs from the claimed distribution. (claim)

(b) 0.05   (c) $\chi_0^2 = 11.071$

(d) Rejection region: $\chi^2 > 11.071$

(e) 3.799   (f) Fail to reject H_0.

(g) There is not enough evidence at the 5% level of significance to conclude that the distribution for people in the United States ages 35–44 differs from the distribution for people ages 25 and older.

4. (a) H_0: The distribution of educational achievement for people in the United States ages 65–74 is 13.4% not a high school graduate, 31.2% high school graduate, 17.2% some college, no degree; 8.8% associate's degree, 19.1% bachelor's degree, and 10.3% advanced degree.

H_a: The distribution of educational achievement for people in the United States ages 65–74 differs from the claimed distribution. (claim)

(b) 0.01

(c) $\chi_0^2 = 15.086$

(d) Rejection region: $\chi^2 > 15.086$

(e) 26.175

(f) Reject H_0.

(g) There is enough evidence at the 1% level of significance to conclude that the distribution for people in the United States ages 65–74 differs from the distribution for people ages 25 and older.

Real Statistics–Real Decisions for Chapter 10 *(page 594)*

1. Reject H_0. There is enough evidence at the 1% level of significance to conclude that the distribution of responses differs from the claimed or expected distribution.

2. (a) $E_{1,1} = 15$, $E_{1,2} = 120$, $E_{1,3} = 165$, $E_{1,4} = 185$, $E_{1,5} = 135$, $E_{1,6} = 115$, $E_{1,7} = 155$, $E_{1,8} = 110$, $E_{2,1} = 15$, $E_{2,2} = 120$, $E_{2,3} = 165$, $E_{2,4} = 185$, $E_{2,5} = 135$, $E_{2,6} = 115$, $E_{2,7} = 155$, $E_{2,8} = 110$

(b) There is enough evidence at the 1% level of significance to conclude that the ages of the victims are related to the type of fraud.

CHAPTER 11

Section 11.1 *(page 604)*

1. A nonparametric test is a hypothesis test that does not require any specific conditions concerning the shapes of populations or the values of population parameters.

A nonparametric test is usually easier to perform than its corresponding parametric test, but the nonparametric test is usually less efficient.

3. When n is less than or equal to 25, the test statistic is equal to x (the smaller number of + or − signs).

When n is greater than 25, the test statistic is equal to

$$z = \frac{(x + 0.5) - 0.5n}{\frac{\sqrt{n}}{2}}.$$

5. Identify the claim and state H_0 and H_a. Identify the level of significance and sample size. Find the critical value using Table 8 (if $n \le 25$) or Table 4 ($n > 25$). Calculate the test statistic. Make a decision and interpret it in the context of the problem.

7. (a) H_0: median $\le$ \$300; H_a: median $>$ \$300 (claim)

(b) 1   (c) 5   (d) Fail to reject H_0.

(e) There is not enough evidence at the 1% level of significance for the accountant to conclude that the median amount of new credit card charges for the previous month was more than \$300.

9. (a) H_0: median $\le$ \$198,000 (claim)

H_a: median $>$ \$198,000

(b) 1   (c) 4   (d) Fail to reject H_0.

(e) There is not enough evidence at the 5% level of significance to reject the agent's claim that the median sales price of new privately owned one-family homes sold in the past year is \$198,000 or less.

11. (a) H_0: median $\ge$ \$3000 (claim); H_a: median $<$ \$3000

(b) −2.05   (c) −1.47   (d) Fail to reject H_0.

(e) There is not enough evidence at the 2% level of significance to reject the institution's claim that the median amount of credit card debt for families holding such debts is at least \$3000.

13. (a) H_0: median $\le$ 30; H_a: median $>$ 30 (claim)

(b) 4   (c) 10   (d) Fail to reject H_0.

(e) There is not enough evidence at the 1% level of significance to support the research group's claim that the median age of Twitter® users is greater than 30 years old.

15. (a) H_0: median $=$ 4 (claim); H_a: median $\ne$ 4

(b) −1.96   (c) −1.90   (d) Fail to reject H_0.

(e) There is not enough evidence at the 5% level of significance to reject the organization's claim that the median number of rooms in renter-occupied units is 4.

17. (a) H_0: median $=$ \$37.06 (claim); H_a: median $\ne$ \$37.06

(b) −2.575   (c) −0.91   (d) Fail to reject H_0.

(e) There is not enough evidence at the 1% level of significance to reject the labor organization's claim that the median hourly wage of computer systems analysts is \$37.06.

19. (a) H_0: The lower back pain intensity scores have not decreased.

H_a: The lower back pain intensity scores have decreased. (claim)

(b) 1   (c) 0   (d) Reject H_0.

(e) There is enough evidence at the 5% level of significance to conclude that the lower back pain intensity scores were lower after the acupuncture.

21. (a) H_0: The SAT scores have not improved.

H_a: The SAT scores have improved. (claim)

(b) 2   (c) 4   (d) Fail to reject H_0.

(e) There is not enough evidence at the 5% level of significance to conclude that the critical reading SAT scores improved.

23. (a) Reject H_0.

(b) There is enough evidence at the 5% level of significance to reject the claim that the proportion of adults who feel older than their real age is equal to the proportion of adults who feel younger than their real age.

25. Hypothesis test results:

Parameter : median of Variable

H_0 : Parameter = 22.55

H_A : Parameter ≠ 22.55

Variable	n	n for test
Hourly wages (in dollars)	14	13

Sample Median	Below	Equal	Above	P-value
26.075	2	1	11	0.0225

$P = 0.0225 < 0.05$, so reject H_0. There is enough evidence at the 5% level of significance to reject the labor organization's claim that the median hourly wage of tool and die makers is $22.55.

27. (a) H_0: median ≤ $638 (claim); H_a: median > $638

(b) 2.33　(c) 1.46　(d) Fail to reject H_0.

(e) There is not enough evidence at the 1% level of significance to reject the organization's claim that the median weekly earnings of female workers is less than or equal to $638.

29. (a) H_0: median ≤ 26 (claim); H_a: median > 26

(b) 1.645　(c) 1.302　(d) Fail to reject H_0.

(e) There is not enough evidence at the 5% level of significance to reject the counselor's claim that the median age of brides at the time of their first marriage is less than or equal to 26 years.

Section 11.2　*(page 615)*

1. If the samples are dependent, use a Wilcoxon signed-rank test. If the samples are independent, use a Wilcoxon rank sum test.

3. (a) H_0: There is no reduction in diastolic blood pressure. (claim)

H_a: There is a reduction in diastolic blood pressure.

(b) Wilcoxon signed-rank test

(c) 10　(d) 17　(e) Fail to reject H_0.

(f) There is not enough evidence at the 1% level of significance to reject the claim that there was no reduction in diastolic blood pressure.

5. (a) H_0: The cost of prescription drugs is not lower in Canada than in the United States.

H_a: The cost of prescription drugs is lower in Canada than in the United States. (claim)

(b) Wilcoxon signed-rank test

(c) 4　(d) 6　(e) Fail to reject H_0.

(f) There is not enough evidence at the 5% level of significance for the researcher to conclude that the cost of prescription drugs is lower in Canada than in the United States.

7. (a) H_0: There is no difference in salaries.

H_a: There is a difference in salaries. (claim)

(b) Wilcoxon rank sum test

(c) ±1.96　(d) −1.94　(e) Fail to reject H_0.

(f) There is not enough evidence at the 5% level of significance to support the representative's claim that there is a difference in the salaries earned by teachers in Wisconsin and Michigan.

9. Reject H_0. There is enough evidence at the 10% level of significance for the engineer to conclude that the gas mileage is improved.

Section 11.3　*(page 623)*

1. The conditions for using a Kruskal-Wallis test are that each sample must be randomly selected and the size of each sample must be at least 5.

3. (a) H_0: There is no difference in the premiums.

H_a: There is a difference in the premiums. (claim)

(b) 5.991　(c) 9.506　(d) Reject H_0.

(e) There is enough evidence at the 5% level of significance to conclude that the distributions of the annual premiums of the three states are different.

5. (a) H_0: There is no difference in the salaries.

H_a: There is a difference in the salaries. (claim)

(b) 6.251　(c) 1.202　(d) Fail to reject H_0.

(e) There is not enough evidence at the 10% level of significance to conclude that the distributions of the annual salaries in the four states are different.

7. Kruskal-Wallis results:

Data stored in separate columns.

Chi Square = 8.0965185 (adjusted for ties)

DF = 2

P-value = 0.0175

Column	n	Median	Ave. Rank
A	6	5	6.75
B	6	8.5	14.5
C	6	5	7.25

$P = 0.0175 > 0.01$, so fail to reject H_0. There is not enough evidence at the 1% level of significance to conclude that the distributions of the number of job offers at Colleges A, B, and C are different.

9. (a) Fail to reject H_0.　(b) Fail to reject H_0.

Both tests come to the same decision, which is that there is not enough evidence to support the claim that there is a difference in the number of days spent in the hospital.

Section 11.4 (page 628)

1. The Spearman rank correlation coefficient can be used to describe the relationship between linear or nonlinear data. Also, it can be used for data at the ordinal level and it is easier to calculate by hand than the Pearson correlation coefficient.

3. The ranks of the corresponding data are identical when r_s is equal to 1. The ranks are in "reverse" order when r_s is equal to -1. The ranks have no relationship when r_s is equal to 0.

5. (a) H_0: $\rho_s = 0$; H_a: $\rho_s \neq 0$ (claim)
   (b) 0.929
   (c) 0.857
   (d) Fail to reject H_0.
   (e) There is not enough evidence at the 1% level of significance to support the claim that there is a correlation between debt and income in the farming business.

7. (a) H_0: $\rho_s = 0$; H_a: $\rho_s \neq 0$ (claim)
   (b) 0.833
   (c) 0.950
   (d) Reject H_0.
   (e) There is enough evidence at the 1% level of significance to conclude that there is a correlation between the oat and wheat prices.

9. Fail to reject H_0. There is not enough evidence at the 5% level of significance to conclude that there is a correlation between science achievement scores and GNI.

11. Reject H_0. There is enough evidence at the 5% level of significance to conclude that there is a correlation between science and mathematics achievement scores.

13. Fail to reject H_0. There is not enough evidence at the 5% level of significance to conclude that there is a correlation between average hours worked and the number of on-the-job injuries.

Section 11.5 (page 637)

1. Answers will vary. *Sample answer:* It is called the runs test because it considers the number of runs of data in a sample to determine whether the sequence of data was randomly selected.

3. Number of runs: 8
   Run lengths: 1, 1, 1, 1, 3, 3, 1, 1

5. Number of runs: 9
   Run lengths: 1, 1, 1, 1, 1, 6, 3, 2, 4

7. n_1 = number of T's = 6
   n_2 = number of F's = 6

9. n_1 = number of M's = 10
   n_2 = number of F's = 10

11. too high: 11; too low: 3

13. too high: 14; too low: 5

15. (a) H_0: The coin tosses were random.
    H_a: The coin tosses were not random. (claim)
    (b) lower critical value = 4
    upper critical value = 14
    (c) 9 (d) Fail to reject H_0.
    (e) There is not enough evidence at the 5% level of significance to support the claim that the coin tosses were not random.

17. (a) H_0: The sequence of leagues of winning teams is random.
    H_a: The sequence of leagues of winning teams is not random. (claim)
    (b) ± 1.96 (c) 1.79 (d) Fail to reject H_0.
    (e) There is not enough evidence at the 5% level of significance to conclude that the sequence of leagues of World Series winning teams is not random.

19. (a) H_0: The microchips are random by gender. (claim)
    H_a: The microchips are not random by gender.
    (b) lower critical value = 8
    upper critical value = 18
    (c) 12 (d) Fail to reject H_0.
    (e) There is not enough evidence at the 5% significance level to reject the claim that the microchips are random by gender.

21. Fail to reject H_0. There is not enough evidence at the 5% level of significance to support the claim that the daily high temperatures do not occur randomly.

23. Answers will vary.

Uses and Abuses for Chapter 11 (page 639)

1. Answers will vary.

2. Sign test $\rightarrow$ z- or t-test
   Paired-sample sign test $\rightarrow$ t-test
   Wilcoxon signed-rank test $\rightarrow$ t-test
   Wilcoxon rank sum test $\rightarrow$ z- or t-test
   Kruskal-Wallis test $\rightarrow$ one-way ANOVA
   Spearman rank correlation coefficient $\rightarrow$ Pearson correlation coefficient

Review Answers for Chapter 11 (page 641)

1. (a) H_0: median ≤ 650 (claim); H_a: median > 650
   (b) 2 (c) 7 (d) Fail to reject H_0.
   (e) There is not enough evidence at the 1% level of significance to reject the bank manager's claim that the median number of customers per day is no more than 650.

3. (a) H_0: median = 2 (claim); H_a: median $\neq 2$
   (b) -1.645 (c) -3.26 (d) Reject H_0.
   (e) There is enough evidence at the 10% level of significance to reject the agency's claim that the median sentence length for all federal prisoners is 2 years.

5. (a) H_0: There is no reduction in diastolic blood pressure. (claim)

H_a: There is a reduction in diastolic blood pressure.

(b) 2   (c) 3   (d) Fail to reject H_0.

(e) There is not enough evidence at the 5% level of significance to reject the claim that there was no reduction in diastolic blood pressure.

7. (a) Independent; Wilcoxon rank sum test

(b) H_0: There is no difference in the total times to earn a doctorate degree by female and male graduate students.

H_a: There is a difference in the total times to earn a doctorate degree by female and male graduate students. (claim)

(c) ± 2.575 (d) -1.357 (*or* 1.357)

(e) Fail to reject H_0.

(f) There is not enough evidence at the 1% level of significance to support the claim that there is a difference in the total times to earn a doctorate degree by female and male graduate students.

9. (a) H_0: There is no difference in the ages of doctorate recipients among the fields of study.

H_a: There is a difference in the ages of doctorate recipients among the fields of study. (claim)

(b) 9.210   (c) 6.741   (d) Fail to reject H_0.

(e) There is not enough evidence at the 1% level of significance to conclude that the distributions of ages of the doctorate recipients in these three fields of study are different.

11. (a) H_0: $\rho_s = 0$; H_a: $\rho_s \neq 0$ (claim)

(b) 0.786   (c) 0.830   (d) Reject H_0.

(e) There is enough evidence at the 5% level of significance to conclude that there is a correlation between overall score and price.

13. (a) H_0: The traffic stops were random by gender.

H_a: The traffic stops were not random by gender. (claim)

(b) lower critical value = 8

upper critical value = 19

(c) 14   (d) Fail to reject H_0.

(e) There is not enough evidence at the 5% level of significance to support the claim that the stops were not random by gender.

Chapter Quiz for Chapter 11   (*page 645*)

1. (a) H_0: There is no difference in the hourly earnings.

H_a: There is a difference in the hourly earnings. (claim)

(b) Wilcoxon rank sum test

(c) ± 1.645 (d) -3.326 (*or* 3.326)   (e) Reject H_0.

(f) There is enough evidence at the 10% level of significance to support the organization's claim that there is a difference in the hourly earnings of union and nonunion workers in state and local governments.

2. (a) H_0: median = 52 (claim); H_a: median $\neq 52$

(b) Sign test   (c) ± 1.96 (d) -2.75 (e) Reject H_0.

(f) There is enough evidence at the 5% level of significance to reject the organization's claim that the median number of annual volunteer hours is 52 hours.

3. (a) H_0: There is no difference in the sales prices among the regions.

H_a: There is a difference in the sales prices among the regions. (claim)

(b) Kruskal-Wallis test

(c) 11.345   (d) 25.957   (e) Reject H_0.

(f) There is enough evidence at the 1% level of significance to conclude that the distributions of the sales prices in these regions are different.

4. (a) H_0: The days with rain are random.

H_a: The days with rain are not random. (claim)

(b) Runs test

(c) lower critical value = 10

upper critical value = 22

(d) 16   (e) Fail to reject H_0.

(f) There is not enough evidence at the 5% level of significance for the meteorologist to conclude that days with rain are not random.

5. (a) H_0: $\rho_s = 0$; H_a: $\rho_s \neq 0$ (claim)

(b) Spearman rank correlation coefficient

(c) 0.829   (d) 0.886   (e) Reject H_0.

(f) There is enough evidence at the 10% level of significance to conclude that there is a correlation between the number of larceny-thefts and the number of motor vehicle thefts.

Real Statistics–Real Decisions for Chapter 11   (*page 646*)

1. (a) Answers will vary.

(b) Answers will vary.

(c) Answers will vary.

2. (a) Answers will vary.

(b) Sign test; You need to use the nonparametric test because nothing is known about the shape of the population.

(c) H_0: median ≥ 4.1; H_a: median < 4.1 (claim)

(d) Fail to reject H_0. There is not enough evidence at the 5% level of significance to support the claim that the median tenure for workers from the representative's district is less than 4.1 years.

3. (a) Wilcoxon rank sum test; You need to use the nonparametric test because nothing is known about the shape of the population.

(b) H_0: There is no difference between the median tenures for male workers and female workers.

H_a: There is a difference between the median tenures for male workers and female workers. (claim)

(c) Fail to reject H_0. There is not enough evidence at the 5% level of significance to support the claim that there is a difference between the median tenures for male workers and female workers.

Cumulative Review for Chapters 9–11 *(page 648)*

1. (a)

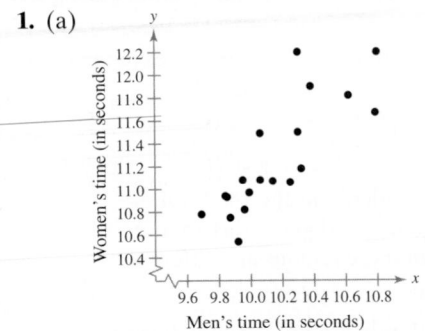

$r \approx 0.815$; strong positive linear correlation

(b) Reject H_0. There is enough evidence at the 5% level of significance to conclude that there is a significant linear correlation between the men's and women's winning 100-meter times.

(c) $\hat{y} = 1.264x - 1.581$

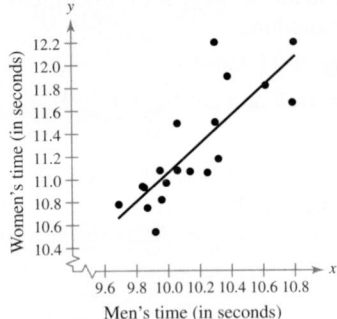

(d) 10.93 seconds

2. There is enough evidence at the 5% level of significance to support the agency's claim that there is a difference in the weekly earnings of workers who are union members and workers who are not union members.

3. There is not enough evidence at the 1% level of significance to reject the company's claim that the median age of people with mutual funds is 50 years.

4. There is enough evidence at the 10% level of significance to reject the claim that the mean expenditures are equal for all four regions.

5. (a) 17,876.15 pounds per acre

(b) 20,148.12 pounds per acre

6. There is not enough evidence at the 10% level of significance to reject the administrator's claim that the standard deviations of reading test scores for eighth grade students are the same in Colorado and Utah.

7. There is enough evidence at the 1% level of significance for the representative to conclude that the distributions of annual household incomes in these regions are different.

8. There is not enough evidence at the 5% level of significance to conclude that the distribution of how much parents intend to contribute to their children's college costs differs from the claimed or expected distributions.

9. (a) 0.733; About 73.3% of the variation in height can be explained by the variation in metacarpal bone length; About 26.7% of the variation is unexplained.

(b) 4.255; The standard error of estimate of the height for a specific metacarpal bone length is about 4.255 centimeters.

(c) $168.026 < y < 190.83$; You can be 95% confident that the height will be between 168.026 centimeters and 190.83 centimeters when the metacarpal bone length is 50 centimeters.

10. There is enough evidence at the 10% level of significance to conclude that there is a correlation between the overall score and the price.

INDEX

Photo Credits

Applet Correlation

Applet	Concept Illustrated	Descriptor	Applet Activity
Random numbers	This applet simulates selecting a random sample from a population by first assigning a unique integer to each experimental unit and then using the random numbers generated to determine the experimental units that will be included in the sample.	This applet generates random numbers from a range of integers specified by the user.	1.3
Mean versus median	The mean and the median of a data set respond differently to changes in the data. This applet investigates how skewedness and outliers affect measures of central tendency.	This applet allows the user to visualize the relationship between the mean and median of a data set. The user may easily add and delete data points. The applet automatically updates the mean and median for each change in the data.	2.3
Standard deviation	Standard deviation measures the spread of a data set. Use this applet to investigate how the shape and spread of a distribution affect the standard deviation.	This applet allows the user to visualize the relationship between the mean and standard deviation of a data set. The user may easily add and delete data points. The applet automatically updates the mean and standard deviation for each change in the data.	2.4
Simulating the stock market	Theoretical probabilities are long run experimental probabilities.	This applet simulates fluctuation in the stock market, where on any given day going up is equally likely as going down. The user specifies the number of days and the applet reports whether the stock market goes up or down each day and creates a bar graph for the outcomes. It also calculates and plots the proportion of days that the stock market goes up during the simulation.	3.1
Simulating the probability of rolling a 3 or 4	Theoretical probabilities are long run experimental probabilities. Use this applet to investigate the relationship between the theoretical and experimental probabilities of rolling a 3 or a 4 as the number of times the die is rolled increases.	This applet simulates rolling a fair die. The user specifies the number of rolls and the applet reports the outcome of each roll and creates a frequency histogram for the outcomes. It also calculates and plots the proportion of 3s and 4s rolled during the simulation.	3.3
Binomial distribution	As the number of samples increases, the estimated probability gets closer to the true value.	This applet simulates values from a binomial distribution. The user specifies the parameters for the binomial distribution (n and p) and the number of values to be simulated (N). The applet plots N values from the specified binomial distribution in a bar graph and reports the frequency of each outcome.	4.2
Sampling distributions	The mean and standard deviation of the distribution of sample means are unbiased estimators of the mean and standard deviation of the population distribution. This applet compares the means and standard deviations of the distributions and assesses the effect of sample size.	This applet simulates repeatedly choosing samples of a fixed size n from a population. The user specifies the size of the sample, the number of samples to be chosen, and the shape of the population distribution. The applet reports the means, medians, and standard deviations of both the sample means and the sample medians and creates plots for both.	5.4

(continued on next page)

Applet	Concept Illustrated	Descriptor	Applet Activity
Confidence intervals for a mean (the impact of not knowing the standard deviation)	Confidence intervals obtained using the sample standard deviation are different from those obtained using the population standard deviation. This applet investigates the effect of not knowing the population standard deviation.	This applet generates confidence intervals for a population mean. The user specifies the sample size, the shape of the distribution, the population mean, and the population standard deviation. The applet simulates selecting 100 random samples from the population and finds the 95% z-interval and 95% t-interval for each sample. The confidence intervals are plotted and the number and proportion containing the true mean are reported.	6.2
Confidence intervals for a proportion	Not all confidence intervals contain the population mean. This applet investigates the meaning of 95% and 99% confidence.	This applet generates confidence intervals for a population proportion. The user specifies the population proportion and the sample size. The applet simulates selecting 100 random samples from the population and finds the 95% and 99% confidence intervals for each sample. The confidence intervals are plotted and the number and proportion containing the true proportion are reported.	6.3
Hypothesis tests for a mean	Not all tests of hypotheses lead correctly to either rejecting or failing to reject the null hypothesis. This applet investigates the relationship between the level of confidence and the probabilities of making Type I and Type II errors.	This applet performs hypotheses tests for a population mean. The user specifies the shape of the population distribution, the population mean and standard deviation, the sample size, and the null and alternative hypotheses. The applet simulates selecting 100 random samples from the population and calculates and plots the t statistic and P-value for each sample. The applet reports the number and proportion of times the null hypothesis is rejected at both the 0.05 level and the 0.01 level.	7.2
Hypothesis tests for a proportion	Not all tests of hypotheses lead correctly to either rejecting or failing to reject the null hypothesis. This applet investigates the relationship between the level of confidence and the probabilities of making Type I and Type II errors.	This applet performs hypotheses tests for a population proportion. The user specifies the population proportion, the sample size, and the null and alternative hypotheses. The applet simulates selecting 100 random samples from the population and calculates and plots the z statistic and P-value for each sample. The applet reports the number and proportion of times the null hypothesis is rejected at both the 0.05 level and the 0.01 level.	7.4
Correlation by eye	The correlation coefficient measures the strength of a linear relationship between two variables. This applet teaches the user how to assess the strength of a linear relationship from a scatter plot.	This applet computes the correlation coefficient r for a set of bivariate data plotted on a scatter plot. The user can easily add or delete points and guess the value of r. The applet then compares the guess to its calculated value.	9.1
Regression by eye	The least squares regression line has a smaller SSE than any other line that might approximate a set of bivariate data. This applet teaches the user how to approximate the location of a regression line on a scatter plot.	This applet computes the least squares regression line for a set of bivariate data plotted on a scatter plot. The user can easily add or delete points and guess the location of the regression line by manipulating a line provided on the scatter plot. The applet will then plot the least squares line. It displays the equations and the SSEs for both lines.	9.2